UNDERSTANDING
Our Universe

STACY PALEN
GEORGE BLUMENTHAL

FOURTH EDITION

NORTON
LOOSE
LEAF

Table of Selected Symbols

Symbol	Meaning/Common Usage in this text
⊙	Sun
⊕	Earth
λ (lambda)	Wavelength
∝	Proportional to
×	Multiplied by
≈	Approximately equal to
°	Degrees
±	Plus or minus
′	Arcminutes
″	Arcseconds
γ (gamma)	Gamma ray
p^+	Proton
e^+	Positron
e^-	Electron
^{2}H	Deuterium
^{4}He	Helium-4
ν (nu)	Neutrino
Ω (omega)	Density ratio
Ω_Λ (omega sub lambda)	Energy density ratio (cosmological constant)
Ω_{mass} (omega sub mass)	Mass density ratio
s	Speed
d	Distance
t	Time
T	Temperature
e	Eccentricity
A	Semimajor axis
P	Period
v	Velocity
a	Acceleration
F	Force
M, m	Mass
R, r	Radius or distance
f	Frequency
L	Angular momentum; luminosity
M_J	Mass of Jupiter
C	Circumference
p	Parallax
E	Energy
z	Redshift
H_0	Hubble constant

Unit Prefixes

Prefix	Name	Factor
n	nano-	10^{-9}
μ	micro-	10^{-6}
m	milli-	10^{-3}
k	kilo-	10^3
M	mega-	10^6
G	giga-	10^9
T	tera-	10^{12}

These prefixes, when appended to a unit, change the size of the unit by the factor given. For example, 1 km (kilometer) is 10^3 meters (m).

Fundamental Physical Constants

Constant	Symbol	Value
Speed of light in a vacuum	c	2.99792×10^8 m/s
Universal gravitational constant	G	6.673×10^{-11} m^3/(kg s^2)
Planck constant	h	6.62607×10^{-34} J s
Electric charge of electron or proton	e	1.60218×10^{-19} C
Boltzmann constant	k	1.38065×10^{-23} J/K
Stefan-Boltzmann constant	σ	5.67040×10^{-8} W/(m^2 K^4)
Mass of electron	m_e	9.10938×10^{-31} kg
Mass of proton	m_p	1.67262×10^{-27} kg

Source: Data from the National Institute of Standards and Technology (http://physics.nist.gov).

Understanding Our Universe

FOURTH EDITION

Understanding Our Universe

FOURTH EDITION

Stacy Palen
WEBER STATE UNIVERSITY

George Blumenthal
UNIVERSITY OF CALIFORNIA—SANTA CRUZ

W. W. NORTON & COMPANY
Independent Publishers Since 1923

W. W. Norton & Company has been independent since its founding in 1923, when William Warder Norton and Mary D. Herter Norton first published lectures delivered at the People's Institute, the adult education division of New York City's Cooper Union. The firm soon expanded its program beyond the Institute, publishing books by celebrated academics from America and abroad. By midcentury, the two major pillars of Norton's publishing program—trade books and college texts—were firmly established. In the 1950s, the Norton family transferred control of the company to its employees, and today—with a staff of five hundred and hundreds of trade, college, and professional titles published each year—W. W. Norton & Company stands as the largest and oldest publishing house owned wholly by its employees.

EDITORS: Erik Fahlgren and Rob Bellinger
PROJECT EDITOR: Taylere Peterson
DEVELOPMENTAL EDITOR: Michael Zierler
EDITORIAL ASSISTANT: Selin Tekgurler
MANAGING EDITOR, COLLEGE: Marian Johnson
PRODUCTION MANAGER: Elizabeth Marotta
MEDIA EDITORS: Michael Jaoui and Rob Bellinger
ASSOCIATE MEDIA EDITOR: Arielle Holstein
MEDIA PROJECT EDITOR: Danielle Belfiore
ASSISTANT MEDIA EDITOR: Jesse Singh
MANAGING EDITOR, COLLEGE DIGITAL MEDIA: Kim Yi
EBOOK PRODUCTION MANAGER: Kate Barnes
MARKETING MANAGER, ASTRONOMY: Ruth Bolster
DESIGN DIRECTOR: Jillian Burr
DESIGNER: Anne DeMarinis
DIRECTOR OF COLLEGE PERMISSIONS: Megan Schindel
COLLEGE PERMISSIONS MANAGER: Bethany Salminen
COLLEGE PERMISSIONS ASSISTANT: Patricia Wong
PHOTO EDITORS: Steph Romeo and Ted Szczepanski
COMPOSITION: Graphic World
ILLUSTRATIONS: Graphic World
MANUFACTURING: LSC Communications—Kendallville

Permission to use copyrighted material is included in the Credits section of this book, which begins on page C-1.

Library of Congress Cataloging-in-Publication Data

Names: Palen, Stacy, author. | Blumenthal, George (George Ray), author.
Title: Understanding Our Universe / Stacy Palen, Weber State University,
   George Blumenthal, University of California–Santa Cruz.
Description: Fourth edition. | New York, NY : W. W. Norton & Company,
   [2021] | Includes index.
Identifiers: LCCN 2020026905 | ISBN 9780393422894 (paperback) | ISBN
   9780393533927 (epub)
Subjects: LCSH: Astronomy—Textbooks. | LCGFT: Textbooks.
Classification: LCC QB43.3 .P35 2021 | DDC 520—dc23
LC record available at https://lccn.loc.gov/2020026905

ISBN 978-0-393-53386-6

W. W. Norton & Company, Inc., 500 Fifth Avenue, New York, NY 10110
wwnorton.com
W. W. Norton & Company Ltd., 15 Carlisle Street, London W1D 3BS
1 2 3 4 5 6 7 8 9 0

Stacy Palen dedicates this book to John Armstrong, with deep gratitude.

George Blumenthal gratefully thanks his wife, Kelly Weisberg,
and his children, Aaron and Sarah Blumenthal, for their support
during this project. He also wants to thank Professor Robert Greenler
for stimulating his interest in all things related to physics.

BRIEF TABLE OF CONTENTS

CONTENTS

PART I   INTRODUCTION TO ASTRONOMY

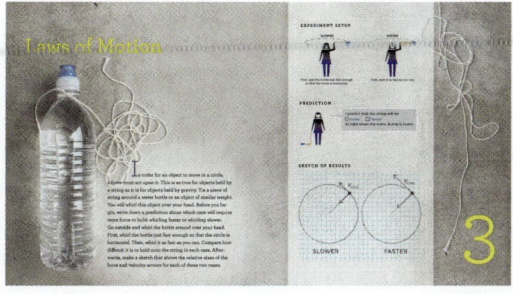

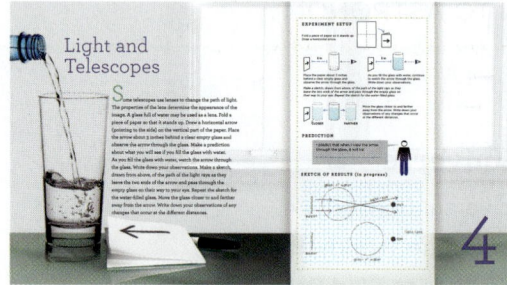

4 Light and Telescopes   70

PART II   THE SOLAR SYSTEM

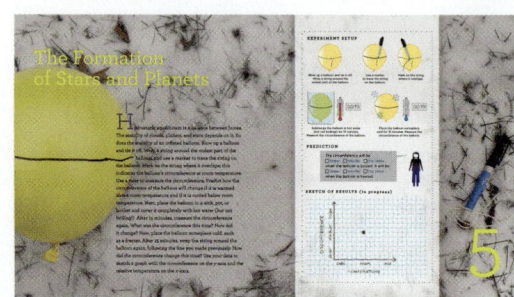

5 The Formation of Stars and Planets   94

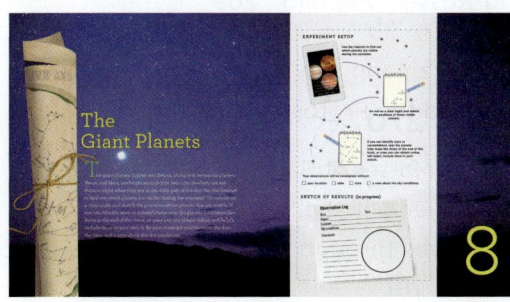

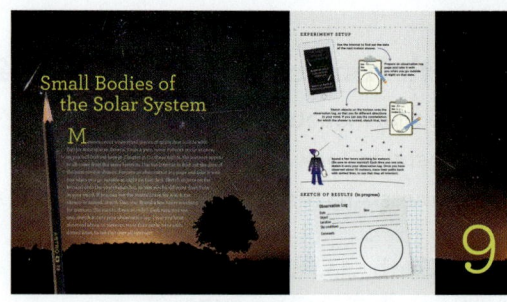

9 Small Bodies of the Solar System   210

PART III   STARS AND STELLAR EVOLUTION

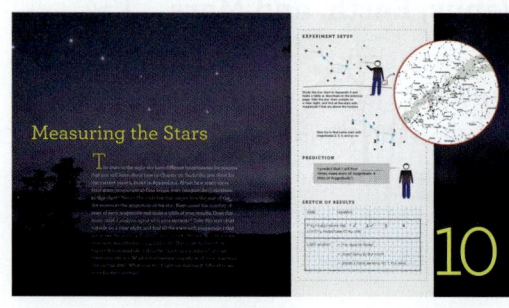

10 Measuring the Stars   240

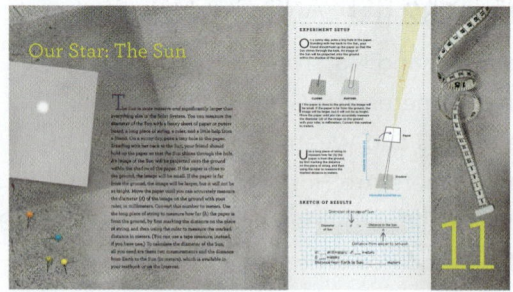

11 Our Star: The Sun   270

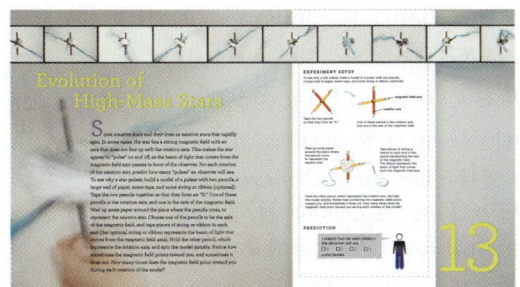

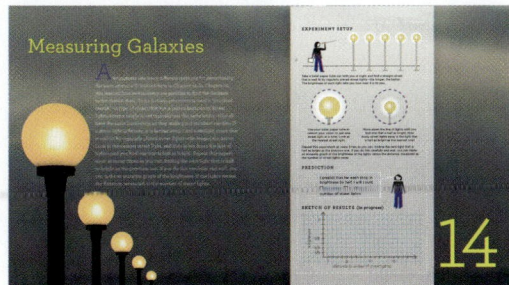

PART IV   GALAXIES, THE UNIVERSE, AND COSMOLOGY

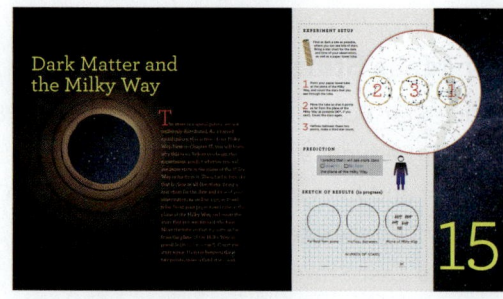

15 Dark Matter and the Milky Way   382

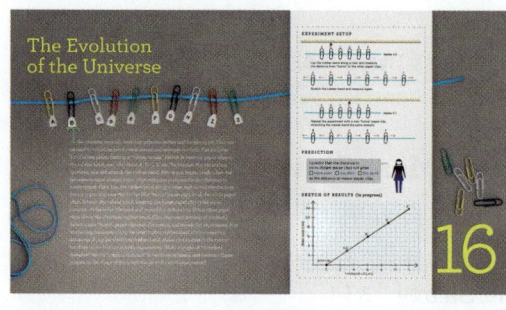

16 The Evolution of the Universe   408

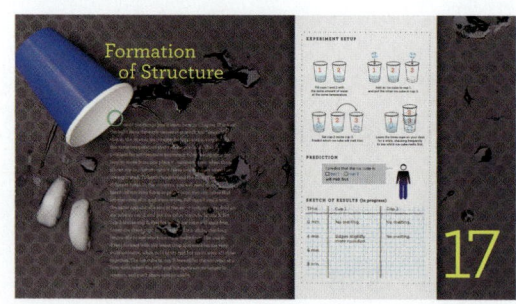

17 Formation of Structure   436

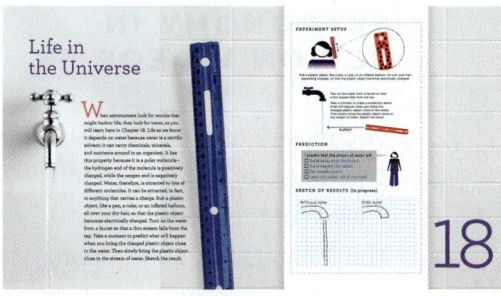

ASTROTOURS

INTERACTIVE SIMULATIONS

ASTRONOMY IN ACTION VIDEOS

Dear Student,

You may wonder why it is a good idea to take a general-education science course. Scientists, including astronomers, have a specific approach to understanding new information. Astronomers "understand" when they can make correct predictions about what will happen next. Astronomers "know" when an idea has been tested dozens, or even hundreds, of times and has stood the test of time.

There are two fundamental goals to keep in mind as you take this course. The first is to understand some basic physical concepts and become familiar with the night sky. The second is to learn to think like a scientist and learn to use the scientific method, or process of science, to answer questions in this course and make decisions about science and technology in your life. We have written the Fourth Edition of *Understanding Our Universe* with these two goals in mind.

Throughout this book, we emphasize the content of astronomy (the masses of the planets, the compositions of stellar atmospheres) as well as *how* we know what we know. The scientific method is a valuable tool that you can carry with you, and use, for the rest of your life.

The most effective way to learn something is to "do" it. Whether you are playing an instrument or a sport or becoming a good cook, reading "how" can only take you so far. The same is true of learning astronomy. The following tools in each chapter help you "do" as you learn:

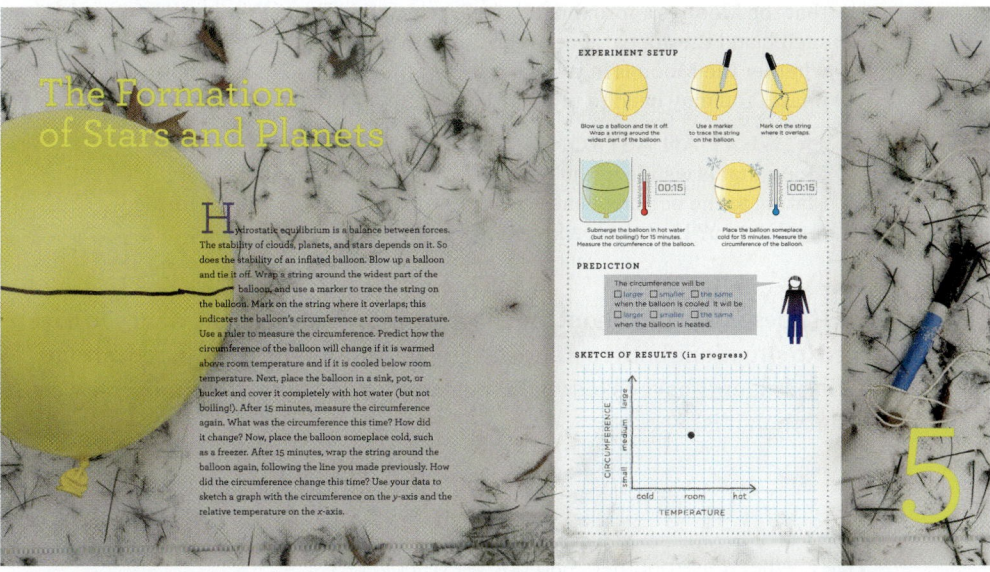

- **Active Learning Figures** open each chapter and ask you to "do" science by setting up an experiment, making either a prediction or an observation, and recording the results. We hope you find that the "answer" isn't the most important part of the activity. Rather, we want the experience of thinking about a physical phenomenon and predicting what will happen next to become a natural way for you to apply your knowledge and understand new concepts.

- To promote active reading, we place **Check Your Understanding** questions at the end of each section of a chapter. These questions act as "speed bumps" so that you will pause and check your comprehension of the material prior to moving on to the next section. These, and the **Questions and Problems** sections at the end of each chapter, are a great way to check whether you have a basic understanding of the material.

> **CHECK YOUR UNDERSTANDING 5.5**
>
> The terrestrial planets are different from the giant planets because when they formed,
> **a.** the inner Solar System was richer in heavy elements.
> **b.** the inner Solar System was hotter than the outer Solar System.
> **c.** the outer Solar System took up a bigger volume, so there was more material to form planets.
> **d.** the inner Solar System was moving faster.

- Within each chapter, the **What an Astronomer Sees** feature helps you understand how astronomers interpret astronomical images, obtaining enormous amounts of information from a single picture. This feature is accompanied by an end-of-chapter question to help you find and interpret visual clues in the universe yourself. Similarly, we have identified all of the figure-based questions at the end of the chapter with an icon; these will also help develop your ability to interpret images and graphs.

> **THINKING ABOUT THE CONCEPTS**
>
> **21.** ★ **WHAT AN ASTRONOMER SEES** Sketch the relative positions of the Earth, Moon, and Sun in Figure 6.1. The caption states that the Moon is approximately in first quarter phase. From your sketch and the appearance of the Earth in this image, determine whether the Moon is in waxing crescent phase or waxing gibbous phase. You may need to review Figure 2.15. ⊛
>
> **22.** ★ **WHAT AN ASTRONOMER SEES** Study the surface of the Moon that is visible in the image in Figure 6.1. Is this surface old (dating to the period of early heavy bombardment) or young (formed much later)? How do you know? ⊛

- **What If** questions throughout the chapter prompt you to apply what you have learned to situations both real and imagined. Many of these are based on questions that students have asked when they wondered, "What if the universe were different than it is?"

> **what if . . .**
>
> What if astronomers had not yet discovered any planets around the 5,000 stars that have been observed so far: Would we be able to conclude (with current technology) that there are no planets around the 100 billion stars in the Milky Way? Why or why not?

Figure 6.1 In December 1968, *Apollo 8* astronauts in orbit photographed Earth rising above the Moon. ★ **WHAT AN ASTRONOMER SEES** An astronomer, viewing this image, will be as awestruck as anyone by the simple fact that humans managed to figure out how to orbit the Moon and take such a picture. Then she will start to look at details. She will notice that the Moon has topography. There are elevation changes, particularly along the horizon, and large and small circular craters that vary in brightness. And she will notice that Earth is very different from the Moon. Studying Earth, she will notice blue water and brown land, as well as swirling cloud formations that indicate an atmosphere with active weather patterns. The Earth is only partially illuminated by the Sun, and that illumination shows that the Sun is located beyond the top of the image and slightly to the right. Because Earth is approximately in quarter phase, the Moon must likewise be approximately in quarter phase. An astronomer may spend considerable time trying to decipher which of the continents are visible on Earth, to figure out whether Earth's North Pole is on the right or the left, deciding eventually that north is on the right. From this information, she can make a mental sketch and figure out that the Moon was just about at first quarter, which is very satisfying.

- **Reading Astronomy News** sections toward the end of each chapter include a news article or press release with questions to help you make sense of how the science is presented. For citizens of the world, recognizing what is credible and questioning what is not are important skills. You make judgments about science, distinguishing between good science and pseudoscience, in order to make decisions in the grocery store, pharmacy, car dealership, and voting booth. You base these decisions on the presentation of information you receive through the media, which is very different from the presentation of information in class. The goal of Reading Astronomy News is to help you build your scientific literacy and your ability to challenge what you hear elsewhere.

- At the very end of each chapter, an **Exploration** activity shows you how to use the concepts and skills you learned in an interactive way. About half of the book's Explorations ask you to use animations and simulations found on the Student Site, while the others are hands-on, paper-and-pencil activities that use everyday objects such as ice cubes or balloons.

We believe that the learn-by-doing approach not only helps you better understand the material, but also makes the material more interesting and, perhaps, fun.

As you learn any new subject, one of the stumbling blocks is often the language of the subject itself. This can be jargon—the specialized words unique to that subject—for example, *supernova* or *Cepheid variable*. But it can also be ordinary words that are used in a special way. As an example, the common word *inflation* usually applies to balloons or tires in everyday life, but economists use it very differently, and astronomers use it differently still. Throughout the book, we have included **Vocabulary Alerts** that point out the astronomical uses of common words to help you recognize how those terms are used by astronomers.

VOCABULARY ALERT

pressure In everyday language, we often use *pressure* interchangeably with the word *force*. Astronomers specifically use *pressure* to mean the force per unit area that atoms or molecules exert as they speed around and collide with each other and their surroundings.

density In everyday language, we use this word in many ways, some of which are metaphorical. Astronomers specifically use *density* to mean "the amount of mass packed into a volume"; denser material contains more mass in the same amount of space. Density has units of kilograms per cubic meter. In practical terms, you are familiar with density by how heavy an object feels for its size: A billiard ball and a tennis ball are roughly the same size, but the billiard ball has greater **mass** and therefore feels heavier because it is denser.

In learning science, there is another potential language issue. The language of science is mathematics, and it can be as challenging to learn as any other language. The choice to use mathematics as the language of science is not arbitrary; nature "speaks" math. To learn about nature, you will also need to speak its language. We don't want the language of

reading Astronomy News

Citizen Scientists Discover Rare Exoplanet
By Ashley Strickland, CNN

K2-288Bb is about twice the size of Earth.

Although NASA's Kepler space telescope ran out of fuel and ended its mission in 2018, citizen scientists have used its data to discover an exoplanet 226 light-years away in the Taurus constellation.

The exoplanet, known as K2-288Bb, is about twice the size of Earth and orbits within the habitable zone of its star, meaning liquid water may exist on its surface. It's difficult to tell whether the planet is rocky like Earth or a gas giant like Neptune.

The planet is in the K2-288 system, which contains a pair of dim, cool M-type stars that are 5.1 billion miles apart, about six times the distance between Saturn and the sun. The brightest of the two stars is half as massive as our sun, and the other star is one-third of the sun's mass. K2-288Bb orbits the smaller, dimmer star, completing a full orbit every 31.3 days.

K2-288Bb is half the size of Neptune or 1.9 times the size of Earth, placing it in the "Fulton gap" between 1.5 and two times the size of Earth. This is a rare size of exoplanet that makes it perfect for studying planetary evolution because so few have been found.

The discovery was announced Monday at the 233rd meeting of the American Astronomical Society in Seattle.

the possible planet transits—or dip in light when a planet passes in front of its star—in the light curve data. Kepler observed other events that could be mistaken for planet transits by a computer.

But the "reboot" of the Kepler mission in 2014 that led to the K2 mission allowed for multiple observation campaigns that brought in even more data. Every three months, Kepler would stare at a different patch of sky.

"Reorienting Kepler relative to the Sun caused miniscule [sic] changes in the shape of the telescope and the temperature of the electronics, which inevitably affected Kepler's sensitive measurements in the first days of each campaign," said study co-author Geert Barentsen, an astrophysicist at NASA's Ames Research Center, in a statement.

Those first three days of data were ignored, and errors were corrected in the rest of the data gathered.

But the scientists couldn't do it alone. There were too many light curves to study on their own.

So the reprocessed, "cleaned-up" light curves were uploaded through the Exoplanet Explorers project on online platform Zooniverse, and the public was invited to "go forth and find us planets," Feinstein said.

In May 2017, citizen scientists began discussing a particular planet candidate, but it

"That's how we missed it—and it took the keen eyes of citizen scientists to make this extremely valuable find and point us to it," Feinstein said.

Follow-up observations were made with multiple telescopes to confirm the exoplanet.

There will be more opportunities for citizen scientists to help discover exoplanets. NASA's latest planet-hunting mission, TESS, will be providing more light curves that are full of potential planets waiting to be found.

Last year at the American Astronomical Society meeting, it was announced that citizen scientists helped discover five planets between the size of Earth and Neptune around star K2-138, the first multiplanet system found through crowdsourcing.

This year, Kevin Hardegree-Ullman, postdoctoral scholar in astronomy at the California Institute of Technology, announced that the Spitzer space telescope followed up on that discovery and discovered a sixth planet, K2-138 g, smaller than Neptune, that orbits the star every 42 days.

"This is only the ninth system discovered containing six or more planets," he said.

EVALUATING THE NEWS

1. What planet detection method was used by the Kepler mission?

2. Why were the first three days of data

EXPLORATION EXPLORING EXOPLANETS   123

exploration Exploring Exoplanets

digital.wwnorton.com/universe4

Visit the Student Site at the Digital Resources page and open the Radial Velocity Simulation in Chapter 5. This applet has a number of different panels that allow you to experiment with the variables that are important for measuring radial velocities. Compare the views shown in the various panels with the colored arrows in the first panel to see where an observer would stand to see the view shown. Start the animation (in the "Animation Controls" panel) and allow it to run while you watch the planet orbit its star from each of the views shown. Stop the animation, and in the "Sample Systems" panel select "Option A."

1. Is Earth's view of this system most nearly like the "side view" or most nearly like the "orbit view"?

2. Is the orbit of this planet circular or elongated?

3. Study the radial velocity graph in the upper right panel. The blue curve shows the radial velocity of the star over a full period. What is the maximum radial velocity of the star?

4. The horizontal axis of the graph shows the "phase," or fraction of the

In the "Sample Systems" window, select "Option B:"

6. What has changed about the orbit of the planet as shown in the views in the upper left panel?

7. When is the planet moving fastest—when it is close to the star or when it is far from the star?

8. When is the star moving fastest—when the planet is close to it or when it is far away?

9. Explain how an astronomer would determine, from a radial velocity graph of the star's motion, whether the orbit of the planet was in a circular or elongated orbit.

Astronomy in Action: Wien's Law

AstroTour: Solar System Formation

Interactive Simulation: Radial Velocity

math to obscure the concepts, so we have placed this book's mathematics in **Working It Out** boxes to make it clear when we are beginning and ending a mathematical argument, so that you can spend time with the concepts in the chapter text and then revisit the mathematics to study the formal language of the argument. Read through a Working It Out box once, then cover the worked example with a piece of paper, and work through the example until you can do it on your own. When you can do this, you will have learned a bit of the language of science. We want you to be comfortable reading, hearing, and speaking the language of science, and we provide you with tools to make it easier.

In addition to learning the language of astronomy, visualizing a process or phenomenon will help you reach a deeper understanding. In addition to the illustrations in the book, many physical concepts are further explained in a series of short **Astronomy in Action** videos, **AstroTour** animations, and new **Interactive Simulations** available in the ebook and on the Student Site. The videos feature one of the authors (and several students) demonstrating physical concepts at work. Each animation is a brief tutorial on a concept or process in the chapter. The simulations allow you to explore topics such as Moon phases, Kepler's laws, and the Hertzsprung-Russell diagram. Your instructor might assign the videos and animations to you, or you might choose to watch them on your own to create a better picture of each concept in your mind.

Finally, every new copy of the book or ebook includes access to *At Play in the Cosmos: The Videogame*. This game invites you to experience the excitement of scientific discovery for yourself. You will apply concepts from the course to fly and repair your ship, make decisions based on scientific data, and complete missions. You can go to the Student Site to download the app.

Astronomy gives us a sense of perspective that no other field of study offers. The universe is vast, fascinating, and beautiful, filled with a wealth of objects that, surprisingly, can be understood using only a handful of principles. By the end of this book, you will have gained a sense of your place in the universe—both how incredibly small and insignificant you are and how incredibly unique and important you are.

Sincerely,
Stacy Palen
George Blumenthal

Dear Instructor,

We wrote this book with a few overarching goals: to inspire students, to use active learning to build scientific literacy, and to create a useful and flexible tool that offers diverse approaches to the content.

As scientists and as teachers, we are passionate about the work we do. We hope to share that passion with students and inspire them to engage in science on their own.

As authors, one way we do this is through the **Active Learning Figures** at the beginning of each chapter. These figures model student engagement and provide an opportunity for students to do an experiment or make an observation on their own, using only everyday objects they can find around the house or dorm room.

Through our own experience, familiarity with education research, and surveys of instructors, we know how students learn and what goals teachers have. We have explicitly addressed many of these goals in this book, sometimes in large, immediately visible ways such as the pedagogical structure, but also through less obvious efforts such as figure-based end-of-chapter exercises that teach students to read astronomical images and graphs, helping them understand important concepts through visualization.

For example, many instructors would like their students to become critical consumers of science news. We have specifically addressed this goal in our **Reading Astronomy News** feature, which presents a news article and a series of questions that guide a student's critical thinking about the article, the data presented, and the sources. For the fourth edition, we have selected new articles for many chapters.

Many instructors want students to develop the ability to understand important concepts through visualization. We address this explicitly by teaching students to make and use spatial models. One example is in Chapter 2 ("Patterns in the Sky— Motions of Earth and the Moon"), where we ask students to use an orange and a lamp to understand the celestial sphere and the phases of the Moon. In nearly every chapter, we use **visual analogies** to compare astronomical concepts to everyday events or objects. Through these analogies, we strive to make the material more interesting, relevant, and memorable. For example, in Chapter 8 ("The Giant Planets"), we show an image of moss flowing downstream from a rock to demonstrate how the magnetospheres of the giant planets are shaped by the solar wind.

Education research shows that the most effective way to learn is by doing. In addition to the Active Learning Figures and Reading Astronomy News features, the **Exploration** activities at the end of each chapter ask students to take the concepts they've learned in the chapter and apply them as they interact with animations and simulations on the Student Site or work through pencil-and-paper activities. Many of these Explorations incorporate everyday objects and can be used either in your classroom or as activities at home.

To learn astronomy, students must also learn the language of science—not just the jargon, but the everyday words we scientists use in special ways. *Theory* is a good example of a word that students think they understand, but their definition is very different from ours. The first time we use an ordinary word in a special way, a **Vocabulary Alert** in the margin calls attention to it, helping to reduce student confusion. This is in addition to the back-of-book Glossary, which includes all the text's boldface words as well as other terms students may be unfamiliar with.

We also believe students should grow more comfortable with the more formal language of science—mathematics. We have placed the math in **Working It Out** boxes so it does not interrupt the flow of the text or get in the way of students' understanding of conceptual material. But we've gone further by beginning with fundamental skills in earlier math boxes and slowly building complexity in math boxes that appear later in the book. We've also worked to remove some of the stumbling blocks that can reduce student confidence by providing calculator hints, references to earlier boxes, and detailed, fully worked examples.

In our overall organization, we have made several efforts to encourage students to engage with the material and build confidence in their scientific skills as they proceed through the book. We introduce some physical principles with a "just-in-time" approach; for example, we cover atomic emission and absorption in Chapter 10 ("Measuring the Stars"), the first time that this level of detail is required to understand astronomical phenomena. Students are not required to flip back several chapters to remember the details. The material on stars is organized to cover the general case first and then to delve into more detail with specific examples. This organization implicitly helps students to understand their place in the universe: Our star is one of many. The Sun is a specific example of a star that can be generalized to learn about all stars because we live in a physical universe in which the same laws apply everywhere. Planets have been organized comparatively to emphasize that science is a process of studying individual examples that lead to collective conclusions. All of these organizational choices were made with the student perspective in mind.

Other items new to the fourth edition of *Understanding Our Universe* include the following:

- Each chapter contains a **What an Astronomer Sees** figure that demonstrates for students how an astronomer can derive a wealth of information from a single image. Each figure is accompanied by an end-of-chapter question and questions in Smartwork that further guide students in developing the skill of interpreting astronomical imagery.

- Multiple times in each chapter, we ask students, "What if?" These **What If** questions often ask students to imagine the universe other than it is or ask them to extrapolate from the information they have learned. These advanced comprehension questions give students the opportunity to experiment with developing and testing hypotheses by asking, "If the universe were not as it is, what else would have to be true?" This feature is accompanied by a "guiding" question in Smartwork that helps students organize their thinking.

- We revised each chapter, updating the science, to reflect the fast pace of astronomical research today. For example, we include a brief discussion of the famous "black hole image" taken by the Event Horizon Telescope, as well as new statistics on gravitational wave detections and exoplanets.

- Two additional **Interactive Simulations**, authored by Stacy Palen, join a growing collection of simulations that pair with the Exploration activities in the text and questions in Smartwork, allowing students to explore topics such as the seasons, the ecliptic, and the habitable zone.

- *Teaching Astronomy by Doing Astronomy* (**tada101.com**) is a blog for introductory astronomy instructors. Stacy Palen writes weekly posts and provide suggestions for adding active learning to classes of any size. Stacy also posts suggested Reading Astronomy News articles, discusses ways to build quantitative literacy, and hosts discussions about recent developments in practical applications of astronomy education research.

- The **Interactive Instructor's Guide** is a searchable online resource that allows instructors to find exactly what they need, right when they need it. Included are discussion points; notes and suggestions for teaching with **AstroTour** animations, **Interactive Simulations**, **Astronomy in Action** videos, in-text learning-by-doing features (such as Active Learning Figures, Reading Astronomy News, and Explorations); worked solutions to all in-chapter and end-of-chapter problems; PowerPoint Lecture Slides; Current Event Update Slides; the Test Bank; and information about incorporating the *Learning Astronomy by Doing Astronomy: Collaborative Lecture Activities* workbook activities into class.

Student engagement is not limited to the classroom or to the text. Norton's online tutorial and homework system, **Smartwork**, allows you to engage students outside the classroom and to easily assess students using interactive, visual content on both tablets and computers. Each of the more than 2,000 questions offers students answer-specific feedback. All questions are tied directly to this text, offering students an ebook link in every problem to motivate reading.

Many textbook problems, including in-chapter Check Your Understanding questions, selected end-of-chapter questions, and versions of the Reading Astronomy News, Exploration, and What an Astronomer Sees questions, are available in the system—all easily findable using the system's filters. In addition, hundreds of ranking, sorting, and labeling tasks challenge students to interact meaningfully with book art, videos, animations, and simulations. Process of Science assignments help walk students through major discoveries and challenges throughout the history of astronomy.

In addition, new Smartwork questions based on the ***Learning Astronomy by Doing Astronomy*, Second Edition** workbook and ***At Play in the Cosmos: The Videogame*** can help you meaningfully integrate these resources into your course. You can easily modify any of the provided questions, answers, and feedback, or you can create your own questions. You can easily set your course up with a range of premade assignments using the pedagogical elements of the book. Smartwork can also integrate directly into your campus learning management system (LMS). Students will sign in to a single website, and their grades will automatically appear in the LMS gradebook.

We approached this text by asking: What do instructors want students to learn, and how can we best help students learn those things? Where possible, we consulted the education research to help guide us, and that guidance has led us down some previously unexplored paths. That research has continued to be useful in this fourth edition, but we continue to draw on another excellent resource—our colleagues who also teach introductory astronomy. We value your input, which often gives us ideas for new approaches, so we hope you will be part of the creative process by sharing your experiences with us on the Teaching Astronomy by Doing Astronomy blog.

Sincerely,
Stacy Palen
George Blumenthal

Ancillaries for Students
smartwork

Steven Desch, Guilford Technical Community College; Violet Mager, Penn State Wilkes-Barre; Todd Young, Wayne State College; David A. Wood, San Antonio College

More than 2,000 questions support *Understanding Our Universe*, Fourth Edition—all with answer-specific feedback, hints, and ebook links. Questions include selected end-of-chapter questions, versions of the Explorations (based on Astro-Tours and new simulations), and Reading Astronomy News. Image-labeling questions based on NASA images allow students to apply course knowledge to images that are not contained in the text. Astronomy in Action video assignments focus on overcoming common misconceptions, while Process of Science assignments take students through the steps of a discovery and ask them to participate in the decision-making process that led to the discovery.

At Play in the Cosmos: The Videogame

Jeff Bary, Colgate University; Adam Frank, University of Rochester; Learning Games Network, University of Wisconsin and Massachusetts Institute of Technology; Gear Learning at UW-Madison's Wisconsin Center for Education Research; Games and Stuff LLC

Every student can experience the excitement of scientific discovery using this groundbreaking and award-winning educational videogame. Using the process of science, students complete missions by flying their ship and choosing the right scientific tools for each task. The game is available in Mac, PC, iOS, and Android formats. Game access is now included *free* with every new print book and every Norton ebook.

A complete videogame instructor's manual provides suggestions for meaningfully integrating the game into your course—for example, as a lab activity or as homework. In addition, new mission-based questions available in Smartwork get students applying what they learned in the game and seeing connections between gameplay and your course.

Norton Digital Resources Page
digital.wwnorton.com/universe4

This site contains:

- Astronomy in Action videos demonstrate the most important concepts in a visual, easy-to-understand, and memorable way.

- Nine Interactive Simulations allow students to manipulate variables and explore topics such as eclipses, seasons, the H-R diagram, and the habitable zone.

- Thirty-one AstroTour animations. These animations, some of which are interactive, use illustrations from the text to help students visualize important physical and astronomical concepts.

Learning Astronomy by Doing Astronomy:
Collaborative Lecture Activities, Second Edition

Stacy Palen, Weber State University, and Ana Larson, University of Washington

Many students learn best by doing. But, for instructors, it can be challenging and time consuming to create activities that use real data, illuminate astronomical concepts, and ask probing questions that get students to confront their misconceptions. In this workbook, the authors draw on their experience teaching thousands of students in many different types of courses (large in-class, small in-class, hybrid, online, flipped, and so forth) to provide 36 field-tested activities that can be used in any classroom today. The activities have been designed to require no special software, materials, or equipment, and to take no more than 50 minutes each to do. Pre-activity and post-activity questions are available in Smartwork. PowerPoint versions of the pre-activity and post-activity questions are available on Norton's Instructor's Site.

Starry Night Planetarium Software
(Norton College Version 8) and Workbook

Steven Desch, Guilford Technical Community College, and Michael Marks, Bristol Community College

Starry Night is a realistic, user-friendly planetarium simulation program designed to allow students in urban areas to perform observational activities on a computer screen. Norton's unique accompanying workbook offers observation assignments that guide students' virtual explorations and help them apply what they've learned from reading their textbook.

Ancillaries for Instructors

Instructor's Manual

Tabitha Buehler, University of Utah

This resource includes brief chapter overviews, suggested classroom discussions points, notes on Active Learning Figures, AstroTour animations, Astronomy in Action videos, Interactive Simulations, Reading Astronomy News, Explorations, and worked solutions to all in-chapter and end-of-chapter Questions and Problems. Also included are notes on teaching with *Learning Astronomy by Doing Astronomy: Collaborative Lecture Activities*, Second Edition.

PowerPoint Lecture Slides

Windsor Morgan, Dickinson College

These ready-to-use lecture slides complement the text by providing summaries of central topics, integrated photos and art, class questions, and links to the animations, simulations, and videos. These lecture slides are editable and are available in Microsoft PowerPoint format. Slides are optimized for use with screen readers, and alt text is included with every image.

Norton also provides a biannual update service—PowerPoint presentations on engaging new topics—that enables instructors to cover new developments in astronomy soon after they occur.

Test Bank

Todd Vaccaro, Ball State University

The Test Bank assesses a set of learning objectives based on the text and consistent with Smartwork online homework and provides over 1,200 multiple-choice and short-answer questions. Now available through Norton Testmaker, assessments can be created from anywhere with an internet connection, without downloading files or installing specialized software. Questions can be filtered by chapter, section, type, difficulty, learning objective, and Bloom's taxonomy (remembering, understanding, applying, analyzing, and evaluating). Assessments can be easily exported to Microsoft Word or Common Cartridge files for import into an LMS. The Test Bank and Testmaker make it easy to construct meaningful and diagnostic exams.

Norton Interactive Instructor's Guide (IIG)

This searchable online resource is designed to help instructors prepare for lecture in real time. It contains the following resources, each tagged by topic, chapter, and learning objective:

- Test Bank
- Instructor's Manual
- Brief chapter overviews and discussion points
- Notes for teaching with Active Learning Figures, Reading Astronomy News, and Explorations
- Astronomy in Action videos
- Interactive Simulations
- AstroTour animations
- Notes for teaching with *Learning Astronomy by Doing Astronomy: Collaborative Lecture Activities*, Second Edition
- Solutions to all in-chapter and end-of-chapter questions and problems

- Lecture PowerPoint slides
- Current Event Update slides
- All art, photos, and tables in JPEG and PPT formats
- Answers to the *Starry Night Workbook* exercises

Resources for Your LMS

Easily add high-quality Norton digital resources to your online, hybrid, or lecture courses. Get started building your course with our easy-to-use integrated resources; all activities can be accessed right within your existing LMS. Norton provides a file containing links to the following resources, organized by chapter:

- Ebook
- Smartwork premade assignments
- Links to Astronomy in Action Videos, Interactive Simulations, and AstroTour animations

Acknowledgments

Thanks so much to Laura Kay, whose work on previous editions was foundational. We missed her input and ideas on this edition, and we wish her the very best in her future endeavors.

Many people collaborated to produce this book. Most of these people worked hard in the background, keeping track of the thousands of small details that no one person could possibly remember. The authors would like to acknowledge the extraordinary efforts of the staff at W. W. Norton: Selin Tekgurler, who coordinated reviewers and kept the project running on time; Taylere Peterson, who shepherded the manuscript through the layout process; Heather Whirlow Cammarn, the copy editor, who made sure that grammar and punctuation survived the multiple rounds of the editing process; and Michael Zierler, who brought his love of science to the developmental editing process. Erik Fahlgren and Rob Bellinger coordinated the Norton team and continued to rein in our excessive use of exclamation points.

Anne DeMarinis and Jillian Burr created the design for the fourth edition. Liz Marotta managed the production. Michael Jaoui, Arielle Holstein, and Jesse Singh worked on the media and supplements, and Ruth Bolster helped get this book into your hands.

And we would like to thank the reviewers, whose input at every stage improved the final product:

Fourth Edition Reviewers

Justin Adkins, *Bluegrass Community and Technical College*
Charles C. Agosta, *Clark University*
Eduardo Araujo-Pradere, *Miami Dade College*
Celso Batalha, *Evergreen Valley College*
Tabitha Buehler, *The University of Utah*
Gene Byrd, *The University of Alabama*
Marianne Caldwell, *Hillsborough Community College*
Philip Chang, *The University of Wisconsin-Milwaukee*
Micol Christopher, *Mt. San Antonio College*
Sophia Cisneros, *University of Denver*
Jess Dowdy, *Abilene Christian University*
Davin Flateau, *University of Cincinnati*

Michael Frey, *Cypress College*
Steven Furlanetto, *University of California–Los Angeles*
Jack Gabel, *Creighton University*
Guillermo Gonzalez, *Ball State University*
David Hedin, *Northern Illinois University*
Patrick Huth, *Community College of Allegheny County*
Preston Johnson, *Allen Community College*
John Kielkopf, *University of Louisville*
Hyun-Chul Lee, *The University of Texas Rio Grande Valley*
Jorge Lopez, *The University of Texas at El Paso*
Naibi Marinas, *University of Florida*
Brian B. McGuinness, *Fairleigh Dickinson University*
Randall L. Milstein, *Oregon State University*

Hon Kie Ng, *Florida State University*
Joshua Ridley, *Murray State University*
Judy Ripley, *Florida SouthWestern State College*
Fatma Salman, *Manchester Community College*
Susan Stolovy, *El Camino College*
Richard J. Wainscoat, *University of Hawaii*
Thulsi Wickramasinghe, *The College of New Jersey*

Previous Editions' Reviewers
Loren Anderson, *West Virginia University*
Jonathan Barnes, *Salt Lake Community College*
Rory Barnes, *University of Washington*
Celso Batalha, *Evergreen Valley College*
Lloyd Black, *Rowan University*
Ann Bragg, *Marietta College*
James S. Brooks, *Florida State University*
Edward Brown, *Michigan State University*
Thomas Brueckner, *University of Central Florida*
Eric Bubar, *Marymount University*
Marianne Caldwell, *Hillsborough Community College*
Karen Castle, *Diablo Valley College–Pleasant Hill*
Kwang-Ping Cheng, *California State University–Fullerton*
James Cooney, *University of Central Florida*
Kelle Cruz, *Hunter College*
Xinyu Dai, *University of Oklahoma*
Noella D'Cruz, *Joliet Junior College*
Steve Desch, *Guilford Technical Community College*
Declan De Paor, *Old Dominion University*
Ethan Dolle, *Northern Arizona University*
Adrienne Dove, *University of Central Florida*
Jess Dowdy, *Abilene Christian University*
Christian Draper, *Utah Valley University*
Hardin Dunham, *Angelo State University*
David Dunn, *Sierra College, Rocklin*
Amy Fredericks, *Long Beach City College*
Michael Frey, *Cypress College*
Robert Friedfeld, *Stephen F. Austin State University*
Jeffrey Gillis-Davis, *University of Hawaii at Manoa*
Karl Haisch Jr., *Utah Valley University*
Joy Moses-Hall, *Pitt Community College*
Steven A. Hawley, *University of Kansas*
Eric R. Hedin, *Ball State University*
Anthony Heinzman, *Victor Valley College*
Sean Hendrick, *Millersville University*
Scott Hildreth, *Chabot College*
Emily S. Howard, *Broward College*
Eric Hsiao, *Florida State University*
Olenka Hubickyj, *Foothill College*
Doug Ingram, *Texas Christian University*
Donald Isenhower, *Abilene Christian University*
C. Renee James, *Sam Houston State University*
Joe Jensen, *Utah Valley University*
Nathan Kaib, *University of Oklahoma*
Kishor Kapale, *Western Illinois University*
Dain Kavars, *Ball State University*

Sean Kemp, *Evergreen Valley College*
Viken Kiledjian, *East Los Angeles College*
Thomas Klee, *Hillsborough Community College*
Kevin Krisciunas, *Texas A&M University*
Ana Larson, *University of Washington*
Lee LaRue, *Texas A&M University, Commerce*
Hector Leal, *The University of Texas Rio Grande Valley*
Hyun-chul Lee, *The University of Texas Rio Grande Valley*
Rod Lee, *Portland Community College*
Lauren Likkel, *University of Wisconsin–Eau Claire*
Duncan Lorimer, *West Virginia University*
Jane H. MacGibbon, *University of North Florida*
Naibi Marinas, *University of Florida*
Norman Markworth, *Stephen F. Austin State University*
Piet Martens, *Georgia State University*
Paul Mason, *The University of Texas at El Paso*
Geoffrey Mathews, *University of Hawaii*
James McAteer, *New Mexico State University*
Ian McLean, *University of California–Los Angeles*
Jo Ann Merrell, *Saddleback College*
Michele Montgomery, *University of Central Florida*
Hon-Kie Ng, *Florida State University*
Christopher Palma, *Pennsylvania State University*
Marina Papenkova, *East Los Angeles College*
Jay Pehl, *Indiana University–Purdue University Indianapolis*
Nicolas A. Pereyra, *University of Texas–Pan American*
Vahe Peroomian, *University of California–Los Angeles*
Barry Rice, *Sierra College*
Dan Robertson, *Monroe Community College–Rochester*
Dwight Russell, *Baylor University*
Ann Schmiedekamp, *Pennsylvania State University–Abington*
Teresa Schulz, *Lansing Community College*
Kent Schwitkis, *El Camino College*
Marc Seigar, *University of Minnesota, Duluth*
Haywood Smith, *University of Florida*
Ulysses J. Sofia, *American University*
Michael Solontoi, *University of Washington*
Thomas Targett, *Sonoma State University*
Greg Taylor, *University of New Mexico*
Don Tendrup, *The Ohio State University*
Tracey Tessier, *University of New Mexico*
Vladimir Tumakov, *Santa Ana College*
Robert Tyson, *The University of North Carolina at Charlotte*
Todd Vaccaro, *St. Cloud University*
Trina Van Ausdal, *Salt Lake Community College*
Nilakshi Veerabathina, *The University of Texas at Arlington*
G. Scott Watson, *Syracuse University*
Amy White, *St. Charles Community College*
Scott Williams, *Angelo State University*
Shannon Willoughby, *Montana State University*
Fred Wilson, *Angelo State University*
Brian Woodahl, *Indiana University–Purdue University Indianapolis*
Laura Woodney, *California State University–San Bernardino*
Kaisa Young, *Nicholls State University*

ABOUT THE AUTHORS

Stacy Palen is an award-winning professor in the physics department and the director of the Ott Planetarium at Weber State University. She received her BS in physics from Rutgers University and her PhD in physics from the University of Iowa. As a lecturer and postdoc at the University of Washington, she taught Introductory Astronomy more than 20 times over four years. Since joining Weber State, she has been very active in science outreach activities ranging from star parties to running the state Science Olympiad. Stacy does research in formal and informal astronomy education and the death of Sun-like stars. She spends much of her time thinking, teaching, and writing about the applications of science in everyday life. She then puts that science to use on her small farm in Ogden, Utah.

George Blumenthal is the director of the UC Berkeley Center for Studies in Higher Education. From 2006 through 2019, he was Chancellor at the University of California–Santa Cruz, where he has been a professor of astronomy and astrophysics since 1972. He received his BS degree from the University of Wisconsin–Milwaukee and his PhD in physics from the University of California–San Diego. As a theoretical astrophysicist, George conducts research that encompasses several broad areas, including the nature of the dark matter that constitutes most of the mass in the universe, the origin of galaxies and other large structures in the universe, the earliest moments in the universe, astrophysical radiation processes, and the structure of active galactic nuclei such as quasars. Besides teaching and conducting research, he has served as Chair of the UC–Santa Cruz Astronomy and Astrophysics Department, has chaired the Academic Senate for both the UC–Santa Cruz campus and the entire University of California system, and has served as the faculty representative to the UC Board of Regents.

Understanding Our Universe

FOURTH EDITION

Our Place in the Universe

The location of the sunset changes throughout the year. This pattern can be used as part of the scientific method—which you will learn about in this chapter—to investigate Earth's orbit.

Three times throughout this course (at the beginning, near the middle, and at the end), go outside to the same location and take a picture of the western horizon just after sunset. Before you begin, write down what you expect to see when you compare these images. When you are finished, compare the images and see if you were right!

EXPERIMENT SETUP

PHOTO 1

PHOTO 2

PHOTO 3

Go to the same location and take three photos of the horizon at a time of day when the Sun is just below the horizon. Be sure that you have a stationary object (like a tree) in each photo. Take a photo at the:

1 beginning of semester
2 middle of semester
3 end of semester

PREDICTION

1
2
3

When I compare the three images, I expect to see:

_____

_____

SKETCH OF RESULTS

1

Historically, the science of **astronomy** focused on measuring positions of and finding patterns among the stars. However, modern astronomy seeks the answers to questions that early astronomers could not even imagine being able to answer: What are the Sun and Moon made of? How do stars shine? How did the universe begin? How will it end? Astronomers seek the answers to these and many other compelling questions. Astronomers, like other scientists, use a set of processes, sometimes collectively called the scientific method, to seek and obtain knowledge. The scientific method often begins by recognizing a pattern in nature. Putting those patterns together to understand how they apply in different places and at different times is a primary goal of science. In this chapter, we begin the study of astronomy by exploring our place in the universe and the way modern science is done.

LEARNING GOALS

①　Relate our place in the universe to larger structures in the universe.

②　Explain how science connects the natural patterns that we observe in our daily lives to the laws that govern the universe as a whole.

③　Describe our astronomical origin.

④　Describe the scientific method.

⑤　Interpret and draw conclusions from graphs.

1.1　Astronomy Creates a Universal Context

Astronomers think of our place in the universe as both a location and a time. Locating Earth in the context of the universe is the first step in learning the science of astronomy.

Our Place in Space

Most people have a home address that consists of a street number, street name, city, state, and country. If we expand our view to include the enormously vast universe, however, our "cosmic address" might include our planet, star, galaxy, galaxy group, galaxy cluster, and galaxy supercluster.

We reside on a planet called Earth, which orbits the Sun under the influence of gravity. The distance of Earth from the Sun is such a useful measurement that astronomers make it a unit all its own— the **Astronomical Unit (AU)**. The Sun is an ordinary, middle-aged star, more massive and luminous than some stars but less massive and luminous than others. (Terms in red signify a "Vocabulary Alert" in the margin of the text.) The Sun is extraordinary only because of its importance to us within our own Solar System. Our Solar System consists of eight planets—Mercury, Venus, Earth, Mars, Jupiter, Saturn, Uranus, and Neptune. It also contains many smaller bodies, including *dwarf planets* (such as Pluto, Ceres, and Eris), *asteroids* (such as Ida and Eros), and *comets* (such as Halley).

The Sun is located about halfway out from the center of the *Milky Way Galaxy*, a disk containing more than a hundred billion stars, along with gas and dust. In the

Figure 1.1 Our place in the universe is given by our cosmic address: Earth, Solar System, Milky Way Galaxy, Local Group, Virgo Supercluster, and Laniakea Supercluster. ★ **WHAT AN ASTRONOMER SEES** In this type of figure, an astronomer will be especially sensitive to the arrows, which show that the figure "zooms out" from panel to panel. While each panel is about the same size on the page, they represent dramatically different sizes in space. This figure is representative, without precision, but an astronomer will know that the Laniakea structure in the last panel is much larger—more than 100,000,000,000,000,000 times larger—than Earth, in the first panel. Learning to work with large numbers and ranges of size is one of the challenges of thinking like an astronomer.

last few decades, astronomers have discovered that most of these stars have planets around them, which suggests that planetary systems are common.

The Milky Way, in turn, is part of a small collection of a few dozen galaxies called the Local Group. The Milky Way Galaxy and the Andromeda Galaxy are the two largest galaxies within the Local Group, and each has several galaxies in orbit around it. The next largest galaxy, the Triangulum Galaxy, is significantly smaller than either the Andromeda Galaxy or the Milky Way Galaxy. Most of the approximately 50 other galaxies are much smaller dwarf galaxies. The Local Group itself is part of a vastly larger collection of thousands of galaxies—a supercluster—called the Virgo Supercluster, which is part of an even larger grouping of galaxies called the Laniakea Supercluster.

With this information in hand, we can now define our cosmic address, illustrated in **Figure 1.1**: Earth, Solar System, Milky Way Galaxy, Local Group, Virgo Supercluster, Laniakea Supercluster. Yet this address is incomplete because the vast structure we just described is only the local universe. The part of the universe that we can see extends much farther and includes *several thousand billion galaxies*—about 10 times as many galaxies as there are stars in the Milky Way.

The Scale of the Universe

Defining your cosmic address in this way immediately leads to one of the first challenges you face when learning astronomy: trying to imagine or describe the size of the universe. A hill may be "big," and a mountain may be "really big." If a mountain is really big, then Earth is "enormous," But we run out of words as we try to describe the size of things larger than Earth. We are going to need to use numbers, both for precision and because we simply run out of words. In order to keep those numbers comprehensible, astronomers often talk about distances by talking about the **light travel time**; how long it would take for light to travel that distance. Why do astronomers choose light for this comparison? Because light travels through space at the fastest possible speed, and that speed is always the same.

To understand how astronomers use light travel time to understand distances, think about traveling in a car at 60 kilometers per hour (km/h). At 60 km/h, you travel 60 km every 60 minutes, so in 1 minute you travel 1 km. In 10 hours, you travel 600 km. To get a feel for the difference between 1 km and 600 km,

what if . . .

What if our Solar System were located in the Andromeda Galaxy: What would our cosmic address be then?

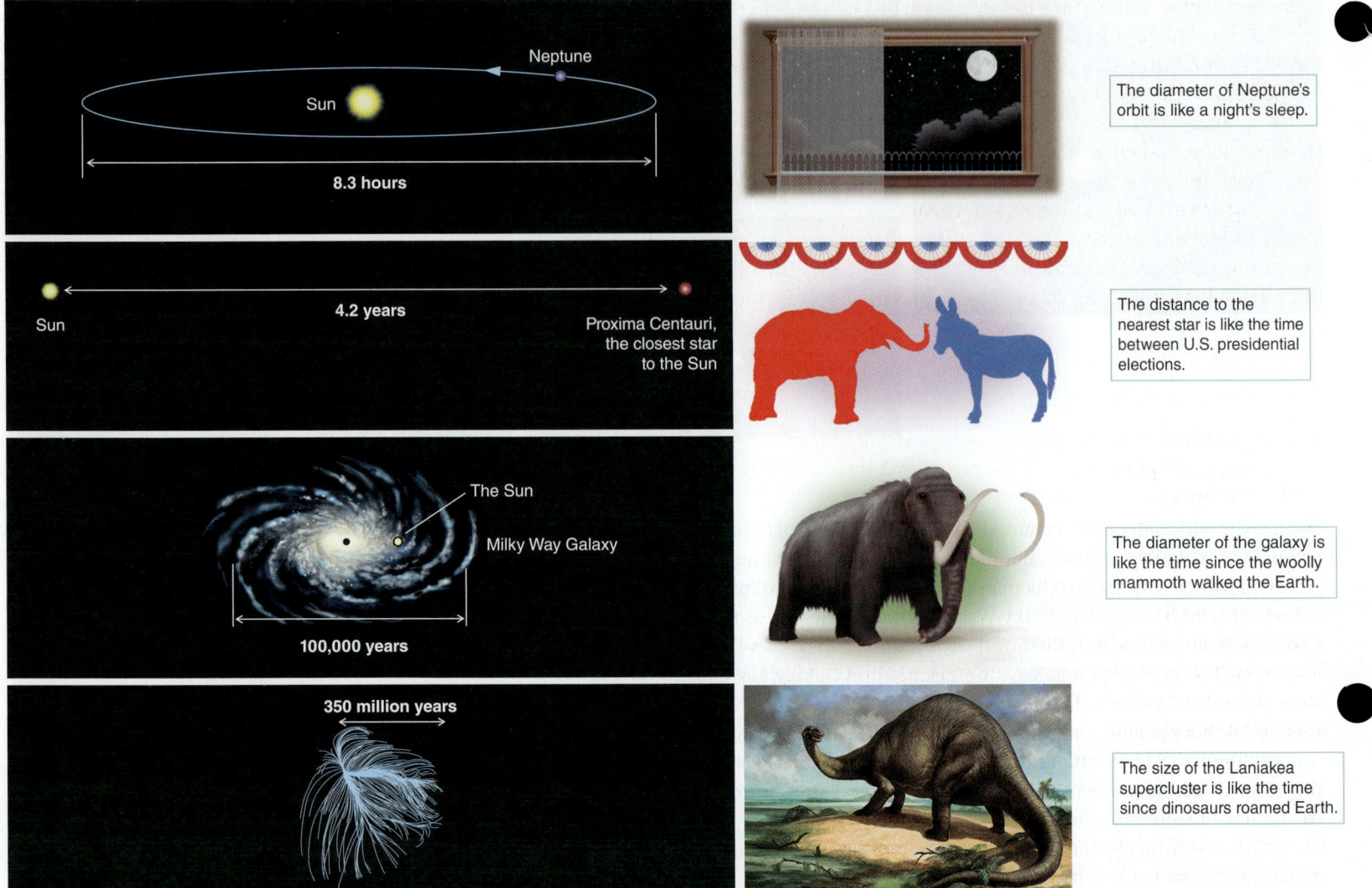

The diameter of Neptune's orbit is like a night's sleep.

The distance to the nearest star is like the time between U.S. presidential elections.

The diameter of the galaxy is like the time since the woolly mammoth walked the Earth.

The size of the Laniakea supercluster is like the time since dinosaurs roamed Earth.

Figure 1.2 Thinking about the time it takes for light to travel between objects helps astronomers to comprehend the vast distances in the observable universe.

you can compare 1 minute (the time to send a brief text) with 10 hours (a long day at work).

In astronomy, the speed of a car is far too slow to be useful. Instead, astronomers use the speed of light. Light travels at 300,000 kilometers per second (km/s). Light can circle Earth (a distance of 40,000 km) in just under one-seventh of a second—about the time it takes you to snap your fingers. Distances within the Solar System are less than a **light-day**, which is the distance light travels in a day. Compare this to the distance to the nearest star, which is measured in **light-years (ly)**, or the distance light travels in a year. Using this way of thinking about distances, **Figure 1.2** progresses outward through the observable universe. Distances in astronomy are so vast that, even traveling at the fastest possible speed, it takes far more than a human lifetime to travel across just one galaxy.

what if . . .

What if Earth formed 4 billion years after the Big Bang, rather than 9.3 billion years after it: How might we expect our planet to be different?

Our Place in Time

The universe began 13.8 billion years ago in an event known as the *Big Bang*. The only chemical elements in the early universe were hydrogen and helium, plus tiny amounts of lithium, beryllium, and possibly boron. Gravity caused clouds of these elements to collapse into stars. In the cores of these stars, light elements like hydrogen combined to form more massive elements, like carbon or silicon. When a star neared the end of its life, it lost much of its material back into space. In some high-energy explosions of stars, even more massive elements were formed. Material from many stars formed large clouds of dust and gas. Those clouds made new stars and planets, such as our Sun and Solar System, which formed 4.5 billion years ago. Therefore, prior "generations" of stars supplied the building blocks for the world you see around you today. Look around you. Everything you see contains atoms of elements that were formed in stars during the first two-thirds of the history of the universe. The Sun and Earth have only been around for the most recent third. Humans are latecomers on Earth, appearing only in the last 0.04 percent of the history of Earth.

An Astronomer's Toolkit

Telescopes often come to mind when we think of astronomers. However, the 21st-century astronomer spends far more time staring at a computer screen than peering through the eyepiece of a telescope. In part, this is because modern astronomers observe many varieties of light, not just the *visible light* that human eyes can see. From the highest-energy *gamma rays* and *X-rays*, through *ultraviolet*, *visible*, and *infrared* radiation, down to the lowest-energy *radio waves*, each type of light carries different information about the universe. **Figure 1.3** shows the Parkes radio telescope as you would see it with your eyes, combined with an image of the Milky Way Galaxy taken with a telescope that detects radio waves, which you could *not* see with your eyes. Modern astronomers use different types of telescopes to observe different kinds of light.

Astronomers use computers to collect and analyze data from telescopes, test and refine physical models of astronomical objects, and prepare reports on the results of their work. Astronomers also use computers to communicate and work with scientists in related fields, such as physics, chemistry, geology, and planetary science, to develop a deeper understanding of physical laws and to make sense of their observations of the distant universe.

In 1957, the Soviet Union launched Sputnik, the first human-made **satellite**, into orbit above the atmosphere. This successful launch ushered in a new era of space exploration (as well as the Space Race), which has given us a new perspective on the universe. Earth's atmosphere blocks much of the light that travels through space. Space telescopes, however, show views hidden from ground-based telescopes by Earth's atmosphere. Astronomers have used space telescopes to make discovery after surprising discovery, all of which were impossible before the space age.

Most of what we know about the Solar System has resulted from the past six decades of exploration since the space age began. Since 1957, 12 people have walked on the Moon (**Figure 1.4**), and dozens of unmanned probes have visited all of the planets. Spacecraft have flown past asteroids, comets, and the Sun. Spacecraft have also landed on Mars, Venus, Titan (Saturn's largest moon), an asteroid, and a comet. One spacecraft brought samples of a comet's tail back to Earth to be studied. In the past few years, spacecraft have begun probing the outer reaches of the Solar System. NASA's *New Horizons* has visited Pluto and Ultima Thule, objects beyond the orbit of Neptune. The spacecraft *Voyager 1* has flown beyond the edge of the Sun's influence on interstellar space and is now farther from Earth than any other human-made object.

Figure 1.3 This image of the Milky Way Galaxy shows dust and gas structures, some of which will form new stars. This is the Milky Way Galaxy as we would see it if our eyes were sensitive to radio waves, shown as a backdrop to the Parkes radio telescope in Australia.

(a)

(b)

Figure 1.4 Exploring the universe has included travels in the Solar System. (a) *Apollo 15* (1971) was the fourth U.S. mission to land on the Moon. Here, astronaut James B. Irwin stands by the lunar rover during an excursion to explore and collect samples from the Moon. (b) Robotic missions have visited all the planets (as well as some other objects). The *Curiosity* rover even took "selfies" on the surface of Mars.

VOCABULARY ALERT

satellite In common language, *satellite* typically refers to a human-made object. Astronomers use this word to describe any object, human-made or natural, that orbits another object.

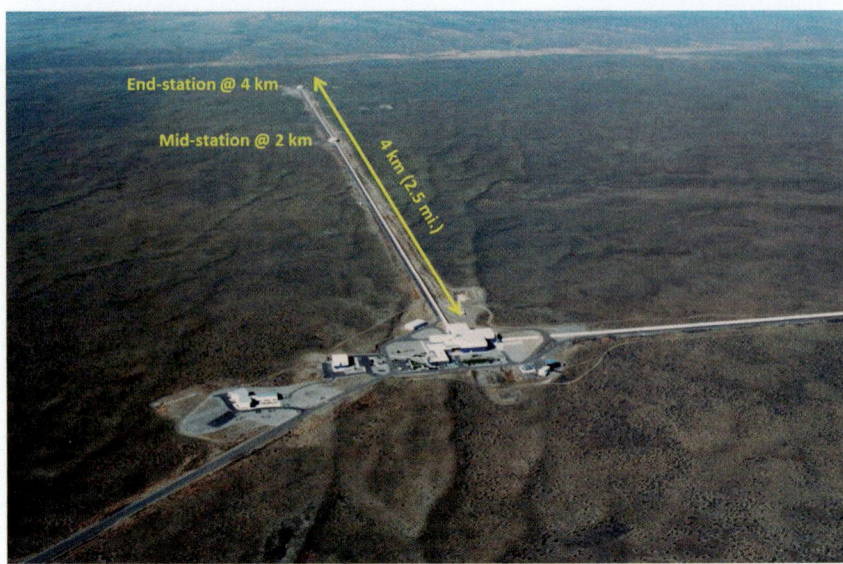

Figure 1.5 The Laser Interferometer Gravitational-Wave Observatory (LIGO) was used to make the first detection of gravitational waves in 2016, a century after they were predicted to exist. The station shown is one of two such instruments that together make up the LIGO facility. The instrument is so large that fitting it into one frame is difficult, and only one "arm" is shown completely here. The arm is so long that perspective distorts the distance between the evenly spaced mid- and end-stations. Gravitational waves are a new way of observing the universe.

Most recently, the detection of gravitational waves from colliding black holes and neutron stars has opened another window on the universe that astronomers are just beginning to explore. Observing gravitational waves requires enormous physics facilities such as the one shown in **Figure 1.5**. Astronomers also often use other large physics facilities like the high-energy particle collider at the European Organization for Nuclear Research (CERN), to learn about interactions among tiny particles. The 21st century is an exciting time to be an astronomer!

CHECK YOUR UNDERSTANDING **1.1**

Rank the following in order from smallest to largest: Sun, Laniakea Supercluster, Earth, Solar System, Local Group, Milky Way Galaxy, Universe.

Answers to Check Your Understanding questions are in the back of the book.

1.2 Science Is a Way of Viewing the World

As scientists, astronomers use the methods, principles, and concepts of science to explore the physical world beyond Earth. Many of these principles and methods are collectively known as the **scientific method**: a systematic, logical process for collecting evidence and using it to test ideas or explanations. Any idea that cannot be tested is not scientific; you choose to accept or reject it on some basis other than evidence.

The Language of Science

You have already seen that scientists often use everyday words, like "massive," in special ways. One source of common confusion is the word "theory." In everyday language, the word *theory* may mean an idea that is little more than a guess: "Do you have a theory about who might have done it?" "My theory is that a third party could win the next election." In everyday language, a theory isn't taken too seriously. "After all," one may say, "it is only a theory."

In stark contrast, scientists use the word **theory** to mean a carefully constructed proposition that takes into account every piece of data as well as our entire understanding of how the world works. A theory has been used to make testable predictions, and all of those predictions have been correct. Every attempt to prove the theory false has failed. A theory such as Einstein's general theory of relativity is a crowning achievement of humankind. Even so, scientific theories are accepted only as long as their predictions are correct. A theory that fails a single test is proved false. In this sense, all scientific knowledge is subject to challenge.

Theories are at the top of the loosely defined hierarchy of scientific knowledge. At the bottom is an *idea*—an untested notion about how something might be. Moving up the hierarchy we come to a *fact*, which is an observation or measurement. For example, the statement that "the radius of Earth is 6,371 km" is a statement of fact. A *hypothesis* is an idea that leads to testable predictions. At the top we reach a *theory*: an idea that has been examined carefully, is consistent with all existing theoretical and experimental knowledge, and makes testable predictions. Ultimately, the success of the predictions is the deciding factor between competing hypotheses.

The term "law" is another that is used differently in everyday language. To scientists, a *law* is a series of observations that scientists can use to make predictions, but a law has no underlying explanation of why the phenomenon occurs. For example, a "law of daytime" might be that the Sun rises and sets once each day. A corresponding "theory of daytime" is that the Sun rises and sets once each day because Earth spins on its axis. Scientists are sometimes sloppy about the way they use these words, and you will sometimes see them used differently than in these formal definitions.

Underlying this hierarchy of knowledge are scientific principles. A scientific principle is a generalization about the universe that guides scientists in the construction of new hypotheses. For example, at the heart of modern astronomy is the cosmological principle. The **cosmological principle** includes the testable assumption that physical laws that apply here and now also apply everywhere and at all times. In other words, the physical laws that act in laboratories also act in the centers of stars or in the hearts of distant galaxies. Each new theory that succeeds in explaining patterns and relationships among objects in the sky adds to our confidence that this principle is correct.

Occam's razor is a guiding principle in science in general. **Occam's razor** states that when we are faced with competing hypotheses that explain all the observations equally well, we should use the one that requires the fewest assumptions. For example, we might hypothesize that atoms are constructed differently in the Andromeda Galaxy than in the Milky Way Galaxy. To accept this hypothesis would require a large number of assumptions about how the atoms in the Andromeda Galaxy may be constructed differently and yet appear to behave identically to atoms in the Milky Way. For example, we might assume that the center of the atom is negatively charged in Andromeda, opposite to that in the Milky Way, where the center of the atom is positively charged. This would require us to form an assumption about where the boundary is between Andromeda-like matter and Milky Way-like matter. We would need to make a further assumption about why atoms on the boundary between the two regions did not destroy each other. And we would need an assumption about *why* atoms in the two regions are constructed so differently, and so on. It is far simpler to assume that the atoms in the Andromeda Galaxy which appear to behave like atoms in the Milky Way Galaxy actually *are* like the atoms in the Milky Way Galaxy.

Principles are not independent of one another. If reasonable experimental evidence is ever found that the cosmological principle is not true, scientists will construct a new description of the universe that takes the new data into account. Until then, it is the hypothesis that has the fewest assumptions, satisfying Occam's razor. To date, the cosmological principle has been repeatedly tested and remains unfalsified.

The Scientific Method

The process of developing and testing hypotheses is often called "the scientific method," as shown in the simplified diagram in **Figure 1.6**. The method often begins with a **fact**—an observation or a measurement. For example, you might observe that the weather changes in a predictable way each year and wonder why that happens. You then create a **hypothesis**, a testable explanation of the observation: "I think that it is cold in the winter and warm in the summer because Earth is closer to the Sun in the summer." You think of a **test**: a prediction of the hypothesis that

what if . . .

What if we found evidence that the cosmological principle is wrong because there is a big region in the universe where the laws of physics are different from the ones on Earth: How should we then think about the laws of physics in yet other parts of the universe?

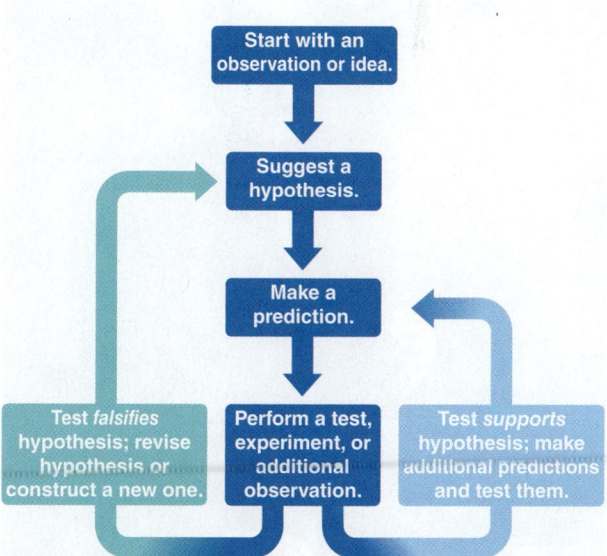

Figure 1.6 The scientific method is the path by which an idea or observation leads to a falsifiable hypothesis. The hypothesis is either accepted as a tested theory or rejected on the basis of observational or experimental tests of its predictions. The process goes on indefinitely as scientists continue to test the hypothesis.

what if . . .

What if a theory is not likely to be falsifiable in many human lifetimes: Might it still be a scientific theory?

Figure 1.7 Einstein helped usher in two different scientific revolutions: relativity and quantum mechanics.

can be measured or observed. In this case, if your hypothesis is correct, then it will be cold everywhere on the planet in January—Australia (in the Southern Hemisphere) should have winter at the same time of year as the United States (in the Northern Hemisphere). You travel to the opposite hemisphere in the winter to test your hypothesis. You find that the seasons are not the same; it is summer there when it is winter back home. Your hypothesis has just been **falsified**, which means that it has been proved incorrect. This is good! It means you now know something you didn't know before. It also means you must revise or completely change your hypothesis to make a new hypothesis that is consistent with the new information you have gained.

Further observations or experiments test the hypothesis repeatedly. A falsifiable, scientific hypothesis or idea does not have to be testable using current technology, but we must be able to imagine an experiment or observation that *could* prove the idea wrong if we could carry it out. Perhaps this process, over time, results in a well-tested idea. Perhaps the idea is rejected in favor of one that better describes the outcome of a test. Notice that there is no end point to this method; every idea undergoes repeated testing over time. Few ideas stand up to this rigorous testing to become accepted as scientific theories.

The scientific method provides the rules for testing whether an idea is false, but it offers no insight into where the idea came from in the first place or how an experiment was designed. Scientists discussing their work use words such as *insight*, *intuition*, and *creativity*. Scientists speak of a beautiful theory in the same way that an artist speaks of a beautiful painting or a musician speaks of a beautiful performance. Science has an aesthetic that is as human and as profound as any found in the arts.

Scientific Revolutions

Limiting the discussion of science to the existing theories fails to convey the dynamic nature of scientific inquiry. Scientists do not have all the answers and must constantly refine their ideas in response to new information and new insights. The vulnerability of knowledge may seem like a weakness. "Gee, you really don't know anything," the cynical person might say. But this vulnerability is actually science's great strength, because it means that science is self-correcting. Wrong ideas are eventually overturned by new information. In science, even our most cherished ideas about the nature of the physical world are subject to challenge by new evidence. Many of history's best scientists earned their status by falsifying a universally accepted idea. This is a powerful motivation for scientists to challenge old ideas constantly, rather than simply accepting what came before.

For example, the classical physics developed by Sir Isaac Newton in the 17th century withstood the scrutiny of scientists for more than 200 years. It seemed that only small details remained to be explained. Yet during the late 19th and early 20th centuries, a series of scientific revolutions completely changed our understanding of the nature of reality. Albert Einstein (**Figure 1.7**) is responsible for some of these scientific revolutions. Einstein's new ideas unified the concepts of mass and energy and destroyed the conventional notion of space and time as separate entities. Einstein showed that Newton's theories were a special case of a far more general and powerful set of physical laws, and Einstein's special and general theories of relativity replaced Newton's mechanics. In practice, it's often far easier (and sufficiently accurate) to use Newton's mechanics to solve problems, so these approaches are still used in "everyday" situations where masses are small and speeds are slow.

Throughout this text, you will encounter many other discoveries and successful ideas that forced scientists to abandon or modify accepted theories. Einstein himself never embraced the view of the world offered by *quantum mechanics*—a second revolution he helped start. Yet quantum mechanics, a statistical description of the behavior of particles smaller than atoms, has held up to challenges for more than 100 years. In science, all authorities are subject to challenge, even Einstein.

CHECK YOUR UNDERSTANDING **1.2**

If we compare our place in the universe with a very distant place, the cosmological principle tells us that

a. the laws of physics are different in each place.

b. some laws of physics are different in each place.

c. all of the laws of physics are the same in each place.

d. some laws of physics are the same in each place, but we don't know about others.

1.3 Astronomers Use Mathematics to Find Patterns

One of the primary advantages of scientific thinking is that it allows us to make predictions. Once a pattern, such as the rising and setting of the Sun, has been observed, scientists can predict what will happen next.

Finding Patterns

Imagine that the natural patterns of life became disrupted. For example, suppose that one day the Sun rose at noon and set at 1:00 P.M., the next day it rose at 6:00 A.M. and set at 10:00 P.M., and the day after that the Sun did not rise at all. Fortunately, the Sun rises, sets, and then rises again at predictable times and in predictable locations. The behavior of the natural world produces patterns in our lives, and these patterns can be used to find out about the nature of the physical world.

Astronomers identify and characterize these patterns and use them to understand the movements of the world around us. The patterns of change in the night sky are some of the most consistent patterns in nature. How will the sky look different a week from now? A month from now? A year from now? As you can see in **Figure 1.8**, patterns of stars in the sky change with the changing of seasons. These patterns of stars can be used to determine when to plant or harvest crops. A familiarity with the night sky has historically been a matter of life or death for human cultures.

Reading Graphs

Astronomers use mathematics to analyze patterns. Mathematical patterns are often shown in graphical form. Reading graphs is a skill that is important not only in astronomy but also in everyday life. Economists, social and political scientists, mortgage brokers, financial analysts, retirement planners, doctors, and scientists all use graphs to evaluate and communicate important information. The graphs you will be working with typically have two axes: the horizontal (x) axis shows the *independent variable*, while the vertical (y) axis shows the *dependent variable*. The independent variable is typically the one the experimenter has control over. Often,

Figure 1.8 Since ancient times, our ancestors have recognized that patterns in the sky change with the seasons. These and other patterns shape our lives. These star maps show the night sky in the Northern Hemisphere during each season.

the independent variable is time: The experimenter chooses the times at which she will make a measurement. The dependent variable *depends* on the independent variable. For example, the experimenter may measure the dependent variable at various (chosen) times.

The relationship between distance and time during a car trip is shown in **Figure 1.9**. Whenever you see a graph, first look at the axes to find out what information is plotted on the graph and in what units. In Figure 1.9, the distance, measured in miles, is plotted against time, measured in minutes. The distance depends on the time at which it is measured.

Trends in plotted data indicate a pattern in the behavior of the system. Study the graph of Figure 1.9. You can see that the data point labeled (a) is plotted at 10 minutes on the horizontal axis and 5 miles on the vertical axis. During those first 10 minutes, the distance grew steadily larger, as the car traveled 5 miles from its starting place. The distance is increasing with time, so the line goes upward.

The slope of a line is calculated by dividing the *rise* (change in the vertical axis) by the *run* (change on the horizontal axis). To get a feel for what's happening with a pattern, you need to understand the slope. A horizontal line has zero slope, which in this case would mean that over some amount of time the distance is not changing (the car is parked); a rising line has positive slope, which in this case would mean that the distance is increasing (the car is moving away from where it started); and a falling line has negative slope, which means that the distance is getting smaller (the car is approaching where it started). A steeper slope means the car is moving faster, either toward or away from where it started. For this type of graph, the slope of the line is equal to the speed of the car.

During the second 10 minutes, between points (a) and (b), the car travels 7 miles. Notice that the line rises more sharply—the slope is steeper than during the first 10 minutes. This steeper slope tells you that the car is going faster.

After 20 minutes, the car begins to slow down. Initially, the distance is still increasing, because the plotted data are still rising. The slope becomes flatter, however, until the line becomes horizontal after point (c), which occurs at 25 minutes. After this time, the car is not moving.

You will encounter graphs often in your life. Always remember to look at these things when you are given a new graph: What is the independent variable? What is the dependent variable? What is the trend of the data (up, down, or flat), and how fast is it changing (steep, shallow, or flat)?

If patterns are the heart of science, and mathematics is the language of patterns, it should come as no surprise that *mathematics is the language of science.* Trying to study science while avoiding mathematics is like trying to study Shakespeare while avoiding the written or spoken word. It cannot be done meaningfully.

Becoming comfortable with mathematics will open up new possibilities for you. Part of the responsibility for making you comfortable with the language of mathematics lies with us, the authors. It is our job to take on the role of translators, using words to express concepts as much as possible, even when these concepts are more concisely and accurately expressed mathematically. When we do use mathematics, we will explain in everyday language what the equations mean and show you how equations express concepts that you can connect to the world. We will also limit the mathematics to a few tools that we would like you to know. These mathematical tools, a few of which are described in **Working It Out 1.1**, enable scientists to convey complex information compactly and accurately.

Your responsibility is to accept the challenge and make an honest effort to work through the mathematical concepts that we use. The mathematics in this book is comparable with what it takes to calculate a sale price, determine if you can afford a mortgage, check your gas mileage, estimate how long it will take to drive to another city, or buy enough food to feed an extra guest or two at dinner.

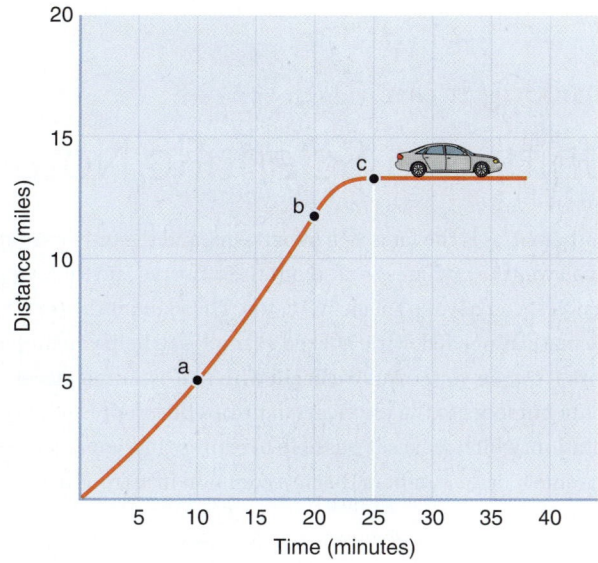

Figure 1.9 This graph of the relationship between distance and time in a car trip shows changes in speed.

CHECK YOUR UNDERSTANDING 1.3

In a graph of distance versus time (like the one in Figure 1.9), a line that sloped from upper left to lower right would mean that

a. the car is approaching.

b. the car is slowing down.

c. the car is moving away but is behind you.

d. the car is speeding up but is behind you.

working it out 1.1

Units and Scientific Notation

Mathematics is the language of patterns, and the universe "speaks" math. If you want to ask questions about the universe and understand the answers, you have to speak math, too. Throughout the text, we present the math as needed, with worked examples to help you understand how to work it out on your own. We begin with two essential pieces: units, which relate numbers to the physical quantities they represent, and scientific notation, which makes it possible to express in a compact manner the extremely large numbers that are common in astronomy.

Units

Scientists use the metric system of units because most metric units are related to each other by multiplying or dividing by 10 and therefore simply involve moving the decimal place to the right or left. Metric measurements also have prefixes that show the relationship of the units. There are 100 *centi*meters (cm) in a meter, 1,000 meters in a *kilo*meter (km), and 1,000 grams in a *kilo*gram (kg). (A more complete listing of metric prefixes is located inside the front cover of this book.) To convert from meters to centimeters, multiply by 100 by moving the decimal place two spaces to the right. For example, 25 meters is $25 \times 100 = 2,500$ cm. But within the English system of units, to convert from yards to inches you must first multiply by 3 to convert from yards to feet and multiply that result by 12 to convert feet to inches.

Though many of the units we use will be unfamiliar, you can start to develop an intuition about them. For example:

1 meter is about 3 feet (ft).

1 cm is a bit less than half an inch; 1 inch is about 2.5 cm.

1 km is approximately two-thirds of a mile.

- **Approximately how many feet are in 2.5 meters?** Because 1 meter is about 3 ft, 2.5 meters is about $3 \times 2.5 = 7.5$ ft. This is about the diameter of the Hubble Space Telescope's primary mirror.

- **Approximately how many centimeters are in a foot?** We can estimate this value in two ways. We know that 1 ft is 12 inches, and 1 inch is about 2.5 cm. Therefore, 1 ft = 12 inches $\times$ 2.5 cm/inch $\approx$ 30 cm. Here is another way to calculate an approximate conversion. Because 1 meter is approximately 3 ft, 1 ft is approximately $\frac{1}{3}$ meter. We know that 1 meter is 100 cm, so there are roughly 33 cm in a foot. If we measure precisely, we find that 1 ft is 30.48 cm, which falls between our two estimates. The telescope used to discover Pluto was just over a foot in diameter.

- **How can you check your own work?** One way to check your own work is to think about whether the answer should be larger or

smaller than the original number. The number of centimeters, for example, should always be larger than the number of meters because there is more than 1 cm in a meter. Often, a quick estimate can also help. For example, 1 ft is about an "order of magnitude" (a power of 10) larger than a centimeter; in other words, there are a few tens—not hundreds—of centimeters in a foot. So if you have converted 1.2 ft to centimeters and calculated an answer of 3,200 cm, you should try again. You can also include the units in every step of the calculation, and at each step, strike out the ones that cancel. For example, if you are multiplying 6 yards by 3 feet per yard, the unit *yards* appears in the numerator and the denominator and divides out.

$$6 \text{ yards} \times \frac{3 \text{ feet}}{1 \text{ yard}} = 18 \text{ feet}$$

This can help you remember if you should multiply or divide by the conversion factor; you want some of the units to cancel.

Scientific Notation

Scientists deal with numbers of vastly different sizes by using scientific notation. Writing out 7,540,000,000,000,000,000,000 in standard notation is tedious and very inefficient. Scientific notation uses the first few digits (the *significant* ones) and counts the number of decimal places to create the condensed form 7.54×10^{21}. Similarly, rather than writing out 0.000000000005, we write 5×10^{-12}. In scientific notation, the exponent on the 10 tells you where to move the decimal point. If it is positive, move the decimal point that many places to the right. If it is negative, move the decimal point that many places to the left.

- **Write the number 13.8×10^9 in standard notation.** Because the exponent of the 10 is 9, and it is positive, we must move the decimal point nine places to the right: 13,800,000,000. This number is the age of the universe in years. Usually, we would say "thirteen point eight billion years," where the word *billion* stands in for "times 10 to the 9."

- **Write the number 2×10^{-10} in standard notation.** The exponent of the 10 is –10. It is negative, so we must move the decimal point 10 places to the left: 0.0000000002. This is about the size of an atom in meters.

Calculator hint: How do you put numbers in scientific notation into your calculator? Most scientific calculators have a button that says EXP or EE. These mean "times 10 to the." So for 4×10^{12}, you would type [4][EXP][1][2] or [4][EE][1][2] into your calculator. Usually, this number shows up in the window on your calculator either just as you see it written in this book or as a 4 with a smaller 12 all the way over on the right of the window.

reading Astronomy News

Breakthrough: Scientists Detect Einstein's Gravity Ripples

by Seth Borenstein, AP Science Writer

This article from the Associated Press was reproduced in small newspapers across the country, because the findings were revolutionary.

WASHINGTON—In an announcement that electrified the world of astronomy, scientists said Thursday that they have finally detected gravitational waves, the ripples in the fabric of space-time that Einstein predicted a century ago.

Astronomers hailed the finding as an achievement of historic proportions, one that opens the door to a new way of observing the universe and the violent collisions that are constantly shaping it. For them, it's like turning a silent movie into a talkie because these waves are the soundtrack of the cosmos in action.

"Until this moment, we had our eyes on the sky and we couldn't hear the music," said Columbia University astrophysicist Szabolcs Marka, a member of the discovery team. "The skies will never be the same."

An all-star international team of astrophysicists used an excruciatingly sensitive, $1.1 billion set of twin instruments known as the Laser Interferometer Gravitational-wave Observatory, or LIGO, to detect a gravitational wave generated by the collision of two black holes 1.3 billion light-years from Earth.

"Einstein would be beaming," said National Science Foundation director France Cordova.

To make the finding easier to comprehend, the scientists converted the wave into sound. At a news conference, they played a recording of a single chirp—the signal they picked up on Sept. 14. It was barely perceptible even when enhanced.

"That's the chirp we've been looking for," said Louisiana State University physicist Gabriela Gonzalez, scientific spokeswoman for the LIGO team. Scientists said they hope to have a greatest hits compilation of the universe in a decade or so.

Some physicists said the finding is as big a deal as the 2012 discovery of the subatomic Higgs boson, known as the "God particle." Some said this is bigger.

"It's really comparable only to Galileo taking up the telescope and looking at the planets," said Penn State physics theorist Abhay Ashtekar, who wasn't part of the discovery team. "Our understanding of the heavens changed dramatically."

Gravitational waves, first postulated by Albert Einstein in 1916 as part of his theory of general relativity, are extraordinarily faint ripples in space-time, the continuum that combines both time and three-dimensional space.

When massive objects like black holes or neutron stars collide, they send gravitational waves across the universe, stretching space-time or causing it to bunch up like a fishing net.

Scientists found indirect proof of gravitational waves in the 1970s by studying the orbits of two [orbiting] stars, and the work was honored as part of the 1993 Nobel Prize in Physics. But now scientists can say they have actually detected a gravitational wave.

"It's one thing to know soundwaves exist, but it's another to actually hear Beethoven's Fifth Symphony," said Marc Kamionkowski, a physicist at Johns Hopkins University who wasn't part of the discovery team. "In this case, we're actually getting to hear black holes merging."

In this case, the crashing of the two black holes stretched and squished Earth so that it was "jiggling like Jell-O," but in a tiny, almost imperceptible way, said David Reitze, LIGO's executive director.

The dual LIGO detectors went off just before 5 a.m. in Louisiana and e-mails started flying. "I went, 'Holy moly,'" Reitze said.

But the finding had to be tested and verified, using conventional telescopes, before the scientists could say with confidence that it was a gravitational wave. They concluded there was less than a 1-in-3.5-million chance they were wrong, he said.

LIGO technically wasn't even operating in full science mode; it was still in the testing phase when the signal came through, Reitze said.

"We were surprised, BOOM, right out of the box, we get one," Reitze said.

Reitze said that given how quickly they found their first wave, scientists expect to hear more of them, maybe even a few per month.

Detecting gravitational waves is so difficult that when Einstein first theorized about them, he figured scientists would never be able to hear them. In fact, the greatest scientific mind of the 20th century came to doubt himself in the 1930s and questioned whether such waves really do exist.

In 1979, the National Science Foundation decided to give money to the California Institute of Technology and the Massachusetts Institute of Technology to come up with a way to detect the waves.

Twenty years later, they started building two LIGO detectors in Hanford, Washington, and Livingston, Louisiana, and they were turned on in 2001. But after years with no luck, scientists realized they had to build a much more sensitive system, which was turned on last September.

Sensitivity is crucial because the stretching and squeezing of space-time by gravitational waves is incredibly tiny. Essentially, LIGO detects waves that pull and compress the entire Milky Way Galaxy "by the width of your thumb," Hanna said.

Each LIGO has two giant perpendicular arms more than 2 miles long. A laser beam is split and travels both arms, bouncing off mirrors to return to the arms' intersection. Gravitational waves stretch the arms to create an incredibly tiny mismatch—smaller than a subatomic particle—in the beams' signature wave curves. That mismatch is what LIGO detects.

A giant team of scientists had to keep the discovery secret until it was time to be announced. The study detailing the research in the journal *Physical Review Letters* had 1,004 authors.

Kip Thorne, the Caltech physicist who co-founded LIGO and has been working on gravitational waves for more than half a century, said he kept the secret even from his wife until just a few days ago. When he heard about the wave, he said, "it was just sort of a sigh of happiness."

EVALUATING THE NEWS

1. Compare this story to the flowchart in Figure 1.6. Which box on the flowchart corresponds to the result described here?

2. How much time passed between Einstein's prediction of gravitational waves and their detection? Place this time in context of the discussion about "falsifiability." Is it necessary for a theory to be falsifiable with current technology at the time when it is formulated?

3. Once the second version of the instrument was built, how much time elapsed between turning it on and detecting gravitational waves? What does this tell you about how often gravitational waves might be detected in the future?

4. Study the graph in **Figure 1.10.** What is the independent variable?

5. In Figure 1.10, what is the dependent variable?

6. How much time is shown in the observations of Figure 1.10? Is that a long or a short amount of time in which to make an observation?

7. The y-axis shows the *strain*. A strain of 10^{-21} tells you that each kilometer of an arm of the instrument stretches by 10^{-21} km as the gravitational wave passes by. If an arm is 4 km long, how far does it stretch all together? The radius of a proton is accepted to be roughly 10^{-18} km. How does the amount of stretch in the arm of the detector compare to the size of a proton?

Source: **Seth Borenstein**, "Breakthrough: Scientists Detect Einstein's Gravity Ripples," from *Standard Examiner*, February 11, 2016. Reprinted with permission of The Associated Press via Wright's Media.

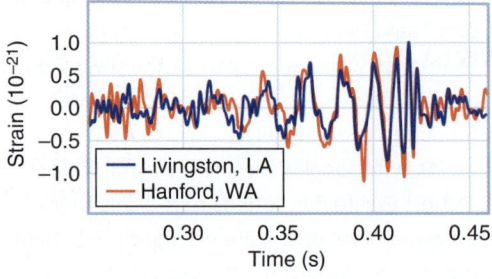

Figure 1.10 The announcement of the discovery of gravitational waves was accompanied by this graph. Even without fully understanding how we detect gravitational waves, you can infer from the very small numbers on the vertical axis that this was an incredibly difficult measurement to make.

SUMMARY

Astronomers employ the scientific method to seek answers to many compelling questions about the universe by studying many different kinds of patterns in the sky. Science is based on objective reality, physical evidence, and testable hypotheses. The great strength of science lies in its ability to continually evolve as new information becomes available.

(1) Earth is a small planet orbiting a more or less average star, halfway out in the disk of a galaxy that is one among billions.

(2) The physical laws that apply on Earth apply everywhere else in the universe.

(3) Two-thirds of the age of the universe passed before the Sun formed 4.5 billion years ago. Humankind has been around for a *very* short amount of time compared to the Sun or the universe.

(4) The scientific method is an approach to learning about the physical world. It includes observation, forming hypotheses, making predictions to enable the testing and refining of those hypotheses, and repeated testing of theories.

(5) Scientists often represent information mathematically, and especially graphically. Reading a graph means looking carefully at the axes and the trends in the data, which can often be found by calculating a slope.

QUESTIONS AND PROBLEMS

TESTING YOUR UNDERSTANDING

1. **T/F:** The Solar System is another name for our galaxy.

2. **T/F:** A scientific theory can be tested by observations and proved to be true.

3. **T/F:** A pattern in nature can reveal an underlying physical law.

4. **T/F** A theory, in science, is a guess about what might be true.

5. **T/F:** A flat slope on a graph means the dependent variable is not changing.

6. The following astronomical events led to the formation of you. Place them in order of their occurrence over astronomical time.
   a. Stars die and distribute heavy elements into the space between the stars.
   b. Hydrogen and helium are made in the Big Bang.
   c. Enriched dust and gas gather into clouds in interstellar space.
   d. Stars are born and process light elements into heavier ones.
   e. The Sun and planets form from a cloud of interstellar dust and gas.

7. The Solar System contains
   a. planets, dwarf planets, asteroids, and galaxies.
   b. planets, dwarf planets, comets, and billions of stars.
   c. planets, dwarf planets, asteroids, comets, and one star.
   d. planets, dwarf planets, one star, and many galaxies.

8. The Sun is part of
   a. the Solar System.
   b. the Milky Way Galaxy.
   c. the universe.
   d. all of the above

9. A light-year is a measure of
   a. distance.
   b. time.
   c. speed.
   d. mass.

10. Earth is 1 astronomical unit from the Sun. Light covers that distance in 8.3 minutes. Mars is 1.5 astronomical units from the Sun. How long does it take for light to get to Mars from the Sun?
    a. 4.15 minutes
    b. 8.3 minutes
    c. 12.45 minutes
    d. 16.6 minutes

11. Which of the following was *not* made in the Big Bang?
    a. hydrogen
    b. lithium
    c. beryllium
    d. carbon

12. The fact that scientific revolutions take place means that
    a. all the science we know is wrong.
    b. the science we know now is more correct than it was in the past.
    c. all knowledge about the universe is relative.
    d. the laws of physics that govern the universe keep changing.

13. Occam's razor states that
    a. the universe is expanding in all directions.
    b. the laws of nature are the same everywhere in the universe.
    c. if two hypotheses fit the facts equally well, we should choose the simpler one.
    d. patterns in nature are really manifestations of random occurrences.

14. According to the cosmological principle,
    a. on a large scale, the universe is the same everywhere at a given time.
    b. the universe is the same from one time to another.
    c. our location is special.
    d. all of the above are correct

15. There are _____ nanometers (10^{-9} meters) in 1 micrometer (10^{-6} meters).
    a. 10
    b. 100
    c. 1,000
    d. 1,000,000

16. Place the following in order from largest to smallest.
    a. micrometer
    b. kilometer
    c. millimeter
    d. megameter

17. In Figure 1.9 ⭐, during the time period from 20 to 25 minutes, the car
    a. is traveling away from the starting point and speeding up.
    b. is traveling away from the starting point and slowing down.
    c. is traveling toward the starting point and speeding up.
    d. is traveling toward the starting point and slowing down.

18. Suppose that you see a graph representing the relationship between the value of a car you own and the time you own it. In this case, _____ would be plotted on the horizontal axis, and _____ would be plotted on the vertical axis.
    a. time; car value
    b. car value; time
    c. dollars; car value
    d. time; dollars

19. In astronomy, we often use a *light curve* to investigate the properties of stars. A light curve shows how the brightness of the object changes as time passes. In this case, _____ would be plotted on the horizontal axis, and _____ would be plotted on the vertical axis.
    a. time; brightness
    b. brightness; time
    c. the name of the star; brightness
    d. time; the name of the star

20. In astronomy, we often use a *light curve* to investigate the properties of stars. A light curve shows how the brightness of the object changes as time passes. Suppose the brightness increased as time passed. In this case, you would expect to see that a line through the data on the graph
    a. was horizontal.
    b. dropped from upper left to lower right.
    c. rose from lower left to upper right.
    d. was vertical.

THINKING ABOUT THE CONCEPTS

21. ★ **WHAT AN ASTRONOMER SEES** In Figure 1.1, the figure repeatedly "zooms out" to show larger and larger segments of space. Compare the steps shown in this figure with the size of the steps shown in Figure 1.2. Each step of "zoom" shown in Figure 1.1 does not show the same increase in size. Why was this figure drawn this way? ⊛

22. Draw a diagram representing your cosmic address. How can you best show the difference in size scales represented in the components of this address? How do these differences in scale compare to the differences in scale in your postal address?

23. Figure 1.2 shows examples of timescales that are analogous to the distance scales in the universe. Pick one of the distances in the figure and think of a different analogy. That is, think of a different activity that takes the same amount of time. ⊛

24. According to Figure 1.2, how long does it take us to know about an event that occurs on the Sun? ⊛

25. The Andromeda Galaxy is 2.5 million light years from the Milky Way Galaxy. If you were to place this distance on Figure 1.2, between which two steps would you place it? ⊛

26. Scientists say that we are "made of stardust." Explain what they mean.

27. What does the word *falsifiable* mean? Give an example of an idea that is *not* falsifiable. Give an example of an idea that *is* falsifiable.

28. Explain how the word *theory* is used differently by a scientist than in everyday language.

29. What is the difference between a *hypothesis* and a *theory*?

30. Astronomers commonly work with scientists in other fields. Use what you have learned about the scientific method to predict what would happen if a discrepancy were found between two theories, each coming from a different field.

31. Suppose you read an article on the Internet that claims that children born under a full Moon become better students than children born at other times.
   a. Is this theory falsifiable?
   b. If so, how could it be tested?

32. A textbook published in 1945 stated that it takes 800,000 years for light to reach us from the Andromeda Galaxy. In this book, we say that it takes 2,500,000 years. What does this tell you about a scientific fact and how our knowledge changes with time?

33. Astrology makes testable predictions. For example, it predicts that the horoscope for your star sign on any day should fit you better than horoscopes for other star signs. Without indicating which sign is which, read the daily horoscopes to a friend, and ask how many of them might describe your friend's experiences on the previous day. Repeat the experiment every day for a week and keep records. Did one particular horoscope sign consistently describe your friend's experiences?

34. Imagine yourself living on a planet orbiting a star in a very distant galaxy. What does the cosmological principle tell you about the way you would perceive the universe from this distant location?

35. Figure 1.8 shows a rough map of the sky as viewed from the Northern Hemisphere in each of the four seasons. Constellations are labeled with their names in blue text. What time of year is it where you are? What constellations would be visible if you went out to view the sky tonight? ⊛

APPLYING THE CONCEPTS

36. The distance to the Sun is 8.3 light-minutes.
   a. If you converted this to light-hours, should you get a larger or a smaller number?
   b. Convert the distance to the Sun from light-minutes to light-hours.
   c. Check your work by comparing the value in light-hours to the value in light-minutes to see if your answer matches your prediction.

37. The slope of a line is calculated by taking the rise (the change in y) and dividing it by the run (the change in x). ⊛
   a. For Figure 1.9, calculate the slope of the line during the first 10 minutes. Your answer should come out in miles per minute, which is the speed of the car.
   b. Check your work by multiplying the slope by 5 minutes to find the distance of the car from the starting point after 5 minutes. Compare this distance to value of the line on the graph at that time.

38. There are 86,400 seconds in a day. The Moon's mass compared to Earth's is 0.0123.
   a. Write 86,400 and 0.0123 in scientific notation.
   b. In each case, check your work by carefully thinking about the exponent of 10, which should be negative if the number is less than 1 and positive if the number is greater than 1.

39. Suppose that you want to travel 100 km at 30 km/h, and you want to know how much time it will take.
   a. Your answer should have units of "hours." Why?
   b. If you multiply the two numbers together, what units will the result have? Are these the units of time?
   c. If you divide the speed by the distance, what units will the result have? Are these the units of time?
   d. If you divide the distance by the speed, what units will the result have? Are these the units of time?
   e. Carry out the operation (multiplication or division) that will give you the units of time that you want for your answer. How long will it take you to drive 100 km at 30 km/h?

40. The average distance from Earth to the Moon is 384,000 km. In the late 1960s, astronauts reached the Moon in about 3 days. How fast (on average) must they have been traveling (in kilometers per hour) to cover this distance in this time? Compare this speed to that of a jet aircraft (800 km/h).
   a. The problem asks for a speed in kilometers per hour, but you have been given a time in days. How many hours are in a day?
   b. Should the number of hours be larger or smaller than the number of days? Do you want to multiply by the number of hours per day, or divide by the number of hours per day to convert 3 days into hours?
   c. The problem is asking for a speed. What are the units of speed?
   d. How should you combine the units of these numbers to find a speed? Multiply or divide them? If divide, which way should you divide them?
   e. Combine the distance in kilometers and the time in hours to find a speed in kilometers per hour.
   f. Compare your answer to the speed of a jet aircraft.

exploration Scaling the Solar System

The Sun is 149.6 million km from Earth. Astronomers use this number so often that they made a new unit to represent it, called the astronomical unit (AU). Just as 1 km is equal to 1,000 meters, 1 AU is equal to 149.6 million km. The distance from the Sun to objects in the Solar System is given in Appendix 2. One of the columns gives this distance in astronomical units.

One way to get a feel for very large numbers is to "scale them" to everyday objects, much like making a map of a large area on a small piece of paper. Imagine the Earth is a speck of dust 1 cm away from the Sun. This sets the scale for a model of the Solar System, in which 1 cm on a piece of paper represents 1 AU in space. On this same scale, Mercury would be located 0.387 cm from the Sun. Using the data from Appendix 2, make a scale map of the Solar System, placing each object at the appropriate scaled distance from the Sun. You may need to tape multiple pieces of paper together to fit all of the objects on your scale map. Label each object on your scale map with its name and its distance from the Sun (in astronomical units).

Patterns in the Sky— Motions of Earth and the Moon

Each month, the Moon changes its appearance in a predictable way. For 1 month, go outside on 10 different dates to the same location and take a picture of the Moon. You will have to think about what time of day to go outside to take these photos. Label each photo with the date and time of the observation. For each photograph, make a sketch of the relative positions of the Sun, Earth, and the Moon when you took the photograph. Think about where you must have been standing on Earth to see that phase of the Moon at that location in your sky, and add a stick figure to your Earth sketch to show where you must have been standing.

EXPERIMENT SETUP

S	M	T	W	T	F	S
				1	2	3
4	5	6	7	8	9	10
11	12	13	14	15	16	17
18	19	20	21	22	23	24
25	26	27	28	29	30	

DAY 10 — Moon Rise 1:55 PM
DAY 13 — Moon Rise 4:14 PM
DAY 16 — Moon Rise 6:07 PM

For 1 month, go outside on 10 different dates to the same location, and take a picture of the Moon. You will have to think about what time of day to go outside to take these photos.

DAY 14

DAY 5

Label each photo with the date and time of the observation.

SKETCH OF RESULTS

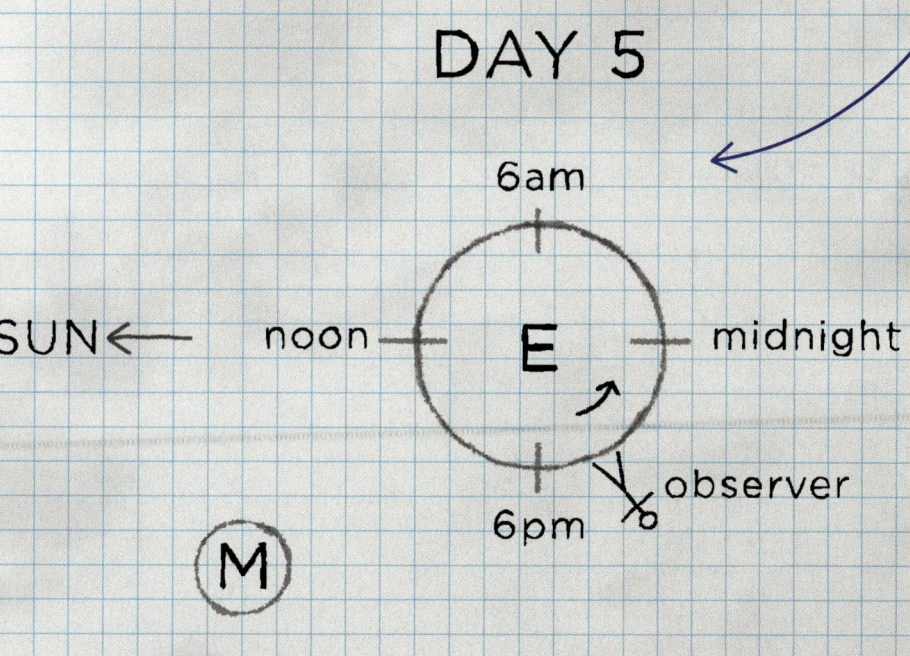

DAY 5

6am

SUN ← noon — E — midnight

6pm

observer

M

2

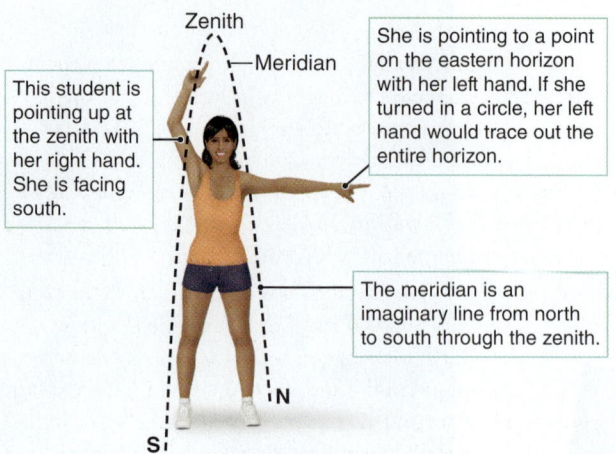

This student is pointing up at the zenith with her right hand. She is facing south.

She is pointing to a point on the eastern horizon with her left hand. If she turned in a circle, her left hand would trace out the entire horizon.

The meridian is an imaginary line from north to south through the zenith.

Figure 2.1 The meridian is a line on the celestial sphere that runs from north to south, dividing the sky into an eastern half and a western half.

▶‖ **AstroTour:** The Earth Spins and Revolves

▶‖ **AstroTour:** The Celestial Sphere and the Ecliptic

🎥 **Astronomy in Action:** Vocabulary of the Celestial Sphere

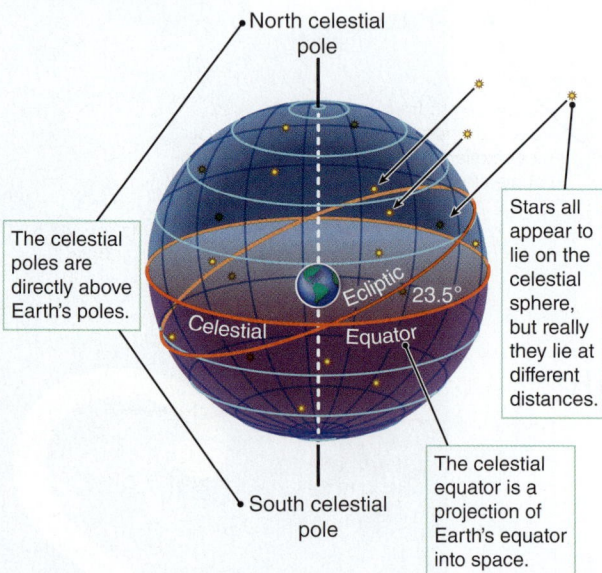

The celestial poles are directly above Earth's poles.

Stars all appear to lie on the celestial sphere, but really they lie at different distances.

The celestial equator is a projection of Earth's equator into space.

Figure 2.2 The celestial sphere is useful for thinking about the appearance and apparent motion of the stars in the sky. This imaginary sphere is centered on Earth and allows us to ignore, temporarily, the complication that stars are not all located at the same distance from Earth. The ecliptic is discussed in Section 2.2.

Through careful observation, our ancestors learned that they could use patterns in the sky to predict the changing length of day, the change of seasons, and the rise and fall of tides. Now we see these patterns with the perspective of centuries of modern science, and we can explain these changes as a consequence of the motions of Earth and the Moon. In this chapter, you will look at patterns in the sky and on Earth, and you will learn about the underlying motions that cause these patterns.

LEARNING GOALS

① Describe how Earth's rotation affects our perception of celestial motions as seen from different places on Earth.

② Visualize how Earth's motion around the Sun determines which stars are visible at night.

③ Explain which seasons are experienced in different locations through the year and why.

④ Connect the motion of the Moon in its orbit around Earth to the phases that are observed and to the spectacle of eclipses.

2.1 Earth Spins on Its Axis

Aristotle (384–322 BCE) and other Greek philosophers knew that Earth was a sphere. However, it was difficult for them to determine that the daily and annual patterns in the sky were caused by Earth's motions, because to an Earth-bound observer, Earth *feels* stationary. Think about traveling in a car down a smooth road; signs on the roadside seems to travel past you from the front of the car to the back. If you had not been in the car when it started moving, it would be very difficult to determine whether you were moving forward or the road was moving backward. As you will learn in this section, Earth's rotation on its axis causes the Sun, Moon, and stars to rise and set each day.

The Celestial Sphere

Imagine that you are outside in a large, flat field, looking up at the sky. Half of the sky will be visible to you, arcing in a dome over your head. The highest point in the sky, directly above your head, is called the **zenith**. Half of the sky is below the horizon, hidden from view by Earth's surface. You can find the **horizon** by pointing your right hand at the zenith and your left hand straight out from your side. Turn in a complete circle. Your left hand has traced out the entire horizon. You can divide the sky into an eastern half and a western half with a line that runs from the horizon at due north through the zenith to the horizon at due south. This imaginary north–south line is called the **meridian**, shown as a dashed line in **Figure 2.1**. Over the course of 1 day, objects will rise in the eastern half of the sky, pass through the meridian, and set in the western half of the sky.

To help visualize the patterns we observe in the Sun and stars, it is useful to change perspective and think of the sky as a huge sphere with Earth at its center. Imagine that the stars and other objects in the sky are painted on the inner surface of this sphere. Astronomers refer to this imaginary sphere, shown in **Figure 2.2**, as

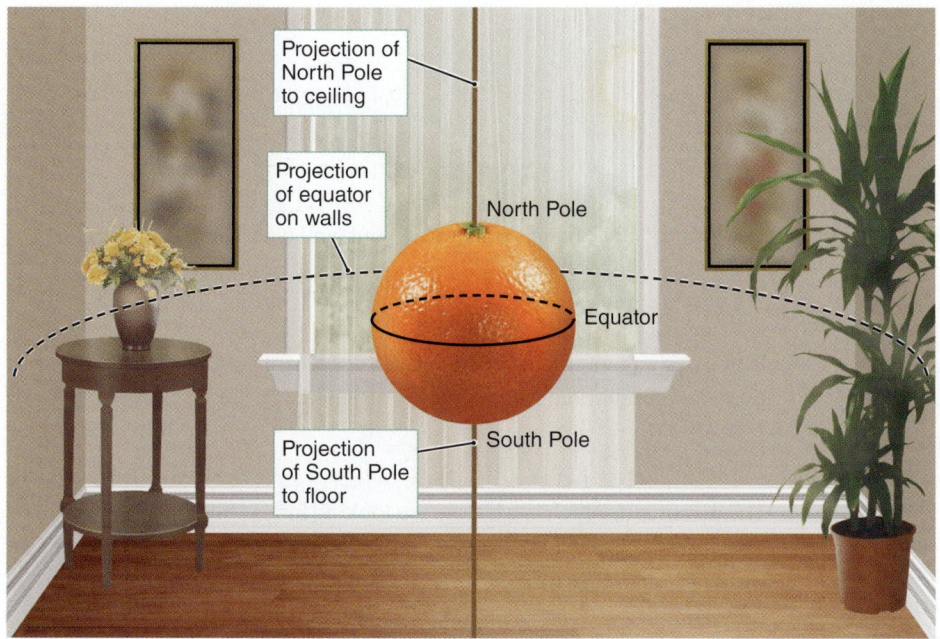

Figure 2.3 You can draw poles and an equator on an orange and imagine these points projected onto the walls of your room. Similarly, we imagine the poles and equator of Earth projected onto the celestial sphere. (Not drawn to scale.)

VOCABULARY ALERT

horizon In common language, the *horizon* is the place where the sky appears to meet the ground. For astronomers, however, it means the circle that is 90° from the zenith, ignoring obstructions, such as mountains. (This is the line you would see if you held your eyes perfectly level and turned all the way around.) A line to the horizon always makes a right angle with a line to the zenith.

day In everyday language, this word means both the time during which the Sun is up in the sky and the length of time it takes Earth to rotate once (from midnight to midnight). The context of the sentence tells us which meaning is intended. Unfortunately, astronomers use this word in both senses as well. You will have to consider the context of the sentence to know which meaning is being used in each instance.

the **celestial sphere**. Each point on the celestial sphere indicates a *direction* in space. Directly above Earth's North Pole is the **north celestial pole (NCP)**. Directly above Earth's South Pole is the **south celestial pole (SCP)**. Directly above Earth's **equator** is the **celestial equator**. The celestial equator divides the sky into a northern half and a southern half. If you point one arm toward the celestial equator and one arm toward the north celestial pole, your arms will always form a right angle, because the north celestial pole is 90° away from the celestial equator. The angle between the celestial equator and the south celestial pole is also 90°. You might want to draw an equator and north and south poles on an orange and visualize those markings projected onto the walls of your room, as shown in **Figure 2.3**. (We will use this orange again throughout this chapter, so don't eat it!)

Take a moment to visualize all these locations in space. You may want to draw a little person on your orange and visualize that person's zenith, meridian, and horizon on the walls of your room. How is the meridian oriented relative to the celestial equator? How is the horizon oriented relative to the celestial equator? When you can answer these questions for the person on your orange and then visualize them from your own perspective, you have oriented yourself to the sky.

Earth Rotates

Earth rotates around an imaginary axis that passes through its center and pokes through the surface at Earth's **North Pole** and Earth's **South Pole**. Imagine you are transported to a point in space far above Earth's North Pole. From this vantage point, you would see Earth rotate counterclockwise once each **day**, as shown in **Figure 2.4**. As an observer on Earth is carried around from west toward east, that observer sees objects in the sky, such as the Sun, *appear* to move in the opposite direction, along an arc in the sky that passes from east to west. The daily path each object takes across the sky is called its *apparent daily motion*. The direction and

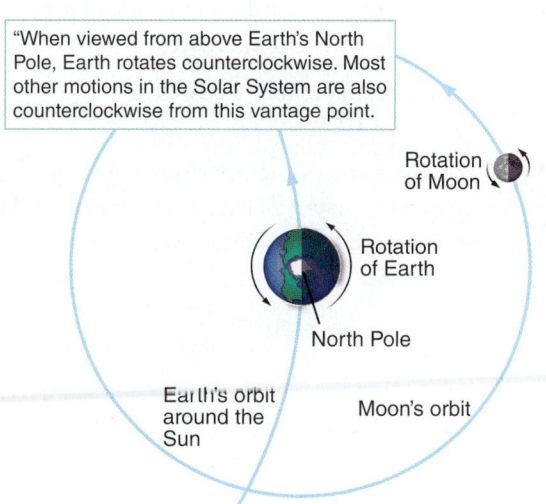

Figure 2.4 When viewed from far above Earth's North Pole, Earth orbits the Sun and rotates on its axis. The Moon also orbits Earth and rotates on its own axis. All of these motions (along with most others in the Solar System) are counterclockwise from this vantage point. (Not drawn to scale.)

working it out 2.1

Manipulating Equations

So far, we have discussed scientific notation and units. Now we want to relate quantities to each other, such as distance and time. In Section 2.1, we mentioned the speed of the surface of Earth as it rotates. How do we know that speed?

How are distance, time, and speed related? If you travel a distance of 100 kilometers (km) in 1 hour, you have traveled at an average speed of 100 km/h. That's how we say it in English. How do we translate this sentence into math? We write the equation:

$$\text{Speed} = \frac{\text{Distance}}{\text{Time}}$$

But that takes up a lot of space, so we abbreviate it by using the letter s to represent speed, the letter d to represent distance, and the letter t to represent time:

$$s = \frac{d}{t}$$

To find the speed of a spot on Earth's equator as Earth rotates, we need the distance that spot travels and the time that it takes. The time it takes to go around once is 1 day, and the distance traveled is Earth's circumference. So the speed is

$$s = \frac{\text{Circumference}}{1 \text{ day}}$$

The circumference of a circle is given by $2 \times \pi \times r$, where r is the radius of the circle. Earth's radius is 6,378 km. To find the circumference, multiply that by 2 and then by π to get 40,070 km. (Does it matter if you multiply first by π and then by 2? If you can't remember the rule, try it and see.)

Now that we have the circumference, we can find the speed:

$$s = \frac{\text{circumference}}{1 \text{ day}} = \frac{40{,}070 \text{ km}}{1 \text{ day}} = 40{,}070 \text{ km/day}$$

But that's not the way we usually write a speed. We usually use kilometers per hour. One day = 24 hours, so (1 day) ÷ (24 hours) is equal to 1. We can always divide or multiply by 1, so let's multiply our speed by (1 day) ÷ (24 hours):

$$s = 40{,}070 \frac{\text{km}}{\text{day}} \times \frac{1 \text{ day}}{24 \text{ hours}}$$

The unit of "day" divides (or cancels) out, and we have

$$s = \frac{40{,}070 \text{ km}}{24 \text{ h}} = 1{,}670 \text{ km/h}$$

How can we manipulate this equation to solve for distance? We start with the original equation:

$$s = \frac{d}{t}$$

The distance d, which is what we are looking for, is "buried" on the right side of the equation. We want to get it all by itself on the left. That's what we mean when we say "solve for." The first thing to do is to flip the equation around:

$$\frac{d}{t} = s$$

We can do this because of the equals sign. It's like saying, "two quarters is equal to 50 cents" versus "50 cents is equal to two quarters." It has to be true either way we say it. We still have not solved for d. To do this, we need to multiply by t. This cancels the t on the bottom of the left side of the equation. But if you do something only to one side of an equation, the two sides are not equal anymore. So whatever we do on the left side of the equation, we also have to do on the right side:

$$\frac{d}{\cancel{t}} \times \cancel{t} = s \times t$$

The two t's on the left side cancel, to give us our final answer:

$$d = s \times t$$

We often write this as

$$d = st$$

For simplicity, the multiplication symbol is left out of the equation. When two terms are written side by side, with no symbols between them, it means you should multiply them.

speed of Earth's rotation can be found from these observations and the radius of Earth, as shown in **Working It Out 2.1**. As Earth rotates, its surface is moving quite fast—about 1,670 kilometers per hour (km/h) at the equator.

To see how the celestial sphere and your view from Earth are related, consider the apparent daily motion of the Sun. Throughout one day, the Sun stays in very nearly the same place on the celestial sphere. As Earth rotates counterclockwise inside the celestial sphere, the Sun appears to rise above the observer's horizon in the east, cross the sky, and set below the horizon in the west. The angle that an observer measures between an object and the point below it on the horizon is called the **altitude**. From most locations on Earth, the Sun's altitude changes throughout the day, from zero, just as it rises, to a maximum as the Sun crosses the observer's meridian, and back to zero again as it sets. The time at which the Sun crosses the meridian is "local astronomical noon." People often say the Sun is "overhead" at noon, but in most locations on Earth, the Sun's altitude is less than 90° as it crosses the meridian. The Sun only passes through the zenith for observers at locations between 23.5° north of the equator and 23.5° south of the equator (the region known as the *Tropics*). Even in those locations, the Sun passes through the zenith only on specific days each year. You have to be in a specific location on a specific day for the Sun to be directly over your head at noon—such as latitude 23.5° north on June 21.

Local astronomical midnight occurs when the Sun is precisely opposite from its position at local noon. When it is noon where you live, Earth has rotated so that you face most directly toward the Sun. Half a day later, at midnight, your location on Earth has rotated so that you face most directly away from the Sun.

The View from the Poles

The apparent daily motions of the stars and the Sun depend on where you live. The apparent daily motions of celestial objects in northern Canada, for example, are quite different from the apparent daily motions seen from an island in the Tropics.

Imagine that you are standing on the North Pole watching the sky, as in **Figure 2.5a**. (Ignore the Sun for the moment and pretend that you can always see stars in the sky.) You are standing where Earth's axis of rotation intersects its surface, which is like standing at the center of a rotating wheel. As Earth rotates, the spot directly above you remains fixed over your head while everything else in the sky appears to revolve in a counterclockwise direction around this spot. If you are standing at the North Pole, the zenith is always in the same place on the celestial sphere, in the direction of the north celestial pole. Objects visible from the North Pole follow circular paths that always have the same altitude. Objects close to the zenith appear to follow small circles, while objects near the horizon follow the largest circles. Figure 2.5b depicts this overhead view. If you have trouble visualizing this, spin your orange and imagine looking at the objects in your room from its "north pole." The view from the North Pole is special because nothing rises or sets each day as Earth turns; from there an observer always sees the *same* half of the celestial sphere.

The view from Earth's South Pole is much the same, but with two major differences. First, the South Pole is on the opposite side of Earth from the North Pole, so the visible half of the sky at the South Pole is precisely the half that is hidden from view at the North Pole. The second difference is that instead of appearing to move counterclockwise around the sky, stars appear to move clockwise around the south celestial pole. Try to visualize this movement. It might help to stand up and rotate around from right to left. As you look at the ceiling, things appear to move in a counterclockwise direction, but as you look at the floor, they appear to be moving clockwise.

VOCABULARY ALERT

altitude In everyday language, *altitude* is the height of an object above the ground, such as an airplane flying overhead. Astronomers use the word to refer to the angle formed between an imaginary line from an observer to an object in the sky and a second line from the observer to the point on the horizon directly below the object.

what if . . .

What if Earth did not rotate at all as it orbited the Sun: What effect might that have on the science of astronomy?

Figure 2.5 An observer (a) standing at the North Pole sees (b) stars moving throughout the night on counterclockwise, circular paths about the zenith. (c) The same half of the sky is always visible from the North Pole.

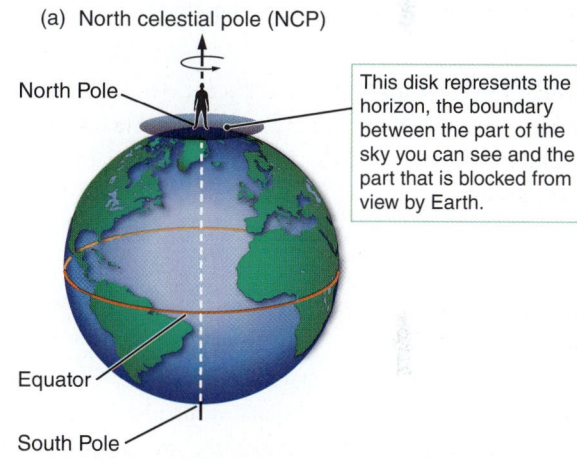

(a) North celestial pole (NCP)

North Pole

This disk represents the horizon, the boundary between the part of the sky you can see and the part that is blocked from view by Earth.

Equator

South Pole

From the North Pole looking directly overhead, the **north celestial pole (NCP)** is at the zenith.

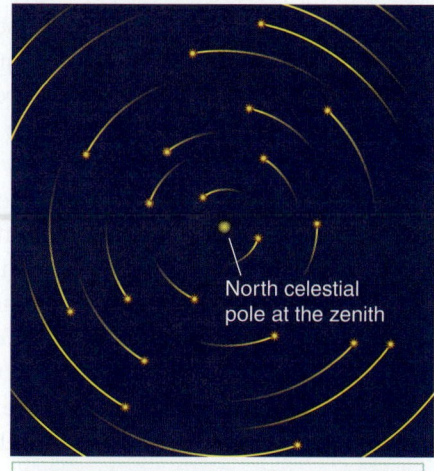

(b)

North celestial pole at the zenith

As Earth rotates, the stars appear to move in a counterclockwise direction around the NCP.

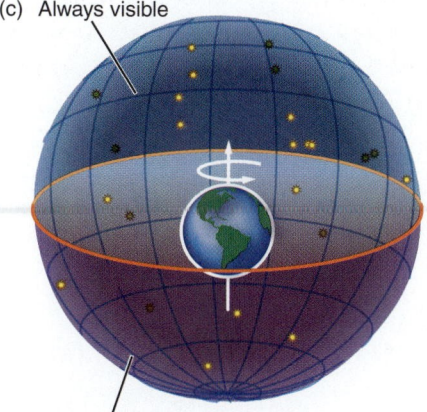

(c) Always visible

Never visible; blocked by Earth

From the North Pole, you always see the same half of the sky.

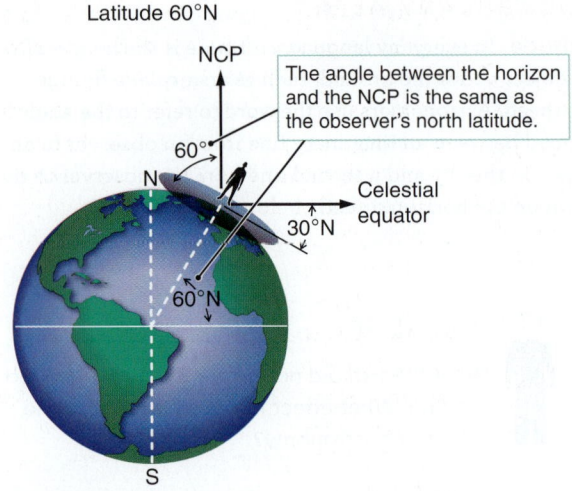

Latitude 60°N

NCP

The angle between the horizon and the NCP is the same as the observer's north latitude.

60°

N

Celestial equator

30°N

60°N

S

Figure 2.6 Our perspective on the sky depends on our location on Earth. The locations of the celestial poles and celestial equator in an observer's sky depend on the observer's latitude. In this case, an observer at latitude 60° north sees the north celestial pole at an altitude of 60° above the northern horizon and the celestial equator at an altitude of 30° above the southern horizon.

▶❙❙ **AstroTour:** The View from the Poles

For any observer on Earth, half of the sky is always below the horizon, as shown in Figure 2.5c; the view of the other half of the sky is blocked by Earth itself. This may be difficult to visualize in the figure, because the celestial sphere is drawn close to Earth so the figure fits on the page. You should picture the celestial sphere very much larger than this to see why an observer on Earth can see fully half of the sky.

The View away from the Poles

Now imagine leaving the North Pole to travel south to lower latitudes. **Latitude** measures how far north or south of the equator you are on the surface of Earth. **Figure 2.6** shows how latitude is measured as the angle between the plane of Earth's equator and your position on Earth's surface. At the North Pole, this angle is 90°. At the equator, it is 0°. So the latitude of the North Pole is 90° north, and the latitude of the equator is 0°. The South Pole is at latitude 90° south.

Your latitude determines the part of the sky that you can see throughout the year. As you move south, the zenith moves away from the north celestial pole, and so the horizon moves as well. At the North Pole, the horizon makes a 90° angle with the north celestial pole, which is at the zenith. At a latitude of 60° north, the horizon is tilted 60° from the north celestial pole. The angle between your horizon and the north celestial pole is equal to your latitude no matter where you are on Earth. At locations other than the poles, the visible half of the sky changes as Earth rotates, because the zenith points at different locations on the celestial sphere as Earth carries the observer around.

The best way to solidify your understanding of the view of the sky at different latitudes is to draw pictures like the one in Figure 2.6. If you can draw a picture like this for any latitude—filling in the values for each angle in the drawing and imagining what the sky looks like from that location—then you will be well on your way to developing a working knowledge of the appearance of the sky. That knowledge will prove useful later when we discuss a variety of phenomena, such as the changing of the seasons.

The apparent motion of the stars about the celestial poles also differs from latitude to latitude, as shown in **Figure 2.7**. For an observer at the North Pole (Figure 2.7a), the celestial equator lies exactly along the horizon. The north celestial pole is at the zenith, and the southern half of the sky is never visible. Stars neither rise nor set; their paths form circles parallel to the horizon.

At other latitudes, the celestial equator intersects the horizon due east and due west. Therefore, a star on the celestial equator rises due east and sets due west. Stars north of the celestial equator rise north of east and set north of west. Stars south of the celestial equator rise south of east and set south of west.

Regardless of where you are on Earth (with the exception of the poles), half of the celestial equator is always visible above the horizon. Therefore, any object located on the celestial equator is visible half of the time—above the horizon for 12 hours each day. Objects that are not on the celestial equator are above the horizon for differing amounts of time. Figure 2.7b and d show that stars in the observer's hemisphere are visible for more than half the day because more than half of each star's path in the sky is above the horizon. In contrast, stars in the opposite hemisphere are visible for less than half the day because less than half of each star's path in the sky is above the horizon.

For example, as seen from the Northern Hemisphere, stars south of the celestial equator are above the horizon for less than 12 hours a day. The farther south

(a) North Pole

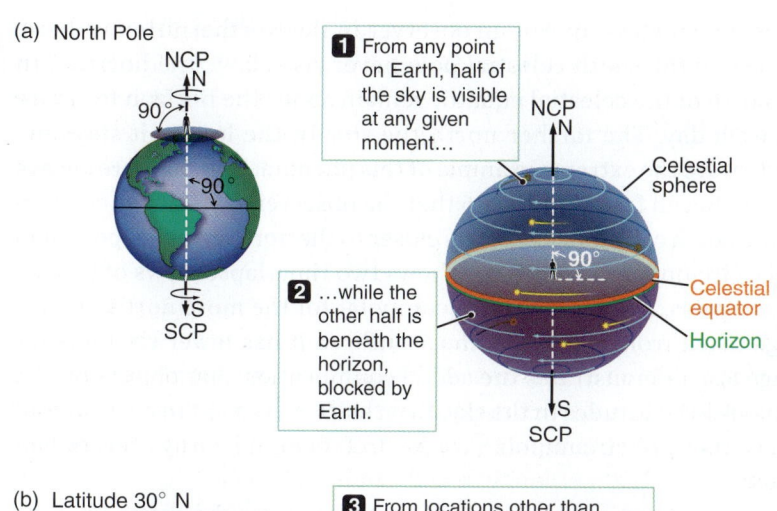

1 From any point on Earth, half of the sky is visible at any given moment…

2 …while the other half is beneath the horizon, blocked by Earth.

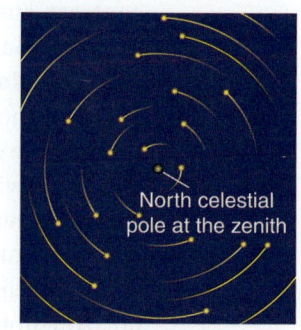

North celestial pole at the zenith

(b) Latitude 30° N

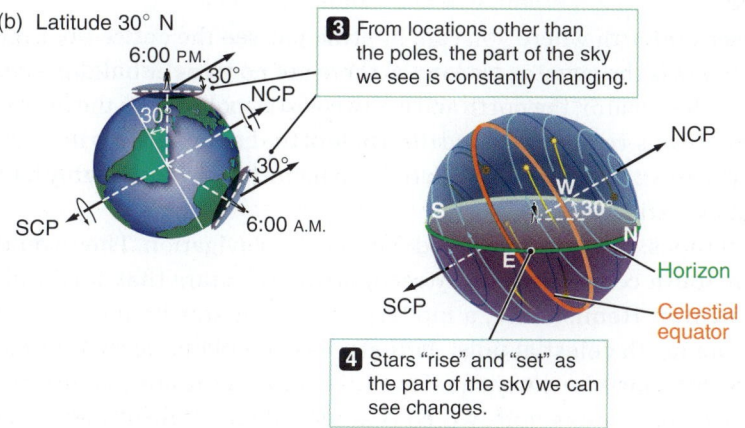

3 From locations other than the poles, the part of the sky we see is constantly changing.

4 Stars "rise" and "set" as the part of the sky we can see changes.

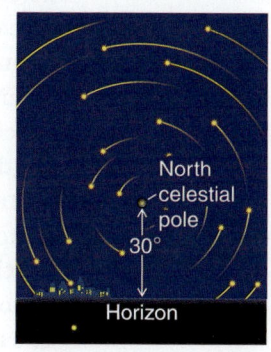

North celestial pole
30°
Horizon

(c) Equator

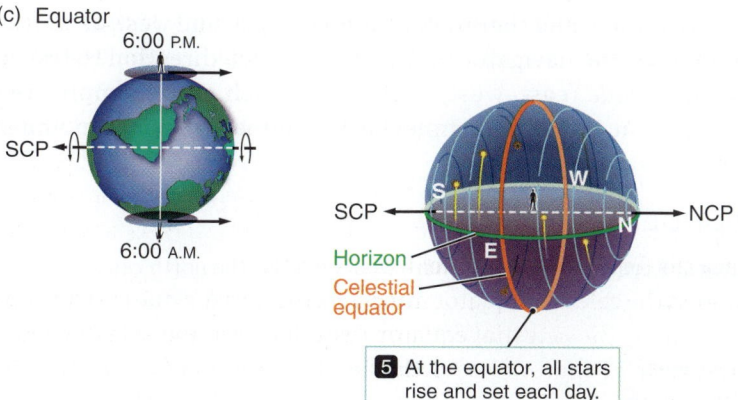

5 At the equator, all stars rise and set each day.

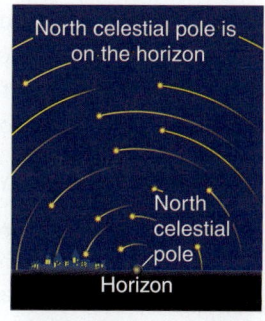

North celestial pole is on the horizon

North celestial pole

Horizon

(d) Latitude 30° S

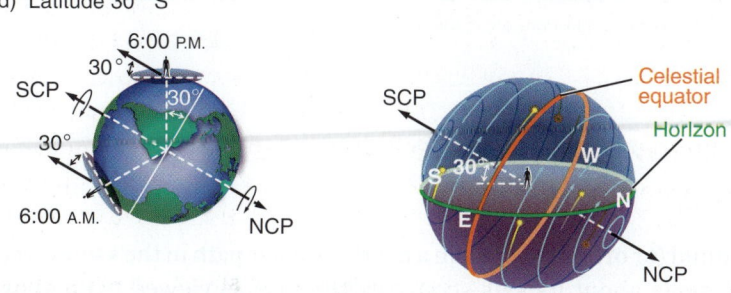

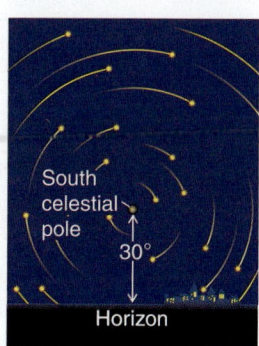

South celestial pole
30°
Horizon

Figure 2.7 The celestial sphere is shown here as viewed by observers at four different latitudes. At all locations other than the poles, stars rise and set as the part of the celestial sphere that we see changes during the day. In (a), the observer is looking straight up and sees the stars making circles around the north celestial pole, which is located at the zenith. In (b), the observer is looking north from a middle latitude (30° north), so she sees some stars circling the north celestial pole, whereas others travel in arcs across the sky from east to west. In (c), the observer is looking north from the equator, so she sees all the stars traveling in arcs across the sky from east to west. In (d), the observer is looking south from a middle latitude (30° south), so she sees some stars circling the South Pole, whereas others travel in arcs across the sky from east to west.

From a location in the Canadian woods, the north celestial pole appears high in the sky and most stars are circumpolar.

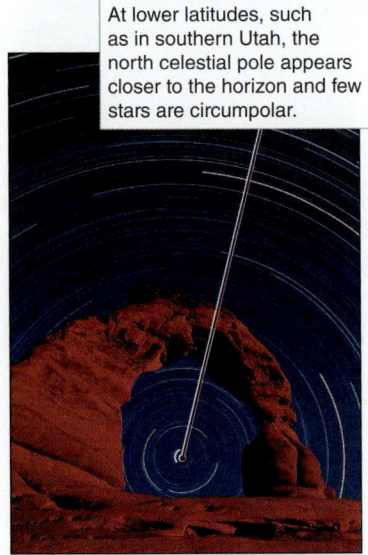

At lower latitudes, such as in southern Utah, the north celestial pole appears closer to the horizon and few stars are circumpolar.

Figure 2.8 Time exposures of the sky show the apparent motions of stars through the night. Note the difference in the circumpolar portion of the sky as seen from the two different latitudes. The observer's latitude determines the number of circumpolar stars, but mountains or other objects on the horizon reduce the number actually seen. In the picture at right, the "floor" of the arch blocks part of the horizon; not all of the circumpolar stars can actually be seen to be circumpolar.

VOCABULARY ALERT

revolve/rotate In everyday language, the words *revolve* and *rotate* are sometimes used synonymously to describe something that spins. Astronomers distinguish between the two terms, using *rotate* to mean that an object spins about an axis through its center and *revolve* to mean that one object orbits another. Earth rotates about its axis (causing our day) and revolves around the Sun (causing our year).

a star is, the less time it stays up. For an observer in the Northern Hemisphere, stars located close to the south celestial pole never rise above the horizon. In contrast, stars north of the celestial equator remain above the horizon for more than 12 hours each day. The farther north the star is, the longer it stays up. **Circumpolar** stars are the extreme example of this phenomenon; they are always above the horizon. Recall from Figure 2.6 that the observer's latitude is equal to the altitude of the north celestial pole. Stars closer to the north celestial pole than this angle will be circumpolar. **Figure 2.8** shows two time-lapse views of the sky from different latitudes. More stars are circumpolar for the more northern latitude. The image taken from Utah is farther south, so it has fewer circumpolar stars. This image also demonstrates the added complication that objects on the horizon often block low altitudes in the sky. Even stars that ought to be visible all the time, because they are circumpolar, might drop behind nearby objects that block the horizon.

The only place on Earth where you can, in principle, see the entire sky over the course of 24 hours is the equator (as long as there are no trees or buildings on the horizon). From the equator, the north and south celestial poles sit on the northern and southern horizons, respectively, and the whole of the heavens passes through the sky each day (Figure 2.7c). (Even though the Sun lights the sky for roughly half this time, the stars are still there.)

Since ancient times, travelers have used the stars for navigation. They would find the north or south celestial poles by recognizing the stars that surround them. In the Northern Hemisphere, a moderately bright star happens to be located close to the north celestial pole. This star is called Polaris, the "North Star." The altitude of Polaris is nearly equal to the latitude of the observer. If you are in Phoenix, Arizona, for example, which is at latitude 33.5° north, then the north celestial pole has an altitude of 33.5°. A navigator who has located the North Star can identify north and therefore also south, east, and west, as well as her latitude. This enables the navigator to determine which direction to travel. Figuring out your *longitude* (east–west location) is much more complicated because of Earth's rotation. Longitude cannot be determined from astronomical observation alone.

CHECK YOUR UNDERSTANDING 2.1

If you were standing at Earth's North Pole, where would you see the north celestial pole?

a. at the zenith

b. on the eastern horizon

c. 23.5° south of the zenith

d. none of the above (the north celestial pole can't be seen from there)

Answers to Check Your Understanding questions are in the back of the book.

2.2 Revolution around the Sun Leads to Changes during the Year

Earth **revolves** around (or orbits) the Sun in a nearly circular path in the same direction that Earth **rotates** about its axis—counterclockwise as viewed from above Earth's North Pole. Because of this motion, the stars in the night sky change throughout the year, and Earth experiences seasons.

Earth Orbits the Sun

Earth's average distance from the Sun is 1.50×10^8 km—this distance is called an astronomical unit (AU). The astronomical unit is useful for measuring distances in the Solar System. One **year** is the time it takes Earth to orbit the Sun once. Because Earth moves a distance over a period of time, it has a speed. This motion is responsible for many of the patterns we see in the sky and on Earth, such as the season-to-season changes in the stars we see overhead. **Figure 2.9** shows that as Earth orbits the Sun, our view of the night sky changes. Six months from now, Earth will be on the opposite side of the Sun from where it is now. The stars that are overhead at midnight 6 months from now are those that are overhead at noon today. Take a moment to visualize this motion. You can again use your orange as Earth and a table lamp with the shade removed, or a flashlight, to represent the Sun. Move the orange around the "Sun" and notice which parts of your room's walls are visible from the "nighttime side" of the orange at different points in the orbit.

Also take a moment to notice the location of your "Sun" relative to the walls of your room for the "observer" on your orange. If we correspondingly note the position of the Sun relative to the stars each day for a year, we find that it traces out a path against the background of the stars called the **ecliptic** (see Figure 2.9), and the plane of Earth's orbit around the Sun is called the **ecliptic plane**. On September 1, the Sun is in the direction of the constellation Leo. Three months later, on December 1, Earth has moved so that the Sun is in the direction of the constellation Scorpius. By six months later, on March 1, Earth is on the opposite side of the Sun from its position in September, and the Sun is in the direction of the constellation Aquarius. The constellations that lie along the ecliptic are called

Figure 2.9 As Earth orbits the Sun, different stars appear in the night sky, and the Sun's apparent position against the background of stars changes. The imaginary circle traced by the annual path of the Sun is called the ecliptic. Constellations along the ecliptic form the zodiac.

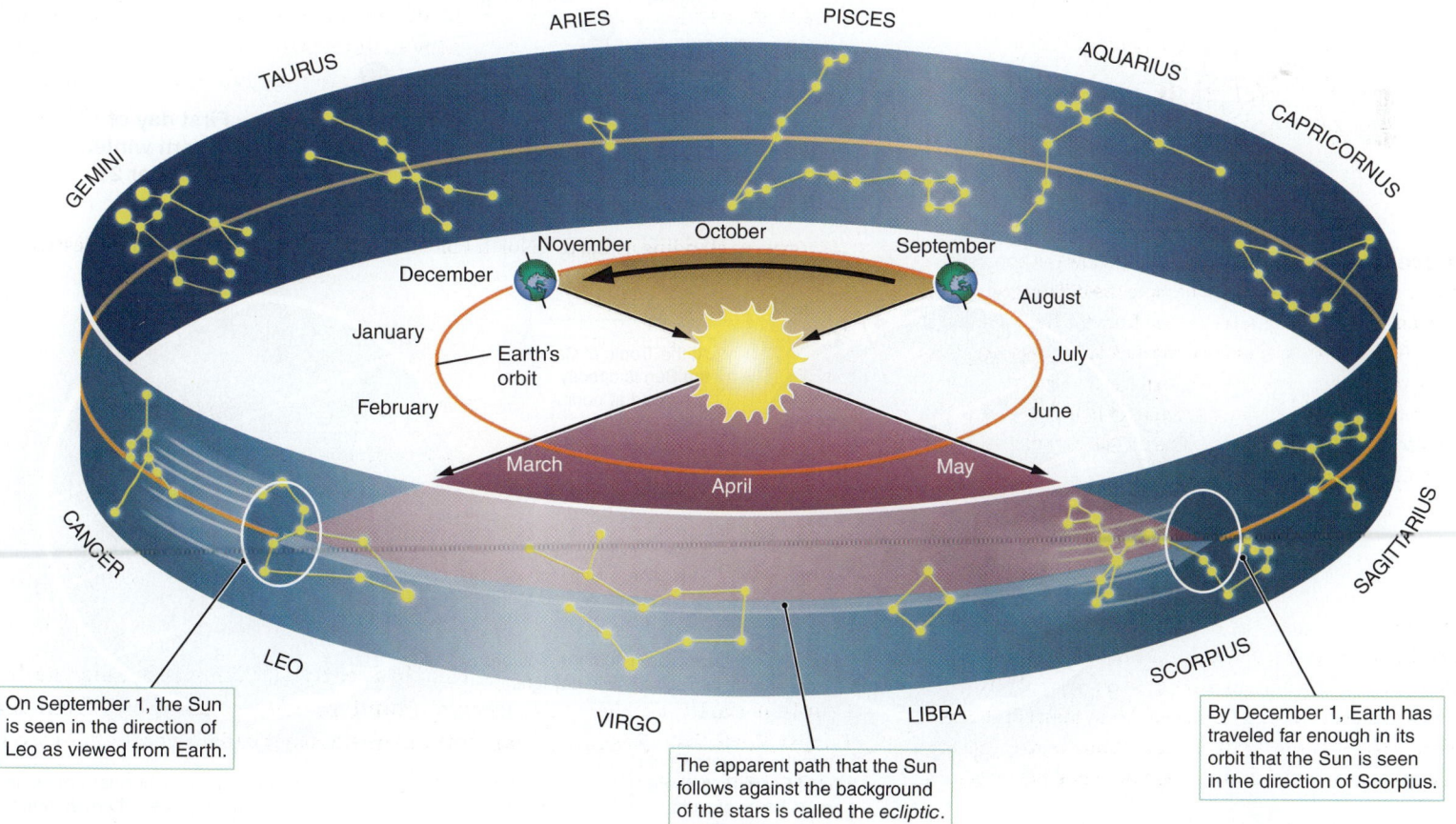

On September 1, the Sun is seen in the direction of Leo as viewed from Earth.

The apparent path that the Sun follows against the background of the stars is called the *ecliptic*.

By December 1, Earth has traveled far enough in its orbit that the Sun is seen in the direction of Scorpius.

the constellations of the **zodiac**. Ancient astrologers assigned special mystical significance to these stars because they lie along the path of the Sun. The constellations of the zodiac are nothing more than random patterns of unrelated distant, bright stars that happen by chance to be located near the ecliptic.

Seasons and the Tilt of Earth's Axis

So far we have discussed the rotation of Earth on its axis and the revolution of Earth around the Sun. To understand why the seasons change, we need to consider the combined effects of these two motions. Many people think that Earth is closer to the Sun in the summer and farther away in the winter and that this changing distance causes the seasons. If this idea were a hypothesis, can it be falsified? It can, because we can make a prediction. If the distance from Earth to the Sun causes the seasons, all of Earth should experience summer at the same time of year. However, the United States experiences summer in June, and Chile experiences summer in December. In modern times, we can directly measure the distance from Earth to the Sun, and we find that Earth is actually closest to the Sun at the beginning of January. We have just falsified this hypothesis (twice!), and we need to come up with another hypothesis that explains *all* the available facts. We will find an answer that fits all the facts by investigating Earth's axial tilt. To visualize what is happening, picture yourself looking at Earth from the outside of Earth's orbit, from a vantage point nearly in the same plane as the orbit.

Earth's axis of rotation is tilted 23.5° from the perpendicular to Earth's orbital plane, as shown in **Figure 2.10**. As Earth orbits the Sun, the north end of this axis always points in the same direction in space, toward Polaris. Because Earth is some-

Figure 2.10 Earth's orbit around the Sun is nearly circular. Here it is shown "from the side," in perspective, and so it looks much more elliptical than it is. This perspective view is a common way to portray orbits of planets and commonly leads people to believe that the orbits are highly elongated. (a) On the first day of northern summer (approximately June 21), the northern end of Earth's axis is tilted toward the Sun, while the southern end is tipped away. (b) Six months later, on the first day of northern winter (approximately December 21), the situation is reversed. This explains why seasons are opposite in the Northern and Southern hemispheres.

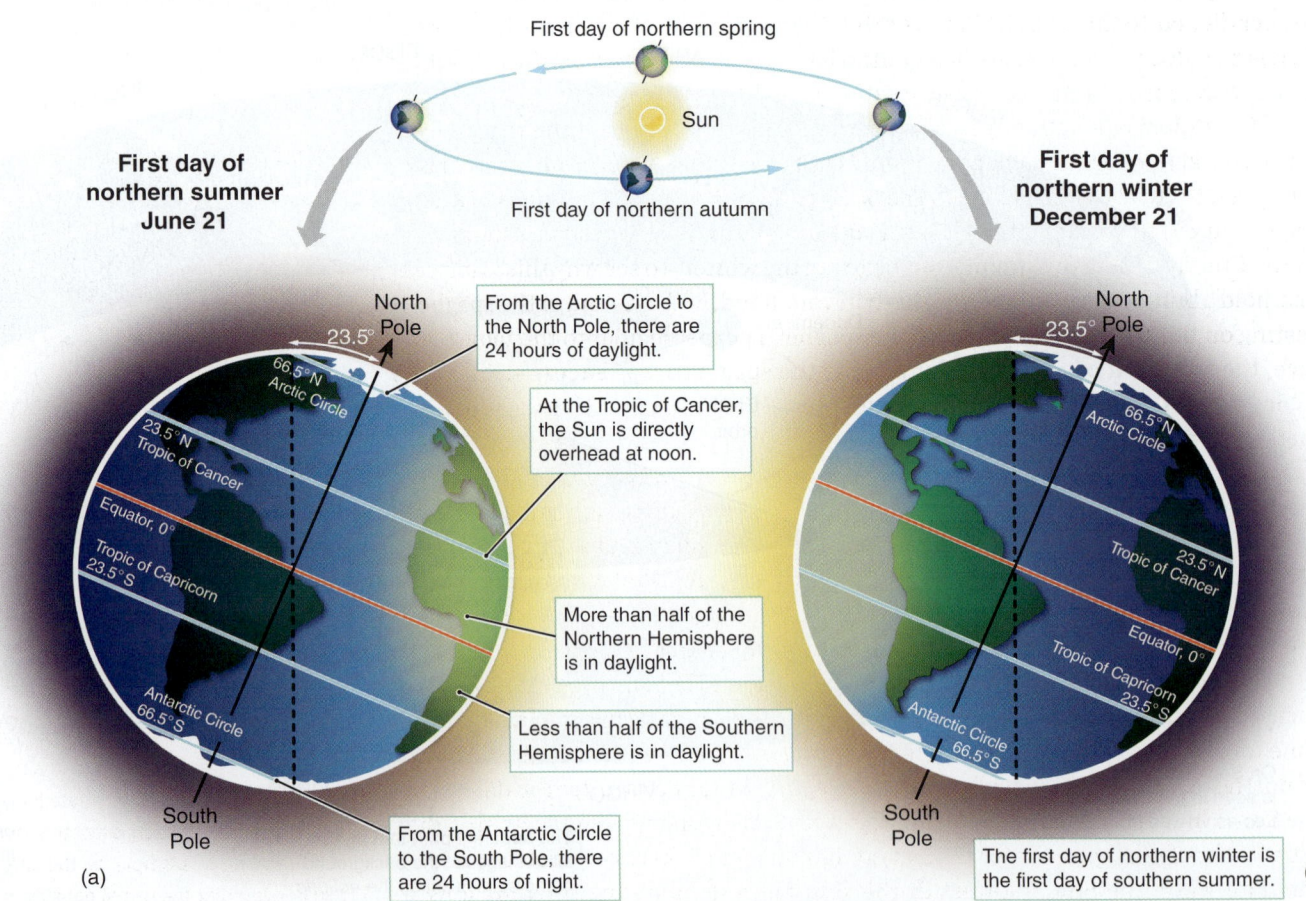

times on one side of the Sun and sometimes on the other side, sometimes Earth's North Pole is tilted toward the Sun and sometimes it is tilted away from the Sun. When Earth's North Pole is tilted toward the Sun, the Sun is *north* of the celestial equator; for observers in the Northern Hemisphere, the Sun is above the horizon more than 12 hours each day. Six months later, when Earth's North Pole is tilted away from the Sun, the Sun is *south* of the celestial equator; for observers in the Northern Hemisphere, the Sun is above the horizon less than 12 hours each day. Take a moment to visualize this tilted axis with your orange and your table lamp. Tilt the orange so that its north pole no longer points at the ceiling but at some distant point in the sky through a wall or window. Keep the north pole dot on your orange pointing in that direction, and "orbit" it around your "Sun." At one point, the tilt will be *toward* the "Sun," whereas halfway around the orbit, the tilt will be *away* from the "Sun."

In the preceding paragraph, we were careful to specify the *Northern* Hemisphere because seasons are opposite in the Southern Hemisphere. Look again at Figure 2.10. On June 21, while the Northern Hemisphere is enjoying the long days and short nights of summer, Earth's South Pole is tilted away from the Sun. As a result, it is winter in the Southern Hemisphere—the days are short and the nights are long there. But on December 21, Earth's South Pole is tilted toward the Sun, so it is summer in the Southern Hemisphere—the days are long and the nights are short there.

To understand how the combination of Earth's axial tilt and its path around the Sun creates seasons, consider a limiting case. If Earth's spin axis were exactly perpendicular to the plane of its orbit, then the Sun would always be on the celestial equator. At every latitude, the Sun would follow the same path through the sky every day, rising due east each morning and setting due west each evening. The Sun would be above the horizon exactly half the time, and days and nights would always be exactly 12 hours long everywhere on Earth. In short, if Earth's axis were exactly perpendicular to the plane of Earth's orbit, there would be no seasons. Because the Sun is not always on the celestial equator, its path through the sky (the ecliptic) is tilted with respect to the celestial equator, as shown in **Figure 2.11**.

The differing length of the night through the year is part of the explanation for seasonal temperature changes, but it is not the whole story. Another important effect relates to the angle at which the Sun's rays strike Earth. The Sun is higher in the sky during the summer than it is during the winter, so sunlight strikes the ground more directly during the summer than during the winter. To see why this is important, hold a bundle of uncooked spaghetti in your hand, with one end of the spaghetti resting on the table. When you hold the spaghetti perpendicular to the table's surface, the spaghetti covers a smaller area. If the spaghetti was energy striking the table, it would be concentrated in a small area. But if you hold the spaghetti at an angle, the strands cover a larger area, and the energy is more spread out. This is exactly what happens with the changing seasons. During the summer, Earth's surface at your location is more nearly perpendicular to the incoming sunlight, so the energy is more intense—more energy falls on each square meter of ground each second. During the winter, the surface of Earth is more tilted with respect to the sunlight, so less energy falls on each square meter of ground each second. This is the main reason why it is hotter in the summer and colder in the winter.

Just as it takes time for a pot of water on a stove to heat up when the burner is turned up and time for the pot to cool off when the burner is turned down, it takes time for Earth to respond to changes in heating from the Sun. The hottest months of northern summer are usually July and August, which come *after* the date when the Sun is highest in the sky, when the days are growing shorter. Similarly, the coldest months of northern winter are usually January and February, which occur *after* the date when the Sun is lowest in the sky, when the days are growing longer.

▶▶ Interactive Simulation: Seasons

Astronomy in Action: The Cause of Earth's Seasons

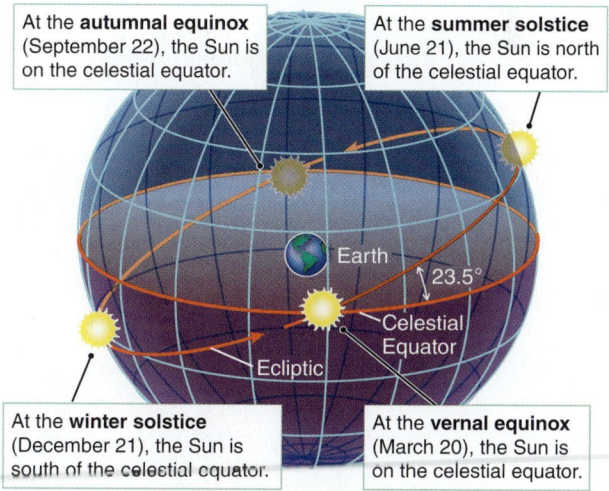

Figure labels:
At the **autumnal equinox** (September 22), the Sun is on the celestial equator.

At the **summer solstice** (June 21), the Sun is north of the celestial equator.

Earth

23.5°

Celestial Equator

Ecliptic

At the **winter solstice** (December 21), the Sun is south of the celestial equator.

At the **vernal equinox** (March 20), the Sun is on the celestial equator.

Figure 2.11 The path of the Sun through the sky is tilted 23.5° from the celestial equator. This establishes four special locations in space: two where the ecliptic crosses the celestial equator, one where the ecliptic is farthest north, and one where it is farthest south. The names for these locations are also the names of the dates when the Sun is located there, relative to the Northern Hemisphere. For example, on the date of the vernal equinox, the Sun is located at the vernal equinox in the sky.

what if . . .

What if humans establish a colony on Mars, where the axial tilt is 25.19°: Considering only this single piece of information, would you expect seasonal variations on Mars to be larger, smaller, or about the same as seasonal variations on Earth? Explain.

Temperature changes on Earth follow the changes in the amount of heating we receive from the Sun.

Together, these two effects—the directness of sunlight and the differing length of the night—mean that there is more heating from the Sun during summer than during winter. The directness of sunlight is the more significant effect.

Marking the Passage of the Seasons

There are four special days during Earth's orbit that mark unique moments in the year. As Earth orbits the Sun, the Sun moves along the ecliptic, which is tilted 23.5° with respect to the celestial equator. The day when the Sun is highest in the sky as it crosses the meridian (recall that the meridian is the line from due north to due south that passes overhead) is called the **summer solstice** (Figure 2.11). On this day, the Sun rises farthest north of east and sets farthest north of west. This occurs each year about June 21, the first day of summer in the Northern Hemisphere. This orientation of Earth and Sun is shown in Figure 2.10a.

Six months later, the North Pole is tilted away from the Sun. This day is the **winter solstice**, shown in Figures 2.10b and 2.11. This occurs each year about December 21, the shortest day of the year and the first day of winter in the Northern Hemisphere. Almost all cultural traditions in the Northern Hemisphere include a major celebration of some sort in late December. These winter festivals celebrate the return of the source of Earth's light and warmth. The days have stopped growing shorter and are beginning to get longer. Spring will come again.

Between these two solstice days, there are days when the Sun lies directly above Earth's equator, so that the entire Earth experiences 12 hours of daylight and 12 hours of darkness. These are called the equinoxes (*equinox* means "equal night"). The **autumnal equinox** marks the beginning of fall, about September 22, which is halfway between summer solstice and winter solstice. The **vernal equinox** marks the beginning of spring, about March 20, which is halfway between winter solstice and summer solstice. Figure 2.11 shows the location of the Sun along the ecliptic on these four special dates. Use your orange and your light source to reproduce these pictures and the motion implied by the arrows. Then use the orange and the light to reproduce the actual motion of Earth around the Sun: the Sun remains still, while Earth orbits around it, with its axis tilted at an angle.

The day does not divide evenly into the year; there is no reason why it should. Historically, this caused problems for people trying to keep a calendar. In modern times, we use the **Gregorian calendar**, which is based on the *tropical year*. The **tropical year** measures the time from one vernal equinox to the next—from the start of Northern Hemisphere spring to the start of the next Northern Hemisphere spring—and is 365.242189 days long. The tropical year has approximately one-quarter "extra" day. To make up for that fraction of a day, nearly every fourth year is made a **leap year**, with an extra day in February. This prevents the seasons from becoming out of sync with the months. The Gregorian calendar also makes other adjustments on longer timescales, to make up for the fact that the fraction of a day is a little bit less than one-quarter.

Seasons and Location

The variation in the length of nighttime with seasons is extreme near Earth's poles. The **Arctic Circle** and the **Antarctic Circle** (see Figure 2.10) are located at 66.5° north and south latitudes, respectively. In the regions from the Arctic Circle to the North Pole and from the Antarctic Circle to the South Pole, the Sun is above the horizon 24 hours a day for part of the year, earning these polar regions the nickname "land of the midnight Sun." There is an equally long period during which the Sun never rises and the nights are 24 hours long. The Sun never rises very high in the sky at these latitudes, so the sunlight is never very direct. Even with the long days at the height of summer, the polar regions remain relatively cool.

In contrast, nighttime at the equator is approximately 12 hours long throughout the year. The Sun passes directly overhead on the first day of spring and the first day of autumn because these are the days when the Sun is precisely on the celestial equator (see Figure 2.11). Sunlight is perpendicular to the ground at the equator on these days. On the summer solstice, the Sun is at its northernmost point along the ecliptic. On this day, and on the winter solstice, the Sun is *farthest* from the zenith at noon, and therefore sunlight is *least* direct at the equator.

The band of latitudes between 23.5° south and 23.5° north is called the **Tropics**. As shown in Figure 2.10, the northern limit of this region is called the Tropic of Cancer, whereas the southern limit is called the Tropic of Capricorn. In the Tropics, the Sun is directly overhead (actually at the zenith) at noon twice during the year.

Earth's Axis Wobbles

When the ancient Egyptian astronomer Ptolemy (Claudius Ptolemaeus) and his associates were formalizing their knowledge of the positions and motions of objects in the sky 2,000 years ago, the Sun appeared in the constellation Cancer on the first day of northern summer and in the constellation Capricornus on the first day of northern winter. Today, the Sun is in Taurus on the first day of northern summer and in Sagittarius on the first day of northern winter (see Figure 2.9). Why has this change occurred? There are *two* motions associated with Earth and its axis. Earth spins on its axis once each day, but the axis also wobbles slowly like the axis of a spinning top, shown in **Figure 2.12a**. The wobble is so slow that it takes about 26,000 years for the north celestial pole to make one trip around a large circle. In Section 2.1, we saw that the north celestial pole currently lies very near Polaris. However, if you could travel several thousand years into the past or future, you would find that the northern sky does not appear to rotate about a point near Polaris, but instead the stars rotate about another point on the path shown in Figure 2.12b. This figure shows the path of the north celestial pole through the sky during one *precession* cycle.

The plane of the celestial equator is perpendicular to Earth's axis. Therefore, as Earth's axis wobbles, the celestial equator must also wobble. You could punch a pencil through a business card, and then wobble the pencil to see this in action. As the celestial equator wobbles, the locations where it crosses the ecliptic—the equinoxes—change as well. During each 26,000-year wobble of Earth's axis, the locations of the equinoxes make one complete circuit around the celestial equator. This change of the position of the equinox, due to the wobble of Earth's axis, is called the **precession of the equinoxes**.

CHECK YOUR UNDERSTANDING 2.2

The tilt of Jupiter's rotational axis with respect to its orbital plane is 3°. If Earth's axis had this tilt, then the seasons on Earth

a. would be much more extreme.

b. would be much less extreme.

c. would be pretty much the same.

d. would occur much differently every year, depending on the alignments.

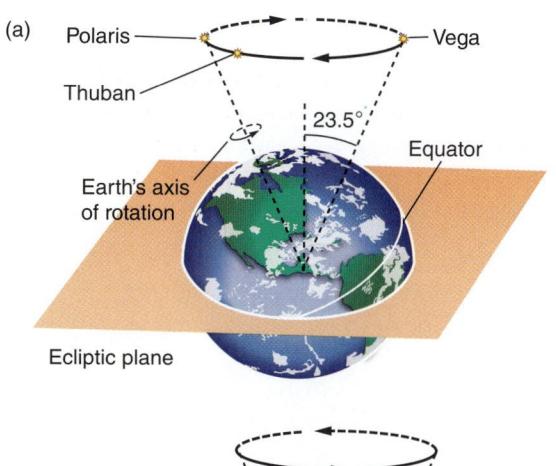

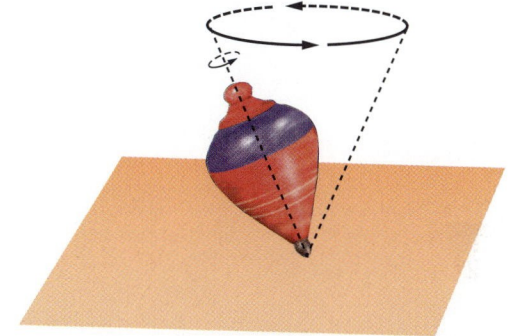

Figure 2.12 (a) Earth's axis of rotation changes orientation in the same way that the axis of a spinning top changes orientation. (b) This precession causes the projection of Earth's rotation axis to move in a circle with a radius of 23.5°, centered on the north ecliptic pole (yellow cross), with a period of 26,000 years. The red cross shows the location of the north celestial pole in the early 21st century. Polaris is currently the pole star. Vega will be the pole star in 14,000 CE, and Thuban was the pole star in 3000 BCE.

2.3 The Moon's Appearance Changes As It Orbits Earth

After the Sun, the Moon is the most prominent object in the sky. Earth and the Moon orbit around each other, and together they orbit the Sun. The Moon takes just over 27 days to orbit Earth. One aspect of the Moon's appearance does not change: We always see the same side from Earth. However, as the Moon orbits, it is illuminated differently, and so its appearance changes constantly. In this section, you will learn how the Moon's appearance changes as it orbits.

The Unchanging Face of the Moon

If you were to go outside next week or next month, or 20 years from now, or 20,000 *centuries* from now, you would still see the same pattern of dark patches on the Moon. (Some of these dark patches are visible in the chapter-opening figure.) This observation is responsible for the common misconception that the Moon does not rotate. In fact, the Moon *does* rotate on its axis—exactly once for each revolution it makes around Earth.

Once again, use your orange to help you visualize this idea. This time, the orange represents the Moon. Use your chair or some other object to represent Earth. Make a mark on the orange, or use a mark you previously made on the orange. First, make the orange "orbit" around the chair while keeping the mark on the orange always facing the same wall of the room. Thus, the orange does not rotate around its axis, and different sides of the orange face toward the chair at different points in the orbit. Next, make the orange orbit the chair, with the mark on the orange always facing the chair. You will have to turn the orange, relative to the walls of the room, to make this happen, which means that the orange is rotating about its axis. By the time the orange completes one orbit, it will have rotated about its axis exactly once. The Moon does exactly the same thing, rotating about its axis once per revolution around Earth, always keeping the same face toward Earth, as shown in **Figure 2.13**. This phenomenon, where the same side of the Moon always faces toward Earth, is called **synchronous rotation** because the revolution and the rotation are synchronized (or "in sync") with each other. The Moon's synchronous rotation occurs because the Moon is not perfectly round but has a bulge on the side closer to Earth. Earth's gravity constantly tugs on this bulge, which causes its *near side* always to fall toward Earth.

The Moon's *far side*, facing away from Earth, is often improperly called the dark side of the Moon. In fact, the far side spends just as much time in sunlight as the near side. The far side is not dark as in "unlit" but dark as in "unknown." Humans did not observe the far side of the Moon until spacecraft orbited the Moon in 1959.

The Changing Phases of the Moon

Unlike the Sun, the Moon has no light source of its own; it shines by reflected sunlight. Like Earth, half of the Moon is always in bright sunlight and half is always in darkness, as depicted in Figure 2.13. From Earth, our view of the illuminated half of the Moon is constantly changing as it orbits, which causes the phases of the Moon. During a *new Moon*, when the Moon is between Earth and the Sun, the side facing away from us is illuminated, so we cannot see the Moon. During a *full Moon*, when Earth is between the Moon and the Sun, the side facing

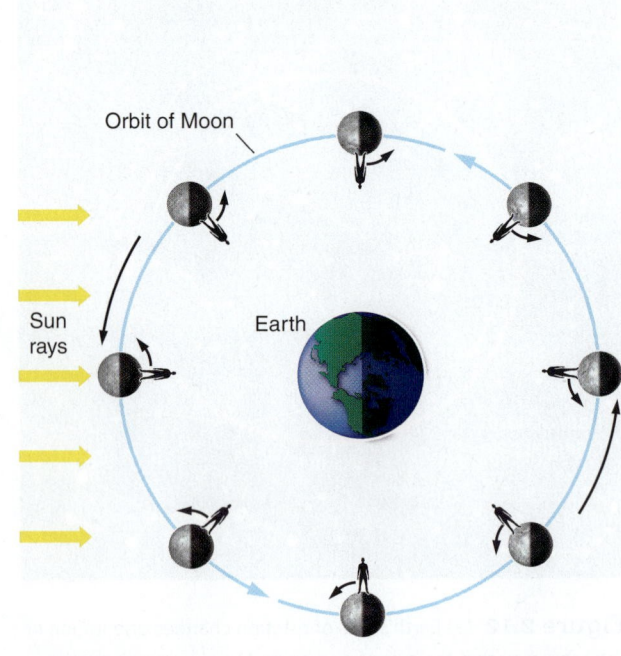

Figure 2.13 The Moon rotates once on its axis for each orbit around Earth—an effect called *synchronous rotation*. This causes the Moon to keep the same side facing toward Earth at all times. Relative to this illustration, the Sun is far to the left of the Earth-Moon system.

toward us is illuminated, so we see a full circle illuminated. The rest of the time, only a part of the illuminated portion can be seen from Earth. Sometimes the Moon appears as a circular disk in the sky. At other times, it is nothing more than a sliver.

To help you visualize the changing phases of the Moon, orient your orange (the Moon), your lamp (the Sun), and your head (Earth), as shown in **Figure 2.14**. Turn off all the other lights in the room, and stand as far from the lamp as you can. Hold the orange slightly above your head so that it is illuminated from one side by the lamp. Move the orange clockwise around you and watch how the appearance of the orange changes. When you are between the orange and the lamp, the face of the orange that is toward you is fully illuminated. The orange appears to be a bright, circular disk. As the orange moves around its circle, you will see a progression of lighted shapes, depending on how much of the bright side of the orange you can see. This progression of shapes mimics the changing phases of the Moon.

The progression of the phases of the Moon is shown in **Figure 2.15**, as viewed from the Northern Hemisphere. The Moon orbits Earth counterclockwise from this vantage point, so the phases proceed in order counterclockwise on this diagram.

The **new Moon** occurs when the Moon is between Earth and the Sun: The far side of the Moon is illuminated, but the near side is in darkness and we cannot see it. At this phase, the Moon is up in the daytime, because it is in the direction of the Sun. A new Moon is never visible in the nighttime sky. It appears close to the Sun

Figure 2.14 You can experiment with illumination effects by using an orange as the Moon, a lamp with no shade as the Sun, and your own head as Earth. As you move the orange around your head, viewing it in different relative locations, you will see that the illuminated part of the orange mimics the phases of the Moon.

▶❚❚ **AstroTour:** The Moon's Orbit: Eclipses and Phases

▶▶ **Interactive Simulation:** Phases of the Moon

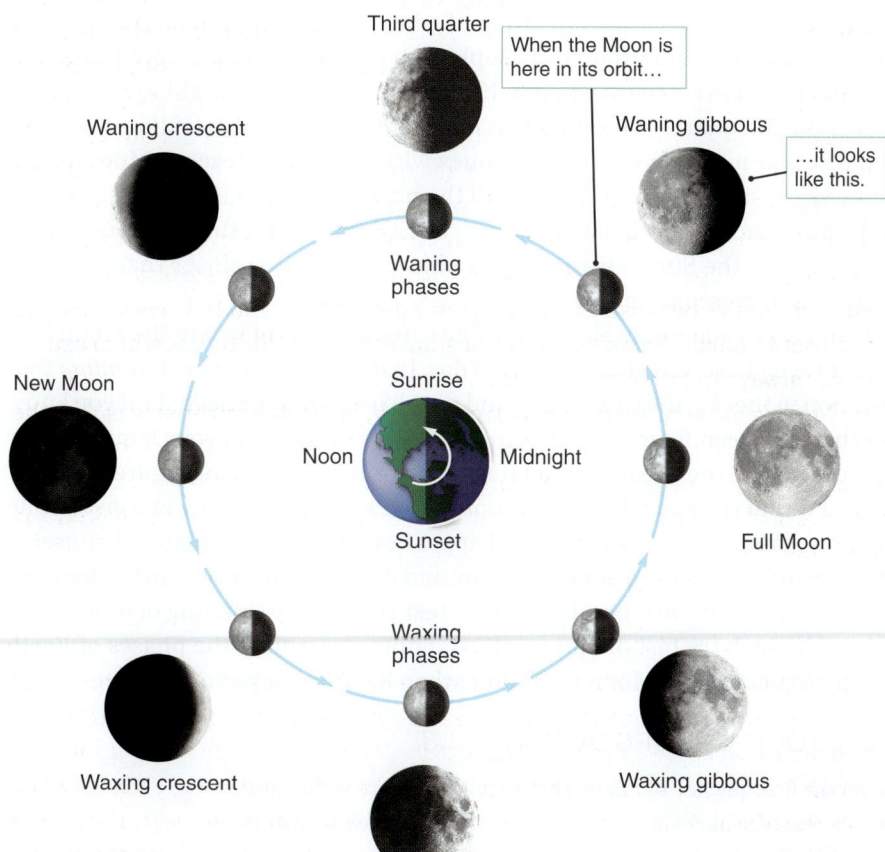

Third quarter

When the Moon is here in its orbit…

Waning crescent

Waning gibbous

…it looks like this.

Waning phases

New Moon

Sunrise

Noon   Midnight

Sunset

Full Moon

Waxing phases

Waxing crescent

Waxing gibbous

First quarter

Figure 2.15 The inner circle of images (connected by blue arrows) shows the Moon as it orbits Earth as seen by an observer far above Earth's North Pole. The outer ring of images shows the corresponding phases of the Moon as seen from an observer in Earth's Northern Hemisphere.

in the sky, so it rises with the Sun at sunrise, crosses the meridian near noon, and sets in the west with the Sun.

A few days later, as the Moon orbits Earth, a sliver of its illuminated half, known as a **crescent**, becomes visible. Because the Moon appears to be "filling out" from night to night at this time, this phase of the Moon is called a **waxing** crescent Moon. (*Waxing* here means "growing in size and brilliance.") During the week that the Moon is in this phase, the Moon is visible east of the Sun. It is most noticeable just after sunset, near the western horizon, when it will be lit on the side of the disk closest to the horizon—the west side.

As the Moon moves farther along in its orbit and the angle between the Sun and the Moon grows, more and more of the near side becomes illuminated, until half of the near side of the Moon is bright—this phase is called **first quarter Moon** because the Moon is one-quarter of the way through its orbit. At this point, because we see only one half of the Moon, and that side is half-illuminated, the Moon appears as a bright half-circle in the sky. The first quarter Moon rises at noon, crosses the meridian at sunset, and sets at midnight.

As the Moon moves beyond first quarter, more than half of the near side is illuminated—the phase is called a waxing **gibbous** Moon. The gibbous Moon waxes until finally we see the entire near side of the Moon—a **full Moon**. The Sun and the Moon are now opposite each other in the sky. The full Moon rises as the Sun sets, crosses the meridian at midnight, and sets in the morning as the Sun rises.

The second half of the Moon's cycle of phases is the reverse of the first half. As the Moon appears to shrink on its way from full to new Moon, it is called **waning**, which means "becoming smaller." When the Moon is waning, the east side appears bright. Just after full Moon, the Moon appears gibbous, but the near side is becoming less illuminated. This phase is called a waning gibbous Moon. A **third quarter Moon** occurs when half the near side is in sunlight and half is in darkness. A third quarter Moon rises at midnight, crosses the meridian near sunrise, and sets at noon. The cycle continues with a waning crescent Moon in the morning sky, rising before the Sun, until the new Moon once more rises and sets with the Sun, and the cycle begins again. Notice that when the Moon is farther than Earth from the Sun, it is in gibbous phases. When it is closer than Earth to the Sun, it is in crescent phases.

Do not try to memorize all the possible combinations of where the Moon is in the sky at each phase and at every time of day. Instead, work on *understanding* that the location in the sky, the time of day, and the phase are all connected—if you know two of these, you can figure out the third. Use your orange and your lamp or draw a picture to follow the Moon around its orbit. From your drawing, figure out what phase you would see and where the Moon would appear in the sky at a given time of day. For example, a full Moon would appear overhead at midnight; at sunset, a full Moon would be on the eastern horizon; and if you see the Moon on the western horizon at sunrise, it must be a full Moon. Test yourself by thinking of other combinations. As an extra test of your understanding, determine the phases of Earth that an astronaut on the Moon would see when looking back at our planet.

Astronomy in Action: Phases of the Moon

CHECK YOUR UNDERSTANDING 2.3

You see the first quarter Moon on the meridian. Where is the Sun?

a. on the western horizon

b. on the eastern horizon

c. below the horizon

d. on the meridian

what if . . .

What if the Moon were a cube rather than a sphere: How would this change the appearance of the Moon's phases?

2.4 Shadows Cause Eclipses

An **eclipse** occurs when the shadow of one astronomical body falls on another. A **solar eclipse** occurs when the Moon passes between Earth and the Sun and casts a shadow on Earth. This shadow is small, compared to the size of Earth, so observers need to be in the right place at the right time to see a solar eclipse. There are three types of solar eclipse: *total*, *partial*, and *annular*. A **total solar eclipse** (**Figure 2.16**) occurs when the Moon completely blocks the disk of the Sun. At any location, a total solar eclipse never lasts longer than 7½ minutes and is usually significantly shorter. Even so, it is one of the most amazing sights in nature. People all over the world travel great distances to see a total solar eclipse. The next one visible in the continental United States will occur on April 8, 2024.

A **partial solar eclipse** occurs when the Moon partially covers the disk of the Sun. An **annular solar eclipse** occurs when the Moon is slightly farther away from Earth in its noncircular orbit, so it appears slightly smaller in the sky. It is centered over the disk of the Sun but does not block the entire disk. A ring is visible around the blocked portion. Observers of any solar eclipse should be very careful to use eye protection or a pinhole camera to make their observations—observing an eclipse without these precautions can cause permanent eye damage!

Figure 2.17a shows the geometry of a solar eclipse, when the Moon's shadow falls on the surface of Earth. Figure 2.17b shows the geometry of a solar eclipse with Earth, the Moon, and the separation between them drawn to the correct scale. Compare this drawing to Figure 2.17a, and you will understand why drawings of Earth and the Moon are rarely drawn to the correct scale. If the Sun were drawn to scale in Figure 2.17b, it would be bigger than your head and located almost 64 meters off the left side of the page.

Figure 2.16 During a total eclipse, the Sun produces a remarkable spectacle, as the solar disk is completely covered by the Moon.

Astronomy in Action: The Earth-Moon-Sun System

(a) Solar eclipse geometry (not to scale)

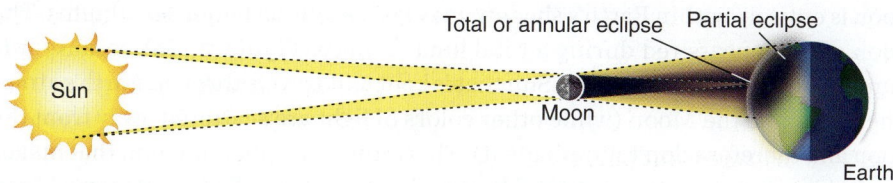

(b) Solar eclipse to scale

(c) Lunar eclipse geometry (not to scale)

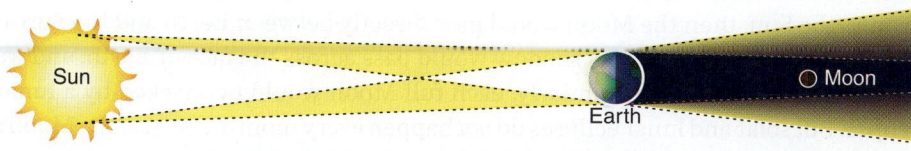

(d) Lunar eclipse to scale

Figure 2.17 (a, b) A solar eclipse occurs when the shadow of the Moon falls on the surface of Earth. (c, d) A lunar eclipse occurs when the Moon passes through Earth's shadow.

(a)

(b)

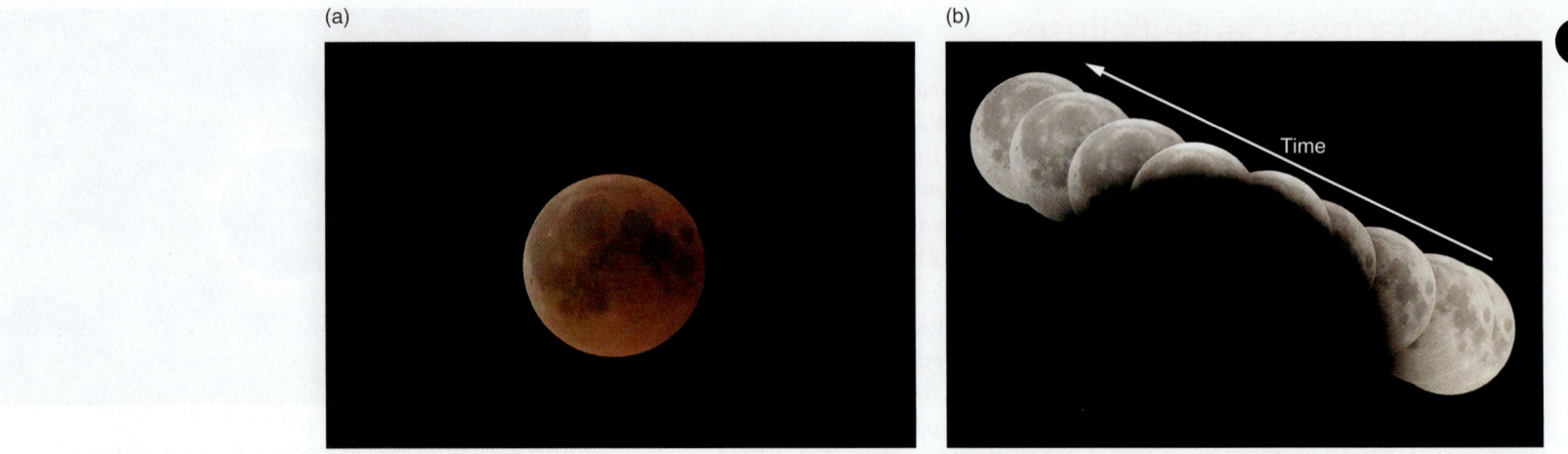

Figure 2.18   (a) During a total lunar eclipse, the Moon often appears blood-red. (b) A time-lapse series of photographs of a partial lunar eclipse clearly shows Earth's shadow. ★ **WHAT AN ASTRONOMER SEES** Astronomers learn to be sensitive to color variations. An astronomer looking at these two images for the first time would immediately conclude from the difference in color between the images that the Moon in the left image is in full lunar eclipse. An astronomer will also notice that in the montage the images of the Moon were carefully aligned, so that in each image Earth's shadow remained stationary. Combining the images in this way makes it possible to see (or measure) the size of Earth's shadow relative to the Moon.

what if . . .

What if Earth had two moons, having the same size and same orbit as each other but 120° apart in that orbit: Would we expect to see more solar eclipses in this case? Would the eclipses for both moons occur during the same eclipse seasons?

A **lunar eclipse** occurs when the Moon is partially or entirely in Earth's shadow. A lunar eclipse is very different in character from a solar eclipse. The geometry of a lunar eclipse is shown in Figure 2.17c and is drawn to scale in Figure 2.17d. Because Earth is much larger than the Moon, Earth's shadow at the distance of the Moon is more than twice the diameter of the Moon. Thus, all observers on the nighttime side of Earth will be able to see a lunar eclipse. A **total lunar eclipse**, when the Moon is entirely within Earth's shadow, may last as long as 1 hour 40 minutes. The Moon often appears red during a total lunar eclipse (**Figure 2.18a**) because it is illuminated by red light from the Sun that is bent as it travels through Earth's atmosphere and hits the Moon (while other colors of light are scattered away from the Moon and therefore don't illuminate it). This is the same phenomenon that makes the sky blue and sunsets red. A total lunar eclipse is often called a blood-red Moon in literature and poetry.

If Earth's shadow incompletely covers the Moon, some of the disk of the Moon remains bright and some of it is in shadow. This is called a **partial lunar eclipse**. Figure 2.18b shows a composite of images taken at different times during a partial lunar eclipse, in which the Moon is nearly completely eclipsed by Earth's shadow.

Imagine Earth, the Moon, and the Sun all sitting on the same flat tabletop. If the Moon's orbit around Earth were in exactly the same plane as the orbit of Earth around the Sun, then the Moon would pass directly between Earth and the Sun at every new Moon. The Moon's shadow would pass across the face of Earth, and we would see a solar eclipse. Similarly, each full Moon would be marked by a lunar eclipse. But solar and lunar eclipses do *not* happen every month, because the Moon's orbit does not lie in *exactly* the same plane as the orbit of Earth. As you can see in **Figure 2.19**, the plane of the Moon's orbit around Earth is inclined by about 5° with respect to the plane of Earth's orbit around the Sun. Most of the time, the Moon is "above" or "below" the line between Earth and the Sun. About twice per year, the orbital planes line up at points called *nodes*, and eclipses can occur. These two times of year are sometimes called *eclipse seasons*.

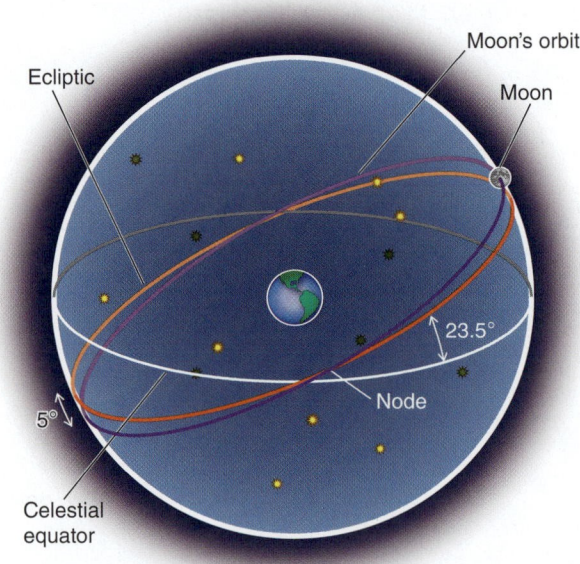

Figure 2.19 The orbit of the Moon is tilted with respect to the ecliptic, so we do not see eclipses every month. If the Sun were located to the left in this figure, its light could still reach the Moon. The Moon would be in full phase but not eclipsed by Earth's shadow.

CHECK YOUR UNDERSTANDING 2.4

If the Moon were in the same orbital plane as Earth's orbit around the Sun, which of the following would happen? (Choose all that apply.)

a. The phases of the Moon would remain unchanged.

b. There would be a solar eclipse every lunar orbit (roughly every month).

c. There would be a lunar eclipse every lunar orbit (roughly every month).

d. The Moon's cycle of phases would not occur.

---

reading Astronomy News

Eclipse 2017: "I need like two more hours of that"

by Jacy Marmaduke and Cassa Niedringhaus, *Coloradoan*

The 2017 total solar eclipse drew people from around the world to sites across the United States. Casper, WY, was particularly popular because the sky was likely to be free of clouds.

CASPER, Wyoming—Two minutes and 26 seconds has never vanished so quickly.

Casper, blessed with one of the longest periods of totality in the nation, lost its sun to a total solar eclipse Monday morning as thousands of onlookers hooted and hollered into the sudden twilight on the Wyoming city's broad, windswept plains.

The eclipse came slowly at first, the moon blacking out the sun bit by bit as eclipse-goers turned collectively skyward.

And then, with minutes to go until totality, the sky dimmed. The air grew chilly as babies cried, dogs barked and cameras clicked.

The light seemed to be falling from the sky.

As the last bit of the sun blinked away, gasping in a final flash of light, cheers rang out. The brilliant white outline that remained seemed to shimmer to the beat of the crowd's cries. A golden-pink sunset burst out from the horizon in every direction. The flame of a hot air balloon flickered in the suddenly-night sky.

Just as quickly as it left, the sun returned to a chorus of exaltation. Onlookers howled themselves hoarse as the sky slipped off its robe of darkness.

"I need like two more hours of that," a woman yelled in the distance.

Thousands from around the world set out for this Western city of 55,000, drawn like moths to the flame of Casper's wide-open landscape, promise of visibility and lengthy period of totality.

Among them were the Meades, a family that recently moved to Superior, Colorado, from Arlington, Virginia. Eric and Julia Meade and their children, 10-year-old Charlotte and 6-year-old Daniel, trekked to Casper to witness the eclipse together. They settled in the grass of Fort Caspar Park, eclipse glasses affixed over their eyes.

"I like that we're making the time to do fun things that happen, while they're kids." Eric said. "... They'll know that we cared enough to take them out and see the fun things that were going on."

Charlotte reached for her father as the sky darkened, and Daniel sat in the grass before hopping around the lawn In excitement.

"The kids were enchanted—it was so mysterious," Julia said. "... I'm feeling really renewed."

Daniel said that just as the sun began to reappear, he saw a shooting star cross the sky.

"Amazing!" he said.

The family shared the lawn with groups from places as far flung as Japan and Germany, as well as a couple celebrating an 80th birthday and a seventh shared eclipse.

Bill and Kath Conolly—Kath celebrated her 80th birthday Sunday—fit Casper into a nine-day trip from British Columbia and through Montana and Wyoming. They've taken their children and grandchildren to eclipses all over the world, and they're scheduled to fly to Argentina's wine country in 2020 for their eighth eclipse.

Of all the places they've been, Casper has charmed them most, they said.

"We have been absolutely dazzled by the Midwest of the United States," Bill said. "We've been to all four corners, Alaska, Puerto Rico, Hawaii, you name it. But the friendliness and the actual beauty of the towns (are incredible)."

The influx of visitors brought a new luster to Wyoming, where a slowdown in oil and gas drilling has stagnated economic growth.

"Our economy in town is not so good," said Jason Laird, manager of Casper's Sterling Hotel. "This event has helped people get through the winter, so to speak."

That's certainly the case for Laird. After lightning struck the hotel he and his wife, Amber, manage in June 2016, an issue with the building's insurance policy forced them to close the business. It cost them their life savings.

Six weeks ago, the Lairds decided to try reopening for the eclipse. Inflated hotel rates—we're talking upward of $1,000 a room—allowed them to fund crucial renovations, carpet-cleaning, pillows and all the other essentials.

All but five of the hotel's 67 rooms were booked as of Sunday night, not including a woman sleeping in the hotel's conference room and a man camping in the parking lot.

Laird's glad so many people chose Casper for its "wide open spaces."

"You feel free here," he said. "It's not congested like a big city."

Casper's downtown had a spit-shined quality as throngs of tourists slid beside old brick storefronts and eclipse vendor tents Sunday. Musicians performed under the wooden awnings of David Street Station, a new events area that has served as a beacon for downtown development.

One of Casper's oldest establishments, the World Famous Wonder Bar, reopened five days ago as the C85 Wonder Bar, swapping cheeseburgers and pool tables for Wagyu tenderloin and quilted leather booths.

In the 1930s, visitors could order a beer at the bar without dismounting from their horses. On Sunday, Justin Timberlake blasted from the speakers of the food court-style sports bar upstairs. The upscale restaurant downstairs was sold out of half a dozen options on its special "Eclipse Menu" and had teemed with visitors all weekend, employees said.

The eclipse proved the state can handle a huge volume of visitors, said Diane Shober, executive director for the Wyoming Office of Tourism.

According to Wyoming Department of Transportation, the state's traffic counts increased by more than 497,000 vehicles between Wednesday and Sunday compared to five-year averages for the same period.

In October, the state will release an economic analysis of eclipse visitor volume and spending. Shober's excited for the results, which she predicts will be "fabulous." But she's more eager for what's next.

"We just want to capitalize on that momentum," she said. "You have to make hay while the sun is shining, and that's our job."

As for Laird, he thanks the eclipse for saving his job. Income from eclipse bookings has allowed the Sterling Hotel, named after his 4-year-old daughter, to stay in business for the foreseeable future.

"I woke up this morning and realized one natural event stopped everything for us," Laird said. "And then, a year later, another brought it all back."

EVALUATING THE NEWS

1. Did you get to see the total solar eclipse of August 21, 2017? If so, where were you? Was the event a total solar eclipse or a partial solar eclipse from your location?

2. Was it possible for anyone on Earth to observe an annular eclipse on August 21, 2017?

3. Make a diagram (not to scale) of the Earth-Moon-Sun system as it would have appeared from far above Earth's North Pole on August 21, 2017.

4. Explain how this natural event changed the town of Casper, Wyoming. What is the local impact of "astronomy tourism"?

SUMMARY

The motions of Earth and the Moon are responsible for many of the repeating patterns that can be observed in the sky and form the basis of our modern calendar. Daily patterns of rising and setting are caused by Earth's rotation about its axis. Annual patterns of the stars in the sky and the passage of the seasons are caused by Earth's revolution around the Sun. The tilt of Earth on its axis changes both the length of daytime and the intensity of sunlight, causing the seasons. The pattern of the phases of the Moon lasts roughly a month and is caused by the Moon's revolution around Earth. Occasionally, special alignments of Earth, the Moon, and the Sun cause eclipses.

① As Earth rotates on its axis, stars follow circular arcs through the sky. For an observer at the equator, these are circular arcs from the eastern horizon to the western horizon. At higher latitudes, some stars move in complete circles around a celestial pole. At Earth's North Pole or South Pole, the observer sees all the stars move in complete circles in the sky.

② As Earth revolves around the Sun, the nighttime side of Earth faces different directions in space, and different stars are visible at night.

③ Seasons occur at different times in the Northern and Southern Hemisphere; when it is winter in the north, it is summer in the south. Because Earth is tilted on its axis, the length of daytime and the intensity of sunlight change throughout the year, causing the seasons.

④ The Moon's motion around Earth causes it to appear to be illuminated differently at different times. When the Moon is farther than Earth from the Sun, it is in gibbous phases. When it is closer than Earth to the Sun, it is in crescent phases. Roughly twice a year, at new or at full Moon, the Moon is exactly in line with Earth and the Sun. At these times, eclipses occur.

QUESTIONS AND PROBLEMS

TESTING YOUR UNDERSTANDING

1. **T/F:** The celestial sphere is not an actual object in the sky.

2. **T/F:** Eclipses happen somewhere on Earth every month.

3. **T/F:** The phases of the Moon are caused by the relative position of Earth, the Moon, and the Sun.

4. **T/F:** If a star rises north of east, it will set south of west.

5. **T/F:** Viewed from the North Pole, all stars in the night sky are circumpolar stars.

6. The Sun, Moon, and stars
   a. change their relative positions over time.
   b. appear to move each day because the celestial sphere rotates about Earth.
   c. rise north or south of west and set north or south of east, depending on their location on the celestial sphere.
   d. always remain in the same position relative to each other.

7. Which stars we see at night is dependent on (select all that apply)
   a. our location on Earth.
   b. Earth's location in its orbit.
   c. the time of the observation.
   d. the motion of stars relative to one another over the course of the year.

8. You see the Moon rising, just as the Sun is setting. What phase is it in?
   a. full
   b. new
   c. first quarter
   d. third quarter
   e. waning crescent

9. Where on Earth can you stand and, over the course of a year, see the entire sky?
   a. only at the North Pole
   b. at either pole
   c. at the equator
   d. anywhere

10. You do not see eclipses every month because
    a. all eclipses happen at night.
    b. the Sun, Earth, and the Moon line up only twice a year.
    c. the Sun, Earth, and the Moon line up only once a year.
    d. eclipses happen randomly and are unpredictable.

11. The tilt of Earth's axis causes the seasons because
    a. one hemisphere of Earth is closer to the Sun in summer.
    b. the days are longer in summer.
    c. the rays of light strike the ground more directly in summer.
    d. both a and b
    e. both b and c

12. On the vernal and autumnal equinoxes,
    a. the entire Earth has 12 hours of daylight and 12 hours of darkness.
    b. the Sun rises due east and sets due west.
    c. the Sun is located on the celestial equator.
    d. all of the above
    e. none of the above

13. We always see the same side of the Moon because
    a. the Moon does not rotate on its axis.
    b. the Moon rotates once each revolution.
    c. when the other side of the Moon is facing toward us, it is unlit.
    d. when the other side of the Moon is facing Earth, it is on the opposite side of Earth.
    e. none of the above

14. You see the Moon on the meridian at sunrise. The phase of the Moon is
    a. waxing gibbous.
    b. full.
    c. new.
    d. first quarter.
    e. third quarter.

15. A lunar eclipse occurs when the _____ shadow falls on the _____.
    a. Earth's; Moon
    b. Moon's; Earth
    c. Sun's; Moon
    d. Sun's; Earth

16. You see the full Moon on the meridian. From this information, you can determine that the time where you are is
    a. noon.
    b. sunrise, about 6:00 A.M.
    c. sunset, about 6:00 P.M.
    d. midnight.

17. What do we call the group of constellations through which the Sun appears to move over the course of a year?
    a. the celestial equator
    b. the ecliptic
    c. the line of nodes
    d. the zodiac

18. If you were standing at Earth's South Pole, which stars would you see rising and setting?
    a. all of them
    b. all the stars north of the Antarctic Circle
    c. all the stars south of the Antarctic Circle
    d. none of them

19. On the summer solstice in the Northern Hemisphere, the Sun
    a. rises due east, passes through its highest point on the meridian, and sets due west.
    b. rises north of east, passes through its highest point on the meridian, and sets north of west.
    c. rises north of east, passes through its highest point on the meridian, and sets south of west.
    d. rises south of east, passes through its highest point on the meridian, and sets south of west.
    e. none of the above

20. In the Tropics,
    a. the Sun is directly overhead twice per year.
    b. the Sun's rays strike Earth precisely perpendicularly twice per year.
    c. the seasons vary less than elsewhere on Earth.
    d. all of the above

THINKING ABOUT THE CONCEPTS

21. Polaris was used for navigation by seafarers such as Columbus as they sailed from Europe to the New World. When Magellan sailed the South Seas, he could not use Polaris for navigation. Explain why.

22. Suppose you are on a plane from Dallas, Texas to Santiago, Chile. On the way there, you realize something amazing. You have just experienced the longest day of the year in the Northern Hemisphere and are about to experience the shortest day of the year in the Southern Hemisphere on the same day! On what day of the year are you flying? How do you explain this phenomenon to the person in the seat next to you?

23. Vampires often appear in popular fiction. These creatures have extreme responses to even a tiny amount of sunlight (the response depends on the author), but moonlight doesn't affect them at all. Is this logical? How is moonlight related to sunlight?

24. We tend to associate certain constellations with certain times of year. For example, we see the zodiacal constellation Gemini in the Northern Hemisphere's winter (Southern Hemisphere's summer) and the zodiacal constellation Sagittarius in the Northern Hemisphere's summer. Refer to Figure 2.9 to explain why we do not see Sagittarius in January or Gemini in June. ⭐

25. Describe the Sun's apparent daily motion on the celestial sphere at the vernal equinox.

26. Why is winter solstice not the coldest time of year?

27. In your study group, two of your fellow students are discussing the phases of the Moon. One argues that the phases are caused by the shadow of Earth on the Moon. The other argues that the phases are caused by the orientation of Earth, the Moon, and the Sun. Explain how the appearance of the Moon at various locations (as shown in Figure 2.15) could be used to falsify one of these hypotheses. ⭐

28. Romance novelists sometimes have the hero ride off into the sunset when the full Moon is overhead. Is this correct? Why or why not? Use Figure 2.15 to create a sketch of the Sun, the Moon, and Earth at full Moon phase that explains your answer. ⭐

29. What is the approximate time of day when you see the full Moon near the meridian? At what time is the first quarter (waxing) Moon on the eastern horizon? Use a sketch to help explain your answers.

30. Assume that the Moon's orbit is circular. Suppose you are standing on the side of the Moon that faces Earth.
    a. How would Earth appear to move in the sky as the Moon made one revolution around Earth?
    b. How would the "phases of Earth" appear to you compared to the phases of the Moon as seen from Earth?

31. Astronomers are sometimes asked to serve as expert witnesses in court cases. Suppose you are called in as an expert witness, and the defendant states that he could not see the pedestrian because the full Moon was casting long shadows across the street at midnight. Is this claim credible? Why or why not?

32. ★ WHAT AN ASTRONOMER SEES Figure 2.18b shows the shadow of Earth, projected onto the Moon over the course of a partial lunar eclipse. From this shadow, what can you determine about the shape of Earth? Many observations like this, taken from many different locations on Earth in different years, show the same shadow shape. From those observations, what can you determine about the shape of Earth? ⭐

33. Explain how you could determine from Figure 2.18b that Earth's shadow is larger than the Moon. Approximately how many times larger than the Moon is Earth's shadow in this image? ⭐

34. From your own home, why are you more likely to witness a partial eclipse of the Sun rather than a total eclipse?

35. Why don't we see a lunar eclipse each time the Moon is full or witness a solar eclipse each time the Moon is new?

APPLYING THE CONCEPTS

36. Working It Out 2.1 shows how to manipulate the equation for speed to find the distance traveled.
    a. Rework this algebraic manipulation to solve for time instead.
    b. Check your work by putting the units for speed (kilometers per hour) and distance (kilometers) into your result and showing that the answer comes out in units of time.

37. Converting between units of time is a skill we will use often in this book. To convert time values from one unit to another, we need to multiply or divide, as we did in Working It Out 2.1 (sometimes by more than one factor to convert from hours to seconds, for example). Perform the unit conversions below, by following these four steps:
    1. Predict: Think about whether the resulting number should be bigger or smaller than the original number.
    2. Decide: Whether to multiply or divide by the conversion factor(s); the number of minutes in an hour, for example.
    3. Act: Multiply or divide by the conversion factor. Include the units so that you can verify that the units cancel out in the way you require.
    4. Check: Compare your result with the given number to check your work.
       a. Convert 65 hours into days.
       b. Convert 1.89 years into days.
       c. Convert 60 seconds into days.

38. The Moon has a circumference of about 11,000 km. Follow the process of Working It Out 2.1 to find the speed at which a point at the equator of the Moon is carried around by the Moon's rotation in kilometers per hour. Don't forget to convert the Moon's rotation period to hours!
    a. The Moon's rotation period and its orbital period are the same. What is the Moon's orbital period, in days?
    b. Make a prediction: do you expect a point on the equator of the Moon to move faster or slower than a point on Earth's equator?
    c. Convert that orbital period to hours.
    d. Calculate the speed of a point on the equator of the Moon.
    e. Check your work by comparing this speed to the speed of a point on Earth's equator (given in Working It Out 2.1). Was your prediction correct?

39. The circumference of a circle is equal to $2 \times \pi \times r$; this relationship can be compactly written as $C = 2\pi r$.
    a. Reorganize this equation to solve for the radius r.
    b. Earth's circumference is 40,070 km. Use your equation from part (a) to find Earth's radius. Check your work by considering whether the radius should be larger or smaller than the circumference.

40. The radius of Earth's orbit is 1.49×10^8 km. Earth takes 1 year to orbit the Sun. Find the speed of Earth in its orbit in kilometers per hour. Check your work by including the units in each step, making sure the result is in the correct units, and by thinking about the size of this number. It should be faster than a jet (about 800 km/h) but slower than the speed of light (300,000 km/h).

exploration Phases of the Moon

digital.wwnorton.com/universe4

In this Exploration, we will be examining the phases of the Moon. Visit the Student Site at the Digital Resources page and open the Phases of the Moon Interactive Simulation in Chapter 2. This simulator animates the orbit of the Moon around Earth, allowing you to control the simulation speed and a number of other parameters.

Begin by starting the animation to explore how it works. Examine all three image frames. The large frame shows the Earth-Moon system, as looking down from far above Earth's North Pole. The upper right frame shows what the Moon looks like to the person on the ground. The lower right frame shows where the Moon appears in the person's sky. Stop the animation, and press "reset" in the upper menu bar.

1. What time of day is this for the person shown on Earth?

_____

_____

_____

_____

2. What phase is the Moon in?

_____

_____

_____

_____

3. Where is the Moon in this person's sky?

_____

_____

_____

_____

Run the animation until the Moon reaches waxing crescent phase.

4. As viewed from Earth, which side of the Moon is illuminated (the left or the right)?

_____

_____

_____

_____

5. The person shown on Earth will observe this waxing crescent Moon either after sunset or before sunrise. At which of these times can the person see the waxing crescent Moon?

_____

_____

_____

_____

Run the animation until the Moon reaches first quarter and the Sun is setting for the person on Earth. (Hint: You may want to slow the animation rate!)

6. How many full days have passed since new Moon?

_____

_____

_____

_____

7. At this instant, where is the first quarter Moon in the person's sky?

_____

_____

_____

_____

8. If an astronaut were standing on the near side of the Moon at this time, what phase of Earth would he see?

_____

_____

_____

_____

Three observations about the phases of the Moon are connected: the location of the Moon in the sky, the time for the observer, and the phase of the Moon. If you know two of these, you can figure out the third. Use the animation to fill in the missing pieces in the following situations:

9. An observer sees the Moon in _____ phase, overhead, at midnight.

10. An observer sees the Moon in third quarter phase, rising in the east, at _____.

11. An observer sees the Moon in full phase, _____, at 6:00 A.M.

Laws of Motion

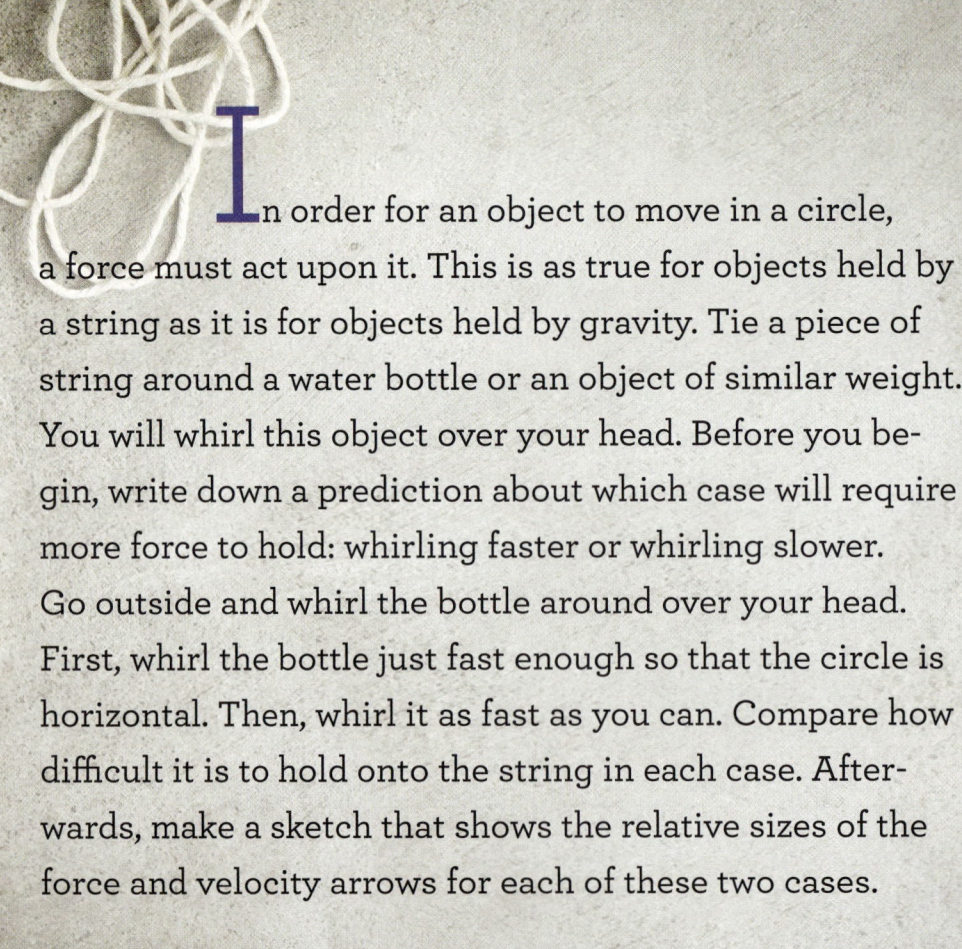

In order for an object to move in a circle, a force must act upon it. This is as true for objects held by a string as it is for objects held by gravity. Tie a piece of string around a water bottle or an object of similar weight. You will whirl this object over your head. Before you begin, write down a prediction about which case will require more force to hold: whirling faster or whirling slower. Go outside and whirl the bottle around over your head. First, whirl the bottle just fast enough so that the circle is horizontal. Then, whirl it as fast as you can. Compare how difficult it is to hold onto the string in each case. Afterwards, make a sketch that shows the relative sizes of the force and velocity arrows for each of these two cases.

EXPERIMENT SETUP

SLOWER

FASTER

First, spin the bottle just fast enough so that the circle is horizontal.

Then, spin it as fast as you can.

PREDICTION

I predict that the string will be
☐ easier ☐ harder
to hold when the water bottle is faster.

SKETCH OF RESULTS

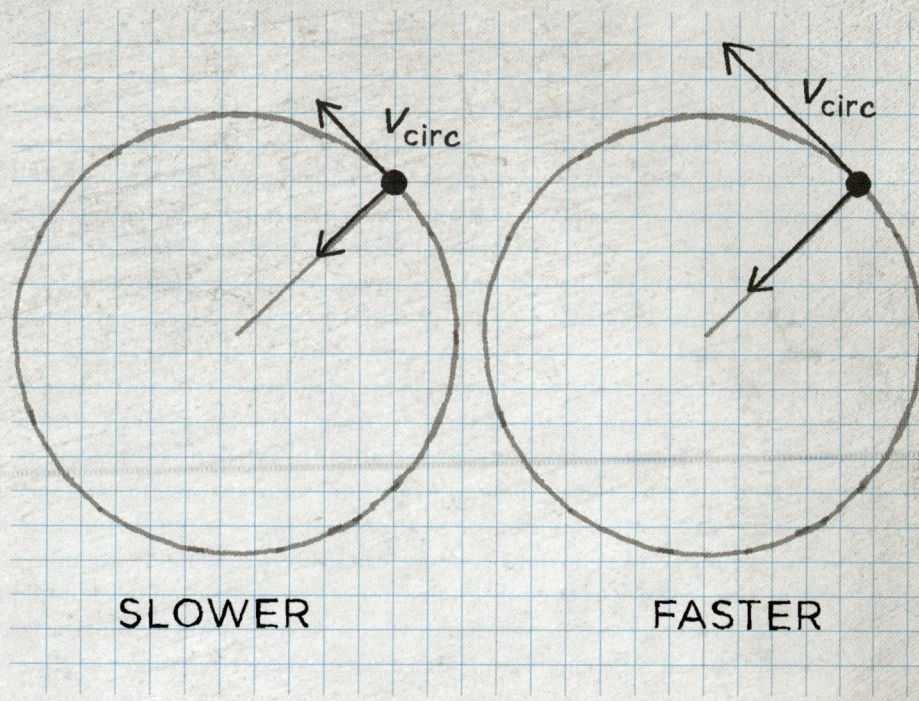

v_{circ}

v_{circ}

SLOWER

FASTER

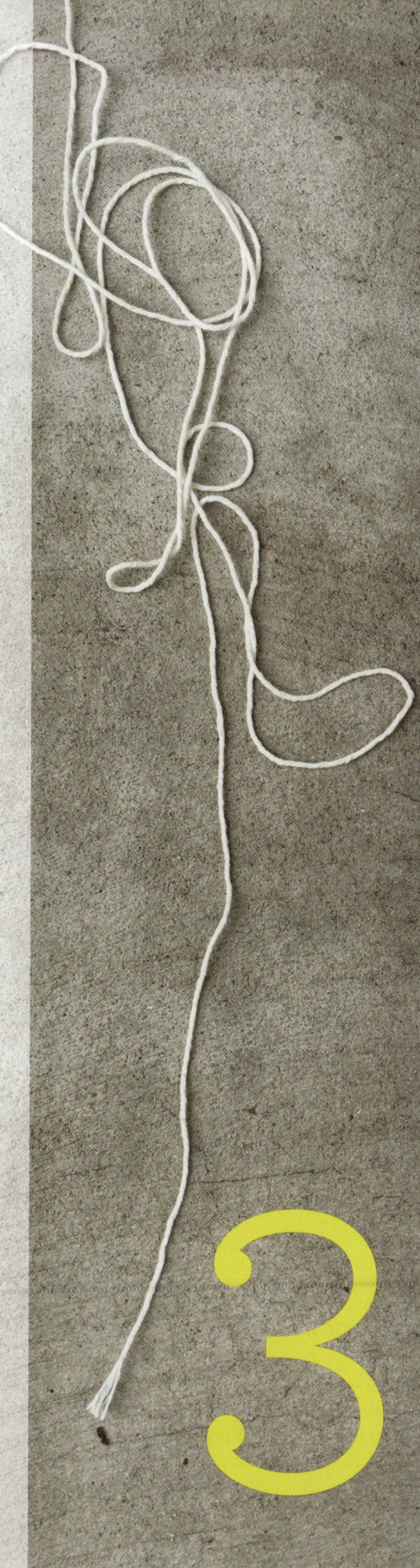

3

In Chapter 2, you learned that Earth orbits the Sun, but not the reason why. Gravity is the force that holds all the planets in orbit. Because the Sun is far more massive than all the other parts of the Solar System combined, its gravity shapes the motions of every object in the Solar System, from the almost circular orbits of some planets to the extremely elongated orbits of comets. Our Solar System is one example of gravity at work.

LEARNING GOALS

(1) Describe each of Kepler's three laws.

(2) State Newton's three physical laws that govern the motion of all objects.

(3) Combine Newton's laws of motion and gravitation to explain why planets remain in orbit.

(4) Identify different frames of reference.

3.1 Astronomers Have Studied the Motions of the Planets since Ancient Times

Astronomy challenges us to think in novel ways, to imagine ourselves in space, looking down at Earth, or even farther away, looking down at the Solar System. The struggle to understand our place in the universe begins with understanding why Earth, the Sun, and the planets move as they do. The history of the progression of ideas about this—from Earth at the center of all things to Earth as a tiny, insignificant rock—is full of heroes and villains and is a wonderful example of the self-correcting nature of science.

Early Astronomy

In ancient times, some astronomers and philosophers hypothesized that the Sun might be the center of the Solar System, but they did not have the tools to test the hypothesis or the mathematical insight to formulate a more complete and testable model. Because we can't feel Earth's motion through space, a **geocentric**—that is, Earth-centered—**model** of the Solar System prevailed. For nearly 1,500 years, most educated people in the West believed that the Sun, the Moon, and the known planets (Mercury, Venus, Mars, Jupiter, and Saturn) all moved in circles around a stationary Earth.

Ancient peoples were aware that planets appear to move in a generally eastward direction among the "fixed stars." They also knew that these planets would sometimes move in **apparent retrograde motion**; that is, they would seem to turn around, move westward for a while, and then return to their normal eastward travel. **Figure 3.1** shows a time-lapse sequence of Mars going through its retrograde "loop." This odd behavior of the five then-known planets—including Venus and Mercury—created a puzzling problem for the geocentric model: if the planets moved in circles around a stationary Earth, the only explanation for the retrograde motion would be that the planets stop, turn around, and go the other way. But why or how could they do that? The Greek astronomer Ptolemy (Claudius Ptolemaeus, 90–168 CE)

VOCABULARY ALERT

model In common language, a *model* is typically a scaled-down, three-dimensional version of a larger object. A model of a car, for example, is as close to the appearance of the real thing as possible, but often it doesn't actually work. In science, a model is a description of a system that accounts for its properties.

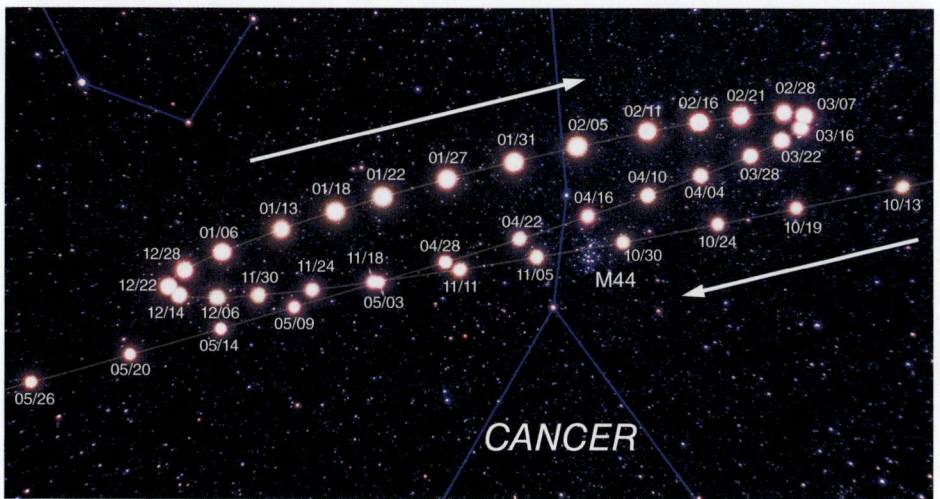

Figure 3.1 This time-lapse photographic series shows Mars as it moves in apparent retrograde motion. ★ **WHAT AN ASTRONOMER SEES** An astronomer will notice from the dates shown that the photographic series begins with Mars on the right side of the image. As time passes, Mars moves toward the left, then completes the loop, and leaves the image on the left, about 7.5 months later. An astronomer will also be charmed to notice that Mars is larger and brighter on the image in the retrograde part of the loop, indicating that Earth and Mars are closer together at that time. This color image clearly shows that some stars are red and some stars (like the ones in the group of stars called M44) are blue.

modified the geocentric model with a complex system of interconnected circles to try to both obtain more accurate results and explain retrograde motion (**Figure 3.2**).

The Copernican Revolution

Nicolaus Copernicus (1473–1543) was not the first person to consider the idea that the Sun was at the center of the Solar System. Aristarchus, for example, proposed this in about 280 BCE. But Copernicus was the first person to develop a mathematical model that made testable predictions about the planetary orbits. It took many decades to actually carry out the observations and test this model, but this work is known as the beginning of the Copernican Revolution. Through the later work of scientists such as Tycho Brahe (1546–1601), Galileo Galilei (1564–1642), Johannes Kepler (1571–1630), and Sir Isaac Newton (1642–1727), the **heliocentric**—or Sun-centered—theory of the Solar System became one of the most well-corroborated theories in all of science.

In 1543, Copernicus published a heliocentric model in a treatise called *De revolutionibus orbium coelestium* ("On the Revolutions of the Heavenly

▶❚❚ **AstroTour:** The Celestial Sphere and the Ecliptic

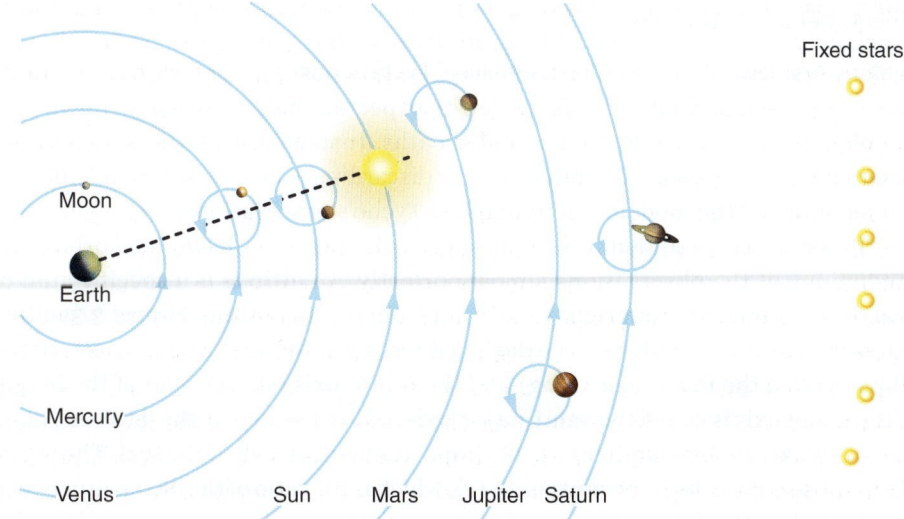

Figure 3.2 Ptolemy's model of the Solar System included a complex system of interconnected circles to explain retrograde motion. While traveling along its larger circle, a planet would at the same time be moving along its smaller circle. At times, the motions would be in opposite directions, creating the observed retrograde motion. In this model, the centers of the circles for Mercury and Venus must always be between Earth and the Sun, as indicated by the dashed line.

(a)

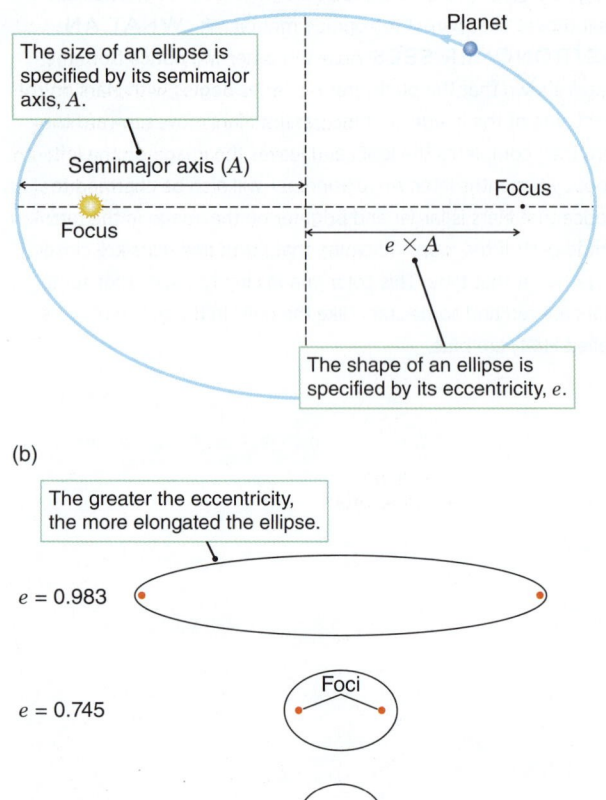

The size of an ellipse is specified by its semimajor axis, A.

Semimajor axis (A)

Focus

Focus

$e \times A$

Planet

The shape of an ellipse is specified by its eccentricity, e.

(b)

The greater the eccentricity, the more elongated the ellipse.

$e = 0.983$

$e = 0.745$ Foci

$e = 0.25$

$e = 0$ When the distance between foci is zero, e is also zero.

Figure 3.3 (a) Planets move in elliptical orbits with the Sun at one focus. The average radius of the orbit is equal to the semimajor axis, and the shape of the orbit is given by its eccentricity, e. (b) Ellipses range from circles to elongated eccentric shapes. All of the planets in the Solar System have eccentricities smaller than 0.25.

what if . . .

What if Earth's orbit were highly elliptical, with Earth being closest to the Sun in northern winter: What factors might cause life to be different in the northern and southern hemispheres in that case?

VOCABULARY ALERT

focus (plural *foci*) In everyday language, this word is used in several ways to indicate directed attention or the place where light is concentrated by a lens. In this mathematical context, it refers to a special point within an ellipse. An ellipse has two of these special points; the distance between them shrinks as the ellipse becomes more circular.

Spheres") that explained retrograde motion much more simply than Ptolemy's modifications to the geocentric model. Imagine that you are traveling in a car and you pass a slower-moving car. From your point of view, the other vehicle appears to move backward when viewed against distant objects on the horizon. Your point of view is formally known as a **frame of reference**: a system within which an observer measures positions and motions using coordinates such as distance and time. In your frame of reference, you are stationary, and you measure all motions relative to you. Copernicus provided a Sun-centered frame of reference for the Sun and its planets, in which the Sun is stationary but all the planets, including Earth, orbit around it. In the Copernican model, the outer planets Mars, Jupiter, and Saturn undergo apparent retrograde motion when Earth overtakes them in their orbits. Likewise, the inner planets Mercury and Venus move in apparent retrograde motion when overtaking Earth. Except for the Sun and the Moon, all Solar System objects exhibit apparent retrograde motion. The size of the effect decreases with increasing distance from Earth. In this frame of reference, retrograde motion is easily understood as an illusion caused by the relative motion between Earth and the other planets.

Combining geometry with observations of the positions of the planets in the sky, including their altitudes and the times they rise and set, Copernicus estimated planet–Sun distances in terms of Earth–Sun distance. These relative distances are remarkably close to those obtained by modern methods. From these observations, Copernicus also found when each planet aligns with Earth and the Sun. He used this information and geometry to figure out how long it took each planet to orbit the Sun. His model made testable predictions of the location of each planet on a given night.

Kepler's Laws

Tycho Brahe made careful measurements of the precise positions of planets in the sky. These measurements were the most comprehensive set of planetary data available at that time. When he died, his assistant, Johannes Kepler, received these records. Kepler used the data to deduce three rules that elegantly and accurately describe the motions of the planets. These three rules are now generally referred to as **Kepler's laws**. These laws are **empirical**: They use prior data to make predictions about future behavior but do not include an underlying theory of why the objects behave as they do.

Kepler's first law. When Kepler compared Tycho's observations with predictions that were based on Copernicus's model, he expected the data would confirm that the planets orbit the Sun along circular paths. Instead, his predictions and the observations disagreed. Rather than simply discarding Copernicus's model, though, Kepler adjusted the model until it matched Tycho's observations.

Kepler discovered that if he replaced circular orbits with *elliptical* orbits, the predictions fit the observations almost perfectly. An **ellipse** is a specific kind of oval. It is symmetric from right to left and from top to bottom. **Figure 3.3a** illustrates the features of ellipses. The dashed lines represent the two main axes of the ellipse, called the major axis (long) and the minor axis (short). Half of the length of the major axis is called the **semimajor axis**, often denoted by the letter A. Along the major axis are two mathematically important points, called the **foci**. The shape of an ellipse is given by its **eccentricity (e)**, which is the ratio of the distance between foci to the length of the major axis. This eccentricity is 0 for a perfect circle and is always less than 1 for any ellipse, as shown in Figure 3.3b. As the two foci approach

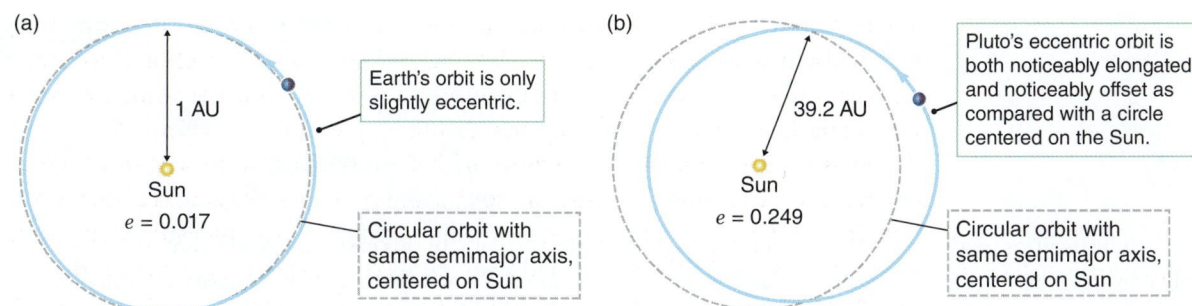

(a) Earth's orbit is only slightly eccentric.

1 AU

Sun
$e = 0.017$

Circular orbit with same semimajor axis, centered on Sun

(b) Pluto's eccentric orbit is both noticeably elongated and noticeably offset as compared with a circle centered on the Sun.

39.2 AU

Sun
$e = 0.249$

Circular orbit with same semimajor axis, centered on Sun

Figure 3.4 When the orbits of (a) Earth and (b) Pluto are compared with circles centered on the Sun, it becomes clear that they are elliptical because the orbits do not precisely overlap the circle. In the case of Pluto, you can also see that the Sun is not at the center of the orbit but instead lies at one focus of the ellipse.

each other, the figure tends toward a circle with eccentricity 0. Correspondingly, as the foci move farther apart, the ellipse becomes more elongated and the eccentricity approaches 1.

Kepler's first law of planetary motion states that the orbit of a planet is an ellipse with the Sun at one focus. There is nothing at the other focus of the ellipse. The semimajor axis of an orbit is equal to the average distance between the planet and the Sun. Most planetary objects in our Solar System have nearly circular orbits with eccentricities close to zero. As shown in **Figure 3.4a**, Earth's orbit is very nearly a circle centered on the Sun, with an eccentricity of 0.017. By contrast, Pluto's orbit, as shown in Figure 3.4b, has an eccentricity of 0.249. The orbit is noticeably elongated, with the Sun offset from center.

Kepler's second law. From Tycho's observations of planetary motions, Kepler found that a planet moves fastest when it is closest to the Sun and slowest when it is farthest from the Sun. Kepler found an elegant way to describe the changing speed of a planet in its orbit around the Sun. **Figure 3.5** shows a planet at six different times in its orbit: t_1, t_2, t_3, t_4, t_5, and t_6. Imagine a straight line connecting the Sun with the planet. As the planet orbits between one time and the next, the line "sweeps out" an area of the ellipse. **Kepler's second law**, also called Kepler's **law of equal areas**, states that the area swept out by a planet during a specific time interval is always the same, regardless of the location of the planet in its orbit. In Figure 3.5, if the three time intervals are equal—that is, if $(t_1 \rightarrow t_2) = (t_3 \rightarrow t_4) = (t_5 \rightarrow t_6)$—then the three areas A, B, and C will be equal as well. In order for that to be true, the planet must be moving faster during the time between t_1 and t_2 than it is during the time between t_5 and t_6.

Kepler's third law. Kepler found that planets on larger orbits not only have further to go in one orbit, but they also travel more slowly. The **period** of an orbit is the time the planet takes to orbit exactly once. Earth orbits the Sun in 1 year, so the period of Earth's orbit is 1 year. Jupiter's average distance from the Sun is 5.2 times larger than Earth's average distance from the Sun. Because a nearly circular orbit's circumference is proportional to its radius, Jupiter must travel 5.2 times farther in its orbit around the Sun than Earth does in its own orbit. If the two planets were traveling at the same speed, Jupiter would complete one orbit in 5.2 years. But Jupiter takes almost 12 years to complete one orbit. Jupiter not only has farther to go but also is moving more slowly than Earth. The farther a planet is from the Sun, the larger the circumference of its orbit and the lower its speed.

Kepler's third law states the mathematical relationship between the period of a planet's orbit (in years) and its semimajor axis (in astronomical units or AU): the period squared is equal to the semimajor axis cubed (**Working It Out 3.1**).

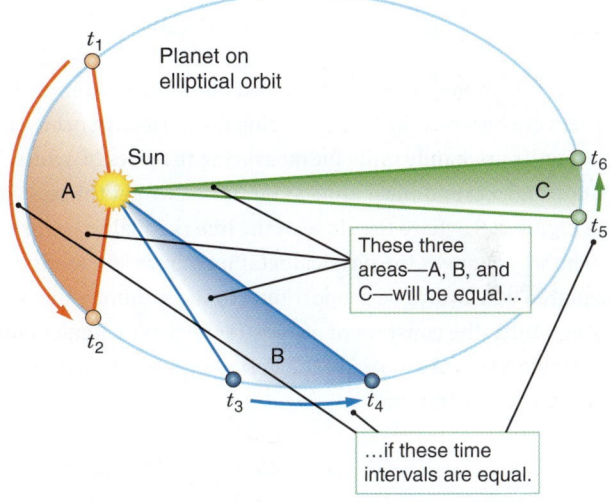

t_1

Planet on elliptical orbit

Sun

A

t_2

B

t_3 t_4

C

t_6

t_5

These three areas—A, B, and C—will be equal…

…if these time intervals are equal.

Figure 3.5 Kepler's second law states that an imaginary line between a planet and the Sun sweeps out an area as the planet orbits such that, if the three time intervals shown are equal, then the areas A, B, and C will also be equal.

▶❙❙ **AstroTour:** Kepler's Laws

what if . . .

What if you read online that "experts have discovered a new planet with a distance from the Sun of 2 AU and a period of 4 years": Use Kepler's third law to argue that this is impossible.

VOCABULARY ALERT

period In everyday language, the word *period* can mean how long a thing lasts. For example, we might talk about the "period of time spent at the grocery store." Astronomers use this word only when talking about repeating intervals, such as the time it takes for an object to complete one orbit.

working it out 3.1

Kepler's Third Law

Squaring a number means that you multiply it by itself, as in $a^2 = a \times a$. *Cubing* a number means that you multiply it by itself twice, as in $a^3 = a \times a \times a$. Kepler's third law states that the square of the period of a planet's orbit measured in years, P_{years}, is equal to the cube of the semi-major axis of the planet's orbit measured in astronomical units, A_{AU}. Translated into math, the law becomes

$$(P_{years})^2 = (A_{AU})^3$$

Here, astronomers use nonstandard units as a matter of convenience. Years are handy units for measuring the periods of orbits, and astronomical units are handy units for measuring the sizes of orbits. When we use years and astronomical units as our units, we get the relationship shown in **Figure 3.6**, where the slope of the line is equal to 1. Our choice of units in no way changes the physical relationship we are studying. The period squared will always be proportional to the semimajor axis cubed; but for other units, the constant of proportionality is no longer equal to 1. For example, we could measure the period in seconds and the semimajor axis in meters and find that

$$(P_{seconds})^2 = (2.9 \times 10^{-19})(A_{meters})^3$$

Suppose that we want to know the average distance of Neptune from the Sun in astronomical units. First, we carefully observe Neptune's position relative to the fixed stars to find out that Neptune takes 165 years to orbit the Sun. Plugging this into Kepler's third law, we find that

$$(165)^2 = (A_{AU})^3$$

To solve this equation, we must first square the left side (which gives 27,225) and then take its cube root. The cube root of $a^3 = a$. For example, $5^3 = 5 \times 5 \times 5 = 125$. So the cube root of 125, which is written as $\sqrt[3]{125}$, is 5.

Calculator hint: Your smartphone likely has a scientific calculator built into it. If you open the calculator app and turn the phone sideways, the calculator turns into a scientific calculator. A scientific calculator usually has a cube root function. It sometimes looks like $x^{1/y}$ and some-times like $^x\sqrt{y}$. You use it by typing the base number, hitting the but-ton, and then typing the root you are interested in (2 for square root, 3 for cube root, and so on). Occasionally, a calculator will instead have a

button that looks like x^y (or y^x). In this case, you need to enter the root as a decimal. For example, if you want to take the square root, you type 0.5 because the square root is denoted by $1/2$. For the cube root, you type 0.333333333 (repeating) because the cube root is denoted by $1/3$.

To find the length of the semimajor axis of Neptune's orbit, we might type 27,225 [$x^{1/y}$] 3. This gives

$$30.1 = A_{AU}$$

The average distance between Neptune and the Sun is 30.1 AU.

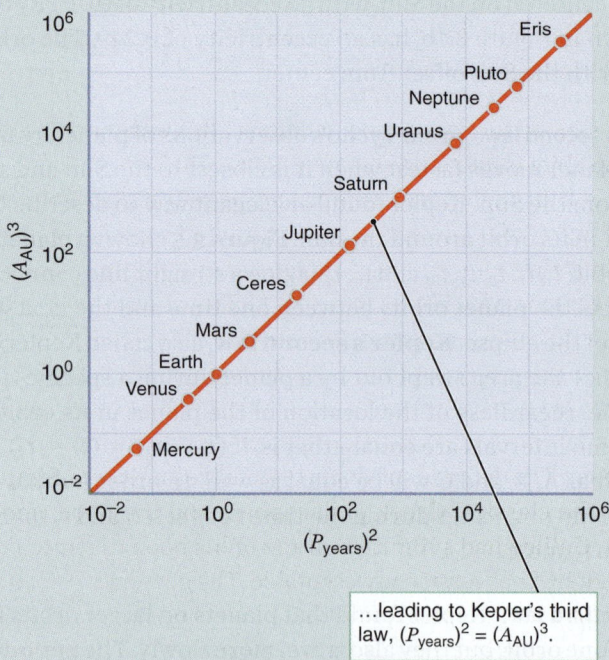

Figure 3.6 A plot of A^3 versus P^2 for the eight planets and three of the dwarf planets in our Solar System shows that they obey Kepler's third law. (Note that by plotting powers of 10 on each axis, we are able to fit both large and small values on the same plot. We will do this frequently.)

3.2 Galileo Was the First Modern Scientist

Galileo Galilei was the first to use a telescope to make significant discoveries about astronomical objects, and much has been written about the considerable danger that Galileo—as he is commonly known—faced as a result of his discoveries. Galileo's telescopes were relatively small, yet they were sufficient for him to observe spots on the Sun, the uneven surface and craters of the Moon, and the large numbers of stars in the band of light in the sky called the Milky Way.

When he turned his telescope to the planet Jupiter, he observed several "stars" in a line near Jupiter. Over time, he saw that there were four of these stars and that their positions changed from night to night. Galileo correctly reasoned that these were moons in orbit around Jupiter. These are the largest of Jupiter's many moons and are referred to as the Galilean moons. This was the first observational evidence that some objects in the sky did not orbit Earth. Galileo also observed that the planet Venus went through phases similar to the Moon's and that the phases correlated with the size of the image of Venus in his telescope (**Figure 3.7**). This is impossible to explain with Ptolemy's model. These observations in particular convinced Galileo that Copernicus was correct to place the Sun at the center of our Solar System.

Galileo's public support for Copernican heliocentricity got him into trouble with the Catholic Church. In 1632, Galileo published *Dialogo sopra i due massimi sistemi del mondo* ("Dialogue Concerning the Two Chief World Systems"). In the *Dialogo*, the champion of the Copernican view of the universe, Salviati, is a brilliant philosopher. The defender of an Earth-centered universe, Simplicio, uses arguments made by the classical Greek philosophers and the pope, and he sounds silly and ignorant. Galileo had submitted drafts of his book to church censors, but the censors found the final version unacceptable. The perceived attack on the pope attracted the attention of the church, and Galileo was eventually placed under house arrest. The book was placed on the Index of Prohibited Works, along with Copernicus's *De revolutionibus*, but it nevertheless traveled across Europe, was translated into other languages, and was read widely by other scientists.

Galileo's work on the motion of objects was at least as important as his astronomical observations, because his approach was different from that of prior natural philosophers who believed that one could understand the universe just by thinking about it. Galileo conducted actual experiments with falling and rolling objects in an excellent application of the scientific method. In one famous experiment, he dropped two objects of different **weights** from the same very tall height and found that they landed at the same time. His work on falling objects demonstrated that gravity on Earth accelerates (changes the speed of) all objects at the same rate,

Figure 3.7 Modern photographs of the phases of Venus show that when we see Venus more illuminated, it also appears smaller, implying that Venus is farther away at that time. In Ptolemy's geocentric model, in contrast, Venus orbits Earth between Earth and the Sun but always on the sunward side of the sky. Gibbous phases could never be observed, and the size of Venus would change in a different way.

VOCABULARY ALERT

weight In everyday language, we often use *weight* and *mass* interchangeably. For example, we often interchange kilograms (a unit of mass) and pounds (a unit of weight), but this is technically incorrect. Mass refers to the amount of stuff in an object—all of its atoms and molecules added together—and weight refers to the force exerted on that object by the planet's gravitational pull. Your weight changes with location, but your mass stays the same no matter what planet you are on.

independent of weight. Galileo's observations and experiments with many types of moving objects, such as carts and balls, led him to disagree with the philosophers about when and why objects continued to move or came to rest. Before Galileo, it was thought that the natural state of an object was to be at rest. But Galileo found that the natural state of an object is to keep doing what it was doing until it is pushed or pulled by another object. Then either its speed or its direction (or both) will change. For example, a ball will just keep rolling until the "push" from the friction of a table slows it down. This idea has implications for not only the motion of carts and balls but also the orbits of planets.

CHECK YOUR UNDERSTANDING 3.2

Suppose that you drop the following objects off of a tall tower. Rank the objects in terms of their acceleration, from smallest to largest.

a. an apple

b. a decorative stone "apple"

c. a solid gold "apple"

3.3 Newton's Laws Govern Motion

The work of Sir Isaac Newton set the standard for what we now refer to as *scientific theory* and *physical law*. Newton proposed three elegant laws that govern the motions of all objects in the universe. These laws connect phenomena on Earth to phenomena in the sky. Newton's laws are essential to our understanding of the motions of the planets and all other celestial bodies. In this section, we will look at each of the three laws in turn.

Newton's First Law: Objects at Rest Stay at Rest; Objects in Motion Stay in Motion

A **force** is a push or a pull. It is possible for two or more forces to oppose one another in such a way that they are perfectly balanced and cancel out. For example, gravity pulls down on you as you sit in your chair. But the chair pushes up on you with an exactly equal force that points in the opposite direction (up). Because these two forces are equal and opposite, they cancel out, and you remain motionless. Forces that cancel out have no effect on an object's motion. When forces add together to produce an effect, we often use the term *net force*, or sometimes just *force*.

Imagine that you are driving a car, and your phone is on the seat next to you. A rabbit runs across the road in front of you, and you press the brakes hard. You feel the seat belt tighten to restrain you. At the same time, your phone flies off the seat and hits the dashboard. You have just experienced what Newton describes in his first law of motion. **Inertia** is the tendency of an object to maintain its state—either of motion or of rest—until it is pushed or pulled by a net force. In the case of the braking car, you did not hit the dashboard because the force of the seat belt on you slowed you down, but the phone went flying because no such force acted upon it.

Newton's first law of motion states that an object in motion tends to stay in motion, in the same direction, until a net force acts upon it; similarly, an object at rest tends to stay at rest until a net force acts upon it. This law followed closely

▶‖ **AstroTour:** Newton's Laws and Universal Gravitation

🎥 **Astronomy in Action:** Velocity, Force, and Acceleration

▶‖ **AstroTour:** Velocity, Acceleration, and Inertia

VOCABULARY ALERT

force In everyday language, the word *force* has many meanings. Astronomers specifically mean a push or a pull.

inertia In everyday language, we think of *inertia* as a tendency to remain motionless. Astronomers and physicists think more generally of inertia as the tendency of matter to resist a change in motion—of an object at rest to remain at rest, and of a moving object to remain in motion.

from Galileo's work on moving objects, and is fundamentally a description of inertia.

Recall from Section 3.1 the concept of a frame of reference. Within a frame of reference, only the relative motions between objects have any meaning. Without external clues, you cannot tell whether you are sitting still or traveling at constant speed in a straight line. For example, in an elevator, once it has accelerated to its normal speed, you cannot tell if it is stationary or moving, nor can you tell how fast you are traveling unless you watch the floor numbers change. Returning to the earlier example, your phone was "at rest" beside you on the front seat of your car, in your reference frame. However, a person standing by the side of the road would measure the phone moving past at the same speed as the car. And people in a car approaching you would see the phone moving quite fast—at the speed they are traveling plus the speed you are traveling! All of these perspectives are equally valid, and all of these speeds of the phone are correct when measured in the appropriate reference frame.

A reference frame moving in a straight line at a constant speed is an **inertial frame of reference**. Any inertial frame of reference is as good as any other—there is no "correct" reference frame. In the frame of reference of the cup of coffee shown in **Figure 3.8a**, *it is at rest in its own frame* even if the car is speeding down the road. (Notice that we use particular colors throughout this text for different quantities. Green arrows are used for acceleration, and red arrows are used for speed or velocity.)

Newton's Second Law: Motion Is Changed by Forces

Suppose that a net force does act. In the previous example, you were traveling in the car, and your motion was slowed when the force of the seat belt acted upon you. Forces change an object's motion—by changing either its speed or its direction. This reflects **Newton's second law of motion**: If a net force acts on an object, then the object's motion changes.

When we speak of "changes in an object's motion," what do we really mean? When you are in the driver's seat of a car, a number of controls are available. On the floor are a gas pedal and a brake pedal. You use these to make the car speed up or slow down. A *change in speed* is one way the car's motion can change. But you also have the steering wheel in your hands. When you are moving down the road and you turn the wheel, your speed does not necessarily change, but the direction of your motion does. A *change in direction* is also a change in motion.

Together, the combined speed and direction of an object's motion is called an object's **velocity**. "Traveling at 50 kilometers per hour (km/h)" indicates speed; "traveling north at 50 km/h" indicates velocity. The rate at which the velocity of an object changes is called **acceleration**. Acceleration tells you how rapidly a change in velocity happens. For example, if you go from 0 to 100 km/h in 4 seconds, you feel a strong push from the seat back as it shoves your body forward, causing you to accelerate along with the car. However, if you take 2 minutes to go from 0 to 100 km/h, the acceleration is so slight that you hardly notice it.

Because the gas pedal on a car is often called the accelerator, some people think *acceleration* always means that an object is speeding up. As used in physics, however, *any* change in motion is an acceleration. Figure 3.8b illustrates the point by showing what happens to the coffee in a coffee cup as the car speeds up, slows down,

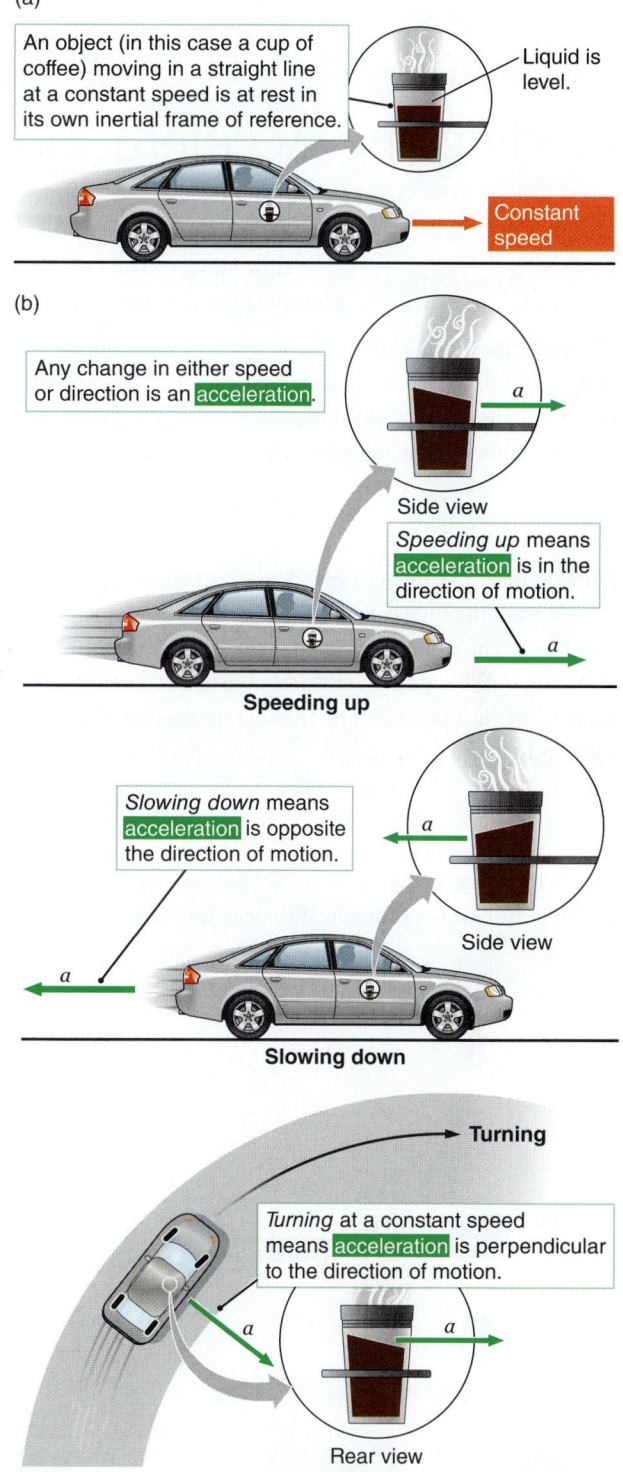

(a)

An object (in this case a cup of coffee) moving in a straight line at a constant speed is at rest in its own inertial frame of reference.

Liquid is level.

Constant speed

(b)

Any change in either speed or direction is an acceleration.

Side view

Speeding up means acceleration is in the direction of motion.

Speeding up

Slowing down means acceleration is opposite the direction of motion.

Side view

Slowing down

Turning

Turning at a constant speed means acceleration is perpendicular to the direction of motion.

Rear view

Figure 3.8 (a) An object moving in a straight line at a constant speed is not accelerating, so the coffee remains level in the cup. (b) When accelerating, the coffee in the cup sloshes forward or backward. (Throughout the text, velocity arrows are shown in red, whereas acceleration arrows are shown in green.) As shown in the bottom panel, even a change in the direction of motion with no change in speed indicates an acceleration that will slosh the coffee in the cup.

working it out 3.2

Finding the Acceleration

The equation for acceleration states

$$\text{Acceleration} = \frac{\text{How much velocity changes}}{\text{How long the change takes to happen}}$$

We can write this more compactly by expressing the change in velocity as $v_2 - v_1$ and the change in time as $t_2 - t_1$, where v_1 is the velocity at time t_1 and v_2 is the velocity at time t_2. Acceleration is commonly abbreviated as a, so our equation can be written mathematically as

$$a = \frac{v_2 - v_1}{t_2 - t_1}$$

For example, if an object's speed goes from 5 meters per second (m/s) to 15 m/s, then the final velocity is $v_2 = 15$ m/s and the initial velocity is $v_1 = 5$ m/s. The change in velocity $(v_2 - v_1)$ is then 15 m/s − 5 m/s = 10 m/s. If that change happens over the course of 2 seconds, then the change in time, $t_2 - t_1$, is 2 seconds, and the average acceleration is 10 m/s divided by 2 seconds:

$$a = \frac{10 \text{ m/s}}{2 \text{ s}} = 5 \text{ m/s/s}$$

This is the same as saying "5 meters per second squared," which is written as 5 m/s² or 5 m s⁻². It means the speed increases by 5 m/s each second.

If we want to know how an object's motion is changing, we need to know the net force acting on the object and the mass of the object. This formula translates into words as "the acceleration is equal to the strength of the net force divided by the mass." Mathematically, it becomes

$$\text{Acceleration} = \frac{\text{Strength of net force}}{\text{Mass}}$$

This is a lot to write every time we want to talk about acceleration, so we usually let a stand for acceleration, F for force, and m for mass:

$$a = \frac{F}{m}$$

Newton's second law is often written as $F = ma$ (to understand how we changed this equation, revisit Working It Out 2.1), giving force as units of mass multiplied by units of acceleration, or kilograms times meters per second squared (kg m/s²). The units of force are named **newtons (N)**, where $1 \text{ N} = 1 \text{ kg m/s}^2$.

This is the mathematical statement of Newton's second law of motion. It says the following three things: (1) when you push on an object, that object accelerates in the direction you are pushing; (2) the harder you push on an object, the more it accelerates; and (3) the more massive the object, the harder it is to accelerate.

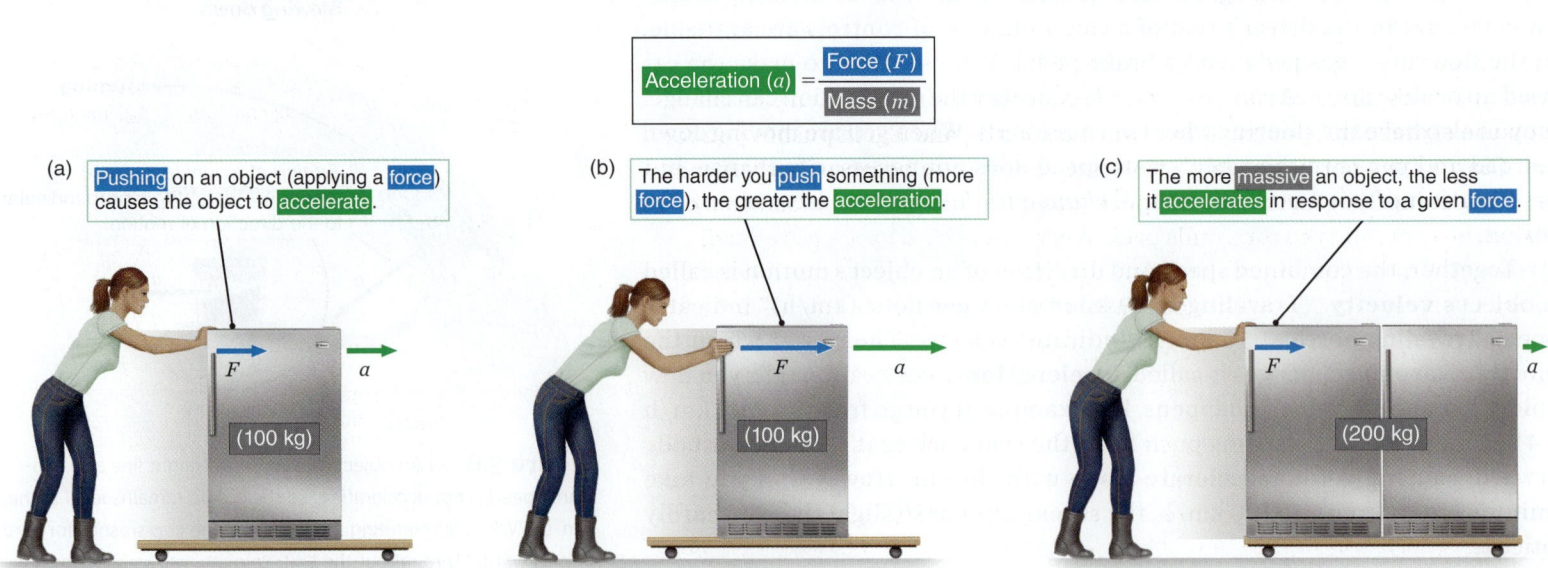

Figure 3.9 Newton's second law of motion says that the acceleration experienced by an object is determined by the force acting on the object, divided by the object's mass. (Throughout the text, force arrows are shown in blue.)

or turns. Slamming on your brakes and going from 100 to 0 km/h in 4 seconds is just as much of an acceleration as going from 0 to 100 km/h in 4 seconds. Similarly, the acceleration you experience as you go through a fast, tight turn at a constant speed is every bit as real as the acceleration you feel when you slam your foot on the gas pedal or the brake pedal. Speeding up, slowing down, turning left, turning right—if you are not moving in a straight line at a constant speed, you are experiencing an acceleration.

According to Newton's second law of motion, net force causes acceleration. The acceleration an object experiences depends on two things. First, as shown in **Figure 3.9a**, the acceleration depends on the strength of the net force acting on the object to change its motion. Push 3 times as hard, and the object experiences 3 times the acceleration (see Figure 3.9b). The acceleration occurs in the direction the net force points. Push an object away from you, and it will accelerate away from you.

The acceleration that an object experiences also depends on its inertia. Some objects, such as the empty box that a refrigerator was delivered in, are easily moved around by humans. The refrigerator that came in the box, however, is not so easily moved, even though it is about the same size. For our purposes, an object's **mass** is interchangeable with its inertia. The greater the mass, the greater the inertia, and thus less acceleration will occur in response to the same net force, as shown in Figure 3.9c. This relationship between acceleration, force, and mass is expressed mathematically in **Working It Out 3.2**.

Newton's Third Law: Whatever Is Pushed, Pushes Back

Imagine you are standing on a skateboard and pushing yourself along with your foot. Each shove of your foot against the ground sends you faster along your way. Why does this happen? Your muscles flex, and your foot exerts a force on the ground. Yet this does not explain why you experience an acceleration. The fact that you accelerate means that as you push on the ground, the ground must be pushing back on you.

Part of Newton's genius was his ability to see patterns in such everyday events. Newton realized that *every* time one object exerts a force on another, a matching force is exerted by the second object on the first. That second force is exactly as strong as the first force but is in exactly the *opposite* direction. When you are riding on the skateboard, you push backward on Earth, and Earth pushes you forward. (Earth does not noticeably accelerate in response because its great mass gives it great inertia.) As shown in **Figure 3.10**, a woman pulling a load on a cart pulls on the rope, and the rope pulls back. A car tire pushes back on the road, and the road pushes forward on the tire. Earth pulls on the Moon, and the Moon pulls on Earth. A rocket engine pushes hot gases out of its nozzle, and those hot gases push back on the rocket, propelling it through space. Because the force of the exhaust on the rocket is larger than the force of gravity on the rocket, the rocket accelerates.

The force pairs in Figure 3.10 are examples of **Newton's third law of motion**, which says that forces always come in pairs, and the forces of a pair are always equal in strength but opposite in direction. The forces in these action-reaction pairs always act on two different objects. The rocket pushes on the gases, and the gases push back on the rocket with the same amount of force. For every force there is *always* an equal and opposite force. This is one of the few times when we can say "always" and really mean it.

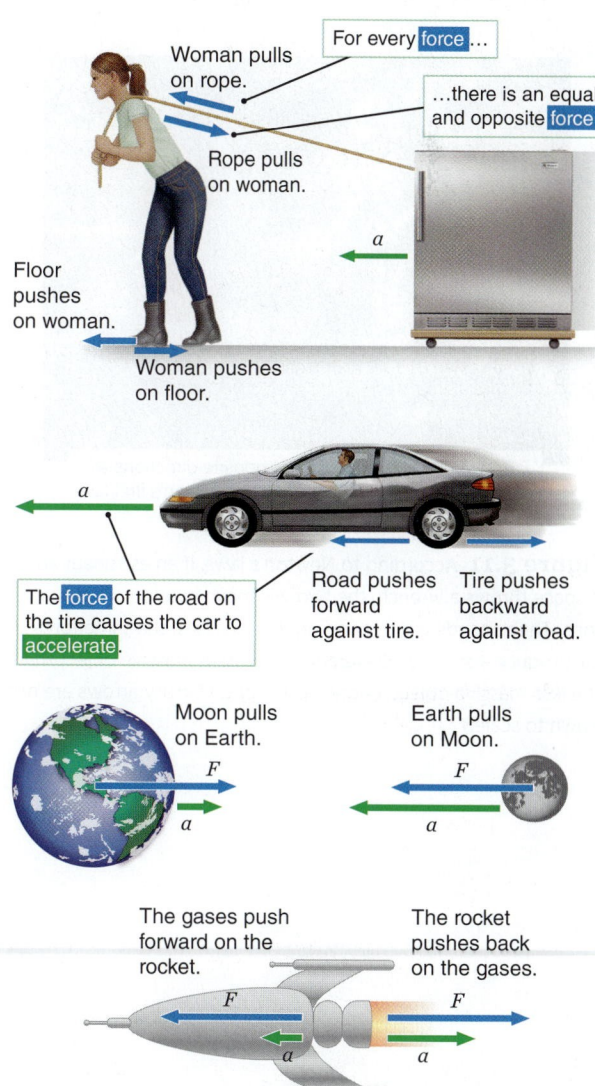

Figure 3.10 Newton's third law states that for every force there is always an equal and opposite force. These opposing forces (action-reaction pairs) always act on two different objects.

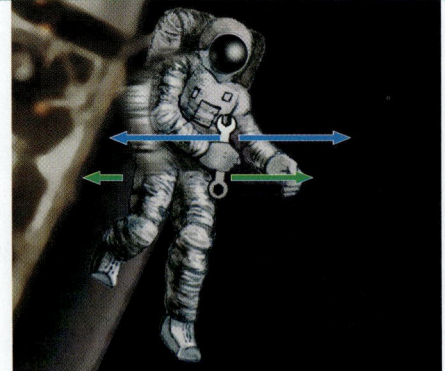

An astronaut adrift in space pushes on a wrench, which, according to Newton's third law, pushes back on the astronaut.

While in contact with each other, the wrench and the astronaut experience accelerations proportional to the inverse of their masses…

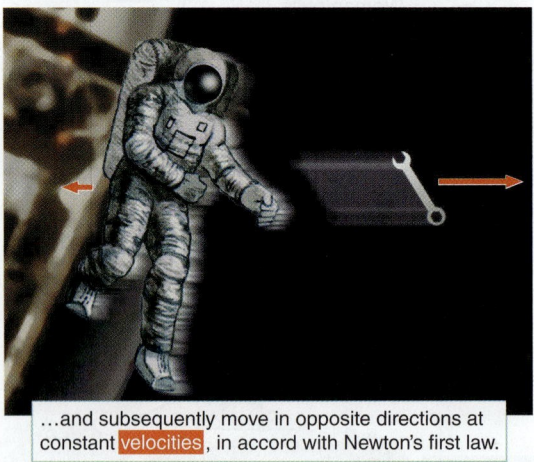

…and subsequently move in opposite directions at constant velocities, in accord with Newton's first law.

Figure 3.11 According to Newton's laws, if an astronaut adrift in space throws a wrench, the two will move in opposite directions. Their speeds will depend on their masses; the same force will produce a smaller acceleration of a more massive object than of a less massive object. (Acceleration and velocity arrows are not drawn to scale.)

Newton's Laws and Motion

To see how Newton's three laws of motion work together, study **Figure 3.11**. An astronaut is adrift in space, motionless with respect to the nearby spacecraft. According to Newton's first law, he's stuck! With no tether to pull on, how can the astronaut get back to the shuttle? Suppose the 100-kg astronaut throws a 1-kg wrench directly away from the shuttle at a speed of 10 m/s. Newton's second law says that to cause the motion of the wrench to change, the astronaut has to apply a force to it in the direction away from the shuttle. Newton's third law says that the wrench must therefore push back on the astronaut with just as much force in the opposite direction. The force of the wrench on the astronaut causes the astronaut to begin drifting toward the shuttle. How fast will the astronaut move? Turn to Newton's second law again. Because the astronaut has more mass, he will accelerate less. The 100-kg astronaut will experience only 1/100 as much acceleration as the 1-kg wrench, and so he will have 1/100 the final velocity. The astronaut will drift toward the shuttle, but only at the leisurely rate of 0.1 m/s.

CHECK YOUR UNDERSTANDING **3.3**

Suppose you are driving a car, and a coffee cup is in the cup holder beside you. Rank, in increasing order, the speed of the coffee cup in the following reference frames. (Represent two equal choices with an "=" between them.)

a. an astronaut on the International Space Station

b. the cup itself

c. an observer at the side of the road

d. an observer in an oncoming car

e. you, the driver

3.4 Gravity Is a Force between Any Two Massive Objects

Newton's work on the motion of objects led to a great insight. Gravity was known before Newton, so he did not "discover" gravity, but Newton realized that the force that caused keys to fall to the floor when dropped also caused the orbits of the Moon around Earth and the planets around the Sun. Gravity is one of the four fundamental forces in nature, and it binds the universe together.

Gravity, Mass, and Weight

What happens when you drop your keys on the floor? They begin at rest and then accelerate downward. This acceleration requires the action of a force, which is commonly known as the **gravitational force**: the mutually attractive force between two objects with mass. The gravitational force on an object attracted by a planet is also typically called the weight of the object.

The acceleration due to the gravitational force—known as gravitational acceleration—near the surface of Earth is usually written as g and has an average value across the surface of Earth of 9.8 m/s². Experiments show that all objects on Earth fall with this same acceleration. Whether you drop a marble or a cannonball, after 1 second it will be falling at a speed of 9.8 m/s, after 2 seconds at 19.6 m/s, and after 3 seconds at 29.4 m/s. (Note that air resistance becomes a factor at higher speeds, but it is negligible for dense, slow objects.)

Newton realized that if all objects fall with the same acceleration, then the gravitational force on an object must be determined by the object's mass. To see why, look back at Newton's second law—acceleration equals force divided by mass, or $a = F/m$. The only way gravitational acceleration can be identical for all objects is if the value of the force divided by the mass is the same for all objects. In other words, make an object twice as massive and you double the gravitational force acting on it. Make an object three times as massive and you triple the gravitational force acting on it. The gravitational force is directly proportional to the mass.

On the surface of Earth, weight is just mass multiplied by Earth's gravitational acceleration, g, which is written mathematically as

$$F_{\text{weight}} = m \times g$$

where F_{weight} is an object's weight in newtons, m is the object's mass in kilograms, and g is Earth's gravitational acceleration in meters per second squared. On Earth, an object with a *mass* of 2 kg has a *weight* of 2 kg $\times$ 9.8 m/s², or 19.6 N. On the Moon, where the gravitational acceleration is 1.6 m/s², the 2-kg mass would have a weight of 2 kg $\times$ 1.6 m/s², or 3.2 N. Although your mass remains the same wherever you are, your weight varies. On the Moon, your weight would be about one-sixth of your weight on Earth. In everyday speech, people equate mass and weight. We often say that an object with a mass of 2 kg "weighs" 2 kg, but as you have just seen, a weight is actually a unit of force.

Newton's Law of Gravity

Newton's next great insight came from applying his third law of motion to gravity. Newton's third law states that for every force there is an equal and opposite force. Drop a 10-kg frozen turkey and it falls toward Earth—but at the same time, Earth falls toward the 10-kg turkey. The reason we do not notice the motion of Earth is that Earth has a lot of inertia. In the time it takes a 10-kg turkey to fall to the ground from a height of 1 km, Earth has "fallen" toward the turkey by a tiny fraction of the size of an atom. Many turkeys may be falling all around Earth simultaneously, however, so the effect of these turkeys will effectively cancel out.

Newton reasoned that if doubling the mass of an object doubles the gravitational force between the object and Earth, then doubling the mass of Earth ought to do the same thing. In short, the gravitational force between Earth and an object must be proportional to the product of the masses of Earth and the object:

$$\text{gravitational force} = \text{something} \times \text{mass of Earth} \times \text{mass of object}$$

If the mass of the object were 2 times greater, then the force of gravity would be 2 times greater. Likewise, if the mass of Earth were 3 times what it is, the force of gravity would have to be 3 times greater as well. If *both* the mass of Earth *and* the mass of the object were greater by these amounts, the gravitational force would increase by a factor of 2 $\times$ 3 = 6 times. Because objects fall toward the center of Earth, we know that gravity is an attractive force acting along a line between the two masses.

If gravity is a force that depends on mass, then there should be a gravitational force between *any* two masses, mass 1 and mass 2 (m_1 and m_2 for short). The gravitational force between them is something multiplied by the product of the masses:

$$\text{gravitational force between two objects} = \text{something} \times m_1 \times m_2$$

We have gotten this far just by combining Galileo's observations of falling objects with (1) Newton's laws of motion and (2) Newton's belief that Earth is a mass just

what if . . .

What if the gravitational force on an object did not depend on the object's mass: How would objects in our world behave differently?

Newton's Law of Gravity: Playing with Proportionality

One of the most useful ways to play with equations is to study proportionalities. In Newton's law of gravity, for example, the force is proportional to each of the masses of the objects involved:

$$F = \frac{Gm_1m_2}{r^2}$$

What does that mean? It means that if the mass of one object is doubled, the force is doubled as well. If the mass of one object is increased by a factor of 2.63147, so is the force. This provides a handy shorthand way of making calculations without actually having to plug G, r, m_1, and m_2 into the equation. Suppose that your mass, m_1, suddenly increased by a factor of 2 (doubled). How would this affect the force of gravity, F, on you; that is, how would it affect your weight? F would increase by a factor of 2, so your weight would double, too. This is somewhat intuitive. If you suddenly had twice as much stuff in your insides, you should certainly weigh twice as much.

We are just using a rule we learned in Working It Out 2.1; namely, whatever you do to one side of the equation, you also have to do to the other side. In this case, we multiplied one of the terms on the right by 2, so we also had to multiply the term on the left by 2.

In these examples, we have been focusing on terms that are directly proportional to each other. Inverse proportions are possible, too. In inverse proportions, doubling one term means the other must be halved. This happens when one term is in the denominator while the other is in the numerator. We saw something like this before when we learned about the relationship between distance (d), time (t), and speed (s):

$$s = \frac{d}{t}$$

Speed and distance are directly proportional to each other, whereas speed and time are inversely proportional. If you traveled twice as far as your friend in the same amount of time, you were going twice as fast. If it took you twice as long to travel the same distance as your friend, however, you were going *half* as fast. Consider our rule about doing the same thing to both sides of the equation. On the right, you multiply the time by 2, but time is in the denominator, so you have effectively multiplied the right side by ½. You must also do this on the left and multiply s by ½. (Notice that two things that are inversely proportional never have the same units.)

Let's return to Newton's law of gravity and examine what happens when we change r, the distance between the objects. How would the gravitational force F between Earth and the Moon change if the distance between them were doubled? F is inversely proportional to r^2

$$F = \frac{Gm_1m_2}{r^2}$$

so doubling r to $2r$ should make F decrease: so far, so good. But what do we do about the square? If we substitute $2r$ for r in the equation, then we end up with $(2r)^2$ in the denominator. This means that the r is squared *and* the 2 is squared: $(2r)^2 = 4r^2$. We have effectively multiplied the right-hand side by ¼. We must do the same on the left-hand side, too, so the force is ¼ as strong as before:

$$\frac{F}{4} = \frac{Gm_1m_2}{4r^2}$$

Doubling the distance reduces the force by a factor of 4. This $1/r^2$ proportionality occurs in many contexts in astronomy. Take a moment to calculate how the force would change if the distance increased by a factor of 3, 5, or 10 and how the force would change if the distance decreased by a factor of ½, ¼, or ¹⁄₁₀. Once you can do this, you will have a tool you can use again and again in the remaining chapters.

like any other mass. But what about that "something" in the two previous expressions? Today we have sensitive instruments that allow scientists to put two masses close to each other in a laboratory, measure the force between them, and determine the value of that something directly. Newton, though, had no such instruments. He had to look elsewhere to continue his exploration of gravity.

Kepler had already thought of another factor that should be important in the interaction between the Sun and the planets. He reasoned that because the Sun is the focal point for planetary orbits, the Sun must be responsible for exerting an influence over the motions of the planets. Kepler speculated that whatever this influence is, it must grow weaker with distance from the Sun. After all, it must surely require a stronger influence to keep the innermost planet, Mercury, whipping around in its tight, fast orbit than it does to keep the outer planets lumbering along their paths around the Sun. Kepler speculated even further. Although he did not know about forces or inertia or gravity, he did know quite a lot about geometry, and geometry alone suggested how this solar "influence" might change for planets progressively farther from the Sun.

Imagine that you have only a certain amount of paint to spread over the surface of a sphere. If the sphere is small, you will get a thick coat of paint. If the sphere is larger, though, the paint has to cover a larger surface area, so you end up with a thinner coat. The surface area of a sphere depends on the square of the sphere's radius. Double the radius of a sphere, and the sphere's surface area increases by 4 times. If you paint this new, larger sphere, the paint must cover 4 times as much area, and the thickness of the paint will be only one-fourth of what it was on the smaller sphere. Triple the radius of the sphere and the sphere's surface area will be 9 times as large, and the coat of paint will be only one-ninth as thick.

The paint in this example describes how Kepler thought about the influence the Sun exerts over the planets. As the influence of the Sun extended farther and farther into space, it would spread out to cover the surface of a larger and larger imaginary sphere centered on the Sun. The influence of the Sun should diminish with the square of the distance from the Sun—a relationship known as an **inverse square law**.

Kepler had an interesting idea but no scientific hypothesis with testable predictions. He lacked an explanation for how the Sun influences the planets and the mathematical tools to calculate how an object would move. Newton had both. If gravity is a force between *any* two objects, then there should be a gravitational force between the Sun and each of the planets. Might this gravitational force be the same as Kepler's "influence"? If so, gravity might behave according to an inverse square law. Newton's expression for gravity came to look like this:

$$\text{gravitational force between two objects} = \text{something} \times \frac{m_1 \times m_2}{(\text{distance between objects})^2}$$

There is still a "something" left in this expression, and that something is a constant: a number that does not change. This constant determines the strength of gravity between objects, and it is the same for *all* pairs of objects. Newton named it the **universal gravitational constant (G)**. It was not until many years later that the actual value of G was first measured. Today, the value of G is known to be $6.673 \times 10^{-11} \, \text{m}^3/(\text{kg s}^2)$.

Putting the Pieces Together: A Universal Law for Gravitation

Newton's **universal law of gravitation** states that gravity is a force between any two objects having mass and has these properties:

1. It is an attractive force, F, acting along a straight line between the two objects.

2. It is proportional to the mass of one object, m_1, multiplied by the mass of the other object, m_2. If we double m_1, F increases by a factor of 2. Likewise, if we double m_2, F increases by a factor of 2. And if both masses double, then F increases by a factor of 4.

3. It is inversely proportional to the square of the distance, r, between the centers of the two objects. If we double r, F decreases by a factor of 4. If we triple r, F falls by a factor of 9 (**Working It Out 3.3**).

These properties are illustrated in **Figure 3.12**. Translated into mathematics, and including the universal gravitational constant, G, the universal law of gravitation is

$$F = G\frac{m_1 m_2}{r^2}$$

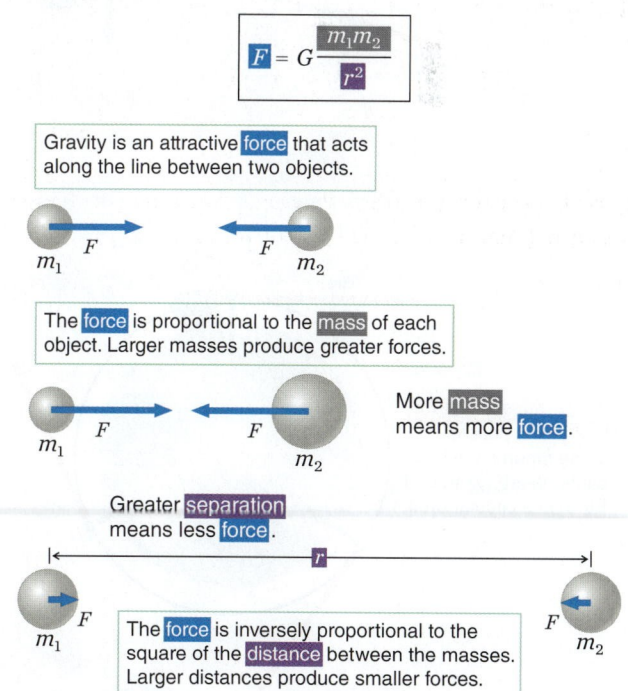

Gravity is an attractive force that acts along the line between two objects.

The force is proportional to the mass of each object. Larger masses produce greater forces.

More mass means more force.

Greater separation means less force.

The force is inversely proportional to the square of the distance between the masses. Larger distances produce smaller forces.

Figure 3.12 Gravity is an attractive force between two objects. The force of gravity depends on the masses of the objects, m_1 and m_2, and the distance, r, between them.

CHECK YOUR UNDERSTANDING **3.4**

Suppose you are transported to a planet with twice the mass of Earth but the same radius as Earth. Your weight would _____ by a factor of _____.

a. increase; 2

b. increase; 4

c. decrease; 2

d. decrease; 4

(a)

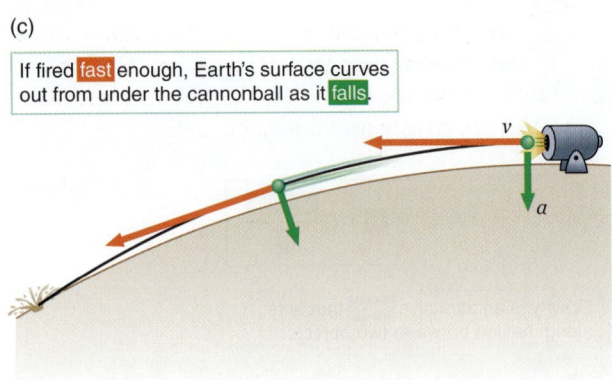

A cannonball travels over the ground as it falls toward Earth.

(b)

If fired faster, it travels farther in the time it takes to fall to the ground.

(c)

If fired fast enough, Earth's surface curves out from under the cannonball as it falls.

(d)

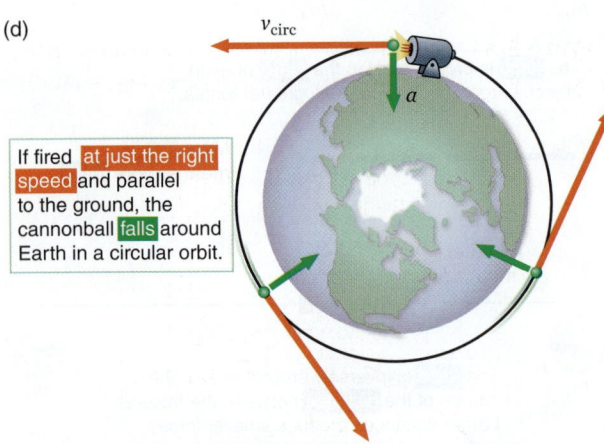

If fired at just the right speed and parallel to the ground, the cannonball falls around Earth in a circular orbit.

Figure 3.13 Newton realized that a cannonball fired at the right speed would fall around Earth in a circle.

3.5 Orbits Are One Body "Falling around" Another

Kepler's laws regarding the motions of planets allowed astronomers to predict the positions of the planets accurately, but these laws did not explain why the planets behaved as they did. Newton's work provided the answer to the question: Why do planets orbit the Sun?

Newton Explains Kepler

Newton used his laws of motion and his proposed law of gravity to calculate the paths that planets should follow as they move around the Sun. When he did so, his calculations predicted the following:

- Planetary orbits should be ellipses with the Sun at one focus.
- Planets should travel faster when they are closer to the Sun.
- The square of the period of the orbit should equal the semimajor axis cubed (in appropriate units).

In short, Newton's universal law of gravitation *predicted* that planets should orbit the Sun in just the way that Kepler's empirical laws described. This was the moment when it all came together. By *explaining* Kepler's laws, Newton found important support for his law of gravitation. Newton argued that the same gravitational force governed the behavior of dropped keys and orbiting planets.

Gravity and Orbits

Newton's laws describe how an object's motion changes in response to forces and how objects interact with each other through gravity. To go from statements about how an object's motion is *changing* to more practical statements about where an object *is*, we must carefully "add up" the object's motion over time. To see how we can do this, let's begin with a "thought experiment"—the same thought experiment that helped lead Newton to his understanding of how planets orbit the Sun.

Drop a cannonball and it falls directly to the ground, just as any mass does. However, if we fire the cannonball from a cannon that is level with the ground, as shown in **Figure 3.13a**, the cannonball still falls to the ground in the same time as before, but while falling it is also traveling *over* the ground, following a curved path that carries it some horizontal distance before it finally lands. As shown in Figure 3.13b, the faster the ball is fired from the cannon, the farther it will go before it hits the ground.

In the real world, this experiment reaches a natural limit. To travel through air, the cannonball must push the air out of its way—an effect we normally refer to as air resistance—which slows it down. But because this is only a thought experiment, we can ignore such real-world complications. Instead, imagine that, having inertia, the cannonball continues along its course until it runs into something. As the cannonball is fired faster and faster, it goes farther and farther before hitting the ground. If the cannonball flies far enough, Earth's surface "curves out from under it," as shown in Figure 3.13c. Eventually, the cannonball is flying so fast that the surface of Earth curves away from the cannonball at exactly the same rate that the cannonball is falling toward Earth. This is the case shown in Figure 3.13d. At this point the cannonball, which always falls *toward the center of Earth*, is literally "falling around the world."

In 1957, the Soviet Union used a rocket to lift Sputnik 1, an object about the size of a basketball, high enough above Earth's upper atmosphere that air resistance ceased to be a concern. With this event, Newton's thought experiment became a matter of great practical importance; Sputnik 1 was moving so fast that it fell around Earth, just as the cannonball did in Newton's mind. Sputnik 1 was the first human-made object to orbit Earth, where an **orbit** is the path of one object that freely falls around another, and "to orbit" is to fall freely around another object.

The concept of orbits also explains why astronauts float freely about the cabin of a spacecraft. It is *not* because they have escaped Earth's gravity; it is Earth's gravity that holds them in their orbit. Instead, the explanation lies in Galileo's early observation that every object falls in just the same way, regardless of its mass. The astronauts and the spacecraft are both moving in the same direction, at the same speed, and are experiencing the same gravitational acceleration, so they fall around Earth together. **Figure 3.14** demonstrates this point. The astronaut is orbiting Earth just as the spacecraft is orbiting Earth. On the surface of Earth our bodies try to fall toward the center of Earth, but the ground gets in the way. We experience our weight when we are standing on Earth because the ground pushes on us hard enough to counteract the force of gravity, which is trying to pull us down. In the spacecraft, however, nothing interrupts the astronaut's fall because the spacecraft is falling around Earth in just the same orbit. The astronaut is in **free fall**: He is falling freely in Earth's gravity.

When one object is falling around another, much more massive object, we say that the less massive object is a **satellite** of the more massive object. Planets are satellites of the Sun, and moons are natural satellites of planets. Newton's imaginary cannonball is a satellite. The spacecraft and the astronauts are independent satellites of Earth that conveniently happen to share the same orbit.

▶▶ **Interactive Simulation:** Planetary Orbits Simulator

what if . . .

What if you jump from a very high tower into a very deep body of water at an amusement park: As you fall, how would your experience compare to that of an astronaut in orbit?

Figure 3.14 A "weightless" astronaut has not escaped Earth's gravity. Rather, the astronaut and spacecraft share the same orbit as they fall around Earth together.

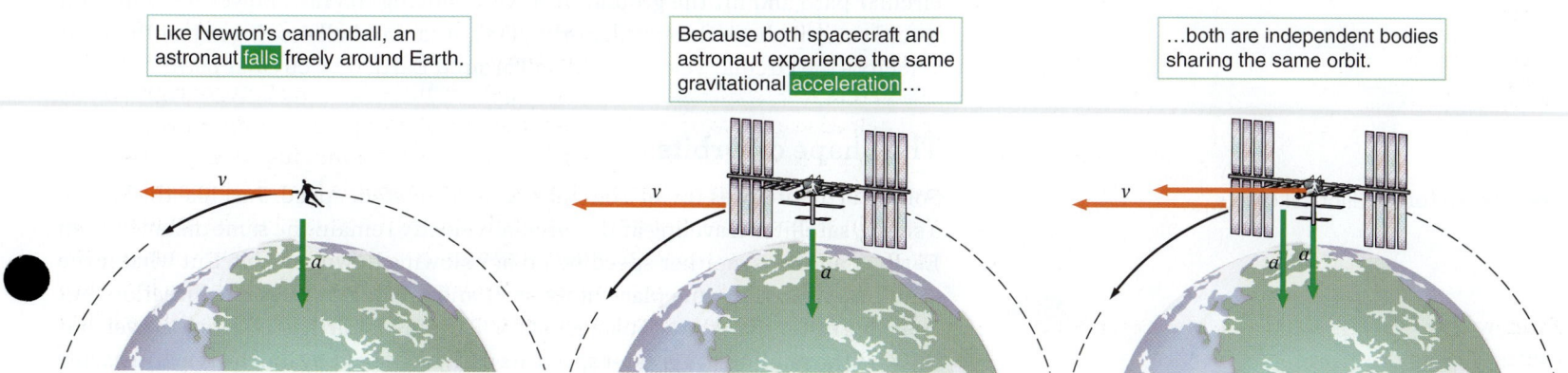

Like Newton's cannonball, an astronaut falls freely around Earth.

Because both spacecraft and astronaut experience the same gravitational acceleration…

…both are independent bodies sharing the same orbit.

Figure 3.15 (a) A string provides the centripetal force that keeps a ball moving in a circle. (We are ignoring the smaller force of gravity that also acts on the ball.) (b) Similarly, gravity provides the centripetal force that holds a satellite in a circular orbit.

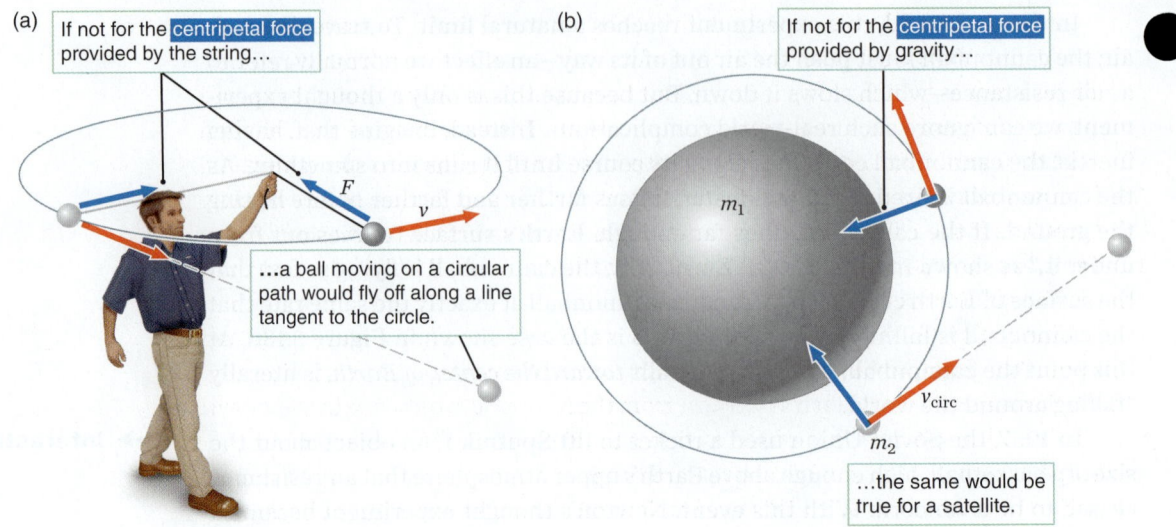

(a) If not for the centripetal force provided by the string…

…a ball moving on a circular path would fly off along a line tangent to the circle.

(b) If not for the centripetal force provided by gravity…

…the same would be true for a satellite.

Centripetal Force and Circular Velocity

If fired fast enough, Newton's cannonball falls around the world; but just how fast is "fast enough"? Newton's orbiting cannonball moves along a circular path at constant speed, a type of motion that is referred to as **uniform circular motion**. You are probably familiar with other examples of uniform circular motion. For example, think about a ball whirling around your head on a string, as shown in **Figure 3.15a**. If you were to let go of the string, the ball would fly off in a straight line in whatever direction it was traveling at the time. The string prevents the ball from flying off by constantly changing the direction the ball is traveling. This central force of the string on the ball is called a **centripetal force**: a force toward the center of a circle. Using a more massive ball, speeding up the ball's motion, or making the string shorter so that the turn is tighter all increase the force needed to keep the ball moving in a circle.

In the case of Newton's cannonball (or a satellite), there is no string to hold the ball in its circular motion. Instead, the force is provided by gravity, as illustrated in Figure 3.15b. For Newton's thought experiment to work, the force of gravity must be just enough to keep the satellite moving on its circular path. Because this force has a specific strength, it follows that the satellite must be moving at a particular speed, which we call its **circular velocity (v_{circ})**. If the satellite were moving at any other velocity, it would not be moving in a circular orbit. This is no different than the cannonball. If the cannonball were moving too slowly, it would drop below the circular path and hit the ground. If it were moving too fast, however, its motion would carry it above the circular orbit. Only a cannonball moving at just the right velocity—the circular velocity—will fall around Earth on a circular path.

The Shape of Orbits

Some Earth satellites travel a circular path at constant speed. Just like the ball on a string, satellites traveling at the circular velocity remain the same distance from Earth at all times, neither speeding up nor slowing down in orbit. But what if the satellite was in the same place in its orbit and moving in the same direction, but traveling faster than the circular velocity? The pull of Earth is as strong as ever, but because the satellite has greater speed, its path is not bent by Earth's gravity sharply enough to hold it in a circle, so the satellite begins to climb above the circular orbit.

▶❙❙ **AstroTour:** Elliptical Orbits

As the distance between the satellite and Earth increases, the satellite slows down. If you throw a ball into the air, as shown in **Figure 3.16a**, Earth's gravity slows the ball down as it climbs higher. The ball keeps climbing, but it slows more and more until its vertical motion stops for an instant and then is reversed; the ball then falls back toward Earth, speeding up along the way. The satellite does the same thing as it moves away from Earth. The farther the satellite is from Earth, the more slowly the satellite moves. The satellite eventually reaches a maximum height on its curving path and then begins falling back toward Earth. As the satellite falls back toward Earth, Earth's gravity speeds it up.

This happens for any object in an elliptical orbit, including a planet orbiting the Sun. According to Kepler's second law, a planet moves fastest when it is closest to the Sun and slowest when it is farthest from the Sun. Now we know why. As shown in Figure 3.16b, planets lose speed as they pull away from the Sun and then gain that speed back as they fall inward toward the Sun.

Newton's laws do more than explain Kepler's laws: They predict orbits beyond Kepler's empirical observations. **Figure 3.17** shows a series of satellite orbits, each with the same point of closest approach to Earth but with different velocities at that point. The greater the speed a satellite has at its closest approach to Earth, the farther the satellite is able to travel from Earth, and the more eccentric its orbit becomes. As long as it remains elliptical, no matter how eccentric, the orbit is called a **bound orbit** because the satellite is gravitationally bound to the object it is orbiting.

In this sequence of faster and faster satellites there comes a point of no return—a point when the satellite is moving so fast that Earth's gravity is unable to reverse the satellite's outward motion, so the satellite coasts away from Earth, never to return. The lowest speed at which this happens is called the **escape velocity** from the orbit, v_{esc}. Once a satellite's velocity at closest approach equals or exceeds v_{esc}, it is in an **unbound orbit**. The object is no longer gravitationally bound to the body that it was orbiting. A comet traveling on an unbound orbit makes only a single pass around the Sun and then is back off into deep space, never to return.

CHECK YOUR UNDERSTANDING **3.5**

Place the following in order from largest to smallest semimajor axis.
a. a planet with a period of 84 Earth days
b. a planet with a period of 1 Earth year
c. a planet with a period of 2 Earth years
d. a planet with a period of 0.5 Earth year

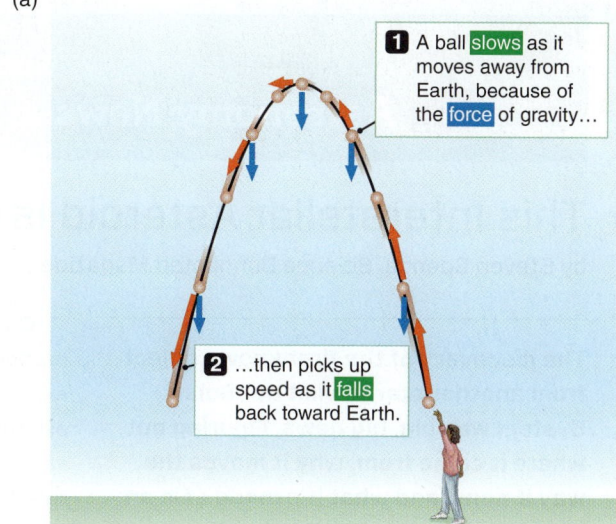

(a)

1 A ball slows as it moves away from Earth, because of the force of gravity…

2 …then picks up speed as it falls back toward Earth.

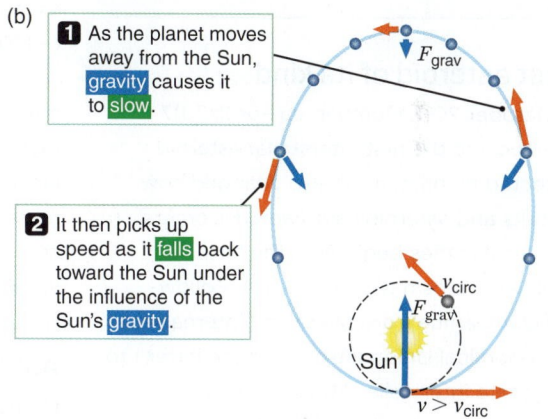

(b)

1 As the planet moves away from the Sun, gravity causes it to slow.

F_{grav}

2 It then picks up speed as it falls back toward the Sun under the influence of the Sun's gravity.

v_{circ}
F_{grav}
Sun
$v > v_{circ}$

Figure 3.16 (a) A ball thrown into the air slows as it climbs away from Earth and then speeds up as it heads back toward Earth. (b) A planet on an elliptical orbit around the Sun does the same thing. (Although no planet has an orbit as eccentric as the one shown here, the orbits of comets can be far more eccentric.)

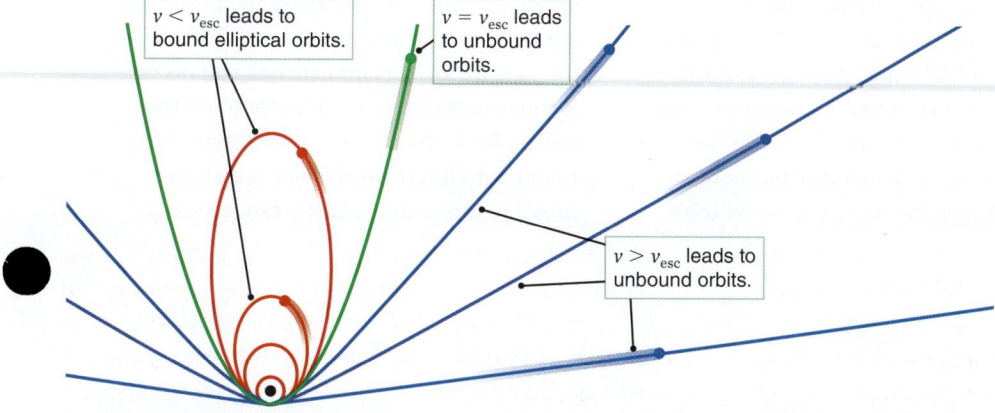

$v < v_{esc}$ leads to bound elliptical orbits.

$v = v_{esc}$ leads to unbound orbits.

$v > v_{esc}$ leads to unbound orbits.

Figure 3.17 A range of different orbits that share the same point of closest approach but differ in velocity at that point. An object's velocity at closest approach determines the orbit shape and whether the orbit is bound or unbound. The dividing line between bound and unbound occurs when the object has a velocity equal to the escape velocity (green line). Objects with higher velocities have excess velocity, and their orbits are not bound (blue lines).

reading Astronomy News

This Interstellar Asteroid Is Accelerating

by Steven Spence, Science Connected Magazine

The discovery of the first known object from another star to visit our Solar System was big, big news. Figuring out where it came from, why it moves the way it does, and what it's made of is an ongoing project.

First asteroid of its kind

In October 2017, Oumuamua—or 1I/2017 U1—became the first interstellar asteroid detected by humans. It also changed how comets and asteroids are named. Comets' technical names begin with the letter C, while asteroids have the letter A. Following Oumuamua's discovery, the International Astronomical Union introduced the letter I to designate interstellar objects.

What do we know about Oumuamua?

Oumuamua, which means "a messenger from afar arriving first" in Hawaiian, was discovered on October 19, 2017, by the University of Hawaii's Pan-STARRS1 telescope. The telescope is part of a NASA-funded program to discover and track asteroids and comets near Earth's orbit. Initially, scientists were not sure what the object was. It was first classified as a comet while details about its orbit and composition were still being worked out. Later, scientists decided to reclassify it as an asteroid due to the lack of visible comet activity. Scientists also could not measure a cyanide gas emission, typically seen in outgassing from comets formed in our solar system.

Oumuamua has a complex rotational motion. It spins on its axis approximately every seven hours and 20 minutes. The asteroid is also unusually shaped. Based on variations in observed brightness and the rotation of the asteroid, models indicate the asteroid is approximately five times as long (~800 meters) as it is in diameter (~160 meters).

Oumuamua has a dull red color. Scientists think its surface was reddened by cosmic rays while it is travelling through interstellar space. The object is dense and rocky or metallic, with little dust or ice detected. No coma of dust and gas from Oumuamua is visible in telescopes, as would be the case with a typical comet.

Asteroid's interstellar origin

Oumuamua is moving on an open hyperbolic trajectory with an eccentricity of 1.23 through our solar system, which is why astronomers were quickly able to identify it as an interstellar object (**Figure 3.18**). Orbital trajectory calculations indicate that the asteroid came in from the direction of Vega in the constellation Lyra. It is exiting the solar system on a trajectory that will take it into the great square of the constellation Pegasus.

Oumuamua is moving too fast (~26 km/s) for the sun to capture it. At perihelion, its nearest approach to the sun, Oumuamua was moving 87.7 km/s. The asteroid was moving so fast then that it could have traveled the average distance from the Earth to the moon in only 73 minutes! For comparison, the Apollo flights to the moon took

approximately three days to go from Earth's orbit to lunar orbit. New Horizons, the spacecraft sent to Pluto, covered the distance in eight hours. By comparison, if we could drive to the moon at a typical speed of 130 km/h (~81 mph), it would take approximately 123 days of non-stop driving to get there.

Oumuamua has sped up

As detailed in a recent paper published in *Nature*, the team under lead author Marco Micheli of the European Space Agency proved that the asteroid isn't moving as it expected. In fact, it passed the orbit of Jupiter earlier than predicted, indicating it accelerated along its trajectory. Oumuamua was approximately 100,000 km farther than could be explained based on purely gravitational interactions. That sounds like a huge distance, but at the asteroid's speed, that difference amounts to passing Jupiter approximately 64 minutes early. For scale, at its average distance from the sun, Jupiter is approximately 774 million km (~43 light minutes) from the sun. A discrepancy of 100,000 km amounts to only 0.01%, which doesn't sound like a lot but is measurable. In space, small percentages can rapidly become a large number in absolute terms.

Astronomers have spent weeks analyzing the discrepancy between Oumuamua's observed trajectory data and the predicted position. In their paper published in *Nature* on June 27, 2018, Marco Micheli and his colleagues reported how they identified the reason for Oumuamua's acceleration. After ruling out effects from "solar-radiation pressure, drag- and friction-like forces,

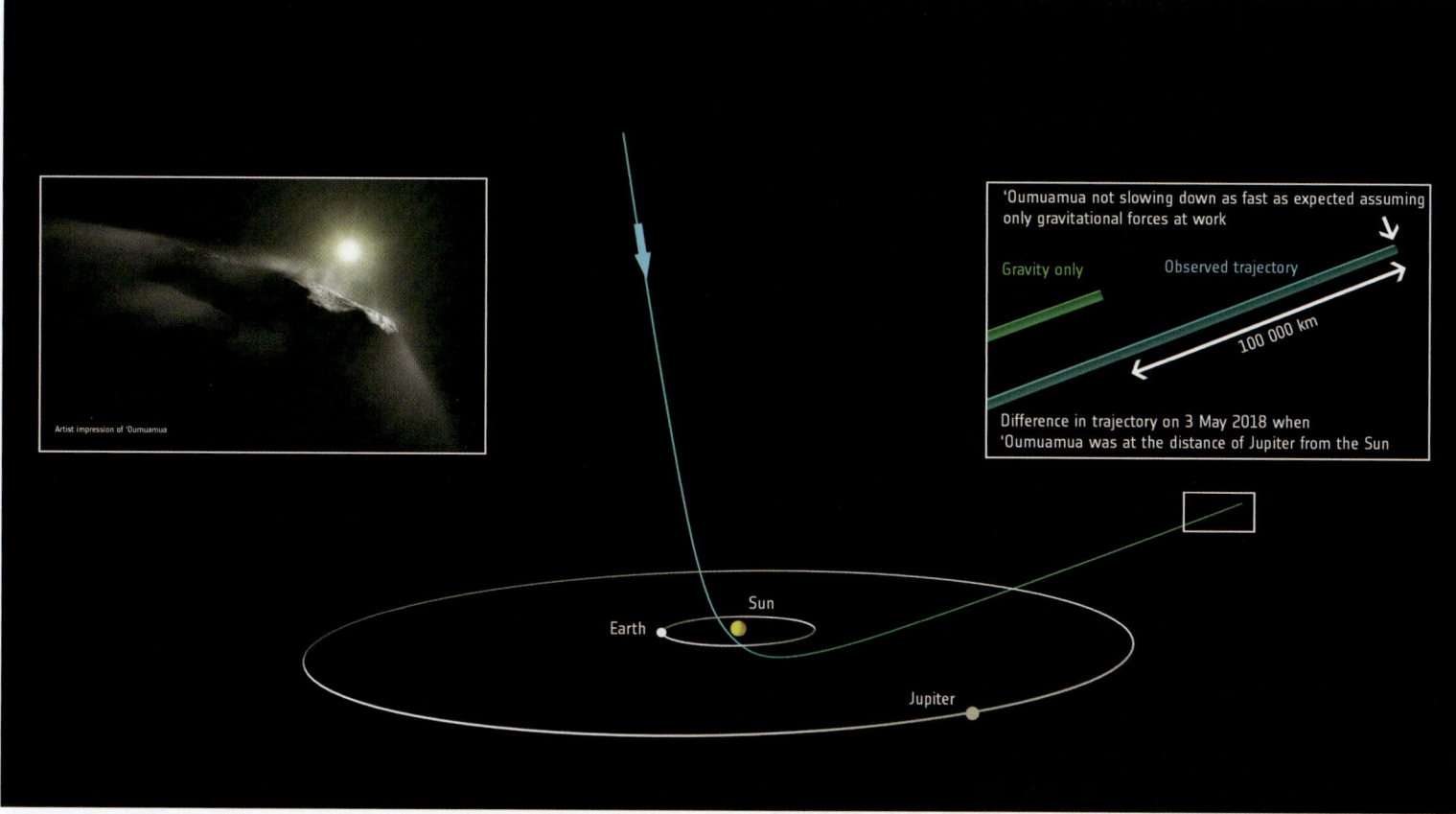

Figure 3.18 The orbit of interstellar object Oumuamua as it passes through the Solar System. The predicted path of Oumuamua and the new course that takes the new measured velocity of the object into account are shown. Credit: European Space Agency.

interaction with solar wind for a highly magnetized object, and geometric effects" due to the shape of the asteroid, they concluded that Oumuamua accelerated due to "comet-like outgassing." This means that Oumuamua vents gases when heated in a similar manner to a comet.

Asteroid leaving the solar system

Oumuamua passed Jupiter's orbit in early May 2018. It will cross Saturn's orbit in January 2019, Uranus's in August 2020, and Neptune's in June 2024. By late 2025 Oumuamua will reach the outer edge of the Kuiper Belt, and then the heliopause boundary—the edge of the solar system—sometime in November 2038.

EVALUATING THE NEWS

1. In the discussion of where 'Oumuamua came from, the author states that it is "moving on an open hyperbolic trajectory." Compare this statement to Figure 3.17. What color of the lines in that figure could correspond to this trajectory?

2. 'Oumuamua's trajectory is described as having an eccentricity of 1.23. But the maximum eccentricity for an ellipse is 1.00. What is the resolution of this apparent contradiction?

3. In the article, 'Oumuamua's speed Is described as fast enough to cover the distance from Earth to the Moon in 73 minutes. The *New Horizons* spacecraft covered the same distance in 8 hours. Approximately how many times faster is 'Oumuamua traveling than *New Horizons*?

4. 'Oumuamua reached the orbit of Jupiter about an hour earlier than expected. Why does this imply that the object has accelerated?

5. The best idea, currently, for how 'Oumuamua is accelerating is that it is due to "comet-like outgassing." Outgassing occurs when a jet of gas shoots out from the object. Use Newton's third law to explain how "outgassing" can cause an object like 'Oumuamua to accelerate.

Source: **Steven Spence**, "This Interstellar Asteroid Is Accelerating," *Science Connected Magazine*, July 21, 2018. Reprinted by permission of Science Connected.

SUMMARY

Early astronomers hypothesized that Earth was stationary at the center of the Solar System. Later astronomers realized that a Sun-centered Solar System was a more straightforward way of describing their observations. Kepler's laws describe the elliptical orbits of planets around the Sun, including details about how fast a planet travels at various points in its orbit. These laws helped Newton to advance science by developing his laws of motion, which govern the motion of all objects (not just orbiting ones). Kepler's laws are explained by Newton's theory of gravity, which describes both how objects fall and how planets are bound to the Sun.

(1) Kepler's first law states that planets move in ellipses. Kepler's second law relates the speed of the planet to different places in its orbit (faster when closer to the Sun, slower when farther from the Sun), so that an imaginary line connecting the planet to the Sun sweeps out equal areas in equal times. Kepler's third law gives an empirically determined relationship between the size of an orbit and the time it takes for the planet to orbit once (the period squared is proportional to the average distance cubed).

(2) Newton's three laws (in shorthand: inertia, $F = ma$, and "every action has an equal and opposite reaction") govern the motion of all objects. Net forces cause accelerations, or changes in motion. Mass is the property of matter that gives it inertia, or resistance to changes in motion. Gravity is one of the fundamental forces of nature and binds the universe together. Gravity is a force between any two objects that is due to their masses. The force of gravity is proportional to the product of the two masses and inversely proportional to the square of the distance between them.

(3) Planets orbit the Sun in bound, elliptical orbits because of gravity. Any circular orbit has a characteristic circular velocity, which is faster if the orbit is smaller, because gravity is stronger when the objects are closer together. A planet on an elliptical orbit speeds up and slows down as it is closer and farther from the Sun, respectively. Orbits become "unbound" when the object reaches escape velocity.

(4) A reference frame is a system in which an observer makes measurements of both space and time. No reference frame is preferred over any other.

QUESTIONS AND PROBLEMS

TESTING YOUR UNDERSTANDING

1. **T/F:** Kepler's three laws explain *why* the planets orbit the Sun as they do.

2. **T/F:** The natural state of objects is to be at rest. This is why a book, given a push across a table, will eventually slow to a stop.

3. **T/F:** To find the period of an object in orbit around the Sun, we only need to know the semimajor axis of the orbit.

4. **T/F:** You are always at rest in your own reference frame.

5. **T/F:** A force is required to keep an object moving at the same velocity.

6. Newton's second law of motion states that
   a. objects have inertia.
   b. the acceleration of an object is proportional to the net force acting on it.
   c. every action has an equal and opposite reaction.
   d. the force of gravity is proportional to the masses of the two objects and inversely proportional to the square of the distance between them.

7. The connection between gravity and orbits enables astronomers to measure the _____ of stars and planets.
   a. distances
   b. sizes
   c. masses
   d. compositions

8. What is the eccentricity of a circular orbit?
   a. 0
   b. 0.5
   c. 1.0
   d. 100 percent

9. Imagine you are walking along a forest path. Which of the following is *not* an action-reaction pair in this situation?
   a. the gravitational force between you and Earth; the gravitational force between Earth and you
   b. your shoe pushes back on Earth; Earth pushes forward on your shoe
   c. your foot pushes back on the inside of your shoe; your shoe pushes forward on your foot
   d. you push down on Earth; Earth pushes you forward

10. A net force must be acting when (choose all that apply)
    a. an object accelerates.
    b. an object changes direction but not speed.
    c. an object changes speed but not direction.
    d. an object changes speed and direction.

11. From Kepler's second law, you can determine that
    a. planetary orbits are ellipses with the Sun at one focus.
    b. the square of a planet's orbital period equals the cube of its semimajor axis.
    c. for every action there is an equal and opposite reaction.
    d. planets move fastest when they are closest to the Sun.

12. Imagine that you are floating in the International Space Station, tossing a bag of dried ice cream to a fellow astronaut. Which of the following are action-reaction pairs? (Choose all that apply.)
    a. you push on the ice cream; the ice cream pushes back on you
    b. Earth pulls on the ice cream; the ice cream pushes on your fellow astronaut
    c. Earth pulls on the space station; the space station pulls on Earth
    d. Earth pulls on you; you pull on Earth

13. Imagine a planet moving in a perfectly circular orbit around the Sun. Is this planet experiencing acceleration?
    a. Yes, because its speed is changing.
    b. Yes, because it is changing its direction of motion all the time.
    c. No, because its speed is not changing.
    d. No, because planets do not experience accelerations.

14. Suppose a piece of rock enters Earth's atmosphere, traveling to your left as a bright streak in the sky. The rock explodes, and a piece of it goes in a direction opposite the motion of the original rock. In your reference frame,
    a. the small piece travels to the left.
    b. the small piece travels to the right.
    c. the small piece falls straight down.
    d. the behavior of the small piece depends on the relative speeds of the small piece and the original rock.

15. Suppose you read on the Web that a new planet has been found. Its average speed in its orbit is 33 kilometers per second (km/s). When it is closest to its star, it moves at 31 km/s, and when it is farthest from its star, it moves at 35 km/s. This story is in error because
    a. the average speed is far too fast.
    b. Kepler's third law says the planet has to sweep out equal areas in equal times, so the speed of the planet cannot change.
    c. planets stay at a constant distance from their stars; they don't move closer or farther away.
    d. Kepler's second law says the planet must move fastest when it is closest to its star, not when it is farthest away from it.
    e. using these numbers, the square of the orbital period will not be equal to the cube of the semimajor axis.

16. Suppose a car can go from 0 to 100 km/h in only 2.0 seconds. This car's acceleration is about
    a. 50 km/h.
    b. 14 m/s^2.
    c. 50 km/(h s).
    d. 200 km.
    e. 0.056 km/h^2.

17. Imagine that you are standing on the (airless) Moon, and you drop four objects, each the size of a bowling ball. Each is made of a different substance: Styrofoam, lead, Bubble Wrap, and pumpkin. In what order do they reach the ground?
    a. lead, pumpkin, Bubble Wrap, Styrofoam
    b. lead, Bubble Wrap, pumpkin, Styrofoam
    c. Styrofoam, lead, Bubble Wrap, pumpkin
    d. none of the above (They all reach the ground at the same time.)

18. Suppose a planet has a mass 100 times bigger than the mass of its moon. How does the force of the planet on the moon compare to the force of the moon on the planet?
    a. The force of the planet on the moon is stronger than the force of the moon on the planet.
    b. The force of the moon on the planet is stronger than the force of the planet on the moon.
    c. Both forces are the same.
    d. Neither force can be measured, so the answer is not known.

19. *Weight* refers to the force of gravity acting on a mass. We often calculate the weight of an object by multiplying its mass by the local acceleration due to gravity. The value of gravitational acceleration on the surface of Mars is 0.377 times that on Earth. If your mass on Earth is 85 kg, your weight on Earth is 830 N ($m \times g = 85$ kg $\times$ 9.8 m/s^2 = 830 N). What would be your approximate mass and weight on Mars?
    a. mass = 830 N; weight = 8,300 N
    b. mass = 85 kg; weight = 830 N
    c. mass = 85 kg; weight = 31 kg
    d. mass = 85 kg; weight = 310 N

20. Suppose a new, Earth-mass planet is discovered around a Sun-like star. This planet is 19 times Earth's distance from the Sun. As a result, the star's gravitational force on this planet is _____ than the Sun's gravitational force on Earth.
    a. 361 times weaker
    b. 19 times weaker
    c. 19 times stronger
    d. 361 times stronger

THINKING ABOUT THE CONCEPTS

21. ★ **WHAT AN ASTRONOMER SEES** In Figure 3.1, Mars appears larger in the part of the loop that is retrograde, when Mars is moving from left to right. Why does Mars appear larger at this time? What conclusion can you make about the cause of the changing size of Venus in Figure 3.7? ★

22. Make an enlarged sketch of the orbit of Venus in the geocentric model shown in Figure 3.2. Label the phase of Venus at four evenly spaced points on the small circle, keeping in mind that, in this model, Venus is always between Earth and the Sun. Which phases in Figure 3.7 cannot be reproduced in this model? ★

23. Ellipses contain two axes, major and minor. Half the major axis is called the semimajor axis. What is especially important about the semimajor axis of a planetary orbit?

24. Suppose that the car in Figure 3.10 is driving past the woman pulling the refrigerator. Are the driver of the car and the woman pulling on the rope in the same reference frame? Explain. ★

25. The distance that Neptune has to travel in its orbit around the Sun is approximately 30 times greater than the distance that Earth must travel, yet it takes nearly 165 years for Neptune to complete one trip around the Sun. Explain why.

26. Figure 3.14 demonstrates that objects of very different masses experience the same gravitational acceleration. How would the situation shown here be different if the space station accelerated more or less than the astronaut? What would happen? ★

27. When riding in a car, we can sense changes in speed or direction through the forces that the car applies on us. Do we wear seat belts in cars and airplanes to protect us from speed or from acceleration? Explain your answer.

28. An astronaut standing on Earth can easily lift a wrench having a mass of 1 kg but not a scientific instrument with a mass of 100 kg. In the International Space Station, the astronaut is quite capable of manipulating both, although the scientific instrument responds much more slowly than the wrench. Explain why.

29. Explain the difference between weight and mass.

30. On the Moon, your weight is different from your weight on Earth. Why?

31. Describe the difference between a bound orbit and an unbound orbit.

32. Two objects are leaving the vicinity of the Sun, one traveling in a bound orbit and the other in an unbound orbit. What can you say about the future of these two objects? Would you expect either of them to return?

33. Suppose astronomers discovered an object approaching the Sun in an unbound orbit. What would that say about the origin of the object?

34. What is the advantage of launching satellites from spaceports located near the equator? Why are satellites never launched toward the west?

35. In 1920, a *New York Times* editor refused to publish an article that was based on rocket pioneer Robert Goddard's paper that predicted spaceflight, saying that "rockets could not work in outer space because they have nothing to push against." The *New York Times* did not retract this statement until July 20, 1969, the date of the *Apollo 11* Moon landing. What was wrong with the editor's logic?

APPLYING THE CONCEPTS

36. Suppose you discover a new dwarf planet in our Solar System with a semimajor axis of 46.4 AU. What is its period (in Earth years)? Check your work by comparing your result to the period of a Solar System object with a similar semimajor axis (see Appendix 2).

37. The average acceleration, *a*, of an object moving in a straight line is equal to the change in velocity divided by the change in time. Suppose that an object accelerates from 0 km/h to 5 km/h and takes 1 hour to do it.
  a. Predict whether you expect the acceleration to be large or small. Has the speed changed very much in an hour? (Is 5 km/h fast or slow?) Think about dropping a ball; would it accelerate much faster or slower than this?
  b. If you divide this change in velocity by this time, what will your units of acceleration be?
  c. Calculate the average acceleration.
  d. Check your work by checking that your units come out as a distance divided by a time squared. Then compare the acceleration you calculated with the acceleration due to gravity (127,000 km/h²). Did you get the answer you expected?

38. A sports car accelerates from 0 km/h to 100 km/h in 4 seconds.
  a. Predict whether you expect this acceleration to be large or small. Would this change in speed happen faster or slower than a similar change in speed for a dropped ball?
  b. Awkwardly, the two times here (in the denominator of the speed and the time elapsed) are in two different units. Convert the speeds from kilometers per hour to meters per second by converting the units, as you learned to do in Chapter 2.
  c. What is the car's average acceleration? Check your work by comparing your result with the acceleration due to gravity (9.8 m/s²).
  d. Suppose the mass of the car is 1,200 kg. Use Newton's second law to calculate the strength of the force on the car.
  e. What supplies the "push" that accelerates the car?

39. Suppose that you are pushing a small refrigerator with a mass of 50 kg on wheels. You push with a force of 100 N.
  a. Solve Newton's second law ($F = ma$) algebraically for the acceleration.
  b. What is the refrigerator's acceleration? Check your work by making sure that the units come out correctly.

40. You are riding along on your bicycle at 20 km/h and eating an apple. You pass a bystander, stationary by the side of the road.
  a. How fast is the apple moving in your frame of reference?
  b. How fast is the apple moving in the bystander's frame of reference?
  c. Whose perspective is more valid?

exploration Newtonian Orbits

digital.wwnorton.com/universe4

In this Exploration, we will use a simulation to explore the Newtonian features of Mercury's orbit. Visit the Student Site at the Digital Resources page and open the Orbits Interactive Simulation in Chapter 3.

Accelerations

To begin exploring the simulation, set parameters for "Mercury" in the "Orbit Settings" panel and then click "OK." Click the "Newtonian Features" tab at the bottom of the control panel. Select "show solar system orbits" and "show grid" under "Visualization Options." Change the animation rate to 0.01, and press the "start animation" button.

Examine the graph at the bottom of the panel.

1. Where is Mercury in its orbit when the acceleration is smallest?

_____

_____

2. Where is Mercury in its orbit when the acceleration is largest?

_____

_____

3. What are the values of the largest and smallest accelerations?

_____

_____

In the "Newtonian Features" graph, mark the boxes for "vector" and "line" that correspond to the acceleration. Checking these boxes will insert an arrow that shows the direction of the acceleration and a line that extends the arrow.

4. To what Solar System object does the arrow point?

_____

Think about Newton's second law.

5. In what direction is the force on the planet?

_____

_____

Velocities

Examine the graph at the bottom of the panel again.

6. Where is Mercury in its orbit when the velocity is smallest?

_____

_____

7. Where is Mercury in its orbit when the velocity is largest?

_____

_____

8. What are the values of the largest and smallest velocities?

_____

_____

Add the velocity vector and line to the simulation by clicking on the boxes in the graph window. Study the arrows carefully.

9. Are the velocity and the acceleration always perpendicular (is the angle between them always 90°)?

_____

_____

10. If the orbit were a perfect circle, what would the angle be between the velocity and the acceleration?

_____

_____

Hypothetical Planet

Use the "Orbit Settings" to change the semimajor axis to 0.8 AU.

11. How does this imaginary planet's orbital motion compare to Mercury's?

_____

_____

_____

Now change the semimajor axis to 0.1 AU.

12. How does this planet's orbital period now compare to Mercury's?

_____

_____

_____

13. Summarize your observations of the relationship between the speed of an orbiting object and the semimajor axis.

_____

_____

_____

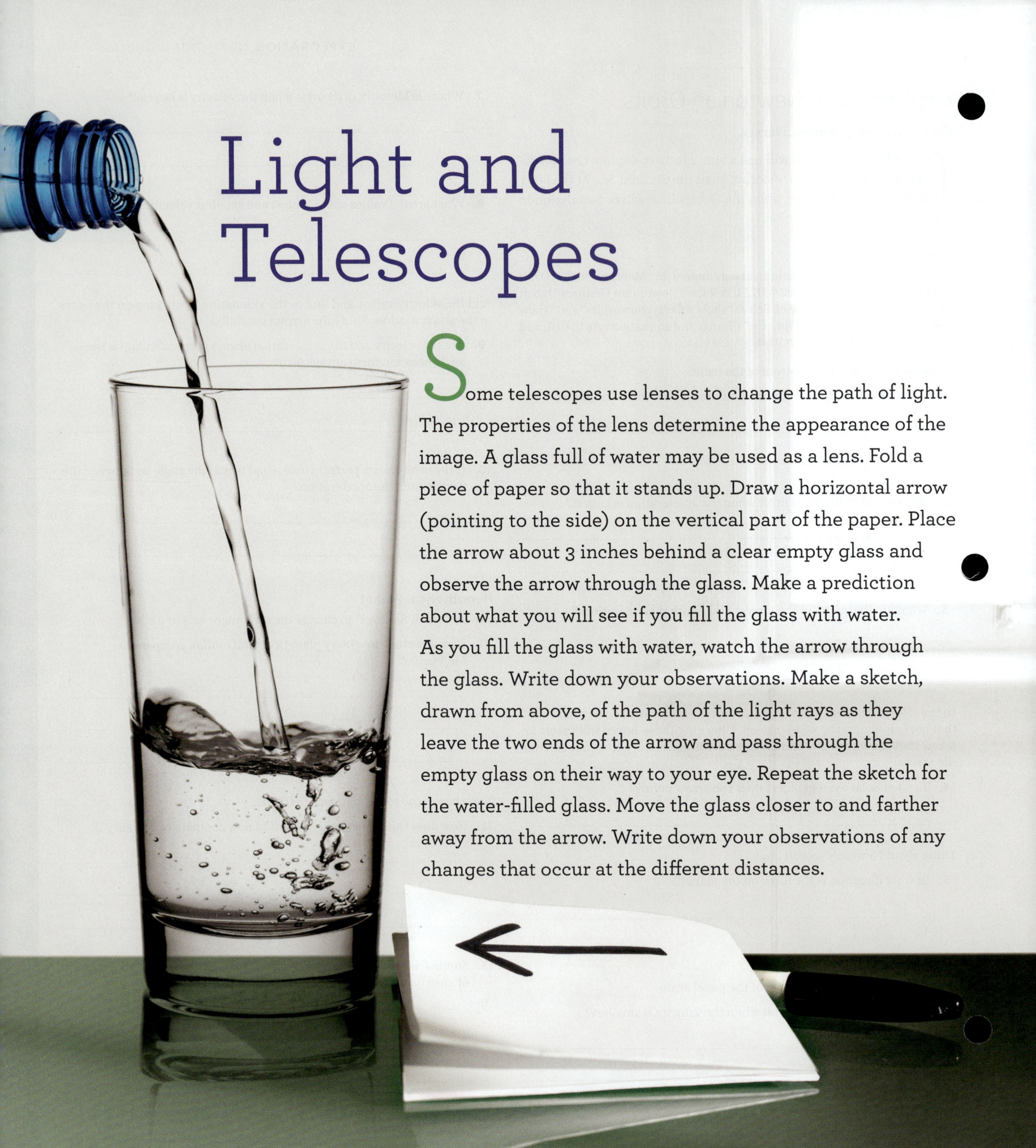

Light and Telescopes

Some telescopes use lenses to change the path of light. The properties of the lens determine the appearance of the image. A glass full of water may be used as a lens. Fold a piece of paper so that it stands up. Draw a horizontal arrow (pointing to the side) on the vertical part of the paper. Place the arrow about 3 inches behind a clear empty glass and observe the arrow through the glass. Make a prediction about what you will see if you fill the glass with water. As you fill the glass with water, watch the arrow through the glass. Write down your observations. Make a sketch, drawn from above, of the path of the light rays as they leave the two ends of the arrow and pass through the empty glass on their way to your eye. Repeat the sketch for the water-filled glass. Move the glass closer to and farther away from the arrow. Write down your observations of any changes that occur at the different distances.

EXPERIMENT SETUP

Fold a piece of paper so it stands up. Draw a horizontal arrow.

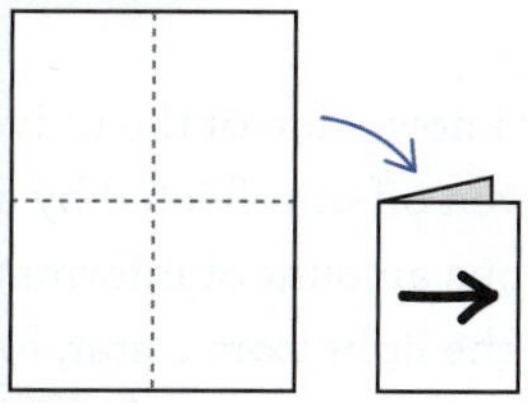

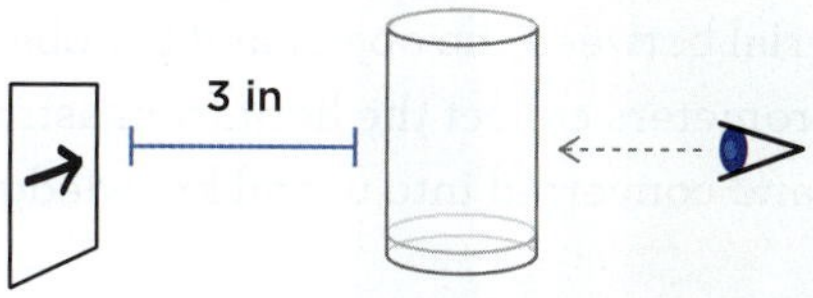

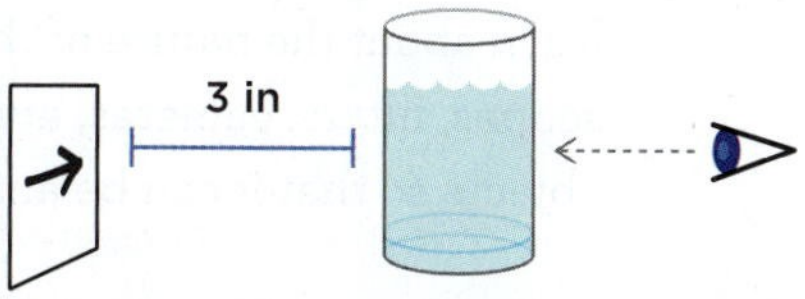

Place the paper about 3 inches behind a clear empty glass and observe the arrow through the glass.

As you fill the glass with water, continue to watch the arrow through the glass. Write down your observations.

Make a sketch, drawn from above, of the path of the light rays as they leave the two ends of the arrow and pass through the empty glass on their way to your eye. Repeat the sketch for the water-filled glass.

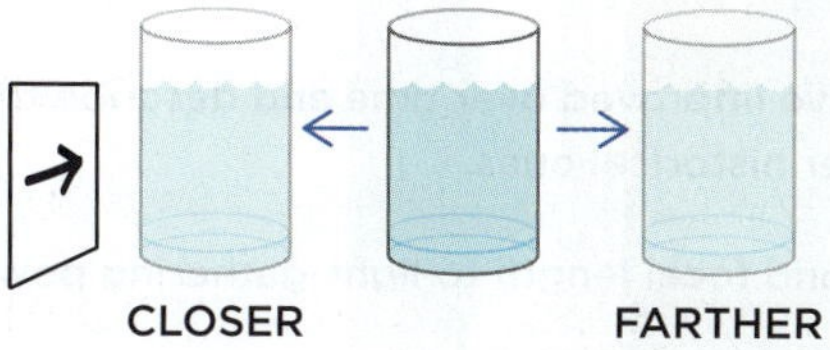

CLOSER   FARTHER

Move the glass closer to and farther away from the arrow. Write down your observations of any changes that occur at the different distances.

PREDICTION

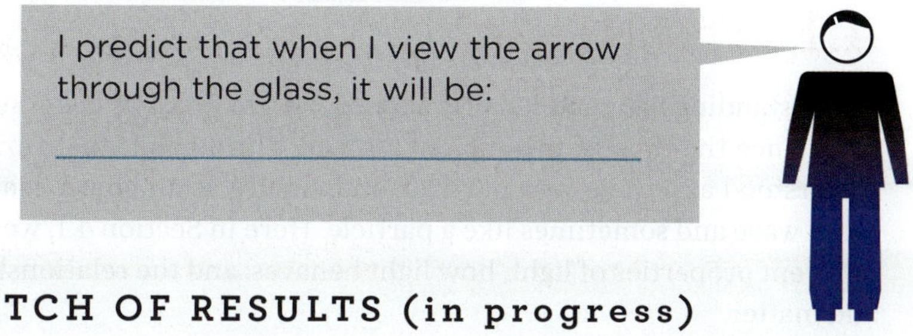

I predict that when I view the arrow through the glass, it will be:

_____

SKETCH OF RESULTS (in progress)

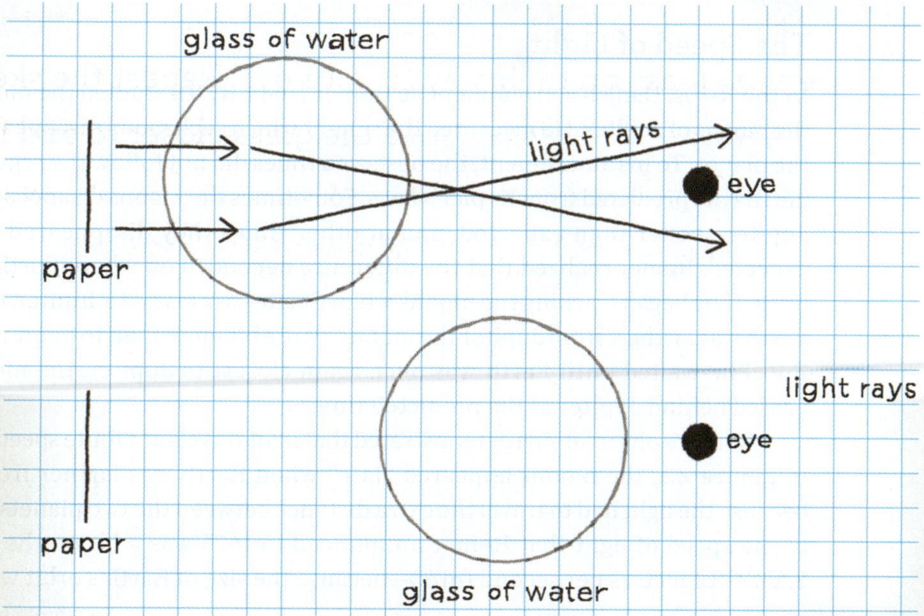

glass of water

light rays

paper   eye

light rays

paper   eye

glass of water

Our knowledge of the universe beyond Earth comes primarily from light given off or reflected by astronomical objects. This light carries an enormous amount of information about the object. When astronomers carefully study the light from a star, for example, they can find out how hot the star is, what it is made of, and how fast and in what direction it is traveling. They can also learn about the nature of the material between the object and the observer. Telescopes, filters, cameras, and spectrometers collect the light from astronomical objects so that it can be analyzed and converted into useful knowledge.

LEARNING GOALS

(1) Compare and contrast the properties of waves with the properties of particles, and give examples of the wave and particle behavior of light.

(2) Describe the electromagnetic spectrum and the types of information that can be obtained by observing light.

(3) Explain how light detectors have improved over time and describe the advantages of modern detectors over historical ones.

(4) Relate a telescope's aperture and focal length to light-gathering power, resolution, and image size.

4.1 What Is Light?

Understanding light and its interactions with matter has been a scientific quest at least since the time of the ancient Greeks. Throughout that time, light has been understood as a wave, as a particle, and, finally, as an object that acts sometimes like a wave and sometimes like a particle. Here in Section 4.1, we will examine the different properties of light, how light behaves, and the relationship between light and matter.

The Speed of Light

In the 1670s, Danish astronomer Ole Rømer (1644–1710) studied the movement of the moons of Jupiter. He measured the time at which each moon disappeared behind the planet. To his amazement, the observed times did not follow the regular schedule that he predicted from Kepler's laws. Sometimes the moons disappeared behind Jupiter sooner than expected, and at other times they disappeared later than expected. Rømer realized that the difference depended on where Earth was in its orbit. If he began tracking the moons when Earth was closest to Jupiter, by the time Earth was farthest from Jupiter the moons were a bit more than $16\frac{1}{2}$ minutes "late." But if he waited until Earth was once again closest to Jupiter, the moons again passed behind Jupiter at the predicted times.

Rømer's observations demonstrated that light travels at a finite speed. As shown in **Figure 4.1**, the moons appeared "late" when Earth was farther from Jupiter because the light had to travel the extra distance between the two planets. The value of the speed of light that Rømer announced in 1676 was a bit on the low side—2.25×10^8 meters per second (m/s)—because the size of Earth's orbit was not well

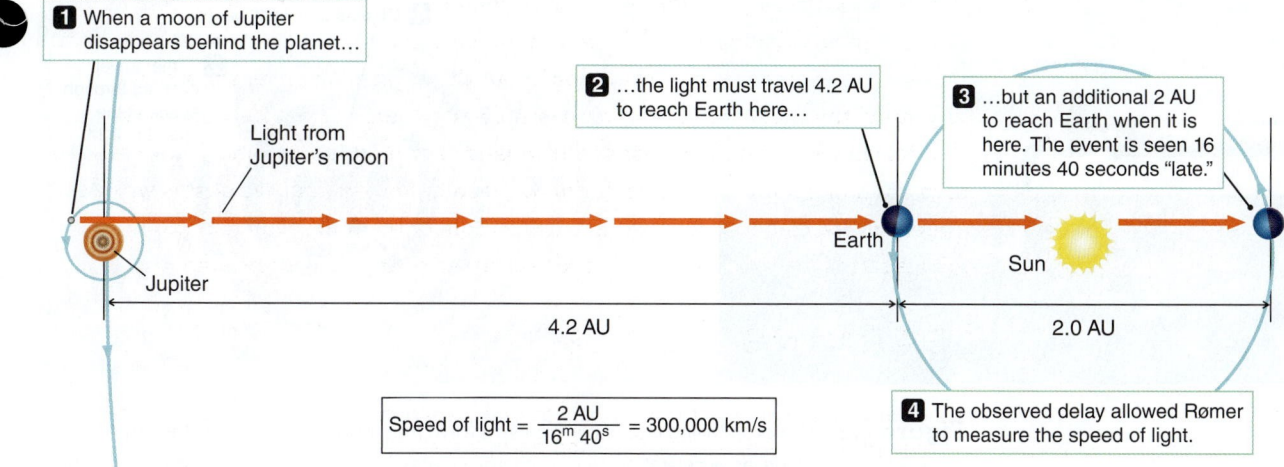

1 When a moon of Jupiter disappears behind the planet…

2 …the light must travel 4.2 AU to reach Earth here…

3 …but an additional 2 AU to reach Earth when it is here. The event is seen 16 minutes 40 seconds "late."

Light from Jupiter's moon

Earth

Sun

Jupiter

4.2 AU

2.0 AU

$$\text{Speed of light} = \frac{2\ \text{AU}}{16^{\text{m}}\ 40^{\text{s}}} = 300{,}000\ \text{km/s}$$

4 The observed delay allowed Rømer to measure the speed of light.

Figure 4.1 Ole Rømer realized that apparent differences between the predicted and observed orbital motions of Jupiter's moons depend on the distance between Earth and Jupiter. He used these observations to measure the speed of light. (Recall that 1 astronomical unit [AU] is the average distance from Earth to the Sun, equal to 1.50×10^8 km.)

known. Modern measurements of the speed of light in a **vacuum** give a value of 2.99792458×10^8 m/s. In this book, we will round up to 3×10^8 m/s (equivalent to 3×10^5 km/s). The speed of light in a vacuum is a fundamental constant, c. Keep in mind, however, that light travels at this speed *only* in a vacuum. The speed of light through a substance such as air or glass is less than c.

The International Space Station moves around Earth at about 28,000 kilometers per hour (km/h), orbiting Earth in about 90 minutes. Light travels almost 40,000 times faster than this and could circle Earth in $\frac{1}{7}$ of a second. Because light is so fast, the travel time of light is a convenient way of expressing cosmic distances. The basic unit is the light-year—the distance light travels in 1 year. Pause for a moment to consider the light-year. Imagine traveling around Earth in $\frac{1}{7}$ of a second. Now try to imagine traveling at that speed for about 8 minutes, which is the time you would need to travel from Earth to the Sun at the speed of light. Now try to imagine traveling at that speed for an entire year. After a year, you would be much farther than the Sun, but you wouldn't be even one-quarter of the way to the next closest star, Proxima Centauri.

As light travels, it carries energy from place to place. **Energy** is the ability of a system to produce a change, and it comes in many forms. **Kinetic energy** is the energy of moving objects. **Thermal energy** is closely related to kinetic energy and is the sum of all the kinetic energy of the moving bits of matter inside a substance. The sum of the energies of all these random motions results in an object's temperature. An object may have **potential energy**, which is energy stored in the system. For example, a stretched rubber band has energy stored within it, which may turn into kinetic energy when the band is released. Energy has units of joules (J). (You may be more familiar with the kilocalorie, which is the unit of energy commonly used for food. One kilocalorie is 4,200 J.)

Energy can change from one type to another, but energy cannot be created or destroyed. When a stretched rubber band is released, the stored potential energy is often converted to kinetic energy. When light from the Sun strikes pavement, the pavement heats up because it gains thermal energy. Light carried that energy from the Sun to the pavement. Rømer knew how long it took for light to travel a given distance and used that to calculate the speed of light, but it would take more than 200 years for physicists to figure out what light actually is.

VOCABULARY ALERT

vacuum In common language, we think of this as a completely empty space, without even an atom in it. Such a perfect vacuum does not exist in nature, so astronomers are comfortable using *vacuum* to mean "a place with hardly anything in it." For the purposes of determining what speed to use for light, the space between the galaxies is considered a vacuum, and so is the space between the stars.

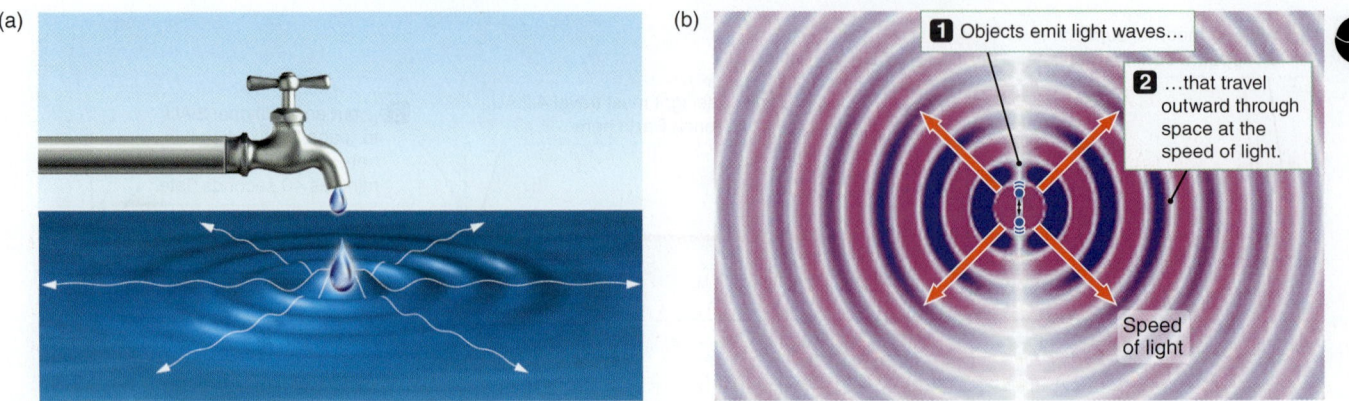

Figure 4.2 (a) A drop falling into water creates a wave that moves across the water's surface. (b) In a similar fashion, an accelerating electric charge creates light waves that move away from the charge at the speed of light.

Light as a Wave

Throughout the late 1800s and early 1900s, the results of light experiments appeared inconsistent. Sometimes light acted like a wave in water, and at other times it acted like a particle—an object that, for the moment, can be thought of as a very, very tiny baseball. Eventually, scientists concluded that whether light acts like a wave or like a particle depends on the type of instrument used to make the observation. In this subsection, we will discuss light's wavelike properties. In the next subsection, we will discuss its particle-like properties.

When a drop of water falls from a faucet into a sink full of water, it causes a disturbance, or wave, like the one shown in **Figure 4.2a**. The wave moves outward as a ripple on the surface of the water. In the same way, a charged particle that moves up and down emits waves that travel away through space. In the case of the charged particle, those waves are light waves, as shown in Figure 4.2b. However, the ripples in the sink are distortions of the water's surface, and they require a **medium**: a substance to travel through. Light waves do not require a medium—they can move through the vacuum of empty space.

Now imagine that a cork is floating in the sink, as in **Figure 4.3a**. The cork remains stationary until the ripple from the dripping faucet reaches it. The rising and falling water causes the cork to rise and fall. This indicates that the wave is carrying energy, because as it passes the cork, it lifts it up, increasing the cork's gravitational potential energy. Light waves similarly carry energy through space and cause electrically charged particles to oscillate up and down, as shown in Figure 4.3b.

Waves are characterized by four quantities—*amplitude, wavelength, frequency,* and *speed*—as illustrated in **Figure 4.4**. The **amplitude** of a wave is the height of the wave above the undisturbed position (Figure 4.4a). For water waves, the amplitude is how far the water is lifted up by the wave. The amplitude of a light wave, however, is related to the brightness of the light. A wave travels at a particular speed, v (Figure 4.4b). The distance from one crest of a wave to the next is the **wavelength**, λ (this is the Greek letter *lambda* in Figure 4.4a, c, and d). The number of wave crests passing a point in space each second is the wave's **frequency**, f. Waves with longer wavelengths have lower frequencies, whereas waves with shorter wavelengths have higher frequencies (Figure 4.4c and d). Frequency is measured in cycles per second, or **hertz (Hz)**.

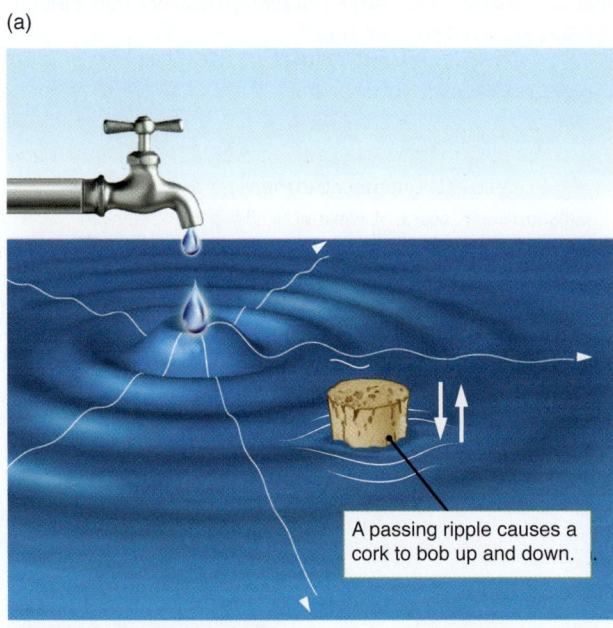

A passing ripple causes a cork to bob up and down.

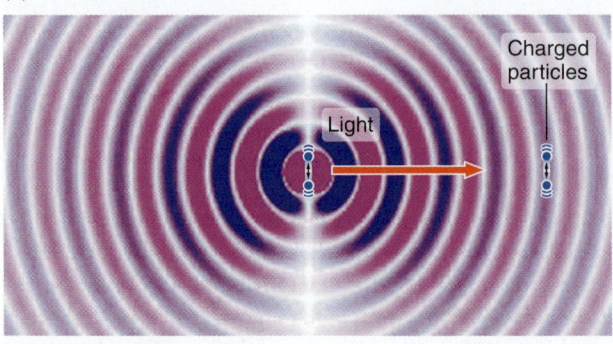

Figure 4.3 (a) When waves moving across the surface of water reach a cork, they cause the cork to bob up and down. (b) Similarly, a passing light wave causes a charged particle to oscillate in response to the wave.

working it out 4.1

Wavelength and Frequency

Radio waves are a type of light with long wavelengths and low frequencies. When you tune in to a radio station at 770 AM, you receive a signal that is broadcast at a frequency of 770 kilohertz (kHz), or 7.7×10^5 Hz. You can use the relationship between wavelength (λ), frequency (f), and the speed of light (c) to calculate the wavelength of the AM signal:

$$\lambda = \frac{c}{f} = \frac{3 \times 10^8 \,\text{m/s}}{7.7 \times 10^5 /\text{s}} = 390 \,\text{m}$$

This AM wavelength is about 4 times the length of an American football field.

The human eye is most sensitive to green light, which has a wavelength of about 500–550 nanometers (nm). You can use the relationship between wavelength (λ), frequency (f), and the speed of light (c) to calculate the frequency of green light. Green light with a wavelength of 520 nm has a frequency of

$$f = \frac{c}{\lambda} = \frac{3 \times 10^8 \,\text{m/s}}{5.2 \times 10^{-7} \,\text{m}} = \frac{5.8 \times 10^{14}}{\text{s}} = 5.8 \times 10^{14} \,\text{Hz}$$

This frequency corresponds to 580 *trillion* (580 million million) wave crests passing by each second.

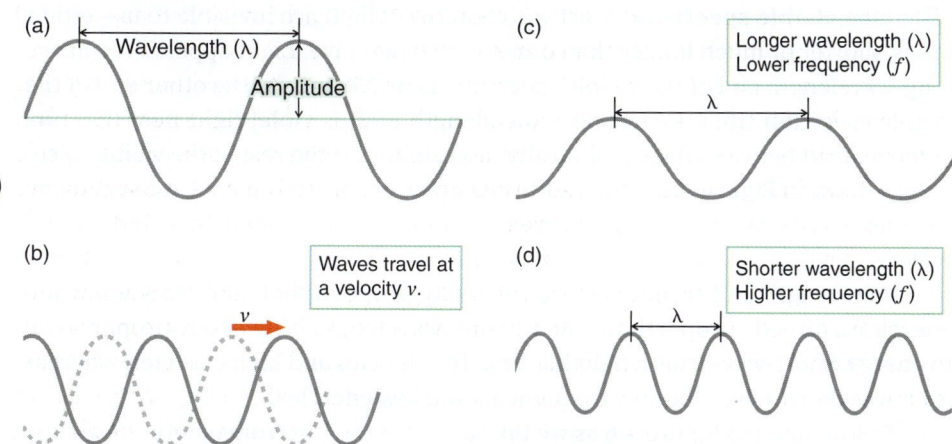

(a) Wavelength (λ)
Amplitude

(b) Waves travel at a velocity v.

(c) Longer wavelength (λ)
Lower frequency (f)
λ

(d) Shorter wavelength (λ)
Higher frequency (f)
λ

Figure 4.4 (a) A wave is characterized by the distance from one peak to the next (wavelength, λ) and the maximum height above the medium's undisturbed state (amplitude) and (b) the speed (v) at which the wave pattern travels. (c) Frequency is also a wave characteristic; it describes the number of waves that pass a location each second. A wave with a longer wavelength has a lower frequency. (d) Conversely, a wave with a shorter wavelength has a higher frequency.

Waves travel a distance of one wavelength each cycle, so the **speed** of a wave (v) can be found by multiplying the wavelength and the frequency: $v = \lambda f$. When we are talking about waves in water, the water itself doesn't travel; it just moves up and down at the same location. Think of a rubber duck in the water; as a wave passes by, the duck moves up and down, but doesn't travel anywhere. The wave, however, does travel from one location to another. In water, the speed of a wave is variable and depends on the density of the medium, among other things, whereas light always moves through a vacuum at the same speed, c. Because light travels at constant speed, its wavelength and frequency are inversely proportional to each other: If the wavelength increases, the frequency decreases. Once the wavelength of a wave of light is measured, the frequency can be found, and vice versa. **Working It Out 4.1** explores this relationship further.

Light as a Particle

Though the wave concept of light is consistent with many observations, there are properties of light that it does not describe well. Scientists working in the late 19th and early 20th centuries discovered that many of the puzzling aspects of light could

what if . . .

What if, like sound, light propagates as a wave through an invisible medium fixed in space: As Earth moves through its orbit around the Sun, how will the measured speed of light change throughout a year?

▶‖ **AstroTour:** Light as a Wave, Light as a Photon

be better understood by thinking of light as a particle. In the particle model, light is made up of massless particles called **photons** (*phot-* means "light," as in *photograph*, and *-on* signifies a particle). Photons travel at the speed of light, and each one carries energy.

The energy of a photon and the frequency of the corresponding wave are directly proportional to each other: higher frequency light waves carry more energy. This relationship connects the particle and wave concepts of light. The constant of proportionality between the energy (E) and the frequency (f) is Planck's constant (h), which is equal to 6.63×10^{-34} joule-seconds (a joule is a unit of energy). Specifically, we write $E = hf$.

The Electromagnetic Spectrum

Light is related to electricity and magnetism, and light waves are sometimes called **electromagnetic waves**. When white light interacts with water droplets and is spread out by wavelength into its component colors, a rainbow like the one shown in **Figure 4.5** is created. This spread of colors is called a **spectrum**. The whole range of different wavelengths of light is collectively referred to as the **electromagnetic spectrum**, and the small part of this spectrum that the human eye can perceive is called the **visible spectrum**. Most wavelengths of light are invisible to us—either much shorter or much longer than our eyes can perceive. Light appears red at the long-wavelength end of the visible spectrum, near 750 nm. At the other end of the visible spectrum (that is, the short-wavelength end) is violet light near 350 nm. Stretched out between the two, literally in a rainbow, is the rest of the visible spectrum, shown in **Figure 4.6**. The traditional order of colors in the visible spectrum, in order of decreasing wavelength, are red, orange, yellow, green, blue, indigo, and violet.

Wavelength and frequency are inversely proportional, and frequency and energy are directly proportional. As a result, wavelength is inversely proportional to energy. Short-wavelength light has high frequencies and high energies, whereas long-wavelength light has low frequencies and low energies.

Follow along in Figure 4.6 as we take a tour of the electromagnetic spectrum, beginning with the shortest wavelengths (on the left) and ending with the longest ones (on the right). The light with the very shortest wavelengths is called **gamma rays**, or sometimes gamma **radiation**. Gamma rays carry very high energy that can penetrate matter easily. **X-rays**, which you may know about from visits to the dentist or doctor, have longer wavelengths than gamma rays. X-ray light has enough energy to penetrate through skin and muscle but is stopped by denser teeth and bone. That's why bones show up on an X-ray image, but skin and muscle do not. **Ultraviolet (UV) radiation** has longer wavelengths than X-rays but shorter wavelengths than visible light. UV light is responsible for sunburns: It has enough energy to penetrate into your skin, but not into muscle and bone.

The visible part of the spectrum—the part you can see with your eyes—is a very small portion of the entire electromagnetic spectrum. Most of the electromagnetic spectrum, and most of the information in the universe, is invisible to the human eye. To detect light outside the visible region, we must use specialized detectors of various kinds that respond to photons with higher or lower energies than the ones that activate the cells in our eyes. In the visible spectrum, short wavelengths are blue, and long wavelengths are red.

For light that is visible to us, the energy of the light is related to its color: Blue photons have more energy than red photons. The energies of green, yellow, and orange photons fall between these two extremes.

Figure 4.5 The visible part of the electromagnetic spectrum is laid out in the colors of this rainbow.

VOCABULARY ALERT

radiation In everyday language, *radiation* is associated with emissions from nuclear bombs or radioactive substances. In some cases, this radiation actually is light in the form of gamma rays. In other cases, it is particles. Astronomers use the word to mean energy carried through space; that is, light. Astronomers often use the two words *light* and *radiation* interchangeably, especially when talking about wavelengths that are not in the visible range.

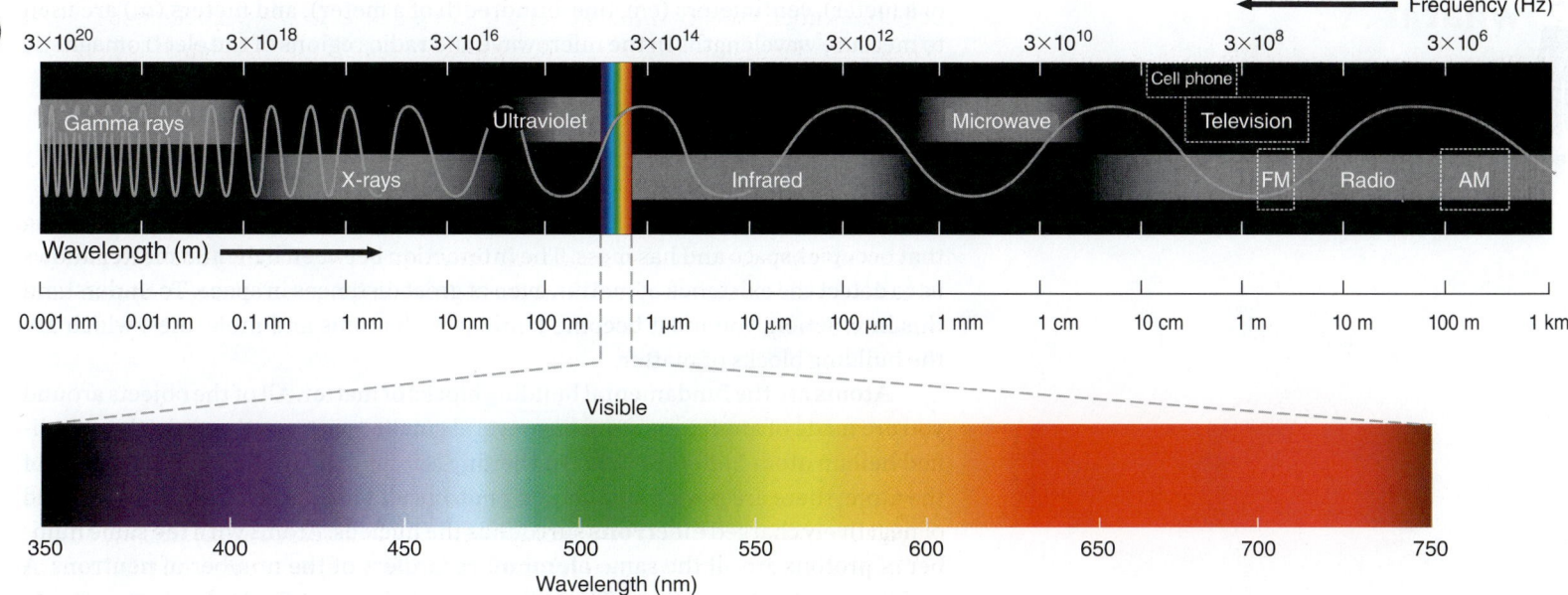

Figure 4.6 By convention, the electromagnetic spectrum is broken into loosely defined regions ranging from gamma rays to radio waves. Throughout the rest of this book, a labeled icon appears below individual astronomical images to identify what part of the spectrum was used to take the image: gamma rays (G), X-rays (X), ultraviolet (U), visible (V), infrared (I), or radio (R). If more than one region of the spectrum was used, multiple labels are highlighted in the icon. ★ **WHAT AN ASTRONOMER SEES** An astronomer looking at this figure will notice that there are two axes shown: frequency is on top and wavelength is on the bottom. She will pause to notice that wavelength is measured in meters and frequency is measured in hertz (Hz). Both scales are logarithmic, so that one tick mark corresponds to a factor-of-10 change in the frequency or wavelength. The visible portion of the spectrum is tiny, compared to the rest of the spectrum, and has been "zoomed" below the main graph. Yellow light marks the middle of this visible part of the spectrum, at about 550 nm, just a bit larger than 3×10^{14} Hz.

A beam of red light can carry just as much energy as a beam of blue light, but the red beam will have more photons than the blue beam, as shown in **Figure 4.7**. In this sense, light is similar to money. Ten dollars is 10 dollars, but it takes a lot more pennies (low-energy photons) than quarters (high-energy photons) to make up 10 dollars.

Light with wavelengths longer than red is called **infrared (IR) radiation**. Some IR radiation can be felt as heat. When you hold your hand next to a hot stove, some of the heat you feel is carried to your hand by infrared radiation emitted from the stove. In this sense, you could think of your skin as a giant infrared eyeball—it is sensitive to infrared wavelengths. Infrared radiation is also used in television remote controls. Night vision goggles detect infrared radiation from warm objects such as animals. Beyond 500 millionths of a meter (5×10^{-4} meter), the light is known as **microwave radiation**. The microwave in your kitchen heats the water in food using light of these wavelengths. The longest-wavelength light, with wavelengths longer than a few centimeters, is called **radio waves**. The FM, AM, television, and cell phone signals used to transmit information around the world are all composed of radio waves.

Astronomers conventionally use *nanometers* when referring to wavelengths at visible and shorter wavelengths. A nanometer (nm) is one-billionth of a meter (10^{-9} meter). One-millionth of a meter is a micrometer (μm), or micron, and it is useful for measuring wavelengths in the infrared. Millimeters (mm, one-thousandth

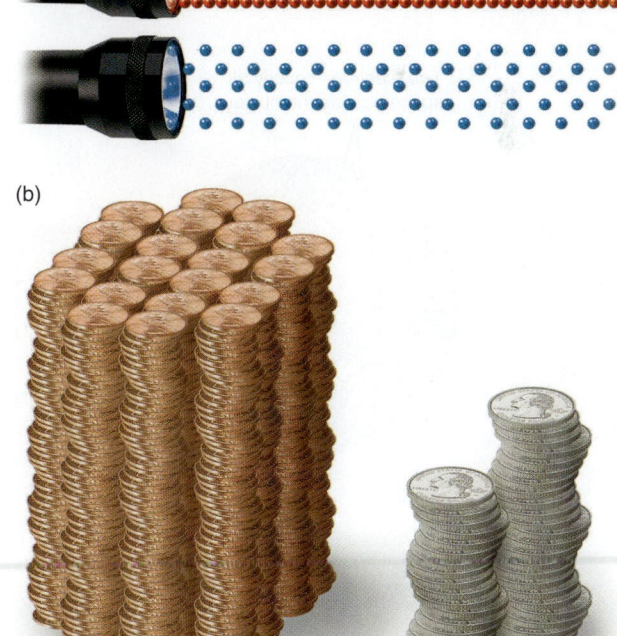

Figure 4.7 (a) A photon of red light carries less energy than a photon of blue light, so it takes more red photons than blue photons to make a beam of a particular intensity. (b) Similarly, pennies are worth less than quarters, so it takes more pennies than quarters to add up to $10.

what if . . .

What if our eyes were sensitive to infrared and X-ray radiation in addition to optical light: How would each of those radiation bands affect our daily life?

of a meter), centimeters (cm, one-hundredth of a meter), and meters (m) are used to measure wavelengths in the microwave and radio regions of the electromagnetic spectrum.

Light and Matter

Light and matter interact with one another. But what is matter? **Matter** is anything that occupies space and has mass. The interaction between light and matter allows us to detect the existence of matter, even at great distances in space. To understand this interaction, you must become familiar with atoms and molecules, which are the building blocks of matter.

Atoms are the fundamental building blocks of matter. All of the objects around you are made of atoms. An atom has several smaller parts, as shown in the simplified helium atom in **Figure 4.8a**. In the tiny but massive **nucleus** at the center of the atom, there are positively charged **protons** and uncharged **neutrons**. A cloud of negatively charged **electrons** surrounds the nucleus. Atoms with the same number of protons are all the same **element**, regardless of the number of neutrons. A helium atom has two protons. The helium atom in Figure 4.8a also has two neutrons in the nucleus, which is typical for helium. A rarer form of helium has two protons but only one neutron. All the types of an element, such as helium, that contain different numbers of neutrons are known as **isotopes**. **Molecules** are two or more atoms bound together by chemical bonds. In a chemical bond in a molecule, two atoms share their electrons with each other. Water (H_2O), depicted in Figure 4.8b, consists of two hydrogen atoms chemically bonded to one oxygen atom. So, a molecule of H_2O contains three atoms and two chemical bonds.

An atom with an equal number of protons and electrons is neutral in charge because the positive charge of the protons is exactly canceled by the negative charge of the electrons. It is common for electrons to be torn loose from the atom and travel through space on their own, leaving a charged atom called an **ion**. Because the electrons are negatively charged, the ion they leave behind is positively charged.

Let us now return to the interactions that occur between light and matter. Recall the cork floating in the sink (Figure 4.3a). The cork remains stationary until the ripple created by the dripping faucet reaches it. The rising and falling water causes the cork to rise and fall. Similarly, light causes charged particles like electrons and protons to oscillate, or move up and down (Figure 4.3b). The reverse is also true: An oscillating electric charge (Figure 4.2b) causes a disturbance that moves outward through space as light.

It takes energy to produce light, and that energy is carried through space by the wave. When light hits matter (atoms, molecules, electrons, and ions) far from the source of the wave, that matter can absorb this energy—a process called **absorption**. Matter also emits energy—a process called **emission**. Absorption and emission of light are at the foundation of our understanding of the universe. We detect light by its interaction with matter, and we detect distant matter by the light that it absorbs or emits.

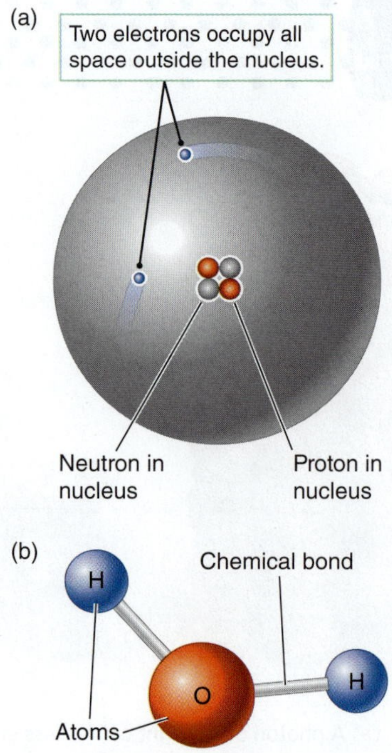

(a)

Two electrons occupy all space outside the nucleus.

Neutron in nucleus

Proton in nucleus

(b)

Chemical bond

H

O

H

Atoms

Figure 4.8 (a) The helium atom has two neutrons, two protons, and two electrons. (b) A molecule, such as a water molecule, consists of several atoms held together by chemical bonds.

CHECK YOUR UNDERSTANDING **4.1**

Rank the following kinds of light in order from longest wavelength to shortest.

a. gamma rays

b. visible light

c. infrared radiation

d. ultraviolet light

e. radio waves

Answers to Check Your Understanding questions are in the back of the book.

4.2 Cameras and Spectrographs Record Astronomical Data

Astronomical observations began with the human eye—information about the overall colors of stars and their brightness in the night sky is apparent even to the "naked" eye, by which we mean the eye unassisted by binoculars or telescopes or filters. The development of lenses and telescopes in the 1600s allowed astronomers to gather more light, multiplying the "light-gathering power" of their eyes. This allowed astronomers to detect fainter and more distant objects, but still the eye was the only available detector, and astronomers relied on descriptions, sketches, and drawings to record their observations. Starting in the 1800s, the development of astronomical detectors, such as cameras, revolutionized astronomy. For the first time, it was possible to record the observation itself, not just a description of it. In this section, you will learn about astronomical detectors, beginning with the human eye.

The Eye

Human eyes respond to light with wavelengths ranging from about 350 nm (deep violet) to 750 nm (far red). **Figure 4.9** shows a simplified schematic of the human eye. The part of the human eye that detects light is called the retina. The individual cells that respond to light falling on the retina are called rods and cones. Cones are located near the middle of the retina at the center of our vision, whereas rods are located away from the center of the retina and are responsible for our peripheral vision. As photons from a star enter the eye, they strike cones at the center of the retina. The cones then send a signal to the brain, which interprets this message as "I see a star." The limit of the faintest stars we can see with our unaided eyes is determined in part by two factors that are characteristic of all detectors of light: integration time and quantum efficiency. Although the human eye is an imperfect analogy for astronomical detectors, it is the detector with which you have the most experience, and so we will use it to give you a sense of how astronomical detectors work.

Integration time is the time interval during which the eye can add up photons—this is analogous to leaving the shutter open on a camera. The brain "reads out" the information gathered by the eye about every 100 milliseconds (ms). Anything that happens faster than that appears to happen all at once. If two images flash on a computer screen 30 ms apart, you will see them as a single image because your eyes will add up (or integrate) whatever they see over an interval of 100 ms. However, if the images occur 200 ms apart, you will see them as separate images. This relatively brief integration time is the most important factor limiting our nighttime vision. Stars that produce too few photons for our eyes to detect in 100 ms are too faint for us to see.

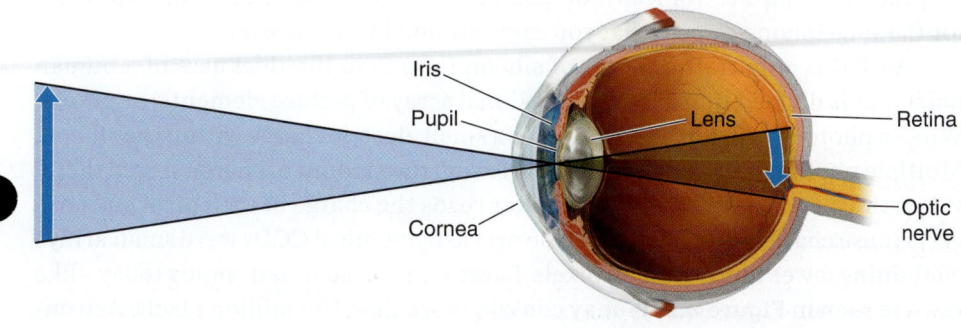

Figure 4.9 A sketch of the human eye shows how the image of the object is oriented upside-down on the retina.

(a)

(b)

G   X   U   V   I   R

Figure 4.10 (a) William Parsons (Lord Rosse, 1800–1867) made this drawing of the galaxy M51 in 1845. (b) John W. Draper photographed the Moon in 1840.

Quantum efficiency describes how many responses occur for each photon received. For the human eye, 10 photons must strike a cone within 100 ms to activate a single response: that is, one signal sent to the brain. So the quantum efficiency of our eyes is about 10 percent. Together, integration time and quantum efficiency determine the rate at which photons must arrive before the brain says, "Aha, I see something."

However, even when we receive a sufficient number of photons in a short-enough time span to see a star, our vision is further limited by the eye's **angular resolution**, which refers to how close two points of light can be to each other before we can no longer distinguish them. The human eye can resolve objects separated by about 1 arcminute ($\frac{1}{60}$ of a degree). This angular distance is about $\frac{1}{30}$ the diameter of the full Moon. This may seem small, but thousands of stars and galaxies may be found within a patch of sky that is 1 arcminute across.

Photographic Plates

Until 1840, the retina of the human eye was the only astronomical detector. Permanent records of astronomical observations were limited to what an experienced observer could sketch on paper while working at the eyepiece of a telescope, as illustrated in **Figure 4.10a**.

In 1840, a chemist named John W. Draper (1811–1882) created the earliest known astronomical photograph, shown in Figure 4.10b. His subject was the Moon. Early photography was slow and very messy, and astronomers were reluctant to use it. In the late 1870s, a faster, simpler process was invented, and astrophotography took off. Astronomers could now create permanent images of planets, nebulae, and galaxies with ease.

The quantum efficiency of most photographic systems used in astronomy was very low—typically 1–3 percent, even poorer than that of the human eye. But unlike the eye, photography can overcome poor quantum efficiency by leaving the shutter open on the camera, increasing the integration time to many hours of exposure. Astronomers were able to record and study objects that were previously not visible. Photography made such an improvement in data collection and storage that, by the middle of the 20th century, the search was on for electronic detectors that could further improve upon photography, which used glass plates or plastic film.

Charge-Coupled Devices

In 1969, scientists at Bell Laboratories invented a remarkable detector called a **charge-coupled device (CCD)**. The CCD soon became the detector of choice in almost all astronomical imaging applications. The output from a CCD is a digital signal that can be sent directly from the telescope to image-processing software, or it can be stored electronically for later analysis. This type of detector is the basis for the smartphone camera that you carry around in your pocket.

A CCD is an ultrathin wafer of silicon (less than the thickness of a human hair) that is divided into a two-dimensional array of picture elements, or **pixels**. When a photon strikes a pixel, it creates a small electric charge within the silicon. Multiple photons will strike a pixel, increasing the amount of charge stored there. When the picture is finished, a computer reads the charge of each pixel and converts these charges into an image. The first astronomical CCDs were small arrays containing fewer than a million pixels. Large CCDs used in astronomy today—like the one seen in **Figure 4.11a**—may contain more than 100 million pixels. Astronomers are currently working on a camera with 3 billion pixels. This is so large that 1,500 high-definition televisions would be required to display a single image. Larger arrays of these cameras, like the one used on the Kepler telescope shown in Figure 4.11b, require ever-faster computing power to keep up with image-processing demands.

The quantum efficiency of CCDs is far superior to that of photographic film. For some digital cameras, in the visible range of wavelengths, this quantum efficiency approaches 90 percent, with an electron produced for nearly every photon that strikes the CCD. This high quantum efficiency decreases the exposure time required to observe faint objects.

The camera in a smartphone takes color pictures with a grid of CCD pixels arranged in groups of three. Each pixel in a group responds only to a particular range of colors—only to red light, for example. This is also true for the pixels in digital image displays. You can see this for yourself if you place a small drop of water on the screen of your smartphone or tablet and turn it on. The water magnifies the grid of pixels so that you can see them individually. This grid degrades the angular resolution of the camera because each spot in the final image requires three pixels of information.

Astronomers make color images in another way. They measure the number of photons that fall on each pixel, without regard to color. Filters in front of the camera allow only light of particular wavelengths to pass through. Astronomers take multiple images of the same object, using a different filter for each image. In the computer, they color each image and then carefully align and overlap them to produce beautiful and informative images. Sometimes the colors are "true"; that is, they are close to the colors you would see if you were actually looking at the object with your eyes. At other times, the colors represent different portions of the spectrum and show the temperature or composition of different parts of the object. Using changeable filters instead of designated color pixels gives astronomers greater flexibility and greater angular resolution.

Spectra and Spectrographs

When astronomers want to know about an object in detail, they pass the light through a prism or diffraction grating. Prisms and gratings disperse the light, spreading it out to create an artificial rainbow, which is recorded by a camera. This arrangement of a camera with a prism or grating is called a **spectrometer**, and the field of study that focuses on the spectra of objects is called **spectroscopy**. Much of what we know about astronomical objects comes from spectroscopy.

For example, from the relative brightness of the different colors in a rainbow, astronomers can find the object's temperature. If the object is brighter in the red part of the spectrum than the blue, the object is relatively cool. Conversely, if it is brighter in the blue part of the spectrum than the red, it is relatively hot. Redder objects are dominated by lower-energy light and have lower temperatures, whereas bluer objects are dominated by higher-energy light and have higher temperatures. Hotter objects not only are bluer but also emit more light of all colors and so shine more brightly than cooler objects. You know this from the burners on an electric stove, which glow more brightly the hotter they are.

Still more information is embedded in the spectrum. Because atoms of different elements interact with different wavelengths of light, they amplify or reduce the spectrum at particular wavelengths. Atoms and molecules produce **emission lines** when they add light to the spectrum and **absorption lines** when they take light away. In either case, as we'll see in more detail in Chapter 10, each

(a)

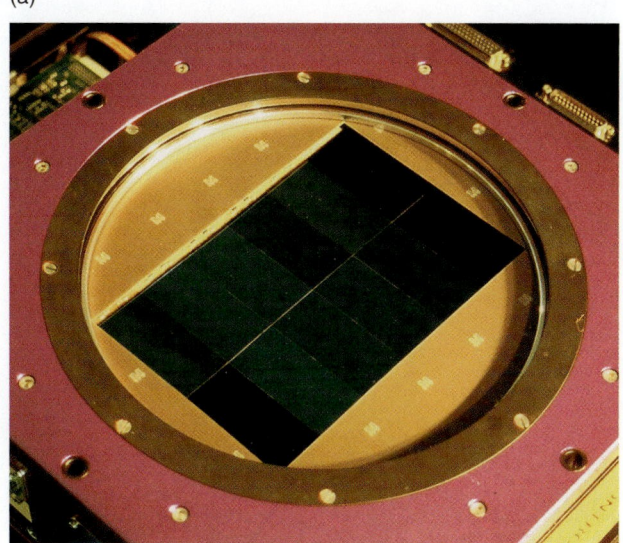

(b)

Figure 4.11 (a) This CCD has approximately 100 megapixels, about 10 times more than a smartphone camera. (b) The Kepler telescope combines 42 CCD cameras into an array.

type of atom or molecule has a unique set of lines, which act like fingerprints that astronomers use to find the composition of an object; that is, to find out which atoms and molecules are present in the object. You will encounter many more applications of spectroscopy throughout the chapters to come.

CHECK YOUR UNDERSTANDING 4.2

CCD cameras are better astronomical detectors than the human eye because (choose all that apply)

a. their quantum efficiency is higher.
b. the integration time can be longer.
c. they can observe at wavelengths beyond the visible.
d. they turn photons into protons.

4.3 Telescopes Collect Light

A telescope is a device for collecting and focusing light. Telescopes collect light in two ways: **refracting telescopes** use lenses, and **reflecting telescopes** use mirrors. The first telescopes were refracting telescopes. But within a few decades, reflecting telescopes were also in use, and now all large modern telescopes are reflectors.

For astronomers to collect light from across the electromagnetic spectrum, they require different types of telescopes in addition to different types of detectors. Radio or infrared telescopes detect low-energy light and are used to study very cool, or even cold, dust. X-ray telescopes, which detect high-energy light, are used to study violently hot gas.

Collecting light from each region of the spectrum presents its own challenges. For example, because warm objects emit infrared light, an infrared telescope requires special shielding, even in space, to keep the entire instrument cool. Otherwise, the instrument itself would emit so much light that its detector would be unable to see any astronomical objects. Gamma-ray and X-ray telescopes, which collect light at the highest energies, require very different systems than those used for the rest of the electromagnetic spectrum because conventional lenses and mirrors do not focus light at these very short wavelengths.

For all telescopes, the "size" of the telescope refers to the diameter of the largest mirror or lens, which determines the light-gathering area. This diameter is called the **aperture**. Thus, a "4.5-meter telescope" has an aperture of 4.5 meters and a primary mirror (or lens) that is 4.5 meters in diameter. An increased aperture increases the light-gathering power of a telescope. This light-gathering power is the ratio of the area of the aperture to the area of a human pupil (typically estimated to be 5 mm). Because both the aperture and the pupil are circular, the light-gathering power (lgp) can be reduced to the ratio of the radii squared $[\text{lgp} = (r_{\text{aperture}}/r_{\text{pupil}})^2]$. Gathering more light is the primary purpose of a telescope.

Refracting Telescopes

As light enters a new medium—for example, as it moves from the air into a glass lens—its speed changes. If the light strikes the surface at an angle, some of the crest of the wave arrives at the surface earlier and some arrives later. This is shown in the sketch of wave crests (red lines) striking a medium at an angle in **Figure 4.12a** and in the photograph of a light ray passing through a medium in Figure 4.12b. Notice that the ray refracts each time the medium changes.

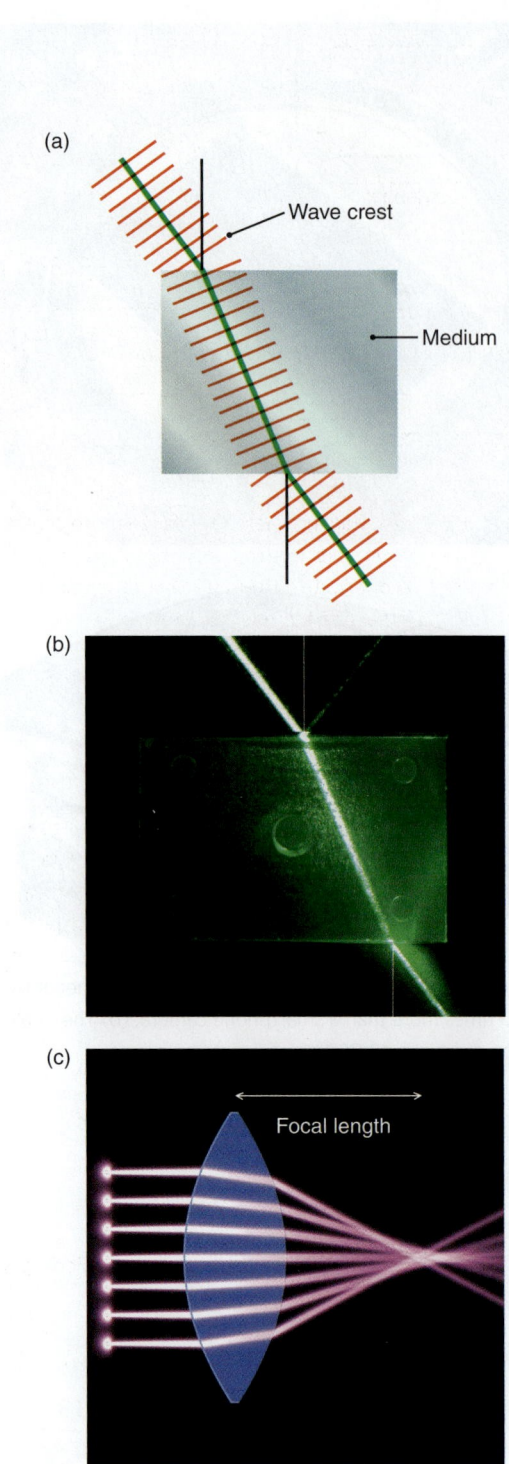

Figure 4.12 (a) When wavefronts enter a new medium, they bend in a new direction, relative to a line perpendicular to the surface (black lines). (b) An actual light ray entering and leaving a medium. Light waves are refracted (bent) when entering a medium with a higher index of refraction. They are refracted again as they reenter the medium with the lower index of refraction. (c) For a curved lens like the one shown, this phenomenon causes the light to focus to a point. This point is in a slightly different location for different wavelengths of light.

The amount of refraction depends on both how the properties of the medium affect the speed of light and the angle at which the light strikes the medium. The ratio of light's speed in a vacuum, c, to its speed in a medium, v, is the medium's **index of refraction (n)**; that is, $n = c/v$. For example, n is approximately 1.5 for typical glass, so the speed of light decreases from 300,000 kilometers per second (km/s) in a vacuum to 200,000 km/s in glass. This index of refraction is different for different materials, and it determines how much light will refract in that medium. Light that arrives at the new medium "head on," so that the crests are parallel to the edge of the medium, will travel straight. As the angle between the crests and the surface grows, the light is refracted more.

A lens is curved, so light at the outer edges strikes the surface at a larger angle than light near the center of the lens. Therefore, light at the outer edges is refracted more than light near the center. This concentrates the light rays entering the lens, bringing them to a sharp focus at a distance referred to as the **focal length**, shown in Figure 4.12c. Flatter, less curved lenses have longer focal length, and lenses with a shorter focal length are more sharply curved. An image is created in the telescope's **focal plane**. **Figure 4.13a** shows the light from two stars passing through a lens of a telescope and converging at the focal plane of the lens. Figure 4.13b shows the same situation for a lens with a flatter, less curved lens, and hence a longer focal length. This longer focal length increases the size and separation of objects in the image, as shown by the positions of the red and blue light rays in the focal plane (compare Figure 4.13a and b). The separation distance is represented by d, as shown in Figure 4.13b.

The larger the area of a telescope's largest lens (the primary lens), the more light-gathering power it has, which makes fainter stars more visible. However, because large lenses are very heavy, there is a practical limit on how large a refracting telescope can be. The largest existing refracting telescope has a lens diameter of slightly more than a meter. There is another serious drawback to refracting telescopes: *chromatic aberration*. Sunlight is made up of all the colors of the rainbow, and each color refracts at a slightly different angle because the index of refraction depends on the wavelength of the light. This is the effect that produces rainbows when sunlight passes through a water droplet. In astronomical applications, this chromatic aberration produces blurry images unless a filter is used to block all but a narrow range of wavelengths. As an example of this blurring, an image of a star will have a colored halo around the star's location; it will not appear as a crisp white point. Manufacturers of quality cameras and telescopes combine two (or more) types of glass into a **compound lens** to correct for chromatic aberration. This adds significantly to the cost and complexity of manufacture.

Reflecting Telescopes

Reflecting telescopes use mirrors to focus light into an image. When light strikes the surface of a mirror, it bounces off the surface and travels in a new direction. Light that strikes the mirror with the wave crests parallel to the surface bounces back along the same path. But light that comes in with the wave crests at an angle to the mirror's surface bounces away with the wave crests at the same angle to the mirror. The sketch of a reflecting telescope in **Figure 4.14** shows how light coming

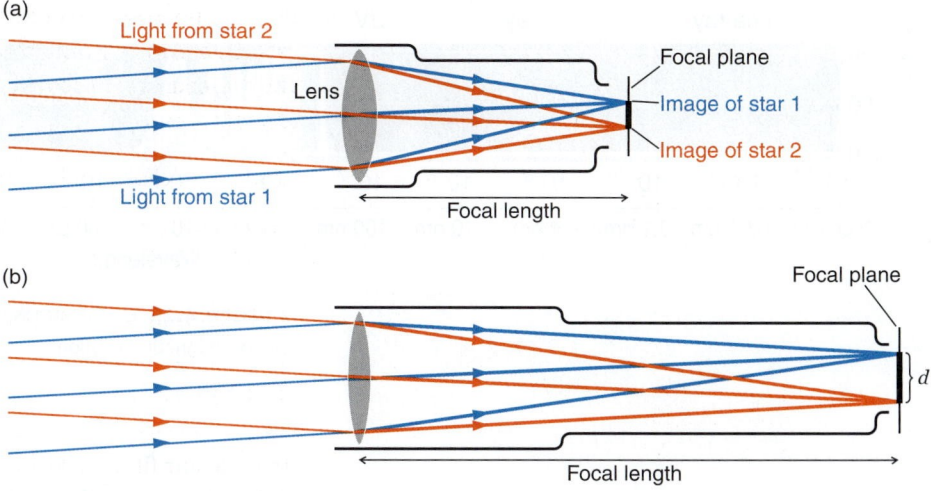

Figure 4.13 (a) A refracting telescope uses a lens to collect and focus light from two stars, forming images of the stars in its focal plane. (b) Telescopes with longer *focal length* spread the images of objects farther apart, producing larger, more widely separated images.

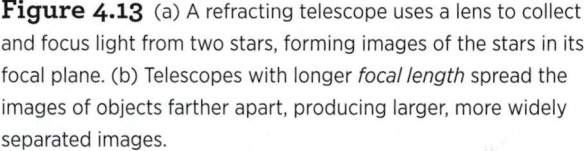

what if . . .

In *The Lord of the Rings*, Gollum's eyes are much larger than the average human eye. Suppose that your eyes, like Gollum's, magically grew to be twice as large as they currently are: How would this affect your vision, both during the day and at night?

▶‖ **AstroTour:** Geometric Optics and Lenses

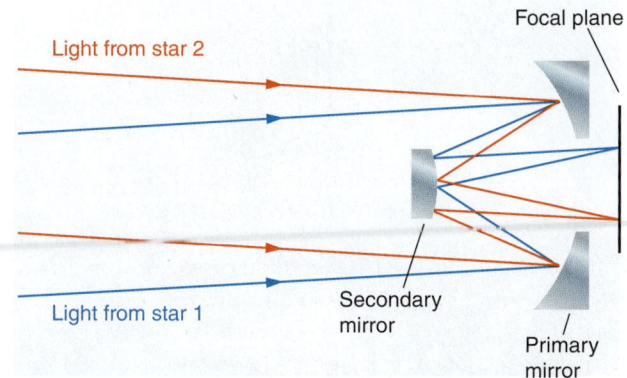

Figure 4.14 Reflecting telescopes use mirrors to collect and focus light. Large telescopes typically use a secondary mirror that directs the light back through a hole in the primary mirror to an accessible focal plane behind the primary mirror.

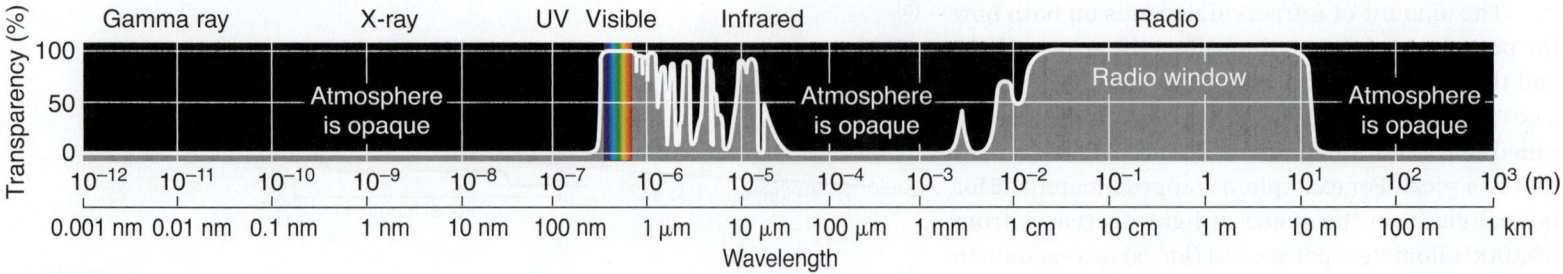

Figure 4.15 Earth's atmosphere is transparent in a few regions of the spectrum but blocks most electromagnetic radiation.

from a star first strikes the *primary mirror* and reflects off it back toward the sky. Typically, a *secondary mirror* then reflects the light back through a hole in the primary mirror to the focal plane.

Reflecting telescopes have several important advantages over refracting telescopes. Because mirrors reflect all colors of light equally, chromatic aberration is no longer a problem. Primary mirrors can be supported from the back, and they can be made thinner and therefore weigh less than large lenses. Very large primary mirrors can be constructed from mosaics of smaller mirrors, which can each be tilted independently to help focus the light. Sometimes, the secondary mirror is angled so that the focal plane is located at the side of the telescope instead of behind the primary mirror. This configuration makes the telescope easier to use because the telescope doesn't need to be lifted off the ground for the user to look through it. Innovations like this one are common, but the fundamental concept of using mirrors to focus the light to a focal plane is always the same.

Radio Telescopes

Not every type of **electromagnetic radiation** reaches the ground. **Figure 4.15** shows the transparency of the atmosphere at different wavelengths of the electromagnetic spectrum; where the white line is high on the graph, the atmosphere is more transparent and lets more light through. There are several atmospheric "windows" where the atmosphere is nearly transparent. The largest of these windows is in the radio part of the spectrum. Because the atmosphere is transparent in this part of the spectrum, we are able to build radio telescopes on the ground rather than in space. Most **radio telescopes** are large, steerable dishes, typically tens of meters in diameter, such as the 100-meter (330-foot) Green Bank Telescope shown in **Figure 4.16a**. The world's largest single-dish radio telescope is the Five-hundred-meter Aperture Spherical radio Telescope (FAST) in China, completed in 2016 (Figure 4.16b). This huge structure (500 meters = 1,640 feet) is too big to steer; it can only observe sources that pass overhead as Earth rotates.

Even though individual radio telescopes are large, they have relatively poor angular resolution compared to optical telescopes. A telescope's angular resolution is determined by the ratio λ/D, where λ is the wavelength of the electromagnetic radiation being observed and D is the telescope's aperture. Recall from section 4.2 that angular resolution describes how close two objects can be and still remain distinguishable. To see fine detail, astronomers want this number to be small. This means that angular resolution is better when λ is small and/or D is large. Radio telescopes have diameters much larger than the apertures of most optical telescopes.

(a)

(b)

Figure 4.16 (a) The Green Bank Telescope is the largest steerable telescope in the world. (b) The FAST radio telescope is the world's largest single-dish telescope.

Increasing D in the denominator of the equation λ/D means that the angular resolution improves. However, the wavelengths of radio waves are much longer than the wavelengths of visible light, and a larger wavelength gives worse angular resolution. A radio telescope may be 10 times larger than an optical telescope, but it observes at a wavelength a million times longer. Radio telescopes are thus hampered by the very long wavelengths they are designed to receive. Consider the huge FAST radio telescope. Its resolution is typically about 3 arcminutes—a little worse than the unaided human eye.

Radio astronomers have developed ways to mathematically combine signals from multiple telescopes to improve resolution. This combination of two (or more) telescopes is called an **interferometer**, and it makes use of the wavelike properties of light. Combining the signals from two radio telescopes gives an angular resolution that is the same as that of a telescope the size of the distance between them. For example, if two 10-meter telescopes are located 1,000 meters apart, the D in λ/D is 1,000, not 10. Usually several telescopes are used in an *interferometric array*. Through the use of very large arrays, astronomers can attain and even exceed the angular resolution possible with optical telescopes.

One of the larger radio interferometric arrays is the Very Large Array (VLA) in New Mexico, shown in **Figure 4.17**. The VLA is made up of 27 individual movable dishes spread out in a Y-shaped configuration with a maximum antenna separation of 36 km. At a wavelength of 10 cm, the VLA can achieve resolutions of less than 1 arcsecond. The Very Long Baseline Array (VLBA) uses 10 radio telescopes spread out over more than 8,000 km from the Virgin Islands in the Caribbean to Hawaii in the Pacific. At a wavelength of 10 cm, this array can reach resolutions better than 0.003 arcsecond. An even larger array of eight telescopes makes up the globe-spanning Event Horizon Telescope. In 2019, this telescope made headlines by producing an image with the unprecedented resolution of 60 microarcseconds (0.000060 arcsecond). This is about the angular size of a tennis ball on the Moon. The telescope operated in the millimeter wavelengths. A radio telescope put into near-Earth orbit as part of a Space Very Long Baseline Interferometer (SVLBI) overcomes even this limit. Future SVLBI projects would extend the baseline to as much as 100,000 km, yielding resolutions far exceeding those of any existing telescope. These resolutions would allow astronomers to see deep into the hearts of galaxies to observe the behavior of objects falling into the supermassive black holes at their centers.

Optical telescopes can also be combined in an array to yield resolutions greater than those of single telescopes, although for technical reasons the individual units cannot be spread as far apart as radio telescopes. The Very Large Telescope (VLT), operated by the European Southern Observatory (ESO) in Chile, combines either four 8-meter telescopes or four movable 1.8-meter auxiliary telescopes (**Figure 4.18**). It has a baseline of up to 200 meters, yielding angular resolution in the milliarcsecond range.

Observing at Other Wavelengths

Earth's atmosphere distorts telescopic images, and molecules such as water vapor in Earth's atmosphere block large parts of the electromagnetic spectrum, so astronomers try to locate their instruments above as much of the atmosphere as possible. Mauna Kea, a dormant Hawaiian volcano and home of the Mauna Kea Observatories (MKO), rises 4,200 meters above the Pacific Ocean. At this altitude, the MKO telescopes sit above 40 percent of Earth's atmosphere; more important, 90 percent of

Figure 4.17 The Very Large Array (VLA) in New Mexico combines signals from 27 different telescopes so that they act as one "very large" telescope.

Figure 4.18 The Very Large Telescope (VLT) operated by the European Southern Observatory in Chile acts as an optical interferometer.

Earth's atmospheric water vapor lies below. Still, for the infrared astronomer, the remaining 10 percent of water vapor is troublesome.

One way to solve the water vapor problem is to make use of high-flying aircraft. The Stratospheric Observatory for Infrared Astronomy (SOFIA), which began operations in 2011, carries a 2.5-meter telescope and works in the far-infrared region of the spectrum. It flies in the stratosphere at an altitude of about 12 km—above 99 percent of the water vapor in Earth's lower atmosphere.

Gaining full access to the complete electromagnetic spectrum, such as the X-ray or gamma-ray regions of the spectrum, requires getting completely above Earth's atmosphere because atmospheric atoms and molecules absorb and redirect this light, so the atmosphere is opaque in these regions of the spectrum. Placing observatories in space enables astronomers to study objects that emit light of these wavelengths. The first astronomical satellite was the British Ariel 1, launched in 1962 to study solar ultraviolet and X-ray radiation and the energy spectrum of primary cosmic rays. Today, orbiting astronomical telescopes cover the electromagnetic spectrum from gamma rays to microwaves.

X-rays and gamma rays are often emitted by very hot gas, such as that produced by an exploding star. These types of high-energy radiation can only be seen by instruments equipped with specialized mirrors and detectors. Unfortunately, the resolution of these telescopes is much more limited than for optical telescopes and radio interferometers.

Astronomers often find it helpful to combine multiple instruments into a single telescope. The Hubble Space Telescope (HST; **Figure 4.19a**), for example, was launched in 1990 and has been collecting ultraviolet, visible, and infrared data for 30 years. The HST has been called a "discovery machine" because it has contributed so much useful data to so many subfields of astronomy. The HST orbits at 600 km above the surface of Earth. But for other astronomical satellites, this is not nearly high enough above Earth's atmosphere for them to function optimally.

The Chandra X-ray Observatory, NASA's X-ray telescope shown in Figure 4.19b, cannot see through even the tiniest traces of atmosphere and therefore orbits more than 16,000 km above Earth's surface. NASA's Spitzer Space Telescope, an infrared telescope, is so sensitive that it needs to be completely free from Earth's own infrared radiation. Spitzer trails tens of millions of kilometers behind Earth as it orbits the Sun. Future space telescopes, including the James Webb Space Telescope—NASA's replacement for the HST—will orbit the Sun at a point that is always farther from the Sun than Earth, on a line that extends from the Sun through Earth.

Resolution and the Atmosphere

As light passes by a sharp edge, the light's waves bend. You have observed this phenomenon as light spreads out to either side after it passes through a doorway into a dark room. This spreading-out phenomenon is called **diffraction**, and it is illustrated in **Figure 4.20a**, where light is shown passing through a gap and traveling on to a screen. The spreading waves from each edge line up on the screen; in some places they combine to make a brighter wave, and in some places they cancel out. A telescope has a circular opening, rather than a vertical gap, so the pattern created by diffraction is a series of concentric circles, as shown in Figure 4.20b. The star in the image no longer appears as a crisp point but rather as a small circle, surrounded by rings caused by diffraction at the edges of the aperture. If there are also crossbars holding up a secondary mirror, you may see diffraction spikes in the image—a

(a)

(b)

Figure 4.19 (a) The Hubble Space Telescope (HST) observes in the ultraviolet, visible, and infrared regions of the electromagnetic spectrum. (b) NASA's Chandra X-ray Observatory, shown in this artist's conception, collects X-rays. The two telescopes are complementary and, along with several others, allow astronomers to observe at all wavelengths of the entire electromagnetic spectrum.

working it out 4.2

Diffraction Limit

The ultimate limit on the angular resolution, θ, of a lens—the diffraction limit—is determined by the ratio of the wavelength of light, λ, to the aperture, D:

$$\theta = 2.06 \times 10^5 \left(\frac{\lambda}{D}\right) \text{arcseconds}$$

The constant, 2.06×10^5, changes the units to arcseconds. An arcsecond is a tiny angular measure found by first dividing a degree by 60 to get arcminutes, and then by 60 again to get arcseconds. To get an idea of the size of an arcsecond, imagine that you hand a tennis ball to your friend and ask her to run 8 miles (almost 13 km) away from you down a straight road and then hold up the tennis ball. The angle you perceive from one side of the tennis ball to the other is approximately 1 arcsecond.

Both λ and D must be expressed in the same units (usually meters). The smaller the ratio of λ/D, the better the resolution. The size of the human pupil can change from about 2 mm in bright light to 8 mm in the dark. A typical pupil size is about 4 mm, or 0.004 meter. Visible (green) light has a wavelength of 550 nm or 5.5×10^{-7} meter. Using these values for the aperture and the wavelength gives

$$\theta = 2.06 \times 10^5 \left(\frac{5.5 \times 10^{-7}\,\text{m}}{0.004\,\text{m}}\right) \text{arcseconds}$$

$$= 28.3 \text{ arcseconds}$$

or about 0.5 arcminute. But, as discussed earlier, the best resolution that the human eye can achieve is about 1 arcminute, and 2 arcminutes is actually more typical. We do not achieve the theoretical maximum resolution with our eyes because the optical properties of the lens and the physical properties of the retina are imperfect.

How does this compare to the resolution of the Hubble Space Telescope, operating in the visible part of the spectrum? Its primary mirror has a diameter of 2.4 meters. If we substitute this value for D and again assume visible light, we have

$$\theta = 2.06 \times 10^5 \left(\frac{5.5 \times 10^{-7}\,\text{m}}{2.4\,\text{m}}\right) \text{arcseconds}$$

$$= 0.047 \text{ arcsecond}$$

This is about 600 times better than the theoretical resolving power of the human eye.

---

bright X centered on a bright star. These are caused as the light bends around the sharp edges of the crossbars.

The amount of blurring depends on the wavelength of the light and the telescope's aperture. The larger the aperture, the smaller the problem posed by diffraction. The best resolution possible for a given aperture is known as the **diffraction limit** (see **Working It Out 4.2**). Larger telescopes have better resolution, so they can distinguish objects that appear closer together. Theoretically,

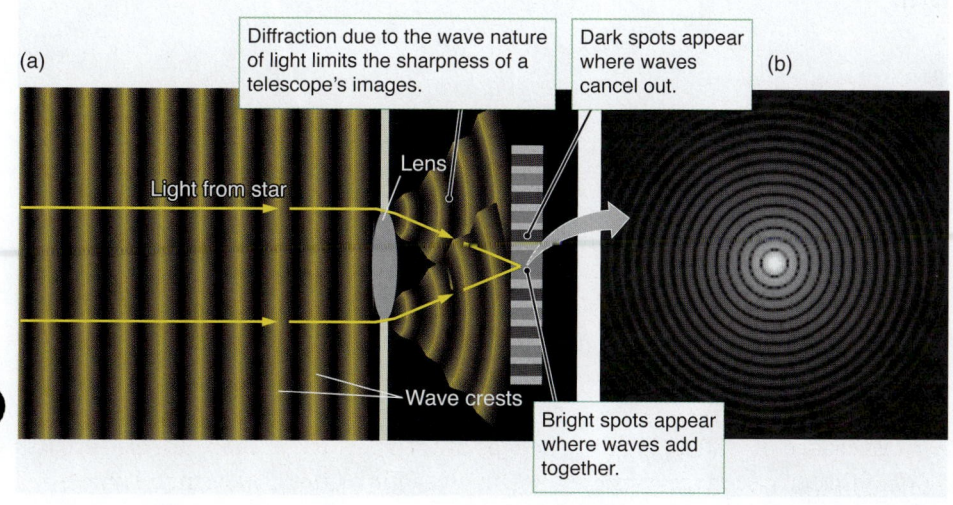

(a)

Diffraction due to the wave nature of light limits the sharpness of a telescope's images.

Dark spots appear where waves cancel out.

(b)

Light from star

Lens

Wave crests

Bright spots appear where waves add together.

Figure 4.20 (a) Light waves from a star are diffracted by the edges of a telescope's lens or mirror. (b) This diffraction causes the stellar image to be blurred, limiting the telescope's ability to resolve objects.

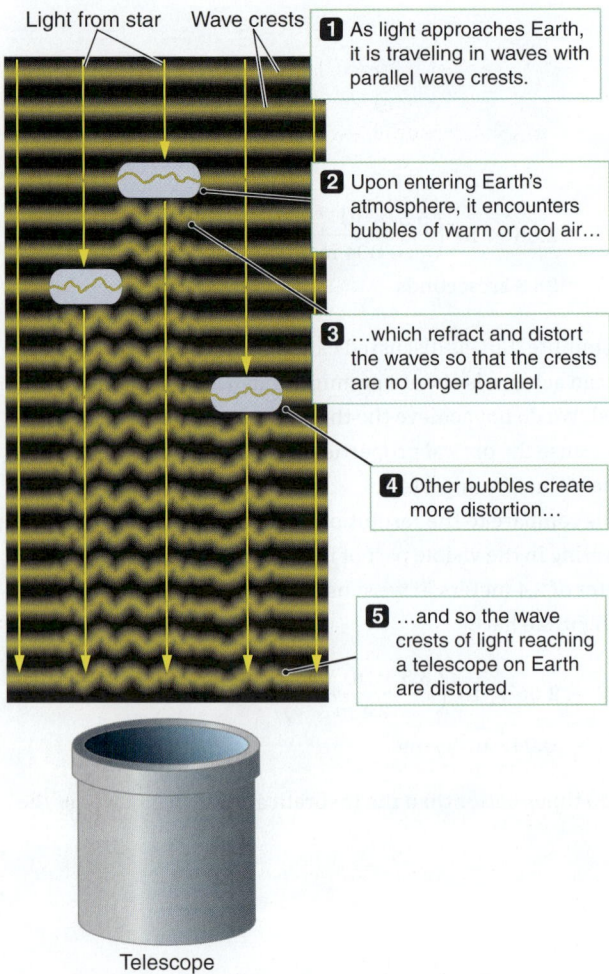

Light from star   Wave crests

1 As light approaches Earth, it is traveling in waves with parallel wave crests.

2 Upon entering Earth's atmosphere, it encounters bubbles of warm or cool air...

3 ...which refract and distort the waves so that the crests are no longer parallel.

4 Other bubbles create more distortion...

5 ...and so the wave crests of light reaching a telescope on Earth are distorted.

Telescope

Figure 4.21 Bubbles of warmer or cooler air in Earth's atmosphere distort the wavefront of light from a distant object.

the 10-meter Keck telescopes located on a mountaintop in Hawaii have a diffraction-limited resolution of 0.0113 arcsecond in visible light, which would allow you to read newspaper headlines 60 km away.

For telescopes with apertures larger than about a meter, Earth's moving atmosphere stands in the way of better resolution. If you have ever looked out across a road on a summer day, you have seen the air shimmer as light is bent this way and that by turbulent bubbles of warm air rising off the hot pavement. This effect is less pronounced when looking overhead, but the twinkling of stars in the night sky is caused by the same phenomenon. Telescopes magnify the angular diameter of an object, but they also magnify the effects of the atmosphere. The limit on the resolution of a telescope on the surface of Earth caused by this atmospheric distortion is called astronomical **seeing**. One advantage of launching telescopes such as the Hubble Space Telescope into space is that they are not hindered by astronomical seeing. But there is another way to correct the problem: *adaptive optics*.

Light from a distant star arrives at the top of Earth's atmosphere with flat, parallel wave crests, as shown in **Figure 4.21**. Earth's atmosphere is filled with small bubbles of air that have slightly different temperatures than their surroundings. Different temperatures mean different densities, and different densities mean each bubble bends light differently. The small air bubbles act as weak lenses, and by the time the waves reach the telescope they are no longer flat, as shown in Figure 4.21. Instead of a tiny disk caused by diffraction, the image in the telescope's focal plane is distorted and swollen. Modern telescopes often use **adaptive optics** to correct this distortion. In an adaptive optics system, at least one optical element (the mirror, typically) adapts to the distortion of the light in real time. Before reaching the telescope's focal plane, the light is reflected off a mirror with a flexible surface. A computer analyzes the incoming light and bends the flexible mirror so that it accurately corrects for the distortion caused by the air bubbles. Examples of images corrected by adaptive optics are shown in **Figure 4.22**. The widespread use of adap-

Uranus on 9 July 2004
The Power of Keck's Adaptive Optics
AO System OFF   AO System ON

2.2 μm

1.6 μm
zoom x2

Figure 4.22 These near-infrared images of Uranus show the planet as seen without adaptive optics (left) and with the technology turned on (right).

tive optics has made the image quality of ground-based telescopes competitive with that of the Hubble Space Telescope.

Light is the primary messenger that astronomers have for exploring the universe, and the vast majority of what is known about the universe beyond Earth has come from studying its properties. Throughout the rest of this book, we will return again and again to the study of light and its detailed properties.

CHECK YOUR UNDERSTANDING 4.3

Match the following properties of telescopes (lettered) with their corresponding definitions (numbered).

a. aperture
b. resolution
c. focal length
d. chromatic aberration
e. diffraction
f. interferometer
g. adaptive optics

1. several telescopes connected to act as one
2. distance from lens to focal plane
3. diameter
4. ability to distinguish objects that appear close together in the sky
5. computer-controlled active focusing
6. rainbow-making effect
7. smearing effect due to a sharp edge

what if . . .

What if human life had developed on a planet containing almost no atmosphere: How might astronomy have developed differently?

reading Astronomy News

NASA's James Webb Space Telescope could study Planet 9

by Bruce Dorminey, contributor, *Forbes* magazine

The James Webb Space Telescope will replace the aging Hubble Space Telescope. This telescope will be much larger and much farther away, and it will be optimized in the infrared. Scientists are busy constructing both the telescope and the cameras that will be attached to it. This article describes progress on part of this telescope, now scheduled to launch in 2021.

NASA's James Webb Space Telescope (JWST)—arguably one of the most anticipated and talked about pieces of space hardware ever conceived—will certainly make observations of the ends of the universe that will likely shake cosmology to its core. But it would also be able to study the large super-earth—dubbed Planet 9—that some astronomers think may lie some

900 AU (or Earth–Sun distances) away; at the outermost fringes of our own solar system.

"If a new planet is found, JWST will be able to fully characterize it," Stefanie Milam, JWST deputy project scientist for planetary science at NASA Goddard Space Flight Center, told me. "Planet 9 is predicted to be fairly large but far, so most ground based facilities [would] barely be able to detect it."

Milam says this would include being able to detect compounds like carbon dioxide in the putative planet's tenuous, icy atmosphere.

Over the life of its $8.8 billion mission, JWST will also make some great, heretofore unobtainable images of Mars and other familiar objects in our own solar system, bridging gaps in solar system science in the process.

The idea is that some six months after launch in October 2018, JWST will be able to

provide near- and mid-infrared observations at least until the end of the next decade. Its observing forte here locally will be monitoring many familiar solar system planets and moons for changes in their atmospheres and surface geology. That is, in ways that can't be achieved from the ground or with current or near-term space missions.

Orbiting the Sun a million miles from Earth, JWST's 6.5-meter primary mirror, Sun shade, and four instruments will offer better sensitivity, spectral resolution, and wavelength coverage than many planetary flyby or orbiter missions. Mars itself will be one of JWST's prime targets.

"JWST will permit instantaneous measurements of the whole observable martian disk at very high spatial resolutions," said Milam, "and at wavelengths not accessible from ground-based observatories."

JWST's other solar system targets will include:

— Asteroids and comets
Unlike flybys or landers, says Milam, JWST will have the opportunity to study multiple comets instead of one target, providing detailed studies of a statistical sample.

— Giant planets – Saturn and Jupiter and the two ice giants Uranus and Neptune. "Uranus and Neptune are by far the least studied planets in our solar system and JWST is ideal for future studies," said Milam.

— Outer planet satellites – all the moons of the planets in our solar system. JWST will allow us to even follow geologic activity on active bodies such as Io or Enceladus, says Milam.

— Rings and small satellites
JWST will not only be able to characterize ring systems across the solar system, says Milam, but also detect new ones; possibly even around Pluto—if they exist.

But JWST will not be able to observe Mercury or Venus, because it can't risk pointing toward the Sun. However, it will be able to observe objects in our solar system beginning at the orbit of Mars on out.

And what sort of serendipitous observations will JWST enable?

Among other things, the detection and characterization of plumes or other geologic activity on natural satellites, says Milam.

"We can also conduct observations as quickly as 48 hours from the time a decision to observe a target is made," said Milam.

EVALUATING THE NEWS

1. Research "Planet 9" online. Has such a planet been found since this article went to press? What is the key observation that led astronomers to predict the existence of "Planet 9"?

2. Why can't ground-based observatories make the same observations as JWST?

3. Consider the overall tone of the article. Would you describe it as objective? Why or why not?

4. Compare the diameters of the James Webb Space Telescope and the Hubble Space Telescope (2.4 m). Are these primary mirror diameters or secondary mirror diameters? What is the advantage of the larger diameter of JWST?

5. JWST cannot observe Mercury or Venus because it can't point toward the Sun. Based on what you learned about orbits in Chapter 3, are there stars that JWST can *never* observe for this reason? Make a sketch to show why or why not.

6. It is common for large projects like JWST to experience delays. Search online to find out whether JWST has launched. If not, when will it launch? If so, what is its current status?

SUMMARY

Light is both a particle and a wave. Whether it is observed to have particle or wave properties depends on the type of observations being made. Visible light is only a tiny portion of the entire electromagnetic spectrum, and studying all the regions of the spectrum provides a range of information about the object being studied. Modern CCD cameras have improved quantum efficiency and lengthened integration times, which allow astronomers to study fainter and more distant objects than were visible with prior detectors. Telescopes are matched to the wavelengths of observation, with different technologies required for each region of the spectrum. The aperture of a telescope both determines its light-gathering power and limits its resolution; larger telescopes are better in both measures.

(1) Waves carry energy through space, and light waves do not require a medium in which to travel. The speed of light in a vacuum is a fundamental constant of nature and is always the same, but light slows when it enters a medium. Photons are particles of light. Light sometimes behaves like a wave (for example, when it refracts) and sometimes behaves like a particle (for example, when it is absorbed by an atom).

(2) The electromagnetic spectrum spans a very large range of wavelengths. Short-wavelength light, like gamma rays, has very high energy. Long-wavelength light, like radio waves, is low in energy. Light carries information about the temperature and composition of objects.

(3) The original light detector was the eye. Over time, astronomers developed better instruments for collecting light from space: from film photography to digital photography using CCD cameras. Spectrographs are specialized instruments that take the spectrum of an object to reveal what the object is made of, along with many other physical properties. Modern detectors are both more responsive to individual photons and can integrate for a longer time than the human eye. In addition, detectors of various types can collect light in regions of the spectrum that cannot be seen by the eye.

(4) Telescopes of larger aperture focus more light on the detector. They therefore collect more light from very faint objects. Also, because the diffraction of the telescope is inversely proportional to the diameter, larger-aperture telescopes have better resolution: Objects that appear close together can still be distinguished from one another. Telescopes with long focal lengths produce more magnified images than telescopes with small focal lengths.

QUESTIONS AND PROBLEMS

TESTING YOUR UNDERSTANDING

1. **T/F:** Light can travel faster than 3×10^8 m/s.

2. **T/F:** The frequency of a wave is related to the energy of the photon.

3. **T/F:** Blue light has a longer wavelength than red light.

4. **T/F:** Blue light has more energy than red light.

5. **T/F:** Visible light is an electromagnetic wave.

6. A light wave does *not* require
   a. a medium.
   b. a speed.
   c. a frequency.
   d. a wavelength.

7. An extremely hot object emits most of its light
   a. at very low energies.
   b. in the radio region.
   c. in the visible region.
   d. at very high energies.

8. When light enters a medium from space,
   a. it slows down.
   b. it speeds up.
   c. it travels at the same speed.
   d. it changes frequency.

9. The amplitude of a light wave is related to
   a. its color.
   b. its speed.
   c. its frequency.
   d. its brightness.

10. Study Figure 4.15. Which of the following can be observed from the ground? (Choose all that apply.) 👁★
    a. radio waves
    b. gamma radiation
    c. ultraviolet light
    d. X-ray light
    e. visible light

11. Which of the following is *not* a property of waves?
    a. speed
    b. wavelength
    c. frequency
    d. mass

12. Which of the following regions of the electromagnetic spectrum has the lowest-energy light?
    a. visible
    b. gamma ray
    c. ultraviolet
    d. radio

13. The advantage of an interferometer is that
    a. the resolution is dramatically increased.
    b. the focal length is dramatically increased.
    c. the light-gathering power is dramatically increased.
    d. diffraction effects are dramatically decreased.
    e. chromatic aberration is dramatically decreased.

14. Suppose that a telescope has a resolution of 1.5 arcseconds at a wavelength of 300 nm. What is its resolution at 600 nm?
    a. 3 arcseconds
    b. 1.5 arcseconds
    c. 0.75 arcsecond
    d. In order to know this, you need to know the diameter of the aperture. Not enough information is given.

15. If the wavelength of a beam of light were halved, how would that affect the frequency?
    a. The frequency would be 4 times larger.
    b. The frequency would be 2 times larger.
    c. The frequency would not change.
    d. The frequency would be 2 times smaller.
    e. The frequency would be 4 times smaller.

16. How does the speed of light in a medium compare to the speed of light in a vacuum?
    a. It's the same, because the speed of light is a constant.
    b. The speed in a medium is always greater than the speed in a vacuum.
    c. The speed in a medium is always less than the speed in a vacuum.
    d. The speed in a medium may be greater or less than the speed in a vacuum, depending on the medium.

17. Astronomers put telescopes in space to
    a. get closer to the stars.
    b. avoid atmospheric effects.
    c. look primarily at radio wavelengths.
    d. improve quantum efficiency.

18. The angular resolution of a ground-based telescope is usually determined by
    a. diffraction.
    b. refraction.
    c. the focal length.
    d. atmospheric seeing.

19. Study Figure 4.15. Gamma-ray telescopes are placed in space because 👁★
    a. gamma rays are too fast to be detected by stationary telescopes.
    b. gamma rays do not penetrate the atmosphere.
    c. the atmosphere produces too many gamma rays and overwhelms the signal from space.
    d. gamma rays are too dangerous to collect in large numbers.

20. The mirror of the James Webb Space Telescope will have a diameter of about 6.5 meters. About how many people of average height could lie, head to foot, across the diameter of this telescope?
    a. 3
    b. 9
    c. 6.5
    d. 12

THINKING ABOUT THE CONCEPTS

21. We know that the speed of light in a vacuum is 3×10^8 m/s. Is it possible for light to travel at a lower speed? Explain your answer.

22. Is the light-year a measure of time, distance, or both?

23. Figure 4.7 compares red light and blue light. Explain how a beam of red light can carry as much energy as a beam of blue light, even though photons of blue light have more energy than photons of red light. 👁★

24. Galileo's telescope used simple single lenses made of one type of glass rather than compound lenses. What is the primary disadvantage of using single lenses, like Galileo's, in a refracting telescope?

25. The largest refracting telescope has an aperture of 1 meter. Why is it impractical to build a larger refractor with, say, twice the aperture?

26. Name and explain at least two advantages that reflecting telescopes have over refracting telescopes.

27. Study Figures 4.13 and 4.14. Is the image produced by a telescope right-side up or upside-down? How do you know? 👁

28. What causes refraction?

29. How do manufacturers of quality refracting telescopes and cameras correct for the problem of chromatic aberration?

30. ★ WHAT AN ASTRONOMER SEES Suppose that you read about a new observation of an object that combines data from the ultraviolet and the radio parts of the spectrum (see Figure 4.6). What approximate wavelength and frequency ranges are included in this observation? Why would astronomers want to combine multiple parts of the spectrum when observing an object?

31. Name two ways in which Earth's atmosphere interferes with astronomical observations.

32. Explain *why* stars twinkle.

33. Explain integration time and how it contributes to the detection of faint astronomical objects.

34. Explain quantum efficiency and how it contributes to the detection of faint astronomical objects.

35. Some people believe that we put astronomical telescopes in orbit because doing so gets them closer to the objects they are observing. Explain what is wrong with this common misconception.

APPLYING THE CONCEPTS

36. You are tuned to 790 on AM radio. This station is broadcasting at a frequency of 790 kHz (7.90×10^5 Hz). What is the wavelength of the radio signal? 👁
    a. This signal is in the radio part of the spectrum. Glance back at Figure 4.6 to check and see what a reasonable number would be for an answer.
    b. Follow along with Working It Out 4.1 to calculate the wavelength.
    c. Check your answer by verifying that it falls in the right range for a radio wavelength.

37. Compare the light-gathering power of a large astronomical telescope (aperture 10 meters) with that of the dark-adapted human eye (aperture 8 mm).
    a. Predict whether the light-gathering power of a telescope should be larger or smaller than that of the human eye.
    b. Calculate the light-gathering power of a large astronomical telescope and of the human eye.
    c. Check your answer by comparing the two systems with a ratio or a fraction.

38. Assume a telescope has an aperture of 1 meter. Calculate the telescope's resolution when observing in the near-infrared region of the spectrum ($\lambda = 1{,}000$ nm). Calculate the resolution in the violet region of the spectrum ($\lambda = 400$ nm). In which region does the telescope have better resolution?
    a. Study the equation for a telescope's resolution given in Working It Out 4.2. What prediction could you make about these two resolutions that would allow you to check your work afterwards?
    b. Calculate both resolutions, as demonstrated in Working It Out 4.2.
    c. Check your work by comparing your answers with your prediction.

39. The resolution of the human eye is about 1.5 arcminutes. What would the aperture of a radio telescope (observing at 21 cm) have to be to have this resolution? Even though the atmosphere is transparent at radio wavelengths, we do not see in the radio region of the electromagnetic spectrum. Using your calculations, explain why humans do not see in the radio region.
    a. Use the equation for a telescope's resolution in Working It Out 4.2 to predict whether the aperture of a radio telescope will need to be larger or smaller than the human eye, given this much larger wavelength.
    b. Calculate the needed aperture to achieve this resolution. This requires rearranging the equation algebraically to solve for *D*.
    c. Check your calculation by comparing your answer to your prediction. (The problem statement hints that your answer will be slightly ridiculous, in the context of the size of a human's head.)

40. The VLBA uses an array of radio telescopes ranging across 8,000 km of Earth's surface from the Virgin Islands to Hawaii.
    a. Calculate the angular resolution of the array when radio astronomers are observing interstellar water molecules at a microwave wavelength of 1.35 cm. Don't forget to make a prediction and check your work.
    b. How does this resolution compare with the angular resolution of two large optical telescopes separated by 100 meters and operating as an interferometer at a visible wavelength of 550 nm? Don't forget to make a prediction and check your work.

exploration Light as a Wave

digital.wwnorton.com/universe4

Visit the Student Site at the Digital Resources page and open the "Light as a Wave, Light as a Photon" AstroTour in Chapter 4. Watch the first section and click through, using the Play button, until you reach Section 2 of 3.

Here we will explore the following questions: How many properties does a wave have? Are any of these related to each other?

Work your way through to the experimental section, where you can adjust the properties of the wave. Watch the simulation for a moment to see how fast the frequency counter increases.

1. Increase the wavelength using the arrow key. What happens to the rate of the frequency counter?

_____

_____

_____

_____

2. Reset the simulation and then decrease the wavelength. What happens to the rate of the frequency counter?

_____

_____

_____

_____

3. How are the wavelength and frequency related to each other?

_____

_____

_____

_____

4. Imagine that you increase the frequency instead of the wavelength. How should the wavelength change when you increase the frequency?

_____

_____

_____

5. Reset the simulation and increase the frequency. Did the wavelength change in the way you expected?

_____

_____

_____

_____

_____

_____

6. Reset the simulation and increase the amplitude. What happens to the wavelength and the frequency counter?

_____

_____

_____

_____

_____

7. Decrease the amplitude. What happens to the wavelength and the frequency counter?

_____

_____

_____

_____

_____

8. Is the amplitude related to the wavelength or frequency?

_____

_____

_____

_____

9. Why can't you change the speed of this wave?

_____

_____

_____

The Formation of Stars and Planets

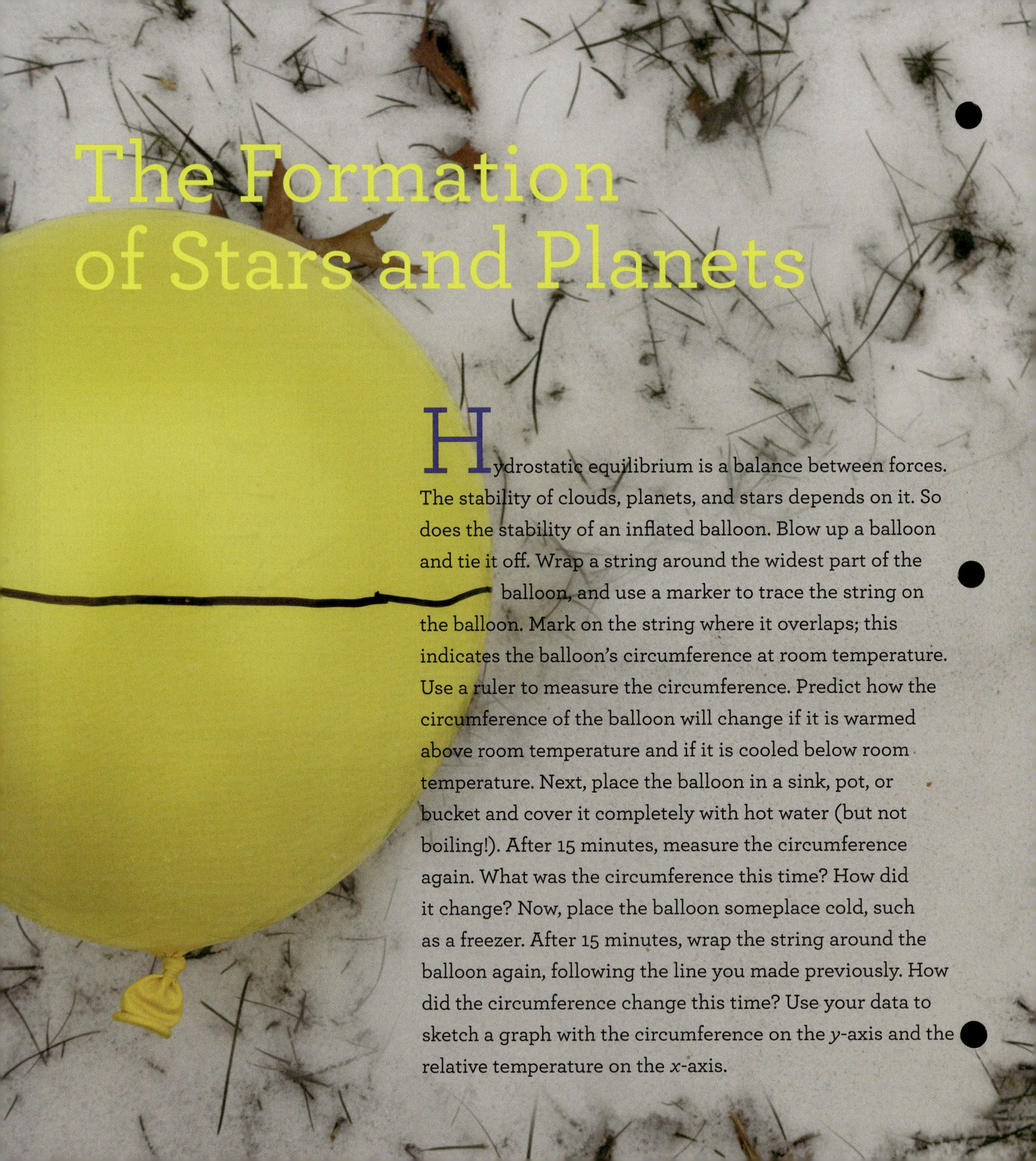

Hydrostatic equilibrium is a balance between forces. The stability of clouds, planets, and stars depends on it. So does the stability of an inflated balloon. Blow up a balloon and tie it off. Wrap a string around the widest part of the balloon, and use a marker to trace the string on the balloon. Mark on the string where it overlaps; this indicates the balloon's circumference at room temperature. Use a ruler to measure the circumference. Predict how the circumference of the balloon will change if it is warmed above room temperature and if it is cooled below room temperature. Next, place the balloon in a sink, pot, or bucket and cover it completely with hot water (but not boiling!). After 15 minutes, measure the circumference again. What was the circumference this time? How did it change? Now, place the balloon someplace cold, such as a freezer. After 15 minutes, wrap the string around the balloon again, following the line you made previously. How did the circumference change this time? Use your data to sketch a graph with the circumference on the y-axis and the relative temperature on the x-axis.

EXPERIMENT SETUP

Blow up a balloon and tie it off. Wrap a string around the widest part of the balloon.

Use a marker to trace the string on the balloon.

Mark on the string where it overlaps.

00:15

Submerge the balloon in hot water (but not boiling!) for 15 minutes. Measure the circumference of the balloon.

00:15

Place the balloon someplace cold for 15 minutes. Measure the circumference of the balloon.

PREDICTION

The circumference will be
☐ larger ☐ smaller ☐ the same
when the balloon is cooled. It will be
☐ larger ☐ smaller ☐ the same
when the balloon is heated.

SKETCH OF RESULTS (in progress)

CIRCUMFERENCE

large
medium
small

cold   room   hot

TEMPERATURE

5

Stars and planets form from clouds of cool dust and gas. Each of the atoms and molecules in a cloud is gravitationally attracted to every other particle. This gravitational pull will cause some clouds to collapse. During the collapse, the clouds may fragment to form multiple stars and then further fragment to form planets. In the past few decades, both stellar astronomers and planetary scientists, working from different perspectives, have arrived at this picture of the early Solar System. However, recent discoveries of exoplanets show that the formation of our Solar System is not typical and challenge this basic picture, furthering the development of a detailed theory of star and planet formation.

LEARNING GOALS

1. Describe the roles that gravity and angular momentum play in the formation of stars and planets.

2. Diagram the process by which dust grains in an accretion disk around a young star stick together to form progressively larger solid objects.

3. Sketch the process that makes Solar System planets orbit the Sun in a plane and revolve in the same direction that the Sun rotates.

4. Explain how temperature at different locations in an accretion disk affects the composition of planets, moons, and other bodies.

5. List the methods that astronomers use to find planets around other stars, and explain how we know that planetary systems around other stars are common.

5.1 Molecular Clouds Are the Cradles of Star Formation

A **star** is a ball of gas, held together by its own gravity, that fuses lighter atoms into heavier ones in its dense core, releasing energy in the process. This energy production causes the outer parts of the star to shine because these parts have been heated to very high temperatures. Stars are often accompanied by **planets**: large, round bodies that travel around the star in individual orbits. In general, a system of planets and other smaller objects surrounding a star is a **planetary system**, and there are many planetary systems in the Milky Way Galaxy. Our **Solar System** is the planetary system that surrounds the Sun. Stars and their associated planets share a common origin in a cloud of dust and gas. We begin our study of star formation by investigating the places where stars form.

Interstellar Clouds

Stars and planets form from large **interstellar clouds** of cool dust and gas located between and among existing stars, like the portion of the Carina Nebula shown in **Figure 5.1**. **Nebula** is the most general term astronomers use for an interstellar cloud of gas and dust. These clouds are held together partly by **self-gravity**, which gravitationally attracts parts of the cloud to each other. The **pressure** of hot gas that occupies the space between the clouds and presses in upon them also contributes to the stability of the clouds. If parts of a cloud have a high enough **density**, the cloud will fragment and collapse to form stars and planets. The dust and gas in these

Figure 5.1 This gorgeous image from the Hubble Space Telescope shows a portion of the Carina Nebula. ★ **WHAT AN ASTRONOMER SEES** An astronomer looking at this image will immediately notice the colors, because color often indicates where different atoms or molecules are present. She will not necessarily know which atoms or molecules are represented by these colors because different astronomers will use a different palette. But even without reading any background on the image, she will know that there is something different about the hazy blue areas and the hazy pink or green ones. An astronomer will recognize and then mostly ignore the diffraction spikes (mentioned in Chapter 4) that form an X around each bright star. She will notice the brown clumps of material that are too dense to see through. These high-density regions indicate that star formation might be happening in this nebula, and this will be confirmed by the small oval protostar in the upper right corner and by the dense "fingers" that stick out in various places. These fingers point the way toward a source of interstellar wind outside the image to the upper right. That wind has eroded away the less dense material around these denser regions and may have triggered star formation by compressing the material at the top of each finger. Each finger is a dense blob of material that creates a wind "shadow" behind it. A new star has just formed at the top of the finger near the middle of the image. An astronomer will identify this new star because of the thin jets of material that are being ejected in opposite directions; new stars sometimes create such jets.

clouds have usually been through several cycles of star formation and stellar death, and so the dust and gas have many different types of atoms that were formed in earlier generations of stars.

As shown in **Figure 5.2**, each part of an interstellar cloud is gravitationally attracted to every other part of the cloud. The sum of all these forces (the *net* force) on each particle points toward the center of the cloud. Self-gravity may be opposed in two ways. In solid objects, like planets, structural strength balances self-gravity; the rocks that make up Earth have molecular bonds that resist being compressed by self-gravity. Self-gravity can also be balanced by the outward force resulting from gas pressure; the hot gases within a star resist being compressed by self-gravity. When the forces are balanced, the object is in **hydrostatic equilibrium**. If the outward force from structural strength or gas pressure is *weaker* than self-gravity, however, then the object is unstable and *contracts*. If the outward force, usually of gas pressure, is *stronger* than self-gravity, then the object is also unstable, but in this case it *expands*. In most interstellar clouds, the internal gas pressure pushing out is much stronger than self-gravity, so the cloud should expand. The much hotter gas surrounding the cloud presses inward on the cloud and helps to hold it together.

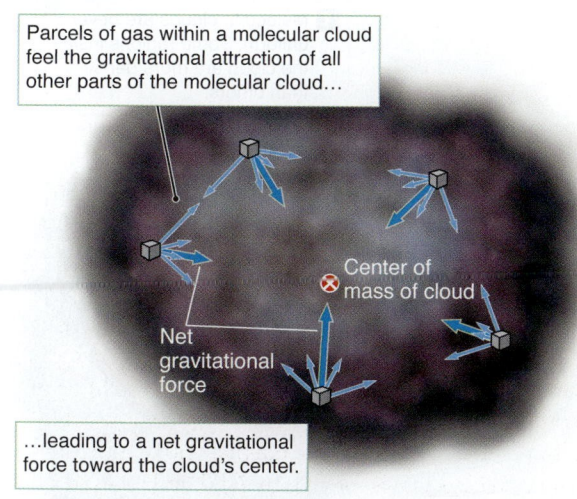

Parcels of gas within a molecular cloud feel the gravitational attraction of all other parts of the molecular cloud...

Center of mass of cloud

Net gravitational force

...leading to a net gravitational force toward the cloud's center.

Figure 5.2 Self-gravity causes a molecular cloud to collapse, drawing parcels of gas toward a single point inside the cloud.

▶❙❙ **AstroTour:** Star Formation

The densest, coolest interstellar clouds are called **molecular clouds** because they are primarily composed of hydrogen molecules—two hydrogen atoms bound together. These clouds also contain dust and other gases. Some molecular clouds are dense enough that their self-gravity overwhelms their internal gas pressure, and they collapse. Other molecular clouds are pushed into a collapse by the explosion of nearby stars or by gravitational interactions with passing stars. These events can trigger star formation in nearby molecular clouds. The collapse of a molecular cloud happens very slowly because several other effects, like interactions with magnetic fields, slow the collapse. However, these effects are temporary, and although gravity is weak, it is also relentless. As the forces that oppose the cloud's self-gravity gradually fade away, the cloud slowly collapses.

Molecular-Cloud Fragmentation

Molecular clouds are never uniform. Some regions are denser and collapse more rapidly than surrounding regions. As these regions collapse, their self-gravity becomes stronger, so they collapse even faster. Slight variations in the density of the gas of individual regions of the cloud lead to some regions becoming very dense and concentrated. Thus, instead of collapsing into a single object, the cloud fragments into a number of very dense **molecular-cloud cores**, as shown in **Figure 5.3**. A single molecular cloud may form hundreds or thousands of molecular-cloud cores, each of which is typically a few light-months in size (a light-month is the distance light travels in 1 month). Some of these cores will eventually form stars.

Because the force of gravity is inversely proportional to the square of the radius of a molecular-cloud core, the gravitational forces grow stronger as a molecular-cloud core shrinks. Suppose a core starts at 4 light-years across. By the time the core has collapsed to half the radius, 2 light-years across, the gravitational attraction that different parts of the core feel toward each other will be 4 times stronger than it was ($2^2 = 4$). When the core is one-fourth as large as when the collapse began, the force of gravity will be 16 times as strong ($4^2 = 16$). As a core collapses, the inward force of gravity increases; as gravity increases, the collapse speeds up; as the collapse speeds up, the gravitational force increases even faster. Eventually, gravity is

Figure 5.3 As a molecular cloud collapses, denser regions within the cloud collapse more rapidly than less dense regions. As this process continues, the cloud fragments into a number of very dense molecular-cloud cores that are embedded within the large cloud. These cloud cores may go on to form stars.

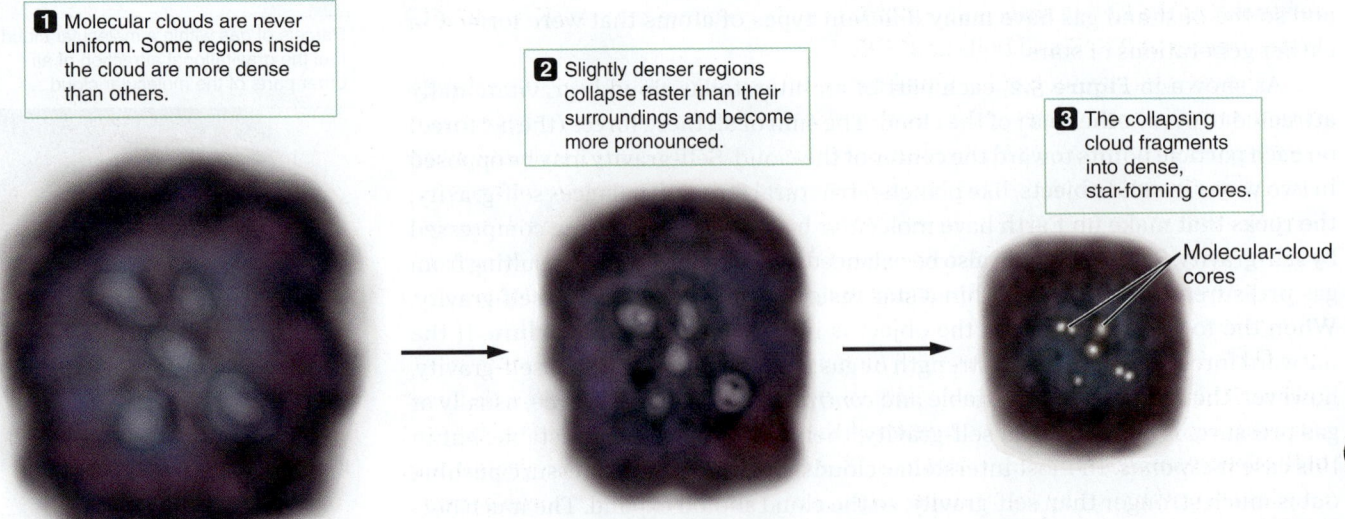

1 Molecular clouds are never uniform. Some regions inside the cloud are more dense than others.

2 Slightly denser regions collapse faster than their surroundings and become more pronounced.

3 The collapsing cloud fragments into dense, star-forming cores.

Molecular-cloud cores

able to overwhelm all the opposing forces. The cloud material is most dense near the center of the cloud core, so the collapse begins there. The inner parts of the cloud core start to fall rapidly inward, allowing the more distant parts of the cloud to fall. The cloud core collapses from the inside out, as shown in **Figure 5.4**. This innermost part of a collapsing molecular-cloud core is called a **protostar**, and it will soon become a star.

CHECK YOUR UNDERSTANDING 5.1

In a collapsing molecular-cloud core, which of the following is true?
- **a.** The strength of self-gravity is decreasing.
- **b.** Gravity is stronger than internal gas pressure.
- **c.** The molecular-cloud core is in hydrostatic equilibrium.
- **d.** The molecular-cloud core density is decreasing.

Answers to Check Your Understanding questions are in the back of the book.

5.2 The Protostar Becomes a Star

Once a molecular-cloud core begins to collapse, several things happen at once. The innermost part of the core eventually becomes a star, while the outer parts may form planets. We will first describe the formation of the star and then explore what becomes of the rest of the dust and gas.

Stars and Protostars

As particles collapse toward the center of the cloud, they move faster and faster. They become more densely packed and begin to crash into each other, causing the particles to move in random directions. The random motion of particles is collectively known as thermal energy. So, gravitational energy is being converted into thermal energy.

When the particles move faster, the thermal energy increases, and the object becomes hotter. Scientists measure the temperature of stars and other objects on the **Kelvin temperature scale**, which has increments known as **kelvins (K)**. These increments are the same size as degrees in the more familiar Celsius scale, but the zero point of the Kelvin scale is at $-273°C$ (that is, absolute zero $= 0\,K = -273°C$). Water freezes at 273 K and boils at 373 K.

The collapse of a molecular-cloud core is an excellent example of the principle of **conservation of energy**, which states that the total amount of energy remains the same in an isolated system. Additional energy cannot be created, and what's there already cannot be destroyed. However, energy can be converted from one type to another. In the collapse of a molecular-cloud core, gravitational energy is transformed into kinetic energy of the gas falling together toward the center, and then the energy is turned into thermal energy, which is the rapid random motion of the atoms and molecules.

Once the temperature of the protostar reaches a few thousand kelvins, the surface begins to radiate energy at visible wavelengths. The hotter it gets, the more energy it radiates, and the wavelengths of the radiation shift toward the blue end of the spectrum (see **Working It Out 5.1**). This is a general property of objects whose spectra are dominated by their temperature. A hypothetical perfect example of such an object would absorb all the electromagnetic radiation it receives, reflecting none.

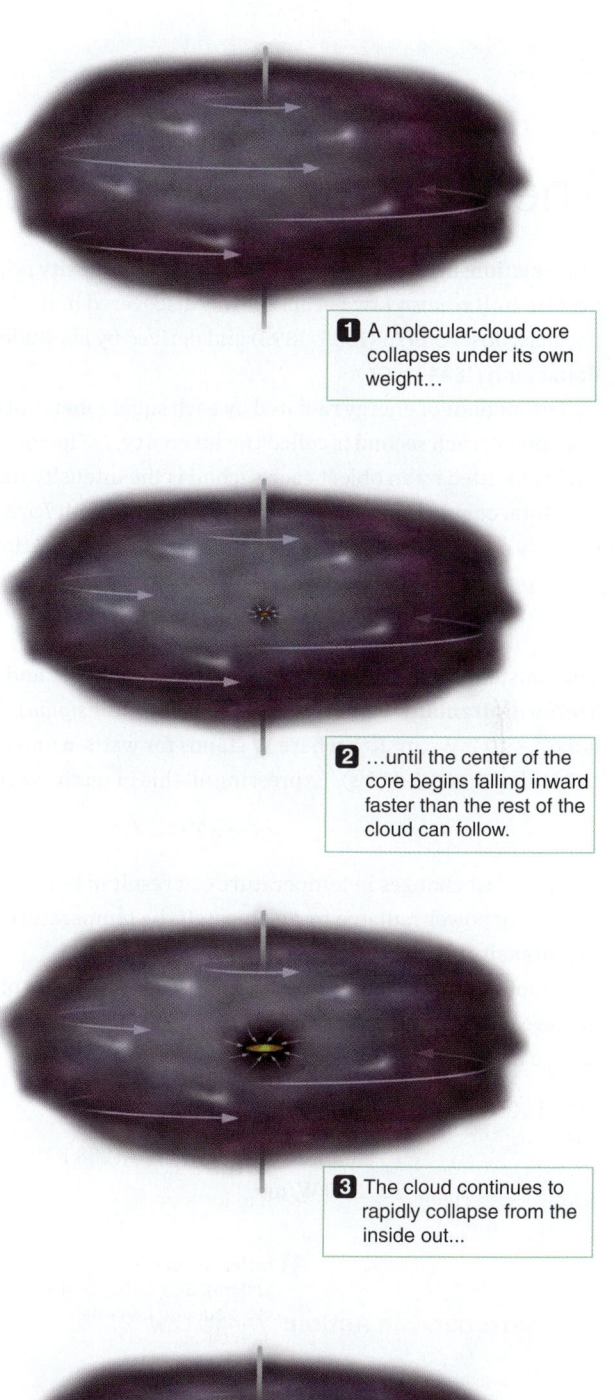

1 A molecular-cloud core collapses under its own weight…

2 …until the center of the core begins falling inward faster than the rest of the cloud can follow.

3 The cloud continues to rapidly collapse from the inside out…

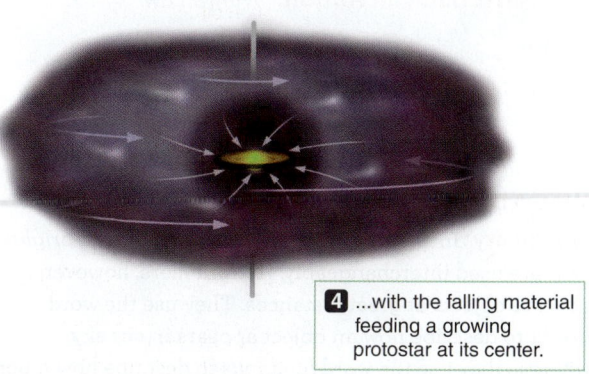

4 …with the falling material feeding a growing protostar at its center.

Figure 5.4 When a molecular-cloud core gets very dense, it collapses from the inside out.

The Stefan-Boltzmann Law and Wien's Law

This relationship between temperature and luminosity is known as the **Stefan-Boltzmann law** because it was discovered in the laboratory by physicist Josef Stefan (1835–1893) and derived by his student, Ludwig Boltzmann (1844–1906).

The amount of energy radiated by each square meter of the surface of an object each second is called the **intensity**, I. The total amount of energy emitted by an object each second is the intensity times the area. This total energy is known as the **luminosity**, L. Both I and L depend strongly on the object's temperature, T. Both are proportional, $\propto$, to the *fourth power* of T:

$$L \propto T^4$$

The constant of proportionality between the intensity and T^4 is the **Stefan-Boltzmann constant**, σ (the Greek letter *sigma*). The value of σ is 5.67×10^{-8} W/(m² K⁴), where W stands for watts, a unit of power equal to 1 joule per second (J/s). Expressing all this in math, we find

$$I = \sigma T^4$$

Even modest changes in temperature can result in large changes in the amount of power radiated by an object. If the temperature triples, then the intensity increases by a factor of 3^4, or 81.

Suppose we want to find the intensity and luminosity of Earth. Earth's average temperature is 288 K, so the amount of energy leaving each square meter of its surface each second is

$$I = \sigma T^4$$
$$I = [5.67 \times 10^{-8} \, \text{W/(m}^2 \, \text{K}^4)](288 \, \text{K})^4$$
$$I = 390 \, \text{W/m}^2$$

To find the luminosity, multiply by the surface area (A) of Earth, given by $4\pi R^2$. The radius of Earth (R) is 6,378,000 meters or 6.378×10^6 meters, so Earth's luminosity is

$$L = I \times A$$
$$L = I \times 4\pi R^2$$
$$L = (390 \, \text{W/m}^2) \, [4\pi (6.378 \times 10^6 \, \text{m})^2]$$
$$L \approx 2 \times 10^{17} \, \text{W}$$

Every second, Earth emits enough energy to power 2,000,000,000,000,000 hundred-watt lightbulbs.

Notice in Figure 5.5 where the peak of each curve lines up along the horizontal axis. As the temperature T increases, the *peak* of the spectrum shifts toward shorter wavelengths. This is an inverse proportion. Translating this into math, and inserting the constant 2.9×10^6 nm K (derived from measurements of this relationship) to fix the units, gives **Wien's law**:

$$\lambda_{\text{peak}} = \frac{2.9 \times 10^6 \, \text{nm K}}{T}$$

Wien's law, pronounced "Veen's law," is named for physicist Wilhelm Wien (1864–1928), who discovered the relationship. In this equation, λ_{peak} is the peak wavelength.

If we insert Earth's average temperature of 288 K into Wien's law, we get

$$\lambda_{\text{peak}} = \frac{2.9 \times 10^6 \, \text{nm K}}{T}$$
$$\lambda_{\text{peak}} = \frac{2.9 \times 10^6 \, \text{nm K}}{288 \, \text{K}}$$
$$\lambda_{\text{peak}} = 10{,}100 \, \text{nm}$$

or slightly more than 10 μm. Thus, Earth's radiation peaks in the infrared region of the spectrum.

Astronomy in Action: Wien's Law

VOCABULARY ALERT

luminosity In everyday language, *luminosity* and *brightness* are used interchangeably. Astronomers, however, observe objects at great distances. They use the word *bright* to describe how an object appears in our sky, whereas they use the word *luminous* to describe how much light the object emits in all parts of the spectrum. As a result, a bright star might be luminous or it might just be very close to Earth, and a faint star might be extremely luminous but very, very far away.

It would shine because it was hot, without any effect from composition or reflection. This hypothetical perfectly absorbing and emitting object is known as a **blackbody**. **Figure 5.5** shows the spectrum of a blackbody at five different temperatures. The vertical axis plots the *intensity*—the energy emitted by every square meter of the object every second. The horizontal axis shows the wavelength, commonly given the symbol λ (the Greek letter *lambda*). Recall from Chapter 4 that longer wavelengths are redder and shorter wavelengths are bluer. As the object's temperature increases, it emits more radiation at every wavelength, so the entire curve is higher. Also notice that the peak, or highest point, of the curve shifts to the left. The wavelength that is directly below this peak on the graph is known, sensibly, as the "peak wavelength" and has the symbol λ_{peak}. Changing one thing about the object produces two effects; increasing the temperature makes the object brighter and bluer.

The surface of a protostar is tens of thousands of times larger than the surface of the Sun, and each square meter of that enormous surface radiates away energy. As a result, the protostar has a very high **luminosity**—the amount of energy it emits from the whole surface every second. A protostar is thousands of times more luminous than our Sun. However, even though the protostar is very luminous, astronomers often cannot see it in visible light. There are two reasons for this. First, most of the protostar's radiation is in the infrared rather than the visible part of the spectrum. Second, the protostar is buried deep in the heart of a dense and dusty molecular cloud. Dust absorbs visible light, especially blue light. Look back at Figure 5.1: You will see that fewer stars are visible in the dark, dense part of the nebula and that all the stars in the image are reddish in color. However, astronomers are able to view these objects in the infrared part of the spectrum because much of the longer-wavelength infrared light from the protostar *is* able to escape through the cloud. In addition, as the dust absorbs visible light, it warms up, and this heated dust also glows in the infrared region.

Sensitive infrared instruments have revolutionized the study of protostars and other young stellar objects by allowing astronomers to peer inside dense clouds of dust and gas. Dark clouds consist of clusters of dense cloud cores, young stellar objects, and glowing dust when viewed in the infrared region. **Figure 5.6** shows infrared pictures of stars forming within columns of gas and dust (shown in visible wavelengths in the larger image) in the Eagle Nebula.

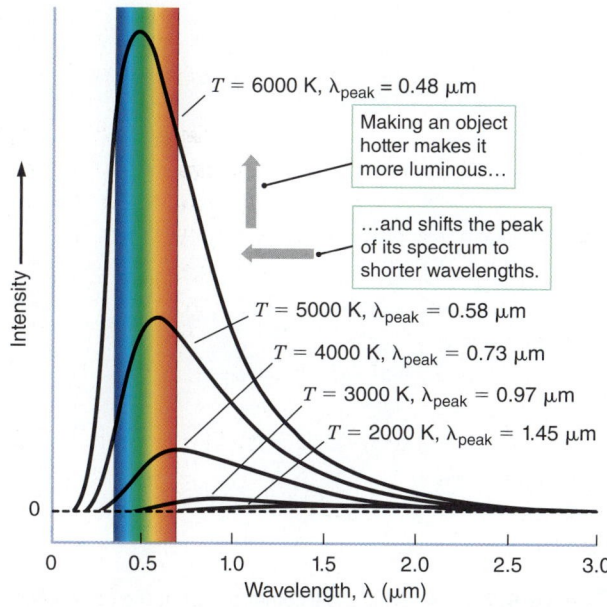

Figure 5.5 This graph shows blackbody spectra emitted by sources with temperatures ranging from 2000 to 6000 K. At higher temperatures, the peak of the spectrum shifts toward shorter wavelengths, and the amount of energy radiated per second from each square meter of the source increases.

Infrared radiation escapes the dusty columns, so infrared images show where new stars are forming.

Figure 5.6 This Hubble Space Telescope (HST) image of the Eagle Nebula shows dense columns of molecular gas and dust illuminated by nearby stars. Infrared images of the same field, also taken with the HST, show young stars forming within these columns. The Eagle Nebula is about 7,000 light-years from Earth. The largest pillar, on the left, is about 4 light-years, or 24 trillion miles, in extent.

A Shifting Balance: The Evolving Protostar

At any given moment, the protostar is in balance: The force from hot gas pushes outward and the force of gravity pulls inward, and these forces exactly oppose each other. However, this balance is dynamic—it is constantly changing. How can an object be in perfect balance and yet be changing at the same time? The spring scale shown in **Figure 5.7a** provides an example of dynamic balance. If sand is slowly poured onto a spring scale, the spring compresses until the downward force of the weight of the sand is exactly balanced by the upward force of the spring. As the weight of the sand increases, the spring is compressed further. The more the spring is compressed, the harder the spring pushes back. The spring and the weight of the sand are always in balance, but this balance is *changing with time* as more sand is added. The weight of the sand is found by measuring the compression of the spring.

In a protostar (Figure 5.7b), the outward pressure of the gas behaves like the spring, whereas material that falls onto the protostar is like the sand. The protostar slowly loses internal thermal energy by radiating it away, but the material that has fallen onto the protostar compresses the protostar and heats it up. The interior becomes denser and hotter, and the pressure rises—just enough to balance the increased weight of the material above it. Dynamic balance is always maintained.

This dynamic balance persists as energy is radiated away and the protostar slowly contracts. Gravitational energy is converted to thermal energy, which heats the core, raising the pressure to oppose gravity. The protostar shrinks and its interior grows hotter, until the center of the protostar is finally hot enough to begin

Figure 5.7 (a) A spring scale comes to rest at the point where the weight of the sand is matched by the upward force of the compressed spring. As sand is added, the location of this balance point shifts. (b) Similarly, the balance between pressure and gravity determines the structure of a protostar. Like the spring scale, the structure of the protostar constantly shifts as additional material falls onto the surface of the protostar and as the protostar radiates away energy (shown by wavy red arrows).

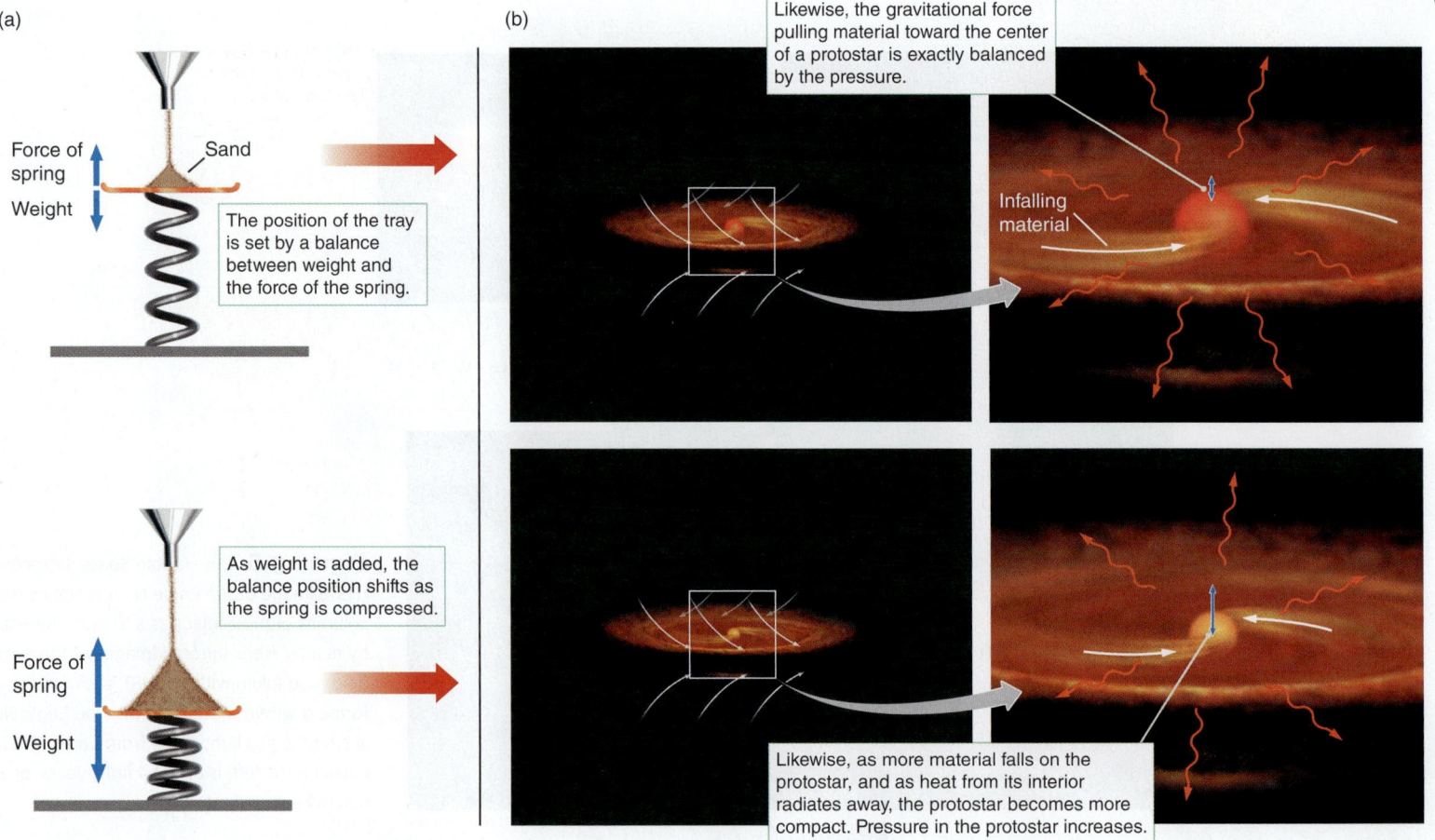

(a)

Force of spring

Sand

Weight

The position of the tray is set by a balance between weight and the force of the spring.

As weight is added, the balance position shifts as the spring is compressed.

Force of spring

Weight

(b)

Likewise, the gravitational force pulling material toward the center of a protostar is exactly balanced by the pressure.

Infalling material

Likewise, as more material falls on the protostar, and as heat from its interior radiates away, the protostar becomes more compact. Pressure in the protostar increases.

turning hydrogen into helium in a process called **hydrogen fusion**. When this change occurs, the protostar becomes a star. Hydrogen fusion is discussed in more detail in Chapter 11. For now, it's enough to know that the process produces energy that increases the pressure inside the star. The newly born star will once again adjust its structure until it is radiating energy away at exactly the rate that energy is being released in its interior. The collapse of a protostar to form a star is shown in **Figure 5.8**.

Some molecular-cloud cores do not form stars. The mass of the cloud core determines whether it will actually become a star. If the mass is greater than about 0.08 times the mass of the Sun (0.08 M_{Sun}), the temperature in the shrinking core will eventually reach 10 million K, and hydrogen fusion will begin. If the mass of the protostar is just under 0.08 M_{Sun}, it will never be hot enough to become a star. Instead, it will be a "failed" star called a **brown dwarf**. A brown dwarf is intermediate between a star and a planet, but it is not quite massive enough to cause sustained hydrogen fusion in its core. The energy emitted from a brown dwarf has the same source as the energy emitted from a protostar: gravitational collapse that turns gravitational energy into thermal energy. As the years pass, a brown dwarf gets gradually smaller and fainter. About 3,000 brown dwarfs have been found in the Milky Way Galaxy since the first one was identified in the mid-1990s.

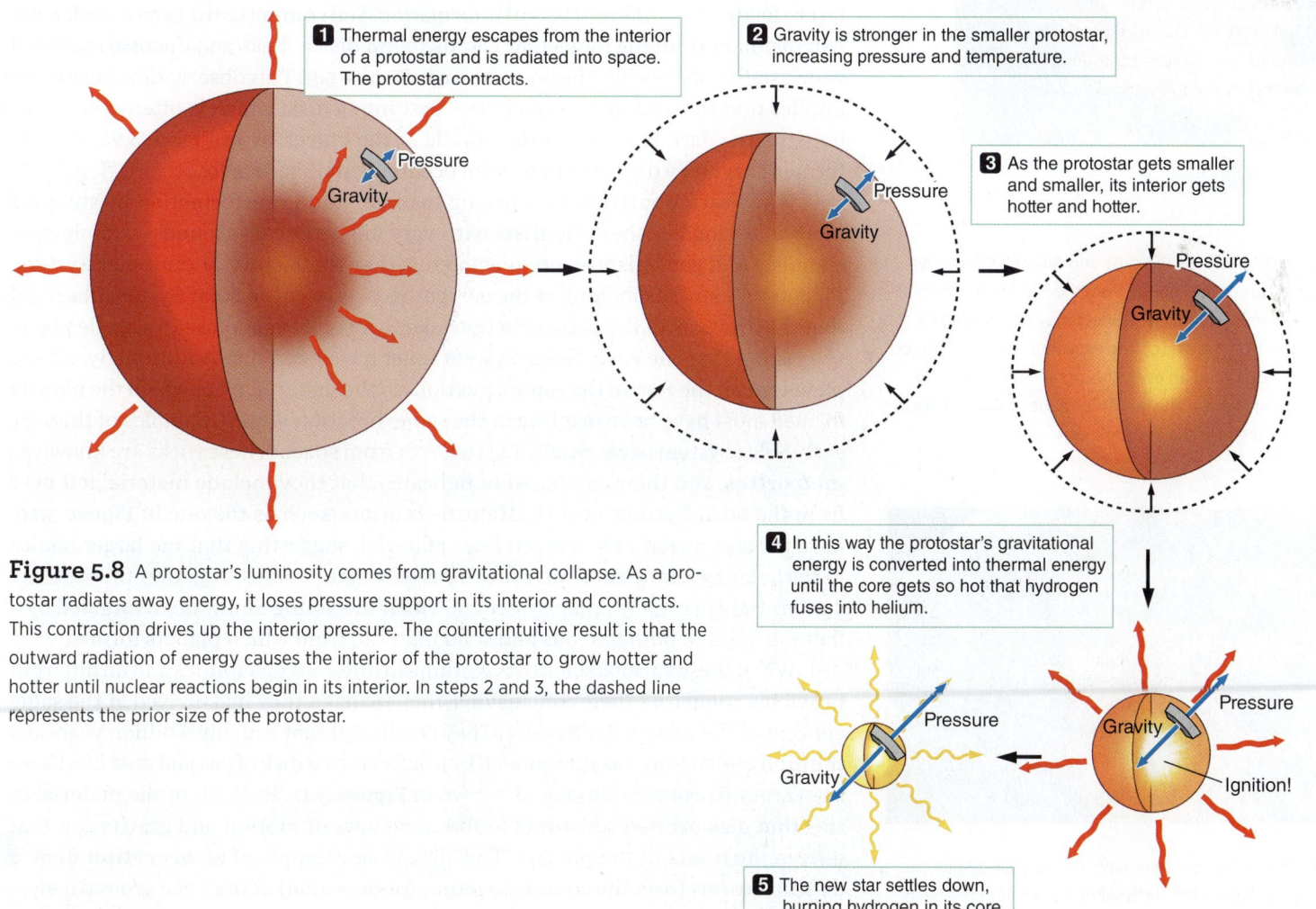

1 Thermal energy escapes from the interior of a protostar and is radiated into space. The protostar contracts.

2 Gravity is stronger in the smaller protostar, increasing pressure and temperature.

3 As the protostar gets smaller and smaller, its interior gets hotter and hotter.

4 In this way the protostar's gravitational energy is converted into thermal energy until the core gets so hot that hydrogen fuses into helium.

5 The new star settles down, burning hydrogen in its core.

Figure 5.8 A protostar's luminosity comes from gravitational collapse. As a protostar radiates away energy, it loses pressure support in its interior and contracts. This contraction drives up the interior pressure. The counterintuitive result is that the outward radiation of energy causes the interior of the protostar to grow hotter and hotter until nuclear reactions begin in its interior. In steps 2 and 3, the dashed line represents the prior size of the protostar.

(a)

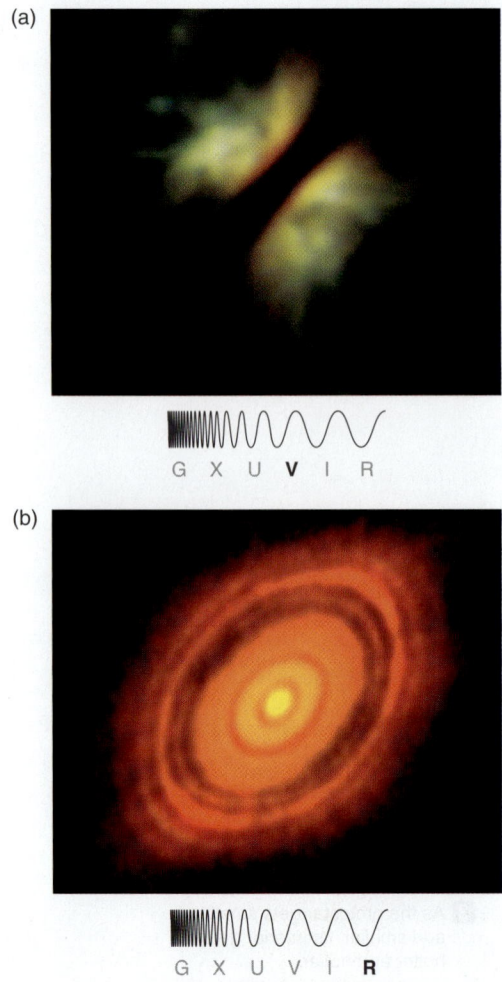

G X U **V** I R

(b)

G X U V I **R**

Figure 5.9 Disks form from the dust and gas around newly formed stars. (a) In this optical image from the Hubble Space Telescope, the disk is seen in silhouette. Planets may be forming or have already formed in this disk. (b) The Atacama Large Millimeter/submillimeter Array obtained an even sharper image of a protoplanetary disk around the star HL Tau. This image shows substructures and possibly planets in the system's dark patches.

Figure 5.10 Meteorites are fragments of the young Solar System that have landed on the surfaces of planets. On the basis of this cross section, the meteorite must have formed from many smaller components that stuck together.

CHECK YOUR UNDERSTANDING 5.2

The two forces that remain in dynamic balance as a protostar collapses are
a. gravity and gravitational energy.
b. gravity and the force exerted by the hot gas.
c. the force exerted by the hot gas and thermal energy.
d. temperature and pressure.

5.3 Planets Form in a Disk around the Protostar

So far we have focused on the innermost portion of the collapsing molecular-cloud core. What happens to the rest of the dust and gas? A piece of dust or a molecule of gas in the disk eventually suffers one of three fates: It travels inward, joining the protostar at its center; it is tossed into interstellar space; or it remains in orbit around the protostar, eventually forming planets and other objects. In this section, we focus on the formation of planets from this dust and gas.

Convergence of Evidence

The theory of star and planet formation is an example of scientists working in different fields, from different sets of information, and coming to the same conclusion. Astronomers studying molecular clouds found disks of gas and dust surrounding young stellar objects like the ones shown in **Figure 5.9**. This observational evidence implies that the dust and gas collapses first into a disk, which is often seen in silhouette as a dark lane across the middle of the object, as in Figure 5.9a. Planets form in this dusty disk, as can be seen occurring in Figure 5.9b.

While astronomers were working to understand star formation by studying molecular clouds, other scientists with very different backgrounds—mainly geochemists and geologists—were piecing together the history of our Solar System. Planetary scientists looking at the current structure of the Solar System observed that the orbits of all the planets in the Solar System lie very close to a single plane, suggesting that the early Solar System must have been flat. Additionally, all the planets orbit the Sun in the same direction, so the material from which the planets formed must have been orbiting in the same direction as well. Samples of the very early Solar System occasionally fall to Earth from space. These rocks are known as **meteorites**, and their composition indicates that they include material left over from the Solar System's youth. Many meteorites, such as the one in **Figure 5.10**, have pebbles mixed with a much finer material, suggesting that the larger bodies in the Solar System grew from collections of smaller bodies. Together these observations imply an early Solar System in which the young Sun was surrounded by a flattened disk of both gaseous and solid material from which planets formed.

When these various scientists working in different disciplines and on different problems compared their conclusions, they realized they had arrived at the same concept of the early Solar System. They concluded that roughly 5 billion years ago the protostellar Sun was surrounded by a flat, orbiting disk of gas and dust like those seen around protostars today, as shown in **Figure 5.11**. Each bit of the material in this thin disk orbited according to the same laws of motion and gravitation that govern the orbits of the planets. This disk is an example of an **accretion disk**: a disk that forms from the coming together (or *accretion*) of material around a massive object. The accretion disk around the forming Sun had only a fraction of the mass of the Sun, but this amount was more than enough to account for the bodies

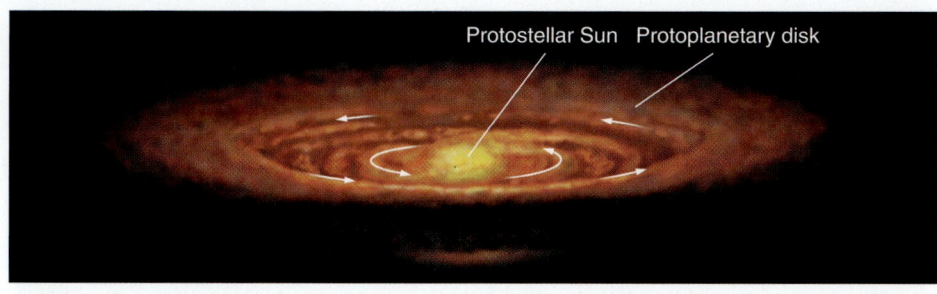

Protostellar Sun   Protoplanetary disk

Figure 5.11 When you consider the young Sun, think of it as being surrounded by a flat, rotating disk of gas and dust that is flared at its outer edge.

that make up the Solar System today. The connection between the formation of stars and the origin and evolution of the Solar System is a wonderful example of the convergence of evidence in science.

▶❙❙ **AstroTour:** Solar System Formation

The Collapsing Cloud and Angular Momentum

What is it about the process of star formation that leads not only to a star but also to a flat, orbiting collection of gas and dust? To understand the answer, you have to know something about angular momentum. The **angular momentum** of a revolving or rotating system depends on both the velocity and distribution of the mass. The angular momentum of an object remains unchanged unless the object is acted on by an external force; this is the principle of **conservation of angular momentum**. For example, the spinning figure-skater shown in **Figure 5.12a**, like any rotating object, has some amount of angular momentum. Unless an external force acts on her, she will always have the same amount of angular momentum.

The amount of angular momentum a spinning object has depends on three things: the mass, the rate of rotation, and how spread out the object is.

1. The mass of the object. If a bowling ball and a basketball are spinning at the same speed, the bowling ball has more angular momentum because it has more mass.

2. How fast the object is rotating. If two basketballs are spinning at different rates, then the faster one has more angular momentum.

3. How spread out the object is. If a spinning skater wants to rotate faster, she brings her arms in tight around her body.

(a)

(b)

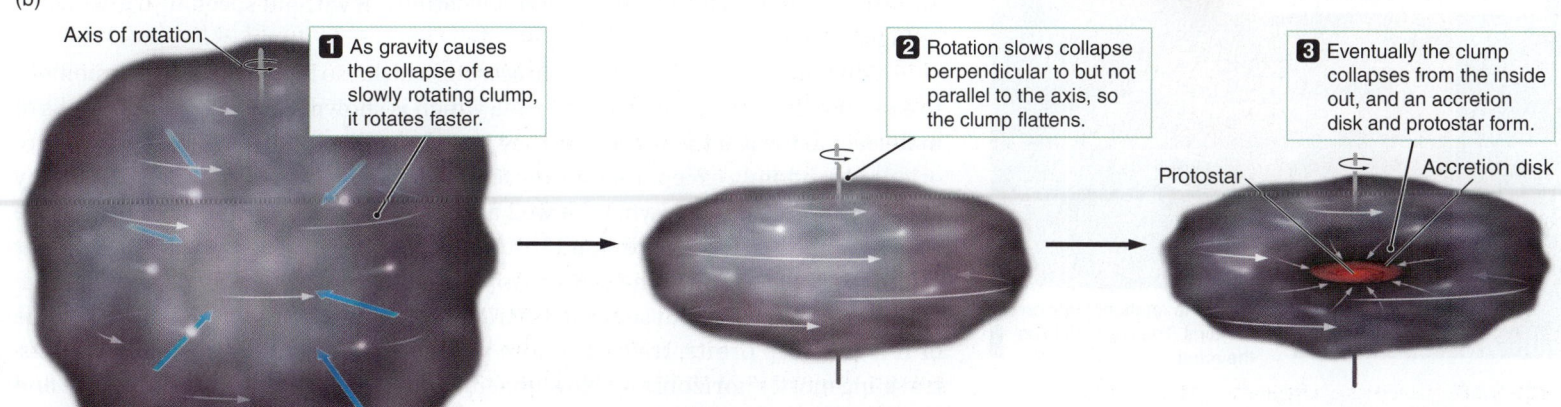

Axis of rotation

1 As gravity causes the collapse of a slowly rotating clump, it rotates faster.

2 Rotation slows collapse perpendicular to but not parallel to the axis, so the clump flattens.

3 Eventually the clump collapses from the inside out, and an accretion disk and protostar form.

Protostar   Accretion disk

Figure 5.12 (a) A figure-skater relies on the principle of conservation of angular momentum to change the speed of her spin. (b) In the same way, a collapsing cloud spins faster as it becomes smaller. Angular momentum is conserved in a collapsing cloud.

Point 3 might be a bit difficult to understand. Let's look at it a couple of additional ways. If twin skaters with the same mass are spinning at the same rate, the one that has her arms out has more angular momentum than the one with her arms held tight to her side. A spread-out object that is rotating slowly has the same angular momentum as a compact object rotating rapidly, if they both have the same mass.

For a small object in a large orbit, like a planet around a star, the relationship between angular momentum and these three factors is often expressed as an equation: $L = m \times v \times r$, where L is the angular momentum, m is the mass of the planet, v is the orbital speed of the planet, and r is the distance of the planet from the star. In order for angular momentum to be conserved, a change in one of the three quantities (mass, rate of spin, or distance of the planet) must be accompanied by a change of another quantity. If the planet gets closer to the star (r gets smaller), it must speed up (v gets bigger). If it moves farther away, it must slow down. That should sound familiar! It's Kepler's second law.

Both a spinning ice-skater and a collapsing interstellar cloud conserve angular momentum. As an ice-skater pulls in her arms to become more compact, she decreases her distribution of mass, so she must spin faster to maintain the same angular momentum. Similarly—as shown in Figure 5.12b—as the cloud that formed our Sun collapsed, it rotated faster and faster.

This description presents a puzzle, though. A typical cloud core is about a light-year across and rotates so slowly that it takes about a million years to complete one rotation. If the Sun formed by collapsing from a cloud like this, its radius decreased dramatically, so it should be spinning so fast that one rotation would take 0.6 second. This is more than 3 million times faster than our Sun actually spins. That rate of rotation is so fast that the Sun would tear itself apart. It appears that angular momentum was *not* conserved in the actual formation of the Sun—but that can't be right, because angular momentum must be conserved. We must be missing something. Where did the angular momentum go?

The Formation of an Accretion Disk

To understand how angular momentum is conserved in the formation of the Solar System, we must think in three dimensions. Imagine that an ice-skater bends her knees, compressing herself downward instead of bringing her arms toward her body. As she does this, she again makes herself less spread out, but her rate of spin does not change because no part of her body has become closer to the axis of rotation. Similarly, a clump of a molecular cloud can flatten out without speeding up, as shown in Figure 5.12b.

Many astronomical objects form accretion disks, so it is worth taking a moment to examine this process in a bit of detail, which is shown in **Figure 5.13**. As blobs of material fall toward the protostar, they travel on elliptical orbits. These orbits are oriented randomly except for one key feature—either they all go clockwise or they all go counterclockwise, when viewed along the axis of rotation. This is what we really mean when we say the cloud "rotates." Imagine yourself in such a cloud, near the edge, looking toward the center. As you watch, all the material is moving from your left to your right, but some of it is traveling upward and some downward. Some of it is on steep orbits, traveling more vertically, and some is on shallow orbits, traveling mostly horizontally. Now imagine that two pieces of material collide and stick together to form a larger piece. Both are on shallow orbits, but one is traveling up and the other is traveling down. What will happen to the new, larger piece? It

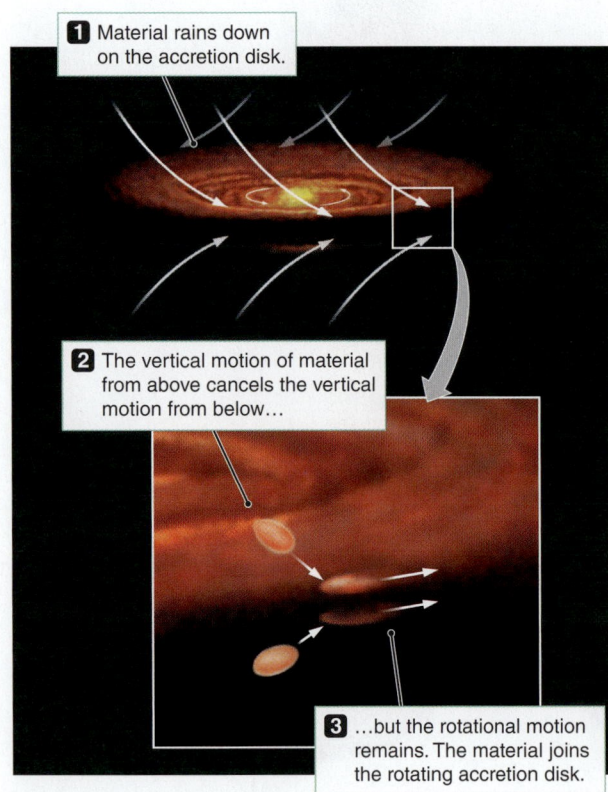

1 Material rains down on the accretion disk.

2 The vertical motion of material from above cancels the vertical motion from below...

3 ...but the rotational motion remains. The material joins the rotating accretion disk.

Figure 5.13 Gas from a rotating molecular cloud falls inward from opposite sides, piling up onto a rotating disk.

will still orbit from left to right, but the upward motion and the downward motion will cancel out, so the orbit will become shallower. Imagine this same scenario for two pieces on steep orbits. Again, the orbit will become less steep. In this way, the upward and downward motions of the material cancel each other out, and the whole cloud flattens out. Over time, a disk is formed in which all the material has very shallow orbits, and all orbits still proceed in the same overall direction—either clockwise or counterclockwise. The angular momentum is unchanged because all the material has maintained its distance from the axis of rotation.

The radius of a rotating accretion disk is thousands of times greater than the radius of the star that will form at its center, so the disk will rotate much more slowly than a protostar would. Much of the angular momentum ends up in its accretion disk rather than in the central protostar. This is why the Sun does not have all the angular momentum that was originally present in the young Sun. Nearly all of the angular momentum of the Solar System (96%) is now in the orbits of the planets. On the basis of these angular momentum considerations, astronomers predicted that accretion disks should be found around young stars long before any disks were actually observed around protostars.

Some of the matter that lands on the accretion disk is ejected back into interstellar space, in the form of jets or other outflows, as seen in **Figure 5.14**. These jets

▶‖ **AstroTour:** Traffic Circle Analogy

what if . . .

What if you observe a close pair of stars forming a binary system: How would you expect the disk around such a pair of stars to differ from the disk around a single star?

(a)

Material moves in toward the protostar in the accretion disk.

Bipolar jets and outflows flow away from the young star and disk.

Wind

Jets

This structure is seen in images of forming stars.

G X U **V** I **R**

(b)

Jets — Star

G X U V **I** R

Figure 5.14 (a) Material falls onto an accretion disk around a protostar and then moves inward, eventually falling onto the star. In the process, some of this material is driven away in powerful jets that stream perpendicular to the disk. Note the nearly edge-on, dark accretion disk surrounding the young star. (b) This infrared Spitzer Space Telescope image shows jets streaming outward from a young, developing star.

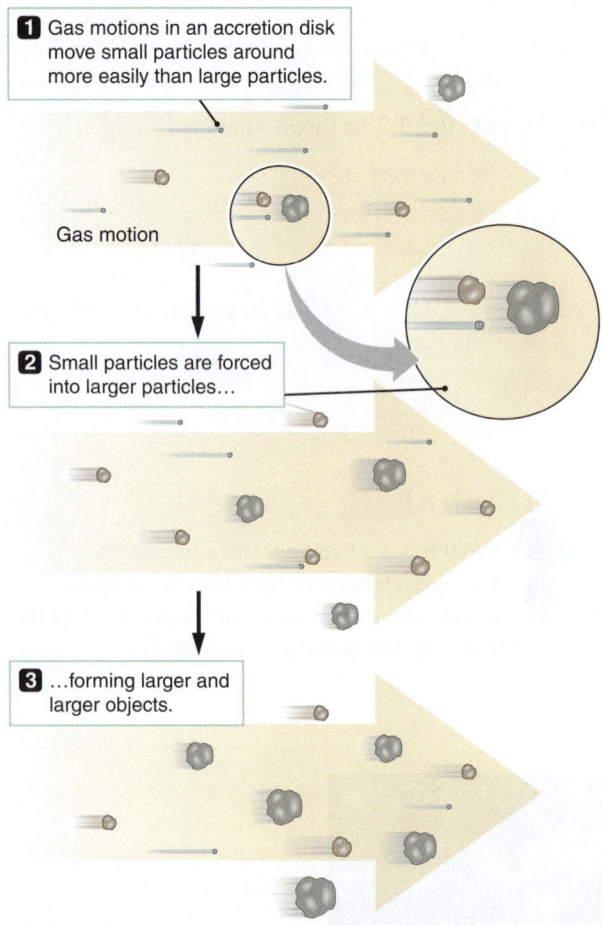

1 Gas motions in an accretion disk move small particles around more easily than large particles.

Gas motion

2 Small particles are forced into larger particles…

3 …forming larger and larger objects.

Figure 5.15 Motions of gas in an accretion disk force smaller particles of dust into larger particles and then make these large particles even larger. This process continues, eventually creating objects many meters in size.

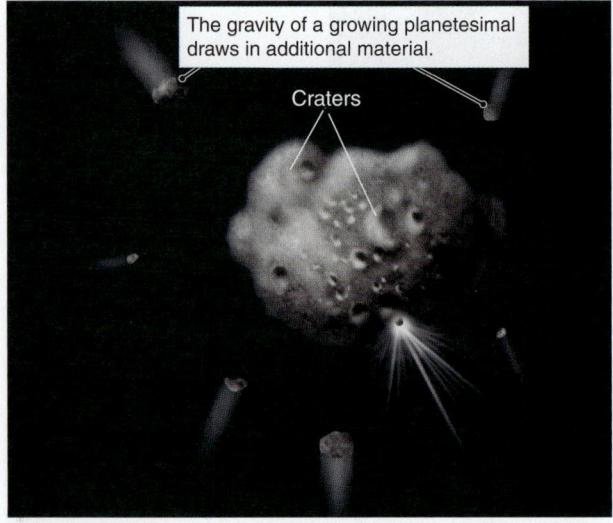

The gravity of a growing planetesimal draws in additional material.

Craters

Figure 5.16 The gravity of a planetesimal is strong enough to capture surrounding material, which causes the planetesimal to grow more rapidly.

are typically bipolar—they come in pairs aligned along an axis. Material swirling in the bipolar jets carries angular momentum away from the accretion disk in the general direction of the poles of the rotation axis. However, a small amount of material is left behind in the disk. It is this leftover material in the disk that forms planets and other objects that orbit the star. Our Sun and Solar System formed from a protostar and disk much like the ones shown in Figures 5.9a and 5.14.

Creation of Planetesimals

How do the small blobs of dust and gas in an accretion disk grow to form planets? First, random motions of material in the accretion disk push smaller grains of solid material toward larger grains, as shown in **Figure 5.15**. As this happens, the smaller grains stick to the larger grains. The "sticking" process among smaller grains is due to the same static electricity that causes dust bunnies to grow under your bed. Starting out at only a few micrometers across, the slightly larger bits of dust grow to the size of pebbles and then to clumps the size of boulders, which are not as easily pushed around by gas. When clumps grow to be objects about 100 meters across, much of the dust and gas has already been collected. The objects are now so far apart that they collide less frequently, and their growth rate slows down.

The process does not yield larger and larger bodies with every collision. When violent collisions occur in an accretion disk, larger clumps break back into smaller pieces. For two large clumps to stick together rather than explode into many small pieces, they must bump into each other very gently; in fact, collision speeds must be about 0.1 meter per second (m/s) or less for colliding boulders to stick together. If your stride is roughly a meter long, then a collision speed of 0.1 m/s would correspond to only one step every 10 seconds. Over a long period of time (tens of millions of years), large bodies do form.

Once the objects are about 100 meters across, they grow by "sweeping up" smaller objects that get in their way. These objects can eventually measure up to several hundred meters across. As the clumps grow to about a kilometer across, a different process becomes important. These kilometer-sized objects are massive enough that their gravity pulls strongly on nearby bodies, as shown in **Figure 5.16**, at which point they are known as **planetesimals** (literally "tiny planets"). Planetesimals grow both by chance collisions with other objects and by pulling in and capturing small objects outside their direct paths. Their growth speeds up, and the larger planetesimals quickly consume most of the remaining bodies in the vicinity of their orbits. The final survivors of this process are large enough to be called planets. As in our Solar System, some planets may be relatively small and others quite large.

Very large objects can capture gas and dust to acquire mini accretion disks of their own. Some of this material may grow into larger bodies in much the same way that material in the accretion disk formed planets. The result is a "mini solar system"—a group of moons all orbiting in the same plane about the planet.

CHECK YOUR UNDERSTANDING **5.3**

Dust grains first begin to grow into larger objects because of
   a. gravity from large planetesimals.
   b. gravity from the central star.
   c. collisions between dust grains.
   d. conservation of angular momentum.

5.4 The Inner and Outer Disk Have Different Compositions

In the Solar System, the inner planets are small and mostly rocky, whereas the outer planets are very large and mostly gaseous. How can we explain this distinct difference between the inner and outer Solar System by using a model of the formation of the stars and planets?

Rock, Metal, and Ice

On a hot summer day, ice melts and water quickly evaporates, but on a cold winter night, even the water in our breath freezes into tiny ice crystals before our eyes. Metals and rocky materials, such as iron, **silicates** (minerals containing silicon and oxygen), and carbon, remain solid even at quite high temperatures. Other materials, such as water, ammonia, and methane, remain in a solid form only if their temperature is very low. Materials like these, which become gases at moderate temperatures, are called **volatile materials** or "volatiles."

The temperature of the accretion disk is lower at points farther from the proto-Sun. Volatiles turn to gas close to the Sun, whereas metals and rocky materials can be found in solid form everywhere in the accretion disk, as shown in **Figure 5.17**. In the inner disk, dust grains are composed almost entirely of metals and rocky materials. Somewhat farther out, some hardier volatiles, such as water ice and some carbon-based substances, can survive in solid form, adding to the materials that make up dust grains. Highly volatile components such as methane, ammonia, and carbon monoxide **ices** and some simple carbon-bearing molecules survive in solid form only in the coldest, outermost parts of the accretion disk, far from the central protostar. The differences in composition of solid materials within the disk are reflected in the composition of the planets formed there. Planets closest to the central star are composed primarily of rock and metals but are deficient in volatiles. Those that form farthest from the central star contain metals and rocky materials, but they also contain large quantities of ices and other volatiles.

Chaotic encounters can change this organization of planetary compositions. Gravitational scattering or interactions with gas in the protoplanetary disk can force some planets to end up far from where they formed, a process called **planet migration**. Uranus and Neptune originally may have formed near the orbits of Jupiter and Saturn but were then driven outward to their current locations by gravitational encounters with Jupiter and Saturn. A planet can also migrate when it gives up some of its orbital angular momentum to the disk material that surrounds

VOCABULARY ALERT

ice In everyday language, *ice* refers specifically to the solid form of water, as we've used it so far in this chapter. The term *dry ice* is also common, which refers to frozen carbon dioxide. To astronomers, *ice* refers to the solid form of any type of volatile material.

chaotic (or chaos) *Chaotic* is commonly used to mean "messy or disorganized." To scientists, a *chaotic* system is one that is very sensitive to tiny changes in the starting conditions: A small change at the beginning can lead to a large change in the later state of a system.

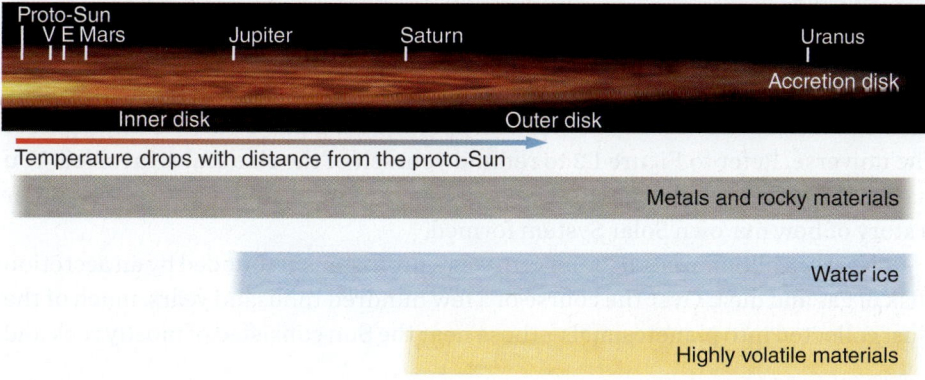

Figure 5.17 Differences in temperature within an accretion disk determine the composition of dust grains that then evolve into planetesimals and planets. The colored bars show where metals and rocky materials, water ice, and volatiles can exist in the accretion disk. Shown here are the proto-Sun and the distances to Venus (V), Earth (E), Mars, Jupiter, Saturn, and Uranus.

it. Such a loss of angular momentum causes the planet to slowly spiral inward toward the central star. We will see examples from other planetary systems when we discuss *hot Jupiters* in Section 5.6.

Atmospheres around Solid Planets

An **atmosphere** is a gaseous layer around a star, moon, or planet that is bound by gravity. Once a solid planet has formed, it may grow an atmosphere by capturing gas from the accretion disk. This must happen quickly because protostars emit fast-moving particles and intense radiation that blow away the accretion disk. Planets like Jupiter have about 10 million years to capture gas to form an atmosphere.

The gas captured by a planet at the time of its formation forms the planet's **primary atmosphere**. Massive planets have more gravity, so they can capture more of the hydrogen and helium gas that makes up the bulk of the accretion disk than can low-mass planets. Massive planets like Jupiter can accumulate atmospheres that are more massive than the solid body at the core of the planet.

A less massive planet may also capture a primary atmosphere, only to lose it later. The gravity of some small planets is too weak to hold low-mass gases such as hydrogen and helium. The atmosphere around a small planet like Earth is a **secondary atmosphere**, which forms later in the life of a planet. Volcanism is one important source of a secondary atmosphere because it releases carbon dioxide, water vapor, and other gases that were trapped inside the planet's interior. In addition, volatile-rich bodies like comets that formed in the outer parts of the disk may be on elliptical orbits. As they travel on these orbits, they fall in toward the new star and sometimes they collide with planets. These collisions may deliver significant amounts of water, carbon-rich compounds, and other volatile materials to planets close to the central star.

CHECK YOUR UNDERSTANDING 5.4

Examine Figure 5.17. During the time shown in the Solar System's formation, why is there no water ice at the distance of the orbit of Mars?

a. The inner disk was too hot.

b. The inner disk was too dense.

c. Earth is the only place where water (in any form) can be found in the Solar System.

d. The outer planets are larger and have more gravity.

5.5 A Case Study: The Solar System

The Solar System is a collection of planets, moons, and other smaller bodies that orbit the star we call the Sun. Enough small bodies occupy particular regions of the Solar System to warrant separate names for these regions, such as the asteroid belt and Kuiper Belt (see **Figure 5.18**). Do not confuse our Solar System with the galaxy or the universe. Our Solar System is a tiny part of our galaxy, which is a tiny part of the universe. Refer to Figure 1.2 to remind yourself of the size scales involved. You have now learned enough about the formation of stars and planets to put together a story of how our own Solar System formed.

Nearly 5 billion years ago, our Sun was a protostar surrounded by an accretion disk of gas and dust. Over the course of a few hundred thousand years, much of the dust collected into planetesimals—those near the Sun consisted of mostly rock and

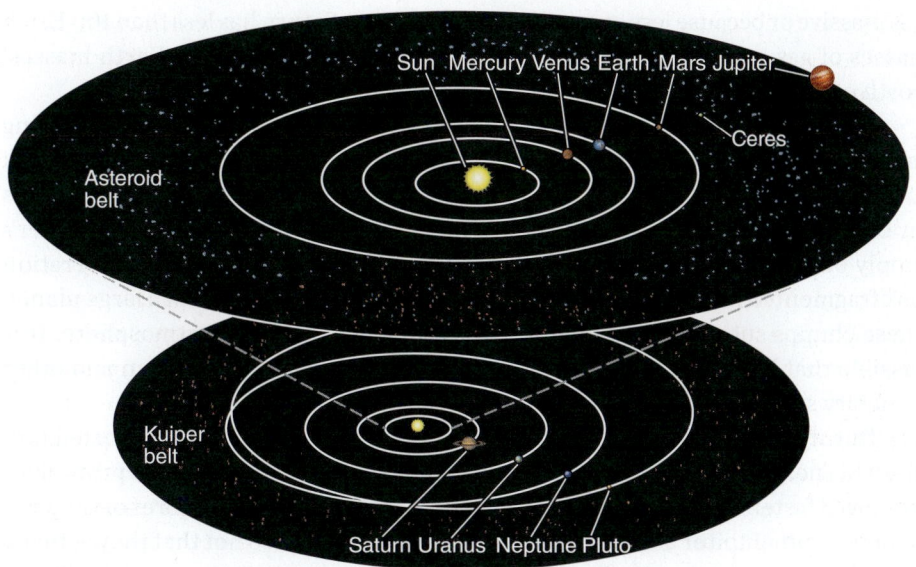

Figure 5.18 Our Solar System includes planets, moons, and other small bodies. Sizes and distances are not to scale in this sketch of the Solar System.

metal, while those farther away also contained large amounts of volatiles. Within the inner few astronomical units (AU) of the disk, about half a dozen rock and metal planetesimals quickly grew to become the dominant masses in their orbits. These larger planetesimals either captured the remaining planetesimals or ejected them from the inner part of the disk. These dominant planetesimals had now become planet-sized bodies with masses ranging between $\frac{1}{20}$ and 1 Earth mass (M_{Earth}). They became the **terrestrial planets** (terrestrial means Earth-like, or rocky). Mercury, Venus, Earth, and Mars are the surviving terrestrial planets. One or two others may have formed in the young Solar System but were later destroyed.

For several hundred million years after the terrestrial planets formed, leftover pieces of debris still in orbit around the Sun continued to smash into their surfaces. Much of this debris may have originated in the outer Solar System, bringing volatiles such as water ice. The scars of these early impacts are still visible on the cratered surfaces of some of the terrestrial planets, such as the surface of Mercury shown in **Figure 5.19**. Collisions continue to happen today but at a much lower rate.

Before the proto-Sun became a true star, gas in the inner part of the accretion disk was still plentiful. During this early period, Earth and Venus accumulated thin primary atmospheres of hydrogen and helium, but these thin atmospheres were soon lost to space. The terrestrial planets later formed the secondary atmospheres that now surround Venus, Earth, and Mars. Mercury's nearness to the Sun and the Moon's small mass prevented these bodies from retaining significant secondary atmospheres.

Beyond 5 AU from the Sun, planetesimals combined to form a number of bodies with masses about 10–20 times that of Earth. These planet-sized objects contained volatiles in addition to rock and metal. Four of these massive bodies became the cores of the **giant planets**: Jupiter, Saturn, Uranus, and Neptune. The giant planets are many times the mass of any terrestrial planet and lack a solid surface. Mini accretion disks formed, capturing large amounts of hydrogen and helium and funneling this material onto the planetary cores.

Jupiter's massive solid core captured and retained the most gas—creating an atmosphere roughly 300 times the mass of Earth. The other planetary cores captured lesser amounts of hydrogen and helium, perhaps because their cores were

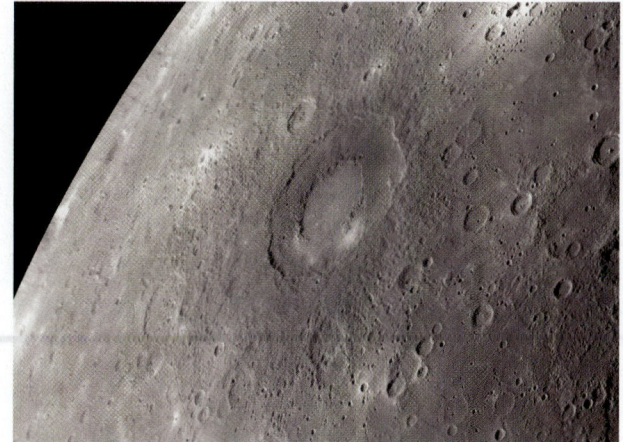

Figure 5.19 Large impact craters on Mercury (and on solid bodies throughout the Solar System) record the final days of the Solar System's youth, when smaller planetesimals rained down on their surfaces.

less massive or because less gas was available to them. Saturn has less than 100 Earth masses of gas, whereas Uranus and Neptune captured only a few Earth masses' worth of gas each.

It may have taken as much as 10 million years for Jupiter to form by capturing gas from the disk. However, some scientists do not think that our accretion disk could have survived long enough to form gas giants such as Jupiter through this process. All the gas may have dispersed in roughly half that time, cutting off Jupiter's supply of hydrogen and helium. An alternative explanation is that the accretion disk fragmented into massive clumps, each of which had the mass of a large planet. These clumps subsequently collapsed to form a planet with its atmosphere. It is possible that both processes played a role in the formation of our own and other planetary systems. This part of the story is still incomplete.

During the formation of the planets, gravitational energy was converted into thermal energy as the individual atoms and molecules that made up the protoplanets moved faster. This conversion warmed the gas surrounding the cores of the giant planets. Proto-Jupiter and proto-Saturn probably became so hot that they actually glowed a deep red color, similar to the heating element on an electric stove. Their internal temperatures may have reached as high as 50,000 K. However, they were never close to becoming stars. As we saw in Section 5.2, a ball of gas must have a mass at least 0.08 times the mass of the Sun for it to become a star. This is about 80 times more massive than Jupiter.

The composition of the moons of the giant planets followed the same trend as the planets that formed around the Sun: The innermost moons formed under the hottest conditions and therefore contained the smallest amounts of volatile material. When Jupiter's moon Io formed, Jupiter was glowing so intensely that it rivaled the distant Sun. The high temperatures created by the glowing planet evaporated most of the volatile substances nearby. Io today contains no water at all. However, water is plentiful on at least three of Jupiter's other large moons—Europa, Ganymede, and Callisto—because these moons formed farther from warm, glowing Jupiter.

Not all planetesimals in the disk became planets. For example, **dwarf planets** orbit the Sun but have not cleared other, smaller bodies from their orbits. Many are not massive enough to be round. Ceres and Pluto are both dwarf planets; their locations are marked in Figure 5.18. More dwarf planets, along with a large number of smaller bodies, are found beyond Pluto's orbit in the **Kuiper Belt**. **Asteroids** are small rocky bodies in orbit around the Sun; most are located in the main **asteroid belt** between Mars and Jupiter. Jupiter's gravity kept the region between Jupiter and Mars so stirred up that planetesimals could not merge to form a large planet.

Planetesimals persist to this day in the frigid outermost part of the Solar System. These objects have retained most of the highly volatile materials found in the original accretion disk. Unlike the crowded inner part of the disk, planetesimals in the outermost parts of the disk were too far apart for large planets to grow. Icy planetesimals in the outer Solar System are relatively pristine samples of the material from which our planetary system formed. The frozen, distant dwarf planets Pluto and Eris are especially large examples of these residents of the outer Solar System.

Even after the initial formation, the Solar System was a remarkably violent and chaotic place. Many objects in the Solar System show evidence of cataclysmic impacts that reshaped worlds. A dramatic difference between the terrains of the northern and southern hemispheres on Mars, for example, may be the result of one or more colossal collisions. Mercury has a crater on its surface from an impact so devastating that it caused the crust to buckle on the opposite side of the planet. In the outer Solar System, Saturn's moon Mimas has a crater roughly one-third the diameter of the moon itself. Uranus suffered a collision violent enough to literally knock the planet on its side; its axis of rotation is tilted to lie almost in its orbital plane.

Not even our own Earth escaped devastation by these cataclysmic events. The Moon may be the result of such a collision. According to the best current hypothesis for the formation of the Moon, the early Solar System included a protoplanet about the same size and mass as Mars. As the newly formed planets were settling into their present-day orbits, this fifth terrestrial planet suffered a grazing collision with Earth and was completely destroyed. The remains of the planet, together with material knocked from Earth's outer layers, formed a huge cloud of debris encircling Earth. For a brief period, Earth may have displayed a magnificent group of rings like those of Saturn. In time, some of this debris coalesced into the single body we know as our Moon. This "impact formation" hypothesis is still an active area of research, as astronomers work to understand all the observations about the Earth-Moon system.

CHECK YOUR UNDERSTANDING 5.5

The terrestrial planets are different from the giant planets because when they formed,

a. the inner Solar System was richer in heavy elements.

b. the inner Solar System was hotter than the outer Solar System.

c. the outer Solar System took up a bigger volume, so there was more material to form planets.

d. the inner Solar System was moving faster.

5.6 Planetary Systems Are Common

In 1995, astronomers announced the first confirmed **exoplanet**, which is a planet orbiting around a star other than the Sun. This planet orbits around a star called 51 Pegasi and is a Jupiter-sized body with a surprisingly small orbit. The International Astronomical Union is a bit more specific about what an exoplanet is. They define an exoplanet as a body that orbits a star other than our Sun and has a mass less than 13 Jupiters.

The Search for Exoplanets

Exoplanet astronomy is advancing quickly. Today, there are more than 4,000 known exoplanets, and new discoveries are occurring almost daily. The first exoplanets were discovered indirectly by observing their gravitational tug on the central star. Astronomers now use several methods to detect and investigate exoplanets. They can directly image exoplanets and have been able to observe the spectra of exoplanets to discover the composition of their atmospheres. We will now look at each of these methods.

The spectroscopic radial velocity method. As a planet orbits a star, the planet's gravity tugs the star around ever so slightly. Astronomers can sometimes detect this motion of the star and infer the mass of the planet and its distance from the star.

(a)

Figure 5.20 The Doppler shift can be seen in waves of all kinds. (a) The waves in front of the moving duck are compressed while the ones behind are stretched out. Waves emitted perpendicular to the direction of travel are neither compressed nor stretched out. (b) Motion of a light or sound source relative to an observer causes waves to be spread out (*redshifted*, or made lower in pitch) or squeezed together (*blueshifted*, or made higher in pitch). Such a change in the wavelength of light or the frequency of sound is called a Doppler shift.

▶‖ **AstroTour:** The Doppler Effect

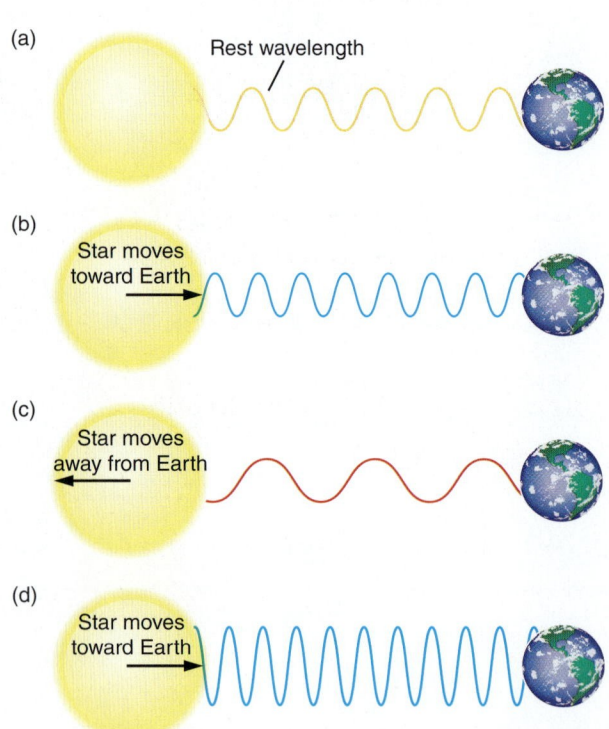

Figure 5.21 Light from an astronomical object will be observed (a) at its rest wavelength, (b) blueshifted, or (c) redshifted, depending on whether the object is (a) at rest, (b) moving toward Earth, or (c) moving away from Earth. (d) If the object is moving faster, the Doppler shift will be more pronounced.

(b)

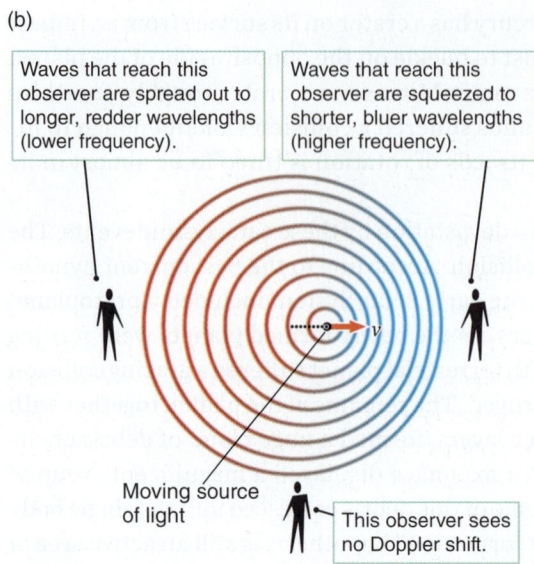

Waves that reach this observer are spread out to longer, redder wavelengths (lower frequency).

Waves that reach this observer are squeezed to shorter, bluer wavelengths (higher frequency).

Moving source of light

This observer sees no Doppler shift.

Have you ever watched a car race on television or in the movies? As a car passes the camera, there is a characteristic change in the frequency of the sound. The approaching car has a higher pitch, and as it passes by, the pitch drops noticeably. If you close your eyes and listen, you have no trouble knowing when the car passed; the change in pitch tells you when it passes the camera. This change in frequency due to motion is known as the **Doppler effect**.

As you learned in Chapter 4, wavelength and frequency are inversely proportional. A higher pitch indicates that the sound waves have shorter wavelengths. A lower pitch indicates that the waves have longer wavelengths. This occurs for any kind of wave, whether light waves, sound waves, or waves in the water. As an object approaches, the waves that it emits "crowd together" in front of the object (**Figure 5.20**). The waves in front of the duck in Figure 5.20a are crowded together (higher frequency) so the wavelength is shorter. Conversely, the waves behind the duck are stretched out (lower frequency) so the wavelength is longer. Waves emitted to the side, perpendicular to the motion, are not shifted. The Doppler effect for light waves is shown schematically in Figure 5.20b.

An object at rest emits light with the **rest wavelength**, as shown in **Figure 5.21a**. If the object moves, the Doppler effect causes a shift in the light emitted. If a star is moving toward you, the light you see is compressed to a shorter (bluer) wavelength, so we say it is **blueshifted** (Figure 5.21b). In contrast, light from a star that is moving away from you is stretched to longer (redder) wavelengths, so we say it is **redshifted** (Figure 5.21c). The faster the star is moving toward or away from you, the larger the shift (compare panels b and d in Figure 5.21). This shift is called the **Doppler shift**, and the size of the Doppler shift depends on the speed of the star.

The Doppler shift provides information only about the **radial velocity** of the object: the part of the motion that is toward you or away from you. A distant star moving *across* the sky, for example, does not move toward or away from you, and so you will not see the Doppler shift. In **Working It Out 5.2**, we explore how to use the Doppler shift to calculate the radial velocity of an astronomical object, or the amount by which its light has shifted from our standpoint.

How does the Doppler shift help us find exoplanets? Let's use our own Solar System as an example. The Sun and Jupiter orbit a common center of gravity that lies just beyond the surface of the Sun, as shown in **Figure 5.22**. (This common center of gravity, which is sometimes called the center of mass, is the location where the effect of one mass balances the other.) The Sun's orbit around this center of mass is much smaller than Jupiter's, but both objects take 11.86 years to orbit once. This means the Sun's speed is much lower than Jupiter's—only 12 m/s, slower than a cougar can run. Imagine that an astronomer on some distant exoplanet points a spectrometer toward the Sun. The astronomer would find that Jupiter tugs the Sun around so that the Sun's radial velocity varies by ±12 m/s, with a period of 11.86 years. From this information, the astronomer would rightly conclude that the Sun has at least one planet with a mass comparable to Jupiter's, but

working it out 5.2

Making Use of the Doppler Shift

Because atoms of different elements interact with different wavelengths of light, they amplify or reduce the spectrum at particular wavelengths. Atoms and molecules produce emission lines when they add light to the spectrum and absorption lines when they take light away. These absorption or emission lines that look something like a bar code instead of a rainbow, and each type of atom or molecule has a unique set of lines. These lines are called *spectral lines*. A prominent spectral line of hydrogen atoms has a rest wavelength, λ_{rest}, of 656.3 nanometers (nm). Suppose that, using a telescope, you measure the wavelength of this line in the spectrum of a distant star and find that instead of seeing the line at 656.3 nm, you see it at an observed wavelength, λ_{obs}, of 659.0 nm. What can we learn from this measurement?

The mathematical form of the Doppler effect shows that the object is moving at a radial velocity (v_r) of

$$v_r = \frac{\lambda_{obs} - \lambda_{rest}}{\lambda_{rest}} \times c$$

$$v_r = \frac{659.0 \text{ nm} - 656.3 \text{ nm}}{656.3 \text{ nm}} \times (3 \times 10^8 \text{ m/s})$$

$$v_r = 1.2 \times 10^6 \text{ m/s}$$

The star is moving with a speed of 1.2×10^6 m/s, or 1,200 kilometers per second (km/s). In addition, you know that the star is moving away from you because the wavelength we observe is longer (and redder) than the rest wavelength. Note that objects traveling away from you have a positive value of v_r.

Now consider our stellar neighbor, Alpha Centauri, which is moving toward us with a radial velocity of -21.6 km/s, or -2.16×10^4 m/s. (Negative velocity means the object is moving toward us; positive means it's moving away from us.) What is the observed wavelength, λ_{obs}, of a magnesium line in Alpha Centauri's spectrum having a rest wavelength, λ_{rest}, of 517.27 nm? First, we need to manipulate the Doppler equation to get λ_{obs} all by itself. Then we can plug in all the numbers.

$$v_r = \frac{\lambda_{obs} - \lambda_{rest}}{\lambda_{rest}} \times c$$

Solve this equation for λ_{obs} to get

$$\lambda_{obs} = \lambda_{rest} + \frac{v_r}{c} \lambda_{rest}$$

Both terms on the right contain λ_{rest}. Factoring it out makes the equation a little more convenient:

$$\lambda_{obs} = \left(1 + \frac{v_r}{c}\right) \lambda_{rest}$$

We are ready to plug in some numbers to solve for the observed wavelength:

$$\lambda_{obs} = \left(1 + \frac{-2.16 \times 10^4 \text{ m/s}}{3 \times 10^8 \text{ m/s}}\right) \times 517.27 \text{ nm}$$

$$\lambda_{obs} = 517.23 \text{ nm}$$

Although the observed Doppler blueshift ($517.23 - 517.27$) is only -0.04 nm, it is easily measured with modern instrumentation.

---

without greater precision, she would be unaware of the other planets in the Solar System. But, spurred on by the excitement of the discovery of Jupiter, the astronomer would improve the sensitivity of her instruments. If the astronomer could measure radial velocities as small as 2.7 m/s, she would be able to detect Saturn, and if the precision extended to radial velocities as small as 0.09 m/s, she would be able to detect Earth.

Astronomy in Action: Doppler Shift

Astronomy in Action: Center of Mass

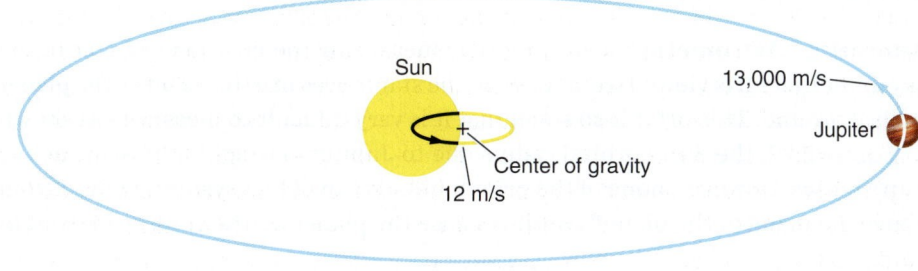

Figure 5.22 Both the Sun and Jupiter orbit around a common center of gravity, which lies just beyond the Sun's surface. Spectroscopic measurements made by an astronomer on an exoplanet would reveal the Sun's radial velocity varying by ±12 m/s over an interval of 11.86 years, which is Jupiter's orbital period. Jupiter travels around its orbit at a speed of 13,000 m/s.

▶▶ **Interactive Simulation:** Radial Velocity

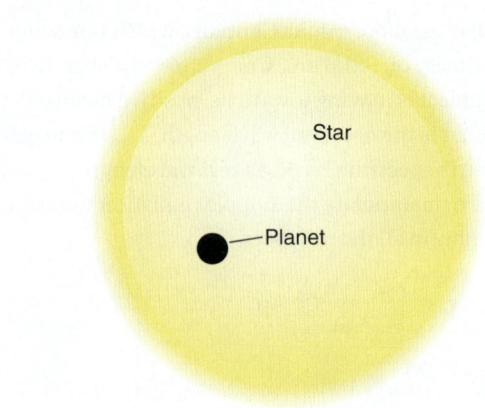

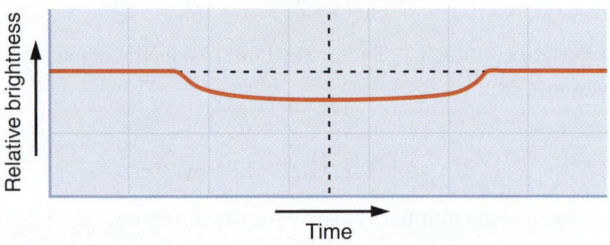

Figure 5.23 As a planet passes in front of a star, it blocks some of the light coming from the star's surface, causing the brightness of the star to decrease slightly. (The brightness decrease has been exaggerated in this illustration.)

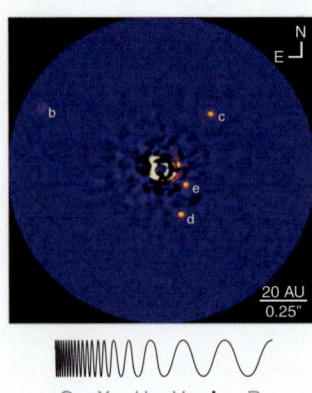

Figure 5.24 An infrared image shows four planets (labeled b, c, d, and e), each with a mass several times that of Jupiter, orbiting the star HR 8799 (hidden behind a mask).

The technique of examining Doppler shifts in the light from stars to detect exoplanets is called the **spectroscopic radial velocity method**. Because this method detects the radial velocity of the star, it cannot be used to find planets whose orbits we observe from "above." In those cases, the star moves around on the sky but does not move toward or away from us; the radial velocity of the star is always zero. Current technology limits the precision of radial velocity instruments to about 0.3 m/s, but it remains the most successful ground-based approach to finding exoplanets. This technique enables astronomers to detect giant planets around solar-type stars. Finding the signal of the Doppler shift in the noise of the observation requires the star to be quite bright in our sky. So this method is limited to nearby stars, within about 160 light-years from Earth. The transit method, discussed next, does not have this limitation.

The transit method. In the **transit method**, a planet passing in front of its parent star blocks a small amount of the light from the star, as illustrated in **Figure 5.23**. For a planet to pass in front of a star from our perspective, Earth must lie nearly in the orbital plane of the planet. This limitation is similar to the one for the spectroscopic radial velocity method, but the limitation here is much more severe. There is another important difference between the radial velocity and transit methods. Whereas the radial velocity method gives us the mass of the planet and its orbital distance from a star, the transit method provides the *size* of a planet. Current ground-based technology limits the sensitivity of the transit method to about 0.1 percent of a star's brightness.

Using the transit method, our aforementioned astronomer could infer the existence of Earth only if the astronomer was located in the plane of Earth's orbit and could detect a 0.009 percent drop in the Sun's brightness. In 2009, NASA launched a solar-orbiting telescope called Kepler with instruments that were able to detect transits of Earth-sized planets. This telescope found more than 4,000 exoplanet candidates in just a few short years. Using other methods, about 2,500 of these candidates have been confirmed to be planets. Many of these are "Earth-like"—they are about Earth's size, with about Earth's mass. A planet must also be in the "habitable zone," the range of distances from the central star that might allow for liquid water, to be considered Earth-like. The discovery of Earth-like planets is an exciting new development in the search for life in the universe. In 2013, the original Kepler mission came to an end due to an equipment failure, but the discovery of new planets is continuing as planet candidates identified by Kepler are confirmed. The Kepler mission was salvaged, running as "K2" until the mission finally, permanently ended in October 2018.

Gravitational lensing. The gravitational field of an unseen planet can bend light like a lens, an effect called **gravitational lensing**. If the planet passes in front of a very distant background star, it causes the star to brighten temporarily while the planet is passing in front of it. Because the effect is small, it is usually called **microlensing**. Like the transit method, lensing can detect Earth-sized planets. This method also provides an estimate of the mass of a planet.

Astrometry. **Astrometry** means precisely measuring the position of a star in the sky. If the system is viewed from "above," the star moves in a tiny orbit as the planet pulls it around. This orbit is so small that it is very difficult to measure—as shown in Figure 5.22, the Sun's orbital radius due to Jupiter is roughly the same as the Sun's radius. However, none of the prior methods can detect systems viewed from above the plane of the planet's orbit because the planet neither passes in front of the star nor causes a shift in its speed along the line of sight. Several space missions, such as the *Gaia* spacecraft, which was launched in 2013 by the European Space Agency, are carrying out observations of this kind.

Direct imaging.   Direct imaging means taking a picture of the planet directly. This technique is conceptually straightforward but is technically difficult because it requires searching for a faint planet in the overpowering glare of a bright star—a challenge far more difficult than looking for a star in the dazzling brilliance of a clear daytime sky. Even when an object is detected by direct imaging, the astronomer must still determine whether the observed object is actually a planet. Suppose we detect a faint object near a bright star. Could it be a more distant star that just happens to be in the line of sight? Future observations could tell if the object shares the bright star's motion through space. But it could also be a brown dwarf rather than a true planet. To distinguish between these two, the astronomer would need to make further observations to determine the object's mass.

Some planets have been discovered by this method using large, ground-based telescopes operating in the infrared region of the spectrum. **Figure 5.24** is an infrared image of four planets orbiting the star HR 8799. Planets can be somewhat easier to image in the infrared region of the spectrum because the star is typically fainter in this region of the spectrum, while the planet emits light in this region. These two effects combine so that, in the infrared region of the spectrum, the planet is relatively brighter compared to the star than it is in the visible-light region of the spectrum.

The first visible-light discovery was made from space while the Hubble Space Telescope was observing Fomalhaut, a bright naked-eye star only 25 light-years away from the Sun. The planet Fomalhaut b (formally named Dagon) is shown in **Figure 5.25**. Fomalhaut b has a mass no more than 3 times that of Jupiter and orbits within a dusty debris ring some 17 billion km from the central star (more than 100 times the distance of Earth from the Sun). A related form of direct observation involves separating the spectrum of a planet from the spectrum of its star to obtain information about the planet directly. Large ground-based telescopes have been able to obtain spectra of the atmospheres of a small number of exoplanets and have found, for example, that they contain carbon monoxide and water.

The most exciting discoveries will probably come with future missions. Future observatories will not only detect Earth-like planets around nearby stars but also measure the planets' physical and chemical characteristics.

Other Planetary Systems

As we noted at the beginning of this section, the search for exoplanets has been remarkably successful. Since the first exoplanet around a solar-type star was confirmed in 1995, astronomers have found thousands of stars with planets, many with multiple planets. **Figure 5.26** shows the distribution of confirmed planets as of October 2019. Many more candidates have yet to be confirmed as actual planets. Such confirmation requires follow-up observations that may take many years. The field is changing so fast that the most up-to-date information can only be found online.

The first discoveries included many **hot Jupiters**, which are Jupiter-type planets orbiting solar-type stars in tight circular orbits that are closer to their parent stars than Mercury is to our own Sun. These planets were among the first to be detected because they are relatively easy targets for the spectroscopic radial

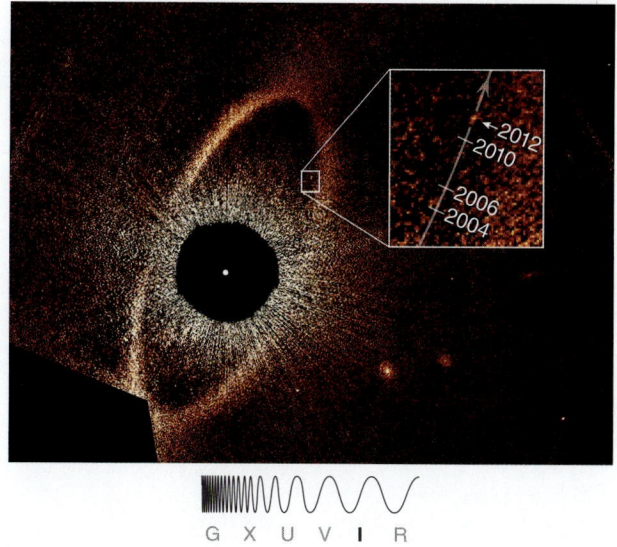

G X U V I R

Figure 5.25 Fomalhaut b (formally named Dagon) is seen here moving in its orbit around Fomalhaut, a nearby star easily visible to the naked eye. The parent star, hidden by an obscuring mask, is about a billion times brighter than the planet, which is located within a dusty debris ring that surrounds the star.

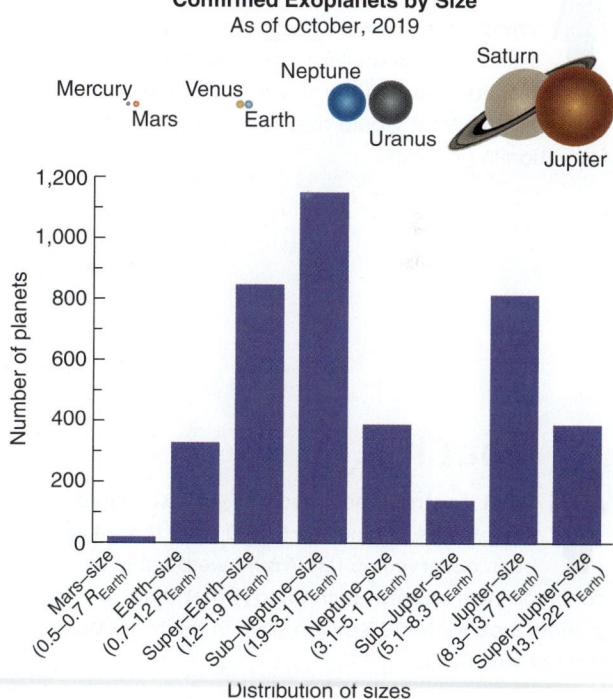

Figure 5.26 Planetary systems have been discovered around hundreds of stars other than the Sun, confirming what astronomers have long suspected—that planets are natural and common by-products of star formation. As time passes and technology improves, more smaller planets will be discovered. This graph shows the distribution of sizes (in Earth radii, R_{Earth}) of 2,325 planet candidates from the Kepler mission.

velocity method. Planets close to the central star have short periods, so they are easier to detect by all methods, because all methods rely on observing the effect of the planet more than once. Planets with longer orbital periods (and therefore larger orbits) can be discovered only when the observations have gone on long enough to observe more than one complete orbit. A close, massive planet also tugs the star very hard, creating large radial velocity variations in the star. Massive planets orbiting close to the central star are more likely to pass in front of the star than those on larger orbits, making them easier targets for the transit method as well. These hot Jupiter systems are easier to find than smaller, more distant planets, although they may turn out to be less common. This is an example of a *selection effect*: A type of object may be less common, but astronomers may find more of them because they are easier to observe.

Many astronomers were surprised by the hot Jupiters because the formation theory at the time (based only on our Solar System) predicted that these giant, volatile-rich planets should not have been able to form so close to their parent stars. Jupiter-type planets should form in the more distant, cooler regions of the accretion disk, where the volatiles that make up much of their composition are able to survive. Astronomers have added planet migration to the theory of system formation. Hot Jupiters form much farther away from their parent stars, but then interact with gas and small bodies in the disk, which transfers angular momentum. This interaction changes the orbit of the hot Jupiter, causing it to move inward or outward, depending on whether it gains or loses angular momentum.

Most of the new planets discovered by Kepler are mini-Neptunes (gaseous planets in the range of roughly 2–10 Earth masses) or super-Earths (rocky planets more massive than Earth but less massive than Neptune). Improvements in technology will enable even smaller planets to be found. Some of the newly discovered planets have highly elliptical orbits, unlike planets in the Solar System. Some orbits of newly discovered planets are highly inclined compared to the plane of rotation of their stars. Yet multiple-planet systems so far discovered tend to reside in flat systems like our own, possibly supporting the accretion disk theory of planet formation. In addition, planets have been found wandering freely through the Milky Way, unattached to any star. These planets may have been ejected from systems as they were forming.

Studies of planetary systems, many unlike our own, challenge some aspects of our understanding of planet formation. Nevertheless, the message conveyed by our discoveries is clear: The formation of planets frequently, and perhaps always, accompanies the formation of stars. The implications of this conclusion are profound. In a galaxy of more than 100 billion stars and a universe of hundreds of billions of galaxies, how many planets, or even moons, with Earth-like conditions might exist? And with all of these Earth-like worlds in the universe, how many might play host to the particular category of chemical reactions that we refer to as "life"? To answer these questions, we must become more familiar with our own Solar System, as we will do in the next few chapters.

what if . . .

What if astronomers had observed all other systems to be similar to our own, with rocky planets closer to the star and gas giants farther away: What would that tell us about star formation in general?

what if . . .

What if astronomers had not yet discovered any planets around the 5,000 stars that have been observed so far: Would we be able to conclude (with current technology) that there are no planets around the 100 billion stars in the Milky Way? Why or why not?

CHECK YOUR UNDERSTANDING 5.6

Planetary systems are probably

a. exceedingly common—nearly every star has planets.
b. common—about half the stars in our galaxy have planets.
c. rare—few stars have planets.
d. exceedingly rare—only one star has planets.

Citizen Scientists Discover Rare Exoplanet

By Ashley Strickland, CNN

K2-288Bb is about twice the size of Earth.

Although NASA's Kepler space telescope ran out of fuel and ended its mission in 2018, citizen scientists have used its data to discover an exoplanet 226 light-years away in the Taurus constellation.

The exoplanet, known as K2-288Bb, is about twice the size of Earth and orbits within the habitable zone of its star, meaning liquid water may exist on its surface. It's difficult to tell whether the planet is rocky like Earth or a gas giant like Neptune.

The planet is in the K2-288 system, which contains a pair of dim, cool M-type stars that are 5.1 billion miles apart, about six times the distance between Saturn and the sun. The brightest of the two stars is half as massive as our sun, and the other star is one-third of the sun's mass. K2-288Bb orbits the smaller, dimmer star, completing a full orbit every 31.3 days.

K2-288Bb is half the size of Neptune or 1.9 times the size of Earth, placing it in the "Fulton gap" between 1.5 and two times the size of Earth. This is a rare size of exoplanet that makes it perfect for studying planetary evolution because so few have been found.

The discovery was announced Monday at the 233rd meeting of the American Astronomical Society in Seattle.

"It's a very exciting discovery due to how it was found, its temperate orbit and because planets of this size seem to be relatively uncommon," said Adina Feinstein, a University of Chicago graduate student in astrophysics and lead author of a paper describing the new planet that was accepted for publication by The Astronomical Journal.

Although all of the data from the Kepler mission was run through an algorithm to determine potential planet candidates, visual manpower was needed to actually look at the possible planet transits—or dip in light when a planet passes in front of its star—in the light curve data. Kepler observed other events that could be mistaken for planet transits by a computer.

But the "reboot" of the Kepler mission in 2014 that led to the K2 mission allowed for multiple observation campaigns that brought in even more data. Every three months, Kepler would stare at a different patch of sky.

"Reorienting Kepler relative to the Sun caused miniscule [sic] changes in the shape of the telescope and the temperature of the electronics, which inevitably affected Kepler's sensitive measurements in the first days of each campaign," said study co-author Geert Barentsen, an astrophysicist at NASA's Ames Research Center, in a statement.

Those first three days of data were ignored, and errors were corrected in the rest of the data gathered.

But the scientists couldn't do it alone. There were too many light curves to study on their own.

So the reprocessed, "cleaned-up" light curves were uploaded through the Exoplanet Explorers project on online platform Zooniverse, and the public was invited to "go forth and find us planets," Feinstein said.

In May 2017, citizen scientists began discussing a particular planet candidate, but it had only two transits, or passes of the planet in front of its star. The scientists needed at least three to mark it as an interesting target.

Feinstein and Makennah Bristow, an undergraduate student at the University of North Carolina Asheville, worked as interns at NASA's Goddard Space Flight Center, searching the data for transits. They had noticed the same system and its two transits.

But the citizen scientists found the third transit hiding in those first few days of data that had been all but forgotten.

"That's how we missed it—and it took the keen eyes of citizen scientists to make this extremely valuable find and point us to it," Feinstein said.

Follow-up observations were made with multiple telescopes to confirm the exoplanet.

There will be more opportunities for citizen scientists to help discover exoplanets. NASA's latest planet-hunting mission, TESS, will be providing more light curves that are full of potential planets waiting to be found.

Last year at the American Astronomical Society meeting, it was announced that citizen scientists helped discover five planets between the size of Earth and Neptune around star K2-138, the first multiplanet system found through crowdsourcing.

This year, Kevin Hardegree-Ullman, postdoctoral scholar in astronomy at the California Institute of Technology, announced that the Spitzer space telescope followed up on that discovery and discovered a sixth planet, K2-138 g, smaller than Neptune, that orbits the star every 42 days.

"This is only the ninth system discovered containing six or more planets," he said.

EVALUATING THE NEWS

1. What planet detection method was used by the Kepler mission?
2. Why were the first three days of data "ignored" by astronomers?
3. What is a "citizen scientist"?
4. Why are citizen scientists needed for a project like the Kepler mission?
5. Look online to see if this project is still available on Zooniverse, and find out if you could participate as a citizen scientist!

SUMMARY

Stars and planets form from clouds of dust and gas. These clouds collapse under their own gravity, sometimes assisted by external events, such as nearby exploding stars. As the clouds collapse, they fragment to form multiple protostars. Conservation of angular momentum produces an accretion disk around the protostars that often further fragments to form multiple planets, as well as smaller objects such as asteroids and dwarf planets, through the gradual accumulation of dust into larger and larger objects. There are multiple methods for finding planets around stars beyond our Sun, and these exoplanets are now thought to be very common. This field of study is evolving very quickly as technology advances.

(1) Gravity pulls clumps of gas and dust together, causing them to compress and heat up. Angular momentum must be conserved, leading to both a spinning central star and an accretion disk that revolves in the same direction as the central star.

(2) Dust grains in the accretion disk first stick together because of collisions and static electricity. As these objects grow, they eventually have enough mass to capture other objects gravitationally. Once this occurs, they begin emptying the space around them. Collisions of these planetesimals can lead to the formation of planets.

(3) As particles orbit the forming star, those on rising tracks impact those on falling tracks. The upward and downward motions cancel, and the cloud of dust and gas flattens into a plane. Conservation of angular momentum determines both the speed and the direction of the revolution of the objects in the forming system.

(4) The temperature is higher near the central protostar. This forces volatile elements, like water, to evaporate and leave the inner part of the disk. Planets in the inner part of the system will have fewer volatiles than those in the outer part of the disk.

(5) Astronomers find planets around other stars using a variety of methods: the radial velocity method, the transit method, gravitational lensing, astrometry, and direct imaging. Although the first exoplanets were not discovered until the 1990s, thousands of exoplanets and exoplanet candidates have been discovered in just the past few years. Astronomers now know of planetary systems of all kinds: systems with many planets, systems that are much like the Solar System, and systems that are completely unlike the Solar System. This great diversity, and the remarkable rate of discovery, implies that planets are very common around stars.

QUESTIONS AND PROBLEMS

TEST YOUR UNDERSTANDING

1. **T/F:** All molecular clouds are held together solely by gravity.

2. **T/F:** Gravity and angular momentum are both important in the formation of planetary systems.

3. **T/F:** A protostar has nuclear reactions inside.

4. **T/F:** Volatile materials are solid only at low temperatures.

5. **T/F:** The Solar System formed from a giant cloud of dust and gas that collapsed under gravity.

6. Figure 5.5 shows a number of curves for objects of different temperatures. Suppose you observe a star with a temperature of 4500 K. What color would the peak wavelength be? ⊙
   a. red
   b. orange
   c. yellow
   d. green
   e. blue

7. Figure 5.5 shows a number of curves for objects of different temperatures. Suppose you observe a star with a temperature of 10,000 K. Where would its spectrum lie on this graph? ⊙
   a. below the $T = 2000$ K curve
   b. between the $T = 3000$ K curve and the $T = 4000$ K curve
   c. between the $T = 4000$ K curve and the $T = 6000$ K curve
   d. above the $T = 6000$ K curve

8. Figure 5.6 shows a molecular cloud in which stars are currently forming. Of the three visible pillars in this image, one is bright and two are dark. From front to back, how are these pillars and the light source arranged? ⊙
   a. colored pillar, light source, black pillars
   b. black pillars, light source, colored pillar
   c. light source, colored pillar, black pillars
   d. colored pillar, black pillars, light source

9. The direction of revolution in the plane of the Solar System was determined by
   a. the plane of the galaxy in which the Solar System sits.
   b. the direction of the gravitational force within the original cloud.
   c. the direction of rotation of the original cloud.
   d. the amount of material in the original cloud.

10. The radial velocity method preferentially detects
   a. more massive planets close to the central star.
   b. small planets close to the central star.
   c. more massive planets far from the central star.
   d. small planets far from the central star.
   e. none of the above (The method detects all of these equally well.)

11. Nuclear reactions require very high _____ and _____.
   a. temperature; density
   b. volume; density
   c. density; area
   d. mass; area
   e. temperature; mass

12. The transit method preferentially detects
    a. large planets close to the central star.
    b. small planets close to the central star.
    c. large planets far from the central star.
    d. small planets far from the central star.
    e. none of the above (The method detects all of these equally well.)

13. Which of the following are true? An "Earth-like" planet
    a. has life on it.
    b. has water on it.
    c. has physical properties similar to Earth's.
    d. orbits a Sun-like star.

14. A planet in the "habitable zone"
    a. is close to the central star.
    b. is far from the central star.
    c. is the same distance from its star as Earth is from the Sun.
    d. is at a distance where liquid water can exist on the surface.
    e. is extremely rare—none has yet been found.

15. Molecular clouds collapse because of
    a. gravity.
    b. angular momentum.
    c. static electricity.
    d. nuclear reactions.

16. Because angular momentum is conserved, an ice-skater who throws her arms out will
    a. rotate more slowly.
    b. rotate more quickly.
    c. rotate at the same rate.
    d. stop rotating entirely.

17. Clumps grow into planetesimals by
    a. gravitationally pulling in other clumps.
    b. colliding with other clumps.
    c. attracting other clumps with opposite charge.
    d. both a and b

18. The terrestrial planets and the giant planets have different compositions because
    a. the giant planets are much larger.
    b. the terrestrial planets are closer to the Sun.
    c. the giant planets are mostly made of solids.
    d. the terrestrial planets have few moons.

19. Which of the following planets still has its primary atmosphere?
    a. Mercury
    b. Earth
    c. Mars
    d. Jupiter

20. Exoplanets have been detected by
    a. the spectroscopic radial velocity method.
    b. the transit method.
    c. the gravitational lensing method.
    d. the direct imaging method.
    e. all of the above methods

THINKING ABOUT THE CONCEPTS

21. Compare the size of our Solar System with the size of the universe.

22. ★ WHAT AN ASTRONOMER SEES In Figure 5.1, identify an unlabeled finger, jet, and diffraction spike. Make a sketch of the image, and label the locations of the features you have identified. ⊙

23. Examine Figure 5.8. Suppose that the universe were different, and in step 2 the shrinking star caused the gravity to decrease, instead of increase, so that the pressure arrow became longer than the gravity arrow. What would this mean for the formation of stars? ⊙

24. Physicists describe certain properties, such as angular momentum and energy, as being "conserved." What does this mean? Do these conservation laws imply that an individual object can never lose or gain angular momentum or energy? Explain your reasoning.

25. How does the law of conservation of angular momentum control a figure-skater's rate of spin?

26. Look under your bed for "dust bunnies." If there aren't any, look under your roommate's bed, the refrigerator, or any similar place that might have some. Once you find them, blow one toward another. Watch carefully and describe what happens as they meet. What happens if you repeat this with another dust bunny? Will these dust bunnies ever have enough gravity to begin pulling themselves together? If they were in space, instead of on the floor, might that happen? What force prevents their mutual gravity from drawing them together into a "bunny-tesimal" under your bed?

27. How does the image of Mercury in Figure 5.19 support the idea that many, many planetesimals were once zooming around the early Solar System? Some of the large craters have smaller craters inside them. Which happened first, a larger crater or the smaller ones? How do you know? What does this tell you about their relative ages? This reasoning becomes extremely important in the next chapter. ⊙

28. What are the two reasons why the inner part of an accretion disk is hotter than the outer part?

29. Why were the four giant planets able to collect massive gaseous atmospheres, whereas the terrestrial planets could not?

30. Explain the fate of the original atmospheres of the terrestrial planets.

31. What happened to all the leftover Solar System debris after the last of the planets formed?

32. Redraw Figure 5.22 looking straight down on the system. Now draw a series of pictures from that same orientation, showing one complete orbit of Jupiter around the Sun. Label the motions of the Sun and Jupiter (toward, away, neither) as they would be viewed by an observer off the page to the right. Are the Sun and Jupiter ever on the same side of the center of mass? ⊙

33. Redraw Figure 5.23, paying close attention to where the line on the graph drops in brightness. Now add three more graphs. In the first, show what happens to the light curve if the planet crosses much closer to the bottom of the star. In the second, show what happens to the light curve if the planet is much larger than the one in Figure 5.23. In the third, show what happens to the light curve if the planet crosses the precise middle of the star, but from top to bottom instead of from side to side. ⊙

34. Early in the discovery of exoplanets, many of the planets that astronomers found orbiting other stars were giant planets with masses more like that of Jupiter than of Earth and with orbits located very close to their parent stars. How did this affect our understanding of the formation of planetary systems?

35. Step outside and look at the nighttime sky. Depending on the darkness of the sky, you may see dozens or hundreds of stars. Would you expect many or very few of those stars to be orbited by planets? Explain your answer.

APPLYING THE CONCEPTS

36. Review Kepler's third law. Suppose a planet has been found around a star with the same mass as the Sun. The planet's orbital period is 200 days.
  a. Make a prediction: If this planet were in our Solar System, would you expect it to be closer or farther from the Sun than Earth is?
  b. Use Kepler's third law to find the semimajor axis of the orbit of this exoplanet.
  c. Check your work by comparing your result to your prediction and to the orbits of Mercury and Mars.
  d. What environmental conditions must this planet experience?

37. Suppose a very young star has a peak wavelength of 0.97×10^{-6} meter. Find the star's temperature.
  a. Make a prediction by reasoning this way: In what region of the spectrum is this peak wavelength? Does this region of the spectrum have shorter or longer wavelengths than those in the visible part of the spectrum? Do you expect to calculate that this star will be hotter or colder than the Sun, which is 5800 K?
  b. Calculate the star's temperature using Wien's law.
  c. Check your work by comparing your answer to your prediction.

38. Suppose a planet has a temperature of 400 K, and it has a radius of 2.0 Earth diameters.
  a. Make a prediction: Compare this temperature to the temperature of Earth, as given in Working It Out 5.1. Is this planet hotter or cooler than Earth? Do you expect it to have a larger or smaller peak wavelength? Do you expect it to emit more light or less?
  b. Use Wien's law to find the peak wavelength of the planet's emission. Does the emission peak in the visible part of the spectrum?
  c. Use the Stefan-Boltzmann law to determine the energy emitted each second from a square meter of the planet's surface (the intensity).
  d. Find the planet's luminosity. Recall that the surface area of a sphere is $4\pi r^2$.
  e. Check your work by comparing your answers to the values for Earth, found in Working It Out 5.1. Do your answers make sense, given the temperature of this planet compared to Earth's?

39. You observe a spectral line of hydrogen at a wavelength of 502.3 nm in a distant star. The rest wavelength of this line is 486.1 nm.
  a. Make a prediction: Compare the observed wavelength to the rest wavelength. Is the observed wavelength longer or shorter than the rest wavelength? Is the star's light redshifted or blueshifted? Is the star moving toward or away from Earth? Compare the difference in wavelength for this star to the difference in wavelength for the star in Working It Out 5.2. Do you expect that this star moves faster or slower than that star?
  b. Use the Doppler effect to find the radial velocity of this star.
  c. Check your work by comparing your answer to Working It Out 5.2. Does the sign (positive or negative) of your answer agree with your prediction? Is the numerical value less than the speed of light 3×10^8 m/s? (Every speed you ever calculate should be less than the speed of light.) Is the speed you have calculated reasonable, when compared to the speed calculated in Working It Out 5.2?

40. Earth tugs the Sun around as it orbits, but it affects the radial velocity of the Sun by only 0.09 m/s. How large a shift in wavelength does this cause in the Sun's spectrum at 575 nm?
  a. Make a prediction by comparing this problem with the problem in Working It Out 5.2.
  b. Calculate the shift in wavelength.
  c. Check your work by comparing both the sign and the size of your answer to the solved problem in Working It Out 5.2.

exploration Exploring Exoplanets

digital.wwnorton.com/universe4

Visit the Student Site at the Digital Resources page and open the Radial Velocity Simulation in Chapter 5. This applet has a number of different panels that allow you to experiment with the variables that are important for measuring radial velocities. Compare the views shown in the various panels with the colored arrows in the first panel to see where an observer would stand to see the view shown. Start the animation (in the "Animation Controls" panel) and allow it to run while you watch the planet orbit its star from each of the views shown. Stop the animation, and in the "Sample Systems" panel select "Option A."

1. Is Earth's view of this system most nearly like the "side view" or most nearly like the "orbit view"?

_____

2. Is the orbit of this planet circular or elongated?

_____

3. Study the radial velocity graph in the upper right panel. The blue curve shows the radial velocity of the star over a full period. What is the maximum radial velocity of the star?

_____

_____

4. The horizontal axis of the graph shows the "phase," or fraction of the period. A phase of 0.5 is halfway through a period. The vertical red line indicates the phase shown in views in the upper left panel. Start the animation to see how the red line sweeps across the graph as the planet orbits the star. The period of this planet is 365 days. How many days pass between the minimum radial velocity and the maximum radial velocity?

_____

5. When the planet moves away from Earth, the star moves toward Earth. The sign of the radial velocity tells the direction of the motion (toward or away). Is the radial velocity of the star positive or negative at this time in the orbit? If you could graph the radial velocity of the planet at this point in the orbit, would it be positive or negative?

_____

_____

_____

In the "Sample Systems" window, select "Option B:"

6. What has changed about the orbit of the planet as shown in the views in the upper left panel?

_____

_____

_____

7. When is the planet moving fastest—when it is close to the star or when it is far from the star?

_____

_____

8. When is the star moving fastest—when the planet is close to it or when it is far away?

_____

_____

9. Explain how an astronomer would determine, from a radial velocity graph of the star's motion, whether the orbit of the planet was in a circular or elongated orbit.

_____

_____

_____

10. Study the Earth view panel at the top of the window. Would this planet be a good candidate for a transit observation? Why or why not?

_____

_____

_____

Terrestrial Worlds in the Inner Solar System

The Moon has many visible surface features that can be used to reveal its history. Use the Internet to find out the date of the next full Moon. Go outside on a clear night within 3 days of the full Moon (either before or after), and sketch the face of the Moon. You may want to look at the Moon through a cardboard tube, such as a toilet paper roll or paper towel roll, or a rolled-up piece of paper. The tube will block some of the ambient light and focus your attention on the surface of the Moon itself, thus helping you to see more detail. Use a map of the Moon to identify at least two mare and two craters and label them on your sketch.

EXPERIMENT SETUP

Use the Internet to find out the date of the next full Moon.

Go outside on a clear night within three days of the full Moon (either before or after), and sketch the face of the Moon.

You may want to look at the Moon through a cardboard tube or a rolled-up piece of paper to block some of the ambient light and focus your attention on the surface of the Moon itself, thus helping you see more detail.

Use a map of the Moon to identify at least two mare and two craters on the surface of the Moon. Label them on your sketch.

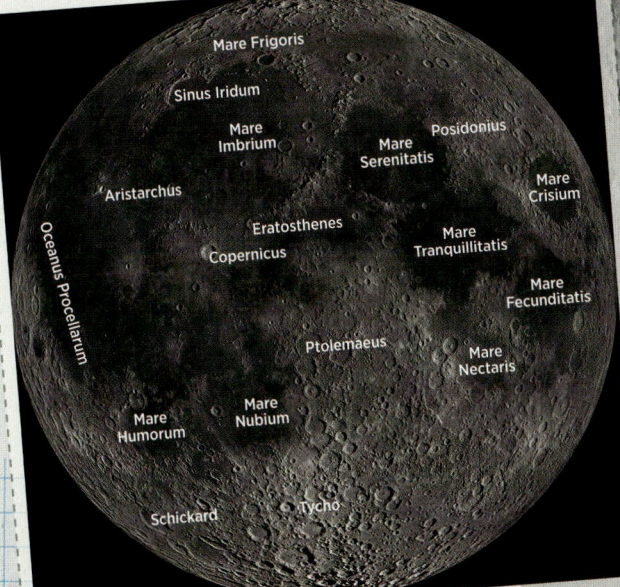

SKETCH OF RESULTS
(in progress)

Mare Crisium

Mare Fecunditatis

6

For most of human history, Earth has seemed completely different from the stars and other planets. Over time, scientists realized that Earth is a planet like any other: small in the astronomical context and changing over the vast timescales of geologic time. The four innermost planets in our Solar System—Mercury, Venus, Earth, and Mars—are known collectively as the *terrestrial planets*. Although the Moon is not a planet—it's Earth's only natural satellite—it has many similarities to the terrestrial planets, so it is useful to explore it at the same time.

When comparing planets, we first compare the basic physical properties, such as distance from the Sun, size, density, and gravitational pull at the surface. These physical properties affect the geological properties and evolution of the planet. For example, a planet with high surface gravity is more likely to retain its atmosphere than a planet with low surface gravity. **Table 6.1** compares the basic physical properties of the terrestrial planets and the Moon. By comparing the terrestrial planets, astronomers have learned that four processes shape these planets: *impact cratering, tectonism, volcanism,* and *erosion*. In this chapter, you'll explore the interiors and surfaces of the Solar System's terrestrial worlds to learn how these four processes shape each of them.

▶❚❚ **AstroTour:** Processes That Shape the Planets

Table 6.1 Comparison of Physical Properties of the Terrestrial Planets and the Moon

Property	Mercury	Venus	Earth	Mars	Moon
Orbital semimajor axis	0.387 AU	0.723 AU	1.000 AU	1.524 AU	384,000 km
Orbital period[†]	0.241 year	0.615 year	1.000 year	1.881 years	27.32 days
Orbital velocity (km/s)	47.4	35.0	29.8	24.1	1.02
Mass (in units of M_{Earth})	0.055	0.815	1.000	0.107	0.012
Equatorial radius (km)	2,440	6,052	6,378	3,397	1,738
Equatorial radius (in units of R_{Earth})	0.383	0.949	1.000	0.533	0.272
Density (relative to water = 1)	5.43	5.24	5.51	3.93	3.34
Rotation period[†]	58.65^d	243.02^d	23^h 56^m	24^h 37^m	27.32^d
Tilt of axis (degrees)[‡]	0.04	177.36	23.45	25.19	6.68
Surface gravity (m/s²)	3.70	8.87	9.78	3.71	1.62
Escape velocity (km/s)	4.25	10.36	11.18	5.03	2.38

[†]The superscript letters d, h, and m stand for days, hours, and minutes of time, respectively.
[‡]A tilt greater than 90° indicates that the planet rotates in a retrograde, or backward, direction.

LEARNING GOALS

① List the four processes involved in the evolution of a terrestrial planet.

② Identify the relative ages of parts of a planet's surface from observations of craters, and explain how radiometric dating can be used to find the ages of rocks.

③ Explain how scientists combine models of planetary interiors and direct observation to determine the internal structure of planets.

④ Describe tectonism and volcanism and the forms they take on each of the terrestrial planets and the Moon.

⑤ Describe how erosion modifies and wears down surface features.

6.1 Impacts Help Shape the Terrestrial Planets

A snapshot taken in December 1968 by *Apollo 8* astronauts looking back at Earth while orbiting the Moon (**Figure 6.1**) allows us to visually compare these two worlds. The most noticeable features on the surface of the Moon are the craters resulting from collisions of the Moon with other rocks in space. Earth's surface has very few visible craters. In this section, you will learn how these craters form and how we can use them to determine the age of a surface. In a later section, you will learn why so few craters remain visible on Earth.

Impacts and Craters

An **impact** is a collision between two bodies. The Solar System is currently dominated by a few large objects in stable orbits and many smaller ones, so the most likely impact with a terrestrial world is from a relatively small object. The space rocks that cause these impacts are referred to by three closely related terms: meteoroids, meteors, and meteorites. **Meteoroids** are small (ranging from tiny, sandlike grains up to 100 meters) cometary or asteroid fragments in space. A meteoroid that enters and burns up in a planetary atmosphere is called a **meteor**. Any meteoroids that survive to hit the ground are known as **meteorites**.

An impact can form a distinctive round surface scar known as an **impact crater** in a process known as **impact cratering**. Large impacts cause a concentrated and sudden release of large amounts of energy, and they tend to leave the largest craters. Planets and other objects orbit the Sun at very high speeds; Earth, for example, orbits at an average speed of about 30 kilometers per second (km/s), equivalent to 67,000 miles per hour (mph).

When an object hits a planet, the energy it releases heats and compresses the surface and ejects material, throwing it far from the resulting impact crater, as shown in **Figure 6.2a**. This material sometimes forms **rays**—bright streaks pointing away from a young crater. Sometimes the ejected material falls back to the surface of the planet with enough energy to cause more scars known as **secondary craters**. Craters on the Moon's surface are often surrounded by strings of smaller secondary craters formed from material thrown out by the original impact, like those shown in Figure 6.2b. Rock at the bottom of the crater is melted, or sometimes even **vaporized**, and forms a smooth floor interior to the crater

Figure 6.1 In December 1968, *Apollo 8* astronauts in orbit photographed Earth rising above the Moon. ★ **WHAT AN ASTRONOMER SEES** An astronomer, viewing this image, will be as awestruck as anyone by the simple fact that humans managed to figure out how to orbit the Moon and take such a picture. Then she will start to look at details. She will notice that the Moon has topography. There are elevation changes, particularly along the horizon, and large and small circular craters that vary in brightness. And she will notice that Earth is very different from the Moon. Studying Earth, she will notice blue water and brown land, as well as swirling cloud formations that indicate an atmosphere with active weather patterns. The Earth is only partially illuminated by the Sun, and that illumination shows that the Sun is located beyond the top of the image and slightly to the right. Because Earth is approximately in quarter phase, the Moon must likewise be approximately in quarter phase. An astronomer may spend considerable time trying to decipher which of the continents are visible on Earth, to figure out whether Earth's North Pole is on the right or the left, deciding eventually that north is on the right. From this information, she can make a mental sketch and figure out that the Moon was just about at first quarter, which is very satisfying.

VOCABULARY ALERT

vaporize In everyday language, to *vaporize* often means "to destroy completely," suggesting even that the stuff the object was made of is not there anymore. Astronomers use this word specifically to mean "turn into vapor." In this case, solid rocks are being turned into gas by the energy of the collision. It takes a lot of energy to do that.

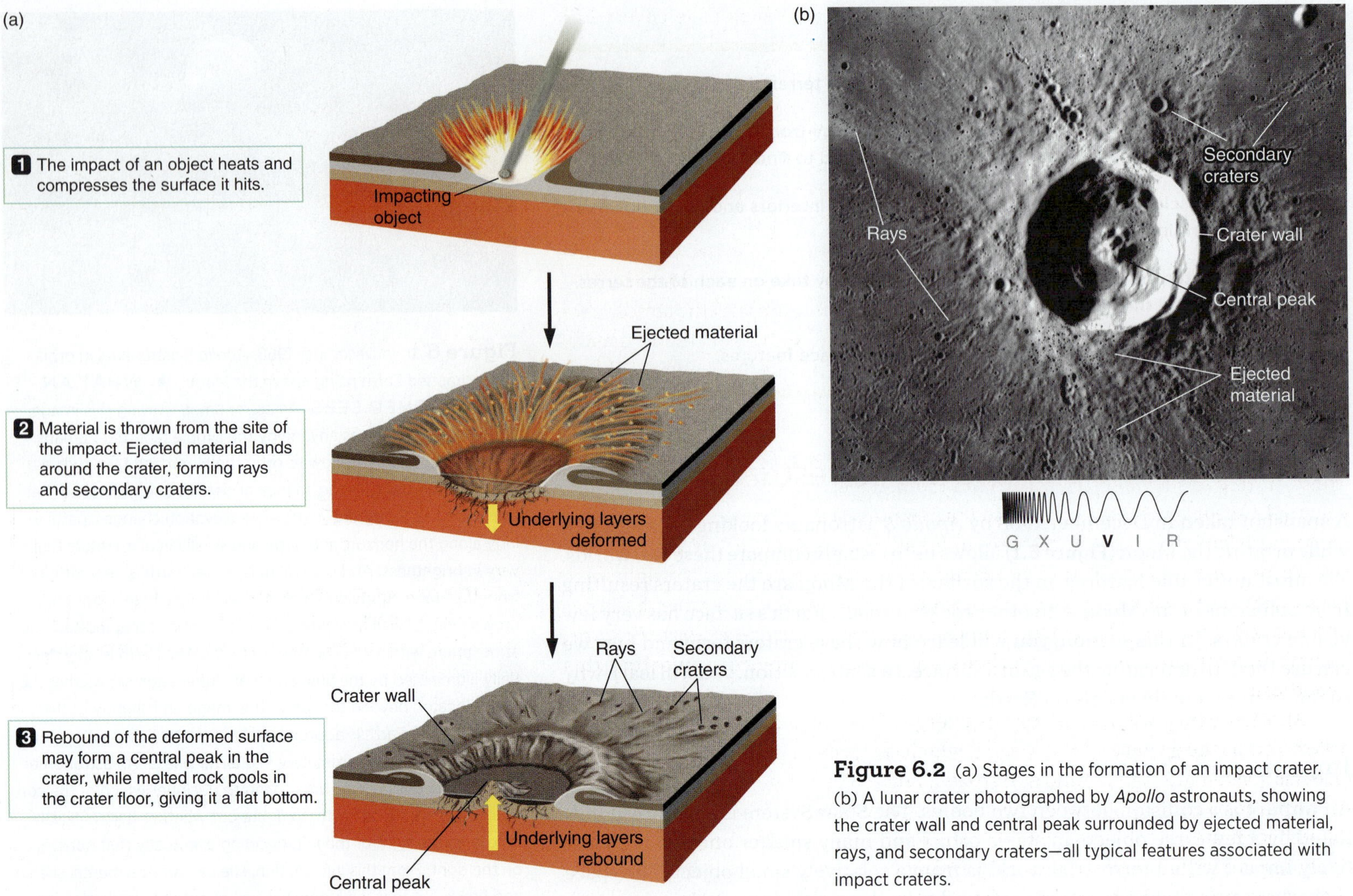

(a)

1 The impact of an object heats and compresses the surface it hits.

Impacting object

Ejected material

2 Material is thrown from the site of the impact. Ejected material lands around the crater, forming rays and secondary craters.

Underlying layers deformed

Rays   Secondary craters

Crater wall

3 Rebound of the deformed surface may form a central peak in the crater, while melted rock pools in the crater floor, giving it a flat bottom.

Underlying layers rebound

Central peak

(b)

Secondary craters

Rays

Crater wall

Central peak

Ejected material

G X U V I R

Figure 6.2 (a) Stages in the formation of an impact crater. (b) A lunar crater photographed by *Apollo* astronauts, showing the crater wall and central peak surrounded by ejected material, rays, and secondary craters—all typical features associated with impact craters.

Figure 6.3 Meteor Crater (also known as Barringer Crater), located in northern Arizona, is an impact crater 1.2 km in diameter. It was formed some 50,000 years ago by the collision of a nickel-iron asteroid fragment with Earth.

walls. The rebound of heated and compressed material may form a central peak or a ring of mountains. All of these features can be seen in the lunar crater in Figure 6.2b. The energy released in an impact can also lead to the formation of new minerals. Because certain minerals form *only* during an impact, finding them on Earth is evidence of ancient impacts on Earth's surface.

One of the best-preserved impact craters on Earth is Meteor Crater in Arizona (**Figure 6.3**). This crater was caused by an impact about 50,000 years ago. From the crater's size and from the remaining pieces of the impacting body, we know that a nickel-iron asteroid fragment about 50 meters across, with a mass of about 300 million kilograms (kg), was traveling at 13 km/s when it hit the upper atmosphere. Approximately half of the original mass was vaporized in the atmosphere before the remainder hit the ground. This collision released about 300 times as much total energy as an atomic bomb. Yet, at only 1.2 km in diameter, Meteor Crater is tiny compared with impact craters seen on the Moon and with more ancient impact scars on Earth.

Impact craters cover the surfaces of Mercury, Mars, and the Moon. On Earth and Venus, by comparison, most impact craters have been obliterated. Fewer than 200 impact craters have been identified on Earth, and about 1,000 have been found on Venus. Nearly all of Earth's craters have been erased by plate tectonics in Earth's ocean basins and erosion on land, while lava flows on Venus have covered its craters.

Look at Table 6.1 at the row labeled "Surface gravity." Note the much higher values for Venus and Earth compared to the other three worlds. This greater surface gravity has enabled Venus and Earth to keep their atmospheres, which partially protect their surfaces from impacts. Because they have weaker surface gravity, Mars, the Moon, and Mercury have lost their atmospheres and do not have such protection. This is another reason for the shortage of craters on Earth and Venus. Rock samples from the Moon show craters smaller than a pinhead, formed by micrometeoroids. In contrast, most meteoroids smaller than 100 meters in diameter that enter Earth's atmosphere either burn up or break up by friction with the atmosphere before they reach Earth's surface. Small meteorites found on Earth are usually pieces of much larger bodies that broke up upon entering the atmosphere. With an atmosphere far thicker than that of Earth, Venus's surface is even better protected against impacts.

Some craters on Mars have a very different appearance from those on the Moon. These craters have smooth flows around them, which look much like the pattern you might see if you threw a rock into mud (**Figure 6.4**). This is evidence that the martian surface rocks contained water or ice at the time of impact. Not all martian craters look like this, so the water or ice must have been concentrated in only some areas, and these icy locations might have changed with time. At the time these craters formed, there may have been liquid water on the surface of Mars. Features resembling canyons and dry riverbeds are further evidence supporting this hypothesis.

Another explanation relies on the energy of the impact itself to create liquid water. Today, the surface of Mars is dry, which suggests that any water on the surface has soaked into the ground and frozen, much like water frozen in the ground in Earth's polar regions. The energy released by an impact could melt this ice, giving the surface material the consistency of wet concrete. When thrown from the crater by the force of the impact, this wet material would hit the surrounding terrain and slide across the surface, forming the flow features around the craters.

Calibrating a Cosmic Clock

Recall that, during the formation of the Solar System, small bodies of rock called "planetesimals" formed. Because many of these planetesimals remained in the early Solar System, every planet experienced a large number of impacts during that early time period, which is known as the *period of heavy bombardment*. All of the terrestrial planets were heavily cratered by the end of the period of heavy bombardment. The number of visible craters on a planet is determined not only by the number of craters that were created but also by the rate at which those craters are destroyed. Geological activity, like tectonics and erosion, erased most of the craters on Earth, Mars, and Venus. The Moon's surface still preserves the scars of craters dating from about 4 billion years ago. The lunar surface has remained essentially unchanged for more than a billion years because the Moon has no atmosphere or surface water and no geological activity. Mercury also has well-preserved craters, although evidence from the *Messenger* mission shows crater floors that are higher on one side than the other—evidence that internal forces lifted the floors unevenly after the craters formed. As time passes, more craters are erased, so planetary scientists use the cratering record to estimate the relative **age** of planetary surfaces—extensive cratering signifies an older planetary surface.

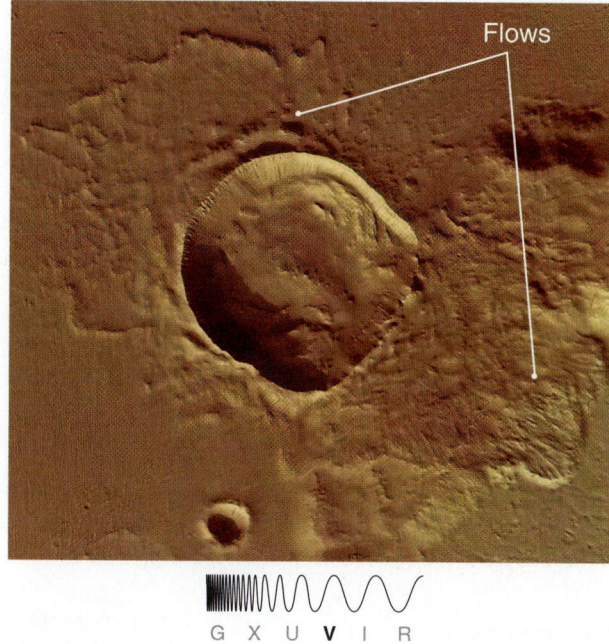

Flows

G X U V I R

Figure 6.4 Some craters on Mars look like those formed by rocks thrown into mud, suggesting that material ejected from the crater contained large amounts of water. This crater is about 20 km across.

what if . . .

What if Earth did not have an atmosphere: Would we expect Earth's surface to have the same cratering as the Moon?

VOCABULARY ALERT

age In everyday language, we usually consider the *age* of a planet to be the length of time since it formed. The age of the surface of a planet can be much younger than that, however, due to volcanism or tectonics moving material from the interior of the planet to the surface, for example. It is important, then, to distinguish between the age of the planet and the age of its surface. This is not as strange as you may think. You are much older than the outermost layer of your skin, which is replaced about every month.

We can use the amount of cratering to measure the *relative* ages of surface features on a planet, because if a planet's surface is very smooth, it must be younger than a surface that has been around long enough to be heavily cratered. In addition, features that lie on top of other features must have happened more recently. For example, if a lava flow has eroded the edge of a crater, the crater must be older than the lava flow. But this reasoning only tells us which surfaces are older than others. It cannot be used to determine how old they actually are, in years. We must use another method to find the age of at least one (and preferably more than one) surface in the Solar System. This "calibrates" the clock; once the actual ages of a few surfaces are known, the relative ages of all other surfaces can be converted to actual ages. The method to find the age of a surface is called "radioisotope dating," and to understand it, you need to learn a little bit about nuclear physics and isotopes.

Recall that an atom is made of a nucleus surrounded by a swirl of electrons. Within the nucleus are two types of particles: protons and neutrons. All atoms of an element have the same number of protons. For example, every helium atom has two protons. Isotopes of an element always have the same number of protons, but they have different numbers of neutrons. For example, a helium atom with two protons and one neutron is an isotope of helium. So is a helium atom with two protons and two neutrons, or two protons and three neutrons. Some isotopes (called "parents") are unstable. These isotopes naturally decay, emitting alpha, beta, or gamma radiation: alpha particles are helium nuclei, beta particles are electrons or their antimatter counterparts, and gamma "particles" are high-energy light. This radioactive decay leaves behind new isotopes or elements (called "daughters"). Over time, the ratio of parent atoms to daughter atoms changes, as the number of parents decreases and the number of daughters grows, as explained in more detail in **Working It Out 6.1**. The process of finding ages from these ratios is called **radiometric dating**. This process is also used, typically with isotopes of carbon, to find the ages of bones on Earth.

Between 1969 and 1976, *Apollo* astronauts and Soviet unmanned probes visited the Moon and brought back samples from nine different locations on the lunar surface. The results from analysis of these samples were surprising. Although smooth areas on the Moon were indeed younger than heavily cratered areas, they were still very old. The oldest, most heavily cratered regions on the Moon date back to about 4.4 billion years ago, whereas most of the smoother parts of the lunar surface are typically 3.1 billion to 3.9 billion years old. As shown in the graph in **Figure 6.5**, the cratering rate dropped abruptly between 4.5 billion and about 3.5 billion years ago. Almost all of the cratering in the Solar System took place within its first billion years.

Figure 6.5 Radiometric dating of lunar samples returned from specific sites by *Apollo* astronauts is used to determine how the cratering rate has changed over time. Cratering records can then be used to tell us the age of other parts of the lunar surface.

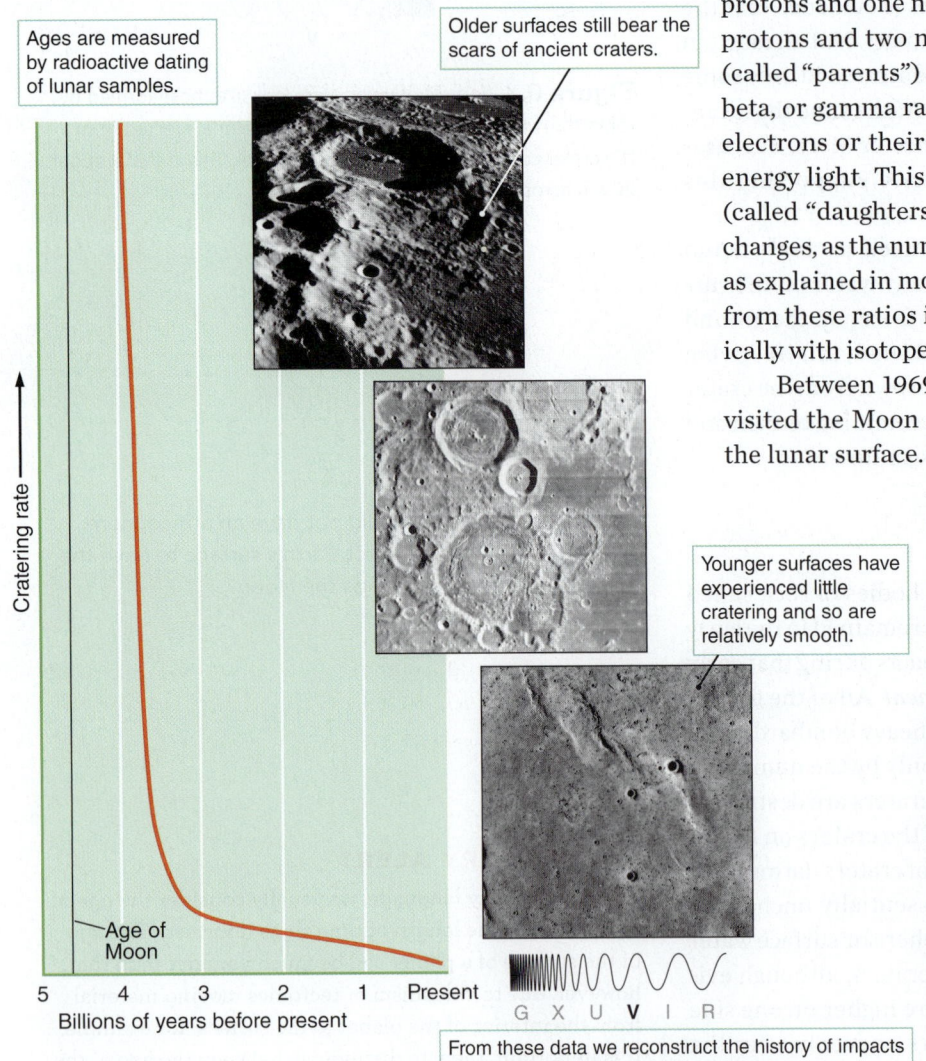

Ages are measured by radioactive dating of lunar samples.

Older surfaces still bear the scars of ancient craters.

Younger surfaces have experienced little cratering and so are relatively smooth.

Cratering rate

Age of Moon

5   4   3   2   1   Present
Billions of years before present

G X U V I R

From these data we reconstruct the history of impacts in the Solar System. Impacts were much more frequent billions of years ago than they are today.

CHECK YOUR UNDERSTANDING 6.1

If crater A is inside crater B, we know that

a. crater A was formed before crater B.

b. crater B was formed before crater A.

c. both craters were formed at about the same time.

d. crater B formed crater A.

e. crater A formed crater B.

Answers to Check Your Understanding questions are in the back of the book.

working it out 6.1

How to Read Cosmic Clocks

A geologist can find the age of a mineral by measuring the relative amounts of a radioactive element, known as a **radioisotope**, and the decay products it turns into. The time that it takes for half of the parent isotopes to decay into daughter isotopes is called the **half-life**. With every half-life that passes, the remaining amount of parent isotopes will decrease by a factor of 2. For example, over three half-lives, the final amount, P_F, of a parent radioisotope will be $\frac{1}{2} \times \frac{1}{2} \times \frac{1}{2} = \frac{1}{8}$ of its original amount, P_O. If we express the number of half-lives more generally as n, then we can translate this relationship into math:

$$\frac{P_F}{P_O} = \left(\frac{1}{2}\right)^n$$

This equation has an exponential expression on the right-hand side. This means that we are raising $\frac{1}{2}$ to the power of n. Recall some of the rules of exponents. For example, if $n = 0$, the right-hand side of this equation will be 1. That makes sense. If no half-lives have gone by, then the final amount will be the same as the original amount, so the ratio of P_F to P_O will be 1. If $n = 1$, then the right-hand side of the equation is $\frac{1}{2}$, and the ratio of the final amount to the initial amount is $\frac{1}{2}$. One half-life has gone by, and half of the parent isotopes remain.

Calculator hint: This kind of exponent is not the same as the "times 10 to the" operation that you learned about in Chapter 1. Because the base of the exponent in this case is not 10, you cannot use the EE or EXP key on your calculator. Instead, you need to use the x^y (sometimes y^x) key. First, type the base ($\frac{1}{2}$, or 0.5) into your calculator, then hit x^y, and then type in your value for n. For example, if you are calculating the fraction of remaining parent isotopes after three half-lives, you would type [0][.][5] [x^y][3][=] to find the answer (try it: you should get $\frac{1}{8}$, or 0.125).

The most abundant form of the element uranium is ^{238}U (pronounced "uranium two-thirty-eight"), which decays through a series of reactions to an isotope of the element lead, ^{206}Pb (pronounced "lead two-oh-six"). It takes 4.5 billion years for half of a sample of ^{238}U to decay through these processes to ^{206}Pb. If we were to find a sample in which half of the atoms were ^{238}U and half of the atoms were ^{206}Pb, we would know that half the uranium atoms had turned to lead, so that

$$\frac{P_F}{P_O} = \frac{1}{2}$$

Compare this to the previous equation, and convince yourself that if $(\frac{1}{2})^n = (\frac{1}{2})$, then $n = 1$. So we find that the sample formed from pure ^{238}U one half-life, or 4.5 billion years, ago.

This equation gives the ratio on the left rather than the number of half-lives, n, but n is what you are looking for! Solving for n in the general case involves logarithms, so it is sometimes easiest to try reasonable numbers in your calculator until you find approximately the right one.

Let's look at an example of finding the age from the fraction of material that has decayed, this time with a different form of uranium (^{235}U) that decays to a different form of lead (^{207}Pb) with a half-life of 700 million years. Suppose that a lunar mineral brought back by astronauts has 15 times as much ^{207}Pb as ^{235}U. If the sample was pure uranium when it formed, this means that $\frac{15}{16}$ of the ^{235}U has decayed to ^{207}Pb, leaving only $\frac{1}{16}$ of the parent element remaining in the mineral sample. The ratio on the left side of the equation is $\frac{1}{16}$, or 0.0625.

$$\frac{1}{16} = \left(\frac{1}{2}\right)^n$$

Since $\frac{1}{16}$ is less than $\frac{1}{2}$, we know that n must be bigger than 1 (more than one half-life has passed). Let's try 2:

$$\text{Does } \frac{1}{16} = \left(\frac{1}{2}\right)^2 ? \text{ No.}$$

Putting that into the calculator gives 0.25—that's too big. Let's try 5:

$$\text{Does } \frac{1}{16} = \left(\frac{1}{2}\right)^5 ? \text{ No.}$$

That gives 0.03125—too small. Try 4:

$$\text{Does } \frac{1}{16} = \left(\frac{1}{2}\right)^4 ? \text{ Yes!}$$

Aha! After $n = 4$ half-lives, the ratio of the remaining material to the original material is 0.0625, or $\frac{1}{16}$. To find out how much time has passed, we multiply the number of half-lives by the length of a half-life: 4×700 million = 2.8 billion years old.

6.2 The Surfaces of Terrestrial Planets Are Affected by Processes in the Interior

While impact cratering is driven by forces external to the planet, two other important processes—tectonism and volcanism—are driven by conditions in the interior of the planet. But how do we know what the interiors of planets are like? The deepest holes ever drilled on Earth, for example, are only about 12 km deep—tiny when compared to Earth's radius of 6,378 km. Even so, scientists have determined a lot about the interior of Earth.

Interior Composition

The composition of Earth's interior can be determined in two different ways. The first approach is to compare Earth's density to the density of various materials. Density is defined as the mass of an object divided by its volume. This is a useful property to measure because it is a property of the material, not the object. The density of a brick remains the same, whether you have the whole brick (more massive, but larger) or half the brick (less massive but smaller). In your experience, you notice density when objects are unusually heavy (or light) for their size. If someone handed you a brick made of Styrofoam or a brick made of lead, you would be surprised. You expect a brick to be made of clay and to have a particular weight for its size.

To find Earth's density, scientists first find the mass of Earth from the strength of Earth's gravity. Dividing the mass of Earth by the volume of Earth gives an average density of about 5,500 kilograms per cubic meter (kg/m^3) (see **Working It Out 6.2**). The average density of rocks that make up the surface, though, is only 2,900 kg/m^3. Because the density of the whole planet is greater than the density of the surface rocks, Earth's interior must contain material that is quite a bit denser than its surface rocks.

The second approach to determining the composition of Earth's interior comes from studying meteorites. Because meteorites and Earth formed at the same time out of similar materials, the overall composition of Earth should resemble the composition of meteorite material, which includes minerals with large amounts of iron. Both of these approaches led planetary scientists to conclude that Earth's interior contains large amounts of iron, which has a density of nearly 8,000 kg/m^3. A similar conclusion can be drawn about Mercury, Venus, and Mars.

Building a Model of Earth's Interior

How do we know about the structure of Earth's interior? It is impossible to drill down into Earth's core to observe Earth's interior structure directly. However, **seismic waves** from earthquakes pass through the interior and can be monitored by instruments placed around the world. As these waves pass through regions of different density, they are blocked or refracted, just like light passing through a lens. Geologists compare observations of these waves with layered models of the interior to determine the size and composition of those interior layers. At any point in Earth's interior, the pressure must be just high enough that the outward forces balance the inward force of the weight of all the material above that point. Otherwise, Earth would expand or contract. This concept of equilibrium provides an additional constraint to the models of Earth's interior.

One type of seismic wave, called a **surface wave**, travels across the surface of a planet, much like a wave on the ocean. Surface waves from earthquakes can

working it out 6.2

The Density of Earth

Density is defined as the mass of an object divided by its volume. The mass of Earth is 5.97×10^{24} kg. Earth's average radius is given in Table 6.1 as 6,378 km, or 6,378,000 meters. If we assume that Earth is a perfect sphere, then we can calculate the volume, V, from the radius, R, using the formula for the volume of a sphere:

$$V = \frac{4}{3}\pi R^3$$

$$V = \frac{4}{3}\pi (6{,}378{,}000 \text{ m})^3$$

$$V = 1.087 \times 10^{21} \text{ m}^3$$

To find the density, we must divide the mass (M) by this volume (V):

$$\text{Density} = \frac{M}{V}$$

$$\text{Density} = \frac{5.97 \times 10^{24} \text{ kg}}{1.087 \times 10^{21} \text{ m}^3}$$

$$\text{Density} = 5{,}490 \text{ kg/m}^3$$

This is the average density of Earth. Because the rocky surface material has a much lower density (2,900 kg/m³) than this average density, the interior density must be much higher to bring the density of the whole Earth up to roughly double the density of the surface rocks.

sometimes be seen rolling across the countryside like ripples on water. These waves cause Earth's surface to heave during an earthquake, buckling roadways and causing buildings to sway. Other, faster seismic waves, called primary and secondary waves, travel *through* Earth. **Primary waves** compress and expand the material in the direction of travel, like sound waves do, whereas **secondary waves** move the material from side to side as they pass, like a wave that travels down a string. In Chapter 4, you learned that light waves bend when they enter a new medium. Seismic waves do the same: they bend if they enter a region with a different density.

Primary waves can travel through solids and liquids, whereas secondary waves can travel only through solids. As shown in **Figure 6.6**, a liquid outer core will create a "shadow" of primary waves (shown in yellow), which bend at the outer edge of the core, and a second shadow is created by the inability of secondary waves (shown in blue) to penetrate it. For more than 100 years, thousands of instruments scattered around the globe have measured the vibrations from earthquakes, volcanic eruptions, and nuclear explosions as these vibrations either pass through Earth or are absorbed and leave a shadow. Scientists have compared these data to the predictions of theoretical models to discover the structure of Earth's interior.

This structure is shown in Figure 6.6. The innermost region of Earth's interior consists of a two-component core—a solid inner core and a liquid outer core. Earth's core consists primarily of iron, nickel, and other dense metals. Outside

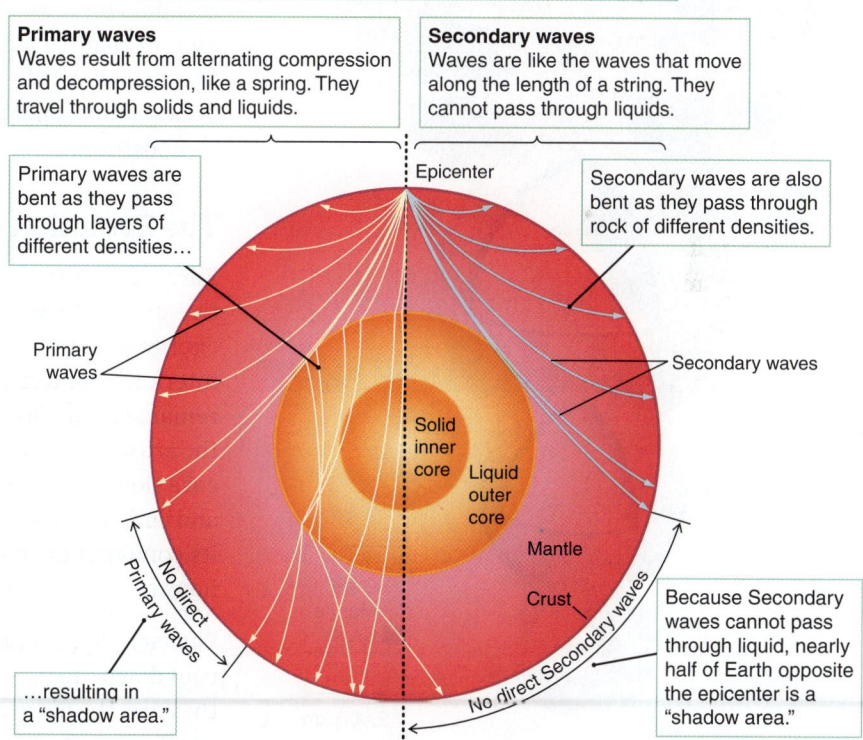

For clarity, each type of wave is shown on one-half of Earth.

Primary waves
Waves result from alternating compression and decompression, like a spring. They travel through solids and liquids.

Secondary waves
Waves are like the waves that move along the length of a string. They cannot pass through liquids.

Primary waves are bent as they pass through layers of different densities…

Epicenter

Secondary waves are also bent as they pass through rock of different densities.

Primary waves

Secondary waves

Solid inner core

Liquid outer core

Mantle

Crust

No direct Primary waves

No direct Secondary waves

…resulting in a "shadow area."

Because Secondary waves cannot pass through liquid, nearly half of Earth opposite the epicenter is a "shadow area."

Figure 6.6 Primary and secondary seismic waves move through the interior of Earth in distinctive ways. Measurements of when and where different types of seismic waves arrive after an earthquake enable scientists to test predictions from detailed models of Earth's interior. Note the "shadow areas" caused both by primary waves (shown in yellow) bending at the outer boundary of the liquid outer core and by the inability of secondary waves (shown in blue) to pass through the liquid outer core.

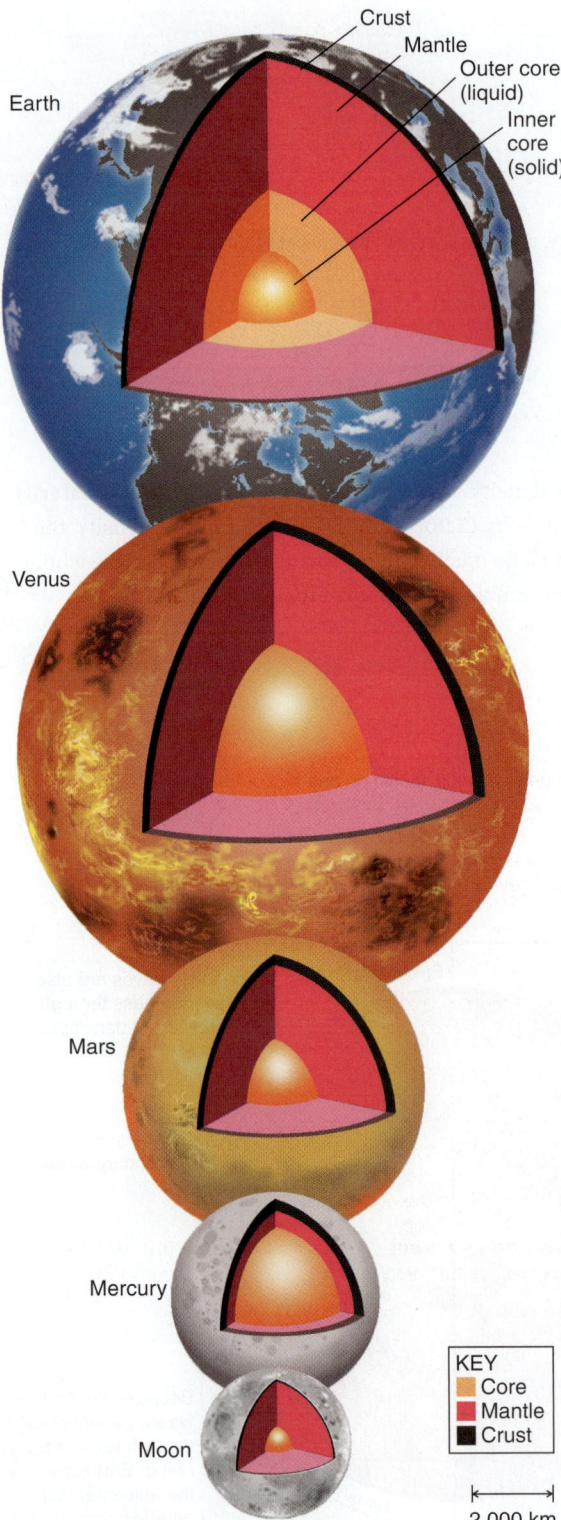

Figure 6.7 These models of the interiors of the terrestrial planets and Earth's Moon show that the thicknesses of the components change from one planet to another. Some fraction of the cores of Mercury, Venus, and Mars is probably liquid.

the core lies a solid portion called the **mantle**, which is made of medium-density materials. Covering the mantle is the **crust**—a thin, hard layer of lower-density materials that is chemically distinct from the interior. We live on top of the crust.

Earth's interior is not uniform. The materials have been separated by density in a process known as **differentiation**. When rock is melted, the denser materials sink to the center and the less dense materials float toward the surface. Today, little of Earth's interior is molten, but the separation of dense materials from lower-density materials shows that Earth was once much hotter, and its interior was liquid throughout.

Instrumentation on the Moon, left behind by Apollo astronauts, has observed "moonquakes," revealing the structure of the Moon's interior. The *InSight* lander on Mars is currently measuring its internal structure. The interior structures of Venus and Mercury have been determined from the effects of their gravitational fields on the orbits of spacecraft. The cut-away models in **Figure 6.7** compare the structures of each of the terrestrial planets and the Moon. The differentiated interiors show that the cores of all the terrestrial worlds, including the Moon, were once molten.

The Moon's Structure and Formation

The Moon has only a tiny core (Figure 6.7) that is composed of material similar to that of Earth's mantle. As explained in Chapter 5, the best explanation of the Moon's origin is that a Mars-sized protoplanet collided with Earth. This explanation accounts for the similar composition of the Moon and Earth's mantle, as well as the Moon's relative lack of water and other volatiles compared to Earth, Venus, and Mars. However, the origin of the Moon is still an active area of astronomical research because this hypothesis does not yet explain every detail of the Moon's geology and composition.

The Evolution of Planetary Interiors

Early in its history, Earth was melted by energy from both collisions and radioactive decay of short-lived radioactive elements. As the surface of Earth radiated energy into space, it cooled rapidly, causing a solid crust to form above a molten interior. A solid crust does not transfer thermal energy well, so the interior of Earth remained hot. Over a long time, though, energy continued to slowly leak through the crust and radiate into space, cooling Earth's interior. The mantle and the inner core slowly solidified. The balance between energy received and energy produced and emitted determines the temperature within any planet. The size of the planet, its composition, and its sources of heating (both internal and external) influence how quickly the core temperature changes.

Cooling. Planets radiate thermal energy from their surfaces, causing the planet to cool down. As you learned in Chapter 4, all objects radiate energy, and the hotter they are, the more energy they radiate. Hotter objects emit more of their energy at shorter wavelengths.

The rate at which a planet cools depends on its radius. Suppose that two rocky planets are the same temperature, but one planet has twice the radius of the other. The larger planet has 4 times the surface area of the smaller one but 8 times the volume. This means the larger planet has 8 times as much mass and 8 times as much thermal energy but only 4 times as much ability to radiate it away. The Moon and a smaller planet like Mercury are less geologically active than larger ones like Venus,

Earth, and Mars because the interiors of the smaller celestial bodies have cooled more since they formed.

Radioactive heating. Many long-lived radioactive isotopes are massive, and so they fall below the crust when the planet is molten. Much of the thermal energy that remains in Earth's interior comes from long-lived radioactive elements, such as uranium-238, thorium-232, and postassium-40, trapped in the mantle. The energy released as these elements decay heats Earth's interior. Equilibrium between this production of energy and the radiation of energy from the surface determines Earth's interior temperature. As time passes, there are fewer remaining radioactive elements to produce energy, and Earth's interior gradually cools.

Tidal heating. Additional sources of energy continue to heat the interior of Earth. One source of continued heating is friction. Earth is repeatedly stretched and squished by the gravity of the Moon and Sun. Friction from these motions heats Earth's interior. This phenomenon, which also occurs on some moons of the outer planets, is called **tidal heating**. It is closely related to more noticeable tidal effects in the oceans of Earth that cause the ocean to advance and retreat along the beach twice each day.

Tidal effects are caused by the change in the strength of gravity across an object. The gravitational force on Earth from the Moon is shown in **Figure 6.8.** The size of the blue arrows indicates the strength of the force. The force of gravity on the part of Earth nearest the Moon is stronger than the force of gravity on the part of Earth farthest from the Moon. This difference in forces stretches Earth and causes it to bulge along the line from Earth to the Moon. The oceans respond dramatically to these differing forces, causing **tides**, when the ocean level moves up and down past its equilibrium position.

The Moon and Sun both contribute to tides on Earth. **Lunar tides** are caused by the gravitational pull of the Moon, which causes Earth to stretch along a line pointing approximately in the direction of the Moon. The rotation of Earth drags the tidal bulge slightly ahead of the position of the Moon, as shown in **Figure 6.9.**

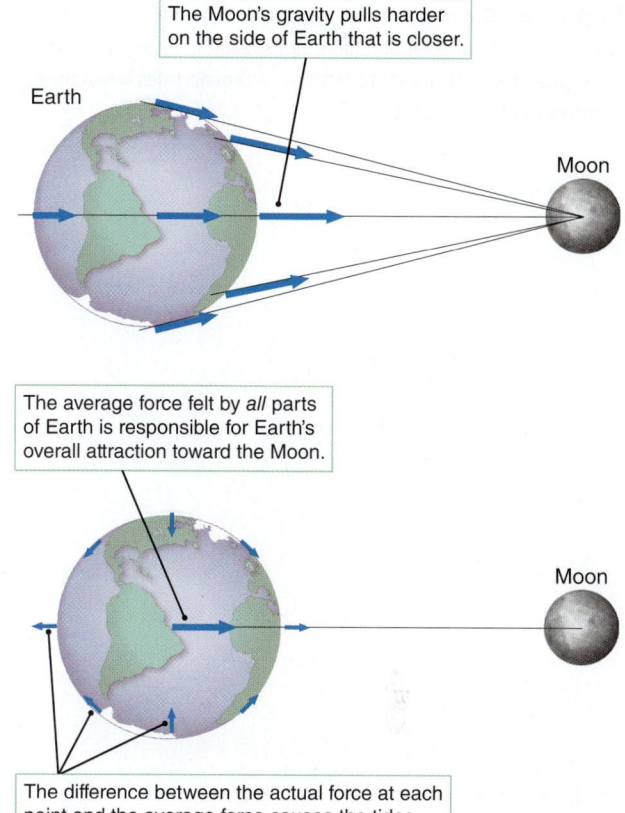

Figure 6.8 Tides stretch Earth along the line between Earth and the Moon but compress Earth perpendicular to this line. In the bottom image, the average force has been subtracted from all the force arrows around the circumference, leaving behind smaller arrows that show the net force at those locations.

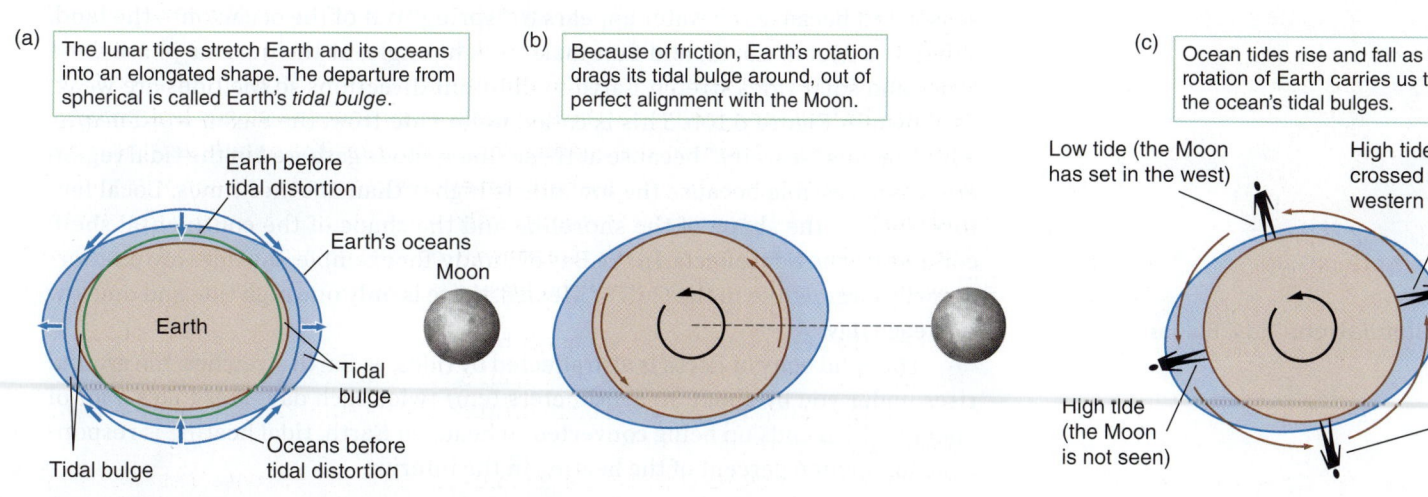

Figure 6.9 (a) Tidal effects exert forces (blue arrows) that pull Earth and its oceans into a tidal bulge. (b) Earth's rotation pulls its tidal bulge slightly out of alignment with the Moon. (c) As Earth's rotation carries us through these bulges, we experience the ocean tides. The magnitude of the tides has been exaggerated in these diagrams for clarity. In the diagrams, the observer is looking down from above Earth's North Pole. Sizes and distances are not to scale.

Figure 6.10 Solar tides are about half as strong as lunar tides. The interactions of solar and lunar tides result in either (a) spring tides when they are added together or (b) neap tides when they partially cancel each other.

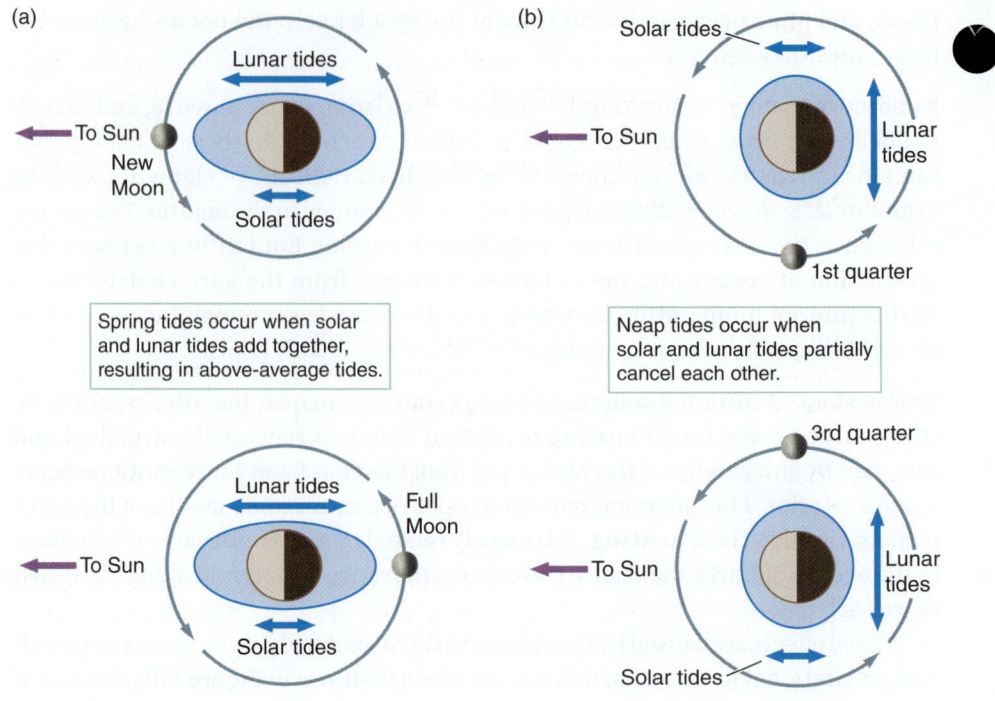

(a)

Spring tides occur when solar and lunar tides add together, resulting in above-average tides.

(b)

Neap tides occur when solar and lunar tides partially cancel each other.

▶❚❚ **AstroTour:** Tides and the Moon

🎥 **Astronomy in Action:** Tides

▶▶ **Interactive Simulation:** Tidal Bulges

Solar tides are caused by the gravitational pull of the Sun, which causes Earth to stretch along a line pointing approximately in the direction of the Sun.

Although the Sun is much more massive than the Moon, it is also much farther away, so solar tides are about half as strong as lunar tides. Depending on the phase of the Moon, the lunar and solar tides either amplify one another or partially cancel each other out. When the Moon, Earth, and Sun are all in a line, at new and full Moon, the lunar and solar tides overlap as shown in **Figure 6.10a**. This creates more extreme tides; the high tides are extra-high and the low tides are extra-low. These extreme tides are called **spring tides**—not because of the season but because the water appears to "spring" out of the ocean onto the land. When the Moon, Earth, and Sun make a right angle, at the quarter phases, the lunar and solar tides stretch Earth in different directions so the tides are weak, as shown in Figure 6.10b. This is called **neap tide** from the Saxon word *neafte*, which means "scarcity," because at these times, foods gathered in the tidal region are less accessible because the low tide is higher than at other times. Local factors, such as the shape of the shoreline and the shape of the continental shelf, complicate the tidal effects. In the Bay of Fundy, for example, the tides are extraordinarily large, while in the Gulf of Mexico, there is only one high tide and one low tide each day.

The solid body of Earth is also affected by tides; as Earth stretches, the ground rises under you by about 30 centimeters (cm) twice each day. This takes a lot of energy, which ends up being converted to heat. On Earth, tidal heating is responsible for about 6 percent of the heating in the interior.

Effects of temperature and pressure on material. Cooling, radioactive heating, and tidal heating determine the temperature in the interior of Earth. This temperature plays an important role, but it's not the only influence—whether a material is solid or liquid depends on pressure as well as temperature. High pressure forces atoms and molecules closer together and makes the material more likely to be a solid. In Earth's outer core, the high temperature dominates over the pressure, allowing the material to exist in a molten state. At the center of Earth, even though the temperature is higher, the pressure is so great that Earth's inner core is solid.

Magnetic Fields

The combination of Earth's rotation about its axis and a liquid, electrically conducting, circulating outer core creates a *magnetic field* around Earth that you may have detected with a compass. Earth's magnetic field is constantly changing in both strength and direction. Studies of this magnetic field provide an additional probe of the interior of Earth, both as it exists now and as it was in the past. Similar observations of the magnetic fields of other planets, while less detailed, can be used to test models of their interiors.

A **magnetic field** is created by moving charges and exerts a force on magnetically reactive objects, such as iron, as shown in **Figure 6.11a**. In this image, small pieces of iron have lined up along magnetic field lines emanating from a bar magnet. Earth has a magnetic field roughly shaped like the magnetic field of a bar magnet. This field is approximately aligned with Earth's rotation axis, as shown in Figure 6.11b. Because the magnetic axis and the rotation axis are not perfectly aligned, the north-pointing end of a compass points at a location in the Arctic Ocean off the coast of northern Canada, rather than at the *geographic* North Pole, about which Earth spins. This is why topographic maps always include a diagram to show you how to correct *magnetic north* (the direction the compass points) to geographic north. Earth's south magnetic pole is off the coast of Antarctica, 2,800 km from Earth's geographic South Pole.

The magnetic pole tends to wander, changing direction as a result of changes in the interior. At the moment, the north magnetic pole is traveling several dozen kilometers per year toward the northwest. If this rate and direction continue, the north magnetic pole will be in Siberia before the end of the century.

Dramatic changes in the magnetic field have occurred over the history of our planet, as geologists discovered when they looked at iron-rich minerals that were once molten but then cooled. When a magnetic material gets hot enough, it loses its magnetization. As the material cools, it again becomes magnetized by any surrounding magnetic field, recording the strength and direction of the field at the time the mineral cooled. From these minerals, geologists have learned that Earth's magnetic field has existed for billions of years, but the north and south magnetic poles switch from time to time. On average, these reversals in Earth's magnetic field take place about every 500,000 years.

Mercury is the only other terrestrial planet with a significant magnetic field today. As with Earth, the planet's rotation and a large iron core, parts of which are molten and circulating, create Mercury's magnetic field. Planetary scientists expected that Venus would have a magnetic field because Venus' mass and distance from the Sun are so similar to Earth's. Venus may lack a magnetic field

(a)

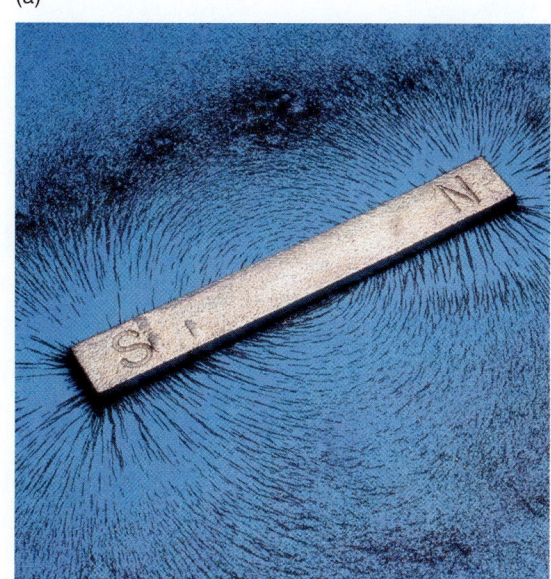

(b)

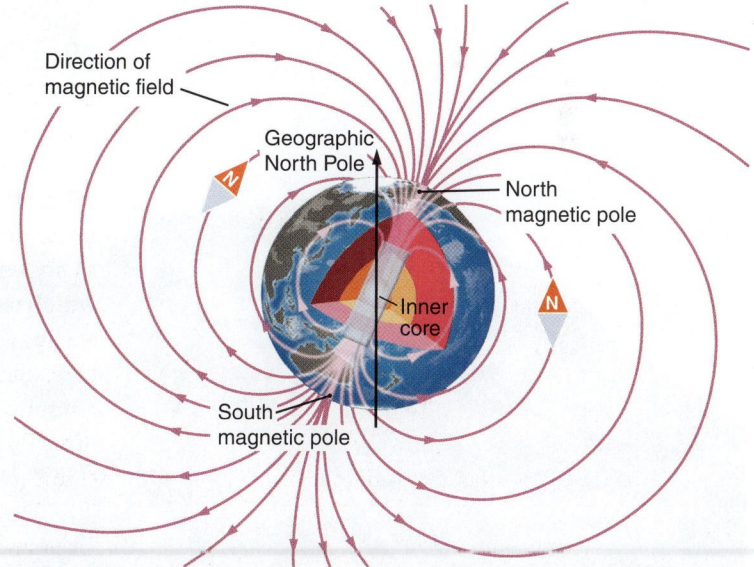

Figure 6.11 (a) Iron filings sprinkled around a bar magnet help us visualize a magnetic field. (b) Earth's magnetic field can be visualized as though it were a giant bar magnet tilted relative to Earth's axis of rotation. Compass needles line up along magnetic field lines and point toward Earth's north magnetic pole.

what if . . .

What if Earth's magnetic field showed signs that it would undergo a reversal within the next hundred years: Would we care whether the magnetic north pole simply migrates to the South Pole or whether the magnetic field disappears entirely and reappears as a reversed field?

▶❙❙ **AstroTour:** Continental Drift

because of its extremely slow rotation (once every 243.0 Earth days), but this explanation is still uncertain.

Mars has a weak magnetic field. The magnetic signature occurs only in the ancient crustal rocks, not in geologically younger rocks. This implies the magnetic field was frozen in place at an earlier time, and Mars no longer has a global magnetic field. The lack of a strong magnetic field today on Mars might be the result of its small core. However, given that Mars is expected to have a partly molten interior and has a rotation rate similar to that of Earth, the lack of a field is still not yet fully understood. The martian lander *InSight* is currently gathering data about the interior of Mars and may help to answer these questions.

Because the Moon is very small and has a small, solid, nonrotating inner core, the magnetic field is weak, and the Moon does not have a global, bar-magnet-like field as Earth does. However, rocks gathered during the Apollo program show that the Moon must have had a strong magnetic field at some time in the past. These rocks may have been magnetized when the Moon had a molten core. But that core was small, so it's difficult to understand how it could have created a strong magnetic field. Another possible explanation is that large impacts might cause temporary magnetic fields, which become frozen into the rocks of the crust as they cool.

CHECK YOUR UNDERSTANDING 6.2

Scientists learn about the interior structure of planets by using (select all that apply)

a. ground-penetrating radar.

b. deep mine shafts.

c. observations of seismic waves.

d. models of Earth's interior.

e. observations of magnetic fields.

f. X-ray observations from satellites.

Figure 6.12 Tectonic processes fold and warp Earth's crust, as seen in the rocks along this roadside cut.

6.3 Planetary Surfaces Evolve through Tectonism

If you have been on a drive through mountainous or hilly terrain, you may have noticed roadcuts like the one shown in **Figure 6.12**. The exposed layers tell the story of Earth through the vast expanse of geologic time. The crust and part of the upper mantle form the **lithosphere** of a planet. **Tectonism** is the faulting or folding or other modification of the outer layer of a planet. Tectonism modifies the lithosphere—warping, twisting, and shifting it to form visible surface features, such as the layers seen in roadcuts. More famously, on Earth tectonism drives the motion of tectonic plates, carrying continents across the surface of Earth. Earth is the only planet in our Solar System with this form of tectonic activity. In this section, you will learn how tectonic processes shape the surface of a planet.

The Theory of Plate Tectonics

As early as 1596, people noticed from studying maps that parts of Earth's continents could be fitted together like pieces of a giant jigsaw puzzle. Later, the layers in the rock and the fossil record on the east coast of South America were found to match those on the west coast of Africa. On the basis of evidence like this, Alfred Wegener (1880–1930) hypothesized that the continents were originally joined in one large landmass that shattered as the continents drifted apart over millions of years. This hypothesis was further developed into the theory known today as **plate tectonics**.

Wegener's idea was met with great skepticism because it was difficult to imagine a mechanism that could move such huge landmasses. In the 1960s, however, studies of the ocean floor provided compelling evidence for plate tectonics. As **Figure 6.13** shows, hot material in ocean rifts rises toward Earth's surface, creating new ocean floor. When this hot material cools to form a type of rock called basalt, it becomes magnetized along the direction of Earth's magnetic field, recording the direction of Earth's magnetic field at that time. Greater distance from the ridge indicates an older ocean floor and an earlier time. Combining this magnetic record with radiometric dates for the rocks showed that the continental plates have moved over long geologic time spans.

More recently, precise surveying techniques and the Global Positioning System (GPS) have confirmed these results more directly. Some areas are being pulled apart by about the length of a pencil each year. Over millions of years of geologic time, such motions become significant. Over 10 million years—a short time by geological standards—15 cm/year becomes 1,500 km, and maps definitely need to be redrawn.

Today, geologists recognize that Earth's crust is composed of a number of **lithospheric plates**. The movement of these plates causes a variety of geological features on our planet, including the continental drift that Wegener hypothesized.

The Role of Convection

Movement of lithospheric plates requires immense forces. These forces are the result of thermal energy escaping from the interior of Earth. The transport of thermal energy by the movement of packets of gas or liquid is known as **convection**. **Figure 6.14a** illustrates the process of convection, which should be familiar to you if you have ever heated a pot of water on a stovetop. Thermal energy from the stove warms water at the bottom of the pot. The warm water expands slightly, becoming less dense than the cooler water above it. The cooler water sinks, displacing the warmer water upward. When the lower-density water reaches the surface, it gives up part of its energy to the air and cools, becomes denser, and sinks back toward the bottom of the pot. Water rises in some locations and sinks in others, forming convection "cells."

Figure 6.14b shows how convection works in Earth's mantle, where radioactive decay provides the heat source. Earth's mantle is not molten, but it is somewhat

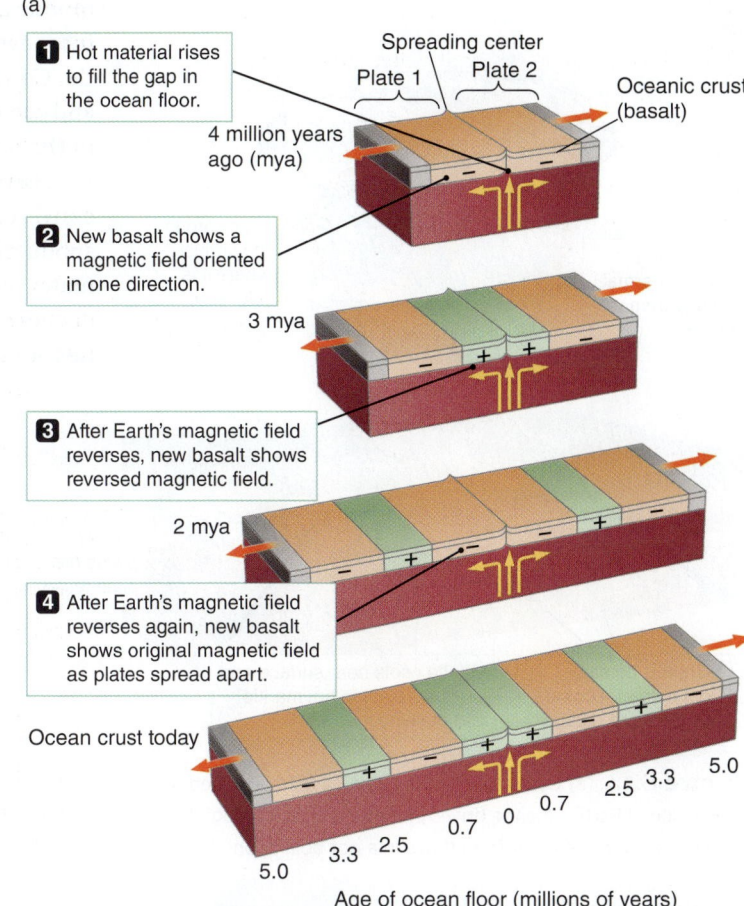

(a)

1. Hot material rises to fill the gap in the ocean floor.
2. New basalt shows a magnetic field oriented in one direction.
3. After Earth's magnetic field reverses, new basalt shows reversed magnetic field.
4. After Earth's magnetic field reverses again, new basalt shows original magnetic field as plates spread apart.

Spreading center
Plate 1 Plate 2
Oceanic crust (basalt)
4 million years ago (mya)
3 mya
2 mya
Ocean crust today

Age of ocean floor (millions of years)

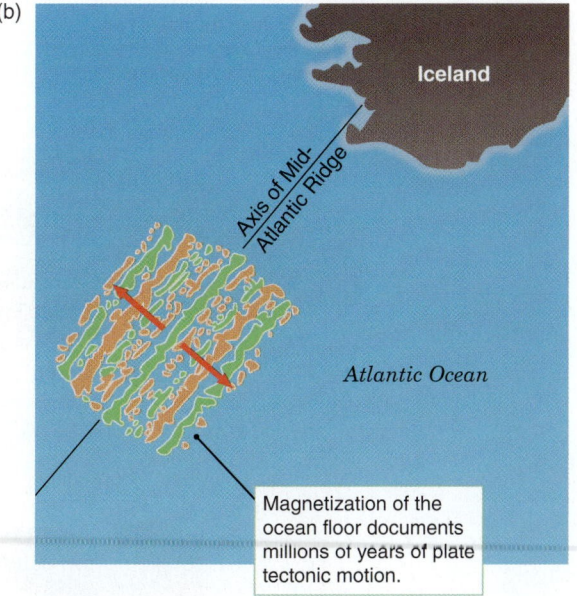

(b)

Iceland

Axis of Mid-Atlantic Ridge

Atlantic Ocean

Magnetization of the ocean floor documents millions of years of plate tectonic motion.

Figure 6.13 (a) As new seafloor is formed at a ridge, the cooling rock becomes magnetized. The magnetized rock is then carried away by tectonic motions. (b) Maps like this one, of banded magnetic structures in the seafloor near Iceland, provide support for the theory of plate tectonics.

(a)

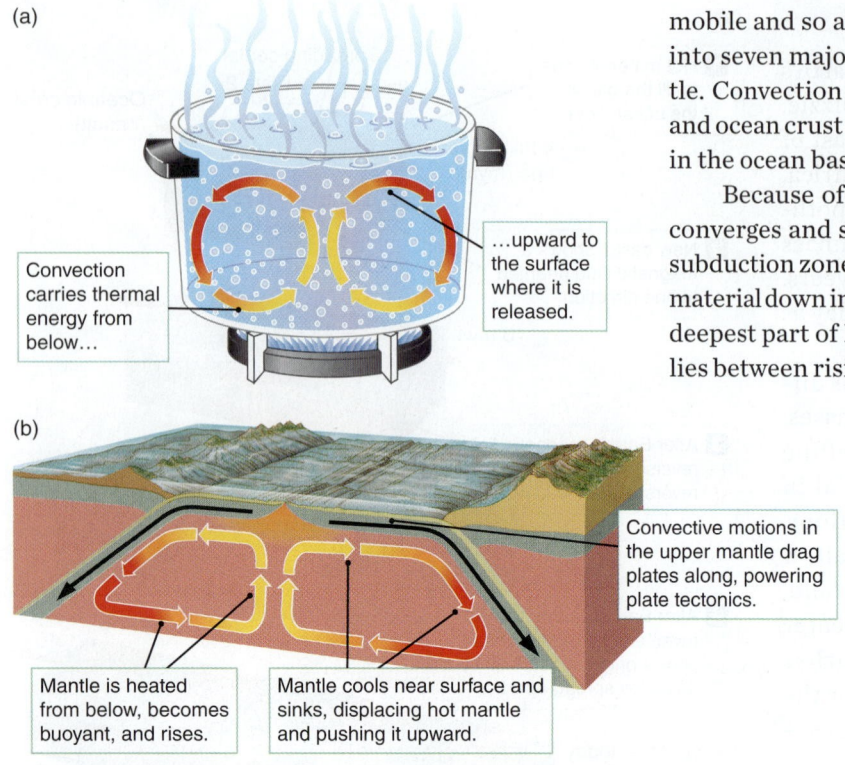

Convection carries thermal energy from below…

…upward to the surface where it is released.

(b)

Convective motions in the upper mantle drag plates along, powering plate tectonics.

Mantle is heated from below, becomes buoyant, and rises.

Mantle cools near surface and sinks, displacing hot mantle and pushing it upward.

Figure 6.14 (a) Convection occurs when a fluid is heated from below. (b) Similarly, convection in Earth's mantle drives plate tectonics, although the timescale and velocities involved are very different from those in a pot boiling on your stovetop.

mobile and so allows convection to take place very slowly. Earth's crust is divided into seven major plates and a half-dozen smaller plates floating on top of the mantle. Convection cells in Earth's mantle drive the plates, carrying both continents and ocean crust along with them. Convection also creates new crust along rift zones in the ocean basins, where mantle material rises up, cools, and slowly spreads out.

Because of plate tectonics, material rises and spreads in some locations but converges and sinks in others, as shown in **Figure 6.15**. In sinking regions, called subduction zones, one plate slides beneath the other, and convection drags the crust material down into the mantle. The Mariana Trench in the western Pacific Ocean—the deepest part of Earth's ocean floor—is a subduction zone. Much of the ocean floor lies between rising and sinking zones, making the ocean floor the youngest portion of Earth's crust. In fact, the oldest seafloor rocks are less than 200 million years old. Compare that with the oldest terrestrial rocks on Earth, which are 3.5 billion to 4 billion years old.

In some places, the plates are colliding with each other and being shoved upward. The highest mountains on Earth, the Himalayas, grow 0.5 meter per century as the Indo-Australian Plate collides with the Eurasian Plate. In still other places, plates meet at oblique angles and slide along past each other. One such place is the San Andreas Fault in California, where the Pacific Plate slides past the North American Plate. A **fault** is a fracture in a planet's crust along which material can slide.

Locations where plates meet tend to be very active geologically. Where plates run into each other, enormous stresses build up. Earthquakes occur when a portion of the boundary between

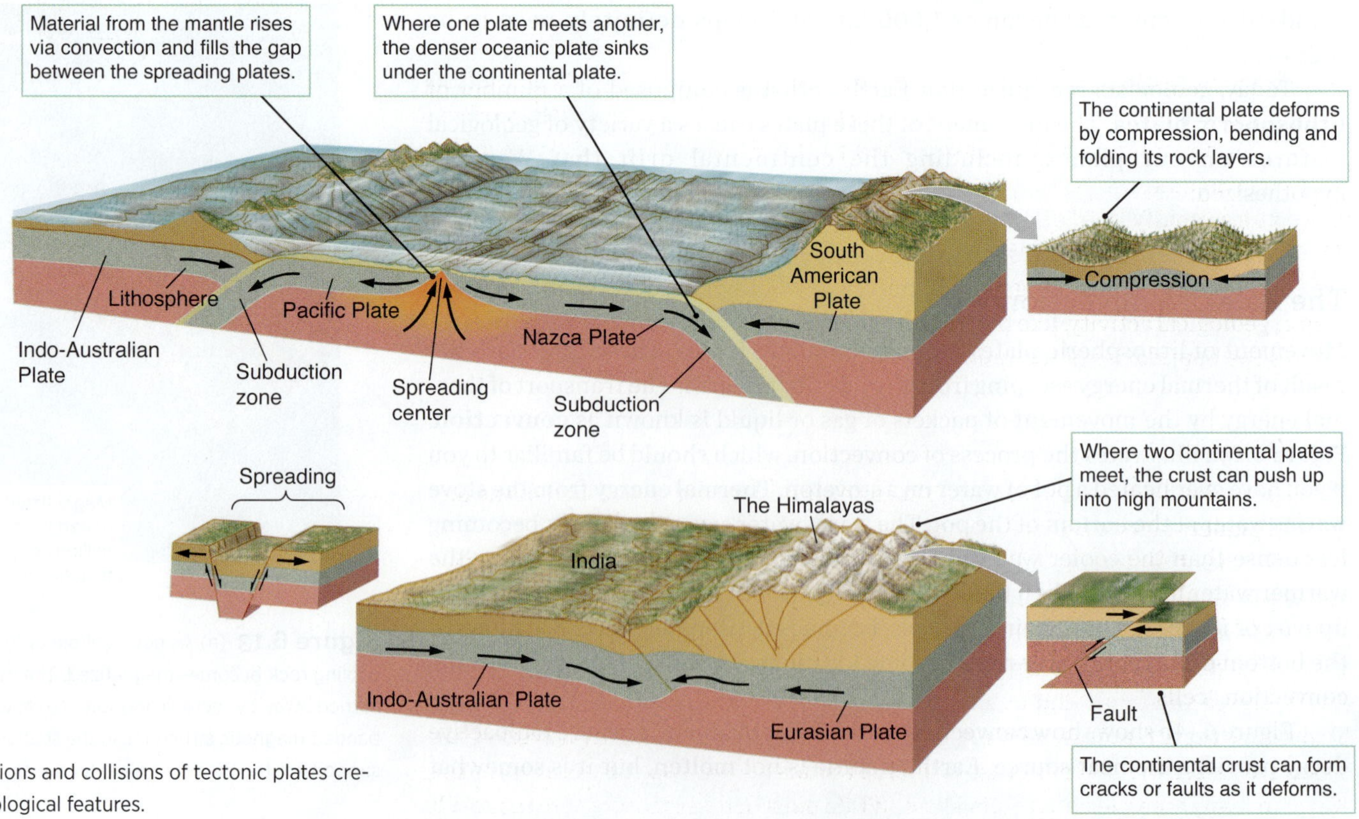

Material from the mantle rises via convection and fills the gap between the spreading plates.

Where one plate meets another, the denser oceanic plate sinks under the continental plate.

The continental plate deforms by compression, bending and folding its rock layers.

Lithosphere

Pacific Plate

South American Plate

Compression

Indo-Australian Plate

Subduction zone

Spreading center

Nazca Plate

Subduction zone

Spreading

Where two continental plates meet, the crust can push up into high mountains.

The Himalayas

India

Indo-Australian Plate

Eurasian Plate

Fault

Figure 6.15 Separations and collisions of tectonic plates create a wide variety of geological features.

The continental crust can form cracks or faults as it deforms.

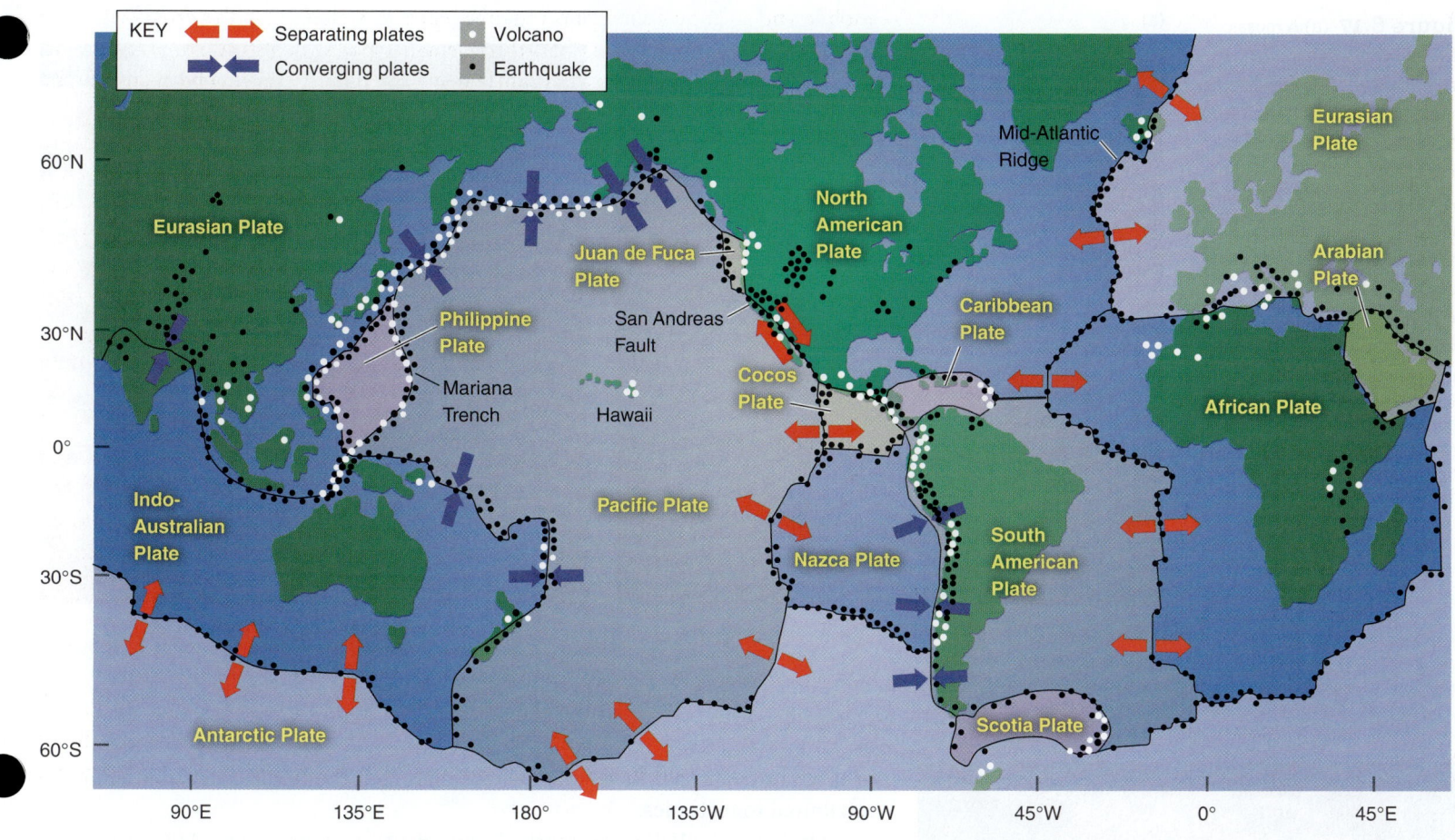

Figure 6.16 Major earthquakes and volcanic activity are often concentrated along the boundaries of Earth's principal tectonic plates.

two plates suddenly slips, relieving the stress. Volcanoes are created when friction between plates melts rock, which is then pushed up through cracks to the surface. Earth also has numerous **hot spots**, such as the Hawaiian Islands, where hot deep-mantle material rises, releasing thermal energy. As plates shift, some parts move more rapidly than others, causing the plates to stretch, buckle, or fracture. These effects are seen on the surface as folded and faulted rocks. Mountain chains are common near converging plate boundaries, where plates buckle and break. A map of geological activity, like the map of earthquakes and volcanoes in **Figure 6.16**, reveals the boundaries of tectonic plates.

Tectonism on Other Planets

Plate tectonics have been observed to occur only on Earth. However, all of the terrestrial planets show evidence of other types of tectonic disruptions.

Like the other terrestrial planets, Mercury was once molten. After the surface of the planet cooled and its crust formed, the interior of the planet continued to cool and shrink. Mercury's crust cracked and buckled in much the same way that a grape skin wrinkles as it shrinks to become a raisin. As a result, Mercury's surface has fractures and faults and numerous cliffs that are hundreds of kilometers long. To produce these features, Mercury's radius must have shrunk slightly (by less than 2 percent) after the planet's crust formed. Fractures have also cut the crust of the Moon in many areas, leaving fault valleys. While most of these features are the result

Figure 6.17 (a) A mosaic of *Viking Orbiter* images shows Valles Marineris, the major tectonic feature on Mars, stretching across the center of the image from left to right. This canyon system is more than 4,000 km long. The dark spots on the left are huge shield volcanoes in the Tharsis region. (b) This close-up perspective view of the canyon wall was photographed by the European Space Agency's *Mars Express* spacecraft.

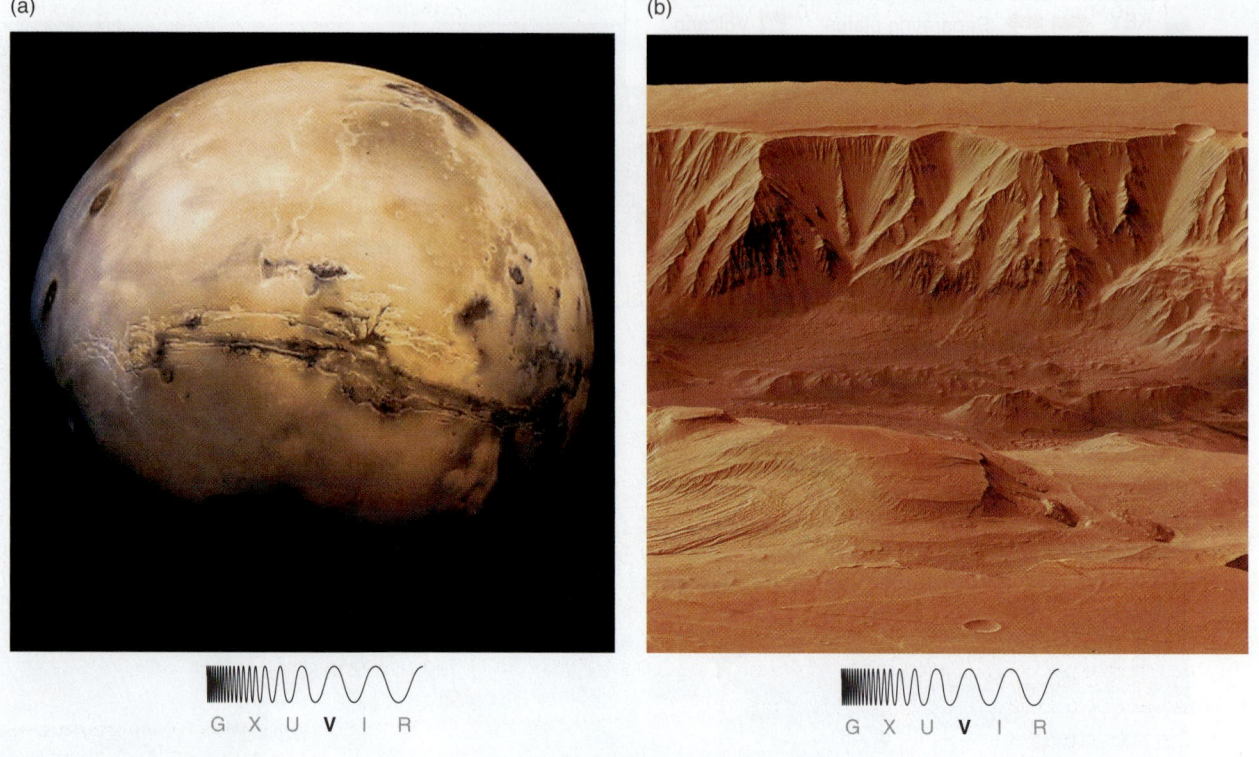

(a) G X U **V** I R

(b) G X U **V** I R

G X U V I **R**

Figure 6.18 The atmosphere of Venus blocks our view of the surface in visible light. This false-color view of Venus is a radar image made by the *Magellan* spacecraft. Bright yellow and white areas are mostly fractures and ridges in the crust. Some circular features seen in the image may be hot spots—regions of mantle upwelling. Most of the surface is formed by lava flows, shown in orange.

of large impacts that have cracked and distorted the lunar crust, some may be attributed to tectonics.

On Mars, the Valles Marineris (**Figure 6.17**) stretches nearly 4,000 km and is nearly 4 times as deep as the Grand Canyon. If located on Earth, this chasm would link San Francisco with New York. Valles Marineris includes a series of massive cracks formed as local forces, perhaps related to mantle convection, pushed the crust upward from below. The surface was not equally supported by the interior everywhere, and unsupported segments fell in. Once formed, the cracks were eroded by wind, water, and landslides, resulting in the structure we see today. Other parts of Mars have faults similar to those on the Moon or Mercury, but cliffs like those on Mercury are absent.

The mass of Venus is 80 percent that of Earth, and its radius is 95 percent of the radius of Earth. Because the two planets are similar in size and mass, the interior of Venus should be very much like the interior of Earth. Convection should be occurring in its mantle. On Earth, mantle convection and plate tectonics release thermal energy from the interior. However, on Venus there is no evidence of plate tectonics. The *Magellan* spacecraft provided the first high-resolution images of the surface of the planet, as shown in **Figure 6.18**. The relative scarcity of impact craters in images like this one indicate that most of Venus' surface is less than 1 billion years old, in turn suggesting a geologically active planet. Circular fractures called *coronae* on the surface of Venus, like the one shown in **Figure 6.19**, range from a few hundred kilometers to more than 2,500 km across. These coronae resemble impact craters, but instead they may be caused by upwelling plumes of hot mantle that have fractured Venus's lithosphere, providing a way for thermal energy to escape the interior. Alternatively, energy may build up in the interior until large chunks of the lithosphere melt and overturn, releasing an enormous amount of energy all at once. Scientists are still trying to figure out why the tectonic processes on Venus and Earth are so different.

CHECK YOUR UNDERSTANDING **6.3**

On which of the following terrestrial objects does plate tectonics occur? (Select all that apply.)

a. Mercury   **b.** Venus
c. Earth   **d.** Moon
e. Mars

6.4 Volcanism Reveals a Geologically Active Planet

Volcanoes release molten rock from the interior onto the surface of a planet. The molten rock originates deep in the lithosphere, where rising convection cells in the mantle, friction from movement in the crust, and concentrations of radioactive elements combine to generate a lot of thermal energy. In this section, we discuss the occurrence of volcanic activity on a planet or moon, which is called **volcanism**. Volcanism not only shapes planetary surfaces but also is a key indicator of a geologically active planet.

Terrestrial Volcanism Is Related to Tectonism

Volcanoes are usually located along plate boundaries and over hot spots. Maps such as the one shown in Figure 6.16 leave little doubt that, on Earth, volcanism is linked to the same forces responsible for plate motions. Places where convection carries hot mantle material toward the surface are frequent sites of eruptions. Iceland, which is one of the most volcanically active regions in the world, sits astride one such place—the Mid-Atlantic Ridge (see Figure 6.16).

A tremendous amount of thermal energy is generated from friction as plates slide under each other, heating rock toward its melting point. However, the weight of a lithospheric plate compresses the material at the base, creating high pressures there. This pressure forces the material to remain solid even at high temperature. As this material seeps through the crust, however, the pressure drops, while the temperature of the rock remains high. Thus, the rock that was solid at the base of a plate may melt as it nears the surface. This molten rock is called **magma**.

Once molten rock reaches the surface of Earth through a volcano, it is known as lava, and it can form many types of structures. Lava flows often form vast sheets, especially if the eruptions come from long fractures. A **shield volcano** forms when very **fluid** lava flows from a single location and spreads out over the surrounding terrain or ocean floor (**Figure 6.20a**). A **composite volcano** (Figure 6.20b) forms when thick lava flows, alternating with explosively generated rock deposits, build a steep-sided structure.

On Earth, volcanism also occurs where convection brings hot material toward the surface in the interiors of lithospheric plates, creating local hot spots like the Hawaiian Islands (Figure 6.20c) and the hot springs and geysers in Yellowstone National Park. Volcanism over hot spots works much like volcanism elsewhere, except that the convective upwelling occurs at a single spot rather than in a line

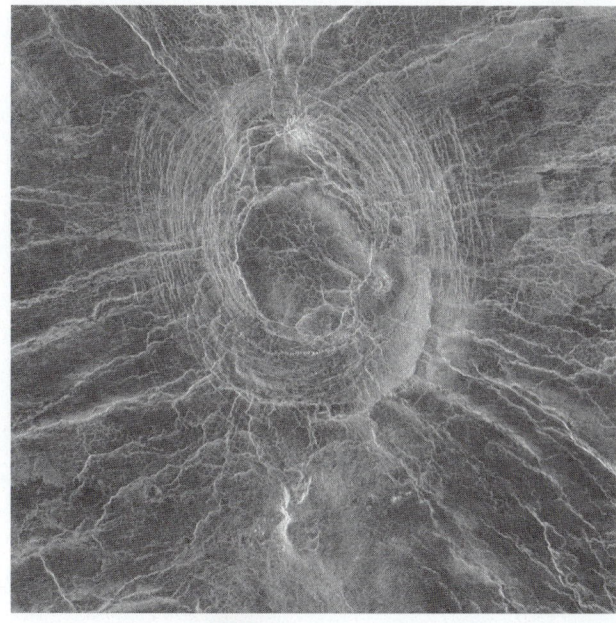

G X U V I **R**

Figure 6.19 Thermal energy escapes from Venus through hot spots like this one, which shows the circular fracture pattern known as a corona.

▶⏸ **AstroTour:** Hot Spot Creating a Chain of Islands

VOCABULARY ALERT

fluid In everyday language, *fluid* is used mainly as a noun, which can refer to either a liquid or a gas. Here, it is used as an adjective, describing how easily a substance flows. If it is very fluid, it flows like water. If it is not very fluid, it flows like honey or tar. Another adjective often used in this context is *viscous*. A viscous substance does not flow easily. Honey is much more viscous than water.

Figure 6.20 Magma reaching Earth's surface commonly forms (a) shield volcanoes, such as Mauna Loa, which have gently sloped sides built up by fluid lava flows, and (b) composite volcanoes, such as Mount Fuji, which have steeply symmetrical sides built up by viscous lava flows. (c) A hot spot forms a series of volcanoes as the plate above it slides by.

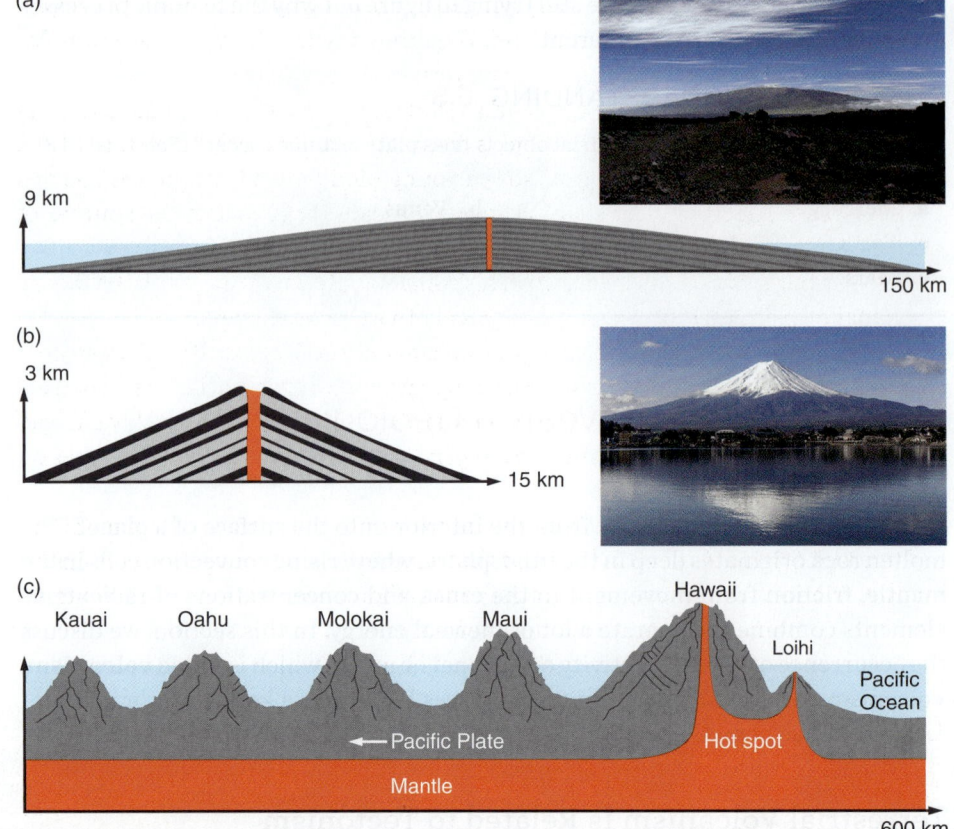

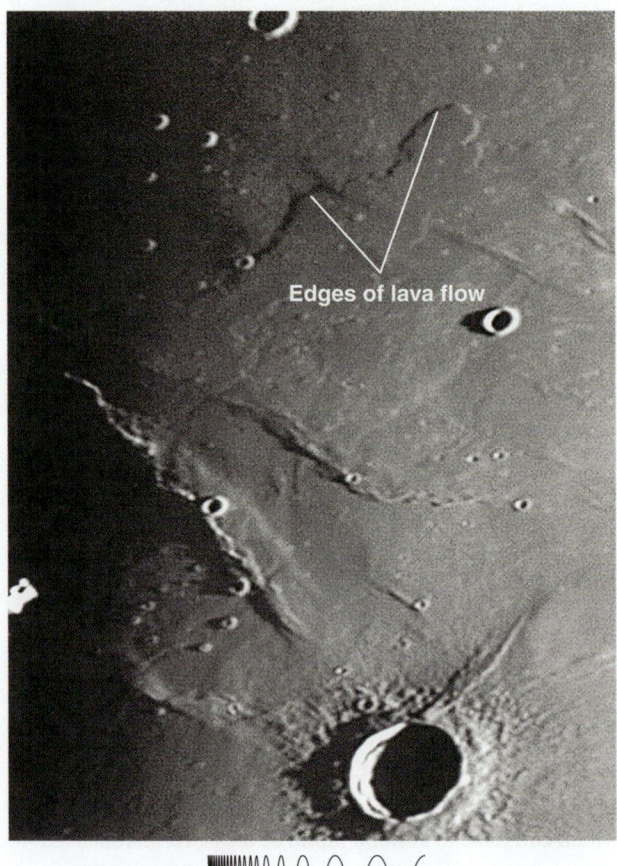

Figure 6.21 Lava flowing across the surface of Mare Imbrium on the Moon must have been relatively fluid to have spread out for hundreds of kilometers in sheets that are only tens of meters thick.

along the edge of a plate. These hot spots force mantle and lithospheric material toward the surface, where it emerges as lava.

The Hawaiian Islands are a chain of shield volcanoes that formed as their lithospheric plate (the Pacific Plate) moved across a hot spot. Each island stops growing when the plate motion carries the island away from the hot spot. Erosion, occurring since each island formed, wears the islands away. Kauai, which is further from the hot spot than the Big Island of Hawaii, is smaller than Hawaii, because erosion has had more time to work on Kauai. Today the Hawaiian hot spot is located off the southeast coast of the Big Island of Hawaii, where it continues to power the active volcanoes. On top of the hot spot, the newest Hawaiian island, Loihi, is forming. Loihi is already a massive shield volcano, rising more than 3 km above the ocean floor. Loihi will eventually break the surface of the ocean and merge with the Big Island of Hawaii—but not for another 100,000 years.

Volcanism in the Solar System

Lunar photographs like the one in **Figure 6.21** show flowlike features in the dark regions of the Moon's surface. These dark regions are called **maria**, Latin for "seas," because some of the first observers to use telescopes thought that these dark areas looked like bodies of water. Maria are vast hardened lava flows, similar to volcanic rocks known as basalts on Earth. Because maria contain relatively few craters, these volcanic flows must have occurred *after* the heavy bombardment ceased. The lava that flowed across the lunar surface must have been relatively fluid to fill low-lying areas such as impact basins. The lava flowed too easily to pile up and form classic shield or composite volcanoes. The volcanic origin was confirmed when the *Apollo* astronauts returned rock samples from the lunar maria. Many of these samples contain gas bubbles typical of volcanic materials (**Figure 6.22**).

Analysis of the lunar rock samples also showed that most of the lunar lava flows were older than 3 billion years. Samples from the heavily cratered terrain of the Moon also originated from magma, so the young Moon must have gone through a molten stage. Together, these observations indicate that most of the sources of heating and volcanic activity on the Moon shut down some 3 billion years ago—unlike on Earth, where volcanism continues. This conclusion is consistent with the theory that smaller planets should cool more quickly than larger planets.

Mercury also shows evidence of past volcanism. *The Mariner 10* and *Messenger* missions revealed smooth plains similar in appearance to lunar maria. These sparsely cratered plains are the youngest areas on Mercury, created when fluid lava flowed into and filled huge impact basins. *Messenger* also found a number of volcanoes on Mercury.

Radar images of Venus reveal a wide variety of volcanic landforms. These include highly fluid flood lavas covering thousands of square kilometers, enormous volcanoes, and riverlike flows of lava thousands of kilometers long. These lavas must have been extremely hot and fluid to flow for such long distances.

Mars has also been volcanically active and may still be. More than half the surface of Mars is covered with volcanic rocks. Hardened lava covers huge regions of Mars, flooding the older, cratered terrain. Most of the vents or long cracks that created these flows are buried under the lava that poured forth from them. There are enormous shield volcanoes on Mars, including the largest mountains in the Solar System. Olympus Mons, standing 27 km high at its peak and 550 km wide at its base (**Figure 6.23**), would tower over Earth's largest mountains (the summit of Mount Everest is 8.8 km above sea level). Most of the largest volcanoes of Mars are shield volcanoes, just like the Hawaiian Islands. Olympus Mons and its neighbors have remained over their hot spots for billions of years, growing ever taller and broader with each successive eruption. Because the hot spot has remained stationary over time, rather than creating a chain of volcanoes, we can conclude that Mars lacks plate tectonics. Volcanic features span nearly the entire history of Mars, from the formation of the crust about 4.4 billion years ago to geologically recent times. "Recent" in this sense could still be more than 100 million years ago; the age of "fresh-appearing" lava flows cannot be determined until rock samples are obtained. Mars could, in principle, experience eruptions today.

Figure 6.22 This rock sample from the Moon, collected by *Apollo 15* astronauts from a lunar lava flow, shows gas bubbles typical of gas-rich volcanic materials. This rock is about 6 × 12 cm.

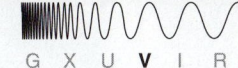

G X U V I R

Figure 6.23 The largest known volcano in the Solar System, Olympus Mons on Mars, is a 27-km-high shield-type volcano, similar to but much larger than Hawaii's Mauna Loa. This oblique view was created from an overhead *Viking Orbiter* image and topographic data provided by the *Mars Orbiter* laser altimeter.

CHECK YOUR UNDERSTANDING **6.4**

Lava flows on the Moon and Mercury created large, smooth plains. We don't see similar features on Earth because

a. there is less lava on Earth.

b. Earth had fewer large impacts in the past.

c. plate tectonics and erosion modify the surface of Earth.

d. Earth is large compared to the size of these plains, so they are not as noticeable.

e. the rotation rate of Earth is much faster than that of either the Moon or Mercury.

what if . . .

What if plate tectonics did not occur on Earth: How would the Hawaiian Islands be different in that case?

6.5 Wind and Water Modify Surfaces

Impacts, tectonism, and volcanism create valleys and mountains on the surface of a terrestrial world. **Erosion**, on the other hand, wears away a planet's surface by a variety of mechanical and chemical actions. Water, wind, and the actions of living organisms create river valleys, wind-sculpted hills, or mountains carved by glaciers. The resulting debris creates river deltas, sand dunes, and piles of rock at the bases of mountains and cliffs. If erosion were the only geological process operating, it would eventually smooth Earth's surface completely. Because Earth is a geologically and biologically active world, however, its surface is constantly changed by processes that create height variations and those that tear them down. In this section, we will examine the evidence for erosion on the terrestrial worlds.

Erosion is most pronounced on planets with water and wind. For example, on Earth, where water and wind are so powerful, most impact craters have been worn down and filled in. Even though the Moon and Mercury have almost no atmosphere and no running water, erosion is still at work. For example, landslides can occur wherever gravity and differences in elevation are present. Although water enhances landslide activity, landslides are seen on Mercury and the Moon.

Weathering

Weathering is the first step in the process of erosion. During **weathering**, rocks are broken into smaller pieces. For example, the pounding waves along shorelines break rocks into beach sand. Other weathering processes include chemical reactions, such as when oxygen in the air combines with iron in rocks to form a type of rust. Radiation from the Sun and from deep space can cause weathering on planets without atmospheres, by slowly decomposing some types of minerals in the top few millimeters of the surface. On the Moon and Mercury, impacts of micrometeoroids also chip away at rocks. One of the most efficient forms of weathering occurs when liquid water runs into crevices and then freezes. The water expands as it freezes, shattering the rock.

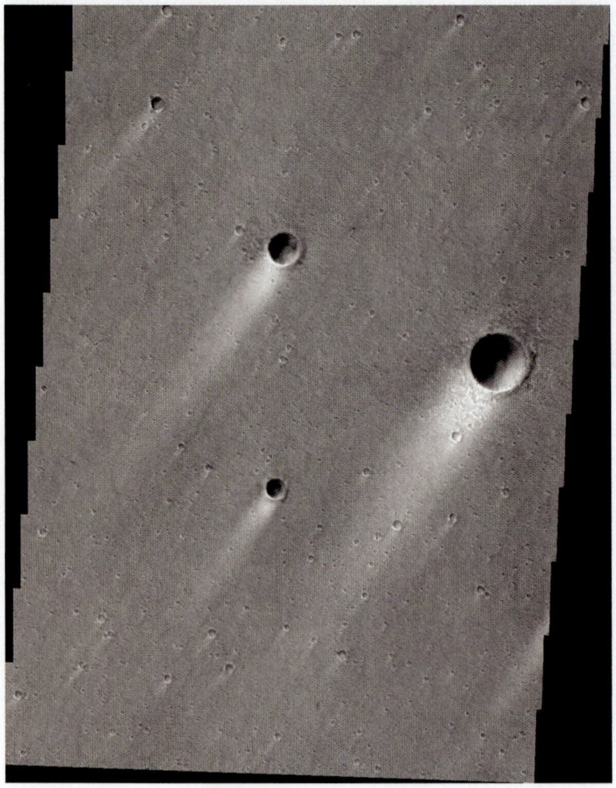

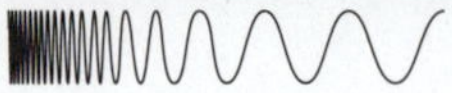

Wind Erosion

Earth, Mars, and Venus all show the effects of wind. Sand dunes, created when wind picks up and deposits small particles, are common on Earth and Mars, and dunes have been identified on Venus. Orbiting spacecraft have also found wind-eroded hills and surface patterns called wind streaks on Mars. These surface patterns appear, disappear, and change in response to winds blowing sediments around hills, craters, and cliffs. They serve as local "wind vanes," as shown in **Figure 6.24**, telling planetary scientists about the direction of local prevailing surface winds. Wind generates dust storms on Mars that can engulf the entire planet.

Water Erosion

Today, Earth is the only planet in the Solar System where the temperature and atmospheric conditions allow extensive liquid surface water to exist. Water is an extremely powerful agent of erosion and dominates erosion on Earth. Every year, rivers and streams on Earth deliver about 10 billion metric tons of sediment into the oceans. Even though there is no longer liquid water on the surface of Mars, at one time water likely flowed across its surface in vast quantities. Features

G X U V I R

Figure 6.24 These wind streaks in Chryse Planitia on Mars were observed by the *Mars Odyssey* thermal emission imaging system. The orientation of these streaks shows that the wind is coming from the upper right and blowing toward the lower left.

resembling water-carved channels on Earth, such as those shown in **Figure 6.25**, suggest the past presence of liquid water on Mars. In addition, many regions on Mars show small networks of valleys that are thought to have been carved by flowing water. Some parts of Mars may even have once contained oceans and glaciers.

The Search for Water in the Solar System

Evidence of water erosion on other planets in our Solar System has implications for the possibility of finding life beyond Earth. Life on Earth requires water as a solvent and as a delivery mechanism for essential chemistry. Because of this, the search for water is central to the search for life in the Solar System and also is an important consideration in the design of future human space travel.

In 2004, NASA sent two vehicles, *Opportunity* and *Spirit*, to search for evidence of water on Mars. *Opportunity* landed inside a small crater. For the first time, rocks on Mars were available for study in the original layers in which they were laid down. Previously, the only rocks that had been studied were those that had been dislodged from their original settings by either impacts or river floods.

The form of the layers at the *Opportunity* site was typical of layered sandy deposits laid down by gentle currents of water. Rover instruments detected a mineral so rich in sulfur that it had almost certainly formed by precipitation from water. Magnified images of the rocks showed "blueberries"—small spherical grains a few millimeters across that probably formed in place among the layered rocks. These are similar to terrestrial features that form by the percolation of water through sediments. Analysis of these blueberries revealed abundant hematite, an iron-rich mineral that forms only in the presence of water. Further observations by the European Space Agency's *Mars Express* and NASA's *Mars Odyssey* orbiters have shown the hematite signature and the presence of sulfur-rich compounds in a vast area surrounding the *Opportunity* landing site. These observations suggest an ancient martian sea larger than the combined area of Earth's Great Lakes and as much as 500 meters deep.

Spirit landed in Gusev, a 170-km-wide impact crater. This site was chosen because it showed signs of ancient flooding by a now-dry river. Scientists hoped that surface deposits would provide further evidence of past liquid water. Surprise, and perhaps some disappointment, followed when *Spirit* revealed that the flat floor of Gusev consisted primarily of basaltic rock. Only when the rover ventured cross-country to some low hills located 2.5 km from the landing site did it find basaltic rocks showing clear signs of having been chemically altered by liquid water.

In August 2012, the Mars rover *Curiosity* landed in Gale, a 150-km crater just south of the equator of Mars. *Curiosity* found evidence of a stream that flowed at a rate of about 1 m/s and was as much as 0.6 meter (2 feet) deep. The streambed is identified by water-worn gravel, shown in **Figure 6.26**. When the rover drilled into a rock, it found sulfur, nitrogen, hydrogen, oxygen, phosphorus, and carbon, together with clay minerals that formed in a water-rich, low-salt environment. Taken together, these pieces of evidence indicate that Mars may have had conditions suitable to support Earth-like microbial life in the distant past.

Where did the water go? Some evaporated into the thin atmosphere of Mars, and at least some of it is locked up as ice in the polar regions, just as the ice caps on Earth hold much of our planet's water. Unlike our own polar caps, those on Mars are a mixture of frozen carbon dioxide and frozen water. Water-ice crystals have been found mixed in with the Martian soil, and NASA's *Phoenix* lander found water ice just a few centimeters beneath surface soils at high northern latitudes

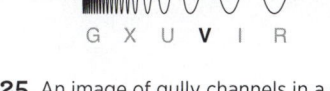

Figure 6.25 An image of gully channels in a crater on Mars taken by the *Mars Reconnaissance Orbiter*. The gullies coming from the rocky cliffs near the crater's rim (out of the image, to the upper left) show meandering and braided patterns similar to those of water-carved channels on Earth.

what if . . .

What if Mars had large quantities of liquid water on its surface throughout its history: How might water erosion have changed the appearance of Olympus Mons or Valles Marineris?

Figure 6.26 This image compares (a) a photograph taken on Mars by NASA's *Curiosity* rover with (b) a photograph of a streambed on Earth. The Mars image shows water-worn gravel embedded in sand, sure evidence of an ancient streambed.

G X U **V** I R   G X U **V** I R

(**Figure 6.27**). However, most of the water on Mars appears to be trapped well below the surface of the planet. Studies of eroding slopes by NASA's *Mars Reconnaissance Orbiter (MRO)* have found evidence of huge quantities of subsurface water ice, not only in the polar areas, as expected, but also at lower latitudes. The history of water on Mars is an area of active study by planetary scientists. Many questions remain about the past, present, and future of water on Mars, including questions about all the factors, like mass, magnetic field strength, and orbital and rotational properties, that have combined to make Mars much different now than it was in the past.

Evidence for water on Venus comes primarily from water vapor in its atmosphere, but color differences between highland and lowland regions hint that water may have been there in the past. This hint of prior water may indicate the presence of granite, which forms in the presence of water.

Although Earth and Mars are the only terrestrial planets that show strong evidence for liquid water at any time in their histories, water ice exists on the Moon and on Mercury today. Some deep craters in the polar regions of both Mercury and the Moon have floors in perpetual shadow. Temperatures in these permanently shadowed areas remain below 180 K. For many years, planetary scientists speculated that ice—perhaps from comets—could be found in these craters. The *Messenger* spacecraft has photographed water ice in the craters at Mercury's north pole.

In 2009, NASA's *Lunar Reconnaissance Orbiter (LRO)* and a companion satellite known as the *Lunar Crater Observation and Sensing Satellite (LCROSS)* were put into lunar orbit to continue the search for possible sources of subsurface water ice. *LCROSS* intentionally crashed the second stage of its Centaur rocket into a lunar polar-region crater, Cabeus. *LRO* collected data from the resulting plume, finding that more than 150 kg of water ice and water vapor were blown out of the crater by the impact. These data revealed the presence of large amounts of water buried beneath the crater's floor. Other evidence of water on the Moon was found by later spacecraft observations. These results are important to scientists and engineers who are thinking about future missions to the Moon and how astronauts will be supplied with water.

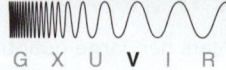

G X U **V** I R

Figure 6.27 Water ice (white region) appears a few centimeters below the surface of Mars in this trench dug by a robotic arm on the *Phoenix* lander. The trench measures about 20 × 30 cm.

CHECK YOUR UNDERSTANDING **6.5**

Erosion is most efficient on planets with
a. tectonism.
b. volcanoes.
c. wind and water.
d. large masses.

reading Astronomy News

The North Pole is mysteriously moving, and the U.S. government finally caught up

Sarah Kaplan, *The Washington Post*

Magnetic north is not where it used to be.

Since 2015, the place to which a compass points has been sprinting toward Siberia at a pace of more than 30 miles (48 kilometres) a year. And this week, after a delay caused by the month-long partial government shutdown in the United States, humans have finally caught up.

Scientists on Monday released an emergency update to the World Magnetic Model, which cellphone GPS systems and military navigators use to orient themselves.

It's a minor change for most of us—noticeable only to people who are attempting to navigate very precisely very close to the Arctic.

But the north magnetic pole's inexorable drift suggests that something strange—and potentially powerful—is taking place deep within Earth. Only by tracking it, said University of Leeds geophysicist Phil Livermore, can scientists hope to understand what's going on.

The planet's magnetic field is generated nearly 2,000 miles (3,200 kilometres) beneath our feet, in the swirling, spinning ball of molten metal that forms Earth's core.

Changes in that underground flow can alter Earth's magnetic field lines—and the poles where they converge.

Consequently, magnetic north doesn't align with geographic north (the end point of Earth's rotational axis), and it's constantly on the move. Records of ancient magnetic fields from extremely old rocks show that the poles can even flip—an event that has occurred an average of three times every million years.

The first expedition to find magnetic north, in 1831, pinpointed it in the Canadian Arctic. By the time the US Army went looking for the pole in the late 1940s, it had shifted 250 miles (400 kilometres) to the northwest.

Since 1990, it has moved a whopping 600 miles (970 kilometres), and it can be found in the middle of the Arctic Ocean, 4 degrees south of geographic north—for the moment.

Curiously, the south magnetic pole hasn't mirrored the peregrinations of its northern counterpart. Since 1990, its location has remained relatively stable, off the coast of eastern Antarctica.

Livermore's research suggests that the North Pole's location is controlled by two patches of magnetic field beneath Canada and Siberia. In 2017, he reported that the Canadian patch seems to be weakening, the result of a liquid iron sloshing through Earth's stormy core.

Speaking at a meeting of the American Geophysical Union in December, he suggested that the tumult far below the Arctic may explain the movement of magnetic field lines above it.

Scientists for the National Oceanic and Atmospheric Administration (NOAA) and the British Geological Survey (BGS) collaborate to produce a new World Magnetic Model—a mathematical representation of the field—every five years. The next update wasn't scheduled until 2020.

But Earth had other plans. Fluctuations in the Arctic were occurring faster than predicted.

By the summer, the discrepancy between the World Magnetic Model and the real-time location of the north magnetic pole had nearly exceeded the threshold needed for accurate navigation, said William Brown, a geomagnetic field modeler for the BGS.

He and his US counterparts worked on a new model, which was nearly ready to be released when much of the US federal government ran out of funding.

Though the British agency was able to publish elements of the new model on its site, NOAA was responsible for hosting the model and making it available for public use. This portion of the model didn't become available until Monday, a week after most NOAA employees were able to go back to work.

Some have speculated that Earth is overdue for another magnetic field reversal—an event that hasn't happened for 780,000 years—and the North Pole's recent restlessness may be a sign of a cataclysm to come.

Livermore was skeptical. "There's no evidence" that the localized changes in the Arctic are a sign of something bigger, he said.

Anyway, magnetic field reversals have typically unfolded over the course of 1,000 years or so—giving plenty of time for even the US federal government to adjust.

This article was originally published by The Washington Post.

EVALUATING THE NEWS

1. The author of the article states that the motion of the magnetic north pole is important to "people who are attempting to navigate very precisely very close to the Arctic." Why is the position of the magnetic north pole important for navigation?

2. Only the north magnetic pole is moving; the south magnetic pole is not. Draw a sketch like the one in Figure 6.11b, and use an arrow to indicate the point about which the metaphorical bar magnet must be turning in order for these two observations to be true.

3. Between 1831 and 1949, what was the average speed of the magnetic pole (in kilometers per year)? Between 1990 and 2015, what was the average speed of the magnetic pole (in kilometers per year)? How many times faster is the pole moving now than it was between 1831 and 1949?

4. Go online and find out what has happened since this textbook was published. Has the Earth's north magnetic pole continued to move so quickly?

SUMMARY

The terrestrial planets in the Solar System include Mercury, Venus, Earth, and Mars. Because Earth's Moon is similar in many ways to these terrestrial planets, it is included in the list of terrestrial worlds. The interiors of terrestrial planets were extremely hot at formation. Tectonism and volcanism are a result of these hot interiors. Over time, the interiors cool, and tectonic and volcanic activities weaken. On Earth, radioactive decay and tidal effects from the Moon continue to heat Earth's interior. Impacts deliver energy and some materials to the surface of a planet, tectonic and volcanic activity deform the surface, and erosion gradually scrubs away the evidence of impact cratering, tectonism, and volcanism. Surface features on the terrestrial planets, such as tectonic plates, volcanoes, mountain ranges, and canyons, are the result of the interplay among these four processes.

① Four processes shape the surfaces of the terrestrial planets: impact cratering, tectonism, volcanism, and erosion. Impact cratering is the result of a direct interaction of an astronomical object with the surface of the planet. Active volcanism and tectonics are the results of a "living" planetary interior: one that is still hot inside. Erosion is a surface phenomenon that results from weathering by wind or water.

② The relative positioning of craters gives their relative ages, with more recent craters found on top of older ones. Crater densities can be used to find the relative ages of regions on a surface, with more heavily cratered regions being older than less cratered ones. Finding the absolute age of a region requires a calibration using radioactive isotopes and their products. These two lines of inquiry are used to find the age of a surface on a terrestrial planet.

③ Earth's interior has been mapped using a combination of models and observation. Theoretical models predict how seismic waves should propagate through the interior. Repeatedly comparing actual observations of seismic waves with model predictions improves the models, so that an accurate picture of the interior of a planet is formed. The interiors of other planets cannot be tested with the same precision by direct observation of seismic waves, but instead verification is obtained by observations of gravitational and magnetic fields. Earth's larger size meant a hotter interior. This, combined with tidal heating from the Moon, means that Earth has had more prolonged geological activity than the other terrestrial planets.

④ Tectonism folds, twists, and cracks the outer surface of a planet. Plate tectonics is unique to Earth, although other types of tectonic disruptions are observed on the other terrestrial planets, such as cracking and buckling on the surface. Volcanism brings magma to the surface, where it flows as lava and reshapes the surface features.

⑤ Weathering by wind and water causes erosion on planetary surfaces. In general, the consequence of erosion is to erase features that are the result of impact cratering, tectonics, and volcanism. Geological evidence of water erosion is one factor in the search for water in the Solar System. This research is important to both the search for extraterrestrial life and the possibilities of human colonization of space.

QUESTIONS AND PROBLEMS

TESTING YOUR UNDERSTANDING

1. T/F: Volcanism has occurred on all the terrestrial planets.

2. T/F: The propagation of seismic waves reveals the structure of Earth's interior.

3. T/F: Large worlds remain geologically active longer than small ones.

4. T/F: Mercury has no volcanoes.

5. T/F: Wind erosion is an important process on Venus.

6. _____, _____, and _____ build up structures on the terrestrial planets, whereas in general, _____ tears them down.
   a. Impacts, erosion, volcanism; tectonism
   b. Impacts, tectonism, volcanism; erosion
   c. Tectonism, volcanism, erosion; impacts
   d. Tectonism, impacts, erosion; volcanism

7. If radioactive element A decays into radioactive element B with a half-life of 20 seconds, then after 40 seconds,
   a. none of element A will remain.
   b. none of element B will remain.
   c. half of element A will remain.
   d. one-quarter of element A will remain.

8. On which of the following worlds is wind erosion negligible? (Select all that apply.)
   a. Mercury   b. Venus
   c. Earth   d. Moon
   e. Mars

9. On which of the following worlds has erosion by wind and water occurred? (Select all that apply.)
   a. Mercury   b. Venus
   c. Earth   d. Moon
   e. Mars

10. Which of the following worlds shows evidence of water? (Select all that apply.)
    a. Mercury   b. Venus
    c. Earth   d. Moon
    e. Mars

11. Of the four processes that shape the surface of a terrestrial world, the one with the greatest potential for future catastrophic rearrangement is
    a. impacts.   b. volcanism.
    c. tectonism.   d. erosion.

12. Geologists can find the relative age of impact craters on a world because
    a. the ones on top must be older.
    b. the ones on top must be younger.
    c. the larger ones must be older.
    d. the larger ones must be younger.
    e. all the features we can see are the same age.

13. Geologists can find the actual age of features on a world by
    a. radioactive dating of rocks retrieved from the world.
    b. comparing cratering rates on one world to those on another.
    c. assuming that all features on a planetary surface are the same age.
    d. both a and b
    e. both b and c

14. Impacts on the terrestrial worlds
    a. are more common than they used to be.
    b. have occurred at approximately the same rate since the formation of the Solar System.
    c. are less common than they used to be.
    d. periodically become more common and then are less common for a while.
    e. never occur any more.

15. Why does Earth have fewer craters than Venus?
    a. Earth's atmosphere provides better protection than that of Venus.
    b. Earth is a smaller target than Venus.
    c. Earth is closer to the asteroid belt.
    d. Earth's surface experiences more erosion.

16. Other than Earth, which of the terrestrial worlds may still be geologically active?
    a. The Moon and Mercury
    b. Mars and Venus
    c. Venus and Mercury
    d. Venus only

17. Spring tides occur only when
    a. the Sun is near the vernal equinox in the sky.
    b. the Moon is in first or third quarter.
    c. the Moon, Earth, and Sun form a right angle.
    d. the Moon is in new or full phase.
    e. the Sun is in full phase.

18. On Earth, one high tide each day is caused by the Moon pulling on that side of Earth. The other is caused by
    a. the Sun pulling on the opposite side of Earth.
    b. the Earth rotating around so that the opposite side is under the Moon.
    c. the Moon pulling the center of Earth away from the opposite side, leaving a tidal bulge behind.
    d. the resonance between the rotation and revolution of the Moon.

19. Scientists know the history of Earth's magnetic field because
    a. the magnetic field hasn't changed since the formation of Earth.
    b. they see how it's changing today and project that back in time.
    c. the magnetic field gets frozen into rocks, and plate tectonics spreads them out.
    d. they compare the magnetic fields on other planets to Earth's.
    e. there are written documents of magnetic field measurements since the beginning of Earth.

20. Water erosion is an important ongoing process on
    a. all the terrestrial worlds.
    b. Earth only.
    c. Earth and Mars only.
    d. Earth, Mars, and the Moon only.
    e. Earth, Mars, and Venus only.

THINKING ABOUT THE CONCEPTS

21. ★ WHAT AN ASTRONOMER SEES Sketch the relative positions of the Earth, Moon, and Sun in Figure 6.1. The caption states that the Moon is approximately in first quarter phase. From your sketch and the appearance of the Earth in this image, determine whether the Moon is in waxing crescent phase or waxing gibbous phase. You may need to review Figure 2.15. ★

22. ★ WHAT AN ASTRONOMER SEES Study the surface of the Moon that is visible in the image in Figure 6.1. Is this surface old (dating to the period of early heavy bombardment) or young (formed much later)? How do you know? ★

23. In discussing the terrestrial planets, why do we include the Moon?

24. Explain how scientists know that rock layers at the bottom of Arizona's Grand Canyon are older than those found on the rim.

25. Suppose that you have two rocks, each containing a radioactive isotope and its decay products. In rock A, there is an equal amount of parent and daughter. In rock B, there is twice as much daughter as parent. Which rock is older? Explain how you know this from the information given.

26. Describe the sources of heating that are responsible for the generation of Earth's magma.

27. Explain why the Moon's core is cooler than Earth's.

28. Compare and contrast tectonism on Venus, Earth, and Mercury.

29. Make a table that shows which of the four processes (impact cratering, volcanism, tectonism, and erosion) occur on each of the five terrestrial worlds (Mercury, Venus, Earth, Mars, and the Moon), both now and in the past.

30. Describe the collision theory of formation of the Moon, including a description of how this theory explains observations of lunar properties.

31. Figure 6.9a shows two tidal bulges on Earth, both caused by the Moon. Compare this figure to Figure 6.10. In your own words, describe why the Moon's gravity causes two tides on Earth—one on the side closest to the Moon and one on the side farthest away. ★

32. Describe and explain the evidence for reversals in the polarity of Earth's magnetic field. Study Figure 6.13a. How often do reversals happen, on average: once every few million years, once in about 10 million years, or once in about 100 million years? ★

33. Why do earthquakes and volcanoes tend to occur near plate boundaries?

34. Does the age of a planetary surface tell you the age of the planet? Why or why not?

35. Explain some of the evidence that Mars once had liquid water on its surface. Why is there no liquid water on Mars today?

APPLYING THE CONCEPTS

36. Suppose you find a piece of ancient pottery and take it to the laboratory of a physicist friend. He finds that the glaze contains radium, a radioactive element that decays to radon and has a half-life of 1,620 years. He tells you that the glaze couldn't have contained any radon when the pottery was being fired but that it now contains three atoms of radon for each atom of radium. Follow Working It Out 6.1 to find the age of the pottery.
    a. To make a prediction, consider reasonable answers: there is still a significant amount of radium left (¼ of the sample), so only a small number of half-lives can have passed. Because the half-life is in thousands of years, your answer should come out in thousands of years, not hundreds or hundreds of thousands.
    b. Calculate the number of half-lives that have elapsed.
    c. Calculate the age of the sample from the number of half-lives and the length of the half-life.
    d. Check your work by verifying that your answer has units of years and that it is in the right numerical range (thousands of years).

37. Archaeological samples are often dated by radiocarbon dating. The half-life of carbon-14 is 5,700 years. Follow along with Working It Out 6.1 and problem 36 to answer these questions.
    a. After how many half-lives will the sample have only ¹⁄₆₄ as much carbon-14 as it originally contained?
    b. How much time will have passed?
    c. Suppose that the daughter of carbon-14 is present in the sample when it forms (even before any radioactive decay happens). Then you cannot assume that every daughter you see is the result of carbon-14 decay. If you did make this assumption, would you overestimate or underestimate the age of a sample?

38. The object that created Arizona's Meteor Crater was estimated to have a radius of 25 meters and a mass of 300 million kg. Follow Working It Out 6.2 to calculate the density of the impacting object, and explain what that may tell you about its composition.
    a. Make a prediction: Should your answer be in hundreds of kilograms per cubic meter (wood), thousands of kilograms per cubic meter (ice, water, or rock), or tens of thousands of kilograms per cubic meter (pure uranium)? (Hint: What kinds of objects potentially made this crater?)
    b. Calculate the density of the impacting object, and draw a conclusion about its composition.
    c. Check your work by verifying that the density (and therefore the composition) you found was reasonable.

39. The east coast of South America and the west coast of Africa are separated by an average distance of about 4,500 km. GPS measurements indicate that these continents are now moving apart at a rate of about 3.75 cm/year. Use the relationship between distance, rate, and time (rate = distance/time) to find out how long ago these two continents separated. You may assume the rate has been constant. Check your work by making sure that the units of your answer are "years."

40. The average temperature of Mars is about 210 K. Review Working It Out 5.1 to answer the following questions.
    a. How much energy per second is emitted from a square meter of Mars? Check your work by verifying that your answer is in watts per square meter.
    b. What is the peak wavelength of radiation from a blackbody with the same temperature as Mars? Check your work by verifying that your answer falls in the infrared region of the electromagnetic spectrum.

exploration Earth's Tides

digital.wwnorton.com/universe4

Visit the Student Site at the Digital Resources page and open the Tidal Bulges Interactive Simulation. Before you start the simulation, examine the setup. You are looking down on Earth from the North Pole.

1. Are the sizes of Earth and the Moon approximately to scale in this image?

_____

_____

2. Is the distance between them to scale in this image?

_____

_____

3. Are the tides shown to scale?

_____

_____

4. Explain why the authors of the simulation made the scaling choices they did.

_____

_____

_____

5. In this position, is the east coast of North America experiencing high or low tide?

_____

_____

Recall from Chapter 2 that Earth rotates counterclockwise when viewed from this vantage point. Click the box that says "Include Effects of Earth's Rotation."

6. What happened to the tidal bulges?

_____

_____

7. Why does Earth's rotation have this effect?

_____

_____

Click the box that says "Play Animation."

8. Over the course of 1 day, how many high tides does the east coast of North America experience?

_____

9. How many low tides does it experience in 1 day?

_____

10. Why are there two high tides? That is, what causes the tide on the side of Earth away from the Moon?

_____

_____

_____

11. Do the tides change as the Moon orbits?

_____

_____

_____

Once the Moon has orbited back to the right side of the window, stop the simulation by clicking "Pause Animation." Click the box that says "Include Sun's Gravity."

12. What happened to the tides when you added the Sun's effect to the simulation?

_____

_____

_____

Now, run the simulation again. Stop the simulation when the Moon is at first quarter. Remove the check mark from the "Include Sun's Gravity" box.

13. What changed about the tides when you removed the Sun's effect?

_____

_____

_____

Run the simulation until the Moon is full. Stop the simulation by clicking "Pause Animation." Click the box that says "Include Sun's Gravity."

14. What changed about the tides when you added the Sun's effect back in?

_____

_____

15. Does the Moon or the Sun dominate the tides on Earth?

_____

_____

Atmospheres of Venus, Earth, and Mars

Throughout the history of Earth, life has modified the atmosphere and the surface, as you will learn here in Chapter 7. In this experiment, you will see that even a small amount of microscopic life can significantly modify its environment. Stretch out a balloon by blowing it up and letting the air out a few times. Combine 1 cup of very warm water, 2 tablespoons of sugar, and 1 packet of active dry yeast in a clean soda bottle, and mix them together until bubbling starts, which is the sign that the yeast are producing CO_2. Follow the instructions to collect your data. After the first measurement, begin to sketch a graph with circumference on the y-axis and time on the x-axis. How do you expect the balloon's size to increase with time? Continue measuring the balloon's circumference until it stops expanding. Graph your results and compare them to your prediction. How long did it take for the balloon to stand upright? Why did the balloon eventually stop expanding?

EXPERIMENT SETUP

Stretch out a balloon by blowing it up and letting the air out a few times. Combine 1 cup of very warm water, 2 tablespoons of sugar, and 1 packet of active dry yeast in a clean soda bottle. Mix them together until bubbling starts.

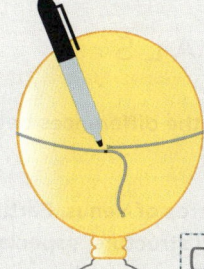

Attach the balloon to the top of the bottle. Note the time, and wait for the balloon to stand upright above the bottle.

Note the time again, and measure the circumference of the balloon. At even time intervals (every few minutes), measure the circumference of the balloon, as you did in Chapter 5. Continue measuring the balloon's circumference until it stops expanding.

 00:00

   00:10 00:12 00:14

PREDICTION

I predict that the circumference of the balloon will stop changing

☐ abruptly.   ☐ smoothly.

SKETCH OF RESULTS (in progress)

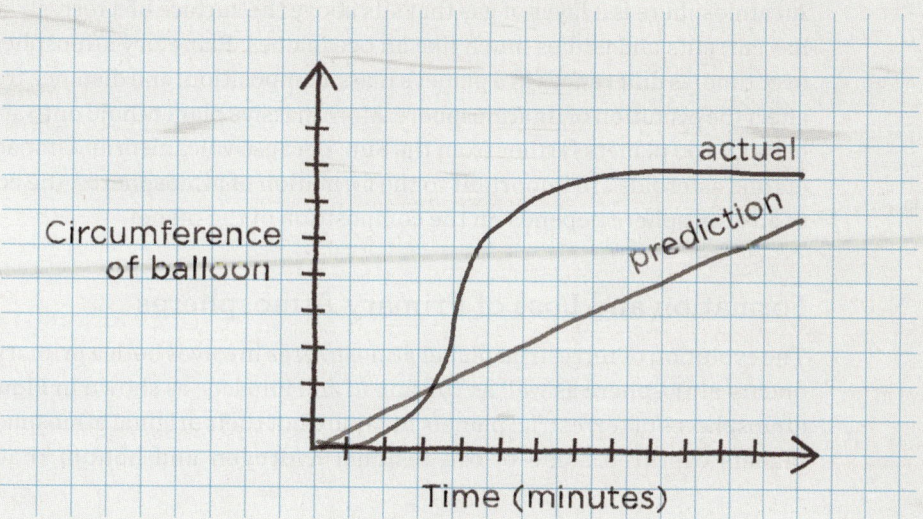

Circumference of balloon

actual

prediction

Time (minutes)

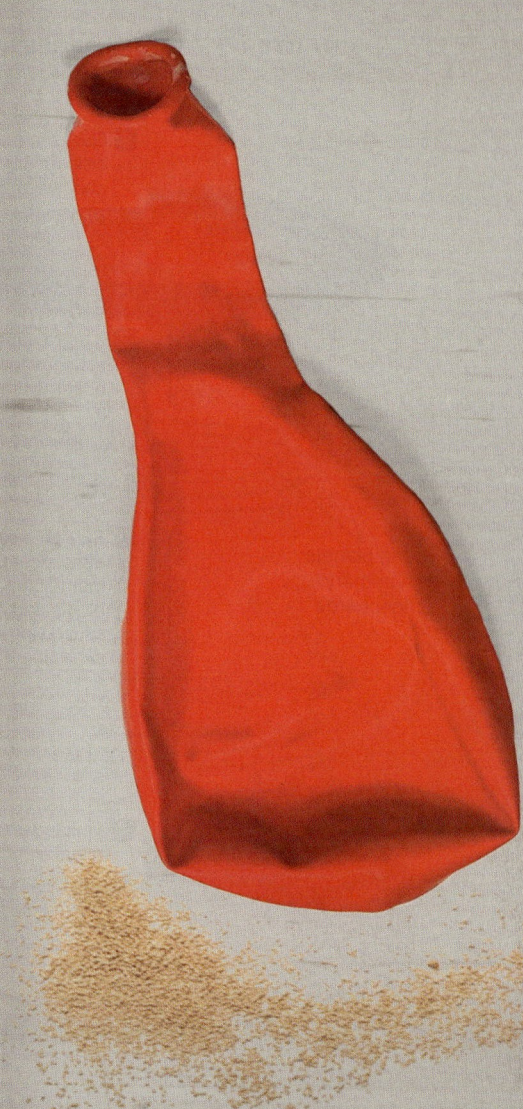

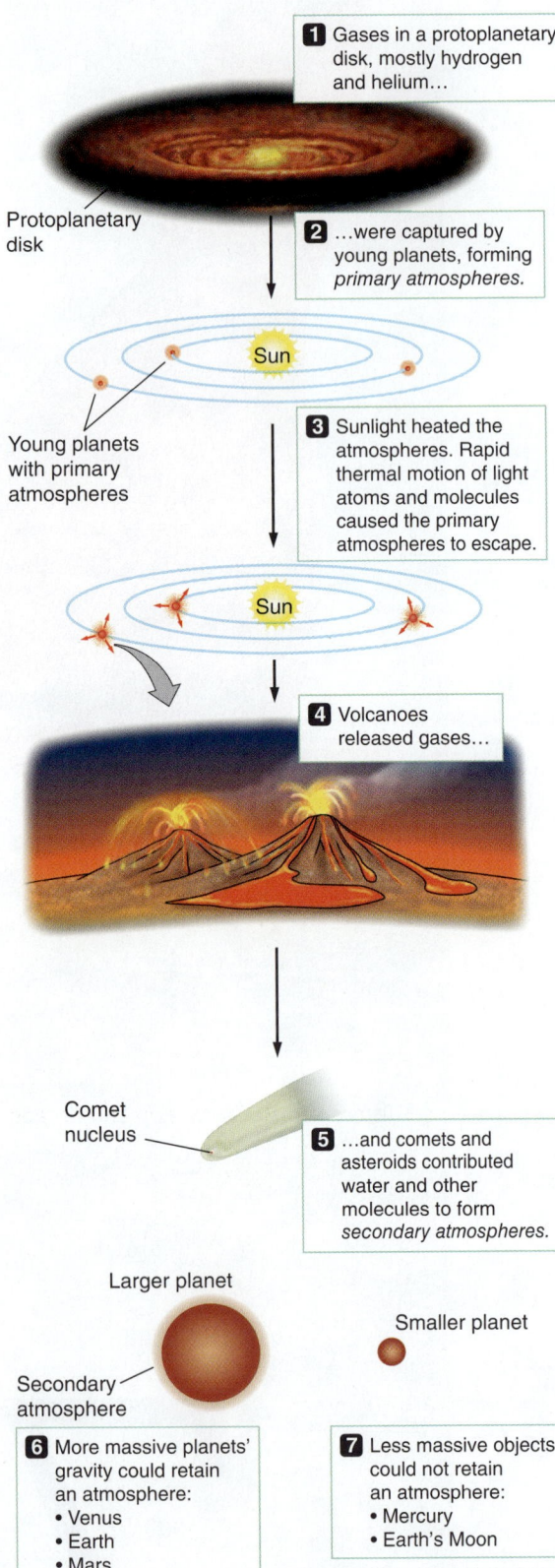

1 Gases in a protoplanetary disk, mostly hydrogen and helium…

Protoplanetary disk

2 …were captured by young planets, forming *primary atmospheres.*

Sun

Young planets with primary atmospheres

3 Sunlight heated the atmospheres. Rapid thermal motion of light atoms and molecules caused the primary atmospheres to escape.

Sun

4 Volcanoes released gases…

Comet nucleus

5 …and comets and asteroids contributed water and other molecules to form *secondary atmospheres.*

Larger planet

Smaller planet

Secondary atmosphere

6 More massive planets' gravity could retain an atmosphere:
• Venus
• Earth
• Mars

7 Less massive objects could not retain an atmosphere:
• Mercury
• Earth's Moon

Figure 7.1 Terrestrial planet atmospheres form and evolve in phases.

Without an atmosphere, Earth would look something like the Moon, and life as we know it would not exist. Earth's thick blanket of atmosphere warms and sustains the climate and determines weather patterns. Water is an important component of these weather patterns. The Sun heats water so that it evaporates from lakes and oceans, moving the water into the atmosphere. Rain and snow deposit the water on mountaintops and plains, where it flows into streams and rivers. Earth's atmosphere also interacts with particles and radiation that flow Earthward from the Sun. Of the five terrestrial bodies that we discussed in Chapter 6, only Venus and Earth have dense atmospheres. Mars has a very low-density atmosphere, and the atmospheres of Mercury and the Moon are so sparse that they can hardly be detected. To understand the origins and evolution of the atmospheres of Venus, Earth, and Mars, we must look back nearly 5 billion years to a time when the planets were just completing their growth from planetesimals.

LEARNING GOALS

1 Explain the origins of and the differences between primary and secondary atmospheres.

2 Explain why the atmospheres of Venus, Earth, and Mars change with altitude and how this leads to a layered structure, especially in Earth's atmosphere.

3 Describe how convection and the Coriolis effect contribute to Earth's weather.

4 Compare the strength of the greenhouse effect on Venus, Earth, and Mars and how this contributes to differences among the atmospheres of these three planets.

5 Describe how Earth's atmospheric composition has been reshaped by life.

6 Describe the evidence that shows Earth's climate is changing.

7.1 Atmospheres Change over Time

An atmosphere is a layer of gas that sits above the surface of a terrestrial planet. It has currents and eddies much like an ocean does. Planetary atmospheres change over time, as differences in a planet's mass, composition, and distance from the Sun affect the evolution of its atmosphere. More massive planets hold onto atmospheres better, as do planets farther from the Sun. Because volcanism and impacts by comets and asteroids are important to the formation of atmospheres, the composition of an atmosphere depends on the composition of the planet.

Formation and Loss of Primary Atmospheres

The evolution of terrestrial planet atmospheres involves both a primary and a secondary atmosphere as well as volcanism and impacts, as shown in **Figure 7.1**. The atmospheres of terrestrial planets today are not their original atmospheres. Young planets captured some of the residual hydrogen and helium that filled the protoplanetary disk surrounding the Sun (see Figure 7.1, steps 1 and 2), forming a **primary atmosphere**. This gas capture continued until the supply of gas ran out, and then this primary atmosphere composed of lightweight atoms and molecules escaped from the planet's gravity and was lost. To understand how this happens, you must learn how particles move within a planetary atmosphere.

Imagine a large box full of air. If the box is left alone for a while, all the particles inside it will collide enough times that, regardless of their mass, they will all have the same average kinetic energy. This is called **thermal equilibrium**. The kinetic energy of a molecule is determined by its mass and its speed (kinetic energy $= \frac{1}{2}mv^2$), so if the particles all have the same average energy, then the lightest molecules must be moving faster than the more massive ones. In a mixture of hydrogen and oxygen at room temperature, for example, the hydrogen molecules travel at about 2,000 meters per second (m/s), while the much more massive oxygen molecules move at only 500 m/s. These are the *average* speeds; some of the molecules will be moving much faster or much slower than average.

Near the surface of a planet, the density of molecules is high, so molecules collide frequently. These collisions tend to slow down the fast molecules and speed up the slow ones. Fast molecules near the ground therefore are likely to slow down and remain near the ground. Higher regions of the atmosphere are less dense, so fast molecules in the upper atmosphere rarely collide with other molecules. Instead, these fast molecules can escape the atmosphere into space. They will escape provided they are heading more or less upward and have a velocity greater than the escape velocity (recall Chapter 3). Putting all of this together, lighter molecules such as hydrogen and helium move faster than more massive molecules such as nitrogen or carbon dioxide, so lighter molecules are more likely to escape the planet's gravity. In the early Solar System, the terrestrial planets lost the hydrogen and helium they had acquired as a primary atmosphere (see Figure 7.1, step 3). This process was likely assisted by collisions between planetesimals, which added energy to the atmosphere of the forming planet.

Formation of Secondary Atmospheres

Earth's primary atmosphere is long gone. The current atmosphere was formed largely by accretion, volcanism, and impacts. In addition, life has played an important role in altering the composition of the atmosphere at least twice during Earth's history. During the planetary accretion process, minerals containing water, carbon dioxide, and other volatile matter collected in Earth's interior. Later, as the interior heated up, these gases were released from the minerals that had held them. Volcanism brought the gases to the surface, where they accumulated and created our **secondary atmosphere**, as shown in step 4 of Figure 7.1.

Impacts by huge numbers of comets and asteroids were another important source of atmospheric gases. As the giant planets of the outer Solar System grew, their gravity perturbed the orbits of comets and asteroids, scattering some of them into the inner Solar System. Some of these objects struck the terrestrial planets, releasing water, carbon monoxide, methane, and ammonia (see Figure 7.1, step 5). On Earth, and likely on Mars as well, most of the water vapor then rained down on the surface and flowed into depressions to form the earliest oceans.

Sunlight also influenced the composition of the secondary atmospheres. Ultraviolet (UV) light from the Sun easily fragments molecules such as ammonia and methane. UV light breaks ammonia into hydrogen and nitrogen atoms. The

▶❙❙ **AstroTour:** Atmospheres: Formation and Escape

VOCABULARY ALERT

perturb *Perturb* is uncommon in everyday usage, although *perturbed* is sometimes used to indicate that someone is upset. Astronomers use it in the context of changes to an object's orbit, often caused by gravitational interactions.

very light hydrogen atoms quickly escape to space, and the much heavier nitrogen atoms remain behind. Pairs of nitrogen atoms then combine to form more massive nitrogen molecules (N_2), which are even less likely to escape into space. Decomposition of ammonia by sunlight was the primary source of molecular nitrogen in the atmospheres of Venus, Earth, and Mars.

Mercury is relatively small, with weak gravity. It is also close to the Sun, so the temperature on the sunlit side is high. As a result, Mercury has lost nearly all of its secondary atmosphere to space, just as it had previously lost its primary atmosphere. Even molecules as massive as CO_2 (which is 44 times more massive than a hydrogen atom) or N_2 (28 times more massive than a hydrogen atom) can escape from a small planet if the temperature is high enough. Because the distance from the Sun to the Moon is much farther than the distance from the Sun to Mercury, the Moon is much cooler than Mercury. But the Moon's mass is so small that molecules can easily escape, even at low temperatures. Thus, both the Moon and Mercury have virtually no atmosphere today because of their small masses and, in the case of Mercury, its proximity to the Sun.

CHECK YOUR UNDERSTANDING **7.1**

Molecules can escape more easily from a _____ planet than a _____ one with the same radius.

a. low mass; high mass

b. high mass; low mass

c. brighter; darker

d. darker; brighter

Answers to Check Your Understanding questions are in the back of the book.

7.2 Secondary Atmospheres Evolve

Although Venus, Earth, and Mars started out with atmospheres of similar composition and comparable quantity, they ended up being very different from one another. All three are volcanically active today or have been volcanically active in their geological past. In addition, all three were subjected to impacts from comets and asteroids that took place in the early Solar System. Their similar geological histories suggest that their early secondary atmospheres might also have been quite similar. However, Earth's secondary atmosphere has changed significantly since it formed—in particular, the development of life has increased the amount of oxygen in the atmosphere. Earth's atmosphere is made up primarily of nitrogen (78 percent) and oxygen (21 percent), with only a trace of carbon dioxide. In contrast, the atmospheres of Venus and Mars today are nearly identical in composition—almost entirely carbon dioxide, with much smaller amounts of nitrogen. The masses of the atmospheres of these planets differ greatly, which in turn causes a large difference in the atmospheric conditions.

To understand why these atmospheres are so different, first recall from Chapter 6 some basic data about these planets: Venus is about 80 percent as massive as Earth. Venus and Earth have similar compositions, and their orbits are less than 0.3 astronomical unit (AU) apart—about one-third of Earth's average distance from the Sun. Mars is also similar in composition, but its mass is only about one-tenth that of Earth, and it is about 1.5 times farther away from the Sun than Earth.

The Effect of Planetary Mass on a Planet's Atmosphere

Despite the similarities in atmospheric composition, Mars and Venus have vastly different *amounts* of atmosphere. The atmosphere of Venus today is more than 2,500 times as massive as the atmosphere of Mars. However, Venus is only 8 times as massive as Mars. Since the two planets have a similar composition, Venus probably had about 8 times as much carbon within its interior to produce its carbon dioxide atmosphere. When their secondary atmospheres formed, the atmosphere of Venus was probably about 8 times more massive than the atmosphere of Mars. Venus has the gravitational pull necessary to hang onto its atmosphere, but since Mars is much smaller, it has much weaker gravity, so most of its secondary atmosphere has escaped. When a planet such as Mars begins to lose its atmosphere to space, the process begins to accelerate. With a thinner atmosphere, there are fewer slow molecules to keep fast molecules from escaping, and the rate of escape increases. This, in turn, leads to even less atmosphere and still faster escape rates. Mars had the added disadvantage of lacking a magnetic field to protect its atmosphere from fast-moving particles emitted by the Sun. These fast-moving particles helped strip away the atmosphere of Mars.

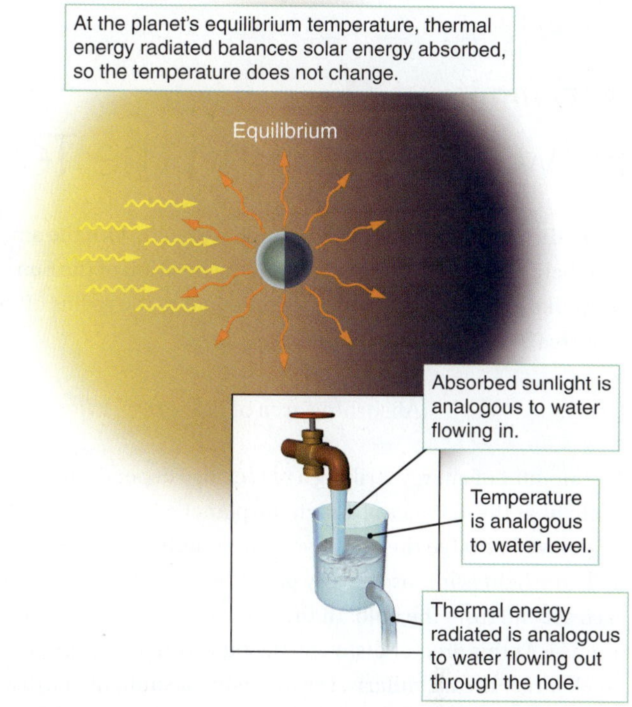

The Atmospheric Greenhouse Effect

Differences in the masses of the atmospheres of Venus, Earth, and Mars have a large effect on their surface temperatures. The temperature of a planet is determined by the balance between the amount of sunlight absorbed and the amount of energy radiated back into space. If the planet's temperature remains constant, then it must radiate as much energy as it absorbs, just as the bucket in **Figure 7.2** fills and empties at the same rate, keeping the water level constant. In **Working It Out 7.1**, we calculate the temperature of a round ball of rock by finding the equilibrium between the amount of energy it receives and the amount of energy it radiates. The results of these calculations are shown in **Figure 7.3**. This calculation gives good results for planets without atmospheres, but for Earth, and especially for Venus, the calculations are far from actual measured values. To match the observed temperatures of Earth and Venus, we must adjust for the **atmospheric greenhouse effect**, which traps solar radiation in the atmosphere.

At the planet's equilibrium temperature, thermal energy radiated balances solar energy absorbed, so the temperature does not change.

Equilibrium

Absorbed sunlight is analogous to water flowing in.

Temperature is analogous to water level.

Thermal energy radiated is analogous to water flowing out through the hole.

Figure 7.2 Planets are heated in part by absorbing sunlight and cooled by emitting thermal radiation into space. If there are no other sources of heating or means of cooling, then the equilibrium between these two processes determines the temperature of the planet.

Figure 7.3 Predicted planetary temperatures, based on the equilibrium between absorbed sunlight and thermal radiation into space, are compared with ranges of observed surface temperatures. Notice that the horizontal axis is a logarithmic scale.

= Range of measured temperatures

= Predicted temperature based on equilibrium model

working it out 7.1

How Can We Find the Temperature of a Planet?

To predict the temperature of a planet, we begin with the amount of sunlight being absorbed. As viewed from the position of the Sun, a planet looks like a circular disk with a radius equal to the radius of the planet, R. The area of the planet that is lit by the Sun is

$$(\text{Absorbing area of planet}) = \pi R^2$$

The amount of energy striking a planet also depends on the intensity of sunlight at the distance at which the planet orbits. Imagine a light bulb on a stage in a large theater. If you are standing on the stage with the light bulb, the light is intense—every part of your body receives a lot of energy each second from the bulb. In the back row of the theater, the light is less intense: At this greater distance, the bulb's energy is spread out over a much larger area. Similarly, the intensity of sunlight at a distance, d, is the luminosity of the Sun, L, divided by the area over which the light is spread, $4\pi d^2$:

$$(\text{Intensity of sunlight}) = \frac{L}{4\pi d^2}$$

A planet does not absorb all the sunlight that falls on it. **Albedo**, a, is the fraction of light that reflects from a planet. The fraction of the sunlight that is absorbed by the planet is 1 minus the albedo. A planet covered entirely in snow would have a high albedo (close to 1), whereas a planet covered entirely by black rocks would have a low albedo (close to 0).

$$(\text{Fraction of sunlight absorbed}) = 1 - a$$

We can now calculate the energy absorbed by the planet each second. Writing this relationship as an equation, we say that

$$\begin{pmatrix} \text{Energy} \\ \text{absorbed} \\ \text{each second} \end{pmatrix} = \begin{pmatrix} \text{Absorbing} \\ \text{area of} \\ \text{planet} \end{pmatrix} \times \begin{pmatrix} \text{Intensity} \\ \text{of} \\ \text{sunlight} \end{pmatrix} \times \begin{pmatrix} \text{Fraction of} \\ \text{sunlight} \\ \text{absorbed} \end{pmatrix}$$

$$= \pi R^2 \times \frac{L}{4\pi d^2} \times (1 - a)$$

The amount of energy that the planet radiates away depends on the surface area of the planet, $4\pi R^2$. According to the Stefan-Boltzmann law (recall Working It Out 5.1), each square meter radiates energy equal to σT^4 every second. Putting these two pieces together gives

$$\begin{pmatrix} \text{Energy} \\ \text{radiated} \\ \text{each second} \end{pmatrix} = \begin{pmatrix} \text{Surface} \\ \text{area of} \\ \text{planet} \end{pmatrix} \times \begin{pmatrix} \text{Energy radiated} \\ \text{per square meter} \\ \text{per second} \end{pmatrix}$$

$$= 4\pi R^2 \times \sigma T^4$$

🎥 **Astronomy in Action:** Changing Equilibrium

▶‖ **AstroTour:** Greenhouse Effect

The atmospheric greenhouse effect in planetary atmospheres and the **conventional greenhouse effect** operate in different ways, although the end results are much the same, so they have similar names. The conventional greenhouse effect occurs in a car on a sunny day when you leave the windows closed. Sunlight pours through the glass, heating the interior and raising the internal air temperature. With all the windows closed, hot air is trapped, and temperatures can climb as high as 180°F (about 80°C). Planetary atmospheres and the interiors of greenhouses are both heated by trapping the Sun's energy, but here the similarities end.

The atmospheric greenhouse effect is illustrated in **Figure 7.4**. Atmospheric gases freely transmit visible light, allowing the Sun to warm the planet's surface. The warmed surface radiates the excess energy in the infrared region of the spectrum. Some atmospheric gases strongly absorb infrared radiation and rerelease it in random directions. Some of this energy warms the atmosphere at ground level, so that convection can play a role in transporting thermal energy to the top of the atmosphere, where it can be more easily radiated to space. However, much of the energy returns to the ground and heats the planet's surface. As a result, the planet is heated by both the Sun and the atmosphere. Because of this warming

For a planet at constant temperature, each second the "energy radiated" must be equal to the "energy absorbed." When we set these two quantities equal to each other, we arrive at the expression

$$\left(\begin{array}{c}\text{Energy radiated}\\\text{each second}\end{array}\right) = \left(\begin{array}{c}\text{Energy absorbed}\\\text{each second}\end{array}\right)$$

$$4\pi R^2 \sigma T^4 = \pi R^2 \frac{L}{4\pi d^2}(1 - a)$$

or

$$T^4 = \frac{L}{16\sigma\pi d^2}(1 - a)$$

This gives us T^4, but we want just T. If we take the fourth root of each side, we get

$$T = \left[\frac{L(1 - a)}{16\sigma\pi d^2}\right]^{1/4}$$

Fortunately, L, σ, and π are known. If we express the distance from the Sun in astronomical units, the equation simplifies to

$$T = 279\,\text{K} \times \left[\frac{1 - a}{d_{\text{AU}}^2}\right]^{1/4}$$

For a blackbody ($a = 0$) at 1 AU from the Sun, the temperature is 279 kelvins (K). For Earth, with an albedo of 0.306 and a distance from the Sun of 1 AU, the temperature is

$$T = 279\,\text{K} \times \left[\frac{1 - 0.306}{1^2}\right]^{1/4}$$

$$T = 255\,\text{K}$$

Calculator hint: To take the fourth root of a number, you can use the x^y (or sometimes y^x) button on your calculator and put in 0.25 for the exponent, or you may have a button labeled $x^{1/y}$, which allows you to put in 4 for y. For example, if you are calculating $3^{1/4}$, you can either type [3][x^y][0][.][2][5] or [3][$x^{1/y}$][4].

Figure 7.3 shows the predicted and actual temperatures of the planets of the Solar System. Overall, there is fairly good agreement between predictions (black dots) and observations (orange bars), so our basic understanding of *why* planets have the temperatures they do is probably close to the mark. The data and predictions for Mercury and Mars agree particularly well.

For Earth and the giant planets, however, the actual temperatures are a bit higher than predicted. For Venus, the actual temperature is much higher than our prediction. As we built our mathematical model for the equilibrium temperatures of planets, we made a number of assumptions. For example, we assumed that the temperature of a planet is the same everywhere. This is clearly not true; for example, we might expect planets to be hotter on the day side than on the night side. We also assumed that a planet's only source of energy is sunlight. Finally, we assumed that a planet is able to radiate energy into space freely as a blackbody.

The discrepancies between our model and the measured temperatures tell us that, for some of these planets, some or all of these assumptions must be incorrect. The question of why these planets are hotter than predicted by this model leads us to a number of new and interesting insights, as described in the main text.

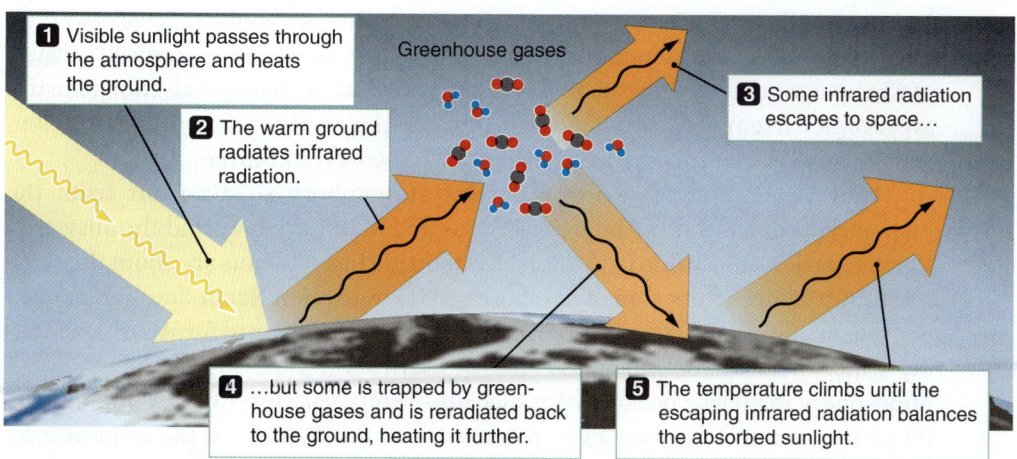

1 Visible sunlight passes through the atmosphere and heats the ground.

Greenhouse gases

2 The warm ground radiates infrared radiation.

3 Some infrared radiation escapes to space…

4 …but some is trapped by greenhouse gases and is reradiated back to the ground, heating it further.

5 The temperature climbs until the escaping infrared radiation balances the absorbed sunlight.

Figure 7.4 In the atmospheric greenhouse effect, greenhouse gases such as water vapor and carbon dioxide absorb infrared radiation and reradiate it in all directions, slowing the transport of energy out of the atmosphere. This causes the temperature of the planet to rise to a new equilibrium.

effect, gases that transmit visible radiation but absorb infrared radiation are known as **greenhouse gases**. Greenhouse gases found in Earth's atmosphere include carbon dioxide and water vapor, along with human-made methane, nitrous oxide, and chlorofluorocarbons (CFCs). The presence of greenhouse gases in an atmosphere increases the surface temperature of a planet.

If greenhouse gases are added to an atmosphere, the planet's surface temperature will rise until the surface becomes sufficiently hot—and therefore radiates enough energy—that the fraction of infrared radiation leaking out through the atmosphere once again balances the absorbed sunlight. Consider the bucket analogy in Figure 7.2. If less water is allowed to flow out of the hole at the bottom, the water level inside the bucket will rise, increasing the pressure, until the amount of water flowing out equals the amount of water flowing in. The system comes to a new equilibrium at a higher water level. Similarly, when greenhouse gases are added that keep energy from leaving a planet, the surface temperature will rise to a new equilibrium level.

Similarities and Differences among the Terrestrial Planets

Even though the atmosphere of Mars is composed almost entirely of carbon dioxide—an effective greenhouse gas—the atmosphere is very thin and contains relatively few greenhouse gas molecules compared to the atmospheres of Venus and Earth. As a result, the atmospheric greenhouse effect is relatively weak on Mars and raises the mean surface temperature by only about 5 K. At the other extreme, on Venus, the massive atmosphere of carbon dioxide and sulfur compounds raises its mean surface temperature by more than 400 K, to about 737 K. At such high temperatures, any water and most carbon dioxide locked up in surface rocks are driven into the atmosphere, further enhancing the atmospheric greenhouse effect. This comparison between Venus and Mars indicates that what really matters is the actual *number* of greenhouse gas molecules in a planet's atmosphere, not the *fraction* they represent. Non-greenhouse-gas atmospheric components (like the nitrogen molecules in Earth's atmosphere) are irrelevant to the greenhouse effect.

The atmospheric greenhouse effect on Earth is greater than it is on Mars but not as severe as it is on Venus—the average global temperature near Earth's surface is about 288 K (15°C). Temperatures on Earth's surface are about 33 K warmer than they would be in the absence of an atmospheric greenhouse effect, mainly because of water vapor and carbon dioxide. Without the natural greenhouse effect, Earth's average global temperature would be −18°C, well below the freezing point of water, leaving us with a world of frozen oceans and ice-covered continents.

Why is the composition of Earth's atmosphere so different from the high-carbon-dioxide atmospheres of Venus and Mars? We may find the answer in Earth's particular location in the Solar System. Earth and Venus are about the same mass, but Venus orbits somewhat closer to the Sun than Earth does. Volcanism poured out large amounts of carbon dioxide and water vapor to form early secondary atmospheres on both planets. Most of Earth's water quickly rained out of the atmosphere to fill vast ocean basins. Because Venus was closer to the Sun, though, its surface temperature was higher than that on Earth. Most of the rainwater on Venus re-evaporated, much as water does in Earth's desert regions. Venus was left with a surface that contained very little liquid water and an atmosphere filled with water vapor. The continuing buildup of both water vapor and carbon dioxide in the atmosphere of Venus led to a runaway atmospheric greenhouse effect that drove up the surface temperature. Ultimately, the surface of Venus became so hot that liquid water could not exist on it.

The early difference between a watery Earth and an arid Venus changed the ways that their atmospheres and surfaces evolved. On Earth, water erosion caused by rain and rivers continually exposed fresh minerals, which then reacted chemically with atmospheric carbon dioxide to form solid carbonates. This reaction removed some of the atmospheric carbon dioxide, burying it within Earth's crust as a component of a rock called limestone. Later, the development of life in Earth's oceans accelerated the removal of atmospheric carbon dioxide. Tiny sea creatures built their protective shells of carbonates, and as they died they built up massive beds of limestone on the ocean floors. Water erosion and the chemistry of life tied up all but a trace of Earth's total inventory of carbon dioxide in limestone beds. Earth's particular location in the Solar System seems to have spared it from the runaway atmospheric greenhouse effect. If all the carbon dioxide now in limestone beds had not been locked up by these reactions, Earth's atmospheric composition would resemble that of Venus or Mars: mostly carbon dioxide. The atmospheric greenhouse effect would be much stronger, and Earth's temperature would be much higher.

Earth today has 100,000 times more water than Venus, more than 97 percent of it in the oceans. What happened to all the water on Venus, which we have just suggested might have caused the runaway greenhouse effect? One possibility is that water molecules high in its atmosphere were broken apart into hydrogen and oxygen atoms by solar UV radiation. The low-mass hydrogen atoms were quickly lost to space, whereas the oxygen atoms migrated downward to the surface, where they were removed from the atmosphere by bonding with surface minerals. The *Venus Express* spacecraft has measured hydrogen and some oxygen escaping from the upper levels of the atmosphere.

CHECK YOUR UNDERSTANDING **7.2**

The _____ of greenhouse gas molecules affects the temperature of an atmosphere.

a. percentage

b. fraction

c. number

d. mass

7.3 Earth's Atmosphere Has Detailed Structure

Now that we understand some of the overall processes that have influenced the evolution of the atmospheres on terrestrial planets, we will look in depth at each of them. We begin with the composition, structure, and weather of Earth's atmosphere, not only because we know it best but also because it will help us better understand the atmospheres of other worlds.

Life and the Composition of Earth's Atmosphere

Earth's atmosphere is about four-fifths nitrogen molecules (N_2) and one-fifth oxygen molecules (O_2). There are also many important minor constituents, such as water vapor and carbon dioxide, the amounts of which vary depending on location and season.

Oxygen is a highly reactive gas. It chemically combines with, or oxidizes, almost any material it touches. The rust (iron oxide) that forms on iron is an example. Earth's atmosphere contains abundant amounts of oxygen, while the atmospheres of other planets do not. A planet that has significant amounts of oxygen in its atmosphere must be constantly replacing the oxygen that is lost through oxidation. On

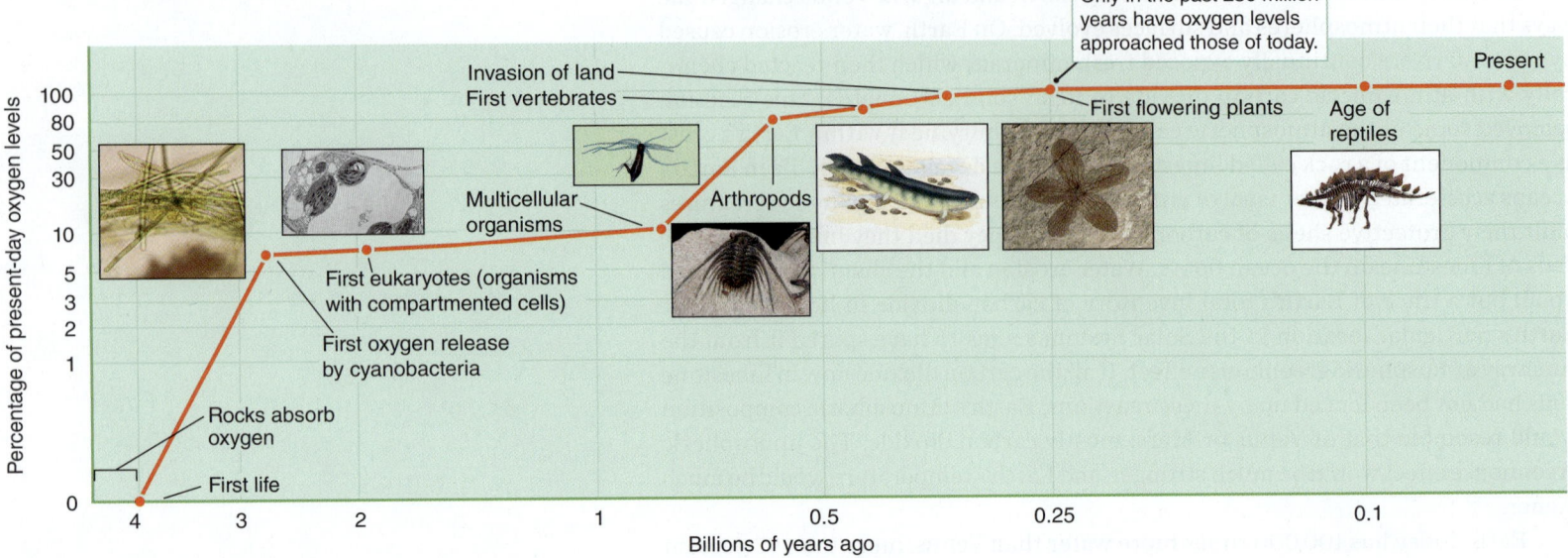

Figure 7.5 The amount of oxygen in Earth's atmosphere has built up over time due to photosynthesis. Notice that the axes are logarithmic, so that equal steps mean a factor of 10 on the *y*-axis and a factor of 2 on the *x*-axis.

what if . . .

What if you could use spectroscopy to detect different kinds of molecules in the atmospheres of planets orbiting other stars: What molecules would be suggestive of life on such a planet and why?

VOCABULARY ALERT

bar *Bar* has many everyday meanings, but scientists often use the bar as a unit of pressure equal to the average atmospheric pressure at sea level on Earth. A millibar (mbar) is one-thousandth of 1 bar and is more commonly used in meteorology and in weather reports.

Earth, plants produce about half of oxygen in our atmosphere. Algae and cyanobacteria produce the rest.

The oxygen concentration in Earth's atmosphere has changed over the history of the planet, as shown in **Figure 7.5**. When Earth's secondary atmosphere first appeared about 4 billion years ago, it had very little oxygen because free oxygen is not found in volcanic gases or comets. Studies of ancient sediments show that, about 2.8 billion years ago, an ancestral form of cyanobacteria (single-celled organisms that contain chlorophyll, which enables them to obtain energy from sunlight) began releasing oxygen into Earth's atmosphere as a waste product. At first, this oxygen combined readily with exposed metals and minerals in surface rocks and soils, so it was removed from the atmosphere as quickly as it formed. Ultimately, the explosive growth of plant life accelerated the production of oxygen, building up atmospheric O_2 concentrations 250 million years ago that approached today's levels. All plants use the energy of sunlight to build carbon compounds out of carbon dioxide and produce oxygen as a waste product in the process called photosynthesis. Earth's atmospheric oxygen content is held in a delicate balance primarily by plants, along with algae and cyanobacteria. If these disappeared, so would nearly all of Earth's atmospheric oxygen, and therefore all animal life—including humans.

The Layers of Earth's Atmosphere

Earth's atmosphere is a blanket of gas several hundred kilometers thick. It has a total mass of approximately 5×10^{18} kilograms (kg), which is less than one-millionth of Earth's total mass. Recall from Chapter 5 that the pressure at any point within a star must be great enough to balance the weight of the overlying layers. The same principle applies to a planetary atmosphere: The atmospheric pressure on a planet's surface must be great enough to hold up the weight of the overlying atmosphere. The atmospheric pressure at Earth's surface produces a force of 100,000 newtons on each square meter (N/m^2). This amount of pressure is called a **bar**, and it is equivalent to about 14.7 pounds pressing on every square inch, or about 130 pounds pressing on the palm of your hand. You are largely unaware of this pressure because the same pressure exists both inside and outside your body, so the force pushing out precisely balances the force pushing in.

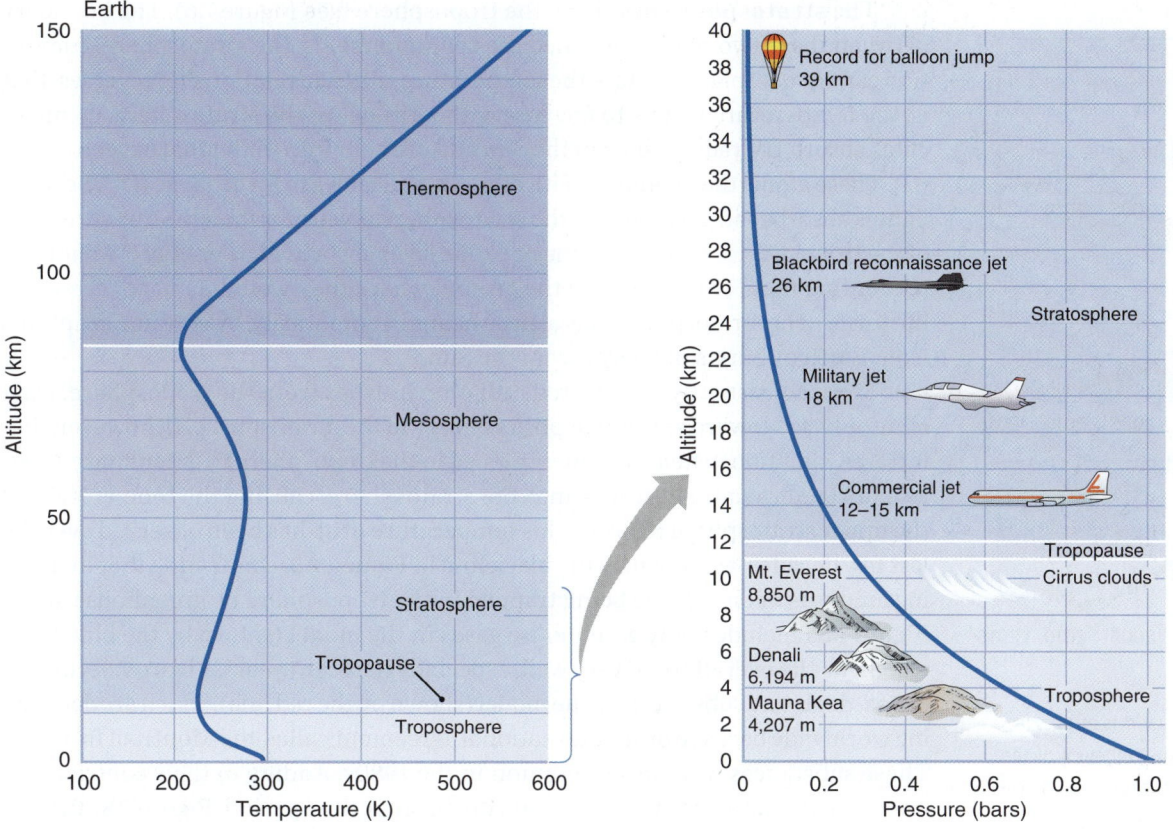

Figure 7.6 (a) Temperature and (b) pressure determine Earth's atmospheric layers. Most human activities are confined to the bottom layers of Earth's atmosphere.

Earth's atmosphere can be divided into layers that are distinguished by changes in temperature and pressure, as shown in **Figure 7.6**. The **troposphere** is the lowest atmospheric layer. The layer in which we live and breathe, it contains 90 percent of the mass of Earth's atmosphere. Within the troposphere, atmospheric pressure and temperature (as well as density) decrease as altitude increases. Convection circulates air between the lower and upper levels of the troposphere, minimizing the temperature difference caused by heating at the bottom and cooling at the top.

Convection also affects the way that water cycles through the environment. Warm air can hold more water vapor than cold air. As warm air is convected upward, it cools until the air can no longer hold all of its water vapor. At that temperature, water vapor begins to condense into tiny droplets or ice crystals. In large numbers, these form clouds. When these droplets combine to form large drops, they fall as rain or snow. This temperature effect causes most of the water vapor in Earth's atmosphere to stay within 2 kilometers (km) of the surface. Because water vapor strongly absorbs infrared light (it is a greenhouse gas), astronomers making infrared observations locate their observatories above as much water vapor as possible. At an altitude of 4 km, the Mauna Kea Observatories (Figure 7.6b) are higher than approximately one-third of Earth's troposphere, but they lie above nine-tenths of the atmospheric water vapor. The water in the atmosphere is often visible as condensed water in the form of clouds and ice, and these clouds can be seen below the observatories in **Figure 7.7**.

Figure 7.7 Observatories are usually located high on mountaintops, above most of Earth's water vapor. This improves access to the infrared portion of the spectrum. Visitors to the Mauna Kea Observatories in Hawaii have a view of the tops of the clouds, stretching away from the mountaintop.

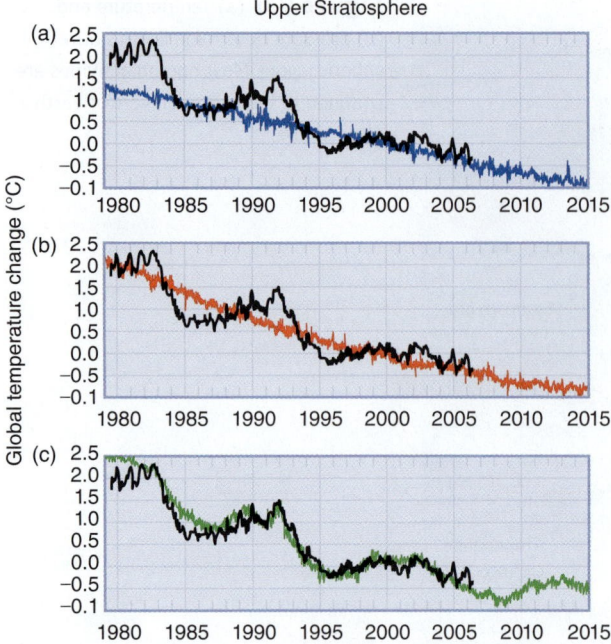

Figure 7.8 Climate scientists incorporate various effects in their models to find the best explanation for the behavior of the temperature of the stratosphere (black line). In this example, climate scientists (a) begin with greenhouse gases (GHG), then (b) add in ozone-depleting substances (ODS), and (c) finally include natural forcing factors, such as volcanoes and the solar cycle. Only when all three are included do the models provide a good explanation for changes in the upper stratosphere. These changes match the predictions that the upper stratosphere should cool when greenhouse gases accumulate in Earth's atmosphere.

VOCABULARY ALERT

plasma In common usage, *plasma* almost always refers to a component of blood (or a type of television). Its scientific meaning, though, refers to a gas in which most of the atoms and molecules have lost some electrons, leaving them with a net positive charge. Because these particles are charged, they interact with electric and magnetic fields.

The **stratosphere** lies above the troposphere (see Figure 7.6). The boundary between these two regions is called the **tropopause.** In the stratosphere, convection cannot take place because the temperature-altitude relationship reverses; that is, the temperature begins to *increase* with altitude. In the stratosphere, O_2 molecules absorb UV radiation from the Sun and are split into individual oxygen atoms (O). These atoms then combine with other O_2 molecules to form ozone (O_3). Because light is absorbed in this process, the stratosphere is warmer than the upper regions of the troposphere. The stratospheric ozone layer absorbs high-energy UV photons, preventing them from reaching the ground where they would damage terrestrial life forms. The stratosphere may also be heated, temporarily, by volcanic eruptions from Earth or bursts of energy from the Sun.

More consistently, as infrared radiation escapes from the troposphere, traveling outward toward space, it deposits energy in the stratosphere. If, however, that infrared radiation instead becomes trapped in the troposphere by greenhouse gases, less infrared radiation escapes into the stratosphere, causing the temperature of the upper stratosphere to drop. This temperature drop has been observed over the last few decades, as shown by the black line in **Figure 7.8**, confirming that increasing amounts of energy are being trapped in the troposphere by greenhouse gases. Figure 7.8a includes only greenhouse gases in the model (shown by the blue line), which fits the overall trend very well, especially in recent years. Figure 7.8b adds in ozone-depleting substances to the model (shown by the red line), which are decreasing worldwide because of an international agreement called the Montreal Protocol. These substances were more common in the 1980s. Adding in their contribution improves the agreement with the data during that time period. Figure 7.8c further adds in natural contributions, such as large volcanic eruptions and variations of the Sun. Using all three contributions together provides the best agreement between predictions from the model (shown by the green line) and observations in the atmosphere, showing that the long-term cooling trend in the upper atmosphere is due to greenhouse gases accumulating in the atmosphere. This is precisely as predicted by climate scientists.

The region above the stratosphere is called the **mesosphere**. It extends from an altitude of about 55 km to 85 km, as shown in Figure 7.6. In the mesosphere there is no ozone to absorb sunlight, so temperatures once again decrease with altitude. The base of the stratosphere and the upper boundary of the mesosphere are the two coldest levels in Earth's atmosphere.

Above the mesosphere, in the **thermosphere**, interactions with the space environment begin to be important. Solar UV radiation and the high-energy particles that stream continually from the Sun, called the solar wind, liberate electrons from the atoms and molecules in the thermosphere, ionizing those atoms and molecules. This once again causes the temperature to increase with altitude, as shown in Figure 7.6. The thermosphere is the hottest part of the atmosphere. Near the top of the thermosphere, at an altitude of 600 km, the temperature can reach 1000 K. However, if you were in a high-altitude balloon at this height, you would not feel warm, because there are so few air particles with so much distance between them that they would rarely hit you and would not transfer much energy to your skin. The gases ionized in the thermosphere form an electrically charged **plasma** layer known as the **ionosphere** (not shown in Figure 7.6). This layer overlaps the thermosphere but also extends farther into space.

Even farther out is Earth's **magnetosphere**, defined by the dominance of Earth's magnetic field (recall Figure 6.11). The magnetosphere's radius is approximately 10 times the radius of Earth, and its volume is greater than 1,000 times the volume of Earth. This large region is filled with charged particles from the Sun that have been captured by Earth's magnetic field. Charged particles can move freely *along* the direction of a magnetic field line but cannot cross magnetic field lines. If a charged particle attempts to move *across* a field line, it experiences a force that causes it to turn in a circle around the field line, as illustrated in **Figure 7.9a**. Notice that the particles are both turning in a circle and moving along a linear path. The result is that they travel in a helical path (shown by red lines). Charged particles in the solar wind are trapped in this region as they try to cross Earth's magnetic field lines.

Earth's magnetic field is pinched together at the north and south magnetic poles, so that the field lines become closer together and the magnetic field is stronger, as shown in Figure 7.9b. Charged particles moving into the pinch feel a force that reflects them back along the magnetic field lines, creating a sort of "magnetic bottle" that contains the charged particles. These charged particles may bounce

Astronomy in Action: Charged Particles and Magnetic Forces

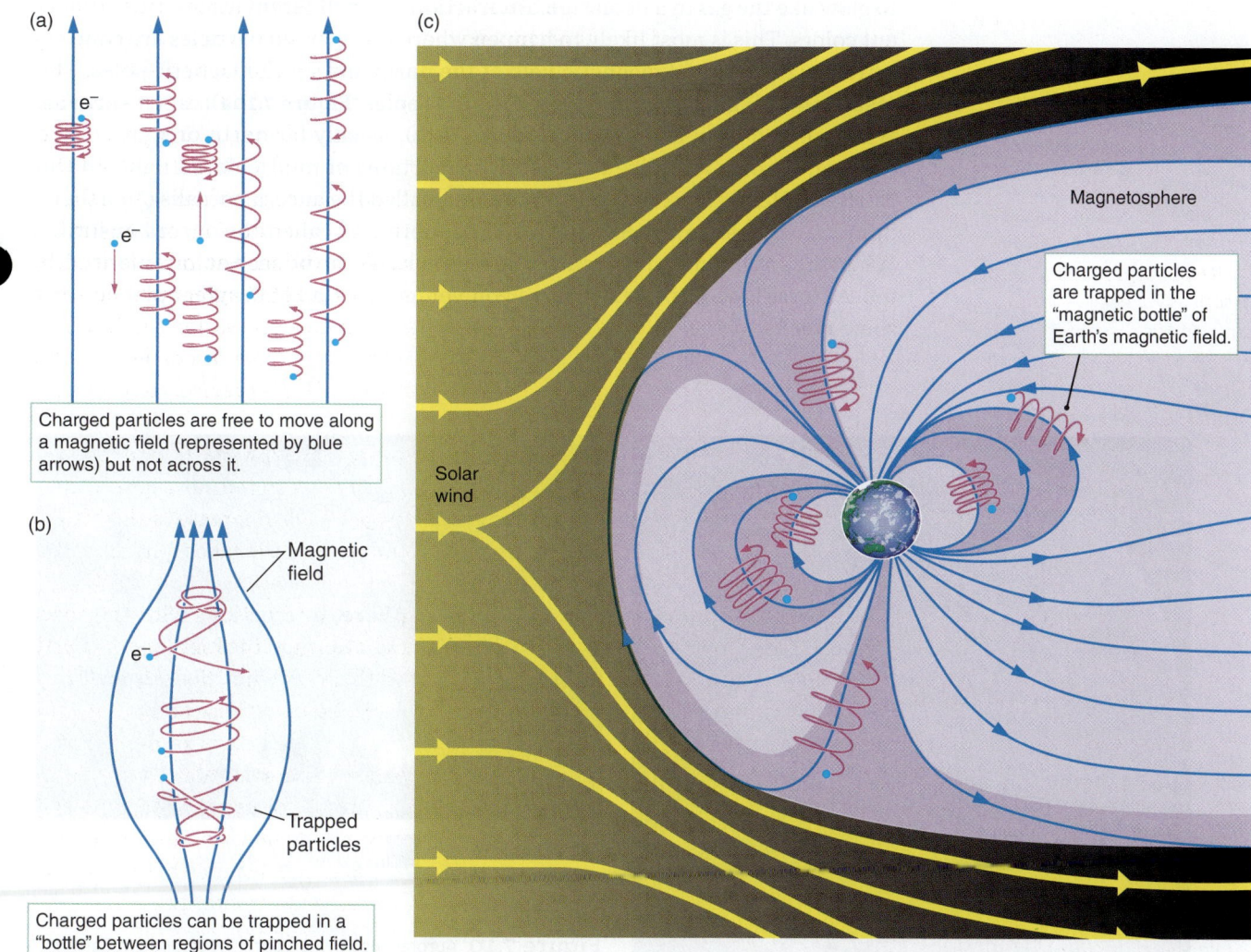

(a)

e^-

e^-

Charged particles are free to move along a magnetic field (represented by blue arrows) but not across it.

(b)

Magnetic field

e^-

Trapped particles

Charged particles can be trapped in a "bottle" between regions of pinched field.

(c)

Magnetosphere

Charged particles are trapped in the "magnetic bottle" of Earth's magnetic field.

Solar wind

Figure 7.9 (a) Charged particles—in this case, electrons—spiral in a uniform magnetic field. (b) When the field is pinched, charged particles can be trapped in a "magnetic bottle." (c) Earth's magnetic field acts like a bundle of magnetic bottles, trapping particles in Earth's magnetosphere. In all these images, the radius of the helical path followed by the charged particle is greatly exaggerated.

back and forth many times in this bottle before an interaction with another particle changes their trajectory.

Most of the charged particles in the solar wind that come toward us are diverted by Earth's magnetic field like a river is diverted around a boulder, as illustrated by the yellow lines in Figure 7.9c. The magnetosphere protects Earth's atmosphere, keeping it from being stripped away by interactions with the solar wind. In addition, Earth's magnetic field deflects dangerous high-energy charged particles, keeping them from reaching Earth's surface. These charged particles can still cause trouble, however. Some regions in the magnetosphere, such as the Van Allen radiation belts, contain especially strong concentrations of energetic charged particles, which can be very damaging to both electronic equipment and humans that pass through them. Disturbances in Earth's magnetosphere caused by changes in the solar wind can affect Earth's magnetic field enough to trip power grids, cause blackouts, and disrupt communications.

Charged particles trapped in the magnetic field eventually collide with atoms such as oxygen, nitrogen, and hydrogen in the upper atmosphere, causing them to glow like the gas in a neon sign. Interactions with different atoms cause different colors. This is most likely to happen where the charged particles are concentrated by the pinched magnetic field at the north and south magnetic poles. The result is glowing rings around these magnetic poles (**Figure 7.10a**), called **auroras**. When viewed from the ground (Figure 7.10b), usually far north or south of the equator, auroras look like eerie, shifting curtains of multicolored light. In the Northern Hemisphere this phenomenon is called the aurora borealis ("northern lights"), whereas in the Southern Hemisphere it is called the aurora australis. When the solar wind is particularly strong, auroras can be seen at lower latitudes, too. Auroras have also been observed on Venus, Mars, all of the giant planets, and some moons.

(a)

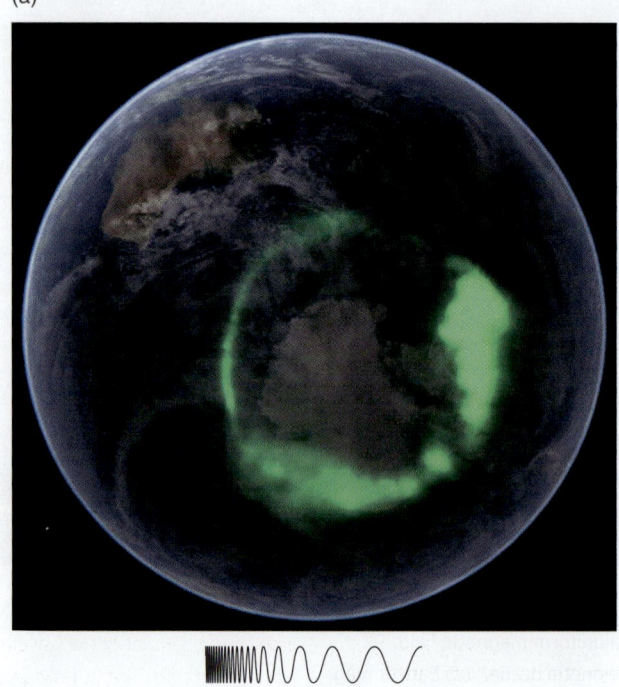

G X U V I R

(b)

G X U V I R

Figure 7.10 Auroras result when particles trapped in Earth's magnetosphere collide with molecules in the upper atmosphere. (a) An auroral ring around Earth's south magnetic pole, as seen from space. (b) Aurora borealis—the "northern lights"—viewed from the ground.

The general structure we have described here is not unique to Earth's atmosphere. The major components—troposphere, stratosphere, and ionosphere—also exist in the atmospheres of Venus, Mars, Saturn's moon Titan, and the giant planets. The magnetospheres of the giant planets are among the largest structures in the Solar System.

Wind

Winds are the natural movement of air, both locally and on a global scale, in response to variations in temperature from place to place. The air is usually warmer in the daytime than at night, warmer in the summer than in winter, and warmer at the equator than near the poles. Large bodies of water, such as oceans, also affect atmospheric temperatures. Heating a gas increases its pressure, which causes it to expand into its surroundings. These pressure differences cause winds. The strength of the winds is governed by the magnitude of the temperature difference.

As air in Earth's equatorial regions is heated by the warm surface, convection causes it to rise. The warmed surface air displaces the air above it, which then has no place to go but toward the poles. This air becomes cooler and denser as it moves toward the poles, so it sinks back down through the atmosphere. It displaces the surface polar air, which is forced back toward the equator, completing the circulation. As a result, the equatorial regions remain cooler and the polar regions remain warmer than they otherwise would be. Air moves between the equator and poles of a planet in a pattern known as **Hadley circulation**, which is shown in **Figure 7.11a**.

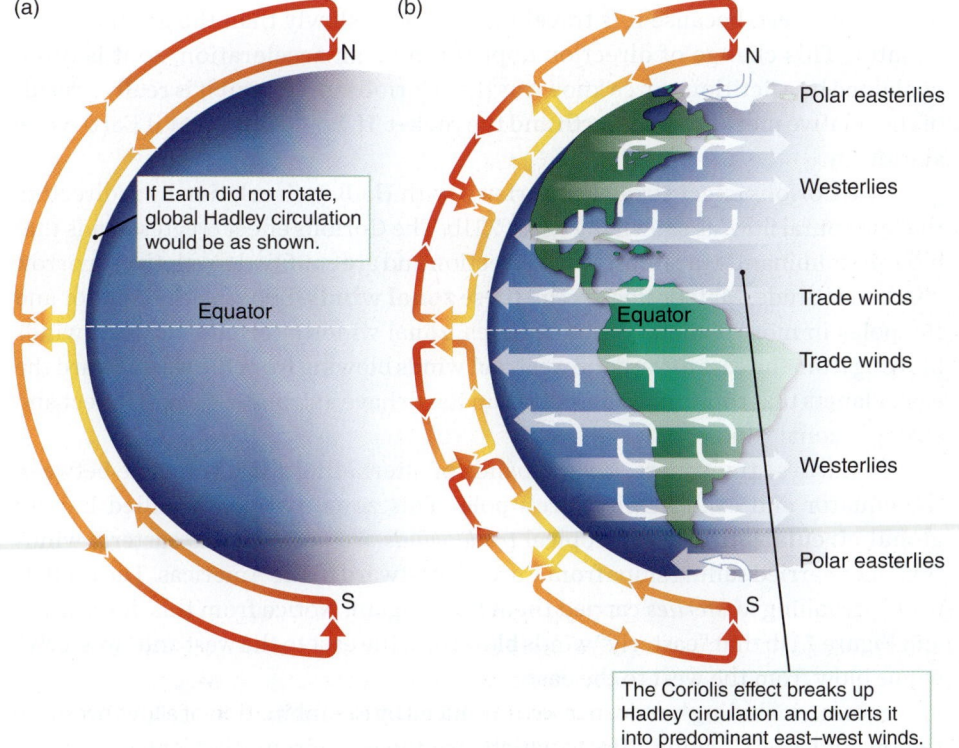

(a)

N

If Earth did not rotate, global Hadley circulation would be as shown.

Equator

S

(b)

N

Polar easterlies

Westerlies

Equator

Trade winds

Trade winds

Westerlies

Polar easterlies

S

The Coriolis effect breaks up Hadley circulation and diverts it into predominant east–west winds.

Figure 7.11 (a) Hadley circulation covers an entire hemisphere. (b) On Earth, Hadley circulation breaks up into smaller circulation cells. The Coriolis effect diverts north–south flow into east–west flow.

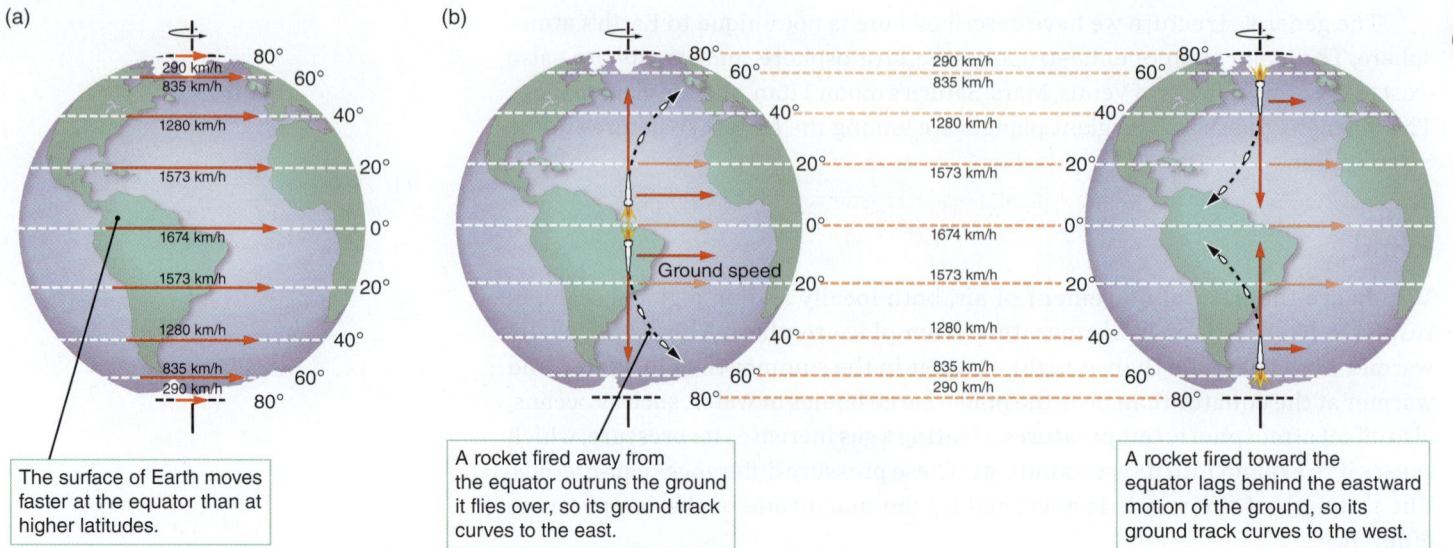

(a) The surface of Earth moves faster at the equator than at higher latitudes.

(b) A rocket fired away from the equator outruns the ground it flies over, so its ground track curves to the east.

A rocket fired toward the equator lags behind the eastward motion of the ground, so its ground track curves to the west.

Figure 7.12 (a) Earth is a solid, so points on the equator must take the same amount of time to rotate once as points at mid-latitudes. To travel a longer distance in the same amount of time, points on the equator must therefore be moving faster than points away from the equator. (b) An object fired perpendicular to the equator will have two components to its speed: the speed parallel to the equator and the speed perpendicular to it. Because the object travels faster parallel to the equator than the ground does, the object's path appears to curve in the direction of rotation. (c) The opposite effect is observed when the object is traveling toward the equator. In this case, the ground is traveling faster in the direction of rotation than the object is, so the object is left behind, and it appears to move along a path that curves away from the direction of rotation.

Earth's rotation also has an effect on winds—and on the motion of any object. This is called the **Coriolis effect** (Figure 7.11b). Objects near the equator move in the direction of rotation (toward the east) faster than objects closer to the poles do, as shown in **Figure 7.12a**. A rocket sitting on the ground at the equator travels to the east at the same speed as the ground (otherwise, it would not keep up with the ground underneath it). Imagine that the rocket is launched directly north from a point on the equator, as shown in Figure 7.12b; its path would appear to curve to the east because the rocket is already traveling to the east faster than other objects farther north—including the ground. If the rocket travels from a northern point toward the equator (Figure 7.12c), the path would appear to curve toward the west, because it is traveling east more slowly than the ground at the equator. This change of direction appears to be an acceleration, so it is often explained by a fictitious force known as the "Coriolis force." But it is really a result of the relative motion of the Earth and the rocket. It would not occur if Earth were stationary.

The Coriolis effect strongly interferes with Hadley circulation by redirecting the horizontal flow, as shown in Figure 7.11b. The Coriolis effect creates winds that blow predominantly in an east–west direction and are confined to relatively narrow bands of latitude. Meteorologists call these **zonal winds**. Between the equator and the poles in most planetary atmospheres, zonal winds alternate between winds blowing from the east toward the west and winds blowing from the west toward the east. Planets that rotate more rapidly than Earth have a stronger Coriolis effect and stronger zonal winds.

In Earth's atmosphere, several bands of alternating zonal winds lie between the equator and each hemisphere's pole. This zonal pattern is called Earth's **global circulation**. The subtropical *trade winds* are more or less easterly winds that once carried sailing ships from Europe westward to the Americas. The midlatitude prevailing *westerlies* carried them home again. Notice from this description and Figure 7.11b that "easterly" winds blow from the east to the west and "westerly" winds blow from the west to the east.

Storms, including hurricanes, are produced by a combination of a low-pressure region and the Coriolis effect, which produces a circulating pattern called **cyclonic motion**. Conversely, high-pressure circulating systems experience **anticyclonic motion** and are generally associated with fair weather.

CHECK YOUR UNDERSTANDING 7.3

If Earth's atmospheric greenhouse effect increases, the temperature of the stratosphere will

a. increase, because there would be more infrared radiation leaving the surface of Earth.

b. increase, because the ultraviolet light would not penetrate as far.

c. decrease, because infrared radiation leaving the surface would be trapped in the lower layers.

d. decrease, because more ultraviolet light would bounce off the top of the stratosphere.

7.4 The Atmospheres of Venus and Mars Differ from Earth's

The atmospheres of Venus, Earth, and Mars are very different. The atmosphere of Venus is very hot and dense compared to that of Earth, whereas the atmosphere of Mars is very cold and thin. Whereas Earth is a lush paradise, perfect for life, the greenhouse effect has turned Venus into a convincing likeness of hell—an analogy made complete by the presence of choking amounts of sulfurous gases, possibly produced by intermittent volcanism. Compared to Venus, the surface of Mars is almost hospitable. For this reason, Mars is the planet of choice as humans consider the colonization of other planets. Understanding why and how these atmospheres are so different helps us understand how Earth's atmosphere may evolve in the future.

Venus

Venus and Earth are so similar in mass, composition, and distance from the Sun that they might be thought of as sister planets. On the basis of these properties, when we used the laws of radiation to predict temperatures for the two planets, we concluded that they should be very similar. But that was before we considered the greenhouse effect and the role of carbon dioxide in trapping infrared radiation from the surface. The atmosphere of Venus is 96 percent carbon dioxide, with a small amount (3.5 percent) of nitrogen and tiny amounts of other gases. This thick blanket of carbon dioxide traps the infrared radiation from Venus, driving up the temperature at the surface of the planet to a sizzling 737 K, which is hot enough to melt lead. This carbon dioxide atmosphere is about 93 times more massive than Earth's atmosphere. As a consequence, the atmospheric pressure at the surface of Venus is 92 times greater than the pressure on Earth's surface. This is enough pressure to crush the hull of a modern nuclear submarine. Venus has no seasons because its axis of rotation is nearly perpendicular to the orbital plane.

As on Earth, the atmospheric temperature of Venus drops as altitude increases throughout the troposphere, reaching a low of about 160 K at the tropopause, as shown in **Figure 7.13**. At an altitude of approximately 50 km, the atmosphere of Venus has an average temperature and pressure similar to those of Earth's atmosphere at sea level. At altitudes between 50 and 80 km, the atmosphere is cool enough for sulfurous oxide vapors to react with water vapor to form clouds of concentrated sulfuric acid (H_2SO_4) droplets. These dense sulfuric acid clouds completely block our view of the surface of Venus, as **Figure 7.14** shows. Throughout the 1960s,

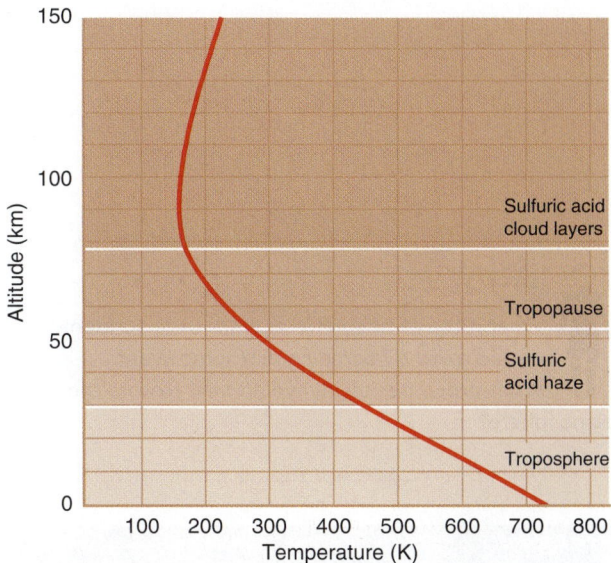

Figure 7.13 The temperature of the atmosphere on Venus primarily decreases as altitude increases, unlike temperatures in Earth's atmosphere, which fall and rise and fall again through the troposphere, stratosphere, and mesosphere, respectively.

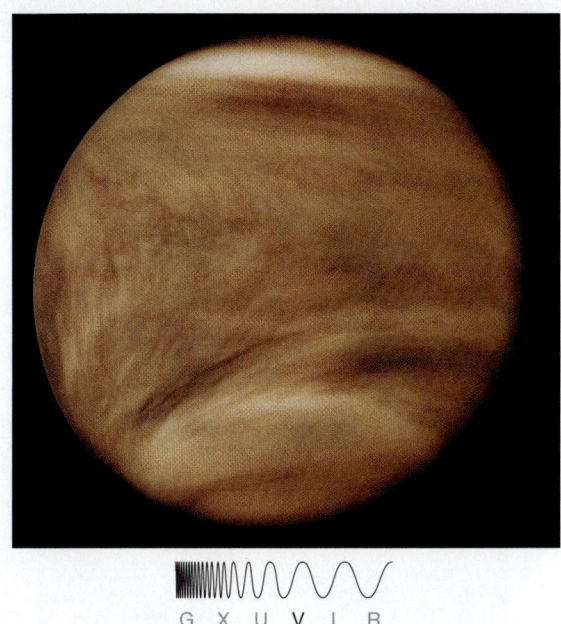

G X U V I R

Figure 7.14 ★ WHAT AN ASTRONOMER SEES

This visible-light image of Venus is an excellent opportunity to describe what an astronomer may conclude from what she does NOT see. In this image, the surface is not visible at all, indicating that the cloud layer is thick and relatively uniform, unlike Earth's. There are no white, fluffy clouds, so if water vapor is present, it is only in small amounts. There are few bands of zonal winds, so an astronomer would conclude that the Coriolis effect is weak on this planet, and thus the planet probably rotates slowly. Phenomena that are absent in astronomical images are sometimes as important as those that are present.

what if . . .

What if the atmosphere of Mars were suddenly ripped away by some catastrophic event: What processes today would help restore the atmosphere?

Figure 7.15 This true-color image of the surface of Mars was taken by the rover *Spirit*. In the absence of dust, the sky's thin atmosphere would appear deep blue. In this image, windblown dust turns the sky pinkish in color.

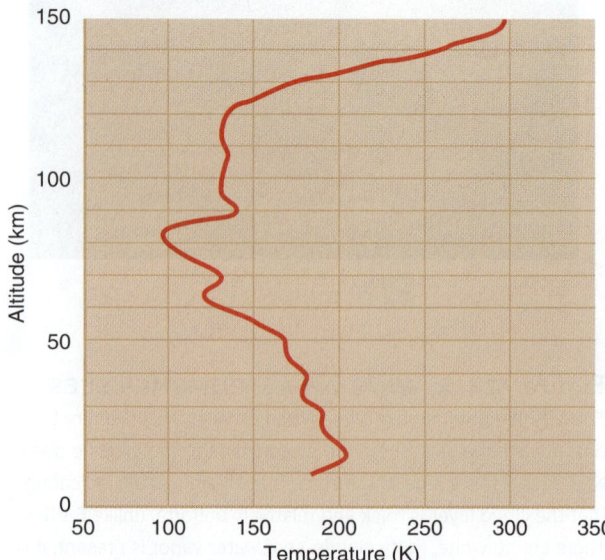

Figure 7.16 Temperature profile of the atmosphere of Mars. Note the differences in temperature and structure between this profile and that of the atmosphere of Venus in Figure 7.13.

radio telescopes and spacecraft with cloud-penetrating radar provided low-resolution views of the surface of Venus. In 1975 the former Soviet Union succeeded in landing cameras on Venus, yielding clear pictures of the surface at individual locations. Finally, in the early 1990s, radar images taken by the *Magellan* spacecraft produced a global map of the surface of Venus (see Figure 6.18).

Unlike the other planets, Venus rotates on its axis in a direction opposite to its motion around the Sun. This is called *retrograde rotation*. (Recall that *retrograde motion* is the apparent movement of a planet in a direction opposite from its normal motion.) Its rotation is also extremely slow—Venus takes 243 Earth days to rotate once on its axis. This slow rotation means that the Coriolis effect is small, so the global circulation is quite close to the Hadley pattern shown in Figure 7.11a. The winds in the upper atmosphere of Venus are very fast, traveling around the planet in only four Earth days. The European Space Agency's *Venus Express* spacecraft observed that these upper-atmosphere winds have been steadily speeding up; the reason for this is not known at present.

Because sunlight cannot easily penetrate the dense clouds, noontime on the surface of Venus is no brighter than a very cloudy day on Earth. The thick atmosphere efficiently retains heat so that the temperature does not drop at night, nor does it drop near the poles. Such small temperature variations at the surface mean there is almost no wind at ground level, so wind erosion is weak compared to wind erosion on Earth and Mars. High temperatures and very light winds near the surface keep the lower atmosphere of Venus free of clouds and hazes, and the strong scattering of light by the dense atmosphere turns any view of distant scenes hazy and bluish. We see the same effect, but to a lesser extent, in our own atmosphere.

Mars

Mars is a stark landscape, reddish in color because of the oxidation of iron-bearing surface minerals. The sky is sometimes a dark blue, but more often it has a pinkish color caused by windblown dust (**Figure 7.15**). The lower density of the Mars atmosphere makes it more responsive than Earth's atmosphere to heating and cooling, so its temperature extremes are greater. The surface near the equator at noontime is a comfortable 20°C—a cool room temperature. At night, however, temperatures typically drop to a frigid −100°C, and during the night at the poles the air temperature can reach −150°C, cold enough to freeze carbon dioxide out of the air in the form of dry-ice frost.

The average atmospheric surface pressure on Mars is equivalent to the pressure at an altitude of 35 km above sea level on Earth, far higher than our highest mountain. Pressures range from a high of 11.5 millibars (mbar) in the lowest impact basins of Mars to a mere 0.3 mbar at the summit of Olympus Mons. Recall that Earth's pressure at sea level is about 1 bar, so the highest pressure on Mars is only 1.1 percent of Earth's pressure at sea level. The temperature profile of the atmosphere of Mars (**Figure 7.16**) is more similar to that of Earth than to that of Venus, with a range of only 100 K up to about 125 km, above which the temperature rises because the upper atmosphere absorbs sunlight.

The inclination of the equator of Mars to its orbital plane (25.2°) is similar to that of Earth (23.4°), so the two planets have similar seasons. The effects on Mars are larger, though, for two reasons: Mars varies more in its annual orbital distance from the Sun than does Earth, and the low density of the atmosphere of Mars makes the atmosphere more responsive to seasonal change. The large daily, seasonal, and

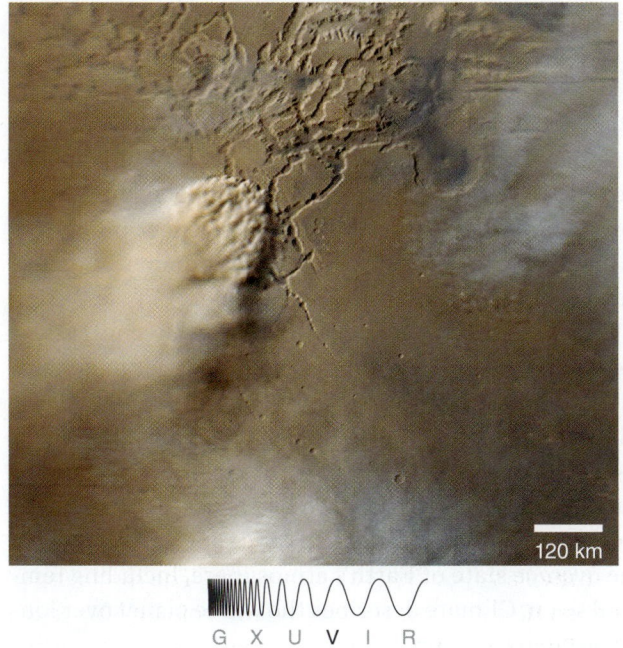

Figure 7.17 A dust storm rages in the canyon lands of Mars.

Figure 7.18 These Hubble Space Telescope images show the development of a global dust storm that enshrouded Mars in September 2001. The same region of the planet is shown in both images.

latitudinal surface temperature differences on Mars often create locally strong winds; some are estimated to be higher than 100 m/s. High winds can stir up huge quantities of dust (**Figure 7.17**) and distribute it around the planet's surface. For more than a century, astronomers have watched the seasonal development of springtime dust storms on Mars. The stronger storms spread quickly and can envelop the entire planet in a shroud of dust within a few weeks, causing the planet's appearance to change dramatically, as shown in **Figure 7.18**. Such large amounts of windblown dust can take many months to settle out of the atmosphere. Seasonal movement of dust from one area to another alternately exposes and covers large areas of dark, rocky surface.

what if . . .

What if humans were to "terraform" Mars, by creating a much thicker atmosphere, among other things: How would this change wind patterns on Mars? Would Mars have zonal winds, like Earth, or be more like Venus? How do you know?

CHECK YOUR UNDERSTANDING 7.4

The difference in surface temperature between Venus, Earth, and Mars is primarily caused by

a. the compositions of their atmospheres.

b. their relative distances from the Sun.

c. the thicknesses of their atmospheres.

d. the times at which their atmospheres formed.

7.5 Earth's Climate Is Changing

Life has changed Earth's atmosphere in the past, and life is changing Earth's atmosphere now. Humans have added vast quantities of carbon dioxide (and other greenhouse gases) to the atmosphere since the industrial revolution. The increasing level of greenhouse gases is increasing the temperature of Earth's surface and oceans. This warming has important consequences for life on Earth.

Atmospheric Carbon Dioxide and Temperature Are Correlated

As mentioned in Section 7.2, Earth and Venus are both significantly warmer than would be expected on the basis of their distances from the Sun. Mars is just slightly warmer than would be expected. By comparing these three planets in our Solar System, we find evidence of how the atmospheric greenhouse effect influences the surface temperature of planets. Mars has a much smaller number of greenhouse gas molecules than Earth, and so its atmosphere holds energy in the system less effectively than Earth's. Mars is thus much cooler. Venus, by contrast, is inhospitably hot due to the extremely large number of greenhouse gas molecules in its atmosphere.

Global changes in greenhouse gases affect climate, which in turn affects weather. To understand observations of climate change, you must first clearly understand the distinction between *climate* and *weather*. The state of Earth's atmosphere at any given time and place is **weather**. Weather is small scale and short term. **Climate** is the term used to define the *average* state of Earth's atmosphere, including temperature, humidity, winds, and so on. Climate describes the whole planet over longer timescales than those involved in weather.

Observations of the temperature and composition of Earth's atmosphere show that Earth's climate is changing. Past temperatures and carbon dioxide levels can be reconstructed from samples of ice in Antarctica, which are obtained by drilling deep down and pulling out a cylinder of ice called an ice core. Climate scientists often show temperature data as the *anomaly*: the difference of the temperature at a given time from a long-term time average over the past few thousand years. A positive anomaly means the global temperature was warmer than the average, whereas a negative anomaly means the temperature was cooler than the average.

Data from ice cores go back 800,000 years to a time long before *Homo sapiens* were here, as shown in **Figure 7.19**. The temperature and carbon dioxide levels over that long time period are tightly correlated: When one rises, so does the other. While correlation does not necessarily mean causation, if there is a physical mechanism that can explain the correlation, then a correlation is strong evidence that one change causes the other. In this case, the discussion of the climates of Venus, Earth, and Mars explains the connection between carbon dioxide and temperature—carbon dioxide in the atmosphere traps infrared radiation, raising the surface temperature of the planet. The more carbon dioxide molecules there are in the atmosphere, the higher the temperature rises. This correlation is exactly what is expected from the atmospheric greenhouse effect.

These ice-core data also show that Earth's climate has lengthy temperature cycles, which can last hundreds of thousands of years and occasionally produce shorter cold periods called *ice ages*. These variations indicate that Earth's atmosphere is extremely sensitive to average global temperature—a drop of only a few degrees can cause an ice age. Earth's climate is a sensitive, chaotic system within which tiny changes in temperature can produce large changes in climate.

Figure 7.19 Temperature anomalies and carbon dioxide levels over the past 800,000 years of Earth's history have been plotted on the same graph to make the similarities and differences easier to see. The right *y*-axis shows temperature data relative to temperatures measured in the mid-20th century, whereas the left *y*-axis shows CO_2 data. The temperature and CO_2 levels are very tightly correlated. The modern CO_2 level is indicated by the large blue arrow. This level is much higher than at any time in the past 800,000 years.

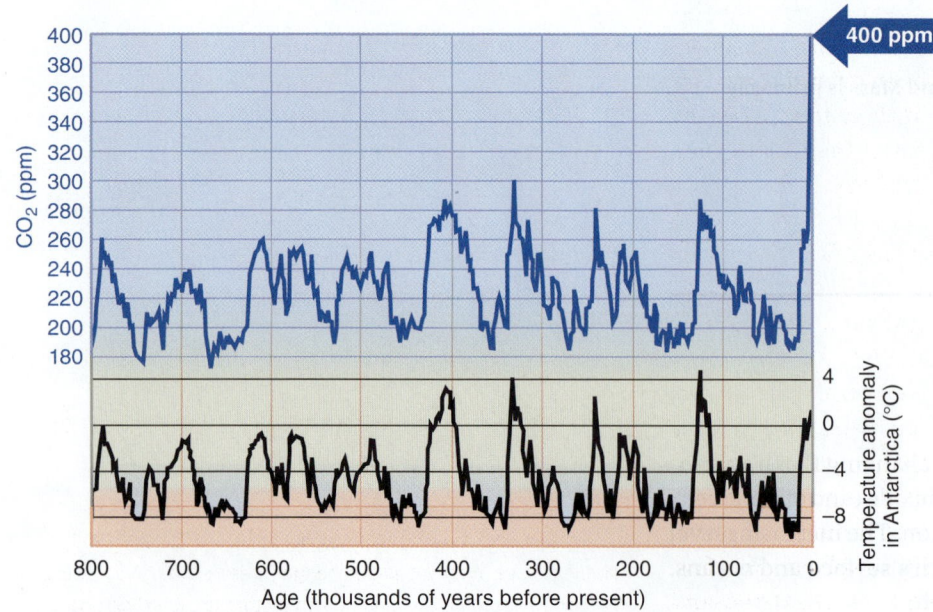

Since the industrial revolution (about 1850), the carbon dioxide level has risen higher than ever before, primarily due to the burning of fossil fuels. At the scale of Figure 7.19, it is easy to see that carbon dioxide has recently risen to the highest levels in 800,000 years, but it is difficult to see what has happened to temperature during the past few hundred years. We need to "zoom in" to see recent history more clearly. Temperature over this time period has been measured in as many ways as scientists can think of, both direct and indirect: from tree rings to ice cores to thermometers to ocean sediment layers. **Figure 7.20** plots temperature measurements since 1850, showing a steady increase in the mean global temperature. This temperature increase correlates with the steady increase in CO_2 levels in the atmosphere. Computer forecasts suggest that this trend represents the beginning of a long-term change caused by the accumulation of anthropogenic (that is, human-made) greenhouse gases. In addition to carbon dioxide and water vapor, anthropogenic greenhouse gases found in Earth's atmosphere include methane, nitrous oxide, and chlorofluorocarbons (CFCs). The presence of these greenhouse gases increases the temperature of Earth.

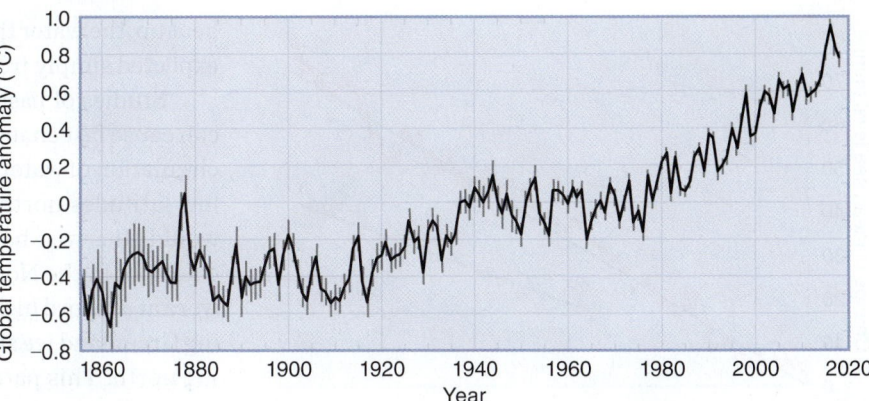

Figure 7.20 Global temperature variations of Earth since 1850 show that the temperature has been rising over this time period, relative to the average temperature from 1951–1980. Vertical lines on the data points are "95% confidence intervals," indicating that researchers are 95% certain that another independent measurement of the temperature would fall within the boundaries of these small vertical lines. Over time, accuracy has improved, and these "error bars" have become smaller. These temperature data were prepared by Berkeley Earth and the Met Office Hadley Centre.

Modeling Earth's Changing Climate

Earth's climate is complex. Small changes in the Sun's energy output, Earth's orbit, or the inclination of Earth's rotation axis affect the climate in a predictable way. Volcanic eruptions, which can produce global sunlight-blocking clouds or hazes, can cause short-term climatic changes. (Recall Figure 7.8, where you saw how some of these factors affect the upper stratosphere.) Long-term interactions between Earth's oceans and its atmosphere, in which oceans act as reservoirs for both heat and carbon dioxide, also contribute. To add to the complexity, Earth's climate is intimately tied to short-term changes in ocean temperatures and currents. We see examples of this connection in the periodic El Niño and La Niña conditions—small shifts in Pacific Ocean temperature that cause much larger global changes in air temperature and rainfall. All these factors must be included to fully model the behavior of the Earth's climate system. **Figure 7.21** shows that, over the past century, natural factors have gradually been overwhelmed by the contribution from **anthropogenic climate change**, such as the release of greenhouse gases into the atmosphere from burning of fossil fuels and other human activities, such as deforestation and land use practices. Simulations that include only natural causes (green shading) fail to fit the data (black line) as well as simulations that include both human activities and natural factors (peach shading). In recent decades, human factors have become dramatically more dominant, as shown in Figure 7.21 by the divergence of the peach and green shaded areas since 1980.

Detailed simulations show that a small addition of greenhouse gases to the atmosphere is extremely likely to have large effects over a relatively short period of about 50–100 years. One prediction is that as Earth warms, sea levels should rise. One cause of sea-level rise is that glaciers and other ice on land melt and flow into the ocean. But a larger factor is that the ocean also expands as it absorbs a lot of heat from the atmosphere. Even during a recent period when atmospheric warming briefly remained flat, the temperature of the oceans continued to rise. As the oceans

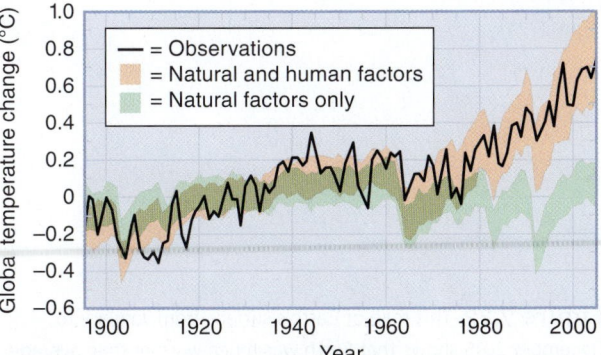

Figure 7.21 Global temperature variations on Earth since 1900 (black line), overlaid with model results that include only natural causes of temperature change (green shading) and those that include both natural and human factors (peach shading).

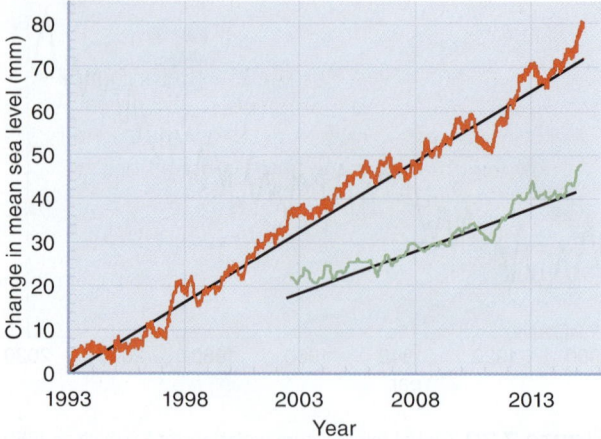

Figure 7.22 The global average absolute sea level has been increasing since the industrial revolution. Since 1993, it has been possible to observe sea level from satellites. Since then, sea level (red line) has risen, on average, 3.2 mm each year. Only 1.9 mm/year of this rise can be attributed to an increase in ocean mass (green line). The other 1.3 mm/year occurs because the oceans expand as they are heated.

heat up, the water they contain expands, causing the sea level to rise above the level expected simply from meltwater, as shown in **Figure 7.22**.

Studies of past climate suggest that changes to the mean global temperature can cause fast changes (occurring over a few decades, rather than centuries) in the circulation of water in the North Atlantic. This circulation brings warm water from low latitudes northward, making the climate nearby far more temperate than it would otherwise be. In 2015, ocean temperature data (**Figure 7.23**) showed a record cold spot in the North Atlantic during the same year that the global temperatures were at a record high. This may have occurred because heavy, cold water from melting Greenland ice shut down the circulation and prevented warm water from coming north. This pattern has persisted at least through 2018. This confirmation of the predictions is extremely worrying, because such a change would precede dramatic changes in the climate in northern Europe.

As an example of how climate scientists test the contribution of different natural and anthropogenic sources of climate change, recall the discussion of the stratosphere from Section 7.3. Climate scientists found the best agreement between simulations and data when they combined multiple contributions: natural contributions, like volcanoes, and anthropogenic contributions, like greenhouse gases and ozone-depleting substances. The physics associated with these changes to Earth's atmosphere are complex. An increase in cloud cover, caused by increased water in the atmosphere, might decrease the amount of sunlight reaching Earth's surface. Ocean currents are critical in transporting thermal energy from one part of Earth to another; predicting how increased temperatures may affect those systems is complicated. The process is so complex that it is still not possible to predict with perfect precision the long-term outcome of the changes that humans are now making to the composition of Earth's atmosphere, although we are improving our predictive accuracy every year. We do know that the anthropogenic CO_2 will remain in the atmosphere for thousands of years.

In a real sense, we are *experimenting* with our atmosphere. We are asking: What happens to Earth's climate if we steadily and rapidly increase the number of greenhouse gas molecules in its atmosphere? The results of this experiment are becoming clear, and scientists have likely underestimated the consequences of inaction. It is important to remember that climate change is just a problem. Mankind has solved global-scale environmental problems before, such as the ozone hole over

Figure 7.23 This map of Earth's surface from January to December 2015 shows that Earth was much warmer than average nearly everywhere, with large regions experiencing the warmest year in recorded history. In the northern Atlantic Ocean, however, a region of the ocean was the coldest it had ever been in recorded history. This may indicate that the circulation system that keeps northern Europe temperate may be shutting down. In 2018 this region was similarly "much cooler than average."

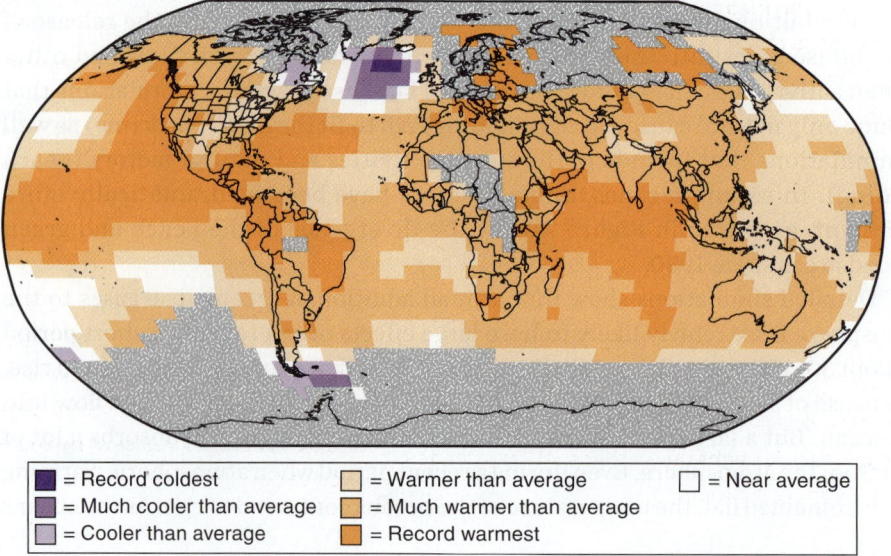

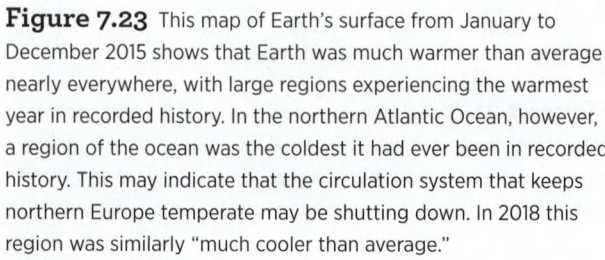

Antarctica, which was caused by the ozone-depleting chemicals mentioned above. This hole is now closing. Acid rain is another global environmental problem that has been addressed successfully through international cooperation. Climate change is not intractable. It also has solutions, if we can muster the will to apply them.

CHECK YOUR UNDERSTANDING 7.5

Over the past 800,000 years, Earth's temperature has closely tracked

a. solar luminosity.

b. oxygen levels in the atmosphere.

c. nitrogen levels in the atmosphere.

d. carbon dioxide levels in the atmosphere.

what if . . .

What if we do nothing to curtail greenhouse gases in Earth's atmosphere and global warming continues unabated for a century: How might those changes affect conditions on Earth's surface?

reading Astronomy News

Curiosity rover sees signs of vanishing Martian atmosphere

by Amina Khan, *Los Angeles Times*

Scientists use many different techniques to reconstruct past events. Here, they use isotopes to investigate the history of the atmosphere of Mars.

Before going incommunicado behind the Sun for a month, NASA's Mars Curiosity rover sent Earth evidence that the Red Planet has lost much of its original atmosphere.

The findings, announced by Jet Propulsion Laboratory [JPL] scientists at the European Geosciences Union meeting in Vienna, bolster the idea that the Martian atmosphere was once much thicker than it is today—and come less than a month after the rover drilled its first rock and found signs that Mars was once hospitable to life.

Curiosity's Sample Analysis at Mars instrument sniffed the Martian atmosphere and counted up the isotopes of argon in the air. Isotopes are heavier and lighter versions of the same element, and when a planet starts to lose its atmosphere, the lighter isotopes in the upper layers are the first to go. So if scientists see fewer of the lighter isotopes than expected, it might mean that there was once much more air there.

The researchers looked at two isotopes of argon: the heavier argon-38 and the lighter argon-36. They found that there was four times as much argon-36 as argon-38—a lower share of the lighter argon than expected based on data from other parts of the Solar System.

Some of that missing argon-36 must have escaped because the top of the atmosphere had started to blow away, they surmised. This could mean Mars' atmosphere was much thicker in the past than what Curiosity picks up today.

The scientists recently used this technique with carbon isotopes, finding similar results. But the new argon measurements provide much firmer evidence than older, less certain data on argon isotopes from the Viking mission in 1976, as well as from Martian meteorites.

Curiosity is experiencing what scientists call "solar conjunction," when the Sun comes between Earth and Mars, blocking communications until May. But Curiosity isn't taking a vacation. The rover has to take weather and radiation measurements every day—part of a detailed list of chores sent by the Mars Science Laboratory team at JPL.

EVALUATING THE NEWS

1. Is the atmosphere being discussed in this article the primary atmosphere or the secondary atmosphere of Mars? How do you know?

2. Explain why the lighter isotopes disappear from the atmosphere first. Where do those isotopes go?

3. Does argon-38 have more, fewer, or the same number of protons as argon-36? How do you know?

4. The article states that the scientists had carried out this experiment with carbon isotopes, finding similar results. Some might argue that this argon experiment was therefore unnecessary and inefficient. Using what you learned in Chapter 1 about the scientific method, argue that the argon experiment should have been done.

5. Sketch the Earth-Mars-Sun system as viewed from above during the "solar conjunction" referenced in the article. Why is *Curiosity* "incommunicado" during the time of the article?

Source: **Amina Khan,** "Curiosity rover sees signs of vanishing Martian atmosphere," from *Los Angeles Times*, April 9, 2013. Used by permission of the *Los Angeles Times*.

SUMMARY

Venus, Earth, and Mars all have significant atmospheres that are different from the atmospheres they captured when the planets first formed. These atmospheres are complex, both in chemical composition and in physical characteristics such as temperature and pressure. Convection drives weather on Earth. The Coriolis effect converts the north-south flow of air, caused by heating near the equator, into east-west circulation. The climates of Venus, Earth, and Mars are all modified by their individual atmospheres. Venus, Earth, and Mars are warmer than they would be from solar illumination alone. The atmospheres of Venus, Earth, and Mars have different amounts of various greenhouse gases, resulting in dramatic differences in atmospheric temperature and pressure. Life has altered Earth's atmosphere several times, most notably in the distant past when photosynthesizing organisms increased the amount of oxygen in the atmosphere and now in modern times as humans drastically increase the amount of greenhouse gases.

(1) The primary atmospheres of the terrestrial planets, composed mostly of hydrogen and helium, were lost early in the history of the Solar System. Secondary atmospheres arose from volcanism and impacts from comets and meteoroids, resulting in atmospheres with different chemical compositions than those of the primary atmospheres. These secondary atmospheres continue to exist today.

(2) The atmospheres of Venus, Earth, and Mars are very different from one another. They have different temperatures, pressures, and compositions. Earth's atmosphere, in particular, has many layers. The layers are determined by variations in temperature and the absorption of solar radiation as they change with altitude throughout the atmosphere.

(3) The Sun heats Earth's surface most strongly near the equator. This heating causes the air to rise in the atmosphere and flow toward the poles, setting up convection cells. These north-south flows are redirected toward the east and west by the Coriolis effect. These motions are the primary cause of weather on Earth.

(4) Naturally occurring greenhouse gases exist on Venus, Earth, and Mars, which increase the average surface temperature of each planet. The amount by which these greenhouse gases raise the temperature of a planet depends on the number of greenhouse gas molecules in the atmosphere. The difference in global temperatures among these planets can be explained in part by their distance from the Sun. However, the different compositions and densities of their atmospheres is a far more significant factor in determining global temperatures.

(5) The oxygen levels in Earth's atmosphere have been enhanced through photosynthesis. The increased amount of oxygen is thought to have been necessary for the development of more advanced life-forms (respiration) and for life to leave the oceans once there was protection by the ozone layer. Forests on land then continued to convert CO_2 into oxygen. Increased oxygen led to the development of other forms of life that continue to modify the atmosphere. Today, human activity is significantly modifying the climate by adding greenhouse gases to the atmosphere.

(6) Indirect measurements of Earth's global temperature provide data going back hundreds of thousands of years. Large variations in global temperature correlate strongly with the number of carbon dioxide molecules in the atmosphere. The current level of greenhouse gases in Earth's atmosphere is higher than any seen in the past 800,000 years and correlates with a subsequent rise in temperature. This increase in greenhouse gases cannot be accounted for by natural causes alone.

QUESTIONS AND PROBLEMS

TESTING YOUR UNDERSTANDING

1. **T/F:** The current atmospheres of the terrestrial planets were formed when the planets formed.

2. **T/F:** Comets and asteroids are one of several sources of the water on Earth.

3. **T/F:** The stratosphere is where most weather happens.

4. **T/F:** All other things being equal, a planet with a high albedo will have a lower temperature than a planet with a low albedo.

5. **T/F:** The average temperature of Earth has risen measurably over the past 100 years.

6. Venus is hot and Mars is cold primarily because
   a. Venus rotates very slowly and in the opposite direction from Mars.
   b. Venus has a much thicker atmosphere.
   c. the atmosphere of Venus is dominated by CO_2, but the atmosphere of Mars is not.
   d. Venus has stronger winds.

7. The atmosphere of Mars is often pinkish because
   a. the atmosphere is dominated by carbon dioxide.
   b. the Sun is at a low angle in the sky.
   c. Mars has no oceans to reflect blue light to the sky.
   d. winds lift dust into the atmosphere.

8. Convection in the _____ causes weather on Earth.
   a. stratosphere
   b. mesosphere
   c. troposphere
   d. ionosphere

9. Auroras are the result of
   a. the interaction of particles from the Sun with Earth's atmosphere.
   b. upper-atmosphere lightning strikes.
   c. the destruction of the upper atmosphere, which leaves a hole.
   d. the interaction of Earth's magnetic field with Earth's atmosphere.

10. The stratospheric ozone layer protects life on Earth from
    a. high-energy particles from the solar wind.
    b. micrometeorites.
    c. ultraviolet radiation.
    d. charged particles trapped in Earth's magnetic field.

11. Hadley circulation is broken into zonal winds by
    a. convection from solar heating.
    b. hurricanes and other storms.
    c. interactions with the solar wind.
    d. the planet's rapid rotation.

12. Earth experiences long-term climate cycles spanning _____ of years.
    a. hundreds of millions
    b. hundreds of thousands
    c. thousands
    d. hundreds

13. Place in chronological order the following steps in the formation and evolution of Earth's atmosphere.
    a. Plant life converts CO_2 to oxygen.
    b. Hydrogen and helium are lost from the atmosphere.
    c. Volcanoes, comets, and asteroids increase the inventory of volatile matter.
    d. Hydrogen and helium are captured from the protoplanetary disk.
    e. Oxygen enables the growth of new life-forms.
    f. Volcanism releases CO_2 from the subsurface into the atmosphere.

14. The oxygen molecules in Earth's atmosphere
    a. were part of the primary atmosphere.
    b. arose when the secondary atmosphere formed.
    c. are the result of life.
    d. are being rapidly depleted by the burning of fossil fuels.

15. Studying climate on other planets is important to understanding climate on Earth because (select all that apply)
    a. the underlying physical processes are the same on every planet.
    b. other planets offer a range of extremes to which Earth can be compared.
    c. comparing climates on other planets helps scientists understand which factors are important.
    d. other planets can be used to test atmospheric models.

16. The Coriolis effect
    a. causes winds to circulate north to south.
    b. causes winds to flow only eastward.
    c. causes winds to flow only westward.
    d. causes winds to circulate east and west.

17. Which of the following statements are true? (Choose all that apply.)
    a. Earth's magnetosphere shields us from the solar wind.
    b. Earth's magnetosphere is essential to the formation of auroras.
    c. Earth's magnetosphere extends far beyond Earth's atmosphere.
    d. Earth's magnetosphere is weaker than Mercury's.

18. On which planet(s) is the greenhouse effect present? (Choose all that apply.)
    a. Mercury
    b. Venus
    c. Earth
    d. Mars

19. The words *weather* and *climate*
    a. mean essentially the same thing.
    b. refer to different size scales.
    c. refer to different timescales.
    d. both b and c

20. Uncertainties in climate science are dominated by
    a. uncertainty about physical causes.
    b. uncertainty about current effects.
    c. uncertainty about past effects.
    d. uncertainty about future effects.

THINKING ABOUT THE CONCEPTS

21. Primary atmospheres of the terrestrial planets were composed almost entirely of hydrogen and helium. Explain why they contained these gases and not others.

22. How were the secondary atmospheres of the terrestrial planets created?

23. Some of Earth's water was released above ground by volcanism. What is another likely source of Earth's water?

24. The force of gravity holds objects tightly to the surfaces of the terrestrial planets. Yet atmospheric molecules are constantly escaping into space. Explain how these molecules are able to overcome gravity's grip. How does the mass of a molecule affect its ability to break free?

25. Use Figure 7.3 to explain why the range of temperatures on Mercury is so much larger than for any other planet. 👁

26. Given that the atmospheres of Venus and Mars are both dominated by carbon dioxide, why is Venus very hot and Mars very cold?

27. In what ways does plant life affect the composition of Earth's atmosphere?

28. What is the principal cause of winds on the terrestrial planets?

29. Why are humans able to get a clear view of the surface of Mars but not Venus?

30. ★ **WHAT AN ASTRONOMER SEES** In Figure 7.14, an astronomer would see evidence that Venus is a slowly rotating planet. Make a sketch of what the clouds in this figure would look like if Venus rotated quickly. Annotate your sketch with arrows to show how the clouds might move relative to one another. 👁

31. In 1975, the former Soviet Union landed two camera-equipped spacecraft on Venus, giving planetary scientists the only close-up views ever seen of the planet's surface. Both cameras ceased to function after an hour. What environmental conditions probably caused them to stop working?

32. What are the two most important characteristics that distinguish one layer of Earth's atmosphere from another?

33. Describe the overall trend in Figure 7.20.

34. Figure 7.19 shows two sets of data on the same graph. The *x*-axis is time, which is the same for both data sets. The *y*-axis on the left is the amount of CO_2 in the atmosphere. The *y*-axis on the right is the temperature anomaly—the difference between the temperature that year and the average temperature over the past 10,000 years. Do both of these lines show the same trend? Does this necessarily mean that one causes the other? What other information do we have that leads us to believe that the increasing CO_2 levels are driving the increasing temperatures? ⊛

35. In Figure 7.23, compare the latitude of England with the latitude of Maine in the United States. Which of these locations is further north? Which of these locations has more temperate weather? Why? What would you expect to occur in this more temperate area, if the circulation of water through the Atlantic shuts down? ⊛

APPLYING THE CONCEPTS

36. A planet with no atmosphere at 1 AU from the Sun would have an average blackbody surface temperature of 279 K if it absorbed all the Sun's electromagnetic energy falling on it (albedo = 0.0). Follow Working It Out 7.1 to find the average temperature of the planet if it had an albedo of 0.1 or 0.9.
   a. Make a prediction: 0.1 and 0.9 are both higher albedos than 0.0. Does this mean a planet with one of these albedos would be more or less reflective than a planet with an albedo of 0.0? Therefore, do you expect temperatures to be higher or lower than 279 K? In which case should the planet be hotter: an albedo of 0.1 or 0.9?
   b. What would the average temperature on this planet be if its albedo were 0.1, typical of a rock-covered surface?
   c. What would the average temperature be if its albedo were 0.9, typical of a snow-covered surface?
   d. Check your work by comparing your answers to 279 K.

37. The Moon's average albedo is approximately 0.12. Follow Working It Out 7.1 to find the average temperature of the Moon.
   a. Make a prediction: Earth has an albedo of 0.3 and a surface temperature of 288 K. Do you expect the Moon's temperature to be hotter or colder than Earth's?
   b. Calculate the Moon's average temperature in kelvins.
   c. Check your work by comparing the Moon's temperature to Earth's temperature.

38. One effect of climate change is that ice melts all over Earth. Ice has a quite high albedo, so it reflects a lot of sunlight. As the ice melts, Earth's albedo drops. Use Working It Out 7.1 to get a feel for the effect of this, by finding the surface temperature of an Earth-sized body with an albedo half of Earth's current albedo and no atmosphere.
   a. Make a prediction: Do you expect the temperature of this Earth-sized body to be higher or lower than the temperature calculated for an (atmosphere-less) Earth in Working It Out 7.1?
   b. Calculate the surface temperature of an Earth-sized body with an albedo half of Earth's current albedo.
   c. Check your work by comparing this result with Earth's (atmosphere-less) surface temperature of 255 K.
   d. Reason further: subtract 255 K from your answer to find out how many degrees the temperature changed when the albedo changed. Compare this answer to the natural effect of the atmosphere, which is to raise the temperature of Earth 33 K. Is it important for climate scientists to include this effect in their work?

39. Suppose you read about a geo-engineering plan to solve the climate crisis by increasing Earth's albedo (this is an actual plan that people talk about!). You can get a rough idea of how much the albedo would have to change by turning around the equation in Working It Out 7.1 to solve for the albedo. This gives $a = 1 - (T/279)^4 d^2$. Assume that you want to counteract the 4°C temperature change that falls in the middle of the predicted temperature increase by 2100, so use 251 K in this equation, along with the semimajor axis of Earth's orbit (1 AU) to find the albedo required to drop Earth's temperature by 4°C.
   a. Make a prediction: Do you expect the albedo to be higher or lower than Earth's current albedo?
   b. Calculate the new albedo required to drop Earth's temperature by 4°C.
   c. Check your work by checking to make sure that your albedo is a number between 0 and 1 and by comparing it to your prediction. Make sense of your answer by comparing it to the albedo of forests, grass, and crops (all about 0.2); the albedo of water (0.1); the albedo of ice (0.4); and the albedo of very thick clouds (0.9). What kinds of changes would need to be made in order to change Earth's albedo to solve this problem?

40. Suppose you read about a plan to solve the climate crisis by moving Earth farther from the Sun in its orbit. Leaving aside the considerable technical challenges of this plan, you decide to figure out "How far are we talking about here?" You can get a rough idea by turning around the equation in Working It Out 7.1 to solve for the distance from the Sun. This gives $d = \sqrt{(1 - a)} /(T/279)^2$. Assume that you want to counteract the 4°C temperature change that falls in the middle of the predicted temperature increase by 2100, so use 251 K in this equation, along with Earth's albedo (0.306) to find the new distance to the Sun in astronomical units.
   a. Make a prediction: do you expect this new distance to be larger or smaller than 1 AU (Earth's current orbital radius)?
   b. Calculate the new orbital distance, using $T = 251$ K and $a = 0.306$.
   c. Check your work by comparing with your prediction. Then, to help you understand this distance, convert the amount Earth would have to be moved (your answer to part b minus Earth's current radius) from astronomical units to kilometers. Compare this to Earth's radius of 6,371 km, and the Moon's orbital radius of 385,000 km. Does this seem like a feasible solution to the climate crisis?

exploration Climate Change

One prediction about climate change is that as the planet warms, ice in the polar caps and in glaciers will melt. This certainly seems to be occurring in the vast majority of glaciers and ice sheets around the planet. It is reasonable to ask whether this actually matters and why. Here you will explore several consequences of the melting ice on Earth.

Experiment 1: Floating Ice

For this experiment, you will need a permanent marker, a translucent plastic cup, water, and ice. Place a few ice cubes in the cup, and add water until the ice cubes float (that is, they don't touch the bottom). Mark the water level on the outside of the cup with the marker, and label this mark so that later you will know it was the initial water level.

1. As the ice melts, what do you expect to happen to the water level in the cup?

_____

_____

_____

Wait for the ice to melt completely, then mark the cup again.

2. What happened to the water level in the cup when the ice melted?

_____

_____

_____

3. Given the results of your experiment, what can you predict will happen to global sea levels when the Arctic ice sheet, which floats on the ocean, melts?

_____

_____

Experiment 2: Ice on Land

For this experiment, you will need the same materials as in experiment 1, plus a paper or plastic bowl. Fill the cup about halfway with water and then mark the water level, labeling it so you know it's the initial level. Poke a hole in the bottom of the bowl, and set the bowl over the cup. Add some ice cubes to the bowl.

4. As the ice melts, what do you expect to happen to the water level in the cup?

_____

_____

Wait for the ice to melt completely, then mark the cup again.

5. What happened to the water level in the cup when the ice melted?

_____

_____

_____

6. In this experiment, the water in the cup is analogous to the ocean, and the ice in the bowl is analogous to ice on land. Given the results of your experiment, what can you predict will happen to global sea levels when the Antarctic ice sheet, which sits on land, melts?

_____

_____

_____

7. The *expansion* of ocean water as it heats is a much larger contributor to sea-level rise than the addition of water from melting ice. What implications does this have for global sea levels as greenhouse gases continue to accumulate and the temperature of the oceans continues to rise?

_____

_____

_____

Experiment 3: Why Does It Matter?

Search online using the phrase "Earth at night" to find a satellite picture of Earth taken at night. The bright spots on the image trace out population centers. In general, the brighter they are, the more populous the area (although there is a confounding factor relating to technological advancement).

8. Where do humans tend to live—near coasts or inland?

_____

_____

Coastal regions are, by definition, near sea level. If both the Arctic and Antarctic ice sheets melted completely, sea levels would rise by 80 meters.

9. How would a sea level rise of about 2 meters (in the range of reasonable predictions) over the next few decades affect the global population? (To help you think about this, remember that a meter is about 3 feet, and one story of a building is about 10 feet.)

_____

_____

The Giant Planets

The giant planets Jupiter and Saturn, along with terrestrial planets Venus and Mars, are bright enough that even city-dwellers can see them at night when they are in the right part of the sky. Use the Internet to find out which planets are visible during the semester. Go outside on a clear night and sketch the positions of the planets that are visible. If you can identify stars or constellations near the planets (star maps like those at the end of this book, or ones you can obtain online, will help!), include them in your sketch. Be sure to record your location, the date, the time, and a note about the sky conditions.

EXPERIMENT SETUP

THE SKY TONIGHT

This evening, Venus, Mars and Jupiter will all be near each other in the night sky.

Use the Internet to find out which planets are visible during the semester.

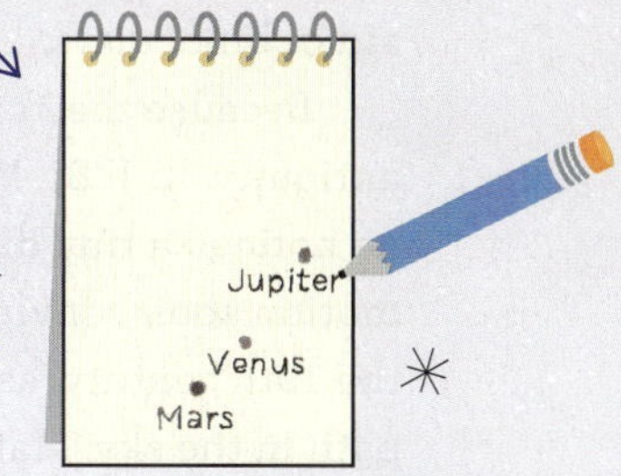

Jupiter
Venus
Mars

Go out on a clear night and sketch the positions of these visible planets.

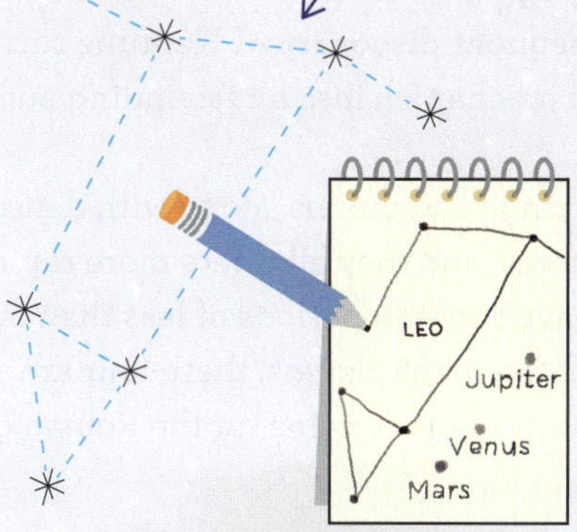

LEO
Jupiter
Venus
Mars

If you can identify stars or constellations near the planets (star maps like those at the end of this book, or ones you can obtain online, will help!), include them in your sketch.

Your observations will be incomplete without:

☐ your location ☐ date ☐ time ☐ a note about the sky conditions.

SKETCH OF RESULTS (in progress)

Observation Log

Date _____ Time _____

Object _____

Location _____

Sky conditions _____

Comments

_____
_____
_____
_____
_____
_____
_____

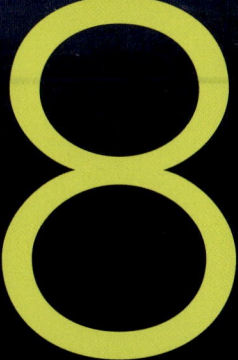

U nlike the rocky planets of the inner Solar System, four worlds in the outer Solar System captured and retained enough material from the Sun's protoplanetary disk to swell to enormous size and mass. Jupiter and Saturn are large enough that the sunlight reflected from them back toward Earth is as bright as the brightest stars. Uranus, which is both smaller and farther away, is slightly brighter than the faintest stars visible on a dark night. Neptune, the outermost planet in our Solar System, cannot be seen without the aid of binoculars.

Because they are so bright, Jupiter and Saturn have been known since antiquity. In 1781, William Herschel was producing a catalog of the sky when he noticed a tiny disk that he thought was a comet. The object's slow nightly motion soon convinced him that it was a new planet—later named Uranus. In the 19th century, astronomers found that Uranus strayed from its predicted path in the sky. Mathematicians applied Newton's and Kepler's laws to Uranus, including the influence of a hypothetical planet in an even larger orbit. In 1846, Johann Gottfried Galle found Neptune in the position predicted by mathematicians. The prediction and subsequent discovery of Neptune turned what looked at first to be a failure of orbital mechanics into a resounding success of the scientific method.

All four giant planets have rings and moons, along with dense cores surrounded by very large atmospheres, and they all rotate more rapidly than Earth. Although they are large, they have rotation periods of less than one Earth day. Because Jupiter is the largest of the giant planets, these four are sometimes called the "Jovian planets" (Jove is another name for the Roman god Jupiter).

LEARNING GOALS

1. List specific differences between the giant planets and the terrestrial planets.

2. Describe how the composition, temperature, and pressure of each giant planet is responsible for how that planet appears to us.

3. Describe how gravitational energy turns into thermal energy and how that affects the atmospheres of the giant planets.

4. Describe the extreme conditions deep within the interiors of the giant planets, and explain how these extreme conditions produce the magnetic fields of the giant planets.

5. Describe the origin and structure of the rings of the giant planets.

8.1 Giant Planets Are Large, Cold, and Massive

As with the terrestrial planets, a lot can be learned about the giant planets by comparing them to each other. Comparative planetology is useful both within planetary groups and between groups. Throughout most of the chapter, we compare giant planets to other giant planets. Still, it is useful to fix in your mind a comparison of

Table 8.1 Comparison of Physical Properties of the Giant Planets

Property	Jupiter	Saturn	Uranus	Neptune
Orbital semimajor axis (AU)	5.20	9.6	19.2	30
Orbital period (Earth years)	11.9	29.5	84.0	164.80
Orbital velocity (km/s)	13.1	9.7	6.8	5.4
Mass (in units of M_{Earth})	317.8	95	14.5	17.1
Equatorial radius (km)	71,490	60,270	25,560	24,300
Equatorial radius (in units of R_{Earth})	11.2	9.5	4.0	3.8
Density (relative to water = 1)	1.33	0.69	1.27	1.64
Rotation period*	$9^h 56^m$	$10^h 39^m$	$17^h 14^m$	$16^h 6^m$
Tilt of axis (degrees)†	3.13	26.7	97.8	28.3
Surface gravity (m/s²)	24.8	10.4	8.7	11.2
Escape velocity (km/s)	59.5	35.5	21.3	23.5

*The superscript letters h and m stand for hours and minutes of time, respectively.
†An axial tilt greater than 90° indicates that the planet rotates in a retrograde, or backward, direction.

at least one giant planet with Earth as a reference point. For example, to understand the size of the giant planets, it is helpful to know that the diameter of Jupiter is 11 times larger than the diameter of Earth, thus yielding a volume into which more than 1,300 Earths would fit, and Jupiter's mass is 318 times as large as Earth's. In this section, we examine the physical properties of the giant planets, which are summarized in **Table 8.1**.

Characteristics of the Giant Planets

Jupiter, the closest giant planet, is more than 5 astronomical units (AU) from the Sun. From Jupiter, the Sun appears as a tiny disk, just 1/27 as bright as it appears from Earth. The Sun appears dim and provides very little warmth at the distance of Jupiter; daytime temperatures hover around 123 kelvins (K) at the cloud tops on Jupiter. At the distance of Neptune, the Sun no longer looks like a disk at all; it appears as a brilliant star about 500 times brighter than the full Moon in our own sky. Daytime on Neptune is equivalent to twilight here on Earth. Daytime temperatures on Neptune's moon Triton are around 37 K.

Jupiter, Saturn, Uranus, and Neptune are enormous compared to their four rocky terrestrial counterparts. Jupiter is the largest of the eight planets; it is more than one-tenth the diameter of the Sun. Neptune is the smallest giant planet, but

Figure 8.1 (a) Occultations occur when a planet, moon, or ring passes in front of a star. (b) As the planet moves (from right to left as seen from Earth), the starlight is blocked. The amount of time that the star is hidden, combined with information about how fast the planet is moving, gives the size of the planet.

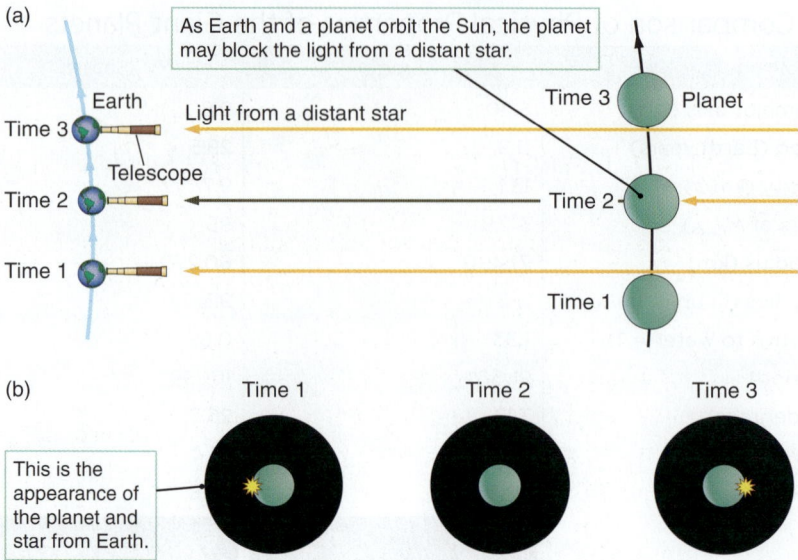

its radius is still almost 4 times the radius of Earth. **Figure 8.1** illustrates the most accurate method for finding the diameter of a planet. As the planet passes in front of a star, the star is eclipsed. This is known as a **stellar occultation**. Because we know the relative speeds of Earth and the giant planets, we can calculate the size of the eclipsing giant planet from the length of time the star's light is blocked. The exact center of a planet rarely passes directly in front of a star, but observations of occultations from several widely separated observatories can provide the geometry necessary to calculate both the planet's size and its shape. Occultations of radio signals transmitted from orbiting spacecraft have also provided accurate measures of the sizes and shapes of planets and their moons, as have images taken by space-craft cameras. An example of this type of calculation is given in **Working It Out 8.1**.

Scientists can measure a planet's mass by observing the motions of its moons. In Chapter 3, you learned how to use Newton's law of gravitation and Kepler's third law to predict the motion of a moon by using the planet's mass. Planetary space-craft now make it possible to measure the masses of planets more accurately. As a spacecraft flies by a planet, its motion is altered by the planet's gravity. By track-ing changes in the spacecraft's path, astronomers can accurately measure the planet's mass.

The giant planets contain 99.5 percent of the mass in the Solar System, not counting the mass of the Sun. All other Solar System objects—terrestrial planets, dwarf planets, moons, asteroids, and comets—are included in the remaining 0.5 per-cent. Jupiter alone is more than twice as massive as all the other planets in the Solar System combined. Even so, its mass is only about a thousandth that of the Sun. The density of a planet can be found from its mass and radius. The densities of the giant planets are low. Saturn's density is so low that it would actually float in water with 70 percent of its volume submerged—if you could find a lake large enough. These low densities indicate that the giant planets are primarily made of gases and liquids, rather than solid rock like the terrestrial planets.

Composition of the Giant Planets

Jupiter and Saturn are composed almost entirely of hydrogen and helium and are therefore known as **gas giants**. Uranus and Neptune are known as **ice giants** because they contain much larger amounts of water and other volatile materials

working it out 8.1

Finding the Diameter of a Giant Planet

Suppose a particular planet is moving at a speed of precisely 25 kilometers per second (km/s) relative to Earth. As the planet moves across the sky toward a background star, you begin to make careful observations, as sketched in Figure 8.1. You note the moment when the planet first eclipses the star, and you wait while the star passes behind the center of the planet and emerges from the opposite side. You find that this event takes exactly 2,000 seconds. The planet has traveled a specific distance (its diameter), at a specific speed, during this time. The equation from Working It Out 2.1 relates speed, distance and time:

$$\text{Speed} = \frac{\text{Distance}}{\text{Time}}$$

Rearranging this equation to solve for distance requires first multiplying both sides by time and then exchanging the right and left sides of the equation:

$$\text{Distance} = \text{Speed} \times \text{Time}$$

Plugging in a speed of 25 km/s and our time of 2,000 seconds gives the distance traveled:

$$\text{Distance} = 25 \text{ km/s} \times 2,000 \text{ s}$$
$$\text{Distance} = 50,000 \text{ km}$$

The planet's diameter is equal to the distance it traveled, or 50,000 km.

than do Jupiter and Saturn. On a giant planet, a relatively shallow atmosphere merges seamlessly into a deep liquid **ocean**, which in turn merges smoothly into a denser liquid or solid core. Although the atmospheres of the giant planets are shallow compared with the depth of the liquid layers below, they are still much thicker than those of the terrestrial planets—thousands of kilometers rather than hundreds. On Jupiter or Saturn, we can see only the tops of a layer of thick **clouds**, the highest of many other layers that lie below. Although a few thin clouds are visible on Uranus, we mostly find ourselves looking deep into the atmosphere. Atmospheric models tell us that thick cloud layers must lie below, but strong scattering of sunlight by molecules in the upper part of the atmosphere prevents us from seeing these lower cloud layers. Neptune displays a few high clouds with a deep, featureless atmosphere showing between them.

Astronomers use the relative amounts of the elements in the Sun as a standard reference, termed solar **abundance**, when thinking about the chemical composition of stars or planets. Hydrogen (H) is the most abundant element in the Sun (and the universe), followed by helium (He). Jupiter's chemical composition is quite similar to that of the Sun, with about a dozen hydrogen atoms for every atom of helium. Jupiter is 98 percent hydrogen and helium. Only 2 percent of its mass is made up of **heavy elements**, which astronomers define as all elements more massive than helium. Atoms of carbon (C), nitrogen (N), sulfur (S), and oxygen (O) have combined with hydrogen in Jupiter's atmosphere to form molecules of methane (CH_4), ammonia (NH_3), hydrogen sulfide (H_2S), and water (H_2O), respectively. More complex molecules are also common. Helium and certain other gases, such as neon and argon, are classified as **inert** gases because they do not combine with other elements or with themselves to make molecules in the atmosphere. Jupiter's liquid core, which contains most of the planet's iron and rocky materials and much of its water, is left over from the original rocky planetesimal around which Jupiter grew. The abundances of the elements in the Sun (and Jupiter) are represented in the astronomer's version of the periodic table of the elements, shown in **Figure 8.2**.

VOCABULARY ALERT

ocean Here on Earth, *ocean* commonly and specifically means a vast expanse of salty liquid water. Astronomers have expanded the definition of *ocean* to mean a vast expanse of any liquid—not necessarily water.

clouds Just as oceans on other worlds are not necessarily bodies of water, the *clouds* that an astronomer discusses are not necessarily clouds of water vapor—remember from Chapter 7 that clouds of sulfuric acid exist on Venus.

abundance In everyday language, *abundance* means a more than adequate supply, or simply "enough." At times, it implies having much more than is needed. Astronomers use *abundance* very specifically to refer to percentages of chemical composition. If you take a sample of an object, and a certain fraction of it is hydrogen, then that fraction is the abundance of hydrogen in your sample. This means that if one star is more abundant in hydrogen than another, it must be less abundant in at least one other chemical element, even if it has more atoms overall. This is because all the fractions must necessarily add together to equal 1.

inert In everyday language, *inert* can mean "immobile" or "sluggish." Scientists use *inert* in its chemical sense to describe an element that generally doesn't form molecules with other elements or itself; that is, it does not react chemically.

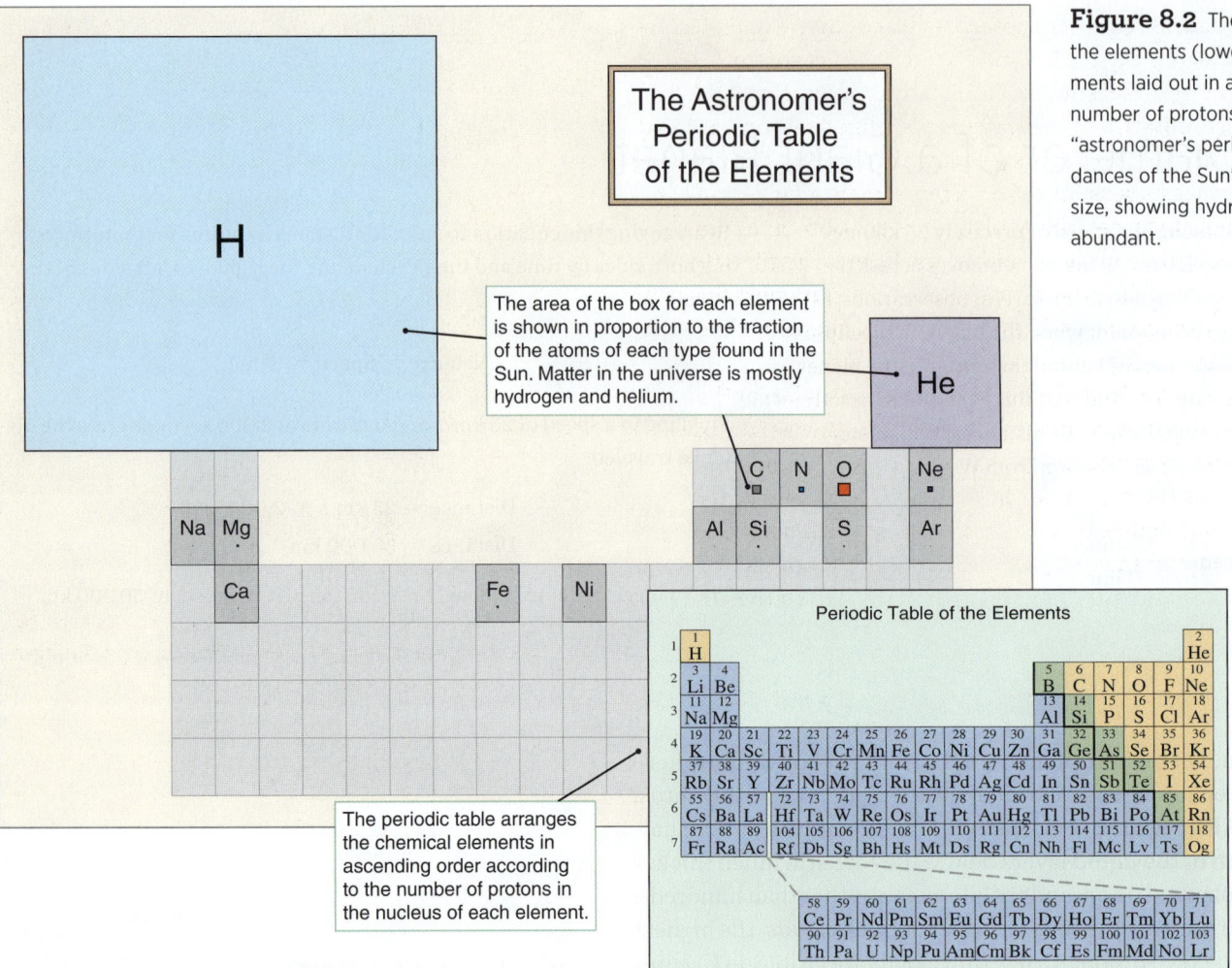

The Astronomer's Periodic Table of the Elements

H

The area of the box for each element is shown in proportion to the fraction of the atoms of each type found in the Sun. Matter in the universe is mostly hydrogen and helium.

He

The periodic table arranges the chemical elements in ascending order according to the number of protons in the nucleus of each element.

Figure 8.2 The traditional periodic table of the elements (lower right) shows the chemical elements laid out in ascending order according to the number of protons in the nucleus of each. But the "astronomer's periodic table" displays the abundances of the Sun's elements in boxes of relative size, showing hydrogen and helium as the most abundant.

The chemical compositions of the giant planets are not all the same. The principal compositional difference among the four giant planets is the abundance of heavy elements each contains. Because of its larger mass, Jupiter had a stronger gravitational pull, and accumulated more hydrogen and helium when it formed than the other planets did. Saturn is more abundant in heavy elements than Jupiter and therefore less abundant in hydrogen and helium. In Uranus and Neptune, heavy elements are more abundant than in either Jupiter or Saturn. These compositional differences support the model of the formation of the Solar System discussed in Chapter 5.

Rotation of the Giant Planets

Giant planets rotate rapidly, as listed in Table 8.1, so their days are short. A day on Jupiter is just under 10 hours long, and a day on Saturn is only a little longer. Neptune and Uranus have rotation periods of 16 and 17 hours, respectively, so the length of their days is between those of Jupiter and Earth. This rapid rotation causes the giant planets to be **oblate**—they bulge at their equators because the inertia of the rotating material near the equator acts to counter gravity. Saturn is significantly oblate; its equatorial diameter is almost 10 percent greater than its polar diameter (**Figure 8.3**). In comparison, the oblateness of Earth is only 0.3 percent. If they did not rotate, the giant planets would be perfectly spherical.

G X U V I R

Figure 8.3 This Hubble Space Telescope (HST) image of Saturn was taken in 1999. The oblateness of the planet is apparent when compared to the white circle. The large moon Titan appears near the top of the disk of Saturn, along with its black shadow.

The intensity of a planet's seasons is determined primarily by the tilt of its axis. Recall from Chapter 2 that Earth's tilt of 23.5° causes our distinct seasons. With a tilt of only 3°, Jupiter has almost no seasons at all. The tilts of Saturn and Neptune are slightly greater than those of Earth, so they have well-defined seasons, as Earth does. The tilt of Uranus is about 98°, so Uranus spins on an axis that lies nearly in the plane of its orbit. This causes its seasons to be extreme; each polar region experiences 42 years of continual sunshine, followed by 42 years of total darkness as the pole first points toward and then away from the Sun. Why is the tilt of Uranus so different than the tilts of most of the other planets? Many astronomers think the planet was "knocked over" by the impact of an Earth-sized planetesimal near the end of its accretion phase.

Uranus is one of only five major Solar System bodies with a tilt larger than 90°. A value greater than 90° indicates that the planet rotates in a clockwise (retrograde) direction when seen from above its orbital plane. Venus, Pluto, Pluto's moon Charon, and Neptune's moon Triton are the only other major bodies that behave this way. This small number of "unusual" bodies is consistent with our ideas about how the Solar System conserved angular momentum when it formed. Each of these objects must have had a unique event in its history to make it different from the others.

CHECK YOUR UNDERSTANDING 8.1

The differences in composition among the giant planets support the model of the formation of stars and planets because the model predicts that

a. the elements were differentiated when the Solar System formed.

b. more massive planets will hold on to more hydrogen and helium.

c. less massive planets will hold on to more hydrogen and helium.

d. planets more distant from the Sun will have more heavy elements.

Answers to Check Your Understanding questions are in the back of the book.

G X U V I R

Figure 8.4 A visible-light image of Jupiter that was taken by the *Cassini* spacecraft while it was on its way to Saturn. The black spot is the shadow of Jupiter's moon Europa.

8.2 The Giant Planets Have Clouds and Weather

When we observe the giant planets through a telescope or in visible-light images from a spacecraft, we typically see a two-dimensional view of the cloud tops. But the atmospheres of the giant planets are unlike anything we have on Earth. The deeper cloud layers on these giant planets are inferred from physical models of the temperature as a function of depth below the cloud tops. In this section, you will learn more about the atmospheres of the giant planets.

Viewing the Cloud Tops

Jupiter is the most colorful of all the giant planets (**Figure 8.4**). Parallel bands, ranging in hue from bluish gray to various shades of orange, reddish brown, and pink, stretch out across its large, pale yellow disk. Convection cells, combined with the Coriolis effect (see Chapter 7), cause this banded pattern. The darker bands are called belts, and the lighter ones are called zones. Belts and zones rotate in opposite directions. The higher, cooler zones are white due to ammonia ice crystals, while the complex chemistry in the lower, warmer belts produces many molecules that in turn cause the brownish color. Many small clouds and storms appear along the edges of, or within, the belts. The most prominent feature is a large, often brick-red oval in Jupiter's southern hemisphere known as the **Great Red Spot**, seen at the lower right in Figure 8.4 and in the close-up image in **Figure 8.5**. At times, having

G X U V I R

Figure 8.5 The Great Red Spot on Jupiter was a bit larger than Earth at the time this *Juno* image was taken in 2017.

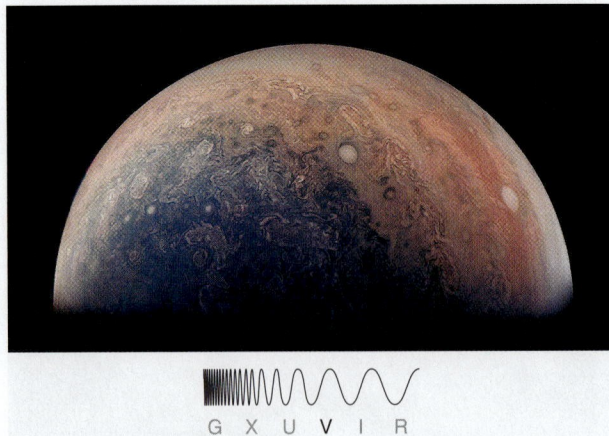

Figure 8.6 The detailed structure of bands, clouds, and storms can be seen even at high latitudes on Jupiter, as shown in this image from *Juno*, taken from an angle close to Jupiter's pole.

Figure 8.7 This *Cassini* image of Saturn was taken in 2004 as the spacecraft approached the planet. This was the last time the whole planet and its rings fit in the camera's field of view. Sunlight is scattered by the cloud-free upper atmosphere, creating a sliver of light in the northern hemisphere.

a length of 25,000 kilometers (km) and a width of 12,000 km, the Great Red Spot could hold two Earths side by side within its boundaries. In 2016, the Great Red Spot shrank to become only as large as one Earth, and it may disappear entirely in a few decades.

The Great Red Spot has been circulating in Jupiter's atmosphere for at least 300 years. It was first seen shortly after the invention of the telescope, and it has since varied unpredictably in size, shape, color, and motion. Observations of small clouds circling the perimeter of the Great Red Spot show that it is an enormous atmospheric whirlpool—a **vortex** (pl. vortices) that swirls in a counterclockwise direction with a period of about a week. Its cloud pattern looks a lot like that of a terrestrial hurricane, but it rotates in the opposite direction, exhibiting anticyclonic rather than cyclonic flow (recall from Chapter 7 that anticyclonic flow indicates a high-pressure system). Comparable whirlpool-like behavior is observed in many of the smaller oval-shaped clouds found elsewhere in Jupiter's atmosphere and in similar clouds observed in the atmospheres of Saturn and Neptune. These vortices are familiar to us on Earth as high- and low-pressure systems, hurricanes, and supercell thunderstorms.

Jupiter is a turbulent, swirling giant with atmospheric currents and vortices so complex that, even after decades of analysis, scientists still do not fully understand the details of how they interact with one another. In a series of time-lapse images, the *Voyager 2* spacecraft observed a number of Alaska-sized clouds being swept into the Great Red Spot. Some of these clouds were carried around the vortex a few times and then ejected, while others were swallowed up and never seen again. Other smaller clouds with structure and behavior similar to that of the Great Red Spot are seen in Jupiter's middle latitudes. The *Juno* spacecraft arrived at Jupiter in 2016 and showed that these clouds are present even at high latitudes (**Figure 8.6**).

Saturn is both farther away than Jupiter and somewhat smaller in radius, so from Earth it appears less than half as large as Jupiter. Like Jupiter, Saturn displays atmospheric bands, but they tend to be wider, and their colors and contrasts are much more subdued than those on Jupiter (**Figure 8.7**). A relatively narrow, meandering band in the mid-northern latitudes encircles the planet in a manner similar to that of our own terrestrial jet stream. Individual clouds on Saturn have been seen only rarely from Earth. On these infrequent occasions, large, white, cloudlike features suddenly erupt in the tropics, spread out longitudinally, and then fade away over a period of a few months. The largest clouds are larger than the continental United States, but many that we see are smaller than terrestrial hurricanes. Close-up views from the *Cassini* spacecraft (**Figure 8.8**) show immense lightning-producing storms in a region of Saturn's southern hemisphere known to mission scientists as "storm alley."

From Earth, even through the largest telescopes, Uranus and Neptune look like tiny, featureless, pale bluish-green disks. Infrared (IR) imaging reveals individual clouds and belts, giving these distant planets considerably more character (**Figure 8.9**). Methane strongly absorbs sunlight, which causes the atmospheres of Uranus and Neptune to appear dark in the near-IR region, allowing the highest clouds and bands to stand out in contrast against the dark background. Methane gas is much more abundant in the atmospheres of Uranus and Neptune than in the atmospheres of Jupiter and Saturn. Like water, methane gas absorbs the longer wavelengths of light—yellow, orange, and red. Absorption of the longer wavelengths leaves only the shorter wavelengths—green and blue—to be scattered from the

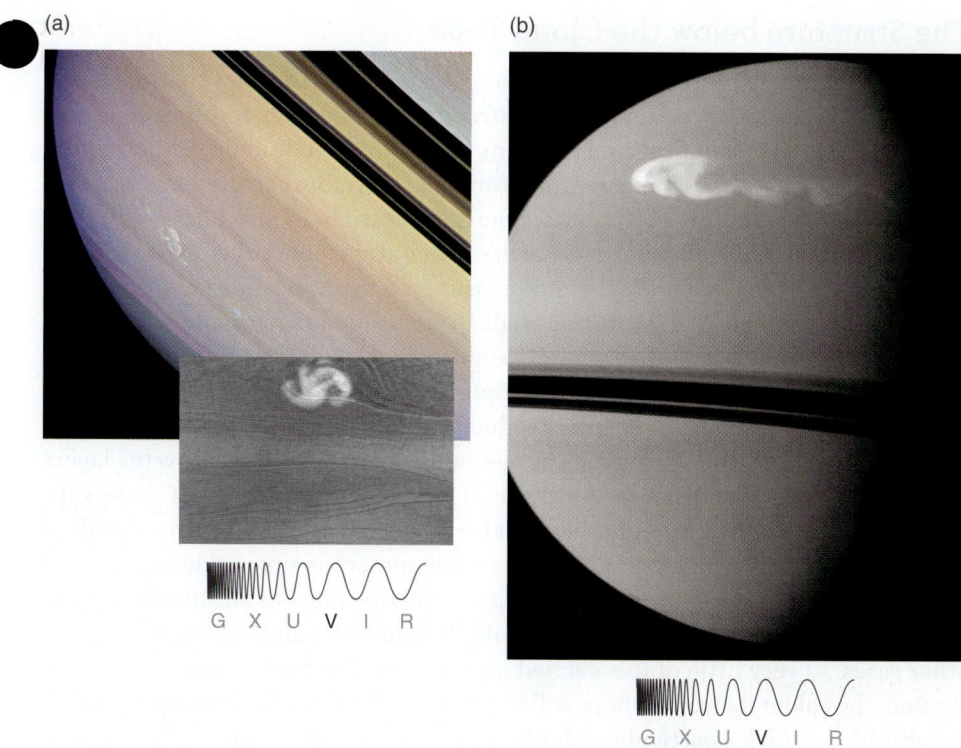

Figure 8.8 Violent storms are known to erupt on Saturn. (a) An enhanced *Cassini* image of an intense lightning-producing storm (left of center) located in Saturn's "storm alley." The inset shows a similar storm on Saturn's night side, illuminated by sunlight reflecting off Saturn's rings. (b) In December 2010, an enormous storm in Saturn's northern hemisphere was discovered by amateur astronomers and subsequently imaged by *Cassini*, as shown here.

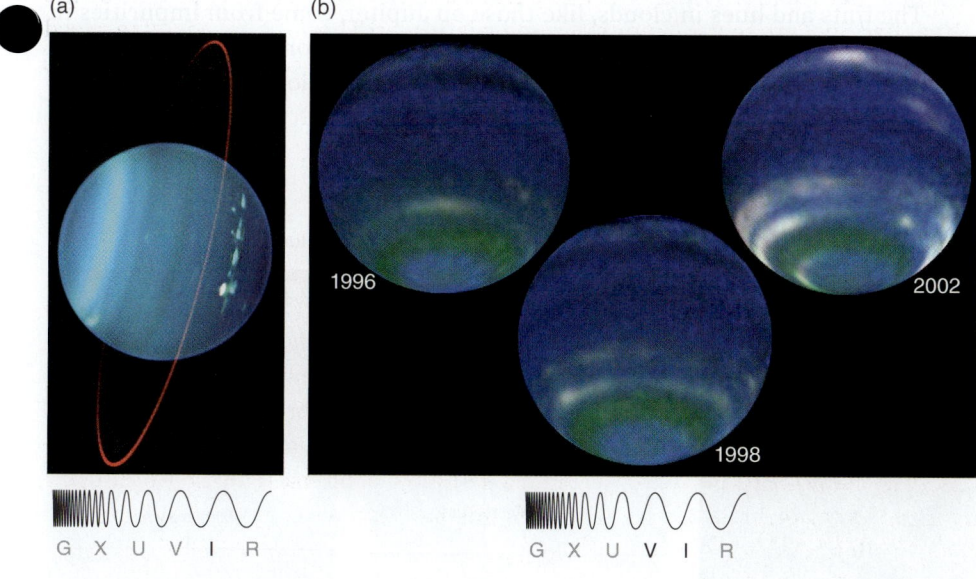

Figure 8.9 (a) The ground-based Keck telescope image of Uranus and (b) the HST images of Neptune were taken at a wavelength of light that is strongly absorbed by methane. The visible clouds are high in the atmosphere. The rings of Uranus show prominently because the brightness of the planet has been subdued by methane absorption. Seasonal changes in cloud formation on Neptune are evident over a 6-year interval.

relatively cloud-free atmospheres of Uranus and Neptune. Thus, Uranus and Neptune appear blue to us in the visible part of the spectrum.

A large, dark, oval feature in Neptune's southern hemisphere, first observed in images taken by *Voyager 2* in 1989, reminded astronomers of Jupiter's Great Red Spot, so they called it the Great Dark Spot. The Neptune feature was gray rather than red, and its length and shape changed more rapidly than did the Great Red Spot. By 1994, it had disappeared. These dark spots come and go; a second was briefly visible in 1994, and a third appeared in 2016.

what if . . .

What if tidal forces caused Jupiter's orbit to shrink, so that it ended up in the same orbit as Venus: How would you expect Jupiter's appearance to change as a result of this?

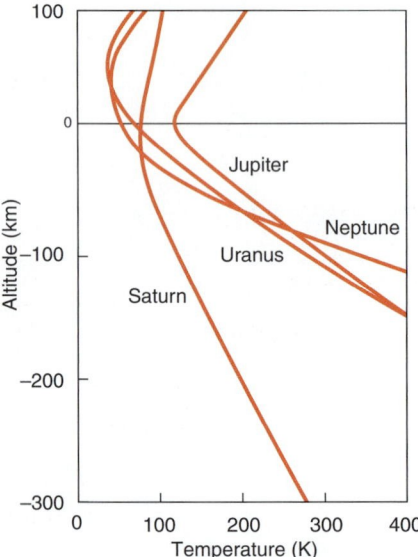

Figure 8.10 The atmospheric temperature changes with height for all four of the giant planets. The arbitrary zero points of altitude are at 0.1 bar for Jupiter and Saturn and at 1.0 bar for Uranus and Neptune. Recall that 1.0 bar corresponds approximately to the atmospheric pressure at sea level on Earth.

The Structure below the Cloud Tops

Although our visual impression of the giant planets is based on views of their cloud tops, atmospheres are three-dimensional structures whose temperatures, densities, pressures, and even chemical compositions vary with altitude and over horizontal distances. As a rule, atmospheric temperature (and therefore density, and pressure) is smaller at higher points in the atmosphere, as shown in **Figure 8.10**.

Water is the only substance in Earth's lower atmosphere that can condense into clouds, but the temperatures and pressures in the atmospheres of the giant planets allow a variety of volatile materials to form clouds surrounded by regions of relatively clear atmosphere. A thin haze above the cloud tops is visible in profile above the **limbs** of the planets. The composition of the haze particles remains unknown, but they may be smoglike products created when ultraviolet (UV) sunlight acts on hydrocarbon gases such as methane. **Figure 8.11** shows how the layers are stacked in the tropospheres of the giant planets. Because each of these substances condenses at a particular temperature and pressure, each forms clouds at a different altitude. Convection carries volatile materials upward along with all other atmospheric gases, and when a particular volatile reaches an altitude with its condensation temperature, most of the volatile condenses and separates from the other gases, so very little of it is carried higher aloft. The farther a planet is from the Sun, the colder its troposphere will be. Distance from the Sun thus determines the altitude at which a particular volatile, such as ammonia or water, will condense to form a cloud layer on each of the planets (see Figure 8.11). If temperatures are too high, some volatiles may not condense at all.

The tints and hues in clouds, like those on Jupiter, come from impurities in the ice crystals, similar to the way that syrups color snow cones. These impurities are often elemental sulfur and phosphorus, as well as various organic materials produced when UV sunlight breaks up hydrocarbons, such as methane, and the fragments recombine to form complex organic compounds.

Figure 8.11

Volatile materials condense at different levels in the atmospheres of the giant planets, leading to chemically different types of clouds at different heights in the atmosphere. As in Figure 8.10, the arbitrary zero points of altitude are at 0.1 bar for Jupiter and Saturn and at 1.0 bar for Uranus and Neptune. Recall that 1.0 bar corresponds approximately to the atmospheric pressure at sea level on Earth.

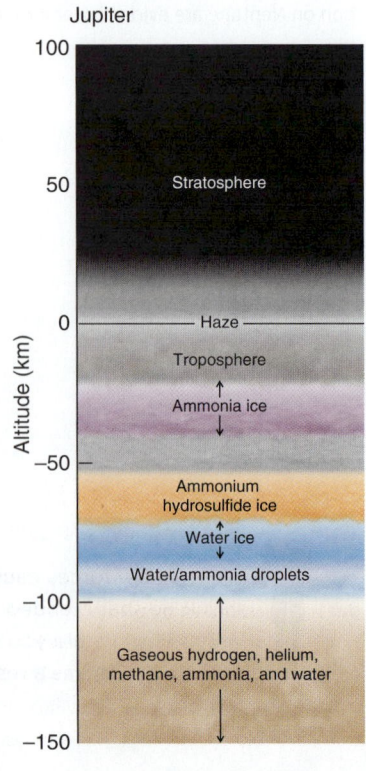

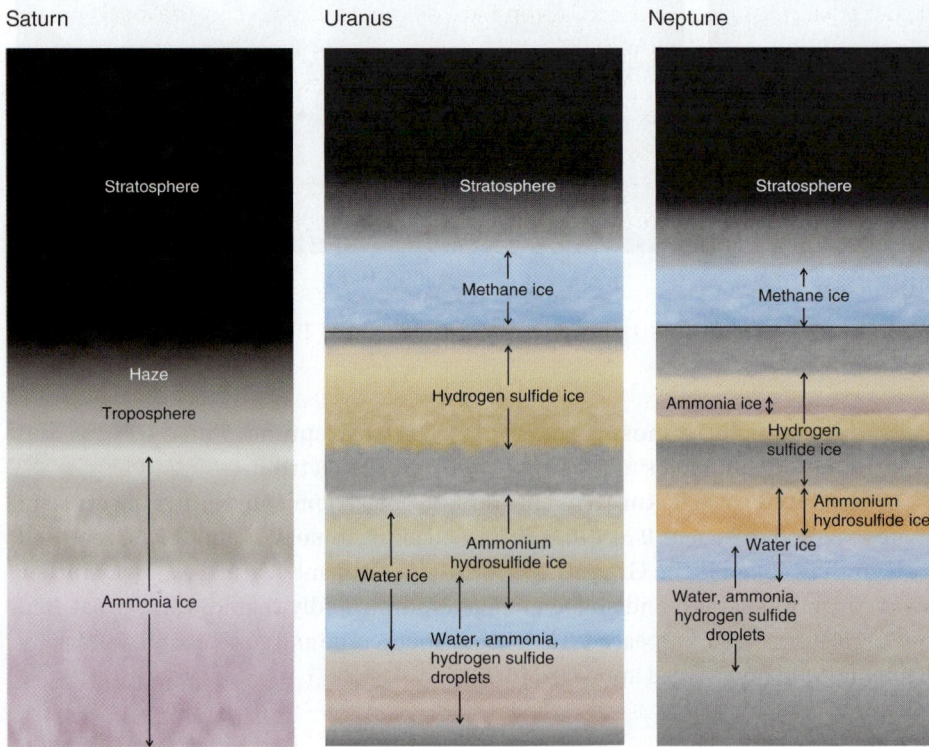

Convection and Weather

The giant planets have much stronger zonal winds than the terrestrial planets. (Recall from Chapter 7 that a zonal wind flows parallel to the equator.) Because the giant planets are farther from the Sun, less thermal energy is available. However, they rotate rapidly, which makes the Coriolis effect very strong. In fact, the Coriolis effect on the giant planets is more important than atmospheric temperature patterns in determining the structure of the global winds.

On the giant planets, the thermal energy that drives convection comes in part from the Sun, but it comes primarily from the hot interiors of the planets themselves. The Coriolis effect shapes that convection into atmospheric vortices. Convective vortices are visible as isolated circular or oval cloud structures, such as the Great Red Spot on Jupiter and the Great Dark Spot on Neptune. As the atmosphere ascends near the centers of the vortices, it expands and cools. Cooling condenses certain volatile materials into liquid droplets, which then fall as rain. As they fall, the raindrops collide with surrounding air molecules, stripping electrons from the molecules and thereby developing tiny electric charges in the air. The cumulative effect of countless falling raindrops can generate an electric charge and a resulting electric field so great that it produces a surge of current and a flash of lightning. A single observation of Jupiter's night side by *Voyager 1* revealed several dozen lightning bolts occurring within an interval of 3 minutes. The strength of these bolts is estimated to be equal to or greater than the "superbolts" that occur in the tops of high convective clouds in Earth's tropics. *Cassini* has imaged the bright dots of lightning flashes in Saturn's atmosphere, and radio receivers on *Voyager 2* picked up lightning static in the atmospheres of both Uranus and Neptune.

Winds on Jupiter and Saturn

How do astronomers measure wind speeds on planets that are so far away, and without entering the atmosphere? As on Earth, clouds are carried by the local winds. By measuring the positions of individual clouds and noting how much they move during an interval of a day or so, astronomers can calculate the local wind speed. However, to find the speed of winds relative to the planet's rotating "surface," astronomers also need to know how fast the planet is rotating.

In the case of the giant planets, there is no solid surface against which to measure the winds. Astronomers instead assume a hypothetical surface—one that rotates as though it were "connected" to the planet's deep interior. The magnetic field of each planet is locked to the conducting liquid layers deep within the planet's interior, so the magnetic field rotates with exactly the same period as that of the deep interior of the planet. Except in the case of Saturn, precise measurement of periodic variations in the radio signals from the giant planets tells us the true rotation periods of these planets. Notably, radio bursts from Saturn have, at different times, implied different rotation periods. Because this phenomenon is not yet understood, astronomers have adopted an average value. Wind speeds on the giant planets are expressed relative to the rotation periods determined from these observations of the magnetic fields.

On Jupiter, the strongest winds are equatorial westerlies, which have speeds of 550 kilometers per hour (km/h), as seen in **Figure 8.12a**. (Recall from Figure 7.11b that westerly winds blow *from*, not toward, the west.) At higher latitudes, the winds alternate between easterly and westerly in a pattern that mimics Jupiter's banded structure of belts and zones. Near a latitude of 20° south, the Great Red

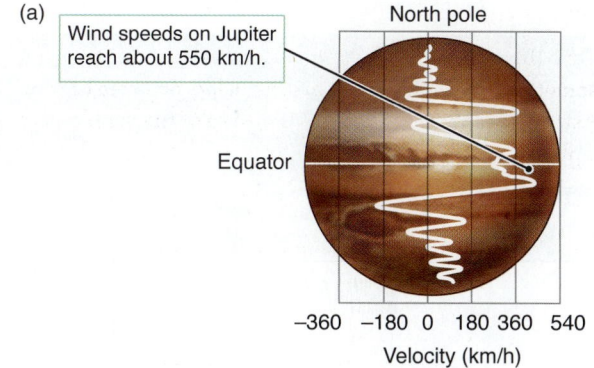

(a) Wind speeds on Jupiter reach about 550 km/h.

North pole
Equator
−360 −180 0 180 360 540
Velocity (km/h)

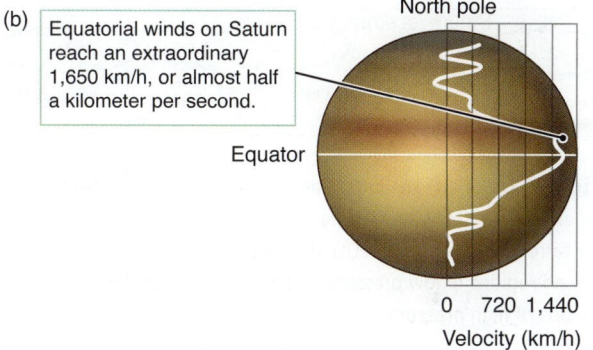

(b) Equatorial winds on Saturn reach an extraordinary 1,650 km/h, or almost half a kilometer per second.

North pole
Equator
0 720 1,440
Velocity (km/h)

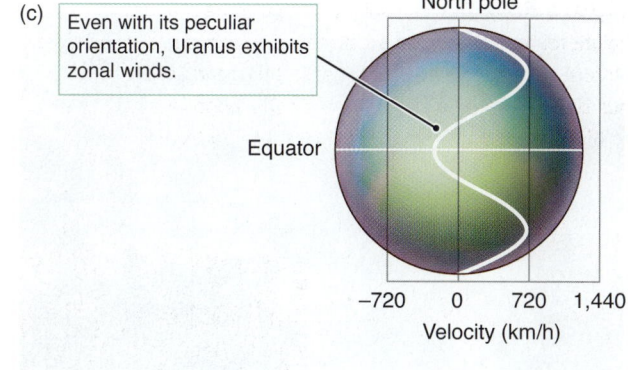

(c) Even with its peculiar orientation, Uranus exhibits zonal winds.

North pole
Equator
−720 0 720 1,440
Velocity (km/h)

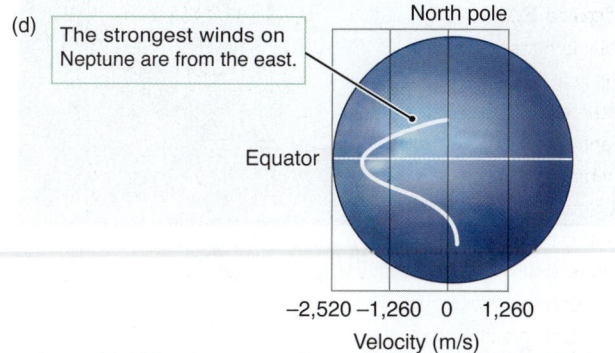

(d) The strongest winds on Neptune are from the east.

North pole
Equator
−2,520 −1,260 0 1,260
Velocity (m/s)

Figure 8.12 Strong winds blow in the atmospheres of the giant planets, driven by powerful convection and the Coriolis effect on these rapidly rotating worlds. The speed of the wind at various latitudes is shown by the white line. When the speed is positive (to the right of 0), the wind is westerly. When the speed is negative (to the left of 0), the wind is easterly.

(a)

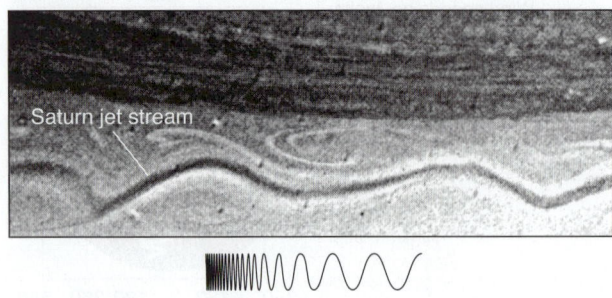

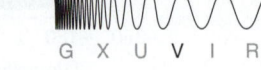

G X U V I R

(b)

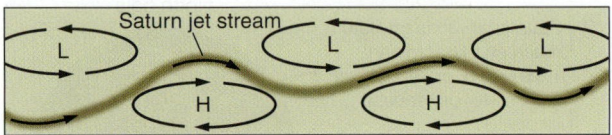

Figure 8.13 (a) The jet stream in Saturn's northern hemisphere seen in this *Voyager* image is similar to jet streams in our terrestrial atmosphere. (b) The jet stream dips equatorward around regions of low pressure and moves poleward around regions of high pressure.

G X U V I R

Figure 8.14 Uranus is approaching equinox in this 2006 HST image. Much of its northern hemisphere is becoming visible. The dark spot in the northern hemisphere (to the right) is similar to but smaller than the Great Dark Spot seen on Neptune in 1989. Uranus is "tipped over," so that the cloud bands are perpendicular to the plane of the Solar System.

Spot vortex lies between a pair of easterly and westerly currents with opposing speeds of more than 200 km/h. The opposing winds at the edge of a zone cause vortices to form at the boundary.

The equatorial winds on Saturn are also westerly, but they are much stronger than those on Jupiter. In the early 1980s, the two *Voyager* spacecraft measured speeds as high as 1,650 km/h. Later, the Hubble Space Telescope (HST) recorded maximum speeds of 990 km/h, and more recently *Cassini* has found them to be intermediate between the *Voyager* and HST measurements. What can be happening here? Saturn's winds appear to decrease with height in the atmosphere, so the *apparent* changes in Saturn's equatorial winds may actually indicate changes in the height of the cloud tops. Alternating easterly and westerly winds also occur at higher latitudes, but unlike Jupiter's case, this alternation seems to bear no clear association with Saturn's atmospheric bands (Figure 8.12b). This is one example of the many unexplained differences among the giant planets.

Saturn's jet stream, at latitude 45° north (**Figure 8.13a**), is a narrow, meandering river of atmosphere that curves around regions of high and low pressure to create alternating crests and troughs (Figure 8.13b). Similarly, Earth's jet stream has high-speed winds that blow generally from west to east but wander toward and away from the poles. Nestled within the crests and troughs of Saturn's jet stream are anticyclonic and cyclonic vortices. These are similar in both form and size to terrestrial high- and low-pressure systems, respectively, which cause alternating periods of fair and stormy weather on Earth.

Winds on Uranus and Neptune

Our knowledge of global winds on Uranus is poorer than that of global winds on the other giant planets (see Figure 8.12c). When *Voyager 2* flew by Uranus in 1986, the few visible clouds were in its southern hemisphere because its northern hemisphere was in complete darkness at the time. The strongest winds observed were 650 km/h westerlies in the middle to high southern latitudes, and no easterly winds were detected. Because Uranus's peculiar orientation makes its poles warmer than its equator, some astronomers had predicted that the global wind system of Uranus might be very different from that of the other giant planets. But *Voyager 2* observed that the Coriolis forces dominate on Uranus as they do on other planets, so the dominant winds on Uranus are zonal, just as they are on the other giant planets.

As Uranus has continued in its orbit, previously hidden regions have become visible (**Figure 8.14**). Observations by HST and ground-based telescopes have shown bright cloud bands in the far north extending more than 18,000 km in length and have revealed wind speeds of up to 900 km/h. As Uranus's long year passes, astronomers will continue to learn much more about the northern hemisphere of the planet.

Each season on Neptune lasts 40 Earth years. On Neptune, the southern hemisphere's summer solstice occurred in 2005, so much of the north is still in darkness. We'll have to wait a while before we can get a good look at its northern hemisphere. As on Jupiter and Saturn, the strongest winds on Neptune occur in the tropics (see Figure 8.12d). They are *easterly* rather than westerly, though, with speeds in excess of 2,000 km/h. Westerly winds with speeds higher than 900 km/h have been seen in Neptune's south polar regions. With wind speeds 5 times greater than those of the fiercest hurricanes on Earth, Neptune and Saturn are the windiest planets known.

8.3 The Interiors of the Giant Planets Are Hot and Dense

Most of what we know about the interiors of the giant planets, such as the state of water in the core, is inferred from terrestrial studies of the behavior of various gases under high pressure. In addition, spacecraft such as *Cassini* and *Juno* have measured densities and magnetic field strengths of the giant planets. Models of the planetary interiors must be consistent with these measurements. **Figure 8.15** shows the structures of the giant planets determined by those models. As a general rule, the material inside the planet forms concentric shells of different materials that increase in density closer to the center of the planet.

The giant planets are massive, so the weight of their outer layers compresses the innermost layers to very high temperatures and densities. For example, the pressure at Jupiter's core is about 45 million bars, about 15 times higher than the pressure at Earth's core. This high pressure heats the fluid to 35,000 K, roughly 6 times higher than the temperature at Earth's core. Central temperatures and pressures of the other, less massive giant planets are lower than those of Jupiter. At the center of each of the giant planets is a dense, liquid core consisting of a very hot mixture of heavy materials such as water, rock, and metals. It may seem strange that water is still liquid at temperatures of tens of thousands of degrees. But, like an enormous pressure cooker, the extremely high pressure in the core prevents water from turning to steam.

Thermal energy from the core drives convection in the atmosphere before it eventually escapes to space as radiation (**Working It Out 8.2**). If energy continually escapes from the interiors of the giant planets, this energy must be replenished. Over the past 4.5 billion years, the giant planets have been shrinking in size, and they continue to do so, which converts gravitational energy into thermal energy. This continual production of thermal energy replaces the internal energy that leaks out of the interior of Jupiter. Jupiter is contracting by only 1 millimeter (mm) or so per year. If it were to continue at this rate, in a billion years Jupiter would shrink by only 1,000 km, a little more than 1 percent of its radius.

At depths of a few thousand kilometers, the atmospheric gases of Jupiter and Saturn are so compressed by the weight of the overlying atmosphere that they turn to liquid. This transition roughly marks the lower boundary of the atmosphere. The difference between a liquid and a highly compressed, very dense gas is subtle, so on Jupiter and Saturn the boundary between the atmosphere and the ocean of liquid that lies below is not sharp. Jupiter's atmosphere is about 20,000 km deep, and Saturn's atmosphere is about 30,000 km deep. At these depths, the pressure climbs

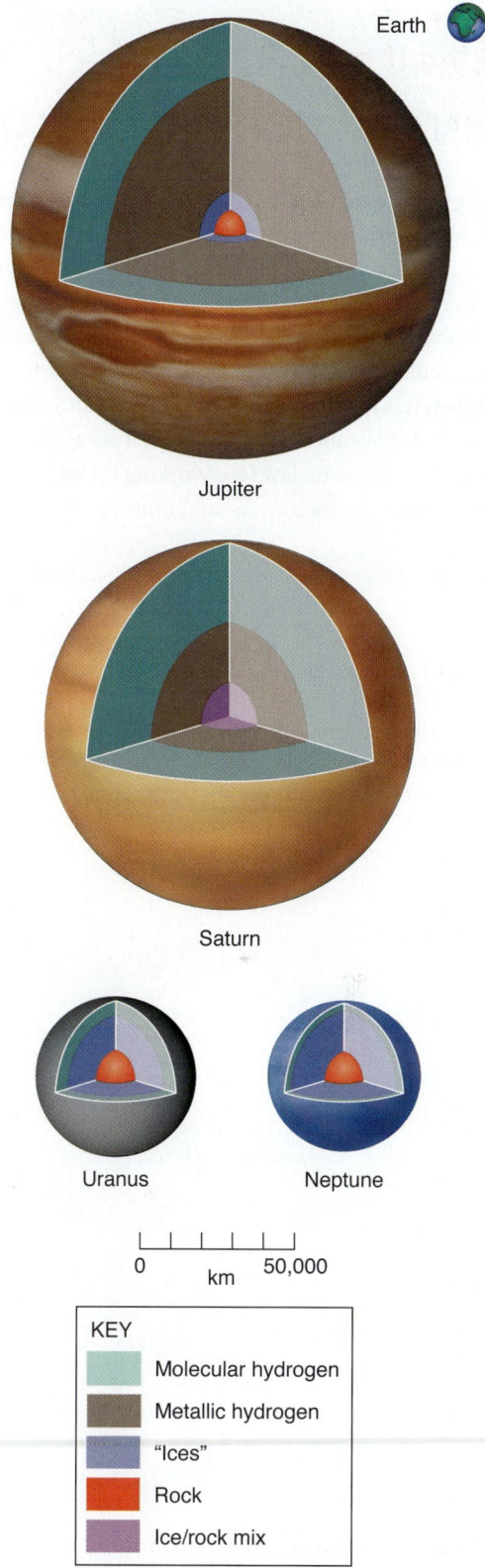

Figure 8.15 The interiors of the giant planets have central cores and outer liquid shells. Only Jupiter and Saturn have significant amounts of the molecular and metallic forms of liquid hydrogen surrounding their cores.

Earth

Jupiter

Saturn

Uranus

Neptune

0   km   50,000

KEY

Molecular hydrogen

Metallic hydrogen

"Ices"

Rock

Ice/rock mix

working it out 8.2

Internal Thermal Energy Heats the Giant Planets

In Chapter 7, we explained the equilibrium between the absorption of sunlight and the radiation of IR light into space, and we saw how the resulting equilibrium temperature is modified by the greenhouse effect on Venus, Earth, and Mars. Yet when we calculate this equilibrium for three of the giant planets, we find that something seems amiss. According to these calculations, the equilibrium temperature for Jupiter should be 109 K. When it is measured, however, we find an average temperature of about 124 K. A difference of 15 K might not seem like much, but according to the Stefan-Boltzmann law (see Working It Out 5.1), the energy radiated by an object depends on its temperature raised to the fourth power.

Applying this relationship to Jupiter, we find

$$\left(\frac{T_{\text{actual}}}{T_{\text{expected}}}\right)^4 = \left(\frac{124}{109}\right)^4 = 1.67$$

Jupiter, therefore, is radiating 1.67 times more energy into space than it absorbs in the form of sunlight. Similarly, the energy escaping from Saturn is about 1.8 times greater than the sunlight that it absorbs. Neptune emits 2.6 times as much energy as it absorbs from the Sun. In contrast, whatever internal energy may be escaping from Uranus is negligible compared with the absorbed solar energy.

what if . . .

The diameter of Jupiter shrinks at a rate of about 1 mm/year: If Jupiter's mass were larger, how would that affect this contraction?

to 2 million bars and the temperature reaches 10,000 K. At this pressure and temperature, hydrogen molecules are battered so violently that their electrons are stripped free, and the hydrogen acts like a liquid metal. In this state, it is called *metallic hydrogen*. Electrical currents in this region may be responsible for the strong magnetic field of Jupiter, although we do not yet have a deep understanding of this process. The first reports of laboratory production of metallic hydrogen came in 2017. These results have been greeted with the usual healthy scientific skepticism, and as of this writing, other scientists are working to repeat the result. Jupiter itself remains the best "laboratory" available for understanding the behavior of metallic hydrogen.

The oceans of liquid hydrogen and helium on the giant planets are tens of thousands of kilometers deep. Uranus and Neptune are less massive than Jupiter and Saturn, have lower interior pressures, and contain a smaller fraction of hydrogen—their interiors probably contain only a small amount of liquid hydrogen, with little or none of it in a metallic state.

Differentiation separates the chemical components inside of a planet (recall Chapter 6). This process has occurred and is still occurring in Saturn, and perhaps in Jupiter too. On Saturn, helium that is mixed in the hydrogen-helium oceans separates into droplets. Because droplets of helium are denser than the liquid around them, they sink toward the center of the planet, converting gravitational energy to thermal energy. This process heats the planet and enriches helium in the core while depleting it in the upper layers. In Jupiter's hotter interior, by contrast, the liquid helium is mostly well-mixed with the liquid hydrogen.

The heavy-element components of the cores of Jupiter and Saturn have masses of about 10–20 Earth masses, while Jupiter and Saturn have total masses of 318 and 95 Earth masses, respectively. The small core mass compared to the planet mass means that the heavy materials contribute little to the average chemical composition. Both Jupiter and Saturn have approximately the same composition as the Sun and the rest of the universe: about 98 percent hydrogen and helium, leaving only 2 percent for everything else.

Uranus and Neptune are about twice as dense as Saturn, so they must contain more heavy-element components than Saturn and Jupiter. Neptune, the densest of the giant planets, is about 1.5 times denser than uncompressed water and only about half as dense as uncompressed rock. These observations indicate that water and other low-density ices, such as ammonia and methane, must be the major compositional components of Uranus and Neptune, along with lesser amounts of silicates and metals. The total amount of hydrogen and helium in these planets is probably limited to no more than 1 or 2 Earth masses, and most of these gases reside in the relatively shallow atmospheres of the planets.

Why do Jupiter and Saturn have so much hydrogen and helium compared with Uranus and Neptune? Why is Jupiter so much more massive than Saturn? The answers may lie both in the time that it took for these planets to form and in the distribution of material from which they formed. The icy planetesimals from which Uranus and Neptune formed were more widely dispersed at their greater distances from the Sun. With more space between planetesimals, their cores would have taken longer to build up. Because of this delay, the cores of Uranus and Neptune were smaller and formed much later than those of Jupiter and Saturn, at a time when most of the gas in the protoplanetary disk had already been blown away. Saturn may have captured less gas than Jupiter both because its core formed somewhat later and because less gas was available at its greater distance from the Sun.

CHECK YOUR UNDERSTANDING 8.3

Deep in the interiors of the giant planets, water is still a liquid even though the temperatures are tens of thousands of degrees above the boiling point of water. This can happen because

a. the giant planets have very little water in the interior.

b. the pressure inside the giant planets is so high.

c. the outer Solar System is so cold.

d. space has very low pressure.

8.4 The Giant Planets Are Magnetic Powerhouses

All of the giant planets have strong magnetic fields: Their field strengths range from 50 to 20,000 times stronger than Earth's when measured at the same distance from the centers of the planets. Field strength decreases with distance, however, so fields at the cloud tops of Saturn, Uranus, and Neptune are comparable in strength to Earth's surface field. Even in the case of Jupiter's exceptionally strong field, the field strength at the cloud tops is only about 15 times that of Earth's surface field. In Jupiter and Saturn, magnetic fields are generated by circulating currents within deep layers of metallic hydrogen (see Figure 8.15). In Uranus and Neptune, magnetic fields arise within deep oceans of liquid water and ammonia made electrically conductive by dissolved salts. The magnetospheres of the giant planets are very large, and they interact with the solar wind (as Earth's magnetosphere does) and with the rings and moons that orbit the giant planets.

The Size and Shape of the Magnetospheres

If we imagine that the magnetic fields are produced by bar magnets, as shown in **Figure 8.16**, we can compare their tilts and their locations in the interior. Saturn's magnetic axis is most closely aligned, in both tilt and location, with its rotation axis, whereas Uranus' magnetic axis is most tilted, at 60°, and Neptune's is offset farthest, by more than half of Neptune's radius. The displacement of the field is primarily toward Neptune's southern hemisphere, thereby creating a field 20 times stronger at the southern cloud tops than at the northern cloud tops. The reason for the unusual geometry of the magnetic fields of Uranus and Neptune remains unknown, but it is not related to the orientations of their rotation axes.

Just as Earth's magnetic field traps energetic charged particles to form Earth's magnetosphere (review Figure 7.9, if necessary), the magnetic fields of the giant planets also trap energetic particles to form magnetospheres of their own. Earth's magnetosphere is tiny in comparison to those of the giant planets. By far, the most colossal of these is Jupiter's. Its radius is 100 times that of the planet itself, roughly 10 times the radius of the Sun. Even the relatively weak magnetic fields of Uranus and Neptune form magnetospheres that are comparable in size to the Sun.

Magnetospheres are influenced by the solar wind. Recall from Chapter 7 that the solar wind supplies some of the particles for a magnetosphere. In addition, the

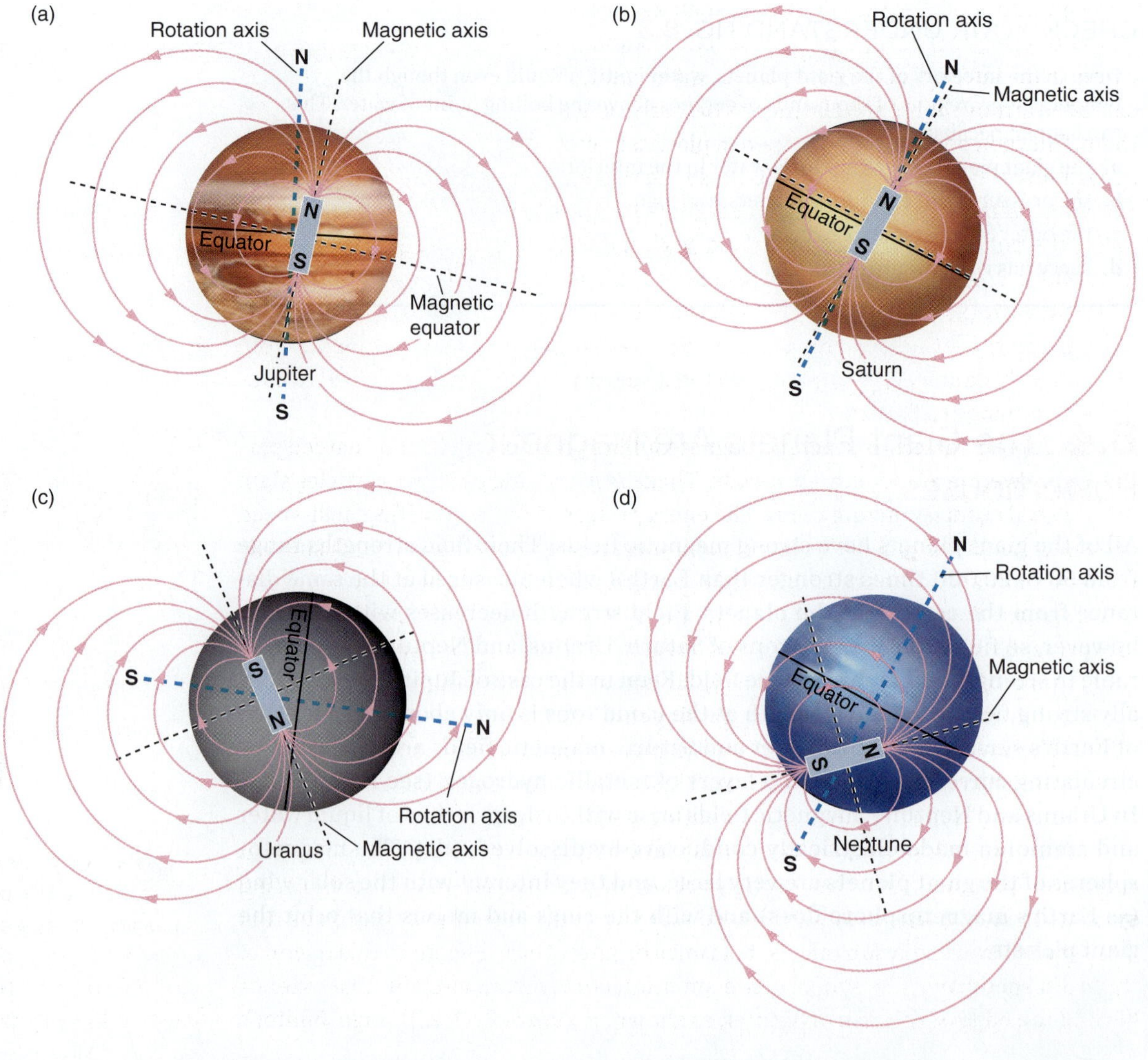

Figure 8.16 The magnetic fields of the giant planets can be approximated by the fields from bar magnets offset and tilted with respect to the rotation axes of the planets. Compare these with Earth's magnetic field, shown in Figure 6.11. (a) Jupiter's magnetic axis is inclined 10° to its rotation axis—an orientation similar to Earth's—but it is offset about a tenth of a radius from the planet's center. (b) Saturn's magnetic axis is located almost precisely at the planet's center and is almost perfectly aligned with the rotation axis. (c) The magnetic axis of Uranus is inclined nearly 60° to its rotation axis and is offset by a third of a radius from the planet's center. (d) The orientation of Neptune's rotation axis is similar to that of Earth, Mars, and Saturn. But Neptune's magnetic axis is inclined 47° to its rotation axis, and the center of this magnetic field is displaced from the planet's center by more than half the radius—an offset even greater than that of Uranus.

pressure of the solar wind pushes on and compresses a magnetosphere, so planets farther from the Sun, where the wind is weaker, have larger magnetospheres. The precise size and shape of a planet's magnetosphere depends on how the solar wind is blowing at any particular time. As illustrated in **Figure 8.17**, the solar wind sweeps the magnetosphere into a tail in much the same way that water sweeps algae downstream of a rock to form a tail. The tail of Jupiter's magnetosphere extends well past the orbit of Saturn. For both Uranus and Neptune, the tilt and the large displacement of the magnetic field from the center of the planet cause the magnetosphere to wobble as the planet rotates. This wobble causes the tail of the magnetosphere to twist like a corkscrew as it trails away.

Charged Particles and Auroras of the Giant Planets

In addition to protons and electrons from the solar wind, the magnetospheres of the giant planets contain large amounts of various elements, such as sodium, sulfur, oxygen, nitrogen, and carbon. These elements come from several sources, including the planets' extended atmospheres and the moons that orbit within them. Charged particles are trapped in planetary magnetospheres. The locations where these particles are trapped are referred to as *radiation belts*. Although Earth's radiation belts are severe enough to worry astronauts, the radiation belts that surround Jupiter are searing in comparison. In 1974, the *Pioneer 11* spacecraft passed through the radiation belts of Jupiter. During its brief encounter, *Pioneer 11* was exposed to a radiation dose of 400,000 rads, or about 1,000 times the lethal dose for humans. Several of the onboard instruments were permanently damaged as a result, and the spacecraft itself barely survived to continue its journey to Saturn.

As Jupiter rotates, it drags its magnetosphere around with it, and charged particles are swept around at high speeds. These fast-moving charged particles slam into neutral atoms, ionizing them. The energy released in the resulting high-speed collisions heats the plasma to extreme temperatures. (Recall that in a plasma the atoms are ionized, so that the positively charged nuclei and the negatively charged electrons are free to move separately.) In 1979, while passing through Jupiter's magnetosphere, *Voyager 1* encountered a region of tenuous plasma with a temperature of more than 300 million K; this is 20 times the temperature at the center of the Sun. *Voyager 1* did not melt when passing through this region because, although each particle was extraordinarily energetic, there were so few of them that the probe passed unscathed through the plasma. This is the same reason that Earth's thermosphere would not feel hot to you, as discussed in Chapter 7.

When fast-moving charged particles interact with magnetic fields, they emit radiation primarily in the radio portion of the spectrum. This is called **synchrotron radiation**. Jupiter has so many trapped high-energy electrons that this synchrotron radiation makes it at times brighter than the Sun in this region of the radio spectrum. The synchrotron emission comes from a region that extends beyond the edge of the disk of Jupiter, as shown in **Figure 8.18**. Although Jupiter's

(a)

(b)

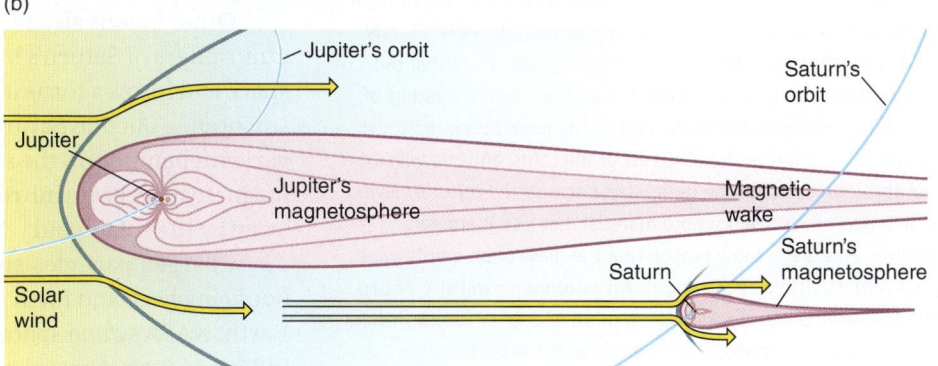

Figure 8.17 (a) Water flowing past a rock sweeps the algae against the rock and into a "tail" pointing in the direction of the water's flow. (b) The solar wind compresses Jupiter's (or any other) magnetosphere on the side toward the Sun and draws it out into a magnetic tail away from the Sun. Jupiter's tail stretches beyond the orbit of Saturn. This drawing is not to scale.

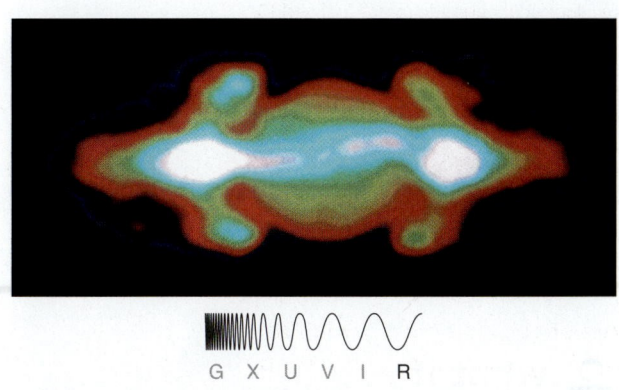

G X U V I R

Figure 8.18 This image from the Very Large Array shows the synchrotron radiation produced by very fast electrons trapped in Jupiter's strong magnetic field. The large extended feature beyond the planet's disk is similar to Earth's Van Allen radiation belt. The gas in Jupiter's atmosphere also produces a fainter radio emission that roughly outlines the planetary disk.

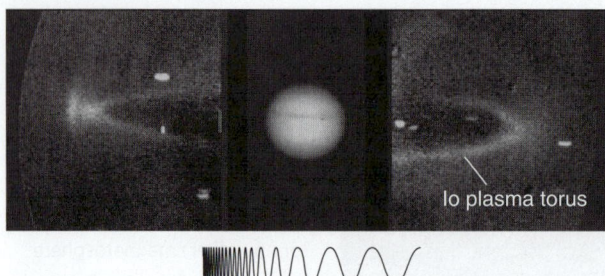

G X U V I R

Figure 8.19 In this image, a faintly glowing torus of plasma surrounds Jupiter (center). The torus is made up of atoms knocked free from the surface of Io by charged particles. Io is the innermost of Jupiter's Galilean moons. (A semitransparent mask was placed over the disk of Jupiter to prevent its light from overwhelming that of the much fainter torus.) ★ **WHAT AN ASTRONOMER SEES** An astronomer will notice that the plasma torus has a radius several times larger than the radius of Jupiter. She will notice that the part of the torus at the right and the left looks brighter than the rest of the torus, and she will know that this is an effect of the geometry; because the torus is viewed at an angle, she is seeing more material there, so it appears brighter. This is the same reason that the glass of an empty wine glass is most visible at the edges. An astronomer will also notice that the camera was tracking the motion of Jupiter across the sky for a considerable amount of time, because the background stars in the image are not points of light but have become smeared into bright streaks.

magnetosphere is much farther away than the Sun, it appears much larger than the Sun in our sky.

Saturn's magnetosphere is also large but much fainter than Jupiter's. Saturn has a strong magnetic field, but pieces of rock, ice, and dust in Saturn's rings absorb magnetospheric particles. With far fewer magnetospheric electrons, there is much less radio emission from Saturn.

Images of the region around Jupiter, taken in the light of emission lines from atoms of sulfur or sodium, show a faintly glowing **torus** (a doughnut-shaped ring) of material supplied by Io, the innermost large moon of Jupiter's (**Figure 8.19**). This is the most intense radiation belt in the Solar System. Io has low surface gravity and violent volcanic activity. Some of the gases erupting from Io's interior escape and become part of Jupiter's radiation belt. Even more material is knocked free of Io's surface and ejected into space after charged particles slam into the moon due to the rotation of Jupiter's magnetosphere.

Other moons also influence the magnetospheres of the planets they orbit. The atmosphere of Saturn's largest moon, Titan, is rich in nitrogen. This gas leaks into space and forms a torus in Titan's wake. The orbit of Titan is sometimes inside and sometimes outside Saturn's magnetosphere, so the density of this radiation belt varies depending on the strength of the solar wind. When Titan is outside Saturn's magnetosphere, any nitrogen molecules lost from Titan's atmosphere are carried away by the solar wind.

Charged particles spiral along the magnetic-field lines of the giant planets, bouncing back and forth between the two magnetic poles just like they do around Earth. NASA's *Juno* spacecraft imaged the bright auroral rings on Jupiter, shown in **Figure 8.20**. Auroral rings like these surround the magnetic poles of the giant

 **what if . . .**

What if you searched for planets around other stars by looking for radio emissions from their magnetospheres: Would this constitute a good way to detect other large planets?

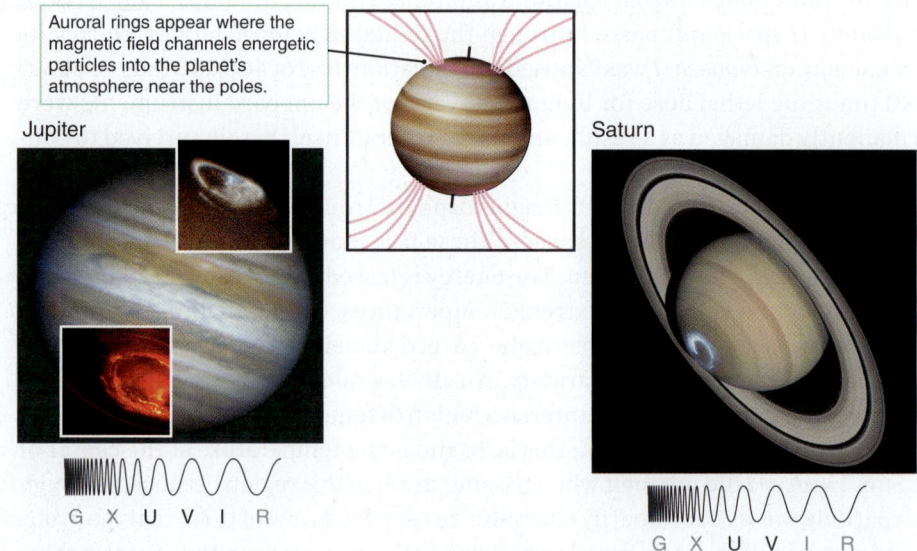

Figure 8.20 NASA's *Juno* spacecraft passed over the south pole of Jupiter to obtain a remarkable IR image of the auroral ring (bottom inset). Views like this cannot be obtained from Earth. The HST took images of auroral rings around Jupiter's north pole and Saturn's south pole. These auroral images were taken in UV light and then superimposed on visible-light images. (High-level haze obscures the UV views of the underlying cloud layers, as the top inset for Jupiter shows.)

planets, just as the aurora borealis and aurora australis ring the north and south magnetic poles of Earth.

Measurements of the magnetic fields of the giant planets indicate what about the interiors?

a. The interiors contain flowing, electrically conductive materials.
b. The interiors are made of giant, solid pieces of magnetic iron.
c. The interiors contain multiple cloud layers, all the way down to the center.
d. The innermost layers of the interior are made of rock.

8.5 Rings Surround the Giant Planets

A planetary ring is a collection of particles—varying in size from tiny grains to house-sized boulders—that orbit individually around the planet, forming a flat disk. Ring systems are found around all of the giant planets but not around terrestrial planets. **Figure 8.21** shows how the ring system of each giant planet varies in size and complexity, with some systems having detailed structures that include numerous small rings. Saturn's E Ring extends well beyond the edge of the image to about 500,000 km from the center of Saturn. Here in Section 8.5, you will learn about orbits of ring particles, ring formation and evolution, and ring composition and structure.

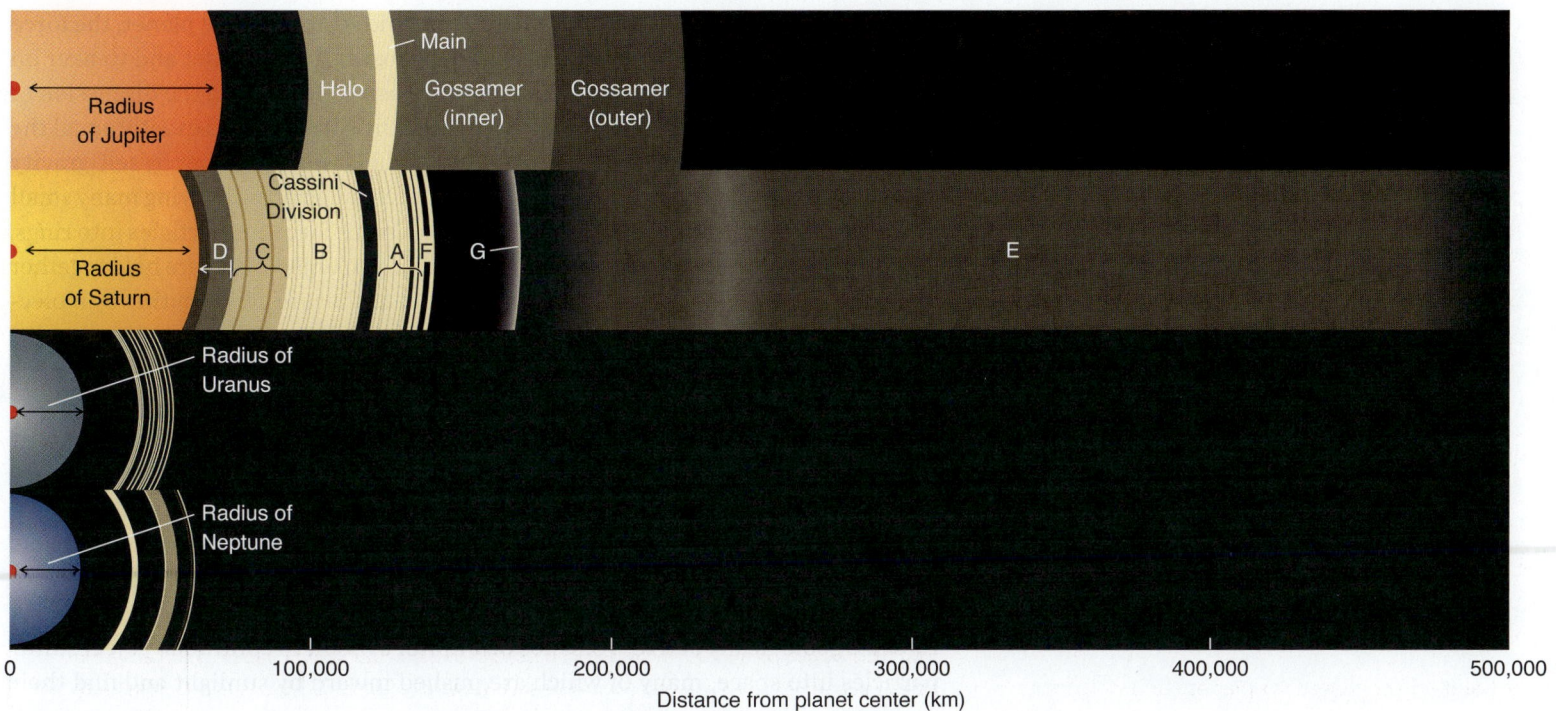

Figure 8.21 The ring systems of the four giant planets vary in size and complexity. Saturn's system, with its broad E Ring, is by far the largest and has the most complex structure in the inner rings.

Orbits of Ring Particles

Kepler's laws (see Chapter 3) dictate the speeds and orbital periods of all ring particles, such that the closest particles move the fastest and have the shortest orbital periods. The orbital periods of particles in Saturn's three bright rings, for example, range from 5 hours 45 minutes at their inner edge to 14 hours 20 minutes at their outer edge. Ring particles all orbit in the same direction, so they have low speeds relative to one another. A particle moving on an upward trajectory will bump into another particle on a downward trajectory and the upward and downward motion will cancel, leaving the particles moving in the same plane. A similar process occurs for particles moving inward and outward, leaving the particles moving at a constant radius. In the densely packed rings of Saturn, collisions force particles into circular orbits in a single plane.

A planet's moons can also influence the orbits of ring particles. A massive moon exerts a significant gravitational tug on a ring particle as it passes by. If this happens to many particles through many orbits, particles will be pulled out of the area, leaving a lower-density gap. This is how Saturn's moon Mimas causes the Cassini Division—a gap in the rings around Saturn (Figure 8.21). Most narrow rings are held in place by pairs of nearby moons, known as **shepherd moons** because of the way they "herd" the flock of ring particles. A shepherd moon just outside a ring robs orbital energy from any particles that drift outward beyond the edge of the ring, causing them to move back inward. A shepherd moon just inside a ring gives up orbital energy to a particle that has drifted too far in, nudging it back outward.

Ring Formation and Evolution

When a moon (or other large body, like an asteroid) orbits a large planet, the force of gravity is stronger on the side of the moon closest to the planet and weaker on the side farther away. This stretches the moon, as you learned in the discussion of tidal forces in Chapter 6. The distance at which the tidal stresses exactly equal the self-gravity is called the **Roche limit**. If a moon that is held together by self-gravity comes within the Roche limit of a giant planet, it is pulled apart, leaving many small particles to spread out and orbit the planet. Collisions force the particles into rings, as described above. The Roche limit does not apply to objects that are held together by other forces—objects like people or bowling balls that are held together by molecular bonds, for example.

Planetary rings do not have the long-term stability of most Solar System objects. Collisions can cause particles at the ring edges to gain or lose energy and drift away, aided by the pressure of sunlight or the solar wind. In some cases, shepherd moons may stabilize the orbits of ring particles and delay the dissipation of the rings, but this is a temporary situation. Most planetary rings eventually dissipate. At least one ring, however, seems immune from this eventual demise: Saturn's moon Enceladus has volcanoes that constantly supply icy particles to Saturn's E Ring (see Figure 8.21), replacing those that drift away. The E Ring will survive for as long as Enceladus remains geologically active. Volcanos on Jupiter's moon Io continually eject sulfur particles into space, many of which are pushed inward by sunlight and find their way into a ring.

Moons can contribute material to rings in other ways as well, as in the case of Jupiter's ring system, shown in **Figure 8.22**. The brightest of Jupiter's rings is a relatively narrow strand only 6,500 km across, consisting of material from the

(a)

(b)

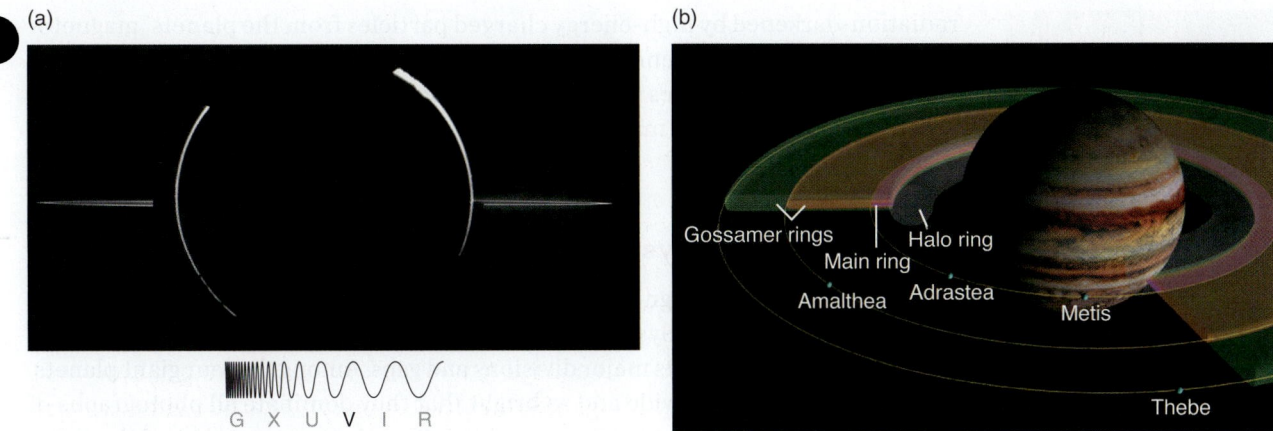

G X U V I R

Figure 8.22 (a) This backlit *Galileo* view of Jupiter's rings also shows the forward scattering of sunlight by tiny particles in the upper layers of Jupiter's atmosphere. (b) A diagram of Jupiter's ring system and the small moons that form the rings.

moons Metis and Adrastea. These two moons orbit in Jupiter's equatorial plane, and the ring they form is narrow. Beyond this main ring, however, are the very different **gossamer rings**, so called because they are extremely tenuous. The gossamer rings are supplied with dust from the moons Amalthea and Thebe. The innermost ring in Jupiter's system, called the halo ring, consists mostly of material from the main ring. As the dust particles in the main ring drift slowly inward toward the planet, they pick up electric charges and are pulled into this rather thick torus by electromagnetic forces associated with Jupiter's powerful magnetic field.

Ring systems have come and gone over the history of the Solar System. Even Earth has probably had several short-lived rings at various times during its long history. Any number of comets or asteroids must have passed close enough to Earth to disintegrate into a swarm of small fragments to create a temporary ring. Because Earth lacks shepherd moons to provide orbital stability, these rings dissipated quickly.

The Composition of Rings

Because much of the material in the rings of the giant planets comes from their moons, the composition of the rings is similar to the composition of the moons. Saturn's bright rings probably formed when a moon or planetesimal came within the Roche limit of Saturn. These rings reflect about 60 percent of the sunlight falling on them. They are made of water ice, though a slight reddish tint tells us they are not made of pure ice but must contain small amounts of other materials, such as silicates. The icy moons around Saturn or the frozen comets of the outer Solar System have similar composition, supporting the hypothesis that similar objects could be the source of this material.

Saturn's rings are the brightest in the Solar System and are the only ones that we know are composed of water ice. In stark contrast, the rings of Uranus and Neptune are among the darkest objects known in our Solar System. They are blacker than coal or soot, and only 2 percent of the sunlight falling on them is reflected back into space. No silicates or similar rocky materials are this dark, so these rings are likely composed of organic materials and ices that have been

what if . . .

What if Earth's Moon had active volcanoes, strong enough to eject particles from the lunar surface: How would the resulting ring be oriented in the sky? Would the ring persist for a long time, or be short-lived? Describe how the night sky would be different when viewed from a dark site.

G X U V I R

Figure 8.23 This *Cassini* image of the rings of Saturn shows many ringlets and minigaps. The cause of most of this structure has yet to be explained in detail.

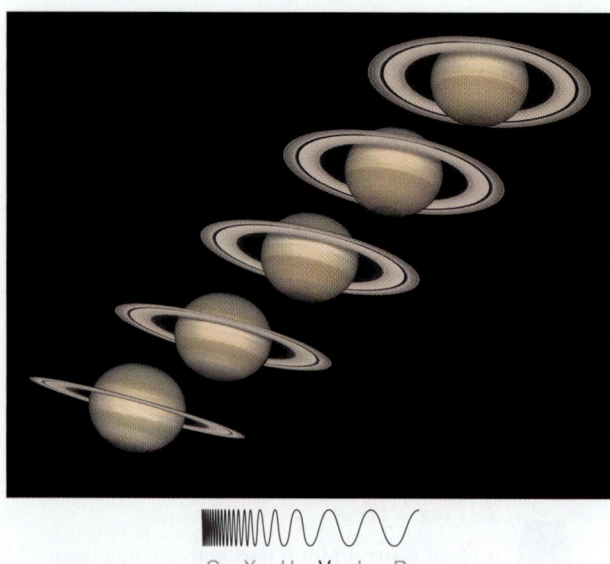

G X U V I R

Figure 8.24 The HST took these images of Saturn over 6 years, showing how the rings appear to change their shape as Saturn moves through its orbit around the Sun. This effect is caused by changes in the position of Earth relative to the rings.

radiation-darkened by high-energy charged particles from the planets' magnetospheres. (Radiation-darkening blackens organic ices such as methane by releasing carbon from the molecules of the ice.) Jupiter's rings are of intermediate brightness, suggesting that they may be rich in dark silicate materials, like the innermost of Jupiter's small moons.

Structure of Ring Systems

Saturn is adorned by a magnificent and complex system of rings, unmatched by any other planet in the Solar System. Figure 8.21 shows the individual components of Saturn's ring system and its major divisions and gaps. Among the four giant planets, only Saturn has rings so wide and so bright that they dominate all photographs of the planet. The outermost bright ring, the A Ring, is the narrowest of the three bright rings. On the A Ring's inner edge, the conspicuous Cassini Division is so wide (4,700 km) that the planet Mercury would almost fit within it. Astronomers once thought that it was completely empty, but images taken by *Voyager 1* show the Cassini Division is filled with material, although it is less dense than the other material in bright rings.

Although Saturn's bright rings are very wide—more than 62,000 km from the inner edge of the C Ring to the outer edge of the A Ring—they are extremely thin. Saturn's bright rings are no more than a hundred meters and probably only a few tens of meters from their lower to upper surfaces. The diameter of Saturn's bright ring system is 10 million times the thickness of the rings. If the bright rings of Saturn were the thickness of a page in this book, six football fields laid end to end would stretch across them.

Saturn's bright rings are not uniform. The A and C Rings contain hundreds, and the B Ring contains thousands, of individual **ringlets**, some only a few kilometers wide (**Figure 8.23**). Each of these ringlets is a narrowly confined concentration of ring particles bounded on both sides by regions of relatively little material.

About every 15 years, the plane of Saturn's rings lines up with Earth, and we view them edge on so that they all but vanish for a day or so in even the largest telescopes. At other times, they are seen more "face-on" as shown in the 6-year montage in **Figure 8.24**. During times when the glare of the rings is absent, astronomers search for undiscovered moons or other faint objects close to Saturn. In 1966, an astronomer was looking for moons when he found weak but compelling evidence for a faint ring near the orbit of Saturn's moon Enceladus. In 1980, *Voyager 1* confirmed the existence of this faint ring—named the E Ring—and found another ring even closer, named the G Ring. The E and G Rings are examples of what astronomers call a **diffuse ring**. In a diffuse ring, particles are far apart, and rare collisions between them can cause their individual orbits to become eccentric, inclined, or both. Because collisions are rare, the particles tend to remain in these disturbed orbits. Diffuse rings spread out horizontally and thicken vertically, sometimes without any obvious boundaries. Diffuse rings are difficult to detect except when lit from behind, when they appear bright because of the strong forward scattering of sunlight by very small ring particles (see Figure 8.22a for an example of Jupiter's rings). In 2009, astronomers using the Spitzer Space Telescope discovered another diffuse ring around Saturn. This dusty ring is thicker than other rings, about 20 times larger than Saturn from top to bottom, and is tilted 27° with respect to the plane of the rest of the rings.

Ring structure among the other giant planets is not as diverse as that of Saturn. Turning back to Figure 8.21, notice that most rings other than Saturn's are quite narrow and often widely spaced. Only a few are diffuse. Gaps between the rings, like the Cassini Division, may appear dark and yet be full of dust. The dust is the right size to scatter sunlight forward instead of reflecting it backward. *Voyager 2* imaged this forward scattering after it passed Uranus and showed that the gaps were not empty. In one of Neptune's four narrow rings, the Adams Ring, the material is clumped together in higher-density segments called **ring arcs (Figure 8.25)**. When first discovered, these ring arcs were a puzzle because collisions should have spread them uniformly around their orbit. Most astronomers now attribute this clumping to gravitational interactions with the moon Galatea. Some parts of Neptune's rings may be unstable: Ground-based images taken 13 years after *Voyager 2* show decay in the ring arcs. One of the ring arcs may disappear entirely before the end of the century.

CHECK YOUR UNDERSTANDING 8.5

Saturn's bright rings are located within the Roche limit of Saturn. This supports the theory that these rings (choose all that apply)

a. were formed from moons torn apart by tidal stresses.

b. formed at the same time that Saturn formed.

c. are relatively recent.

d. are temporary.

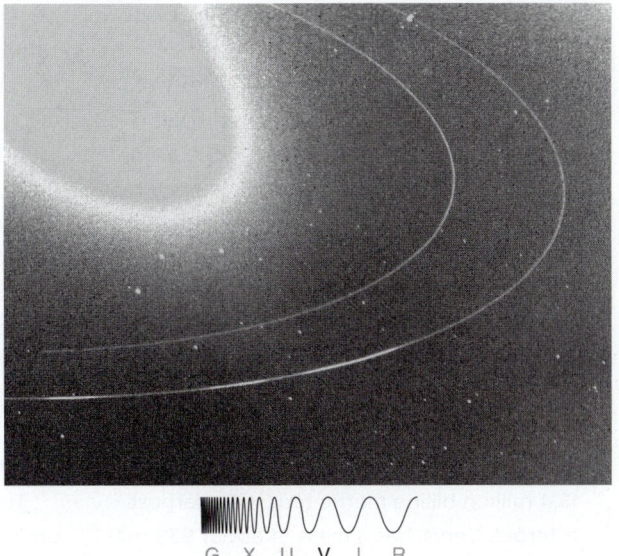

Figure 8.25 Neptune itself is very much overexposed in this *Voyager 2* image of the two brightest arcs in Neptune's Adams Ring.

---

reading Astronomy News

Saturn's Rings May Be Younger Than the Dinosaurs

By Charles Q. Choi, *Space.com*, January 17, 2019

While scientific knowledge is vast, it is not complete. Even the moons and rings in our own Solar System constantly surprise us.

Saturn has not always had rings — the planet's haloes may date only to the age of dinosaurs, or after it, a new study finds.

The age of Saturn's rings has long proven controversial. Some researchers had thought the iconic features formed along with the planet about 4.5 billion years ago from the icy rubble left in orbit around it after the formation of the solar system. Others suggested the rings are very young, perhaps originating after Saturn's gravitational pull tore apart a comet or an icy moon.

One way to solve this mystery is to weigh Saturn's rings. The rings were initially made of bright ice, but over time have become contaminated and darkened by debris from the outer reaches of the solar system. A few years back, NASA's Saturn-orbiting Cassini mission determined that the rings are only about 1 percent impure. If scientists could weigh Saturn's rings, they could estimate the amount of time it would take for them to accumulate enough contaminants to get 1 percent impure and thus calculate their age, lead study author Luciano Iess, a planetary scientist at the Sapienza University of Rome, told Space.com.

Iess and his colleagues relied on more Cassini data. Before the spacecraft plunged to its death into Saturn's atmosphere in September 2017, it coasted between the planet and its rings and let their gravitational pulls tug it around. The strength of a body's gravity depends on its mass, and by analyzing how much Cassini was pulled one way or the other during the "grand finale" phase of its mission, the mission team could measure the gravity and mass of both Saturn and its rings.

During six of Cassini's crossings between Saturn and its rings at altitudes about 1,615 miles to 2,425 miles (2,600 to 3,900 kilometers) above the planet's clouds, scientists monitored the radio link between the spacecraft and Earth. Much as how an ambulance siren sounds higher pitched as the vehicle drives toward you and lower pitched as it moves away, the radio signals would lengthen in wavelength as their source moved away Earth and shorten as their source moved toward it — an effect called the Doppler shift.

"I'm astonished by the fact that we were able to measure the velocity of a distant spacecraft 1.3 billion kilometers [807 million miles] away from Earth with an accuracy that is a hundredth or a thousandth the speed of a snail — a few hundreds of millimeters per second," less said.

Previous estimates based on data from the Voyager flybys of Saturn suggested the rings' mass was about 28 million billion metric tons. The new data from Cassini now suggests the rings' mass is only about 15.4 million billion metric tons. (The largest asteroid, Ceres, has a mass of about 939 million billion metric tons.)

All in all, the researchers suggest the rings formed between 10 million to 100 million years ago. In comparison, the age of dinosaurs ended about 66 million years ago.

Cassini's grand finale also revealed key details about the internal structure of Saturn. For example, it found that jet streams seen around Saturn's equator — the strongest measured in the solar system, with winds of up to 930 mph (1,500 km/h) — extend to a depth of at least 5,600 miles (9,000 km), rotating a colossal amount of mass around the planet about 4 percent faster than the layer below it.

"The discovery of deeply rotating layers is a surprising revelation about the internal structure of the planet," Cassini project scientist Linda Spilker at NASA's Jet Propulsion Laboratory in Pasadena, California, who did not participate in the study, said in a statement. "The question is, What causes the more rapidly rotating part of the atmosphere to go so deep, and what does that tell us about Saturn's interior?"

The new findings also suggest that Saturn's rocky core is about 15 to 18 times the mass of Earth, similar to prior estimates.

The scientists detailed their findings online Jan. 17 in the journal Science.

EVALUATING THE NEWS

1. The basic experiment described here is to measure the mass of the rings of Saturn as Cassini moved outward from Saturn toward the rings. If the rings had no mass, how would you expect them to affect Cassini's orbit as Cassini approached them: Would it speed up, slow down, or neither? If the rings were very massive, what effect would you expect as Cassini approached them: Would it speed up, slow down, or neither?

2. If Cassini were approaching Earth and being accelerated by the gravity of Saturn's rings, would the radio signals from Cassini be redshifted, blueshifted, or neither (see Chapter 5)?

3. The rings have been measured to be 15.4 million billion metric tons, but they were predicted to be 28 million billion metric tons. Was the measured Doppler shift due to the rings larger or smaller than the expected shift? How do you know?

4. The article states that "If scientists could weigh Saturn's rings, they could estimate the amount of time it would take for them to accumulate enough contaminants to get 1 percent impure and thus calculate their age." Why does the total mass of the rings matter in this analysis?

5. What is the most likely origin for the contaminants that darken the rings?

Source: **Charles Q. Choi**, "Saturn's Rings May Be Younger Than the Dinosaurs," *Space.com*, January 17, 2019. Reprinted with permission of EnVeritas Group.

SUMMARY

The giant planets are much larger and less dense than the terrestrial planets, consist primarily of low-mass elements rather than rock, and are much colder on their surfaces. Because of their rapid rotation and the Coriolis effect, zonal winds are very strong on these planets. Volatiles become ices at various heights in these atmospheres, leading to a layered cloud structure. Jupiter, Saturn, and Neptune are still shrinking, and as they do, their gravitational energy is being converted to thermal energy, heating both the cores and the atmospheres from the inside. Uranus does not seem to have a significant heat source inside. Temperatures and pressures in the cores of the giant planets are very high, leading to novel states of matter, such as metallic hydrogen. All four giant planets have ring systems, which are temporary, created from and maintained by moons also in orbit around these planets.

(1) Jupiter and Saturn are made up mostly of hydrogen and helium—a composition similar to that of the Sun. Uranus and Neptune contain larger amounts of "ices," such as water, ammonia, and methane, than do Jupiter and Saturn. These compositions set them apart from the terrestrial planets. In addition, all four giant planets are much larger than Earth.

(2) We see only atmospheres on the giant planets because solid or liquid surfaces, if they exist, lie deep below the cloud layers. Clouds on Jupiter and Saturn are composed of various kinds of ice crystals colored by impurities. The change in temperature and pressure with depth in the atmosphere produces convection cells that cause a banded appearance. Uranus and Neptune have relatively few clouds, so their atmospheres appear more uniform and are predominantly colored blue by the properties of methane.

(3) The ongoing collapse of the giant planets converts gravitational energy to thermal energy. This process heats most of the giant planets from within, producing convection. Powerful convection and the Coriolis effect drive high-speed winds in the upper atmospheres of all of the giant planets.

④ The interiors of the giant planets are very hot and very dense because of the high pressures exerted by the overlying atmosphere. Within Jupiter and Saturn, these conditions force hydrogen into the unusual state known as metallic hydrogen, which leads to strong magnetic fields. In Uranus and Neptune, the magnetic fields are produced by the motions of electrically conductive dissolved salts in deep oceans of liquid water and ammonia. All four giant planets have large magnetospheres that interact with the solar wind.

⑤ All four giant planets are surrounded by rings, many of which trace their origin either to the collisions of moons or to individual moons that strayed within the Roche limit of their home planet and were torn apart by tidal forces. Planetary rings form a series of concentric bands that lie in a flat disk.

QUESTIONS AND PROBLEMS

TESTING YOUR UNDERSTANDING

1. **T/F:** Uranus has extreme seasons because its poles are nearly in the plane of the Solar System.

2. **T/F:** All the giant planets have clouds and belts.

3. **T/F:** Water never forms visible clouds in the atmospheres of giant planets.

4. **T/F:** Storms on giant planets last much longer than storms on Earth.

5. **T/F:** The cores of the giant planets are all similar.

6. The interiors of the giant planets are heated by gravitational contraction. We know this because
   a. the cores are very hot.
   b. the giant planets radiate more energy than they receive from the Sun.
   c. the giant planets have strong magnetic fields.
   d. the giant planets are mostly atmosphere.

7. As depth increases, _____ and _____ increase, which causes changes in the chemical composition of clouds in giant planet atmospheres.

8. Zonal winds on the giant planets are stronger than those on the terrestrial planets because
   a. the giant planets have more thermal energy.
   b. the giant planets rotate faster.
   c. the moons of the giant planets provide additional pull.
   d. the moons of the giant planets feed energy to their planet through the magnetosphere.

9. Volcanoes on Enceladus affect the E Ring of Saturn by
   a. pushing the ring around.
   b. stirring the ring particles.
   c. supplying ring particles.
   d. dissipating the ring.

10. Imagine a giant planet, very similar to Jupiter, that was ejected from its solar system at formation. (These objects exist and are probably numerous, although their total number is still uncertain.) This planet would almost certainly still have (choose all that apply)
    a. a magnetosphere.
    b. thermal energy.
    c. auroras.
    d. rings.

11. Stellar occultations are the most accurate way to measure the _____ of a Solar System object.
    a. mass
    b. density
    c. temperature
    d. diameter

12. The elemental compositions of Jupiter and Saturn are most similar to those of
    a. Uranus and Neptune.
    b. the terrestrial planets.
    c. their moons.
    d. the Sun.

13. Uranus and Neptune are different from Jupiter and Saturn in that
    a. Uranus and Neptune have a higher percentage of ices in their interiors.
    b. Uranus and Neptune have no rings.
    c. Uranus and Neptune have no magnetic field.
    d. Uranus and Neptune are closer to the Sun.

14. The Great Red Spot on Jupiter is
    a. a surface feature.
    b. a "storm" that has been going on for more than 300 years.
    c. caused by the interaction between the magnetosphere and Io.
    d. about the size of North America.

15. The different colors of clouds on Jupiter are primarily a result of
    a. temperature.
    b. composition.
    c. motion of the clouds toward or away from us.
    d. all of the above

16. Consider a northward-traveling particle of gas in Jupiter's northern hemisphere. According to the Coriolis effect, which way will this particle of gas turn?
    a. in the direction of rotation
    b. opposite the direction of rotation
    c. north
    d. south

17. Metallic hydrogen is *not*
    a. a metal that acts like hydrogen.
    b. hydrogen that acts like a metal.
    c. common in the cores of giant planets.
    d. a result of high temperatures and pressures.

18. Which of the giant planets has the most extreme seasons?
    a. Jupiter
    b. Saturn
    c. Uranus
    d. Neptune

19. Currents in the metallic hydrogen core of Jupiter may produce Jupiter's
    a. magnetic field.
    b. belts and zones.
    c. colors.
    d. giant storms.

20. The rings of Saturn periodically disappear and reappear when
  a. observed in the direction of the Sun.
  b. the Sun has set on Saturn.
  c. the rings dissipate.
  d. viewed edge-on.

THINKING ABOUT THE CONCEPTS

21. Jupiter's chemical composition is more like that of the Sun than Earth's is, yet both planets formed from the same protoplanetary disk. Explain why they are different today.

22. As shown in Figure 8.2, astronomers take the unusual position of lumping together all atomic elements other than hydrogen and helium into a single category, which they call heavy elements. Why is this a reasonable position for astronomers to take? ★

23. None of the giant planets are truly round. Explain why they have a flattened appearance.

24. The Great Red Spot is a long-lasting atmospheric vortex in Jupiter's southern hemisphere. Winds rotate counterclockwise around its center. Is the Great Red Spot cyclonic or anticyclonic? Is it a region of high or low pressure? Explain.

25. What is the source of color in Jupiter's clouds? Uranus and Neptune, when viewed through a visible-light telescope, appear distinctly bluish green in color. What are the two reasons for their striking appearance?

26. Compare the atmospheric layers shown in Figure 8.11 to the temperature profiles in Figure 8.10. At what temperature does hydrogen sulfide ice form? ★

27. What drives the zonal winds in the atmospheres of the giant planets?

28. Jupiter, Saturn, and Neptune all radiate more energy into space than they receive from the Sun. Does this violate the law of conservation of energy? What is the source of the additional energy?

29. What creates metallic hydrogen in the interiors of Jupiter and Saturn, and why do we call it metallic?

30. Jupiter's core is thought to consist of rocky material and ices, all in a liquid state at a temperature of 35,000 K. How can materials such as water be liquid at such high temperatures?

31. ★ **WHAT AN ASTRONOMER SEES** In Figure 8.19, the astronomer concludes that the camera tracked Jupiter's motion across the sky while the image was being taken. The stars, therefore, formed bright streaks in the background as the camera moved. Did Jupiter move in an up-down direction or in a right-left direction, according to the orientation of this photo? ★

32. What creates auroras in the polar regions of Jupiter and Saturn?

33. Will the particles in Saturn's bright rings eventually stick together to form one solid moon orbiting at the average distance of all the ring particles? Explain your answer.

34. Explain the mechanism that creates gaps in Saturn's bright ring system.

35. Astronomers believe that most planetary rings eventually dissipate. Explain why the rings do not last forever. Name one ring that might continue to exist indefinitely, and explain why it could survive when others might not.

APPLYING THE CONCEPTS

36. An old astronomy joke goes: "Saturn is less dense than water, so if you could find a bathtub large enough, it would float . . . and leave a ring." Using the radius and mass information in Table 8.1, refer back to Working It Out 6.2 to calculate the density of Saturn.
  a. Make a prediction: Should your density be larger or smaller than the density of water (1,000 kg/m³)?
  b. Convert the mass of Saturn given in Table 8.1 (which is in units of Earth masses) to kilograms by multiplying by the mass of Earth in kilograms.
  c. Follow Working It Out 6.2 to calculate Saturn's density.
  d. Check your work by comparing the density that you calculate to the density of water.

37. Uranus occults a star at a time when the relative motion between Uranus and Earth is 23.0 km/s. An observer on Earth sees the star disappear for 37 minutes 2 seconds and notes that the center of Uranus passes directly in front of the star. Follow Working It Out 8.1 to find the diameter of Uranus.
  a. Make a prediction: On the basis of Table 8.1, in what range do you expect to find the diameter of a giant planet like Uranus: thousands of kilometers, tens of thousands of kilometers, or hundreds of thousands of kilometers?
  b. On the basis of the observations, calculate the diameter of Uranus. (Notice that you will have to convert the time into seconds.)
  c. Check your work by verifying that your answer has units of kilometers and comparing to the radius of Uranus given in Table 8.1.

38. Suppose that you plan to observe Neptune during an upcoming occultation. At that time, Neptune will be moving at 18 km/s relative to Earth. How long will the occultation last?
  a. Make a prediction: In problem 37, Uranus occulted a star for about 37 minutes. Do you expect your answer to be much longer, much shorter, or last about the same amount of time?
  b. Adapt the strategy in Working It Out 8.1 to solve for the time the occultation will last.
  c. Calculate the time the occultation will last.
  d. Check your work by comparing your answer to your prediction.

39. The equilibrium (expected) temperature for Saturn should be 82 K, but instead we find an actual temperature of 95 K. Follow Working It Out 8.2 to find out how much more energy is Saturn radiating into space than it absorbs from the Sun.
  a. Make a prediction: Do you expect your answer to be more or less than 1?
  b. Calculate how many times more energy Saturn is radiating than it receives.
  c. Check your answer by comparing your result both to 1 and to the results for Jupiter in Working It Out 8.2.

40. Neptune radiates into space 2.6 times as much energy as it absorbs from the Sun. Its equilibrium (expected) temperature is 47 K. What is its actual temperature?
  a. Make a prediction: Do you expect Neptune to be hotter or colder than its equilibrium temperature?
  b. Rearrange the equation in Working It Out 8.2 to solve for the actual temperature.
  c. Calculate the true temperature of Neptune.
  d. Check your answer by comparing your result to your prediction.

exploration Estimating Rotation Periods of Giant Planets

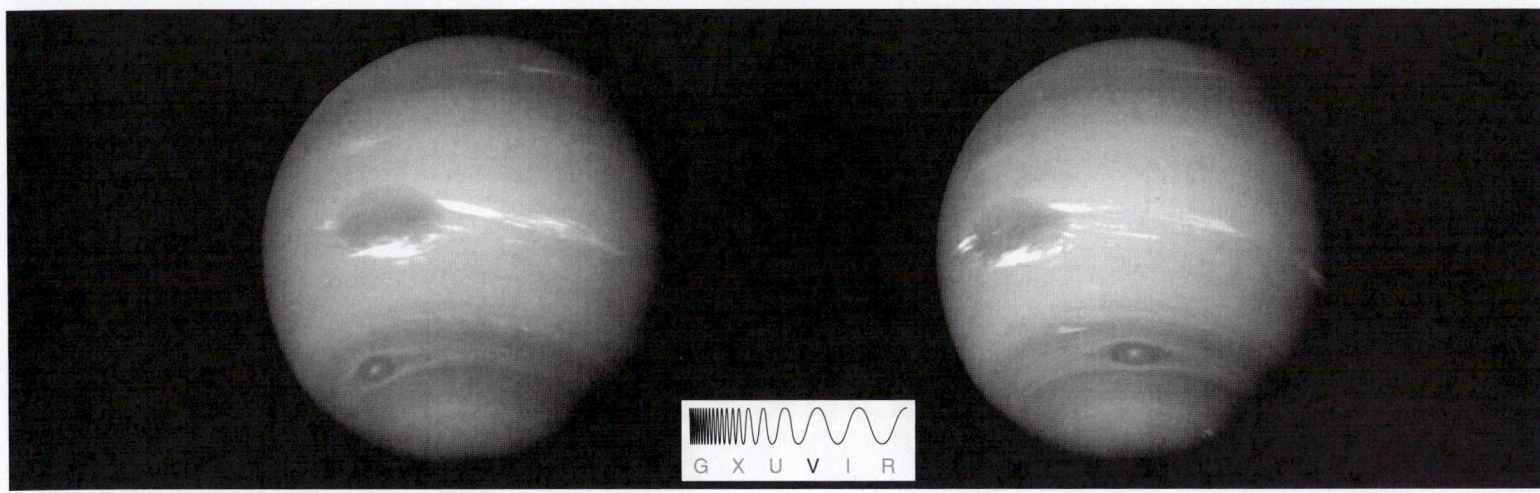

G X U V I R

Figure 8.26 Two images of Neptune, taken 17.6 hours apart by *Voyager 2*.

Neptune rotates from left to right (from west to east) in the two images in **Figure 8.26**. The image on the left was taken first, whereas the image on the right was taken 17.6 hours later. During this time, the Great Dark Spot completed nearly one full rotation. The small storm at the bottom of the image of the planet completed slightly more than one rotation. You would be very surprised to see this result for locations on Earth.

We can find the rotation period of the smaller storm in Figure 8.26 by equating two ratios. First, use a ruler to find the distance in millimeters (1 mm = 0.1 centimeter [cm]) from the left limb of the planet to the small storm (as shown in **Figure 8.27a**). Do this for both images in Figure 8.26. Estimate the radius of the circle traveled by the storm by measuring from the limb to a line through the center of the planet (as shown in Figure 8.27b).

1. Estimate the radius of the circle the small storm makes around Neptune (in millimeters) by making the appropriate measurements on Figure 8.26. Then calculate the circumference of the circle, $2\pi R$ (in millimeters).

_____

_____

Because the small storm has rotated *more than* one time, the total distance it has traveled is the circumference of the circle plus the distance between its location in the two images, which can be measured directly.

2. Add those two numbers together to get the total distance traveled (in millimeters) between these images.

_____

_____

Now we are ready to find the rotation period. The ratio of the rotation period, T, to the time elapsed, t, must be equal to the ratio of the circumference of the circle around which it travels, C (in millimeters), to the total distance traveled, D (in millimeters).

$$\frac{T}{t} = \frac{C}{D}$$

3. What is the small storm's rotation period? (To check your work, note that your answer should be less than 17.6 hours. Why?)

_____

_____

(a)

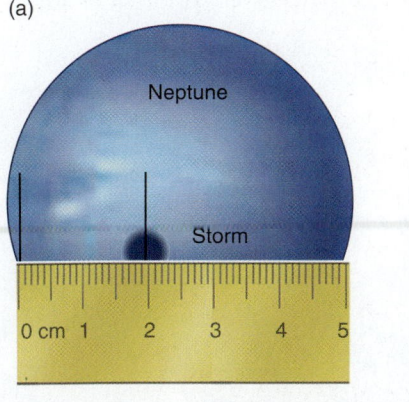

(b)

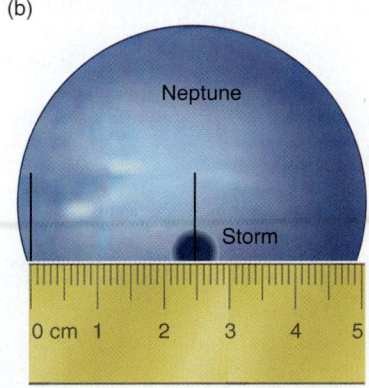

Figure 8.27 (a) How to measure the position of the storm. (b) How to measure the radius of the circle the storm has traveled.

Small Bodies of the Solar System

Meteors occur when small pieces of space dust collide with Earth's atmosphere. Several times a year, many meteors occur at once, as you will find out here in Chapter 9. On these nights, the meteors appear to all come from the same location. Use the Internet to find out the date of the next meteor shower. Prepare an observation log page and take it with you when you go outside at night on that date. Sketch objects on the horizon onto the observation log, so that you fix different directions in your mind. If you can see the constellation for which the shower is named, sketch that, too! Spend a few hours watching for meteors. (Be sure to dress warmly!) Each time you see one, sketch it onto your observation log. Once you have observed about 10 meteors, trace their paths back with dotted lines, to see that they all intersect.

EXPERIMENT SETUP

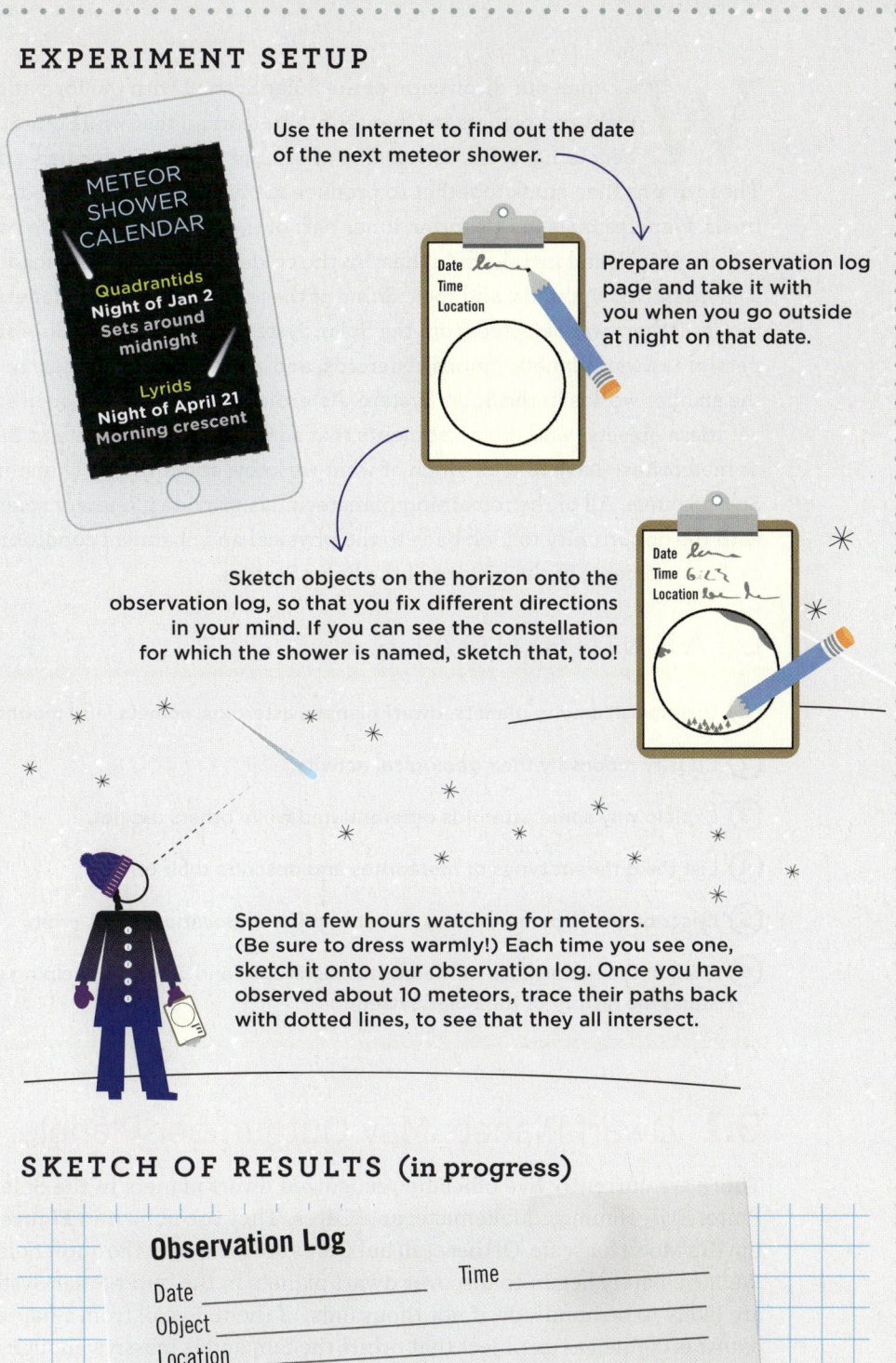

Use the Internet to find out the date of the next meteor shower.

METEOR SHOWER CALENDAR

Quadrantids
Night of Jan 2
Sets around midnight

Lyrids
Night of April 21
Morning crescent

Prepare an observation log page and take it with you when you go outside at night on that date.

Sketch objects on the horizon onto the observation log, so that you fix different directions in your mind. If you can see the constellation for which the shower is named, sketch that, too!

Spend a few hours watching for meteors. (Be sure to dress warmly!) Each time you see one, sketch it onto your observation log. Once you have observed about 10 meteors, trace their paths back with dotted lines, to see that they all intersect.

SKETCH OF RESULTS (in progress)

Observation Log

Date _____ Time _____
Object _____
Location _____
Sky conditions _____

Comments

9

Moon

Pluto

Eris

Haumea

Makemake

Ceres

Figure 9.1 This montage of the five known dwarf planets and Earth's Moon (for scale) reveals that, in addition to being different sizes, they have different shapes and compositions. The images of the Moon, Pluto, and Ceres are photographs. The images of Eris, Haumea, and Makemake are artistic representations based on the known properties of each object, such as shape and albedo.

We began our discussion of the Solar System with the formation of stars and planets in Chapter 5. You learned that while our Sun was becoming a star, tiny grains of primitive material formed a disk. These grains then stuck together to produce many small bodies called planetesimals. Planetesimals in the hotter, inner part of the Solar System were composed mostly of rock and metal, while those in the colder, outer part were made up of ice, organic compounds, and rock. Some of these objects became planets and moons; others were ejected from the Solar System, or fell into the Sun; still others persist as dwarf planets, moons, asteroids, and comets. Dwarf planets are among the smaller worlds in the Solar System. Asteroids and comets are even smaller, yet these objects—and their fragments that survive the atmosphere to fall to Earth as meteorites—have told us much of what we know about the early history of the Solar System. All of the remaining planetesimals provide planetary scientists with the opportunity to look back to the physical and chemical conditions of the earliest moments in the history of the Solar System.

LEARNING GOALS

1. Distinguish among planets, dwarf planets, asteroids, comets, and moons.

2. Classify moons by their geological activity.

3. Explain why some asteroids differentiated while others did not.

4. List the different types of meteorites and describe their origins.

5. Describe the appearance of a comet at different locations in its orbit.

6. Explain the importance of meteorites, asteroids, and comets in helping us understand the history of the Solar System.

9.1 Dwarf Planets May Outnumber Planets

There are currently five officially recognized dwarf planets in the Solar System: Pluto, Eris, Haumea, Makemake, and Ceres. They are shown in **Figure 9.1**, with Earth's Moon for scale. Of these, all but Ceres are located in the outer Solar System. We have barely begun to discover dwarf planets in the outer Solar System; there are likely to be hundreds, if not thousands, of them. Recall from Chapter 5 that a *planet* is defined as an object that orbits the Sun and is massive enough to (1) pull itself into a round shape and (2) clear the area around its orbit, so that there are no other comparable objects sharing the orbit. Dwarf planets are objects that are large enough to meet criterion 1, but they fail to meet criterion 2 because they have not cleared the region near their orbits. At least 100 known objects in our Solar System are potential dwarf planets, but because they are located beyond the orbit of Pluto, their shapes have not yet been measured well enough for definitive classification.

Eris, Haumea, and Makemake

Eris has a radius of 1,100 kilometers (km); as shown in Figure 9.1, it is about the same size as Pluto but is more massive. Eris has a relatively large moon, called Dysnomia. The orbit of Eris is inclined at an angle of about 44° to the ecliptic plane (the plane of Earth's orbit) and is also highly eccentric, as shown in **Figure 9.2**. The orbit of Eris carries it from 37.8 astronomical units (AU) out to 97.6 AU, with an orbital period of 562 Earth years. Eris is currently near the most distant point in its orbit, making it the most remote object known in the Solar System. (The eccentric orbits of some other Solar System bodies will eventually carry them farther from the Sun.) Eris is highly reflective, so the surface must be covered in pristine ice. Observations of the spectrum indicate this ice is probably mostly methane, which will evaporate to form an atmosphere when Eris comes closest to the Sun in the year 2257.

Haumea and Makemake are both smaller and have slightly larger orbits than Pluto. Haumea has two known moons—Hi'iaka and Namaka. While it has enough mass to be round, Haumea spins so rapidly on its axis that its shape is distorted into a flattened ellipsoid with an equatorial radius that is approximately twice its polar radius, as shown in Figure 9.1. This is the most distorted shape of any known planet, dwarf or otherwise. One moon has been found orbiting around Makemake. We know less about this dwarf planet than about Pluto, Haumea, or Eris.

Ceres: Dwarf Planet among the Asteroids

With a diameter of about 975 km, Ceres is larger than most moons but smaller than any planet. Ceres is a dwarf planet because it is round (see Figure 9.1), but it has not cleared its surroundings. Ceres is located in the asteroid belt, the region between Mars and Jupiter that contains most of the Solar System's asteroids. Ceres contains about a third of the total mass of the main asteroid belt. Ceres rotates on its axis with a period of about 9 hours, typical of many asteroids. In 2015, NASA's *Dawn* spacecraft arrived at Ceres, revealing a dwarf planet with significant geologic activity.

Cryovolcanism is similar to terrestrial volcanism but is driven by subsurface low-temperature volatiles such as water and nitrogen rather than molten rock. *Dawn* spotted a cryovolcano on Ceres, along with evidence that there may have been prior cryovolcanoes that have been flattened out by gravity over hundreds of millions of years. Because these volcanoes are made of ice rather than rock, they flow downward slowly over time, much like a blob of honey gradually spreads out in the bottom of a mug.

As the spacecraft approached Ceres, a mysterious bright spot was resolved into two, and then many (**Figure 9.3a**). These spots are located in the bottom of Occator Crater, one of the youngest craters on the dwarf planet. These bright spots are thought to be the result of salts that were brought to the surface by dissolved brine in an impact event. The water evaporated, leaving the bright salt behind. The 34-km-wide Haulani Crater (Figure 9.3b) has landslides on the crater's limb and

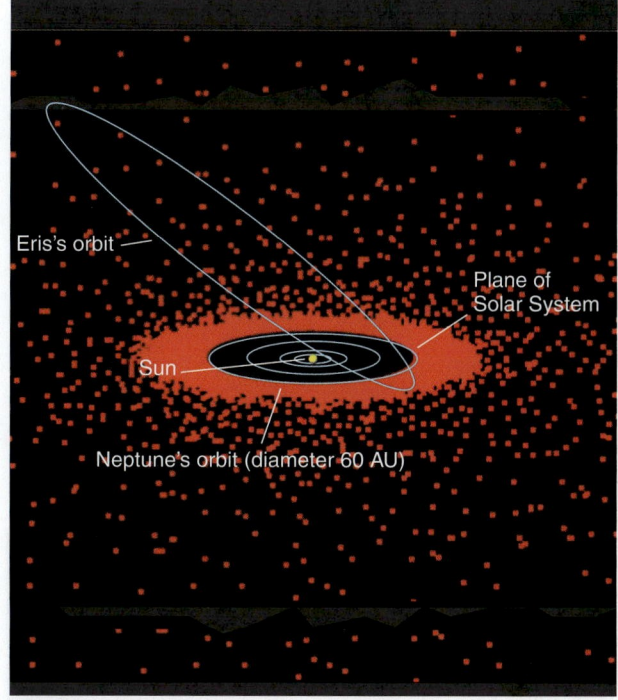

Figure 9.2 The orbit of Eris is both highly eccentric and highly inclined to the rest of the Solar System. The red dots indicate objects in the spherical Oort Cloud, which surrounds the Solar System.

Figure 9.3 (a) NASA's *Dawn* mission arrived at Ceres in 2015, revealing bright spots in the bottom of Occator Crater. (b) This image of the 34-km-wide Haulani Crater shows landslides on the crater's limb, as well as smooth material and a central peak on the crater floor. Blue indicates a younger surface on Ceres. The color in this image is enhanced.

(a)

(b)

G X U V I R

G X U V I R

what if . . .

What if, in 2006, instead of demoting Pluto to a dwarf planet, the International Astronomical Union had redefined a planet to be any round astronomical object smaller than a star: What kinds of objects would then qualify as planets?

G X U V I R

Figure 9.4 ★ WHAT AN ASTRONOMER SEES

NASA's *New Horizons* visited Pluto briefly in 2015, taking images that surprised astronomers around the world. Astronomers expected to see a cratered ice world that had long been geologically dead. However, the pale heart-shaped region toward the lower right is very smooth, indicating to an astronomer that Pluto's core has been hot in the very recent past. The enhanced colors in this image indicate regions of different composition, further indicating that the surface has been modified over time. If Pluto had differentiated and then frozen in place, it should have the same composition over the whole surface. An astronomer would zoom in and notice ice mountains and other features that show that Pluto has had active geology. This realization about Pluto was so surprising that the *New Horizons* mission was extended and the spacecraft was repurposed to visit another distant Solar System object.

blue streaks radiating outward from the central crater peak. These color variations indicate that the surface of Ceres has a different composition than the subsurface layer. The color in this image is enhanced, and blue indicates young surfaces on Ceres. As much as a quarter of its mass exists in the form of a water-ice mantle that surrounds a rocky inner core. This may explain the unusual lack of large craters on the surface of Ceres. An impact large enough to leave a large crater would also allow significant amounts of watery slush to flow on the surface, partially filling in the crater and leaving a slight depression.

Extraordinary Pluto

Pluto's orbit is 248 Earth years long and is elliptical, so that Pluto was closer to the Sun than Neptune was from 1979 to 1999. Pluto's orbit is inclined, however, so that it never actually crosses Neptune's orbit. Pluto has five known moons, all of which are the same age as Pluto, based on the density of their surface craters. This implies that Pluto and its moons were all formed at the same time by a collision in the Kuiper Belt. The largest moon is Charon, about half the size of Pluto; Pluto itself is only two-thirds as large as our Moon. Pluto and Charon are a tidally locked pair: each has one hemisphere that always faces the other object. The total mass of the Pluto-Charon system is 1/400 that of Earth. Pluto and Charon each have a rocky core surrounded by a water-ice mantle. Pluto's surface is an icy mixture of frozen water, carbon dioxide, nitrogen, methane, and carbon monoxide, but Charon's surface is primarily water ice. Pluto has a thin atmosphere of nitrogen, methane, and carbon monoxide. These gases freeze out of the atmosphere and accumulate on the surface when Pluto is more distant from the Sun and is therefore colder.

In 2015, NASA's *New Horizons* spacecraft zipped past Pluto, taking pictures and other data for a very brief time, and then headed into the outer reaches of the Solar System. It took 18 months to send all the data and images back to Earth, but the wait was worthwhile. With the very first images, Pluto was revealed to be completely different than what astronomers expected (**Figure 9.4**). Pluto lies far from the Sun, it is small, and it does not have a large neighbor to cause tidal heating. How, then, has the interior remained hot enough for geological activity to continue to create large smooth areas in recent times? One explanation is that Pluto has more radioactive material in its core than other objects of similar size (Ceres or Charon, for example). This radioactive material heats the core and causes convection cells that continually bring hot material to the surface. But why would Pluto have more radioactive materials than other objects? Scientists continue to analyze the imagery and data from *New Horizons* for answers to this question and many other new and unexpected observations, including the discovery of cryovolcanism and hydrocarbon clouds on Pluto.

CHECK YOUR UNDERSTANDING 9.1

Pluto differs significantly from the eight Solar System planets in that (choose all that apply)

a. it is farther from the Sun than any of these planets.

b. it has a different composition than any of these planets.

c. its orbit is chaotic.

d. it is not round.

e. it has not cleared its orbit.

Answers to Check Your Understanding questions are in the back of the book.

9.2 Moons as Small Worlds

The giant planets have many moons in orbit around them, ranging in size from small boulders to planet-sized objects. A selection of moons of the Solar System are shown in **Figure 9.5**. Some moons have remained the same since their formation during the early history of the Solar System, while others are even more geologically active than Earth. We can categorize the geological activity of moons as follows: (1) definitely active today, (2) probably or possibly active today, (3) active in the past but not today, and (4) inactive. We will organize our tour of these moons according to similarities in their geological histories as evidenced by their surface features. Comparing these worlds in this way allows us to draw conclusions about how they formed and the physical or geological principles governing their evolution. As we discuss each moon in some detail, please refer back to Figure 9.5.

Geologically Active Moons

Geologically active moons show evidence of ongoing geologic activity, such as volcanoes or tectonism. These moons require an active source of energy to maintain geologic activity into the present time. This energy source is often tidal heating, as described in Chapter 6. For example, because Io's orbit is elliptical, Jupiter's

Figure 9.5 The major moons of the Solar System are shown to scale, along with Mercury and Pluto for comparison. Both Ganymede and Titan are larger than Mercury.

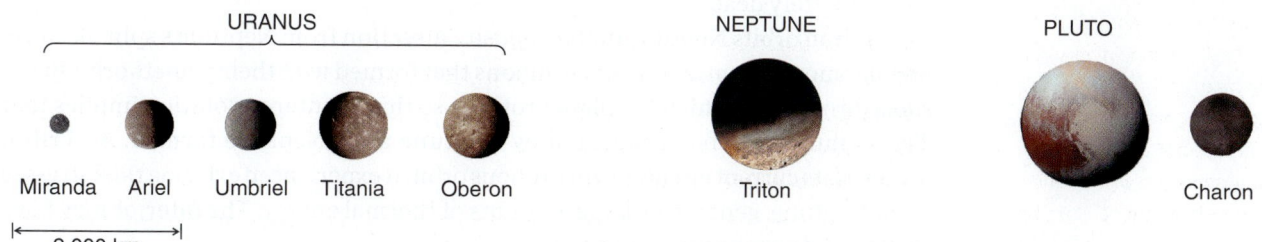

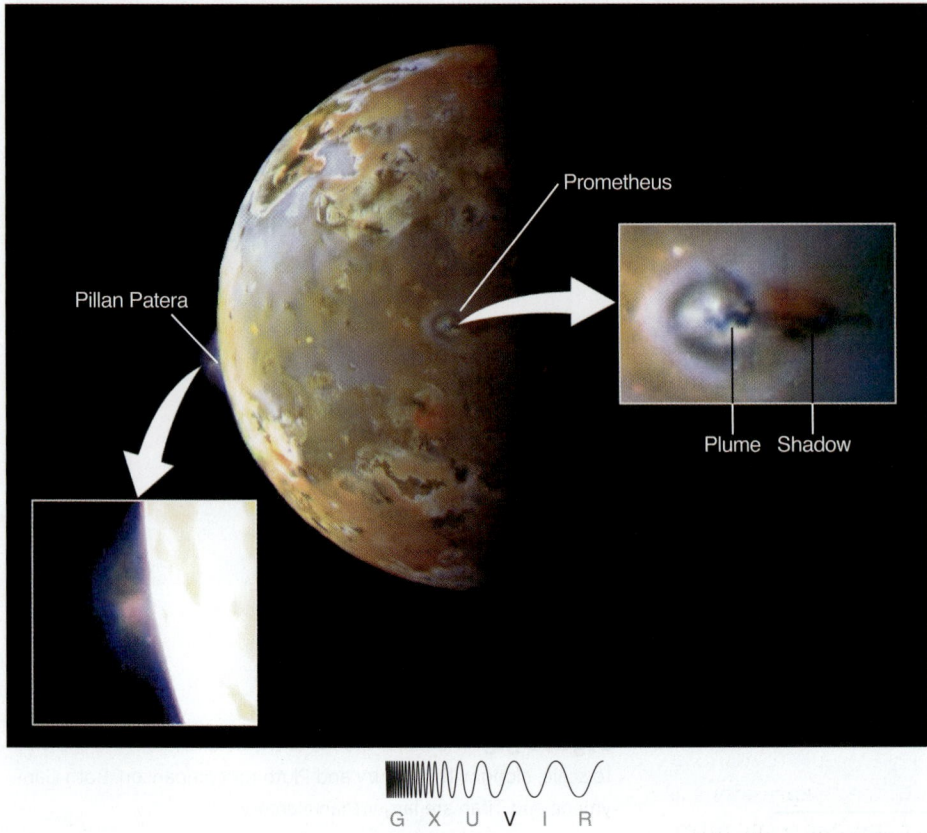

Pillan Patera

Prometheus

Plume   Shadow

G X U V I R

Figure 9.6 When this image of Io was obtained by *Galileo*, two volcanic eruptions could be seen at once. The plume of Pillan Patera rises 140 km above the limb of the moon on the left, while the shadow of a 75-km-high plume can be seen to the right of the vent of Prometheus, near the moon's center.

gravitational pull flexes Io's crust and generates enough thermal energy to melt parts of it, powering the most active volcanism known in the Solar System (**Figure 9.6**). Lava flows and volcanic ash bury impact craters as quickly as they form, so no impact craters have been observed on the surface. Images taken by *Voyager* and *Galileo* reveal plains, irregular craters, and flows, all related to the eruption of mostly silicate magmas onto the surface of the moon. Explosive eruptions from more than 150 active volcanoes (sometimes several of these simultaneously) send debris hundreds of kilometers above Io's surface. Huge structures have multiple summit craters, showing a long history of repeated eruptions followed by collapse of the partially emptied magma chambers. These repeated, stationary eruptions, along with the fact that volcanoes on Io are randomly distributed, allow us to infer that plate tectonics does not occur on Io.

Because of its active volcanism, Io's mantle has turned "inside-out" more than once in the past, with material from inside the moon repaving the surface. Volatiles such as water and carbon dioxide escaped into space long ago. Differentiation caused heavier materials to sink to the interior to form a core. Sulfur and various sulfur compounds, as well as silicate magmas, are constantly being recycled to form the complex surface and wide variety of colors we see today. Images of Io show tall mountains—some nearly twice the height of Mount Everest.

Saturn's moon Enceladus is heated by both tidal heating and radioactive decay within the moon's rocky core, creating a liquid ocean, 10 km deep, that is buried beneath 5–10 km of ice crust. Internal energy heats ice at the bottom of the crust and drives it to the surface, creating cryovolcanoes. Active cryovolcanic plumes (**Figure 9.7a**) are energetic enough to overcome the moon's low gravity, sending tiny ice crystals into space to replace particles continually lost from Saturn's E Ring (recall Figure 8.21). The entire surface shows a wide variety of ridges, faults, and smooth plains. Some impact craters appear softened, perhaps by the viscous flow of ice, like the flow that occurs in the bottom layers of glaciers on Earth. Parts of the moon have no craters, indicating recent resurfacing. Terrain near the south pole of Enceladus is cracked and twisted (Figure 9.7b). The cracks are warmer than their surroundings, confirming that the moon is warmer on the inside. All of this evidence of tectonic processes is unexpected for a small (500-km) body. It is a mystery that Enceladus is so active whereas Mimas, a neighboring moon of about the same size and also subject to tidal heating, appears to be completely dead.

Triton orbits Neptune in the opposite direction from Neptune's spin. Because angular momentum is conserved, moons that formed *with* their planets orbit in the *same* direction in which the planet rotates, so this counter-revolution implies that Triton must have been captured by Neptune after Neptune formed. As Triton achieved its current circular, synchronous orbit, it experienced extreme tidal stresses from Neptune, generating large amounts of thermal energy. The interior may have melted and may be differentiated.

(a) (b)

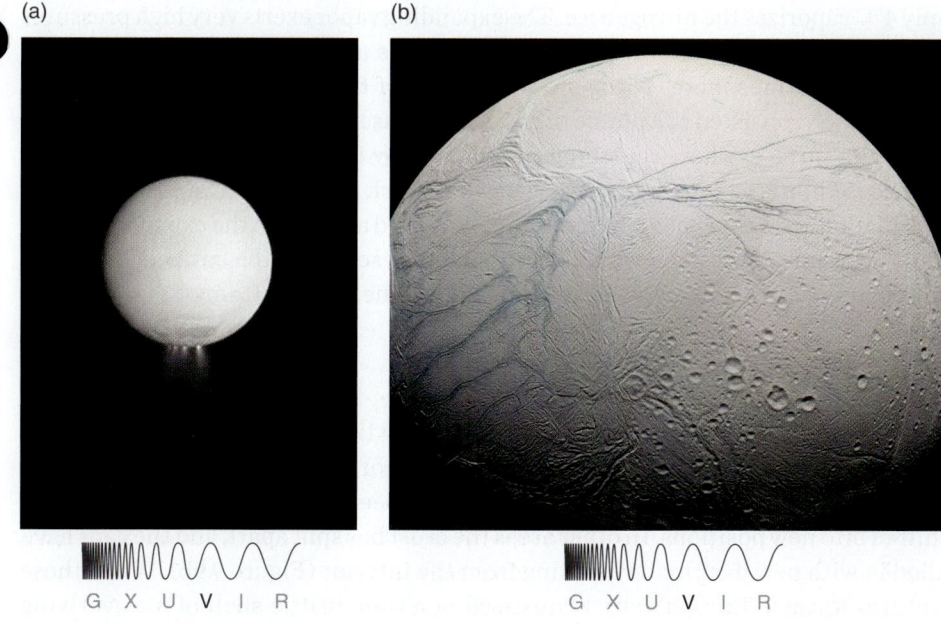

G X U V I R   G X U V I R

Figure 9.7 *Cassini* images of Enceladus show evidence of cryovolcanism. (a) Cryovolcanic plumes in the south polar region spew ice particles into space. In this image, there are two light sources. The surface of the moon is illuminated by sunlight reflected from Saturn while the plumes are backlit by the Sun itself, located almost directly behind Enceladus. (b) The twisted and folded surface of deformed ice cracks near the moon's south pole (shown blue in false color) is warmer than the surrounding terrain and has been found to be the source of cryovolcanism.

Most of Triton is covered with smooth volcanic plains, so the surface is geologically young. Part of Triton is covered with terrain that looks like the skin of a cantaloupe (**Figure 9.8a**), with irregular pits and hills that may be caused by slushy ice emerging onto the surface from the interior. Irregularly shaped depressions as wide as 200 km (Figure 9.8b) formed when mixtures of water, methane, and nitrogen ice melted in the interior of Triton and erupted onto the surface, much as rocky magmas erupted onto the lunar surface and filled impact basins on Earth's Moon (see Chapter 6).

Cryovolcanism also occurs on Triton. Triton has a thin atmosphere and a surface composed mostly of ices and frosts of methane and nitrogen at a temperature of about 38 K. Clear nitrogen ice creates a localized greenhouse effect, trapping solar energy beneath the ice to raise the temperature. A temperature increase of

Figure 9.8 *Voyager 2* images of Triton show evidence of cryovolcanism. (a) This mosaic shows various terrains on the Neptune-facing hemisphere of Triton. "Cantaloupe terrain" is visible at the top; its lack of impact craters indicates that this area is geologically younger than the bright, cratered terrain at the bottom. (b) This irregular basin on Triton has been partly filled with frozen water, forming a relatively smooth ice surface. The state of New Jersey could just fit within the basin's boundary.

(a) (b)

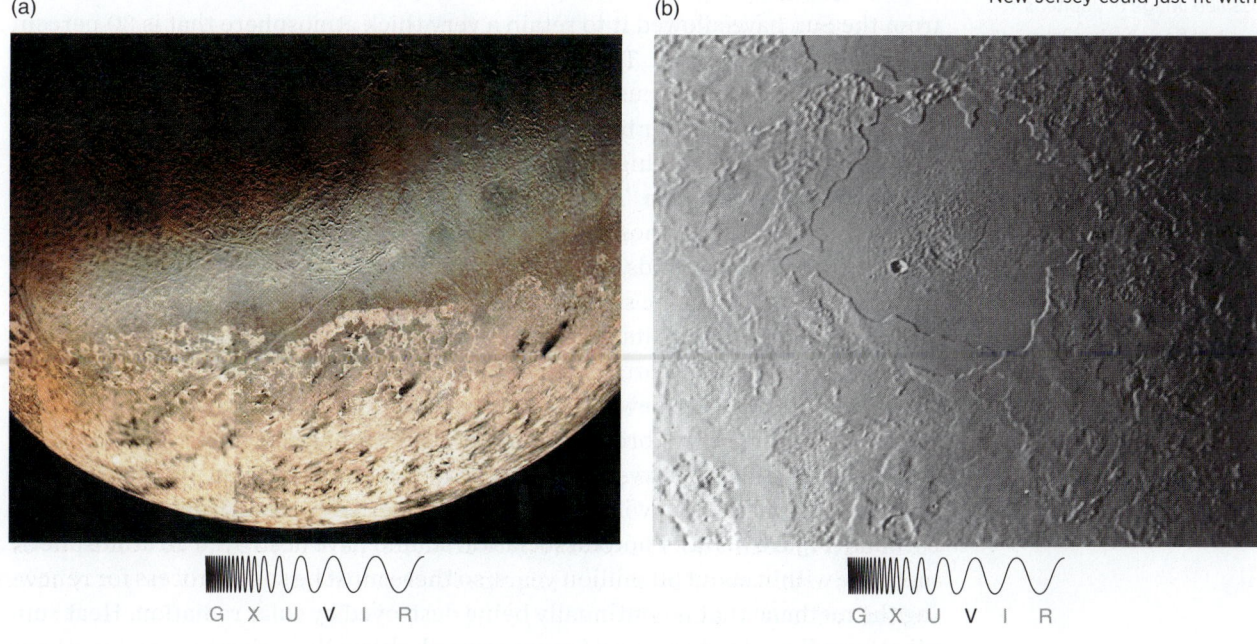

G X U V I R   G X U V I R

(a)

G X U V I R

(b)

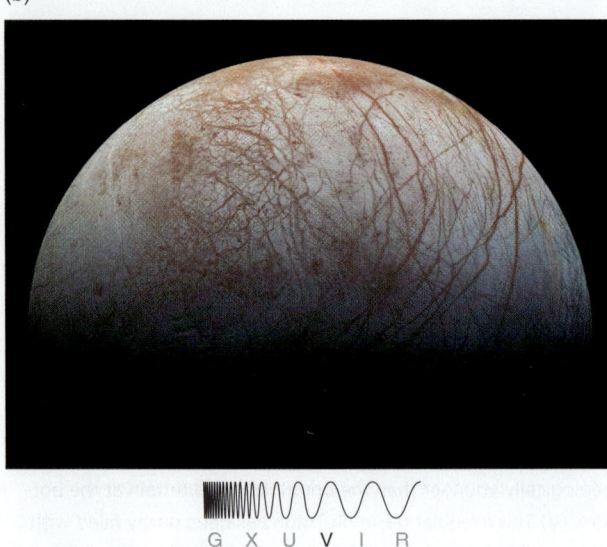

G X U V I R

Figure 9.9 (a) A high-resolution *Galileo* image of Jupiter's moon Europa shows where the icy crust has been broken into slabs that, in turn, have been rafted into new positions. These areas of chaotic terrain are characteristic of a thin, brittle crust of ice floating atop a liquid or slushy ocean. The area shown in this image measures approximately 25 × 50 km. (b) This larger image of Europa was produced from a mosaic of color images from *Galileo*. The image shows many long reddish-brown fractures in Europa's surface that may contain hints about Europa's geologic past, as well as the chemistry of the ocean under the ice.

only 4°C vaporizes the nitrogen ice. The expanding vapor exerts very high pressures beneath the ice cap. Eventually, the ice ruptures and the gas explodes into the low-density atmosphere. *Voyager 2* found four of these active cryovolcanoes on Triton. Each consisted of a plume of gas and dust as much as 1 km wide, rising 8 km above the surface, where the plume was caught by upper atmospheric winds and carried for hundreds of kilometers downwind. Dark material, perhaps silicate dust or radiation-darkened methane ice grains, is carried along with the expanding vapor into the atmosphere, from which it subsequently settles to the surface and forms dark patches streaked out by local winds, as seen near the bottom of Figure 9.8a.

Possibly Active Moons

Like Io, Europa experiences continually changing tidal stresses from Jupiter that generate heat and possibly volcanism. Regions of chaotic terrain, as shown in **Figure 9.9a**, are places where the icy crust has been broken into slabs that have shifted into new positions. In other areas the crust has split apart, and the gaps have filled in with new dark material rising from the interior (Figure 9.9b). When these features formed, Europa's crust consisted of a thin, brittle shell of ice overlying either liquid water or slushy ice. The lightly cratered and thus geologically young surface indicates that Europa may still possess a global ocean 100 km deep that contains more water than all of Earth's oceans. Supporting this conclusion, the Hubble Space Telescope has observed a water geyser erupting from the icy surface. If these observations are confirmed by future observations, Europa will be reclassified as a geologically active moon.

Europa's ocean should be salty with dissolved minerals. *Galileo*'s observations of Europa's variable magnetic field support this hypothesis. The variability indicates that the interior of Europa contains an electrically conducting fluid. Europa's ocean may also contain an abundance of organic material. The conditions that create and support life—liquid water, heat, and organic material—could all be present in Europa's oceans, making Europa a high-priority target in the search for extraterrestrial life.

Saturn's largest moon, Titan, is bigger than Mercury and has a composition of about 45 percent water ice and 55 percent rocky material. Titan's mass and distance from the Sun have allowed it to retain a very thick atmosphere that is 30 percent denser than that of Earth. Titan's atmosphere, like Earth's, is mostly nitrogen. As Titan differentiated, various ices, including methane (CH_4) and ammonia (NH_3), emerged from the interior to form an early atmosphere. The ammonia and methane were broken apart by high-energy ultraviolet photons from the Sun—a process called **photodissociation**. Through this process, ammonia is likely the source of the nitrogen in Titan's atmosphere. Methane breaks into fragments that recombine to form organic compounds, including hydrocarbons. These compounds form tiny particles, creating organic smog much like the air over Beijing on a bad day. This gives Titan's atmosphere its characteristic orange hue (**Figure 9.10a**).

The *Cassini* spacecraft, which began orbiting Saturn in 2004, gave astronomers the first close-up views of Titan's surface (Figure 9.10b). Haze-penetrating infrared imaging shows broad regions of dark and bright terrain (Figure 9.10c and d). Radar imaging shows irregularly shaped features in Titan's northern hemisphere that appear to be widespread lakes of methane, ethane, and other hydrocarbons (**Figure 9.11a**). Photodissociation should have destroyed all atmospheric methane within about 50 million years, so there must be some process for renewing the methane that is continually being destroyed by solar radiation. Heat supplied by radioactive decay could cause cryovolcanism that releases "new" methane

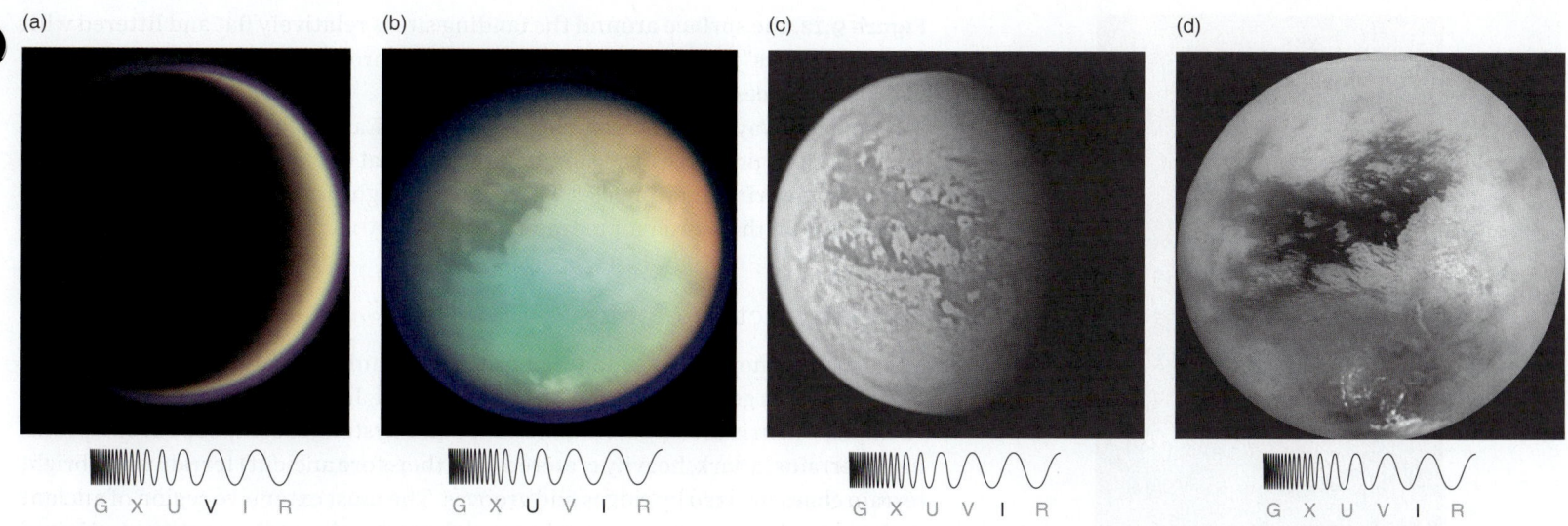

(a)   (b)   (c)   (d)

G X U V I R   G X U V I R   G X U V I R   G X U V I R

Figure 9.10 *Cassini* took images of Saturn's largest moon, Titan, in several regions of the electromagnetic spectrum. (a) Visible-light imaging shows Titan's orange atmosphere, which is caused by the presence of organic smoglike particles. (b) This infrared- and ultraviolet-light combined image shows surface features and a bluish atmospheric haze caused by scattering of ultraviolet sunlight by small atmospheric particles. Bright clouds in Titan's lower atmosphere are seen near the moon's south pole. (c) Infrared imaging penetrates Titan's smoggy atmosphere and reveals surface features. (d) This infrared image covers the same general area of Titan as seen in (b). The large dark area, called Xanadu, is about the size of the contiguous United States. Methane ice clouds are visible at the bottom of the image, near Titan's south pole.

from underground. To date, there is no direct evidence of active cryovolcanism on Titan, but the presence of abundant atmospheric methane and methane lakes, as well as regional changes in albedo, strongly suggests that Titan is indeed geologically active. (Recall from Chapter 7 that albedo measures how much light reflects from a surface.)

Titan has terrains reminiscent of those on Earth (Figure 9.11b), with networks of channels, ridges, hills, and flat areas that may be dry lake basins. The near absence of impact craters on the surface indicates recent erosion. These features suggest a cycle in which methane rain falls to the surface, washes the ridges free of dark hydrocarbons, and then collects in drainage systems that empty into low-lying liquid methane pools. An infrared camera photographed a reflection of the Sun from such a lake surface. The type of reflection observed proves that the lake contains a liquid and is not frozen or dry.

Cassini carried a probe, named *Huygens*, that plunged through Titan's atmosphere, measuring composition, temperature, pressure, and winds and taking pictures as it descended. *Huygens* confirmed the presence of nitrogen-bearing organic compounds in the clouds. These are key components in the production of proteins found in terrestrial life. Once on Titan's surface, *Huygens* took pictures and made physical and compositional measurements. The surface was wet with liquid methane, which evaporated as the probe—heated to 2000 K during its passage through the atmosphere—landed in the frigid soil. The surface was also rich with other organic compounds. As shown in

Figure 9.11 (a) Features commonly associated with terrestrial lakes, such as islands, bays, and inlets, are clearly visible in this false-color radar image of Titan. (b) The surface of Titan, viewed from the *Huygens* probe during its descent to the surface, has dark drainage patterns that resemble river systems on Earth.

(b)

(a)

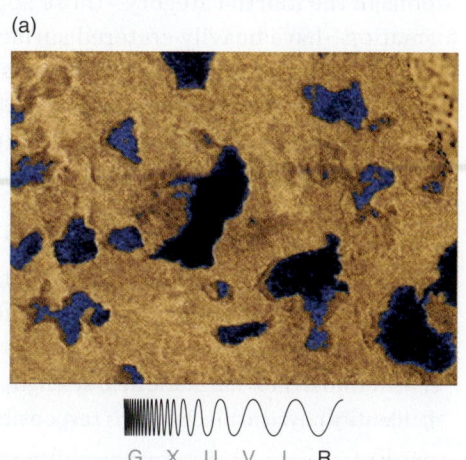

G X U V I R

G X U V I R

G X U V I R

Figure 9.12 This view of the surface of Titan, obtained from *Huygens*, shows a relatively flat surface littered with water-ice "rocks." The dark "soil" is probably made up of hydrocarbon ices.

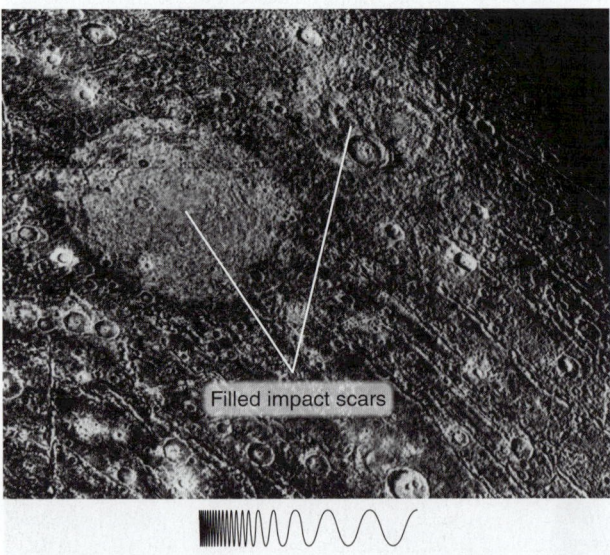

Filled impact scars

G X U V I R

Figure 9.13 This *Voyager* image shows filled impact scars on Jupiter's moon Ganymede. Filled impact scars form as viscous flow smooths out structures left by impacts on icy surfaces.

Figure 9.12, the surface around the landing site is relatively flat and littered with rounded "rocks" of water ice. The dark "soil" is probably a mixture of water and hydrocarbon ices.

In many ways, Titan resembles a primordial Earth, albeit at much lower temperatures. The presence of organic compounds that could be biological precursors in the right environment makes Titan another high-priority target for continued exploration in the search for extraterrestrial life.

Formerly Active Moons

Some moons show clear evidence of past ice volcanism and tectonic deformation but no current geological activity. The surface of Jupiter's moon Ganymede, for example, which is the largest moon in the Solar System, is composed of two prominent terrains: a dark, heavily cratered (and therefore ancient) terrain and a bright terrain characterized by ridges and grooves. The most extensive region of ancient dark terrain includes a semicircular area about the size of the contiguous United States on the leading hemisphere. Parallel ripples occurring in many dark areas are among Ganymede's oldest surface features. They may represent surface deformation from internal processes, or they may be relics of impact cratering.

Impact craters on Ganymede range up to hundreds of kilometers in diameter, and the larger craters are proportionately shallower. The icy crater rims slowly slump, like a lump of soft clay. The craters are seen as bright, flat, circular patches found principally in the moon's dark terrain (**Figure 9.13**) and are thought to be scars left by early impacts onto a thin, icy crust overlying water or slush.

In Chapter 6, you learned how planetary surfaces can be fractured by faults or folded by compression resulting from movements initiated in the mantle. On Ganymede, fracturing and faulting may have completely deformed the icy crust, destroying all signs of older features, such as impact craters, and creating the bright terrain. The energy that powered Ganymede's early activity was liberated during a period of differentiation when the moon was very young. Once the differentiation process was complete, that source of internal energy ran out and geological activity ceased.

Ganymede is not the only formerly active moon. Several other large moons in the outer Solar System have also been active in the past, including Saturn's moons Tethys and Dione and Uranus's moons Miranda and Ariel.

Inactive Moons

Moons in the fourth category—those apparently not active at any time since their formation—have heavily cratered surfaces and show no modifications other than those caused by a long history of impacts. This category includes Jupiter's Callisto, Uranus's Umbriel, and a large assortment of irregular moons. Irregular moons have orbits that are highly inclined or very elliptical or that revolve in the opposite direction from the planet's spin.

CHECK YOUR UNDERSTANDING **9.2**

Classifying moons according to their geology allows us to (choose all that apply)

a. compare and contrast their features.

b. explain how they formed.

c. determine their magnetic field strengths.

d. identify physical mechanisms responsible for their evolution.

9.3 Asteroids Are Pieces of the Past

An asteroid is a primitive planetesimal that did not become part of the accretion process that formed planets. Some asteroids form their own small systems: moons have now been found around nearly 200 asteroids. At least five asteroids are known to have two moons, and at least one asteroid has a ring around it. Because asteroids are part of our Solar System, many of them move quickly enough across the sky that their motion is noticeable over a few hours. Detecting an unexpected asteroid is exciting and often inspires amateur astronomers to continue their observations. The work of amateur astronomers has added significantly to the body of knowledge about asteroids, including orbital parameters, rotation periods, and colors.

Asteroid Groups

Asteroids are found throughout the Solar System, although most, such as Ceres and Sylvia, are located in the main **asteroid belt** between the orbits of Mars and Jupiter, as shown in **Figure 9.14**. There are several other groups of asteroids, which are divided according to their orbital characteristics. Trojan asteroids (Trojans) share a planet's orbit and are held in place by interactions with the planet's gravitational field. Three other groups are defined by their varied relationships to the orbits of Earth and Mars: Apollo asteroids cross the orbits of both Earth and Mars, Aten asteroids cross Earth's orbit but not that of Mars, and Amor asteroids cross the orbit of Mars but not that of Earth (Figure 9.14). Each of these three groups is named for a prototype that is representative of the group.

Asteroids whose orbits bring them within 1.3 AU of the Sun are called **near-Earth asteroids**. These asteroids, along with a few icy comet nuclei, are known collectively as **near-Earth objects (NEOs)**. Astronomers estimate that between 500 and 1,000 NEOs have diameters larger than 1 km. Collisions with these large objects are geologically important and can dramatically alter life on Earth, as explained in Chapter 6.

Asteroid Composition

The planetesimals that formed our Solar System's planets and moons have been so severely modified by planetary processes that nearly all information about their original physical condition and chemical composition has been lost. By contrast, asteroids constitute an ancient and far more pristine record of what the early Solar System was like. Asteroids continue to collide with one another today and produce small fragments of rock and metal, some of which crash to Earth's surface as meteorites. If you visit a planetarium or science museum, you may find on display a meteorite that you can touch. The meteorite may be older than Earth itself and might even contain tiny grains of material that predate the formation of our Solar System.

Asteroids account for only a tiny fraction of the total mass in the Solar System. If all of the asteroids were combined into a single body, it would be only about one-third the mass of Earth's Moon. There are many more small asteroids than large ones. Most are too small for their self-gravities to have pulled them into spherical shapes. Some are very elongated, suggesting they are either fragments of larger bodies or the result of haphazard collisions between smaller bodies.

Most asteroids are relics of rocky or metallic planetesimals that formed in the region between Mars and Jupiter. Although early collisions between these planetesimals created several bodies large enough to differentiate, Jupiter's gravitational tug prevented them from forming a single planet.

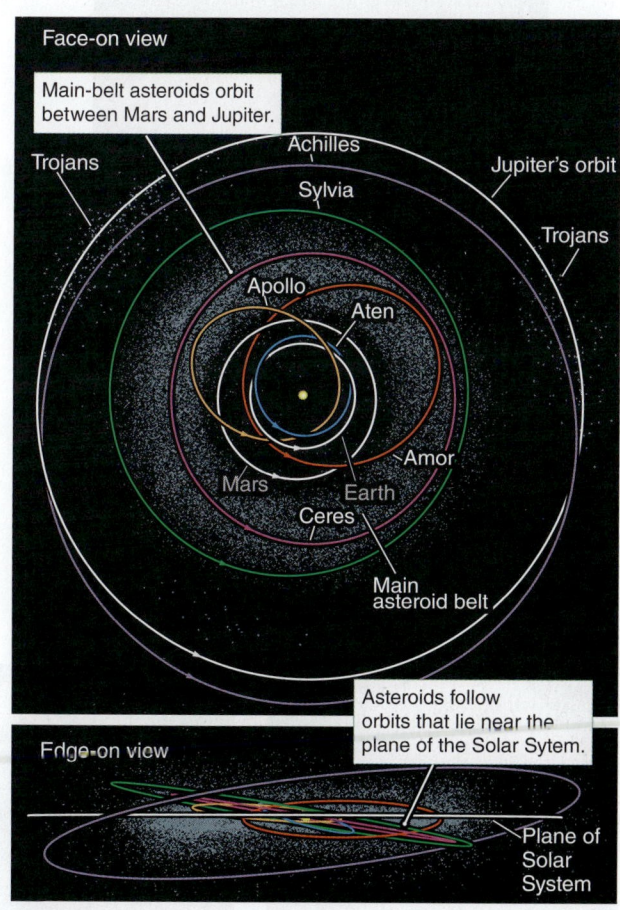

Figure 9.14 Blue dots show the locations of known asteroids at a single point in time. The orbits of Aten, Amor, Achilles, Sylvia, and Apollo (prototype members of some groups of asteroids) are shown.

---

what if . . .

What if you were in charge of NASA's manned space program and had the funds to send humans to one of the moons in the outer Solar System: Which moon do you think would provide the highest scientific reward?

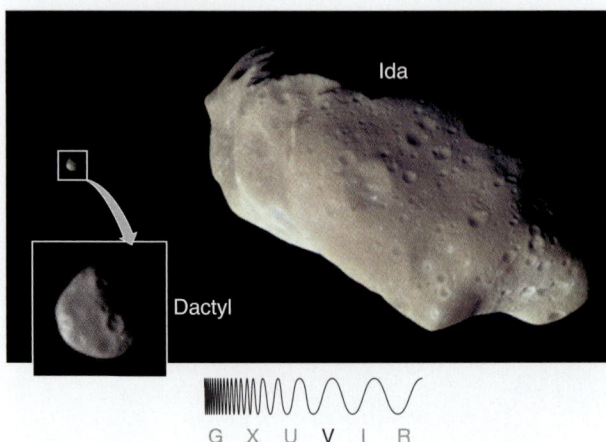

G X U V I R

Figure 9.15 The *Galileo* spacecraft imaged the asteroid Ida with its tiny moon, Dactyl (also shown enlarged in the inset).

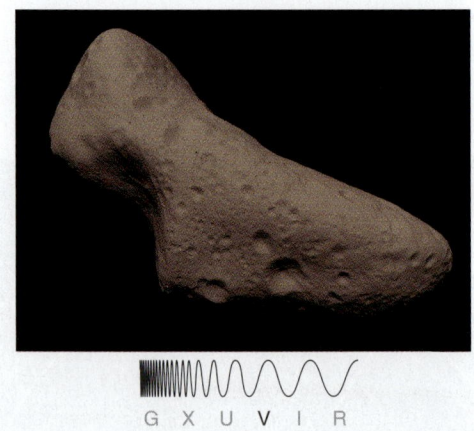

G X U V I R

Figure 9.16 This image of the asteroid Eros was produced via high-resolution scanning of the asteroid's surface by the *NEAR Shoemaker* spacecraft's laser range finder. *NEAR Shoemaker* became the first spacecraft to land on an asteroid when it was gently crashed onto the surface of Eros.

Figure 9.17 *Mars Reconnaissance Orbiter* and *Mars Express* imaged the two tiny moons of Mars: (a) Deimos and (b) Phobos.

(a)   (b)

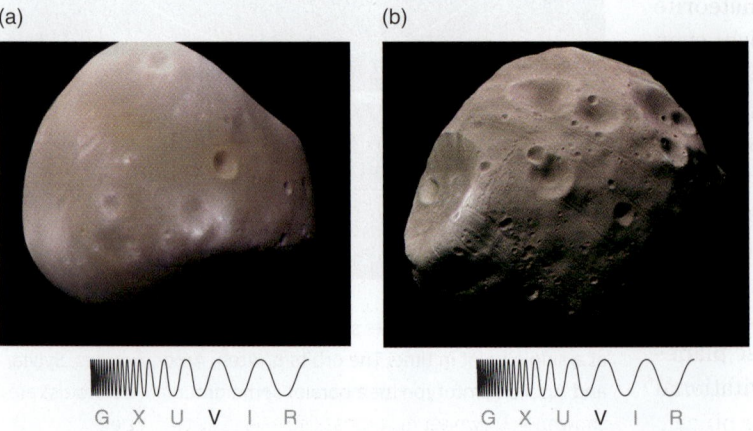

G X U V I R   G X U V I R

Astronomers have measured the masses of some asteroids by noting the effect of their gravity on spacecraft passing nearby. Knowing the mass and the size of asteroids enables us to determine their densities, which range between 1.3 and 3.5 times the density of water. Those at the lower end of this range are "rubble piles" with large gaps between the fragments, and they are considerably less dense than the meteorite fragments they create. This is what we would expect of objects that were assembled from smaller objects and then suffered a history of violent collisions.

Visits to Asteroids

In 1991, the *Galileo* spacecraft passed by the asteroid Gaspra and obtained images of its surface. Gaspra is cratered and irregular in shape, about $11 \times 12 \times 19$ km in size. Faint, groovelike patterns may be fractures from the impact that chipped Gaspra from a larger planetesimal. Distinctive colors imply that Gaspra is covered with a variety of rock types.

Later in its mission, *Galileo* returned to the main belt, passing close to the asteroid Ida (**Figure 9.15**) and its moon Dactyl. *Galileo* flew so close to Ida that its cameras could see details as small as 10 meters across—about the size of a small house. Ida is 54 km long, ranging from 15 to 24 km in diameter. Ida's craters indicate that the surface is about 1 billion years old, at least twice the age estimated for Gaspra. Like Gaspra, Ida is fractured. The fractures indicate that these asteroids must be made of relatively solid rock. (A loose pile of rubble won't "crack.") This supports the idea that some asteroids are chips from larger, solid objects.

In early 2000, the *NEAR Shoemaker* spacecraft entered orbit around the asteroid Eros to begin long-term observations (**Figure 9.16**). Eros is one of more than 5,000 known objects whose orbits bring them within 1.3 AU of the Sun. Eros is roughly a cylinder 34 km long and 11 km in diameter. Like Gaspra and Ida, Eros shows a surface with grooves, rubble, and impact craters, including a crater 8.5 km across. The scarcity of smaller craters suggests that its surface is younger than Ida's. After a year of observing, the spacecraft landed on the asteroid's surface. Chemical analyses confirmed that the composition of Eros is comparable to that of primitive meteorites.

In November 2005, the Japanese spacecraft *Hayabusa* made contact with the asteroid Itokawa. It collected samples and then returned to Earth in 2010. In addition to being a significant engineering accomplishment, this mission was important because the dust was a pristine sample from a specific, well-characterized asteroid. *Hayabusa2* has collected samples from another asteroid, as well as deploying four small rovers that hop around on its surface. This spacecraft will return to Earth in December 2020.

Usually, scientists cannot match a meteorite sample to a specific asteroid. However, in 2011, NASA's *Dawn* spacecraft visited the asteroid Vesta and found that it is a leftover intact protoplanet that formed within the first 2 million years of the condensation of the first solid bodies in the Solar System. It has an iron core and is differentiated, so it is more like the planets than like other asteroids. Vesta's composition matches the composition from a peculiar group of meteorites. A collision that created one of the two large impact basins in the south polar region of Vesta blasted material into space that then landed on Earth as these meteorites. This transfer of material from one object to another is a useful way to gather detailed information about objects in the Solar System.

The appearance and spectra of Deimos and Phobos (**Figure 9.17**), the tiny moons of Mars, are similar to those of some asteroids. Many scientists think these moons may be asteroids that have been captured by the gravity of Mars. But are Deimos and Phobos really asteroids? Some scientists argue that it is unlikely Mars could have captured two asteroids, proposing instead that Deimos and Phobos must somehow have evolved together with Mars. Another possibility is that Mars and its moons were all parts of a much larger body that was fragmented by a collision early in Mars's history. A visit to these moons is required to sort out these possibilities.

CHECK YOUR UNDERSTANDING 9.3

If an asteroid is not spherical, then

a. it is made of iron.

b. its mass is low.

c. it is made of rocky material.

d. it is very young.

9.4 Comets Are Clumps of Ice

Comets are icy planetesimals that formed from primordial material. These icy planetesimals range in size from a few dozen meters to several hundred kilometers across and are composed of ice, organic compounds, and dust grains. They spend most of their lives adrift in the frigid outer reaches of the Solar System. Most are much too small and far away to be seen, so no one really knows how many there are. Estimates for the number of these icy bodies in our Solar System range as high as a trillion (10^{12})—more than all the stars in the Milky Way Galaxy. Comets put on a show only when they come close enough to the Sun to show the effects of solar heating; then they are called **active comets**.

Homes of Comets

We know where comets come from by observing their orbits as they pass through the Solar System. Comets fall into two distinct groups, based on their origin. These groups are named for scientists Gerard Kuiper (1905–1973) and Jan Oort (1900–1992). The Kuiper Belt and the Oort Cloud are both enormous reservoirs of icy planetesimals, some of which follow highly eccentric orbits that carry them into the inner Solar System, where they become fully fledged comets.

The **Kuiper Belt** is disk-shaped and aligned with the Solar System. The Kuiper Belt begins at about 30 AU from the Sun, near the orbit of Neptune, and extends outward to about 50 AU (**Figure 9.18**). Some of the icy planetesimals known as **Kuiper Belt objects (KBOs)** are large enough to be dwarf planets; all of the known dwarf planets except Ceres are KBOs. Although brightness and approximate distance can be measured, the sizes of KBOs are difficult to determine with precision because the compositions and therefore albedos of KBOs are uncertain. The physical properties of these very distant objects, such as shapes and surface features, are also uncertain. Some KBOs have moons, and at least one has five moons. More than 1,000 KBOs have been discovered. On the basis of statistics for known KBOs, astronomers estimate that the Kuiper Belt contains tens of thousands of KBOs, most of them smaller than the known Kuiper Belt objects.

In 2019, the *New Horizons* spacecraft, originally sent to Pluto, arrived at another KBO a billion miles farther away than Pluto. This KBO, called Arrakoth, brought its own surprises, including an unusual shape (**Figure 9.19**). This "snowman" shape indicates that Arrakoth was originally two objects that gently collided and stuck together. Scientists are currently debating the origin and evolution of this object.

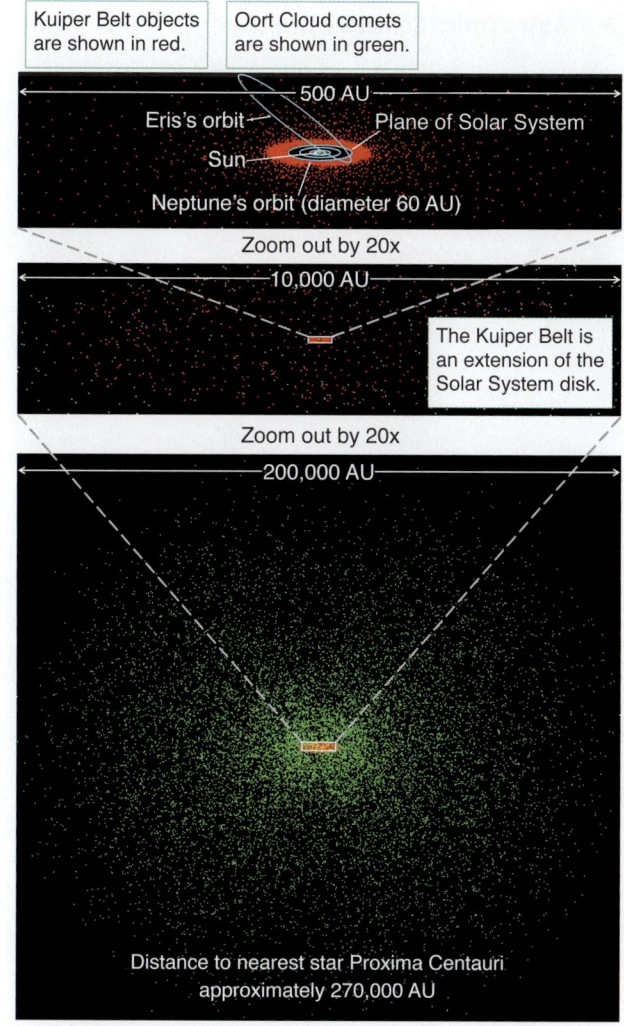

Figure 9.18 Most comets near the inner Solar System (shown in red) populate an extension to the disk of the Solar System called the Kuiper Belt. The spherical Oort Cloud is far larger and contains many more comets (shown in green).

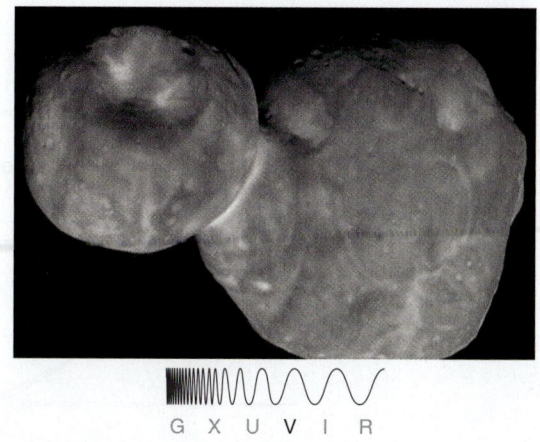

Figure 9.19 After its visit to Pluto, NASA's *New Horizons* spacecraft was redirected to visit a second Kuiper Belt object. Arrokoth presented its own set of surprises for scientists to puzzle over.

▶❙❙ **AstroTour:** Cometary Orbits

They wonder, for example, whether the bright circular patches are impact craters, places where the surface has subsided, or have some other origin.

Unlike the disk-shaped Kuiper Belt, the **Oort Cloud** is a spherical distribution of icy planetesimals that are too distant to be seen by even the most powerful telescopes. Astronomers infer the size and shape of the Oort Cloud from the orbits of comets that originate from it. These bodies approach the Sun from all directions and from as far away as 100,000 AU—more than one-third of the way to the nearest stars.

Orbits of Comets

The lifetime of a comet nucleus depends on how frequently it passes by the Sun and how close it comes. There are about 400 known **short-period comets**, which by definition have periods less than 200 years. Additionally, each year astronomers discover about six new **long-period comets**, whose orbital periods are greater than 200 years. About 3,000 long-period comets have been observed.

The KBOs are packed closely enough to interact with each other from time to time. When this happens, one object gains energy while the other loses it (overall energy is conserved). The "winner" may gain enough energy to be sent into an orbit that reaches far beyond the boundary of the Kuiper Belt, while the "loser" falls inward toward the Sun. This is the origin of short-period comets. Short-period comets tend to orbit the Sun in the same direction that the planets orbit (prograde motion) and to have orbits in the ecliptic plane, so they frequently pass close enough to a planet for its gravity to change the comet's orbit around the Sun. As these

Figure 9.20 Orbits of a number of comets (colored lines) are shown in face-on and edge-on views of the Solar System. Populations of (a) short-period comets and (b) long-period comets have very different orbital properties. Comet Halley, which appears in both diagrams for comparison, is a short-period comet.

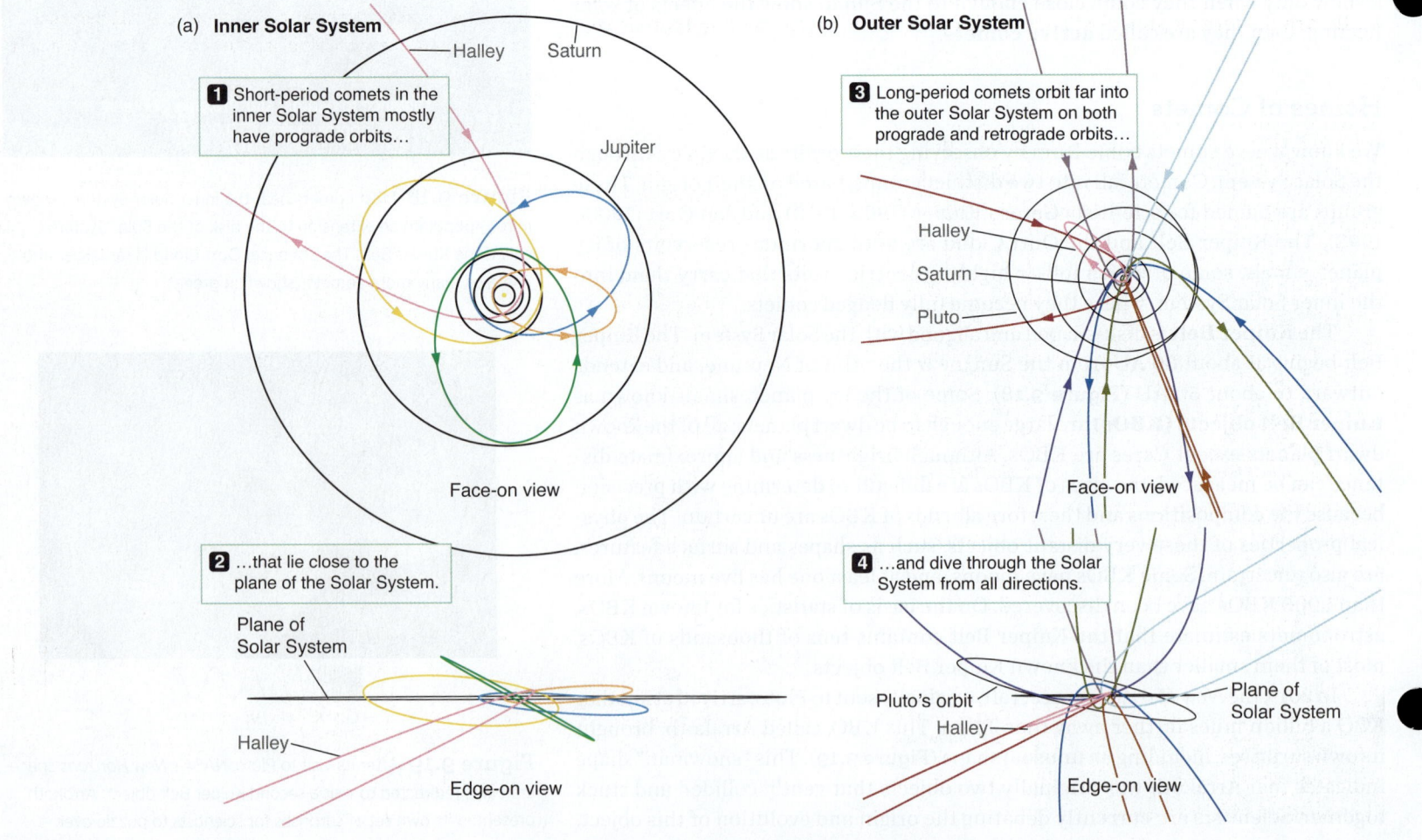

comets fall in toward the Sun, they are forced into their current short-period orbits by these gravitational encounters.

Approximately every 5 million to 10 million years, a star passes "nearby"—within about 100,000 AU of the Sun. The gravitational force from these stars may stretch the orbit of an Oort Cloud planetesimal, which will then pass much closer to the Sun in a future orbit. Similarly, the gravitational attraction from huge clouds of dense interstellar gas can also stir up the Oort Cloud. This is the origin of long-period comets. Long-period comets come into the inner Solar System from all directions; some show prograde motion, whereas others orbit the Sun in the opposite direction from planetary orbits (retrograde motion). More than 600 long-period comets have well-determined orbits. Some have orbital periods of hundreds of thousands or even millions of years. Almost all their time is spent in the Oort Cloud in the frigid, outermost regions of the Solar System. Because of their very long orbital periods, these comets have made at most one appearance throughout the course of recorded history. **Figure 9.20** illustrates the differences between the orbits of short-period and long-period comets.

Anatomy of an Active Comet

As an icy planetesimal approaches the Sun, sunlight heats its surface, vaporizing ices that stream away (**Figure 9.21a**). These gases carry dust particles along with them. The gases and dust form a nearly spherical atmospheric cloud around the planetesimal called the **coma**. The remaining planetesimal, called the comet **nucleus**, and the inner part of the coma are sometimes referred to collectively as the comet's **head**. Long streamers of dust, gas, and ions called the **tails** (Figure 9.21b) are the largest and most spectacular part of a comet, although there is very

Figure 9.21 (a) The principal components of a fully developed active comet are the nucleus, the coma, and two types of tails called the dust tail and the ion tail. Together, the nucleus and the coma are called the head. (b) Comet Hale-Bopp, the great comet of 1997. The ion tail is blue in this image, whereas the dust tail is white.

(a)

Ion tail

Dust tail

Head

Sun

Nucleus

Coma

The much larger coma and tail are material driven off the nucleus by solar heating.

Comet nuclei are "dirty snowballs."

(b)

G X U V I R

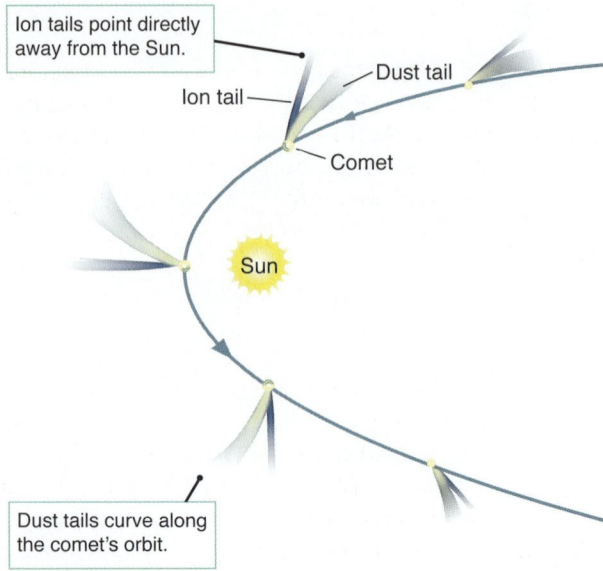

Ion tails point directly away from the Sun.

Dust tail

Ion tail

Comet

Sun

Dust tails curve along the comet's orbit.

Figure 9.22 The orientations of the dust and ion tails change as the comet orbits. The ion tail points directly away from the Sun, whereas the dust tail curves along the comet's orbit.

Figure 9.23 (a) In 2007, Comet McNaught was the brightest comet to appear in decades, but its true splendor was visible only to observers in the Southern Hemisphere. (b) Months after its closest approach to the Sun in 2007, Comet Holmes suddenly became a half-million times brighter within hours, turning into a naked-eye comet with an angular diameter larger than that of the full Moon. Comet Holmes favored Northern Hemisphere observers.

little material in them. The density of gas in a comet's tail is no more than a few hundred molecules per cubic centimeter (cm^3). This is much less than the density of Earth's atmosphere, which at sea level contains more than 10^{19} molecules/cm^3. Dust particles in the tail are typically about 1 micrometer (μm) in diameter, roughly the size of smoke particles.

Active comets have two different types of tails, called *ion tails* and *dust tails*, as shown in Figure 9.21. Many particles in the comet's coma are ionized atoms or molecules. Because they are electrically charged, the solar wind pushes on these particles, rapidly accelerating them away from the Sun to speeds of more than 100 kilometers per second (km/s)—far greater than the orbital velocity of the comet itself. The ionized particles are swept into a long, wispy structure known as the **ion tail**. Ion tails are usually very straight and point from the head of the comet directly away from the Sun.

Dust particles also have a net electric charge and feel the force of the solar wind. In addition, sunlight itself exerts a force on cometary dust. But dust particles are much more massive than individual ions, so they are accelerated more gently and do not reach the high relative speeds that the ions do. As a result, the **dust tail** often curves gently away from the head of the comet as the dust particles are gradually pushed from the comet's orbit in the direction away from the Sun.

Figure 9.22 shows the tails of a comet at various points in its orbit. Remember that both types of tails always point *away from the Sun*, regardless of which direction the comet is moving. As the comet approaches the Sun, its two tails trail behind its nucleus. But the tails extend *ahead* of the nucleus as the comet moves away from the Sun.

Tails vary greatly from one comet to another. Some comets display both types of tails simultaneously; others, for reasons that we do not yet understand, produce no tails at all. A tail often forms as a comet crosses the orbit of Mars, because at this distance heating from the Sun warms the comet enough to drive gas and dust away from the nucleus.

Most naked-eye comets develop a coma and then form an extended tail as they approach the inner Solar System. Comet McNaught in 2007 was such a comet (**Figure 9.23a**), but there are exceptions. Comet Holmes was a very faint telescopic object when it reached its closest point to the Sun just beyond the orbit of Mars. Then, several months later, as it was well on its way outward toward Jupiter's orbit, the comet suddenly became a half-million times brighter in just 42 hours. Comet Holmes became a bright naked-eye comet in the Northern Hemisphere skies for several months (Figure 9.23b). Astronomers remain puzzled over the cause of this dramatic brightening. Explanations range from a meteoroid impact to a sudden (but unexplained) buildup of subsurface gas.

Generally, short-period comets provide a less spectacular display than long-period comets.

(a)

G X U V I R

(b)

G X U V I R

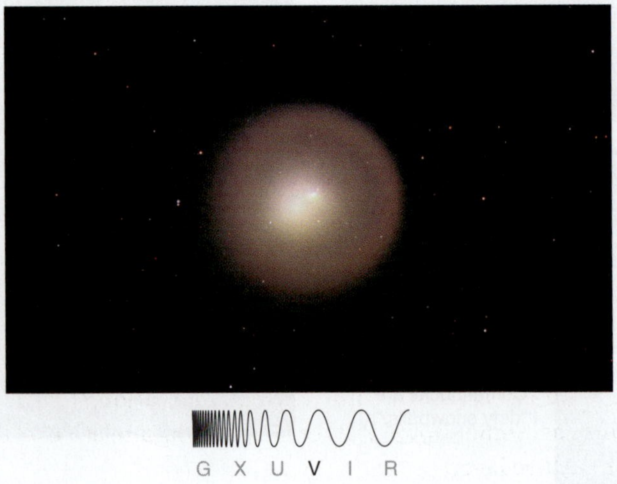

The nuclei of short-period comets are eroded from repeated heating by the Sun, and as the volatile ices are driven away, some of the dust and organic compounds are left behind on the surface of the nucleus. The buildup of this covering slows down cometary activity and makes their solar approaches less spectacular. In contrast, long-period comets are relatively pristine. More of their supply of volatile ices remains close to the surface of the nucleus, so they can produce a truly magnificent show. However, most long-period comets never pass close enough to either Earth or the Sun to become bright enough to attract much public attention.

When will the next bright comet come along? On average, a spectacular comet appears about once per decade, but it is all a matter of chance. It might be many years from now—or it could happen tomorrow.

Visits to Comets

Comets provide an engineering challenge for spacecraft designers. We seldom have enough advance knowledge of a comet's visit or its orbit to mount a successful mission to intercept it. The closing speed between an Earth-launched spacecraft and a comet can be extremely high. Observations must be made very quickly, and there is a danger of high-speed collisions with debris from the nucleus. Despite these challenges, nearly a dozen spacecraft have rendezvoused with comets, including a five-spacecraft armada sent to Comet Halley by the Soviet, European, and Japanese space agencies in 1986. Much of what we know about comet nuclei and the innermost parts of the coma comes from data sent back by these missions.

The Soviet *Vega 1* and *Vega 2* and the European *Giotto* spacecraft entered the coma of Comet Halley when they were still nearly 300,000 km from its nucleus. We learned that the dust from Comet Halley was a mixture of light organic substances and heavier rocky material, and the gas was about 80 percent water and 10 percent carbon monoxide with smaller amounts of other organic molecules. The surface of the comet's nucleus is among the darkest known objects in the Solar System, which means that it is rich in complex organic matter that must have been present as dust in the disk around the young Sun—perhaps even in the interstellar cloud from which the Solar System formed. As the three spacecraft passed close by the nucleus, they observed jets of gas and dust (**Figure 9.24**) from several small fissures that covered only about a tenth of the surface. The material in these jets moved away from the comet's surface at speeds of up to 1 km/s, far above the escape velocity.

In 2001, NASA's *Deep Space 1* spacecraft flew within 2,200 km of the nucleus of Comet Borrelly. Its tar-black surface is also among the darkest seen on any Solar System object and showed no evidence of water ice.

In 2004, NASA's *Stardust* spacecraft flew within 235 km of the nucleus of Comet Wild 2 (pronounced "vilt 2"). This comet is a newcomer to the inner Solar System. A close encounter with Jupiter in 1974 perturbed its orbit, bringing this relatively pristine body in from its orbit between Jupiter and Uranus. The nearly spherical nucleus of Wild 2 is about 5 km across. At least 10 gas jets were active, some of which carried large chunks of surface material. (A few particles as large as a bullet penetrated the outer layer of the spacecraft's protective shield.) The surface of Wild 2 is covered with features that may be impact craters modified by ice sublimation, small landslides, and erosion by jetting gas. Some show flat floors, suggesting a relatively solid interior beneath a porous surface layer.

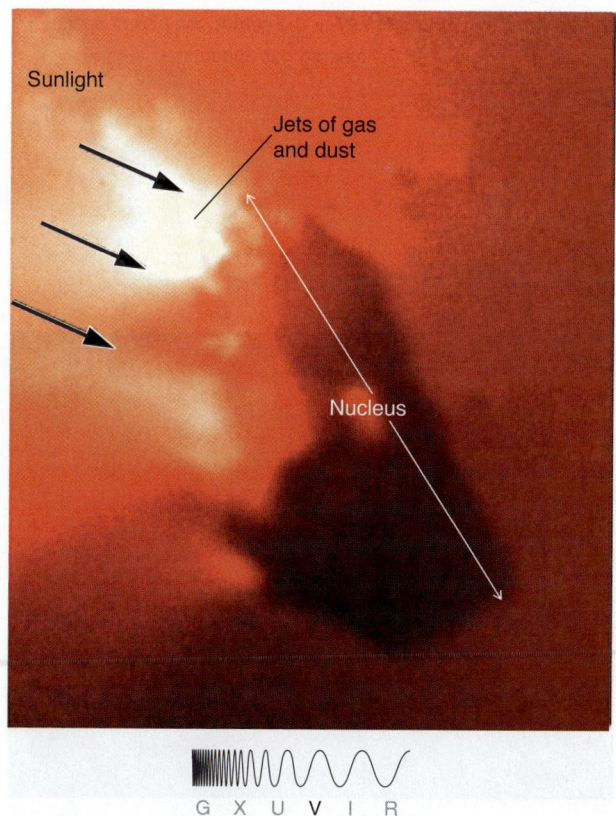

Figure 9.24 The nucleus of Comet Halley was imaged by the *Giotto* spacecraft in 1986. Here it is shown in false color to emphasize details.

(a)

(b)

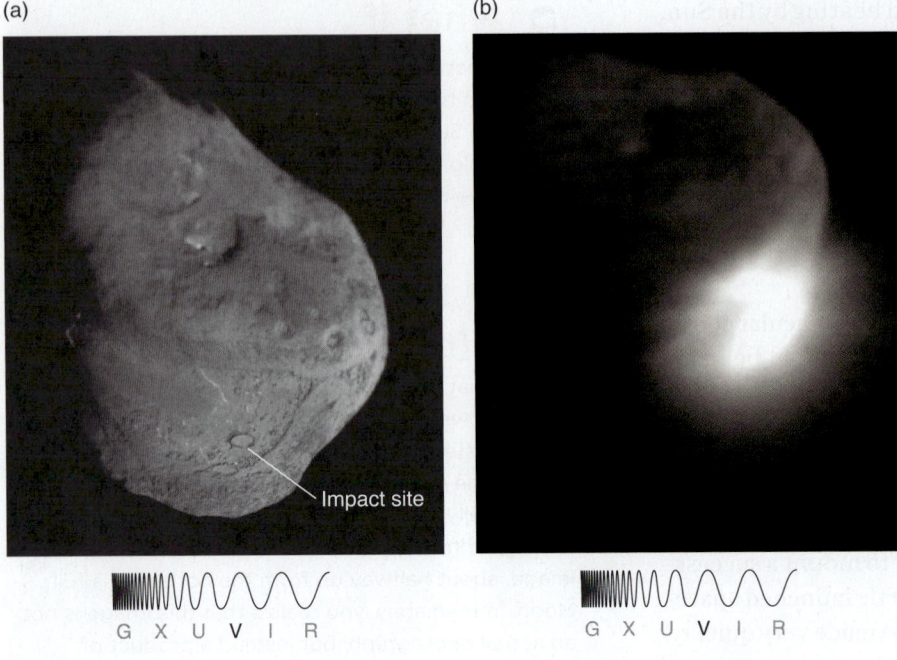

G X U V I R   G X U V I R

Figure 9.25 (a) The surface of the nucleus of Comet Tempel 1 was imaged just before impact by the *Deep Impact* projectile. The impact occurred between the two 370-meter-diameter craters located near the bottom of the image. The smallest features appearing in this image are about 5 meters across. (b) Sixteen seconds after the impactor struck the comet, the parent spacecraft took this image of the ejected gas and dust.

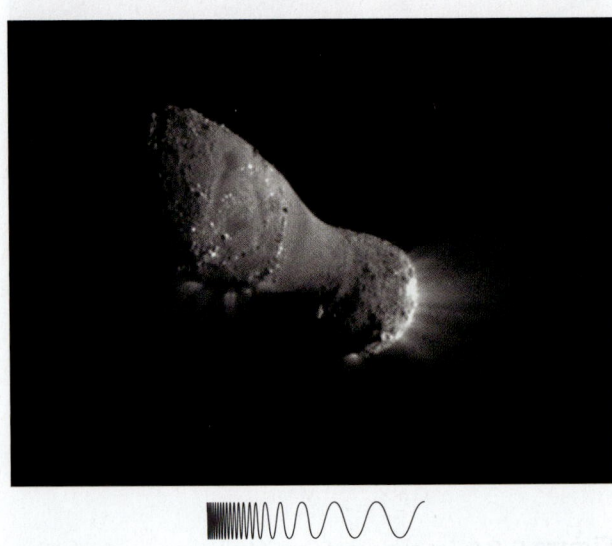

G X U V I R

Figure 9.26 This image of Comet Hartley 2 taken by the *EPOXI* spacecraft shows two distinct surface types.

In 2005, NASA's *Deep Impact* spacecraft launched a 370-kilogram (kg) impacting projectile into the nucleus of Comet Tempel 1 at a speed of more than 10 km/s. The impact sent 10,000 tons of water and dust flying off into space at speeds of 50 meters per second (m/s; **Figure 9.25**). A camera mounted on the projectile snapped photos of its target until it was vaporized by the impact. Observations of the event were made both locally by *Deep Impact* and back on Earth by a multitude of orbiting and ground-based telescopes.

Water, carbon dioxide, hydrogen cyanide, iron-bearing minerals, and a host of complex organic molecules were identified. The comet's outer layer is composed of fine dust with the consistency of talcum powder. Beneath the dust are layers made up of water ice and organic materials. One surprise for scientists was the presence of well-formed impact craters, which had been absent in close-up images of Comets Borrelly and Wild 2. Why some comet nuclei have fresh impact craters and others have none remains an open question.

In late 2010, the *EPOXI* spacecraft flew past Comet Hartley 2 (**Figure 9.26**), imaging not only jets of dust and gas, indicating a remarkably active surface, but also an unusual separation of rough and smooth areas that have very different natures. The "waist"—the narrow part at the middle—is a smooth inactive area where ejected material has fallen back onto the cometary nucleus. Carbon dioxide jets shoot out from the rough areas. It is unclear whether this unusual shape is a result of how Comet Hartley 2 formed 4.5 billion years ago or is due to more recent evolution of the comet. Further observations by the Herschel Space Observatory showed that the water on this comet has the same ratio of deuterium to hydrogen as the water in Earth's oceans. (A deuterium nucleus is a hydrogen nucleus with an extra neutron.). This is strong evidence that some of Earth's water came from the Kuiper Belt. Comets from the Oort Cloud have a different ratio of hydrogen isotopes, and so they have been ruled out as the source of Earth's water.

In 2014, the European Space Agency's *Rosetta* spacecraft arrived at Comet 67P/Churyumov-Gerasimenko (67P). **Figure 9.27** shows an image of the comet's surface, with the shadow of the spacecraft visible in the lower part of the image. The spacecraft carried a small lander, known as *Philae*. This was the first mission to rendezvous with a comet, follow it as it passed the Sun, and also launch a lander to the surface of the comet. Unfortunately, *Philae* landed in a dark crevice; its solar panels could not collect energy for electrical power, and it could not communicate with *Rosetta*. Detailed analysis of data from *Rosetta* has shown that comets are primordial; they formed in the early Solar System, rather than resulting from collisions later on. This means that studies of cometary material can tell us about the conditions in the early Solar System directly. *Rosetta* also found that the ingredients of life (complex organic molecules) were present on the comet. This finding has direct implications for our understanding of how life arose on early Earth. Finding these complex organic molecules in the primordial material of the Solar System supports the hypothesis that comets may have delivered the necessary ingredients for life to Earth. The Rosetta mission ended in 2016, when Rosetta gently (at 3.2 km/h) touched down on the comet, turned off its radio transmitter and stopped broadcasting.

9.5 Comet and Asteroid Collisions Still Happen Today

Almost all hard-surfaced objects in the Solar System still bear the scars of a time when tremendous impact events were common. Certainly, Earth experienced large impacts throughout its history, such as the one that ended the era of the dinosaurs 65 million years ago. Although such impacts are far less frequent than they once were, they still happen today.

The Tunguska River flows through a remote region of western Siberia. In summer 1908, the region was blasted with the energy equivalent of 2,000 Hiroshima atomic bombs. **Figure 9.28** shows a map of the region, along with a photograph of the devastation caused by the blast. Eyewitness accounts detailed the destruction of dwellings, the incineration of reindeer (including one herd of 700), and the deaths of at least five people. Although trees were burned or flattened for more than 2,150 square kilometers—an area greater than metropolitan New York City—no crater was left behind! The Tunguska event was the result of a tremendous high-altitude explosion that occurred when a small body hit Earth's atmosphere, ripped apart, and formed a fireball before reaching Earth's surface. Recent expeditions to the Tunguska area have recovered resin from the trees blasted by the event. Chemical traces in the resin suggest that the impacting object may have been a stony asteroid.

On February 12, 1947, yet another planetesimal struck Earth, this time in the Sikhote-Alin region of eastern Siberia. Composed mostly of iron, the object had an estimated diameter of about 100 meters and broke into a number of fragments before hitting the ground, leaving a cluster of craters and widespread devastation. Witnesses reported a fireball brighter than the Sun and sound that was heard 300 km away.

In February 2013, a known NEO about half the size of an American football field passed so close to Earth that it came within the orbit of man-made satellites. This near miss was uneventful, and the object simply continued on its way. However, in an unrelated event on the same day, a previously unknown meteoroid estimated to have a radius of about

Figure 9.27 shows details of the surface, as well as the shadow of the spacecraft near the bottom of the image. The shadow is about 50 meters long.

Figure 9.27 The *Rosetta* spacecraft followed Comet 67P as it passed near the Sun. This image shows details of the surface, as well as the shadow of the spacecraft near the bottom of the image. The shadow is about 50 meters long.

Figure 9.28 A large region of forest near the Tunguska River in Siberia was flattened in 1908 by a small asteroid or comet exploding in Earth's atmosphere.

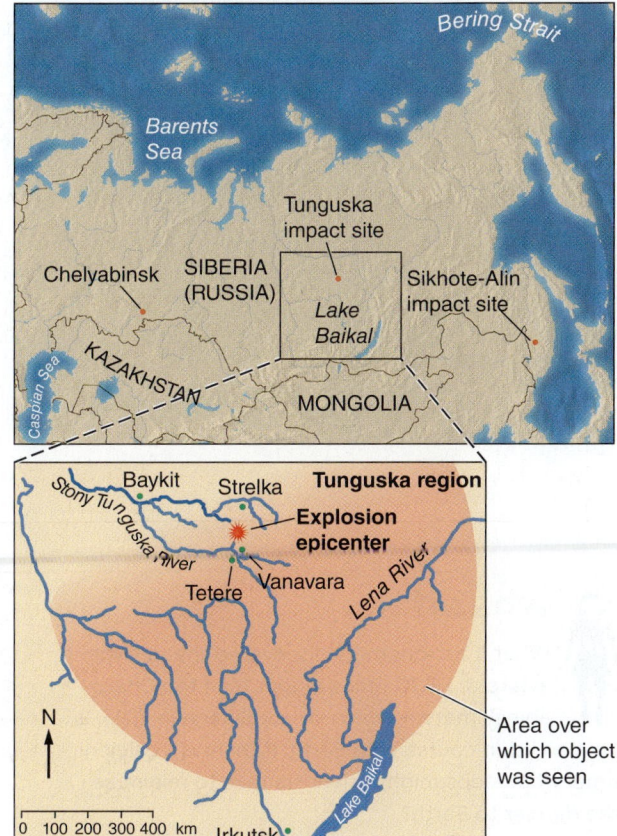

working it out 9.1

Finding the Radius of a Meteoroid

How do astronomers know how big the Chelyabinsk meteoroid was? The energy released by the meteor has been estimated from the strength of the shock wave to be the equivalent of about 2 minutes of worldwide energy usage. Converting that number to more ordinary units gives about 1.8×10^{15} joules (J).

The speed at which the object was traveling can be found from video of the object's motion through the sky: 18.6 km/s. The energy of motion (the kinetic energy) of a moving object is related to both its mass and its speed. The kinetic energy is given by

$$E_K = \frac{1}{2}mv^2$$

where E_K is the kinetic energy in joules, m is the mass in kilograms, and v is the magnitude of the velocity (the speed) in meters per second. We now have a relationship between the estimated energy (which we know), the speed (which we know), and the mass (which we do not know, but which is related to the size). Solving this equation for mass requires first multiplying both sides by 2 and then dividing both sides by v^2:

$$2E_K = mv^2$$

$$\frac{2E_K}{v^2} = m$$

$$m = \frac{2E_K}{v^2}$$

In the last step, we flipped the equation left to right. Plugging in the values given earlier for E_K and v gives

$$m = \frac{2(1.8 \times 10^{15} \text{ J})}{(18.6 \times 10^3 \text{ m/s})^2}$$

$$m = 10{,}000{,}000 \text{ kg}$$

To find the radius, we must use the fact that the density of an object is equal to the mass divided by the volume. Therefore, the volume of an object can be found by dividing the mass by the density. In this case, we assume the meteoroid was spherical, so it has a volume of $\frac{4}{3}\pi r^3$, where r is the radius:

$$\text{Volume} = \frac{\text{Mass}}{\text{Density}}$$

$$\frac{4}{3}\pi r^3 = \frac{m}{\text{Density}}$$

Because this meteorite was made of rock, we know the density was about 2,500 kg/m³. Solving for r gives

$$r^3 = \frac{3m}{(4\pi) \times \text{Density}}$$

$$r = \sqrt[3]{\frac{3m}{(4\pi) \times \text{Density}}}$$

$$r = \sqrt[3]{\frac{3 \times 10{,}000{,}000 \text{ kg}}{(4\pi) \times 2{,}500 \text{ kg/m}^3}}$$

$$r = \sqrt[3]{950 \text{ m}^3}$$

$$r = 9.8 \text{ meters}$$

This is precisely how scientists estimated the radius of the Chelyabinsk meteoroid.

what if . . .

What if astronomers discovered that a large asteroid, 1,000 times bigger than the Chelyabinsk meteor which struck Siberia in 2013, is on a collision course with Earth: If there is sufficient time, what steps might scientists take to minimize the danger to Earth?

10 meters (**Working It Out 9.1**) exploded over Chelyabinsk, Russia (**Figure 9.29a**), damaging thousands of buildings in six cities (Figure 9.29b) and injuring more than 1,000 people. Several rock fragments of meteoritic composition have been recovered, and detailed studies of this object and its origins are ongoing.

These impacts are sobering events. It is highly improbable that a populated area on Earth will experience a collision with a large asteroid within our lifetimes. Comets and smaller asteroids, however, are less predictable. There may be as many as 10 million asteroids larger than a kilometer across, but only about 130,000 have well-known orbits, and most of the unknown objects are too small to see until they come very close to Earth. Although the probability of a collision between a small asteroid and Earth is quite small, the consequences could be catastrophic.

Accordingly, NASA has been given a congressional mandate to catalog all NEOs and to scan the skies for those that remain undiscovered.

Comets present a more serious problem. Half a dozen unknown long-period comets enter the inner Solar System each year. If one happens to be on a collision course with Earth, we might not notice it until just a few weeks or months before impact. Although this has become a favorite theme of science-fiction disaster stories, Earth's geological and historical record suggests that impacts by large cometary bodies are infrequent events.

Collisions also still occur on other planets. Comet Shoemaker-Levy 9 passed so close to Jupiter in 1992 that tidal stresses broke it into two dozen major fragments, which subsequently spread out along its orbit. The fragments took one more orbit around the planet, and throughout a week in 1994, the string of fragments crashed into Jupiter. The impacts occurred just behind the limb of the planet, where they could not be observed from Earth, but fortunately the *Galileo* spacecraft was able to image some of the impacts. Immense plumes rose from the impacts to heights of more than 3,000 km above the cloud tops. Sulfur and carbon compounds released by the impacts formed Earth-sized scars in the atmosphere that persisted for months.

CHECK YOUR UNDERSTANDING 9.5

Large impacts of asteroids and comets with Earth are

a. impossible.
b. infrequent.
c. unimportant.
d. unknown.

(a)

(b)

Figure 9.29 (a) In February 2013, a meteoroid entered the atmosphere over Russia, creating a fireball that eyewitnesses said was brighter than the Sun. (b) The shock wave from the meteor damaged buildings, and more than 1,000 people were injured by flying glass and other debris.

9.6 Meteorites Are Remnants of the Early Solar System

Recall from Chapter 6 that when a meteoroid enters Earth's atmosphere, it produces an atmospheric phenomenon called a meteor, and any fragment of the meteoroid that survives to land on a planet's surface is called a meteorite. In this section, you will learn more about meteorites and what can be learned from them about the early Solar System.

Meteors and Meteor Showers

Comet nuclei that repeatedly enter the inner Solar System disintegrate within a few hundred thousand years as a result of their repeated passages by the Sun. Asteroids have much longer lives but are still being slowly broken into smaller pieces due to occasional collisions with each other. Disintegration of comet nuclei and asteroid collisions create most of the debris that fills the inner part of the Solar System. About 100,000 kg of this meteoritic debris is swept up by Earth every day, and particles smaller than 100 μm eventually settle to the ground as fine dust.

Bits of dust and other debris from a comet nucleus tend to spread out along the orbit of the original comet nucleus (**Figure 9.30**). When Earth passes through this concentration of cometary debris, the result is a **meteor shower**—a period of time with more meteors than usual. More than a dozen comets have orbits that come close enough to Earth's orbit to produce annual meteor showers. Because the

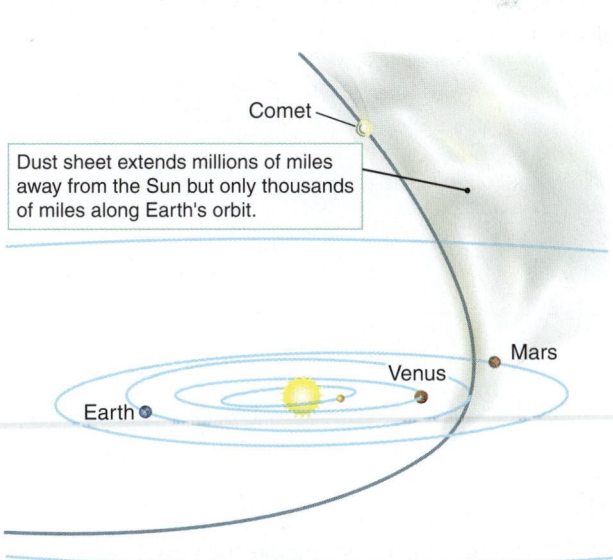

Figure 9.30 Comets leave debris in their path as they orbit the Sun. As Earth travels around in its orbit, it occasionally intersects this debris, causing meteor showers.

Comet

Dust sheet extends millions of miles away from the Sun but only thousands of miles along Earth's orbit.

Mars
Venus
Earth

(a)

(b)

> Meteors in a shower appear to move away from a point called the radiant.

> Actually the meteor paths are parallel. The radiant is a vanishing point.

G X U V I R

Figure 9.31 (a) Meteors appear to stream away from the radiant of the Leonid meteor shower. (b) Such streaks are actually parallel paths that appear to emerge from a vanishing point, as in our view of these railroad tracks.

meteoroids in a shower are all in similar orbits, they all enter our atmosphere moving in the same direction—the paths through the sky are parallel to one another. Therefore, all the meteors appear to originate from the same point in the sky (**Figure 9.31a**), just as the parallel rails of a railroad track appear to vanish to a single point in the distance (Figure 9.31b). This apparent point of origin is the shower's **radiant**, and this is used in the name of the meteor shower. For example, the Perseid meteor shower has a radiant in the constellation Perseus.

Fragments of asteroids are much denser than cometary meteoroids. The fall of a 10-kg meteoroid can produce a fireball so bright that it lights up the night sky more brilliantly than the full Moon. Such a large meteoroid may create a sonic boom heard hundreds of kilometers away or explode into multiple fragments as it nears the end of its flight. Some glow with a brilliant green color caused by metals in the original meteoroid.

Types of Meteorites

If an asteroid fragment is large enough—about the size of your fist—it can become a meteorite. Thousands of meteorites reach the surface of Earth every day, but only a tiny fraction of these are ever found and identified. Antarctica offers the best meteorite hunting in the world. Meteorites are no more likely to fall in Antarctica than anywhere else, but in Antarctica they are far easier to distinguish from their surroundings because in many places the *only* stones to be found on top of the ice are meteorites. Because Antarctica is actually very dry (annual precipitation averages about 16 cm), Antarctic meteorites also tend to show little weathering or contamination from terrestrial dust or organic compounds, making them excellent specimens for study. Scientists compare meteorites to rocks found on Earth and the Moon and contrast their structure and chemical makeup with those of rocks studied by spacecraft that have landed on Mars and Venus. Meteorites are also compared with asteroids and other objects on the basis of the colors of sunlight they reflect and absorb.

Meteorites are grouped into three categories (stony, iron, and stony-iron) according to their materials and the degree of differentiation they experienced within their parent bodies. More than 90 percent of meteorites are **stony meteorites** (**Figure 9.32a** and b), which are similar to terrestrial silicate rocks. A stony meteorite is characterized by the thin coating of melted rock that forms as it passes through the atmosphere. Many stony meteorites contain **chondrules** (see Figure 9.32a), once-molten droplets that rapidly cooled to form crystallized spheres ranging in size from sand grains to marbles. Stony meteorites containing chondrules are called **chondrites**. Conversely, stony meteorites without chondrules are called **achondrites** (see Figure 9.32b). **Carbonaceous chondrites**, chondrites that are rich in carbon, are thought to be the building blocks of the Solar System. Indirect measurements suggest that these meteorites are about 4.56 billion years old—consistent with all other measurements of the time that has passed since the Solar System was formed.

Iron meteorites (see Figure 9.32c) are the easiest to recognize. The surface of an iron meteorite has a melted and pitted appearance generated by frictional heating as it streaked through the atmosphere. Despite their recognizable appearance, many are never found, either because they land in water or simply because no one happens to recognize them. The Mars rover *Opportunity* has discovered a

Figure 9.32 Cross sections of several kinds of meteorites: (a) a chondrite (a stony meteorite with chondrules), (b) an achondrite (a stony meteorite without chondrules), (c) an iron meteorite, and (d) a stony-iron meteorite.

G X U V I R

Figure 9.33 A basketball-sized iron meteorite was found lying on the surface of Mars by the Mars exploration rover *Opportunity*.

handful of iron meteorites on the surface of Mars (**Figure 9.33**). Both their appearance—typical of iron meteorites found on Earth—and their position on the smooth, featureless plains made them instantly recognizable.

The final category of meteorites, known as **stony-iron meteorites** (see Figure 9.32d), consist of a mixture of rocky material and iron-nickel alloys. Stony-iron meteorites are relatively rare.

Meteorites and the History of the Solar System

Meteorites come from asteroids, which in turn come from planetesimals containing both stone and iron. During the growth of the terrestrial planets, large amounts of thermal energy were released as larger planetesimals accreted into smaller objects, and these bodies were heated further as radioactive elements inside them decayed. Despite this heating, some planetesimals never reached the high temperatures needed to melt their interiors—instead, they simply cooled off, and they have since remained pretty much as they were when they first formed. These planetesimals are known as **C-type asteroids**, and they are composed of primitive material that astronomers believe is essentially unmodified since the origin of the Solar System almost 4.6 billion years ago.

Some planetesimals, however, did melt and differentiate, with denser matter such as iron sinking to their center. Lower-density material—such as compounds of calcium, silicon, and oxygen—floated toward the surface and combined to form a mantle and crust of silicate rock. **S-type asteroids** are pieces of the mantles and crusts of such differentiated planetesimals. They are chemically more similar to igneous rocks found on Earth than to C-type asteroids, because they were hot enough at some point to lose their carbon compounds and other volatile materials to space. Similarly, **M-type asteroids** (from which iron meteorites come) are fragments of the iron- and nickel-rich cores of one or more differentiated planetesimals that shattered into smaller pieces during collisions with other planetesimals. Slabs cut from iron meteorites show large, interlocking crystals characteristic of iron that cooled very slowly from molten metal. Rare stony-iron meteorites may come from the transition zone between the stony mantle and the metallic core of such a planetesimal.

Some types of meteorites fail to follow the patterns just discussed. Whereas most achondrites have ages in the range of 4.5 billion to 4.6 billion years, some members of this group are less than 1.3 billion years old and are chemically and physically similar to the soil and atmospheric gases that NASA's lander instruments have measured on Mars. The similarities are so strong that most planetary scientists believe these meteorites are pieces of Mars that were knocked into space by asteroidal impacts. This means researchers have pieces of another planet that they can study in laboratories here on Earth.

Sometimes findings are controversial. In 1996, a NASA research team announced that the meteorite ALH84001 showed possible physical and chemical evidence of past life on Mars. The team's extraordinary conclusions have been challenged, and the debate continues even today.

If pieces of Mars have reached Earth from its orbit almost 80 million km beyond our own, we might expect that pieces of our companion Moon would have found their way to Earth as well. Indeed, meteorites of another group bear striking similarities to rock samples returned from the Moon. Like the meteorites from Mars, these are chunks of the Moon that were blasted into space by impacts and later fell to Earth.

The story of how planetesimals, asteroids, and meteorites are related is one of the great successes of planetary science. Scientists have assembled a wealth of information about this diverse collection of objects to piece together a picture of how planetesimals grow, differentiate, and then shatter in subsequent collisions.

CHECK YOUR UNDERSTANDING 9.6

The three types of meteorites come from different parts of their parent bodies. Stony-iron meteorites are rare because

a. they are hard to find.
b. only a small amount of a parent body has *both* stone and iron.
c. there is very little iron in the Solar System.
d. the magnetic field of the Sun attracts the iron.

reading Astronomy News

NASA won't launch a mission to hunt deadly asteroids

by Tim Fernholz, *Quartz*, July 5, 2019

Large space missions are not simply a matter of science. They are also a matter of politics.

NASA says it can't afford to build a space telescope considered the fastest way to identify asteroids that might impact the Earth with terrible consequences.

A 2015 law gave the space agency five years to identify 90% of near-Earth objects larger than 140 meters in diameter, which could devastate cities, regions and even civilization itself if they were to impact the planet. NASA isn't going to meet that deadline, and scientists believe they have so far only identified about a third of the asteroids considered a threat.

Researchers at NASA's Jet Propulsion Laboratory, led by principal investigator Amy Mainzer, developed a proposal for a space telescope called NEOCam that would use infrared sensors to find and measure near-

Earth objects [NEO]. The National Academy of Sciences [NAS] issued a report this spring concluding that NEOCam was the fastest way to meet the asteroid-hunting mandate. But NASA will not approve the project to begin development.

"The Planetary Defense Program at NASA does not currently have sufficient funding to approve development of a full space-based NEO survey mission as was proposed by the NEOCam project," a NASA spokesperson told Quartz this week.

The agency said it was prioritizing funding for ground-based telescopes looking for asteroids, though the NAS report concluded that they would not fulfill its mandate. The agency is also funding the Double Asteroid Redirection Test mission (DART), which will pilot the technologies needed to do something about any threatening near-Earth objects. Still, the agency said the infrared telescope proposed for NEOCam "could be ready for any future flight mission development effort."

Scientists say that NEOCam is caught in a game of "cosmic chicken" between Congress and NASA. NASA managers are reluctant to prioritize planetary defense over scientific missions, while Congress has yet to appropriate specific funding for an asteroid survey. The NAS report recommended that "missions meeting high-priority planetary defense objectives should not be required to compete against missions meeting high-priority science objectives."

"Although highly unlikely, being taken by surprise by a catastrophic asteroid impact that could have been detected would be an epic failure in the history of science," MIT planetary scientist Richard Binzel told Quartz. "We now have a capability to know what's out there, meaning we have no excuse for an ongoing lack of knowledge."

Binzel says that the onus is now on Congress to increase NASA funding for planetary defense by $40 million annually, which would allow the agency to develop the spacecraft and eventually launch it in the years ahead.

NASA, however, did not ask for that plus-up in the White House's budget proposal to Congress this year. Instead, the agency has asked for $1.6 billion in new funding to support a human return to the moon that could cost as much as $30 billion over the next five years. In contrast, the total cost of the NEOCam mission is estimated to be $500 to $600 million. But NASA funding is a fairly low priority in Washington, as Congress has yet to agree on even a broad outline for 2020 funding.

Lawmakers haven't been enthusiastic about funding the lunar return, but asteroid hunting might be an easier sell. A national poll of Americans taken in May found 68% supported missions to find asteroids that might impact Earth, while just 23% thought a return to the moon would be a good idea.

EVALUATING THE NEWS

1. The NEOCam was designed as an infrared camera. Why would this be the optimal part of the electromagnetic spectrum for observing near-Earth objects?

2. NASA has been tasked with the mission to identify most of the near-Earth objects greater than 140 meters in diameter. Compare this diameter to the diameter of the Chelyabinsk meteoroid (see Working It Out 9.1; note that the radius of the meteoroid is given there). Has NASA been asked to find objects like the Chelyabinsk meteoroid?

3. The energy of a meteoroid impact is proportional to the mass of the meteoroid (see Working It Out 9.1). The smallest near-Earth object that NASA has been tasked to find is roughly 340 times more massive than the Chelyabinsk meteoroid. Suppose the Chelyabinsk meteoroid had been one of these larger objects. How many times more energy would it have delivered on impact?

4. The Chelyabinsk meteoroid property damages have been estimated at $33 million. Estimate the property costs if the meteoroid had been 140 meters in diameter, by multiplying your answer from question 3 by $33 million.

5. The cost of the NEOCam mission is about $600 million. Compare this to the cost of the damage from one near-Earth object, which you have just estimated. Explain why, in 2015, Congress tasked NASA with the mission of finding nearly all of the near-Earth objects larger than 140 meters in diameter.

6. The reporter implies that NASA may be using the NEOCam mission in a game of "cosmic chicken"; what does he mean by that?

Source: **Tim Fernholz**, "NASA won't launch a mission to hunt deadly asteroids," from *Quartz*, July 5, 2019. Republished with permission of Quartz Media, Inc. Permission conveyed through Copyright Clearance Center, Inc.

SUMMARY

Pluto, Eris, Haumea, Makemake, and Ceres are classified as dwarf planets, rather than planets, because although they have enough mass to be spherical, they have not cleared their orbits. The moons of the outer Solar System are composed of rock and ice. A few moons are geologically active, with interiors warmed by gravitational interaction with planets, but most are dead. Asteroids are small Solar System bodies made of rock and metal. Some asteroids cross Earth's orbit and are potentially dangerous. Comets come from small, icy planetesimals that reside in the frigid regions beyond the planets. Comets that venture into the inner Solar System are warmed by the Sun, often producing an atmospheric coma and a long tail. Very large asteroids or comets striking Earth create enormous explosions that can dramatically affect terrestrial life. Meteoroids are small fragments of asteroids and comets. When a meteoroid enters Earth's atmosphere, frictional heat causes the air to glow, producing a phenomenon called a meteor. A meteoroid that survives to reach a planet's surface is called a meteorite.

(1) Small bodies in the Solar System include dwarf planets, moons, asteroids, comets, Kuiper Belt objects, Oort Cloud objects, and meteoroids. These can be distinguished from one another on the basis of their location in the Solar System, their composition, their size and shape, and the properties of their orbits. Dwarf planets are round but have not cleared their orbits of other bodies. Moons are gravitationally bound to objects other than the Sun. Asteroids are rocky bodies primarily found between the orbits of Mars and Jupiter. Comets are icy bodies on elliptical orbits that approach the Sun closely enough to cause some of the ice to vaporize into a tail. Kuiper Belt objects are found outside the orbit of Pluto. Oort Cloud objects occupy a sphere around the entire Solar System. Meteoroids are much smaller rocks or particles in orbit around the Sun.

(2) The moons of the outer planets can be classified as geologically active, possibly active, formerly active, and inactive. Surface features can provide evidence of both current and past geological activity.

(3) The collisions that created massive asteroids also provided sufficient heat to make them molten. This allowed the materials in these asteroids to differentiate. Less massive asteroids did not become molten and thus remain undifferentiated.

④ Meteorites are fragments of asteroids that did not burn up in the atmosphere and thus survived to reach the ground. Different types of meteorites come from different types of asteroids or from different parts of differentiated asteroids. Some meteorites are fragments of larger objects (for example, Mars) that were blasted into space during an impact.

⑤ As a comet orbits the Sun, its appearance changes. Near the Sun, the icy body develops a coma and two tails. Both the size and the orientation of its two tails vary with its location in its orbit.

⑥ Meteorites, asteroids, and comets provide samples of material from times throughout the history of the Solar System.

QUESTIONS AND PROBLEMS

TESTING YOUR UNDERSTANDING

1. T/F: There are no dwarf planets interior to Jupiter's orbit.

2. T/F: Most large moons of the outer Solar System are, or once were, geologically active.

3. T/F: Asteroids are mostly rock and metal; comets are mostly ice.

4. T/F: Comet tails always point directly away from the Sun.

5. T/F: Major impacts on Earth don't happen anymore.

6. Pluto is classified as a dwarf planet because
  a. it is not round.
  b. it orbits the Sun too far away to be considered a planet.
  c. it has company in its orbit.
  d. it is made mostly of ice.

7. Io is an example of a moon that
  a. is definitely active today.
  b. is probably or possibly active today.
  c. was active in the past but not today.
  d. was not active at any time since its formation.

8. Asteroids are
  a. small rock and metal objects orbiting the Sun.
  b. small icy objects orbiting the Sun.
  c. small rock and metal objects found only between Mars and Jupiter.
  d. small icy bodies found only in the outer Solar System.

9. Aside from their periods, short- and long-period comets differ because
  a. short-period comets orbit prograde, whereas long-period comets have either prograde or retrograde orbits.
  b. short-period comets formed with less ice, whereas long-period comets contained more.
  c. short-period comets do not develop ion tails, whereas long-period comets do.
  d. short-period comets come closer to the Sun at closest approach than long-period comets do.

10. Congress tasked NASA with searching for near-Earth objects because
  a. they might impact Earth, as others have in the past.
  b. they are close by and easy to study.
  c. they are moving fast.
  d. they are scientifically interesting.

11. *Rosetta* found that comets are primordial. This means that they formed
  a. with the Solar System.
  b. while dinosaurs roamed Earth.
  c. in collisions of larger bodies.
  d. recently in deep space.

12. What is the source of meteors we see during a meteor shower?
  a. near-Earth asteroids
  b. dust left over from the original disk of dust and gas around the Sun
  c. comet debris
  d. dust left over from the formation of Earth's moon

13. A meteoroid is found _____, a meteor is found _____, and a meteorite is found _____.
  a. in space; in the atmosphere; on the ground
  b. on the ground; in space; between Mars and Jupiter
  c. between Mars and Jupiter; in the atmosphere; on the ground
  d. between Mars and Jupiter; in the atmosphere; elsewhere in the Solar System

14. On average, a bright comet appears about once each decade. Statistically, this means that
  a. one will definitely be observed every tenth year.
  b. one will definitely be observed in each 10-year period.
  c. exactly 10 comets will be observed each century.
  d. about 10 comets will be observed each century.

15. During a meteor shower, all meteors trace back to a single region in the sky known as the radiant. This happens because
  a. the meteors originate from the same point in the atmosphere.
  b. all the meteors are traveling in the same direction relative to Earth.
  c. all the meteors burn up at the same altitude.
  d. all the meteors come from the direction of the Sun.

16. A differentiated asteroid was definitely
  a. formed beyond the orbit of Jupiter.
  b. too small to become round.
  c. made mostly of ice.
  d. once hot enough to be molten.

17. As a comet leaves the inner Solar System, its ion tail always points
  a. back along the orbit.
  b. forward along the orbit.
  c. toward the Sun.
  d. away from the Sun.

18. Meteorites can tell us about (select all that apply)
  a. the early composition of the Solar System.
  b. the composition of asteroids.
  c. other planets.
  d. the Oort Cloud.

19. From the following, select the ways in which Titan resembles early Earth. (Choose all that apply.)
    a. It has a thick atmosphere.
    b. Its atmosphere is mostly nitrogen.
    c. It has liquid water on the surface.
    d. It has terrain similar to Earth's.
    e. It is rich in organic compounds.

20. The smooth, heart-shaped region on Pluto surprised astronomers because
    a. most things in space are round or disk-shaped.
    b. the composition of the surface was unexpected.
    c. they expected to see lots of craters all over Pluto.
    d. it hints at the existence of world-changing aliens.

THINKING ABOUT THE CONCEPTS

21. ★ **WHAT AN ASTRONOMER SEES** In Figure 9.4, the large heart-shaped feature is observed to be smooth, and astronomers take this to be evidence that Pluto must have recently been geologically active. Explain why a smooth surface implies recent geologic activity. 👁

22. Study Figure 9.25a. Compare the surface near the impact site with the surface in the top half of the image. Which of these regions has the younger surface? How do you know? 👁

23. Compare the dust and ion tails shown in Figure 9.22. When the comet is far from the Sun, are the two tails closer together or farther apart than when it is near the Sun? 👁

24. Figures 9.1 and 9.5 compare various Solar System objects to Earth's Moon. Are all moons smaller than planets or dwarf planets? 👁

25. Explain the process that drives volcanism on Jupiter's moon Io.

26. Describe cryovolcanism, and explain its similarities and differences with respect to terrestrial volcanism.

27. Europa and Titan may both be geologically active. What is the evidence for this?

28. Discuss evidence supporting the idea that Europa might have a subsurface ocean of liquid water.

29. Titan contains abundant amounts of methane. Why does this require an explanation? What process destroys methane in this moon's atmosphere?

30. Describe differences between the Kuiper Belt and the Oort Cloud as sources of comets.

31. Kuiper Belt objects (KBOs) are actually comet nuclei. Why don't they display tails?

32. Explain the importance of meteorites, asteroids, and comets in studying the history of the Solar System.

33. If collisions of comet nuclei and asteroids with Earth are rare events, why should we be concerned about the possibility of such a collision?

34. Make a table of the types of meteorites, a distinguishing characteristic of each type, and the origin of each type.

35. What are the differences between a comet and a meteor in terms of their size, distance, and how long they remain visible?

APPLYING THE CONCEPTS

36. The Sikhote-Alin meteoroid, mentioned in Section 9.5, had a diameter of 100 m. Follow Working It Out 9.1 in reverse to find the mass of this rocky meteoroid.
    a. Make a prediction: Reread the description of the effects of this meteoroid and the one that struck Chelyabinsk in 2013. Do you expect from this description that this meteoroid is likely more or less massive than the Chelyabinsk meteoroid?
    b. Calculate the mass of the Sikholte-Alin meteoroid.
    c. Check your work by comparing your answer to the mass of the Chelyabinsk meteoroid given in Working It Out 9.1.

37. NASA has been assigned to find all near-Earth objects larger than 70 meters in radius. Follow Working It Out 9.1 backward to find the mass of the smallest of these objects.
    a. Make a prediction: Compare the radius of 70 meters with the radius of the Chelyabinsk meteoroid from Working It Out 9.1 (about 10 meters). Do you expect the mass of the smallest near-Earth object to be larger or smaller than the mass of the Chelyabinsk meteoroid? Do you expect the difference to be large or small?
    b. Improve your prediction by thinking a little harder: How does the mass depend on the radius? Is it proportional to the radius, to the radius squared, or to the radius cubed?
    c. Calculate the mass of a rocky meteoroid that is 70 meters in radius.
    d. Check your answer by comparing the mass you have calculated to the mass of the Chelyabinsk meteoroid given in Working It Out 9.1.

38. Assume that one of the smallest near-Earth objects that NASA finds (70 meters in radius) enters Earth's atmosphere at the same speed as the Chelyabinsk meteoroid. Follow Working It Out 9.1 backward to determine how much energy this object will release on impact. (Hint: If you have found the mass of the meteoroid in the previous question, you may begin there.)
    a. Make a prediction: Do you expect this energy to be more or less than the amount of energy deposited by the Chelyabinsk meteoroid? Do you expect the difference to be large or small?
    b. Calculate how much energy would be released in an impact by a meteoroid that is 70 meters in radius.
    c. Check your work by comparing your answer to your prediction.

39. The dwarf planet Ceres is located in the asteroid belt. The semimajor axis of its orbit is 2.768 AU. Refer to Working It Out 3.1 to find the period of its orbit.
    a. Make a prediction: Since Ceres is farther from the Sun than Earth is, do you expect the period to be larger or smaller than 1 year?
    b. Calculate the period of the orbit.
    c. Check your work by comparing your result with your prediction.

40. The dwarf planet Makemake has a mass of 4.18×10^{21} kg and a radius of 750 km. What is the density of Makemake?
    a. Make a prediction: On the basis of Makemake's location in the outer Solar System, do you expect it to be a rocky or an icy body? What density range do you then expect for your answer?
    b. Refer to the latter half of Working It Out 6.2 to find the density of Makemake, and determine whether it is likely to be made of rock or of ice.
    c. Check your work: Is the density of Makemake consistent with your expectation based on its location in the Solar System?

exploration Comparative Dwarf Planetology

Much of astronomy consists of gathering information about individual objects and then comparing and contrasting them to find similarities and differences. We used this approach for the planets of the Solar System, in which case it is called *comparative planetology*. The first step is always to gather together the fundamental information about the objects in one place, so that we can compare objects with one another and find patterns (see Tables 6.1 and 8.1).

Make a table of information (similar to the tables in Chapters 6 and 8) about the dwarf planets discussed in this chapter. In some cases, all the information is not yet available for some of these dwarf planets. In such a case, mark the relevant cell in the table with an N/A for "not available." You might wish to add rows to the table that are of interest to you, such as the number of moons, for example, or the year of discovery. Check online to see if any new dwarf planets have been discovered, and if so, include data about those dwarf planets as well.

To help you get started, the partial table in **Figure 9.34** shows how you might lay out the rows and columns. Be sure to include units of measurement for all of the data. Then answer the following questions.

1. Compare the masses of these dwarf planets with the mass of the Moon, listed in Table 6.1. Are these dwarf planets more or less massive than the Moon?

   _____

   _____

2. Compare the densities of these dwarf planets with the densities of the terrestrial planets (Table 6.1) and the giant planets (Table 8.1). From this comparison, what can you determine about the compositions of these dwarf planets?

   _____

   _____

   _____

3. Compare the surface gravities of these dwarf planets with the surface gravity of Earth. About how much less would you weigh on Eris than you do on Earth?

   _____

   _____

4. Compare Pluto's axial tilt to the axial tilts of the terrestrial planets (Table 6.1) and the giant planets (Table 8.1). Which planet will have seasons most similar to Pluto's?

   _____

   _____

   _____

5. Compare the orbital velocities of these dwarf planets. How does the orbital velocity change as the orbital radius grows larger? Use concepts from Chapter 3 (that is, Kepler's laws) to explain this trend.

   _____

   _____

   _____

6. Add a row to your table for the orbital inclinations of these dwarf planets. How does the inclination of the orbit change as the orbital radius grows larger? Use concepts from Chapter 5 to explain this trend.

   _____

   _____

   _____

7. Thinking about Kepler's third law, predict what a graph of the semimajor axis cubed (in astronomical units) on the *y*-axis and the orbital period (in years) on the *x*-axis should look like for these dwarf planets. Make this graph. Did you predict correctly how it would look?

   _____

   _____

Comparison of the Properties of the Dwarf Planets

Property	Pluto	Ceres	Eris
Semimajor axis (AU)			
Orbital period (Earth years)			
Orbital velocity (km/s)			
Mass (Earth masses)			

Figure 9.34 A table of information about the dwarf planets is useful for comparison. This example shows how the rows and columns might be laid out.

Measuring the Stars

The stars in the night sky have different brightnesses for reasons that you will learn about here in Chapter 10. Study the star chart for the current season, found in Appendix 4. About how many more faint stars (magnitude 4) than bright stars (magnitude 1) are there in this chart? Notice the scale bar that shows how the size of the dot relates to the magnitude of the star. Next, count the number of stars of each magnitude and make a table of your results. Does this more careful analysis agree with your estimate? Take this star chart outside on a clear night, and find all the stars with magnitude 1 that are above the horizon. This will orient you to the sky. Now find some stars with magnitudes 2, 3, 4, and so on. If you can find no stars fainter than magnitude 3, then the "limiting magnitude" of your observing site is 3. What is the limiting magnitude of your observing site on this date? What sources of light are making it difficult to see stars fainter than this?

EXPERIMENT SETUP

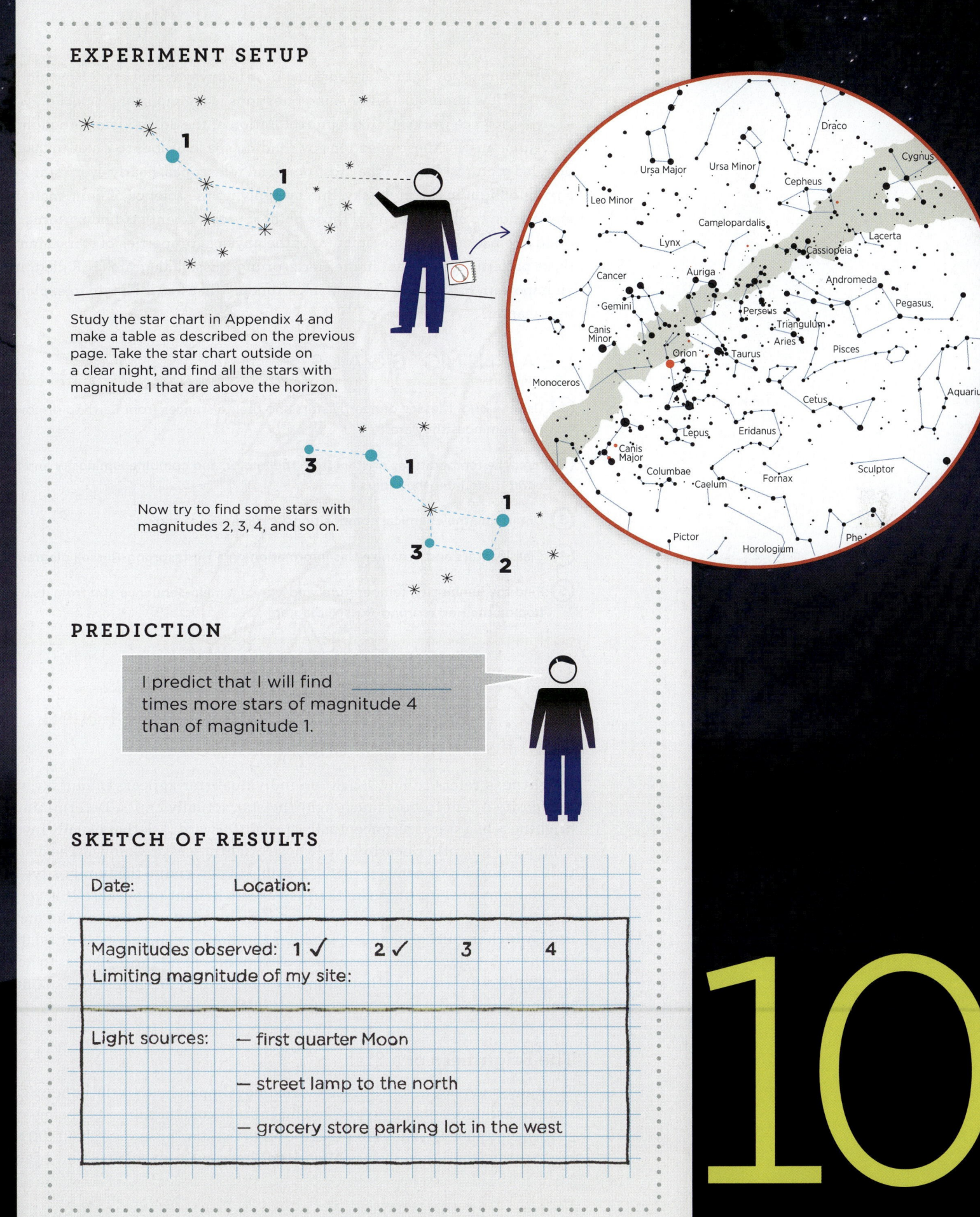

Study the star chart in Appendix 4 and make a table as described on the previous page. Take the star chart outside on a clear night, and find all the stars with magnitude 1 that are above the horizon.

Now try to find some stars with magnitudes 2, 3, 4, and so on.

PREDICTION

I predict that I will find _____ times more stars of magnitude 4 than of magnitude 1.

SKETCH OF RESULTS

Date: _____   Location: _____

Magnitudes observed: 1 ✓   2 ✓   3   4

Limiting magnitude of my site: _____

Light sources: — first quarter Moon

— street lamp to the north

— grocery store parking lot in the west

10

Humans, by nature, are curious. How far away is that star? How big is it? How luminous is it? Asking questions is an important aspect of how science works. Unlike our exploration of the Solar System, though, we cannot answer these questions by sending space probes to a star to take detailed pictures. Even to the most powerful telescopes, nearly every star is just a point of light in the night sky. But, by applying the science of light, matter, and motion to observations of those point sources, scientists find patterns and build a remarkably detailed picture of the physical properties of stars. Many of these patterns are evident in the Hertzsprung-Russell diagram (H-R diagram), an important organizing diagram in astronomy that we will begin to explore in this chapter.

LEARNING GOALS

1. Use the brightnesses of nearby stars and their distances from Earth to discover how luminous they are.

2. Infer the temperatures of stars from their color, and combine luminosity and temperature to infer the radius.

3. Determine the chemical compositions and masses of stars.

4. Classify stars and organize this information on a Hertzsprung-Russell diagram.

5. Find the luminosity, temperature, and size of a main-sequence star from its position on the Hertzsprung-Russell diagram.

10.1 The Luminosity of a Star Can Be Found from Its Brightness and Distance

Brightness refers to how bright an individual star appears in our sky, while *luminosity* refers to how much light the star actually emits. Determining the brightness of a star is a conceptually straightforward task that usually involves comparing it to other nearby stars whose brightnesses are known. The distance to a star must be known before it is possible to determine the luminosity—does the star appear faint because it emits very little light or because it is very far away? Finding the distance is somewhat involved, depending on the method used, which in turn depends on whether the star is relatively near or relatively far. In this section, you will learn how astronomers find the distances to nearby stars and how to combine that information with the star's brightness to find the luminosity.

The Brightness of a Star

Two thousand years ago, the Greek astronomer Hipparchus classified the brightest stars he could see as being "of the first magnitude" and the faintest as being "of the sixth magnitude." We know now that the brightest stars he saw are about 100 times brighter than the dimmest stars. **Magnitude** has come to mean a measure of a star's brightness in the sky, but note that the magnitude of the object *decreases* as its brightness *increases*. It is also an example of logarithmic behavior, so that an object of magnitude 2 is much more than twice as bright as an object of magnitude 4. Each decrease of 2.5 magnitudes corresponds to an increase in brightness by a factor of 10; an object with a magnitude of $M = 1$ is 100 times brighter than an object with a magnitude of $M = 6$. Hipparchus himself must have had typical eyesight, because an average person under dark skies can see stars only as faint as 6th magnitude. Modern telescopes can see much "**deeper**" than this. The Hubble Space Telescope can detect stars as faint as 30th magnitude—4 billion times fainter than what the naked eye can see.

Objects that are brighter than 1st magnitude in this system have smaller magnitudes; that is, magnitudes less than 1 or even negative. For example, Sirius, the brightest star in the sky in visible wavelengths, has a magnitude of -1.46. Venus, at magnitude -4.4, is bright enough to cast shadows. The magnitude of the full Moon is -12.7, and that of the Sun is -26.7. Thus the Sun is 14.0 magnitudes (about 400,000 times) brighter than the full Moon. Again, take note: The magnitude scale is backward, so that larger numbers mean fainter stars.

A star is often observed through an optical **filter** that lets through only a small range of wavelengths. The brightness of the star generally varies depending on the filter used, so astronomers use special symbols to represent magnitudes through these filters. Two of the most common filters are a blue filter (B) and a "visual" or yellow-green filter (V). The term *visual* is used because yellow-green light roughly corresponds to the range of wavelengths to which our eyes are most sensitive.

The magnitude of a star, as we have discussed it, is called the star's **apparent magnitude**, because it is the brightness of the star as it *appears* to us in our sky. Stars are found at different distances from us, so a star's apparent magnitude does not tell us how much light it actually emits. To find out how much light it emits—its luminosity—we must know the distance from us to the star. Then we can put it on a scale with all other stars of known distances, and we can calculate how bright they *would* be if they were all located at the same distance from us. This is called the **absolute magnitude**: the brightness that would be measured for each star if every star were the same distance from Earth. That reference distance is 32.6 light-years, or 10 parsecs, as will be explained shortly.

Astronomers Use Parallax to Measure Distances to Nearby Stars

Hold up your finger in front of you, quite close to your nose. View it with your right eye only and then with your left eye only. Each eye sends a slightly different image to your brain, and so your finger *appears* to move back and forth relative to the background behind it. Now hold up your finger at arm's length, and blink your right eye, then your left eye. Your finger appears to move much less. The way your brain combines the information from your two eyes to determine the distance to an object is called **stereoscopic vision**. **Figure 10.1a** shows an overhead view of the experiment you just performed with your finger. The left eye sees the blue pencil nearly directly between the objects on the bookshelf, but the right eye sees the blue pencil to the left of both objects. The apparent position of the pink pencil also varies. Because the pink pencil is closer to the observer, its apparent

VOCABULARY ALERT

deep In everyday language, *deep* has many meanings, such as referring to depth in the ocean or to a profound idea. Astronomers use *deep* to refer to an object's distance—how deep it is in space. Because distant objects are typically fainter, *deep* and *faint* are often related and used somewhat interchangeably.

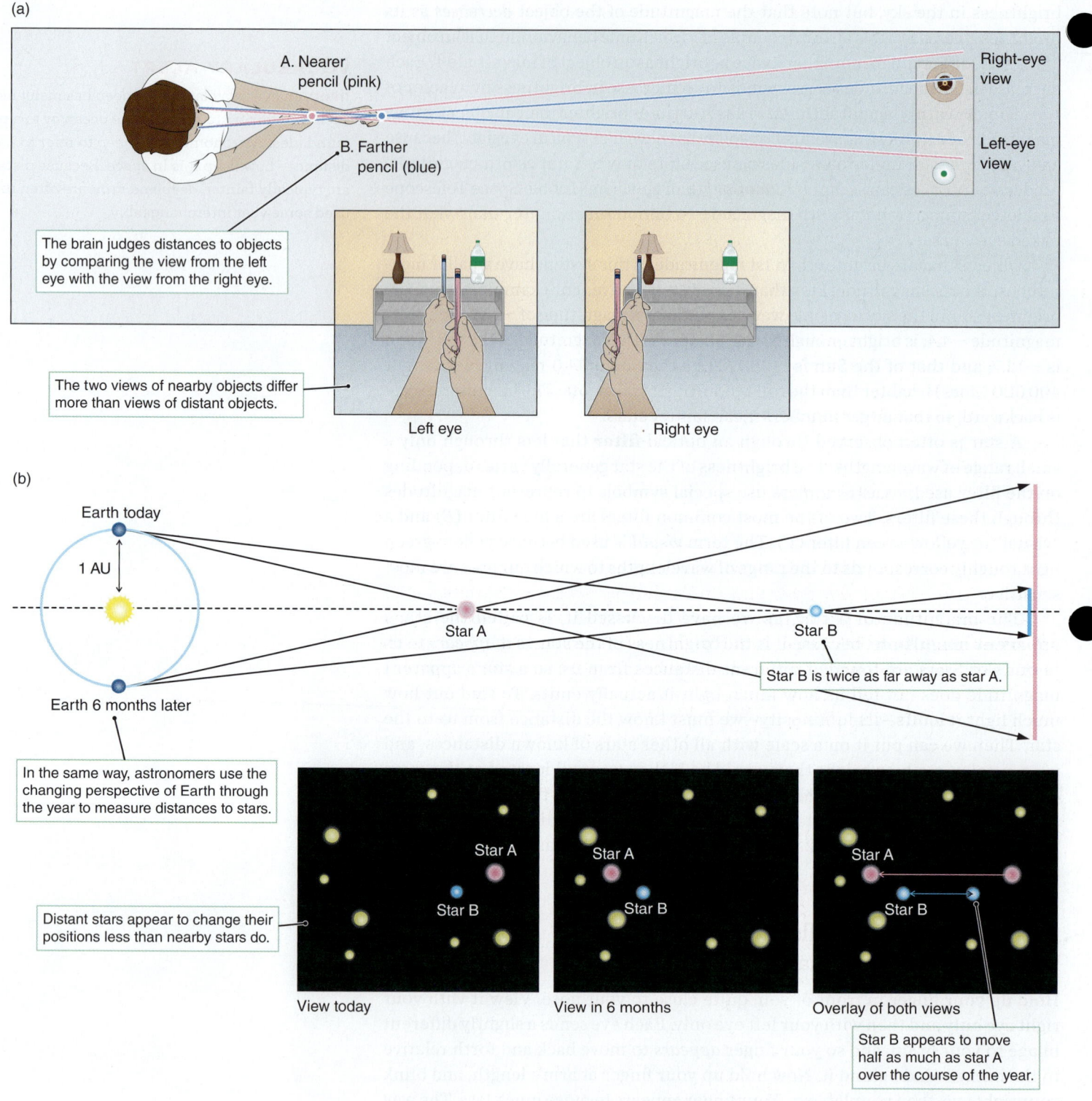

Figure 10.1 (a) Stereoscopic vision allows you to judge the distance to an object by comparing the view from each eye. (b) Similarly, comparing views from different places in Earth's orbit allows us to determine the distance to stars. As Earth moves around the Sun, the apparent positions of nearby stars change more than the apparent positions of more distant stars. (The diagram is not to scale.) This is the starting point for measuring the distances to nearby stars.

position changes more than the apparent position of the blue pencil—it moves from the right of the blue pencil to the left of the blue pencil, so it must have shifted farther.

Stereoscopic vision allows you to judge the distances of objects as far away as 10 meters, but beyond that it is of little use. Each of your two eyes has an identical view of a mountain several kilometers away—all you can determine is that the mountain is too far away for you to judge its distance stereoscopically. The distance over which our stereoscopic vision works is limited by the separation between our two eyes—about 6 centimeters (cm). If you could separate your eyes by several meters, you could judge the distances to objects that were about half a kilometer away.

Although we cannot literally take our eyes out of our heads and hold them apart at arm's length, we can "take a step to the side" to see that the apparent movement of an object is larger when measured across a longer baseline. If you took a photograph from each location, you could compare the photographs to determine which objects were closer and which were farther away. Similarly, if an astronomer takes a picture of the sky tonight and then waits 6 months to take another picture, the distance between the two locations is the diameter of Earth's orbit (2 astronomical units [AU]) (Figure 10.1b). This long baseline yields very powerful stereoscopic vision. This illustration mimics the illustration in Figure 10.1a, viewing the observer from "above." The change in position of Earth over 6 months is like the distance between the right eye and the left eye in Figure 10.1a. The nearby (pink and blue) stars are like the pink and blue pencils, while the distant yellow stars are like the objects on the bookcase. Because of the shift in perspective as Earth orbits the Sun, nearby stars appear to shift their positions. The pink star, which is closer, appears to move farther than the more distant blue star. Over the course of a full year, the nearby star appears to move and then move back again with respect to distant background stars, returning to its original position 1 year later. We can determine the distance to the star using the amount of this shift, the distance from Earth to the Sun, and geometry.

Figure 10.2a shows the same configuration as Figure 10.1b. Look first at star A, the closer star. When Earth is at the top of the figure, it forms a right triangle

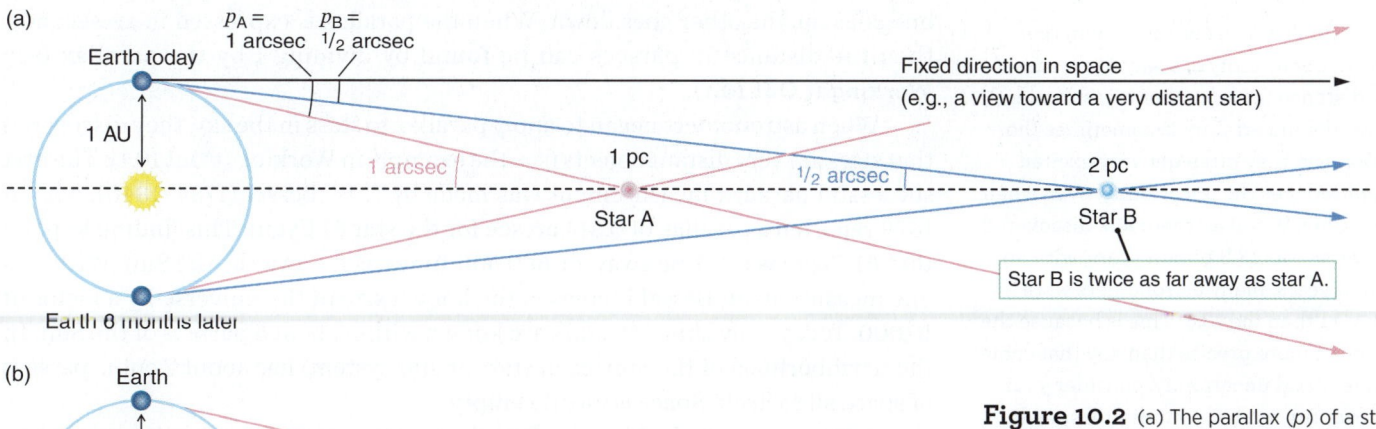

Figure 10.2 (a) The parallax (p) of a star is inversely proportional to its distance. More distant stars have smaller parallax angles. (The diagram is not to scale.) (b) Earth, the Sun, and the target star form a right triangle when Earth is in the best location to measure the parallax angle of the target star.

what if . . .

What if you measure the distances to all the visible stars in the constellation Sagittarius: Would you expect the distances to be similar or very different for these stars, and what would that imply?

VOCABULARY ALERT

uncertainty An uncertain distance does not mean that the distance is unknown. Uncertainty is a way of expressing how precisely the distance (or any other measurement) is known. For scientists, the uncertainty is sometimes the most important number, and they often get very excited when a new experiment reduces the uncertainty, even when it does not change the value. When astronomers discovered that the age of the universe was 13.8 billion years, with an uncertainty of 0.1 billion years, many astronomers were more excited about the 0.1 than the 13.8. This is because the measurement was so much more precise than any that came before it. To better understand uncertainty, consider your speed while driving down the road. If your digital speedometer says 10 kilometers per hour (km/h), you might actually be traveling 10.4 km/h or 9.6 km/h. The precision of your speedometer is limited to the nearest 1 km/h, but that doesn't mean you don't have any idea about your speed. You are certainly not traveling 100 km/h, for example.

with the Sun and star A at the other corners, as shown in Figure 10.2b. (Remember that a right triangle is one with a 90° angle in it.) The short leg of the triangle is the distance from Earth to the Sun, which is 1 AU. The long leg of the triangle is the distance from the Sun to star A. The small angle at the star A corner of the triangle measures the change in the apparent position of the star. This change in position, measured as an angle, is known as the **parallax** of the star. The apparent motion of a star across the sky, described earlier for Figure 10.1b, is equal to twice the parallax.

The parallax of a real star is tiny, much smaller than any of the angles shown in Figure 10.2. Parallax angles are so small that we need special units for them. Just as an hour on the clock is divided into minutes and seconds, a degree that measures an angle can be divided into *arcminutes* and *arcseconds*. An **arcminute (arcmin)** is $\frac{1}{60}$ of a degree, and an **arcsecond (arcsec)** is $\frac{1}{60}$ of an arcminute, and therefore 1/3,600 of a degree. An arcsecond is an important astronomical unit, and it can be difficult to imagine how small it is. Imagine a person standing directly in front of you, uncomfortably close. To look from one of their eyes to the other, you need to turn your own eyes through a (small) angle. An arcsecond is the size of the angle that your eyes would have to move if that person were 10.8 kilometers (km; equivalent to 6.71 miles) away. An arcsecond is really small.

The distances to real stars are large, and in this book we normally use units of light-years to describe them. As we saw in Chapter 4, 1 light-year is the distance that light travels in 1 year—about 9.5 trillion km. We use this unit because it is the unit you are most likely to see in a newspaper article or a popular book about astronomy. When astronomers discuss distances to stars and galaxies, however, they often use the **parsec (pc)**, which is short for *parallax arcsecond*: 1 parsec is the distance to a star that has a parallax of 1 arcsecond. This unit was invented because it makes the mathematics easy when calculating parallax distances. One parsec is equal to 3.26 light-years.

More distant stars make longer and skinnier triangles and have smaller parallax angles. Returning to Figure 10.2, note that star B is twice as far away as star A, so its right triangle is longer and skinnier than the triangle for star A. Because the triangle is longer and skinnier, star B's parallax is half the parallax of star A. Stars that are farther away have smaller parallax angles than stars that are closer to us. In other words, the parallax of a star is inversely proportional to its distance: When one goes up, the other goes down. When the parallax is expressed in arcseconds, then the distance in parsecs can be found by dividing 1 by the parallax (see **Working It Out 10.1**).

When astronomers began to apply parallax to stars in the sky, they discovered that stars are very distant objects (see the example in Working It Out 10.1). The first successful parallax measurement was made by F. W. Bessel (1784–1846), who in 1838 reported a parallax of 0.314 arcsec for the star 61 Cygni. This finding implied that 61 Cygni was 3.2 pc away, or 660,000 times as far away as the Sun. With this one measurement, Bessel increased the known size of the universe by a factor of 10,000. Today, only about 60 stars are known within about 5 parsecs of the Sun. In the neighborhood of the Sun, each star (or star system) has about 7 cubic parsecs of space all to itself. Space is mostly empty.

Astronomers worked hard to find the parallax of known stars for more than 150 years. But most stars were so far away that this motion relative to background stars was too small to measure with ground-based telescopes. Knowledge of our stellar neighborhood took a tremendous step forward during the 1990s, when the Hipparcos satellite measured the positions and parallax angles of 120,000 stars.

working it out 10.1

Parallax and Distance

Recall from earlier chapters that "inversely proportional" means that on one side of the equation a variable is in the numerator, while on the other side a different variable is in the denominator. The relationship between distance (d) and parallax (p) is an inverse proportion:

$$p \propto \frac{1}{d} \quad \text{or} \quad d \propto \frac{1}{p}$$

The parsec has been adopted by astronomers because it makes the relationship between distance and parallax easier than using light-years:

$$\left(\begin{array}{c}\text{Distance measured}\\\text{in parsecs}\end{array}\right) = \frac{1}{\left(\begin{array}{c}\text{Parallax measured}\\\text{in arcseconds}\end{array}\right)}$$

or

$$d = \frac{1}{p}$$

Notice that the proportionality sign has turned into an equals sign: You don't have to remember any constants if the distance is in parsecs and the parallax is in arcseconds.

Suppose that the parallax of a star is measured to be 0.5 arcsec. To find the distance to the star, we substitute that into the parallax equation for p:

$$d = \frac{1}{0.5}$$

$$d = 2 \text{ pc}$$

Suppose that the parallax of a star is measured to be 0.01 arcsec. What is its distance in light-years? First, we find its distance in parsecs:

$$d = \frac{1}{0.01}$$

$$d = 100 \text{ pc}$$

Then, we convert to light-years by remembering that a parsec is 3.26 light-years:

$$d = 100 \text{ pc} \times \frac{3.26 \text{ light-years}}{1 \text{ pc}}$$

$$d = 100 \times 3.26 \text{ light-years}$$

$$d = 326 \text{ light-years}$$

The star closest to us after the Sun is Proxima Centauri. Located at a distance of 4.22 light-years, Proxima Centauri is a faint member of a system of three stars called Alpha Centauri. What is this star's parallax? First, we must convert from light-years back to parsecs:

$$d = 4.22 \text{ light-years} \times \frac{1 \text{ pc}}{3.26 \text{ light-years}}$$

$$d = 1.29 \text{ pc}$$

Then we find the parallax from the distance:

$$d = \frac{1}{p}$$

Solve for p to get:

$$p = \frac{1}{d}$$

Then insert our value for the distance in parsecs:

$$p = \frac{1}{1.29}$$

$$p = 0.77 \text{ arcsec}$$

This star has a parallax of only 0.77 arcsec. Because Proxima Centauri is the closest star to us, every other star we observe will have a smaller parallax than this.

The accuracy of each Hipparcos parallax measurement is about ±0.001 arcsec. Because of this observational **uncertainty**, measurements of the distances to stars are likewise uncertain. For example, a star with a Hipparcos parallax of 0.004 ± 0.001 arcsec really has a parallax between 0.003 arcsec and 0.005 arcsec. This gives a corresponding distance range of 200–330 pc from Earth. Since 2014, the European Space Agency's *Gaia* satellite has been using parallax to measure the positions and distances to many more stars (more than 1 billion astronomical objects all together) to construct the largest and most precise three-dimensional map of the Milky Way Galaxy and its neighborhood ever made. Methods other than parallax—to be discussed later—are used for more distant stars.

Figure 10.3 The brightness of a visible star in our sky depends on both its luminosity—how much light it emits—and its distance from us. When brightness and distance are measured, luminosity can be calculated.

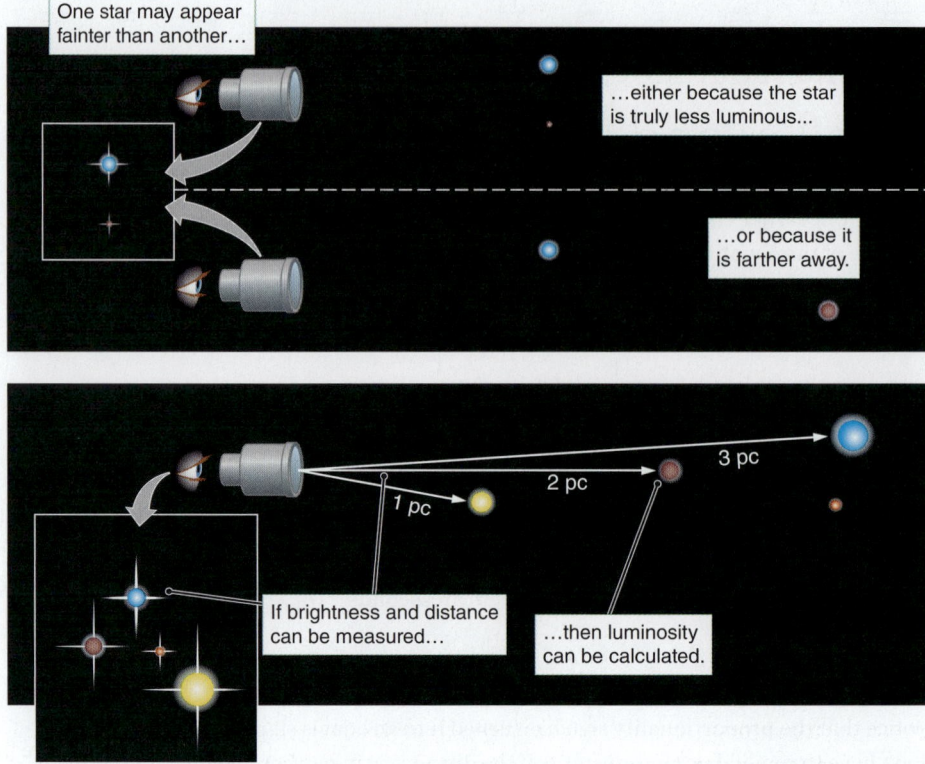

Finding Luminosity

Although the brightness of a star is directly measurable, it does not immediately tell us much about the star itself. As illustrated in **Figure 10.3**, an apparently bright star in the night sky may in fact be intrinsically dim but close by. Conversely, a faint star may actually be very luminous, but because it is very far away, it appears faint to us. We can specifically say that its apparent brightness, which measures the starlight that reaches us, is inversely proportional to the square of our distance from the star. If we know this distance, we can then use our measurement of the star's apparent brightness to find its luminosity. Two everyday concepts—stereoscopic vision and the fact that objects appear brighter when closer—have given us the tools we need to measure the distances and therefore the luminosities of the closest stars. Once the vast distances are known, it is clear that stars are not merely faint points of light in the night sky; they are extraordinarily powerful beacons located at great distances.

The range of possible luminosities for stars is very large. The Sun provides a convenient basis for measuring the properties of stars, including their luminosities. (We compare stars to the Sun so often that properties of the Sun have a special subscript: for example, L_{Sun} is the luminosity of the Sun.) The most luminous stars are more than a million times the luminosity of the Sun. The least luminous stars have luminosities less than $1/10{,}000\, L_{Sun}$. The most luminous stars are therefore more than 10 billion times more luminous than the least luminous stars. Very few stars are near the upper end of this range of luminosities, and the vast majority of stars are far less luminous than our Sun. **Figure 10.4** shows the relative number of stars compared to their luminosity in solar units.

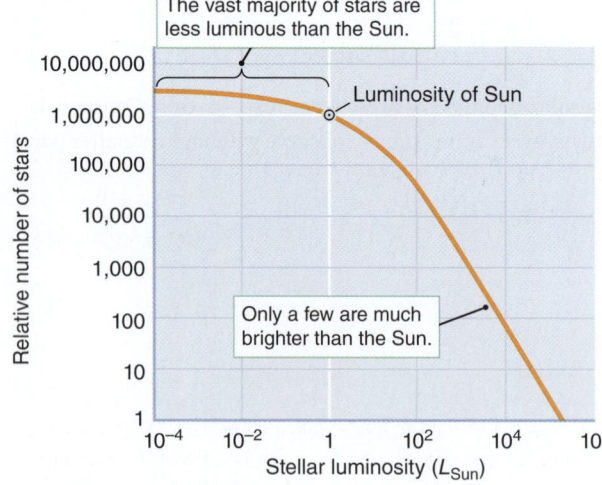

Figure 10.4 The distribution of the luminosities of stars is plotted here logarithmically, so that increments are in powers of 10. For example, for every million stars with a luminosity equal to the Sun, there are only about 100 with a luminosity 10,000 times greater.

CHECK YOUR UNDERSTANDING   **10.1**

It is necessary to know _____ in order to find the luminosity of a star.
  a. both the distance and the brightness
  b. either the distance or the brightness

Answers to Check Your Understanding questions are in the back of the book.

10.2 Radiation Tells Us the Temperature, Size, and Composition of Stars

Studying the details of the light from stars yields an enormous amount of information. In this section, we focus on how astronomers determine temperature, radius, and composition of stars from the colors in their spectra.

Wien's Law Revisited: The Color and Surface Temperature of Stars

Look back at Working It Out 5.1 to refresh your understanding of the Stefan-Boltzmann law (which states that, among same-sized objects, the hotter objects are more luminous) and Wien's law (which states that hotter objects are bluer). In this section, we will use these two laws to measure the temperatures and sizes of stars. We will also develop a more detailed understanding of emission lines mentioned in Chapters 4 and 5, and we will use that to obtain information about the composition of stars.

Wien's law shows that hotter objects emit bluer light. Stars with especially hot **surfaces** are blue; stars with especially cool surfaces are red; and stars like our Sun, which emit almost equal amounts of red, yellow, and blue, are white. (The Sun appears yellow in Earth's sky because the red and blue light are bent and scattered by Earth's atmosphere.) If you measure the wavelength at which a star's spectrum is brightest (or "peaks"), then Wien's law will tell you the temperature of the star's surface. The color of a star tells us only about the temperature at the surface, because this layer is giving off most of the radiation that we see. Stellar interiors are far hotter than stellar surfaces.

In practice, it is usually unnecessary to obtain a complete spectrum of a star to determine its temperature. Instead, astronomers often measure the colors of stars by comparing the brightness at two different, specific wavelengths. From a pair of pictures of a group of stars, each taken through a different filter (*B* and *V* filters, for example, as described in the previous section), astronomers can find an approximate value of the surface temperature of every star in the picture—perhaps hundreds or even thousands—all at once. When we do, we find there are many more cool stars than hot stars. We also discover that most stars are cooler than the Sun.

The Stefan-Boltzmann Law and Finding the Sizes of Stars

Stars are so far away that the vast majority of them cannot be imaged as more than point sources. The size of a star must be inferred from other measurements, such as the temperature and the luminosity. The temperature of a star can be found

▶‖ **AstroTour:** Stellar Spectrum

VOCABULARY ALERT

surface In everyday language, we don't use *surface* to refer to a layer within a gaseous body. But here, astronomers mean the part of the star that gives off most of the radiation that we see. A star's surface is not solid like the surface of a terrestrial planet, and stars usually have more layers outside of this "surface." The Sun's corona, for example, lies above the photosphere, which is the surface that we see.

(a)

(b)

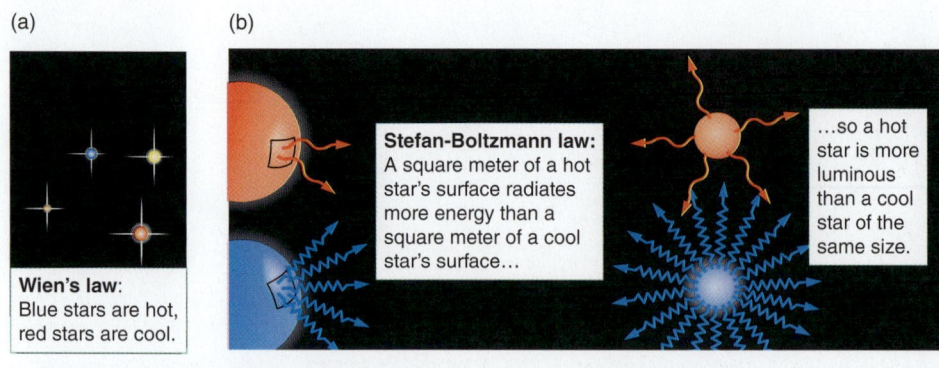

(c)

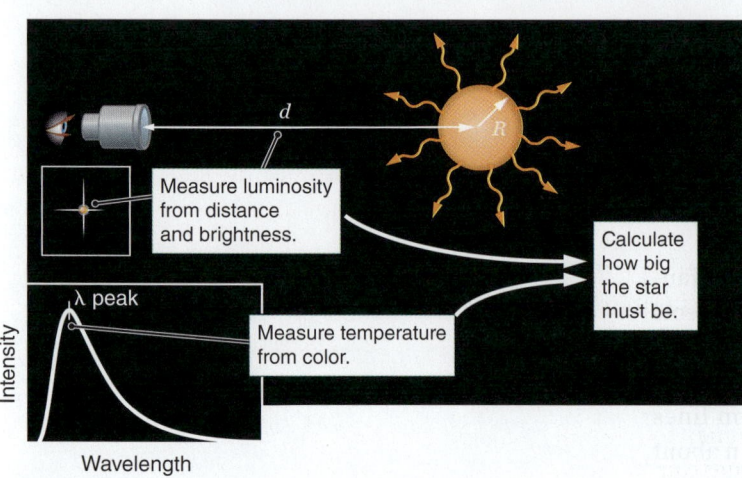

Figure 10.5 (a) The temperature of a star can be found from its color by using Wien's law. (b) The luminosity depends on both the temperature and the size of the star. (c) Once the temperature and the luminosity are known, the size of the star can be calculated from the Stefan-Boltzmann law.

directly, either from Wien's law, as illustrated in **Figure 10.5a**, or from another method. The luminosity of a star can be found from its brightness and its distance, as we discussed in the previous section.

The temperature of a star is one factor that influences its luminosity. The size of the star is also important. If a large star and a small star are the same temperature, they will emit the same energy from every equal-sized patch of surface, but the large star has more patches, so it is more luminous altogether. Conversely, if two stars are the same size, the hotter one will be more luminous than the cooler one. This is an application of the Stefan-Boltzmann law, shown in Figure 10.5b.

Because the luminosity, the temperature, and the radius of the star are all related, combining the luminosity with the temperature allows us to determine the radius of the star, as shown in Figure 10.5c. Recall that we carried out a calculation relating these three quantities in Working It Out 5.1. In that case, Earth was the example, but the laws of physics remain the same for planets and for stars.

The relationship between luminosity, temperature, and radius has been used to estimate the radii of thousands of stars. The radius of the Sun, written as R_{Sun}, is about 700,000 km. The smallest stars, called white dwarfs, have radii that are only about 1 percent of the radius of the Sun ($R = 0.01 R_{Sun}$). The largest stars, called red supergiants, can have radii more than 1,000 times that of the Sun. There are many more small stars than large stars. Most stars are smaller than the Sun.

Atomic Energy Levels

So far, we have concentrated on what we can learn about stars by applying our understanding of blackbody radiation. However, the spectra of stars are not smooth, continuous blackbody spectra. Instead, when we pass the light of stars through a prism, we see dark and bright lines at specific wavelengths in their spectra. These lines have been used to determine much of what we know about the universe. To understand these lines, we must first know how light interacts with matter. You learned a little about light and matter in Chapter 4, but here in Chapter 10, we will explore more about how they interact.

Electrons in an atom can have only particular energies. These energies are known as "energy states" of the atom. We can imagine these energy states as a set of shelves in a bookcase, as shown in **Figure 10.6a**. The energy of an atom might correspond to the energy of one shelf or to the energy of the next shelf, but the energy of the atom will never be found between the two states, just as a book will never be found floating between two shelves. Astronomers keep track of the allowed states of an atom using energy level diagrams, as shown in Figure 10.6b, where each energy level is like a shelf on the bookcase. Both of these metaphors (the bookcase and the energy level diagram) are simplifications of a three-dimensional system: Atoms are three-dimensional objects formed of a very small, dense nucleus surrounded by a cloud of electrons. Changing the energy state of an atom changes the shape and size of the electron cloud that surrounds the nucleus, loosely sketched in Figure 10.6c.

The lowest possible energy state for a system (or part of a system) such as an atom is called the **ground state**. When the atom is in the ground state, the electron

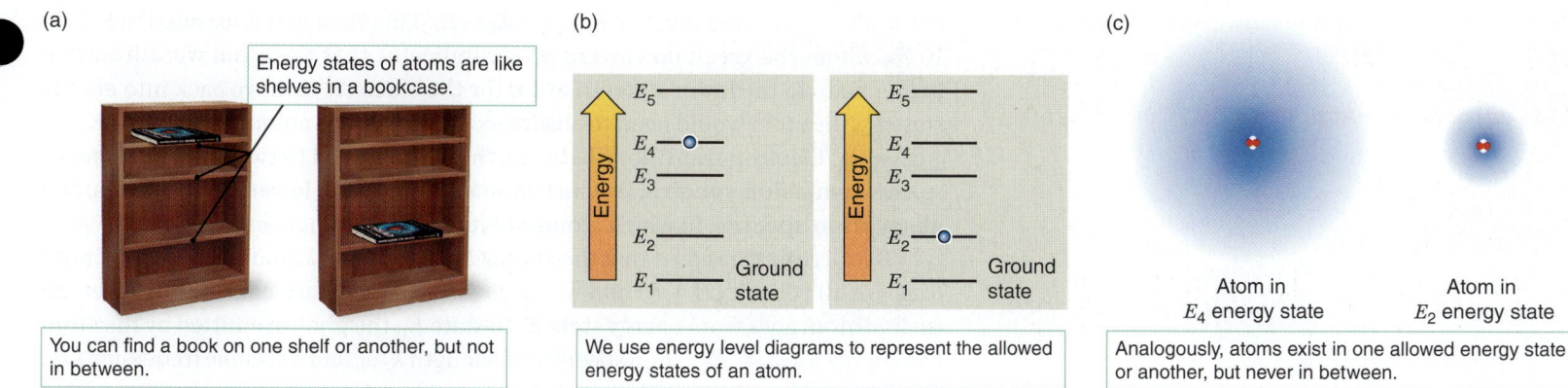

(a)

Energy states of atoms are like shelves in a bookcase.

You can find a book on one shelf or another, but not in between.

(b)

We use energy level diagrams to represent the allowed energy states of an atom.

(c)

Atom in E_4 energy state

Atom in E_2 energy state

Analogously, atoms exist in one allowed energy state or another, but never in between.

Figure 10.6 (a) Energy states of an atom are analogous to shelves in a bookcase. You can move a book from one shelf to another, but books can never be placed between shelves. (b) Energy level diagrams are used to represent the different amounts of energy that an atom can have. (c) Atoms exist in one allowed energy state or another but never in between. There is no level below the ground state.

has its minimum energy. It can't give up any more energy to move to a lower state, because there isn't a lower state. An atom will remain in its ground state forever unless it absorbs energy from outside the system.

Energy levels above the ground state are called **excited states**. An atom in an excited state can transition to the ground state by getting rid of the "extra" energy all at once. It does this when the electron emits a photon. The atom goes from one energy state to another, but it never has an amount of energy in between. Money is similarly *quantized*, as shown in **Figure 10.7a**. If you have a penny, a nickel, and a dime, you have 16 cents. If you give away the nickel, you are left with 11 cents. You never had an intermediate amount of money, such as 12 cents or 13.6 cents. You had 16 cents, and then you had 11 cents. Atoms do not accept and give away money to change energy states, but they do accept and give away photons. Atoms falling from a higher-energy state with energy E_2 to a lower-energy state, E_1, lose an amount of energy exactly equal to the difference in energy levels, $E_2 - E_1$. Therefore, the energy

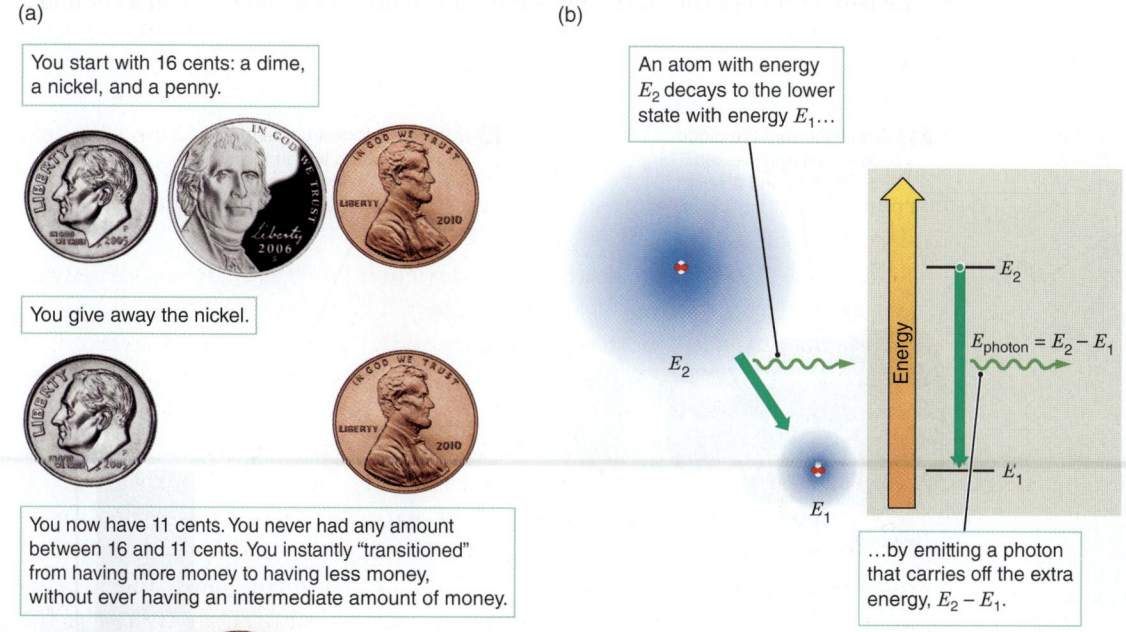

(a)

You start with 16 cents: a dime, a nickel, and a penny.

You give away the nickel.

You now have 11 cents. You never had any amount between 16 and 11 cents. You instantly "transitioned" from having more money to having less money, without ever having an intermediate amount of money.

(b)

An atom with energy E_2 decays to the lower state with energy E_1...

E_2

E_1

$E_{photon} = E_2 - E_1$

E_2

E_1

...by emitting a photon that carries off the extra energy, $E_2 - E_1$.

Figure 10.7 (a) The quantized energy associated with transitions between energy states is analogous to individual coins in a handful. (b) Similarly, an atom can give up photons with only specific energies. A photon with energy $E_2 - E_1$ is emitted when an atom in the higher-energy state decays to the lower-energy state.

▶️‖ **AstroTour:** Atomic Energy Levels and the Bohr Model

🎥 **Astronomy in Action:** Emission and Absorption

▶️‖ **AstroTour:** Light as a Wave, Light as a Photon

of the photon emitted must be $E_{photon} = E_2 - E_1$. This change is illustrated in Figure 10.7b, where the green downward arrow indicates that the atom went from the higher state to the lower state. In order for the electron to move back into energy state E_2, the atom would need to absorb exactly the same amount of energy, $E_{photon} = E_2 - E_1$. Electron transitions between these states lead to two different types of spectra: **emission spectra**, in which atoms are falling to lower-energy states, and **absorption spectra**, in which atoms are jumping to higher-energy states.

Recall from Chapter 4 that the energy E, wavelength λ, and frequency f of photons are all related ($E = hf$ and $\lambda = c/f$, where h and c are constants). When an excited atom goes from energy state E_2 to state E_1, the photon emitted by the atom has energy $E_{photon} = E_2 - E_1$, a specific wavelength $\lambda_{2 \to 1}$, and a specific frequency $f_{2 \to 1}$. Therefore, these emitted photons have a specific color, and every photon emitted in any transition from E_2 to E_1 will have this same color. Because of this connection between energy and wavelength, the energy level structure of an atom determines the wavelengths of the photons it emits, and hence the color of the light that the atom gives off. An atom can emit photons with energies corresponding *only* to the difference between two of its allowed energy states.

An atom sitting in its ground state will remain there forever unless it absorbs just the right amount of energy to kick it up to an excited state. In general, an atom absorbs energy from a photon, or it collides with another atom (or perhaps a free electron) and absorbs some of the other atom's energy. Atoms moving from a lower-energy state E_1 to a higher-energy state E_2 can *absorb* only the amount of energy $E_2 - E_1$, whether it comes in the form of photons or collisions.

Imagine a cloud of hot gas consisting of atoms with only two energy states, E_2 and E_1, as shown in **Figure 10.8**. Because the gas is hot, the atoms are continually zooming around and colliding, thus getting kicked up from the ground state E_1 into the higher-energy state E_2. Any atom in the higher-energy state quickly drops to the lower energy state and emits a photon in a random direction. This emitted light contains only photons with the specific energy $E_{photon} = E_2 - E_1$. In other words, all of the light coming from the cloud is the same color. If passed through a slit and a

Figure 10.8 A hot cloud of gas containing atoms with two energy states, E_1 and E_2 (left), emits photons with energy $E_2 - E_1$. When these photons pass through an astronomical instrument (middle), they appear in the spectrum (right) as a single *emission line*.

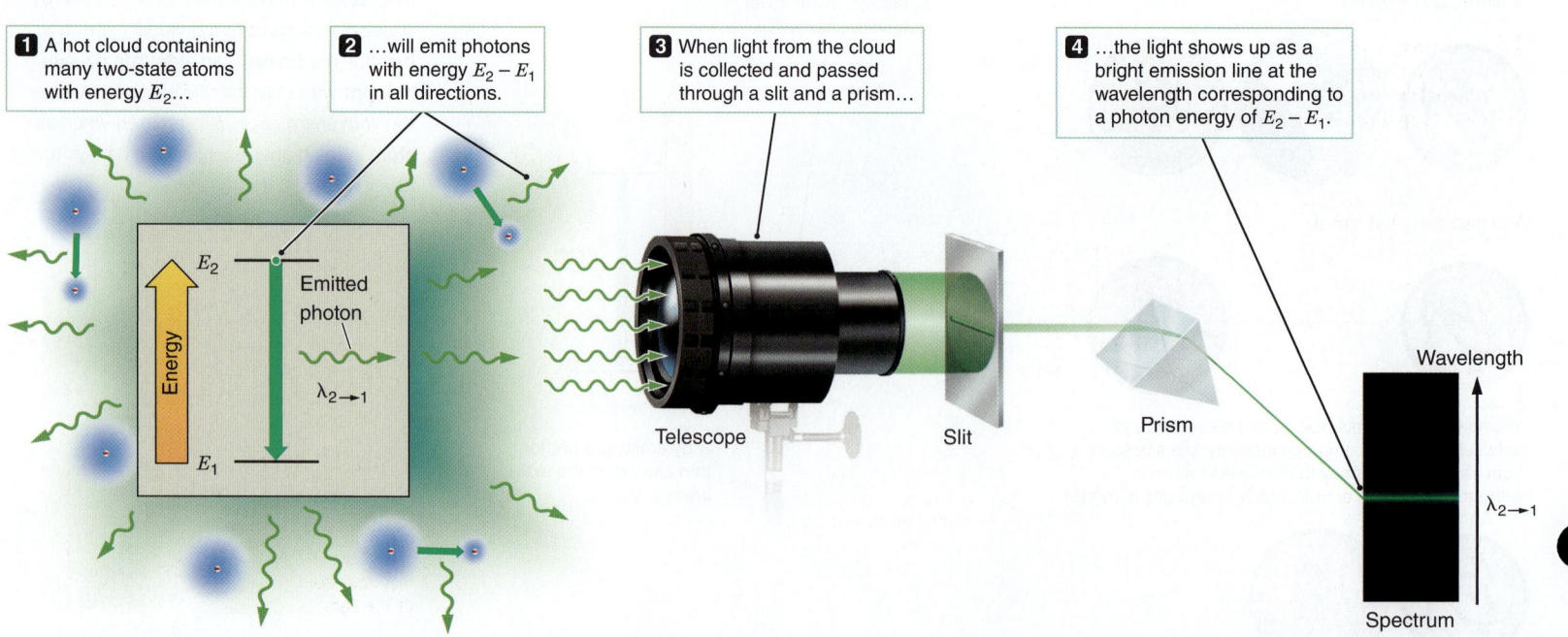

1 A hot cloud containing many two-state atoms with energy E_2...

2 ...will emit photons with energy $E_2 - E_1$ in all directions.

3 When light from the cloud is collected and passed through a slit and a prism...

4 ...the light shows up as a bright emission line at the wavelength corresponding to a photon energy of $E_2 - E_1$.

E_2

Energy

Emitted photon

$\lambda_{2 \to 1}$

E_1

Telescope

Slit

Prism

Wavelength

$\lambda_{2 \to 1}$

Spectrum

prism, it forms a single bright line of one color, called an *emission line.* This is how some neon signs work: Each color in a "neon" sign comes from a different gas (not necessarily neon) trapped inside the glass tubes. A spectrum like the one shown to the right in Figure 10.8 is an emission spectrum, identifiable because the continuum spectrum is missing; it is dominated by a narrow emission line of a particular color. Real objects have atoms with more than two energy states; therefore, emission spectra of real objects have more than one emission line.

When an object shines because it is hot, it emits all the colors of light, because the particles are traveling at a wide range of speeds, colliding from all directions and losing energy in random amounts. This produces a blackbody spectrum. In **Figure 10.9a**, we imagine viewing a hot filament through a spectrometer. Because all the colors are present, the spectrometer creates a complete rainbow (a complete spectrum) on a detector. Notice that the green light shows up in the same place on the detector here as it did in Figure 10.8. All light of the same color has the same wavelength, so a system like this will always place the green light at the same location on the detector.

When the white light passes through a cool cloud of gas, however, some photons will be absorbed. Imagine that, as in the hot cloud of gas shown in Figure 10.8, the atoms in this cool cloud have only two energy states. Almost all of the photons will pass through the cloud of gas unaffected because they do not have the right

Figure 10.9 (a) When passed through a prism, white light produces a spectrum containing all colors. (b) If light of all colors passed through a cloud of atoms with only two possible states, photons with energy $E_2 - E_1$ would be absorbed, leading to the dark absorption line in the spectrum.

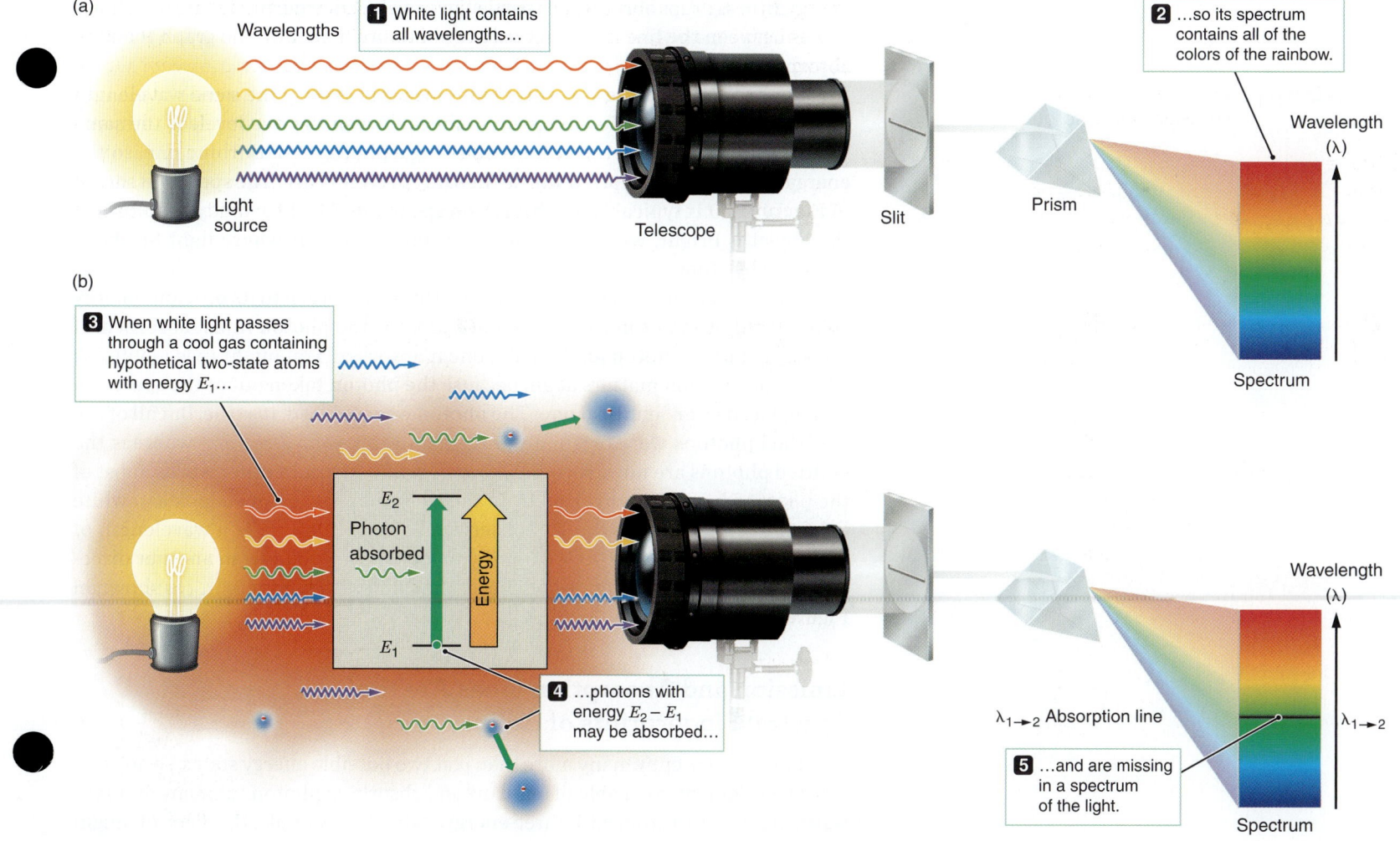

Figure 10.10 Absorption lines in the spectrum of a star may be viewed two ways: (a) The camera attached to the telescope captures an image of the "rainbow" with dark lines where absorption has occurred. (b) Astronomers measure the brightness at each wavelength and make a graph that shows the shape of the absorption line in more detail.

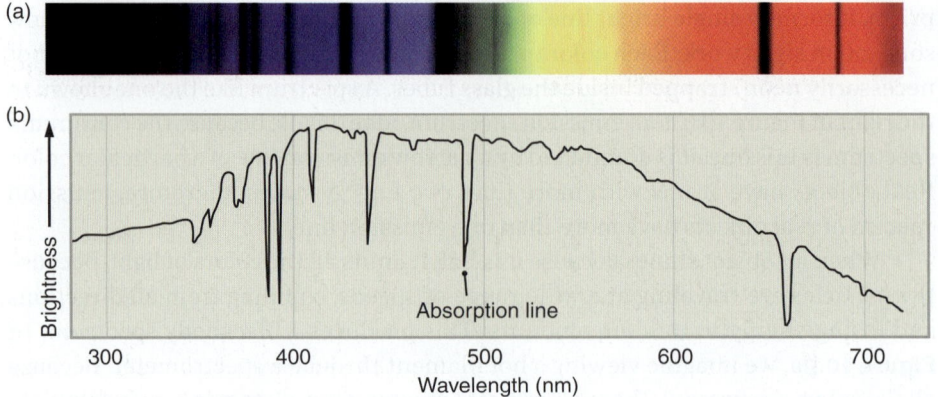

▶❚❚ **AstroTour:** Atomic Energy Levels and Light
Emission and Absorption

amount of energy ($E_2 - E_1$) to be absorbed by atoms of the gas, but photons with *just* the right amount of energy will be absorbed. As a result, they will be *missing* from the spectrum. We will see a sharp, dark line at the wavelength corresponding to this energy, as shown in Figure 10.9b. This process is called *absorption*, and the dark line is called an *absorption line*. **Figure 10.10** shows such absorption lines in the spectrum of a star. The spectrum is shown in two different ways here: as a rainbow with light missing, and then again as a graph of the brightness at every wavelength. Comparing the top and bottom versions of the spectrum, you can see that the brightness drops abruptly at a particular wavelength where there are dark lines. Places between the lines are brighter and therefore higher on the graph than the absorption lines.

For any element, the absorption lines occur at exactly the same wavelength as the emission lines. The energy difference between the two levels is the same whether the electron in the atom is emitting a photon or absorbing one, so the energy of the photon involved will be the same in either case. The spectrum shown in Figure 10.10 is typical of an absorption spectrum: The blackbody spectrum of the object is bright, with dark lines superimposed on it where light has been absorbed by atoms.

When an atom absorbs a photon, it may quickly return to its previous energy state, emitting a photon with the same energy as the photon it just absorbed. If the atom emits a photon just like the one it absorbed, you might reasonably ask why the absorption matters at all, because the photon taken out of the spectrum was replaced by an identical one. The photon was replaced, it's true, but all of the absorbed photons were originally traveling in the *same direction*, whereas the emitted photons are now traveling in *random directions*. In other words, most of the photons have been diverted from their original paths. If you look at a white light *through* the cloud, you will observe an absorption line at a wavelength of $\lambda_{1\rightarrow2}$, as shown in Figure 10.9b. But if you look at the cloud from another direction, you will observe an emission line at the same wavelength, $\lambda_{2\rightarrow1}$, as shown in Figure 10.8.

Emission and Absorption Lines Are the Spectral Fingerprints of Atoms

Real atoms can occupy many more than just two possible energy states, so an atom of a given element is capable of emitting and absorbing photons at many different wavelengths. In an atom with three energy states, for example, the electron might

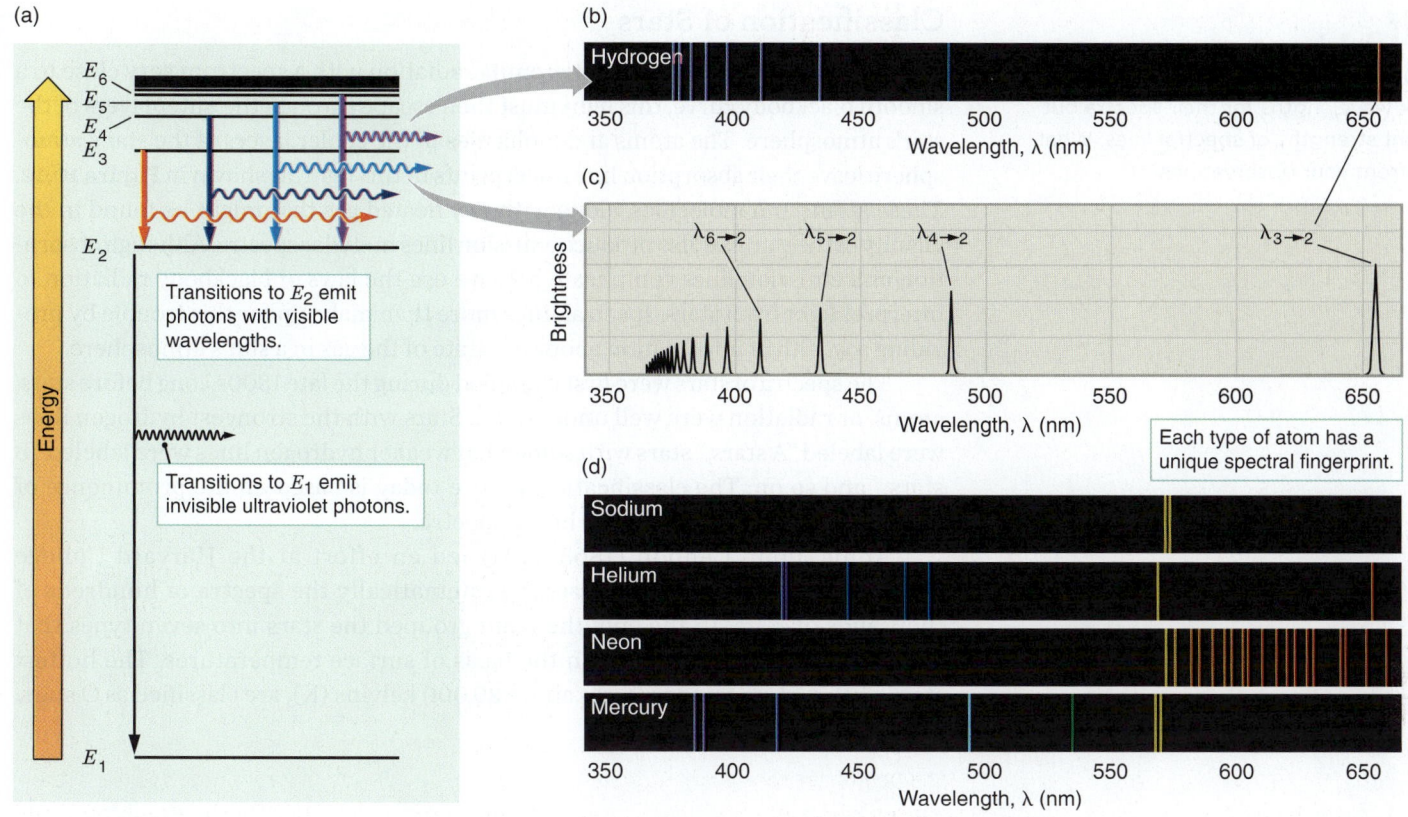

Figure 10.11 (a) Electrons make transitions between the energy states of the hydrogen atom. Transitions from higher levels to level E_2 emit photons in the visible part of the spectrum. (b) The light from a hydrogen lamp produces an emission spectrum. This is the image a camera would produce if you took a picture of the spectrum. (c) The brightness at every wavelength can be measured to produce a graph of the brightness of spectral lines versus their wavelength. (d) Emission spectra from several other types of gases.

jump from state 3 to state 2, or from state 3 to state 1, or from state 2 to state 1. Its spectrum will have three distinct emission lines. Each element's atoms will produce a unique set of lines. Every atom of hydrogen, for example, has the same energy states available to it, so all hydrogen atoms have the same emission and absorption lines. **Figure 10.11a** shows the energy level diagram of hydrogen. Figure 10.11b shows the spectrum of emission lines for hydrogen in the visible part of the spectrum. Figure 10.11c displays this same information as a graph. Each different element has a unique set of available energy states and therefore a unique set of wavelengths at which it can emit or absorb radiation. These unique sets of wavelengths serve as unmistakable spectral "fingerprints" for each element. Figure 10.11d shows the sets of emission lines from several different kinds of atoms.

If the spectral lines of hydrogen, helium, carbon, oxygen, or any other element are present in the light from a distant object, then that element is present in that object. The **strengths** of various absorption lines tell us not only what kinds of atoms are present in the gas but also the abundance of each. However, we must take great care in interpreting spectra to account properly for the temperature and density of the gas in the atmosphere of a star. This type of analysis is how Cecilia Payne-Gaposchkin, in 1925, figured out that stars are made mostly of hydrogen and helium. Typically, more than 90 percent of the atoms in the atmosphere of a star are identified as hydrogen, while helium accounts for most of what remains. All other elements are present only in very small amounts.

The spectral lines also tell us about other physical properties of stars, such as pressure and magnetic-field strength. In addition, by making use of the Doppler shift (see Chapter 5), we can measure rotation rates, motions within the atmosphere, expansion and contraction of the star, "winds" driven away from stars, and other dynamic properties of stars.

VOCABULARY ALERT

strength In this context, *strength* means how bright the emission line is or how faint the absorption line is. More atoms interact with more photons, so they either add more light (in the case of emission) or remove more light (in the case of absorption), making a stronger line. A strong emission line may rise high above the rest of the spectrum, add light to a wide region of the spectrum, or both.

what if . . .

What if you observe two stars that have very similar peak wavelengths for their spectra but very different strengths of spectral lines: What can you conclude from your observations?

Classification of Stars

Although the hot "surface" of a star emits radiation with a spectrum very close to a smooth blackbody curve, this light must then escape through the outer layers of the star's atmosphere. The atoms and molecules in the cooler layers of the star's atmosphere leave their absorption line fingerprints in this light, as shown in **Figure 10.12**. These atoms and molecules, along with any heated gas that might be found in the vicinity of the star, can also produce emission lines in stellar spectra. Although absorption and emission lines complicate how we use the laws of blackbody radiation to interpret light from stars, spectral lines more than make up for this trouble by providing a wealth of information about the state of the gas in a star's atmosphere.

The spectra of stars were first classified during the late 1800s, long before stars, atoms, or radiation were well understood. Stars with the strongest hydrogen lines were labeled "A stars," stars with somewhat weaker hydrogen lines were labeled "B stars," and so on. The classification we use today is based on the prominence of particular absorption lines seen in the spectra.

Annie Jump Cannon (1863–1941) led an effort at the Harvard College Observatory to examine and classify systematically the spectra of hundreds of thousands of stars. In the end, the team grouped the stars into seven types that were subsequently reordered on the basis of surface temperatures. The hottest stars, with surface temperatures above 30,000 kelvins (K), are classified as O stars.

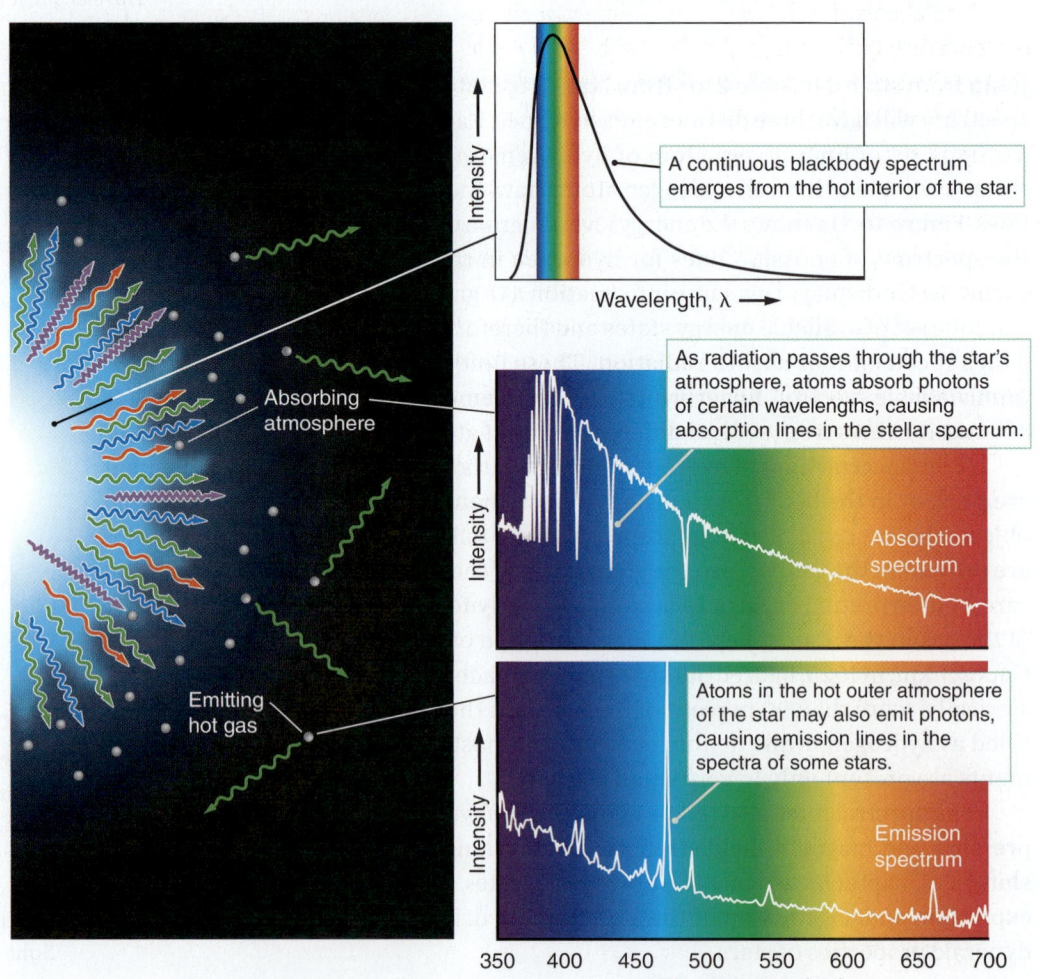

Figure 10.12 Absorption and emission lines both appear in the spectra of stars. The blackbody spectrum is the light emitted from a hot object, just because it is hot (recall Chapter 5). As that light passes through a gas, some of it is absorbed, producing an absorption spectrum. Hot gas also emits light and produces emission lines in the spectra of some stars.

A continuous blackbody spectrum emerges from the hot interior of the star.

As radiation passes through the star's atmosphere, atoms absorb photons of certain wavelengths, causing absorption lines in the stellar spectrum.

Atoms in the hot outer atmosphere of the star may also emit photons, causing emission lines in the spectra of some stars.

Absorbing atmosphere

Emitting hot gas

Absorption spectrum

Emission spectrum

Intensity

Wavelength, λ

350 400 450 500 550 600 650 700
Wavelength, λ (nm)

O stars have weak absorption lines, even from hydrogen and helium. The coolest stars—M stars—have temperatures as low as 2800 K. M stars show myriad lines from many different types of atoms and molecules. The complete sequence of **spectral types** of stars, from hottest to coolest, is O, B, A, F, G, K, M. Spectra of stars of different types are shown in **Figure 10.13**. The boundaries between spectral types are imprecise. A hotter-than-average G star is very similar to a cooler-than-average F star. Brown dwarfs, described as "failed stars" in Chapter 5, are even cooler than M stars and have classifications L and T.

Astronomers divide the main spectral types into subclasses by adding numbers to the letter designations, so that each star is classified by a letter followed by a number. For example, the hottest B stars are called B0 stars, slightly cooler B stars are called B1 stars, and so on. The coolest B stars are B9 stars, which are only slightly hotter than A0 stars. The Sun is a G2 star. Figure 10.13 shows that the brightest part of the spectrum shifts from the left (blue) end to the right (red) end, as the temperature decreases. For example, compare a B5 star with a surface temperature of 15,400 K with a K5 star at 4350 K. This is a visual statement of Wien's Law: that hotter objects are bluer.

The absorption lines in stellar spectra change with temperature as well. The temperature of the gas in the atmosphere of a star affects the state of the atoms in that gas, which in turn affects the energy level transitions available to absorb radiation. In O stars, the temperature is so high that most atoms have had one or more electrons stripped from them by energetic collisions within the gas. Few transitions are available in the visible part of the electromagnetic spectrum, so the visible spectrum of an O star is relatively featureless. At lower temperatures, there are more atoms that can absorb light in the visible part of the spectrum, so the visible spectra of cooler stars are more complex than the spectra of O stars, as shown in Figure 10.13.

Most absorption lines have a temperature at which they are strongest. For example, absorption lines from hydrogen are most prominent at temperatures of about 10,000 K, which is the surface temperature of an A star. (Recall that spectral-type A stars were classified as "A" because they are the stars with the strongest lines of hydrogen in their spectra.) At the very lowest stellar temperatures, atoms in the atmosphere of the star form molecules. Molecules such as titanium oxide (TiO) are responsible for much of the absorption in the atmospheres of cool M stars.

Because different spectral lines are formed at different temperatures, these absorption lines can be used to measure a star's temperature directly. The temperatures of stars measured in this way agree extremely well with the temperatures of stars measured by use of Wien's law, again confirming that the physical laws that apply on Earth apply to stars also.

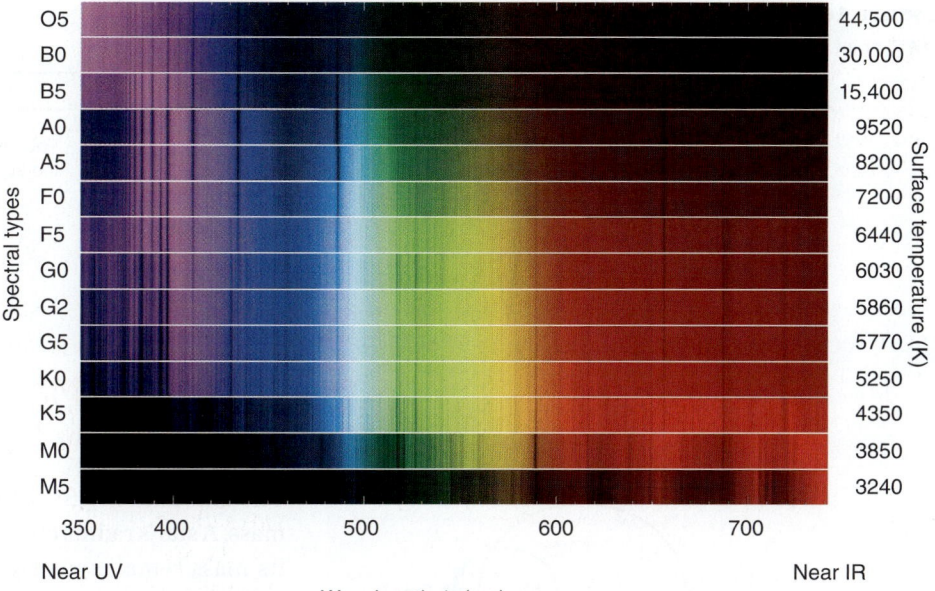

Figure 10.13 Spectra of stars with different spectral types are shown, ranging from hot blue O stars to cool red M stars. Hotter stars are more luminous at shorter wavelengths. The dark lines are absorption lines. Our Sun is a G2 star.

CHECK YOUR UNDERSTANDING 10.2

If a star has very weak hydrogen lines and is blue, what does that most likely mean?
a. The star is too hot for hydrogen lines to form.
b. The star has very little hydrogen.
c. The star is too cold for hydrogen lines to form.
d. The star is moving too fast to measure the lines.

Figure 10.14 The center of mass of two objects is the "balance" point on a line joining the centers of the two masses.

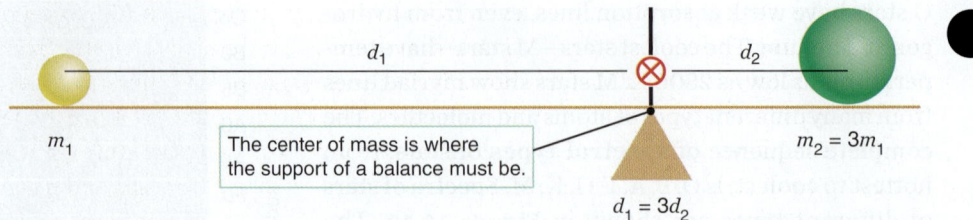

The center of mass is where the support of a balance must be.

m_1

$m_2 = 3m_1$

$d_1 = 3d_2$

10.3 The Mass of a Star Can Be Determined in Some Binary Systems

The amount of light from a star and the star's size are not good measures of its mass. A star's radius changes throughout its lifetime, as does its temperature, while its mass remains nearly constant. The mass, therefore, does not determine the radius or the temperature of the star. However, the mass does determine how the star interacts with objects around it. If we replaced the Sun with a more massive star, that star would exert a larger gravitational force on Earth, no matter how large its radius or how low its temperature. When astronomers are trying to determine the masses of astronomical objects, they almost always wind up looking for the effects of gravity.

In Chapter 3, you learned that Kepler's laws of planetary motion are the result of gravity, and you learned that the orbit of a planet can be used to measure the mass of the Sun. Similarly, Newton's version of Kepler's laws can be used to find the mass of a star when two stars orbit each other. About half of the higher-mass stars in the sky are actually systems consisting of several stars orbiting under the influence of their mutual gravity. Most of these are **binary stars** in which two stars orbit each other. Most stars are single, however, so their masses cannot be found this way.

Binary Stars Orbit a Common Center of Mass

The **center of mass** is the balance point of a system. If two objects were sitting on a seesaw in a gravitational field, the support of the seesaw would have to be directly under the center of mass for the objects to balance, as shown in **Figure 10.14**. In a binary system, the two stars orbit the center of mass, a point in space that is seldom located inside either star but usually somewhere in between.

When Newton applied his laws of motion to the problem of orbits, he found that two objects must move in elliptical orbits around each other, and that their common center of mass lies at one focus shared by both of the ellipses, as shown in **Figure 10.15**. The center of mass, which lies along the line between the two objects, remains stationary. The two objects will always be found on exactly opposite sides of the center of mass.

Because the orbit of the less massive star is larger than the orbit of the more massive star, the less massive star has farther to go than the more massive star. But it must cover that distance in the same amount of time, so the less massive star must be moving *faster* than the more massive star. The velocity of a star in a binary system is inversely proportional to its mass.

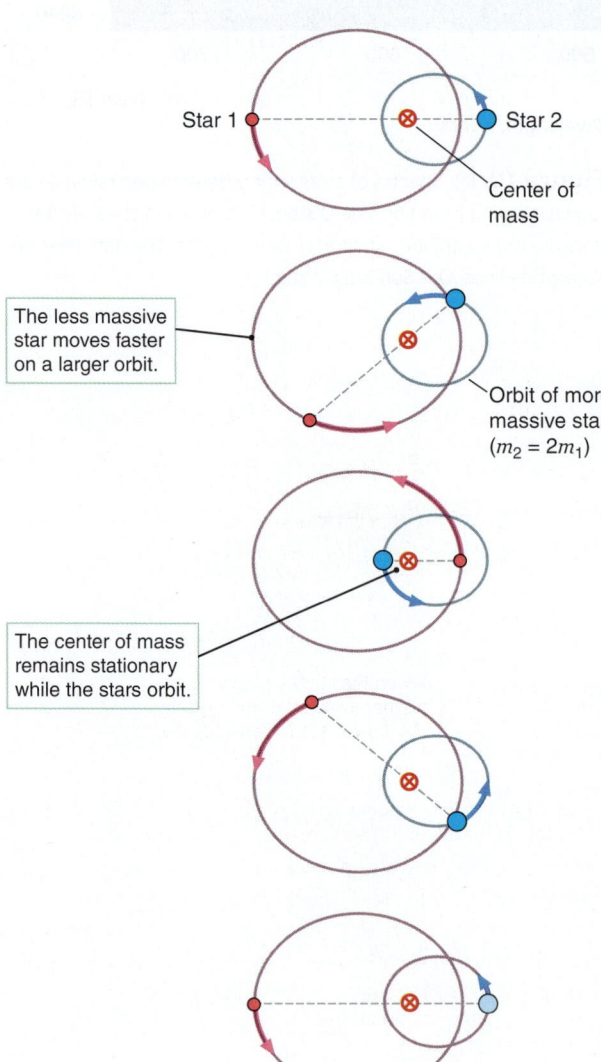

Star 1   Star 2

Center of mass

The less massive star moves faster on a larger orbit.

Orbit of more massive star ($m_2 = 2m_1$)

The center of mass remains stationary while the stars orbit.

Figure 10.15 In a binary star system, the two stars orbit on elliptical paths about their common center of mass. In this case, star 2 has twice the mass of star 1. The eccentricity of the orbits is 0.5. There are equal time steps between the frames.

Imagine that you are watching a binary star as shown in **Figure 10.16a**. As seen from above, two stars orbit the common center of mass. The less massive star (star 1) must complete its orbit in the same time that the more massive star does. Because the less massive star has farther to go around the center of mass, it must move more quickly. In this view from above, no determination of the Doppler shift (recall Chapter 5) can be made because all the motion is in the plane of the sky, and none is toward or away from the observer.

When the system is edge-on to the observer, however, the observer can take advantage of the Doppler shift to find out about the motion. Observations of the spectrum of the combined system (Figure 10.16b) show that the spectral lines of the stars shift back and forth as the stars move toward and away from the observer. When star 2 approaches, star 1 recedes. The light coming from star 2 will be shifted

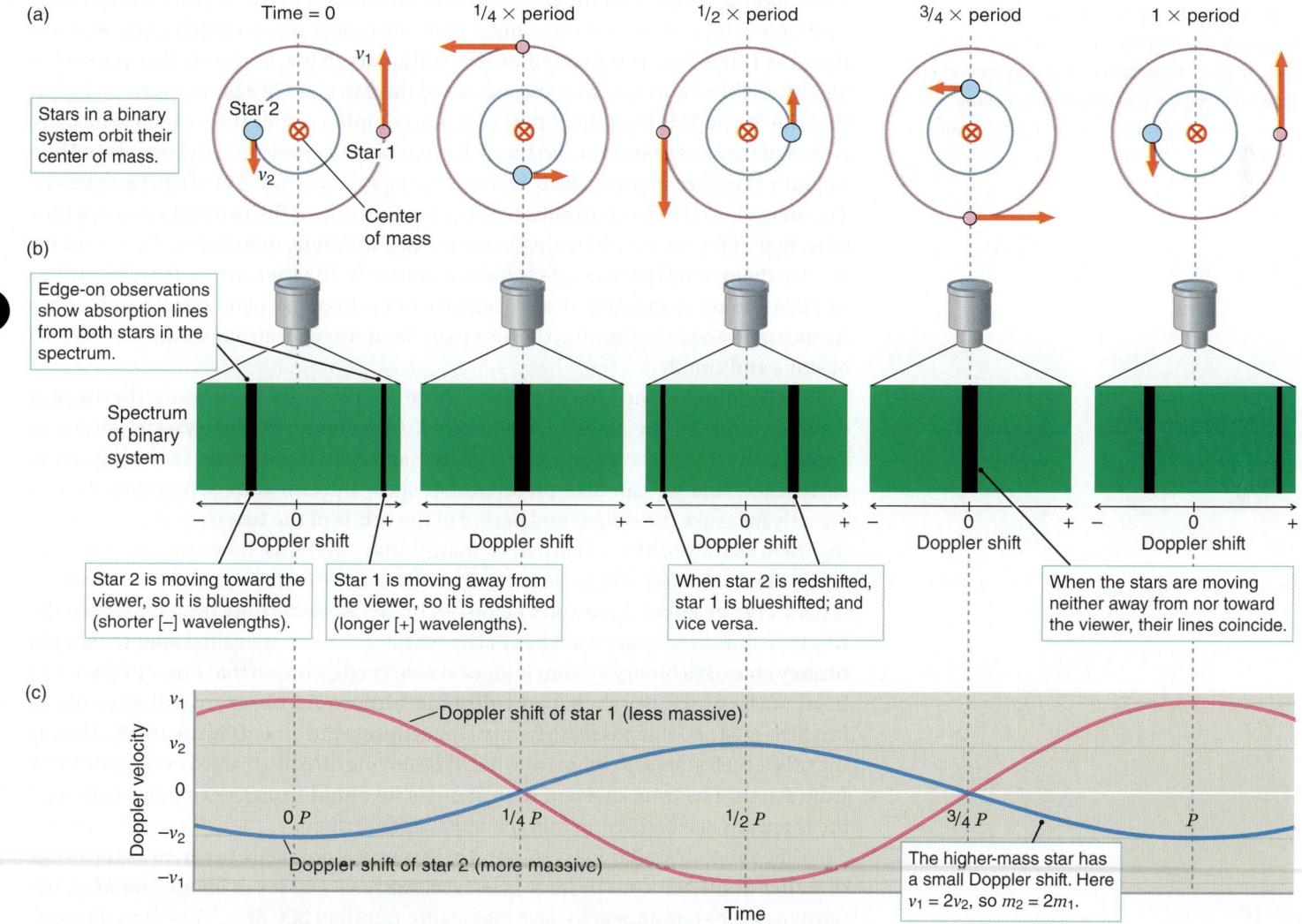

Figure 10.16 (a) The view from "above" the binary system shows that both stars orbit a common center of mass. (b) The spectrum of the combined system (seen edge-on) shows that the spectral lines of each star shift back and forth. (c) Graphing the Doppler shifts of stars 1 and 2 versus time reveals that star 2 has half the maximum Doppler shift, so star 2 is twice as massive as star 1. P is the period of the orbit.

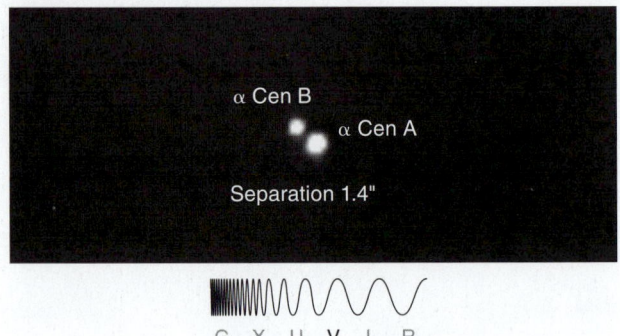

Figure 10.17 ★ **WHAT AN ASTRONOMER SEES**
The two stars of this visual binary are resolved. These stars are two components of Alpha Centauri, the nearest star system to the Sun. When looking at images of stars, astronomers know that the brighter stars appear larger on the image. So in this image, α Cen A is significantly brighter than α Cen B. Knowing that these two stars are part of a system of stars, an astronomer will further conclude that they are the same distance away. She will then know that α Cen A is not only brighter but also more luminous than α Cen B.

what if . . .

What if three stars were orbiting about their common center of mass, all in the same plane: Could you still use measured velocities to determine the orbital radii, and would Kepler's third law give you the mass of the system?

to shorter wavelengths by the Doppler effect as it approaches, so the light will be blueshifted, and the light coming from star 1 will be shifted to longer wavelengths as it recedes, so the light will be redshifted. Halfway around the orbit, the situation will be reversed: lines from star 2 will be redshifted, and lines from star 1 will be blueshifted.

The less massive star has a larger orbit—and consequently moves more quickly—than the more massive star. Comparing the maximum Doppler shift for star 2 with the maximum Doppler shift for star 1 (Figure 10.16c) gives the *ratio* of the masses of the two stars. That is, we can find in Figure 10.16 that star 2 is 2 times as massive as star 1, but we can't determine the actual mass of either star from these observations alone.

Kepler's Third Law and Total Mass of a Binary System

In Chapter 3, we ignored the complexity of the motion of two objects around their common center of mass; we assumed that one object was so much more massive than the other that it remained nearly stationary. Now, however, this very complexity enables us to measure the masses of the two stars in a binary system. Recall that the "period" is the time it takes for one complete orbit. If we can measure the period of the binary system and the average separation between the two stars, then Kepler's third law gives us the total mass in the system: the sum of the two masses. The analysis in the previous subsection gives the ratio of the two masses, so we now have two different relationships between two different unknowns. This is all we need to determine the mass of each star separately. In other words, if we know that star 2 is 2 times as massive as star 1, and we know that star 1 and star 2 together are 3 times as massive as the Sun, then we can calculate separate values for the masses of star 1 and star 2.

Depending on the type of system, there are two ways to measure the average distance between the stars and the period. In a **visual binary** system, shown in **Figure 10.17**, the system is close enough to Earth, and the stars are far enough from each other, that we can take pictures that show the two stars separately. We can directly measure the shapes and period of the orbits of the two stars, just by watching them as they orbit each other. In many binary systems, however, the two stars are so close together and so far away from us that we cannot actually see the stars separately. We know these stars belong to binary systems only because we see the Doppler shift in the spectral lines of the two stars; these are called **spectroscopic binary** stars. If a binary system is viewed nearly edge-on, so that one star passes in front of the other, it is called an **eclipsing binary**. An observer will see a dip in brightness as one star passes in front of (or eclipses) the other (**Figure 10.18**). During the orbit, each star has a moment when it is moving directly toward or directly away from Earth. The total speed of each star can be found from the Doppler shift, and these speeds can be used to find the mass, as described.

The range of stellar masses found in this way is not nearly as great as the range of stellar luminosities. The least massive stars have masses of about 0.08 M_{Sun}; the most massive stars appear to have masses greater than 200 M_{Sun}. Why does the mass of a star have any limits? These limits are determined solely by the physical processes that go on deep in the interior of the star. You will learn in the chapters ahead that a minimum stellar mass is necessary to ignite the nuclear furnace that keeps a star shining, but the furnace can run out of control if the stellar mass is too great. Thus, although the most luminous stars are 10^{10} (10 billion) times more luminous

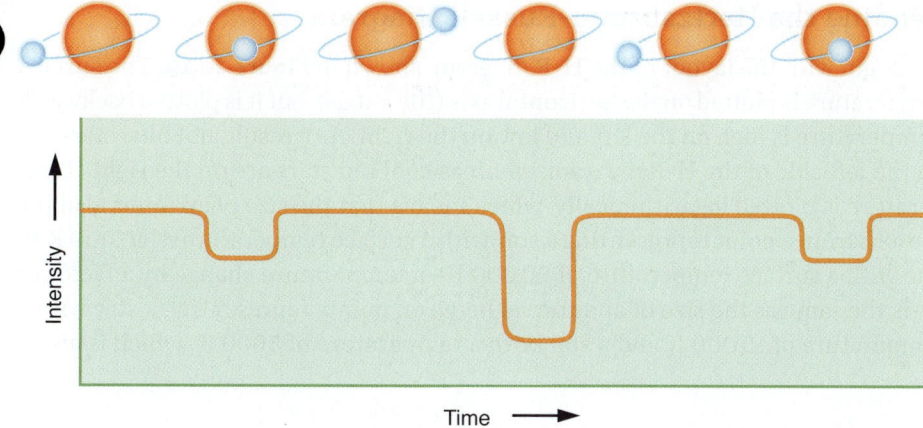

Figure 10.18 In an eclipsing binary system, the system is viewed nearly edge-on, so that the stars repeatedly pass behind one another, blocking some of the light. Even though the blue star here is smaller, it is significantly more luminous because its temperature is higher. When the blue star passes in front of the larger, cooler star, less light is blocked than when it passes behind the red star. The shape of the dips in the light curve of an eclipsing binary can reveal information about the relative size and surface brightness of the two stars.

than the least luminous ones, the most massive stars are only about 10^3 (a thousand) times more massive than the least massive stars.

CHECK YOUR UNDERSTANDING **10.3**

To find the masses of both stars in a binary system, you must find the _____ of each star, the _____ of the orbit, and the average distance between the stars.

a. temperature; period

b. speed; magnitude

c. speed; period

d. temperature; size

10.4 The Hertzsprung-Russell Diagram Is the Key to Understanding Stars

We have come a long way in our effort to measure the physical properties of stars. However, just knowing some of the basic properties of stars does not mean that we understand stars. The next step involves looking for patterns in the properties we have determined. The first astronomers to take this step were Ejnar Hertzsprung (1873–1967) and Henry Norris Russell (1877–1957). In the early part of the 20th century, Hertzsprung and Russell studied the properties of stars independently; each plotted the luminosities of stars versus their surface temperatures. The resulting plot is referred to as the Hertzsprung-Russell diagram, or simply the **H-R diagram**. Astronomers use the H-R diagram to track the evolution of stars, from birth to death. In this section, we take a first look at this important diagram and the way stars are organized within it.

▶❚❚ **AstroTour:** H-R Diagram

what if . . .

What if Hertzsprung and Russell had made their diagram a plot of stellar radius versus temperature: How would the main sequence appear in such a diagram?

▶▶ **Interactive Simulation:** H-R Diagram

Reading the Hertzsprung-Russell Diagram

We begin with the layout of the H-R diagram, shown in **Figure 10.19**. The surface temperature is plotted on the horizontal axis (the x-axis), but it is plotted backward: Temperature is high on the left and low on the right. As a result, hot blue stars are on the left side of the H-R diagram, whereas cool red stars are on the right. Temperature is plotted logarithmically, which means that the size of an interval along the axis from a point representing a star with a surface temperature of 40,000 K to one with a surface temperature of 20,000 K—a temperature change by a factor of 2—is the same as the size of an interval between points representing a star with a temperature of 10,000 K and a star with a temperature of 5000 K, which is also a

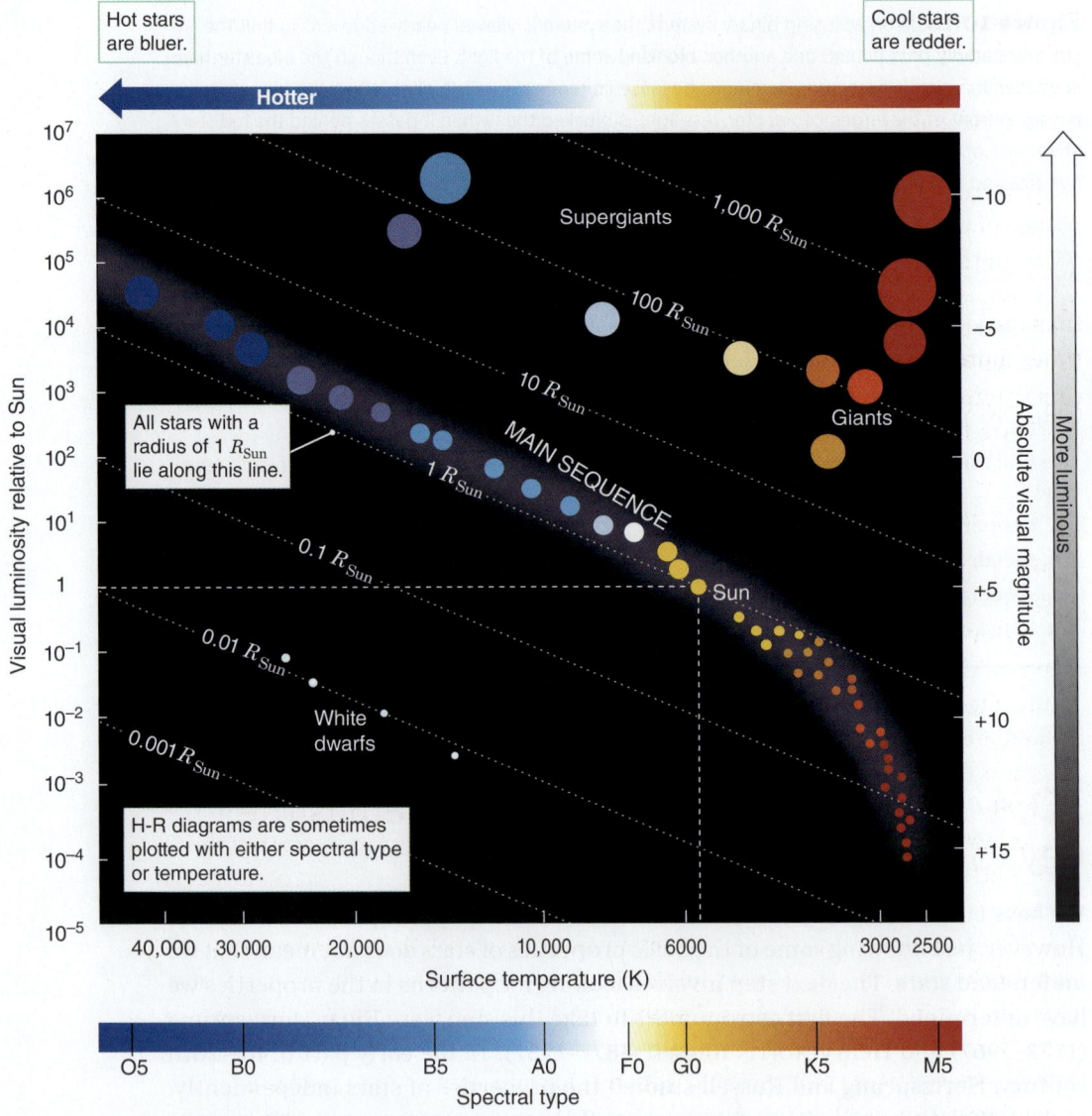

Figure 10.19 The Hertzsprung-Russell (H-R) diagram is used to plot the properties of stars. More luminous stars are at the top of the diagram. Hotter stars are on the left. Stars of the same radius (R) lie along the dotted lines moving from upper left to lower right.

temperature change by a factor of 2. The temperature axis is sometimes labeled with another characteristic that corresponds to temperature, such as the color or the spectral type, as shown at the bottom of Figure 10.19.

Along the vertical axis (the *y*-axis), we plot the luminosity of stars—the total amount of energy that a star radiates each second. More luminous stars are toward the top of the diagram, and less luminous stars are toward the bottom. Like temperature, luminosities are plotted logarithmically, in this case with each step along the left *y*-axis corresponding to a multiplicative factor of 10. To understand why the plotting is done this way, recall that the most luminous stars are 10 billion times more luminous than the least luminous stars, yet all of these stars must fit on the same plot. Sometimes the luminosity axis is labeled with the absolute visual magnitude instead of luminosity, as shown on the right *y*-axis.

Recall from Section 10.2 that the temperature, the luminosity, and the radius are all related. Because each point on the H-R diagram is specified by a surface temperature and a luminosity, we can find the radius of a star at that point as well. A star in the upper right corner of the H-R diagram is very cool, so each square meter of its surface radiates a small amount of energy. This star is also extremely luminous, however, so it must be huge to account for the feeble amount of radiation coming from each square meter of its surface. Conversely, a star in the lower left corner of the H-R diagram is very hot, which means that a large amount of energy is coming from each square meter of its surface. This star has a very low luminosity, however, so it must be very small. Starting in the lower left corner of the H-R diagram and then moving up and to the right takes you to larger and larger stars. Moving down and to the left takes you to smaller and smaller stars. All stars with the same radius lie along slanted lines across the H-R diagram, as shown by dotted lines in Figure 10.19.

The Main Sequence

Figure 10.20 shows more than 4 million nearby stars plotted on an H-R diagram. The data are based on observations obtained by the *Gaia* satellite. There are so many stars here that plotting them all as individual dots leads to **confusion**. The colors in the middle of the diagram show the density of stars in that area; yellow means higher density, and red means lower density. A quick look at this diagram immediately reveals a remarkable fact, one that was first discovered in the original diagrams of Hertzsprung and Russell. About 90 percent of the stars in the sky lie in a well-defined region running across the H-R diagram from lower right to upper left, known as the **main sequence**. On the left end of the main sequence are the O stars: hotter, larger, and more luminous than the Sun. On the right end of the main sequence are the M stars: cooler, smaller, and fainter than the Sun.

If you know that a star lies on the main sequence and has a particular temperature (both of which can be determined from the star's spectrum), then you also know its luminosity and radius. Combining the luminosity with the brightness yields the distance to the main-sequence star. This method of determining distances to main-sequence stars from their spectra, luminosity, and brightness is called **spectroscopic parallax**. Despite the similarity between the names, this method is very different from the *parallax* method that uses geometry, discussed earlier in this chapter. Spectroscopic parallax is useful at much larger distances than the geometric method of parallax, although it is less precise.

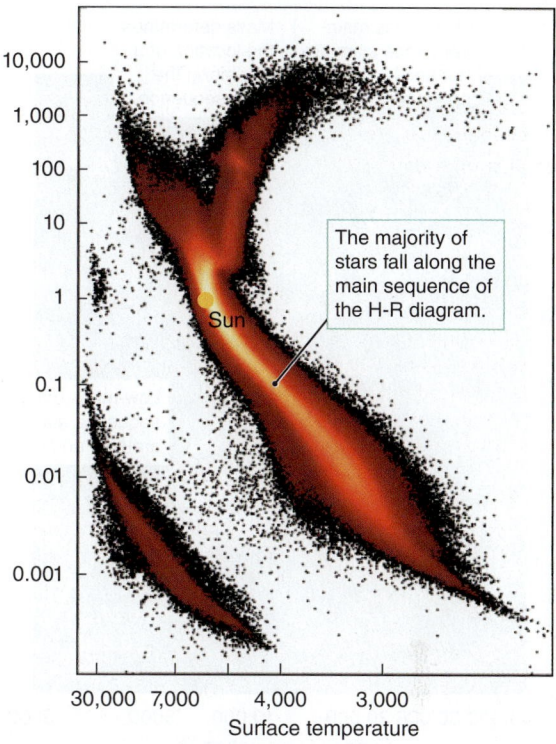

The majority of stars fall along the main sequence of the H-R diagram.

Sun

30,000 7,000   4,000   3,000
Surface temperature

Figure 10.20 An H-R diagram for more than 4 million stars plotted from data obtained by the *Gaia* satellite clearly shows the main sequence. Most of the stars lie along this band running from the upper left of the diagram toward the lower right.

VOCABULARY ALERT

confusion In common language, confusion means the state of being bewildered, or unclear. In science, the word "confusion" is closely related to this everyday term. "Confusion" means that the data points or the stars or the leaves on the tree overlap so that you can't tell them apart. Your colleagues will then be "confused" about which object you might be talking about.

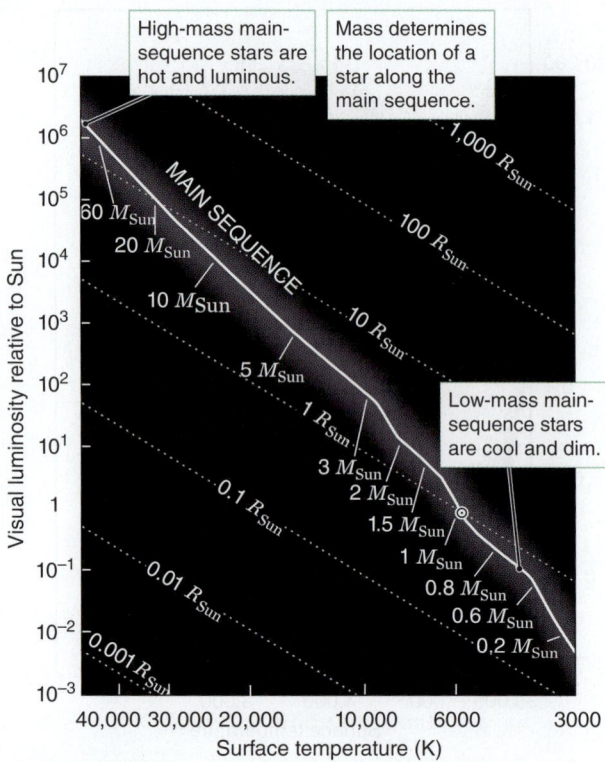

Figure 10.21 The main sequence of the H-R diagram is a sequence of masses.

From a combination of observations of binary star masses, parallax, luminosity measurements, and mathematical models of the interiors of stars, astronomers have determined that stars of different masses lie on different parts of the main sequence. If a main-sequence star is less massive than the Sun, it will be smaller, cooler, redder, and less luminous than the Sun; it will be located to the lower right of the Sun on the main sequence. Conversely, if a main-sequence star is more massive than the Sun, it will be larger, hotter, bluer, and more luminous than the Sun; it will be located to the upper left of the Sun on the main sequence. The mass of a main-sequence star determines where on the main sequence the star will lie, as shown in **Figure 10.21**.

For main-sequence stars with similar chemical compositions, the mass alone determines all of the star's other characteristics. The mass of a main-sequence star determines the star's radius, its surface temperature, luminosity, internal structure, lifetime, evolutionary path, and final fate. Therefore, a star's mass is its most important characteristic.

Stars Not on the Main Sequence

Although 90 percent of stars are main-sequence stars, some stars are found in the upper right portion of the H-R diagram, well above the main sequence. These stars must be large, cool giants with radii hundreds or thousands of times the radius of the Sun. At the other extreme are stars found in the lower left corner of the H-R diagram. These stars are tiny, comparable to the size of Earth. Their small surface areas explain why they have such low luminosities despite having such high temperatures.

Stars that lie off the main sequence on the H-R diagram can be identified by their luminosities (determined by their distances) or by slight differences in their spectral lines. The density and temperature of gas in a star's atmosphere affect the width of a star's spectral lines. In general, hotter stars have broader lines. Puffed-up red giant stars above the main sequence have lower densities and lower temperatures compared to main-sequence stars. When using the H-R diagram to estimate the distance to a star by the spectroscopic parallax method, astronomers must know whether the star is on, above, or below the main sequence in order to find the star's luminosity. Stars both on and off the main sequence have a property called **luminosity class**, which tells us the *size* of the star. Supergiant stars, which are the largest stars that we see, are luminosity class I, bright giants are class II, giants are class III, subgiants are class IV, main-sequence stars are class V, and white dwarfs are class WD. The Sun is a G2V star. Luminosity classes I through IV lie above the main sequence, whereas class WD falls below and to the left of the main sequence. The labels in Figure 10.19 show the approximate locations of some of these classes of stars on the H-R diagram. Thus, the complete spectral classification of a star includes both its spectral type, which tells us temperature and color, and its luminosity class, which indicates size.

The existence of the main sequence, together with the fact that the mass of a main-sequence star determines where on the main sequence it will lie, is a grand pattern that points to the possibility of a deep understanding of what stars are and how they shine. The existence of stars that do *not* follow this pattern raises yet more questions. In the coming chapters, you will learn that the main sequence tells us what stars are and how they work, and that stars off the main sequence tell us how stars form, how they evolve, and how they die.

CHECK YOUR UNDERSTANDING 10.4

Suppose you are studying a star with a luminosity of 100 L_{Sun} and a surface temperature of 4000 K. According to the H-R diagram, this star is a

a. main-sequence star.
b. giant red star.
c. white dwarf.
d. giant blue star.

reading Astronomy News

Mystery of nearby SS Cygni star system finally resolved

by John P. Millis, for redOrbit.com – Your Universe Online

Astronomers are still finding out about stars. Sometimes we are lucky enough to get new data that help settle a long-standing issue.

In 1990, the Hubble Space Telescope measured the distance to a nearby star system known as SS Cygni, composed of a low-mass main-sequence star and a compact object known as a white dwarf—a stellar remnant about the mass of our Sun, but compressed to the size of Earth.

The distance measured by Hubble puzzled scientists, as the measured brightness of the system was considerably higher than expected. If correct, it would call into question the mechanisms by which a white dwarf interacts with a nearby companion.

"If SS Cygni was actually as far away as Hubble measured, then it was far too bright to be what we thought it was, and we would have had to rethink the physics of how systems like this worked," noted James Miller-Jones from the Curtin University campus of the International Centre for Radio Astronomy Research (ICRAR).

Miller-Jones and other astronomers have used two of the most powerful radio telescope networks in the world—the VLBA [Very Long Baseline Array] and EVLA

[Expanded Very Large Array]—to measure the distance to SS Cygni, attempting to resolve the dilemma created by the Hubble result. The team used a method known as parallax, whereby the system is observed at various points during Earth's orbit around the Sun, and then the position of the system is measured against the fixed, distant background.

"If you hold your finger out at arm's length and move your head from side to side, you should see your finger appear to wobble against the background. If you move your finger closer to your head, you'll see it starts to wobble more. We did the exact same thing with SS Cygni—we measured how far it moved against some very distant galaxies as Earth moved around the Sun," noted Miller-Jones.

"The wobble we were detecting is the equivalent of trying to see someone stand up in New York from as far as away as Sydney."

The team found that SS Cygni is about 372 light-years from Earth, considerably closer than previous measurements made using the Hubble Space Telescope.

"The pull of gas off a nearby star onto the white dwarf in SS Cygni is the same process that happens when neutron stars and black holes are orbiting with a nearby companion,

so a lot of effort has gone in to understanding how this works," explained Miller-Jones.

"Our new distance measurement has solved the puzzle of SS Cygni's brightness, it fits our theories after all."

EVALUATING THE NEWS

1. In the second paragraph, the article states that "the distance measured" was a puzzle, because "the measured brightness of the system was considerably higher than expected." Did the author really mean "brightness" or "luminosity" here? How can you tell?

2. Evaluate the author's description of parallax. Is his explanation understandable to an average reader and also correct?

3. Given the distance of 372 light-years, calculate the measured parallax. (This is the opposite calculation of the one the astronomers did.) Is the analogy of the person in New York as viewed from Sydney approximately correct?

4. Sketch an H-R diagram. Label the locations of the two stars of SS Cygni.

5. How do the studies described in this article reflect what you learned about the scientific method in Chapter 1?

SUMMARY

Finding the distances to stars is a difficult but important task for astronomers. Parallax and spectroscopic parallax are two of the methods that astronomers use to determine distances to stars. Combining the brightness with the distance yields the luminosity. Combining the luminosity and the temperature yields the radius. Careful study of the light from a star, including its spectral lines, gives the temperature, size, and composition. Study of binary systems gives the mass of stars of various spectral types, which we can extend to all stars of the same spectral type. The H-R diagram shows the relationship among the various physical properties of stars. The major determining factor in all the properties of a star is its mass.

(1) The luminosity of a star is found by combining stellar distance from Earth with the brightness of the star in the sky. The luminosity is the total energy emitted from the star each second.

(2) The temperature of a star is determined by its color, with blue stars being hotter and red stars being cooler. Combining this temperature information with the luminosity of the star gives the radius.

(3) Spectral lines carry a great deal of information about a star: temperature, composition, and mass, the motions of individual stars, and indirectly the stellar radius.

(4) The H-R diagram is a key to understanding stellar properties. Temperature increases to the left, so that hotter stars lie on the left side of the diagram, while cooler stars lie on the right. Luminosity increases vertically, so that the most lumious stars lie near the top of the diagram. Ninety percent of stars lie along the main sequence.

(5) The mass of a main-sequence star is the fundamental determining factor of all of the star's other characteristics: luminosity, temperature, and size. The main sequence on the H-R diagram is a sequence of masses.

QUESTIONS AND PROBLEMS

TESTING YOUR UNDERSTANDING

1. **T/F:** Red stars have cooler surfaces than blue stars.

2. **T/F:** The brightness of a star in the sky tells you its luminosity.

3. **T/F:** An atom can emit or absorb a photon of any wavelength.

4. **T/F:** If star A has a parallax angle larger than that of star B, then star A is farther away than star B.

5. **T/F:** The mass of an isolated star must be inferred from the star's position on the H-R diagram.

For problems 6–8, star A is blue and star B is red. They have equal luminosities, but star A is twice as far away as star B.

6. Which star appears brighter in the night sky?
   a. star A
   b. star B

7. Which star is hotter?
   a. star A
   b. star B

8. Which star is larger?
   a. star A
   b. star B

9. If a star has very strong hydrogen absorption lines, which of the following is true?
   a. The temperature is right for hydrogen to make lots of transitions.
   b. Hydrogen is abundant in the star because hydrogen is absorbing at these wavelengths.
   c. Hydrogen is depleted in the star because hydrogen is not emitting at these wavelengths.

10. Two stars have equal luminosities, but star A has a much larger radius than star B. What can you say about these stars?
   a. Star A is hotter than star B.
   b. Star A is cooler than star B.
   c. Star A is farther away than star B.
   d. Star A is brighter in our sky than star B.

11. Most stars are
   a. cool with a low luminosity.
   b. hot with a low luminosity.
   c. cool with a high luminosity.
   d. hot with a high luminosity.

12. Suppose an atom has energy levels at 1, 3, 4, and 4.3 (in arbitrary units). Which of the following is *not* a possible energy for an emitted photon?
   a. 2
   b. 1
   c. 1.3
   d. 2.3

13. An eclipsing binary system has a primary eclipse (star A is eclipsed by star B) that is deeper (more light is removed from the light curve) than the secondary eclipse (star B is eclipsed by star A). What does this tell you about stars A and B?
   a. Star A is hotter than star B.
   b. Star B is hotter than star A.
   c. Star B is larger than star A.
   d. Star B is moving faster than star A.

For problems 14–16, suppose you are studying a main-sequence star of 10 solar masses. Consult the H-R diagram in Figure 10.21 to answer these questions. ⊙

14. This star's luminosity is roughly
   a. 0.01 L_{Sun}.
   b. 1 L_{Sun}.
   c. 100 L_{Sun}.
   d. 10,000 L_{Sun}.

15. This star's temperature is roughly
   a. 1000 K.
   b. 10,000 K.
   c. 20,000 K.
   d. 100,000 K.

16. This star's radius is roughly
   a. 0.1 R_{Sun}.
   b. 1 R_{Sun}.
   c. 10 R_{Sun}.
   d. 100 R_{Sun}.

17. If a star is found *directly* above the Sun on the H-R diagram, what can you conclude about its size?
   a. It is larger than the Sun.
   b. It is smaller than the Sun.
   c. It is exactly the same size as the Sun.
   d. You would need to know if it was on the main sequence to answer this question.

18. If a star is found *directly* above the Sun on the H-R diagram, what can you conclude about its luminosity?
   a. It is more luminous than the Sun.
   b. It is less luminous than the Sun.
   c. It is exactly the same luminosity as the Sun.
   d. You would need to know if it was on the main sequence to answer this question.

19. If a star is found *directly* to the right of the Sun on the H-R diagram, what can you conclude about its temperature?
   a. It is hotter than the Sun.
   b. It is cooler than the Sun.
   c. It is exactly the same temperature as the Sun.
   d. You would need to know if it was on the main sequence to answer this question.

20. If a star has the same mass as the Sun, what can you conclude about its temperature?
   a. It is hotter than the Sun.
   b. It is cooler than the Sun.
   c. It is the same temperature as the Sun.
   d. You would need to know if it was on the main sequence to answer this question.

THINKING ABOUT THE CONCEPTS

21. Make a table from the information in this chapter; show each property discussed and the methods of finding that property of an individual star. For example, one row might read:

Property	Method
Composition	Analyze the lines in the star's spectrum.

22. The light from some stars passes through dust in our galaxy before it reaches us, making these stars appear dimmer and redder than they actually are. How does this phenomenon affect parallax? How does it affect spectroscopic parallax?

23. What would happen to our ability to measure stellar parallax if we were on the planet Mars? What about Venus or Jupiter?

24. Albireo, a star in the constellation Cygnus, is a visual binary system whose two components can be seen easily with even a small amateur telescope. Viewers describe the brighter star as "golden" and the fainter one as "sapphire blue."
   a. What does this description tell you about the relative temperatures of the two stars?
   b. What does it tell you about their respective sizes?

25. Very cool stars have temperatures of about 2500 K and emit black-body spectra with peak wavelengths in the red part of the spectrum. Do these stars emit any blue light? Why or why not?

26. ★ **WHAT AN ASTRONOMER SEES** Suppose that Figure 10.17 showed a third star in the system of stars, with the third star being much smaller than the other two in the image. What could you conclude about that third component's brightness, distance, and luminosity? ⊙

27. You obtain a spectrum of an object in space. The spectrum consists of a number of sharp, bright emission lines. Is this object a cloud of hot gas, a cloud of cool gas, or a star?

28. Look carefully at the spectrum in Figure 10.10. What about this spectrum tells you that there is a white light source? What tells you there is a cool cloud of gas? Suppose the cool cloud of gas were located behind the white light source. How would this spectrum be different? ⊙

29. Study Figure 10.11. What are the differences between the ultraviolet photon and the red one? In this schematic diagram, all the photons leave the atom traveling to the right. Is that true for a real batch of hydrogen atoms in space? ⊙

30. Explain why the stellar spectral types (O, B, A, F, G, K, M) are not in alphabetical order. Also, explain the sequence of temperatures defined by these spectral types.

31. In Figure 10.13, there is an absorption line at about 410 nm that is weak for O stars and weak for G stars but very strong in A stars. This particular line comes from the transition from the second excited state of hydrogen up to the sixth excited state. Why is this line weak in O stars? Why is it weak in G stars? Why is it strongest in the middle of the range of spectral types? ⊙

32. Other than the Sun, the only stars whose mass we can measure directly are those in binary systems. Explain why.

33. In Figure 10.15, two stars orbit a common center of mass. ⊙
   a. Explain why star 2 has a smaller orbit than star 1.
   b. Re-sketch this picture for the limiting case where star 1 has a very low mass, perhaps close to that of a planet.
   c. Re-sketch this picture for the limiting case where both stars 1 and 2 have the same mass.

34. Suppose you discover a red star with a radius of 100 R_{Sun}. Use Figure 10.19 to argue whether this is a main-sequence star. ⊙

35. Compare the temperature, luminosity, and radius of stars at the lower left and upper right of the H-R diagram (for example, Figure 10.19). ⊙

APPLYING THE CONCEPTS

36. Sirius, the brightest star in the sky, has a parallax of 0.379 arcsec. You want to know how far away it is.
  a. Make a prediction: Compare this parallax to the parallax of Proxima Centauri calculated in Working It Out 10.1. Do you expect the distance to Sirius to be larger or smaller than the distance to Proxima Centauri? Do you expect the distance to Sirius to be less than a parsec, a few parsecs, or hundreds of parsecs? Given that there are 3.26 light-years in a parsec, do you expect the distance in light-years to be larger or smaller than the distance in parsecs?
  b. Follow Working It Out 10.1 to calculate the distance to Sirius in parsecs. Convert the answer to light-years.
  c. Check your work by comparing your answer to your prediction.

37. Betelgeuse (in Orion) has a parallax of 0.00763 ± 0.00164 arcsec, as measured by the Hipparcos satellite. You want to know the distance to Betelgeuse (in light-years) and the uncertainty in that measurement.
  a. Make a prediction: Do you expect the distance to Betelgeuse to be larger or smaller than the distance to Proxima Centauri given in Working It Out 10.1?
  b. Calculate the distance to Betelgeuse using the average value of the parallax (the first number).
  c. Calculate the largest reasonable distance (using the "−" of the ± 0.00164 arcsec) and the smallest reasonable distance (using the "+" of the ± 0.00164 arcsec). Half the difference between the largest and smallest reasonable distances is your uncertainty. Find the uncertainty in the distance, and express the distance with its uncertainty in the form distance $\pm$ uncertainty.
  d. Check your work by comparing your calculated distance to your prediction.

38. Barnard's star is located 5.978 light-years from Earth. You want to know its parallax.
  a. Make a prediction: Compare this distance to the distance to Proxima Centauri given in Working It Out 10.1. Do you expect the parallax of Barnard's star to be larger or smaller than the parallax of Proxima Centauri?
  b. Follow Working It Out 10.1 to calculate the parallax of Barnard's star (don't forget to convert from light-years to parsecs first!)
  c. Check your work by comparing your result with your prediction.

39. Suppose Figure 10.2 included a third star, C, located 4 times as far away as star A. You want to know its parallax. ⊛
  a. Make a prediction: Distance and parallax are inversely related, so if C is farther than A, will it appear to move more or less than star A? Will the parallax be larger or smaller? Consider the same analysis for star B.
  b. Follow Working It Out 10.1 to find the parallax angle for star C.
  c. Check your work by comparing your answer to your prediction.

40. Suppose you see a red pencil jump from side to side by half a degree as you blink back and forth between your eyes, and you also see a blue pencil that moves only one-third of a degree.
  a. Make a prediction: Is an object that moves one-third of a degree closer or farther than one that moves half a degree? Is it closer or farther than one that moves one-quarter of a degree?
  b. Calculate the relative distances for the red and blue pencils by using a ratio. You don't need the actual distances; just determine how many times farther or closer the blue pencil is than the red one.
  c. Check your work by comparing your answer to your prediction.

exploration The Hertzsprung-Russell Diagram

digital.wwnorton.com/universe4

Visit the Student Site at the Digital Resources page and open the Hertzsprung-Russell Diagram Interactive Simulation in Chapter 10. This simulation allows you to compare stars on the H-R diagram in two ways. You can compare an individual star (marked by a red ✕) to the Sun by varying its properties in the box on the left half of the window. Or you can compare groups of the nearest and brightest stars. Play around with the controls for a few minutes to familiarize yourself with the simulation.

Let's begin by exploring how changes to the properties of the individual star change its location on the H-R diagram. First, click the reset button.

Decrease the temperature of the star by dragging the temperature slider to the left. Notice that the luminosity remains the same. Because the temperature has decreased, each square meter of star surface must be emitting less light. What other property of the star changes in order to keep the total luminosity of the star constant?

Predict what will happen when you slide the temperature slider all the way to the right. Now do it. Did the star behave as you expected?

1. As you move to the left across the H-R diagram, what happens to the radius?

_____

_____

2. As you move to the right across the H-R diagram, what happens to the radius?

_____

_____

Click "reset," and experiment with the luminosity slider.

3. As you move up on the H-R diagram, what happens to the radius?

_____

_____

_____

4. As you move down on the H-R diagram, what happens to the radius?

_____

_____

_____

Click "reset" again and then predict how you would have to move the slider bars to move your star into the red giant portion of the H-R diagram (the upper right). Adjust your slider bars until the star is in that area. Were you correct?

5. How would you have to adjust the slider bars to move the star into the white dwarf area of the H-R diagram?

_____

_____

_____

_____

Click the reset button, and let's explore the right side of the window. Add the nearest stars to the graph by clicking the box. Using what you learned above, compare the temperatures and luminosities of these stars to the Sun (marked by the ✕).

6. Are the nearest stars generally hotter or cooler than the Sun?

_____

7. Are the nearest stars generally more or less luminous than the Sun?

_____

Add the brightest stars in the sky to the plot. Compare these stars to the Sun.

8. Are the brightest stars generally hotter or cooler than the Sun?

_____

9. Are the brightest stars generally more or less luminous than the Sun?

_____

10. How do the temperatures and luminosities of the brightest stars in the sky compare to the temperatures and luminosities of the nearest stars? Does this information support the claim in the chapter that there are more low-luminosity stars than high-luminosity stars? Explain.

_____

_____

_____

_____

Our Star: The Sun

The Sun is more massive *and* significantly larger than everything else in the Solar System. You can measure the diameter of the Sun with a heavy sheet of paper or poster board, a long piece of string, a ruler, and a little help from a friend. On a sunny day, poke a tiny hole in the paper. Standing with her back to the Sun, your friend should hold up the paper so that the Sun shines through the hole. An image of the Sun will be projected onto the ground within the shadow of the paper. If the paper is close to the ground, the image will be small. If the paper is far from the ground, the image will be larger, but it will not be as bright. Move the paper until you can accurately measure the diameter (d) of the image on the ground with your ruler, in millimeters. Convert this number to meters. Use the long piece of string to measure how far (h) the paper is from the ground, by first marking the distance on the piece of string, and then using the ruler to measure the marked distance in meters. (You can use a tape measure, instead, if you have one.) To calculate the diameter of the Sun, all you need are these two measurements and the distance from Earth to the Sun (in meters), which is available in your textbook or on the Internet.

EXPERIMENT SETUP

On a sunny day, poke a tiny hole in the paper. Standing with her back to the Sun, your friend should hold up the paper so that the Sun shines through the hole. An image of the Sun will be projected onto the ground within the shadow of the paper.

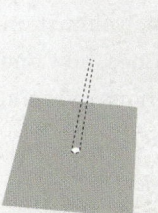 

CLOSER **FARTHER**

If the paper is close to the ground, the image will be small. If the paper is far from the ground, the image will be larger, but it will not be as bright. Move the paper until you can accurately measure the diameter (d) of the image on the ground with your ruler, in millimeters. Convert this number to meters.

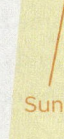

Sun

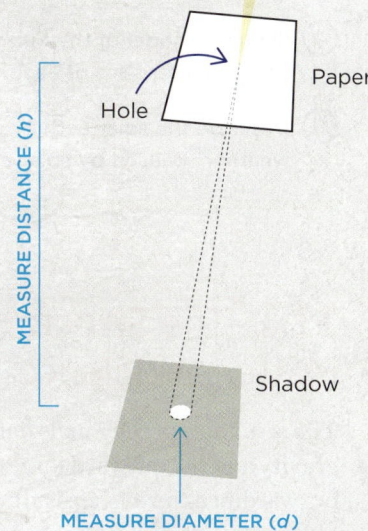

Paper

Hole

MEASURE DISTANCE (h)

Use a long piece of string to measure how far (h) the paper is from the ground, by first marking the distance on the piece of string, and then using the ruler to measure the marked distance in meters.

Shadow

MEASURE DIAMETER (d)

SKETCH OF RESULTS

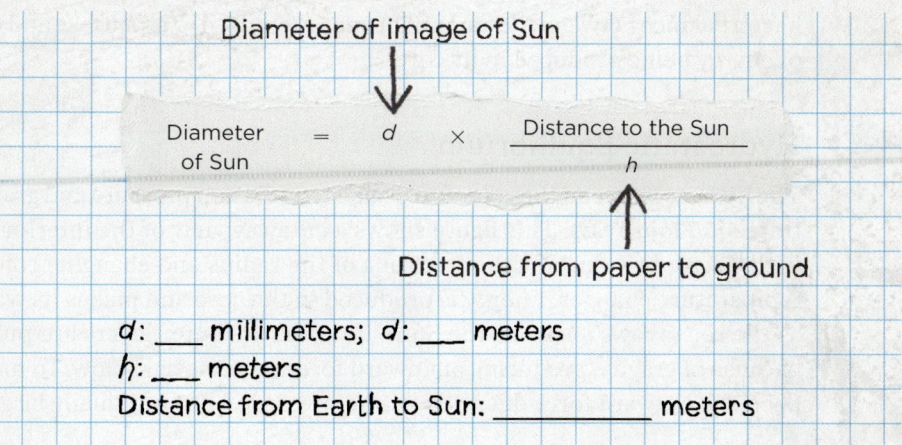

Diameter of image of Sun

↓

$$\text{Diameter of Sun} = d \times \frac{\text{Distance to the Sun}}{h}$$

↑

Distance from paper to ground

d: ___ millimeters; d: ___ meters
h: ___ meters
Distance from Earth to Sun: _____ meters

11

The Sun is a middle-aged, main-sequence star of spectral type G2. As stars go, it may be fairly average, but it is awesome on a human scale. The Sun's mass is more than 300,000 times that of Earth, and its diameter is more than 100 times that of Earth. The Sun produces more energy in a second than human technology would produce in half a million years. The Sun is the only star we can study at close range, so much of the detailed information that we know concerning main-sequence stars, which are 90 percent of the stars in the sky, has come from studying the Sun. To understand the Sun, we must understand how it avoids collapse despite the pull of gravity, how it shines, how energy moves from the core to the surface, and the details of how it changes over time.

LEARNING GOALS

(1) Describe the balance between the forces that determine the structure of the Sun.

(2) Explain how mass is converted into energy in the Sun's core and why fusion happens only in the core and not in the outer layers.

(3) List the different ways that energy moves outward from the Sun's core toward its surface.

(4) Sketch a model of the Sun's interior and atmospheric layers and describe important characteristics of each.

(5) Describe the solar activity cycles of 11 and 22 years, and explain how "space weather" caused by solar activity affects Earth and human technology.

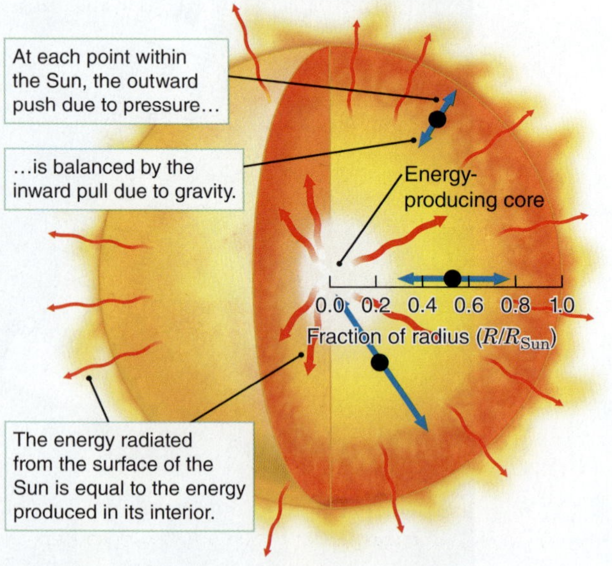

11.1 The Structure of the Sun Is a Matter of Balance

The structure of the Sun is determined by the balance between the inward force of gravity and the outward force due to the pressure of radiation produced inside. The outward pressure is a result of energy finding its way outward from deep in the Sun's interior. To understand this summary statement, you need to know how these forces are produced and how they continually change to balance each other. There is a second balancing act to consider. To maintain a constant size, the Sun must also have a balance between the amount of energy emitted at its surface and the amount of energy being produced in its core.

Hydrostatic Equilibrium

The balance between the solar forces due to radiation pressure and gravity is illustrated in **Figure 11.1**. This figure shows a cutaway view of the interior of the Sun, with a scale bar marking the fractions of the radius and changing colors to show temperature changes. Energy is produced in the core and makes its way outward, as the red arrows show. In the Sun's interior, the outer layers are pulled inward because of gravity, producing an inward force on material below. To maintain balance, the outward force due to radiation pressure must be equally large. If gravity

At each point within the Sun, the outward push due to pressure…

…is balanced by the inward pull due to gravity.

Energy-producing core

0.0 0.2 0.4 0.6 0.8 1.0
Fraction of radius (R/R_{Sun})

The energy radiated from the surface of the Sun is equal to the energy produced in its interior.

Figure 11.1 The structure of the Sun is determined by the balance between the outward force due to radiation pressure and the inward force of gravity. Energy generated in the Sun's core is equal to energy radiated from its surface. The scale bar shows the fraction of the Sun's radius. This could be interpreted as a percentage: 0.2 is 20 percent of the way out from the very center of the Sun.

were greater than the force due to radiation pressure, the Sun would collapse. If the force due to radiation pressure were greater than gravity, the Sun would blow itself apart. At every point within the Sun's interior, the pressure must be just enough to hold up the weight of all the layers above that point. This balance between the effects of radiation pressure and gravity is known as *hydrostatic equilibrium*. It is the same balance we saw in protostars in Chapter 5.

In a gas, higher pressure means higher density and/or higher temperature. **Figure 11.2** shows how conditions vary inside the Sun. Toward the center of the Sun, the pressure increases, and as it does, the density and temperature of the gas increase as well. The energy-producing core occupies only the innermost 20 percent of the Sun.

Nuclear Fusion

Stars like the Sun are remarkably stable objects. Models of stellar evolution indicate that the luminosity of the Sun is increasing with time, but very, very slowly. The Sun's luminosity about 4.5 billion years ago was about 70 percent of its current luminosity. To remain in balance, the Sun must produce enough energy in its interior to replace the energy lost from its surface. This energy balance tells us how much energy must be produced each second in the interior of the Sun. The amount of energy produced by the Sun each second is truly astronomically large at 3.85×10^{26} watts (W). Knowing this fact leads immediately to the question: What process can produce this much energy this quickly? (In other words, what makes the Sun shine?) This was one of the most basic questions pioneers of stellar astrophysics sought to answer. Many possible explanations were tested, and failed. The answer finally came from theoretical physics and the laboratories of nuclear physicists as they began to understand the atomic nucleus.

The nuclei of most hydrogen atoms consist of a single proton. Nuclei of all other atoms are built from a mixture of protons and neutrons. Most helium nuclei, for example, consist of two protons and two neutrons. Remember that protons have a positive electric charge, whereas neutrons have no charge. Because like charges repel, and the closer they are the stronger the repulsive force, all of the protons in an atomic nucleus electrically repel each other with a tremendous force. The nuclei of atoms should fly apart because of electric repulsion—but they don't. The **strong nuclear force**, an attractive force between particles inside the nucleus, overcomes the repulsion and holds the nucleus together. However, the strong nuclear force acts only over very short distances of about twice the diameter of the proton.

A very large amount of energy is required to tear a nucleus apart, much more than is required to take an electron out of an atom. Conversely, when an atomic nucleus is formed, a larger amount of energy is released. This energy release is often referred to as "energy production," which is technically not correct. The energy is not "produced"; rather, it is released from another form. Still, the term "energy production" is so common that we will use it here. **Nuclear fusion**—the process of combining two less massive atomic nuclei into a more massive atomic nucleus—occurs when atomic nuclei are brought close enough together for the strong nuclear force to overcome the force of electric repulsion, as illustrated in **Figure 11.3**. Many kinds of nuclear fusion can occur in stars. In main-sequence stars like the Sun, the primary energy production process is the fusion of hydrogen into helium—a process called **hydrogen burning**.

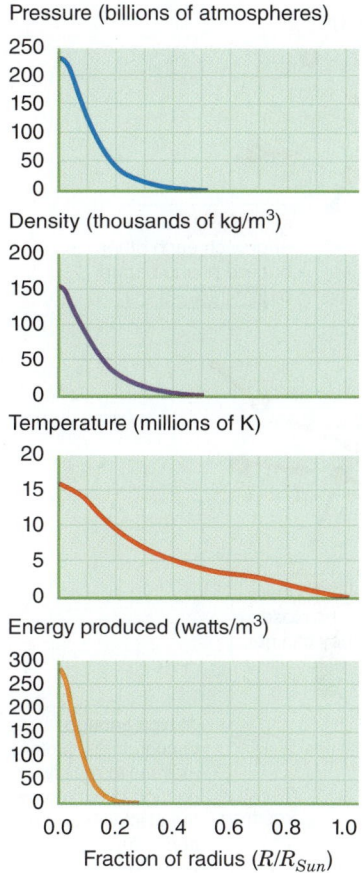

Figure 11.2 Pressure, density, and temperature increase toward the center of the Sun. The energy produced by the Sun is generated only in the Sun's core.

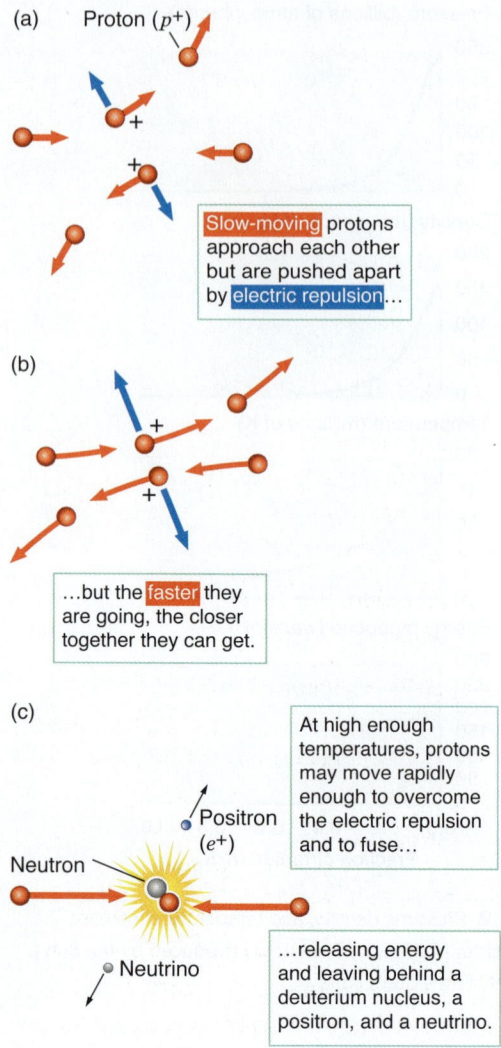

(a) Proton (p^+)

Slow-moving protons approach each other but are pushed apart by electric repulsion...

(b)

...but the faster they are going, the closer together they can get.

(c)

At high enough temperatures, protons may move rapidly enough to overcome the electric repulsion and to fuse...

Positron (e^+)

Neutron

Neutrino

...releasing energy and leaving behind a deuterium nucleus, a positron, and a neutrino.

Figure 11.3 (a) Atomic nuclei, such as the atomic nucleus of hydrogen, which is a single proton, are positively charged and electrically repel each other. (b) The faster two nuclei are moving toward each other, the closer they will get before veering away. (c) At the temperatures and densities found in the centers of stars, nuclei are energetic enough to overcome this electric repulsion, so fusion takes place. (A positron is a small, positively charged particle.)

The energy released in nuclear reactions comes from the conversion of mass into energy. The exchange rate between mass and energy is given by Einstein's equation, $E = mc^2$, in which E is energy, m is mass, and c^2 is the speed of light (3.00×10^8 meters per second, m/s) squared. For any nuclear reaction, we can determine the mass that is turned into energy by calculating the mass that is lost. To find this lost mass, we subtract the mass of the outputs from the mass of the inputs. In hydrogen burning, the inputs are four hydrogen nuclei, and the output is a helium nucleus. The mass of four separate hydrogen nuclei is greater than the mass of a single helium nucleus, so when hydrogen fuses to make helium, energy is released.

Energy is produced in the Sun's innermost region, the core, where conditions are the most extreme. The density is about 150 times the density of water (which is 1,000 kilograms per cubic meter [kg/m^3]), and the temperature is about 15 million kelvins (K). The atomic nuclei have tens of thousands of times more kinetic energy than atoms at room temperature. As illustrated in Figure 11.3c, under these conditions, atomic nuclei slam into each other hard enough to overcome the electric repulsion, allowing the strong nuclear force to act.

In hotter and denser gases, collisions happen more frequently. For this reason, the rate of nuclear fusion reactions is extremely sensitive to the temperature and density of the gas, which is why these energy-producing collisions are concentrated in the Sun's core. Half of the energy produced by the Sun is generated within the inner 9 percent of the Sun's radius. This region represents less than 0.1 percent of the volume of the Sun. Over the lifetime of the Sun, about 10 percent of the hydrogen will be fused into helium. The rest of the hydrogen is too far from the core to become hot enough or dense enough for fusion to occur. It is possible to estimate the lifetime of the Sun by combining information about the fraction of the Sun that will be involved in nuclear fusion with the energy released both from the Sun and from a fusion event, as shown in **Working It Out 11.1**.

Hydrogen burning is the most significant source of energy in main-sequence stars. There are three reasons that this is so. First, hydrogen is the most abundant element in the universe, so it is the most abundant source of nuclear fuel. Second, hydrogen burning is also the most efficient nuclear process to convert mass into energy. Third, and most important, hydrogen is the easiest type of atom to fuse. Hydrogen nuclei—protons—have an electric charge of +1. The electric barrier that must be overcome to fuse hydrogen is the repulsion of one proton against one other proton. Any other type of fusion requires even higher temperatures and pressures to overcome the repulsion. To fuse carbon, for example, the repulsion of six protons in one carbon nucleus pushing against the six protons in another carbon nucleus must be overcome. The repulsion between two carbon nuclei is $6 \times 6 = 36$ times stronger than that between two hydrogen nuclei. Carbon nuclei would need to be traveling much faster than hydrogen nuclei in order to overcome the barrier. The average energy of the carbon atoms, which is related to the temperature, would need to be much higher than the average energy of the hydrogen atoms. Therefore, hydrogen fusion occurs at a much lower temperature than any other type of nuclear fusion.

The Proton-Proton Chain

In the core of the Sun, hydrogen burns primarily through a process called the **proton-proton chain**. It consists of the three steps depicted in **Figure 11.4a**. Notice that the first two steps are shown twice, along the top and bottom of part (a). These steps must occur twice, with different sets of the same type of particles, which then combine in the third step. Each step produces particles and/or energy in the form

working it out 11.1

How Much Longer Will the Sun "Live"?

Like all stars, the Sun's lifetime is limited by the amount of fuel available to it. We can calculate how long the Sun will live by comparing the mass involved in nuclear fusion with the amount of mass available. Converting four hydrogen nuclei (protons) into a single helium nucleus results in a loss of mass. The mass of a hydrogen nucleus is 1.6726×10^{-27} kg. So, four hydrogen nuclei have a mass of 4 times that, or 6.6904×10^{-27} kg. The mass of a helium nucleus is 6.6447×10^{-27} kg, which is less than the mass of the four hydrogen nuclei. The amount of mass lost, m, is

$$m = 6.6904 \times 10^{-27}\ \text{kg} - 6.6447 \times 10^{-27}\ \text{kg} = 0.0457 \times 10^{-27}\ \text{kg}$$

We can move the decimal point to the right and change the exponent on the 10 to rewrite the mass lost as 4.57×10^{-29} kg.

Using Einstein's equation $E = mc^2$ (and the fact that 1 kg m²/s² is 1 joule [J]), we can see that the energy released by this mass-to-energy conversion is

$$E = mc^2 = (4.57 \times 10^{-29}\ \text{kg})(3.00 \times 10^8\ \text{m/s})^2 = 4.11 \times 10^{-12}\ \text{J}$$

Each reaction that takes four hydrogen nuclei and turns them into a helium nucleus releases 4.11×10^{-12} J of energy, which doesn't seem like very much. But consider how small an atom is. Fusing a single gram of hydrogen into helium releases about 6×10^{11} J, which is equivalent to the energy obtained by burning 100 barrels of oil. In a sense, what this number tells you is how astonishingly big the Sun really is. For the Sun to produce as much energy as it does, it must convert roughly 600 billion kg of hydrogen into helium every second (and 4 billion kg of matter is converted to energy in the process). It has been doing this for the past 4.6 billion years. But how do we know how long the Sun will last?

Only about 10 percent of the Sun's total mass will ever be involved in fusion. Ten percent of the mass of the Sun is

$$M_{\text{fusion}} = (0.1)(M_{\text{Sun}})$$
$$M_{\text{fusion}} = (0.1)(2 \times 10^{30}\ \text{kg})$$
$$M_{\text{fusion}} = 2 \times 10^{29}\ \text{kg}$$

That is the amount of fuel the Sun has available. How long will it last? The Sun consumes hydrogen at a rate of 600 billion kilograms per second (kg/s), and there are 3.16×10^7 seconds in a year, so each year the Sun consumes

$$M_{\text{year}} = (600 \times 10^9\ \text{kg/s})(3.16 \times 10^7\ \text{s/year})$$
$$M_{\text{year}} = 2 \times 10^{19}\ \text{kg/year}$$

If we know how much fuel the Sun has available for fusion (2×10^{29} kg), and we know how much the Sun burns each year (2×10^{19} kg/year), then we can divide the amount by the rate to find the lifetime of the Sun:

$$\text{Lifetime} = \frac{M_{\text{fusion}}}{M_{\text{year}}} = \frac{2 \times 10^{29}\ \text{kg}}{2 \times 10^{19}\ \text{kg/year}}$$

$$\text{Lifetime} = 1 \times 10^{10}\ \text{years} = 10 \times 10^9\ \text{years} = 10\ \text{billion years}$$

When the Sun was formed, it had enough fuel to power it for about 10 billion years. At 4.6 billion years old, the Sun is nearly halfway through its total life span, and it will continue fusing hydrogen into helium for about 5.4 billion more years, give or take a few hundred million.

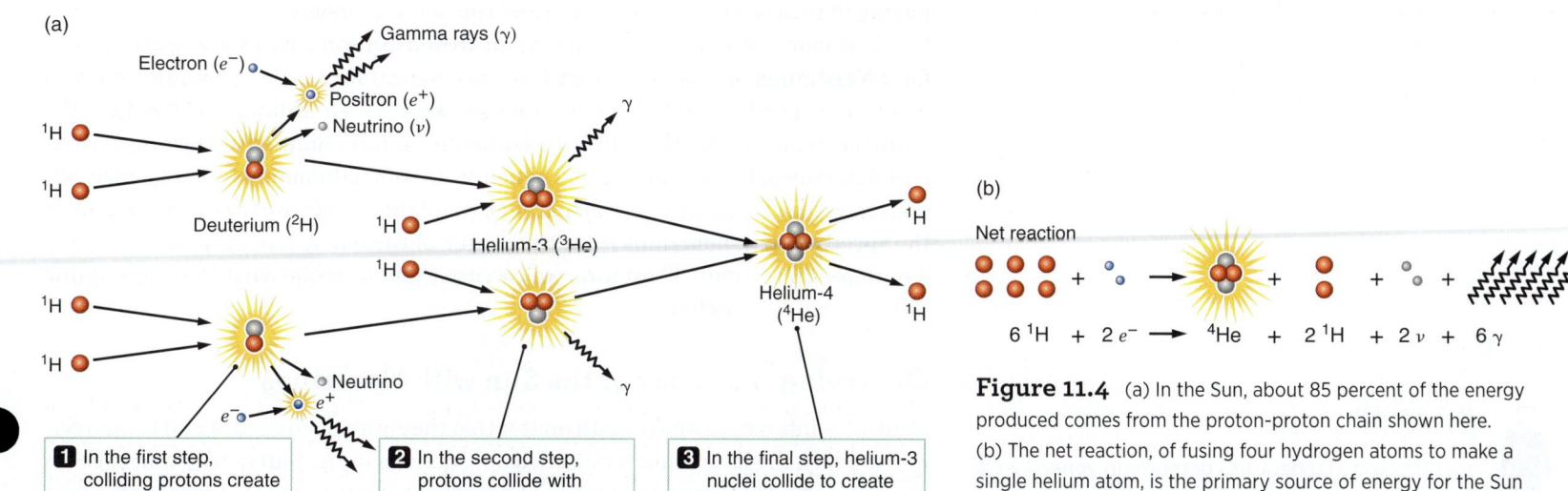

(a)

1 In the first step, colliding protons create deuterium (²H).

2 In the second step, protons collide with deuterium nuclei to produce helium-3 (³He).

3 In the final step, helium-3 nuclei collide to create helium-4 (⁴He).

(b)

Net reaction

$$6\,^1\text{H} + 2\,e^- \longrightarrow\ ^4\text{He} + 2\,^1\text{H} + 2\,\nu + 6\,\gamma$$

Figure 11.4 (a) In the Sun, about 85 percent of the energy produced comes from the proton-proton chain shown here. (b) The net reaction, of fusing four hydrogen atoms to make a single helium atom, is the primary source of energy for the Sun and all other main-sequence stars. The energy is released in the form of gamma rays, neutrinos, and kinetic energy.

▶❚❚ **AstroTour:** The Solar Core

▶▶ **Interactive Simulation:** Proton-Proton Chain

what if . . .

What if you had a neutrino detector in your classroom: How would you expect the signal of solar neutrinos to vary between night and day?

of light. Once these steps were understood in detail, astronomers were able to search for the products of this fusion reaction to test the proposal that the Sun shines because of nuclear fusion. We will begin by following the creation of the helium nucleus, and then go back to find out what happens to the other products of the reaction.

In the first step, two hydrogen nuclei fuse. During this process, one of the protons turns into a neutron. Two particles are emitted, conserving energy and charge: a positively charged particle, called a positron, and a neutral particle, called a neutrino. Neutrinos are commonly designated by the Greek letter nu (ν). Energy is also emitted as light in the form of gamma rays. The new nucleus has a proton and a neutron. It is still hydrogen because it has only one proton. Because it has more than the normal number of neutrons, we say it is an isotope of hydrogen; different isotopes of an element have the same number of protons but different numbers of neutrons. This particular isotope is called deuterium (^{2}H).

In the second step of the proton-proton chain, another proton slams into the deuterium nucleus, forming the nucleus of a helium isotope, ^{3}He, consisting of two protons and one neutron. The energy released in this step is carried away as a gamma-ray photon.

In the third and final step of the proton-proton chain, two ^{3}He nuclei collide and fuse, producing an ordinary ^{4}He nucleus and ejecting two protons in the process. These protons may go on to interact with others in additional reactions. The energy is released in this step as the kinetic energy (energy of motion) of the helium nucleus and the ejected protons. Overall, four hydrogen nuclei have combined to form one helium nucleus, as summarized in Figure 11.4b.

Let's go back and look at what happens to the other products of the reaction. In step 1, when two protons collided, a positron was produced. A **positron** is a type of particle known as antimatter. An antimatter particle has the same mass as a corresponding matter particle but the opposite value of other properties, such as charge. The positron (e^+) is the antimatter counterpart of the electron (e^-). When a particle of matter (such as an electron) and a particle of antimatter (such as a positron) meet, they annihilate each other. The total mass is converted to energy in the form of gamma-ray photons. The positrons and electrons inside the Sun collide and annihilate each other, and the emitted photons heat the surrounding gas. The gamma rays emitted in step 2 similarly heat the surrounding gas. Energy, carried by photons, moves very slowly outward from the hot core, so the fact that we see light coming from the surface of the Sun today indicates that the Sun was producing energy in the core more than a hundred thousand years ago.

The neutrino emitted in step 1 of the proton-proton chain has a very different fate. **Neutrinos** are particles that have no charge, very little mass, and travel at nearly the speed of light. They interact so weakly with ordinary matter that the neutrino escapes from the Sun without further interactions with any other particles. The core of the Sun lies buried beneath 700,000 kilometers (km) of dense, hot matter, yet the Sun is nearly transparent to neutrinos. Because they travel at nearly the speed of light, neutrinos from the center of the Sun arrive at Earth after only 8 minutes 20 seconds. Therefore, we can use them to probe what the Sun is doing today—if we can catch them.

Observing the Heart of the Sun with Neutrinos

Neutrinos interact so weakly with matter that they are extremely difficult to observe. Fortunately, the large number of nuclear reactions in the Sun means that the Sun produces a truly enormous number of neutrinos. As you read this sentence, about 400 trillion solar neutrinos have passed through your body. This happens even at night, because neutrinos easily pass through Earth. With so many neutrinos about, a neutrino detector does not have to detect a very large percentage of them to be effective.

The first apparatus designed to detect solar neutrinos was built 1,500 meters underground, within the Homestake Mine in Lead, South Dakota. Astronomers filled a tank with 100,000 gallons of dry-cleaning fluid, primarily composed of chlorine. When a neutrino strikes a chlorine atom, there is a small probability that the chlorine atom will turn into an argon atom. Over the course of 2 days, roughly 10^{22} solar neutrinos pass through the Homestake detector. On average, only *one* of these neutrinos interacts with a chlorine atom within the fluid to form an atom of argon. Even so, this instrument produces a measurable number of argon atoms over time.

The Homestake experiment detected these argon atoms—evidence of neutrinos from the Sun, confirming that nuclear fusion powers the Sun. As with many good experiments, however, these results raised new questions. After astronomers' initial joy, they noticed that there seemed to be only one-third as many solar neutrinos as predicted. The difference between the predicted and measured number of solar neutrinos was called the **solar neutrino problem**.

The solar neutrino problem, in turn, led particle physicists to suspect that their understanding of the neutrino itself was incomplete. The neutrino was long thought to have zero mass, like photons, and to travel at the speed of light. But if neutrinos actually do have a tiny amount of mass, then neutrinos should oscillate—alternate back and forth—among three different kinds of neutrinos, called *electron, muon,* and *tau neutrinos*. Early neutrino experiments could detect only the electron neutrino, so the other two types of neutrinos (muon and tau neutrinos) went undetected. It is not surprising that scientists did not think of this ahead of time. The more familiar particles in nature—electrons, protons, and neutrons—do not behave this way.

Since Homestake began operating in 1965, more than two dozen additional neutrino detectors have been built, each using different reactions to detect different kinds of neutrinos. Experiments at high-energy physics labs, nuclear reactors, and neutrino telescopes around the world have shown that neutrinos *do* have a nonzero mass, and this work has uncovered evidence of neutrino oscillations. The Nobel Prize in Physics for 2015 was awarded jointly to Takaaki Kajita and Arthur B. McDonald for the discovery of neutrino oscillations.

Solving the solar neutrino problem is a good example of how science works. The Homestake experiment revealed a gap in our understanding. Theories of particle physics suggested possible resolutions, and more sophisticated experiments tested which one of the competing hypotheses was correct. Through this process, the solar neutrino problem led to a deeper knowledge of basic physics. Perhaps this helps explain why scientists are often excited to find out they are wrong. Sometimes it means that completely new knowledge is right around the corner.

CHECK YOUR UNDERSTANDING 11.1

The structure of the Sun is determined by both the balance between the forces due to _____ and gravity and the balance between energy generation and energy _____.

a. pressure; production

b. pressure; loss

c. ions; loss

d. solar wind; production

Answers to Check Your Understanding questions are in the back of the book.

what if . . .

What if a typical photon produced in the Sun's center took 50,000 rather than 100,000 years to escape from the Sun's surface: What could be different about this hypothetical Sun?

11.2 Energy in the Sun's Core Moves through Radiation and Convection

In Chapter 6, we saw that although we cannot travel deep inside Earth to find out how it is structured, we are able to build a model of its interior. Similarly, we can create a model of the Sun's interior by using our knowledge of the balance of forces and energy within the Sun and our understanding of how energy moves from one place to another. Waves traveling through the Sun, as well as some surface distortions, allow us to test and refine our model of its interior.

Energy Transport

Some of the energy released by hydrogen burning in the core of the Sun is carried directly into space by neutrinos, but most of the energy heats the particles in the solar interior and then gradually moves outward through the Sun to the surface. This **energy transport** is a key determinant of the Sun's structure. In any system, energy transport can occur by conduction, convection, or radiation; in all three cases, energy moves from a hotter region to a colder one. Conduction is important primarily in solids. For example, when you pick up a hot object, your fingers are heated by conduction. Because the Sun is made of gas, energy in the Sun is not transported by conduction, but rather it moves by convection and radiation through different zones, as shown in **Figure 11.5a**. First, energy moves outward through the

(a)

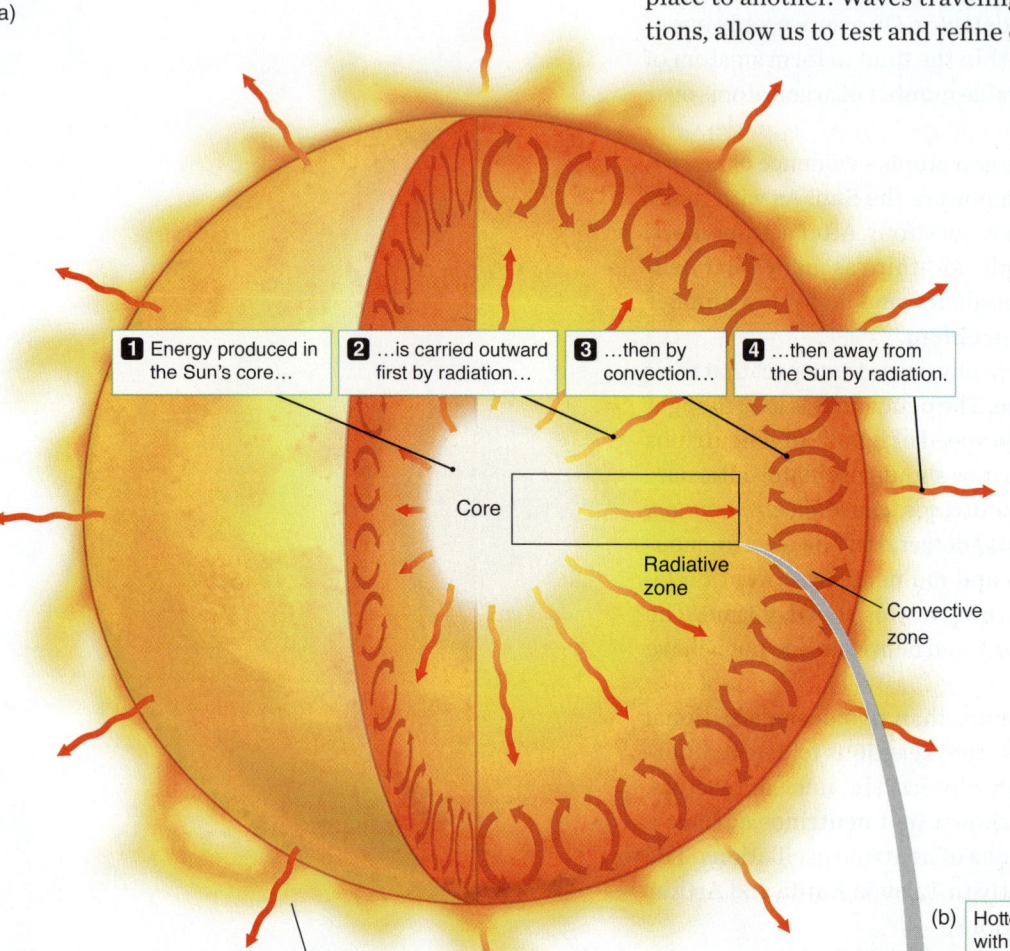

1 Energy produced in the Sun's core…

2 …is carried outward first by radiation…

3 …then by convection…

4 …then away from the Sun by radiation.

Core

Radiative zone

Convective zone

Energy radiated from the Sun's surface

(b) Hotter regions are more crowded with photons than cooler regions…

Hotter region

Cooler region

…making it more likely that photons will move from hotter to cooler regions than in the reverse direction.

Net energy transfer

As a result, radiation carries energy outward from the hot, energy-generating core of the Sun.

Figure 11.5 (a) The interior structure of the Sun is divided into radiative and convective zones on the basis of where energy is produced and how it is transported outward. (b) More radiation flows from the hotter regions to the cooler regions, carrying energy outward from the center of the Sun.

inner layers of the Sun by radiation in the form of photons. Next, energy moves by convection in parcels of gas. Finally, energy radiates from the Sun's surface as light (photons).

The details of energy transport from the center of the Sun outward depend on decreasing temperature and density at larger distances from the core. Near the core, radiation transfers energy from hotter to cooler regions via photons, which carry the energy with them. Consider a hotter region of the Sun located next to a cooler region, as shown in Figure 11.5b. The hotter region contains more (and more energetic) photons than the cooler region. More photons will move by chance from the hotter, more crowded region to the cooler, less crowded region than in the reverse direction. A net transfer of photons and photon energy occurs from the hotter region to the cooler region, and so radiation carries energy outward from the Sun's core.

The rate at which energy moves from one point to another by radiation depends on how freely photons can move from one point to another within a star. The degree to which matter blocks the flow of photons through it is referred to as **opacity**. The opacity of a material depends on many things, including the density of the material, its composition, its temperature, and the wavelength of the photons moving through it.

Radiative transfer is efficient in regions with low opacity. The **radiative zone** is the region in the inner part of the Sun where the opacity is relatively low, and radiation carries the energy produced in the core outward through the star. This radiative zone extends about 70 percent of the way out toward the surface of the Sun. Even though this region's opacity is low enough for radiation to dominate over convection as an energy transport mechanism, photons still travel only a short distance within the region before being absorbed, emitted, or deflected by matter, much like a beach ball being batted about by a crowd of people, as illustrated in **Figure 11.6**. Each interaction reduces the energy of the photon and sends it in an unpredictable direction—not necessarily toward the surface of the star. The distances between interactions are so short that, on average, it takes the energy of a gamma-ray photon about 170,000 years to find its way to the outer layers of the Sun. Opacity holds energy within the interior of the Sun and lets it seep away only slowly. As it travels, the gamma-ray photon gradually becomes converted to lower-energy photons, emerging as optical and infrared radiation from the surface.

From a peak of 15 million K in the core of the Sun, the temperature falls to about 2 million K at the outer margin of the radiative zone. At this cooler temperature, the opacity is higher, so radiation is less efficient in carrying energy from one place to another. The energy that is flowing outward through the Sun "piles up" against this edge of the radiative zone. Beyond this region, the temperature drops off very quickly.

Farther from the core of the Sun, radiative transfer becomes inefficient and the temperature changes quickly. Convection takes over, which transports energy by moving packets of material. These packets of hot gas, like hot air balloons, become buoyant and rise up through the lower-temperature gas above them, carrying energy with them. Just as convection carries energy from the interior of planets to their surfaces or from the Sun-heated surface of Earth upward through Earth's atmosphere, convection also plays an important role in the transport of energy outward from the interior of the Sun. The solar **convective zone** (see Figure 11.5) extends from the outer boundary of the radiative zone to just below the visible surface of

VOCABULARY ALERT

opacity *Opacity* is related to the word *opaque*. In everyday language, when something is opaque, you can't see through it at all. The term *opacity* allows for more subtle variations. A gas can have a low opacity, which means it is mostly transparent and allows photons to pass through, or a high opacity, which means it is mostly opaque and prevents photons from passing straight through.

(a)

(b)

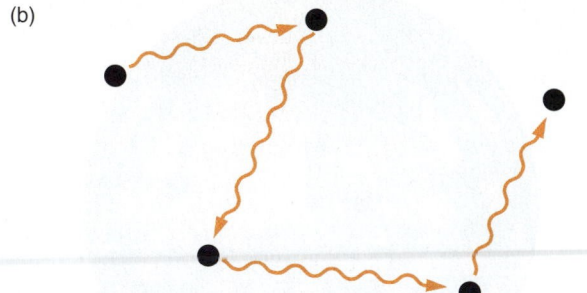

Figure 11.6 (a) When a crowd of people play with a beach ball, the ball moves randomly, sometimes toward the front of the crowd, sometimes toward the back. It often takes a ball a long time to move from one edge of the crowd to the other. (b) Similarly, it takes a long time for a photon to make its way out of the Sun, because interactions with matter make the photon move randomly.

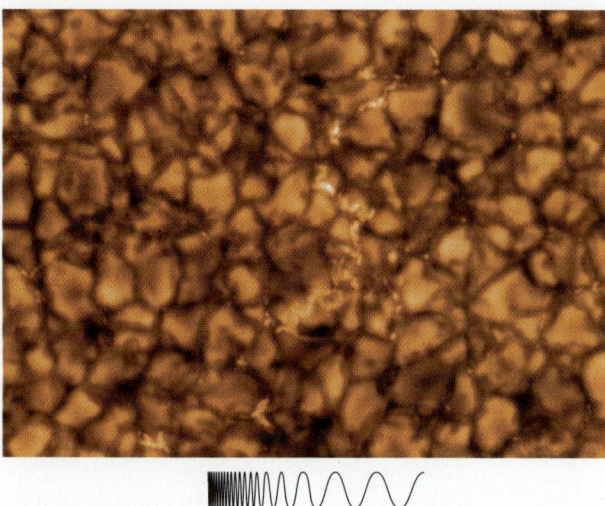

G X U V I R

Figure 11.7 The top of the convective zone shows the bubbling of the surface caused by rising and falling packets of gas.

Astronomy in Action: Random Walk

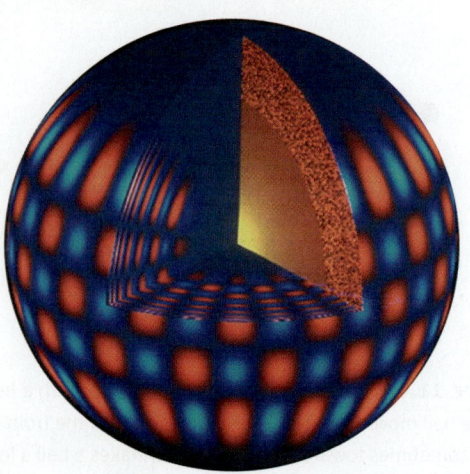

Figure 11.8 The interior of the Sun vibrates as helioseismic waves move through it. This figure shows one particular mode of the Sun's vibration. Red shows regions where gas is traveling inward; blue shows regions where gas is traveling outward.

the Sun, where evidence of convection can be seen in the "bubbling" of that surface (**Figure 11.7**).

In the outermost layers of the Sun, radiation again takes over as the primary mode of energy transport, and it is radiation that transports that energy through space to Earth and elsewhere.

Helioseismology

In Chapter 6, we found that models of Earth's interior predict how density and temperature change from place to place in the interior. These differences affect seismic waves traveling through Earth, bending the paths that they travel. To test and refine our models of Earth's interior, we compared measurements with the predictions.

The same method can be applied to the Sun. Detailed observations of the surface of the Sun show that the Sun vibrates or rings, something like a bell that has been struck. Unlike a well-tuned bell—which vibrates primarily at one frequency—the vibrations of the Sun are very complex; many different frequencies of vibrations occur simultaneously, causing some parts of the Sun to bulge outward (the blue areas in **Figure 11.8**) and some to draw inward (the red areas in Figure 11.8). Some waves are amplified and some are suppressed, depending on how they overlap as they travel through the Sun. Just as geologists use seismic waves from earthquakes to probe Earth's interior, solar physicists use the surface waves on the Sun to probe the solar interior. They study these waves using the Doppler effect (see Chapter 5), which distinguishes between parts of the Sun that move toward the observer and parts that move away. The science that uses surface waves to study the Sun is called **helioseismology**.

To detect the disturbances of helioseismic waves on the surface of the Sun, astronomers must measure Doppler shifts of less than 0.1 m/s while detecting changes in brightness of only a few parts per million at any given location on the Sun. Tens of millions of different wave motions are possible within the Sun. Some waves travel around the circumference of the Sun, providing information about the density of the upper convective zone. Other waves travel through the interior of the Sun, revealing the density structure near the Sun's core. Still others travel inward toward the center of the Sun until they are bent by the changing solar density and return to the surface.

All of these wave motions are going on at the same time, so sorting out this jumble requires computer analysis of long, unbroken strings of solar observations from several sources. The Global Oscillation Network Group (GONG) is a network of six solar observation stations spread around the world, enabling astronomers to observe the surface of the Sun approximately 90 percent of the time. Other satellites and ground-based networks add to the data stream.

Scientists compare the strength, frequency, and wavelengths of helioseismic data against predicted vibrations calculated from models of the solar interior. This technique provides a powerful test of models of the solar interior. For example, some scientists had proposed that the solar neutrino problem might be solved if we had overestimated the amount of helium in the Sun. This explanation was ruled out by analysis of the waves that penetrate to the core of the Sun. Helioseismology also showed that the value for opacity used in early solar models was too low. This realization led astronomers to recalculate the location of the bottom of the convective zone. Both theory and observation now put the base of the convective zone at 71.3 percent of the way out from the center of the Sun, and uncertainty in this number is less than half a percent.

Working back and forth between observation and theory has enabled astronomers to probe the otherwise inaccessible interior of the Sun. We now know that the energy is produced by nuclear fusion deep in the core and that it moves outward by radiation to the bottom of the convection zone. Then it travels outward by convection to the surface. We also know how the temperature, density, and pressure change with radius and how these factors change the opacity at different distances from the center. This kind of collaboration between theory and observation is essential to observational sciences like astronomy. Even though it is usually not possible to sample directly or to set up controlled experiments, the combined power of theory and observation provides answers to many questions about objects we cannot sample directly.

CHECK YOUR UNDERSTANDING 11.2

As energy moves out from the Sun's core toward its surface, it first travels by _____, then by _____, and then by _____.

a. radiation; conduction; radiation

b. conduction; radiation; convection

c. radiation; convection; radiation

d. radiation; convection; conduction

11.3 The Atmosphere of the Sun

Beyond the convective zone lie the outer layers of the Sun, which are collectively known as the Sun's atmosphere. These concentric layers, shown in **Figure 11.9**, include the *photosphere*, the *chromosphere*, and the *corona*. We can observe these layers of the Sun directly using telescopes, satellites, and even pinhole cameras. (A safety notice: Do *not* look directly at the Sun, especially through an unfiltered telescope. You can cause permanent damage to your vision.) Observations of the Sun's

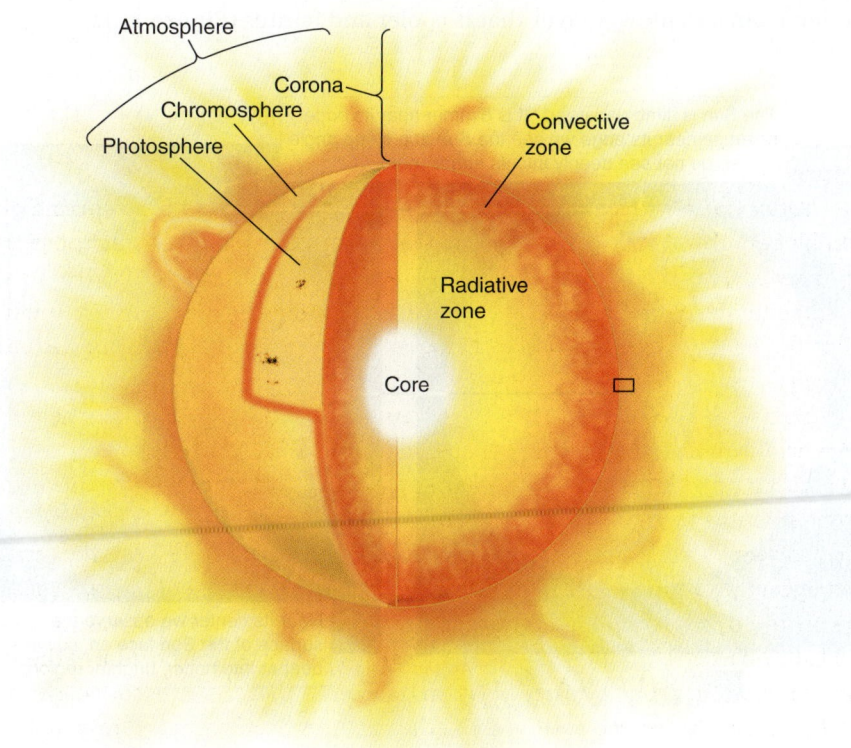

Figure 11.9 The components of the Sun's atmosphere are located above the convective zone and include the photosphere, the chromosphere, and the corona.

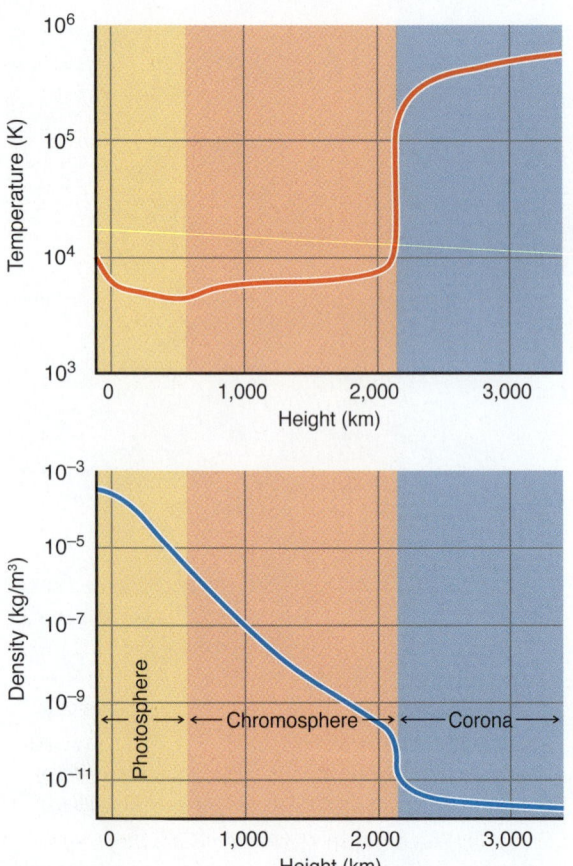

Figure 11.10 The temperature and density of the Sun's atmosphere change abruptly at the boundary between the chromosphere and the corona. The height on these graphs is measured from left to right across the small black box in Figure 11.9.

atmosphere are important because activity in the Sun's atmosphere has consequences both for human-built infrastructure (power grids and satellites in orbit around Earth) and for Earth-bound observers.

Observing the Sun

The Sun is a large ball of gas; unlike Earth, it has no solid surface. Its apparent surface is like a fog bank on Earth. Imagine watching a person walking into a fog bank. After the person disappears from view, you would say he is inside the fog bank, even though he never passed through a definite boundary. The apparent surface of the Sun is similar. Light from the Sun's surface can escape directly into space, so we can see it. Light from below the Sun's surface cannot escape directly into space, so we cannot see it.

The Sun's apparent surface is called the solar **photosphere**. (*Photo* means "light"; the photosphere is the sphere that light comes from.) There is no instant when you can say that you have suddenly crossed the surface of a fog bank, and by the same token there is no instant when we suddenly cross the photosphere of the Sun. The photosphere is about 500 km thick. The Sun appears to have a well-defined surface and a sharp outline when viewed from Earth because 500 km does not look very thick when viewed from a distance of 150 million km. As you can see in the graphs in **Figure 11.10**, the temperature increases sharply across the boundary between the chromosphere and the corona, while the density (and pressure) fall sharply across the same boundary. The density and pressure in the corona are much lower than the density and pressure in Earth's atmosphere.

The Sun appears fainter near its edges than near its center, an effect known as **limb darkening**. Limb darkening is an artifact of the structure of the Sun's photosphere. Near the edge of the Sun you are looking through the photosphere at a steep angle. As a result, you do not see as deeply into the interior of the Sun as when you are looking near the center of the Sun's disk. The light from the limb of the Sun comes from a shallower layer that is cooler and fainter (**Figure 11.11**).

Figure 11.11   (a) When viewed in visible light, the Sun appears to have a sharp outline, even though it has no true surface. The center of the Sun appears brighter, whereas the limb of the Sun is darker—an effect known as limb darkening. The tiny black dot in the middle of the white circle is the planet Mercury transiting the Sun. (b) Looking at the middle of the Sun allows us to see deeper into the Sun's interior than we can by looking at the edge of the Sun. Because higher temperature means more luminous radiation, the middle of the Sun appears brighter than its limb.

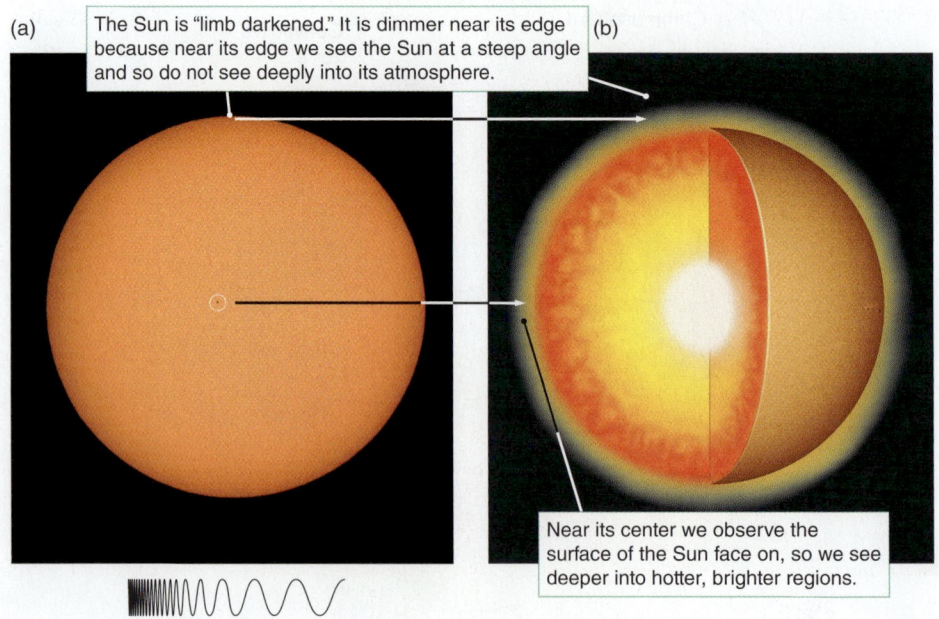

(a)

The Sun is "limb darkened." It is dimmer near its edge because near its edge we see the Sun at a steep angle and so do not see deeply into its atmosphere.

(b)

Near its center we observe the surface of the Sun face on, so we see deeper into hotter, brighter regions.

G X U **V** I R

The Solar Spectrum

In the Sun's atmosphere, the density of the gas drops very rapidly with increasing altitude (see Figure 11.10). All visible solar phenomena take place in the Sun's atmosphere. Most of the radiation from below the Sun's photosphere is absorbed by matter and reemitted at the photosphere as a blackbody spectrum.

As we examine the structure of the Sun in more detail, however, we see that this simple description is incomplete. Light from the solar photosphere must escape through the upper layers of the Sun's atmosphere, which puts its fingerprints on the spectrum that we observe. As photospheric light travels upward through the photosphere, atoms in the cooler gas absorb the light at distinct wavelengths, forming absorption lines (**Figure 11.12**). Absorption lines from more than 70 elements have been identified. Analysis of these lines forms the basis for much of our knowledge of the solar atmosphere, including the composition of the Sun. This is also the starting point for our understanding of the atmospheres and spectra of other stars.

The Sun's Outer Atmosphere: The Chromosphere and Corona

Moving upward through the Sun's photosphere, the temperature falls from 6600 K at the photosphere's bottom to 4400 K at its top. At this point, the trend reverses and the temperature slowly begins to climb, rising to about 6000 K at a height of 1,500 km above the top of the photosphere (see Figure 11.10). This region of increasing temperature is called the **chromosphere** (**Figure 11.13**). The reason for the chromosphere's temperature reversal with increasing height is not well understood, but it may be caused by magnetic fields throughout the region.

The chromosphere was discovered in the 19th century during observations of total solar eclipses, like the one shown in Figure 11.13b. The chromosphere is seen most clearly at the solar limb as a source of emission lines, especially a particular hydrogen line that is produced when the electron falls from the third energy state to the second energy state. This line is known as the Hα line (the "hydrogen alpha" line). The deep red color of the Hα line is what gives the chromosphere its name; *chromo* means "color." The element helium was discovered in 1868 from a spectrum of the chromosphere of the Sun nearly 30 years before it was recognized on Earth; helium is named after *helios*, the Greek word for "Sun."

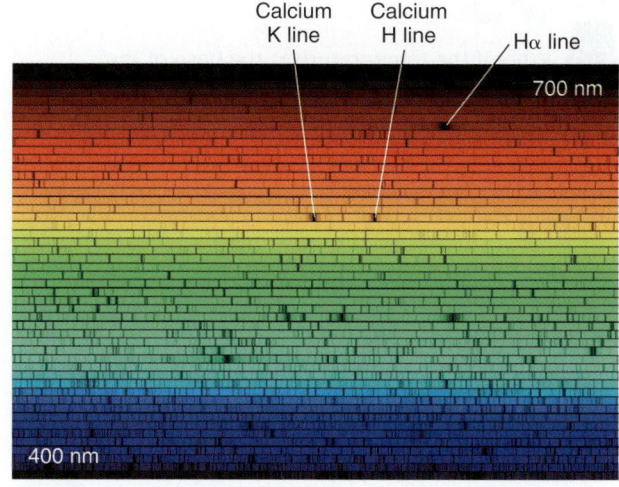

Calcium K line   Calcium H line   Hα line

700 nm

400 nm

Figure 11.12 ★ **WHAT AN ASTRONOMER SEES**
This high-resolution spectrum of the Sun stretches from 400 nanometers (nm) in the lower left corner to 700 nm in the upper right corner and shows black absorption lines. An astronomer will know that this spectrum was produced by passing the Sun's light through a prism-like device and then cutting and folding the single long spectrum (from blue to red) into rows so that it will fit in a single image taken by a camera. She will notice the particularly strong absorption lines, which show up as dark blotches. The one near the top in the red part of the spectrum is particularly noticeable, and an astronomer who looks at a lot of spectra will recognize this line by its color: It is the hydrogen alpha (Hα) line, marking the transition of electrons from the third down to the second energy state of hydrogen. She might also recognize the strong "calcium H" and "calcium K" lines in the orange part of the spectrum. These are recognizable from the combination of their color, their relative strength, and their nearness to each other. The Sun's spectrum is crowded with absorption lines, and an astronomer will immediately know that the outer layers of the Sun are cooler than the layers deep down, because atoms in those outer layers are absorbing energy as it makes its way out from the Sun.

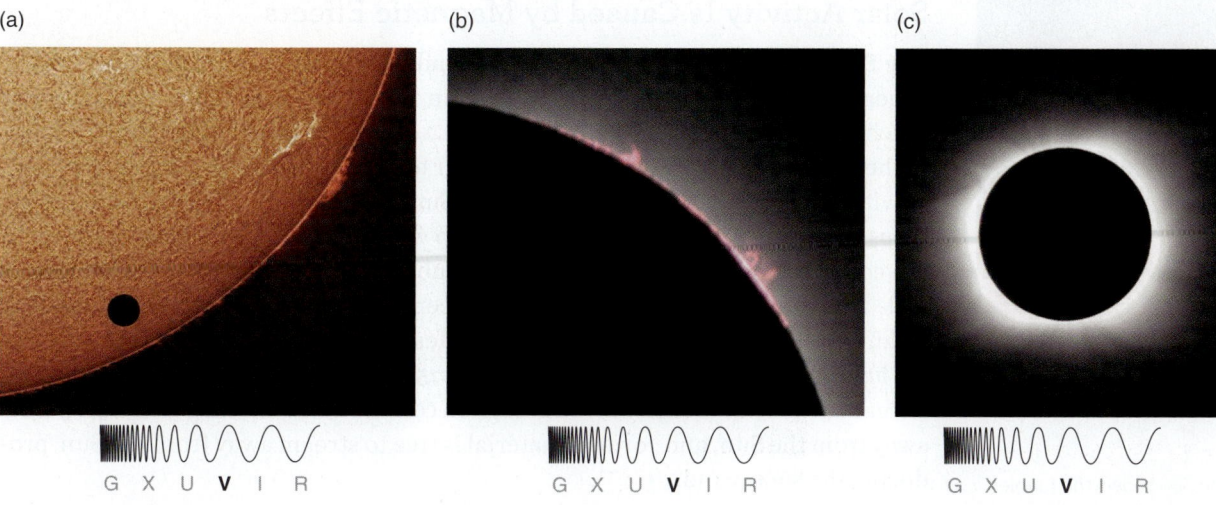

(a)   (b)   (c)

G X U V I R   G X U V I R   G X U V I R

Figure 11.13 (a) This image of the Sun, taken in hydrogen alpha (Hα) light during a transit of Venus, shows structure in the Sun's chromosphere. (The planet Venus is seen in silhouette against the disk of the Sun.) (b) The chromosphere can also be seen during a total eclipse. (c) This eclipse image shows the Sun's corona, consisting of million-kelvin gas that extends for millions of kilometers beyond the surface of the Sun.

what if . . .

What if the temperature at the base of the corona were 10 million K rather than 1 million K: How would the light from this region be different in this hypothetical Sun?

At the top of the chromosphere, across a transition region that is only about 100 km thick, the temperature suddenly soars (see Figure 11.10), while the density abruptly drops. In the **corona**, the hot, outermost part of the Sun's atmosphere, temperatures reach 1 million to 2 million K. The corona is thought to be heated by magnetic fields in much the same way the chromosphere is, but why the temperature changes so abruptly at the transition between the chromosphere and the corona is not at all clear. This is an active area of research today. Because it is so hot, the corona is also a strong source of X-rays. There is so much energy in these X-ray photons that many electrons are stripped away from nuclei, leaving atoms in the corona highly ionized.

The corona is visible during total solar eclipses, when the majority of the light from the Sun is blocked by the Moon. The corona stretches several solar radii beyond the Sun's surface (Figure 11.13c). Spacecraft such as the *Solar and Heliospheric Observatory* (*SOHO*) create artificial eclipses by placing a disk between the Sun and the camera. This makes it possible to study the corona even when there is not a total solar eclipse. These observations help scientists understand how the Sun affects Earth's atmosphere.

CHECK YOUR UNDERSTANDING 11.3

The temperature and density change abruptly at the interface between
a. the radiative zone and the photosphere.
b. the photosphere and the chromosphere.
c. the chromosphere and the corona.
d. the corona and space.

11.4 The Atmosphere of the Sun Is Very Active

The best-known features on the surface of the Sun are relatively dark blemishes in the solar photosphere, called **sunspots**. Sunspots come and go over time. These spots are associated with regions of increased high-energy radiation, loops of material, and explosions that fling particles far out into the Solar System. Long-term patterns have been observed in the variations of sunspots, revealing that the magnetic field of the Sun is constantly changing.

Solar Activity Is Caused by Magnetic Effects

The Sun's magnetic field causes virtually all of the structure in the Sun's atmosphere. High-resolution images of the Sun reveal **coronal loops** that make up much of the Sun's lower corona (**Figure 11.14**). This texture is the result of loops in the magnetic field. The corona itself is far too hot to be held in place by the Sun's gravity, but over most of the surface of the Sun, coronal gas is confined by magnetic loops with both ends firmly anchored deep within the Sun. The magnetic field in the corona acts almost like a network of rubber bands that coronal gas is free to slide along but cannot cross. About 20 percent of the surface of the Sun is covered by an ever-shifting pattern of **coronal holes**. These regions are dark in extreme UV and X-ray images of the Sun (**Figure 11.15**), indicating that they are cooler and lower in density than their surroundings. In coronal holes, the magnetic field points away from the Sun, and coronal material is free to stream away from the Sun, producing the solar wind.

G X **U** V I R

Figure 11.14 A close-up image of the Sun shows the tangled structure of coronal loops.

Recall that the solar wind is responsible for shaping the magnetospheres of planets (see Chapters 7 and 8) and for blowing the tails of comets away from the Sun (see Chapter 9). The relatively steady part of the solar wind consists of lower-speed flows with velocities of about 350 km/s and higher-speed flows with velocities up to about 700 km/s. These higher-speed flows originate in coronal holes. Depending on their speed, particles in the solar wind take about 2–5 days to reach Earth.

The solar wind has been detected by spacecraft as far as 100 astronomical units (AU) from the Sun. But the solar wind does not go on forever. The farther it gets from the Sun, the more it spreads out. Just like radiation, the density of the solar wind follows an inverse square law. At a distance of about 100 AU from the Sun, the solar wind is probably no longer powerful enough to push the gas and dust that lie between stars in the Milky Way. There the solar wind stops abruptly, "piling up" against the pressure of the interstellar medium. **Figure 11.16** shows the region of space over which the wind from the Sun holds sway.

The boundary at which the solar wind piles up against the interstellar medium is one definition of the edge of the Solar System. The *Voyager 1* spacecraft, launched in 1977, is now crossing this boundary. Determining precisely when *Voyager 1* arrived at this boundary has been difficult, with several reports that were later retracted. In 2013, *Voyager 1* entered a region called the "magnetic highway," detected by a change in energetic particles. In this region, the Sun's magnetic field lines are connected to interstellar magnetic field lines, allowing charged particles to zoom in or out of the Solar System. This is one way to define a transition region from inside to outside the Solar System. But some specialists contend that only a change in the *direction* of the magnetic field will serve as a definitive indicator that the spacecraft has reached interstellar space. This change in direction will indicate that the spacecraft has entered a region where the magnetic field lines can no longer be traced back to the Sun.

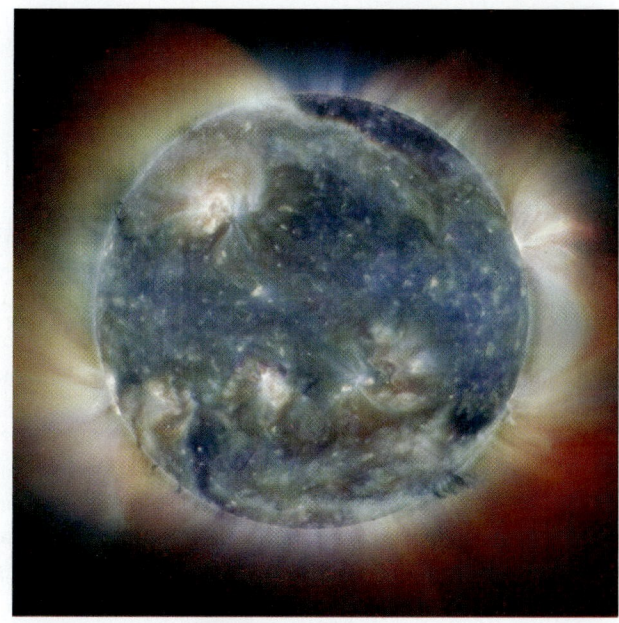

G **X** U V I R

Figure 11.15 X-ray images of the Sun show a very different picture of our star than do images taken in visible light. The brightest X-ray emission comes from the base of the Sun's corona, where gas is heated to temperatures of more than 1 million K. This heating is most powerful above magnetically active regions of the Sun. The dark areas are coronal holes—regions where the Sun's corona is cooler and has lower density.

Astronomy in Action: Inverse Square Law

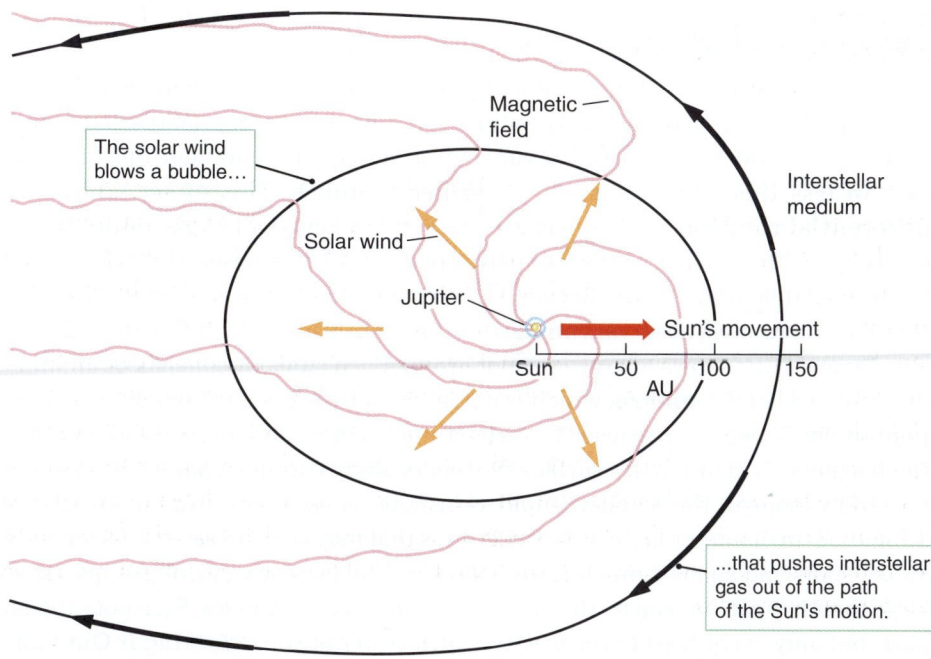

Figure 11.16 The solar wind streams away from the Sun for about 100 AU, until it finally piles up against the pressure of the interstellar medium through which the Sun is traveling. In this figure, the Sun travels to the right, on its orbit around the center of the Milky Way Galaxy (shown by the red arrow). Jupiter's orbit is shown for scale, as the small blue circle.

Sunspots and Temperature

Sunspots are about 1500 K cooler than their surroundings. What does this tell us about their brightness? Think back to the Stefan-Boltzmann law in Chapter 5. The energy emitted from every square meter of a blackbody every second (the "intensity") is proportional to the fourth power of the temperature, T. The constant of proportionality is the Stefan-Boltzmann constant, σ, which has a value of 5.67×10^{-8} W/(m² K⁴). We write this relationship as

$$I = \sigma T^4$$

How much less energy comes out of a sunspot than out of the rest of the Sun? Let's take round numbers for the temperatures of a typical sunspot and of the surrounding photosphere: 4500 K and 6000 K, respectively. We can set up two equations:

$$I_{spot} = \sigma T_{spot}{}^4$$

and

$$I_{surface} = \sigma T_{surface}{}^4$$

We could solve each of these separately and then divide the value of I_{spot} by the value of $I_{surface}$ to find out how much fainter the sunspot is, but it is much easier to solve for the *ratio* $I_{spot}/I_{surface}$. We divide the left side of one equation by the left side of the other equation and do the same with the right sides. Anything that is the same in both equations will divide out.

$$\frac{I_{spot}}{I_{surface}} = \left(\frac{\sigma T_{spot}{}^4}{\sigma T_{surface}{}^4} \right)$$

The constant σ divides out. Because both terms on the right are raised to the fourth power, we can divide them first and then take them to the fourth power:

$$\frac{I_{spot}}{I_{surface}} = \left(\frac{T_{spot}}{T_{surface}} \right)^4$$

Plugging in our values for T_{spot} and $T_{surface}$ gives

$$\frac{I_{spot}}{I_{surface}} = \left(\frac{4500\,K}{6000\,K} \right)^4$$

$$\frac{I_{spot}}{I_{surface}} = 0.32$$

Multiplying both sides by $I_{surface}$ gives

$$I_{spot} = 0.32 I_{surface}$$

So the amount of energy coming from a square meter of sunspot every second is about one-third as much as the amount of energy coming from a square meter of surrounding surface every second. In other words, the sunspot is about one-third as bright as an equal area of the surrounding photosphere. This is still extremely bright—if you could cut out the sunspot and place it elsewhere in the sky, it would be brighter than the full Moon.

Sunspots and Changes in the Sun

Early telescopic observations of sunspots made during the 17th century led to the discovery that the Sun rotates. Like Saturn, the Sun rotates more rapidly at its equator than it does at higher latitudes; sunspots at the equator complete one full rotation in less time than sunspots at higher latitudes. This effect, known as **differential rotation**, occurs because the Sun is a large ball of gas rather than a solid object. The Sun has an average period of about 27 days as seen from Earth and 25 days relative to the stars. Because Earth orbits the Sun in the same direction that the Sun rotates, observers on Earth see a slightly longer rotation period.

Figure 11.17 shows a large sunspot group. Each sunspot consists of an inner dark core called the **umbra**, which is surrounded by a less dark region called the **penumbra**. Sunspots are caused by magnetic fields thousands of times greater than the magnetic field at Earth's surface. Sunspots always occur in pairs that are connected by loops in the magnetic field. Sunspots range in size from a few tens of kilometers in diameter up to complex groups that may contain several dozen individual spots and span as much as 150,000 km. The largest sunspot groups are so large that they can be seen by the naked eye, using eclipse glasses. Sunspots appear dark, but only in contrast to the brighter surface of the Sun (**Working It Out 11.2**).

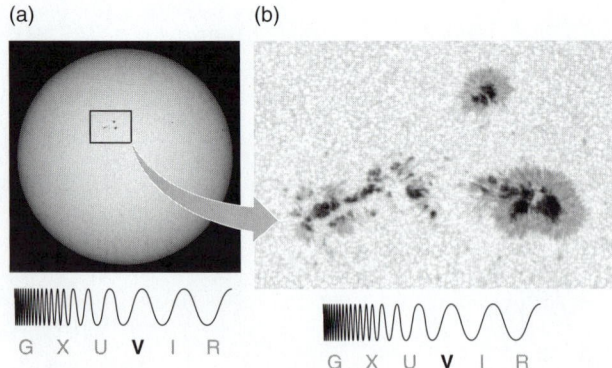

(a)   (b)

G X U V I R   G X U V I R

Figure 11.17 (a) This image from the Solar Dynamics Observatory (SDO), taken in 2010, shows a large sunspot group. Sunspots are magnetically active regions that are cooler than the surrounding surface of the Sun. (b) The close-up shows a very high-resolution view of a group of sunspots: The dark *umbra* is surrounded by the lighter *penumbra*. The solar surface around the sunspot bubbles with separate cells of hot gas. The smallest features are about 100 km across.

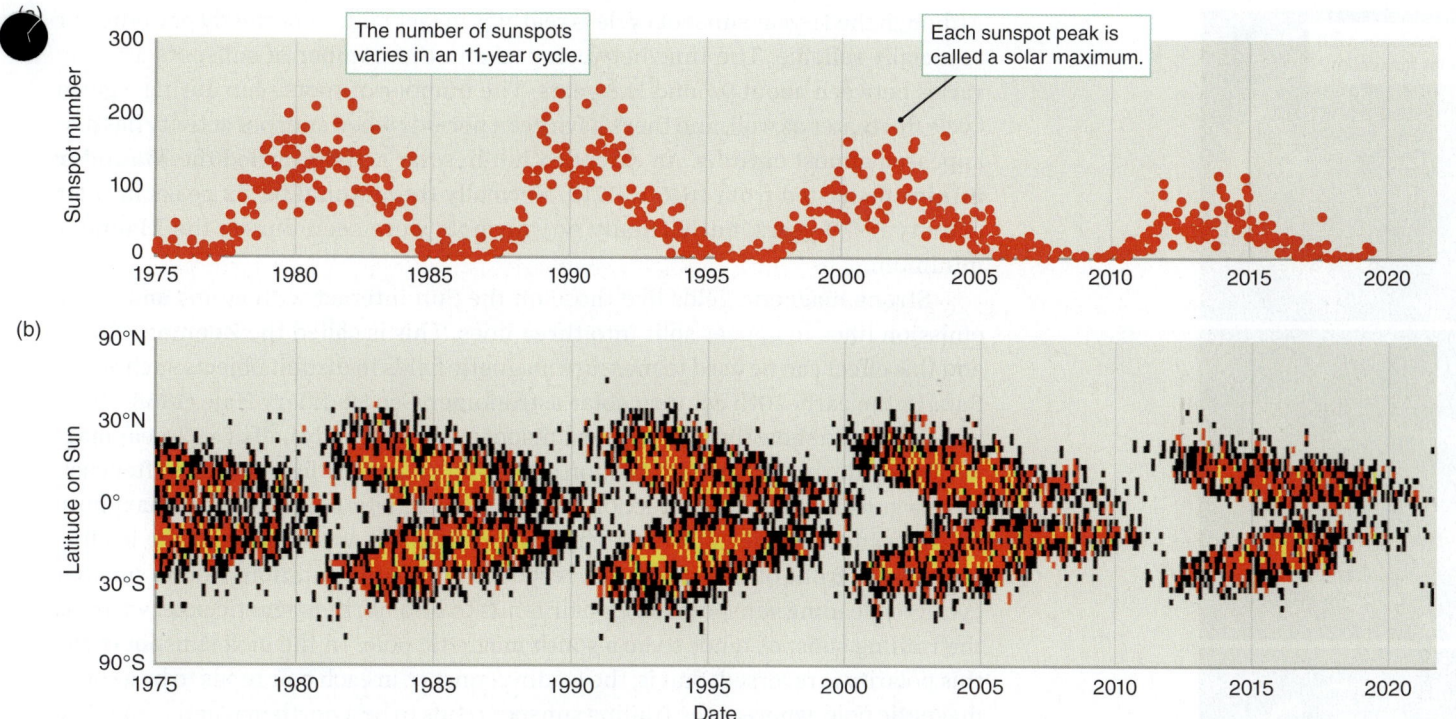

Figure 11.18 (a) The number of sunspots varies with time, as shown in this graph of the past few solar cycles. (b) The solar butterfly diagram shows the fraction of the Sun covered at each latitude. The data are color coded to show the percentage of the strip at that latitude that is covered in sunspots at that time: black, 0–0.1 percent; red, 0.1–1.0 percent; yellow, greater than 1.0 percent.

Individual sunspots do not stay around for very long. Although sunspots occasionally last 100 days or longer, half of all sunspots come and go in about 2 days, and 90 percent are gone within 11 days. The number of sunspots on the Sun varies in an 11-year pattern called the **sunspot cycle**. **Figure 11.18a** shows data for several recent cycles. Over 11 years, both the number and the location of sunspots change. At the beginning of a cycle, sunspots appear at solar latitudes approximately 30° north and south of the solar equator. Over the following years, sunspots are found closer to the equator as their number increases to a maximum known as the **solar maximum** and then declines. As the last few sunspots approach the equator, sunspots reappear at middle latitudes and the next cycle begins. Figure 11.18b shows this pattern; this diagram is often referred to as the sunspot "butterfly diagram."

Telescopic observations of sunspots date back 400 years. Records of sunspot observations by Chinese astronomers date back to nearly 2,000 years before that. **Figure 11.19** shows the historical record of sunspot activity since 1600.

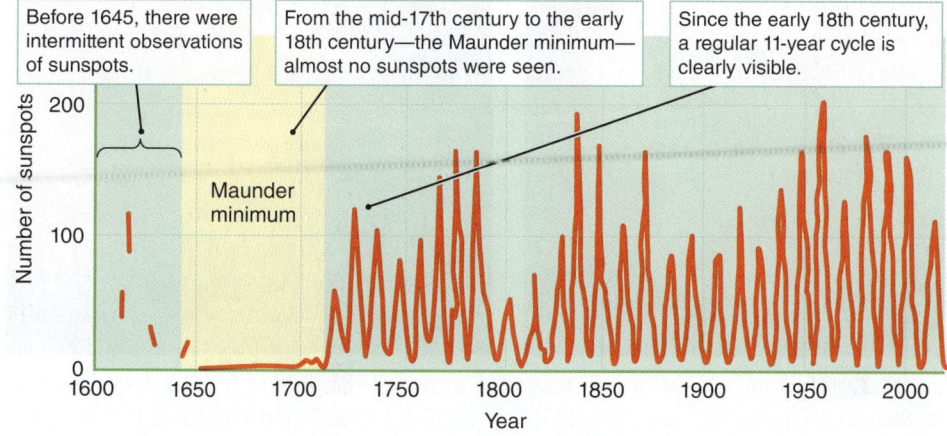

Figure 11.19 Sunspots have been observed for hundreds of years. In this plot, the 11-year cycle in the number of sunspots (half of the 22-year solar magnetic cycle) is clearly visible. Sunspot activity varies greatly over time. The period from the middle of the 17th century to the early 18th century, when almost no sunspots were seen, is called the Maunder minimum.

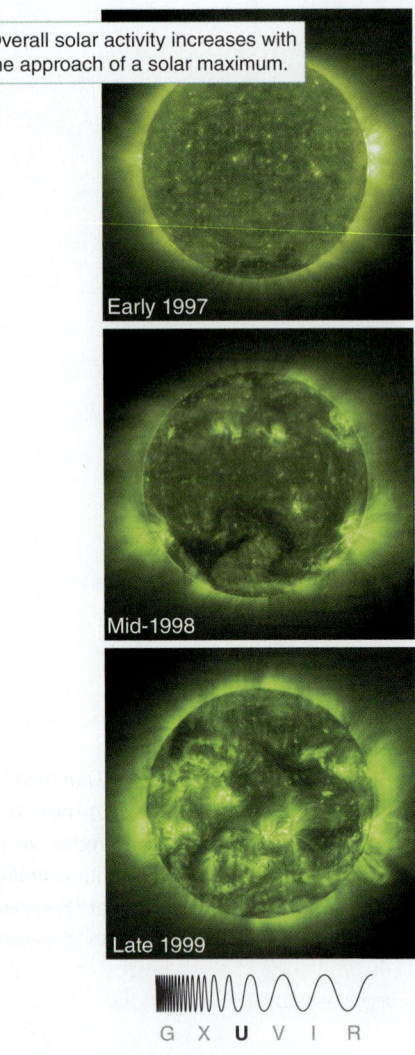

Overall solar activity increases with the approach of a solar maximum.

Early 1997

Mid-1998

Late 1999

G X **U** V I R

Figure 11.20 This time sequence of *Solar and Heliospheric Observatory* (*SOHO*) images shows the increase in activity on the Sun as solar maximum approaches.

Although the 11-year sunspot cycle is real, it is in fact neither perfectly periodic nor especially reliable. The time between peaks in the number of sunspots actually varies between about 9.7 and 11.8 years. The number of spots seen during a given cycle fluctuates as well, and there have been periods when sunspot activity has disappeared almost entirely. An extended lull in solar activity, called the **Maunder minimum**, lasted from 1645 to 1715. Normally there would be six peaks in solar activity in 70 years, but virtually no sunspots were seen during the Maunder minimum.

Strong magnetic fields like those on the Sun interact with atoms and cause emission lines to appear split into three lines. This is called the **Zeeman effect**, and this effect can be used to measure magnetic fields in distant objects such as the Sun. In the early 20th century, solar astronomer George Ellery Hale (1868–1938) was the first to show that the 11-year sunspot cycle is actually half of a 22-year magnetic cycle during which the direction of the Sun's magnetic field reverses after each 11-year sunspot cycle. The direction of the Sun's magnetic field flips at the maximum of each sunspot cycle. Sunspots tend to come in pairs, with one spot (the leading sunspot) in front of the other with respect to the Sun's rotation. In one sunspot cycle, the leading sunspot in each pair tends to be a north magnetic pole, whereas the trailing sunspot tends to be a south magnetic pole. In the next sunspot cycle, this polarity is reversed; that is, the leading sunspot in each pair tends to be a south magnetic pole, whereas the trailing sunspot tends to be a north magnetic pole. The transition between these two magnetic polarities occurs near the peak of each sunspot cycle.

Sunspots are only one of several phenomena that follow the Sun's 22-year cycle of magnetic activity. The solar maximum is a time of intense activity, as can be seen in the ultraviolet images of the Sun in **Figure 11.20**. Sunspots are often accompanied by a brightening of the solar chromosphere that is seen most clearly in emission lines such as Hα. These bright regions are known as solar **active regions**. The magnificent loops arching through the solar corona, shown in **Figure 11.21**, are called solar **prominences**. They are magnetic tubes of relatively cool (5000–10,000 K) gas extending through the million-kelvin gas of the corona. These prominences are anchored in the active regions. Although most prominences are relatively quiet, others can erupt out through the corona, towering a million kilometers or

(a)   (b)

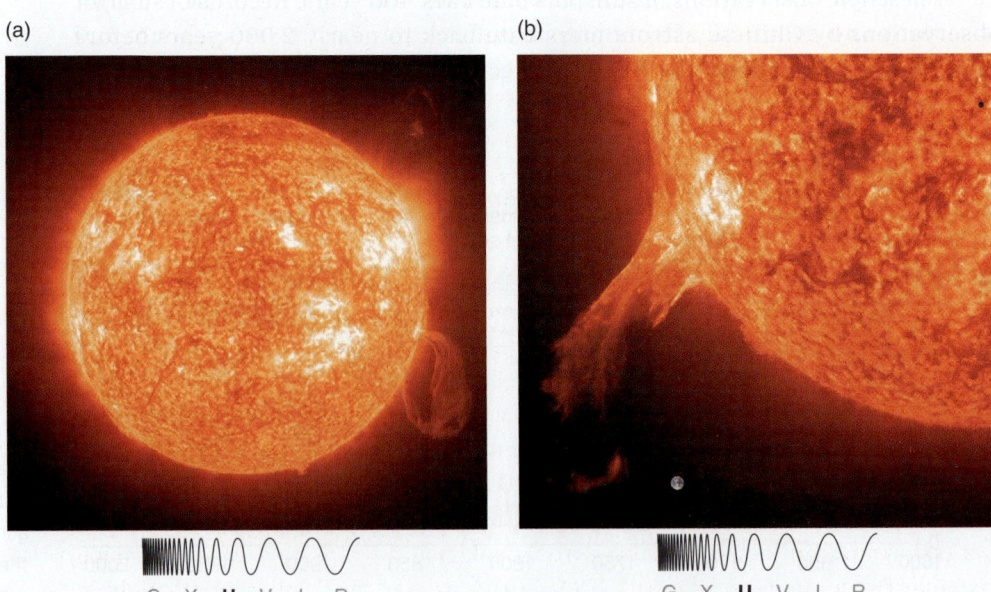

G X **U** V I R   G X **U** V I R

Figure 11.21 (a) Solar prominences are magnetically supported arches of hot gas that rise high above active regions on the Sun. (b) Large prominences are very large indeed, as shown in this close-up view of the base of a large prominence. Earth is shown to scale.

more over the surface of the Sun and ejecting material into the corona at speeds of 1,000 km/s.

Figure 11.22 shows **solar flares** erupting from two sunspot groups. Solar flares are the most energetic form of solar activity. They are violent eruptions in which enormous amounts of magnetic energy are released over the course of a few minutes to a few hours. Solar flares can heat gas to temperatures of 20 million K, and they are the source of intense X-ray and gamma-ray radiation. Hot plasma (consisting of ions and free electrons) moves outward from flares at speeds that can reach 1,500 km/s. Magnetic effects can then accelerate subatomic particles to almost the speed of light. Such events, called **coronal mass ejections** (**Figure 11.23**), send powerful bursts of energetic particles outward through the Solar System. Coronal mass ejections occur about once per week during the minimum of the sunspot cycle and as often as several times per day near the maximum of the cycle.

The Effects of Solar Activity on Earth

On average, Earth receives 1,361 watts per square meter (W/m²) from the Sun. As you can see in **Figure 11.24**, satellite measurements of the amount of radiation coming from the Sun show that this value can vary by as much as 0.2 percent over periods of a few weeks as dark sunspots in the photosphere and bright spots in the chromosphere move across the disk. However, the increased radiation from active regions on the Sun more than makes up for the reduction in radiation from sunspots. On average, the Sun is about 0.1 percent brighter during the peak of a solar cycle than it is at its minimum.

Solar activity affects Earth in many ways. Solar active regions are the source of most of the Sun's extreme ultraviolet and X-ray emissions, energetic radiation that heats Earth's upper atmosphere and, during periods of increased solar activity, causes Earth's upper atmosphere to expand. When this happens, the swollen upper atmosphere can significantly increase the atmospheric drag on spacecraft orbiting at relatively low altitudes, such as that of the Hubble Space Telescope, causing their orbits to **decay**. Periodic boosts are necessary to keep the Hubble Space Telescope in its orbit. In the past, these boosts were provided by the space shuttle servicing missions, which ceased in 2009. The Hubble Space Telescope will likely remain in orbit for a few decades, but over time the orbit will decay and eventually the telescope will reenter Earth's atmosphere and burn up.

Earth's magnetosphere is the result of the interaction between Earth's magnetic field and the solar wind. Recall from Chapter 7 that this interaction causes the aurora in Earth's atmosphere. Increases in the solar wind, especially coronal mass ejections, can cause magnetic storms that disrupt power grids and cause large regional blackouts. Coronal mass ejections that are emitted in the direction of Earth also hinder radio communication and navigation, and they can damage sensitive satellite electronics, including those of communication satellites. In addition, energetic particles accelerated in solar flares pose one of the greatest dangers to human exploration of space.

The *Solar and Heliospheric Observatory* (*SOHO*) spacecraft is a joint mission between NASA and the European Space Agency. The *SOHO* orbits in lockstep with Earth at a location approximately 1,500,000 km (0.01 AU) from Earth and almost directly in line between Earth and the Sun. The *SOHO* carries 12 scientific instruments that monitor the Sun and measure the solar wind upstream of Earth. Through the Solar Dynamics Observatory (SDO), NASA studies the solar magnetic field in

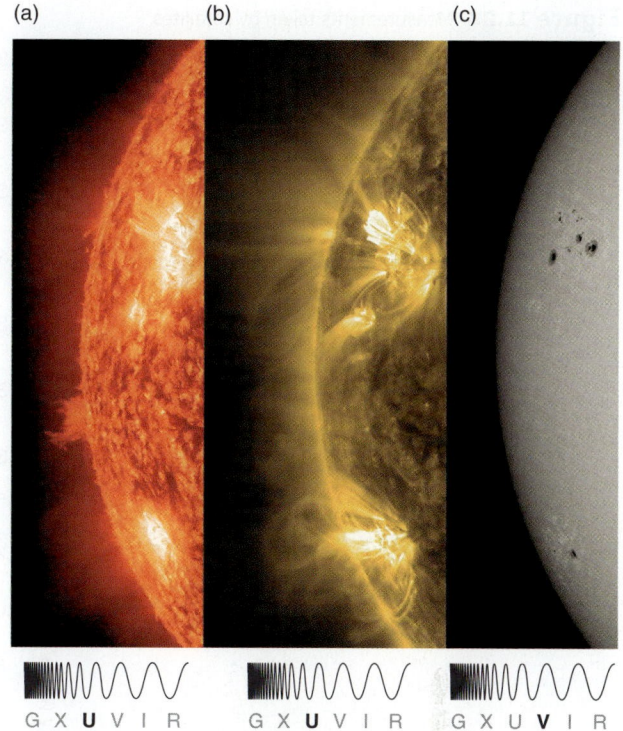

(a)   (b)   (c)

G X **U** V I R   G X **U** V I R   G X **U** V I R

Figure 11.22 The Solar Dynamics Observatory (SDO) captured these images of active regions of the Sun that produced solar flares in August 2011. (a) Activity near the surface at 60,000 K is visible in extreme ultraviolet light (along with a prominence rising up from the Sun's edge). (b) Viewed at other ultraviolet wavelengths, many looping arcs and plasma heated to about 1 million K become visible. (c) The dark spots in this visible-light image are the magnetically intense sunspots that are the sources of all the activity.

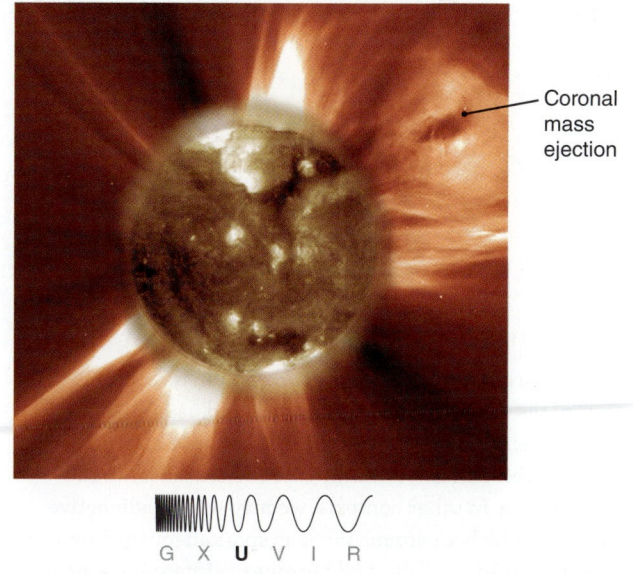

Coronal mass ejection

G X **U** V I R

Figure 11.23 This image from the *SOHO*, composed from multiple instruments, shows a coronal mass ejection (upper right), with a simultaneously recorded ultraviolet image of the Sun superimposed.

Figure 11.24 Measurements taken by satellites above Earth's atmosphere show that the amount of light from the Sun changes slightly over time.

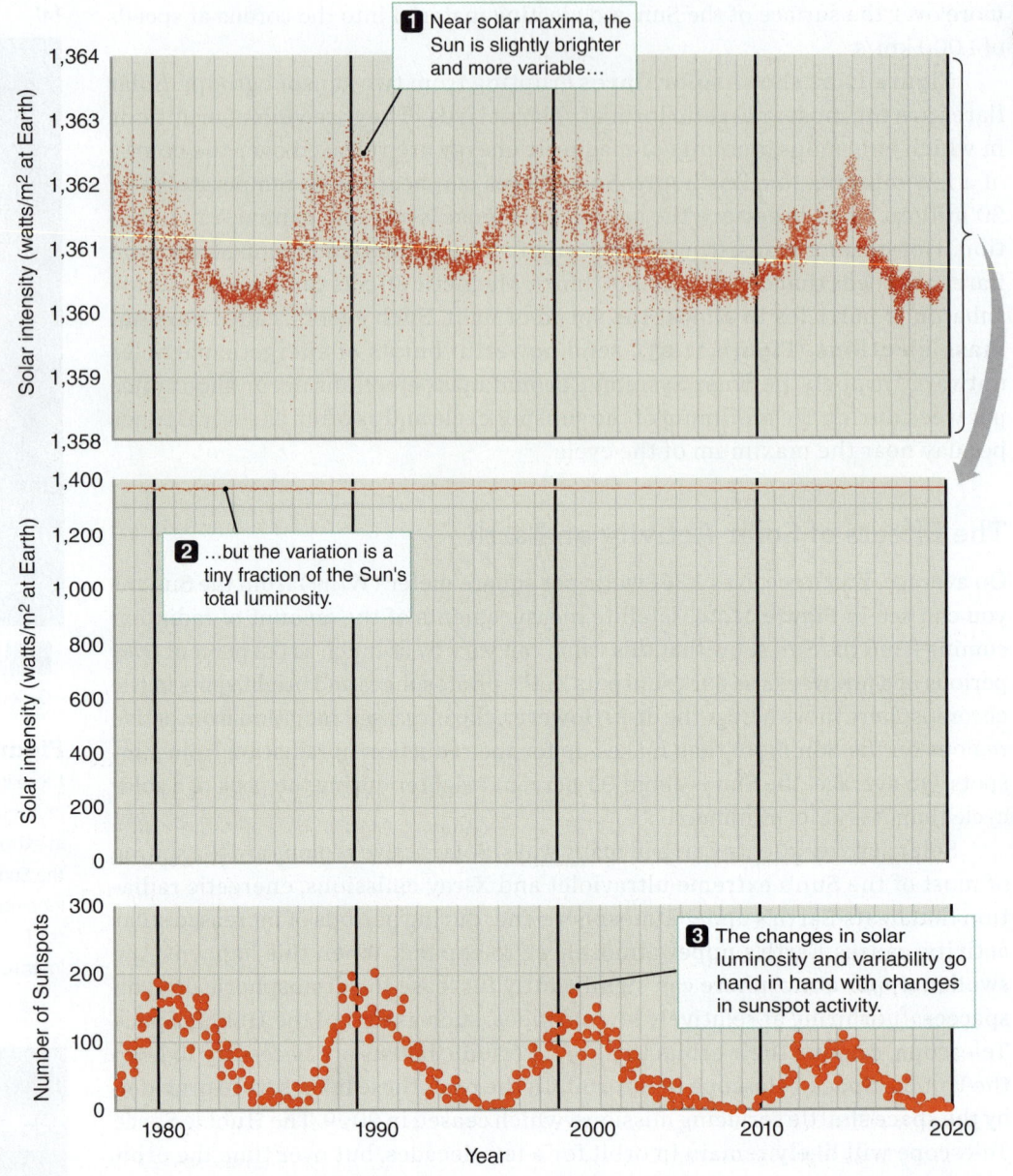

1 Near solar maxima, the Sun is slightly brighter and more variable…

2 …but the variation is a tiny fraction of the Sun's total luminosity.

3 The changes in solar luminosity and variability go hand in hand with changes in sunspot activity.

VOCABULARY ALERT

decay In everyday language, *decay* often describes a process of decomposition via microbes (and usually implies a bad smell). Here we mean that the satellite loses energy, and its orbit shrinks, until the satellite crashes into the atmosphere. In other contexts, we may mean radioactive decay, in which an atomic nucleus spontaneously emits energy and/or particles and becomes a different element.

order to predict when major solar events will occur, rather than simply responding after they happen. Detailed observations from the SDO help astronomers understand the complex nature of the solar atmosphere.

Solar activity certainly affects Earth's upper atmosphere, and we might imagine that it could affect weather patterns as well, although a mechanism to connect the two is unknown. Solar physics models suggest that observed variations in the Sun's luminosity could account for differences of only about 0.1 K in Earth's average temperature—much less than the effects due to the ongoing buildup of carbon dioxide in Earth's atmosphere. However, triggering the onset of an ice age may require a drop in global temperatures of only about 0.2–0.5 K, so the search for a definite link between solar variability and changes in Earth's climate persists.

CHECK YOUR UNDERSTANDING 11.4

Sunspots, flares, prominences, and coronal mass ejections are all caused by

a. magnetic activity on the Sun.

b. the interaction of the Sun's magnetic field and the interstellar medium.

c. the interaction of the solar wind and Earth's magnetic field.

d. the interaction of the solar wind and the Sun's magnetic field.

what if . . .

What if the Sun had only a very small magnetic field on the surface: How might that change the appearance of the Sun?

---

reading Astronomy News

New sunspots potentially herald increased solar activity

by Karen Fox and Lina Tran, NASA's Goddard Space Flight Center

The start of a new solar cycle is an exciting time for solar scientists, because the exact moment is unpredictable.

On May 29, 2020, a family of sunspots—dark spots that freckle the face of the sun, representing areas of complex magnetic fields—sported the biggest solar flare since October 2017. Although the sunspots are not yet visible (they will soon rotate into view over the left limb of the sun), NASA spacecraft spotted the flares high above them.

The flares were too weak to pass the threshold at which NOAA's Space Weather Prediction Center (which is the U.S. government's official source for space weather forecasts, watches, warnings and alerts) provides alerts. But after several months of very few sunspots and little solar activity, scientists and space weather forecasters are keeping their eye on this new cluster to see whether they grow or quickly disappear. The sunspots may well be harbingers of the sun's solar cycle ramping up and becoming more active.

Or they may not. It will be a few more months before we know for sure.

As the sun moves through its natural 11-year cycle, in which its activity rises and falls, sunspots rise and fall in number, too. NASA and NOAA track sunspots in order to determine, and predict, the progress of the solar cycle—and ultimately, solar activity. Currently, scientists are paying close attention to the sunspot number as it's key to determining the dates of solar minimum, which is the official start of Solar Cycle 25. This new sunspot activity could be a sign that the sun is possibly revving up to the new cycle and has passed through minimum.

However, it takes at least six months of solar observations and sunspot-counting after a minimum to know when it's occurred. Because that minimum is defined by the lowest number of sunspots in a cycle, scientists need to see the numbers consistently rising before they can determine when exactly they were at the bottom. That means solar minimum is an instance only recognizable in hindsight: It could take six to 12 months after the fact to confirm when minimum has actually passed.

This is partly because our star is extremely variable. Just because the sunspot numbers go up or down in a given month doesn't mean it won't reverse course the next month, only to go back again the month after that. So, scientists need long-term data to build a picture of the sun's overall trends through the solar cycle. Commonly, that means the number we use to compare any given month is the average sunspot number from six months both backward and forward in time—meaning that right now, we can confidently characterize what October 2019 looks like compared to the months before it (there were definitely fewer sunspots!), but not yet what November looks like compared to that.

On May 29, at 3:24 a.m. EST, a relatively small M-class solar flare blazed from these sunspots. Solar flares are powerful bursts of radiation. Harmful radiation from a flare cannot pass through Earth's atmosphere to physically affect humans on the ground, however—when intense enough—they can disturb the atmosphere in the layer where GPS and communications signals travel. The intensity of this flare was below the threshold that could affect geomagnetic space and below the threshold for NOAA to create an alert.

Nonetheless, it was the first M-class flare since October 2017—and scientists will be watching to see if the sun is indeed beginning to wake up.

EVALUATING THE NEWS

1. This article mentions the Space Weather Prediction Center. Why is such a center necessary to the economic health of the United States? In other words, why do we fund such a center?

2. What is the connection between sunspots and solar flares?

3. In your own words, explain why a single large flare is not enough information to determine whether the Sun is entering a more active part of its cycle.

4. We have not discussed the details of flare types in this text. The article describes the flare as a "relatively small M-class" flare. Use the internet to learn more about flare classification. What is the smallest class? What is the largest? Where do M-class flares fit into this classification scheme?

5. Look up what's happening on the Sun now. Did this large flare indicate that the solar cycle was "ramping up and becoming more active?"

Source: **Karen Fox and Lina Tran**, "New Sunspots Potentially Herald Increased Solar Activity," NASA's Goddard Space Flight Center, June 1, 2020. Via Science X at Phys.org.

SUMMARY

The forces due to pressure and gravity balance each other in hydrostatic equilibrium, maintaining the Sun's structure. Nuclear reactions converting hydrogen to helium are the source of the Sun's energy. As hydrogen fuses to form helium in the core of the Sun, neutrinos are emitted. Neutrinos are elusive, almost massless particles that interact only very weakly with other matter. Observations of neutrinos confirm that nuclear fusion is the Sun's primary energy source. Energy created in the Sun's core moves outward to the surface, first by radiation and then by convection. The Sun has multiple concentric layers, each with a characteristic density, temperature, and pressure. The temperature of the Sun's atmosphere ranges from about 6600 K near the bottom to as much as 2 million K at the top. Material streaming away from the Sun's corona creates the solar wind. Sunspots are photospheric regions that are cooler than their surroundings, and they reveal 11- and 22-year cycles in solar activity. Solar storms can disrupt power grids and damage satellites.

(1) The outward pressure of radiation produced inside the Sun balances the inward pull of gravity at every point. This balance is dynamically maintained. An energy balance is also maintained, with the energy produced in the core of the Sun balancing the energy lost from the surface.

(2) When four hydrogen atoms become one helium atom, some mass is lost. This mass is turned into energy, nearly all of which leaves the Sun either as photons (light) or as neutrinos. Fusion is confined to the core of the Sun because elsewhere in the Sun the temperature and pressure are too low.

(3) Energy moves outward through the Sun by radiation and by convection.

(4) The interior of the Sun is divided into zones that are defined by how energy is transported in that region. This model of the interior of the Sun has been tested by helioseismology, in much the same way that the model of Earth's interior has been tested by seismology. The outer layers of the Sun (the photosphere, chromosphere, and corona) are distinguished by changes in temperature and pressure.

(5) Activity on the Sun follows a cycle that peaks every 11 years but takes 22 full years for the magnetic field to reset completely. This activity, particularly the ejections of mass from the corona, produces auroras and affects human technology. Changes in the space around Earth that are caused by effects like outflows from the Sun are often called "space weather."

QUESTIONS AND PROBLEMS

TESTING YOUR UNDERSTANDING

1. **T/F:** Hydrostatic equilibrium is the balance between energy production and loss.

2. **T/F:** Six hydrogen nuclei are involved in the proton-proton chain, although only four of them wind up in the helium nucleus.

3. **T/F:** Photons travel outward through the Sun because the innermost part of the Sun is more crowded with photons than the outer parts.

4. **T/F:** Neutrinos make it out of the Sun faster than photons. This means they travel faster than photons.

5. **T/F:** Sunspots can be comparable in size to Earth.

6. The physical model of the Sun's core has been confirmed by observations of
   a. neutrinos and seismic vibrations.
   b. sunspots and solar flares.
   c. neutrinos and positrons.
   d. sample returns from spacecraft.
   e. sunspots and seismic vibrations.

7. Ultimately, the Sun's energy comes from
   a. the mass of hydrogen nuclei.
   b. gravitational collapse.
   c. residual heat of formation.
   d. the slowing of its rotation rate.

8. Radiation transports energy by moving _____, whereas convection transports energy by moving _____.
   a. neutrinos; matter
   b. neutrinos; light
   c. light; matter
   d. matter; light

9. The solar wind
   a. makes a perfectly spherical bubble around the Solar System.
   b. does not interact with the magnetic field of the Sun.
   c. creates a teardrop-shaped bubble around the Solar System.
   d. creates a wind pressure much stronger than a wind on Earth.

10. Sunspots peak every _____ years, a consequence of the _____-year magnetic solar cycle.
   a. 11; 22
   b. 22; 11
   c. 5.5; 11
   d. 22; 44

11. Hydrostatic equilibrium inside the Sun means that
   a. energy produced in the core equals energy radiated from the surface.
   b. radiation pressure balances the weight of outer layers pushing down.
   c. the Sun absorbs and emits equal amounts of energy.
   d. the Sun is collapsing.

12. The energy that is emitted from the Sun is produced
   a. at the interface between the chromosphere and the photosphere.
   b. at the top of the convection zone.
   c. in the core, by nuclear fusion.
   d. at the surface.

13. In the proton-proton chain, four hydrogen nuclei are converted to a helium nucleus. This does not happen spontaneously on Earth because the process requires
   a. vast amounts of hydrogen.
   b. very high temperatures and pressures.
   c. hydrostatic equilibrium.
   d. very strong magnetic fields.

14. The solar neutrino problem pointed to a fundamental gap in our knowledge of
   a. nuclear fusion.
   b. neutrinos.
   c. hydrostatic equilibrium.
   d. magnetic fields.

15. Sunspots appear dark because
   a. they have very low density.
   b. magnetic fields absorb most of the light that falls on them.
   c. they are cooler than their surroundings.
   d. they are regions of very high pressure.

16. Sunspots change in number and location during the solar cycle. This phenomenon is connected to
   a. the rotation rate of the Sun.
   b. the temperature of the Sun.
   c. the magnetic field of the Sun.
   d. the tilt of the axis of the Sun.

17. Suppose an abnormally large amount of hydrogen suddenly burned in the core of the Sun. Which of the following would be observed first?
   a. The Sun would become brighter.
   b. The Sun would swell and become larger.
   c. The Sun would become bluer.
   d. The Sun would emit more neutrinos.

18. The solar corona has a temperature of more than 1 million K; the photosphere has a temperature of only about 6000 K. Why isn't the corona much, much brighter than the photosphere?
   a. The magnetic field traps the light.
   b. The corona emits only X-rays.
   c. The photosphere is closer to us.
   d. The corona has a much lower density.

19. The Sun rotates once every 25 days relative to the stars. The Sun rotates once every 27 days relative to Earth. Why are these two numbers different?
   a. The stars are farther away.
   b. Earth is smaller.
   c. Earth moves in its orbit during this time.
   d. The Sun moves relative to the stars.

20. Some engineers and physicists have been working to solve the world's energy supply problem by constructing power plants that would convert hydrogen to helium. Our Sun seems to have solved this problem. On Earth, what is the major obstacle to this solution?
   a. It is difficult to achieve the temperatures and pressures of the interior of the Sun.
   b. There is not enough hydrogen.
   c. The helium shortage is driving up prices.
   d. All the hydrogen is locked up in water already.

THINKING ABOUT THE CONCEPTS

21. The Sun's stability depends on hydrostatic equilibrium and energy balance. Describe how both of these work.

22. Explain how hydrostatic equilibrium acts as a safety valve to keep the Sun at its constant size, temperature, and luminosity.

23. In Figure 11.3, two protons are shown interacting at various speeds.
   a. What is different about parts (a) and (b)?
   b. Why are the blue arrows larger in (b) than in (a)? What do these blue arrows represent? ⊙

24. Radiative transfer takes photons from hot regions to cool regions, on average. Why?

25. Describe nuclear fusion and how it relates to the Sun's source of energy.

26. Two of the three atoms in a molecule of water (H_2O) are hydrogen. Why are Earth's oceans not fusing hydrogen into helium, transforming our planet into a star?

27. Sunspots are dark splotches on the Sun. Why are they dark?

28. Imagine that the Sun was more massive than it is. How would this change the gravitational force in the center of the Sun? How would this change in gravitational force affect the fusion rate in the core? How would the Sun's luminosity change as a result?

29. Figure 11.4 diagrams the proton-proton chain. In this figure, a number of squiggly arrows are pointing away from the interactions. What do these squiggly arrows represent? Are they waves or particles or both? Where does the energy come from that is turned into these objects? ⊛

30. The proton-proton chain is often described as "fusing four hydrogens into one helium," but actually six hydrogen nuclei are involved in the reaction. Why don't we include the other two nuclei in our description?

31. On Earth, nuclear power plants use *fission* to generate electricity. A heavy element like uranium is broken into many smaller atoms, where the total mass of the fragments is less than the original atom. Explain why fission could not be powering the Sun today.

32. ★ WHAT AN ASTRONOMER SEES In Figure 11.12, an astronomer would identify hydrogen and calcium lines immediately. Explain how these lines may have been identified the first time, and how experience might lead an astronomer to be able to instantly recognize them. ⊛

33. Discuss the "solar neutrino problem" and how this problem was solved.

34. What technique do you find in common between how we probe the internal structure of the Sun and how we probe the internal structure of Earth?

35. How is the fate of the Hubble Space Telescope tied to solar activity?

APPLYING THE CONCEPTS

36. Imagine that the Sun were twice as luminous as it is now, so that it burned twice as much hydrogen into helium each second. How would this affect the lifetime of the Sun?
    a. Make a prediction: Should the Sun have a longer or a shorter lifetime if it burns its fuel faster?
    b. Follow Working It Out 11.1 to calculate the lifetime of the Sun if it were twice as luminous.
    c. Check your work by comparing your answer to your prediction.

37. Suppose that 30 percent of the Sun's mass could be involved in nuclear fusion instead of 10 percent. How would this affect the lifetime of the Sun?
    a. Make a prediction: Should the Sun have a longer or a shorter lifetime if more of it is involved in nuclear fusion?
    b. Follow Working It Out 11.1 to calculate the lifetime of the Sun if 30 percent of the mass were involved in nuclear fusion.
    c. Check your work by comparing your answer to your prediction.

38. From the H-R diagram at the end of Chapter 10 (see Figure 10.21), you can determine that a star that with a mass of 0.6 M_{Sun} has a luminosity of about 1/10 the luminosity of the Sun (0.1 L_{Sun}). This luminosity means that the star must be burning 1/10 as much mass as the Sun each year. How will this affect the lifetime of the star?
    a. Make a prediction: This star has less mass than the Sun, but it burns it much more slowly. Compare the two fractions in the problem statement. Do you expect the star to last longer or die more quickly than the Sun?
    b. Follow Working It Out 11.1 to calculate the lifetime of this star.
    c. Check your work by comparing your answer with your prediction.

39. A sunspot has a temperature of approximately 5000 K. What does this tell us about the energy per square meter per second coming from the spot (I_{spot}) relative to that coming from the rest of the photosphere ($I_{surface}$)?
    a. Make a prediction: Do you expect this sunspot to be bright or faint compared to the example sunspot in the Working It Out box? So then do you expect that $I_{spot}/I_{surface}$ will be closer or farther from 1 than 0.32?
    b. Follow Working It Out 11.2 to compare the energy per square meter per second coming from the spot (I_{spot}) to the same quantity coming from the rest of the photosphere ($I_{surface}$); that is, calculate the ratio $I_{spot}/I_{surface}$.
    c. Check your work by comparing your answer to your prediction.

40. A sunspot appears only 70 percent as bright as the photosphere ($I_{spot} = 0.7 \times I_{surface}$). The photosphere has a temperature of approximately 6000 K. What does this tell us about the temperature of the sunspot?
    a. Make a prediction: Do you expect this sunspot to be hotter or colder than the example sunspot in the Working It Out box?
    b. Work backward through Working It Out 11.2 to calculate the temperature of the sunspot.
    c. Check your work by comparing your answer to your prediction.

exploration The Proton-Proton Chain

digital.wwnorton.com/universe4

The proton-proton chain powers the Sun by fusing hydrogen into helium. As a by-product, several different particles are produced, which eventually produce energy. The process has multiple steps, and this Exploration is designed to explore these steps in detail, hopefully to help you keep them straight.

Visit the Student Site at the Digital Resources page and open the Proton-Proton animation Interactive Simulation in Chapter 11.

Press play, and watch the animation all the way through once.

Press reset and then play again, and pause the animation after the first collision. Two hydrogen nuclei (both positively charged) have collided to produce a new nucleus with only one positive charge.

1. Which particle carried away the other positive charge?

_____

_____

2. What is a neutrino? Did the neutrino enter the reaction, or was the neutrino produced in the reaction?

_____

_____

_____

Compare the interaction on the top with the interaction on the bottom.

3. Did the same reaction occur in each instance?

_____

_____

Press play again, and pause the animation after the second collision.

4. What two types of nuclei entered the collision? What type of nucleus resulted?

_____

_____

_____

5. Was charge conserved in this reaction, or was it necessary for a particle to carry charge away?

_____

_____

6. What is a gamma ray? Did the gamma ray enter the reaction, or was it produced by the reaction?

_____

_____

_____

Press play again, and allow the animation to run to the end.

7. What nuclei enter the final collision? What nuclei are produced?

_____

_____

_____

8. In chemistry, a catalyst is a reaction helper. It facilitates the reaction but is not used up in the process. Are there any nuclei that act like catalysts in the proton-proton chain?

_____

_____

_____

Make a table of inputs and outputs. Which of the particles in the final frame of the animation were inputs to the reaction? Which were outputs? Fill in your table with these inputs and outputs.

9. Which of the outputs are converted into energy that leaves the Sun as light?

_____

_____

10. Which of the outputs could become involved in another reaction immediately?

_____

_____

_____

11. Which of the outputs is likely to stay in that form for a very long time?

_____

_____

_____

Evolution of Low-Mass Stars

As stars begin the final stages of their evolution, they become both brighter and redder, as you will learn here in Chapter 12. Stars are red, white, or blue, depending on the temperature of their outer layers. Predict whether you will see more red, white, or blue stars in the night sky. These colors can be subtle; some of the most prominent stars in the night sky are red giants or supergiants, but many people do not notice the color until it is pointed out to them! On a clear night, take a star chart out with you, and carefully observe the stars. When you find a red one, identify it with your star chart, and make a log of your observation, including the other bright stars of the constellation in which you found it. While not all red stars are red giants or supergiants, the vast majority of red stars are far too faint to see. If you see a bright red star in the night sky, it is most likely a dying star that is quickly passing through the giant or supergiant phase. While you are observing, carefully note the colors of other stars that you see.

EXPERIMENT SETUP

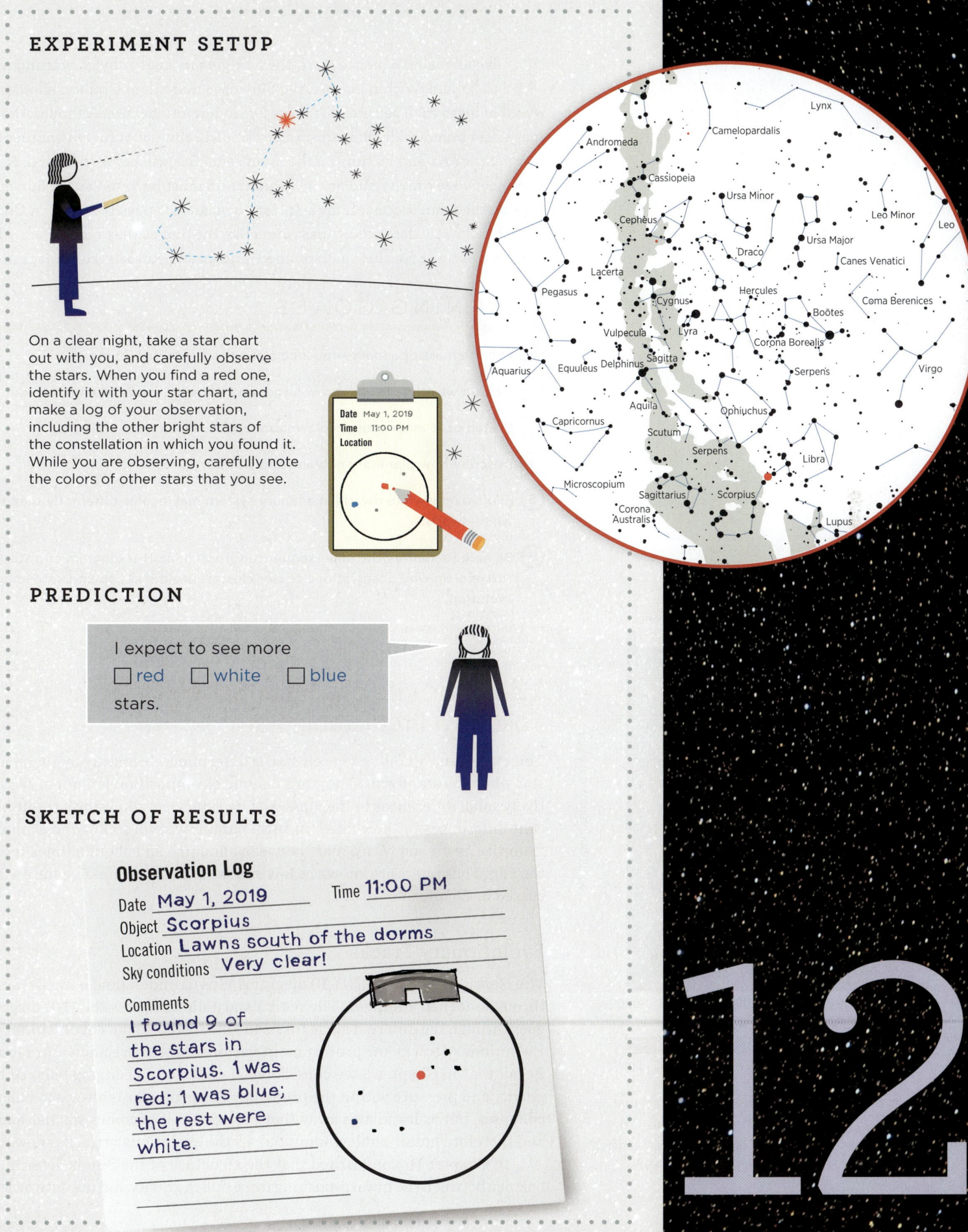

On a clear night, take a star chart out with you, and carefully observe the stars. When you find a red one, identify it with your star chart, and make a log of your observation, including the other bright stars of the constellation in which you found it. While you are observing, carefully note the colors of other stars that you see.

Date May 1, 2019
Time 11:00 PM
Location

PREDICTION

I expect to see more
☐ red   ☐ white   ☐ blue
stars.

SKETCH OF RESULTS

Observation Log

Date May 1, 2019   Time 11:00 PM
Object Scorpius
Location Lawns south of the dorms
Sky conditions Very clear!

Comments
I found 9 of the stars in Scorpius. 1 was red; 1 was blue; the rest were white.

12

Like all main-sequence stars, the Sun gets its energy by converting hydrogen to helium. Within its core, the Sun loses more than 4 billion kilograms (kg) of mass each second as it fuses hydrogen into helium. Although the Sun may seem immortal by human standards, it will run out of fuel in another 5.4 billion years or so, and its time on the main sequence will come to an end. New balances between gravity and energy production must be found as the Sun evolves beyond the main sequence, until at last no balance is possible. Here in Chapter 12, we will follow the fate of low-mass stars like the Sun as they run out of hydrogen, and we will trace out their path on the Hertzsprung-Russell (H-R) diagram.

LEARNING GOALS

① Use the mass of a main-sequence star to estimate its lifetime.

② Explain why the Sun will grow larger and more luminous when it runs out of fuel.

③ Sketch post-main-sequence evolutionary tracks on an H-R diagram.

④ Describe how planetary nebulae and white dwarfs form.

⑤ Explain how white dwarfs in a binary system may evolve differently than solitary white dwarfs.

⑥ Be able to identify the main-sequence turnoff in an H-R diagram of a star cluster and explain why observations of star clusters are the key to understanding stellar evolution.

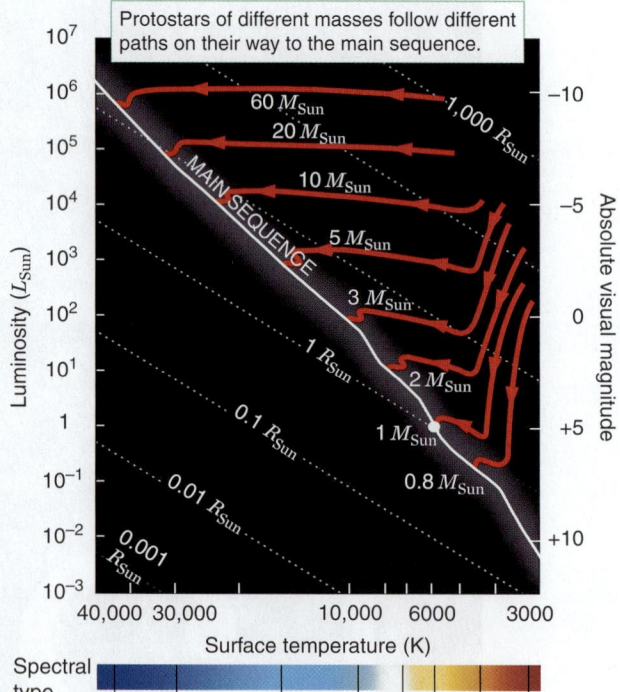

Figure 12.1 The H-R diagram can be used as a tool to show how a star evolves, meaning how it changes over time. In this figure, the pre-main-sequence tracks of protostars with different masses show that they follow different paths on their way to the main sequence.

12.1 The Life of a Main-Sequence Star Follows a Predictable Path

The evolutionary course of each star is determined primarily by the mass of the star when the star forms. The star's chemical composition is a minor factor. Relatively small differences in the masses of two stars can sometimes result in significant differences in their fates. In this chapter, we will follow the evolution of a "Sun-like" star; one with a mass between about 0.4 and about 8 times the mass of the Sun. These stars are known as **low-mass stars**. More massive stars will be discussed in Chapter 13.

Evolutionary Tracks

The Hertzsprung-Russell (H-R) diagram helps us understand how all stars change throughout their lifetimes. The path a star follows across the H-R diagram as it goes through the different stages of its life is called the star's **evolutionary track**. Evolutionary tracks for protostars of various masses are shown in **Figure 12.1**. Recall that in Chapter 5 we considered the idea of a changing balance between gravity and pressure within the protostar. The protostar is always in balance as it collapses, but as it radiates away thermal energy, it becomes smaller and denser and therefore hotter but less luminous, as the evolutionary tracks show.

In Chapter 11, you learned that the structure of the Sun is determined by a balance between the inward-pushing force due to gravity and the outward-pushing force due to the pressure of radiation moving outward from the core. When the Sun formed, about 90 percent of its atoms were hydrogen atoms. Most other atoms in the Sun were helium. Because helium is more massive than hydrogen, this means that the *mass* of the Sun was 70 percent hydrogen. Since reaching the main sequence, the Sun has produced energy by converting hydrogen into helium via the proton-proton chain (see Figure 11.4). This energy released by nuclear fusion maintains the outward pressure within the Sun. However, the temperature, density, and pressure of the core of a main-sequence star must continually shift in response to the changing core composition as hydrogen is converted to helium.

Compared to the events that follow, stars evolve slowly while they are on the main sequence. Between the time the Sun was born (about 4.6 billion years ago) and the time it will leave the main sequence (about 5.4 billion years from now), its luminosity will roughly double (remember that the luminosity of a star is the total energy emitted every second). Most of this change will occur during the last billion years of its life on the main sequence. The evolutionary track of a star on the main sequence is essentially a dot; the star remains in nearly the same location for its entire main-sequence lifetime.

The mass of a star determines the main-sequence lifetime because the mass and luminosity of main-sequence stars are related, as shown in **Figure 12.2**. More mass means stronger gravity; stronger gravity means higher temperature and pressure in the interior; higher temperature and pressure mean faster nuclear reactions; and faster nuclear reactions mean a more luminous star and a faster rate of fuel consumption. More massive stars are therefore more luminous and burn their fuel more quickly than less massive stars. **Table 12.1** lists the main-sequence lifetimes of stars of different spectral types and masses. (Recall that you learned about spectral types in Chapter 10.) Notice here again that more massive stars have *shorter* lifetimes. **Working It Out 12.1** demonstrates how to work with this relationship between mass, luminosity, and lifetime.

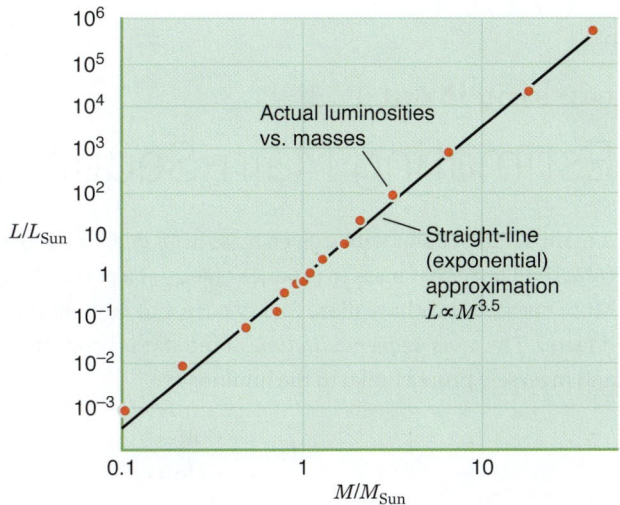

Figure 12.2 The mass and luminosity of main-sequence stars are related such that L is proportional to $M^{3.5}$. The exponent (3.5) is an average value over the wide range of main-sequence star masses.

Helium Ash in the Center of the Star

While a low-mass star is on the main sequence, the helium "**ash**" produced in the core does not fuse to form heavier elements. At the temperature found at the center of these stars, atomic collisions are not energetic enough to overcome the electric repulsion between helium nuclei. This nonburning helium ash does not build up evenly throughout the interior of a star. Because the temperature and pressure are highest at the center of a main-sequence star, hydrogen burns most rapidly there.

Table 12.1 Main-Sequence Lifetimes

Spectral Type	Mass (M_{Sun})	Luminosity (L_{Sun})	Main-Sequence Lifetime (years)
O5	60	500,000	3.6×10^5
B5	5.9	480	1.2×10^8
A5	2.0	12.3	1.8×10^9
F5	1.4	2.6	4.3×10^9
G2 (our Sun)	1.0	1.0	1.0×10^{10}
G5	0.92	0.8	1.2×10^{10}
K5	0.67	0.32	2.7×10^{10}
M5	0.21	0.008	4.9×10^{11}

VOCABULARY ALERT

ash In everyday language, *ash* specifically refers to the gray, dusty product of a fire, accumulating as the fire burns and collecting, for example, at the bottom of a fireplace. In this book, we use *ash* metaphorically to refer to the products of fusion, which collect in the core of the star.

working it out 12.1

Estimating Main-Sequence Lifetimes

A main-sequence star with a given mass available for hydrogen fusion (M_{MS}) will burn that mass to produce energy at a particular rate (L_{MS}). More massive stars have more fuel to burn, but more luminous stars burn it faster. The *main-sequence lifetime* of the star is proportional to the mass and inversely proportional to the luminosity:

$$\text{Lifetime} \propto \frac{M_{MS}}{L_{MS}}$$

We can put this relationship in quantitative terms by comparing a main-sequence star with the Sun. In practice, this means: (a) putting the mass of the star in solar masses (we write M_{MS}/M_{Sun}), (b) expressing its luminosity in solar luminosities (we write L_{MS}/L_{Sun}), and (c) introducing a constant of proportionality, 10 billion (1.0×10^{10}) years, which is the computed lifetime (in years) of the Sun:

$$\text{Lifetime} = (1.0 \times 10^{10}) \times \frac{M_{MS}/M_{Sun}}{L_{MS}/L_{Sun}} \text{ years}$$

Relatively small differences in the masses of stars result in large differences in their main-sequence luminosities. One method for estimating luminosities of main-sequence stars is known as the **mass-luminosity relationship**, $L \propto M^{3.5}$, which is based on observed

luminosities of stars of known mass (see Figure 12.2). As done previously, we can express this relationship relative to the Sun's mass and luminosity:

$$\frac{L_{MS}}{L_{Sun}} = \left(\frac{M_{MS}}{M_{Sun}}\right)^{3.5}$$

Substituting the mass-luminosity relationship into the lifetime equation gives a way to determine the lifetime from the mass of the star:

$$\text{Lifetime} = (1.0 \times 10^{10}) \times \frac{M_{MS}/M_{Sun}}{(M_{MS}/M_{Sun})^{3.5}}$$

$$= (1.0 \times 10^{10}) \times \left(\frac{M_{MS}}{M_{Sun}}\right)^{-2.5} \text{ years}$$

As an example, let's look at a main-sequence K5 star. According to the masses listed in Table 12.1 (where the corresponding lifetimes are based on more detailed models than our calculations here), a K5 star has a mass that is about equal to 0.67 times that of the Sun:

$$\text{Lifetime}_{K5} = (1.0 \times 10^{10}) \times (0.67)^{-2.5} = 2.7 \times 10^{10} \text{ years}$$

This is 27 billion years. Thus, a K5 star has a main-sequence lifetime 2.7 times longer than the Sun's. Even though the K5 star starts out with less fuel than the Sun, it burns that fuel much more slowly, so it lives longer. This is true across all stellar masses: low-mass stars live longer than high-mass stars.

As a result, helium accumulates more rapidly at the center of the star than elsewhere.

The chemical composition inside a star like the Sun changes throughout its main-sequence lifetime. When the Sun formed, it had a uniform composition of about 70 percent hydrogen and 30 percent helium by mass, as shown in **Figure 12.3**. As hydrogen was fused into helium, the helium fraction in the core of the Sun climbed. Today, roughly 5 billion years later, only about 35 percent of the mass *in the core* of the Sun is hydrogen. The chemical composition changes most rapidly at the center of a star, where nuclear fusion occurs, and less rapidly further out.

 **what if . . .**

What if Jupiter were 10 times more massive than it is, thus qualifying as a small brown dwarf: How would Jupiter's appearance in the night sky be different?

The Lowest-Mass Stars

Brown dwarfs occupy the dividing region between the planets with the largest mass and stars with the smallest mass. Brown dwarfs have masses less than 0.08 M_{Sun}, which is about 75 times the mass of Jupiter. Temperatures of brown dwarfs range from about 500 K to about 2800 K, so they glow a deep red in color. Because they are not massive enough to fuse hydrogen into helium, brown dwarfs do not become main-sequence stars.

The majority of stars in our galaxy are red dwarf stars with masses lower than about 0.4–0.5 M_{Sun}, but they are small and difficult to detect. All of these stars that have ever formed are still on the main sequence, because they have main-sequence lifetimes longer than the 13.8-billion-year lifetime of the universe.

CHECK YOUR UNDERSTANDING 12.1

Which of the following stars will have the longest lifetime?
a. a star one-tenth as massive as the Sun
b. a star one-fifth as massive as the Sun
c. the Sun
d. a star 5 times as massive as the Sun
e. a star 10 times as massive as the Sun

Answers to Check Your Understanding questions are in the back of the book.

12.2   A Star Runs Out of Hydrogen and Leaves the Main Sequence

Eventually, a star exhausts all of the hydrogen fuel in its core, leaving the innermost core composed entirely of helium ash. As thermal energy leaks out of the helium core into the surrounding layers of the star, no more energy is generated within the core to replace it. The balance between pressure and gravity is broken. The star's life on the main sequence has come to an end, and its further evolution depends on temperature changes in the core, which govern fusion reactions.

Electron-Degenerate Matter in the Helium Core

All of the matter you directly experience, such as books and tables and chairs, is mostly *empty space*, because each atom that composes that matter is mostly empty space except for the tiny volume occupied by the nucleus and the electrons. The same is true for the matter within the Sun. At the enormous temperatures within the Sun, the gas is completely ionized—a mixture of electrons and atomic nuclei all flying about freely. This gas is still mostly empty space, and the electrons and atomic nuclei fill only a tiny fraction of the volume.

When a low-mass star like the Sun exhausts the hydrogen at its center, the situation changes dramatically. As gravity begins to win its shoving match against pressure, the helium core is crushed to an ever-smaller size and an ever-greater density. But there is a limit to how dense the core can get: two electrons cannot occupy the same state at the same time. This limits the number of electrons that can be packed into a given volume of space at a given pressure. As the matter is compressed further and further, it finally reaches this limit. The space is now effectively "filled" with electrons that are smashed tightly together. This matter is so dense that a single cubic centimeter (about the size of a standard six-sided die) has a mass of more than 1,000 kg (about the mass of a small car). Matter that has been compressed to the point at which electrons are packed as closely as possible is called **electron-degenerate** matter. The presence of the degenerate core triggers a chain of events that will dominate the evolution of the 1-M_{Sun} star (this notation means

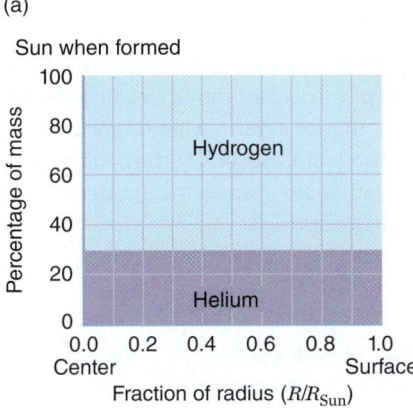

(a)

Sun when formed

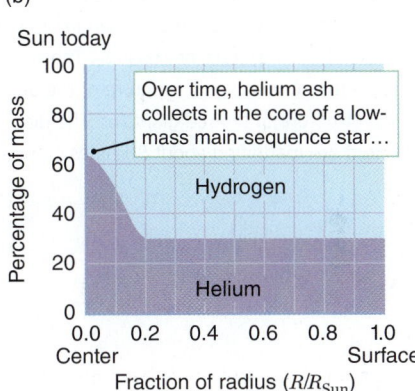

(b)

Sun today

Over time, helium ash collects in the core of a low-mass main-sequence star...

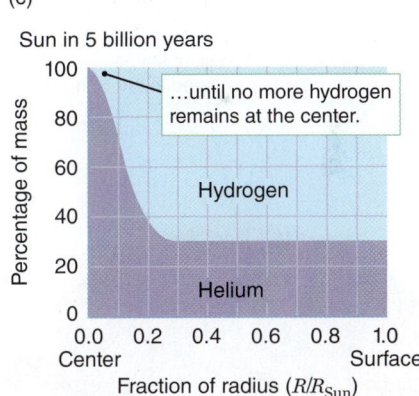

(c)

Sun in 5 billion years

...until no more hydrogen remains at the center.

Figure 12.3 The chemical composition of the Sun is plotted here as a percentage of mass against distance from the center of the Sun. (a) When the Sun formed 5 billion years ago, about 30 percent of its mass was helium and 70 percent was hydrogen throughout. (b) Today the material at the center of the Sun is about 65 percent helium and 35 percent hydrogen. (c) The Sun's main-sequence life will end in about 5 billion years, when all of the hydrogen at the center of the Sun is gone.

what if . . .

What if the Earth were squeezed to a hundredth of its current radius, making the Earth's density that of a gas having degenerate electrons: How would that affect the orbit of the moon and what would it do to the acceleration of gravity on Earth's surface?

Figure 12.4 The size and structure of the Sun (a) are compared with the size and structure of a star near the top of the *red giant branch* of the H-R diagram (b). The left-hand diagrams in (a) and (b) compare the sizes of the two stars. While the radius of the star increases by a factor of 50 during this change of evolutionary state, the mass of the star does not change. The right-hand diagrams compare the core size and structure of the Sun (a) with the core size and structure of the red giant (b). The diagrammatic images on the right are magnified about 50 times compared to those on the left.

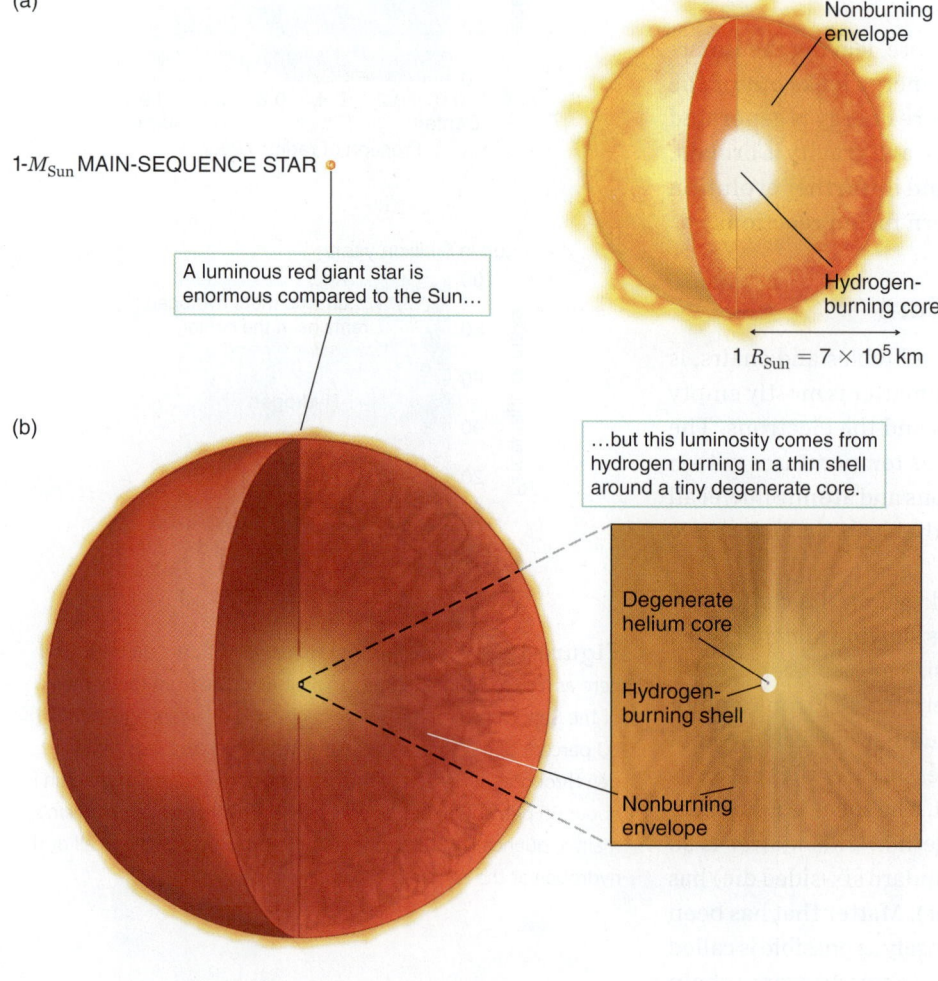

(a)

1-M_{Sun} MAIN-SEQUENCE STAR

A luminous red giant star is enormous compared to the Sun...

Nonburning envelope

Hydrogen-burning core

$1\ R_{Sun} = 7 \times 10^5$ km

(b)

...but this luminosity comes from hydrogen burning in a thin shell around a tiny degenerate core.

Degenerate helium core

Hydrogen-burning shell

Nonburning envelope

1-M_{Sun} RED GIANT STAR

$50\ R_{Sun} = 3.5 \times 10^7$ km

a star with the mass of one Sun) for the next 50 million years after the hydrogen runs out. Degenerate matter behaves much differently than normal matter, as you will see.

Hydrogen Shell Burning

Once a low-mass star exhausts the hydrogen at its center, nuclear burning pauses in the core. Because energy production has ceased, the inward force of gravity causes the core to begin to collapse and shrink. The more compact the core becomes, the higher the temperature rises, as gravitational potential energy is converted to thermal energy. This heat moves outward into the region surrounding the core, where the temperature and pressure in a shell of hydrogen rise high enough for fusion to occur. Astronomers call this **hydrogen shell burning** because the star's hydrogen now burns only in a shell surrounding the core, not in the core itself. This hydrogen-burning shell produces helium ash that sinks to the core below, increasing the core's mass.

As more and more helium ash piles up on the degenerate core from the hydrogen-burning shell, the core *gets smaller*. This is one of the ways that degenerate matter differs from normal matter: More massive objects are smaller than less massive objects. (This is noticeably not true for cows, for example!) Another striking difference is that degenerate matter does not expand when the temperature increases. So, unlike the outer layers of a star, degenerate matter cannot expand as a normal gas does when the pressure increases.

A degenerate core means stronger gravity in the shell around the core, stronger gravity means higher pressure, and higher pressure means faster nuclear burning in the shell, producing greater and greater amounts of energy. This increase in energy generation heats the overlying layers of the star, causing them to expand to form a cool, luminous giant star. As illustrated in **Figure 12.4**, the internal structure of the main-sequence star (Figure 12.4a) changes as the star evolves to a *red giant* (Figure 12.4b). A red giant star fuses hydrogen in a shell around a degenerate helium core and is both larger and redder than it was on the main sequence. The giant has a luminosity hundreds of times the luminosity of the Sun and a radius of more than 50 solar radii (50 R_{Sun}). Yet the core is far more compact than the core of the Sun, and much of the star's mass is concentrated into a volume that is only a few times the size of Earth.

As the star becomes larger, it becomes more luminous because there is more surface area to emit light. The enormous surface allows it to cool very efficiently. Even though its interior grows hotter and its luminosity increases, the surface temperature of the star begins to drop because energy can escape more quickly from the larger surface. This temperature drop is directly connected to the color of the star, and so the star becomes redder.

Evolution of the Star on the H-R Diagram

As the star evolves away from the main sequence, the H-R diagram provides a tool to keep track of the changing luminosity and surface temperature of the star. As soon as the star exhausts the hydrogen in its core, it leaves the main sequence and begins to move upward and to the right on the H-R diagram, growing more luminous but cooler. As the star continues to evolve, it grows larger and cooler. But after a time, its progress to the right on the H-R diagram ceases. The surface layers regulate how much radiation can escape from the star, thus preventing it from becoming any cooler.

The path that a star follows on the H-R diagram as it leaves the main sequence is like a tree "branch" growing out of the "trunk" of the main sequence, as shown in **Figure 12.5**. Astronomers refer to this track as the **red giant branch** of the H-R diagram.

As the star leaves the main sequence, the changes in its structure occur slowly at first, but then the star moves up the red giant branch faster and faster. It takes

▶▶ **Interactive Simulation:** H-R Diagram

Figure 12.5 As a star moves up the red giant branch in the H-R diagram, the luminosity of the star grows faster and faster. This is because the burning of hydrogen to helium in a shell surrounding a degenerate helium core is a cycle that feeds on itself.

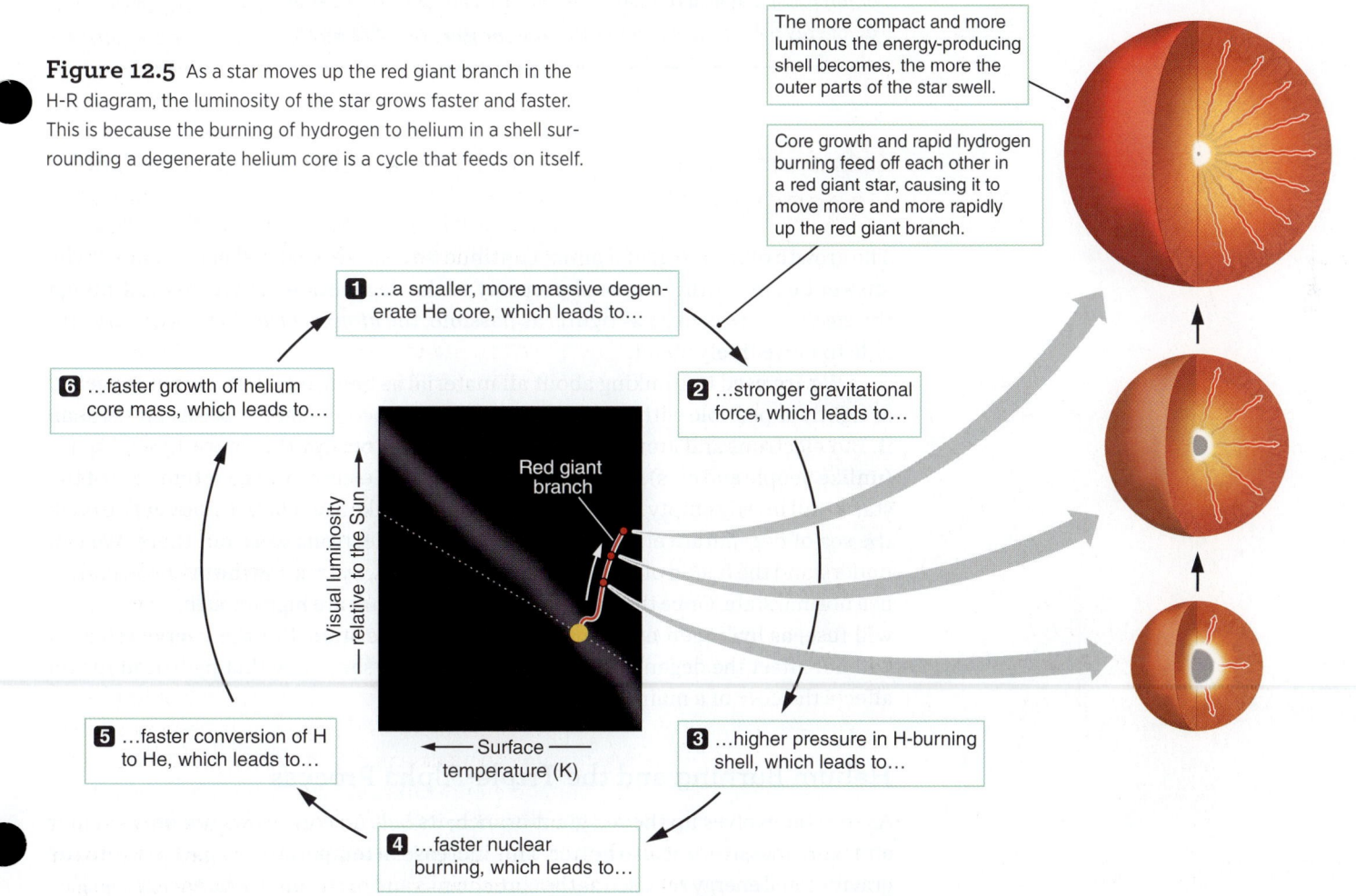

The more compact and more luminous the energy-producing shell becomes, the more the outer parts of the star swell.

Core growth and rapid hydrogen burning feed off each other in a red giant star, causing it to move more and more rapidly up the red giant branch.

1 ...a smaller, more massive degenerate He core, which leads to...

6 ...faster growth of helium core mass, which leads to...

2 ...stronger gravitational force, which leads to...

Red giant branch

Visual luminosity — relative to the Sun →

← Surface → temperature (K)

5 ...faster conversion of H to He, which leads to...

3 ...higher pressure in H-burning shell, which leads to...

4 ...faster nuclear burning, which leads to...

what if . . .

What if the Sun becomes a red giant: How might that affect the planets in our Solar System?

roughly 1 billion years for a star like the Sun to climb up the red giant branch. During the first half of this time period, the star's luminosity increases to about 10 times the luminosity of the Sun ($10\ L_{Sun}$). During the second half of this time, the star's luminosity skyrockets to almost $1,000\ L_{Sun}$.

The electron-degenerate helium core of the star grows in mass—but not in radius—as hydrogen is converted to helium in the hydrogen-burning shell and the helium ash adds to the core. The increasing mass of the ever-more-compact helium core increases the force due to gravity in the heart of the star. Stronger gravity means higher pressure, and higher pressure accelerates nuclear burning in the shell. Faster nuclear reactions in the shell convert hydrogen into helium more quickly, so the mass of the core grows more rapidly. The star has come full circle in a cycle (Figure 12.5) that feeds on itself. Increasing core mass leads to ever-faster burning in the shell, and the faster hydrogen burns in the shell, the faster the core mass grows. As a result, the star's luminosity increases at an ever-higher rate.

CHECK YOUR UNDERSTANDING **12.2**

When the Sun runs out of hydrogen in its core, it will become larger and more luminous because

a. it starts fusing hydrogen in a shell around a helium core.

b. it starts fusing helium in a shell and hydrogen in the core.

c. infalling material rebounds off the core and puffs up the star.

d. energy balance no longer holds, and the star just drifts apart.

12.3 Helium Begins to Burn in the Degenerate Core

The growth of the red giant cannot continue forever, so what will happen next? The answer lies in another unusual property of the degenerate helium core: Although the *electrons* are packed as tightly as possible, the *atomic nuclei* in the core are still able to move freely about.

We are used to thinking about all material as being equal: If a room is packed as tightly as possible with cats, we'd be surprised if people could move freely through it. But electrons and atomic nuclei may overlap to occupy the same physical space (unlike people and cats). To an atomic nucleus, the electron-degenerate core of the star is still mostly empty space. The nuclei behave like a normal gas, moving through the sea of degenerate electrons almost as if the electrons were not there. We can understand the fusion of helium, which comes next, by treating the nuclei as matter in a normal state. Once the pressure and temperature are high enough, these nuclei will fuse, as hydrogen nuclei do in main-sequence stars. But the energy released will not affect the degenerate electron core in the same way that hydrogen fusion affects the core of a main-sequence star.

Helium Burning and the Triple-Alpha Process

As the star evolves up the red giant branch, its helium core grows not only smaller and more massive, but also hotter. This increase in temperature is partly due to the gravitational energy released as the core shrinks and partly due to the energy released by the ever-faster pace of hydrogen burning in the surrounding shell. The thermal

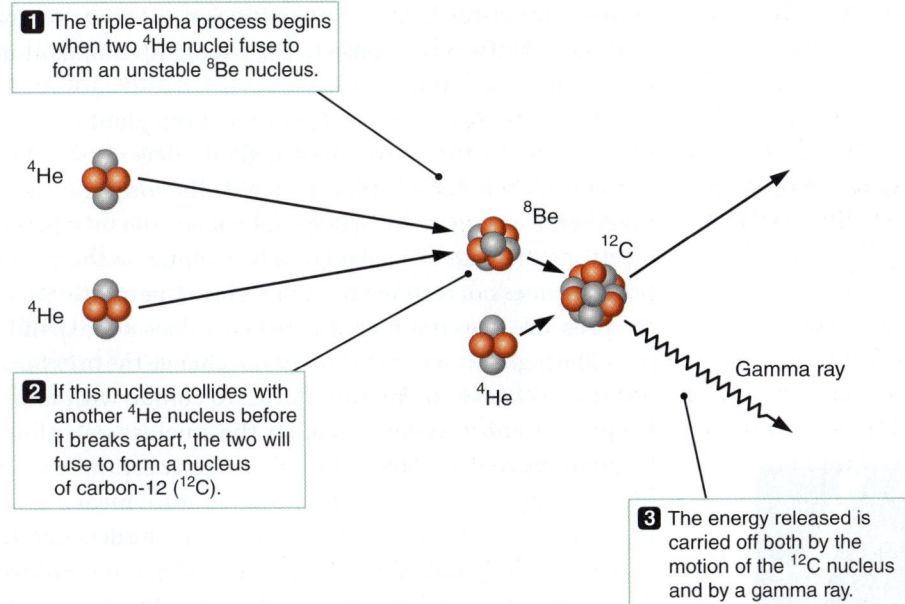

1 The triple-alpha process begins when two ^{4}He nuclei fuse to form an unstable ^{8}Be nucleus.

Figure 12.6 The triple-alpha process produces a stable nucleus of carbon-12 from three helium-4 nuclei.

^{4}He

^{4}He

^{8}Be

^{12}C

Gamma ray

2 If this nucleus collides with another ^{4}He nucleus before it breaks apart, the two will fuse to form a nucleus of carbon-12 (^{12}C).

^{4}He

3 The energy released is carried off both by the motion of the ^{12}C nucleus and by a gamma ray.

motions of the atomic nuclei in the core become more and more energetic. Eventually, at a temperature of about 10^8 K, the collisions among helium nuclei in the core become energetic enough that helium nuclei are slammed together hard enough for the strong nuclear force to act, and helium burning begins.

Helium burns in a two-stage process referred to as the **triple-alpha process**, which is illustrated in **Figure 12.6**. First, two helium-4 nuclei (^{4}He) fuse to form a beryllium-8 nucleus (^{8}Be) consisting of four protons and four neutrons. The ^{8}Be nucleus is extremely unstable. Left on its own, it would break apart after only about a trillionth of a second. But if, in that short time, it collides with another ^{4}He nucleus, the two nuclei will fuse into a stable nucleus of carbon-12 (^{12}C) consisting of six protons and six neutrons. The triple-alpha process takes its name from the fact that it involves the fusion of three ^{4}He nuclei, which are referred to as alpha particles. Other nuclear reactions between heavier atoms and ^{4}He nuclei can result in the formation of oxygen, neon, and magnesium.

Helium Flash and the Horizontal Branch

During the next phase of the star's evolution, the helium in the core begins burning. Degenerate material is a very good conductor of thermal energy, so any differences in temperature within the core rapidly even out. As a result, when helium burning begins at the center of the core, the energy released quickly heats the entire core. Within a few minutes, the entire core is burning helium by the triple-alpha process.

In a normal gas like the air around you, the pressure of the gas comes from the random thermal motions of the atoms. Increasing the temperature of a normal gas increases the random motions of the atoms, so the pressure of the gas increases. If the helium core of a red giant star were a normal gas, the increase in temperature from helium burning would increase the pressure. The core of the star would expand; the temperature, density, and pressure would decrease; nuclear reactions would slow; and the star would settle down into a new balance between gravity and pressure. These are exactly the sorts of changes that are steadily occurring within the

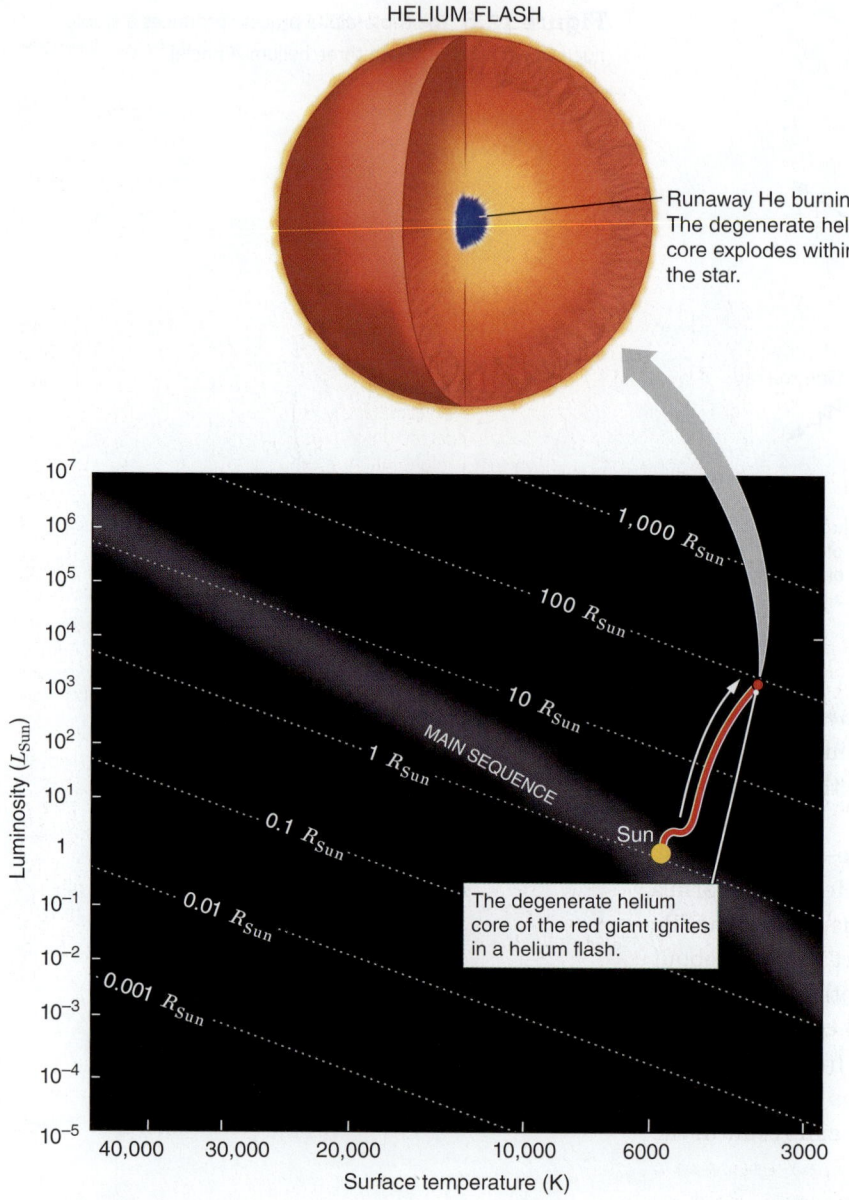

HELIUM FLASH

Runaway He burning: The degenerate helium core explodes within the star.

The degenerate helium core of the red giant ignites in a helium flash.

Figure 12.7 A low-mass star travels a complex path on the H-R diagram at the end of its life. The first part of that path takes it up the red giant branch to a point where helium ignites in a helium flash.

core of a main-sequence star like the Sun as the structure of the star shifts in response to the changing composition of the core.

However, the degenerate core of a red giant is not a normal gas. The pressure in a red giant's degenerate core comes from how densely the electrons in the core are packed together. Heating the core does not change the number of electrons that can be packed into its volume, so the core's pressure does not respond to changes in temperature. Since the pressure does not increase, the core does not expand.

The higher temperature does not change the pressure, but it does cause the helium nuclei to collide with more frequency and greater force, so the nuclear reactions become more vigorous. More vigorous reactions mean higher temperature, and higher temperature means even more vigorous reactions. Helium burning in the degenerate core runs wildly out of control as increasing temperature and increasing reaction rates feed each other. As long as the degeneracy pressure from the electrons is greater than the thermal pressure from the nuclei, this feedback loop continues.

For stars less than $2\,M_{Sun}$, within seconds of helium ignition, the thermal pressure increases until it is no longer smaller than the degeneracy pressure. At this point, the helium core literally explodes in what astronomers call a **helium flash**, illustrated in **Figure 12.7**. The explosion is contained deep within the star, where it cannot be directly observed. The drama is over within a few hours because the helium-burning core is no longer degenerate, and the star is on its way toward a new equilibrium. Stars between $2\,M_{Sun}$ and $8\,M_{Sun}$ will not undergo a helium flash but instead will just begin stable helium burning.

The tremendous energy released during the helium flash goes into fighting gravity and "puffing up" the core. After the helium flash, the core is much larger, so the force of gravity within it and the surrounding shell is much smaller. Weaker gravity means less force pushing down on the core and the shell, which means lower pressure. Lower pressure, in turn, slows the nuclear reactions. The net result is that, after the helium flash, core helium burning keeps the core of the star puffed up, but the slower energy production in the shell means that the star becomes less luminous than it was as a red giant. Helium burning in the core does not cause the star to grow more luminous.

The star takes about 100,000 years to settle into stable helium burning. It then spends about 100 million years burning helium into carbon in a nondegenerate core while hydrogen burns to helium in a surrounding shell. The star is roughly a hundred times less luminous than it was when the helium flash occurred, because the outer layers of the star are not puffed up as much as they were when the star was a red giant. The star shrinks, and its surface temperature climbs.

At this point in their evolution, low-mass stars with chemical compositions similar to that of the Sun lie on the H-R diagram just to the left of the red giant branch. Stars that have higher concentrations of heavy elements (that is, heavier than helium) than the Sun tend to distribute themselves farther away from the red giant branch along a nearly horizontal line on the H-R diagram. This stage of stellar evolution takes its name from this horizontal band. The star is now referred to as a **horizontal branch** star, shown in **Figure 12.8**.

CHECK YOUR UNDERSTANDING 12.3

Degenerate matter is different from ordinary matter because

a. degenerate matter does not interact with other particles.

b. degenerate matter has no mass.

c. degenerate matter does not interact with light.

d. degenerate matter objects get smaller as they get more massive.

12.4 The Low-Mass Star Enters the Last Stages of Its Evolution

The evolution of a star like our Sun from the main sequence through its helium flash and onto the horizontal branch is fairly well understood. Just as our understanding of the interior of the Sun comes from computer models of the physical conditions within our local star, our understanding of the evolution of a red giant comes from computer models that look at the changes in structure as the star's degenerate helium core grows. These models show that any star with a mass of about 1 M_{Sun} will follow the march from main sequence to helium flash and then drop down onto the horizontal branch. But when we try to use computer models to understand what happens next, making predictions gets a bit trickier. We just noted that differences in chemical composition between stars significantly affect where they fall on the horizontal branch. From this point on, small changes in the properties of a star—mass, chemical composition, the strength of the star's magnetic field, or even the rate at which the star is rotating—can lead to noticeable differences in how the star evolves.

With this caution in mind, we follow the most likely sequence of events awaiting our Sun, the prototypical example of a 1-M_{Sun} star with solar composition.

Moving Up the Asymptotic Giant Branch

The star's life on the horizontal branch is much shorter than its life on the main sequence. There are two indications that the star consumes fuel faster than when it is on the main sequence. First, the star is more luminous, so it is consuming fuel more rapidly. Second, helium is a much less efficient nuclear fuel than hydrogen, so the star has to burn more helium to achieve the same luminosity as when it burned hydrogen.

The temperature at the center of a horizontal branch star is not high enough for carbon to burn, so carbon ash builds up in the heart of the star. Gravity once again begins to win as the nonburning carbon ash core is crushed by the

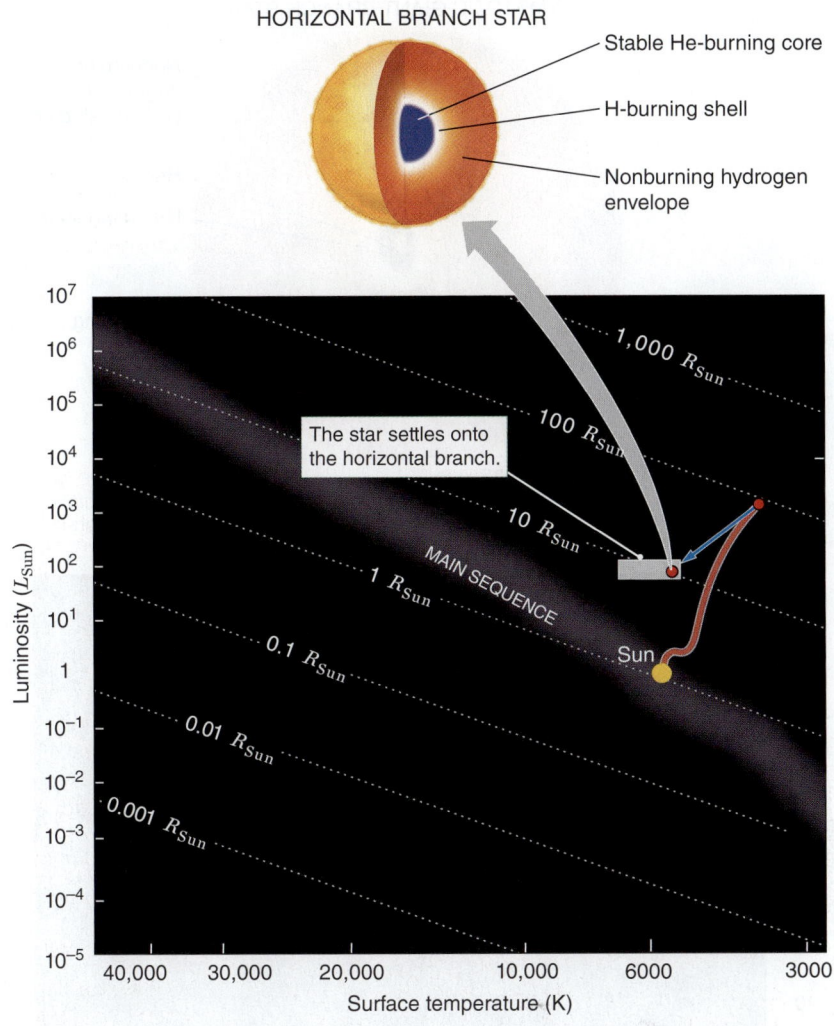

Figure 12.8 A low-mass star travels a complex path on the H-R diagram at the end of its life. The second part of that path takes the star down from the red giant branch onto the horizontal branch.

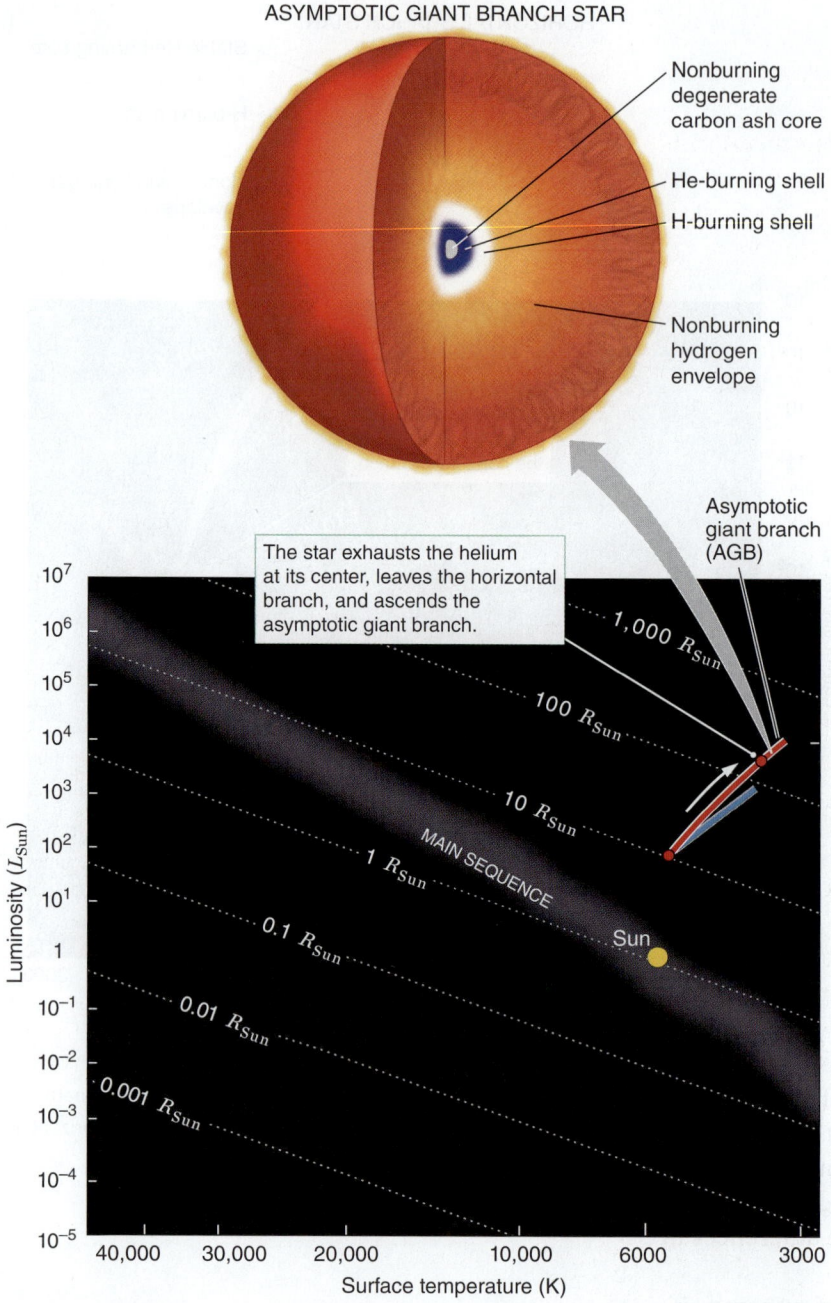

ASYMPTOTIC GIANT BRANCH STAR

- Nonburning degenerate carbon ash core
- He-burning shell
- H-burning shell
- Nonburning hydrogen envelope

Asymptotic giant branch (AGB)

The star exhausts the helium at its center, leaves the horizontal branch, and ascends the asymptotic giant branch.

Figure 12.9 A low-mass star travels a complex path on the H-R diagram at the end of its life. The third part of that path takes the star up from the horizontal branch onto the asymptotic giant branch (AGB).

weight of the layers of the star above it. The electrons become packed together so tightly that the core becomes an electron-degenerate carbon core, with physical properties much like those of the degenerate helium core at the center of a red giant.

The strength of gravity in the inner parts of the star increases, which in turn drives up the pressure, which speeds up the nuclear reactions, which causes the degenerate core to grow more rapidly—you have heard this story before. The internal changes occurring within the star are similar to the changes that took place at the end of the star's main-sequence lifetime, and the path the star follows as it leaves the horizontal branch echoes that earlier phase of evolution as well. Just as the star accelerated up the red giant branch as its degenerate helium core grew, the star now leaves the horizontal branch and once again begins to grow larger, redder, and more luminous as its degenerate carbon core grows. As you can see in **Figure 12.9**, the path that the star follows, known as the **asymptotic giant branch (AGB)** of the H-R diagram, roughly parallels the path it followed as a red giant, approaching the red giant branch as the star grows more luminous. An AGB star burns helium and hydrogen in nested concentric shells surrounding a degenerate carbon core, as the star moves once again up the H-R diagram.

Stellar Mass Loss

Building on our analogy between AGB stars and red giants, you might imagine that the next step in the evolution of an AGB star should be a "carbon flash" when carbon burning begins in the star's degenerate core. A carbon flash never happens, though. Before the temperature in the carbon core becomes high enough for carbon to burn, the star loses its gravitational grip on itself and expels its outer layers into interstellar space.

Red giant and AGB stars are enormous. When our Sun becomes an AGB star, its outer layers will swell to the point that they engulf the orbits of some of the innermost planets, possibly including Earth. When a star expands to such a size, the gravitational force at its surface is only 1/10,000 as strong as the gravity at the surface of the present-day Sun. It takes little extra energy to push surface material away from the star. **Stellar mass loss**—the loss of mass from the outer layers of the star as it evolves—actually begins when the star is still on the red giant branch; by the time a 1-M_{Sun} main-sequence star reaches the horizontal branch, it may have lost 10–20 percent of its total mass. As the star ascends the asymptotic giant branch, it loses another 20 percent or even more of its total mass. And by the time it is well up this branch, the star may have lost more than half of its original mass.

Mass loss on the asymptotic giant branch can be spurred on by the star's unstable interior. The extreme sensitivity of the triple-alpha process to temperature in the core can lead to episodes of rapid energy release, which provide the extra kick needed to expel material from the star's outer layers. Even stars that are initially quite similar can behave very differently when they reach this stage in their evolution.

The Post–Asymptotic Giant Branch Star

Toward the end of an AGB star's life, mass loss itself becomes a runaway process. When a star loses a bit of mass from its outer layers, the weight pushing down on the underlying layers of the star is reduced. Without this weight holding them down, the outer layers of the star puff up even larger than they were before. The star, which is now both less massive and larger, is even less tightly bound by gravity, so even less energy is needed to push its outer layers away. Much of the remaining mass of the star is ejected into space, typically at speeds of 20–30 kilometers per second (km/s).

A tiny, very hot, electron-degenerate carbon core is left behind, surrounded by a thin envelope in which hydrogen and helium are still burning. This star is now somewhat less luminous than when it was at the top of the asymptotic giant branch, but it is still much more luminous than a horizontal branch star. The remaining hydrogen and helium in the star rapidly burn to carbon, and as more and more of the mass of the star ends up in the carbon core, the star shrinks and becomes hotter and hotter. Over the course of only 30,000 years or so after the beginning of runaway mass loss, the star moves very rapidly from right to left across the top of the H-R diagram, as shown in **Figure 12.10**. During this phase, the star expels its outer layers to become a white dwarf.

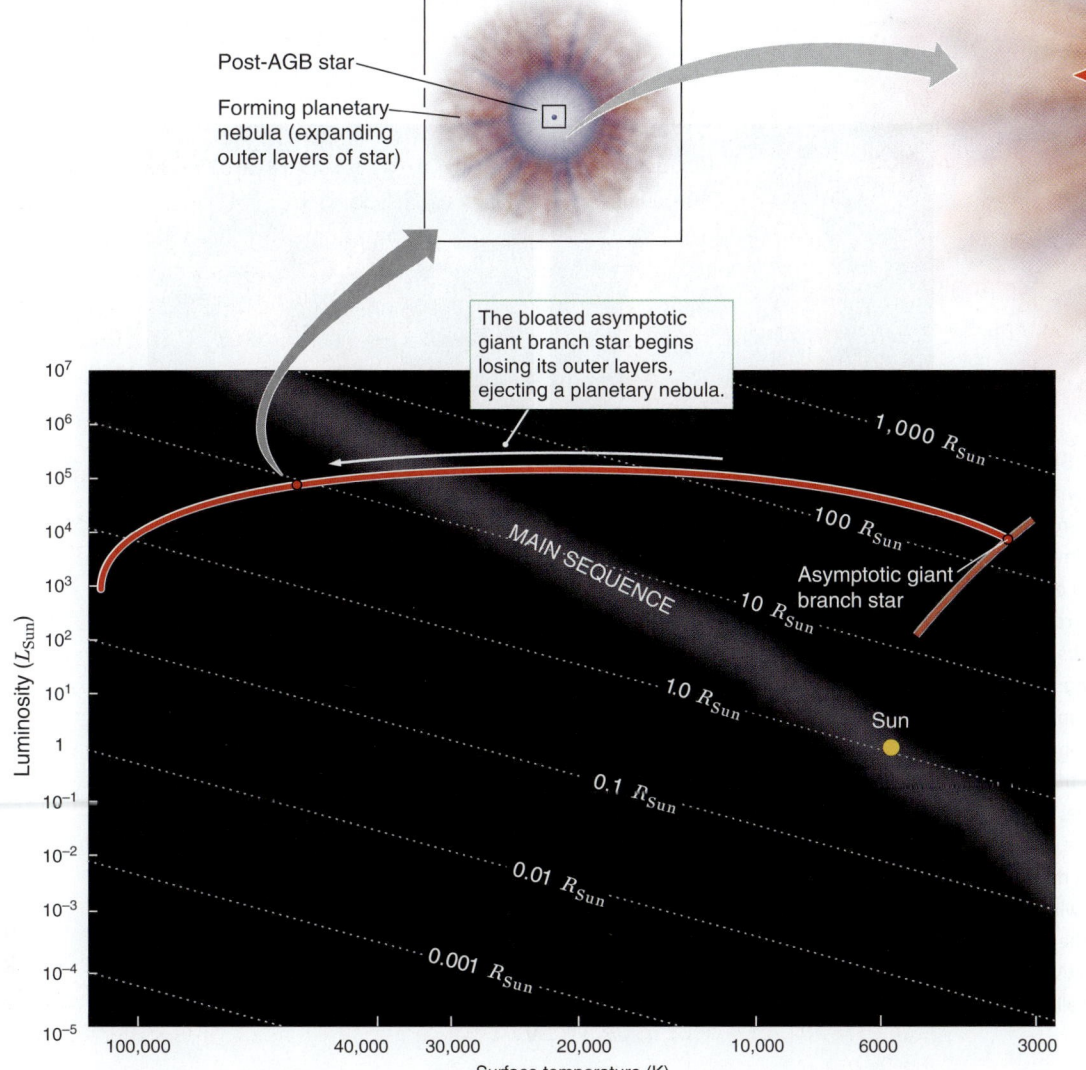

Figure 12.10 At the end of the AGB star's life, it ejects most of its mass in a planetary nebula, becoming a post-AGB star.

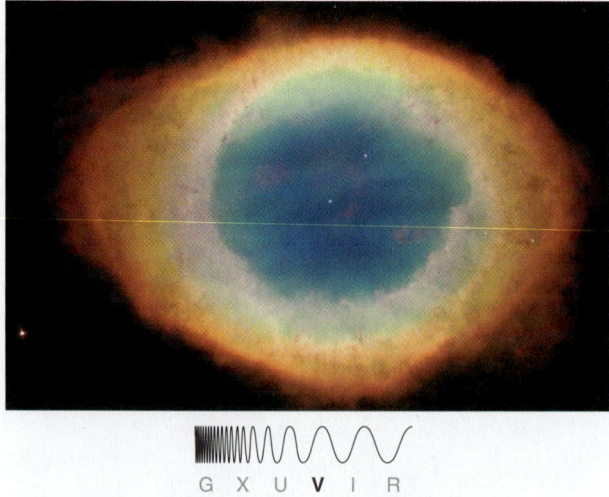

G X U V I R

Figure 12.11 At the end of its life, a low-mass star ejects its outer layers and may form a planetary nebula consisting of an expanding shell of gas surrounding the white-hot remnant of the star. Astronomers initially thought these objects looked like planets. However, a Hubble Space Telescope image of the Ring Nebula shows the remarkable and complex structure of this expanding shell of gas.

Planetary Nebulae

What has happened to the mass ejected by the AGB star? This mass will pile up in a dense, expanding shell. If you were to look at such a shell through a small telescope, you would see a round or perhaps oblong patch of light, perhaps with a hole and a dot in the middle (**Figure 12.11**). In forming this nebula, a lot of mass was lost nearly all at once. Then the mass loss ceased, resulting in a hollow shell around the central star. When these glowing shells were first observed in small telescopes, they looked round, like planets, but fuzzy, like nebulae; they were named **planetary nebulae**. But there is nothing planetary about them. A planetary nebula consists of the remaining outer layers of a star, which were ejected into space as a dying gasp at the end of the star's ascent of the asymptotic giant branch. Not every star forms a planetary nebula. Stars more massive than about 8 M_{Sun} pass through the post-AGB stage too quickly. Stars with insufficient mass take too long, so their envelope dissipates before they can illuminate it. Some astronomers think that our own Sun will not retain enough mass during its post-AGB phase to form a planetary nebula.

Planetary nebulae can be dazzling in appearance. Four examples are shown in **Figure 12.12**: (a) Spirograph Nebula, (b) Cat's Eye Nebula, (c) Butterfly Nebula, and (d) Eskimo Nebula. The structure of a planetary nebula tells of eras when mass loss was slower or faster and of times when mass was ejected primarily from the

Figure 12.12 ★ WHAT AN ASTRONOMER SEES

An astronomer looking at these four images will be reminded that planetary nebulae come in a wide variety of shapes that reveal the details of the history of mass loss from the central star. In general, material that is further from the central star was emitted earlier, although sometimes later material travels faster and overtakes earlier emissions. An astronomer will know that the colors are due to atom- or ion-specific emission lines, and she will recognize that the colors can be used not only to find out what atoms and ions are present but also to further figure out what kinds of photons are reaching that area. She will further be distracted by trying to think about the objects in three dimensions. For example, if we could view the Butterfly Nebula in panel (c) along its axis, from a direction near the top right of the image, it might look much like the Spirograph Nebula in panel (a). An astronomer will also know that bright spots might mean more material—or they might mean that the material is better illuminated. Figuring out what's happening with a planetary nebula requires sorting out all of these effects.

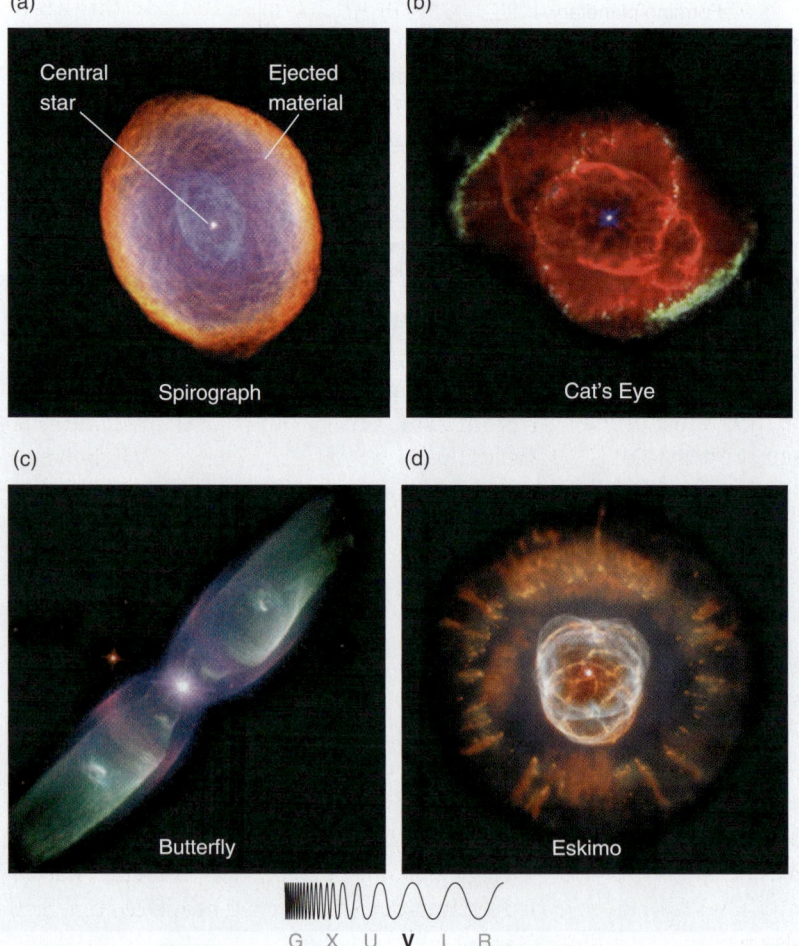

G X U V I R

star's equator or its poles. The colors come from emission lines due to particular atoms and ions.

Mass loss from giant stars carries the chemical elements enriching the stars' outer layers off into interstellar space. Planetary nebulae often show an overabundance of elements such as carbon, nitrogen, and oxygen compared to the outer layers of the Sun. These elements are by-products of nuclear burning. Once this chemically enriched material leaves the star, it mixes with interstellar gas, increasing the abundance of these elements in the universe.

White Dwarfs

Within 50,000 years or so, a post-AGB star burns all of the fuel remaining on its surface, leaving nothing behind but a cinder—a nonburning ball of carbon, oxygen, and small amounts of some other elements. In the process, the star falls down the left side of the H-R diagram, becoming smaller and fainter, as shown in **Figure 12.13**. Within a few thousand years the burned-out core shrinks to about the size of Earth, at which point it has become electron-degenerate and can shrink no further. This remnant of a low-mass star is called a **white dwarf**. The white dwarf continues to radiate energy away into space. As it does so it cools, just like the heating coil on an electric stove once it is turned off. The cooling white dwarf moves down and to the right on the H-R diagram, following a line of constant radius. The white dwarf may remain very hot for 10 million years or so, but its tiny size means the luminosity may now be only one-thousandth that of our Sun. All of the known white dwarfs are too faint to be seen with the naked eye or even with binoculars.

At the far left of its evolutionary track, the white dwarf may have a surface temperature of 100,000 K or more. According to Wien's law (recall Chapter 10 and Working It Out 5.1), most of the light from a star at those kinds of temperatures is in the high-energy ultraviolet (UV) part of the spectrum. The intense UV light heats and ionizes the expanding shell of gas that was recently ejected by the star, causing it to glow.

Summary of Evolution of Low-Mass Stars

Figure 12.14 recaps the evolution of a solar-type, 1-M_{Sun} main-sequence star through its final existence to a 0.6-M_{Sun} white dwarf. The star leaves the main sequence, climbs the red giant branch, falls to the horizontal branch, climbs back up the asymptotic giant branch, takes a left across the top of the diagram while ejecting a planetary nebula, and finally falls to its final resting place in the bottom of the diagram. This process is representative of the fate of low-mass stars. Although all low-mass stars form white dwarfs at the end points of their evolution, the exact path a low-mass star follows from core hydrogen burning on the main sequence to white dwarf depends on details particular to that star.

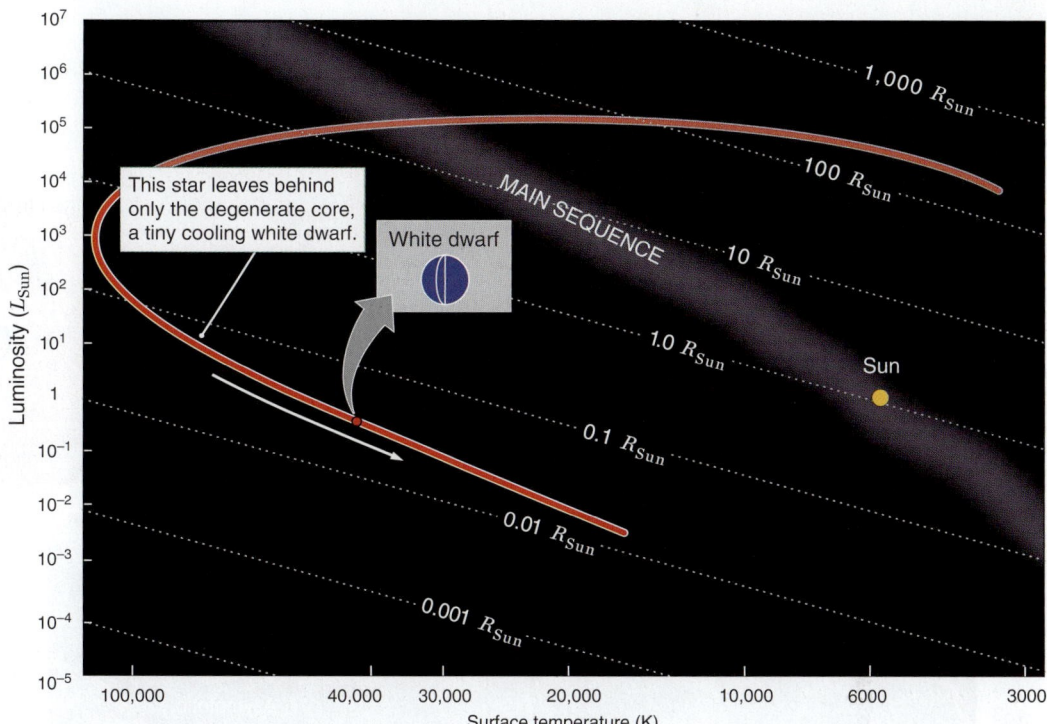

Figure 12.13 Low-mass stars finally leave behind a white dwarf—a small, hot, dense object that slowly fades from view.

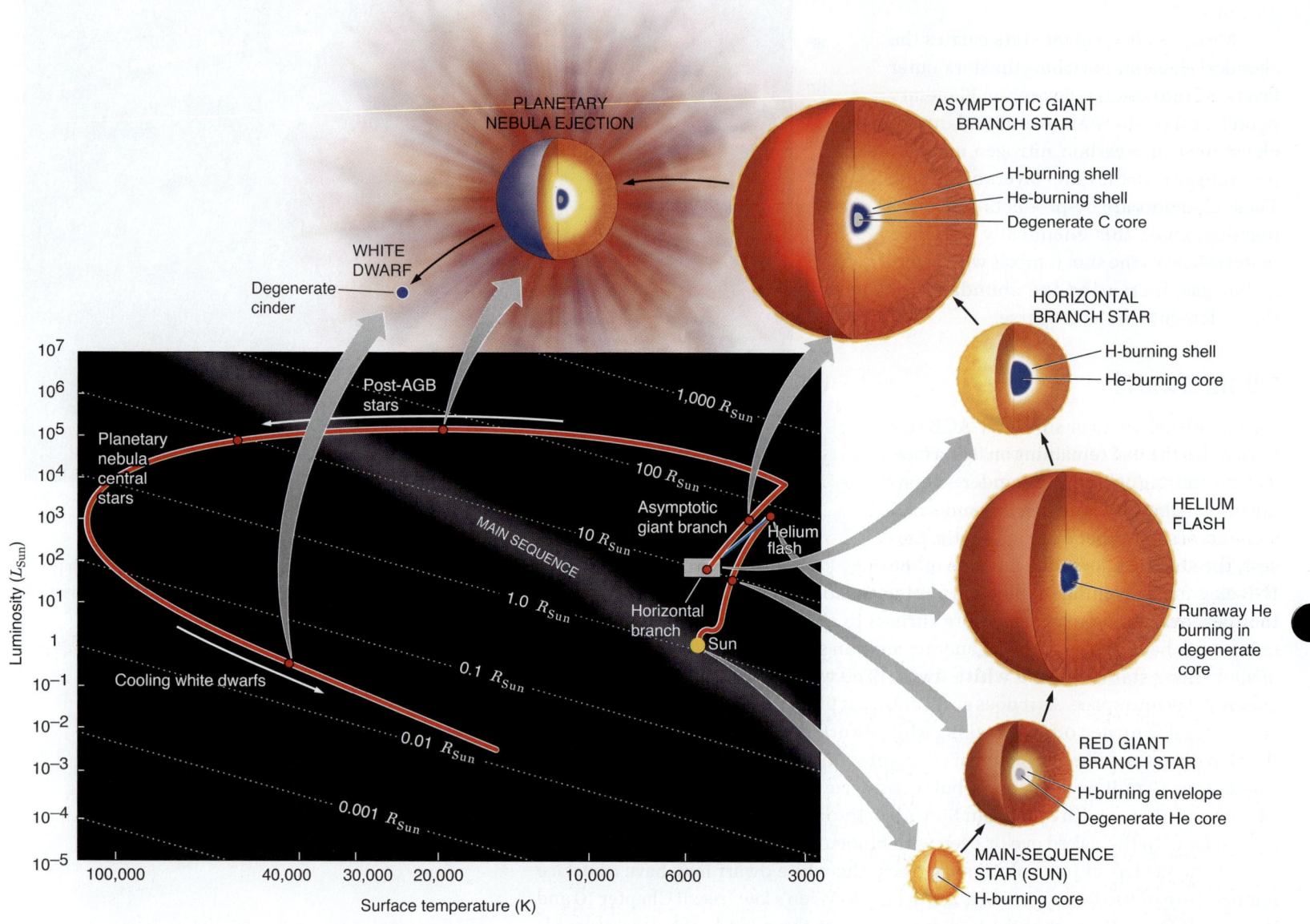

Figure 12.14 This H-R diagram summarizes the stages in the post-main-sequence evolution of a 1-M_{Sun} star.

Some 6 billion or so years from now, our Sun will become a white dwarf that will fade as it radiates its thermal energy away into space. This superdense ball—with a density of a ton per teaspoonful—actually began its life billions of years earlier as a cloud of interstellar gas much less dense than the vacuum in the best vacuum chamber on Earth.

Once it leaves the main sequence, the Sun will travel the path from red giant to white dwarf in less than one-tenth the time that it spent on the main sequence, steadily burning hydrogen to helium in its core. Stars spend most of their luminous lifetimes on the main sequence, which is why most of the stars that we see in the sky are main-sequence stars. In the end, though, white dwarfs will constitute the final resting place for the vast majority of stars that have been or ever will be formed.

CHECK YOUR UNDERSTANDING **12.4**

Place the following steps of the evolution of a low-mass star in order from earliest to latest.

a. The star moves onto the horizontal branch.

b. The white dwarf cools.

c. A clump forms in a giant molecular cloud.

d. The star moves onto the red giant branch.

e. The star moves onto the asymptotic giant branch.

f. A protostar forms.

g. The star sheds mass, producing a nebula.

h. Hydrogen fusion begins.

i. A helium flash occurs.

12.5 Binary Stars Sometimes Share Mass, Resulting in Novae and Supernovae

The evolution of low-mass stars becomes more complicated when we look at the many stars that are members of binary systems. While both members of a binary pair are on the main sequence, they usually have little effect on each other. But in some cases, if the separation between the stars is small and one star is more massive than the other, their evolution may become linked. In this section, we will trace the steps from binary star through to the end point of the evolution: nova or Type Ia supernova.

The Flow of Mass from an Evolving Star to Its Companion

Think for a moment about how gravity would affect you if you were to travel in a spacecraft from Earth toward the Moon. When you are still near Earth, the pull of Earth's gravity on you is far stronger than that of the Moon. As you move away from Earth and closer to the Moon, the interaction with Earth weakens, and the pull of the Moon's gravity on you becomes stronger. Eventually, you reach an intermediate zone where these interactions are equally strong. If you continue beyond this point, the interaction with the Moon grows stronger until you are firmly in the grip of the Moon.

A diagram of this effect looks like two teardrops, joined at the pointed ends, as shown in **Figure 12.15a**. These regions are called Roche lobes, and in this figure, they surround two stars. When one star swells up, its outer layers may cross from one Roche lobe into another, and any material that crosses this line no longer belongs to the first star. Depending on the speed and direction of the material, some is lost to the surrounding space, but some can be pulled toward the companion.

Evolution of a Close Binary System

The best way to understand how mass transfer affects the evolution of stars in a binary system is to apply what we have learned from the evolution of single low-mass stars to each star. Figure 12.15a shows a close binary system consisting of two low-mass stars of somewhat different mass. The more massive of the two stars is "star 1," and the less massive of the two is "star 2." This is an ordinary binary system, and each of these stars is an ordinary main-sequence star for most of the system's lifetime.

More massive stars evolve more rapidly. Therefore, star 1 will be the first to use up the hydrogen in its core and begin to evolve off the main sequence (Figure 12.15b).

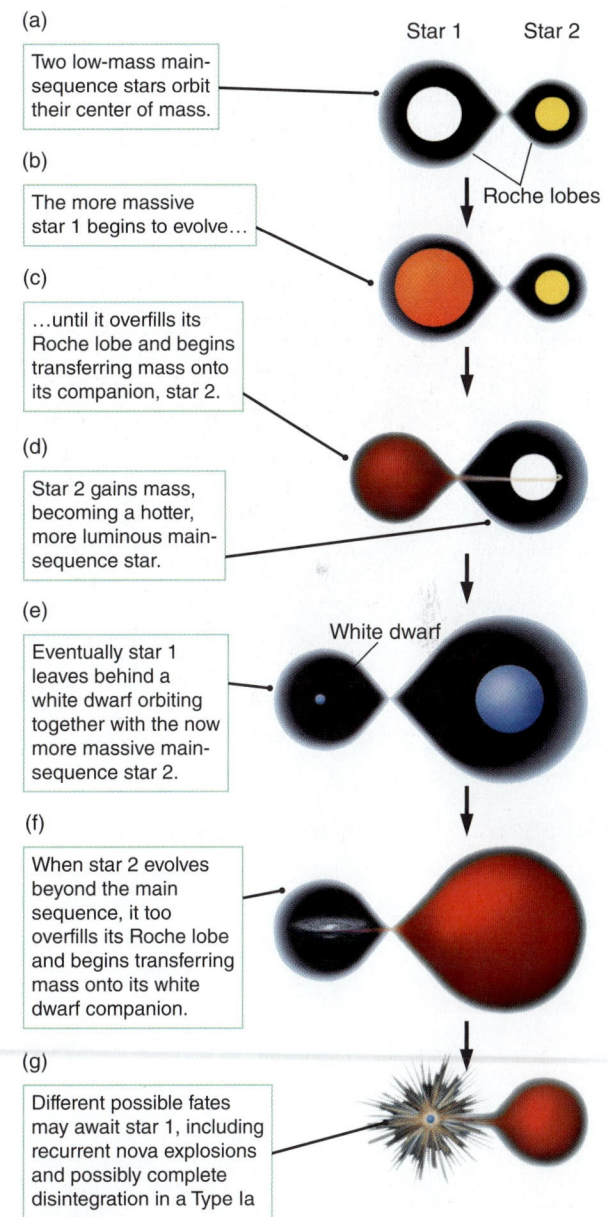

(a)

Star 1   Star 2

Two low-mass main-sequence stars orbit their center of mass.

(b)

Roche lobes

The more massive star 1 begins to evolve…

(c)

…until it overfills its Roche lobe and begins transferring mass onto its companion, star 2.

(d)

Star 2 gains mass, becoming a hotter, more luminous main-sequence star.

(e)

White dwarf

Eventually star 1 leaves behind a white dwarf orbiting together with the now more massive main-sequence star 2.

(f)

When star 2 evolves beyond the main sequence, it too overfills its Roche lobe and begins transferring mass onto its white dwarf companion.

(g)

Different possible fates may await star 1, including recurrent nova explosions and possibly complete disintegration in a Type Ia supernova.

Figure 12.15 A compact binary system consisting of two low-mass stars passes through a sequence of stages as the stars evolve and mass is transferred back and forth.

what if . . .

What if star 1 and star 2 are very widely separated, so that their Roche lobes are very large? In this case, star 1 may not get large enough to fill its Roche lobe. How would the evolution of this binary system be different from that of the system in Figure 12.15?

If the stars are close enough to each other, star 1 will grow to overfill its Roche lobe, and material will transfer to star 2 (Figure 12.15c). This exchange of material between the two stars is called **mass transfer**. The structure of star 2 changes to accommodate its new status as a higher-mass star (Figure 12.15d). If we plotted star 2's position on the H-R diagram during this period, we would see it move up and to the left along the main sequence, becoming larger, hotter, and more luminous.

A number of interesting things can happen at this point. For example, the transfer of mass between the two stars can result in a sort of "drag" that causes the orbits of the two stars to shrink, bringing the stars closer together and further enhancing mass loss. The two stars can even reach the point where they are effectively two cores sharing the same extended envelope of material.

Star 1, because it is losing mass to star 2, never grows large enough to move to the top of the H-R diagram as a red giant. It continues to evolve, though, burning helium in its core on the horizontal branch, proceeding through a stage of helium shell burning, and finally losing its outer layers and leaving behind a white dwarf. Figure 12.15e shows the binary system after star 1 has completed its evolution. All that remains of star 1 is a white dwarf orbiting around star 2, its bloated main-sequence companion.

The Second Star Evolves

Figure 12.15f picks up the evolution of the binary system as star 2 begins to evolve off the main sequence. Like star 1 before it, star 2 grows to fill its Roche lobe; material from star 2 begins to transfer into star 1's Roche lobe. Because the white dwarf is so small, the infalling material generally misses the star. Instead of landing directly on the white dwarf, the infalling mass forms an accretion disk around the white dwarf, similar to the accretion disk that forms around a protostar. As in the process of star formation, the accretion disk serves as a way station for material that is destined to find its way onto the white dwarf but starts out with too much angular momentum to hit the white dwarf directly.

A white dwarf has a mass comparable to that of the Sun but a size comparable to that of Earth. A large mass and a small radius mean strong surface gravity. A kilogram of material falling from space onto the surface of a white dwarf releases 100 times more energy than a kilogram of material falling from the outer Solar System onto the surface of the Sun. All of this kinetic energy is turned into thermal energy. The spot where the stream of material from star 2 hits the accretion disk is heated to millions of kelvins, where it glows in the far ultraviolet and X-ray parts of the electromagnetic spectrum.

The infalling material accumulates on the surface of the white dwarf, where it is compressed by the enormous gravitational pull of the white dwarf to a density close to that of the white dwarf itself. As more and more material builds up on the surface of the white dwarf, the white dwarf shrinks (just like the core of a red giant shrinks as it grows more massive). The density increases, and at the same time the release of gravitational energy drives up the temperature of the white dwarf. Since the infalling material is from the outer, unburned layers of star 2, it is composed mostly of hydrogen.

Once the temperature at the base of the white dwarf's surface layer of hydrogen reaches about 10 million K, this hydrogen begins to burn explosively. Energy released by hydrogen burning drives up the temperature, and the higher temperature drives up the rate of hydrogen burning. This runaway reaction is much like the runaway helium burning that takes place during the helium flash, except now there are no overlying layers of the star to keep things contained. An explosion called a

nova (pl. novae) (Figure 12.15g) occurs that blows part of the layer covering the white dwarf out into space at speeds of thousands of kilometers per second.

Roughly 50 novae occur in our galaxy each year. About 10 of these can be detected with telescopes, but very few become visible to the naked eye. The rest are blocked from our view by dust in the disk of our galaxy. Novae reach their peak brightness in only a few hours, and for a brief time they can be several hundred thousand times more luminous than the Sun. Although the brightness of a nova sharply drops in the weeks after the outburst, it can sometimes still be seen for years. During this time, the glow from the expanding cloud of ejected material is caused by the decay of radioactive isotopes created in the explosion.

The explosion of a nova does not destroy the underlying white dwarf star. Afterward, the binary system is in much the same configuration as before; that is, material from star 2 is still pouring onto the white dwarf. This cycle can repeat itself many times, as material builds up and ignites again and again on the surface of the white dwarf. In most cases, outbursts are separated by thousands of years, so most novae have been seen only once in historical times. Some novae, however, erupt every decade or so.

A Stellar Cataclysm

A white dwarf cannot have a mass greater than 1.4 M_{Sun} and remain stable. This value is referred to as the **Chandrasekhar limit**, named for Subrahmanyan Chandrasekhar (1910–1995), who derived it. Above this mass, even the pressure supplied by degenerate electrons is no longer enough to balance gravity, so the white dwarf will collapse. A white dwarf that is accumulating mass likely does not quite reach the Chandrasekhar limit. As the star reaches about 1.38 M_{Sun}, core pressures and temperatures rise enough to ignite carbon and begin a simmering phase that holds off the runaway thermonuclear process for a while. Once the temperature reaches about 1.0×10^8 K, the runaway carbon burning involves the entire white dwarf. Within about a second, the entire white dwarf is consumed in the resulting explosion. In this instant, 100 times more energy is liberated than will be given off by the Sun over its entire 10-billion-year lifetime on the main sequence. Runaway fusion reactions convert a large fraction of the mass of the star into elements such as iron and nickel, and the explosion blasts the shards of the white dwarf into space at top speeds in excess of 20,000 km/s, enriching the interstellar medium with these heavier elements. The explosion is known as a **Type Ia supernova**.

How might a white dwarf gain mass and explode in this way? There are three options. First, it is possible that, in the binary system discussed earlier, star 2 eventually will simply go on to form a white dwarf, leaving behind a stable binary system consisting of two white dwarfs, as in **Figure 12.16a**. These two white dwarfs may eventually merge. If the sum of their masses is greater than 1.4 M_{Sun}, the resulting merged star will explode. This explosion destroys both of the original white dwarfs. The amount of mass involved in the explosion may range from 1.4 to 2.8 M_{Sun}. As many as 80% of Type Ia supernovae may be of this type.

Second, star 2 may lose mass to the white dwarf while it is still a main-sequence star, as in Figure 12.16b. Through millions of years of mass transfer from star 2 onto the white dwarf, and possibly through countless nova outbursts, the white dwarf's mass slowly increases—until it approaches the Chandrasekhar limit and subsequently explodes. In this case, the white dwarf's mass is almost exactly 1.4 M_{Sun}, every time. Only one object of this type has so far been observed, from a pulse of blue light emitted, when the white dwarf exploded and heated up star 2.

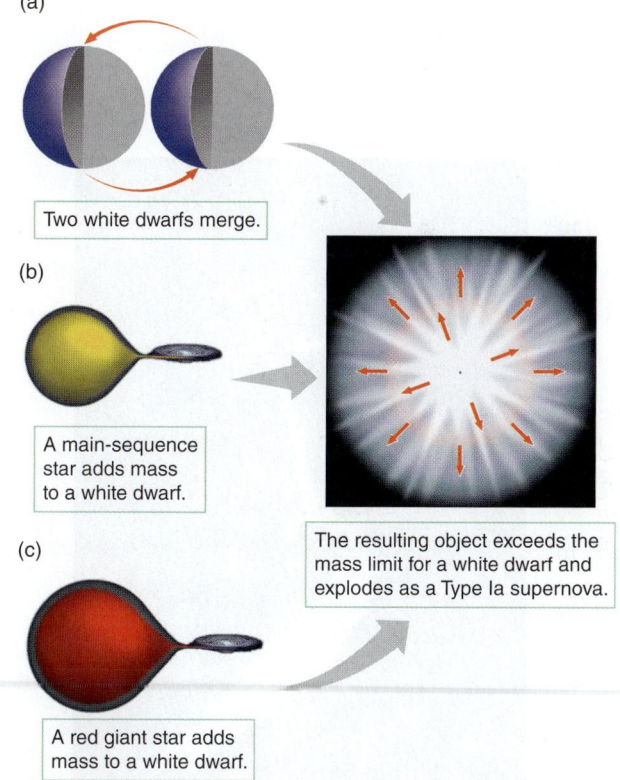

(a)

Two white dwarfs merge.

(b)

A main-sequence star adds mass to a white dwarf.

(c)

A red giant star adds mass to a white dwarf.

The resulting object exceeds the mass limit for a white dwarf and explodes as a Type Ia supernova.

Figure 12.16 A Type Ia supernova results when a white dwarf exceeds the mass limit. This may happen because (a) two white dwarfs merge, (b) mass from a main-sequence companion falls onto the white dwarf, increasing its mass to the limit, or (c) mass from an evolved companion falls onto the white dwarf, increasing its mass to the limit.

what if . . .

What if there were a nearby binary system with a white dwarf accreting material from a star with 0.8 M_{Sun}: Should we worry about being too close to a nearby supernova?

Third, star 2 may evolve off the main sequence to become a red giant, filling its Roche lobe. The material from this star flows onto the white dwarf over millions of years, as shown in Figure 12.16c. As in the above case, the white dwarf approaches the Chandrasekhar limit and subsequently explodes. In this case, the white dwarf's mass is almost exactly 1.4 M_{Sun}, every time. A red giant companion has never yet been observed in a Type Ia supernova.

Type Ia supernovae occur in a galaxy the size of the Milky Way about once a century. For a brief time they can shine with a luminosity billions of times that of our Sun, possibly outshining the galaxy itself. These objects are particularly useful to astronomers because their luminosities can be approximately determined from a careful study of their light curves. Because this type of supernova happens during the explosion of an object with a mass between 1.4 and 2.8 M_{Sun}, the total energy involved will vary by at most a factor of 2. Therefore, the luminosity of Type Ia supernovae should vary by no more than a factor of 2. Even with the uncertainty in the underlying mechanism, Type Ia supernovae are the best standard candle we have for measuring distances to very distant galaxies: Combining the luminosity found from the shape of the light curve of a Type Ia supernova with the apparent brightness gives the distance to the host galaxy.

CHECK YOUR UNDERSTANDING 12.5

There are many steps involved when a low-mass star in a binary system becomes a nova or supernova. Place the following steps in order from earliest to latest.

a. Star 1 (the more massive star) begins to evolve off the main sequence.

b. A white dwarf orbits a more massive main-sequence star.

c. Two low-mass main-sequence stars orbit each other.

d. Star 2 gains mass, becoming hotter and more luminous.

e. Star 2 fills its Roche lobe and begins transferring mass to the white dwarf.

f. Star 1 fills its Roche lobe and begins transferring mass to star 2.

g. The white dwarf becomes either a nova or a supernova.

h. Star 1 becomes a white dwarf.

12.6 Star Clusters Are Snapshots of Stellar Evolution

Stars live for a very long time compared to humans, so it may not be obvious how astronomers are able to determine how stars evolve, or change, over time. Imagine that you took many photographs of people of different ages. From these snapshots, you would be able to reconstruct how a human ages, without ever watching a single individual go through the process. Astronomers use a similar logic to determine how stars evolve. In Chapter 5, you saw that when an interstellar cloud collapses, it fragments into pieces, creating many stars of different masses bound together by gravity. These large groups of stars born at the same time, out of the same cloud of dust and gas, are called **star clusters**. We see many such star clusters around us today, containing anywhere from a few dozen to millions of stars. Because all the stars in a cluster form at nearly the same time, all the stars in the cluster are the same age. By looking at stars of similar mass, say 3-M_{Sun}, in different clusters, we see these 3-M_{Sun} stars at different ages and can reconstruct how a 3-M_{Sun} star changes over time.

Figure 12.17 shows the observed H-R diagram for the real cluster 47 Tucanae. Notice that the main sequence is cut off—it does not extend to the upper left. Instead,

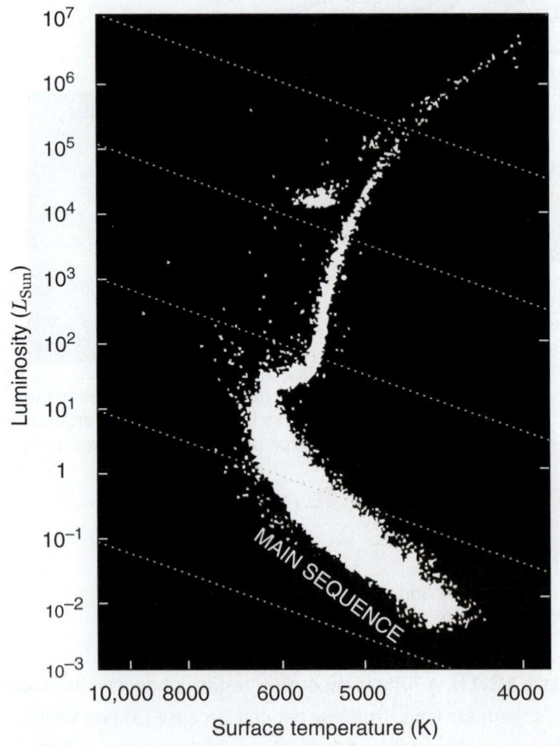

Figure 12.17 The observed H-R diagram of stars in the cluster 47 Tucanae shows a distinct end to the main sequence, where a branch into the upper right of the diagram begins.

the main sequence "turns off" to the right, onto a long branch that extends to the upper right of the diagram. The point at which the main sequence ends is called the **main-sequence turnoff**. Recall from Chapter 10 that the main sequence is a sequence of masses, with low-mass stars in the lower right and high-mass stars in the upper left. Stars above the main-sequence turnoff on the main sequence were more massive, so the pressure in the core was higher, causing fusion to occur faster. These stars were therefore more luminous and, importantly, had shorter lifetimes. The farther a star is up the main sequence, the more massive it is, and the shorter its lifetime on the main sequence. We can use this information to find the ages of clusters and, therefore, of the stars in them.

Figure 12.18 shows H-R diagrams of a simulated cluster of 40,000 stars at several different ages. In Figure 12.18a, stars of all masses are located on the **zero-age main sequence**, showing where they begin their lives as main-sequence stars. (Not all stars reach the main sequence at the same time. For clarity, we have left out the stars in these simulations that have not yet reached the main sequence.)

The more massive a star is, the less time it spends on the main sequence. After only 4 million years (Figure 12.18b), all stars with masses greater than about 20 M_{Sun}

Figure 12.18 Hertzsprung-Russell diagrams of star clusters are snapshots of stellar evolution. These are H-R diagrams of a simulated cluster of 40,000 stars of solar composition seen at different times after the birth of the cluster. Note the progression of the main-sequence turnoff to lower and lower masses. Very young stars, still moving onto the main sequence, have been omitted for clarity.

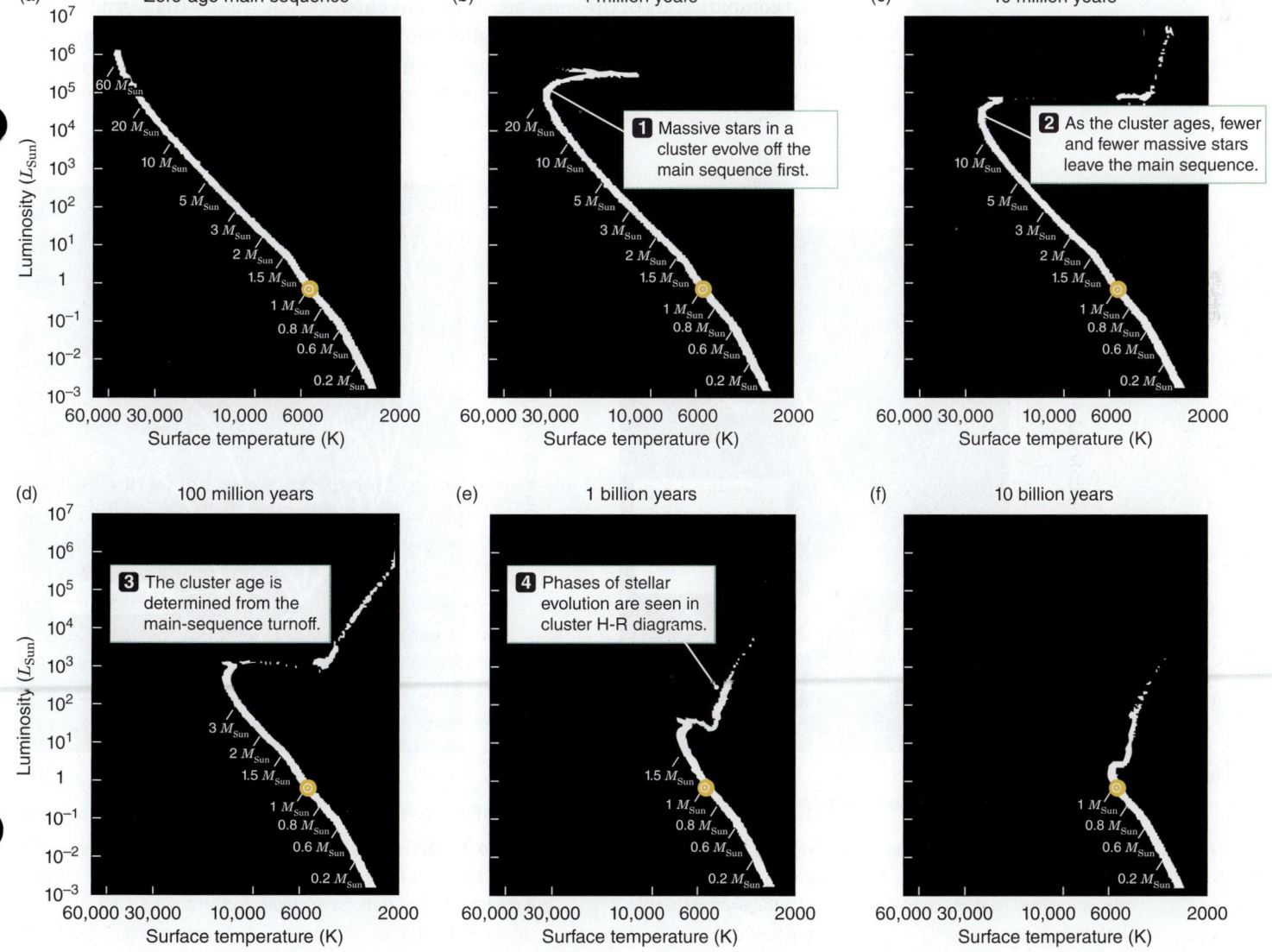

have evolved off the main sequence and are now spread out across the top of the H-R diagram. As time goes on, stars of lower mass evolve off the main sequence, and the main-sequence turnoff moves toward the bottom right along the main sequence. By the time the cluster is 10 million years old (Figure 12.18c), only stars with masses less than about 15 M_{Sun} remain on the main sequence.

As a cluster ages further (Figure 12.18d and e), we see the details of all stages of stellar evolution. By the time the star cluster is 10 billion years old (Figure 12.18f), stars with masses of only 1 M_{Sun} are beginning to die. Stars slightly more massive than this are seen as giants of various types. Note how few giant stars are present in any of the cluster H-R diagrams. This indicates that stars pass quickly through this phase of evolution, in comparison to their main-sequence lifetimes.

Figure 12.19 again shows the observed H-R diagram of the cluster 47 Tucanae, along with a theoretical calculation of the H-R diagram for a 12-billion-year-old cluster. The fact that the predictions of models agree so well with H-R diagrams of observed star clusters is strong support for these theories of stellar evolution. **Figure 12.20** overlays the observed H-R diagrams for several real star clusters. NGC 2362 has a high turnoff, so it is a young cluster. Its complement of massive, young stars shows it to be only a few million years old. In contrast, NGC 752 is quite old; its low main-sequence turnoff indicates a cluster age of about 7 billion years.

Studies of evolved clusters can tell us how individual stars evolve, because we can compare stars of the same age in various clusters. The theory that explains *why* all this change occurs is called "stellar evolution." This theory of stellar evolution is one of the most robust in science. From it, we learn that stars with low masses

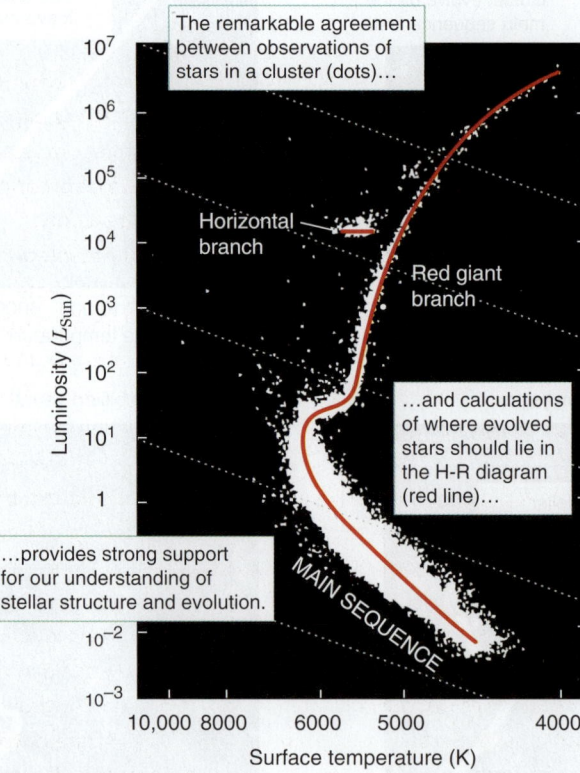

Figure 12.19 The observed H-R diagram of stars (dots) in the cluster 47 Tucanae agrees remarkably well with the theoretical calculation (red line) of the H-R diagram of a 12-billion-year-old cluster.

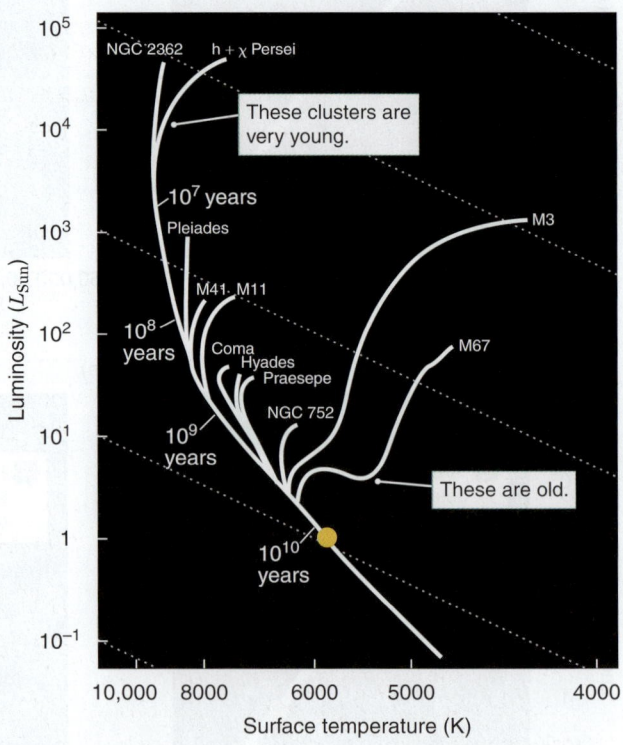

Figure 12.20 Clusters having a range of different ages are overlaid on this H-R diagram. The ages associated with the different main-sequence turnoffs are indicated.

evolve quite differently than those with high masses once they leave the main sequence. Here in Chapter 12, we have concentrated on the stellar evolution of low-mass stars—stars with masses of less than about 8 M_{Sun}—leaving our discussion of stars of higher mass for Chapter 13. These high-mass stars undergo evolution that is quite different from that of low-mass stars.

CHECK YOUR UNDERSTANDING 12.6

Very young star clusters have main-sequence turnoffs
a. nowhere. All the stars have already turned off in a young cluster.
b. at the top left of the main sequence.
c. at the bottom right of the main sequence.
d. in the middle of the main sequence.

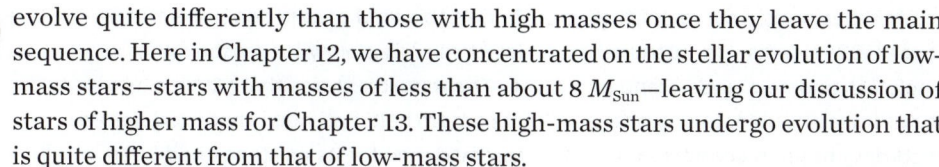

reading Astronomy News

Scientists May Be Missing Many Star Explosions

by Space.com Staff

Even stellar explosions sometimes are undetected!

Some of the brightest stellar explosions in the galaxy may be flying under astronomers' radar, a new study suggests.

Researchers using observations from a Sun-studying satellite detected four novae—exploding stars not quite as bright or dramatic as supernovae. The scientists were able to follow the explosions in intricate detail over time, including before the novae reached maximum brightness.

While other astronomers had discovered all four novae before, two of them escaped detection until after they had reached peak luminosity, the study revealed. This fact suggests that many other stellar explosions, even some that are incredibly bright, may be occurring unnoticed, researchers said.

"So far, this research has shown that some novae become so bright that they could

have been easily detected with the naked eye by anyone looking in the right direction at the right time, but are being missed, even in our age of sophisticated professional observatories," study lead author Rebekah Hounsell, a graduate student at Liverpool John Moores University (LJMU) in England, said in a statement.

The new observations are also allowing scientists to study nova explosions in unprecedented detail, according to researchers.

Hounsell and her colleagues analyzed measurements from an instrument aboard the U.S. Department of Defense's Coriolis satellite. The instrument, called the Solar Mass Ejection Imager (SMEI), was designed to detect disturbances in the solar wind. SMEI maps out the entire sky during its 102-minute orbit around Earth.

The researchers found that SMEI was also detecting star explosions, or novae. Novae occur when small, extremely dense stars called white dwarfs suck up gas from

a nearby companion star, igniting a runaway thermonuclear explosion.

Unlike supernovae, novae do not result in the destruction of their stars. Stars can go nova repeatedly.

SMEI detected four novae, including one confirmed repeater called RS Ophiuchi, which is found about 5,000 light-years away in the constellation Ophiuchus. RS Ophiuchi may ultimately die in a supernova explosion—one of the brightest, most dramatic events in the universe—researchers said.

Ground-based instruments missed the peak flare-up of two of these four novae, according to researchers. That suggests that space-based instruments like SMEI might be needed to pick up many novae, after which their progress can be tracked with telescopes on the ground, researchers said.

"Two of the novae observed by SMEI have confirmed that even the brightest novae may be missed by conventional ground-based observing techniques," said co-author Mike Bode, also of LJMU.

The researchers reported their results in a recent issue of the *Astrophysical Journal*.

The new observations are giving astronomers key insights into the earlier days of novae, revealing a great deal about how they start and evolve, researchers said.

"The SMEI's very even cadences and uniformly exposed images allow us to sample the sky every 102 minutes and trace the entire evolution of these explosions as they brighten and dim," said co-author Bernard Jackson of the University of California, San Diego.

The new observations have revealed, for example, that three of the explosions faltered significantly before regaining strength and proceeding. Such a "pre-maximum halt" had been theorized before, but evidence for its existence had been inconclusive, researchers said.

Because SMEI performs a survey of the entire sky every 102 minutes, the instrument could also help astronomers understand a wide variety of transient objects and phenomena, according to the research team.

"[This] work has shown how important all-sky surveys such as SMEI are and how their [data] sets can potentially hold the key to a better understanding of many variable objects," Bode said.

EVALUATING THE NEWS

1. What kind of stellar explosion is being discussed in this article? Why do you think ground-based observatories missed these novae?

2. Which figure in this chapter corresponds most closely to the astronomical events being discussed in this article?

3. Can this kind of explosion occur in an isolated star? Explain your reasoning.

4. The reporter states that "Such a 'pre-maximum halt' had been theorized before, but evidence for its existence had been inconclusive." Think back to Chapter 1. Is the reporter using the word *theory* correctly here?

5. These observations provide a picture of the same region of the sky every 102 minutes. Why is that an advantage in studying novae?

6. Why might the astronomers quoted in the article think that RS Ophiuchi will ultimately explode as a supernova?

Source: **Space.com Staff**, "Scientists May Be Missing Many Star Explosions," from Space.com, December 27, 2010. Reprinted with permission of EnVeritas Group.

SUMMARY

All stars eventually exhaust their nuclear fuel as it is turned to helium ash in the cores of main-sequence stars. Less massive stars exhaust their fuel more slowly and have longer lifetimes than more massive stars. After exhausting its hydrogen, a low-mass star similar to our Sun leaves the main sequence and swells to become a red giant, with a helium core made of electron-degenerate matter. The red giant burns helium via the triple-alpha process until the core ignites in a helium flash, and the star then moves onto the horizontal branch. A horizontal branch star accumulates carbon ash in its core and then moves up the asymptotic giant branch when helium fusion in the core ceases. In their dying stages, some stars eject their outer layers to form planetary nebulae. All low-mass stars eventually become white dwarfs, which are very hot but very small.

The transfer of mass within some binary systems can lead to a nuclear explosion. A nova occurs when hydrogen collects and ignites on the surface of a white dwarf in a binary system. If the mass of a white dwarf approaches $1.4\ M_{Sun}$, whether by merger with another white dwarf or by accretion of mass from a companion, the entire star explodes in a Type Ia supernova.

Hertzsprung-Russell diagrams of clusters give snapshots of stellar evolution. The location of the main-sequence turnoff indicates the age of the cluster.

(1) More massive stars have shorter lifetimes.

(2) When a star runs out of hydrogen in the core, it will begin to burn hydrogen in a shell around the core. This will cause it to swell onto the red giant branch of the H-R diagram.

(3) After leaving the main sequence, low-mass stars follow a convoluted path along the H-R diagram that includes the red giant branch, the horizontal branch, the asymptotic giant branch, and a path across the top and then down to the bottom of the diagram.

(4) A planetary nebula forms as a low-mass star loses mass, eventually leaving behind a white dwarf.

(5) Solitary white dwarfs simply cool. White dwarfs in binary systems may exchange mass and become novae or supernovae.

(6) The main-sequence turnoff point of a star cluster is located at the top left of the main sequence. This turnoff can be used to find the age of a star cluster. We can know the ages of stars in star clusters, so we may put the observations of stars of similar mass in evolutionary order.

QUESTIONS AND PROBLEMS

TESTING YOUR UNDERSTANDING

1. **T/F:** According to the mass-luminosity relationship, if a star is twice as massive, then it is twice as bright.

2. **T/F:** Even though their masses are the same, a red giant of 1 M_{Sun} can have a radius 50 times as large as the Sun's.

3. **T/F:** Some types of material get smaller as the mass increases.

4. **T/F:** More than one kind of nuclear fusion can occur in low-mass stars.

5. **T/F:** A solitary low-mass star can sometimes become a supernova.

6. All the stars in a star cluster are or have the same
   a. mass.
   b. luminosity.
   c. age.
   d. radius.

7. As a star leaves the main sequence, its position on the H-R diagram moves
   a. up and to the left.
   b. up and to the right.
   c. down and to the left.
   d. down and to the right.

8. Over time, the main-sequence turnoff of a star cluster moves _____ on the H-R diagram.
   a. up and to the left
   b. up and to the right
   c. down and to the left
   d. down and to the right

9. Planetary nebulae form when
   a. the dust and gas that make planets disperses.
   b. the dust and gas that make planets is collected around a protostar.
   c. a star loses mass at the end of its life, forming a cloud of dust and gas.
   d. a planet explodes at the end of its life, forming a cloud of dust and gas.

10. An accretion disk around a white dwarf in a binary system
    a. eventually forms planets.
    b. slowly disperses.
    c. feeds material onto the white dwarf.
    d. holds all of the mass of the star that sheds it.

11. As a protostar moves onto the main sequence to become a low-mass star, the evolutionary track is nearly vertical until the very end of the process. During this vertical portion,
    a. the temperature and the luminosity both decrease.
    b. the luminosity decreases, but the temperature remains nearly constant.
    c. the temperature decreases, but the luminosity remains nearly constant.
    d. the temperature and the luminosity both remain nearly constant.

12. The helium abundance in the outer 20 percent of the Sun
    a. remains nearly the same for its entire main-sequence lifetime.
    b. gradually increases over its main-sequence lifetime.
    c. gradually decreases over its main-sequence lifetime.
    d. remains the same for a long time and then increases abruptly before it leaves the main sequence.

13. When the Sun becomes a red giant, its _____ will decrease and its _____ will increase.
    a. density; luminosity
    b. density; temperature
    c. temperature; density
    d. density; rotation

14. As a low-mass star dies, it moves across the top of the H-R diagram because
    a. it is heating up.
    b. we see deeper into the star.
    c. the dust and gas around it heat up.
    d. a new fuel source is tapped.

15. A star cluster with a main-sequence turnoff at 1 M_{Sun} is approximately
    a. 10 million years old.
    b. 100 million years old.
    c. 1 billion years old.
    d. 10 billion years old.

16. A low-mass star might become a nova or supernova if and only if
    a. it is in a close binary system.
    b. it is really hot.
    c. it is really dense.
    d. it has a lot of heavy elements in it.

17. A white dwarf is located in the lower left of the H-R diagram. From this information alone, you can determine that
    a. it is very massive.
    b. it is very dense.
    c. it is very hot.
    d. it is very bright.

18. A red giant is located in the upper right of the H-R diagram. From this information alone, you can determine that
    a. it is very massive.
    b. it is very dense.
    c. it is very hot.
    d. it is very luminous.

19. The evolution of a star is primarily determined by its
    a. mass.
    b. composition.
    c. neighbors or lack of them.
    d. density.

20. In a white dwarf merger that results in a Type Ia supernova, the maximum mass involved is 2.8 M_{Sun} because
    a. 2.8 M_{Sun} is the maximum mass of a white dwarf.
    b. each white dwarf must separately have mass less than 1.4 M_{Sun}.
    c. stars more massive than 2.8 M_{Sun} die by a different process.
    d. stars more massive than 2.8 M_{Sun} do not form white dwarfs.

THINKING ABOUT THE CONCEPTS

21. Why are most nearby stars low-mass, low-luminosity stars?

22. Is it possible for a star to skip the main sequence and immediately begin burning helium in its core? Explain your answer.

23. What is the primary reason that the most massive stars have the shortest lifetimes? (Note: The answer can be expressed in just a few words.)

24. Describe some possible ways in which a star might increase the temperature within its core while at the same time lowering its density.

25. Suppose a main-sequence star suddenly started burning hydrogen at a faster rate in its core. How would the star react? Discuss changes in size, temperature, and luminosity.

26. Astronomers typically say that the mass of a newly formed star determines its destiny from birth to death. However, there is a frequent environmental circumstance for which this statement is not true. Identify this circumstance, and explain why the birth mass of a star might not fully account for its destiny.

27. Is it fair to assume that stars do not change their structure while on the main sequence? Why or why not?

28. When a star runs out of hydrogen in its core, Figure 12.5 shows that it moves up and to the right on the H-R diagram. Why does the star move up? Why does it move to the right? ⊙★

29. When a star leaves the main sequence, its luminosity increases tremendously. What does this increase in luminosity imply about the amount of time the star has left or the amount of time the star spends on any subsequent part of its evolutionary path?

30. When it is compressed, ordinary gas heats up, but degenerate gas does not. Why, then, does a degenerate core heat up as the star continues burning the shell around it?

31. Why is a horizontal branch star (which burns helium at a high temperature) less luminous than a red giant branch star (which burns hydrogen at a lower temperature)?

32. As an AGB star evolves into a white dwarf, it runs out of nuclear fuel, and one might argue that the star should cool off and move to the right on the H-R diagram. Why does the star move instead to the left, as shown in Figure 12.10? ⊙★

33. ★ WHAT AN ASTRONOMER SEES The Eskimo Nebula in Figure 12.13 looks somewhat like a face, surrounded by a fur hood. In this nebula, which material was most likely emitted first from the central star: the "face" or the "fur hood"? How do you know? ⊙★

34. The intersection of the Roche lobes in a binary system is the equilibrium point between the two stars, where the gravitational attraction from both stars is equally strong and opposite in direction. Is this an example of a stable or unstable equilibrium? Explain.

35. In Latin, *nova* means "new." Novae, as we now know, are not new stars. Explain how novae might have gotten their name and why they are really not new stars.

APPLYING THE CONCEPTS

36. What is the main-sequence lifetime of an F5 star?
    a. Make a prediction: Compare the mass of an F5 star with the mass of the Sun. Do you expect an F5 star to have a lifetime longer or shorter than the lifetime of the Sun?
    b. Follow Working It Out 12.1 to calculate the lifetime of an F5 star.
    c. Check your work by comparing the result of your calculation with your prediction. Then compare with the lifetime of an F5 star given in Table 12.1.

37. What is the main-sequence lifetime of a star with a mass of 0.5 M_{Sun}?
    a. Make a prediction: Do you expect this star to have a longer or shorter lifetime than the Sun?
    b. Follow Working It Out 12.1 to calculate the lifetime of a 0.5-M_{Sun} star.
    c. Check your work by comparing your answer to the main-sequence lifetime of the Sun and to those of stars in Table 12.1.

38. For most stars on the main sequence, luminosity scales with mass at a rate proportional to $M^{3.5}$. How does this affect the luminosity of a star with a mass of 2.0 M_{Sun}?
    a. Make a prediction: Should a star with twice the mass of the Sun be more or less luminous than the Sun?
    b. Calculate the luminosity of a 2.0 M_{Sun} star, in terms of the luminosity of the Sun (see Working It Out 12.1).
    c. Check your work by comparing your answer to your prediction and to the luminosities of stars in Table 12.1.

39. For most stars on the main sequence, luminosity scales with mass at a rate proportional to $M^{3.5}$. How does this affect the luminosity of a star with a mass of 0.50 M_{Sun}?
    a. Make a prediction: Should a star with half the mass of the Sun be more or less luminous than the Sun?
    b. Calculate the luminosity of a 0.5 M_{Sun} star, in terms of the luminosity of the Sun (see Working It Out 12.1).
    c. Check your work by comparing your answer to your prediction, and by comparing your answer to the luminosities of stars in Table 12.1.

40. Suppose you are studying a star cluster, and you find that it has no main-sequence stars with surface temperatures hotter than 6000 K. What does this tell you about the age of this cluster?
    a. Make a prediction: What is a reasonable range for the ages of stellar clusters? Is 100 years old an appropriate answer? How about 100 trillion years?
    b. Find the age of the cluster by studying Figure 12.20. ⊙★
    c. Check your work by making sure your answer is reasonable. Think further about the surface temperature and age of the Sun. Does your answer make sense?

exploration Evolution of Low-Mass Stars

digital.wwnorton.com/universe3

The evolution of a low-mass star, as discussed here in Chapter 12, corresponds to many twists and turns on the H-R diagram. In this exploration, we return to the Hertzsprung-Russell Diagram Interactive Simulation to investigate how these twists and turns affect the appearance of the star.

Visit the Student Site on the Digital Resources page and open the Hertzsprung-Russell Diagram Interactive Simulation in Chapter 12. The box labeled "Comparison to Sun" shows an image of both the Sun and the test star. Initially, these two stars have identical properties: the same temperature, the same luminosity, and the same size.

Examine the box labeled "Test Star Properties." This box shows the temperature, luminosity, and radius of a test star located at the X in the H-R diagram. Before you change anything, answer questions 1–4.

1. What is the temperature of the test star?

2. What is the luminosity of the test star?

3. What is the radius of the test star?

4. What do you predict will happen to the temperature, luminosity, and radius of the test star if it moves up and to the right on the H-R diagram?

As a star leaves the main sequence, it moves up and to the right on the H-R diagram. Grab the cursor (the X on the H-R diagram) and move it up and to the right.

5. What changes about the image of the test star next to the Sun?

6. What is the test star's temperature? What property of the image of the test star indicates that its temperature has changed?

7. What is the test star's luminosity?

8. What is the test star's radius?

9. Ordinarily, the hotter an object is, the more luminous it is. In this case, the temperature has decreased, but the luminosity has increased. How can this be?

The star then moves around quite a lot in that part of the H-R diagram. Look at the H-R diagrams in Chapter 12 (Figures 12.7–12.10), and then use the cursor to approximate the motion of the star as it moves up the red giant branch, back down and onto the horizontal branch, and then back to the right and up the asymptotic giant branch.

10. Are the changes you observe in the image of the star as dramatic as the ones you observed for question 5?

11. What is the most noticeable change in the star as it moves through this portion of its evolution?

Next, the star begins moving across the H-R diagram to the left, maintaining almost the same luminosity.

12. Predict how the temperature, luminosity, and radius of the star will change as it moves across the top of the diagram toward the left.

Drag the cursor across the top of the H-R diagram to the left, and study what happens to the image of the star in the "Comparison to Sun" box.

13. What changed about the star as you dragged it across the H-R diagram?

14. How does its size now compare to that of the Sun?

Finally, the star drops to the bottom of the H-R diagram and then begins moving to the right.

15. Predict how the temperature, luminosity, and radius of the star will change as it drops to the bottom of the H-R diagram and moves to the right.

Move the cursor toward the bottom of the H-R diagram, where the star becomes a white dwarf.

16. What changed about the star as you dragged it down the H-R diagram?

17. How does its radius now compare to that of the Sun?

To solidify your understanding of stellar evolution, press the reset button and then move the star from main sequence to white dwarf several times. This will help you remember how this part of a star's life appears on the H-R diagram.

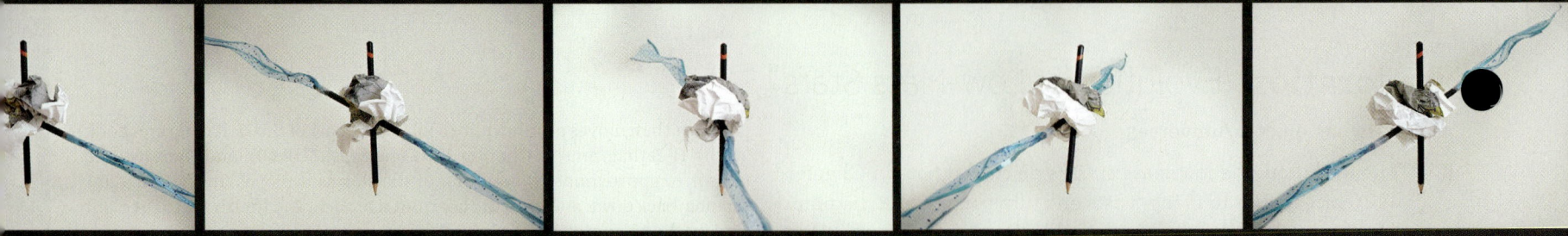

Evolution of High-Mass Stars

Some massive stars end their lives as neutron stars that rapidly spin. In some cases, the star has a strong magnetic field with an axis that does not line up with the rotation axis. This makes the star appear to "pulse" on and off, as the beam of light that comes from the magnetic field axis passes in front of the observer. For each rotation of the neutron star, predict how many "pulses" an observer will see. To see why a star pulses, build a model of a pulsar with two pencils, a large wad of paper, some tape, and some string or ribbon (optional). Tape the two pencils together so that they form an "X." One of these pencils is the rotation axis, and one is the axis of the magnetic field. Wad up some paper around the place where the pencils cross, to represent the neutron star. Choose one of the pencils to be the axis of the magnetic field, and tape pieces of string or ribbon to each end (the optional string or ribbon represents the beam of light that comes from the magnetic field axis). Hold the other pencil, which represents the rotation axis, and spin the model quickly. Notice how sometimes the magnetic field points toward you, and sometimes it does not. How many times does the magnetic field point toward you during each rotation of the model?

EXPERIMENT SETUP

To see why a star pulses, build a model of a pulsar with two pencils, a large wad of paper, some tape, and some string or ribbon (optional).

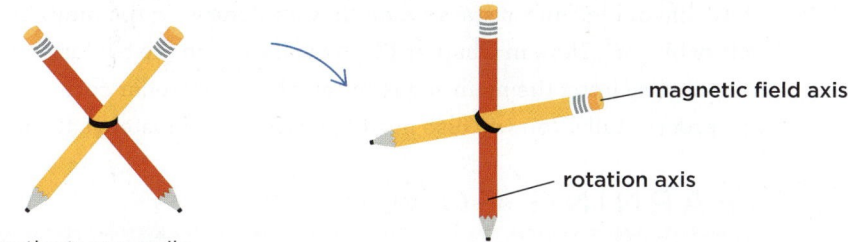

magnetic field axis

rotation axis

Tape the two pencils so that they form an "X."

One of these pencils is the rotation axis, and one is the axis of the magnetic field.

Wad up some paper around the place where the pencils cross, to represent the neutron star.

Tape pieces of string or ribbon to each end of the pencil representing the axis of the magnetic field. The ribbon represents the beam of light that comes from the magnetic field axis.

Hold the other pencil, which represents the rotation axis, and spin the model quickly. Notice how sometimes the magnetic field points toward you, and sometimes it does not. How many times does the magnetic field point toward you during each rotation of the model?

PREDICTION

I predict that for each rotation the observer will see

☐ 0   ☐ 1   ☐ 2   ☐ 4

pulse/pulses.

13

So far in our discussion of the lives of stars (Chapters 11 and 12), we have concentrated on what happens to low-mass stars like our Sun. High-mass stars have masses greater than about 8 solar masses (8 M_{Sun}). More mass means stronger gravitational force on the inner parts of the star. Greater force means higher pressure; higher pressure means faster reaction rates; faster reaction rates mean more energy released each second; and finally, more energy release each second means greater luminosity. These stars are thousands or even millions of times more luminous than the Sun. These high luminosities mean that the stars burn fuel faster and do not last as long: only millions of years, as opposed to billions of years for stars like the Sun. The most dramatic difference in evolution comes at the end of the life of high-mass stars, when they explode with the luminosity of an entire galaxy of stars. Here in Chapter 13, you will explore the evolution of high-mass stars as they leave the main sequence and become stellar corpses. To understand this process fully, you will also need to learn about Einstein's theories of relativity.

LEARNING GOALS

1. Describe how the evolution of high-mass stars differs from the evolution of low-mass stars.

2. List the stages of evolution of high-mass stars, and explain the origin of chemical elements heavier than iron.

3. Discuss some of the implications of special and general relativity.

4. Describe the key properties of supernovae, neutron stars, and black holes. Identify how black holes can be observed.

13.1 High-Mass Stars Follow Their Own Path

In Chapter 12, we followed low-mass stars along their evolutionary path (see Figure 12.14 to review the process). High-mass stars evolve differently as they leave the main sequence, following different paths through the Hertzsprung-Russell (H-R) diagram, because a different structure develops in the interior of the star.

Leaving the Main Sequence

 **Interactive Simulation:** CNO Cycle

At the very high temperatures in the center of a high-mass star, nuclear reactions other than the proton-proton chain (see Figure 11.4) occur. For example, in some high-mass stars, hydrogen (H) is fused into helium (He) via the **carbon-nitrogen-oxygen (CNO) cycle**. In this process, carbon is a **catalyst**—a substance that helps the reaction proceed but is not used up in the reaction. The CNO cycle consumes hydrogen and turns it into helium, as shown in **Figure 13.1**. In the core of a high-mass star, the nuclear reactions increase dramatically with temperature, so that the innermost core generates a lot of energy, while the layer above the core generates much less. This situation is exactly right for convection to occur in the core, and so high-mass stars have convective, hydrogen-burning cores.

As the high-mass star runs out of hydrogen in its core, the weight of the overlying star compresses the core. Yet long before the core becomes electron-degenerate, the pressure and temperature become high enough for helium burning.

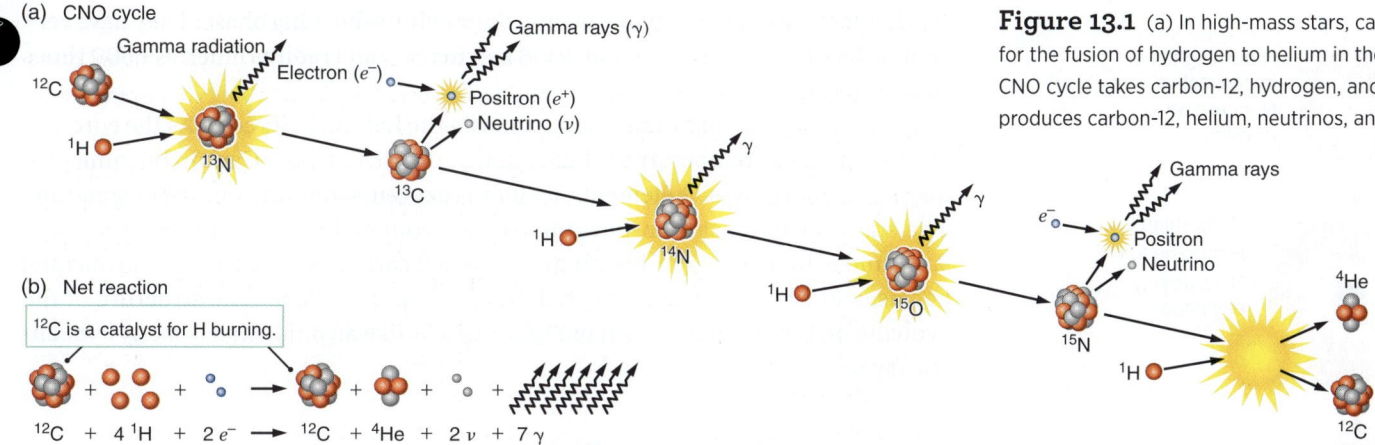

(a) CNO cycle

Figure 13.1 (a) In high-mass stars, carbon serves as a catalyst for the fusion of hydrogen to helium in the CNO cycle. (b) The CNO cycle takes carbon-12, hydrogen, and electrons as inputs and produces carbon-12, helium, neutrinos, and gamma rays.

(b) Net reaction

^{12}C is a catalyst for H burning.

$$^{12}C + 4\,^{1}H + 2\,e^{-} \longrightarrow {}^{12}C + {}^{4}He + 2\,\nu + 7\,\gamma$$

Because the core is not made of degenerate matter, the structure of the star responds to the increase in temperature, but its luminosity changes relatively little. The star makes a fairly smooth transition from hydrogen burning to helium burning.

The high-mass star now burns helium in its core and hydrogen in a surrounding shell, like a low-mass horizontal branch star. It increases in size while its surface temperature decreases. The result, illustrated in **Figure 13.2**, is that the star moves to the right on the H-R diagram, leaving the main sequence. Stars of more than

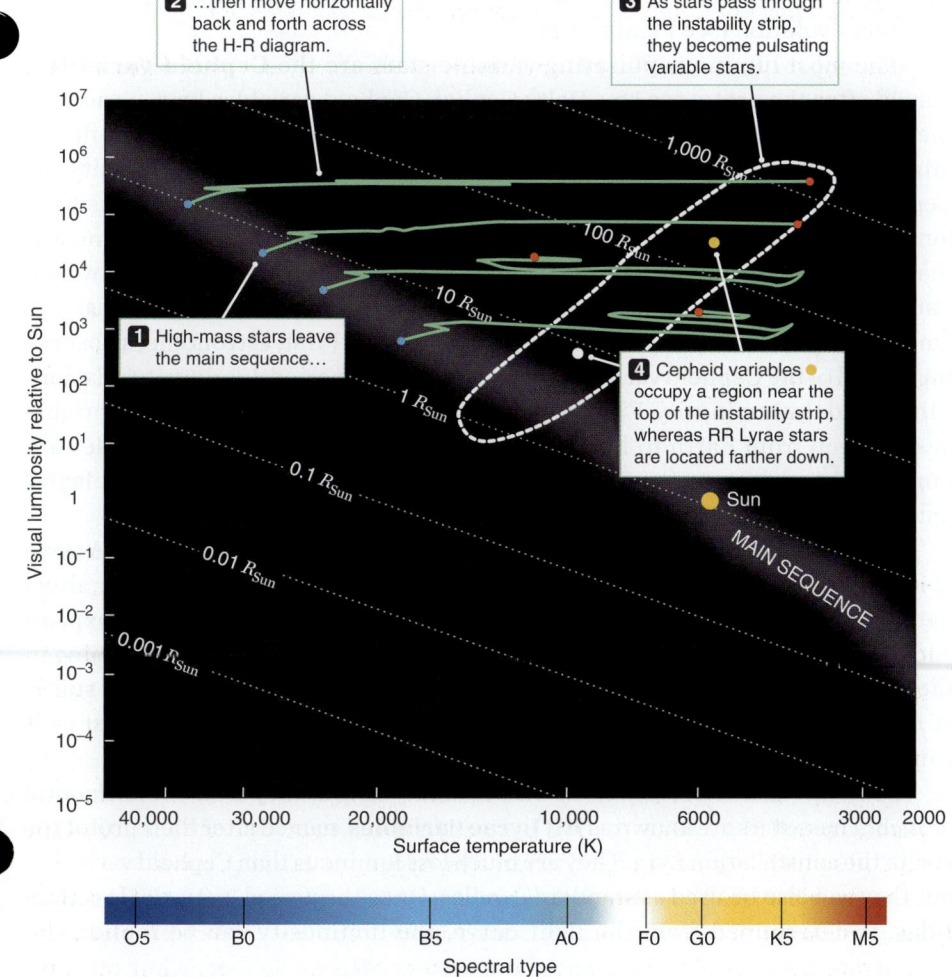

Figure 13.2 When massive stars leave the main sequence, they move horizontally across the H-R diagram. Pulsations occur as the stars cross the instability strip, causing the stars to vary in brightness with a period that is related to their luminosity.

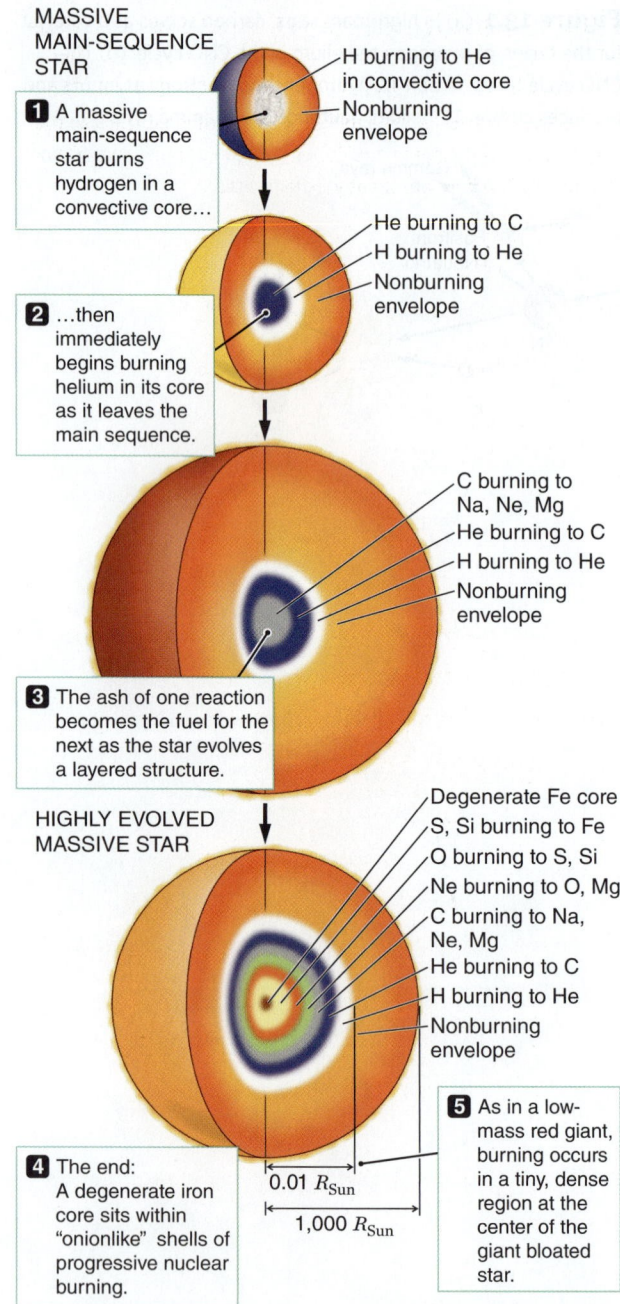

MASSIVE MAIN-SEQUENCE STAR

— H burning to He in convective core
— Nonburning envelope

1 A massive main-sequence star burns hydrogen in a convective core...

— He burning to C
— H burning to He
— Nonburning envelope

2 ...then immediately begins burning helium in its core as it leaves the main sequence.

— C burning to Na, Ne, Mg
— He burning to C
— H burning to He
— Nonburning envelope

3 The ash of one reaction becomes the fuel for the next as the star evolves a layered structure.

HIGHLY EVOLVED MASSIVE STAR

— Degenerate Fe core
— S, Si burning to Fe
— O burning to S, Si
— Ne burning to O, Mg
— C burning to Na, Ne, Mg
— He burning to C
— H burning to He
— Nonburning envelope

4 The end: A degenerate iron core sits within "onionlike" shells of progressive nuclear burning.

$0.01\ R_{Sun}$

$1,000\ R_{Sun}$

5 As in a low-mass red giant, burning occurs in a tiny, dense region at the center of the giant bloated star.

Figure 13.3 As a high-mass star evolves, it builds up a layered structure like that of an onion; progressively advanced stages of nuclear burning are found deeper and deeper within the star. Note the change in scale for the bottom image.

$10\ M_{Sun}$ become red supergiants during their helium-burning phase. They have very cool surface temperatures (about 4000 kelvins [K]) and radii as much as 1,500 times that of the Sun.

Eventually, the high-mass star exhausts the helium in its core. As the core collapses, it reaches temperatures high enough to burn carbon. Carbon burning produces even more massive elements, including oxygen, sodium, neon, and magnesium. The star now has a carbon-burning core surrounded by a helium-burning shell surrounded by a hydrogen-burning shell. When carbon is exhausted, neon burning begins; and when neon is exhausted, oxygen begins to burn. The structure of the evolving high-mass star, shown in **Figure 13.3**, is like an onion; it has many concentric layers.

Stars on the Instability Strip

As a star evolves, it may become a **pulsating variable star**, which does not maintain a steady balance between the inward-pushing force of gravity and the outward-pushing force due to the pressure of energy escaping from the core, known as radiative pressure. As a result, the star repeatedly grows larger and smaller. Larger stars have more surface area, so they emit more light and have a higher luminosity than smaller stars of similar temperature. As a pulsating variable star's size varies, this effect causes the star's luminosity to repeatedly vary as well. The time it takes a star to complete one brightness cycle is known as the period. These pulsating variable stars occupy an area of the H-R diagram known as the **instability strip**. The star may pass through the instability strip more than once during its post-main-sequence evolution (see Figure 13.2).

The most luminous pulsating variable stars are the **Cepheid variables**, named after the prototype star Delta Cephei. Cepheid variables have periods in the range from 1 to 100 days. Astronomer Henrietta Leavitt (1868–1921) determined that there is a relationship between the period and the luminosity for Cepheid variable stars. Short-period Cepheid variables are less luminous than long-period Cepheid variables. This **period-luminosity relationship** means that Cepheid variables are an example of a **standard candle**: an object whose luminosity can be determined independently from its distance. Therefore, astronomers can find the distance to any Cepheid variable. First, astronomers observe the period of the Cepheid variable. Then they use the period-luminosity relationship to find the luminosity. Finally, they combine this luminosity with the brightness as seen from Earth to find the distance to the star. Cepheid variables are luminous enough that astronomers can use them to find the distance to galaxies outside of the Milky Way.

Thermal energy powers the pulsations of stars like Cepheid variables. A cycle of ionization and deionization and reionization of helium atoms in the star alternately traps and releases thermal energy, causing the star to repeatedly expand and contract (**Figure 13.4**). These pulsations do not affect the nuclear burning in the star's interior, but they do affect the light escaping from the star. The star is at its brightest and bluest while it expands and at its faintest and reddest as it contracts.

Some low-mass stars also enter the instability strip as they cross the horizontal branch. These stars are known as **RR Lyrae variables**, named after their prototype star in the constellation Lyra. They are much less luminous than Cepheid variables, but they can also be used as standard candles. Once the period, typically less than 1 day, is determined from the light curve, the luminosity can be found. The

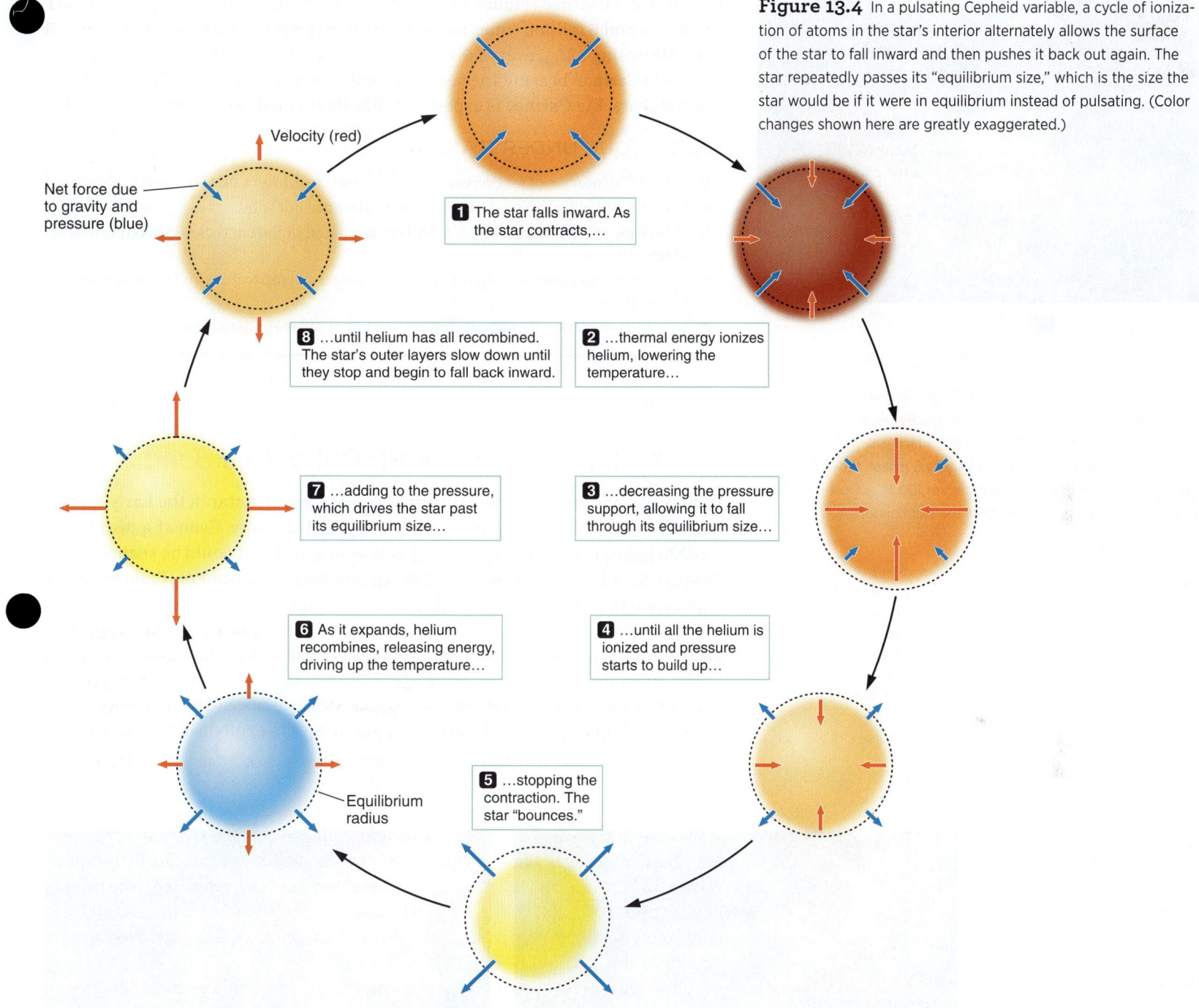

Figure 13.4 In a pulsating Cepheid variable, a cycle of ionization of atoms in the star's interior alternately allows the surface of the star to fall inward and then pushes it back out again. The star repeatedly passes its "equilibrium size," which is the size the star would be if it were in equilibrium instead of pulsating. (Color changes shown here are greatly exaggerated.)

Velocity (red)

Net force due to gravity and pressure (blue)

1 The star falls inward. As the star contracts,...

8 ...until helium has all recombined. The star's outer layers slow down until they stop and begin to fall back inward.

2 ...thermal energy ionizes helium, lowering the temperature...

7 ...adding to the pressure, which drives the star past its equilibrium size...

3 ...decreasing the pressure support, allowing it to fall through its equilibrium size...

6 As it expands, helium recombines, releasing energy, driving up the temperature...

4 ...until all the helium is ionized and pressure starts to build up...

Equilibrium radius

5 ...stopping the contraction. The star "bounces."

instability strip also intersects the main sequence around spectral type A, and many type A stars do show variability.

The pressure of intense radiation on the gas at the surface of a massive star overcomes the star's gravity and causes winds with speeds of about 3,000 kilometers per second (km/s) and mass loss ranging from about 10^{-7} to $10^{-5} M_{Sun}$ per year. These numbers may sound tiny, but over millions of years, this mass loss becomes significant. Type O stars with masses of 20 M_{Sun} lose about 20 percent of their mass while on the main sequence, and possibly more than 50 percent over their entire

what if . . .

What if a Cepheid variable star with a mass of 5 M_{Sun} is in orbit around a white dwarf with a mass of 1 M_{Sun}: What evolutionary steps might both of these stars have taken for the system to reach this state?

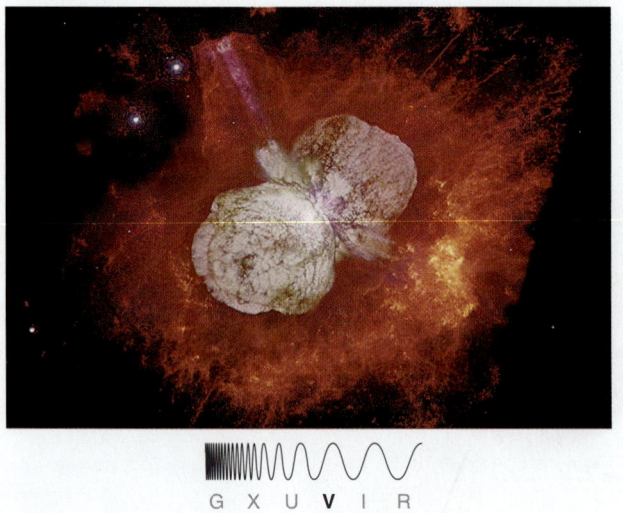

G X U V I R

Figure 13.5 This Hubble Space Telescope image shows an expanding cloud of dusty material ejected by the luminous blue variable star Eta Carinae. The star itself, which is largely hidden by the surrounding dust, has a luminosity 5 million times that of the Sun and a mass probably in excess of 100 M_{Sun}. Dust is created when volatile material ejected from the star condenses.

lifetimes. Eta Carinae (**Figure 13.5**), a binary star more than 100 times as massive as the Sun and 5 million times as luminous, is an extreme example. Eta Carinae is currently losing 1 M_{Sun} every 1,000 years. However, during a 19th century eruption when Eta Carinae became the second-brightest star in the sky, it shed 10 M_{Sun} in only 20 years. Eta Carinae is expected to one day explode violently.

CHECK YOUR UNDERSTANDING 13.1

Why does the interior of an evolved high-mass star have layers like an onion?
  a. Heavier atoms sink to the bottom because stars are not solid.
  b. Before the star formed, heavier atoms accumulated in the centers of clouds because of gravity.
  c. Heavier atoms fuse closer to the center because the temperature and pressure are higher there.
  d. Different energy transport mechanisms occur at different densities.

Answers to Check Your Understanding questions are in the back of the book.

13.2 High-Mass Stars Go Out with a Bang

In 1987, astronomers observed the explosion of a massive star in the Large Magellanic Cloud (LMC), a companion galaxy to the Milky Way. Even at a distance of 170,000 light-years, Supernova 1987A was so bright that it could be seen with the naked eye, and it dazzled sky gazers in the Southern Hemisphere (**Figure 13.6**). A high-mass star had come to the end of its life and exploded. The explosion was more luminous than all the other stars in the host galaxy! By comparing two images of a galaxy, one before the explosion (Figure 13.6a) and one after the explosion (Figure 13.6b), you can identify the exploding star. Observations like those in **Figure 13.7** are helping astronomers understand how remnants from a supernova change over time. Here in Section 13.2, you will learn how a star comes to explode in such a dramatic fashion.

(a)   (b)

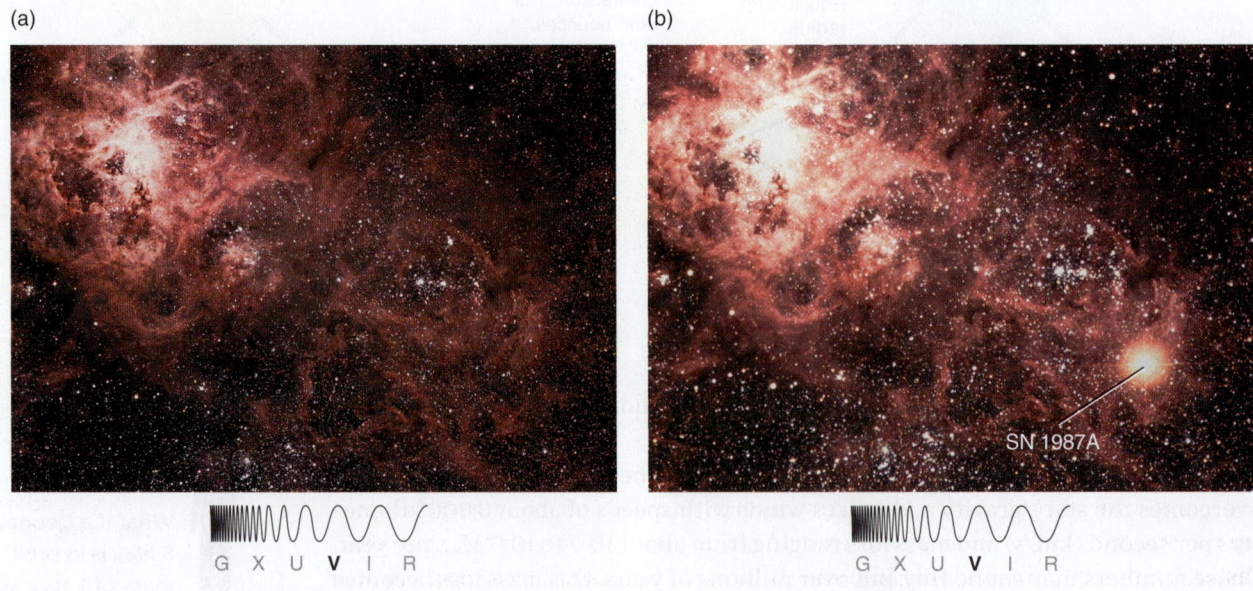

G X U V I R   G X U V I R

Figure 13.6 Supernova 1987A (SN 1987A) was a supernova that exploded in a small companion galaxy of the Milky Way called the Large Magellanic Cloud (LMC). These images show the LMC (a) before the explosion and (b) while the supernova was near its peak.

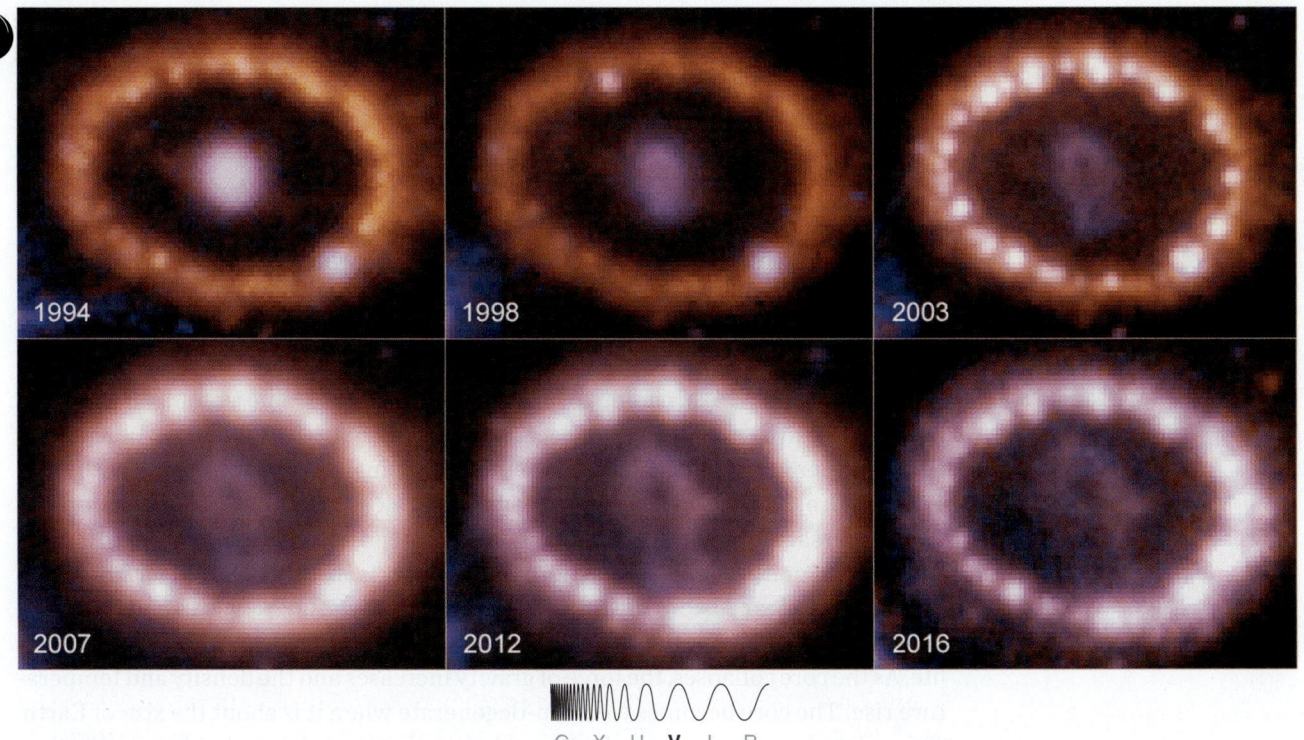

G X U **V** I R

Figure 13.7 ★ **WHAT AN ASTRONOMER SEES** In general, astronomical objects change very slowly, so it is a little bit unusual that an astronomer has the chance to see such dramatic changes in celestial objects. This series of Hubble Space Telescope images shows how the innermost region of the supernova explosion SN1987a changed from 1994 to 2016. The ring of gas was produced late in the life of the star, at least 20,000 years before it exploded, and then was lit up by a blast of ultraviolet light from the explosion, causing it to glow for at least 30 years. An astronomer will notice that in 1994, seven years after the supernova occurred, the material leaving the star is still very compact, near the center of the ring. By 2012, that material had blasted out along an axis, nearly reaching the ring. She might look up the scale of the image to find that, in 2016, the central structure measured more than half a light year across. From images like these, astronomers can observe the evolution of the supernova remnant, which has an organized structure reminiscent of some planetary nebulae.

The Final Days in the Life of a Massive Star

Maintaining the balance in a star between the inward-pushing force of gravity and the outward-pushing force due to radiative pressure is somewhat like trying to keep a leaky ball inflated. The larger the leak, the more rapidly air must be pumped into the ball to keep it inflated. A star that is burning hydrogen or helium, such as the one on the left side of **Figure 13.8**, is like a ball with a slow leak. At the temperatures of hydrogen or helium burning, energy leaks out of the interior of the star primarily by radiation and convection, but the outer layers of the star act like a thick, warm blanket. As a result, much of the energy is kept in the star, so nuclear fuels do not need to burn very fast to support the star against gravity.

Once carbon burning begins, this balance shifts. Energy is carried away from the core primarily by neutrinos rather than radiation and convection. Recall from Section 11.1 that neutrinos escape from a star easily. In a high-mass star, once cooling by neutrinos becomes significant, the outer layers of the star fall inward, driving up the star's density and temperature and forcing nuclear reactions to run even faster. Carbon burning supports the star for about a thousand years, oxygen burning lasts about a year, and silicon burning lasts only about a day. A star that is burning silicon does not emit much more light than it did when it was burning helium. However, during this phase, neutrinos cool the core and carry about 200 million

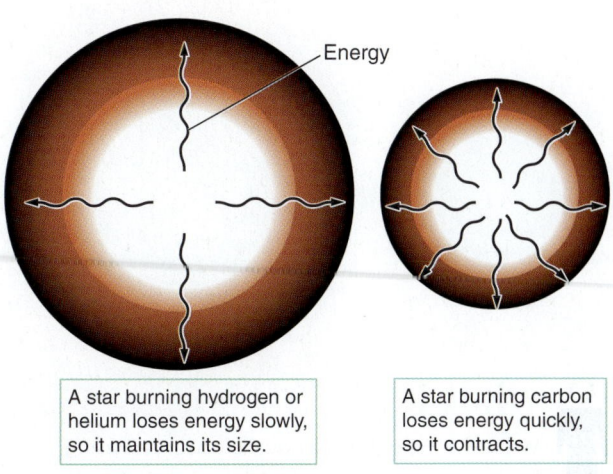

Energy

A star burning hydrogen or helium loses energy slowly, so it maintains its size.

A star burning carbon loses energy quickly, so it contracts.

Figure 13.8 If energy leaves a star faster than it can be replaced, the energy balance is disrupted, and the star begins to contract.

times more energy away from the star each second than when it was burning helium. Therefore, the *total* amount of energy emitted is much higher than when the star was burning helium.

The Core Collapses and the Star Explodes

In Figure 13.3, we saw that an evolving high-mass star builds up its onionlike structure as hydrogen burns to helium; helium burns to carbon; carbon burns to sodium (Na), neon (Ne), and magnesium (Mg); neon burns to oxygen and magnesium; oxygen burns to sulfur (S) and silicon (Si); and sulfur and silicon burn to iron (Fe). After silicon burning, the end comes suddenly and dramatically. Many types of nuclear reactions occur up to this point, forming almost all of the stable isotopes of elements less massive than iron. However, because iron does not *release* energy when it fuses (it *absorbs* energy instead), the chain of nuclear fusion reactions stops with iron. When hydrogen is fused into helium, energy is released, maintaining the temperature necessary to keep the reaction going; once the reaction begins, it is self-sustaining. For iron, once the reaction starts, energy is absorbed. No longer supported by thermonuclear fusion, the iron core of the massive star begins to collapse.

Figure 13.9 shows the stages a high-mass star passes through at the end of its life. As the core collapses, the force of gravity increases and the density and temperature rise. The core becomes electron-degenerate when it is about the size of Earth. The weight bearing down on the iron ash core is too great to be held up by electron degeneracy pressure (step 1 in Figure 13.9). As the core collapses, the core temperature climbs to temperatures of more than 10 billion K, while the density exceeds 10^{10} kilograms per cubic meter (kg/m^3)—10 times the density of a white dwarf.

At these temperatures, the nucleus of the star is filled with photons so energetic that they can break iron nuclei apart (step 2 in Figure 13.9). This process, called *photodisintegration*, absorbs thermal energy and begins tearing apart the nuclei built up by nuclear fusion. At the same time, the density of the core is so great that electrons are forced into atomic nuclei, where they combine with protons to produce neutrons and neutrinos (step 3 in Figure 13.9). Both this process and photodisintegration absorb much of the energy that was holding up the dying star. Neutrinos take still more energy with them as they leave the star. The collapse of the core accelerates, reaching a speed of 70,000 km/s, or almost one-fourth the speed of light, on its inward fall (step 4 in Figure 13.9). All of these events together take place in less than a second.

As material in the collapsing core exceeds the density of an atomic nucleus, the strong nuclear force actually becomes repulsive. About half of the collapsing core suddenly slows its inward fall. The remaining half slams into the innermost part of the star at a significant fraction of the speed of light and "bounces," sending a tremendous shock wave back out through the star (step 5 in Figure 13.9).

Over the next second or so, almost 20 percent of the material in the core is converted into neutrinos. Most fly outward through the star, but at these phenomenal densities, not even neutrinos pass through with complete freedom. The dense material behind the expanding shock wave traps a few tenths of a percent of the neutrinos. The energy of these trapped neutrinos drives the pressure and temperature in this region higher, inflating a bubble of extremely hot gas and intense radiation around the core of the star (step 6 in Figure 13.9). The pressure of this bubble adds to the shock wave moving outward through the star. Within about a minute, the shock wave has pushed its way out through the helium shell within the star. Within a few hours, it reaches the surface of the star itself, heating the stellar

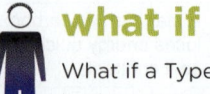

what if . . .

What if a Type II supernova is observed using both neutrinos and optical light: In what order would you expect to see these signals arrive at Earth?

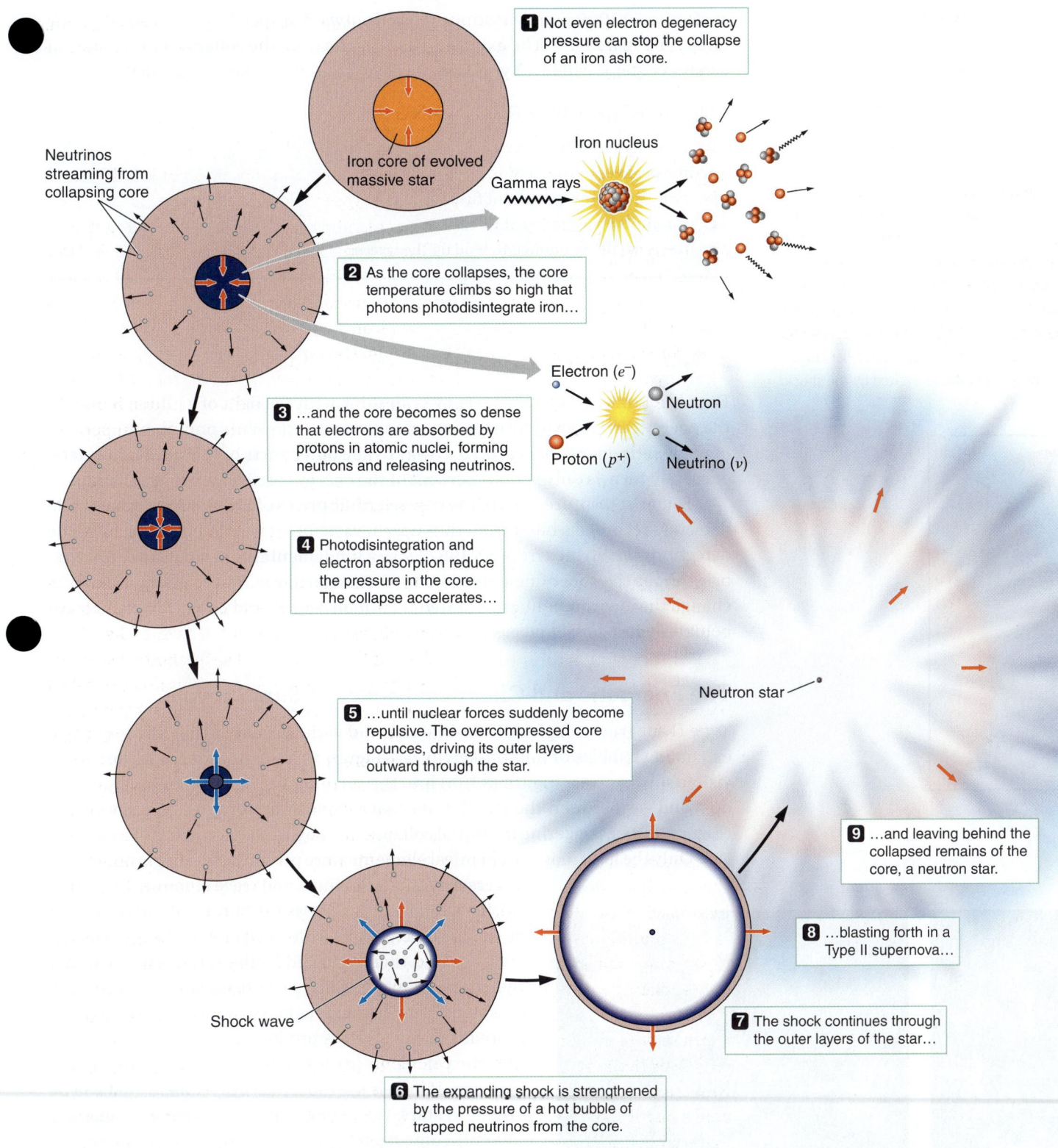

1 Not even electron degeneracy pressure can stop the collapse of an iron ash core.

Iron core of evolved massive star

Iron nucleus

Gamma rays

Neutrinos streaming from collapsing core

2 As the core collapses, the core temperature climbs so high that photons photodisintegrate iron…

Electron (e^-)

Neutron

Proton (p^+)

Neutrino (ν)

3 …and the core becomes so dense that electrons are absorbed by protons in atomic nuclei, forming neutrons and releasing neutrinos.

4 Photodisintegration and electron absorption reduce the pressure in the core. The collapse accelerates…

Neutron star

5 …until nuclear forces suddenly become repulsive. The overcompressed core bounces, driving its outer layers outward through the star.

9 …and leaving behind the collapsed remains of the core, a neutron star.

8 …blasting forth in a Type II supernova…

7 The shock continues through the outer layers of the star…

Shock wave

6 The expanding shock is strengthened by the pressure of a hot bubble of trapped neutrinos from the core.

Figure 13.9 A high-mass star goes through several stages at the end of its life as its core collapses and the star explodes as a Type II supernova. Recall that we use red arrows to represent velocity, and blue arrows to represent force. In this figure, the black arrows indicate neutrinos escaping from the collapsing core.

🎥 **Astronomy in Action:** Type II Supernova

Figure 13.10 The Cygnus Loop is a supernova remnant—an expanding interstellar blast wave caused by the explosion of a massive star thousands of years ago. (a) Gas in the interior of the cloud, with a temperature of millions of kelvins, glows in X-rays. (b) Visible light comes from the edges of the expanding bubble: locations where the expanding blast wave pushes through denser gas in the interstellar medium. (c) A Hubble Space Telescope image of a location where the blast wave is hitting an interstellar cloud. The colors indicate different types of atoms excited by the shock.

(a)

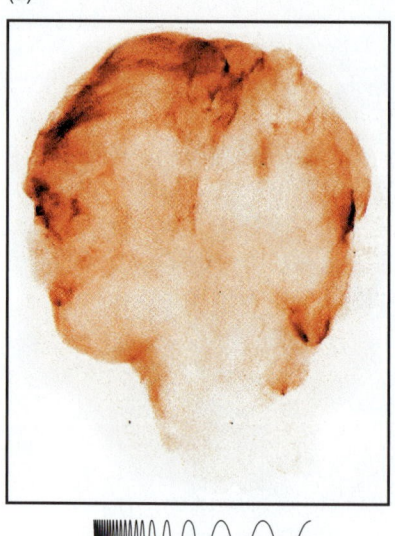

G **X** U V I R

(b)

(c)

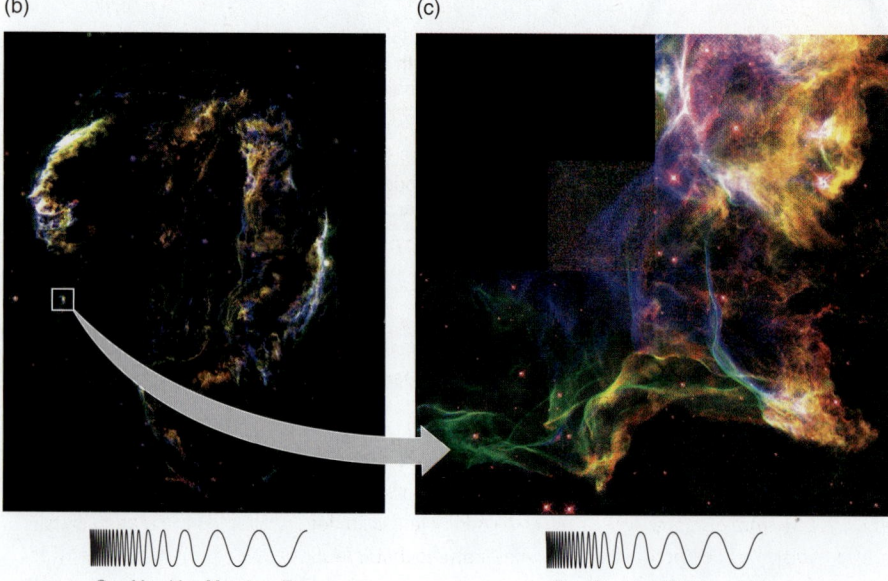

G **X** U V I R   G **X** U **V** I R

surface to 500,000 K and blasting material outward at speeds of up to 30,000 km/s (step 7 in Figure 13.9). The explosion that results from the collapse and rebound of a massive star is called a **Type II** supernova (steps 8 and 9 in Figure 13.9).

CHECK YOUR UNDERSTANDING 13.2

Iron fusion cannot support a star because
a. iron oxidizes too quickly.
b. iron absorbs energy when it fuses.
c. iron emits energy when it fuses.
d. iron is not dense enough to hold up the layers.

13.3 Supernovae Change the Galaxy

For a brief time, a Type II supernova can shine with the light of a billion Suns. Yet the energy carried away by light from a supernova represents only about 1 percent of the kinetic energy being carried away by the outer parts of the star. This kinetic energy, in turn, is only about 1 percent of the energy carried away by neutrinos.

Scientists captured one of the true scientific prizes of SN 1987A when neutrino detectors recorded a burst from the supernova before it was detected visually. The detection of neutrinos from SN 1987A was a fundamental confirmation of our theories about the role of neutrinos in Type II supernovae. These energetic events change the environment around them, both physically and chemically, and leave behind shells of dust and gas and dense cores.

The Energetic and Chemical Legacy of Supernovae

Type II supernova explosions leave a rich and varied legacy to the universe. Huge expanding bubbles of million-kelvin gas (**Figure 13.10a**) glow in X-rays and drive visible shock waves (Figure 13.10b) into the surrounding interstellar medium (the dust and gas between the stars). Supernova explosions compress nearby clouds (Figure 13.10c), triggering the initial collapse that begins the formation of new stars.

Only the least massive chemical elements were present at the beginning of the universe: hydrogen, helium, and trace amounts of lithium (Li), beryllium (Be), and possibly boron (B). All the rest of the chemical elements were formed in stars through nuclear reactions and then returned to the interstellar medium when the stars exploded. Nuclear fusion in the core of low-mass stars forms elements as massive as carbon and oxygen, whereas high-mass stars produce elements as massive as iron. Most elements in the periodic table are more massive than iron. If iron is the most massive element that can be formed in stars, then where do these even more massive elements come from?

Under normal circumstances, electric repulsion keeps positively charged atomic nuclei far apart. Extreme temperatures are needed to slam nuclei together with enough kinetic energy to overcome this electric repulsion. Free neutrons, however, have no net electric charge, so there is no electric repulsion to prevent them from simply running into an atomic nucleus. Under normal conditions, free neutrons are rare. Under the conditions of a Type II supernova, however, free neutrons are produced in very large numbers. These are easily captured by massive atomic nuclei and later decay to become protons, forming elements more massive than iron.

Nuclear physics predicts the abundances of the elements. These predictions agree with abundances that have been measured in the Solar System and in the atmospheres of stars, as shown in **Figure 13.11**. As predicted, less massive elements are far more abundant than more massive elements. An exception to this pattern is the decrease in the abundances of the light elements lithium, beryllium, and boron. Nuclear burning easily destroys these elements, and they are not produced by the common reactions involved in burning hydrogen and helium. Conversely, carbon, nitrogen, and oxygen are big winners in the most common process of helium burning, so they are more abundant. More massive elements are progressively built up from less massive elements. The spike in the abundances of elements near iron is evidence of processes that favor these tightly bound nuclei. Even the sawtooth pattern in the abundances of even- and odd-numbered elements can be explained as a consequence of the way atomic nuclei form in stars.

Our understanding of the interiors of dying stars is confirmed by the chemical composition of the Earth under our feet. Our own bodies are formed of atoms that were made in stars. Our growing understanding of the chemical evolution of the universe and our connection to it is one of the triumphs of modern astronomy.

Neutron Stars and Pulsars

Let's now return to the remains of the star that has exploded (see step 9 in Figure 13.9). The core has collapsed to the point where it has about the same density as the nucleus of an atom. Stars with initial mass less than about 25 M_{Sun} will leave cores with mass less than about 2.2 M_{Sun}. The collapse of these cores is halted when *neutrons* are packed as tightly together as the rules of physics allow. This neutron-degenerate core is called a **neutron star**. It is roughly the size of a small city, with a diameter of about 20 km. That volume is packed with a mass between 1.1 M_{Sun} and 2.2 M_{Sun}. At a density of about 10^{18} kg/m³, the neutron star is a billion (10^9) times denser than a white dwarf and a thousand trillion (10^{15}) times denser than water.

If the original massive star was part of a close binary system, then the neutron star will have a companion, much like the white dwarf binary systems discussed in Section 12.5 (see Figures 12.15 and 12.16). **Figure 13.12** illustrates an **X-ray binary**, a binary

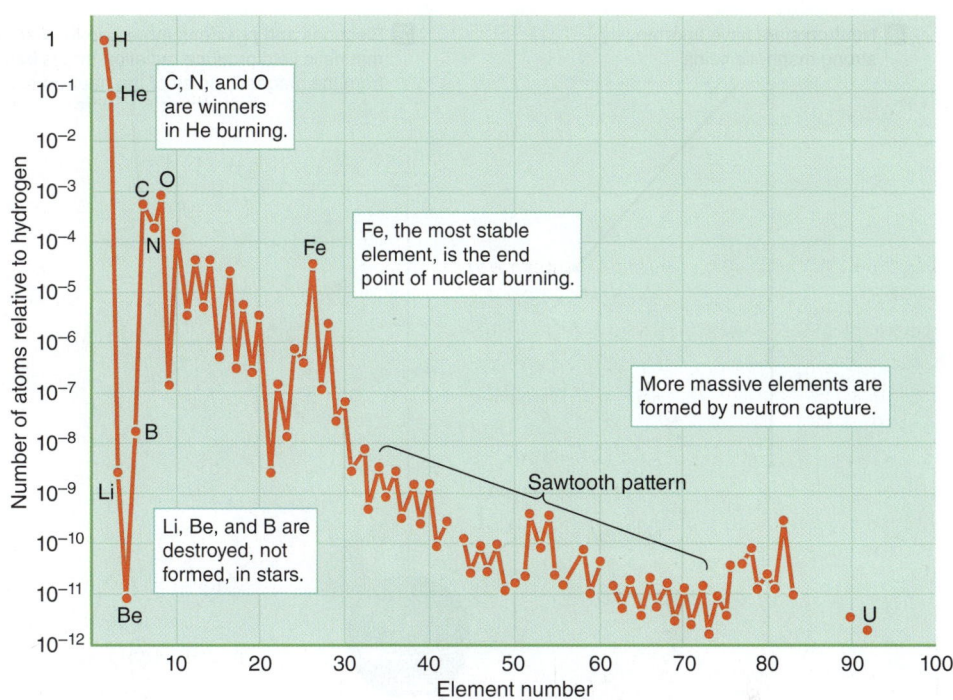

Figure 13.11 In this graph, relative abundances of different elements in the Solar System are plotted against each element's number. These abundances are compared to the number of hydrogen atoms, which is set to 1. So, for example, there are roughly one-tenth as many helium atoms as hydrogen atoms. As the atomic number increases, the number of protons (and typically also neutrons) in the nucleus increases, so the nuclei are more massive. The pattern of abundances is a direct result of the production of atomic nuclei in stars.

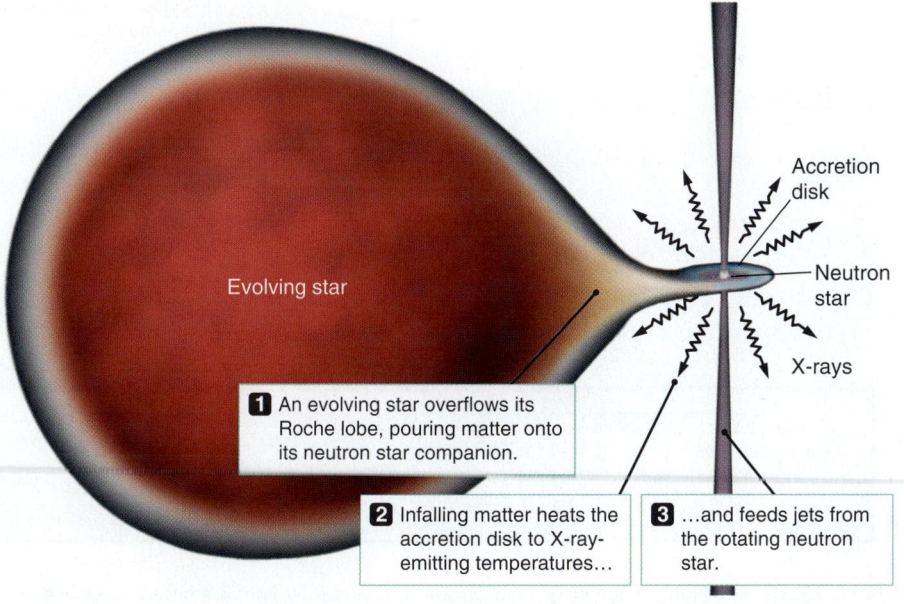

Figure 13.12 X-ray binaries are systems consisting of a normal evolving star and a white dwarf, neutron star, or black hole. As the evolving star overflows its Roche lobe, mass falls toward the collapsed object. The gravitational well of the collapsed object is so deep that when the material hits the accretion disk, it is heated to such high temperatures that it radiates away most of its energy as X-rays.

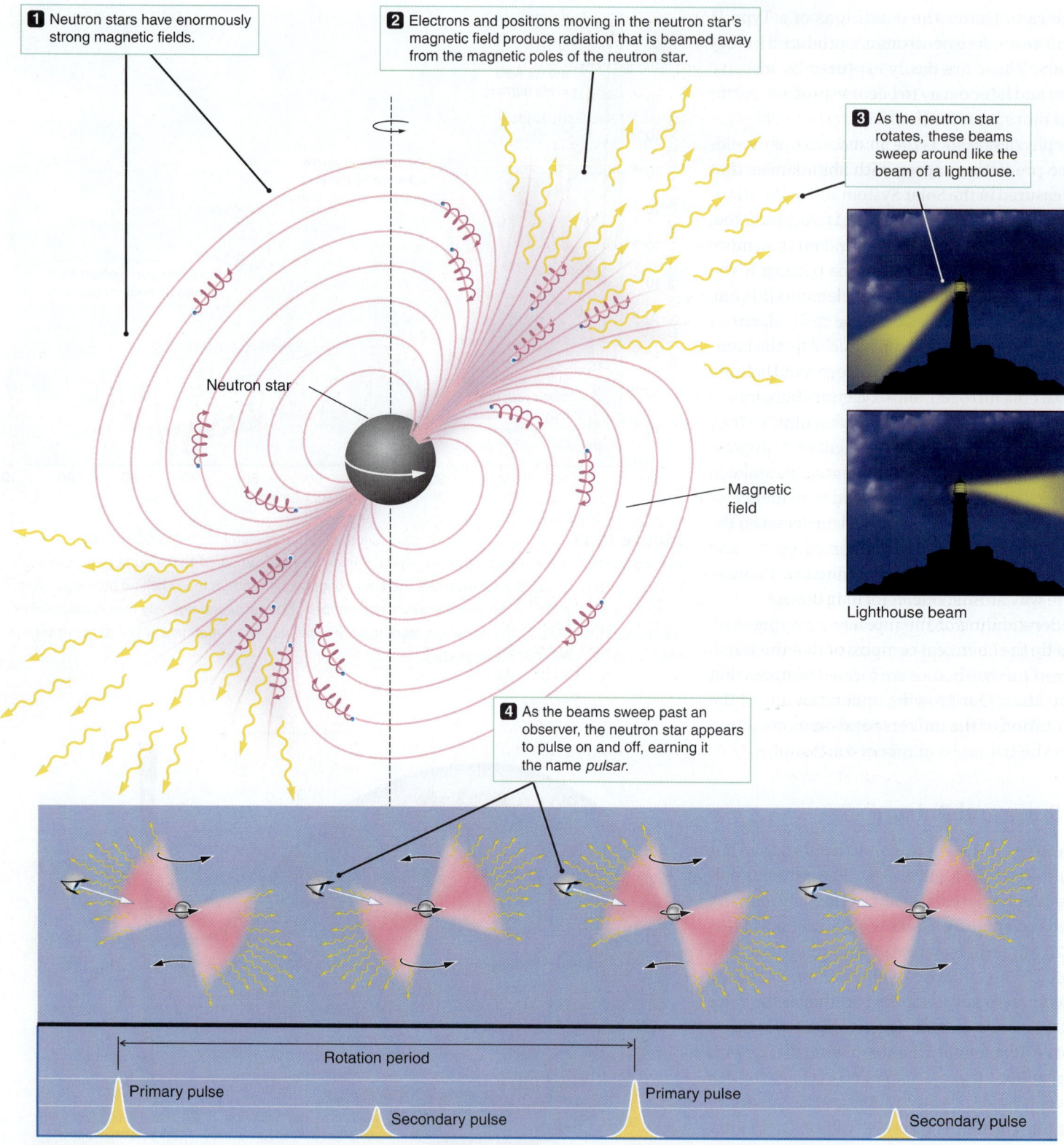

1 Neutron stars have enormously strong magnetic fields.

2 Electrons and positrons moving in the neutron star's magnetic field produce radiation that is beamed away from the magnetic poles of the neutron star.

3 As the neutron star rotates, these beams sweep around like the beam of a lighthouse.

Lighthouse beam

Neutron star

Magnetic field

4 As the beams sweep past an observer, the neutron star appears to pulse on and off, earning it the name *pulsar*.

Rotation period

Primary pulse

Secondary pulse

Primary pulse

Secondary pulse

Figure 13.13 As a highly magnetized neutron star rotates rapidly, light is emitted, much like the beams from a rotating lighthouse lamp. From our perspective, as these beams of radio emission sweep past us, the star appears to pulse on and off, earning it the name *pulsar*.

system in which mass from an evolving star spills over onto a collapsed companion such as a white dwarf, neutron star, or black hole. (Black holes will be discussed in Sections 13.4 and 13.6.) As the lower-mass star evolves and overfills its Roche lobe, matter falls toward the accretion disk around the neutron star, heating it to millions of kelvins and causing it to glow brightly in the X-ray region of the electromagnetic spectrum (see Figure 4.6). X-ray binaries sometimes develop powerful jets of material that are perpendicular to the accretion disk and carry material away at speeds close to the speed of light.

Binary systems like this can undergo a process very similar to the "nova" process that occurs with lower-mass stars. Material may fall from the main-sequence star onto the more evolved star, adding hydrogen to the surface, which fuses into helium. The helium accumulates until it explodes in a brief burst, typically increasing the X-ray luminosity by a factor of about 100. The system releases a lot of energy in the form of X-rays, so it is called an **X-ray burster**.

As the core of the original massive star collapses, it spins faster, because angular momentum is conserved (see Chapter 5). A main-sequence O star rotates perhaps once every few days. As a neutron star, it might rotate tens or even hundreds of times each second. The collapsing star concentrates the magnetic field of the original star to strengths trillions of times greater than the magnetic field at Earth's surface. A neutron star has a magnetosphere just like Earth and several other planets do, except that the neutron star's magnetosphere is vastly stronger and is whipped around many times a second by the spinning star. Just like in planets, in stars the magnetic axis is often not aligned with the rotation axis.

Electrons and positrons move along the magnetic-field lines and are "funneled" by the field toward the magnetic poles of the system. The particles produce beams of radiation along the magnetic poles of the neutron star as shown in **Figure 13.13**. As the neutron star rotates, these beams sweep through space, much like the rotating beams of a lighthouse. The neutron star appears to flash on and off with a regular period equal to the period of rotation of the star (or half the rotation period, if we see both beams). These rotating neutron stars are known as **pulsars**. Each one has a distinct period, which is gradually slowing down at a unique rate. As of this writing, more than 2,000 pulsars are known, and more are being discovered all the time. Most of these pulsars are radio pulsars, with a few also emitting in other regions of the electromagnetic spectrum. For example, X-ray pulsars occur when accretion onto the magnetic poles from a companion star causes periodic outbursts of X-ray emission.

The Crab Nebula: Remains of a Stellar Cataclysm

In 1054, Chinese astronomers noticed a "guest star" in the direction of the constellation Taurus. The new star was so bright that it could be seen during the daytime for 3 weeks, and it did not fade away altogether for many months. This star was a fairly typical Type II supernova. Today, an expanding cloud of debris from this explosion occupies this place in the sky—forming an object called the Crab Nebula (**Figure 13.14a**).

The Crab Nebula has filaments of glowing gas expanding away from the central star at 1,500 km/s. These filaments contain anomalously high abundances of helium and other, more massive chemical elements—the products of the nuclear reactions that took place in the supernova and its progenitor star.

The Crab Pulsar at the center of the nebula flashes 60 times a second: first with a main pulse associated with one of the "lighthouse" beams, then with a fainter

Astronomy in Action: Pulsar Rotation

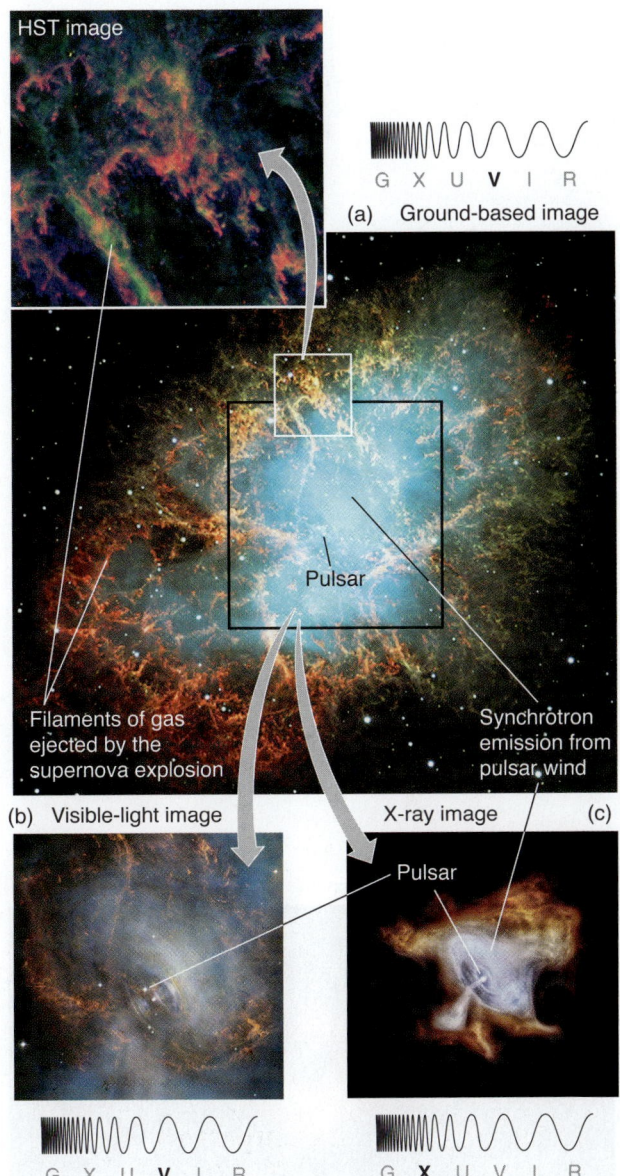

Figure 13.14 The Crab Nebula is the remnant of a supernova explosion witnessed by Chinese astronomers in 1054. (a) The object we see today is an expanding cloud of "shrapnel" from that earlier cataclysm. The spinning pulsar at the heart of the Crab Nebula sends off a "wind" of electrons and positrons moving at close to the speed of light. Inset: An expanded view of the filaments obtained by the Hubble Space Telescope (HST). The synchrotron radiation from these particles is shown (b) in visible light and (c) in X-rays.

what if . . .

What if you could observe the Crab Nebula a million years from now: How do you think the nebula would appear, and what will be the fate of the rotating pulsar?

secondary pulse associated with the other beam. As the Crab Pulsar spins 30 times a second, it whips its powerful magnetosphere around with it. At a distance from the pulsar about equal to the radius of the Moon, material in the magnetosphere must move at almost the speed of light to keep up with this rotation. Like a tremendous slingshot, the rotating pulsar magnetosphere flings particles away from the neutron star in a wind moving at nearly the speed of light. This wind fills the space between the pulsar and the expanding shell. The Crab Nebula is like a big balloon, filled with a mix of very fast particles and strong magnetic fields. The energy that accelerates these particles is exactly equal to the energy lost as the pulsar's rotation slows down. Images of the Crab Nebula, such as those in Figure 13.14b and c, show this bubble as an eerie glow from synchrotron radiation (see Chapter 8) released as the particles spiral around the magnetic field.

CHECK YOUR UNDERSTANDING 13.3

We observe a pulsar "pulsing" because
a. its spin axis crosses our line of sight.
b. it spins.
c. it has a strong magnetic field.
d. its magnetic axis crosses our line of sight.

13.4 Einstein Moved beyond Newtonian Physics

As you may recall from Chapter 12, white dwarfs have a maximum mass of about 1.4 M_{Sun}. The electron degeneracy pressure, which results from electrons being packed as closely together as possible, can only hold up this much mass. Similarly, if the mass of a neutron star exceeds about 2.2 M_{Sun}, then gravity will be stronger than the neutron degeneracy pressure. Stars with initial masses more than about 25 M_{Sun} will leave cores that are more than 2.2 M_{Sun}. These more massive cores shrink quickly under their own gravity, in turn increasing that gravitational pull. Eventually, gravity is so strong that the speed required to escape exceeds the speed of light. The object is now called a **black hole**: an object with gravity so strong that even light cannot escape it. We can, however, observe how it affects the light and matter nearby, as in **Figure 13.15**, in which astronomers observed the bright ring that forms when light is bent by the intense gravity of a black hole. Black holes are so strange, so far from the common understanding of reality, that the laws of Newtonian physics (see Chapter 3) cannot be used to describe them. Here in Section 13.4, you will learn how the constant speed of light led to the formation of the special theory of relativity, which changed our understanding of space and time.

The Speed of Light in Vacuum

Implicit in Newton's laws is the idea that every observer inhabits a reference frame, in which they measure distances and speeds. Observers in different reference frames will measure different speeds. Imagine that you are sitting in a red car traveling at 50 miles per hour (mph) down the highway, as shown in **Figure 13.16a**, and you throw a ball with a speed of 25 mph out the window at an oncoming car, also traveling at 50 mph. An observer standing by the side of the road watches the entire event. In your reference frame, the red car is stationary; that is, the car stays the same distance from you all the time (otherwise, you would not stay inside it!). So, in your reference frame, you would measure the ball traveling at 25 mph (Figure

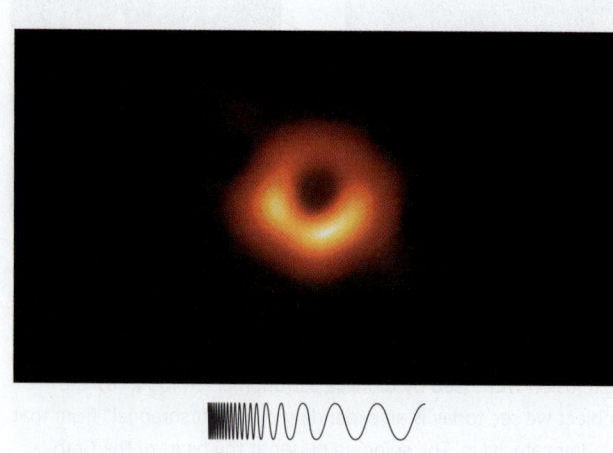

Figure 13.15 This image of the black hole at the center of the galaxy M87 is the first image of a black hole ever obtained. The intense gravity of this 6.5 billion solar mass black hole bends light around it to form a bright ring with a hole in the middle that is about 2.5 times the size of the event horizon of the black hole.

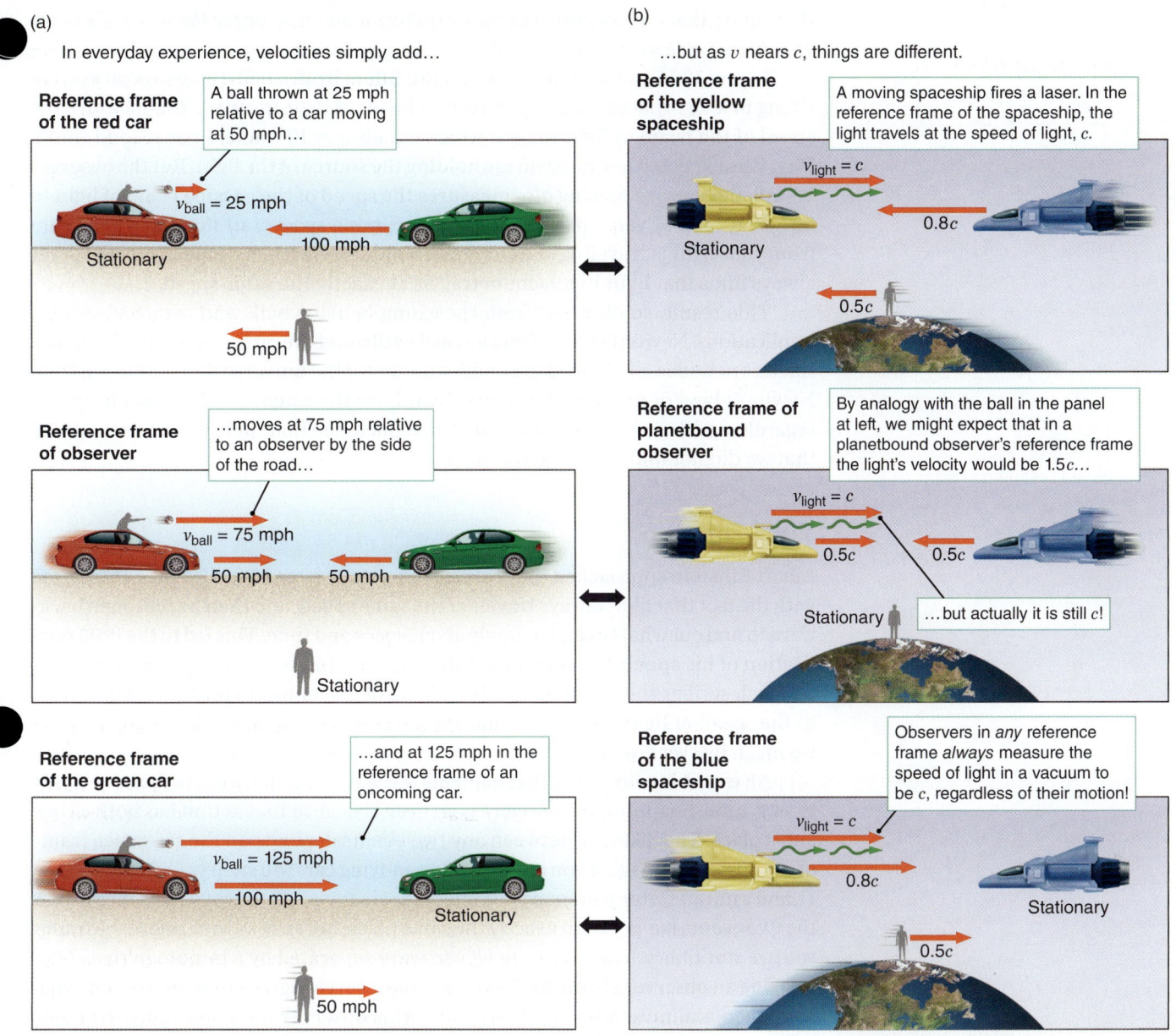

(a)

In everyday experience, velocities simply add…

Reference frame of the red car

A ball thrown at 25 mph relative to a car moving at 50 mph…

$v_{ball} = 25$ mph

100 mph

Stationary

50 mph

Reference frame of observer

…moves at 75 mph relative to an observer by the side of the road…

$v_{ball} = 75$ mph

50 mph   50 mph

Stationary

Reference frame of the green car

…and at 125 mph in the reference frame of an oncoming car.

$v_{ball} = 125$ mph

100 mph

Stationary

50 mph

(b)

…but as v nears c, things are different.

Reference frame of the yellow spaceship

A moving spaceship fires a laser. In the reference frame of the spaceship, the light travels at the speed of light, c.

$v_{light} = c$

0.8c

Stationary

0.5c

Reference frame of planetbound observer

By analogy with the ball in the panel at left, we might expect that in a planetbound observer's reference frame the light's velocity would be 1.5c…

$v_{light} = c$

0.5c 0.5c

Stationary   …but actually it is still c!

Reference frame of the blue spaceship

Observers in *any* reference frame *always* measure the speed of light in a vacuum to be c, regardless of their motion!

$v_{light} = c$

0.8c

Stationary

0.5c

Figure 13.16 The rules of motion that apply in our daily lives (a) break down when speeds approach the speed of light (b). The fact that light itself always travels at the same speed for any observer is the basis of special relativity. (Note that relativity also affects the relative speeds of the two spacecraft.)

13.16a, top). To the observer standing by the road, however, the ball is moving at 75 mph (25 mph from throwing it plus 50 mph from the motion of the car; Figure 13.16a, middle). To passengers in the oncoming car moving at 50 mph, the speed of the ball is 125 mph, because the ball approaches them at 75 mph, and they approach the ball at 50 mph (Figure 13.16a, bottom). The speed of the ball depends on who is measuring it and how the observer, the car, and the ball are moving relative to one another. The velocities are added together (with appropriate positive or negative signs to account for direction) to find the velocity of one object relative to another. This dependence of the speed on the reference frame is *Galilean relativity,* and you are familiar with it in everyday life.

During the closing years of the 19th century and the early years of the 20th century, the results of laboratory experiments with light puzzled physicists. They expected that the speed of light should differ from one observer to the next as a result of the observer's motion, just like the speed of a ball in our example. Instead, they found that *all observers measure exactly the same value for the speed of a beam of light, regardless of their motion!*

Figure 13.16b demonstrates how light differs from a ball. Imagine that you are riding in a fast spaceship and you shine a beam of light forward. You measure the speed of the beam of light to be c, which equals 3×10^8 meters per second (m/s). That is as expected because you are holding the source of the light. But the observer on a planet you are passing *also* measures the speed of the passing beam of light to be 3×10^8 m/s. Even a passenger in an oncoming spacecraft finds that the beam from your light is traveling at exactly c in her own reference frame. Every observer always finds that light in a vacuum travels at exactly the same speed, c.

This result, so different from the example using balls and cars, had serious implications. Newton's laws of motion had withstood every experimental challenge. But this new idea challenged the understanding of the universe that developed from Newton's laws of motion. How could light have the same speed for *all* observers, regardless of their own velocity? Scientists concluded that there must be something that we did not understand about light.

Time Dilation

Albert Einstein approached the problem differently from his colleagues. He started with the fact that light always travels at the same speed, and then he reasoned backward to find out what that must imply about space and time. This led to the 1905 publication of his **special theory of relativity**, sometimes called "special relativity," which describes the counterintuitive effects of traveling at constant speeds close to the speed of light. To understand special relativity, we must first explore what we mean by *event*, *space*, and *time*.

An **event** is something that happens at a particular location in space at a particular time. Snapping your fingers is an event because that action has both a time and a place. The distance between any two events depends on the reference frame of the observer. Imagine you are sitting in a moving car. You snap your fingers (event 1), and a minute later you snap your fingers again (event 2). In your reference frame, the two events happened at exactly the same place, because in your reference frame, *you* are stationary. The events, however, were separated by a minute in *time*. Now imagine an observer sitting by the road. This observer agrees that the second event happened a minute after the first, but to this observer the two events were also separated from each other in space. In this "Newtonian" view, the distance between two events depends on the motion of the observer, but the time between the two events does not.

Einstein carried out a thought experiment, called the "boxcar experiment," as described in **Working It Out 13.1**. In this experiment, light travels farther when measured by the observer outside the moving boxcar than by an observer inside. In Galilean relativity, we would conclude that the light must travel faster when measured by the outside observer. But the speed of light is always the same, for any observer. The only way out of this contradiction is to conclude that *the passage of time is different for each observer*. For moving observers, the time is stretched out, so that each second is longer, a phenomenon known as **time dilation**. The observer inside the boxcar measures a smaller number of seconds than the observer outside the boxcar. So when they divide the shorter distance by the smaller time, the speed of light is c. The observer outside the box divides a longer distance by a larger time, to find that the speed of light is c. In our everyday lives, the march of time seems unchanging and absolute, but Einstein discovered that time flows differently for different observers.

working it out 13.1

The Boxcar Experiment

The special theory of relativity is a *very* counterintuitive idea, but it is so central to our modern understanding of the universe that it is worth wrestling with a bit. In the boxcar experiment, observer 1 is in a boxcar of a train moving to the right. Observer 1 has a lamp, a mirror (mounted to the roof of the boxcar), and a clock. Observer 2 is standing on the ground outside.

Figure 13.17a shows the experimental setup as seen by observer 1, who is stationary with respect to the clock. At time t_1, event 1 happens: the lamp gives off a pulse of light. The light bounces off a mirror a distance l meters away. At time t_2, event 2 happens: the light arrives at the clock. The time between events 1 and 2 is the distance the light travels ($2l$ meters) divided by the speed of light: $t_2 - t_1 = 2l/c$.

Now let's look from the perspective of observer 2, who is standing on the ground outside the train (Figure 13.17b). In *this* observer's reference frame, he is stationary and the boxcar is moving at speed v. In observer 2's reference frame, the clock *moves* to the right between the two events, so the light has *farther to go*. (If you do not see this, use a ruler to measure the total length of the light path in Figure 13.17b and compare it with the total length of the light path in Figure 13.17a.) The time between the two events ($t_2 - t_1$) is longer for observer 2 than for observer 1 because the distance traveled is *longer* than $2l$ meters.

The two events are the *same two events*, regardless of the reference frame from which they are observed. But because the speed of light is the same for all observers, there *must* be more time between the two events when viewed by observer 2. The seconds of a moving clock are stretched. Moving clocks must run slow, and the passage of time must depend on an observer's reference frame.

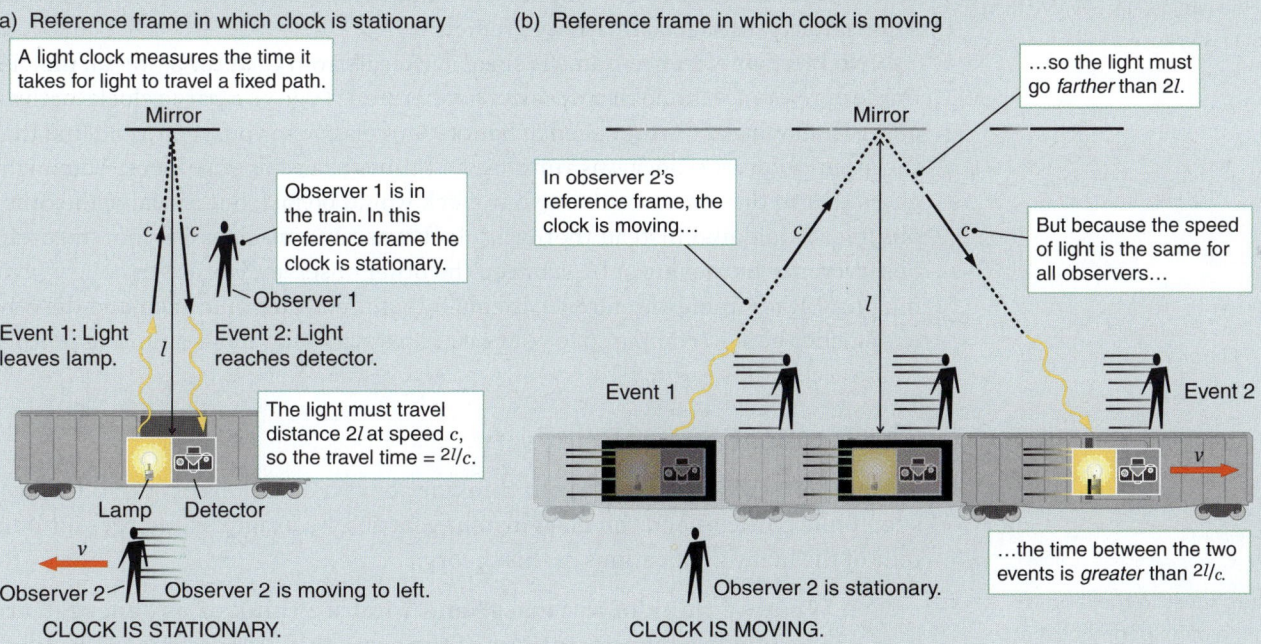

(a) Reference frame in which clock is stationary

A light clock measures the time it takes for light to travel a fixed path.

Mirror

Observer 1 is in the train. In this reference frame the clock is stationary.

Observer 1

Event 1: Light leaves lamp.

Event 2: Light reaches detector.

The light must travel distance $2l$ at speed c, so the travel time $= 2l/c$.

Lamp   Detector

Observer 2   Observer 2 is moving to left.

CLOCK IS STATIONARY.

(b) Reference frame in which clock is moving

In observer 2's reference frame, the clock is moving…

Mirror

…so the light must go *farther* than $2l$.

But because the speed of light is the same for all observers…

Event 1   Event 2

Observer 2 is stationary.

…the time between the two events is *greater* than $2l/c$.

CLOCK IS MOVING.

Figure 13.17 The "tick" of a light clock is different when seen in two different reference frames: (a) stationary and (b) moving. As Einstein's thought experiment demonstrates, if the speed of light is the same for every observer, then moving clocks *must* run slow.

By the time Einstein finished working out the implications of his insight, he had given time and space equal footing and reshaped the Newtonian three-dimensional universe into a four-dimensional one called **spacetime**. Spacetime is a combination of the three dimensions of space and the one dimension of time. Events occur at specific locations within this four-dimensional spacetime, but how this spacetime is divided into what we perceive as "space" and what we perceive as "time" depends on our reference frame.

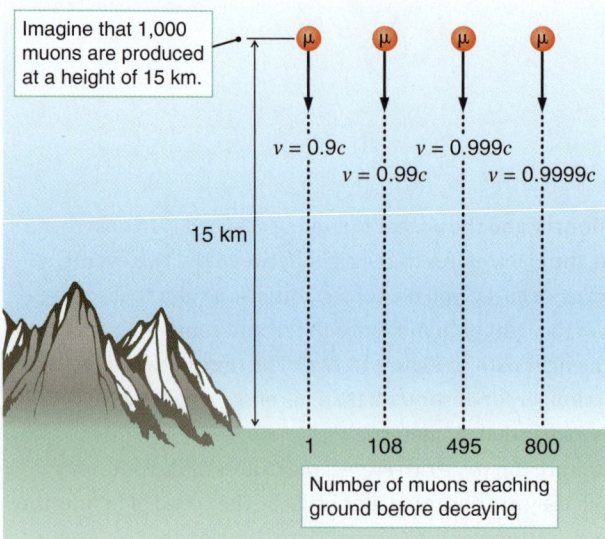

Figure 13.18 Muons (μ) created by cosmic rays high in Earth's atmosphere decay long before reaching the ground if they are not traveling at nearly the speed of light. Here we show what happens to 1,000 muons produced at an altitude of 15 km for a variety of speeds. Faster muons have slower clocks, so more of them survive long enough to reach the ground—many more than would be expected simply due to the faster speed.

what if . . .

What if you could measure mass in the laboratory with great precision: In an experiment that mixed two chemicals together inside a closed container, you observe that the chemical reaction produces light. Would you expect the mass of the container full of chemicals to change? Explain why.

what if . . .

What if we send an astronaut to Mars and back: Would you expect the returning astronaut to have aged differently than someone who remained on Earth? Explain your thinking.

Einstein did not "disprove" Newtonian physics. We were not wasting our time in Chapter 3 when we studied Newton's laws of motion, because at speeds much less than the speed of light, Einstein's equations become the same equations that describe Newtonian physics. In our everyday lives, we experience a Newtonian world because we never encounter speeds approaching that of light. Even the fastest spacecraft ever made by humans (the *Parker Solar Probe*) will reach a top speed of only $0.00064c$ (or 691,000 km/h). Only when objects approach the speed of light do observations begin to depart noticeably from the predictions of Newtonian physics. When great velocities cause an effect different from what Newtonian physics predicts, we say a **relativistic** effect has occurred.

But can it possibly be true? Can time really pass differently for objects traveling at different speeds or is that just how it "seems"? Scientists are motivated to test special relativity because if a scientist can prove it wrong, they will instantly become more famous than Einstein. Many experiments have been conducted to test this theory, which has passed all of these tests. For example, as illustrated in **Figure 13.18**, fast particles called muons are produced 15 km up in Earth's atmosphere when high-energy particles strike atmospheric atoms or molecules. Muons have very short lives, and decay very rapidly into other particles. This happens so quickly that virtually all muons should have decayed long before traveling the 15 km to reach Earth's surface. However, time dilation slows the muons' clocks, so the particles have longer lifetimes. They travel farther during their longer lifetime and reach the ground. The faster they are traveling, the more of them reach the ground. This is exactly as special relativity predicts.

No inertial reference frame is special. (Recall from Chapter 3 that an inertial frame of reference is one that does not accelerate.) If you compared clocks with an observer moving at $^9/_{10}$ the speed of light ($0.9c$) relative to you, you would find that the other observer's clock was running 0.44 times as fast as your clock. You might guess that to the other observer, your clock would be fast; but actually, the other observer would find instead that *your* clock was running slow. To you, the other observer may be moving at $0.9c$; but to the other observer, *you* are the one who is moving. Either frame of reference is equally valid, so you would each find the other's clock to be slow compared to your own. This symmetry holds as long as neither frame accelerates.

The Implications of Relativity

Today, special relativity shapes our thinking about the motions of both the tiniest subatomic particles and the most distant galaxies. As examples, we present only four of the many implications of this theory:

1. **What we think of as "mass" and what we think of as "energy" are actually two manifestations of the same thing.** According to Einstein's equation $E = mc^2$, even a *stationary* object has an intrinsic "rest" energy that equals the mass (m) of the object multiplied by the speed of light (c) squared. The speed of light is a very large number. This relationship between mass and energy means that a single tablespoon of water has a rest energy equal to the energy released in the explosion of more than 300,000 tons of TNT. Recall that you used this relationship between mass and energy when learning about the nuclear physics that makes stars shine (see Section 11.1).

2. **The speed of light is the ultimate speed limit.** As an object's speed approaches the speed of light, it requires more and more energy to accelerate it even a little bit. Because the energy required to accelerate increases with

speed, there is not enough energy in the entire universe to accelerate even one electron to the speed of light. Although getting the electron arbitrarily close to that number—0.99999999999999... × c —is no problem (at least in principle), there is not enough energy available to accelerate the electron beyond that to the speed of light. The necessary conclusion is that nothing travels faster than light. Despite hundreds of examples in science fiction books, shows, and movies, there is no way to travel faster than light.

3. **Time passes more slowly in a moving reference frame.** For moving objects, the seconds are stretched out by time dilation. This effect is important in particle physics, such as the muons we discussed earlier. But it is also important to future space travel. This effect is represented (and misrepresented) in science fiction books and movies on a regular basis.

4. **An object is shorter (in the direction of motion only) than it is at rest.** Moving objects are compressed in the direction of their motion, a phenomenon referred to as **length contraction**. A meter stick moving at 0.9c is only 0.44 meter long. Both time dilation and length contraction are important effects in the strong gravitational fields near neutron stars and black holes.

CHECK YOUR UNDERSTANDING **13.4**

Imagine that two spaceships travel toward each other. One spaceship travels to the right at a speed of 0.9c, and the other travels to the left at 0.9c. The pilot of the spaceship traveling right shoots a yellow laser at the spaceship traveling left. The pilot of the spaceship traveling left observes

a. blue light traveling at c.
b. blue light traveling at 1.9c.
c. yellow light traveling at c.
d. yellow light traveling at 1.9c.

13.5 Gravity Is a Distortion of Spacetime

Our exploration of special relativity began with the observation that the speed of light is always the same, regardless of the motion of an observer or the source. This phenomenon changed the way people understood space and time. But special relativity applies only in the "special" circumstance in which no acceleration occurs. The more general case applies when acceleration DOES occur; this is known as general relativity. One of the mind-bending consequences of general relativity is that four-dimensional spacetime is itself warped and distorted by the masses it contains. Here in Section 13.5, you will learn about the general theory of relativity, which describes how mass affects space and time.

Free Fall and Free Float

According to special relativity, any inertial reference frame is as good as any other. As long as there is no acceleration, there is no way to distinguish between sitting in an enclosed spaceship floating stationary in deep space (**Figure 13.19a**) and sitting in an enclosed spaceship traveling through our galaxy at a constant speed of 0.9999c (Figure 13.19b). As long as nothing accelerates either spaceship, the laws of physics are exactly the same inside both spacecraft, and so too are the experiences of the people inside.

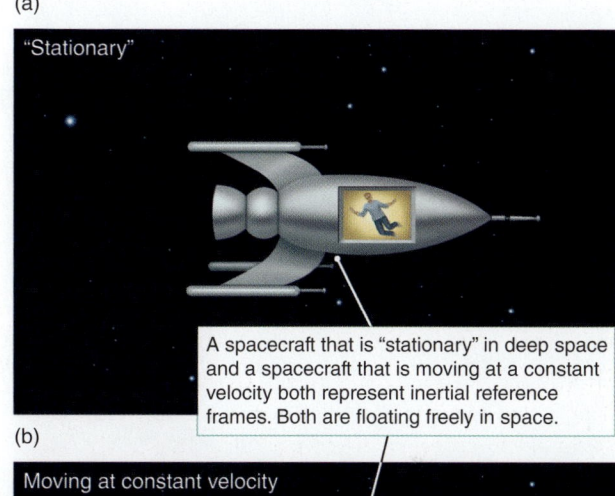

(a) "Stationary"

A spacecraft that is "stationary" in deep space and a spacecraft that is moving at a constant velocity both represent inertial reference frames. Both are floating freely in space.

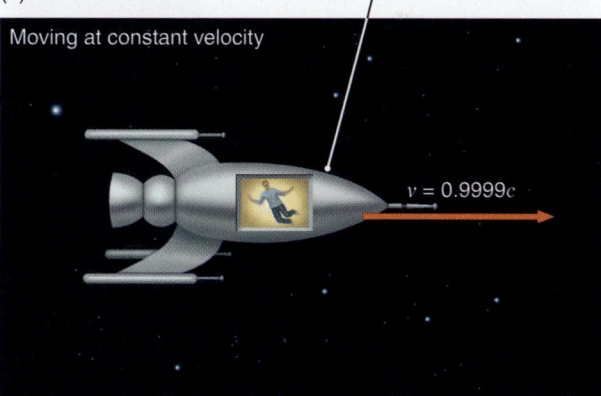

(b) Moving at constant velocity

$v = 0.9999c$

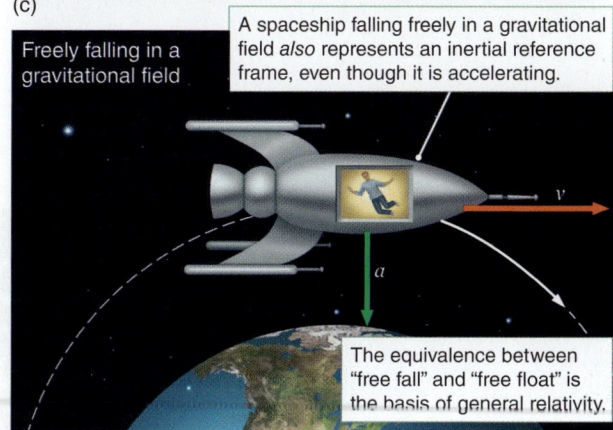

(c) Freely falling in a gravitational field

A spaceship falling freely in a gravitational field *also* represents an inertial reference frame, even though it is accelerating.

a

v

The equivalence between "free fall" and "free float" is the basis of general relativity.

Figure 13.19 According to *special* relativity, there is no difference between (a) a reference frame that is floating "stationary" in space and (b) one that is moving through the galaxy at constant velocity. According to *general* relativity, there is no difference between these inertial reference frames and (c) an inertial reference frame that is falling freely in a gravitational field. Free fall is the same as free float, as far as the laws of physics are concerned.

what if . . .

What if you are orbiting Earth in the International Space Station and you throw a ball out the door: From your perspective, once you let go of the ball, is it accelerating? What about from the perspective of someone on the surface of Earth?

The natural path that an object will follow through spacetime is referred to as the object's **geodesic**. In the absence of a gravitational field, an object's geodesic is a straight line, which is clear from Newton's first law: An object will move at a constant speed in a constant direction, unless acted on by an external force. However, the shape of spacetime becomes distorted in the presence of mass, so an object's geodesic becomes curved into an orbit. Rather than thinking of gravity as a "force" that "acts on" objects, it is more accurate to say that *gravitation results from the shape of spacetime that objects move through*. The warping of spacetime leads directly to the gravity that holds you to Earth. The **general theory of relativity**, or sometimes just "general relativity," describes how mass distorts spacetime and is another of Einstein's great contributions to science.

Consider an astronaut inside a spaceship orbiting Earth, as shown in Figure 13.19c. This spaceship is accelerating, because it is constantly changing direction as it orbits Earth. Still, in this case, the astronaut has no way to tell the difference between being inside the spaceship as it falls around Earth and being inside a spaceship floating through interstellar space. Even though its velocity is constantly changing as it falls, the spaceship is following its geodesic at constant speed. This is why an astronaut can place a wrench in the air next to him, and then pick it up again later. Both the astronaut and the wrench are following the geodesic in free fall, precisely the same as the spaceship does. The wrench just "floats" there and doesn't fall to the deck of the spacecraft.

Imagine you are in a box inside a spaceship that is accelerating through deep space at a rate of 9.8 meters per second per second (m/s²) in the direction of the arrow shown in **Figure 13.20**. The floor of the box pushes on you to overcome your inertia (as measured by your inertial mass) and causes you to accelerate at 9.8 m/s², so you feel as though you are being pushed into the floor of the box. Now imagine instead that you are sitting in a closed box on the surface of Earth. The floor of the box pushes on you to keep you from following your geodesic downward. Again, you feel as though you are being pushed into the floor of the box, because your gravitational mass is attracted to the gravitational mass of Earth. This statement that an acceleration can't be distinguished from a gravitational field is called the **equivalence principle**. The equivalence principle further means that your inertial mass (the one that responds to a push or a pull) is the same as your gravitational mass (the one that responds to gravity).

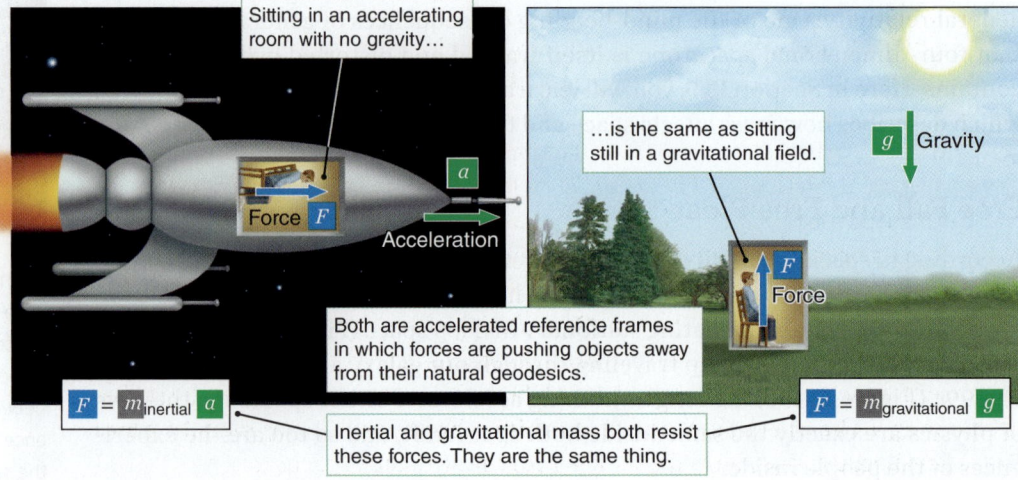

Figure 13.20 According to the equivalence principle, sitting in a spaceship accelerating at 9.8 m/s² feels the same as sitting still on Earth.

There is no difference between sitting in an armchair in a spaceship with an acceleration of 9.8 m/s² and sitting in an armchair on the surface of Earth reading this book. In the first case, the force of the spaceship is pushing you away from your "floating" straight-line geodesic through spacetime. In the second case, Earth's surface is pushing you away from your curved "falling" geodesic through a spacetime that has been distorted by the mass of Earth. An acceleration is an acceleration, regardless of whether you are being accelerated on a straight-line geodesic through deep space or being accelerated on a "falling" geodesic in the gravitational field of Earth.

There is an important caveat to the equivalence principle. In an accelerated reference frame such as an accelerating spaceship, the same acceleration is experienced everywhere. In contrast, the curvature of space by a massive object changes from place to place. Tidal forces are one result of increased curvature of space closer to a gravitating body. A more careful statement of the equivalence principle is that the effects of gravity and acceleration are indistinguishable *locally*; that is, as long as we restrict our attention to small enough regions of spacetime that changes in gravity from one part of the region to another can be ignored, the equivalence principle holds.

Spacetime as a Rubber Sheet

The general theory of relativity describes how mass distorts the geometry of spacetime. Imagine the surface of a tightly stretched, flat rubber sheet, like a very large trampoline. A marble rolled across the sheet will travel in a straight line. On this sheet, Euclidean geometry (the geometry of everyday life) applies. You may remember some rules of Euclidean geometry: the circumference of a circle is equal to 2π times its radius, r; the angles of a triangle add up to 180°; lines that are parallel anywhere are parallel everywhere.

Now imagine placing a bowling ball in the middle of the rubber sheet, creating a "well," as in **Figure 13.21**. The sheet will be stretched and distorted. If you roll a marble across the sheet, its path curves (Figure 13.21a). If you draw a circle around the bowling ball, the radius has to go down into the well, so the circumference is less than $2\pi r$ (Figure 13.21b). If you draw a triangle, the angles add up to more than 180° (Figure 13.21c). The surface of the sheet is no longer flat, and Euclidean geometry no longer applies.

Similarly, mass stretches out spacetime, changing the distance between any two locations or events. You can imagine, at least in principle, stretching a rope all the way around the circumference of a circular orbit around the Sun, approximately along Earth's orbit, and then comparing the length of that rope with the length of a rope taken from the orbit to the center of the Sun. If the Sun were not there, the space would be flat, and you would expect to find that the circumference of the orbit is equal to 2π times the radius of the orbit, as it is for a circle drawn on a flat piece of paper. If you carried out this experiment, however, you would find that the radius is "too long": The circumference of the orbit is shorter (by 10 km) than 2π times the radius because the radius dips down into the stretched spacetime.

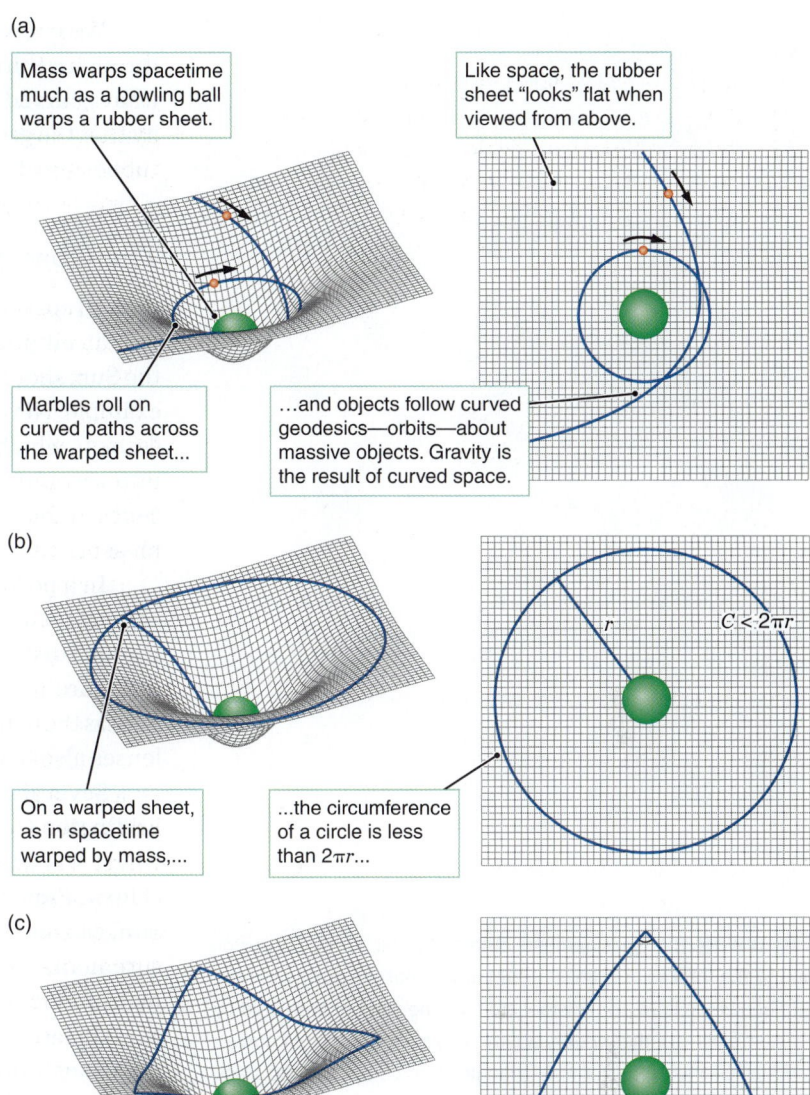

(a)

Mass warps spacetime much as a bowling ball warps a rubber sheet.

Like space, the rubber sheet "looks" flat when viewed from above.

Marbles roll on curved paths across the warped sheet...

...and objects follow curved geodesics—orbits—about massive objects. Gravity is the result of curved space.

(b)

On a warped sheet, as in spacetime warped by mass,...

...the circumference of a circle is less than $2\pi r$...

r $C < 2\pi r$

(c)

...and triangles have angles that sum to more than 180°.

Figure 13.21 Mass warps the geometry of spacetime in much the same way that a bowling ball warps the surface of a stretched rubber sheet. This distortion of spacetime has many consequences: for example, (a) objects follow curved paths or geodesics through curved spacetime, (b) the circumference of a circle around a massive object is less than 2π times the radius of the circle, and (c) angles in triangles do not add up to exactly 180°.

We can visualize how a rubber sheet with a bowling ball on it is stretched through a third spatial dimension, but it is impossible for most people to visualize what a curved four-dimensional spacetime would "look like." Yet experiments verify that the geometry of our four-dimensional spacetime is distorted much like the rubber sheet, whether or not we can easily picture it.

The Observable Consequences of General Relativity

Curved spacetime does have observable consequences. General relativity predicts that an elliptical orbit, in which the planet swings in closer and then farther from the Sun, should *precess*; that is, the long axis should slowly change its direction. For example, Newton's law of gravity predicts the major component of Mercury's precession, which is due to the influence of other planets. But even after this is taken into account, there remains a very small component, equal to 43 arcseconds per century, that cannot be explained by Newton's laws alone. This discrepancy is easily measurable, but it was not explained until Einstein developed general relativity.

In a positively curved space (curved like the surface of a sphere) the real-life equivalent of the triangle with more than 180° is easier to visualize. As light travels through distorted spacetime, its path bends to follow the geodesic, just like the lines in Figure 13.21c are bent by the curvature of the sheet. This bending of the light path as the light passes through bent spacetime is called gravitational lensing because lenses also bend light paths. But how could you ever measure this?

Several months before the total solar eclipse of 1919, Sir Arthur Stanley Eddington (1882–1944) measured the positions of stars in the direction of the sky where the eclipse would occur. He then repeated the measurements during the eclipse. **Figure 13.22** shows how the light from distant stars curved as it passed the Sun, causing the measured positions of the stars to shift outward. From his measurements, Eddington concluded that the stars appeared farther apart in his second measurement than in his first, as predicted by general relativity. During the eclipse, the triangle formed by Earth and any two stars contained more than 180°—just like the triangle on the surface of our rubber sheet. More recently, gravitational lensing has been used to search for unseen massive objects adrift in space; namely, planets ejected from their star systems and more massive objects like low-luminosity galaxies or the dark matter that surrounds them. The gravity from these objects noticeably (and predictably) distorts the light from background objects.

Mass distorts time as well. Deep within the gravitational field of a massive object, clocks run more slowly from the perspective of a distant observer. This effect is called **general relativistic time dilation**. Suppose a light is attached to a clock sitting on the surface of a neutron star. The light is timed so that it flashes once a second. Because time near the surface of the star is dilated, an observer far from the neutron star perceives the light to be pulsing with a lower frequency—less than once a second. Now suppose there is an emission-line source on the surface of the neutron star. Because time is running slowly on the surface of the neutron star, the light that reaches the distant observer will have a lower frequency than when it was emitted. A lower frequency means a longer wavelength, so the light

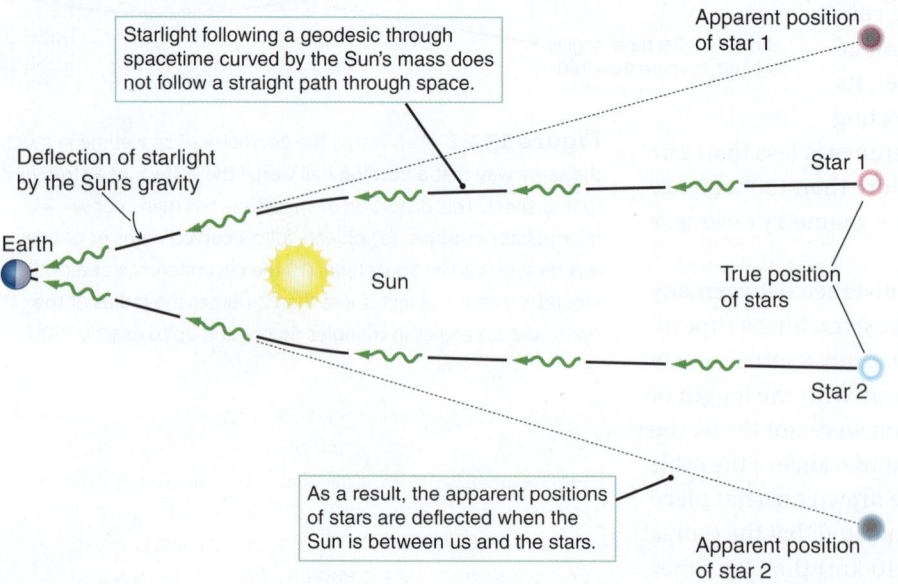

Figure 13.22 Measurements obtained by Sir Arthur Eddington during the total solar eclipse of 1919 found that the gravity of the Sun bends the light from distant stars by the amount predicted by Einstein's general theory of relativity. This is an example of gravitational lensing. Note that the "triangle" formed by the solid black lines connecting Earth and the two stars contains more than 180°, just like the triangle in Figure 13.21c.

Starlight following a geodesic through spacetime curved by the Sun's mass does not follow a straight path through space.

Deflection of starlight by the Sun's gravity

Earth

Sun

Apparent position of star 1

Star 1

True position of stars

Star 2

As a result, the apparent positions of stars are deflected when the Sun is between us and the stars.

Apparent position of star 2

from the source will be seen at a longer, redder wavelength than the wavelength at which it was emitted.

The reddening of light as it climbs out of a gravitational well, shown in **Figure 13.23**, is called the **gravitational redshift** because the wavelengths of light from objects deep within a gravitational well are shifted to longer, redder wavelengths. The effect of gravitational redshift is similar to the Doppler redshift. In fact, there is no way to tell the difference between light that has been redshifted by gravity and light that has been Doppler redshifted.

Bringing this phenomenon a bit closer to home, gravitational effects mean that a clock on the top of Mount Everest runs faster, gaining about 80 nanoseconds (80 billionths of a second) a day compared with a clock at sea level. The difference between an object on the surface of Earth and an object in orbit is even greater. Satellites in orbit travel quickly enough that special relativistic effects are measurable. Even after an allowance for the slowing due to special relativity, the clocks on the satellites that make up the Global Positioning System (GPS) run faster than clocks on the surface of Earth. If the satellite clocks and your GPS receiver did not correct for this, then the position your GPS receiver reported would be in error by up to half a kilometer within a single hour. The fact that the GPS works is actually a strong experimental confirmation of two predictions of general relativity; namely, gravitational redshift and general relativistic time dilation.

There is a direct prediction of general relativity that remained unverified until 2016 because the technological advances required to verify it were imposing. If you strike the surface of a rubber sheet, accelerating it downward, waves will move away

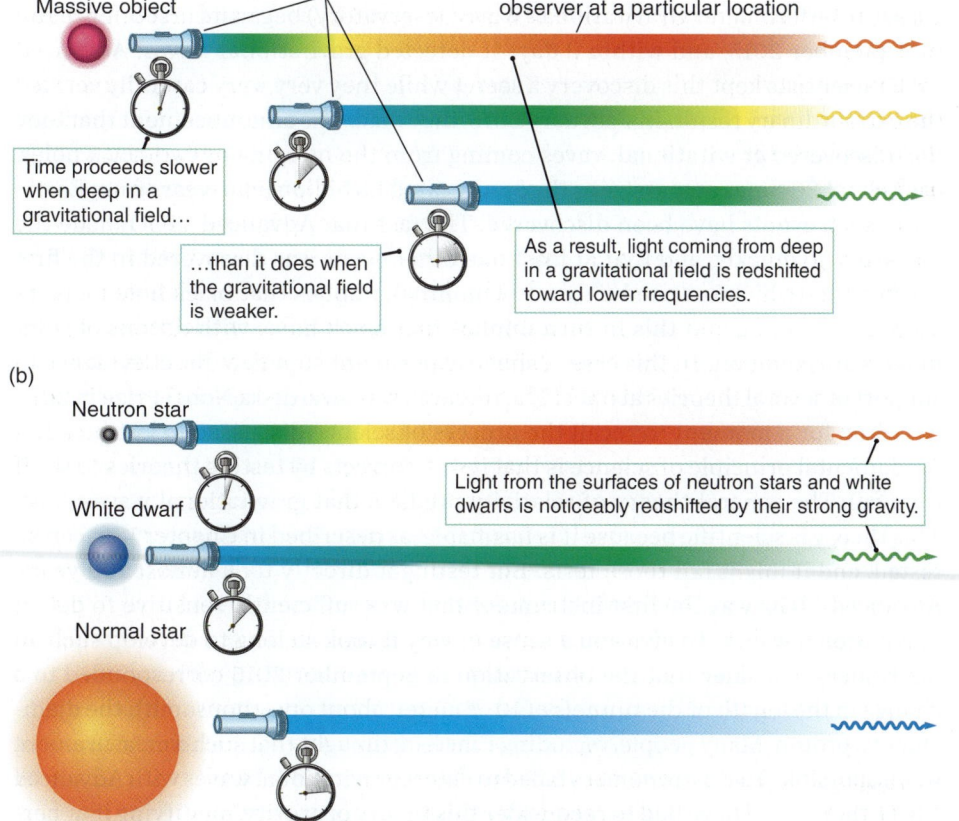

Figure 13.23 Time passes more slowly near massive objects because of the curvature of spacetime. To a distant observer, therefore, light from near a massive object will have a lower frequency and longer wavelength. (a) The closer the source of radiation is to the object or (b) the more massive and compact the object is, the greater the gravitational redshift will be.

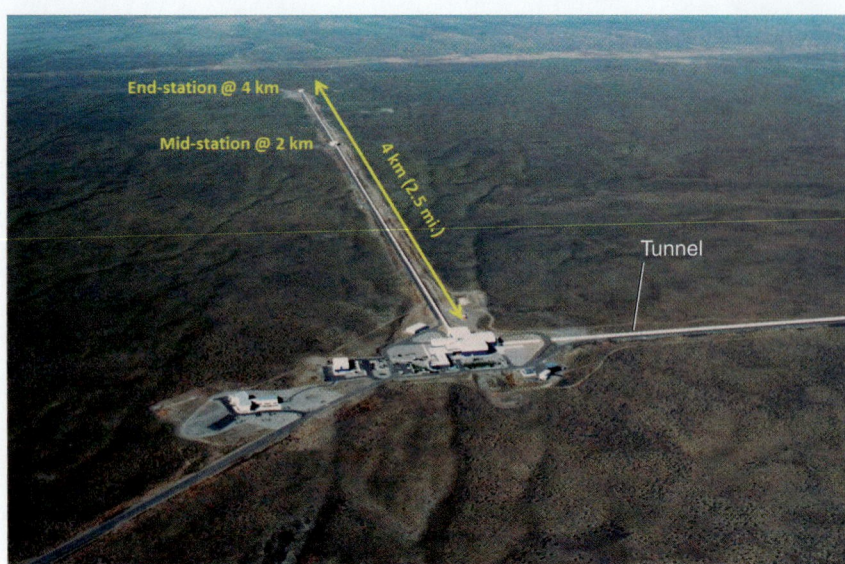

Figure 13.24 One installation of the Advanced LIGO instrument. One arm can be seen in its entirety; perspective makes the more distant 2 km look much smaller than the nearer 2 km. The second arm extends to the right from the building complex. It too is 4 km long.

from where you struck it, something like ripples spreading out over the surface of a pond. Similarly, the equations of general relativity predict that if you accelerate the fabric of spacetime (for example, with the merging of two black holes), then waves in spacetime, called **gravitational waves**, will move outward at the speed of light. These gravitational waves alternately stretch and compress spacetime as they pass through it. They are like electromagnetic waves in some respects. Accelerating an electrically charged particle gives rise to an electromagnetic wave. Accelerating a massive object gives rise to gravitational waves.

General relativity predicts that the orbits of binary neutron stars should lose energy, which will be carried away as gravitational waves. In 1974, astronomers discovered a binary system of two neutron stars, one of which is a pulsar. Using the pulsar as a precise clock, astronomers accurately measured the orbits of both stars. The orbits are gradually losing energy at the rate predicted by general relativity. The results were later duplicated with a similar binary pair with smaller orbits. These measurements very strongly suggested that gravitational waves exist, but this evidence is indirect.

Over many decades, astrophysicists developed the gravitational wave observatory to detect gravitational waves. One such observatory involved two installations, one in Washington State and one in Louisiana. Each installation uses mirrors and lasers to independently measure the changes in the length of a pair of perpendicular tunnels, shown in **Figure 13.24**. These changes are caused by the stretching and squishing of spacetime by gravitational waves as they pass by. Advanced LIGO (Laser Interferometer Gravitational-Wave Observatory) began its first science run in September 2015, and within 2 days it detected gravitational waves! Advanced LIGO scientists kept this discovery a secret while they very, very carefully verified the extraordinary result. In February 2016, they made the announcement that they had discovered gravitational waves coming from the merging of two black holes, each about 30 times as massive as the Sun, located 1.3 billion light-years from Earth. More such events have been discovered. The fact that Advanced LIGO made the discovery so quickly, and that at least one other event was discovered in the first science run (which spanned only a few months), implies that black hole mergers are very common, and this in turn implies that black holes with dozens of solar masses are common. In this case, a single experiment supplied direct evidence in support of several theories at once. The research was awarded a Nobel Prize in 2017.

Stop for a moment to recall the process of science discussed in Chapter 1. A fundamental principle of science is that it self-corrects by testing theories to see if they fail. The general theory of relativity predicts that gravitational waves exist. This theory is scientific because it is falsifiable, as described in Chapter 1—it can be tested, and it might fail those tests. But testing it directly took almost 100 years. Advanced LIGO was the first instrument that was sufficiently sensitive to detect gravitational waves. To give you a sense of why it took so long to develop such an instrument, consider that the observation in September 2015 corresponded to a change in the length of the tunnels of 10^{-18} meter, about one-thousandth the diameter of a proton. Many people, including Einstein, thought that such a measurement was impossible. Had astronomers failed to detect gravitational waves with Advanced LIGO, they would have had to reconsider this theory of gravity, modifying it or perhaps discrediting it all together. As you learn about parts of the universe where your

what if . . .

What if you wanted to detect gravitational waves from a binary system: What properties of the binary system would make it more likely that you could detect the gravitational radiation?

working it out 13.2

Finding the Schwarzschild Radius

The mass of a black hole (M_{BH}) determines the Schwarzschild radius (R_S):

$$R_S = \frac{2GM_{BH}}{c^2}$$

where G is the universal gravitational constant (6.673×10^{-11} m³/kg s²) and c is the speed of light. Inserting values for the constants and expressing the mass in terms of the mass of the Sun gives

$$R_S = 3 \text{ km} \frac{M_{BH}}{M_{Sun}}$$

This relationship shows that the Schwarzschild radius is proportional to the mass, expressed in solar masses. The constant of proportionality is 3 km. A 1-M_{Sun} black hole has a Schwarzschild radius of about 3 km. To put that in perspective, at average walking pace, it would take about 40 minutes to walk 3 km. A 3-M_{Sun} black hole has a Schwarzschild radius of about 9 km, and a 5-M_{Sun} black hole has a Schwarzschild radius of about 15 km.

intuition can no longer guide your understanding, it will be important to keep this particular strength—the self-correcting nature of science through experiment—in mind.

CHECK YOUR UNDERSTANDING 13.5

Which of the following are possible consequences of distortions of spacetime caused by mass?

a. time dilation
b. gravity
c. length contraction
d. tidal forces
e. gravitational lensing
f. precession

13.6 Black Holes Are a Natural Limit

A heavy object on the surface of a rubber sheet causes a funnel-shaped distortion that is analogous to the distortion of spacetime by a mass. Now imagine the funnel is **infinitely** deep—it gets narrower as we go deeper, but it has no bottom. This is the rubber-sheet analog to a black hole. The mathematics describing the shape of a black hole fail in the same way that the mathematical expression $1/x$ fails when x = 0. Such a mathematical anomaly is called a **singularity**. Black holes are singularities in spacetime.

Properties of Black Holes

We can never actually "see" the singularity that is the core of a black hole. As the distance from a black hole decreases, the escape velocity increases, until the velocity reaches the speed of light—at this distance from the center of the black hole, even light cannot escape. The radius where the escape velocity equals the speed of light is called the **Schwarzschild radius**, named for physicist Karl Schwarzschild (1873–1916), and it is proportional to the mass of the black hole (**Working It Out 13.2**).

VOCABULARY ALERT

infinite In everyday language, *infinite* sometimes just means "really, really big" or "really, really far." But astronomers mean *infinite* in a mathematical sense—a quantity without limit or boundary. If we say something is infinite, there is no beyond.

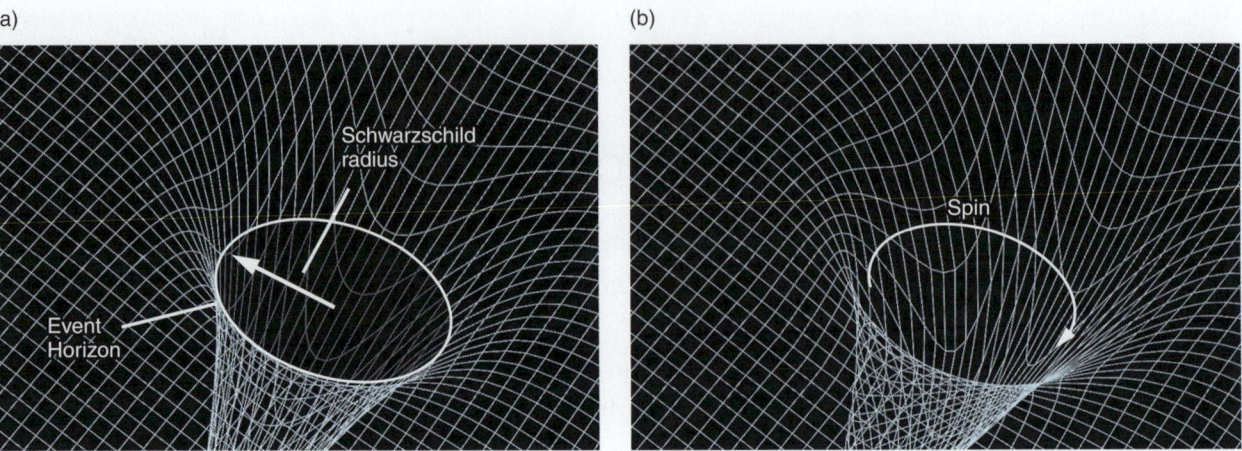

(a)   (b)

Schwarzschild
radius

Spin

Event
Horizon

Figure 13.25 (a) A black hole's size is determined by the Schwarzschild radius and the corresponding event horizon. This image is a two-dimensional analogy for a black hole. In reality, the event horizon is a sphere. (b) If the object that formed the black hole was spinning, its angular momentum is conserved, and the black hole twists the spacetime around it.

The sphere around the black hole at this distance is called its **event horizon**. **Figure 13.25a** shows a rubber-sheet analog to a black hole, with the Schwarzschild radius and the event horizon labeled. A black hole with a mass of 1 M_{Sun} has a Schwarzschild radius of about 3 km. A black hole with the mass of Earth would have a Schwarzschild radius of only about a centimeter.

A black hole has only three properties: mass, electric charge, and angular momentum. The mass of a black hole determines the Schwarzschild radius. The electric charge of a black hole is the net electric charge of the matter that fell into it. The angular momentum of a black hole twists spacetime around it (Figure 13.25b). Apart from these three properties, all information about the material that created the black hole is lost. Nothing of its former composition, structure, or history survives.

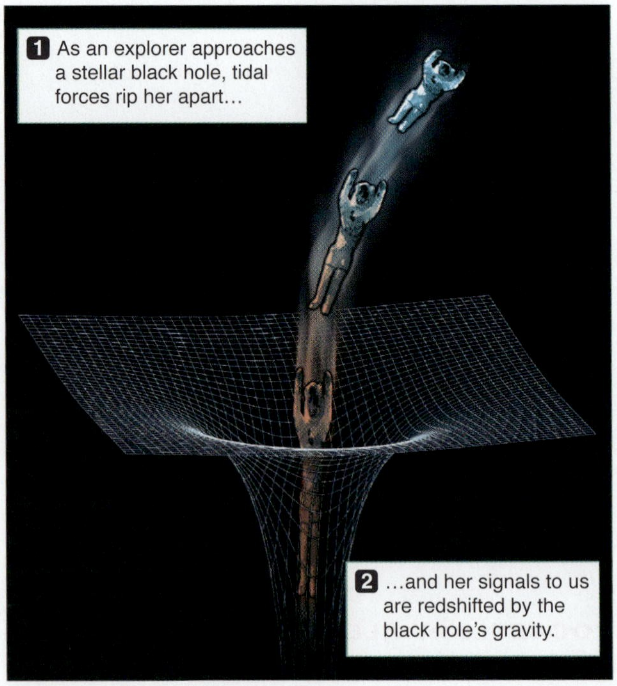

Imagine that an adventurer journeys into a previously created black hole, as illustrated in **Figure 13.26**. This adventurer is carrying a clock with her that flashes a signal back to you once each second. Since she carries this clock with her, it is a record of her clock time in her local inertial reference frame. From our perspective outside the black hole, we would see our adventurer fall toward the event horizon; but as she did, the time between flashes from her clock will grow longer and longer as she approaches the event horizon, until at the event horizon, a last flash would never arrive. Our adventurer would approach the event horizon, but from our perspective she would never quite make it. Yet the adventurer's experience would be very different. From her perspective, there would be nothing special about the event horizon at all. She would fall past the event horizon and on, deeper into the black hole's gravitational well.

Unfortunately, our intrepid explorer would have been torn to shreds long before she reached the black hole. Near the event horizon of a 3-M_{Sun} black hole, the difference in gravitational acceleration between our explorer's feet and her head would be about a billion times larger than her weight. Her feet would be accelerating a billion times faster than her head, and she would be stretched out uncomfortably far, into a thin string (astronomers call this "spaghettification"). This is not an experiment we would ever want to perform. Although scientific theories must produce testable predictions, not all individual predictions have to be tested directly.

1 As an explorer approaches a stellar black hole, tidal forces rip her apart…

2 …and her signals to us are redshifted by the black hole's gravity.

Figure 13.26 An adventurer falling into a black hole would be "spaghettified" by the extreme tidal forces.

"Seeing" Black Holes

It is impossible to directly observe light from a black hole, because the light cannot escape. There are other ways to detect black holes, however, by looking for objects that are massive but extremely small. The X-ray binary star system Cygnus X-1 provides one example. This system contains a normal B0 supergiant star and an unseen, compact source that sometimes emits X-rays. The radio emission from Cygnus X-1 flickers rapidly, changing in as little as 0.01 second. This means that the source must be smaller than the distance that light travels in 0.01 second, or 3,000 km. This is smaller than Earth. The B0 supergiant star has a mass of about 30 M_{Sun} and an orbital period of 5.6 days. Newton's version of Kepler's third law can be used to find that the mass of the unseen compact source must be at least 6 M_{Sun}. The companion is too small to be a normal star, yet it is much more massive than the largest possible white dwarf or neutron star. Such an object can only be a black hole. X-rays are emitted from Cygnus X-1 when material from the B0 supergiant falls onto an accretion disk surrounding the black hole, as illustrated in **Figure 13.27**.

Dozens of other good candidates for stellar-mass black holes have been discovered. Although this evidence that these systems contain black holes is circumstantial, the arguments that lead to this conclusion seem airtight. As we will discuss in Chapter 14, supermassive black holes with masses millions of times larger than the mass of the Sun have been found in the centers of galaxies (see also Figure 13.15). In the Milky Way, for example, stars orbiting close to this central black hole move so quickly in such small orbits that a black hole of more than a million solar masses must be located there.

The gravitational wave observations that confirmed general relativity also provided direct evidence that black holes exist in binary systems and that those systems can collapse, merging two black holes into one. As shown in **Figure 13.28**,

Figure 13.27 This artist's rendering of the Cygnus X-1 binary system shows material from the B0 supergiant being pulled off and falling onto an accretion disk surrounding the black hole, thereby producing X-ray emission.

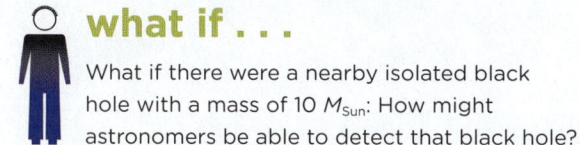

what if . . .

What if there were a nearby isolated black hole with a mass of 10 M_{Sun}: How might astronomers be able to detect that black hole?

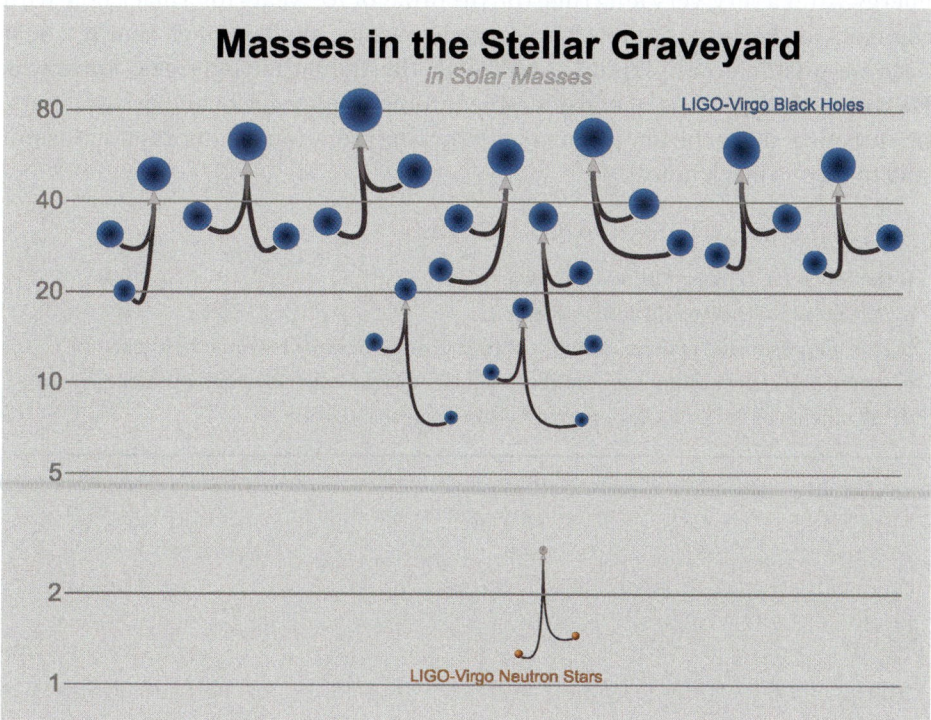

Figure 13.28 This diagram shows all of the merging black holes and neutron stars detected in the first and second observing runs of LIGO-Virgo collaboration. In each case, two smaller objects merged to become one larger object. The masses of the objects before and after the merger can be read from the left side of the graphic.

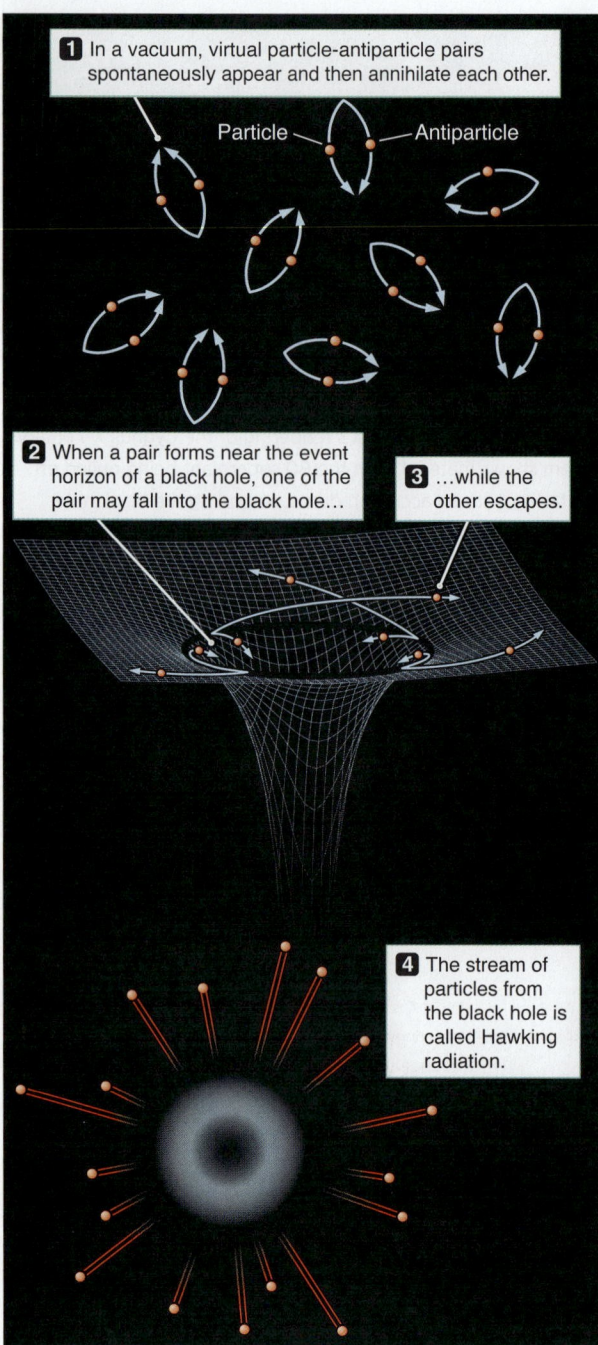

1 In a vacuum, virtual particle-antiparticle pairs spontaneously appear and then annihilate each other.

Particle — Antiparticle

2 When a pair forms near the event horizon of a black hole, one of the pair may fall into the black hole…

3 …while the other escapes.

4 The stream of particles from the black hole is called Hawking radiation.

Figure 13.29 In the vacuum of empty space, particles and antiparticles are constantly being created and then annihilating each other. Near the event horizon of a black hole, however, one particle may cross the horizon before it recombines with its partner. The remaining particle leaves the black hole as light.

roughly a dozen merging systems have been detected so far (one of these systems was composed of two neutron stars, rather than black holes). This detection rate indicates that binary black holes are likely to be far more common than astronomers previously expected—so common that they might very well be a significant portion of the mass in the universe.

The most energetic explosions in the universe may be related to black holes. **Gamma-ray bursts (GRBs)** are intense bursts of gamma rays, followed by a weaker afterglow. Because GRBs happen all over the sky and are not concentrated in the plane of the Milky Way, they must be so distant that they lie outside our own galaxy. Scientists think that those of short duration—less than 2 seconds—are probably the result of two dense objects merging—either two neutron stars or a neutron star and a black hole. Longer-duration GRBs are produced during a supernova event when the rapidly spinning star collapses to form a dense object such as a neutron star or a black hole. Scientists have confirmed the connection between merging black holes and GRBs, using simultaneous gravitational wave and gamma-ray observations of a single object.

Can black holes ever go away? In 1974, Stephen Hawking realized that black holes should be sources of radiation. Somewhat surprisingly, quantum mechanics tells us that even "empty" space is full of activity. In the ordinary vacuum of empty space, particle-antiparticle pairs appear and then, within 10^{-21} second, annihilate each other. This process is ongoing, everywhere, all the time. If this happens near the event horizon of a black hole, as shown in **Figure 13.29**, then one of the particles might fall into the black hole while the other particle escapes. Hawking showed that this process will produce a blackbody spectrum from the region near the event horizon and that the effective temperature of this spectrum would increase as the black hole became smaller through this "evaporation" process. One way to think about this is that the energy to make the escaping particles has been stolen from the warped spacetime of the black hole, leaving it with less energy. After a very, very long time (on the order of 10^{61} years for a black hole with the mass of the Sun), the black hole would become small enough that it would become unstable and explode. Although the light that emerges, known as **Hawking radiation**, is of considerable interest to physicists and astronomers, in a practical sense the low intensity of Hawking radiation means it is not a useful way to "see" a black hole.

CHECK YOUR UNDERSTANDING **13.6**

If the Sun were replaced by a 1-M_{Sun} black hole,

a. the Solar System would collapse into it.

b. the planets would remain in their orbits, but smaller objects would be sucked in.

c. small objects would remain in their orbits, but larger objects would be sucked in.

d. all objects in the Solar System would remain in their orbits.

reading Astronomy News

Ligo is back and it's on the hunt for gravitational waves again

by Matt Burgess, *Wired*

Any new discovery holds the potential for more, future discoveries. Scientists around the world hailed the discovery of gravitational waves as one of the greatest scientific achievements of mankind. Then they immediately started looking ahead, to other discoveries that might be made.

The hunt for gravitational waves is back on. Physicists at the helm of the Laser Interferometer Gravitational-wave Observatory (Ligo) have restarted the machines after work to improve them was completed.

In January, the two Ligo machines—located in Hanford, Washington and Livingston, Louisiana—stopped collecting data on their continuing hunt for gravitational waves for a scheduled round of repairs and upgrades.

While the improvements varied across each of the machines, in one of them the work has increased the laser power within the machine's interferometer and the other has seen new components installed in its vacuum system.

Overall, the changes and updates make Ligo more likely to be able to detect more of the mysterious waves. "The Livingston detector is now sensitive enough to detect a merger from as far away as 200 million parsecs (660 million light-years)," Peter Fritschel the director for LIGO at MIT said. "This is about 25 per cent farther than it could 'see' in the first observing run."

The "modest" improvements should allow scientists to find more mergers of black holes deep in space. "We should add to our knowledge of the black hole population in the universe," Fritschel explains. As well as clearing up some mysteries about black holes, it is hoped the sensitive wave detecting machines will identify gravitational waves from neutron stars.

In February, the Ligo machines and its team of international researchers confirmed the existence of gravitational waves—proving Albert Einstein's 100-year-old theory to be correct. The waves, which are ripples in the curvature of spacetime that travel outwards from their source, were detected after emerging from the collision of two black holes 1.3 billion years ago.

The breakthrough discovery was followed in June with the detection of gravitational waves for a second time. Waves spotted this second time were sent through spacetime when two black holes, eight and 14 times the mass of the Sun, collided. When it happened, 1.4 billion years ago, a giant spinning black hole 21 times the mass of the Sun was created—the gravitational waves in the discovery were said to be stronger than those in the first.

In both instances, the waves were spotted for the first time in 2015 but only confirmed after rigorous checking. Previously, claims of discovery had been proved false.

The latest round of upgrades and repairs will be tested as the machines collect data for a six-month period. However, with increasing technological improvements more gravitational waves are likely to be found in the coming years.

A paper in the *Nature* journal predicted Ligo will be so sensitive within four years that it will be capable of detecting 1,000 black hole collisions each year.

EVALUATING THE NEWS

1. LIGO scientists are predicting they will detect 1,000 mergers of black holes each year. About how many events should they detect over 4 months? Why did they not detect this many in the first science run from September to December 2015?

2. The article states that the detector is now sensitive enough to detect a neutron star merger "from as far away as 200 million parsecs (660 million light-years). This is about 25 percent farther than it could 'see' in the first observing run." How far could LIGO see these mergers in its first observing run?

3. The first detection of gravitational waves from merging black holes occurred 1.3 billion years after the black holes merged. How far away are these black holes?

4. Go online and look for recent LIGO results. Since this article in 2016, how many events has LIGO detected?

5. Go online and look for recent LIGO progress. How close is it to full design sensitivity?

Source: **Matt Burgess**, "Ligo is Back and It's on the Hunt for Gravitational Waves Again," Wired.co.uk, December 2, 2016. Matt Burgess / Wired © The Condé Nast Publications Ltd.

SUMMARY

As high-mass stars evolve, their interiors form concentric shells of progressive nuclear burning. Once they leave the main sequence, they may pass through the instability strip and become pulsating variable stars. The chain of nuclear fusion reactions consists of increasingly shorter stages of burning, resulting in the production of more massive elements up to iron. High-mass stars eventually explode as Type II supernovae, which eject newly formed massive elements into interstellar space. Some high-mass stars leave behind neutron stars, and the most massive high-mass stars leave behind black holes. In the environment surrounding black holes, relativistic effects become important. Time runs more slowly, and objects deep in a black hole's gravitational well appear redshifted to an external observer. A black hole's mass determines its Schwarzschild radius: the boundary from which light cannot escape.

(**1**) The larger masses of high-mass stars allow them to fuse heavier elements than those produced in low-mass stars. This leads to a very different and more violent death for high-mass stars that leaves massive cores behind.

(**2**) Evolving high-mass stars leave the main sequence as they burn heavier elements. Once an iron core is produced, the star becomes unstable and the core collapses, heating the material to cause photodisintegration of the iron nuclei and the merging of protons and electrons into neutrons. The outer layers bounce off the dense core and produce a shock wave that travels outward. This shock wave causes neutrons to penetrate atomic nuclei and form more massive elements, which are present on Earth today.

(**3**) The speed of light in a vacuum is the same for all observers. This is the basis of relativity, which has several other profound implications: mass and energy are equivalent; when traveling very fast or in a strong gravitational field, time runs slower; and when traveling very fast or in a strong gravitational field, lengths are shorter. The development of relativity changed the way that scientists think about the natural world.

(**4**) The supernova explosion that ends the life of a massive star leaves behind a neutron star or a black hole. These explosions are remarkably luminous, often brighter than the entire host galaxy. A supernova remnant, the expanding debris cloud, seeds the surrounding space with heavier elements and may trigger new star formation in nearby clouds. Neutron stars are the most common result and are very dense, city-sized stars. Pulsars are neutron stars whose radiating magnetic poles periodically aim at us, becoming brighter for a brief time. The extreme density of a black hole warps space such that light cannot escape it. The mathematical singularity at the center of a black hole is still a mystery to science. Black holes have been detected in binary star systems and through observation of gravitational waves. They may be destroyed after a very, very long time by evaporation through Hawking radiation.

QUESTIONS AND PROBLEMS

TESTING YOUR UNDERSTANDING

1. **T/F:** The end result of the CNO cycle is that four hydrogen nuclei become one helium nucleus.

2. **T/F:** When iron fuses into heavier elements, it produces energy.

3. **T/F:** Electrons and protons can combine to become neutrons.

4. **T/F:** A supernova can be as bright as its entire host galaxy.

5. **T/F:** A pulsar changes in brightness because its size pulsates.

6. Place in order the stages of nuclear burning in evolving high-mass stars.
   a. helium
   b. neon
   c. oxygen
   d. silicon
   e. hydrogen
   f. carbon

7. An astronaut who fell into a black hole would be stretched because
   a. the gravity is so strong.
   b. the gravity changes dramatically over a short distance.
   c. time is slower near the event horizon.
   d. black holes rotate rapidly, dragging spacetime with them.

8. Elements heavier than iron originate
   a. in the Big Bang.
   b. in the cores of low-mass stars.
   c. in the cores of high-mass stars.
   d. in the explosions of high-mass stars.

9. X-ray binaries are similar to another type of system we have studied. This system is
   a. the Solar System.
   b. progenitors of Type Ia supernovae.
   c. progenitors of Type II supernovae.
   d. progenitors of planetary nebulae.

10. The fact that the speed of light is a universal constant forces us to completely rethink classical physics. This is an example of
    a. scientists always being completely wrong.
    b. the self-correcting nature of science.
    c. a theory becoming a hypothesis and then becoming a law.
    d. the universe changing with time.

11. In a high-mass star, hydrogen fusion occurs via
- **a.** the proton-proton chain.
- **b.** the CNO cycle.
- **c.** gravitational collapse.
- **d.** spin-spin interaction.

12. The layers in a high-mass star occur roughly in order of
- **a.** atomic number.
- **b.** decay rate.
- **c.** atomic abundance.
- **d.** spin state.

13. Pulsations in a Cepheid variable star are controlled by
- **a.** the spin.
- **b.** the magnetic field.
- **c.** the ionization state of helium.
- **d.** the gravitational field.

14. Eta Carinae is an extreme example of
- **a.** a massive star.
- **b.** a rotating star.
- **c.** a magnetized star.
- **d.** a high-temperature star.

15. When photodisintegration starts in a star, a process begins that *always* results in a
- **a.** supernova.
- **b.** neutron star.
- **c.** black hole.
- **d.** pulsar.

16. Supernova remnants
- **a.** are viewable at all wavelengths.
- **b.** are viewable only at a few emission lines.
- **c.** are never seen in radio waves.
- **d.** have colors because the moving gas emits Doppler-shifted emission lines.

17. Marisol is standing in a small, windowless room. The force on her feet suddenly ceases. Which of the following conclusions can she draw?
- **a.** She is in an elevator, which is now accelerating downward.
- **b.** She is now in space, freely falling around Earth.
- **c.** The rocket she was in has stopped accelerating upward.
- **d.** Any of the above could be true.

18. If you could draw a very large circle in space near a black hole, its circumference
- **a.** would equal $2\pi r$.
- **b.** would be greater than $2\pi r$.
- **c.** would be less than $2\pi r$.
- **d.** can't be determined from the information given.

19. An object that passes near a black hole
- **a.** always falls in because of gravity.
- **b.** is sucked in because of vacuum pressure.
- **c.** is pushed in by photon pressure.
- **d.** is deflected by the curvature of spacetime.

20. Gravitational waves are an important confirmation of
- **a.** Galilean relativity.
- **b.** special relativity.
- **c.** general relativity.
- **d.** Newton's theory of gravitation.

THINKING ABOUT THE CONCEPTS

21. Write a brief obituary of a star that produced a supernova. Besides summarizing the important moments of its life, be sure to mention what the star leaves behind.

22. Why does the core of a high-mass star not become degenerate, as in the cases of low-mass stars?

23. For what two reasons does each post-helium-burning cycle for high-mass stars (carbon, neon, oxygen, silicon, and sulfur) become shorter than the preceding cycle?

24. In Figure 13.2, the locations of both Cepheid variables and RR Lyrae stars are shown on the H-R diagram. Use the diagram to explain why Cepheid variables are more useful than RR Lyrae stars for finding the distance to distant galaxies. 👁

25. Advanced LIGO uses two instruments, one in Louisiana and one in Washington State. Explain how this configuration can be used to rule out other sources of vibration (for example, passing trucks or minor earthquakes) that might cause a change in the location of the mirrors.

26. Identify and explain two important ways in which supernovae influence the formation and evolution of new stars.

27. Neutrinos from SN 1987A were received in the detector on February 23, 1987. About 3 hours later, the supernova was detected in optical light. Why did this time delay occur?

28. Why can the accretion disk around a neutron star release so much more energy than the accretion disk around a white dwarf?

29. ★ **WHAT AN ASTRONOMER SEES** All of the images in Figure 13.7 are the same size. Study the bumps in the large bright ring. Do these bumps change location over the two decades shown here, or do they simply change brightness? Does the ring change size during this time? 👁

30. An astronomer sees a redshift in the spectrum of an object. With no other information available, can she determine whether this is an extremely dense object (gravitational redshift) or one that is receding from us (Doppler redshift)? Explain your answer.

31. Suppose that, in a speeding spaceship, the travelers are playing soccer with a perfectly spherical soccer ball. What is the shape of the ball according to observers outside the spacecraft?

32. Explain why more muons reach the ground traveling at $0.9999c$ than at $0.9c$.

33. Suppose astronomers discover a $3\text{-}M_{Sun}$ black hole located a few light-years from Earth. Should they be concerned that its tremendous gravitational pull will lead to our planet's untimely demise?

34. Study Figure 13.23. If you could watch a star falling into a black hole, how would the color of the star change as it approached the event horizon? 👁

35. Why are we unaware of the effects of special and general relativity in our everyday lives here on Earth?

APPLYING THE CONCEPTS

36. Follow Working It Out 11.1 to use Einstein's mass-energy equivalence formula ($E = mc^2$) to verify that 5.88×10^{13} joules (J) of energy are released from fusing 1.00 kg of helium via the triple-alpha process.

37. Figure 13.11 shows the relative abundance of the elements. Is this a log or a linear plot? Explain what it means that oxygen lies on the y-axis at 10^{-3}. ✦

38. You want to know the Schwarzschild radius of a black hole with the mass of Jupiter.
   a. Make a prediction: Should this black hole have a Schwarzschild radius larger or smaller than that of an Earth-mass black hole (centimeters)? Should this black hole have a Schwarzschild radius larger or smaller than that of a Sun-mass black hole (3 km)?
   b. Use the first equation in Working It Out 13.2 to find the Schwarzschild radius of a black hole with the mass of Jupiter. (The mass of Jupiter can be found in Appendix 2.)
   c. Check your work by comparing your answer to your prediction.

39. You want to know the Schwarzschild radius of a 6-M_{Sun} black hole.
   a. Make a prediction: Working It Out 13.2 gives the Schwarzschild radius for a 3-M_{Sun} black hole as 9 km. Do you expect the Schwarzschild radius of a 6-M_{Sun} black hole to be larger or smaller than 9 km? About how much larger or smaller?
   b. Use the second equation in Working It Out 13.2 to calculate the Schwarzschild radius of a 6-M_{Sun} black hole.
   c. Check your work by comparing your answer to your prediction.

40. Suppose that a black hole has a Schwarzschild radius of 6 km. You want to know how massive this black hole is.
   a. Make a prediction: Should this black hole be more or less massive than the Sun? By about how much?
   b. Calculate the mass, in solar masses, of a black hole with a Schwarzschild radius of 6 km.
   c. Check your work by comparing your answer to your prediction.

exploration The CNO Cycle

digital.wwnorton.com/universe4

Nuclear reactions are quite complex, and they usually involve many steps. In the Chapter 11 Exploration, you investigated the proton-proton chain. In this Exploration, you will study the CNO cycle, which is even more complex. Visit the Student Site at the Digital Resources page and open the CNO Cycle Animation Interactive Simulation in Chapter 13.

First, press "Play" and watch the animation all the way through. Press "Reset" to clear the screen, and then press "Play" again, allowing the animation to proceed past the first collision before pressing "Pause."

1. Which atomic nuclei are involved in this first collision?

_____

2. What color is used to represent the proton (hydrogen nucleus)?

_____

3. What is represented by the blue wiggle?

_____

_____

4. What atomic nucleus is created in the collision?

_____

5. The resulting nucleus is not the same type of element as either of the two that entered the collision. Why not?

_____

Press "Play" again, and then press "Pause" as soon as the yellow ball and the dashed line appear.

6. Is this a collision or a spontaneous decay?

_____

7. What is represented by the yellow ball?

_____

8. What is represented by the dashed line?

_____

9. The resulting nucleus has the same number of particles in the nucleus (13), but it is a different element. What happened to the proton that was in the nitrogen nucleus but is not now in the carbon nucleus?

_____

_____

Proceed past the next two collisions, to ^{15}O.

10. Study the pattern that is forming. When a blue ball comes in, what happens to the number of protons and neutrons in the nucleus, and the type of the nucleus (that is, what happens to the "12" and the "C" or the "14" and the "N")?

_____

11. What is emitted in these collisions?

_____

Proceed until ^{15}N appears.

12. Is this a collision or a spontaneous decay?

_____

13. Which previous reaction is this most like?

_____

Now proceed to the end of the animation.

14. After this last collision, a line is drawn back to the beginning. This tells you what type of nucleus the upper red ball represents. What is this nucleus?

_____

15. How many nuclear particles are not accounted for by that upper red ball? (Hint: Don't forget the ^{1}H that came into the collision.) These nuclear particles must be in the nucleus represented by the bottom red ball.

_____

16. Carbon has six protons, whereas nitrogen has seven. How many protons are in the nucleus represented by the bottom red ball?

_____

17. How many neutrons are in the nucleus represented by the bottom red ball?

_____

18. What element is represented by the bottom red ball?

_____

19. What is the net reaction of the CNO cycle? That is, what nuclei are combined and turned into the resulting nucleus?

_____

_____

20. Why do we not consider ^{12}C part of the net reaction?

_____

Measuring Galaxies

Astronomers use many different methods for determining distance, as you will find out here in Chapter 14. In Chapter 10, you learned how astronomers use parallax to find the distance to the closest stars. To go further, astronomers need a "standard candle"—a type of object that has a known luminosity. Street lights along a single street typically use the same bulbs—they all have the same luminosity, so they make good standard candles. If a street light is fainter, it is farther away. Find a straight street that is well-lit by regularly placed street lights—the longer, the better. Look at the nearest street light, and then move down the line of lights until you find one that is half as bright. Repeat this experiment as many times as you can, finding the next light that is half as bright as the previous one. If you do this carefully and well, you can make an accurate graph of the brightness of the lights versus the distance, measured as the number of street lights.

EXPERIMENT SETUP

Take a toilet paper tube out with you at night and find a straight street that is well-lit by regularly placed street lights—the longer, the better. The brightness of each light tells you how near it is to you.

Use your toilet paper tube to restrict your vision to just one street light at a time. Look at the nearest street light.

Move down the line of lights until you find one that is half as bright. How many street lights away is the light that is half as bright as the nearest one?

Repeat this experiment as many times as you can, finding the next light that is half as bright as the previous one. If you do this carefully and well, you can make an accurate graph of the brightness of the lights versus the distance, measured as the number of street lights away.

PREDICTION

I predict that for each drop in brightness by half, I will count ☐ the same ☐ a different number of street lights.

SKETCH OF RESULTS (in progress)

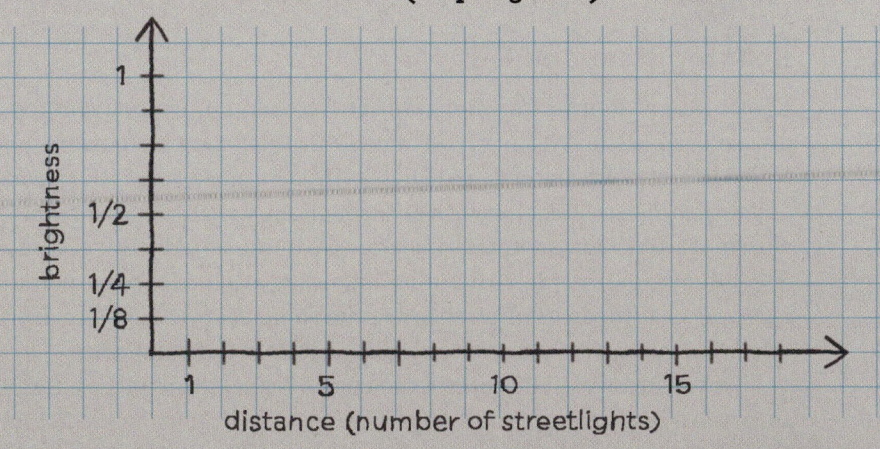

14

In 1920, the astronomers Harlow Shapley (1885–1972) and Heber D. Curtis (1872–1942) debated the nature of observed "spiral and elliptical nebulae." Shapley believed that our galaxy was the entire universe and therefore these nebulae must be inside the Milky Way. Curtis believed these objects were separate from the Milky Way and very distant. While this "Great Debate" did not resolve the issue at the time, it set the stage for the work of Edwin P. Hubble (1889–1953), whose discoveries about the distances and velocities of galaxies fundamentally changed our understanding of the universe. We now know that Curtis was correct: These spiral and elliptical objects are separate galaxies of stars, dust, and gas, like the Milky Way. Here in Chapter 14, you will learn about the types of galaxies and their properties and composition.

LEARNING GOALS

(1) Describe the classifications of galaxies.

(2) Explain how astronomers measure vast distances.

(3) Combine measurements of distance and velocity to show that the universe is expanding.

(4) Explain why scientists think that most—perhaps all—large galaxies have supermassive black holes at their centers.

14.1 Galaxies Come in Many Sizes and Shapes

A **galaxy** is a gravitationally bound grouping of stars, dust, and gas that is recognizable as an object distinct from its surroundings. Our galaxy, the Milky Way, is one of trillions of galaxies that fill a vast observable universe. These galaxies come in many shapes and a wide range of sizes, from ~100 parsecs (pc) to hundreds of thousands of parsecs across. (Recall that a parsec is 3.26 light-years.) All are remarkably distant. The nearest large spiral galaxy is roughly half a million times farther from the Milky Way than the nearest star is from the Sun. Astronomers learn about distant galaxies by using all of the tools available to them, such as imagery, spectroscopy, computer simulation, and reason. The study of galaxies is a young field, and even with all these tools, scientists are just beginning to understand galaxies in detail. In this section, you will learn about the different types of galaxies in the universe.

Relating Near and Far

Look again at Figure 1.1, where we set Earth in a universal context: It orbits an unremarkable star within an enormous spiral galaxy, which itself is part of a group, which is part of a supercluster, which is, in turn, part of an even larger supercluster of galaxies. These structures exist throughout the entire observable universe. As with most other astronomical objects, we cannot actually travel to distant galaxies to conduct experiments, so our understanding of galaxies rests on the *cosmological principle* that the laws of physics are the same everywhere. Gravity, for example, works the same way in distant galaxies as it does here on Earth.

As we stressed in Chapter 1, the cosmological principle is a testable scientific theory. An important prediction of the cosmological principle is that the conclusions we reach about our universe should be the same, whether we observe it from

what if . . .

What if the universe were homogeneous but not isotropic: Would that mean that we live in a special place in the universe?

the Milky Way or from a galaxy billions of light-years away. One test of the cosmological principle is to see whether our universe is a **homogeneous** universe that is the same in every direction.

It is not easy to directly determine if the universe is homogeneous. We cannot ever hope to travel to other galaxies to see whether conditions are the same. However, we can compare light arriving from nearby and distant locations in the universe. For example, we can look at the distribution of galaxies in distant space and ask whether that distribution is similar to the distribution of nearby galaxies.

In addition to predicting that the universe is homogeneous, the cosmological principle requires that all observers measure the same properties of the universe, regardless of the *direction* in which they are looking. If the universe is the same in all directions, then it is **isotropic**. This prediction of the cosmological principle is straightforward to test directly, by looking in different directions. For example, if galaxies were lined up in rows, observers would measure very different properties in different directions, depending on whether they were looking along the rows or diagonally across them. Such a universe would not be isotropic. Notice that, in this case, the universe would still be homogeneous. In most instances, isotropy goes hand in hand with homogeneity, but the cosmological principle requires them both.

The isotropy and homogeneity of the distribution of galaxies in the universe are predictions of the cosmological principle that we can test directly. All of our observations show that the properties of the universe are the same regardless of the direction in which we look—the universe is isotropic. On very large scales, the universe appears homogeneous as well. The cosmological principle has withstood these tests and forms the basis for the study of distant galaxies and of the universe itself.

The Shapes of Galaxies

Imagine taking a handful of different coins and throwing them into the air, as shown in **Figure 14.1a**. All of these objects are flat and circular. When you look at the objects falling through the air, however, they do not appear to be the same. Some coins appear face-on, and they look circular. Others appear edge-on, and they look like thin lines. Most coins are seen from an angle between these two extremes. Even if this image of many coins was the only information you had, you could use it to figure out the three-dimensional shape of a coin—flat and circular.

Astronomers use a similar method to discover the true three-dimensional shapes of galaxies. Figure 14.1b shows a set of galaxies seen from various viewing angles, from face-on to edge-on. We can infer that, just like the coins in Figure 14.1a, these particular galaxies are disk-shaped and are randomly oriented on the sky. Other galaxies are more egg-like, while still others are irregularly shaped.

Galaxies that are elliptical in three dimensions, something like an egg, are called **elliptical galaxies**; several examples are shown in **Figure 14.2**. These galaxies have a circular or elliptical outline on the sky. As with the coins tossed in the air, the appearance of an elliptical galaxy in the sky does not necessarily tell us its true shape. For example, a galaxy might actually be shaped like a jelly bean, but if

(a)

(b)

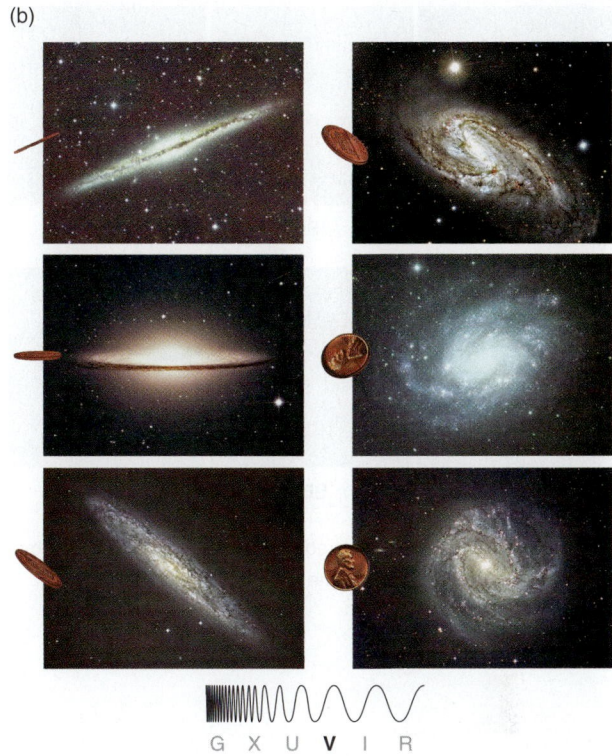

G X U V I R

Figure 14.1 (a) A handful of coins thrown in the air provides a helpful analogy for the difficulties in identifying the shapes of certain types of galaxies. We see some face-on, some edge-on, and most somewhere in between. (b) Disk-shaped galaxies seen from various perspectives, or viewing angles. The variety of orientations we see for galaxies corresponds to the range of perspectives for the coins in (a).

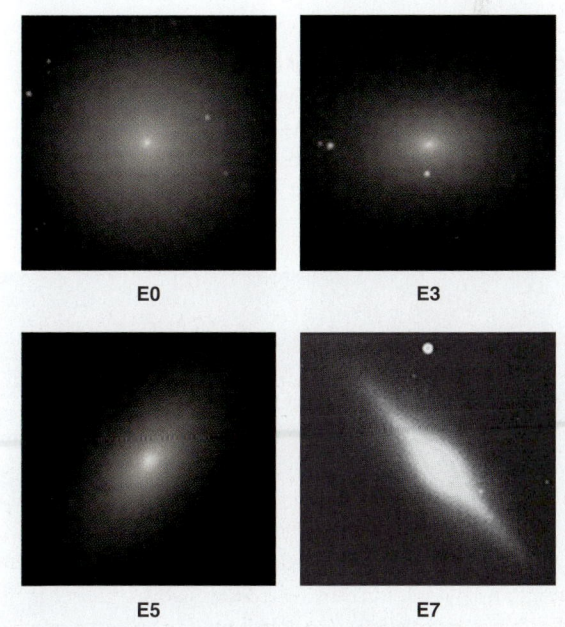

E0   E3

E5   E7

Figure 14.2 Elliptical galaxies are classified on the basis of how elliptical they appear to be from our point of view. The most circular are E0 galaxies, while the most elliptical are classified as E7.

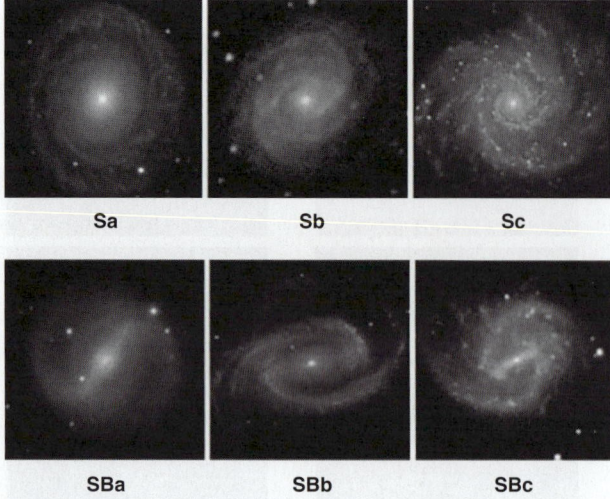

Sa   Sb   Sc

SBa   SBb   SBc

Figure 14.3 Spiral and barred spiral galaxies are classified by the openness of the arms and the prominence of the bulge.

Astronomy in Action: Galaxy Shapes and Orientation

Figure 14.4 (a) The irregular band of dust in the plane of the Milky Way obscures our view toward the galactic center. (b) Similarly, the dust in the plane of the nearly edge-on spiral galaxy M104 is seen as a dark, obscuring band in the midplane of the galaxy.

(a)

G X U **V** I R

we happen to see it end-on, it will look round like a gumball. Elliptical galaxies are divided into numbered subtypes ranging from nearly spherical (E0) to quite flattened (E7). Elliptical galaxies have few young stars. Many, if not most, elliptical galaxies contain small rotating disks at their centers.

Spiral galaxies, designated with an initial S, are characterized by a flattened, rotating disk rich with dust and gas. The spiral arms that give these galaxies their name lie in this disk. In addition to disks and arms, spiral galaxies have central bulges, which extend above and below the disk. Roughly half of all galaxies with spiral arms have a bulge that is bar-shaped when viewed from above the plane of the galaxy; these are known as **barred spiral galaxies** (designated SB). Both spiral and barred spiral galaxies are subdivided into types a, b, and c, according to the prominence of the central bulge and how tightly the spiral arms are wound, as shown in **Figure 14.3**. For example, Sa and SBa galaxies have the largest bulges and display tightly wound and smooth spiral arms. Both Sc and SBc galaxies have small central bulges and more loosely wound spiral arms, often very knotty in appearance. Our own Milky Way Galaxy is a loosely wound barred spiral galaxy (SBc).

The distinction between spiral and elliptical galaxies is not always clear. Some galaxies are a combination of the two types, having stellar disks but no spiral arms, so that the disk is smooth in appearance, like an elliptical galaxy. These intermediate types are called **S0 galaxies**, or "lenticular" galaxies, because they are shaped like a lens. Elliptical and S0 galaxies share other similarities. For example, both contain small rotating disks at their centers, and neither produces many new stars.

Galaxies that fall into none of these classes are called **irregular galaxies** (designated Irr). As their name implies, irregular galaxies are often without symmetry in shape or structure. Often, irregular galaxies are the result of the collision of two or more galaxies.

Differences among Galaxies

In addition to shape, there are other important distinctions between spiral and elliptical galaxies. These distinctions affect the way a galaxy looks and indicate the galaxy's past and future evolution.

Gas and Dust. Most spiral galaxies contain large amounts of dust and cold, dense gas concentrated in the midplanes of their disks. Just as the dust in the disk of our own galaxy is visible on a clear summer night as a dark band slicing the Milky Way in two (**Figure 14.4a**), the dust in an edge-on spiral galaxy appears as a dark band running along the midplane of the disk (Figure 14.4b). The cold gas that accompanies the dust in spiral galaxies is cold enough that the hydrogen atoms are neutral and the electrons are in the ground state. Emission from these atoms occurs because of a change in the alignment of the electron and the

(b)

G X U **V** I R

proton, and this emission can only be seen in radio observations. In contrast, elliptical galaxies contain large amounts of very hot gas that emits strongly in the X-ray region of the spectrum.

The difference in shape between elliptical and spiral galaxies offers some insight into why the gas in elliptical galaxies is hot, while in spiral galaxies it is cold. Recall from Chapter 5 that the conservation of angular momentum causes a rotating cloud to form a disk of gas around a star. Spiral galaxies also rotate, so the conservation of angular momentum causes the cold gas to form a disk around the bulge of a spiral galaxy. In contrast, elliptical galaxies do not have an overall rotation, so the gas does not settle into a disk. In an elliptical galaxy, the only place that cold gas could collect is at the center. However, the density of stars in the center of elliptical galaxies is so high that evolving stars and Type Ia supernovae continually reheat the gas, preventing most of it from cooling off.

Color. The colors of spiral and elliptical galaxies tell us a great deal about their star formation histories. Stars form from dense clouds of cold gas. Because the gas we see in elliptical galaxies is very hot, active star formation is not taking place in those galaxies today. The reddish colors of elliptical and S0 galaxies confirm that little or no star formation has occurred there for quite some time. The stars in these galaxies are an older population of low-mass stars. Conversely, the bluish colors of the disks of spiral galaxies confirm that stars are forming in the cold molecular clouds contained within the disk. Even though *most* of the stars in a spiral disk are old, the massive young stars are so luminous that their blue light dominates, and the disks appear blue. Most irregular galaxies are also forming new young stars. Some irregular galaxies (known as "starburst galaxies") are currently forming stars at prodigious rates, given their relatively small sizes, and so they are quite blue in color.

Size. Only subtle differences in color and concentration of stars exist between large and small galaxies, making it difficult for us to distinguish which are large and which are small. Even when a smaller, nearby spiral galaxy is seen next to a larger, distant spiral, it can be hard to tell which is which by appearance alone. Astronomers have been able to determine that galaxies range in size from ~100 pc to hundreds of thousands of parsecs. There is no distinct size difference between elliptical and spiral galaxies; about half of both types of galaxies fall within a similar range of sizes.

Luminosity. Galaxies range in luminosity from about a million up to a trillion solar luminosities (10^6 to 10^{12} L_{Sun}). Although the most luminous elliptical galaxies are more luminous than the most luminous spiral galaxies, there is considerable overlap in the range of luminosities among all types. The relationship between luminosity and size among the different types of galaxies is not straightforward; a galaxy twice as large as another will not necessarily be twice as luminous.

Mass. Recall that mass is the single most important parameter in determining the properties and evolution of a star. In contrast, differences in mass do not lead to obvious differences among galaxies. Still, galaxies that have relatively low luminosity (less than 1 billion L_{Sun}) are called **dwarf galaxies**, because the brightness indicates the number of stars, and therefore the amount of stellar mass. Galaxies more than 1 billion times as luminous as the Sun are called **giant galaxies**. Only elliptical and irregular galaxies come in both types; among spiral and S0 galaxies, we find only giants. It is relatively easy to tell the difference between a dwarf elliptical galaxy and a giant elliptical galaxy (**Figure 14.5**). Giant elliptical galaxies have a much higher density of stars, which are more centrally concentrated than stars in dwarf elliptical galaxies.

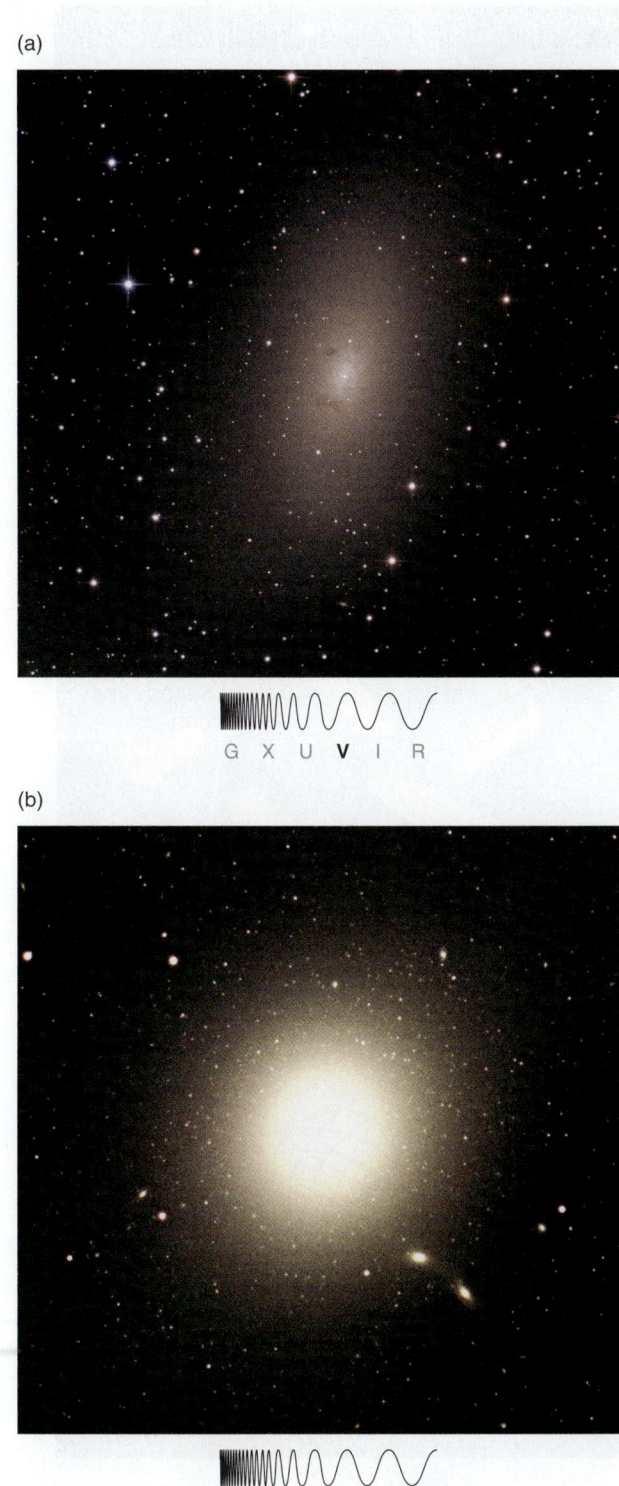

(a)

G X U V I R

(b)

G X U V I R

Figure 14.5 (a) Dwarf elliptical galaxies differ in appearance from (b) giant elliptical galaxies. Giant elliptical galaxies are more centrally concentrated than dwarf elliptical galaxies.

Figure 14.6 Elliptical galaxies take their shape from the disordered orbits of most of the stars in the galaxy. The colored lines superimposed on the galaxy represent the complex, irregular orbits of its stars.

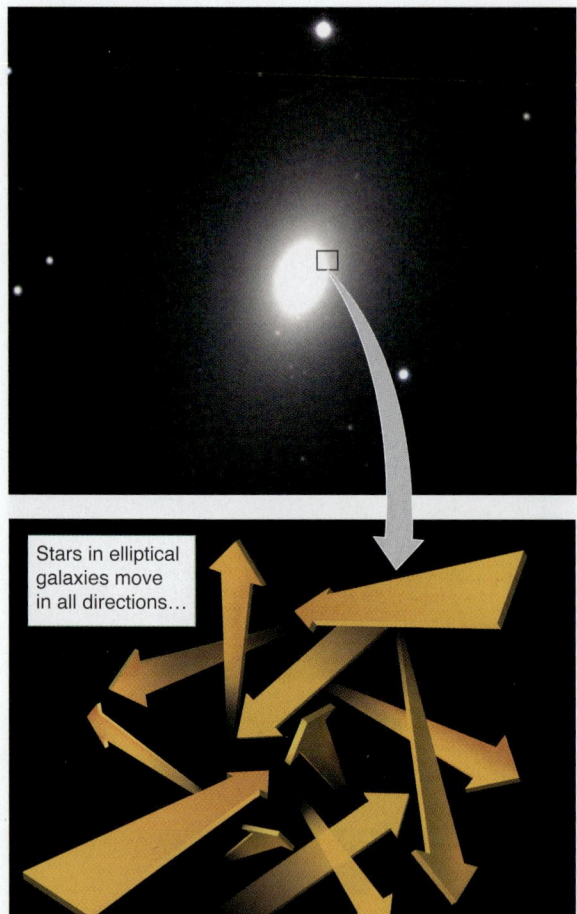

Stars in elliptical galaxies move in all directions...

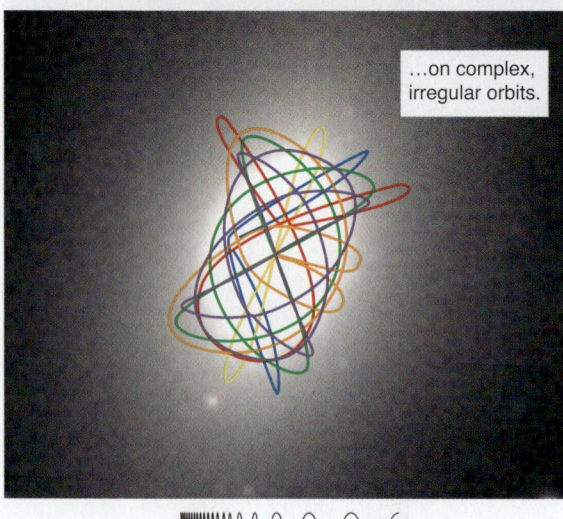

...on complex, irregular orbits.

G X U V I R

Stellar Motions Give Galaxies Their Shapes

A galaxy is not a solid object like a coin but rather a collection of stars, gas, and dust. Unlike planets in our Solar System, which move in the same direction on nearly circular orbits around the Sun, stars in an elliptical galaxy follow orbits with a wide range of shapes, oriented randomly, as shown in **Figure 14.6**. The orbits of stars in elliptical galaxies are more complex than the orbits of planets because the gravitational field within an elliptical galaxy does not come from a single central object. Taken together, the variation in shape and orientation of all these stellar orbits give an elliptical galaxy its shape.

If the stars in an elliptical galaxy are moving in truly random directions, the galaxy will have a spherical shape. However, if stars tend on average to move faster in one direction than in others, the galaxy will be more spread out in that direction, giving it an elongated (elliptical) shape. These differences in stellar orbits cause some elliptical galaxies to be round and others to be elongated. Orbital speeds are also a factor. The faster the stars are moving, the more spread out the galaxy is.

The components of a spiral galaxy (in this case a barred spiral) are shown in **Figure 14.7**. Its defining feature is that it has a flattened, rotating disk. Like the planets of our Solar System, most of the stars in the disk of a spiral galaxy travel in the same direction around a concentration of mass at the center of the galaxy, taking tens of millions of years to orbit once. The stellar orbits in a spiral galaxy's central bulge, however, are quite different from those in the galaxy's disk. As with elliptical galaxies, the gravitational field within the bulge does not come from a single object, and the stars therefore follow a wide variety of orbits. The bulges of

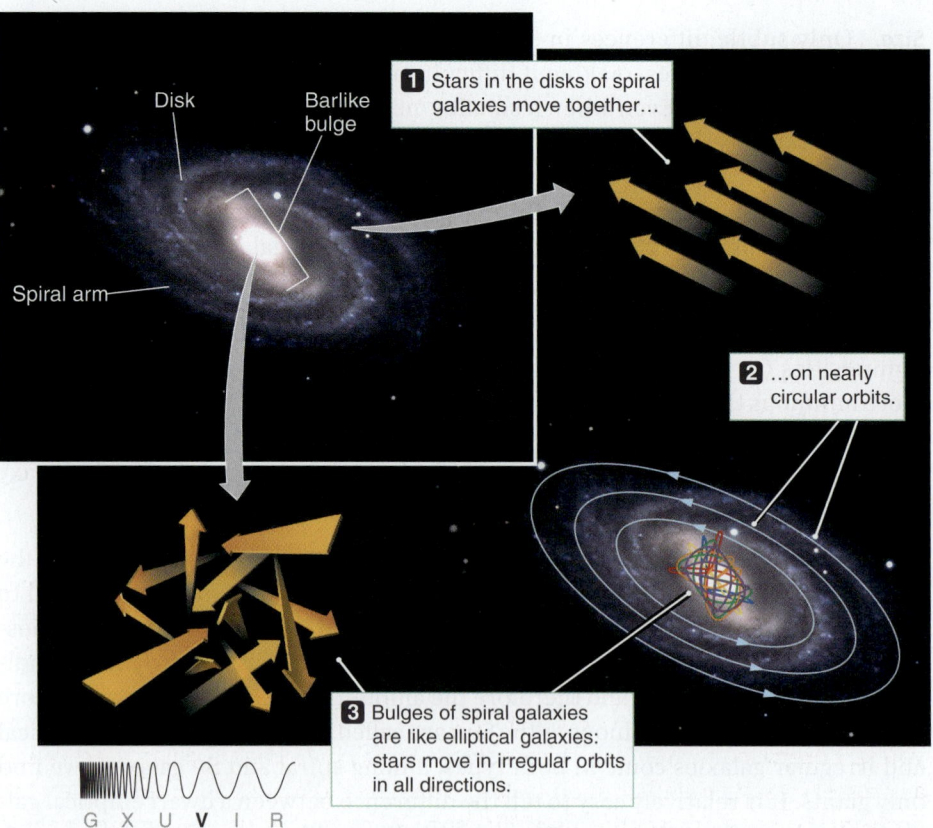

G X U V I R

Figure 14.7 The components of a barred spiral galaxy include a disk, spiral arms, and a barlike bulge. The orbits of stars in the rotating disk and the elliptically shaped bulge are different.

spiral galaxies (but not barred spirals) are thus roughly spherical in shape. Surrounding the entire spiral galaxy is a halo—a spherical distribution of stars, clusters of stars, and dark matter (which will be discussed in Chapter 15). Objects in the halo have orbits in random orientations around the center of the galaxy.

CHECK YOUR UNDERSTANDING **14.1**

If a galaxy is quite elongated but has no dark lane running through it, then its classification is most likely

a. E5.

b. Sa.

c. SBc.

d. Irr.

Answers to Check Your Understanding questions are in the back of the book.

14.2 There Are Many Ways to Measure Astronomical Distances

Studies of galaxies in the 1920s revolutionized the way we see the universe. The discovery that galaxies are separate from the Milky Way and located at great distances from us led astronomers to understand that the universe is much larger than previously thought. Astronomers also realized that there are far more stars than anyone had imagined. In this section, you will learn how astronomers determine the sizes and distances of galaxies.

Finding the Distance to Globular Clusters

A **globular cluster**, such as M92 (**Figure 14.8a**), is a sphere of stars that contains between tens of thousands and a million stars. Many such clusters can be seen through small telescopes. The motions of stars within a globular cluster are much like the motions of stars within an elliptical galaxy. However, globular clusters are much smaller, and the stars are closer together. A typical globular cluster consists of 500,000 stars packed into a volume of space with a radius of only 15 light-years—much more crowded than the average density of stars within the Milky Way.

The Milky Way has more than 150 cataloged globular clusters, but dust in the disk likely hides many more from view. About one-fourth of the globular clusters in the Milky Way reside in or near the disk. The rest occupy the halo of the Milky Way. Globular clusters are very luminous, with luminosities ranging from a low of about 1,000 L_{Sun} to a high of about 1 million L_{Sun}. Most lie outside the dusty disk, so they can easily be seen at great distances. To find the distance to a globular cluster, we need to look at the properties of the stars they contain.

Figure 14.8b shows the Hertzsprung-Russell (H-R) diagram for the stars in M92. The diagram's main-sequence turnoff occurs for stars with masses of about 0.8 M_{Sun}, which corresponds to a main-sequence lifetime of close to 13 billion years. This age is typical for globular clusters, making them the oldest known objects in our galaxy. Globular clusters formed when the universe and our galaxy were very young. Compared to globular-cluster stars, our Sun, at about 5 billion years old, is a relatively young member of our galaxy. In an H-R diagram of an old cluster, the horizontal branch crosses the instability strip, so globular clusters contain Cepheid variables and RR Lyrae stars, which make good standard candles to measure distance, as discussed in Chapter 13.

(a)

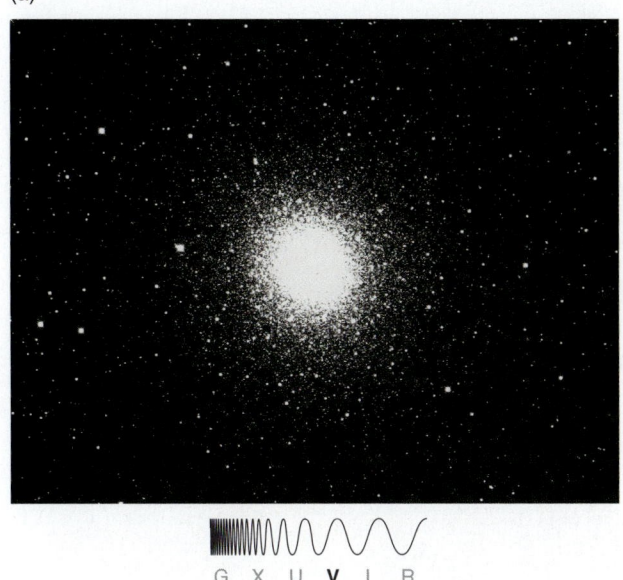

G X U **V** I R

(b)

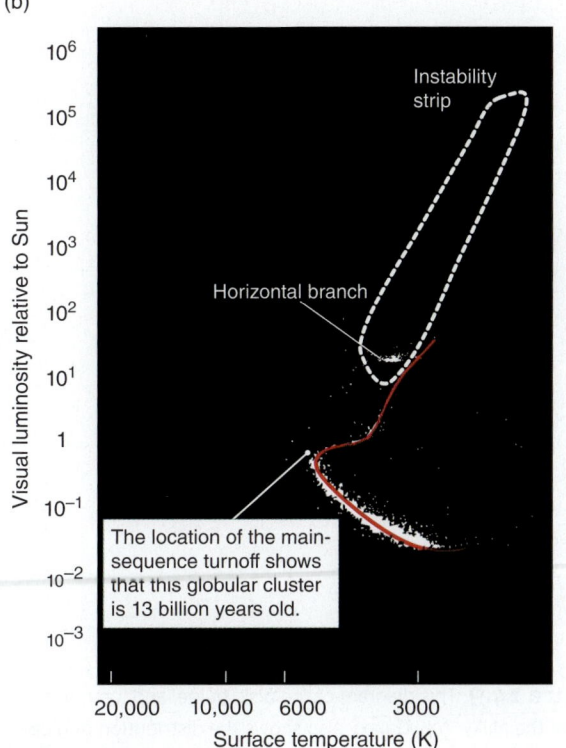

Figure 14.8 (a) The globular cluster M92 and (b) a Hertzsprung-Russell (H-R) diagram of the stars it contains. The main-sequence turnoff indicates the age of the cluster.

RR Lyrae stars have a distinctive light curve; they change with time in a distinctive way. This makes them easy to find in a globular cluster. Harlow Shapley used the period-luminosity relationship to find the luminosities of RR Lyrae stars in globular clusters. He then combined these luminosities with measured brightnesses to determine the distances to globular clusters. Finally, Shapley cross-checked his results by noting that the clusters he thought were farther away also tended to appear smaller in the sky, as expected.

Finding the Size of the Milky Way

The size of the Milky Way and the distance to the center of the galaxy cannot be measured directly, in part because dust, gas, and stars obscure our view through the disk. Astronomers determine the size of the Milky Way from observations of the distance and direction of globular clusters. Using these measurements, Shapley made a three-dimensional map of globular clusters. This map showed that globular clusters occupy a roughly spherical region of space with a diameter of about 300,000 light-years. These globular clusters trace out the halo of the Milky Way Galaxy, as shown in **Figure 14.9**, which reflects the modern view of the globular-cluster distribution. Globular clusters orbit the gravitational center of the galaxy, so the center of the distribution of globular clusters is the same as the gravitational center of the galaxy. The distance from the Sun to this center of distribution is about 27,000 light-years, so Earth is roughly halfway between the center and the edge of the disk of the Milky Way. The distance to the edge of the disk is determined by looking at pulsating variable stars that are not in globular clusters.

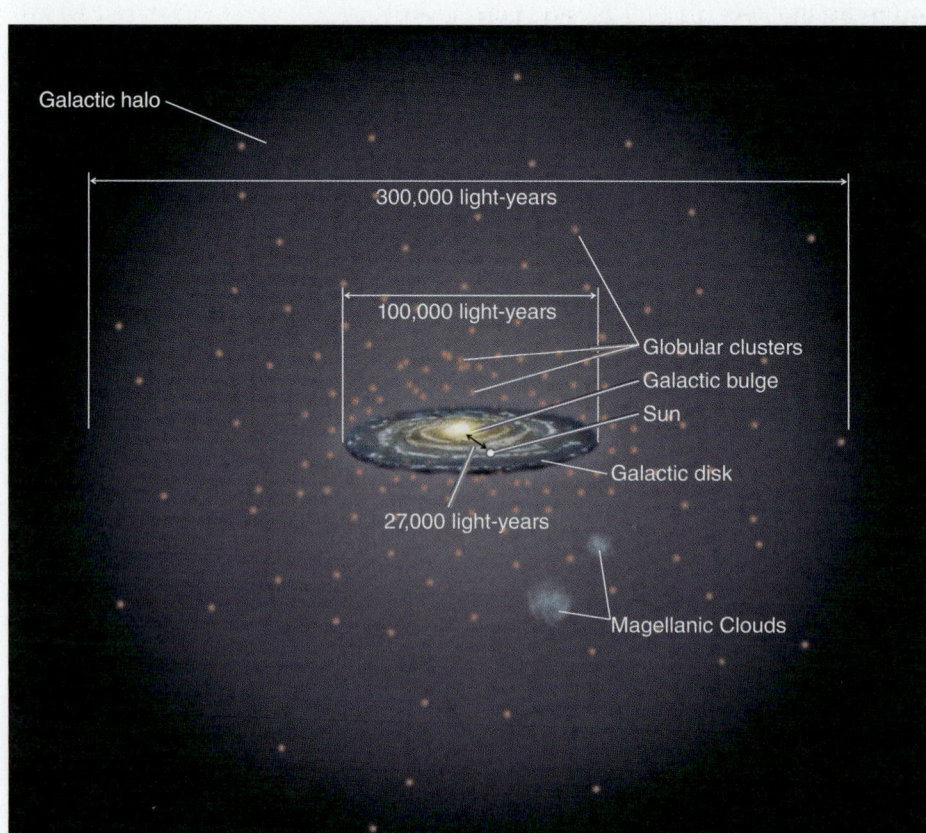

Figure 14.9 This diagram of the disk, bulge, and luminous halo of the Milky Way Galaxy also shows the distribution of globular clusters, two companion galaxies to the Milky Way (the Large and Small Magellanic Clouds), and the location of the Sun within the Milky Way disk.

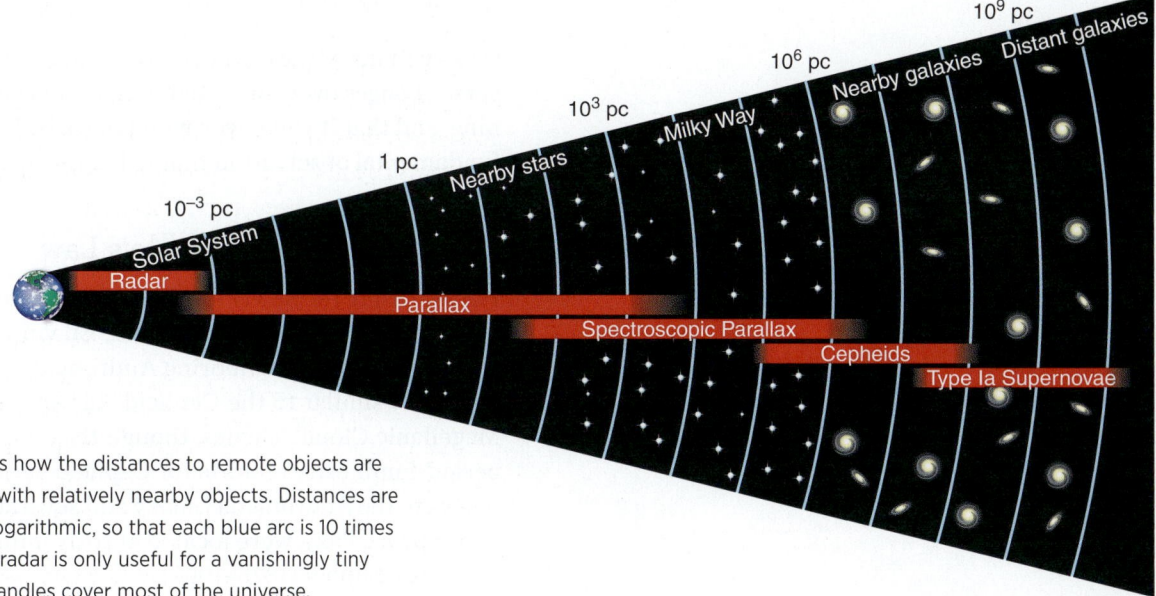

Figure 14.10 The distance ladder indicates how the distances to remote objects are measured through a series of steps beginning with relatively nearby objects. Distances are given in parsecs (pc). Notice that the scale is logarithmic, so that each blue arc is 10 times farther than the one before it. This means that radar is only useful for a vanishingly tiny portion of the universe, and distant standard candles cover most of the universe.

Finding Distances to Other Galaxies

The **distance ladder** relates distances of remote objects on a variety of scales, as illustrated in **Figure 14.10**. Within the Solar System, distances are measured using radar and signals from space probes. Once the distance to the Sun is known, parallax (as discussed in Chapter 10) gives distances to nearby stars, allowing astronomers to find their luminosities and plot them on the H-R diagram. Astronomers use parallax to calibrate the diagram, so that for the next rung of the distance ladder, they need only a main-sequence star's temperature (relatively easy to measure from its spectrum) to find its luminosity. This information enables them, in turn, to estimate the star's distance by comparing its *apparent* brightness with its luminosity. This process for finding the distance to stars is known as *spectroscopic parallax*.

Moving farther out, astronomers measure the distance to relatively nearby galaxies using standard candles, such as Cepheid variables, which can be used to accurately measure distances as far away as 30 million parsecs (30 megaparsecs [Mpc]). Type Ia supernovae are used as standard candles (as described in Chapter 12) to determine even larger distances. The peak luminosities of these Type Ia supernovae can be determined from careful study of the way the light increases and decreases again. Therefore, astronomers can both observe their brightness and estimate their luminosity—everything they need to determine the distance. Type Ia supernovae are so bright (**Figure 14.11**) that modern telescopes can observe them almost to the edge of the observable universe.

CHECK YOUR UNDERSTANDING 14.2

The size of the Milky Way is determined from studying _____ in globular clusters.

a. standard candles
b. velocities
c. standard models
d. expansions

what if . . .

What if someone discovers a systematic error in all spectroscopic parallax measurements, so that the distances to stars measured in this way are 10% larger than previously thought: How might that affect our understanding of the distances to other galaxies?

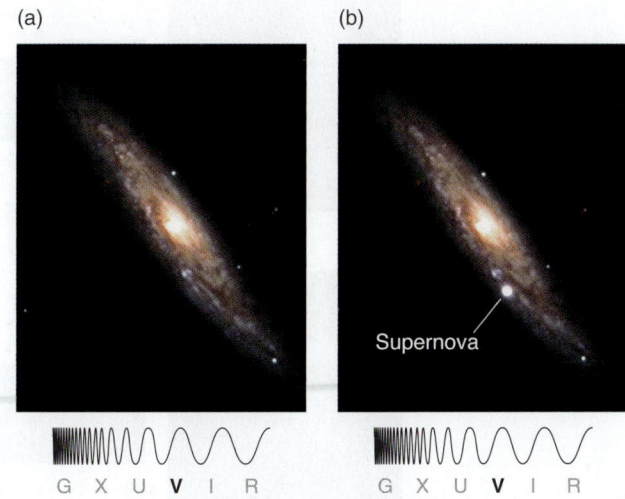

(a)   (b)

Figure 14.11 Galaxy NGC 3877 is shown (a) before and (b) after the explosion of a Type Ia supernova. Type Ia supernovae are extremely luminous standard candles. This galaxy is 15.5 Mpc from Earth.

14.3 We Live in an Expanding Universe

Observations of the distances and motions of galaxies demonstrated that the universe changes over time—that it contains large-scale motions, that it had a beginning, and that it will have an end of sorts. In this section, you will learn about the fundamental observation that led to the discovery that the universe is expanding.

The Discovery of Hubble's Law

In 1925, using the newly finished 100-inch telescope on Mount Wilson, high above the then small city of Los Angeles, Edwin Hubble was able to find some variable stars in the large neighboring Andromeda Galaxy. He recognized that these stars were very similar to the Cepheid variable stars in the Milky Way and the nearby Magellanic Cloud galaxies, though they appeared much fainter. Hubble used the period-luminosity relation for Cepheid variable stars (see Chapter 13) to find distances to the Andromeda Galaxy and several others. These galaxies, similar in size to our own galaxy, were located at truly immense distances.

Vesto Slipher (1875–1969) at Lowell Observatory in Flagstaff, Arizona, obtained spectra of these galaxies. Unsurprisingly, Slipher's galaxy spectra looked like the spectra of collections of stars with a bit of glowing interstellar gas mixed in. However, the emission and absorption lines in these spectra were shifted from the locations of lines in spectra created in the laboratory. The galaxy spectral lines were shifted to longer, or redder, wavelengths, as shown in **Figure 14.12**.

Hubble interpreted Slipher's redshifts as Doppler shifts, and he concluded that almost all of the galaxies in the universe are moving away from the Milky Way. Recall that if an object is moving away, the Doppler shift causes the observed wavelengths of emitted spectral lines to shift toward the red end of the spectrum. The wavelength of a line emitted by a stationary object in the laboratory is called the rest wavelength of the line and is written as λ_{rest}. The redshift of a galaxy's spectral line is written as z. Objects with higher values of z are moving away more quickly.

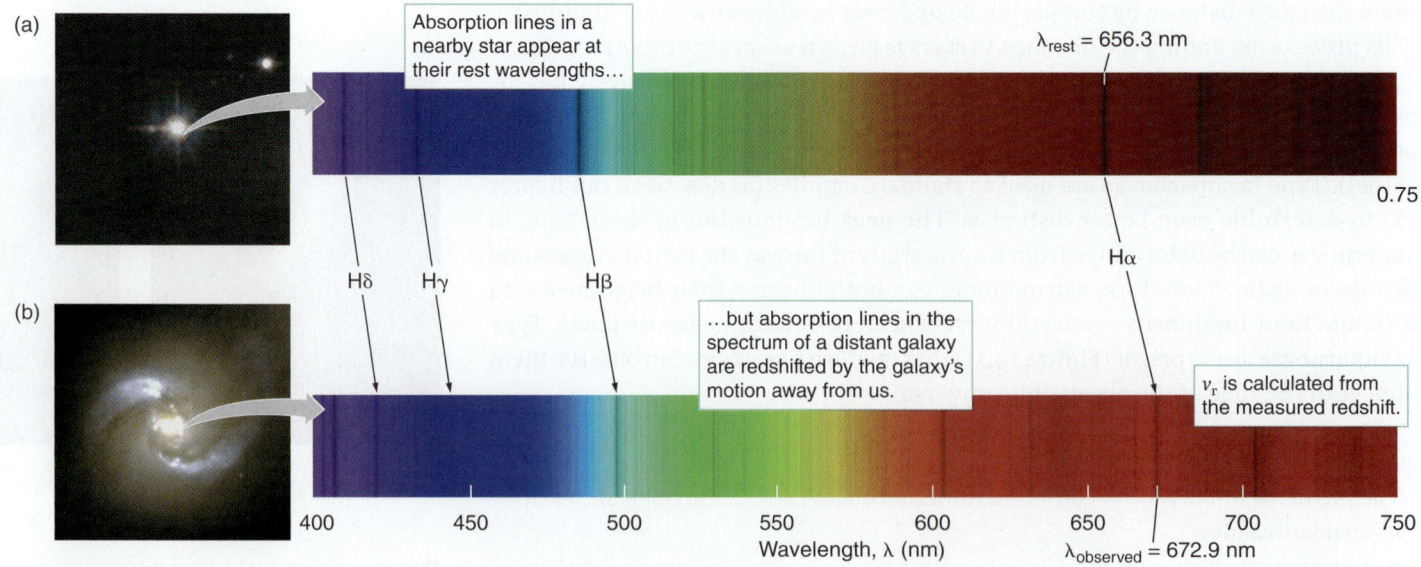

Figure 14.12 (a) The spectrum of a star in our galaxy shows absorption lines, which in this case lie at the rest wavelength. (b) A distant galaxy, shown with its spectrum at the same scale as that of the star, has lines that are redshifted to longer wavelengths. v_r is recession velocity.

When Hubble combined Slipher's measurements of galaxy **recession** velocities with his own estimates of the distances to these galaxies, he made one of the most important discoveries in the history of astronomy. Hubble found that *the recession velocity of a galaxy is proportional to the distance of that galaxy*. A galaxy at a distance of 30 Mpc from Earth moves away from us twice as fast as a galaxy at a distance of 15 Mpc from Earth. This relationship between distance and recession velocity has become known as **Hubble's law**.

The **Hubble constant** is the constant of proportionality between the recession velocity and the distance. Its mathematical symbol is H_0 (which astronomers pronounce as "H naught"), so that $v_r = H_0 d$. Another way to say this is that a graph of v_r versus d should be a straight line, with the slope equal to H_0.

Figure 14.13 plots the measured recession velocities of galaxies against their measured distances. The points lie along a line on the graph, showing that velocity and distance are proportional to each other. There is a small amount of scatter around this line, which is due in part to the "peculiar" motion of individual galaxies as they are tugged around by the gravity of other nearby galaxies. The slope of this line is equal to the proportionality constant H_0. This constant is one of the most important numbers in astronomy, and many astronomical careers have been dedicated to trying to determine its value precisely and accurately. Today, we have measured the Hubble constant to an accuracy of a few percent. The value is likely to be further refined in the years to come. In this text, we use a value of 70 kilometers per second per megaparsec or (km/s)/Mpc as an approximation to the best current value. **Working It Out 14.1** demonstrates how to find the distance to a galaxy using Hubble's Law.

VOCABULARY ALERT

recession In everyday language, *recession* is associated almost entirely with economics. Astronomers use it to refer to an object that is receding, or moving away. The recession velocity is therefore the velocity at which an object moves away.

▶‖ **AstroTour:** Hubble's Law

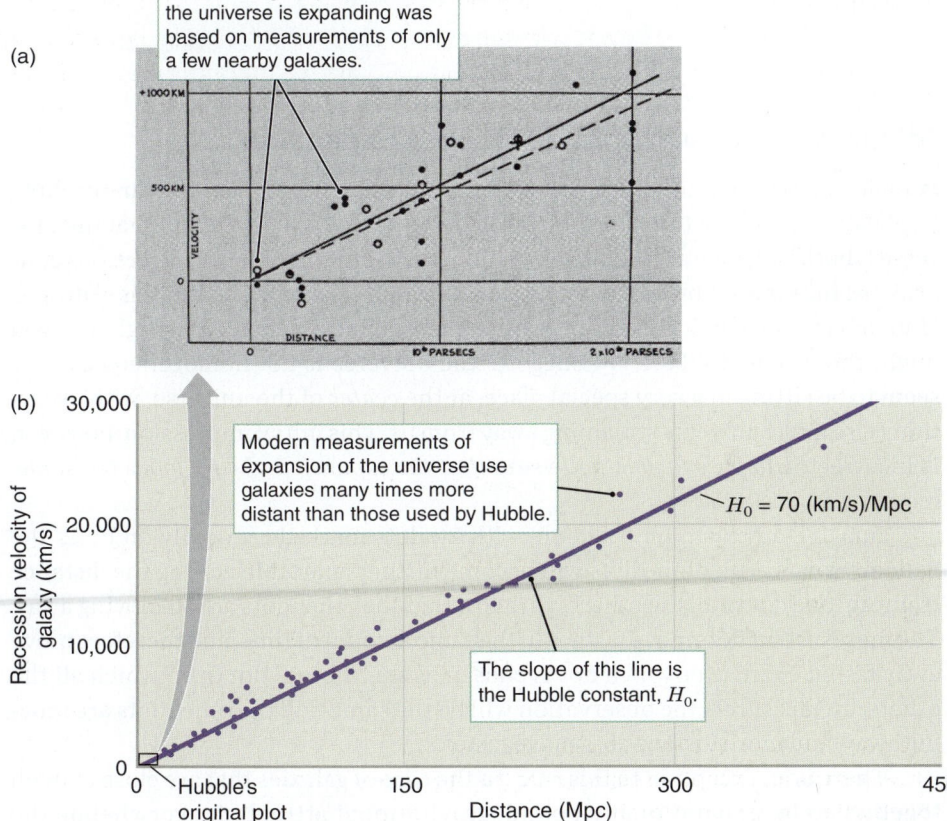

(a)

Hubble's original proposal that the universe is expanding was based on measurements of only a few nearby galaxies.

(b)

Modern measurements of expansion of the universe use galaxies many times more distant than those used by Hubble.

$H_0 = 70$ (km/s)/Mpc

The slope of this line is the Hubble constant, H_0.

Hubble's original plot

Recession velocity of galaxy (km/s)

Distance (Mpc)

Figure 14.13 (a) Hubble's original figure illustrating that more distant galaxies are receding faster than less distant galaxies. (b) Modern data extend much further and show that recession velocity is proportional to distance far into the early universe.

working it out 14.1

Redshift: Calculating the Recession Velocities and Distances of Galaxies

The Doppler equation you learned for spectral lines (in Working It Out 5.2) showed that

$$v_r = \frac{\lambda_{obs} - \lambda_{rest}}{\lambda_{rest}} \times c$$

The fraction in front of the c is equal to z, the redshift. Substituting for this fraction, we get

$$v_r = z \times c$$

(Note: This correspondence works only for velocities much slower than the speed of light.)

Suppose a hydrogen line is seen in the spectrum of a distant galaxy. In the laboratory, this hydrogen line has a measured rest wavelength of 122 nanometers (nm). If the observed wavelength of the hydrogen line is 124 nm, then its redshift is

$$z = \frac{\lambda_{obs} - \lambda_{rest}}{\lambda_{rest}}$$

$$z = \frac{124 \text{ nm} - 122 \text{ nm}}{122 \text{ nm}}$$

$$z = 0.016$$

We can now calculate the recession velocity from this redshift:

$$v_r = z \times c = 0.016 \times 300,000 \text{ km/s} = 4,800 \text{ km/s}$$

How far away, though, is our distant galaxy? This is where Hubble's law and the Hubble constant, $H_0 = 70$ (km/s)/Mpc, come in. Hubble's law relates a galaxy's recession velocity to its distance and can be expressed mathematically as $v_r = H_0 \times d_G$, where d_G is the distance to a galaxy measured in millions of parsecs (that is, in megaparsecs). We can divide through by H_0 to get

$$d_G = \frac{v_r}{H_0}$$

$$d_G = \frac{4,800 \text{ km/s}}{70 \text{ (km/s)/Mpc}} = 69 \text{ Mpc}$$

From a measurement of the wavelength of a hydrogen line, we have learned that the distant galaxy is approximately 69 Mpc away.

All Observers See the Same Hubble Expansion

Hubble's law is a remarkable observation about the universe that has far-reaching implications. For one thing, Hubble's law helps us test the prediction that the universe is both isotropic and homogeneous. Observations in different directions confirm that the same Hubble's law applies in every direction, so the universe is isotropic. Hubble's law also indicates that the universe is homogeneous. At first glance, you might think that Hubble's law suggests the universe is not homogeneous, as we seem to be sitting in a very special place: at the *center* of the universe, with everything else in the universe streaming away from us. This initial impression, however, is incorrect. *Hubble's law actually says that the expansion always looks the same, regardless of location.*

Imagine an expanding balloon, with 20 dots marked randomly on it. As the balloon expands, all of the dots will become farther apart. Measuring the distance from any one dot to all the others will show that the other dots are all moving away. The more distant dots move farther in the same amount of time, and therefore move away faster. No matter which dot is chosen as the "home" dot from which all the others are measured, the observation will be the same—all the other dots are moving away, and more distant dots move faster.

There is an exception to this rule. In the case of galaxies that are close enough together to be gravitationally bound, gravitational attraction overwhelms the

what if . . .

What if Hubble had discovered that the recession velocity increases as the square of the distance, so that $v_r = H_0 d^2$: What would that imply about Earth's location in the universe?

expansion of space. For example, the Andromeda Galaxy and the Milky Way are approaching each other at about 300 km/s, so light from the Andromeda Galaxy is blueshifted. This unusual velocity is caused by the gravitational interaction between the two galaxies, which is stronger than the expansion of space. The fact that gravitational or electromagnetic forces can overwhelm the expansion of space also explains why the Solar System is not expanding, and neither are you.

Hubble's Law Maps the Universe in Space and Time

Hubble's law is a practical tool for measuring distances to extremely remote objects. Once the value of H_0 is measured from a sample of galaxies, the distance to any other galaxy can be found by measuring its redshift and using $v_r = H_0 d$ to calculate its distance. Hubble's law does more than place galaxies in space: it also places galaxies in time. Remember that light travels at a huge but finite speed. When we look at the Sun, we see it as it existed 8.3 minutes ago. When we look at the center of our galaxy, the picture we see is about 27,000 years old. When looking at a distant object, astronomers speak of its **look-back time**—the time it has taken for the light from that object to reach the telescope. In the distant universe, look-back times become very large. The distance to a galaxy with redshift $z = 0.1$ is about 1.4 billion light-years, if the constant $H_0 = 70$ (km/s)/Mpc is used, so the look-back time to that galaxy is 1.4 billion years. As we look at objects with greater and greater redshifts, we see increasingly younger stages of our universe. The most distant observed objects have a look-back time of nearly 13.8 billion years, the age of the universe.

CHECK YOUR UNDERSTANDING **14.3**

If galaxy A has a recession velocity of 2,500 km/s while galaxy B has a recession velocity of 5,000 km/s, then

a. galaxy A is 4 times as far away as galaxy B.
b. galaxy A is twice as far away as galaxy B.
c. galaxy B is twice as far away as galaxy A.
d. galaxy B is 4 times as far away as galaxy A.

14.4 A Supermassive Black Hole Exists at the Heart of Most Galaxies

Probing the centers of galaxies is difficult because there are so many stars and so much dust and gas in the way that we cannot get a clear picture of the center, even for nearby galaxies. An understanding of what lies at the heart of massive galaxies did not initially come from studying nearby galaxies but instead from observing distant objects in the universe. From these studies, astronomers found that most galaxies have a supermassive black hole in the center.

The Discovery of Quasars

In the late 1950s, radio surveys detected a number of bright, compact objects that at first seemed to have no optical counterparts. Improved radio positions revealed that these radio sources coincided with faint, very blue, starlike objects. Unaware of the true nature of these objects, astronomers called them "radio stars." Obtaining spectra of the first two radio stars was a laborious task, requiring 10-hour

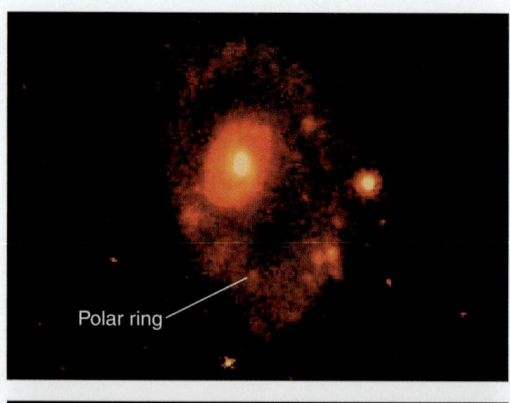

Polar ring

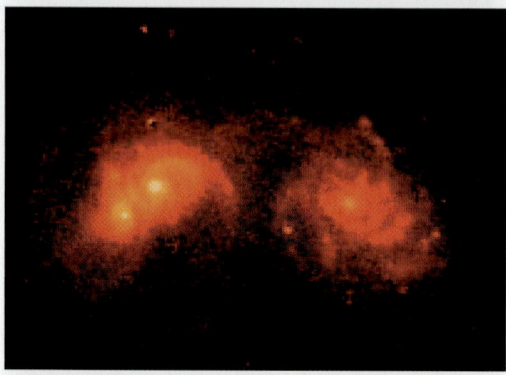

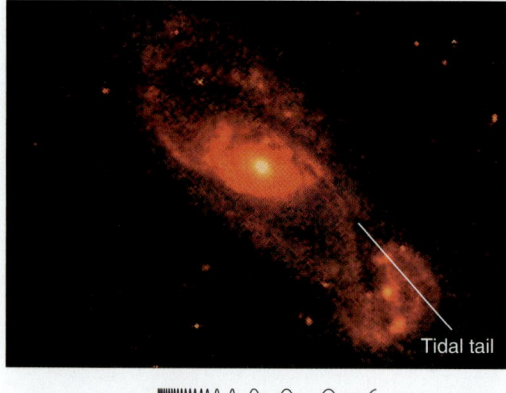

Tidal tail

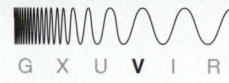

G X U V I R

Figure 14.14 The environments around quasars, which are found in the centers of galaxies, often show evidence of interactions with other galaxies, such as polar rings, disrupted structure, and tidal tails.

exposures with the cameras on the telescopes. Astronomers were greatly puzzled by the results. Rather than displaying the expected absorption lines characteristic of blue stars, the spectra showed only a single pair of emission lines, which were broad—indicating very rapid motions within these objects—and which did not seem to correspond to the lines of any known substances.

For several years, astronomers believed they had discovered a new type of star, until Maarten Schmidt realized that these broad spectral lines were the highly red-shifted lines of ordinary hydrogen. The Hubble law indicates that a high redshift implies a great distance from us. These "stars" were not stars at all. They were extraordinarily luminous objects at enormous distances. These "quasi-stellar radio sources" were dubbed **quasars**. Today we know that quasars result from extreme activity in the nuclei of galaxies, which often results from interactions with other galaxies, as seen in **Figure 14.14**. Together, quasars and their less luminous but still active cousins are called **active galactic nuclei (AGNs)**.

Quasars are phenomenally powerful, with luminosities that range between a trillion and a thousand trillion (10^{12} to 10^{15}) Suns. They are also far away—the closest really bright quasar, 3C 273, is about 750 Mpc away. Billions of galaxies are closer to us than the nearest quasar. Recall that the distance to an object also tells us the amount of time that has passed since the light from that object left its source. Because we see quasars only at great distances, we know that they are quite rare in the universe at this time but were once much more common. This discovery of the changing occurrence of quasars was one of the first pieces of evidence to demonstrate that the universe has evolved over time.

Active Galactic Nuclei Are the Size of the Solar System

The enormous radiated power and mechanical energy of active galactic nuclei are astonishing on their own, but they are made even more spectacular because all of this power emerges from a region about a light-day or so across—comparable in size to our own Solar System. How can we make such a claim? For one thing, quasars and other AGNs at the centers of galaxies appear only as **unresolved** points of light, even in our most powerful telescopes. For more evidence, we turn not to the sky but to the halftime show at a local football game.

Figure 14.15 illustrates a problem faced by every director of a marching band. When the players in a band are clustered at the center of the field, the notes sound clear and crisp in the stands. But as the band spreads out across the field, its sound begins to get mushy. This is not because the marchers are poor musicians. Rather, it is because sound travels at a finite speed. On a cold, dry day, sound travels at a speed of about 330 meters per second (m/s). At this speed, sound travels from one end of the football field to the other in approximately one-third of a second. Even if every musician on the field played a note at exactly the same instant, an observer in the stands would hear the nearby instruments first but would have to wait longer for the sound from the far end of the field to arrive.

If the band is spread from one end of the field to the other, then the beginning of a note will be smeared out over about one-third of a second, which is the time it takes for sound to travel from one end of the field to the other. If the band were spread out over two football fields, it would take about two-thirds of a second for the sound from the most distant musicians to arrive at your ear. If the marching band were spread out over a kilometer, then it would take roughly 3 seconds—the time it takes sound to travel a kilometer—to hear a crisply played note start and stop. Even without looking, it would be easy to tell whether the band was in a tight group or spread out across the field.

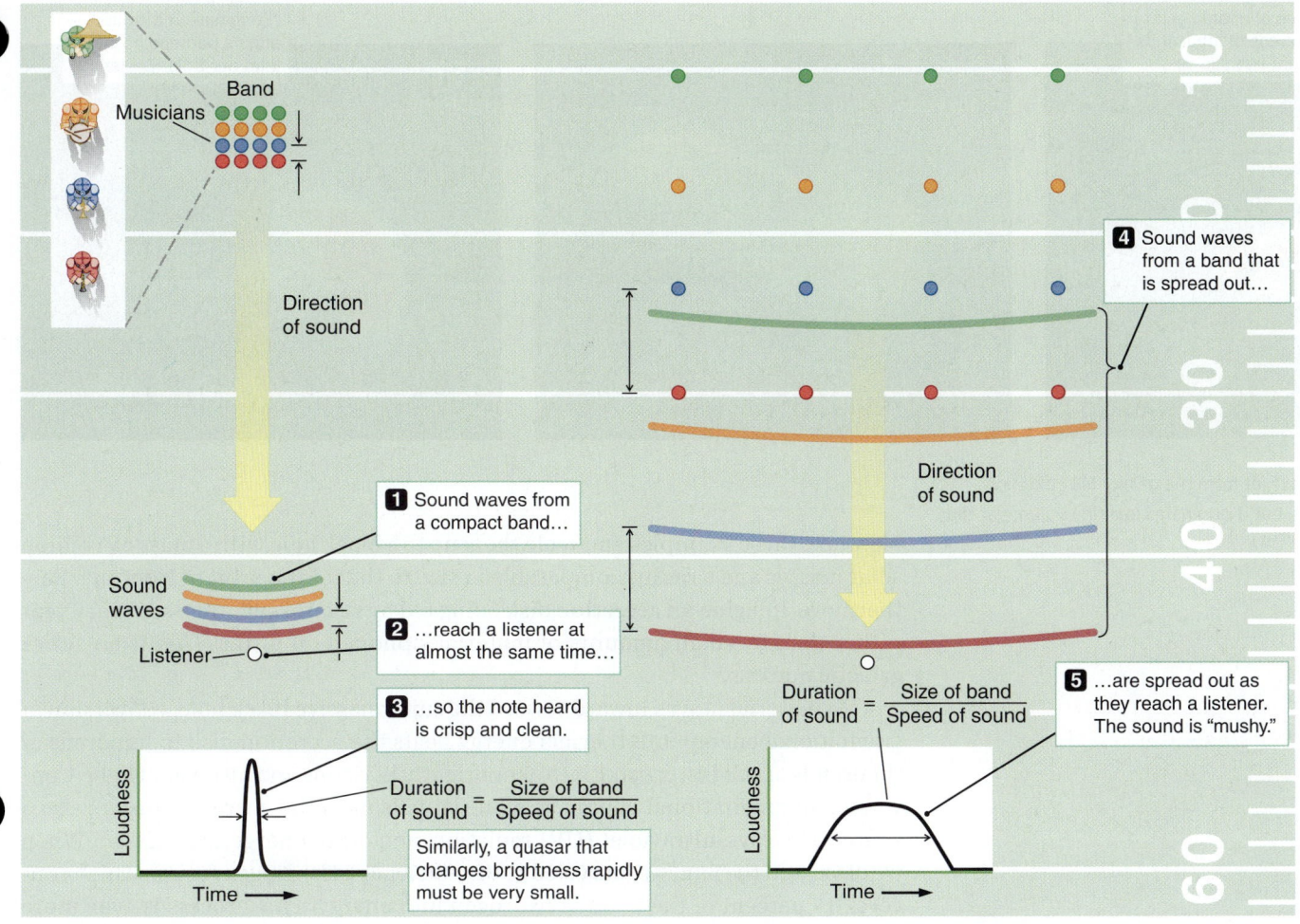

Figure 14.15 The sound produced by a marching band spread out across a field will not be crisp. Similarly, AGNs must be very small to explain their rapid variability.

This principle also applies to the light we observe from active galactic nuclei. Quasars and other AGNs change their brightness dramatically over the course of only a day or two—and in some cases as briefly as a few hours. This rapid variability sets an upper limit on the size of an AGN, just as hearing clear music from a marching band tells us that the band musicians are close together. The AGN powerhouse must therefore be no more than a light-day or so across, because if it were larger, the light would take longer to vary. From this line of reasoning, astronomers determined that an AGN has the light of up to 10,000 galaxies pouring out of a region of space that would come close to fitting within the orbit of Neptune.

Supermassive Black Holes and Accretion Disks

The observation that AGNs emit so much light from such a small region of space implies that these galaxies contain **supermassive black holes**—black holes with masses from thousands to tens of billions of solar masses. Violent accretion disks around these black holes are thought to power AGNs. You have learned about accretion disks several times in this text. Accretion disks surround young stars, providing the raw material for planetary systems. Accretion disks around white dwarfs, fueled by material torn from their bloated evolving companions, lead to novae and Type Ia supernovae. Accretion disks around neutron stars and stellar-mass black holes a few kilometers across are seen as X-ray binary stars.

Astronomy in Action: Size of Active Galactic Nuclei

▶❚❚ **AstroTour:** Active Galactic Nuclei

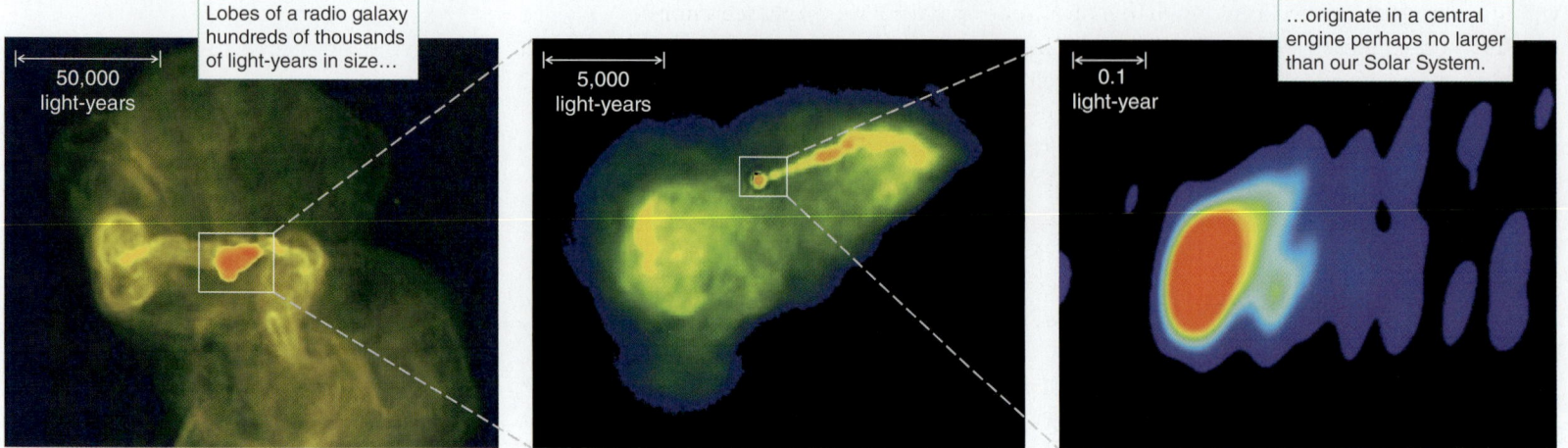

Lobes of a radio galaxy hundreds of thousands of light-years in size…

50,000 light-years

5,000 light-years

0.1 light-year

…originate in a central engine perhaps no larger than our Solar System.

Figure 14.16 The visible jet from the galaxy M87 extends more than 30,000 parsecs, but it originates in a tiny volume at the heart of the galaxy.

Now take these examples and scale them up to a black hole with a mass of a billion solar masses and a radius comparable in size to that of the orbit of Neptune. Furthermore, imagine an accretion disk being fed by several solar masses every year rather than by small amounts of material siphoned off a star. *That* is an active galactic nucleus.

As material moves inward toward the supermassive black hole, conversion of gravitational energy into thermal energy heats the accretion disk to hundreds of thousands of kelvins, causing it to glow brightly in visible and ultraviolet light. Conversion of gravitational energy as material falls onto the accretion disk is also a source of X-rays, ultraviolet (UV) radiation, and other energetic emission. When we discussed the Sun, we marveled at the efficiency of hydrogen fusion, which converts 0.7 percent of the mass of hydrogen into energy. This process is even more efficient: Approximately 15 percent of the mass of infalling material around a supermassive black hole is converted to luminous energy. The rest of that mass falls into the black hole itself, causing it to grow even more massive.

The interaction of the accretion disk with magnetic fields creates powerful jets of material that emerge perpendicular to the disk (**Figure 14.16**) and emit radio waves. Throughout, twisted magnetic fields accelerate charged particles, such as electrons and protons, to relativistic speeds, producing synchrotron radiation (see Chapter 8). Gas in the accretion disk or in nearby clouds orbiting the central black hole at high speeds produces emission lines that are smeared out by the Doppler effect into the broad lines seen in many AGN spectra. This accretion disk surrounding a supermassive black hole is the "central engine" that produces AGNs. An outer torus, or "doughnut," of dust and gas plays a somewhat different role. Located far from the inner turmoil of the accretion disk, and far larger than the central engine, some of the outer torus is ionized by UV light from the AGN. This outer torus may obscure our view of the central engine in different ways, depending on our viewing angle. Viewing similar objects from different angles gives these objects a very different appearance, accounting for the different types of AGNs.

what if . . .

What if you observed two AGNs in the process of merging, with their nuclei quite close together: Would you expect to observe gravitational waves from this system?

Normal Galaxies and Active Galactic Nuclei

The essential elements of an AGN are a central engine (an accretion disk surrounding a supermassive black hole) and a source of fuel (gas and stars flowing onto the accretion disk). Without a source of matter falling onto the black hole, an AGN would no longer be active. If we were to look at such an object, we would see a normal (not active) galaxy with a supermassive black hole sitting in its center.

Only a small percentage of present-day galaxies contain AGNs as luminous as the host galaxy. When the universe was younger, there were many more high-luminosity AGNs than exist today, because all the nearby matter hadn't fallen onto the accretion disk yet. If our understanding of AGNs is correct, then all the supermassive black holes that powered those dead AGNs should still be around. The number of AGNs in the past, combined with ideas about how long a given galaxy remains in an active phase, implies that many—perhaps even *most*—normal galaxies today contain supermassive black holes. This is a somewhat startling prediction that can be tested.

A concentration of mass at the center of a galaxy should have surrounding stars orbiting close to it. The central region of such a galaxy would be much brighter in the presence of such a mass than if stars alone were responsible for the gravitational field. Stars feeling the gravitational pull of a supermassive black hole in the center of a galaxy should also orbit at very high velocities and therefore show very large Doppler shifts. Astronomers have found evidence of this sort in every normal galaxy with a substantial bulge in which they have conducted a careful search. The masses inferred for these black holes range from 10,000 M_{Sun} to 5 billion M_{Sun}. The mass of the supermassive black hole seems to be related to the mass of the bulge of the galaxy in which it is found. Most large galaxies probably contain supermassive black holes. This prediction is confirmed by observations.

Apparently, the only difference between a normal galaxy and an active galaxy is whether the supermassive black hole at its center is being fed at the time we see that galaxy. The rarity of present-day galaxies with very luminous AGNs does not indicate which galaxies have the potential for AGN activity. Rather, it indicates which galaxy centers are being lit up at the moment. If we were to drop a large amount of gas and dust directly into the center of any large galaxy, this material would fall inward toward the central black hole, forming an accretion disk and a surrounding torus. This process would change the nucleus of this galaxy into an AGN.

Galaxies do not exist in isolation, and they often interact. Interactions between galaxies can pull galaxies into distorted shapes in which stars and gas are drawn out into sweeping arcs and tidal tails, as shown in the montage of interacting galaxies in **Figure 14.17**. Galaxies that show evidence of recent interactions with other galaxies are more likely to house AGNs in their centers, because interactions can cause gas far from the center of a galaxy to fall inward to provide fuel for an AGN. During mergers, a significant fraction of a cannibalized galaxy might wind up in the accretion disk. Hubble Space Telescope images of quasars, like those in Figure 14.14, often show that quasar host galaxies are tidally distorted or are surrounded by other visible matter that is probably still falling into the galaxies. The most violent forms of AGN activity were common in the early universe because that was when galaxies were forming, and matter was drawn in by the gravity of newly formed galaxies.

Figure 14.17 ★ WHAT AN ASTRONOMER SEES

This gallery of interacting galaxies shows the distorted shapes that sometimes form, as the gravitational interaction warps and twists the galaxies. Interactions and mergers often disrupt the dust and gas in a galaxy, triggering many stars to form at once. An astronomer will notice the color differences, from one region to another. Young stars are hot and blue, and so very blue regions often indicate recent star formation. An astronomer will look for dark lanes of dust and gas, as well as reddish glow from old stars. Tidal tails of stars, dust and gas are often left behind as the galaxies accelerate toward each other, showing the direction each galaxy came from. An astronomer looking at images like these sees evidence of a dynamic, ever-changing universe.

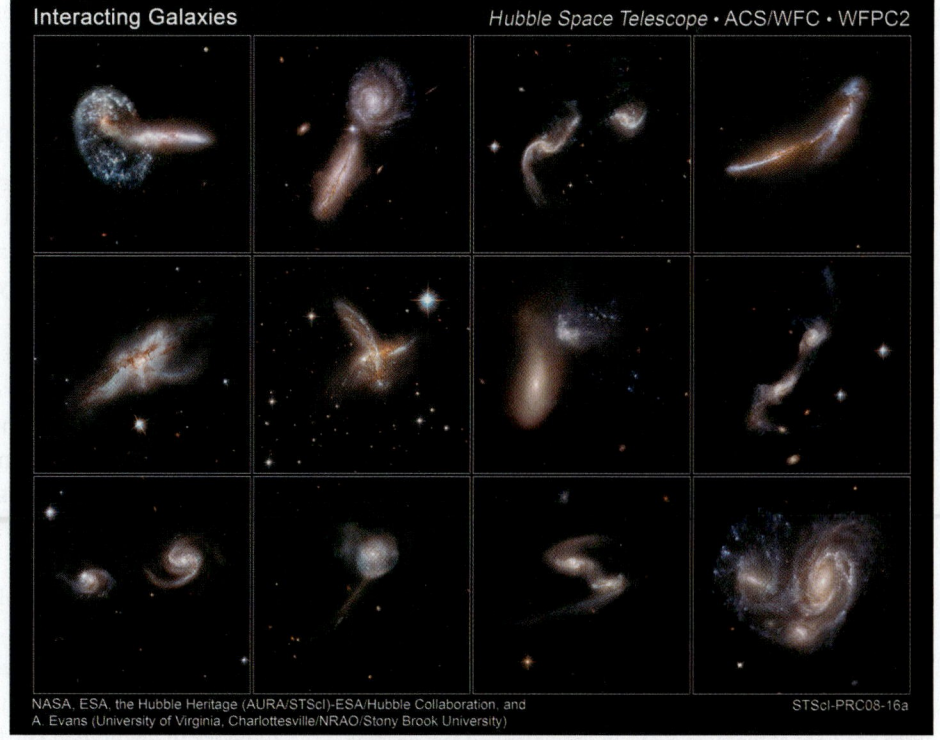

Interacting Galaxies — *Hubble Space Telescope* • ACS/WFC • WFPC2

NASA, ESA, the Hubble Heritage (AURA/STScI)-ESA/Hubble Collaboration, and A. Evans (University of Virginia, Charlottesville/NRAO/Stony Brook University)

STScI-PRC08-16a

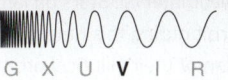

G X U V I R

(a)

(b)

(c)

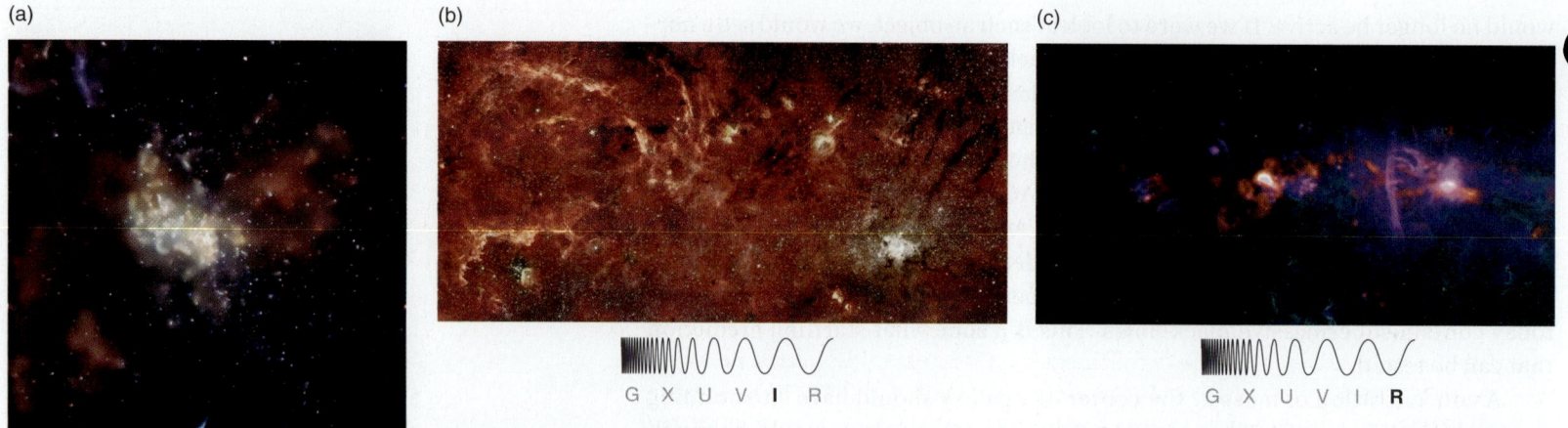

G X U V I **R**

G X U V I **R**

G X U V I **R**

Figure 14.18 (a) An X-ray view of the Milky Way's central region shows the active source, Sgr A*, as the brightest spot at the middle of the image. Lobes of superheated gas (shown in red) are evidence of recent explosions near Sgr A*. (b) This wide infrared view (890 light-years across) shows hundreds of thousands of stars. The bright white spot at the lower right marks the galaxy's center, home of a supermassive black hole. (c) Radio observations reveal wispy molecular clouds (purple) glowing from strong synchrotron emission. Cold dust (20–30 K) associated with molecular clouds is shown in orange. Diffuse infrared emission appears in blue-green. The galactic center (Sgr A*) lies within the bright area to the right of center.

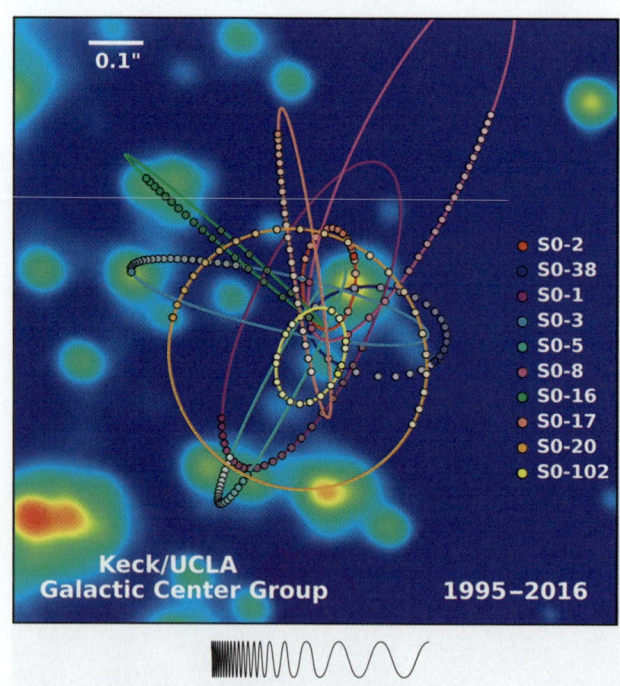

0.1"

- ● S0-2
- ○ S0-38
- ● S0-1
- ● S0-3
- ● S0-5
- ● S0-8
- ● S0-16
- ● S0-17
- ● S0-20
- ● S0-102

Keck/UCLA Galactic Center Group 1995–2016

G X U V I **R**

Figure 14.19 Orbits of several stars (labeled with SO and a number) within 0.1 light-year (about 6,000 astronomical units [AU]) of the Milky Way's center. The scale bar shows an angular size of 0.1 arcseconds (0.1") on the image; this entire image spans about 1 arcsecond. The Keplerian motions of these stars reveal the presence of a 4-million-M_{Sun} supermassive black hole at the galaxy's center. That black hole lies within the ellipses of all of these orbits, but emits no light of its own. Colored dots show the measured positions of the stars over a 21-year interval. Stars on large orbits have not completed an entire orbit during this time, so they only have positions marked on a portion of their elliptical orbit.

Interactions and mergers must have been much more prevalent when the universe was younger, which explains the larger number of AGNs that existed in the past. However, this process is still at work today.

Our understanding of AGNs is far from complete. For example, we do not know why one quasar can be a powerful radio source while another, identical in all other respects, emits no radio waves that we can detect, even with the most sensitive radio telescopes. Also, we still cannot predict how long an outburst of AGN activity will last or how often galaxies will undergo episodes of AGN activity. Any large galaxy might be only a chance encounter away from containing an AGN.

The Milky Way Contains a Supermassive Black Hole

There is evidence for a supermassive black hole at the center of our own galaxy. Dense clouds of dust and gas hide, though, our visible-light view of the Milky Way's center. Fortunately, infrared, radio, and some X-ray radiation passes through dust (**Figure 14.18**). The X-ray view (Figure 14.18a) shows the location of a strong radio source called Sagittarius A* (abbreviated Sgr A*), which astronomers believe lies at the exact center of the Milky Way. The infrared image (Figure 14.18b) cuts through the dust to reveal the crowded, dense core of the galaxy containing hundreds of thousands of stars. Radio observations (Figure 14.18c) reveal synchrotron emission from wisps and loops of material distributed throughout the region. This is similar to the synchrotron emission seen from AGNs but at far lower intensity.

The motions of stars closest to the Sgr A* source suggest a central mass very much greater than that of the few hundred stars orbiting there. Stars less than 0.1 light-year from the galaxy's center follow Kepler's laws. The closest stars studied are only about 0.01 light-year from the center—so close that their orbital periods are only about a dozen years. The positions of these stars change noticeably over time, and we can see them speed up as they whip around what can only be a supermassive black hole at the focus of their elliptical orbits (**Figure 14.19**). Using Kepler's third law, we estimate that the black hole at the center of our own galaxy is a relative lightweight, having a mass of "only" about 4 million M_{Sun}.

Clouds of interstellar gas at the center of the Milky Way are heated to millions of degrees by shock waves from supernova explosions and colliding stellar winds blown outward by young massive stars. Superheated gas produces X-rays, and the Chandra X-ray Observatory has detected more than 9,000 X-ray sources within the central region of the galaxy. These include frequent, short-lived X-ray flares near Sgr A*, which provide direct evidence that matter falling toward the supermassive black hole fuels the energetic activity at the galaxy's center.

The Fermi Gamma-ray Space Telescope has observed bubbles that extend about 25,000 light-years above and below the plane of the Milky Way. The bubbles may have formed after a burst of star formation a few million years ago produced massive star clusters near the center of the galaxy. If some of the gas formed stars and about 2,000 M_{Sun} of material fell into the supermassive black hole, enough energy could have been released to power the bubbles. More recently, faint gamma-ray signals were observed that look like jets coming from the center, within the bubbles, shown in **Figure 14.20**. If these jets are originating from material falling into the supermassive black hole, activity might be even more recent, maybe 20,000 years ago, as observed from the distance of Earth. Some astronomers predict that gas clouds are heading toward the center and will soon be accreted by the black hole. Nevertheless, at this time, this activity is not as intense as that seen in active galaxies with central black holes. The inner Milky Way is a reminder that our galaxy was almost certainly "active" in the past and could become active once again.

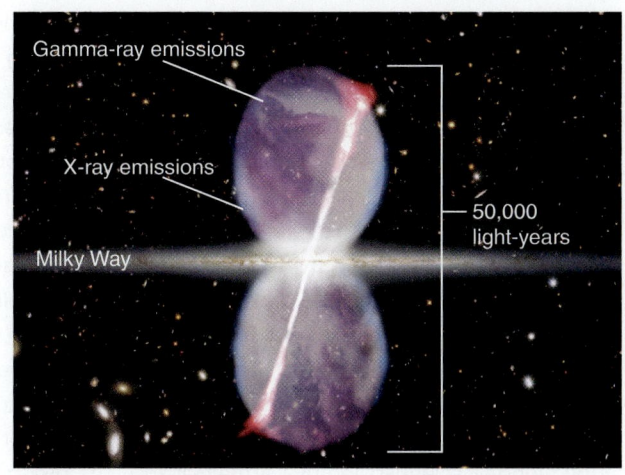

Figure 14.20 The Fermi Gamma-ray Space Telescope observed gamma-ray bubbles (purple) extending about 25,000 light-years above and below the galactic plane. In this artist's conceptual view from outside of the galaxy, the gamma-ray jets (pink) are tilted with respect to the bubbles, which might imply that the accretion disk around the black hole is tilted as well.

CHECK YOUR UNDERSTANDING 14.4

Evidence of a supermassive black hole at the center of the Milky Way comes from (choose all that apply)

a. direct observations of stars that orbit it.

b. X-rays from material that is falling in.

c. strong radio emission from the region of the accretion disk.

d. the abundance of dust in the galaxy.

---

reading Astronomy News

The Milky Way Ate Another Galaxy 10 Billion Years Ago—And We Now Know More about This Cosmic Cannibalism Event

by Hannah Osborne, *Newsweek*

Interactions between galaxies, even galaxy mergers, are common. A single galaxy like the Milky Way may experience more than one merger event, each of which may trigger an episode of star formation.

Ten billion years ago, the Milky Way gobbled up another galaxy, creating the cosmic structure our solar system sits in today. By analyzing the age distribution of stars in the Milky Way's inner halo, scientists have now been able to create a better picture of our galaxy's formation.

They found that many of the stars in the inner halo are up to 10 billion years old, pro-viding more evidence of the timing of the merger—and helping identify the Milky Way's original stars that were there at its onset, 13.5 billion years ago.

Galaxies are not in fixed positions in space. They move around over time, occasionally smashing into one another.

Evidence of a massive merger between the Milky Way and the Gaia Enceladus galaxy came in 2018 when scientists used data from the European Space Agency's Gaia satellite to show a vast number of stars appeared to be out of place. In a letter published by *Nature*, the team said the inner halo of the Milky Way is "dominated by debris" from another galaxy. This galaxy was found to be about a quarter of the size of the Milky Way.

However, when this huge collision took place has been debated.

In a study published in *Nature Astronomy* on Monday, scientists led by Carme Gallart, from Spain's Institute of Astrophysics of the Canary Islands, looked at the stars in the Milky Way's inner halo and created cosmological simulations to find the age limit for when the merger took place.

Their findings showed that most of the stars in the inner halo are up to 10 billion years old—suggesting this is the point when the merger took place. By being able to put a constraint on the timing of the merger, the team was also able to identify the Milky Way's original stars.

Concluding, they say the simulations provide a "clear picture of the formation" of the Milky Way: "In this picture, a primitive Milky Way had been forming stars over a period of some [three billion years] when a smaller galaxy, which had been forming stars on a simi-

lar timescale but was less chemically enriched owing to its lower mass, was accreted into it."

They said heat from the merger helped create the halo-like configuration we see today, and that a large supply of gas "ensured the maintenance of a disk-like configuration, with the thick disk continuing to form stars at a substantial rate."

Gallart told *Newsweek*: "Finding the date of the merger helps to understand what are its effects in the Milky Way. For example, in our study we see that, after the merger, the rate at which stars form in the Milky Way disk go up, and thus we can infer that this merger contributed to star formation in the disk of the Milky Way."

This is not the only time the Milky Way has merged with other galaxies. It is thought that throughout its history it has consumed many other, smaller galaxies.

Effects of mergers are not really noticeable on a small scale, Gallart said. "The distance between stars in a galaxy are so huge . . . that the two galaxies intermix, change their global shape, more star formation may happen in one and maybe the other—the small one—stops forming stars."

"But the individual stars in each galaxy don't collide, don't really notice the force of the event in a way that affect their individual evolution, or the evolution of the planetary systems that may be attached to them."

At the moment, the Milky Way is colliding with the Canis Major [d]warf galaxy. Our solar system is believed to have entered the Milky Way during a merger with the Sagittarius dwarf galaxy.

Eventually, the Milky Way will merge with our largest, closest neighbor, Andromeda. When this happens, around 4.5 billion years from now, the Milky Way as we know it will cease to exist.

EVALUATING THE NEWS

1. The article states that "many of the stars in the inner halo are up to 10 billion years old." Why does this piece of information determine the time of the merger?

2. In the article, Gallart states that, after the collision, the rate of star formation went up. How might we know that? In what part of the colliding galaxies is this star formation likely to be taking place?

3. How might heat help to create a halo configuration of the Milky Way?

4. Did this merger occur before or after the formation of the Solar System? Did the merger have any effect on the Sun and its planets?

5. Why are stars in a galaxy not affected by a galaxy merger?

Source: **Hannah Osborne**, "The Milky Way Ate Another Galaxy 10 Billion Years Ago—And We Now Know More about This Cosmic Cannibalism Event," from *Newsweek*, July 22, 2019. Reprinted with permission of Newseek.

SUMMARY

Galaxies are classified on the basis of their shapes and the types of orbits of their stars. We live in the disk of the Milky Way, a barred spiral galaxy that is 100,000 light-years across. The Sun is about 27,000 light-years from the center of our galaxy. Variable stars of known luminosity yield the distances to globular clusters, which enable us to measure the size of the Milky Way's extended halo. These methods of measuring distance form the basis of a "distance ladder," which is used by astronomers to determine the distances to objects throughout the universe. The observation of galaxies led to a revolution in our thinking about the universe by showing that the universe is expanding. Most galaxies have a supermassive black hole at the center, which may become an active galactic nucleus (AGN) if gas accretes onto it. At the center of the Milky Way is a massive black hole, which produces rapid orbital velocities in the nearby stars.

(1) Spiral galaxies are distinguished by their flat disk and spiral arms. The stars in this disk all orbit the center of the galaxy in the same direction. Elliptical galaxies are roughly egg-shaped, and the stars orbit in all directions. Irregular galaxies are galaxies that fit neither of these classifications, usually because they are interacting with another galaxy.

(2) The distances to globular clusters can be found from the luminosity of variable stars within them. Because these globular clusters are evenly distributed around the center of the Milky Way, the center of the distribution is located at the galaxy's center. Other standard candles, such as supernovae, are used to determine distances even farther out in the universe. Each method for measuring distance builds on the methods for measuring smaller distances, forming a "distance ladder" to the most distant places in the universe.

3 Combining the distances to galaxies with measurements of their velocities proved especially useful, leading to the discovery that the universe is expanding. This discovery allows astronomers to find the distances to very distant galaxies by measuring their recession velocities and then applying Hubble's law.

4 Observations of distant galaxies, such as quasars, reveal extremely luminous, compact sources near the centers of galaxies. These active galactic nuclei are best explained as supermassive black holes surrounded by an accretion disk and a torus of dust and gas. Stars orbiting the centers of nearby galaxies have such high velocities that there must also be a supermassive black hole at the center of each of these galaxies. Most (perhaps all) large galaxies contain a supermassive black hole at the center.

QUESTIONS AND PROBLEMS

TESTING YOUR UNDERSTANDING

1. **T/F:** Galaxies are sometimes difficult to classify because their orientation affects their appearance.

2. **T/F:** Elliptical galaxies are elongated along only one axis.

3. **T/F:** The orbits of stars in the bulge of a spiral galaxy are disordered.

4. **T/F:** The expansion of space causes a change in the spectrum of a galaxy, similar to the change that the Doppler shift causes in the spectrum of a star.

5. **T/F:** Active galactic nuclei surround some supermassive black holes.

6. The orbits of stars in the bulges of spiral galaxies most closely resemble the orbits of
   a. planets in the Solar System.
   b. stars in the disk of a spiral galaxy.
   c. stars in an elliptical galaxy.
   d. objects in the Kuiper Belt.

7. Large galaxies have sizes in the range of
   a. hundreds of parsecs.
   b. tens of thousands of parsecs.
   c. hundreds of thousands of parsecs.
   d. millions of parsecs.

8. The orbits of stars in elliptical galaxies
   a. are all in the same direction.
   b. have random orientations.
   c. are all circular.
   d. have many different objects at the focus.

9. Why are globular clusters *uniquely* useful in determining the size of the galaxy?
   a. They are large.
   b. They are bright.
   c. They are evenly distributed around the center.
   d. They consist mostly of old stars and have been there a long time.

10. In order for a variable star to be useful as a standard candle, its luminosity must be related to its
    a. period of variation.   b. mass.
    c. temperature.   d. radius.

11. The distances to galaxies used to establish Hubble's law are found from
    a. radar.
    b. parallax.
    c. spectroscopic parallax.
    d. standard candles, such as supernovae.

12. The Hubble constant is found from
    a. the slope of the line fit to the data in Hubble's law.
    b. the y-intercept of the line fit to the data in Hubble's law.
    c. the spread in the data in Hubble's law.
    d. the inverse of the slope of the line fit to the data in Hubble's law.

13. If the universe were not expanding, the relationship between the velocity of a galaxy and its distance (as in the Hubble law plot) would
    a. be horizontal.
    b. follow a downward trend.
    c. first go up and then flatten out.
    d. start out flat and then fall.

14. In every direction that astronomers look, they see the same Hubble law, with the same slope. This is an example of
    a. isotropy.   b. homogeneity.
    c. hydrostatic equilibrium.   d. energy balance.

15. Rank the following galaxies in order of increasing distance from the Milky Way.
    a. a galaxy with a recession velocity of $+200$ km/s
    b. a galaxy with a recession velocity of $+400$ km/s
    c. a galaxy with a recession velocity of -50 km/s
    d. a galaxy with a recession velocity of $+500$ km/s

16. Supermassive black holes
    a. are extremely rare. There are only a handful in the universe.
    b. are completely hypothetical.
    c. occur in most, perhaps all, large galaxies.
    d. occur only in the space between galaxies.

17. Supermassive black holes
    a. gradually consume their host galaxies.
    b. are "fed" by disturbed gas when galaxies interact.
    c. are a recent development in the universe's history.
    d. are an as-yet-untested hypothesis.

18. The light that comes to us from active galactic nuclei comes from a volume about the size of
    a. Earth.
    b. the Solar System.
    c. a globular cluster.
    d. the bulge of the Milky Way.

19. Emission lines from a spinning accretion disk around a black hole in an active galactic nucleus
    a. are always redshifted.
    b. are always blueshifted.
    c. are both redshifted and blueshifted.
    d. depend on the viewing angle.

20. The jets of active galactic nuclei have strong radio emissions because
    a. they are highly redshifted.
    b. they are cold.
    c. they have strong magnetic fields.
    d. they are very compact.

THINKING ABOUT THE CONCEPTS

21. Why is it so difficult for astronomers to get an overall picture of the central region of our galaxy?

22. What do astronomers mean by *standard candle*? Why are stars in globular clusters so useful as standard candles in determining the size of the galaxy?

23. Describe the distribution of globular clusters within the galaxy, and explain what that implies about the size of the galaxy and our distance from its center.

24. How do we know that the stars in globular clusters are among the oldest stars in our galaxy?

25. What are the main differences in the gas out of which the stars in globular clusters and stars in the disk of the galaxy formed?

26. Sketch the galaxy shown in Figure 14.4b. Label the disk, the bulge, and the halo. ⊛

27. How does gas temperature differ between elliptical and spiral galaxies?

28. The nearest bright quasar is about 750 Mpc away. Why do we not see any that are closer?

29. Contrast the size of a typical AGN with the size of our own Solar System. How do we know how big an AGN is?

30. Describe what must be happening at the centers of galaxies that contain AGNs.

31. Why must we use X-ray, infrared, and radio observations to probe the center of our galaxy?

32. Why is the Milky Way Galaxy not expanding along with the rest of the universe?

33. Suppose that the universe is not expanding but is contracting. Sketch what Figure 14.13b should look like in that case. Now, suppose that it was doing neither but is static. Sketch Figure 14.13b for the case of a static universe. ⊛

34. Explain the evidence for a supermassive black hole at the center of the Milky Way. How does the mass of the supermassive black hole at the center of our galaxy compare with that found in most other spiral galaxies?

35. ★ **WHAT AN ASTRONOMER SEES** Figure 14.17 shows 12 different pairs of interacting galaxies. How many of these interactions are between two spiral galaxies? How many are between two ellipticals? How many are between an elliptical and a spiral? ⊛

APPLYING THE CONCEPTS

36. Like the stars near the center of the Milky Way, the Sun also orbits around the Milky Way. The Sun completes one orbit in approximately 230 million years. How many orbits has the Sun completed since its formation 4.6 billion years ago?
    a. Make a prediction: Consider roughly how much larger 4.6 billion is than 230 million. Do you expect the Sun to have made more or less than one orbit? Do you expect the Sun to have made tens, hundreds, or thousands of orbits?
    b. Calculate the number of orbits by comparing the total time since the Sun formed with the time for one orbit.
    c. Compare your answer to your prediction.

37. Suppose that the observed wavelength of a hydrogen line in a galaxy is 122.5 nm. The rest wavelength of this hydrogen line is 122 nm. How fast is this galaxy traveling?
    a. Make a prediction: In Working It Out 14.1, the hydrogen line was shifted by 2 nm. Should your answer for this problem be faster or slower than the speed of the galaxy in Working It Out 14.1?
    b. Calculate the recession velocity of this galaxy.
    c. Check your work by comparing your answer to your prediction.

38. Suppose that a galaxy has a recession velocity of 2,000 km/s. You want to know the distance to this galaxy.
    a. Make a prediction: How should the distance to this galaxy compare to the distance to the galaxy in Working It Out 14.1?
    b. Calculate the distance to this galaxy.
    c. Check your work by comparing your answer to the distance found in Working It Out 14.1.

39. Suppose that a galaxy has a recession velocity of 2,000 km/s. What will be the observed wavelength of the hydrogen line that has a rest wavelength of 122 nm?
    a. Make a prediction: Should this line have shifted more or less than the 2 nm that the line shifted in Working It Out 14.1?
    b. Calculate the observed wavelength.
    c. Check your work by comparing your answer to the observed wavelength in Working It Out 14.1.

40. Suppose that a galaxy is located 140 Mpc from the Milky Way. You want to know the recession velocity of this galaxy.
    a. Make a prediction: Is this galaxy nearer or farther than the galaxy in Working It Out 14.1? Should the recession velocity be larger or smaller than the recession velocity of that galaxy? About how much larger or smaller?
    b. Calculate the recession velocity of this galaxy.
    c. Check your work by comparing your answer to your prediction.

exploration Galaxy Classification

Galaxy classification sounds simple, but it can become complicated when you actually attempt it. **Figure 14.21**, taken by the Hubble Space Telescope, shows a small portion of the Coma Cluster of galaxies. The Coma Cluster contains thousands of galaxies, each containing billions of stars. Some of the objects in this image (the ones with a bright cross) are foreground stars in the Milky Way. Some of the galaxies in this image are far behind the Coma Cluster. Working with a partner, you will classify the 20 or so brightest galaxies in this cluster.

First, make a map by laying a piece of paper over the image and numbering the 20 or so brightest (or largest) galaxies in the image. Copy this map so that you and your partner each have a list of the same galaxies.

Separately, classify the galaxies (label them galaxy 1, galaxy 2, and so forth). If it is a spiral galaxy, what is its subtype: a, b, or c? If it is an elliptical galaxy, how elliptical is it? Make a table that contains the galaxy number, the type you have assigned, and any comments that will help you remember why you made that choice.

When you are done classifying, compare your list with your partner's. Now comes the fun part! Argue about the classifications until you agree—or until you agree to disagree.

If you find this activity interesting and rewarding, astronomers can use your help: Go to https://www.zooniverse.org/projects/zookeeper/galaxy-zoo/ to get involved in a "citizen science" project to classify galaxies, some of which have never been viewed before by human eyes.

1. Which galaxy type was easiest to classify?

   _____

2. Which galaxy type was hardest to classify?

   _____

3. What makes it hard to classify some of the galaxies?

   _____

   _____

4. Which galaxy type did you and your partner agree about most often?

   _____

5. Which galaxy type did you and your partner disagree about most often?

   _____

6. Discuss with your partner ways to refine your classification technique. How could you eliminate some of the disagreement in your classifications?

   _____

   _____

Figure 14.21 This Hubble Space Telescope image of the Coma Cluster shows a diversity of shapes.

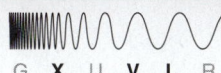

G X U V I R

Dark Matter and the Milky Way

The stars in a spiral galaxy are not uniformly distributed. As a barred spiral galaxy, this is true of our Milky Way. Here in Chapter 15, you will learn why this is so. Before you begin this experiment, predict whether you will see more stars in the plane of the Milky Way or far from it. Then, find a dark sky that is clear in all directions. Bring a star chart for the date and time of your observation, as well as a paper towel tube. Point your paper towel tube at the plane of the Milky Way, and count the stars that you see through the tube. Move the tube so that it points as far from the plane of the Milky Way as possible (90°, if you can!). Count the stars again. Halfway between these two points, make a third star count.

EXPERIMENT SETUP

Find as dark a site as possible, where you can see lots of stars. Bring a star chart for the date and time of your observation, as well as a paper towel tube.

1 Point your paper towel tube at the plane of the Milky Way, and count the stars that you see through the tube.

2 Move the tube so that it points as far from the plane of the Milky Way as possible (90°, if you can!). Count the stars again.

3 Halfway between these two points, make a third star count.

PREDICTION

I predict that I will see more stars ☐ close to ☐ far from the plane of the Milky Way.

SKETCH OF RESULTS (in progress)

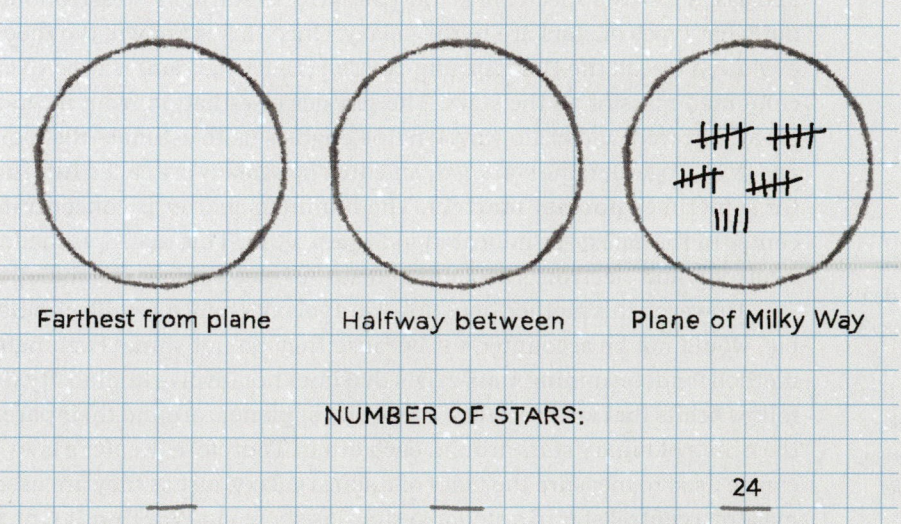

Farthest from plane   Halfway between   Plane of Milky Way

NUMBER OF STARS:

_____   _____   24

15

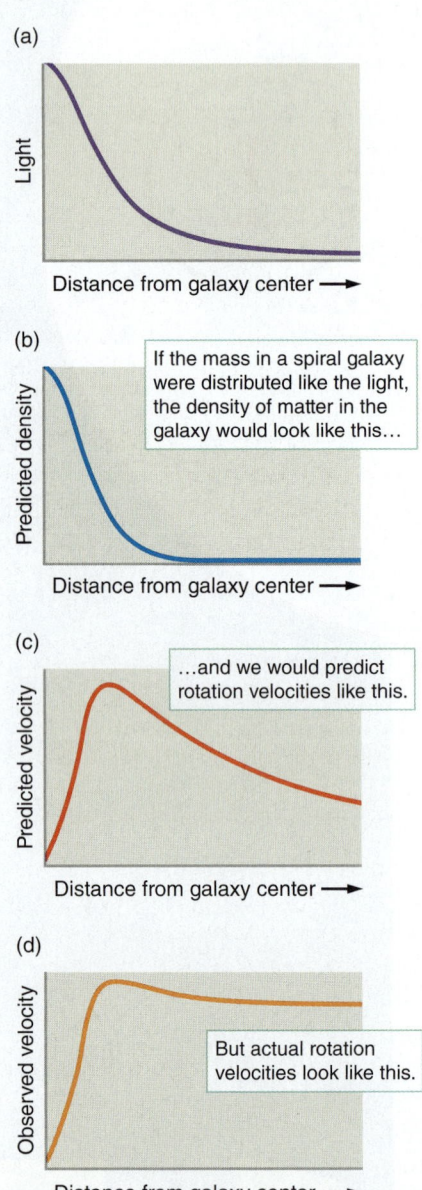

(a)

Light

Distance from galaxy center ⟶

(b)

Predicted density

If the mass in a spiral galaxy were distributed like the light, the density of matter in the galaxy would look like this…

Distance from galaxy center ⟶

(c)

Predicted velocity

…and we would predict rotation velocities like this.

Distance from galaxy center ⟶

(d)

Observed velocity

But actual rotation velocities look like this.

Distance from galaxy center ⟶

Figure 15.1 (a) The profile of visible light in a typical spiral galaxy drops off with distance from the center. (b) The mass density of stars and gas located at a given distance from the galaxy's center resembles the light profile. (c) If stars and gas accounted for the entire mass of the galaxy, then the galaxy's predicted rotation curve would be as shown here. This rotation curve is similar to the one for the Solar system; it is sometimes called "Keplerian." (d) The rotation curves actually observed for galaxies, however, are flatter—more like the one shown here.

W e live in a universe full of galaxies of many sizes and types, visible in our most powerful telescopes all the way to the edge of the observable universe. Yet we cannot see much of this universe in the night sky. Galaxies are made primarily of dark matter—a form of matter that interacts gravitationally with ordinary matter but does not absorb or emit light, so it is invisible. Our own galaxy is mostly made of dark matter, and studying dark matter in the Milky Way has helped us better understand dark matter elsewhere. Connecting studies of the Milky Way to the study of other galaxies has also helped us understand how stars form and evolve within galaxies. The Milky Way has also provided clues to how galaxies form in the universe. Here in Chapter 15, you will learn about dark matter and about the Milky Way.

LEARNING GOALS

① Describe the evidence for the existence of dark matter in galaxies.

② Explain why the arms of spiral galaxies are sites of star formation.

③ Explain how differences in the age and chemical composition of groups of stars tell us about the history of star formation and the evolution of chemical composition in the Milky Way.

④ List the clues that the Milky Way provides about galaxy formation.

15.1 Galaxies Are Mostly Dark Matter

Astronomers measure the mass of galaxies in two different ways. Curiously, the results of these measurements did not agree for any galaxy, and the discrepancy between these two sets of findings was the first convincing evidence for dark matter.

Finding the Mass of a Galaxy

One way to find the mass of a galaxy is to add up the mass of the visible stars, dust, and gas. A galaxy's spectrum is due primarily to starlight, so astronomers can find out what types of stars are in the galaxy. Once this is known, the theory of stellar evolution (recall the Hertzsprung-Russell [H-R] diagram) is used to calculate the combined mass of all the stars. The physics of radiation from interstellar gas at X-ray, infrared, and radio wavelengths enables us to estimate the mass of the gas and dust. Together, the stars, gas, and dust in a galaxy are called **luminous matter** (or sometimes **normal matter**). The luminous matter is concentrated near the center of the galaxy, as indicated in **Figure 15.1a**. This matter is made of protons, neutrons, and electrons that emit and absorb electromagnetic radiation.

However, this method does not give a galaxy's total mass. Black holes, for example, would not be accounted for because they do not shine. Fortunately, another method for determining mass exists that does not involve luminosity. Stars in disks follow orbits that are much like the orbits of planets around their parent stars and the orbits of binary stars around each other. Therefore, Kepler's laws (Chapter 3) can be used to measure the mass of a spiral galaxy, just as they are used to find the mass of a central star in a stellar system, as shown in **Working It Out 15.1**.

working it out 15.1

Finding the Mass of a Galaxy

The Sun is located about 27,000 light-years (1.7×10^9 astronomical units, AU) from the center of the Milky Way. By dividing the circumference of the Sun's orbit by its velocity around the center, you could find that the Sun takes 230 million years to orbit the center of the Milky Way. From this information, we can find the mass of the Milky Way that lies interior to the orbit of the Sun. To find the mass of a binary system, we use Newton's version of Kepler's third law (see Working It Out 3.1), which relates the total mass of an orbiting system ($m_1 + m_2$), in units of solar masses, to the orbital period (P) in years and the average radius of the orbit (A) in astronomical units:

$$m_1 + m_2 = \frac{A_{\mathrm{AU}}{}^3}{P_{\mathrm{years}}{}^2}$$

In this case, the mass of the Sun is tiny compared to the mass of the galaxy interior to the Sun's orbit, so $m_1 + m_2$ is essentially equal to m_1, the mass of

the Milky Way Galaxy interior to the Sun's orbit. Plugging in the values for the orbit of the Sun gives

$$m_1 = \frac{A_{\mathrm{AU}}{}^3}{P_{\mathrm{years}}{}^2}$$

$$m_1 = \frac{(1.7 \times 10^9 \, \mathrm{AU})^3}{(230 \times 10^6 \, \mathrm{years})^2}$$

$$m_1 = 9.3 \times 10^{10} \, M_{\mathrm{Sun}}$$

The mass of the Milky Way, interior to the Sun's orbit, is 93 billion solar masses. The Sun is not at the edge of the disk of the Milky Way but rather is located partway out. Some mass is located farther from the center of the Milky Way than the Sun, so the total mass of the Milky Way is larger than the number we've just calculated. Still, this calculation indicates the vast amount of mass contained within the Milky Way.

Observations of Dark Matter

Astronomers would also like to know how the mass is distributed in a galaxy. They hypothesized that the mass is distributed in the same way as the light. Because the light of all galaxies, including spiral galaxies, is highly concentrated toward the center, they proposed that nearly all the mass of a spiral galaxy was contained in its center (Figure 15.1b). This situation is much like the Solar System, where nearly all the mass is in the Sun, at the Solar System's center. This hypothesis predicts fast orbital velocities near the center of the spiral galaxy and slower orbital velocities farther out (Figure 15.1c). The disk is often said to "rotate," even though the disk is not a solid object like a wheel, so these orbital velocities are often called "rotational velocities" of the disk at that location. The stars all travel in the same direction around the center of the galaxy, like cars on a racetrack. A graph showing how the rotational velocity of a galaxy's disk changes with distance from the galaxy's center is called a **rotation curve**.

To test this prediction, astronomers used the Doppler effect to measure orbital motions of stars, gas, or dust at various distances from the galaxy's center. From these data, they graphed the orbital velocity versus distance, as shown in Figure 15.1d.

Vera Rubin (1928–2016) pioneered this work on galaxy rotation rates. She discovered that, contrary to earlier predictions (Figure 15.1c), the rotation velocities of spiral galaxies remain about the same from about halfway to the edge of the disk all the way out to the most distant measured parts of the galaxies (Figure 15.1d). Observations of clouds of neutral hydrogen show that the rotation curves remain flat even well outside the extent of the visible disks. The hypothesis that mass and light are distributed in the same way is completely *wrong*.

(a)

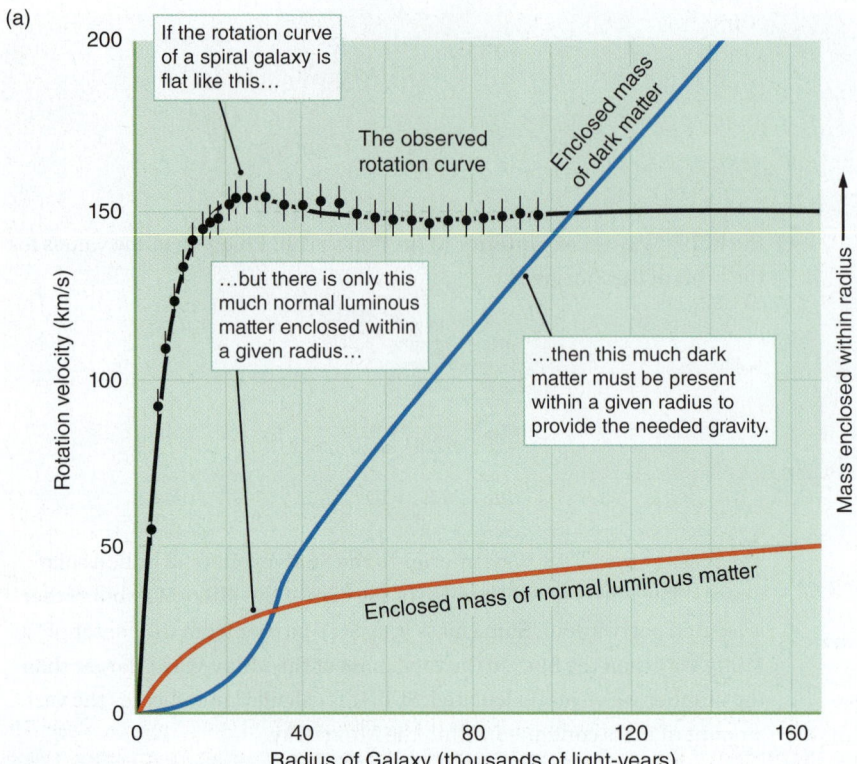

(b)

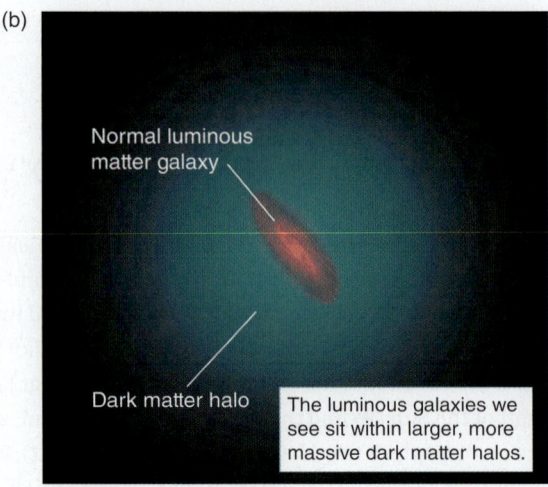

Figure 15.2 (a) We can use the flat rotation curve of the spiral galaxy NGC 3198 to determine the total mass within a given radius. Notice that the normal mass that can be accounted for by stars and gas is only part of the needed mass. Extra dark matter is needed to explain the rotation curve. (b) In addition to the matter we can see, galaxies must be surrounded by halos containing a large amount of dark matter.

The logical next step is to turn the question around and ask, "What mass distribution would cause this unexpected rotation curve?" **Figure 15.2a** shows the result of such a calculation for the spiral galaxy NGC 3198. The black line shows the measured speed of rotation *at* a particular radius (velocity increases vertically on the graph). The red line shows how much luminous mass is observed to be inside a particular radius (mass increases vertically on the graph). To produce a rotation curve like the one shown in black, this galaxy must have a second component consisting of matter that does not show up in our census of stars, gas, and dust. This material, which does not interact with light and therefore reveals itself only by the influence of its gravity, is called **dark matter**. The blue line shows how much dark matter must be inside a particular radius in order to provide enough mass to make the galaxy rotate as it does.

The rotation curves of the inner parts of spiral galaxies match predictions that are based on their luminous matter, indicating that the inner parts of spiral galaxies are mostly luminous matter. However, rotation curves in the outer parts of the galaxies do not match the predictions that are based on luminous matter, indicating that the outer parts of spiral galaxies are mostly dark matter. Astronomers currently estimate that as much as 95 percent of the total mass in some spiral galaxies consists of a greatly extended **dark matter halo** (Figure 15.2b)—far larger than the visible spiral portion of the galaxy located at its center. This is a startling statement. A spiral galaxy shines only from the inner part of a much larger distribution of mass that is dominated by some type of matter that is invisible to all wavelengths of light.

What about elliptical galaxies? Again, astronomers compared the luminous mass with the gravitational mass. Recall from Figure 14.6 that stars in an elliptical galaxy orbit the center in all directions. Therefore, Kepler's laws cannot be used to measure the gravitational mass in an elliptical galaxy, because there is no organized movement of stars. Instead, astronomers determine the gravitational mass statistically from the variety of speeds of individual stars. Like spiral galaxies, elliptical galaxies are mostly made of dark matter. Another avenue of exploration confirms this result. An elliptical galaxy's ability to hold on to its hot gas depends on its mass: If the galaxy is not massive enough, the hot atoms and molecules will escape into intergalactic space. To find the mass of an elliptical galaxy, first we find the total amount of luminous hot gas from X-ray images, such as the blue and purple halo seen in **Figure 15.3**. Next we calculate the mass that is needed to hold onto that gas. We then compare the luminous mass with the gravitational mass. The amount of dark matter is the difference between the observed luminous mass and the amount of mass that is needed to hold on to the hot gas.

Some elliptical galaxies contain up to 20 times as much mass as can be accounted for by their stars and gas alone, so they must be dominated by dark matter, just like spiral galaxies. As with spiral galaxies, the luminous matter in elliptical galaxies is more centrally concentrated than is the dark matter. The inner parts of galaxies (where luminous matter dominates) transition smoothly to the outer parts (which are dominated by dark matter). Some galaxies may contain less dark matter than others; on average, however, about 90 percent of the total mass in a typical galaxy is dark matter.

The Milky Way Is Mostly Dark Matter

As in all other spiral galaxies, there is compelling evidence that dark matter dominates the Milky Way. Let's see how we know this.

Clouds of cold neutral hydrogen are located throughout the disk of the Milky Way. These clouds emit radiation at a wavelength of 21 cm, in the radio region of the spectrum. This radiation was predicted in the 1940s and observed in the 1950s. Measurements of the Doppler shift of the 21-cm line from neutral hydrogen indicate whether the neutral hydrogen is moving toward us or away from us, and how quickly. Because of its long wavelength, 21-cm radiation freely penetrates the dust between the stars, enabling us to see neutral hydrogen throughout our galaxy. The velocities of neutral interstellar hydrogen measured from 21-cm radiation all over the galaxy are plotted versus the direction of observation in **Figure 15.4**.

G **X** U V I R

Figure 15.3 In this combined visible-light and X-ray image of elliptical galaxy NGC 1132, the false-color blue and purple halo is X-ray emission from hot gas surrounding the galaxy. The hot gas extends well beyond the visible light from stars.

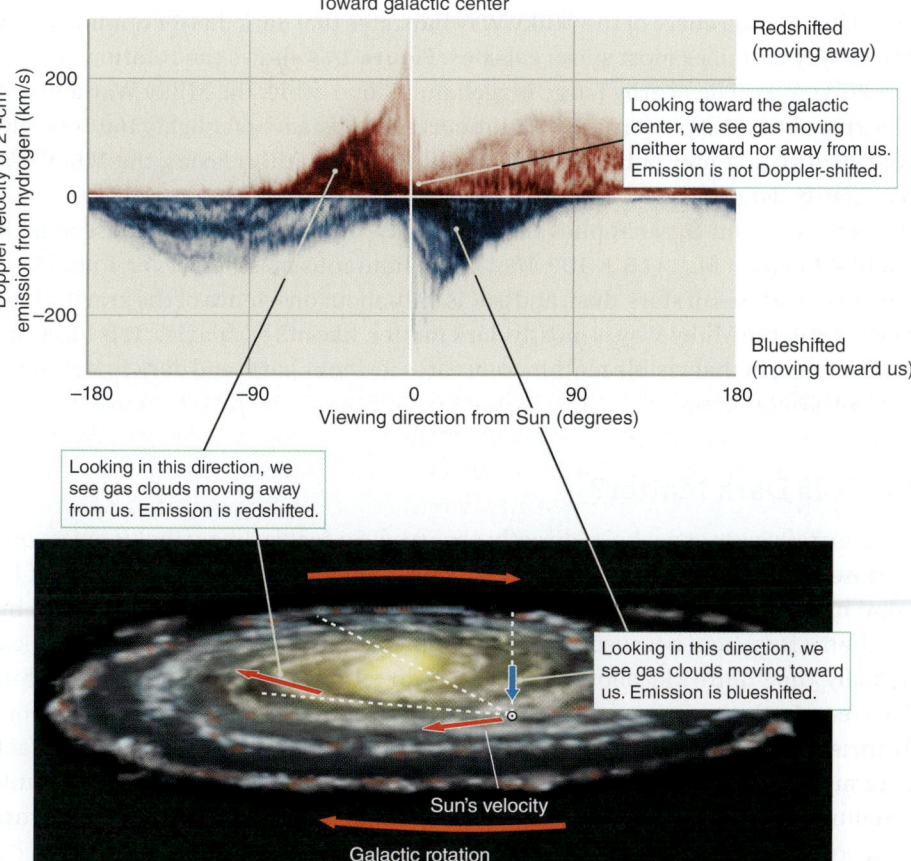

Figure 15.4 Doppler velocities are measured from observations of 21-cm emission from interstellar clouds of neutral hydrogen. These velocities vary between redshift and blueshift as we look around in the plane of our galaxy. Notice that, from our perspective within the Solar System (sight lines are shown as dashed white lines), we see the clear signature of a rotating disk when looking on either side of the galactic center.

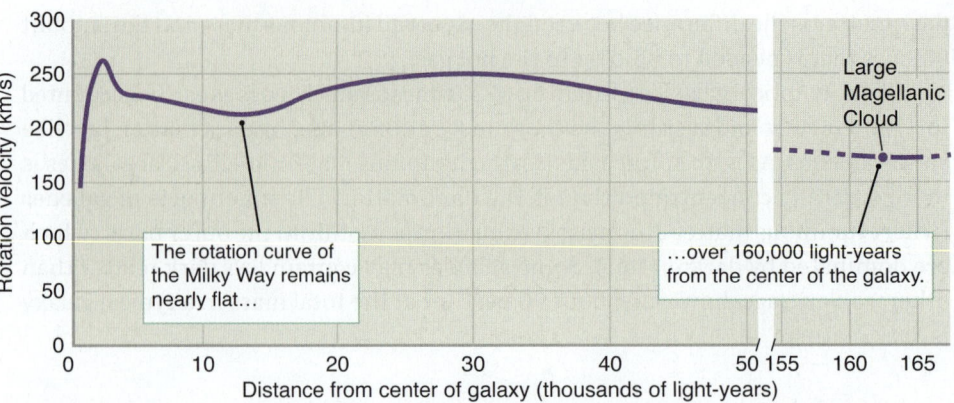

Figure 15.5 A plot showing rotation velocity versus distance from the center of the Milky Way. The most distant point comes from measurements of the orbit of the Large Magellanic Cloud. The nearly flat rotation curve indicates that dark matter dominates the outer parts of our galaxy.

Looking toward the center of the galaxy, but slightly on one side of the center, hydrogen clouds are moving toward Earth, so they are blueshifted. On the other side of the galaxy's center, clouds are moving away from Earth and are redshifted. This is the pattern of the rotation velocity of a disk. In other directions, the velocities we see are complicated by our moving vantage point within the disk and so are more difficult to interpret at a glance. Even so, observed velocities of neutral hydrogen enable us to determine the shape and location of the spiral arms in the disk of our galaxy. These observed velocities also allow us to measure our galaxy's rotation curve.

The rotation curve of the Milky Way indicates that dark matter dominates the Milky Way as it does most spiral galaxies. **Figure 15.5** shows the rotation curve of the Milky Way. The nearby Large Magellanic Cloud orbits the Milky Way and provides the outermost point in the rotation curve, at a distance of roughly 160,000 light-years from the center of the Milky Way. Like other spiral galaxies, the Milky Way has a fairly flat rotation curve.

This rotation curve implies that the Milky Way's gravitational mass must be about 1.3 trillion M_{Sun} ($1.3 \times 10^{12}\ M_{Sun}$). The luminous mass, however, found from adding the masses of stars, dust, and gas, is only about one-tenth of the gravitational mass. Thus, the Milky Way is mostly dark matter, like other galaxies. It is like other galaxies, too, in that visible matter dominates its inner parts and dark matter dominates its outer parts.

What Is Dark Matter?

▶‖ **AstroTour:** Dark Matter

Dark matter has not yet been directly observed. So far, we know that it exists only by observing its effect on other objects. This is analogous to an observation of the wind in the trees outside your window. You do not observe the wind directly, but you know it is there because of the way it affects the objects you *can* see. You can actually know quite a lot about the wind (its direction and its speed) from watching the trees. In the same way, we know quite a lot about dark matter (how much of it there is, and how it is distributed) from watching spiral galaxies rotate. But what is dark matter? A number of suggestions are under investigation, including Jupiter-sized objects, numerous black holes, large numbers of white dwarf stars, and exotic unknown elementary particles. These candidates can be lumped into two categories, called MACHOs and WIMPs.

Dark matter candidates such as small main-sequence M stars, planets, white dwarfs, neutron stars, or black holes are collectively referred to as **MACHOs**, which stands for "massive compact halo objects." Because these objects orbit the center of a galaxy, they behave much like the comets you learned about in Chapter 9. When they are far from the center of a galaxy, they are moving very slowly. They speed up as they fall toward the center, and then they slow again as they retreat back to the halo. MACHOs accumulate in the halo for the same reason that comets accumulate in the Oort Cloud: that is where they are moving most slowly, so that is where they spend the most time.

If the dark matter in the Milky Way's halo consists of MACHOs, there must be a lot of these objects, and they must each exert gravitational force but emit almost no light. Because they have mass, MACHOs should gravitationally deflect light according to Einstein's general theory of relativity, a phenomenon called gravitational lensing (see Section 13.5). If a MACHO passed in front of a distant star, the star's light would be deflected and perhaps, if the geometry were just right, focused by the intervening MACHO as it passed across our line of sight, as illustrated in **Figure 15.6a**. Because gravity affects all wavelengths equally, such lensing events should look the same in all colors, ruling out other causes of variability.

We would be remarkably lucky if such an event occurred just as we were observing a single distant star. In order to look for MACHOs, astronomers monitored the stars in the Large and Small Magellanic Clouds (two of the small companion

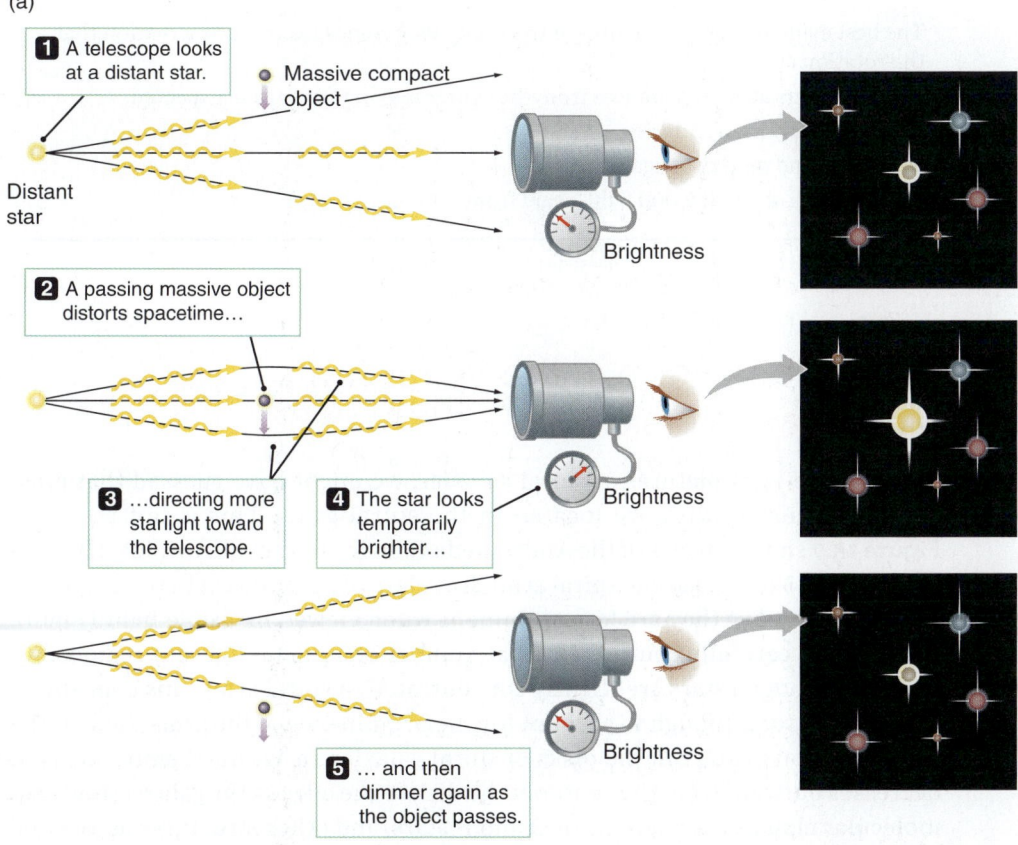

(a)

1 A telescope looks at a distant star.

Massive compact object

Distant star

Brightness

2 A passing massive object distorts spacetime…

3 …directing more starlight toward the telescope.

4 The star looks temporarily brighter…

Brightness

5 … and then dimmer again as the object passes.

Brightness

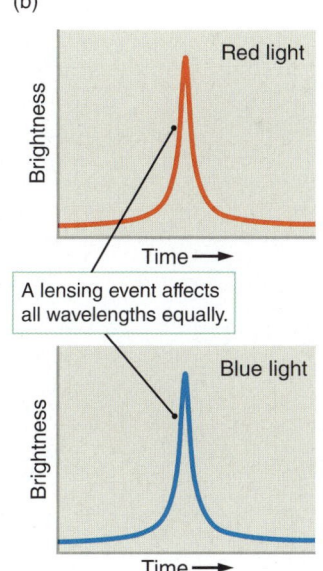

(b)

Red light

Brightness

Time →

A lensing event affects all wavelengths equally.

Blue light

Brightness

Time →

Figure 15.6 (a) The light from a distant star is affected by a compact object crossing our line of sight. (b) The observed light curves of a real star experiencing a lensing event have a distinctive shape and are the same in all colors of light.

what if . . .

What if we could directly observe WIMPs from the galactic halo passing through Earth? The Sun revolves around the galaxy at 220 kilometers per second (km/s), while Earth moves around the Sun at 30 km/s. How would the WIMP signal vary throughout the year due to these motions?

galaxies of the Milky Way Galaxy), observing tens of millions of stars for several years. They found a number of examples of events of the sort shown in Figure 15.6b, but not nearly enough to account for the amount of dark matter in the halo of our galaxy. Thus, it was concluded that the dark matter in the Milky Way is probably *not* composed primarily of MACHOs—at least not this type of MACHO, which has roughly the mass of a planet.

It is too early to tell whether black holes like the ones that provided the first gravitational wave detection (discussed in Chapter 13) could make up the mass that is missing. The black holes are massive (in the range of about 30 solar masses) and compact, so they would be an exotic type of "MACHO." These black holes are much more massive than the black holes that scientists were looking for in lensing experiments, so there would be fewer of them, spread more thinly across the halo. They might be too rare to have had a lensing event in the surveys. Some scientists have argued that it is plausible that these black holes make up dark matter, but too little is known about their number or distribution to be able to draw conclusions yet.

This leaves the exotic unknown elementary particles commonly known as **WIMPs**, which stands for "weakly interacting massive particles." These particles are predicted to be something like neutrinos; they would barely interact with ordinary matter yet would have some mass. Experiments are under way at the Large Hadron Collider and on the International Space Station to discover such particles, and additional experiments are being done to detect such particles from our galactic halo as they pass through Earth. There is a long way to go before the claim could be made that the mystery of dark matter is solved.

CHECK YOUR UNDERSTANDING 15.1

The best evidence for dark matter in the Milky Way comes from the observation that the rotation curve

a. shows stars at great distances from the center moving faster than expected.

b. rises swiftly in the interior.

c. falls off and then rises again.

d. has a peak at about 2,000 light-years from the center.

Answers to Check Your Understanding questions are in the back of the book.

15.2 Stars Form in the Spiral Arms of a Galaxy's Disk

From visible-light pictures of spiral galaxies, we might have guessed that most stars in a galaxy's disk are located in the spiral arms. This is not the case. **Figure 15.7** shows images of the Andromeda Galaxy taken in ultraviolet (UV) and visible light. Notice that the spiral arms are relatively prominent in the UV image (Figure 15.7a), but they are less prominent when viewed in visible light (Figure 15.7b). If we carefully count the actual number of stars in different regions, we find that although stars are slightly concentrated in spiral arms, this concentration is not strong enough to account for the prominence of the arms. In fact, the concentration of stars in the disks of spiral galaxies varies quite smoothly as it decreases outward from the center of the disk to the edge of the galaxy. However, molecular clouds, associations of O and B stars, and other structures associated

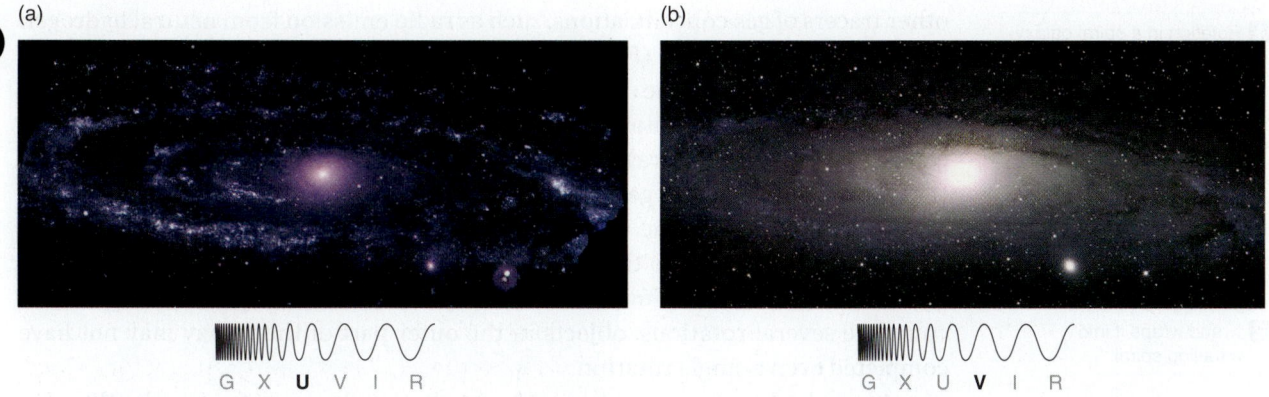

Figure 15.7 The Andromeda Galaxy is shown in (a) UV light and (b) visible light. The spiral arms are most prominent in UV light because they are dominated by young, hot stars and contain emissions from interstellar clouds that are ionized by the radiation from these young, hot stars.

with star formation are all concentrated in spiral arms. Spiral arms look so prominent when viewed in blue or UV light because they contain significant concentrations of young, massive, luminous stars.

As you learned in Chapter 5, stars form when dense interstellar clouds become so massive and concentrated that they begin to collapse under the force of their own gravity. If stars form within spiral arms, then clouds of interstellar gas must pile up and compress in spiral arms. There are many ways to trace the presence of gas in the spiral arms of galaxies. Pictures of face-on spiral galaxies, such as the one featured in **Figure 15.8a**, show dark lanes where clouds of dust block starlight. These lanes provide one of the best tracers of spiral arms. Spiral arms also show up in

what if . . .

What if astronomers observed only K- and M-type stars in the spiral arms of galaxies? What would you expect to find out about the concentration of dust and gas in these arms?

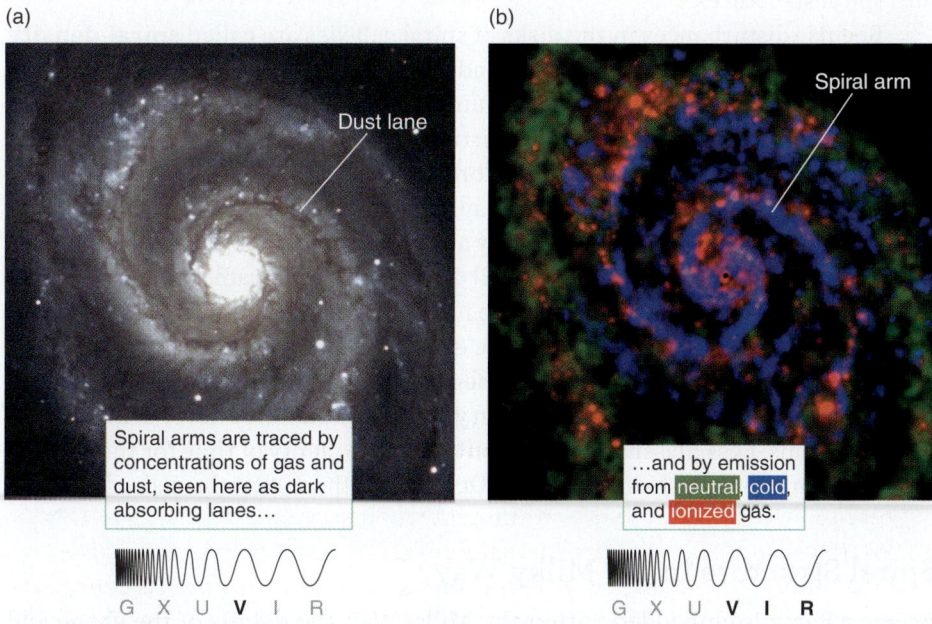

Figure 15.8 These two images of a face-on spiral galaxy show the spiral arms. (a) This visible-light image also shows dust absorption. (b) This image of 21-cm emission shows the distribution of neutral interstellar hydrogen, CO emission from cold molecular clouds, and Hα emission from ionized gas.

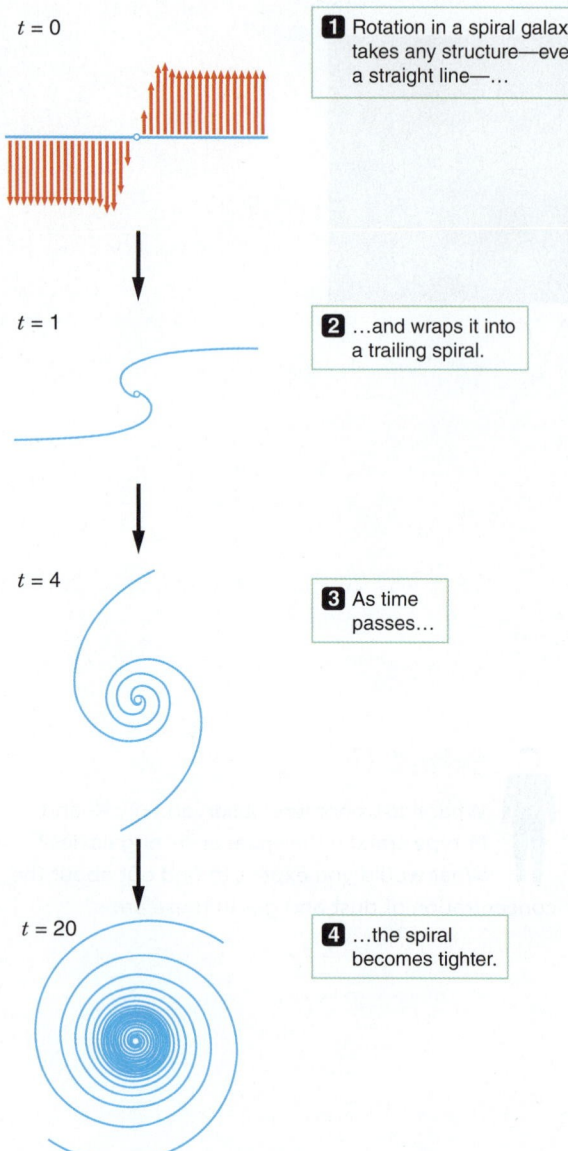

$t = 0$

1 Rotation in a spiral galaxy takes any structure—even a straight line—…

$t = 1$

2 …and wraps it into a trailing spiral.

$t = 4$

3 As time passes…

$t = 20$

4 …the spiral becomes tighter.

Figure 15.9 The differential rotation of a spiral galaxy will naturally take even an originally linear structure and wrap it into a progressively tighter spiral as time (t) passes.

other tracers of gas concentrations, such as radio emission from neutral hydrogen or from carbon monoxide (Figure 15.8b).

Any disturbance in the disk of a spiral galaxy will cause a spiral pattern because the disk rotates. Disks do not rotate like a solid body. Instead, material close to the center takes less time to travel around the galaxy than material farther out, and so the inner part of the disk gets ahead of the outer part. **Figure 15.9** illustrates the point: In the second frame, you can see that the outer part of the line is trailing behind the inner part. As the galaxy rotates, a straight line through the center becomes a spiral. In the time it takes for objects in the inner part of the galaxy to complete several rotations, objects in the outer part of the galaxy may not have completed even a single rotation.

A spiral galaxy can be disturbed by gravitational interactions with other galaxies or a burst of star formation, for example. However, a single disturbance will not produce a *stable* spiral-arm pattern. Spiral arms produced from one disturbance will wind themselves up completely in two or three rotations of the disk and then disappear. But repetitive disturbances sustain spiral structure indefinitely. When the bulge in the center of a spiral galaxy is elongated (as seems to be the case for most spiral galaxies), then the bulge gravitationally disturbs the disk, pushing on the clouds of dust and gas within it. As the disk rotates through this disturbance, repeated episodes of star formation occur, and stable spiral arms form.

A spiral structure can also be created by star formation. Regions of star formation release energy into their surroundings through UV radiation, stellar winds, and supernova explosions. This energy compresses clouds of gas and triggers more star formation. Typically, many massive stars form in the same region at about the same time, and their combined mass outflows and supernova explosions occur over only a few million years, creating large, expanding bubbles of hot gas. These bubbles concentrate the gas into dense clouds at their edges, causing more star formation. Rotation of the disk bends the resulting strings of star-forming regions into spiral structures.

Regular disturbances in the disks of spiral galaxies are called **spiral density waves**: regions of greater mass density and increased pressure in the galaxy's interstellar medium. These waves move around a disk in the pattern of a two-armed spiral. Because they are waves, it is the disturbance that moves around the disk, not the material. As the matter in the disk orbits, material passes *through* the spiral density waves. The stars in the arm today are not the same stars that were in the arm 20 million years ago. This is roughly analogous to a traffic jam on a busy highway. The cars in the jam are changing all the time, yet the traffic jam persists as a place of higher density—where there are more cars than usual.

A spiral density wave has very little effect on stars, but it does compress the gas that flows through it. Stars form in the resulting compressed gas. Massive stars have such short lives (typically 10 million years or so) that they never drift far from the spiral arms. Less massive stars, in contrast, have plenty of time for their orbits to carry them far from their place of birth, filling in the rest of the disk.

Spiral Structure in the Milky Way

Because Earth is embedded within the Milky Way, the details of the shape and structure of the Milky Way are actually more difficult to determine than for other galaxies. Comparing observations of the Milky Way with observations of more distant galaxies improves our understanding of the Milky Way. Astronomers also use observations of globular clusters within the Milky Way to determine the size of the Milky Way and our location within it. The European Space Agency's *Gaia* mission is in the process of creating a precise three-dimensional map of astronomical objects and their motions throughout the Milky Way. This giant observational database will help astronomers further develop ideas about the origin, structure, and evolutionary history of the Milky Way.

Figure 15.10a shows the Milky Way as a luminous band stretching across the night sky. Compare this image to the image in Figure 15.10b, which shows an edge-on spiral galaxy. Simply by comparing the similarities between these two images, you might determine that the Milky Way is a spiral galaxy and that we are viewing it edge-on, from inside the disk. Confirming this hunch and finding further information about the size and shape of the Milky Way require more detailed observations in the visible, infrared, and radio regions of the electromagnetic spectrum.

(a)

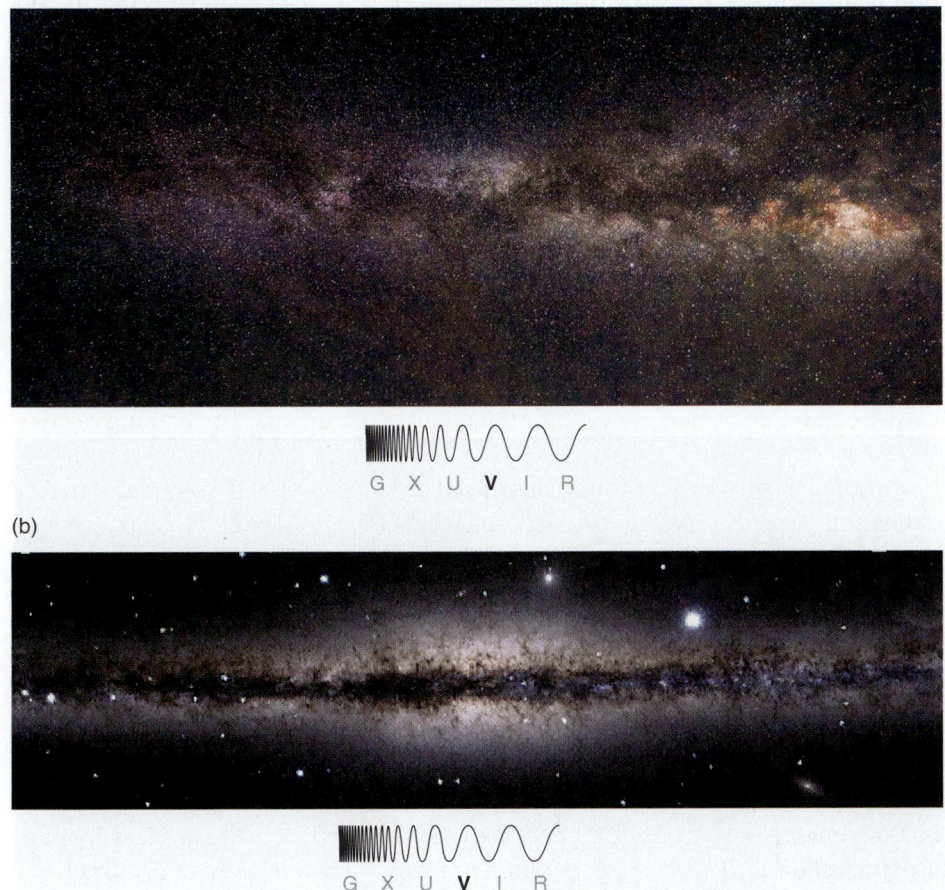

G X U **V** I R

(b)

G X U **V** I R

Figure 15.10 ★ WHAT AN ASTRONOMER SEES Most astronomical objects have relatively simple shapes: a sphere, an ellipse, a disk, or a combination of these shapes. Because of this small set of shapes to choose from, astronomers can often compare two-dimensional images of different objects to determine whether they are the same shape in three dimensions. In (a), an astronomer would note the prominent dark lanes caused by interstellar dust that obscures the light from more distant stars. She would identify the brighter yellow area to the right of the center of the image as the galactic center of the Milky Way. A comparison of this image to the image in (b), in which the disk of the edge-on spiral galaxy NGC 891 greatly resembles the Milky Way, would convince an astronomer that the Milky Way is a spiral galaxy and the Solar System is set within the disk.

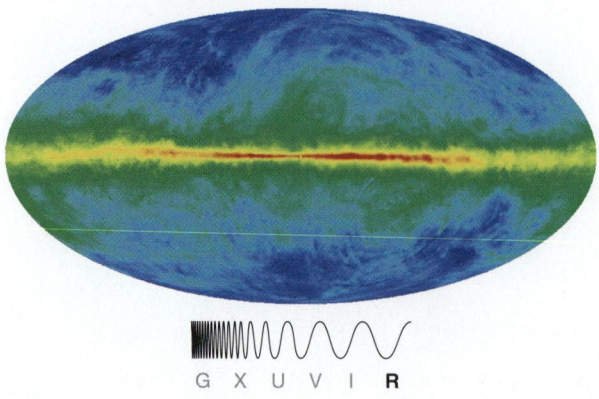

GXUVI **R**

Figure 15.11 This radio image of the sky taken at a wavelength of 21 cm shows clouds of neutral hydrogen gas throughout our galaxy. Because radio waves penetrate interstellar dust, 21-cm observations are a crucial method for probing the structure of our galaxy.

Figure 15.11 shows the full sky, as mapped in 21-cm radiation from neutral hydrogen. By 1952, astronomers had the first maps of the neutral hydrogen in the Milky Way and other galaxies, tracing out the distribution of the gas in these galaxies. The maps showed spiral structure in the other galaxies and suggested spiral structure in the Milky Way. At about the same time, observations of ionized hydrogen gas in visible light showed two spiral arms with concentrations of young, hot O and B stars. These observations together confirmed that the Milky Way is a spiral galaxy.

In the 1990s, some hints emerged that the Milky Way is a barred spiral, with an elongated bulge. This was later confirmed using infrared observations from the Spitzer Space Telescope. These observations of the distribution and motion of stars toward the center of the galaxy showed a substantial bar with a modest bulge at the center (**Figure 15.12**). Two major spiral arms connect to the ends of the central bar. There are several smaller arm segments, including the Orion Spur, which contains the Sun and Solar System. The Milky Way is a giant barred spiral that is more luminous than the average spiral. If viewed from the outside, the Milky Way would look much like the barred spiral galaxy M109, shown in **Figure 15.13**.

CHECK YOUR UNDERSTANDING **15.2**

Spiral arms are like traffic jams because they both
a. last only a short time.
b. are composed of objects that enter and leave constantly.
c. are typically the result of collisions.

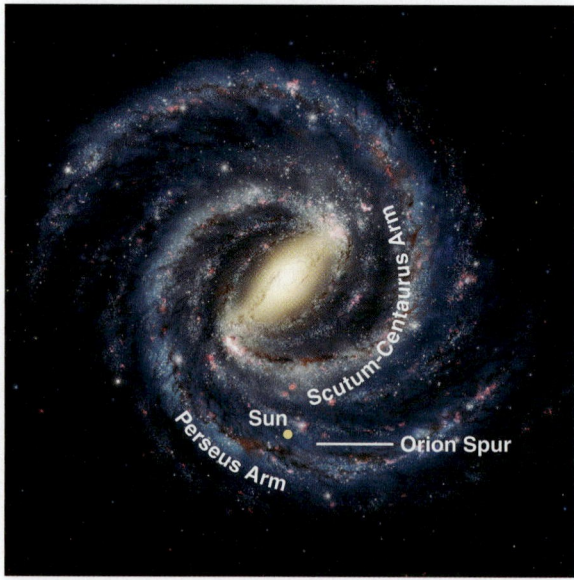

Figure 15.12 Infrared and radio observations contribute to an artist's model of the Milky Way Galaxy. The galaxy's two major arms (Scutum-Centaurus and Perseus) are attached to the ends of a thick central bar. The Sun and Solar System are located within the Orion Spur, which is situated between the two major arms.

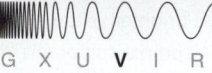

GXUVI R

Figure 15.13 From the outside, the Milky Way would look much like this barred spiral galaxy, M109.

15.3 Components of the Milky Way Reveal Its Evolution

Our Sun is a middle-aged star located among other middle-aged stars that orbit around the galaxy within the disk. Near the Sun, however, are other stars—most much older—that are a part of the halo of the Milky Way and whose orbits are carrying them *through* the disk. Using the ages, chemical abundances, and motions of nearby stars, we can differentiate between disk and halo stars to learn more about the galaxy's structure.

Age and Chemical Composition of Stars

Most globular clusters in the Milky Way are in the halo. With ages of up to 13 billion years, they are among the oldest objects known; in fact, no young globular clusters have been observed. In contrast, **open clusters**, like the one shown in **Figure 15.14**, are much less tightly bound collections of a few dozen to a few thousand stars that are found in the disk of a spiral galaxy. As with globular clusters, the stars within an open cluster all formed in the same region at about the same time, and so their ages can be found from the main-sequence turnoff (recall Section 12.6). Different open clusters have a wide range of ages. Some contain the very youngest stars known, while others contain stars somewhat older than the Sun. Because open clusters are loosely bound together, tidal forces from nearby objects easily disrupt them, and they do not survive long in the disk of our galaxy. The oldest open clusters are several billion years younger than the youngest globular clusters.

The difference in ages between globular and open clusters indicates that stars in the halo formed first, but this epoch of star formation did not last long. Star formation in the disk started later but has been continuing ever since. Star formation processes in the massive, compact globular clusters also must have been much different from the processes in the less massive, more scattered open clusters.

Stars add heavy elements to the interstellar medium as part of a cycle in which material is taken from the interstellar medium to form stars, processed through the life and death of a star, and then lost to the interstellar medium again, as shown in **Figure 15.15**. When the universe was very young, only the least massive (light) elements existed. All elements more massive than boron must have formed in the cores of stars. For this reason, the abundance of heavy elements in the interstellar medium provides a record of all the star formation that has taken place up to the

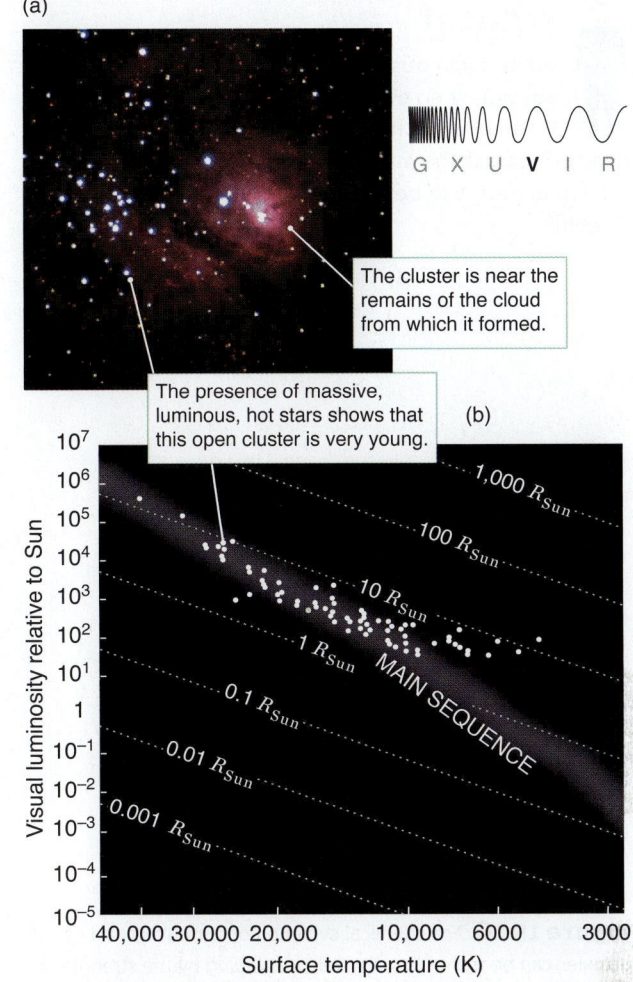

(a)

The cluster is near the remains of the cloud from which it formed.

The presence of massive, luminous, hot stars shows that this open cluster is very young.

(b)

Figure 15.14 (a) The open star cluster NGC 6530 is located 5,200 light-years away, in the disk of our galaxy. (b) The H-R diagram of stars in this cluster indicates that it has an age of a few million years or less, because the cluster does not show a main-sequence turnoff.

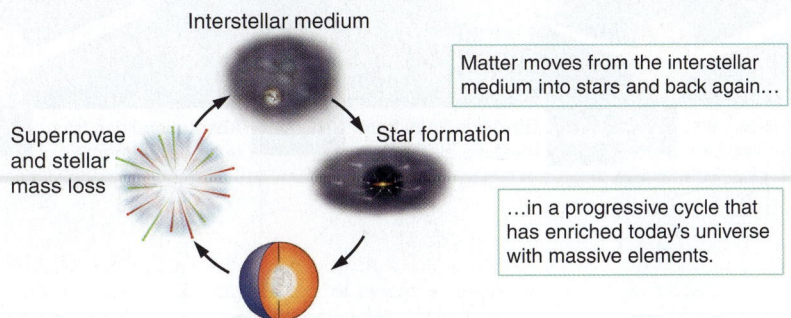

Interstellar medium

Matter moves from the interstellar medium into stars and back again...

Star formation

Supernovae and stellar mass loss

...in a progressive cycle that has enriched today's universe with massive elements.

Figure 15.15 As subsequent generations of stars form, live, and die, they enrich the interstellar medium with heavy elements—the products of the formation of new elements in stars.

what if . . .

What if you observe a nearby star that shows absolutely no evidence of elements heavier than helium in its spectrum: What could you conclude about the mass of this star, and how fast do you expect it to be moving relative to the solar system?

Figure 15.16 The chemical evolution of our galaxy and other galaxies can be traced in many ways, including by the strength of emission lines from interstellar dust and gas and absorption lines in the spectra of stars.

present time. Gas that is rich in heavy elements has gone through a great deal of stellar processing, whereas gas that is poor in heavy elements has not.

In turn, the abundance of heavy elements in the atmosphere of a star provides a snapshot of the chemical composition of the interstellar medium *at the time the star formed.* (In main-sequence stars, material from the core does not mix with material in the atmosphere, so the amounts of chemical elements inferred from the spectra of a star are the same as the amounts in the interstellar gas from which the star formed.) As illustrated in **Figure 15.16**, the chemical composition of a star's atmosphere reflects the cumulative amount of star formation that has occurred up to the moment the star formed.

Stars in globular clusters, among the earliest stars to form, contain only very small amounts of heavy elements. Some globular-cluster stars contain only 0.5 percent as much of these heavy elements as our Sun has. This relationship between age and abundances of heavy elements is evident throughout much of the galaxy. Lower abundances of heavy elements characterize not just globular-cluster stars but all the stars in our galaxy's halo. Within the disk, the chemical evolution of the Milky Way has continued as generations of stars have further enriched the interstellar medium with the products of their nuclear fusion processes. Older disk stars are typically poorer in heavy elements than younger disk stars, because they formed earlier in the history of the galaxy, before the heavier elements were available in the dust and gas. Similarly, older stars in the outer parts of our galaxy's bulge are poorer in heavy elements than younger stars in the disk.

Star formation is generally more active in the inner parts of the Milky Way than in the outer parts because interstellar gas is denser in the inner parts. If such activity has continued throughout the history of our galaxy, we might predict heavy elements to be more abundant in the inner parts of our galaxy than in the outer parts. Observations of chemical abundances in the interstellar medium, based both on interstellar absorption lines in the spectra of stars and on emission lines in

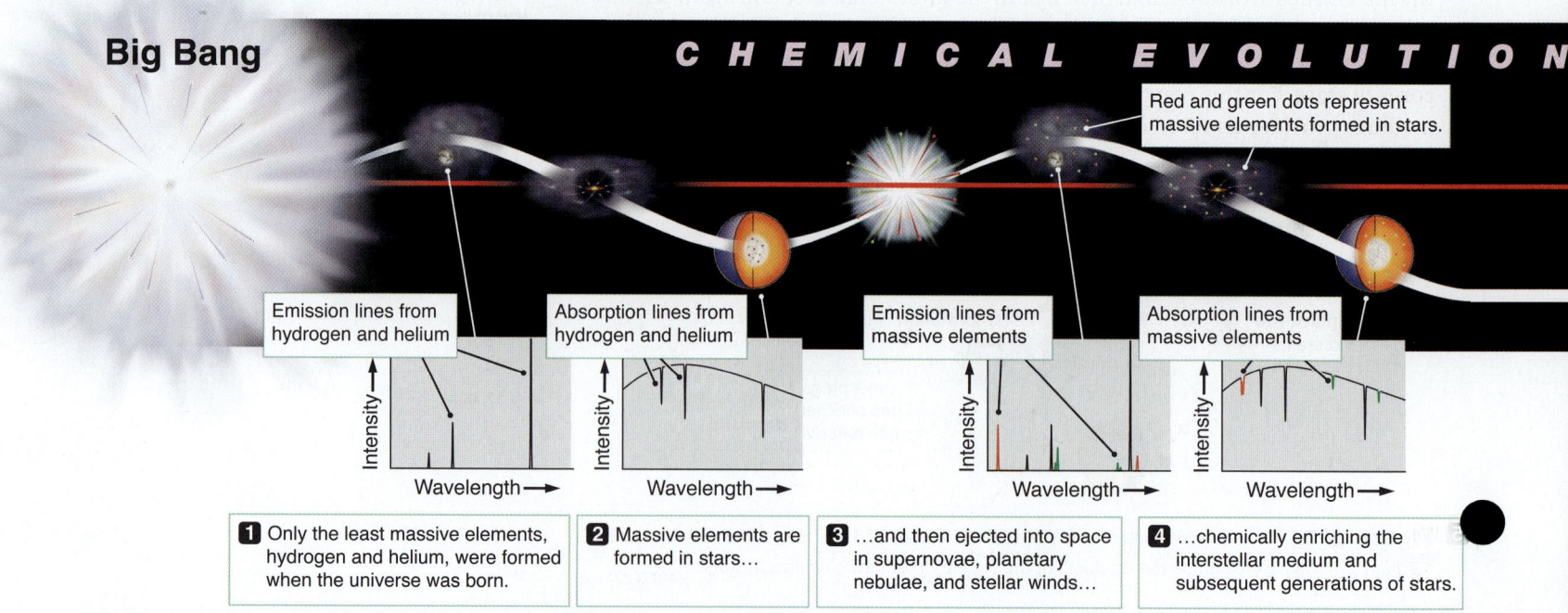

Big Bang

CHEMICAL EVOLUTION

Red and green dots represent massive elements formed in stars.

Emission lines from hydrogen and helium

Intensity

Wavelength

Absorption lines from hydrogen and helium

Intensity

Wavelength

Emission lines from massive elements

Intensity

Wavelength

Absorption lines from massive elements

Intensity

Wavelength

1 Only the least massive elements, hydrogen and helium, were formed when the universe was born.

2 Massive elements are formed in stars...

3 ...and then ejected into space in supernovae, planetary nebulae, and stellar winds...

4 ...chemically enriching the interstellar medium and subsequent generations of stars.

glowing clouds of gas known as H II regions, confirm this prediction. The composition of stars also confirms this prediction. Other galaxies also have more abundant heavy elements in the inner parts than in the outer parts.

The basic idea that higher abundances of heavy elements should correspond to the more active star formation in the inner galaxy seems correct, but the chemical composition of the interstellar medium at any location depends on several factors. New material falling into the galaxy might affect the amounts of heavy elements in the interstellar medium. As illustrated in **Figure 15.17**, chemical elements produced in the inner disk might be blasted into the halo by the energy of massive stars, only to fall back onto the disk elsewhere. Past interactions with other galaxies might have mixed gas from those other galaxies with our own. The variations of chemical abundances within the Milky Way and other galaxies—and what these variations indicate about the history of star formation and the formation of elements—remain active topics of research.

Although the details are complex, several important lessons can be learned from patterns in the amounts of heavy elements in the galaxy. The first is that even the very oldest globular-cluster stars contain *some* amount of heavy chemical elements. This implies that globular-cluster stars and other halo stars were not the first stars to form. At least one generation of massive, short-lived stars lived and died, ejecting heavy elements into space, before even the oldest-known globular clusters formed. The second is that we find no disk stars with exceptionally low amounts of heavy elements, even though every star less massive than about $0.8\,M_{Sun}$ that ever formed is still around today. The gas in the plane of the Milky Way must have seen a significant amount of star formation before it settled into the disk of the galaxy and made stars.

These variations in chemical abundances from place to place in the Milky Way tell us a lot about the history of our galaxy and a lot about the origin of the material that we are made from. It is important to remember, however, that even a

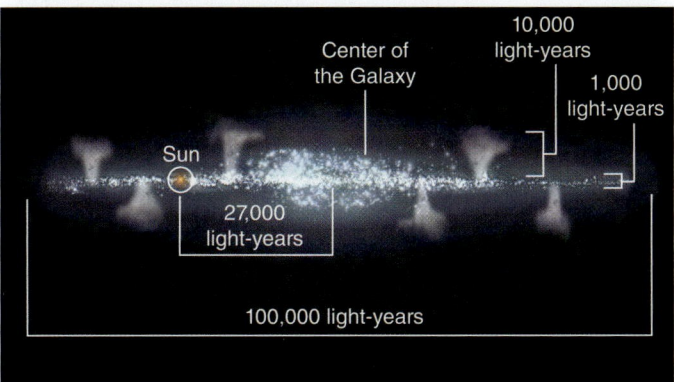

Figure 15.17 In the "galactic fountain" model of the disk of a spiral galaxy, gas is pushed away from the plane of the galaxy by energy released by young stars and supernovae and then falls back onto the disk.

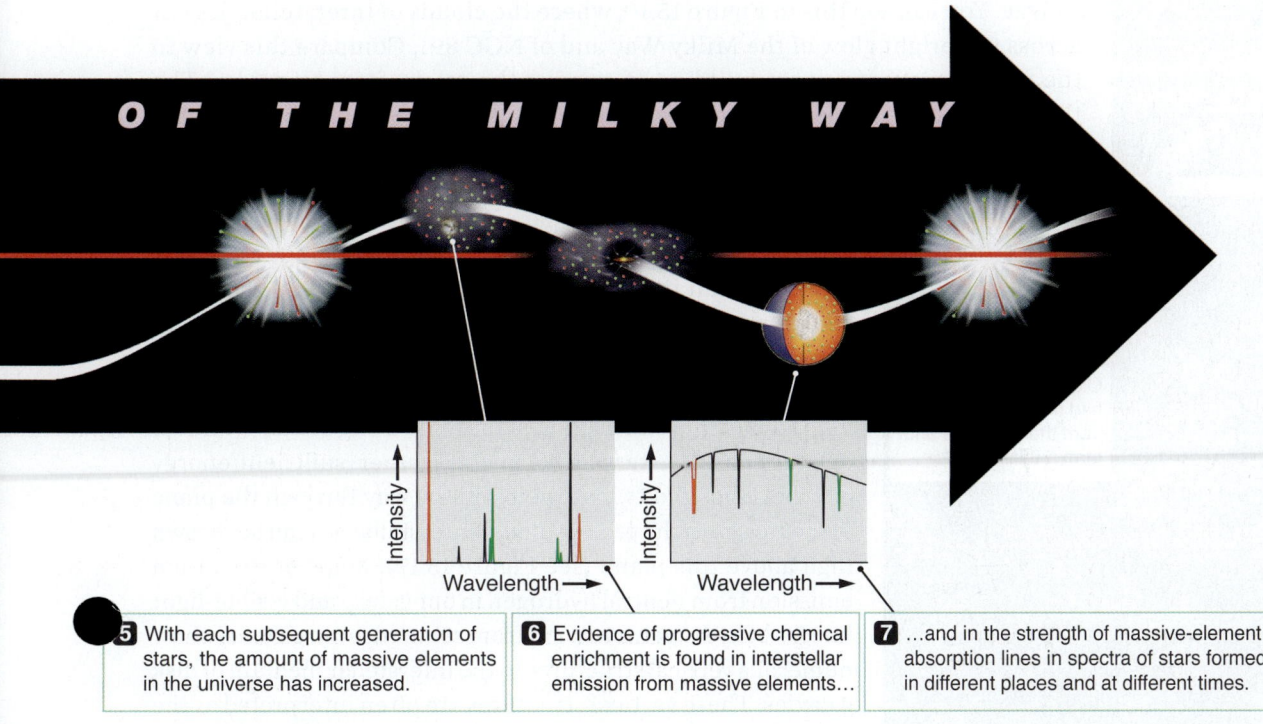

OF THE MILKY WAY

5 With each subsequent generation of stars, the amount of massive elements in the universe has increased.

6 Evidence of progressive chemical enrichment is found in interstellar emission from massive elements…

7 …and in the strength of massive-element absorption lines in spectra of stars formed in different places and at different times.

chemically "rich" star like the Sun, which is made of gas processed through approximately 9 billion years of previous generations of stars, is composed of less than 2 percent heavy elements. Luminous matter in the universe is still dominated by hydrogen and helium that formed long before the first stars.

Looking at a Cross Section through the Disk

The youngest stars in our galaxy are most strongly concentrated in the galactic plane, defining a disk more than 100,000 light-years across but only about 1,000 light-years thick, which is very thin by comparison. The older population of disk stars, distinguishable by lower abundances of heavy elements, has a much "thicker" distribution: about 12,000 light-years thick. This thick distribution of old disk stars extends beyond the boundaries of **Figure 15.18**, which shows how the population of stars changes with distance from the galactic plane. The youngest stars are concentrated closest to the plane of the galaxy because this is where the molecular clouds are. Older stars make up the thicker parts of the disk. There are two hypotheses for the origin of this thicker disk. One suggests that these stars formed in the midplane of the disk long ago but have since been kicked up out of the plane of the galaxy, primarily by gravitational interactions with massive molecular clouds (as shown in Figure 15.18). The other hypothesis suggests that these stars were acquired from the merging process that formed our galaxy.

A rotating cloud of gas naturally collapses into a thin disk when gas falling from one direction (above the disk) collides with gas falling from the other direction (below the disk). The same process applies to clouds of gas that are pulled by gravity toward the midplane of the disk of a spiral galaxy. Although stars are free to pass back and forth from one side of the disk to the other, cold, dense clouds of interstellar gas settle into the central plane of the disk. These clouds appear as concentrated dust lanes that lie in the middle of the disks of spiral galaxies, like tomato sauce that lies between the cheese and the crust on a pizza. You can see this in Figure 15.10, where the clouds of interstellar gas cut across the bright glow of the Milky Way and of NGC 891. Compare this view to the mental picture you are building of viewing the galaxy from "outside." The thin lane slicing through the middle of the disk is the place where new stars are found and are continuing to form.

The interstellar medium is a dynamic place—energy from star-forming regions can shape it into impressively large structures. We mentioned earlier that energy from regions of star formation can impose an interesting structure on the interstellar medium, clearing out large regions of gas in the disk of a galaxy. Many massive stars forming in the same region can blow "chimneys" out through the disk of the galaxy via a combination of supernova explosions and strong stellar winds. If enough massive stars are formed together, sufficient energy may be deposited to blast holes all the way through the plane of the galaxy. In the process, dense interstellar gas can be thrown high above this plane (see Figure 15.17). Maps of the 21-cm emission from neutral hydrogen in our galaxy and visible-light images of hydrogen emission from some edge-on galaxies show numerous vertical structures in the interstellar medium of disk galaxies. These vertical structures are often interpreted as the "walls" of chimneys.

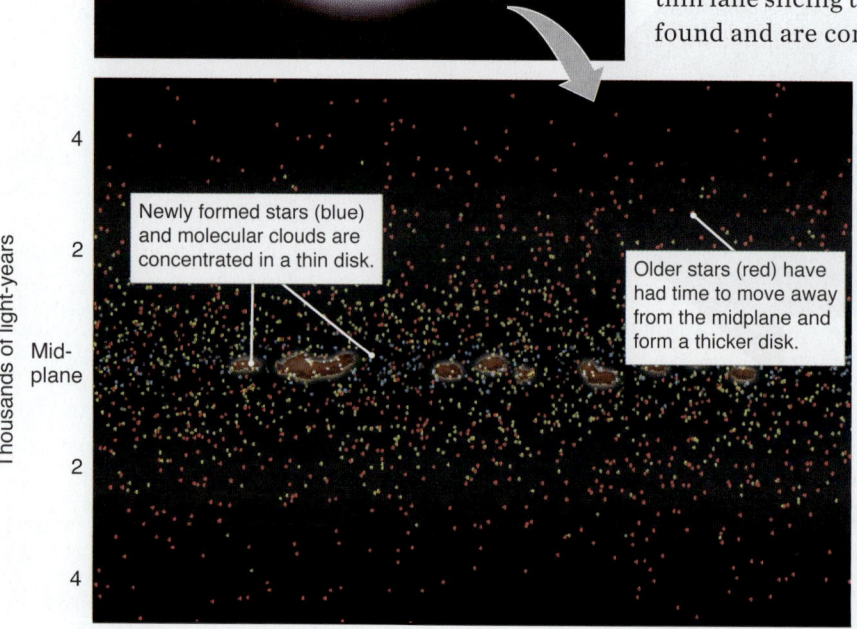

Figure 15.18 A vertical profile of the disk of the Milky Way Galaxy. Gas and young stars are concentrated in a thin layer in the center of the disk. Older populations of stars make up the thicker portions of the disk.

Newly formed stars (blue) and molecular clouds are concentrated in a thin disk.

Older stars (red) have had time to move away from the midplane and form a thicker disk.

Thousands of light-years

4
2
Mid-plane
2
4

Other Halo Components

The globular clusters in the galactic halo tell astronomers a great deal about the history of star formation in the halo. Yet globular clusters account for only about 1 percent of the total mass of stars in the halo. As halo stars fall through the disk of the Milky Way, some (such as Arcturus, the third brightest star in our sky) pass close to the Sun, providing a sample of the halo that can be studied at closer range.

The Sun and most of the stars near it are disk stars. Astronomers can distinguish nearby halo stars in two ways. First, most halo stars are poorer in heavy elements than disk stars. Second, halo stars are moving much faster, relative to the Sun, than disk stars. The disk stars and the Sun all orbit the center of the galaxy in the same direction, and those near the Sun travel at nearly the same speed. In contrast, halo stars orbit the center of the galaxy in random directions, so the relative velocity between the halo stars and the Sun tends to be high. These stars are known as high-velocity stars.

By studying the orbits of high-velocity stars, astronomers have determined that the halo has two separate components: an inner halo, which includes stars up to about 50,000 light-years from the center, and an outer halo that extends far beyond that. The stars in the outer halo have a lower fraction of heavier elements, implying that they formed very early and therefore are very old. Many of them are moving in the opposite direction to the rotation of the galaxy. This suggests that the outer halo may have its origins in a merger with a small dwarf galaxy long ago, lending support to the idea that the Milky Way formed from the merger of multiple galaxies.

X-ray observations suggest there is also a halo of hot (about 2 million K) gas surrounding the Milky Way. This gas halo may extend for about 300,000 light-years from the galactic center and may contain as much mass as all the stars in the Milky Way. Even so, this hot halo does not have nearly enough mass to account for the pull of gravity due to the dark matter that is hypothesized to exist around the Milky Way.

Magnetic Fields and Cosmic Rays

The interstellar medium of the Milky Way is laced with a measurable magnetic field that is wound up and strengthened by the rotation of the galaxy's disk. This interstellar magnetic field is 100,000 times weaker than Earth's magnetic field. Charged particles spiral around magnetic field lines, moving along the field rather than across it. Conversely, magnetic fields cannot freely escape from a cloud of gas containing charged particles. The dense clouds of interstellar gas in the midplane of the Milky Way, shown in **Figure 15.19**, anchor the galaxy's magnetic field to the disk. These magnetic fields trap charged particles known as **cosmic rays**, which are charged particles, such as electrons, that travel close to the speed of light. Despite their name, cosmic rays are not a form of electromagnetic radiation; instead, they are particles with mass. (They were named before their true nature was known.)

Cosmic rays are continually hitting Earth. Most cosmic-ray particles are protons, but some are nuclei of helium, carbon, and other elements. A few are high-energy electrons and other subatomic particles. Cosmic

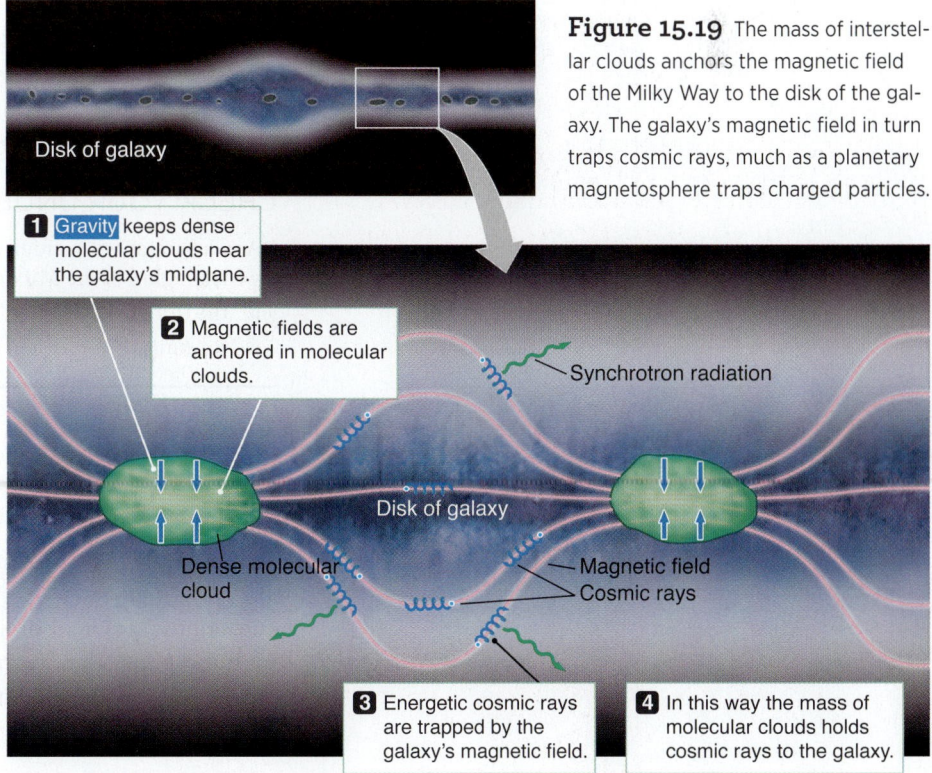

Figure 15.19 The mass of interstellar clouds anchors the magnetic field of the Milky Way to the disk of the galaxy. The galaxy's magnetic field in turn traps cosmic rays, much as a planetary magnetosphere traps charged particles.

Disk of galaxy

1 Gravity keeps dense molecular clouds near the galaxy's midplane.

2 Magnetic fields are anchored in molecular clouds.

Synchrotron radiation

Disk of galaxy

Dense molecular cloud

Magnetic field
Cosmic rays

3 Energetic cosmic rays are trapped by the galaxy's magnetic field.

4 In this way the mass of molecular clouds holds cosmic rays to the galaxy.

(a)

(b)

Figure 15.20 (a) In this artist's sketch, the Pierre Auger Observatory in Argentina is an array of stations designed to catch the particles that shower from collisions of cosmic rays with the upper atmosphere. (b) Each station in the array is equipped with its own particle collectors, carefully protected from the elements.

rays span an enormous range in particle energy. We can observe the lowest-energy cosmic rays, with energies as low as about 10^{-11} joules (J), by using interplanetary spacecraft. These energies correspond to the energy of a proton moving at a velocity of a few tenths the speed of light. In contrast, the most energetic cosmic rays are 10 trillion (10^{13}) times as energetic as the lowest-energy cosmic rays. To get a better sense of just how much energy we're talking about, consider this: If you were to drop your copy of *Understanding Our Universe* (this irreplaceable textbook) from a second-story window, it would hit the ground with the same energy as a *single* high-energy cosmic-ray proton! These high-energy cosmic rays are measured using the showers of elementary particles that they cause when crashing through Earth's atmosphere. These particle showers have been observed by special telescopes such as the High Resolution Fly's Eye Observatory in Utah, or the Pierre Auger Observatory in Argentina (**Figure 15.20**).

Astronomers hypothesize that most cosmic rays are accelerated to high energies by the shock waves produced in supernova explosions. The very highest-energy cosmic rays are as much as 100 million times more energetic than any particle ever produced in a particle accelerator on Earth. These extremely high energies make them much more difficult to explain than those with lower energies.

The disk of our galaxy glows from synchrotron radiation (see Section 8.4) produced by cosmic rays (mostly electrons) spiraling around the galaxy's magnetic field. Such synchrotron radiation is seen in the disks of other spiral galaxies as well, telling us that they, too, have magnetic fields and populations of energetic cosmic rays. Even so, the cosmic rays with the highest energies are moving much too fast to be confined to our galaxy. Any such cosmic rays formed in the Milky Way soon stream away from the galaxy into intergalactic space. It is likely that some of the energetic cosmic rays reaching Earth originated in energetic events outside our galaxy.

The total energy of all the cosmic rays in the galactic disk can be estimated from the energy of the cosmic rays reaching Earth. The strength of the interstellar magnetic field can be measured in a variety of ways, including the effect that it has on radio waves passing through the interstellar medium. These measurements indicate that, in our galaxy, the magnetic-field energy and the cosmic-ray energy are about equal to each other. Both are comparable to the energy present in other energetic components of the galaxy, including the motions of interstellar gas and the total energy of electromagnetic radiation within the galaxy.

CHECK YOUR UNDERSTANDING **15.3**

A nearby star with a high velocity and low heavy element abundance is most likely from which region of the Milky Way Galaxy?

a. the thin disk

b. the thick disk

c. the bulge

d. the halo

15.4 The Milky Way Offers Clues about How Galaxies Form

A fundamental goal of stellar astronomy is to understand the life cycle of stars, including how stars form from clouds of interstellar gas. Galactic astronomy has the same basic goal. That is, astronomers would like very much to have a complete and well-tested theory of how our galaxy formed. The distribution of stars of different ages, with different amounts of heavy elements, is one clue. Additional clues come from studying other galaxies at different distances (and therefore different ages), their supermassive black holes, and their merger history.

The properties of globular clusters and high-velocity stars in the halo are also important clues in understanding how galaxies form. These objects must have been among the first stars formed that still exist today. They are not concentrated in the disk or bulge of the galaxy, so they must have formed from clouds of gas well before those clouds had settled into the galaxy's disk. The observations that globular clusters are very old and that the youngest globular cluster is older than the oldest disk stars confirm this idea. The presence of small concentrations of heavy elements in the atmospheres of halo stars indicates that at least one generation of stars must have lived and died *before* the formation of the halo stars we see today. Astronomers are looking for stars from that first generation.

Galaxies do not exist in isolation. The vast majority of galaxies are parts of gravitationally bound collections of galaxies. The smallest and most common of these are called **galaxy groups**. A galaxy group contains as many as several dozen galaxies, most of which are dwarf galaxies (see Chapter 14), in a space between 4 million and 6 million light-years across. The Milky Way is a member of the Local Group, first identified by Edwin Hubble in 1936. There are about 50 members of the Local Group, including two large barred spiral galaxies, the Milky Way and the Andromeda Galaxy, which account for nearly 98 percent of the mass of the Local Group. The third largest galaxy, Triangulum, is a nonbarred spiral with about one-tenth the mass of the Milky Way Galaxy. The Local Group also includes a few elliptical and irregular galaxies (**Figure 15.21**) and at least 30 smaller dwarf galaxies. Many of these smaller dwarf galaxies were unknown until very recently because of their low luminosity, and there are likely more to be discovered. Most of the dwarf galaxies in the group (but not all) are satellites of the Milky Way or Andromeda galaxies. The Local Group interacts with a few nearby groups.

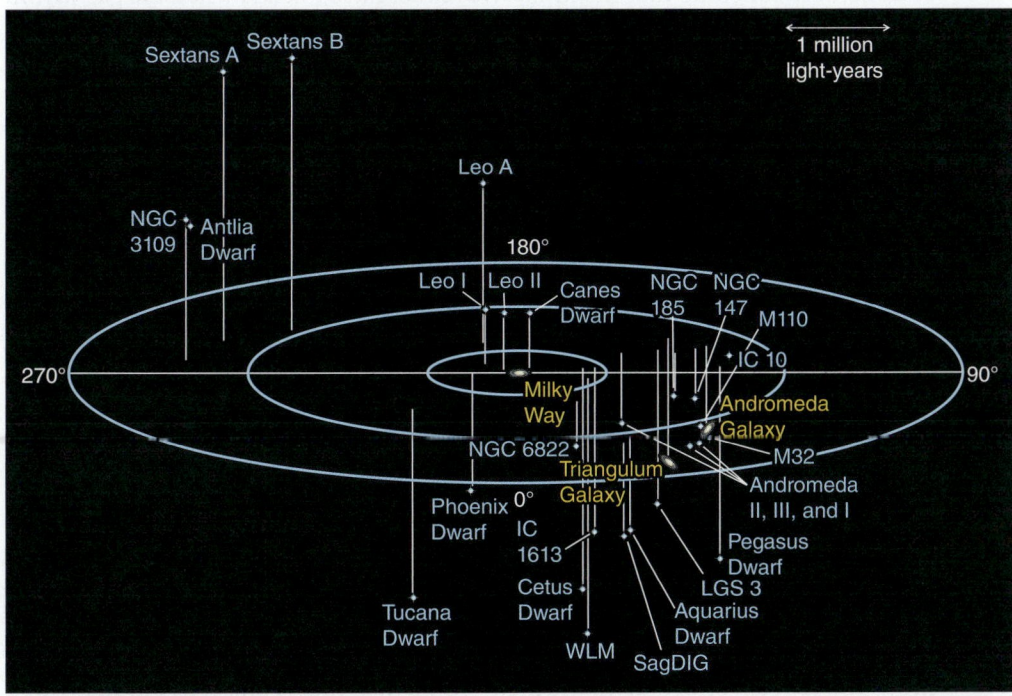

Figure 15.21 A graphical map showing the locations of some of the members of the Local Group of galaxies. Most are dwarf galaxies. Spiral galaxies are shown in yellow.

Figure 15.22 Our Milky Way is surrounded by more than 20 dwarf companion galaxies, the largest among them being the Large and Small Magellanic Clouds. The Magellanic Clouds were named for Ferdinand Magellan (c. 1480–1521), who headed an early European expedition that ventured far enough into the Southern Hemisphere to see them.

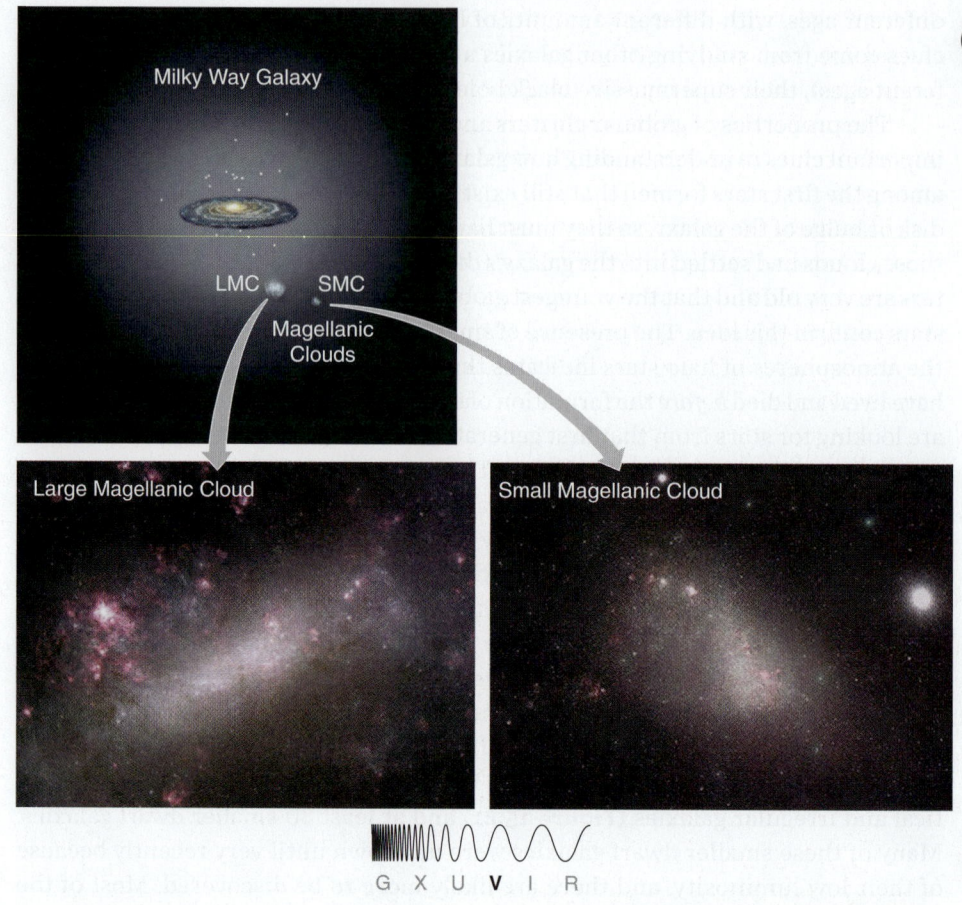

From the properties of globular clusters and high-velocity stars, the presence of the central supermassive black hole, and the number of nearby dwarf galaxies, astronomers conclude that the Milky Way must have formed when the gas within a huge "clump" of dark matter collapsed into a large number of small protogalaxies. Some of these smaller clumps are still around today in the form of small, satellite dwarf galaxies near our own. The largest among them are the Large Magellanic Cloud and Small Magellanic Cloud (**Figure 15.22**), which are easily seen by the naked eye in the Southern Hemisphere and appear much like detached pieces of the Milky Way. Another companion, the elliptical Sagittarius Dwarf, is now plowing through the disk of the Milky Way on the other side of the bulge. Astronomers have observed streams of stars from the Sagittarius Dwarf and some other dwarf galaxies that are being tidally disrupted by the Milky Way. At some point, these dwarf galaxies will become incorporated into the Milky Way—an indication that our galaxy is still growing by cannibalizing other galaxies.

The dwarf galaxies are the lowest-mass galaxies observed, and they are dominated by an even larger fraction of dark matter than are other known galaxies. These ultrafaint dwarf galaxies offer clues to the formation of the Local Group. In addition, observations of the motions and speeds of the dwarf galaxies about the Milky Way will lead to improved estimates of the amount of dark matter in the Milky Way itself.

Protogalaxies merged to form the barred spiral galaxy we call the Milky Way. In this process, stars were formed in the halo, in the bulge, and in the disk. The first stars to form ended up in the halo—some in globular clusters, but many not.

Gas that settled into the disk of the Milky Way quickly formed several generations of stars. A dense, concentrated mass quickly grew into the supermassive black hole at the center of the galaxy. The details of this process are still sketchy, but computer simulations indicate that so much mass was concentrated in this small region that almost any sequence of events would have led to the formation of a massive black hole.

The Milky Way offers many clues about the way galaxies form, but much of what we know of the process comes from looking beyond our local system. Images of distant galaxies (which we see as they existed billions of years ago), as well as observations of the glow left behind from the early stages of the universe itself, provide equally important pieces of the puzzle. In Chapter 16, we turn our attention to the evolution of the universe.

CHECK YOUR UNDERSTANDING 15.4

How do galaxies like the Milky Way form?

a. Individual stars form first and then gather together to form galaxies.

b. A single cloud of gas collapses to form a single large galaxy.

c. A large cloud fragments, forming small protogalaxies that later merge to become a large galaxy.

d. Galaxies like the Milky Way formed at the beginning of the universe and do not change.

---

reading Astronomy News

Dark matter particles won't kill you. If they could, they would have already

by Lisa Grossman, *Science News*

A lack of mysterious deaths from hypothetical 'macros' suggests dark matter is small and light.

The fact that no one seems to have been killed by speeding blobs of dark matter puts limits on how large and deadly these particles can be, a study posted July 18 at arXiv.org suggests.

"In the last 30 years, if someone had died of this, we would have heard of it," says physicist Glenn Starkman of Case Western Reserve University in Cleveland.

Physicists think the invisible dark matter must exist because they can see its gravita-tional effects on visible matter throughout the cosmos. But no one knows what it's actually made of. Among the leading candidates are weakly interacting massive particles, or WIMPs, but scientists have hunted for them for decades with no success (*SN: 6/23/18, p. 13*).

So physicists are turning to other theor-etical candidates (*SN Online: 4/9/18*). Starkman and colleagues focused on mac-roscopic dark matter, or macros, first pro-posed by physicist Edward Witten in the 1980s (*SN Online: 10/7/13*). If they exist, macros would be made up of subatomic particles called quarks, just like ordinary matter, but combined in a way never before observed.

Theoretically, macros could have almost any size and mass. And because dark matter doesn't interact with regular matter, there would be nothing to stop these particles from zipping around unimpeded. So Starkman—along with Case Western physicist Jagjit Singh Sidhu and physicist Robert Scherrer of Vanderbilt University in Nashville—decided to do a gut check using human flesh as a dark matter detector.

If a macro as small as a square microm-eter zipped through your body at hyper-sonic speed, it would deposit about as much energy in your body as a typical metal bul-let, the team calculated. But the damage it caused would be different from that of a bullet: A macro would heat the cylinder of tissue in its wake to about 10,000,000° Celsius—vaporizing the tissue and leaving a path of plasma.

"It's like if you were in Star Wars, and a Jedi hit you with their lightsaber, or someone shot you with their phaser [gun]," Starkman says.

There would be nothing you could do to shield yourself from such a macro strike. Still, there's no reason to worry, Starkman says. Considering there have been no reports of anyone suddenly suffering a mysterious lightsaber wound, the researchers concluded that if macros exist, they have to be smaller than a micrometer and heavier [see Editor's Note] than about 50 kilograms.

"The odds of dying from this are less than 1 in 100 million," Starkman says.

As wacky as this might sound, physicist Katherine Freese thought these calculations were worth doing. "This study is fun," says Freese of the University of Michigan in Ann Arbor. "Looking for macros in already existing detectors, such as the human body, is a good idea." Though she wasn't involved in the macro research, she and colleagues did a similar thought experiment with WIMPs in 2012. "But weak interactions are so weak as to be harmless" to human bodies.

Next, Starkman and Sidhu plan to look for macro tracks in slabs of granite, which would appear as cylinders of black obsidian running straight through the rock. They're starting with a cemetery near the Case Western campus.

Editor's Note: This story was updated July 26, 2019, to correct the size constraint put on macroscopic dark matter. According to an update from the authors, a macro would have to be heavier, not lighter, than 50 kilograms.

EVALUATING THE NEWS

1. This study proposes a dark matter candidate called "macros." Are these macros more similar to WIMPs or to MACHOs? Justify your answer.

2. Imagine that these macros were actually prevalent in the universe and were a significant component of dark matter. How would life on Earth be different than it is today?

3. The article discusses a "macro as small as a square micrometer." A human hair has a diameter of 50 micrometers, whereas a carbon filament has a diameter of about 6 micrometers. Are these macros large enough to see?

4. As a macro passed through your body, it would vaporize the tissue in its path. Would this hole be large enough to see? If not, would there be any other way to know that you had just had this interaction?

5. ". . . the researchers concluded that if macros exist, they have to be smaller than a micrometer and heavier than about 50 kilograms . . ." The mass of a grown human being is in the range of 70 kilograms. For the sake of visualization, consider a human hair (50 times larger in diameter than the size limit for the largest possible macro) with a mass of 50 kilograms. What would happen if your hair had this density?

Source: **Lisa Grossman**, "Dark matter particles won't kill you. If they could, they would have already," from *Science News*, July 25, 2019; updated July 26, 2019. Used with permission.

SUMMARY

Most of the mass of a galaxy is dark matter, which interacts with light very weakly, if at all. The form of this matter is not yet known. The Doppler velocities of radio lines show that the rotation curve of the Milky Way is flat, like those of other galaxies, so most of the mass of our galaxy is in the form of dark matter. The arms of spiral galaxies are regions of intense star formation, and the concentration of young stars makes the arms more visible. Disturbances to the disk can lead to spiral arms, but persistent spiral arms are the result of density waves. The chemical composition of the Milky Way has evolved with time, and there must have been a generation of stars before the formation of the oldest halo and globular-cluster stars we see today. Star formation is still actively occurring in the disk of our galaxy, leading to complex structures within the disk. The Milky Way formed from a collection of smaller protogalaxies that collapsed out of a halo of dark matter.

(1) Dark matter is identified by its gravitational interaction with ordinary matter—it has never been observed directly. The Milky Way's flat rotation curve is the most compelling evidence for a significant amount of dark matter in the Milky Way.

(2) Stars form in spiral arms where there is an abundance of gas and dust. The Milky Way has a disk and is therefore a spiral galaxy. A sketch of the spiral structure would include an elongated bulge with a spiral arm coming from each end, and a few arm fragments.

(3) The ages of stellar clusters can be determined from the main-sequence turnoff. Younger stars are abundant in heavier elements because previous generations of stars enriched the galaxy in these heavier elements as they died.

(4) The Milky Way provides many clues to understand galaxy formation. The properties of stars in the halo, including stars in globular clusters and high-velocity stars, indicate that these stars formed from clouds of gas well before those clouds settled into the galaxy's disk. The dwarf satellites and other neighbors in the Local Group are evidence that the Milky Way is growing by absorbing other galaxies.

QUESTIONS AND PROBLEMS

TESTING YOUR UNDERSTANDING

1. **T/F:** Dark matter is found by comparing the amount of light from stars with the amount of light from dust and gas.

2. **T/F:** The ages of globular clusters put a lower limit on the age of the Milky Way.

3. **T/F:** The Milky Way contains almost no dark matter.

4. **T/F:** Stars that are more distant from the center of a galaxy orbit much more slowly than those closer to the center.

5. **T/F:** The disk of the Milky Way is threaded with magnetic field lines.

6. Dark matter was first detected because the _____ for the amount of gravity that could be present due to the luminous mass.
   a. speed of galaxies through space was too high
   b. speed of galaxies through space was too low
   c. rotation speed of galaxies was too high
   d. rotation speed of galaxies was too low

7. In spiral galaxies, stars form predominantly in
   a. the arms of the disk.   b. the halo.
   c. the bulge.   d. the bar.

8. The rotation curves of galaxies tell us that galaxies are mostly
   a. stars.   b. dust and gas.
   c. dark matter.   d. black holes.

9. MACHOs are most commonly found in the halo because they
   a. move most quickly there, so they are easier to see in motion.
   b. move most quickly there, so their Doppler shift is larger.
   c. move most slowly there, so they spend more time there.
   d. are easier to see far from the disk of the Milky Way.

10. The flat rotation curves of spiral galaxies imply that the distribution of mass resembles
   a. the Solar System; most mass is concentrated in the center.
   b. a wheel; the density remains the same as the radius increases.
   c. the light distribution of the galaxy; a large concentration occurs in the middle, but significant mass exists quite far out.
   d. an invisible sphere much larger than the visible galaxy.

11. In general, older stars have lower _____ than younger stars.
   a. orbital speeds
   b. abundance of heavy elements
   c. luminosities
   d. rotation rates

12. The Milky Way formed when the gas within a clump of dark matter collapsed into a large number of
   a. globular clusters.
   b. high-velocity stars.
   c. supermassive black holes.
   d. small protogalaxies.

13. Radio emission reveals that the Milky Way is
   a. an elliptical galaxy.
   b. an irregular galaxy resulting from a collision.
   c. a spiral with three arms.
   d. a barred spiral with two major arms.

14. In the disk of the Milky Way, stars are _____ and dust and gas are more _____ than in the halo.
   a. younger; diffuse   b. older; diffuse
   c. older; dense   d. younger; dense

15. Looking toward the galactic center, we see no redshift or blueshift. This tells us
   a. the center of the galaxy is motionless.
   b. the Sun and the center are not moving toward or away from each other.
   c. the Milky Way is stationary with respect to other galaxies.
   d. the Sun is stationary with respect to the center of the galaxy.

16. In general, the Milky Way
   a. has the same chemical composition as time passes.
   b. has more abundant hydrogen as time passes.
   c. has more abundant heavy elements as time passes.
   d. has less abundant heavy elements as time passes.

17. Cosmic rays are
   a. a form of electromagnetic radiation.
   b. high-energy particles.
   c. high-energy dark matter.
   d. high-energy photons.

18. In the Hubble scheme for classifying galaxies, what type of galaxy is the Milky Way?
   a. elliptical
   b. spiral
   c. barred spiral
   d. irregular

19. The youngest stars in the Milky Way are
   a. in the core.
   b. in the bulge.
   c. in the disk.
   d. in the halo.

20. Why are most of the Milky Way's satellite galaxies so difficult to detect?
   a. They are very faint.
   b. Their dark matter halos obscure them.
   c. The halo of the Milky Way obscures the view.
   d. They are very massive, and light does not escape them.

THINKING ABOUT THE CONCEPTS

21. What is the difference between the way dark matter interacts with electromagnetic radiation and the way normal matter does?

22. Contrast the rotation curve for a galaxy containing only normal matter and that for a galaxy containing a large amount of dark matter.

23. How does a spiral galaxy's dark matter halo differ from its visible spiral component?

24. To find dark matter in a galaxy, you must be able to measure both how fast it is spinning and how the brightness drops with radius. Would a face-on spiral galaxy (like the one shown in Figure 15.8) be a good candidate to study if you were trying to find dark matter? Why or why not? ⊙

25. To find dark matter in a galaxy, you must be able to measure both how fast it is spinning and how the brightness drops with radius. Would an edge-on spiral galaxy (like the one shown in Figure 15.10b) be a good candidate to study if you were trying to find dark matter? Why or why not? ⊙

26. Explain how the temperature of a hot gas can be used to find out about the gravitational pull of a galaxy.

27. Describe the spiral arms in a galaxy, and explain at least one of the mechanisms that create them. Explain why star formation in spiral galaxies takes place mostly in the spiral arms.

28. ★ **WHAT AN ASTRONOMER SEES** On the basis of images like Figure 15.10a, explain the logic that leads us to determine that the Milky Way is a spiral galaxy. How would these images be different if we lived in an elliptical galaxy? ⊙

29. How do 21-cm radio observations reveal the rotation of our galaxy? What does the rotation curve of our galaxy say about the presence of dark matter in the galaxy?

30. Explain how astronomers know that open clusters are younger than globular clusters.

31. What does the abundance of a star's heavy elements tell us about the age of the star?

32. Where do we find the youngest stars in our galaxy?

33. Halo stars are found in the vicinity of the Sun. What observational evidence distinguishes them from disk stars?

34. Can a cosmic ray travel at the speed of light? Why or why not?

35. What is the origin of the Milky Way's satellite galaxies? What has been the fate of most of the Milky Way's satellite galaxies? Why are most of the Milky Way's satellite galaxies so difficult to detect?

APPLYING THE CONCEPTS

36. Working It Out 15.1 requires that the average radius of the orbit be given in astronomical units, in order that the period comes out in years. However, most distances within galaxies are given in light-years or parsecs. Remind yourself how to convert between light-years, parsecs, and astronomical units. The table inside the back cover gives the conversion factors.
  a. Make a prediction: Is a parsec bigger or smaller than a light-year? Do you expect the number of parsecs to be bigger or smaller than the number of light-years? Is an astronomical unit bigger or smaller than a light-year? Do you expect the number of astronomical units to be bigger or smaller than the number of light-years?
  b. Convert 160,000 light-years into parsecs. To determine whether you need to multiply or divide by 3.262, consider your prediction about whether the answer should be larger or smaller than 160,000.
  c. Convert 160,000 light-years into astronomical units.
  d. Check your work by comparing your answers to your predictions.

37. How far does a star at a distance of 33,000 light-years from the center of the Milky Way travel in one orbit around the center of the Milky Way? Recall that the circumference of a circle is $2\pi R$ and that stars in the disk travel on roughly circular orbits.
  a. Make a prediction: Should the circumference of the orbit be larger or smaller than 33,000 light-years? By about how much?
  b. Calculate the circumference of the orbit.
  c. Check your work by comparing your answer to your prediction.

38. Building on problem 37, what is the period of a star at a distance of 40,000 light-years from the center of the Milky Way?
  a. Make a prediction: Compare this star's rotation velocity to the rotation velocity of the Sun (27,000 light-years from the center of the Milky Way) in Figure 15.5. Is this star closer to or farther from the center of the Milky Way than the Sun? Will it take more or less time to orbit once? ⊙
  b. Solve the equation distance = rate × time for "time." Then calculate the time it takes the star to travel the circumference of the orbit at its rotation velocity.
  c. Check your work by comparing your answer to the period of the Sun's orbit, given in Working It Out 15.1.

39. Building on problems 36 and 38, you have solved for the period of the orbit in years, and you have the semimajor axis in astronomical units. Follow Working It Out 15.1 to find the mass of the Milky Way interior to this star's orbit.
  a. Make a prediction: Do you expect your answer to be larger or smaller than the answer for the Sun in Working It Out 15.1?
  b. Calculate the mass of the Milky Way interior to this star's orbit.
  c. Check your work by comparing your answer to your prediction.

40. Follow Working It Out 15.1 (and problems 36–39) to find the mass of the Milky Way interior to a star orbiting at 50,000 light-years from the center of the Milky Way.
  a. Make a prediction: Do you expect this answer to be larger or smaller than the answer for the Sun in Working It Out 15.1? Larger or smaller than your answer to problem 39?
  b. Calculate the mass of the Milky Way interior to a star orbiting at 50,000 light-years from the center of the Milky Way.
  c. Check your work by comparing your answer to your prediction.

exploration Dark Matter in a Galaxy

The data in **Table 15.1** have been taken from the rotation curve of a spiral galaxy. Astronomers have found that most of the stars in this galaxy lie within a radius of $r = 1.64 \times 10^{20}$ meters (m) and that the total mass within this radius is $M = 1.54 \times 10^{41}$ kilograms (kg). In the absence of dark matter, the stars outside this radius should orbit in the same way that planets orbit the Sun. In this activity, you will test that hypothesis by comparing the calculated speed for an object orbiting, like a planet, to the measured speed of actual stars in the galaxy.

Table 15.1 Rotation Velocities and Derived Masses for a Spiral Galaxy

Orbital Radius (m)	Calculated Speed (m/s)	Measured Speed (m/s)	Gravitational Mass (kg)	Percent Missing Mass (%)
1.85×10^{20}	2.36×10^{5}	2.47×10^{5}	1.69×10^{41}	9
2.75×10^{20}	1.93×10^{5}	2.40×10^{5}	2.37×10^{41}	35
3.18×10^{20}	1.80×10^{5}	2.37×10^{5}	2.68×10^{41}	43
4.26×10^{20}	1.55×10^{5}	2.25×10^{5}	3.23×10^{41}	52
6.48×10^{20}	1.26×10^{5}	2.47×10^{5}	5.93×10^{41}	74

1. The "calculated speed" assumes that the mass of the galaxy is concentrated toward the center of the galaxy. You have seen images of spiral galaxies. Does this seem like a reasonable assumption? Explain why or why not.

_____

_____

_____

2. Study the numbers in the column labeled "calculated speed." How does the calculated speed change as the radius increases? What would you expect a graph of these data to look like?

_____

_____

_____

_____

3. Study the numbers in the column labeled "measured speed." How does the calculated speed change as the radius increases? What would you expect a graph of these data to look like?

_____

_____

_____

4. Make a graph of calculated speed versus orbital radius, and smoothly connect the data points to make a curve. Label this curve "calculated speed." *On the same graph*, add a second curve of measured speed versus orbital radius.

5. Compare the curves you have made. Do they support or refute the hypothesis that stars in a galaxy orbit in the same manner as planets around a star? Why or why not?

_____

_____

_____

6. The gravitational mass has been calculated for you from the measured speed. This mass has then been compared to the luminous mass in the final column of Table 15.1. How does the amount of missing mass change as the radius increases? If you could extend these data out to even further radii, would you expect the percentage of missing mass to increase, decrease, or stay the same?

_____

_____

_____

7. Explain how observations like these led to the discovery of dark matter.

_____

_____

_____

The Evolution of the Universe

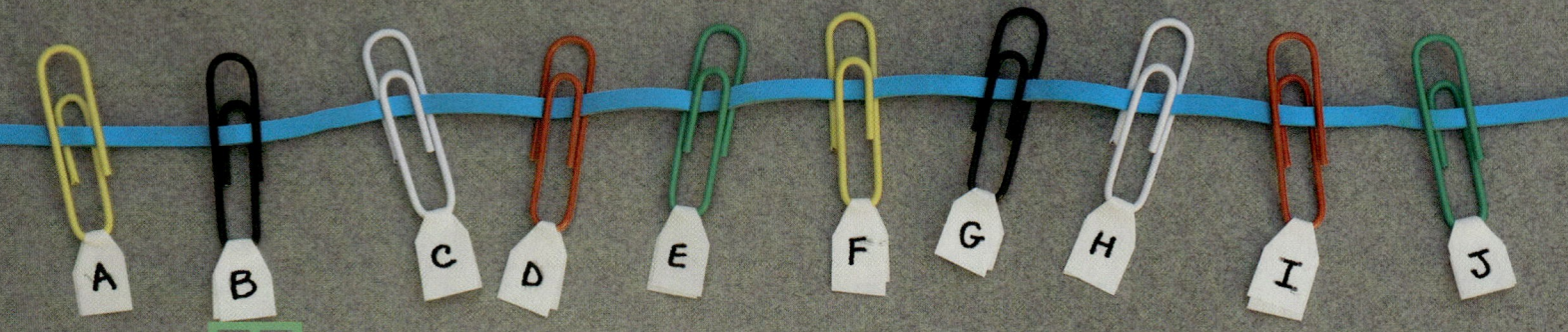

The universe expands, carrying galaxies farther and farther apart. This can be hard to visualize, but a one-dimensional example can help. Cut a rubber band in one place, making a "rubber string." Attach at least six paper clips to the rubber band, and label them A, B, C, D, etc. To simulate the expanding universe, you will stretch the rubber band. Before you begin, predict how the distance to more distant paper clips will grow compared to the distance to nearer paper clips. Lay the rubber band along a ruler, and record the distance from a paper clip near the center (the "home" paper clip) to all the other paper clips. Stretch the rubber band, keeping the home paper clip at the same location on the ruler. Measure and record the distance to all the other paper clips along the stretched rubber band. (You may need the help of a friend.) Select a new "home" paper clip near the center, and repeat the experiment, first measuring distances along the unstretched rubber band, then measuring distances along the stretched rubber band. Make sure to stretch the rubber band the same amount in both experiments. Make a graph of "stretched distance" versus "original distance" for each experiment, and compare these graphs. Is the shape of the graph the same for each experiment?

EXPERIMENT SETUP

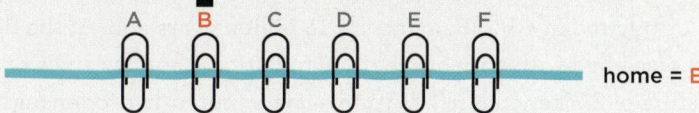

home = B

Lay the rubber band along a ruler and measure
the distance from "home" to the other paper clips.

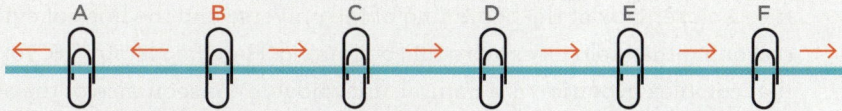

Stretch the rubber band and measure again.

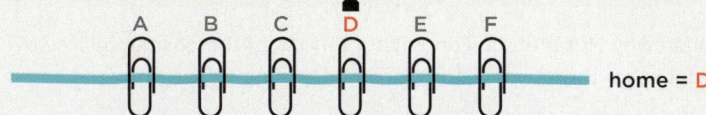

home = D

Repeat the experiment with a new "home" paper clip,
stretching the rubber band the same amount.

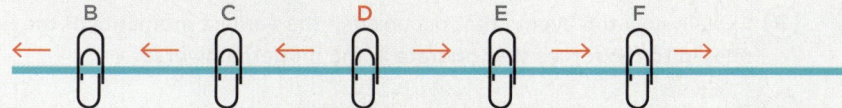

PREDICTION

I predict that the distance to
more distant paper clips will grow
☐ more than ☐ less than ☐ the same
as the distance to nearer paper clips.

SKETCH OF RESULTS (in progress)

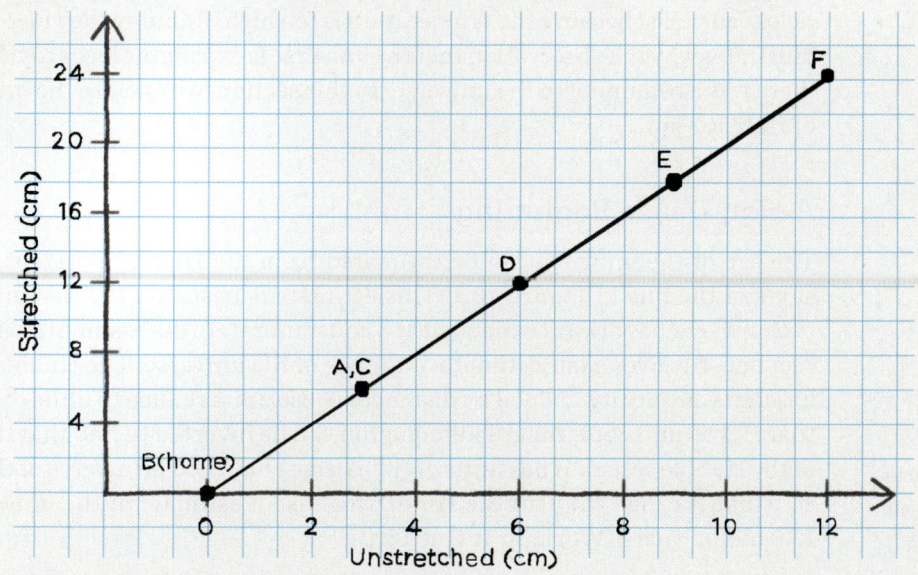

16

Cosmology is the study of the universe on the very grandest of scales, including its nature, origin, evolution, and ultimate destiny. The universe originated in a Big Bang nearly 14 billion years ago. At the Big Bang, the universe was filled with very hot thermal radiation (that has since cooled to a temperature of 2.7 kelvins [K]). Within a few minutes, hydrogen and helium were produced. The universe that emerged from the Big Bang was incredibly uniform, wholly unlike today's universe of galaxies, stars, and planets. In this chapter, we take a closer look at the beginning of the universe and the lines of evidence and reasoning that help us explore this beginning. Here in Chapter 16, you will find that complex evolution is a natural, unavoidable consequence of the action of physical laws in our universe.

LEARNING GOALS

(1) Explain why Hubble's law and observations of the cosmic microwave background indicate that there was a hot, dense beginning to the universe.

(2) Describe the proposed connection between the accelerating universe and dark energy.

(3) Explain how the events that occurred in the earliest moments of the universe are related to the forces that operate in the modern universe.

(4) Describe how inflation solves both the flatness problem and the horizon problem.

16.1 Hubble's Law Implies a Hot, Dense Beginning

 **▶❚❚ AstroTour:** Hubble's Law

Recall that Hubble's law shows that the universe is expanding. Imagine watching a video of the universe: Time is moving forward and the galaxies move apart. Now reverse the video, and run it backward in time. As you watch, the separation between the galaxies shrinks as the universe becomes younger. The universe gets increasingly hot and dense as time rolls backward in the video, until atoms and even particles can't exist because the temperature is too high. Running the video backward illustrates why the observation that the universe is expanding leads to the idea that there was a beginning to the universe. In this section, we explore the implications of Hubble's law.

A Hot, Dense Beginning

The age of the universe can be estimated from the Hubble constant (H_0)—the slope of the line in **Figure 16.1**. This estimate is based on the assumption that the universe has always expanded at a constant rate. This assumption is not correct but still gives a fair estimate of the age of the universe. The Hubble constant has units of velocity divided by distance. These units reduce to units of 1/time, so that H_0 has units of 1/time. This equation can be inverted to find that the inverse of the Hubble constant has units of time (time = $1/H_0$). The inverse of the Hubble constant is called the **Hubble time**, which is an estimate of the universe's age: 13.8 billion years (**Working It Out 16.1**).

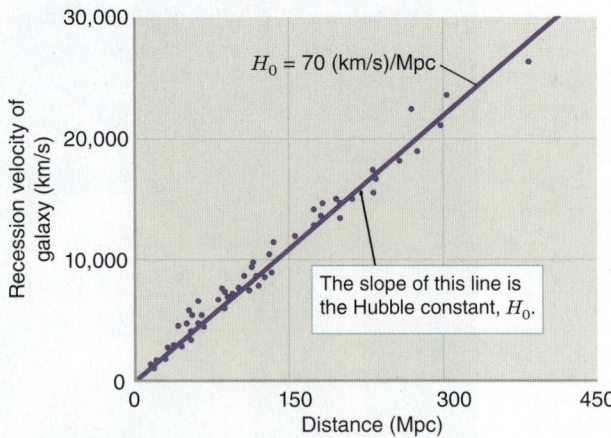

Figure 16.1 The slope of the line in the graph of the Hubble law is the Hubble constant. The inverse of this constant is called the Hubble time and is an estimate of the age of the universe.

(Figure labels: Recession velocity of galaxy (km/s); H_0 = 70 (km/s)/Mpc; The slope of this line is the Hubble constant, H_0. Distance (Mpc))

Expansion and the Age of the Universe

We can use Hubble's law to estimate the age of the universe. Consider two galaxies located 30 megaparsecs (Mpc) from each other (that is, $d_G = 9.3 \times 10^{20}$ kilometers [km]; see **Figure 16.2**). If these two galaxies are moving apart from each other at constant speed, then at some time in the past they must have been together in the same place at the same time. According to Hubble's law (Figure 16.1), and assuming that $H_0 = 70$ kilometers per second per megaparsec or (km/s)/Mpc, the distance between these two galaxies is increasing at the rate

$$v_r = H_0 \times d_G$$

$$v_r = 70 \ (\text{km/s})/\text{Mpc} \times 30 \ \text{Mpc}$$

$$v_r = 2{,}100 \ \text{km/s}$$

Knowing the speed at which they are traveling, we can calculate the time it took for the two galaxies to become separated by 9.3×10^{20} km:

$$\text{Time} = \frac{\text{Distance}}{\text{Speed}} = \frac{9.3 \times 10^{20} \ \text{km}}{2{,}100 \ \text{km/s}} = 4.4 \times 10^{17} \ \text{s}$$

Dividing by the number of seconds in a year (about 3.2×10^7) gives

$$\text{Time} = 1.4 \times 10^{10} \ \text{yr}$$

In other words, *if* expansion of the universe has been constant, two galaxies that today are 30 Mpc apart started out at the same place about 14 billion years ago.

Now let's do the same calculation with two galaxies that are 60 Mpc (19×10^{20} km) apart. These two galaxies are twice as far apart, but the distance between them is increasing twice as rapidly:

$$v_r = H_0 \times d_G = (70 \ \text{km/s})/\text{Mpc} \times 60 \ \text{Mpc} = 4{,}200 \ \text{km/s}$$

Therefore,

$$\text{Time} = \frac{\text{Distance}}{\text{Speed}} = \frac{19 \times 10^{20} \ \text{km}}{4{,}200 \ \text{km/s}} = 4.5 \times 10^{17} \ \text{s}$$

Dividing by the number of seconds in a year gives

$$\text{Time} = 1.4 \times 10^{10} \ \text{yr}$$

Thus, these galaxies also took about 14 billion years to reach their current locations. We could do this calculation again and again for any pair of galaxies in the universe today. (Small differences in the intermediate steps of these calculations are due to rounding and are not significant to the argument.) The most precise measurements give a result of 13.8 billion years.

If we work out the example using words instead of numbers, we can see why the answer is always the same. Because the velocity we are calculating comes from Hubble's law, velocity equals the Hubble constant multiplied by distance. Writing this out as an equation, we get

$$\text{Time} = \frac{\text{Distance}}{\text{Velocity}}$$

$$\text{Time} = \frac{\text{Distance}}{H_0 \times \text{Distance}}$$

Distance divides out to give

$$\text{Time} = \frac{1}{H_0}$$

This time, equal to $1/H_0$, is known as the *Hubble time*. This estimate assumes that the universe has always expanded at the same rate.

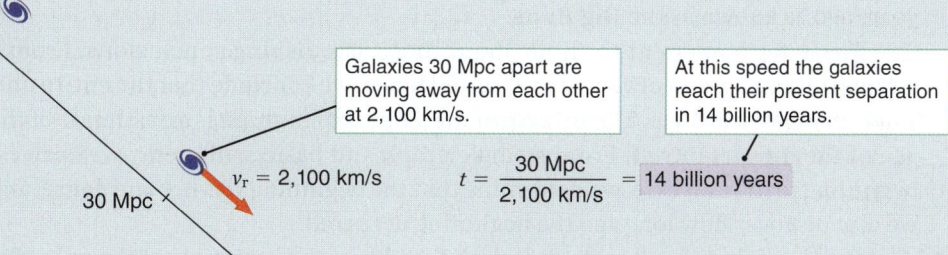
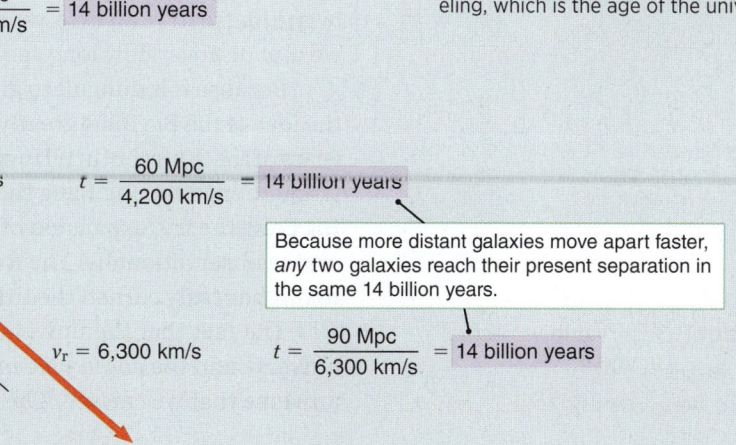

Galaxies 30 Mpc apart are moving away from each other at 2,100 km/s.

At this speed the galaxies reach their present separation in 14 billion years.

30 Mpc

$v_r = 2{,}100$ km/s $t = \dfrac{30 \ \text{Mpc}}{2{,}100 \ \text{km/s}} = 14$ billion years

60 Mpc

$v_r = 4{,}200$ km/s $t = \dfrac{60 \ \text{Mpc}}{4{,}200 \ \text{km/s}} = 14$ billion years

Because more distant galaxies move apart faster, *any* two galaxies reach their present separation in the same 14 billion years.

90 Mpc

$v_r = 6{,}300$ km/s $t = \dfrac{90 \ \text{Mpc}}{6{,}300 \ \text{km/s}} = 14$ billion years

Figure 16.2 These three galaxies have three different apparent speeds. They are also located at three different distances from the Milky Way. Dividing each distance by each speed gives the same elapsed time: about 14 billion years. This is the amount of time that they have been traveling, which is the age of the universe.

Figure 16.3 Looking backward in time, the distance between any two galaxies is smaller and smaller, until the entire universe is concentrated together at the same point: the Big Bang.

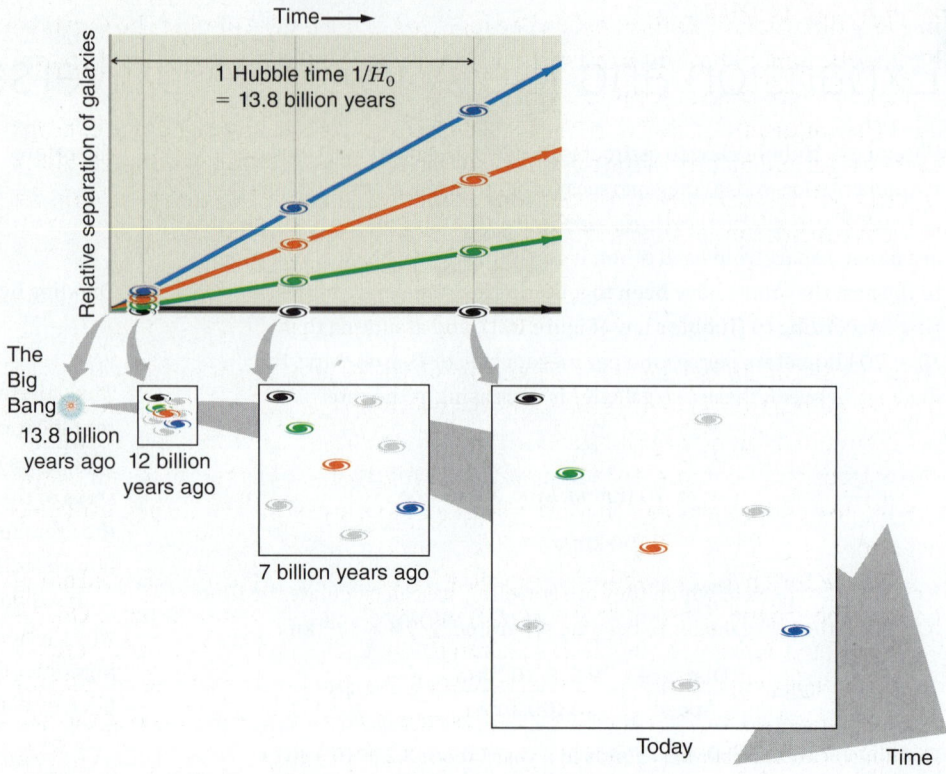

what if . . .

What if the universe were contracting, rather than expanding, with a negative Hubble constant: How would this affect the Doppler shift of other galaxies?

 **Astronomy in Action:** Observable vs. Actual Universe

If the universe has always expanded at the same rate, then 6.9 billion years ago, when the universe was half its current age, all of the galaxies in the universe were half as far apart as they are now (**Figure 16.3**). About 12.4 billion years ago, all of the galaxies in the universe were about a tenth as far apart. A little less than one Hubble time ago—13.8 billion years ago—there was almost no space between the particles that constitute today's universe. All matter as well as energy in the universe was incredibly dense. The universe has cooled since then, so it was much hotter than it is today in its expanded state. This hot, dense beginning, 13.8 billion years ago, is known as the **Big Bang**.

Pause for a moment to think about these astonishing conclusions. From a straightforward graph of velocity versus distance, we conclude that the entire universe changes over time. The inverse of the slope of this graph is an estimate of the age of the universe itself. From a single graph and basic arithmetic, we derive a beginning to the universe, we determine that the beginning was hot and dense, and we also discover how long ago the beginning occurred.

Because it is difficult to grapple with the idea of the beginning of the universe, the idea of the Big Bang greatly troubled many astronomers in the early and middle years of the 20th century. However, further observations and discoveries have only strengthened the Big Bang theory. As you will see, all the major predictions of the Big Bang theory (expansion of the universe is only one of them) have been corroborated observationally. The Big Bang theory is now so well supported by evidence that it has truly earned the title of scientific theory.

The fact that the universe has a finite age creates a distinction between the universe and the *observable universe*. The **observable universe** is the part of the universe that we can see. The observable universe extends 13.8 billion light-years in every direction. This limit exists because that is the length of time the universe has been around. The light from more distant regions has not yet had time to travel to us, and so we cannot see it yet.

The implications of Hubble's law are striking and have changed our concept of the origin and history of the universe in which we live. At the same time, Hubble's law has pointed to many new questions about the universe. To address them, we need to consider precisely what the term *expanding universe* means.

 **Astronomy in Action:** Expanding Balloon Universe

Carried Along by the Expansion

At this point in our discussion, you may be picturing the expanding universe as a cloud of debris from an explosion flying outward through surrounding space. This is a common depiction of the Big Bang in movies and television shows. However, this depiction is completely incorrect. The Big Bang is not an explosion in the usual sense of the word, and there is no surrounding space into which the universe expands.

Recall the chapter-opening experiment, in which the observer is carried along on the rubber band. The paper clips do not move relative to the rubber band: Observation of their recession velocities measures how much the rubber band stretches between paper clips. Similarly, observation of the recession velocities of galaxies measures how much the space between them has expanded, not how fast the galaxies are moving through space.

The Shape and Size of the Universe

The overall shape of the universe may be curved or flat, and if it is curved, it may have positive or negative curvature. Flat space has the type of geometry you learned in school. For example, parallel lines remain parallel no matter where you measure them. You have already learned about an example of positively curved space, near a massive object such as a black hole. The geometric rules of positively curved space apply when we consider positions or distances on the curved surface of Earth. In positively curved space, parallel lines converge. For example, lines of longitude are parallel to each other at the equator but converge to a point at the North Pole. Negatively curved space is like the shape of the center of a Pringles potato chip, and it has geometric rules opposite to those on the surface of Earth. That is, parallel lines diverge away from each other in negatively curved space.

As you will learn in Section 16.5, current observations indicate that our universe is flat on the largest scales, to a quite high degree of precision. (It is so flat, in fact, that its very flatness is a problem!) From the cosmological principle, we may reason that if the universe is flat in the part of the universe we can see, it is flat everywhere. If that is true, then two parallel lines would remain parallel forever and never converge or diverge. This is another way to say that the universe is infinite. An infinite universe extends forever in every direction, so that it has no edges. This means that it also has no center, because in order to find the center of something, you first find the edges, and then find a point halfway between them.

We do not know for certain that the universe is exactly flat, because measurements are not infinitely precise. Therefore, we also do not know if the universe is truly infinite or just very much larger than the part of it that is observable. However, if the universe is infinite, there is certainly no outside, so it is not expanding "into" anything.

 **Astronomy in Action:** Infinity and the Number Line

As a tiny creature, you are more or less confined to the surface of Earth. This surface is finite, yet you could walk around it and around it and around it, and never find an edge. The universe as a whole object might be similarly finite, yet without boundaries. If you traveled far enough through the universe in one direction, you might find yourself back where you started. In this type of universe, there is nothing for the universe to expand into; there is no "outside." We have no observational evidence that our universe curves around itself in this way. We may never know if our universe is flat everywhere or if it eventually curves back upon itself in this way.

In an infinite universe, there is no center, because there are no edges. What about the finite, unbounded universe? This universe has no center, because, again, there are no edges. Just as there is no central point on the surface of Earth, there is no center of a finite, unbounded universe. Both of these "centerless" cases fit nicely with the cosmological principle, which can be interpreted as "no place is special." If there were a center to the universe, that would be a special location, perhaps with its own unique physics. If there were an edge to the universe, that too would be a special location, and the rules of physics might very well be different there.

Wherever you are in the universe today, you are sitting at the site of the Big Bang. The Big Bang took place *everywhere*. The entire universe formed at once, and the space between locations in the universe has been stretching ever since, carrying the galaxies along in the expansion. Hubble's law is an observable consequence of the expanding space between galaxies.

Redshift and the Expansion of the Universe

We have seen that the galaxies are not moving through space; rather, the space between them is growing. If the galaxies are stationary, why is their light redshifted? Although it is true that the distance between galaxies is increasing as a result of the expansion of the universe and that we can use the *equation* for Doppler shifts to express galaxy redshifts in terms of velocity, these redshifts are not due to Doppler shifts.

Recall the rubber-sheet analogy that we used when discussing black holes in Section 13.5. In that case, we imagined space as a two-dimensional rubber sheet. **Figure 16.4** uses this rubber-sheet analogy to explain why the increasing distance between galaxies causes the light to be redshifted. If a series of bands on the rubber sheet represent the crests of a light wave, we can see what happens to the wave as the sheet is stretched. As light comes from distant galaxies, the space through which the light travels is stretching, and the light also becomes "stretched out." The farther light travels through expanding space, the longer (redder) its wavelength becomes. Light from distant galaxies is redder than that from nearby ones, because the light has traveled farther through expanding space to reach us. The redshift of light from distant galaxies is therefore a direct measure of how much the universe has expanded since the time the light left its source.

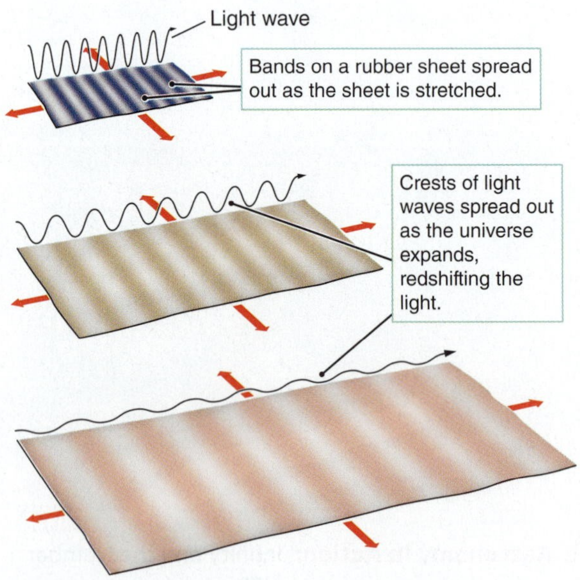

Light wave

Bands on a rubber sheet spread out as the sheet is stretched.

Crests of light waves spread out as the universe expands, redshifting the light.

Figure 16.4 Bands drawn on a rubber sheet represent the positions of the crests of a light wave in space. As the rubber sheet is stretched—that is, as the universe expands—the wave crests get farther apart. The light is redshifted.

CHECK YOUR UNDERSTANDING 16.1

Hubble's law and the Big Bang theory are related because
a. Hubble's law is an observation that led to the development of the theory.
b. Hubble's law was observed to test the hypothesis of the Big Bang.
c. Hubble's law is a hypothesis that eventually became the Big Bang theory.
d. Hubble's law is a fact, and the Big Bang theory is an untested hypothesis.

Answers to Check Your Understanding questions are in the back of the book.

16.2 The Cosmic Microwave Background Confirms the Big Bang

We live in a time when astronomers are finding real, testable answers to cosmological questions using the empirical methods of science. What evidence supports the Big Bang, apart from Hubble's law? One important piece of evidence comes from direct observations of the early universe across the entire sky. Here in Section 16.2, you will learn about the cosmic microwave background, one of the major confirming observations of the theory that the universe had a beginning.

Radiation from the Big Bang

In the late 1940s, Ralph Alpher (1921–2007), Robert Herman (1914–1997), and George Gamow (1904–1968) made a prediction that was based on the conditions of the early universe. When the universe was very young and small, it consisted of an extraordinarily hot, dense gas. This hot, dense, early universe was filled with blackbody radiation. As the universe expanded and cooled, the wavelength of the radiation became longer and longer. As illustrated in **Figure 16.5**, doubling the wavelength of the photons in a blackbody spectrum by stretching space in the universe is equivalent to halving the temperature of the object emitting the blackbody spectrum. Alpher, Herman, and Gamow concluded that the radiation from the early universe should still be visible today as a blackbody spectrum with a temperature of about 5–10 K.

Early attempts to detect this radiation were unsuccessful. However, in the early 1960s, two physicists at Bell Laboratories, Arno Penzias (b. 1933) and Robert Wilson (b. 1936), were trying to bounce radio signals off Echo 1, a newly launched satellite. This seems trivial today, when we routinely use smartphones that communicate directly with the Global Positioning System (GPS) and other satellites.

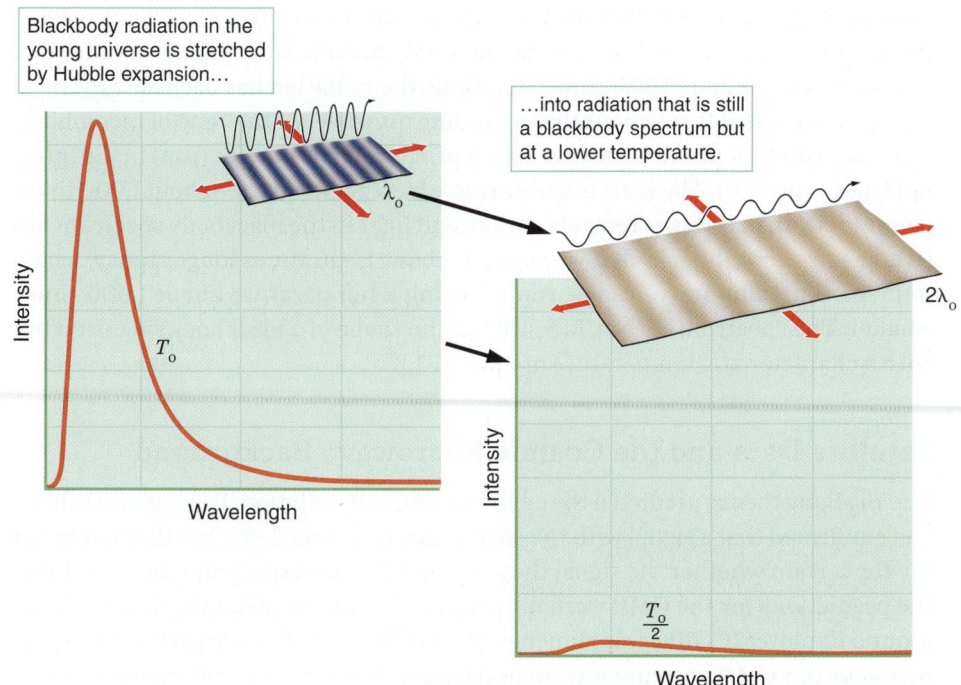

Figure 16.5 As the universe expanded, blackbody radiation left over from the hot young universe was redshifted to longer wavelengths. (Recall that λ [lowercase lambda] represents wavelength.) Redshifting a blackbody spectrum is equivalent to lowering its temperature.

Figure 16.6 Robert Wilson (left) and Arno Penzias (right) are shown next to the Bell Labs radio telescope antenna with which they discovered the cosmic microwave background. This antenna in Holmdel, New Jersey, is now a U.S. National Historic Landmark.

But at the time, the effort pushed technology to its limits. Penzias and Wilson needed a very sensitive microwave telescope to detect the faint signals that bounced off the satellite. It had to be so sensitive, in fact, that any signals coming from the telescope itself might wash out the signals they were trying to detect. Wilson and Penzias, pictured with their radio telescope in **Figure 16.6**, worked tirelessly to eliminate all possible interference from their instrument. This work included such endless and menial tasks as keeping the telescope free of bird droppings. Even so, Penzias and Wilson found that they still detected a faint microwave signal when they pointed the telescope at the sky. Eventually, they came to accept that the faint signal was coming from the sky itself, which glows faintly in microwaves.

In the meantime, a physicist named Robert Dicke (1916–1997) and his colleagues at Princeton University also predicted a hot early universe, arriving independently at the same basic conclusions that Alpher, Herman, and Gamow had reached 20 years earlier. When Dicke and colleagues heard of the faint microwave signal from the sky, they interpreted it as the radiation left behind by the hot early universe. The strength of the detected signal was consistent with the glow from a blackbody with a temperature of about 3 K, close to the predicted value. Penzias and Wilson's results, published in 1965, reported the discovery of the glow from the Big Bang. Penzias and Wilson shared The Nobel Prize in Physics 1978 for this remarkable discovery.

The microwave radiation from the early universe is called **cosmic microwave background (CMB) radiation**. The early universe was hot enough that all of the atoms were ions, with their liberated electrons zipping freely about. Free electrons in such conditions interact strongly with radiation, blocking its progress. As illustrated in **Figure 16.7a**, the universe was opaque because it was so hot and dense.

As the universe expanded, the gas within it cooled. By the time the universe was several hundred thousand years old, the temperature had dropped to about 3000 K. This temperature is low enough that hydrogen and helium nuclei can combine with electrons to form neutral atoms in a process called **recombination**. Hydrogen atoms block radiation much less effectively than free electrons, so when recombination occurred, the universe suddenly became transparent to radiation, as illustrated in Figure 16.7b. Since that time, the radiation has been able to travel largely unimpeded throughout the expanding universe. At the time of recombination, the radiation peaked at a wavelength of about 1 micrometer (μm) in the infrared (see Figure 4.6). The scale of the universe has expanded to be about 1,000 times larger today than it was at recombination, and light in the blackbody spectrum has stretched so that it peaks at a wavelength about 1,000 times longer today, which puts it in the microwave range, representing a temperature about 1,000 times smaller. The spectrum of the CMB still has the shape of a blackbody spectrum but with a characteristic temperature of only 2.73 K.

Satellite Data and the Cosmic Microwave Background

The Big Bang theory predicted the existence of CMB radiation. Penzias and Wilson had confirmed that a signal with the correct strength was there, but they could not say for certain whether the signal they saw had the right spectrum. Decades later, the predictions for the CMB were fully tested. A satellite called the Cosmic Background Explorer (COBE) was launched in 1989. The COBE made precise measurements of the CMB at many wavelengths, from 100 μm to 1 centimeter (cm). In

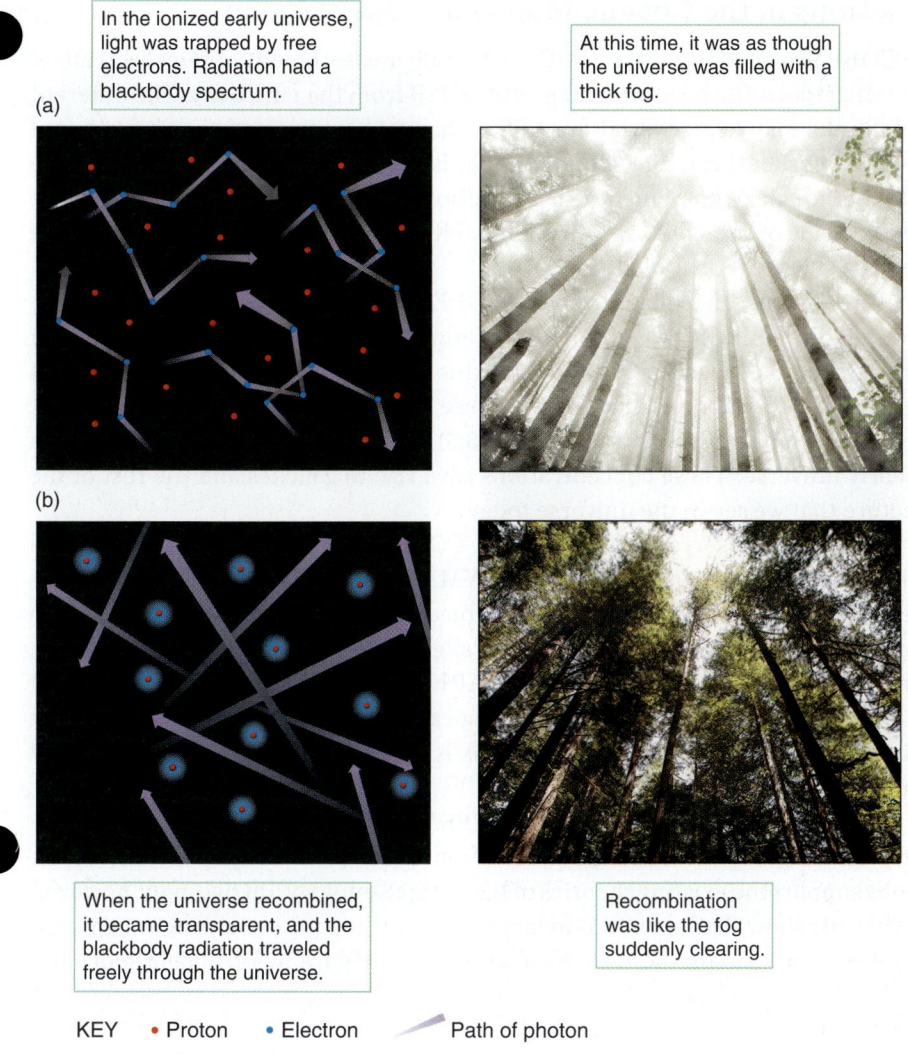

In the ionized early universe, light was trapped by free electrons. Radiation had a blackbody spectrum.

At this time, it was as though the universe was filled with a thick fog.

When the universe recombined, it became transparent, and the blackbody radiation traveled freely through the universe.

Recombination was like the fog suddenly clearing.

KEY   • Proton   • Electron   Path of photon

Figure 16.7 The cosmic microwave background radiation originated at the moment the universe became transparent. (a) Before recombination, the universe was like a foggy day, except that the "fog" was a sea of hydrogen atoms. Radiation interacted strongly with free electrons and so could not travel far. The trapped radiation had a blackbody spectrum. (b) When the universe recombined, the fog cleared and this radiation was free to travel unimpeded.

January 1990, hundreds of astronomers gathered in a large conference room in Washington, DC, at the winter meeting of the American Astronomical Society to hear the COBE team present its first results. Security surrounding the new findings had been tight, so the atmosphere in the room was electric. The tension did not last long; the presentation of a single graph brought the room's occupants to their feet in a spontaneous ovation.

The data shown on that graph are reproduced in **Figure 16.8**. The small dots in the figure are the COBE measurements of the CMB at different frequencies. The uncertainty (the "error bar") in each measurement is far less than the thickness of the line, indicating that the observation is extremely precise. The line in the figure, which runs perfectly through the data points, is a blackbody spectrum with a temperature of 2.73 K. The agreement between theoretical prediction and observation is truly remarkable. The observed spectrum so perfectly matches the one predicted by Big Bang cosmology that there can be no real doubt we are seeing the residual radiation left behind from the hot, dense beginning of the early universe. John Mather (b. 1946) and George Smoot (b. 1945) were awarded The Nobel Prize in Physics 2006 for their work on the COBE.

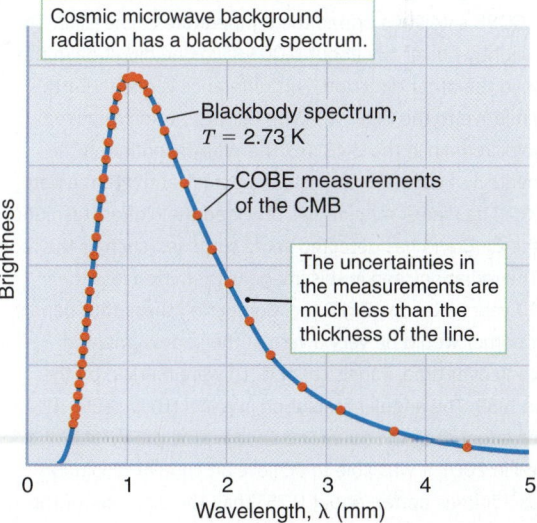

Cosmic microwave background radiation has a blackbody spectrum.

Blackbody spectrum, $T = 2.73$ K

COBE measurements of the CMB

The uncertainties in the measurements are much less than the thickness of the line.

Brightness

Wavelength, λ (mm)

Figure 16.8 The spectrum of the CMB, as measured by the Cosmic Background Explorer (COBE) satellite, is shown by the red dots. The uncertainty in the measurement at each wavelength is much less than the size of a dot. The line running through the data is a blackbody spectrum with a temperature of 2.73 K.

Variations in the Cosmic Microwave Background

The COBE provided us with more than a single measurement of the spectrum of the CMB. **Figure 16.9a** shows a map of the CMB from the entire sky. The different colors in the map depict variations of less than 0.1 percent in the temperature of the CMB. Most of this range of temperature is due to the Sun's motion around the center of the galaxy, combined with the motion of the galaxy itself. Together, these cause one side of the sky to appear slightly blueshifted (warmer) and the opposite side to be slightly redshifted (cooler).

If we subtract this effect from the COBE map, and if we subtract microwave sources in the Milky Way, only slight variations in the CMB remain, as shown in Figure 16.9b. The brighter (red) parts of this image are only about 0.001 percent brighter than the fainter (blue) parts. These tiny fluctuations in the CMB result from gravitational redshifts (see Section 13.5) caused by concentrations of mass in the early universe. These concentrations gave rise to galaxies and the rest of the structure that we see in the universe today.

Subsequent observations support the COBE findings. Beginning in 2001, the Wilkinson Microwave Anisotropy Probe (WMAP) satellite made more precise measurements of the CMB. These maps are color-coded so that red is brighter and blue is fainter; this means that the red spots are slightly warmer while the blue spots are slightly colder. Figure 16.9c shows the WMAP higher-resolution map, which enabled astronomers to determine several cosmological parameters. For example, the value of the Hubble constant we use in this book is the value inferred from the WMAP experiment.

In 2013, the European Space Agency (ESA) Planck mission went one step further, producing maps in even higher resolution. Overall, these highly detailed observations support the existing theories of Big Bang cosmology. Intriguingly, however, early results show that variations at large scales are not quite as strong as expected from the theoretical predictions. Also, as shown in Figure 16.9d, a "cold spot" first

Figure 16.9 ★ WHAT AN ASTRONOMER SEES

(a) The COBE satellite mapped the temperature of the CMB. The CMB is slightly hotter (by about 0.003 K) in one direction in the sky than in the other direction. This difference is due to Earth's motion relative to the CMB. (b) With Earth's motion removed, tiny ripples remain in the CMB. (c) The WMAP confirmed the fundamentals of cosmological theory at small and intermediate scales. (d) The Planck mission has provided the highest resolution yet of the CMB and has detected some surprises, such as the "cold spot." The radiation seen in this image was emitted less than 400,000 years after the Big Bang. Comparing these four panels, an astronomer would be very aware of the improvement in technology over time. Panels (a) and (b) were imaged by the COBE in 1989. The angular resolution in panel (b) is about 10°, 20 times larger than the full moon. By the time the WMAP launched in 2001, it was able to achieve an angular resolution that was 33 times better: about 0.25° (half the diameter of the full Moon). An astronomer would notice this distinct improvement in panel (c). Planck launched in 2009, and panel (d) has an angular resolution of about 5 arcminutes (0.083°). An astronomer will notice that large patterns remain the same, even when more detail is apparent, which lends confidence to the accuracy and interpretation of the images.

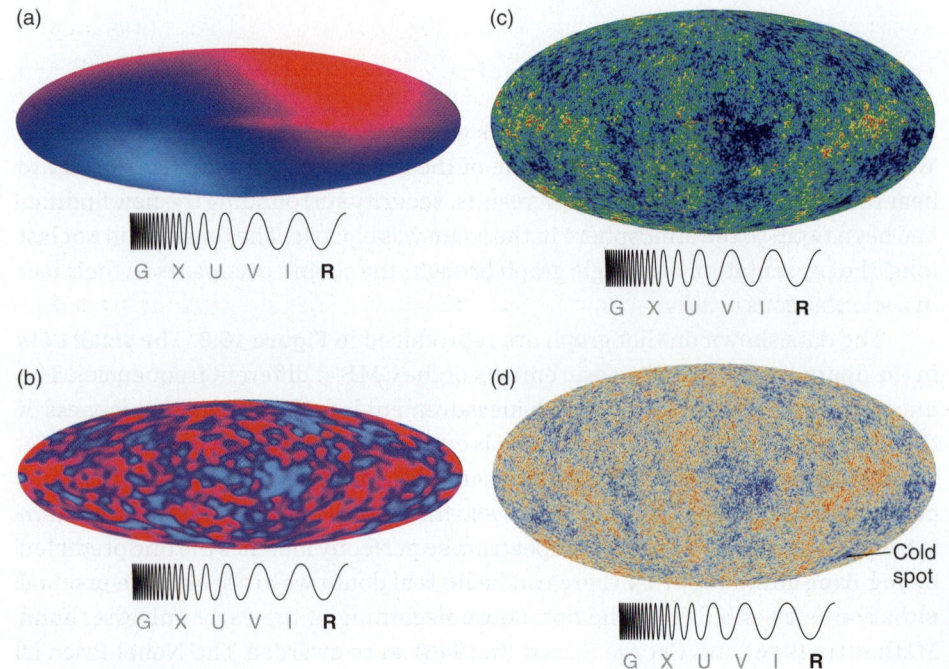

detected in the WMAP data was confirmed to be real. This was exciting: There is something new to figure out, and details may need to be added to existing theories to account for these improved observations.

Because the Big Bang theory is so fundamental to our contemporary understanding of the universe, it is one of the most challenged theories in all of science. Astronomers, particle physicists, and other scientists are continually making predictions from the theory and then testing those predictions. So far, the theory has resisted all efforts to falsify it. As we will see in Chapter 17, there are other confirmed predictions of the Big Bang beyond the CMB.

CHECK YOUR UNDERSTANDING 16.2

The CMB and the Big Bang theory are related because

a. the CMB is an observation that led to the development of the theory.

b. the Big Bang theory predicted the existence of the CMB.

c. the Big Bang is a hypothesis that eventually became known as the CMB.

d. the CMB is a fact, and the Big Bang theory is an untested hypothesis.

16.3 The Expansion of the Universe Is Speeding Up

The simplest approach to the history of the universe is to assume that the expansion occurs at a constant rate, neither speeding up nor slowing down. We made this assumption in Section 16.1. But is this assumption true? The universe contains mass and therefore gravity that pulls galaxies together. It makes sense that this mass might slow the expansion of the universe. Does it? How could we know? Astronomers asked those very questions, and the answers were surprising.

Gravity and the Expansion of the Universe

To see how gravity affects the expansion of the universe, recall the motion of projectiles. The fate of a projectile fired straight up from the surface of a planet depends on the planet's mass and radius. If the planet is massive enough, it will slow the projectile, stop it, and pull it back down. However, if the planet is not massive enough, the projectile will slow down but still escape to space. The size of the planet is also important. If the planet has a small radius, then the mass is more densely packed than in a large planet of equal mass. As a result, the projectile begins closer to the center of mass of the small planet, so the gravitational pull is stronger, and the planet can pull the projectile down. But a planet with a large radius has lower density, so the projectile at the surface starts farther from the center of the planet, the gravitational pull is weaker, and the projectile may escape. Whether the projectile escapes depends on both the planet's mass and the distance between the projectile and the center of the planet.

Just as the mass of the planet slows the climb of a projectile, the mass in the universe slows the expansion of the universe. The fate of the universe is determined by the universe's mass and the separation between masses. If the mass is packed closely together, so that the average density is high, expansion of the universe will slow, then stop, and then reverse; and the universe will collapse. On the other hand, if the mass is very spread out, so that the average density is low, the universe will expand forever.

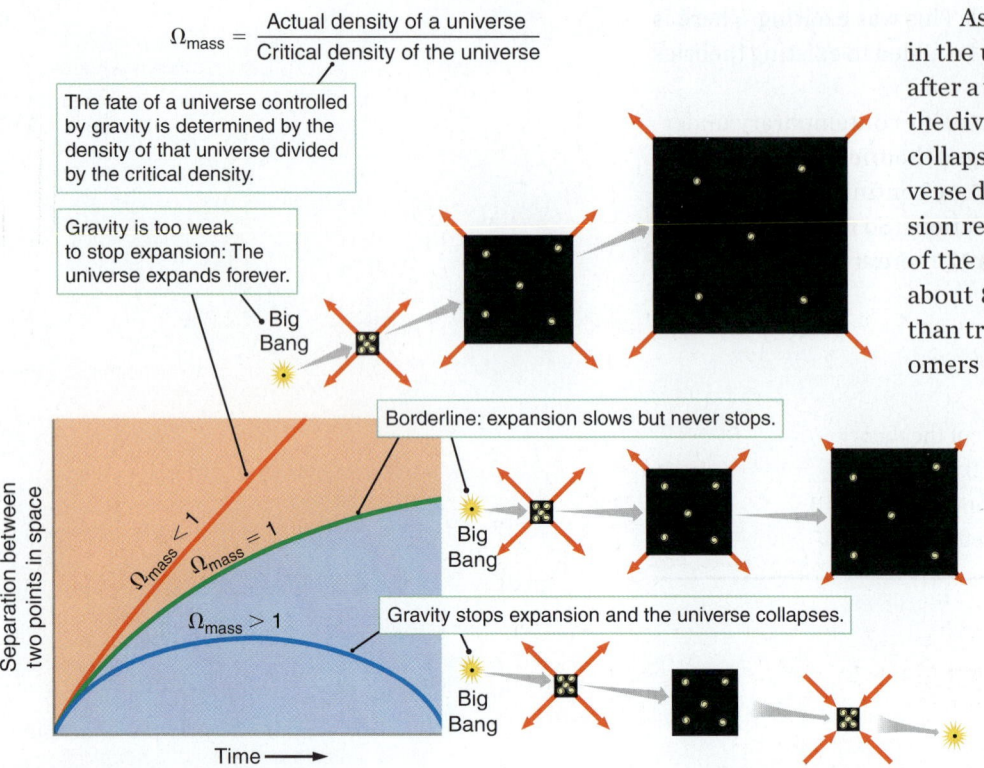

$$\Omega_{mass} = \frac{\text{Actual density of a universe}}{\text{Critical density of the universe}}$$

The fate of a universe controlled by gravity is determined by the density of that universe divided by the critical density.

Gravity is too weak to stop expansion: The universe expands forever.

Big Bang

$\Omega_{mass} < 1$

$\Omega_{mass} = 1$

$\Omega_{mass} > 1$

Borderline: expansion slows but never stops.

Big Bang

Gravity stops expansion and the universe collapses.

Big Bang

Separation between two points in space

Time →

Figure 16.10 There are three possible fates of the universe, which are based on the critical density of the universe and the average density of the universe today. It can expand forever ($\Omega_{mass} < 1$); expansion can slow but never stop ($\Omega_{mass} = 1$); or expansion can stop and then the universe will collapse ($\Omega_{mass} > 1$).

Astronomers define a **critical density** for which the mass in the universe would cause it to just barely coast to a stop after a very, very long time. This critical density determines the dividing line between two possible fates of the universe: collapse or expansion forever. The critical density of a universe depends on the speed of the expansion; a faster expansion requires a higher density to stop it. Current measures of the expansion of our universe give a critical density of about 8×10^{-27} kilogram per cubic meter (kg/m³). Rather than trying to keep track of such awkward numbers, astronomers talk about the *ratio* of the actual density of the universe to this critical density. This ratio is called Ω_{mass} (pronounced "omega mass"). If Ω_{mass} is larger than 1, the universe has a density higher than the critical density, and the universe will collapse. If Ω_{mass} is less than 1, the universe has a density lower than the critical density, and the universe will expand forever. The dividing line, where Ω_{mass} equals 1, corresponds to a universe in which the expansion slows down but never quite stops. The expansion of different possible mass-dominated universes is shown in **Figure 16.10**, which plots the expansion of the universe versus time for different values of Ω_{mass}.

Until the closing years of the 20th century, most astronomers thought that gravity was all there was to the question of expansion and collapse. Researchers carefully measured the masses of galaxies and collections of galaxies in the hope that this would reveal the density and therefore the fate of the universe. The luminous matter gives a value for Ω_{mass} of about 0.02. Galaxies contain about 10 times as much dark matter as normal matter, so adding in the dark matter in galaxies pushes the value of Ω_{mass} up to about 0.2. When we include the mass of dark matter *between* galaxies, Ω_{mass} could increase to 0.3 or a bit higher. By this accounting, there is only about one-third as much mass in the universe as is needed to stop the universe's expansion. Many astronomers were convinced that the expansion would slow down due to gravity but would likely not slow to zero. They looked for evidence that the expansion was slowing down and found out they were wrong.

The Accelerating Universe

If the expansion of the universe is slowing down, then the expansion was faster in the past. Objects that are very far away (so that we see them as they were long ago) were moving faster, so they should have larger redshifts than the local Hubble law would predict.

During the 1990s, astronomers tested this prediction. They compared the measured brightness of Type Ia supernovae in very distant galaxies with the brightness that was expected on the basis of the redshifts of those galaxies. The findings of these studies were startling. Rather than showing that the expansion of the universe is slowing, the data indicated that the expansion is *speeding up*. To describe this increasing expansion rate, people often say, "The universe is accelerating." This does not mean that the universe is zooming through space faster and faster like a car along a road, because that would imply a space outside the universe, which is impossible. Instead, "accelerating" means that the expansion is happening faster

VOCABULARY ALERT

critical In everyday language, *critical* often means "indispensable." Here, astronomers use *critical* in the sense of a turning point or boundary between two cases.

and faster. In order for the expansion rate to increase, a force that is stronger than gravity—a previously unknown force—must be acting. And it must be repulsive: pushing galaxies apart. Naturally, astronomers repeated the experiment many times, in different ways, and continued to improve their certainty in the result. Results from the WMAP experiment early in the 21st century confirmed the result independently, and The Nobel Prize in Physics 2011 was awarded to the original discoverers of the acceleration: Saul Perlmutter, Brian P. Schmidt, and Adam G. Riess. The origins of the repulsive force remain unknown, but astronomers have dubbed the source of the acceleration "**dark energy**."

The idea of a repulsive force that opposes gravity was not new. Nearly a century earlier, more than a decade before Hubble discovered the expansion of the universe, Einstein developed the general theory of relativity. At that time, Einstein believed that the universe was static—that it neither expands nor collapses. When Einstein first used general relativity to determine the structure of the universe, he was greatly troubled. The theory clearly indicated that any universe containing mass could not sit still, any more than a ball can hang motionless in the air. Einstein inserted a "fudge factor," called the **cosmological constant**, into his equations. The cosmological constant acts as a repulsive force that opposes gravity. If it has just the right value, the cosmological constant can lead to a static universe.

When Hubble announced his discovery that the universe is expanding, Einstein realized his mistake. Einstein could have predicted that the universe must either be expanding or contracting with time but instead forced his equations to comply with conventional wisdom. He called the introduction of the cosmological constant the "biggest blunder" of his scientific career. Ironically, the discovery of an accelerating universe decades later restored the cosmological constant to general relativity. A repulsive force is just what is needed to describe a universe that is accelerating. Today, we write the cosmological constant as Λ (uppercase lambda). The fraction of the critical density provided by the cosmological constant is written as Ω_Λ (pronounced "omega lambda"). Because the universe's expansion is accelerating, we know Ω_Λ is not 0.

While a universe is young and densely packed, gravity dominates over the effect of dark energy. As a universe expands, gravity gets weaker because the mass spreads out. The cosmological constant (because it *is* a constant) becomes increasingly important as Ω_{mass} declines. Unless gravity is able to turn the expansion around, the cosmological constant wins in the end, causing the expansion to continue accelerating forever. Even if the density of the universe is greater than the critical density, so that Ω_{mass} is greater than 1, a large enough Ω_Λ could overwhelm gravity and make the universe expand forever. **Figure 16.11** shows plots of the expansion versus time that are similar to those shown in Figure 16.10, but now we have included the effects of Ω_Λ. If Ω_{mass} is sufficiently large, a universe will collapse back on itself, regardless of the value of Ω_Λ. If a universe expands forever, however, its future will depend on Ω_Λ. Current observations indicate that Ω_{mass} is 0.3 and Ω_Λ is 0.7, so that the sum $\Omega_{mass} + \Omega_\Lambda$ is very, very close to 1. This means that the total density of the universe is very close to the critical density, implying that the universe is flat and will never collapse.

Space has distinct physical properties of its own; as you have learned, it can be curved. Space can also have a nonzero energy even in the total absence of matter. This is another way to think about dark energy: It is the energy of empty space. The cosmological constant is an example of dark energy that accelerates the expansion of the universe. But while the cosmological constant does not change, other versions of dark energy can evolve over time. Dark energy is a very active area of study,

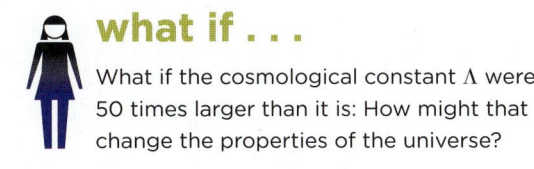

what if . . .

What if the cosmological constant Λ were 50 times larger than it is: How might that change the properties of the universe?

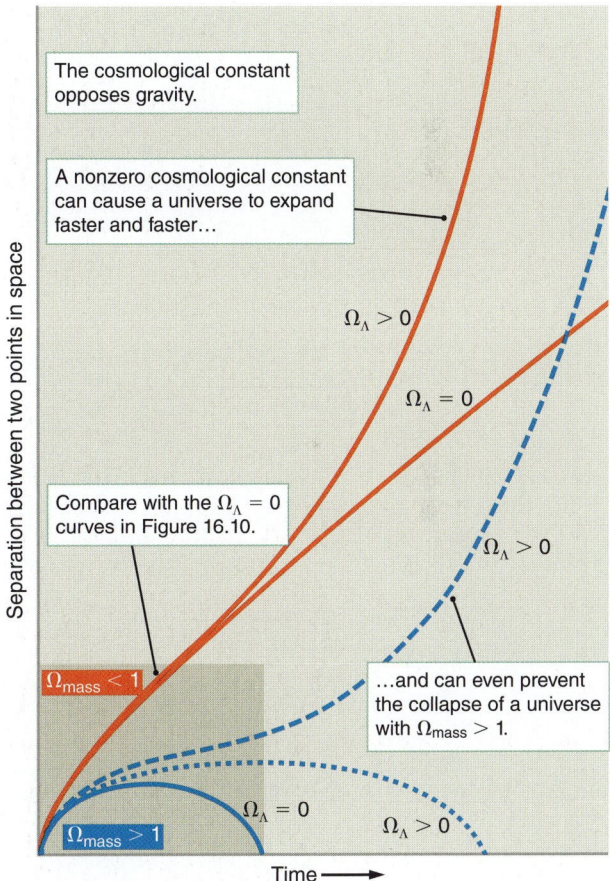

Figure 16.11 The distance between two points in the universe changes with time. This is shown here for cosmologies with and without a cosmological constant. The $\Omega_\Lambda = 0$ curves at lower left are the same curves as those in Figure 16.10, in which Ω_Λ was set equal to 0. If there is enough mass in a universe, gravity could still overcome the cosmological constant and cause that universe to collapse. Any universe without enough mass to eventually collapse will instead end up expanding at an ever-increasing rate because of a nonzero cosmological constant.

as scientists work to figure out where this energy comes from, its form and properties, and whether the amount of it changes over time.

CHECK YOUR UNDERSTANDING 16.3

Dark energy has been hypothesized to solve which problem?
a. The universe is expanding.
b. The CMB is too smooth.
c. The expansion of the universe is accelerating.
d. Stars orbit the centers of galaxies too fast.

16.4 The Earliest Moments of the Universe Connect the Very Largest Size Scales to the Very Smallest

To understand the earliest moments of the universe, we must study the universe on the smallest possible scales. Particle physics is the study of subatomic particles, which are smaller than atoms. In this section, you will learn a little about particle physics and what it has to say about the earliest moments of the universe, when atoms had not yet formed.

The Standard Model

There are four fundamental forces in nature—the electromagnetic force, strong nuclear force, weak nuclear force, and gravity—and everything in the universe is a result of their action. The **electromagnetic force**, which includes both electric and magnetic interactions, acts on charged particles like protons and electrons. This force governs not only chemistry but also light. The *strong nuclear force* binds together the protons and neutrons in the nuclei of atoms and governs nuclear reactions like those in the cores of stars. The **weak nuclear force** governs the decay of a neutron into a proton, an electron, and an antineutrino. Finally, *gravity*, which plays a major role in astronomy, governs how matter affects the geometry of spacetime.

In the standard model, forces between particles, like the attraction between a proton and an electron, are caused by the exchange of carrier particles. Charged particles act like baseball players engaged in an endless game of catch. As baseball players throw and catch baseballs, they experience forces. Similarly, charged particles "throw" and "catch" an endless stream of "virtual photons." The force that results from the average of all these exchanges produces the electromagnetic force that we observe in chemistry. The theory of the standard model is one of the most accurate, well tested, and precise branches of physics. As of this writing, not even the tiniest measurable difference between the predictions of the theory and the outcome of an experiment has been found.

In **electroweak theory**, the electromagnetic force and the weak nuclear force are combined into a single "electroweak force" that requires three carrier particles. In the 1980s, physicists identified these carrier particles in laboratory experiments, thus confirming the essential predictions of this theory.

The strong nuclear force is described by a third theory in which particles such as protons and neutrons are composed of more fundamental building blocks, called

quarks. These quarks are bound together by the exchange of another type of carrier particle.

At high temperatures, the strong nuclear, weak nuclear, and electromagnetic forces are indistinguishable because the different carrier particles have such high energy. In the very early universe, the temperature was sufficiently high that these forces could not be distinguished; they were unified into one force. The *Higgs field* is responsible for causing different kinds of carrier particles to become distinguishable. As the universe cooled, the carrier particles (and also other particles like electrons) gained mass because of the Higgs field. The carrier particles became distinguishable, and so the forces also began to be distinguishable. The Higgs field is mediated by the *Higgs boson,* an elementary particle. The existence of this particle was predicted in 1964 and finally detected at the Large Hadron Collider in 2012. Peter Higgs and François Englert shared The Nobel Prize in Physics 2013 for their work predicting this particle.

Together, these theories of particle physics form the **standard model** of particle physics. Excluding gravity, the standard model explains all the observed interactions of matter and has made many predictions that have been confirmed by laboratory experiments. How does gravity fit into this scheme? Unlike electromagnetism, there is no complete theory of gravity alone involving the exchange of particles. By itself, gravity describes only the bending of spacetime. Work is ongoing to find out how gravity was unified with the other three forces at the earliest moments of the universe. Should this ever be explained, it will be a "theory of everything" that explains all known physics phenomena—a grand achievement of humankind.

A Universe of Particles and Antiparticles

In the standard model, every particle in nature has an **antiparticle** that is identical in mass, but opposite in charge, to the particle. For example, the positron emitted during the proton-proton chain, discussed in Section 11.1, is the antiparticle of an electron. If a particle-antiparticle pair meet, the two particles will annihilate each other, and their energy will be carried away by two photons. The reverse process is also possible: Two photons may collide to produce a particle and its antiparticle by a process called **pair production**.

In the extremely hot early universe, protons and antiprotons formed and annihilated one another. Slightly later, before the universe was roughly 100 seconds old, the temperature was above a billion kelvins, and electron-positron pairs were formed and annihilated one another. The creation and annihilation reached an equilibrium determined strictly by temperature. At this time the universe was filled with photons, electrons, and positrons, as illustrated in **Figure 16.12a**.

As the universe expanded, the photons no longer had enough energy to create particle pairs, so the particles and antiparticles that filled the early universe annihilated each other and were not replaced. When this happened, every proton should have been annihilated by an antiproton, and a positron should have annihilated every electron. This was almost the case, but not quite. Instead, for every 10 billion positrons, there were 10 billion and *one* electrons in the early universe. Similarly, for every 10 billion antiprotons (and antineutrons), there were 10 billion and one protons (and the same for neutrons). This tiny excess of electrons, protons, and neutrons over their antiparticle partners meant that when particle-antiparticle pairs finished annihilating each other, there were electrons, protons, and neutrons remaining—enough to account for all of these subatomic particles in all the atoms in the universe today (Figure 16.12b).

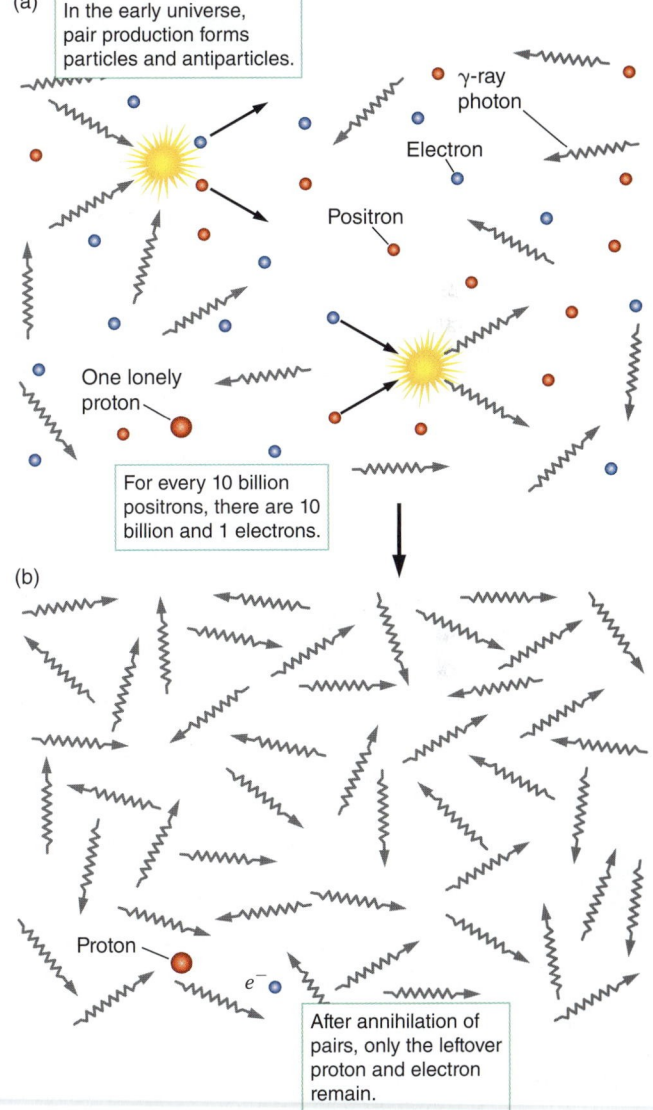

(a) In the early universe, pair production forms particles and antiparticles.

γ-ray photon

Electron

Positron

One lonely proton

For every 10 billion positrons, there are 10 billion and 1 electrons.

(b)

Proton

e^-

After annihilation of pairs, only the leftover proton and electron remain.

Figure 16.12 (a) A swarm of electrons, positrons, and photons filled the very early universe. For every 10 billion positrons, there are 10 billion and one electrons. Before this time, an era of proton-antiproton annihilation left this small piece of universe with only one proton and no antiprotons. (b) After the electrons and positrons annihilate, only the one electron is left, along with the remaining proton and many photons.

what if . . .

What if dark matter particles have dark matter antiparticles associated with them: Is our universe consistent with there having been an excess of dark matter particles over dark matter antiparticles, just as for ordinary matter?

If the standard model of particle physics were a complete description of nature, then the imbalance of one part in 10 billion between matter and antimatter would not have been present in the early universe. The symmetry between matter and antimatter would have been exact. No matter would have survived into today's universe, and we would not exist. The fact that you are reading this page demonstrates that something more needs to be added to the model.

Several competing ideas seek to explain why the amounts of matter and antimatter were not equal in the early universe. **Grand unified theories (GUTs)** explain *how* the electromagnetic force, weak nuclear force, and strong nuclear force joined together into a single force in the early universe. When the universe was very young (younger than about 10^{-35} second) and very hot (hotter than about 10^{27} K), enough energy was available for particles associated with this unified force to be freely created. Perhaps further exploration of this unification will explain the "extra" protons and electrons that persist today. Grand unified theories are only one possible explanation, and there are several of them. As technology advances, it may become possible to determine which explanation is the correct one.

Toward a Theory of Everything

Even earlier in the universe than the time when the three forces were unified, when the universe was younger than about 10^{-43} second, its density was incomprehensibly high. The *observable* universe was so small that 10^{60} of them would have fit into the volume of a single proton. Under these extreme conditions, general relativity can no longer describe spacetime. This era in the history of the universe—just after the Big Bang, when the entire universe must be described with quantum mechanics—is referred to as the **Planck era**.

To understand the Planck era, we need to combine general relativity and our understanding of particles into a single theoretical framework unifying all four of the fundamental forces—a **theory of everything (TOE)**.

Physicists are currently grappling with what a TOE might look like. A successful TOE would determine which of the possible GUTs is correct and would explain the nature of dark matter and dark energy. A successful TOE would also explain the how, when, and why of an early, rapid expansion known as inflation. One contender for the title is **string theory**, in which elementary particles are viewed not as points but as tiny vibrating loops called "strings." In this concept, different elementary particles are like different "notes" on vibrating loops of string.

String theory requires a universe with one dimension of time and nine dimensions of space. Whereas the three familiar spatial dimensions spread out across the vastness of our universe, the other six spatial dimensions predicted by string theory wrap tightly around themselves. These six spatial dimensions extend no further today than they did a brief instant after the Big Bang.

To get a feel for the concept of nine dimensions, imagine living within a thin sheet of paper that extends billions of light-years in two directions but was far smaller than an atom in the third. In such a universe, you would easily be aware of length and width—you could move in those directions at will. But you could not move in the third dimension and so might not even recognize that it was there. Perhaps the only inkling of the true nature of space would be that in order to explain the results of particle physics experiments, you would have to assume that particles extended into a third, unseen dimension.

String theory is only a pale shadow of the sort of well-tested theories that we have made use of throughout this book. In some respects, string theory is no more than a promising idea providing direction to theorists searching for a TOE. Thus, calling it a "theory" is an unfortunate use of the term *theory*. "String hypothesis" or "string idea" is more consistent with the definitions in Section 1.2.

We may never be able to build particle accelerators with high enough energies to search directly for the most fundamental particles predicted by a TOE. Fortunately, some progress may be made by studying the ultimate particle accelerator: the Big Bang itself.

The Cooling Universe

To understand the Big Bang, we have been discussing progressively higher energies occurring at progressively earlier times. Let's review all we've learned by organizing the events from the beginning, following along with **Figure 16.13**. Time begins at the bottom of the figure, at the Big Bang.

In the first 10^{-43} second after the Big Bang, the physics of elementary particles and the physics of spacetime were the same, and a TOE applied. As the universe expanded and cooled, gravity separated from the unified force, and spacetime took on the properties described by general relativity. As the universe expanded, the temperature decreased, so less energy was available to create particle-antiparticle pairs. When the particles governing GUT interactions could no longer form, the strong force split off from the others. Some of those GUT particles remained, while the particle-antiparticle pairs continued to annihilate one another. As a result, the universe ended up with more matter than antimatter.

Next, the electromagnetic and weak nuclear forces split out, leaving these two forces independent of each other. All four fundamental forces of nature that govern today's universe were now separate. At one 10-trillionth of a second, the temperature of the universe had fallen to 10^{16} K. It was another full minute or two before the universe cooled below 1 billion K and pair production could no longer occur.

Although the universe was now too cool for pair production, it was still hot enough for nuclear reactions to take place. These nucleosynthesis reactions formed low-mass elements, including helium, lithium, beryllium, and possibly boron, but could not create more massive elements. These atoms enriched the universe and therefore provide a further test of the Big Bang theory, as will be discussed in Chapter 17.

The nuclear reactions had come to an end by the time the universe was 5 minutes old, when the temperature of the universe had dropped below about 800 million K. The density of the universe was only one-tenth that of water. Normal matter in the universe consisted of atomic nuclei, electrons, and photons. The universe expanded and cooled for the next several hundred thousand years, until finally the temperature dropped so far that electrons were able to combine with atomic nuclei to form neutral atoms. This was the era of recombination, which marks the moment

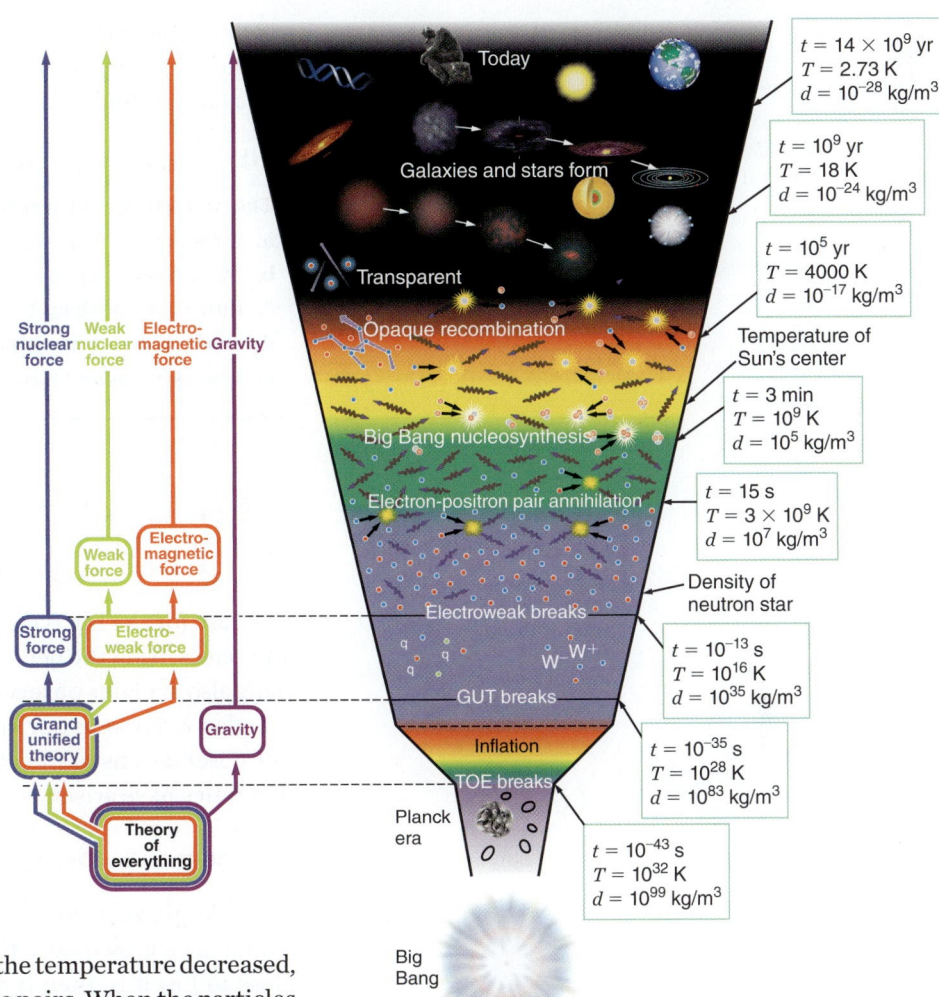

Figure 16.13 As the universe expanded and cooled after the Big Bang, it went through a series of phases determined by what types of particles (for example, exotic W or q particles or electron-positron pairs) could be created freely at that temperature. Later, the structure of the universe was set by the gravitational collapse of material to form galaxies and stars and by the chemistry made possible by elements formed in stars.

at which the universe became transparent and light could move freely through it for the first time. We see this moment directly when we look at the cosmic microwave background.

16.5 Inflation Solves Several Problems in Cosmology

The case for the Big Bang is compelling. The Big Bang explains observations at all size scales, from the behavior of particles to the expansion of the universe. The Big Bang also explains observations at all times, from the earliest observable moments to today. Even so, as our knowledge of the expansion of the universe has grown and our observations of the cosmic microwave background have improved, a number of puzzles have arisen.

The Flatness Problem

The first puzzle is known as the **flatness problem**: Any deviation from flatness would have grown over time, so for the present-day value of $\Omega_{mass} + \Omega_\Lambda$ to be as close to 1 as it is, then it must have been equal to 1 all the way out to at least the tenth decimal place (0.0000000001) when the universe was 1 second old. At even earlier times, it had to be much flatter still. The flatness problem is that the universe is so very flat that it seems unlikely. Something about the early universe must have *forced* $\Omega_{mass} + \Omega_\Lambda$ to have a value incredibly close to 1, but what?

The Horizon Problem

The second problem is that the CMB is surprisingly smooth. After the discovery of the CMB in the 1960s, many observers turned their attention to mapping this background glow. At first they were reassured as result after result showed that the temperature of the CMB is remarkably constant in every direction in the sky. Yet over time, this strong confirmation of the Big Bang theory became a challenge. The CMB is not just smooth—it is *too* smooth.

To understand this, think about carefully adding a tablespoon of cold water to a larger volume of hot water, as illustrated in **Figure 16.14**. At the instant of addition, there is a large temperature difference between the cold water and the hot water. Over time, energy moves from the hot water to the cold water, and the entire cup full of water eventually reaches thermal equilibrium; that is, it comes to the same temperature. However, if the cup of water somehow grows much larger before the temperature has evened out, there will be cold spots and hot spots that grow as the cup of water grows. Similarly, in the early universe, conditions fluctuated in unpredictable ways. Some regions were hotter than others. There was not enough time for the whole universe to come to thermal equilibrium before it was too large for the energy to travel between every location. But the CMB is uniform, showing

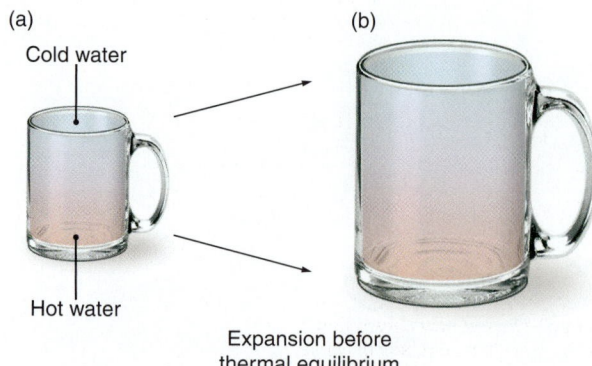

(a) Cold water

Hot water

(b) Expansion before thermal equilibrium

Figure 16.14 (a) When cold water is carefully added to hot water, initially the cold and hot water are at very different temperatures. (b) If the cup grows, the cold spot persists.

that the universe *did* come to thermal equilibrium. The fact that the CMB is so uniform is referred to as the **horizon problem**: Even parts of the universe that were far apart ("over the horizon") from one another came to thermal equilibrium before the CMB was emitted. How could that happen?

Inflation: Early, Rapid Expansion

In the early 1980s, Alan Guth (b. 1947) offered a solution to the flatness and horizon problems. Guth suggested that the universe has not always expanded at the same rate but that it started out much more compact than uniform expansion would predict. Then, for a brief time, the young universe expanded at a rate *far* in excess of the speed of light. This rapid expansion of the universe is called **inflation**. Very early, between about 10^{-35} second and 10^{-33} second after the Big Bang, the space between every two points in the universe increased by a factor of at least 10^{30} and perhaps very much more. This is an absurdly short period of time—about a billionth of the time it takes light to cross the nucleus of an atom. And it is a very large increase in the size of space; for comparison, a fine grain of sand is about a factor of 10^{30} smaller than the current observable universe. Inflation does not violate the rule that nothing can travel *through* space at greater than the speed of light, because space itself was expanding.

To understand how inflation solves the flatness problem, imagine that you are an ant living in the two-dimensional, positively curved universe defined by the surface of a golf ball, as shown in **Figure 16.15**. Now imagine that this golf-ball universe suddenly grew to the size of Earth. The curvature of the universe would no longer be apparent. An ant (or person) walking around on the surface of Earth

VOCABULARY ALERT

inflation In everyday language, *inflation* has two meanings. The increase in the cost of goods due to the decrease in the value of money is known as inflation. Also, the blowing up of a balloon is inflation. Here astronomers are using *inflation* in the second sense, as an analogy to the behavior of the universe. Keep in mind, however, that the universe is not a balloon; that is, the universe does not inflate into anything.

Figure 16.15 If a round, lumpy golf ball were suddenly inflated to the size of Earth, it would seem extraordinarily flat and smooth to an ant on its surface. Similarly, after undergoing inflation, any universe would seem both extremely flat and extremely smooth, regardless of the exact geometry and irregularities it started with. Notice that the ant does not expand with the golf ball, just as galaxies do not expand with the universe—gravity keeps galaxies from expanding.

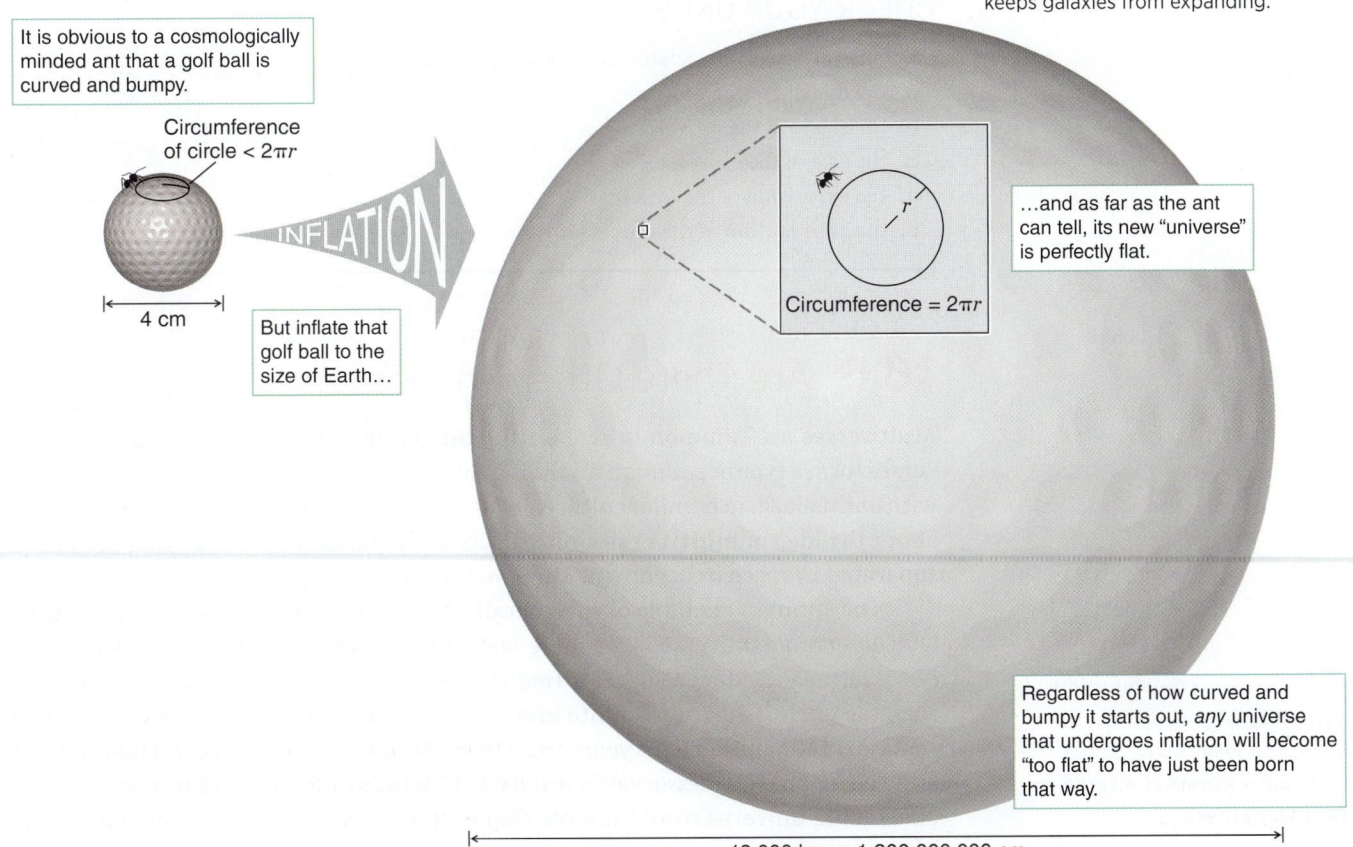

It is obvious to a cosmologically minded ant that a golf ball is curved and bumpy.

Circumference of circle < $2\pi r$

4 cm

But inflate that golf ball to the size of Earth…

INFLATION

Circumference = $2\pi r$

…and as far as the ant can tell, its new "universe" is perfectly flat.

Regardless of how curved and bumpy it starts out, *any* universe that undergoes inflation will become "too flat" to have just been born that way.

13,000 km = 1,300,000,000 cm

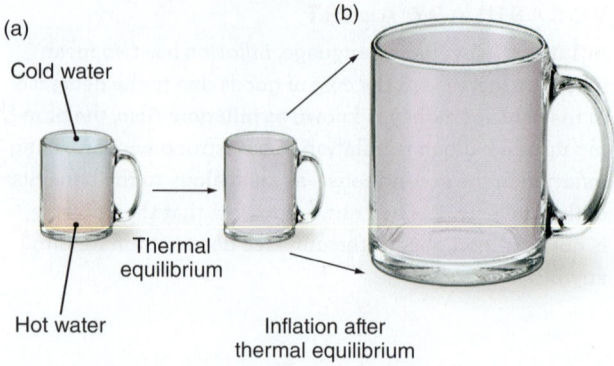

Figure 16.16 (a) If cold water is added to a small mug of hot water, the system comes to thermal equilibrium quickly. (b) If the cup (and the liquid) then proceeds to grow rapidly, the system stays in thermal equilibrium because it had time to reach equilibrium before the inflation started.

what if . . .

What if inflation never happened: What could you conclude about conditions right after the Big Bang?

what if . . .

What if we observed a galaxy a few hundred megaparsecs away that appeared to be absolutely identical to the Milky Way: Could we conclude that we are seeing a parallel universe, or is something more bizarre taking place?

would find it hard to prove that Earth is not flat. If inflation occurred, the universe after inflation would be extraordinarily flat (that is, with $\Omega_{mass} + \Omega_\Lambda$ extraordinarily close to 1) *regardless* of what the geometry of the universe was before inflation. Because the universe was inflated by a factor of at least 10^{30}, $\Omega_{mass} + \Omega_\Lambda$ immediately after inflation must have been equal to 1 to within one part in 10^{60}, which is flat enough for $\Omega_{mass} + \Omega_\Lambda$ to remain close to 1 today. If inflation occurred, then today's universe is not flat by chance. It is flat because *any* universe that underwent inflation would be flat.

What about the horizon problem? Consider again adding cold water to hot water, illustrated in **Figure 16.16**. If the cup was small when the cold water was added, the temperature of all of the fluid in the cup would rapidly equalize. Inflating the cup (and the fluid inside) would preserve (at a new larger scale) the thermal equilibrium already established in the cup. Similarly, the universe prior to inflation was so much more compact that there was time for the variations in temperature to smooth out before the CMB was emitted.

An early era of inflation in the history of the universe offers a handy way of solving the horizon and flatness problems, but it seems quite remarkable that the universe should have undergone a period during which it expanded at such an "astronomical" rate. Because of this rapid expansion, the universe must also have cooled dramatically. As inflation ended, the universe must somehow have rapidly reheated to a very high temperature. The cause of inflation, and its effects on the universe, lies in the fundamental physics that governed the behavior of matter and energy at the earliest moments of the universe. While the existence of an inflationary epoch is difficult to test, it is not impossible, and astronomers are currently devising ways to test whether inflation occurred in the early universe. No firm evidence has yet been found to support the concept of inflation.

CHECK YOUR UNDERSTANDING 16.5

The flatness problem relates to (choose all that apply)
a. the density of the universe.
b. the shape of the universe.
c. the size of the universe.
d. the expansion of the universe.
e. the acceleration of the universe.

16.6 Are There Other Universes?

Multiverses are common in science fiction and in popular science books. Multiverses form a type of common mythology about alternative realities that is explored with enthusiasm in popular culture. In addition, many cosmologists think seriously about the idea of **multiverses**, or collections of parallel universes that are either separated in space or occupying the same space as ours.

The simplest example of such parallel universes is illustrated in **Figure 16.17**. Our *observable* universe is a sphere with a radius of 13.8 billion light-years. The observational evidence suggests that the geometry of space is flat. If this is true, then the universe is truly infinite in size. In this universe, there might be an inhabited world 100 billion light years from here. Those inhabitants would look around and see a spherical observable universe 13.8 billion light years in radius. Their observable universe would not overlap with ours; we would have no galaxies in common. In a truly infinite universe, there must be an infinite number of similar spheres. These spheres constitute parallel universes that are simply too far away for us ever to be able to observe them. Furthermore, they are accelerating away from us, so we will *never* be able to observe them.

What are these other parallel universes like? Because of the cosmological principle, on large scales each of these observable universes should look similar to our own, although details may be different. There are no more than 10^{118} particles in the observable universe, and there are only so many ways that this finite number of particles can be arranged. So in a truly infinite universe there must be an infinite number of observable universes exactly like ours, with an exact copy of you reading an identical version of *Understanding Our Universe*. On average, the distance to an observable universe identical to our own is about $10^{10^{118}}$ Mpc. Yes, that's 10 raised to the power 10^{118}. There must also be a universe in which the only difference from ours is that your copy of *Understanding Our Universe* has an extra period at the end of this sentence.. In general, other universes would be different from our own.

Other types of multiverses include those in which a universe undergoes eternal inflation, with no beginning or end to the inflation. If such a universe exists, then small fluctuations in the universe may cause some regions to expand more slowly than the rest of the universe. As a result, such a region may form a bubble (or parallel universe) whose inflating phase will soon end. In this scenario, we live inside such a region, and our Big Bang is just the condensation of our "local" bubble within the eternally inflating universe. This type of multiverse neatly answers the question of what there was before the Big Bang. Because the universe has been inflating and will continue to inflate forever, there is no beginning or end. Our own bubble separated from the rest of the universe at a time we call the Big Bang, but other bubbles are constantly separating and beginning their own big bangs.

In one version of multiverses derived from quantum physics, every time an event occurs, the universe spawns multiple universes in which each possible outcome of the event exists. While cosmologists consider this at the level of particle interactions, it is more simply explained at the human scale: You made a decision about breakfast this morning. That event caused two universes: one in which you did have breakfast and one in which you didn't. Which one are you in while you read this book? It is fun to wonder what is happening for the "other you."

Is the idea of parallel universes, or multiverses, really science? Throughout this book we have emphasized that any legitimate scientific theory must be testable and ultimately falsifiable. There is considerable debate within the scientific community as to whether models such as eternal inflation can be truly falsified. Are there tests capable of proving these multiverse ideas to be wrong? Possibly. The eternal inflation model is difficult to test, because we will never directly observe its parallel universes (although some scientists interpret the latest data about the cosmic microwave background as hinting at this type of universe). But if we obtain a theory of everything that predicts eternal inflation, and if that theory of everything is itself falsifiable, then there will be a connection between eternal inflation and observation.

CHECK YOUR UNDERSTANDING 16.6

Parallel universes are
a. an observation.
b. a law.
c. a hypothesis.
d. a theory.

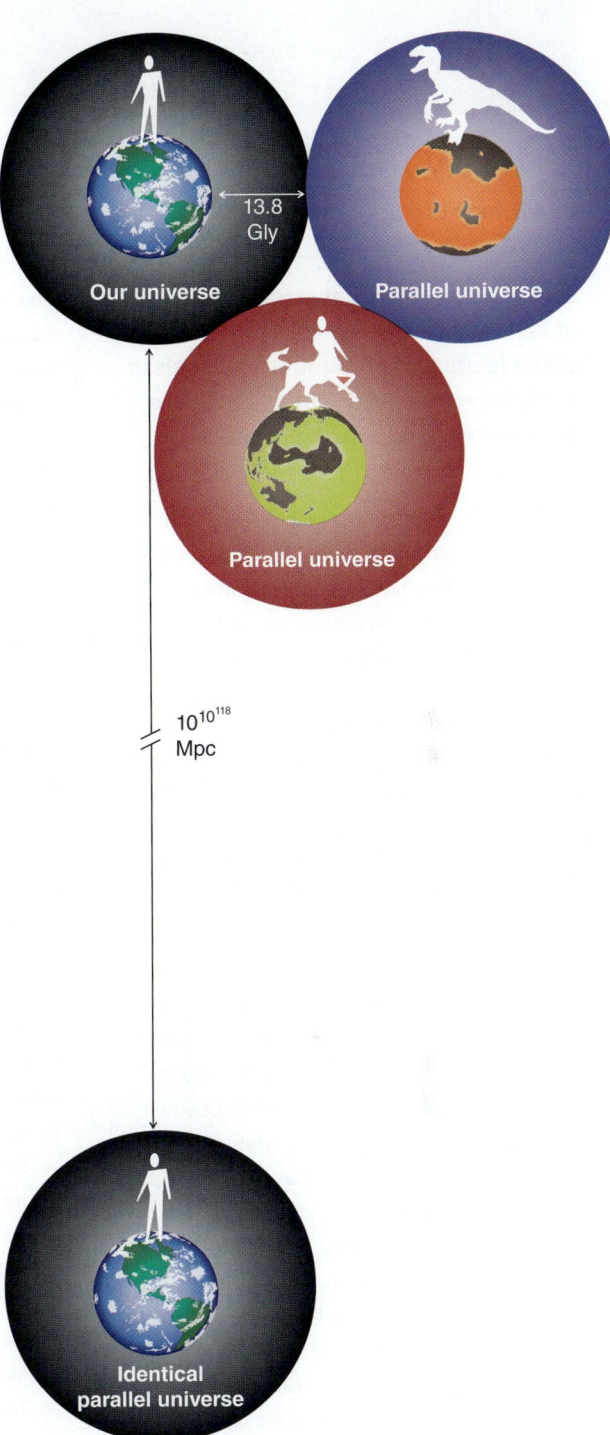

Figure 16.17 The observable universe is a sphere with a radius equal to the distance light has traveled since the Big Bang (13.8 billion light-years, or 13.8 giga-light-years [Gly]). If the universe is infinite, there must be an infinite number of similar spheres. The rules of probability dictate that some of these are exactly like our own.

Gravity may have chased light in the early universe

by Michael Brooks, *New Scientist*

One of science's great strengths is that it is self-correcting. Whenever an unusual claim is made, independent scientists seek ways to test it. These tests are the basis for scientific progress.

It's supposed to be the most fundamental constant in physics, but the speed of light may not always have been the same. This twist on a controversial idea could overturn our standard cosmological wisdom.

In 1998, Joao Magueijo at Imperial College London proposed that the speed of light might vary, to solve what cosmologists call the horizon problem. This says that the universe reached a uniform temperature long before heat-carrying photons, which travel at the speed of light, had time to reach all corners of the universe.

The standard way to explain this conundrum is an idea called inflation, which suggests that the universe went through a short period of rapid expansion early on—so the temperature evened out when the cosmos was smaller, then it suddenly grew. But we don't know why inflation started, or stopped. So Magueijo has been looking for alternatives.

Now, in a paper to be published 28 November in *Physical Review*, he and Niayesh Afshordi at the Perimeter Institute in Canada have laid out a new version of the idea—and this one is testable. They suggest that in the early universe, light and gravity propagated at different speeds.

If photons moved faster than gravity just after the big bang, that would have let them get far enough for the universe to reach an equilibrium temperature much more quickly, the team say.

A Testable Theory

What really excites Magueijo about the idea is that it makes a specific prediction about the cosmic microwave background (CMB). This radiation, which fills the universe, was created shortly after the big bang and contains a "fossilized" imprint of the conditions of the universe.

In Magueijo and Afshordi's model, certain details about the CMB reflect the way the speed of light and the speed of gravity vary as the temperature of the universe changes. They found that there was an abrupt change at a certain point, when the ratio of the speeds of light and gravity rapidly went to infinity.

This fixes a value called the spectral index, which describes the initial density ripples in the universe, at 0.96478—a value that can be checked against future measurements. The latest figure, reported by the CMB-mapping Planck satellite in 2015, places the spectral index at about 0.968, which is tantalizingly close.

If more data reveals a mismatch, the theory can be discarded. "That would be great—I won't have to think about these theories again," Magueijo says. "This whole class of theories in which the speed of light varies with respect to the speed of gravity will be ruled out."

But no measurement will rule out inflation entirely, because it doesn't make specific predictions. "There is a huge space of possible inflationary theories, which makes testing the basic idea very difficult," says Peter Coles at Cardiff University, UK. "It's like nailing jelly to the wall."

That makes it all the more important to explore alternatives like varying light speeds, he adds.

John Webb of the University of New South Wales in Sydney, Australia, has worked for many years on the idea that constants may vary, and is "very impressed" by Magueijo and Afshordi's prediction. "A testable theory is a good theory," he says.

The implications could be profound. Physicists have long known there is a mismatch in the way the universe operates on its smallest scales and at its highest energies, and have sought a theory of quantum gravity to unite them. If there is a good fit between Magueijo's theory and observations, it could bridge this gap, adding to our understanding of the universe's first moments.

"We have a model of the universe that embraces the idea there must be new physics at some point," Magueijo says. "It's complicated, obviously, but I think ultimately there will be a way of informing quantum gravity from this kind of cosmology."

EVALUATING THE NEWS

1. This model predicts that "the ratio of the speeds of light and gravity rapidly went to infinity" at an early time in the universe. According to this model, then, did light travel faster or slower than gravity at that early time?

2. Explain how the difference between the speed of light and the speed of gravity would solve the horizon problem.

3. The model makes a testable prediction that the initial density ripples in the universe would have a "spectral index" of 0.96478. Current measurements of this value are 0.968. Calculate the percent difference (the difference between measurement and prediction, divided by the prediction). One of the scientists responsible for the model is quoted as saying that if the data and the model fail to match, "that would be great." Explain his reasoning—why would he be satisfied if his model is a "failure"?

4. What is it about inflation that makes Peter Coles describe it as "like nailing jelly to a wall"?

5. Consider the "theory of everything" described in this chapter. How does this relate to the theories of the universe described in the last two paragraphs of the news article?

SUMMARY

Our universe has been expanding since the Big Bang, which occurred nearly 14 billion years ago. The Big Bang happened everywhere. It is not an explosion spreading out from a single point. The universe has no center and no edge, although the observable universe has limits. Observed redshifts of distant galaxies result from the increasing space between them. Observations of the cosmic microwave background independently confirm that the universe had a hot, dense beginning. Both gravity and dark energy determine the fate of the universe. Observations indicate that, rather than slowing down, the expansion of the universe is accelerating. The best explanation for this phenomenon is dark energy. During the very earliest moments in the universe, the four fundamental forces of nature were unified. The very early universe may have gone through a brief but dramatic period of exceptionally rapid expansion, known as inflation. If this is true, inflation would explain both the flatness and the uniformity of the universe we see today. Ideas about multiple universes have been proposed, but none is yet directly testable.

(1) Hubble's law indicates that the universe was once very hot and very dense and has since expanded to become cooler and less dense. This hot, dense beginning is known as the Big Bang. The Big Bang the-

ory predicts that we should be able to observe the radiation from a few hundred thousand years after the Big Bang. This radiation should have the same spectrum in every direction. Later observations of the cosmic microwave background confirm precisely this prediction.

(2) Recent observations suggest that the expansion of the universe is speeding up, implying that there must be a force acting to increase the expansion rate. This force comes from dark energy: the energy of empty space. Because the rate of expansion of the universe is increasing, the universe will likely expand forever.

(3) During the earliest moments of the universe, the four forces—originally unified—split off, each becoming separate at a different time. Modern physicists are searching for the combined theory of all four forces: the theory of everything.

(4) Inflation is an early epoch of rapid expansion, followed by reheating, and is proposed to solve the horizon and flatness problems in cosmology. The pre-inflation universe was much smaller and so was able to come to thermal equilibrium, solving the horizon problem. As the universe rapidly expanded, variations were smoothed out, solving the flatness problem. Strong observational confirmation of inflation remains elusive.

QUESTIONS AND PROBLEMS

TESTING YOUR UNDERSTANDING

1. **T/F:** The cosmological constant makes the expansion of the universe accelerate.

2. **T/F:** In our universe, $\Omega_{mass} + \Omega_{\Lambda}$ is as close to 0 as we can measure.

3. **T/F:** Inflation is the theory that the universe is expanding today.

4. **T/F:** The early universe was filled with almost equal numbers of matter and antimatter particles.

5. **T/F:** Cosmological redshift causes a change in the spectrum of a galaxy similar to the change that the Doppler shift causes in the spectrum of a star.

6. Astronomers can get a rough estimate of the age of the universe from Hubble's law by
   a. finding the slope.
   b. finding the inverse of the slope.
   c. squaring the slope.
   d. subtracting the value of the slope from 1.

7. The Hubble constant is found from the
   a. slope of the line fit to the data in Hubble's law.
   b. *y*-intercept of the line fit to the data in Hubble's law.
   c. spread in the data in Hubble's law.
   d. inverse of the slope of the line fit to the data in Hubble's law.

8. A page of this book is
   a. bounded and finite.
   b. bounded and infinite.
   c. unbounded and finite.
   d. unbounded and infinite.

9. The *observable* universe is finite, with a radius of nearly 14 billion light years. This is because
   a. the age of the universe is finite.
   b. the universe itself is finite.
   c. we are at the center of the universe.
   d. light only travels so far before it fades away.

10. Some galaxies are observed to have recession velocities greater than the speed of light. This does not violate Einstein's special theory of relativity because
    a. Einstein's special theory of relativity only applies on Earth.
    b. galaxies are so massive that they warp the space around them, and the light gets redshifted as it emerges from the gravitational well.
    c. young galaxies have no elements in them, so the redshift measurement is very imprecise.
    d. the shift in the wavelength of the light from the galaxies occurs as the light travels through space, not as it leaves the galaxy.

11. Which of the following are properties of the young universe revealed by the cosmic microwave background? (Select all that apply.)
    a. It was hot.
    b. It was cold.
    c. It was dense.
    d. It was diffuse.
    e. It was uniform on large scales.
    f. It was "clumpy" on large scales.

12. Place the following forces in order of their "freeze-out" in the first moments after the Big Bang.
    a. gravity
    b. strong nuclear force
    c. weak nuclear force
    d. electromagnetic force

13. Which of the following adjectives describe the Big Bang? (Choose all that apply.)
    a. hot
    b. dense
    c. loud
    d. tiny
    e. vast
    f. slow
    g. fast

14. Which of the following adjectives describe the early universe, as observed through the CMB? (Choose all that apply.)
    a. hot
    b. cold
    c. dense
    d. diffuse
    e. uniform on large scales
    f. uniform on small scales
    g. rapidly expanding
    h. slowly expanding

15. If a universe is dominated by dark energy, it will
    a. expand forever.
    b. expand and then contract.
    c. expand but gradually slow down.
    d. remain static.

16. If a universe is dominated by matter, it will
    a. expand forever.
    b. expand and then contract.
    c. expand but gradually slow down.
    d. remain static.

17. A positron is related to an electron in that it has all the same properties except
    a. opposite mass.
    b. opposite charge.
    c. opposite spin.
    d. how it interacts with light.

18. The flatness problem states that
    a. the ratio of matter to antimatter is too large.
    b. the sum of Ω_{mass} and Ω_Λ is too close to 1.
    c. the cosmic microwave background is too bumpy.
    d. the cosmic microwave background is too uniform.

19. The cosmic microwave background looks like the spectrum of a blackbody at the low temperature of 2.73 K because
    a. this was the temperature of the universe at the Big Bang.
    b. the light has been redshifted since the Big Bang.
    c. the light has been reddened by dust since the Big Bang.
    d. Earth is receding slowly from the Big Bang.

20. Why is the cosmological redshift in Hubble's law distinct from the redshift arising from the Doppler effect?
    a. The redshift happens as the photon moves through expanding space.
    b. The redshift happens as the galaxy moves through space.
    c. It is used for the light, not for sound.
    d. The two are not distinct; they are the same.

THINKING ABOUT THE CONCEPTS

21. (a) Early in the 20th century, astronomers discovered that most galaxies are moving away from the Milky Way (that is, they are redshifted). What was the significance of this discovery? (b) Edwin Hubble later made an even more important discovery: The speed with which galaxies are receding is proportional to their distance. Why was this among the more important scientific discoveries of the 20th century?

22. ★ **WHAT AN ASTRONOMER SEES** Figure 16.9 shows four different views of the cosmic microwave background. Even without knowing any specifics, an astronomer would recognize that technology had improved between the times that images b, c, and d were taken. How would an astronomer know that the images were different because of a technology improvement, rather than because of a change in the cosmic microwave background? 👁★

23. Figure 16.4 shows light that is traveling from left to right, which is being stretched by the Hubble expansion. This stretch is shown as both a thin wavy line that stretches and as bands of light and dark on the sheet that stretch. Imagine that light was traveling from the back of the sheet to the front, while the same Hubble expansion occurred. According to the cosmological principle, would the effect on this light be the same or different than the effect on the light traveling from left to right? 👁★

24. As applied to the universe, what is the meaning of *critical density*?

25. What set of circumstances would cause an expanding universe to reverse its expansion and collapse?

26. Describe what astronomers mean by dark energy and the observational evidence that suggests dark energy exists. If the universe is being forced apart by dark energy, why isn't our galaxy, Solar System, or planet being torn apart?

27. What is the flatness problem, and why has it been a problem for cosmologists?

28. During the period of inflation, the universe may have briefly expanded at 10^{30} (a million trillion trillion) or more times the speed of light. Why did this ultrarapid expansion not violate Einstein's special theory of relativity, which says that neither matter nor communication can travel faster than the speed of light?

29. List the four fundamental forces in nature. Of these, which one depends on electric charge?

30. Figure 16.11 shows a number of different outcomes for the universe, based on the values of Ω_{mass} and Ω_{Λ}. Of the five cases shown, do most show a universe expanding forever, or do most show a universe that halts and collapses? ⭐

31. Consider the term *string theory* in light of the discussion in Section 1.2. How does the use of the term *theory* in this instance vary from the definition in Chapter 1?

32. As the sensitivity of our instrumentation increases, we are able to look ever farther into space and, therefore, ever further back in time. However, we can see no further back in time than the era of recombination. Explain why.

33. What is meant by *Hubble time*?

34. As the universe expands after the Big Bang, we know that galaxies are not actually flying apart from one another. What is really happening? Why is this distinction important?

35. What is the origin of the CMB? Why is it significant that the CMB displays a blackbody spectrum? What is the significance of the tiny brightness variations that are observed in the CMB?

APPLYING THE CONCEPTS

36. Consider a galaxy located 60 Mpc from the Milky Way. You want to find the rate at which the distance between this galaxy and the Milky Way is increasing.
  a. Make a prediction: How do you expect the recession velocity (v_r) to compare to the recession velocity of a galaxy 30 Mpc away, as found in Working It Out 16.1? Should this galaxy be moving away faster or slower than 2,100 km/s? About how many times faster or slower?
  b. Follow Working It Out 16.1 to calculate the recession velocity, v_r.
  c. Check your answer by comparing your result to your prediction.

37. Consider a galaxy that is receding from the Milky Way at 1,000 km/s. You want to find the distance to this galaxy.
  a. Make a prediction: Compare this recession velocity with the recession velocity for the galaxy in Working It Out 16.1. Do you expect this galaxy to be closer or farther than 30 Mpc away? Roughly how much closer or farther?
  b. Use the Hubble's law equation in Working It Out 16.1 to calculate the distance to this galaxy, d_G, in megaparsecs.
  c. Check your answer by comparing your result to your prediction.

38. Consider a galaxy located 20 Mpc from the Milky Way. You want to know the distance in kilometers.
  a. Make a prediction: Compare this distance to 30 Mpc, which is equal to 9.3×10^{20} km, as stated in Working It Out 16.1. Do you expect to find a larger or smaller number of kilometers for your answer? About how much larger or smaller?
  b. Calculate the distance to this galaxy in kilometers.
  c. Check your work by comparing your answer to your prediction.

39. Consider two galaxies located 3.1×10^{20} km apart, with recession velocity 700 km/s. You want to find out how long ago the two galaxies began moving apart.
  a. Make a prediction: Should your answer be larger or smaller than the 1.4×10^{10} years calculated in Working It Out 16.1? Should the difference be large or small?
  b. Follow Working It Out 16.1 to calculate how long ago the two galaxies began moving apart.
  c. Check your work by comparing your answer to your prediction. If you are surprised, you may want to read Working It Out 16.1 more carefully.

40. The Hubble time is found by taking the inverse of Hubble's constant and converting units appropriately. In 2019, a new study was released that indicates the value of Hubble's constant is 69.8 (km/s)/Mpc.
  a. Make a prediction: do you expect a universe with a Hubble constant of 69.8 (km/s)/Mpc to be older or younger than one with a Hubble constant of 70 (km/s)/Mpc? Will this difference in age be large or small?
  b. Calculate the Hubble time for a Hubble constant of 69.8 (km/s)/Mpc.
  c. Check your work by comparing your answer to your prediction.
  d. Evaluate: Is the result of this 2019 study revolutionary or refining? That is, does it make a large change in the way we think about the age of the universe or a small change?

exploration Hubble's Law for Balloons

The expansion of the universe is extremely difficult to visualize, even for professional astronomers. In this Exploration, you will use the surface of a balloon to get a feel for how an "expansion" changes distances between objects. Throughout this Exploration, you must remember to think of the surface of the balloon as a two-dimensional object, much like the surface of Earth is a two-dimensional object for most people. The average person can move east or west, or north or south, but into Earth and out to space are not options. For this Exploration, you will need a balloon, 11 small stickers, a marker, a piece of string, and a ruler. A partner is helpful as well. **Figure 16.18** shows some of the steps in this process.

Blow up the balloon partially, and do not tie it shut. Stick the 11 stickers on the balloon (these represent galaxies) and number them. Galaxy 1 is the reference galaxy.

Measure the distance between the reference galaxy and each of the other numbered galaxies. The easiest way to do this is to use your piece of string.

Lay it along the balloon between the two galaxies and then measure the string. Record these data in the "Distance 1" column of a table like the example shown. on the following page.

Simulate the expansion of your balloon universe by *slowly* blowing up the balloon the rest of the way. Have your partner count the number of seconds it takes you to do this, and record this number in the "Time Elapsed" column of the table (each row has the same time elapsed, because the expansion occurred for the same amount of time for each galaxy). Tie the balloon shut. Measure the distance between the reference galaxy and each numbered galaxy again. Record these data in the "Distance 2" column of the table.

Subtract the first measurement from the second. Record the difference in the table.

Divide this difference, which represents the distance traveled by the galaxy, by the time it took to blow up the balloon. A distance divided by a time gives you an average speed.

Make a graph of velocity (on the *y*-axis) versus distance 2 (on the *x*-axis) to get "Hubble's law for balloons." You may want to roughly fit a line to these data to clarify the trend.

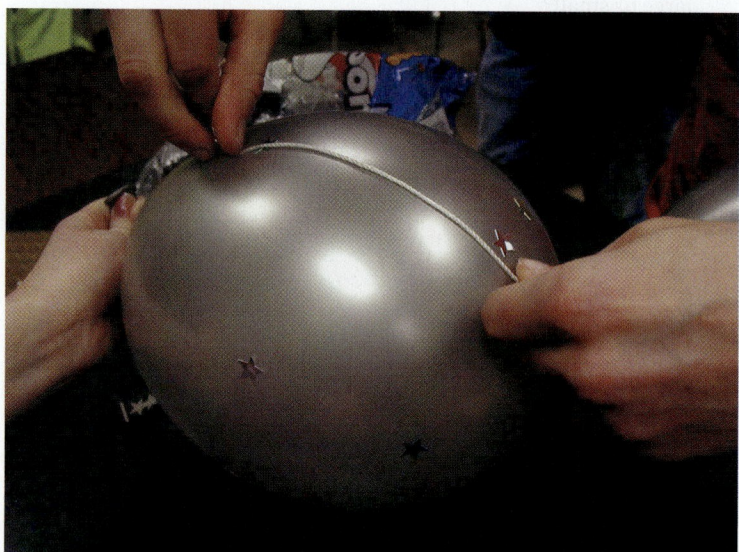
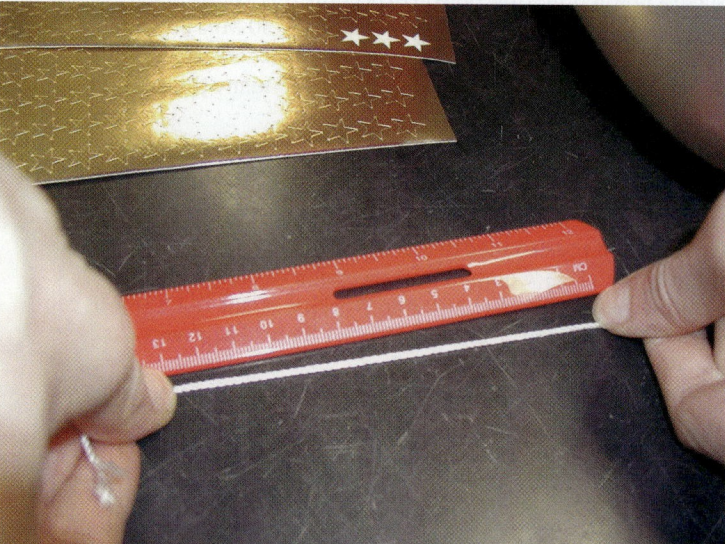

Figure 16.18 The easiest way to measure the distance around a curved balloon is to use a string.

Galaxy Number	Distance 1	Distance 2	Difference	Time Elapsed	Velocity
1 (reference)	0	0	0		0
2					
3					
4					
5					
6					
7					
8					
9					
10					
11					

1. Describe your data. If you fit a line to them, would it be horizontal, trend upward, or trend downward?

_____

_____

_____

2. Was there anything special about your reference galaxy? Was it different in any way from the others?

_____

_____

3. If you had picked a different reference galaxy, would the trend of your line be different? If you are not sure about the answer to this question, get another balloon and try it!

_____

_____

_____

4. The expansion of the universe behaves similarly to the movement of the galaxies on the balloon. We don't want to carry the analogy too far, but there is one more thing to think about. In your balloon, you probably have some areas that expanded less than others because the material was thicker—there was more "balloon stuff" holding it together. How is this similar to some places in the actual universe?

_____

_____

_____

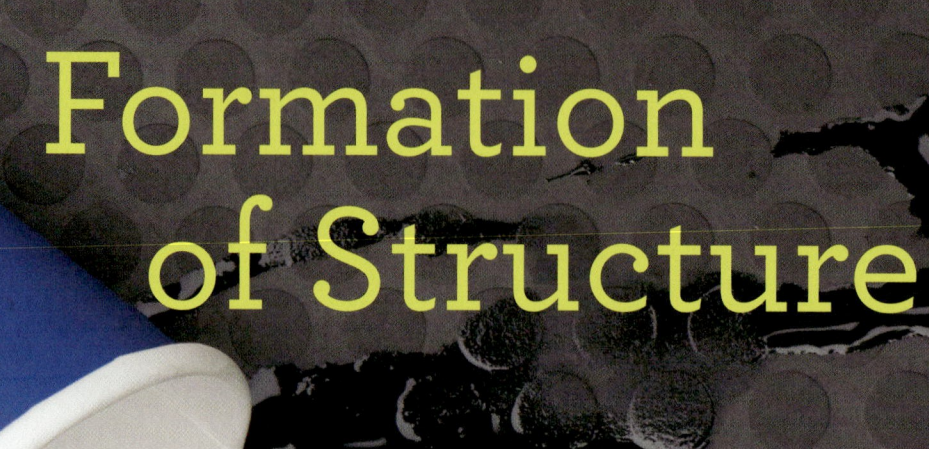

Formation of Structure

The light from the early universe is much too "smooth"; that is, the cosmic microwave background is very nearly the same temperature everywhere you look. This is a problem for astronomers because it takes a long time for heat to move from one place to another—even when those places are in contact—and it takes much longer when they are separated. To better understand the flow of heat at different times in the universe, you will need three identical stacking foam or paper cups, two ice cubes that are the same size, and some water. Fill cups 1 and 2 with the same amount of water at the same temperature. Add an ice cube to cup 1, and put the other ice cube in cup 3. Set cup 2 inside cup 3. Predict which ice cube will melt first. Leave the three cups on your desk for a while, checking frequently to see which ice cube melts first. The one in direct contact with the water (cup 1) resembles the very early universe, when cold spots and hot spots were all close together. The ice cube in cup 3 resembles the universe at a later time, when the cold and hot spots are no longer in contact, and can't share energy easily.

EXPERIMENT SETUP

Fill cups 1 and 2 with the same amount of water at the same temperature.

Add an ice cube to cup 1, and put the other ice cube in cup 3.

Set cup 2 inside cup 3. Predict which ice cube will melt first.

Leave the three cups on your desk for a while, checking frequently to see which ice cube melts first.

PREDICTION

I predict that the ice cube in
☐ cup 1   ☐ cup 3
will melt first.

SKETCH OF RESULTS (in progress)

Time	Cup 1	Cup 3
0 min.	No melting.	No melting.
4 min.	Edges slightly more rounded.	No melting.
6 min.		
8 min.		

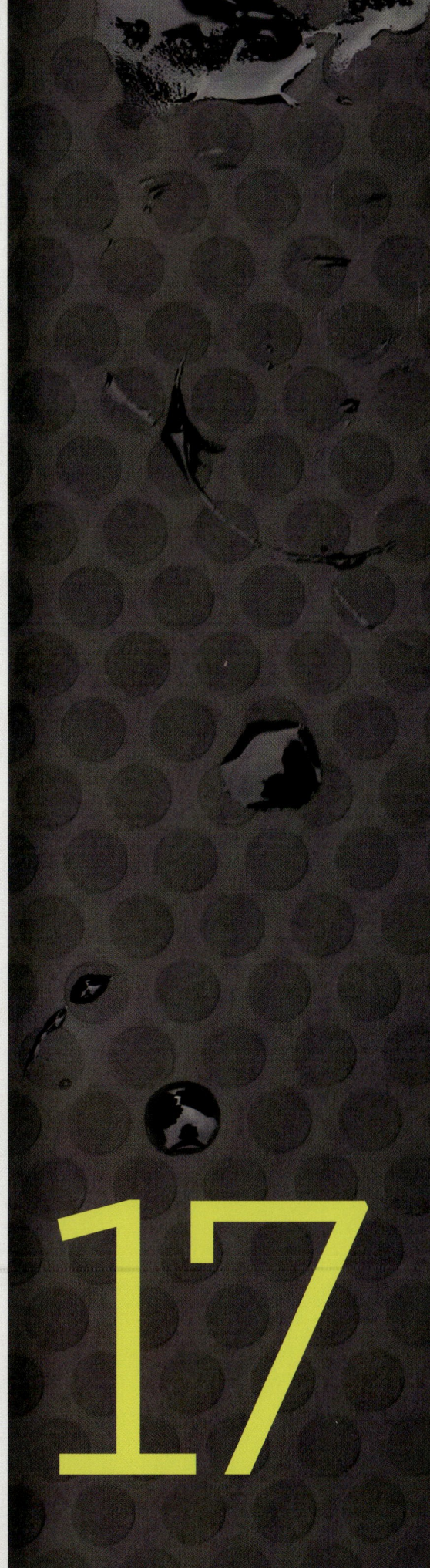

17

The universe that emerged from the Big Bang was incredibly uniform, but the modern universe has structure. Planets, stars, moons, and smaller bodies are parts of galaxies, and galaxies are grouped into larger and larger structures. Cosmologists study how the universe evolved from the uniformity of the earliest times to the nonuniform structure of today. Large-scale structure is a natural consequence of the physical laws that operate in our evolving universe. Here in Chapter 17, we take a closer look at the nature of the universe and how its structure has evolved over time, and we contemplate its ultimate fate.

LEARNING GOALS

1. Describe the distribution of galaxies in the universe.

2. Explain how galaxies formed in the early universe and the role that dark matter plays in galaxy formation.

3. Distinguish between the formation of the first stars and the formation of stars today.

4. Place in order the steps that led to large-scale structure in the universe.

17.1 Galaxies Form Groups, Clusters, and Larger Structures

Galaxies are not distributed evenly throughout the universe. Instead, they are clumped together into structures of varying sizes.

Types of Galaxy Structures

The vast majority of galaxies belong to gravitationally bound collections of galaxies. Galaxy groups are the smallest and most common galaxy collections (see Chapter 15). They contain up to several dozen galaxies, most of them dwarf galaxies. Recall that the Milky Way is a member of the Local Group, which consists of three large spiral galaxies—Milky Way, Andromeda, and Triangulum—along with more than 30 dwarf galaxies and a few elliptical and irregular galaxies in a volume of space roughly 2 megaparsecs (Mpc) in diameter (recall Figure 15.21). Most of the galaxy mass (both luminous and dark matter) in the Local Group resides in the Milky Way and the Andromeda Galaxy.

Galaxy clusters are gravitationally bound collections of hundreds to thousands of galaxies, typically 3–10 Mpc (about 10 million to 30 million light-years) across. **Figure 17.1** shows the Local Group's position relative to two well-known galaxy clusters, the Virgo Cluster and the Coma Cluster. In a cluster, as in a group, dwarf galaxies outnumber giant galaxies. However, the giant galaxies still contain more *mass* than the population of dwarf galaxies. Spiral galaxies are common in most systems, whereas giant elliptical galaxies are prevalent in only about one-fourth of galaxy clusters. The Virgo Cluster, located 17 Mpc from the Local Group, is an example of a cluster containing mostly spiral galaxies. The more distant Coma Cluster is dominated by giant elliptical and S0 galaxies.

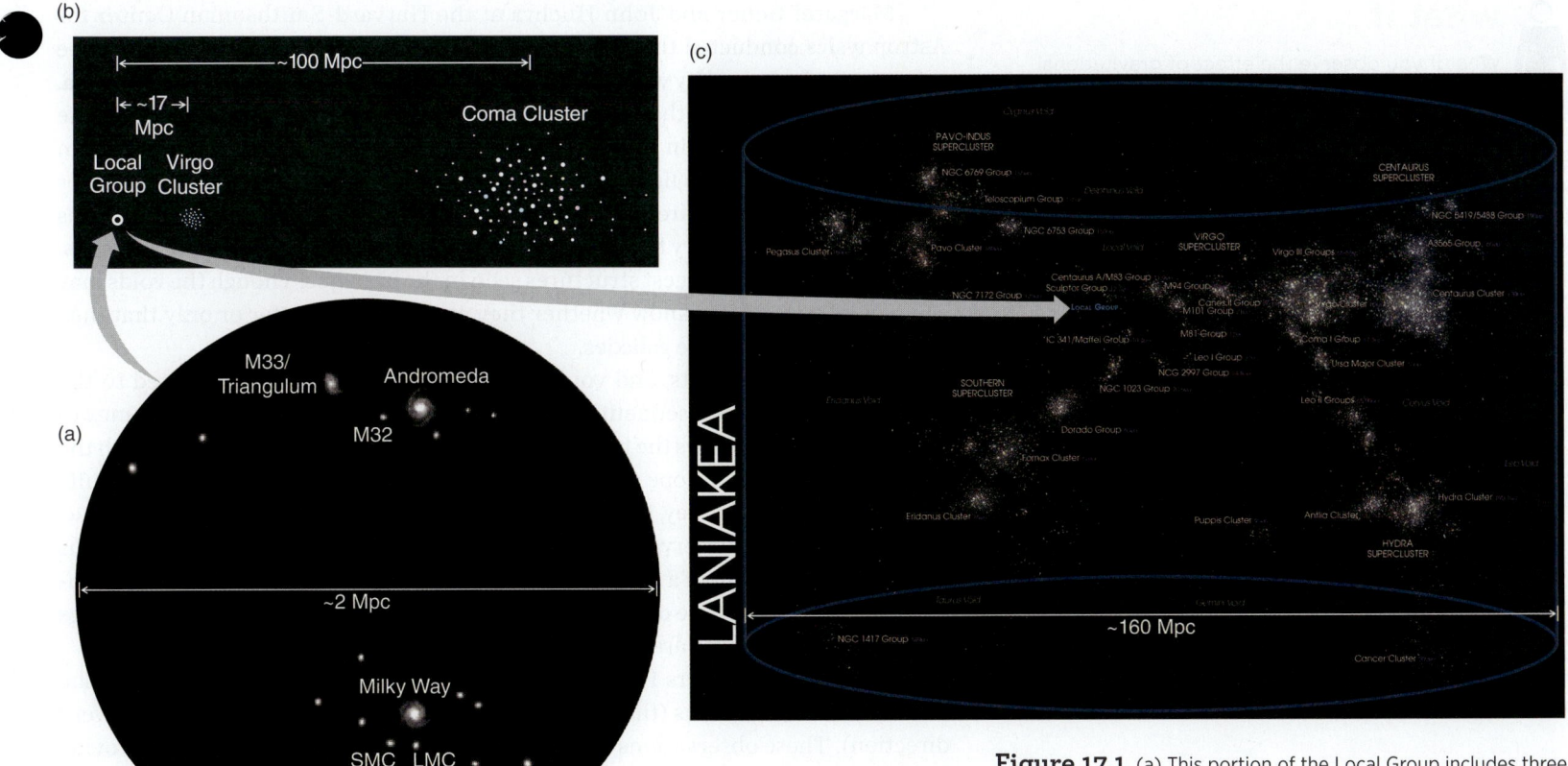

Figure 17.1 (a) This portion of the Local Group includes three spiral galaxies and many smaller dwarf galaxies. LMC is the Large Magellanic Cloud; SMC is the Small Magellanic Cloud. (b) The Local Group is 17 Mpc from the Virgo Cluster and 100 Mpc from the Coma Cluster. (c) These clusters are in turn part of a much larger structure called Laniakea.

Galaxy clusters and galaxy groups bunch together to form enormous **superclusters**, which contain tens of thousands or even hundreds of thousands of galaxies and typically span regions of space more than 30 Mpc in size. Our Local Group is part of the Virgo Supercluster, which also includes the Virgo Cluster. In turn, the Virgo Supercluster joins with four other superclusters to form the even larger Laniakea Supercluster, as shown in Figure 17.1. The Laniakea Supercluster contains more than 100,000 galaxies, and it takes light more than 520 million years to travel from one side of the supercluster to the other. As we will learn, there are even larger structures in the universe. First, let's look at how astronomers map structure in the universe.

Mapping Galaxy Structures

Hubble's law of the expansion of the universe is a powerful tool for mapping the distribution of galaxies, groups, clusters, and superclusters in space. The distance to a galaxy can be obtained by measuring the redshift from a single observation of the galaxy's spectrum. We now know the redshifts to more than 1 million galaxies, which means we know the approximate distances to these galaxies, too. From this information, we can develop a map of the structure of the universe on the largest of scales.

Margaret Geller and John Huchra at the Harvard-Smithsonian Center for Astrophysics conducted the first large redshift survey in 1986 and presented the astronomical community with a "slice of the universe," as displayed in **Figure 17.2a**. The observations show that clusters and superclusters are linked in an intricate network of relatively thin structures known as filaments and walls that can span several hundred million light-years in length. Filaments are stringlike structures made of galaxies. Walls are concentrations of galaxies that surround large regions of space that contain very few galaxies. These enclosed spaces are known as **voids**. Voids are some of the largest structures seen in the universe. Though the voids may seem empty, we do not know whether they are empty of matter or only that they have very few observable galaxies.

The walls, filaments, and voids seen in Figure 17.2a are not limited to the "nearby" universe. Subsequent surveys have looked at much larger volumes of space. Figure 17.2b shows the results of a galaxy redshift survey conducted with the Anglo-Australian Telescope at the Siding Spring Observatory in Australia. Results from a more recent survey, the Sloan Digital Sky Survey, are shown in Figure 17.2c. For as far out as our observations can currently measure, the universe has a porous structure reminiscent of a sponge or a pile of soap bubbles. Take a moment to visualize and sketch this type of structure. Together, galaxies and the larger groupings in which they are found are referred to as **large-scale structure**.

Recall from Chapters 1 and 14 that the cosmological principle states that the universe is homogeneous (the same everywhere) and isotropic (the same in every direction). These observations of the large-scale structure of the universe provide direct observational evidence for the cosmological principle, because they show

(a)

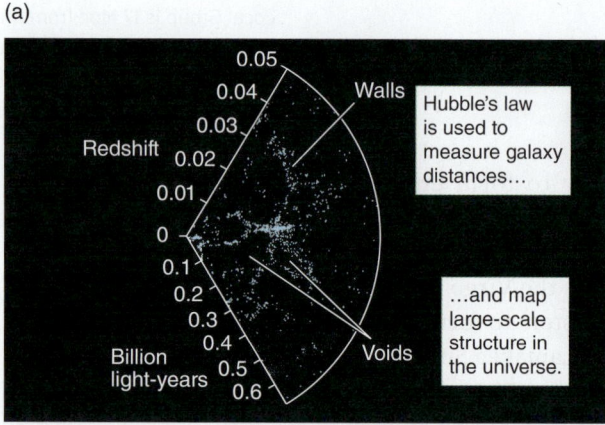

(b)

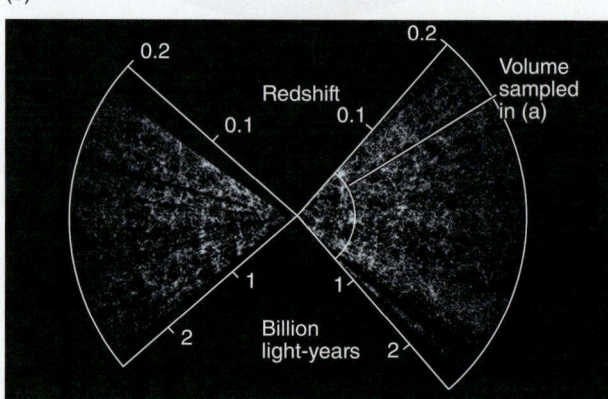

(c)

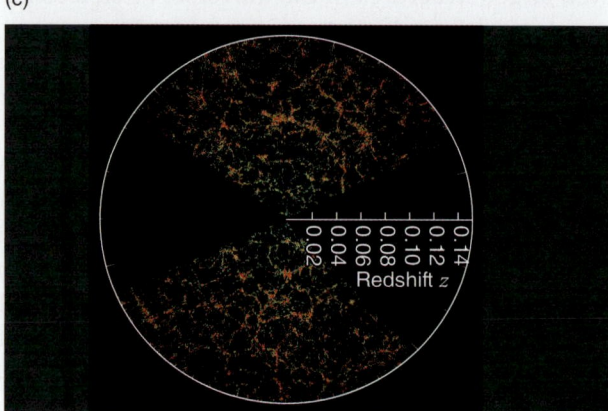

Figure 17.2 Redshift surveys use Hubble's law to map the universe. (a) In 1986, the Harvard-Smithsonian Center for Astrophysics redshift survey, called "A Slice of the Universe," was the first to show that clusters and superclusters of galaxies are part of even larger-scale structures. (b) The Two-degree-Field (2dF) Galaxy Redshift Survey, completed in 2003, shows similar structures at even larger distances, out to about 2.5 billion light-years. (c) The 2008 Sloan Digital Sky Survey map of the universe extends nearly as far—to a distance of almost 2 billion light-years. Shown here is a sample of 67,000 galaxies colored according to the ages of their stars: The redder, more strongly clustered points show galaxies that are made of older stars. The redshift scale is shown on all three images, so that you might easily compare the relative sizes of the maps.

that on this very largest scale, the structure of the universe is the same everywhere and in every direction. If this were not true, the cosmological principle would have been proved false. If, for example, the top half of the multicolored image of Figure 17.2c showed a uniform distribution of galaxies while the bottom half of the image showed walls and voids, then the universe would not be isotropic. All conclusions based on the cosmological principle would have been called into doubt. As it is, however, observations support that the cosmological principle is a valid assumption.

Observing Components of Galaxy Structures

Understanding how clusters form and evolve requires observations beyond the visible. Clusters are bright in the X-ray region of the spectrum (**Figure 17.3a**), indicating that they are rich in hot gas. From the amount of visible, luminous mass that we see, there is not enough gravity to keep this hot gas from escaping. Like stars in the disks of galaxies, cluster galaxies orbit the center of mass of the cluster much faster than can be accounted for by the luminous mass we observe (see **Working It Out 17.1**). From these observations, we can infer the existence of large amounts of dark matter. This dark matter generates other effects that can be observed directly. For example, the total luminous and dark mass in clusters acts as a gravitational lens for light coming from galaxies in the background. Gravitational lenses warp the images of background galaxies into distorted arcs. One such

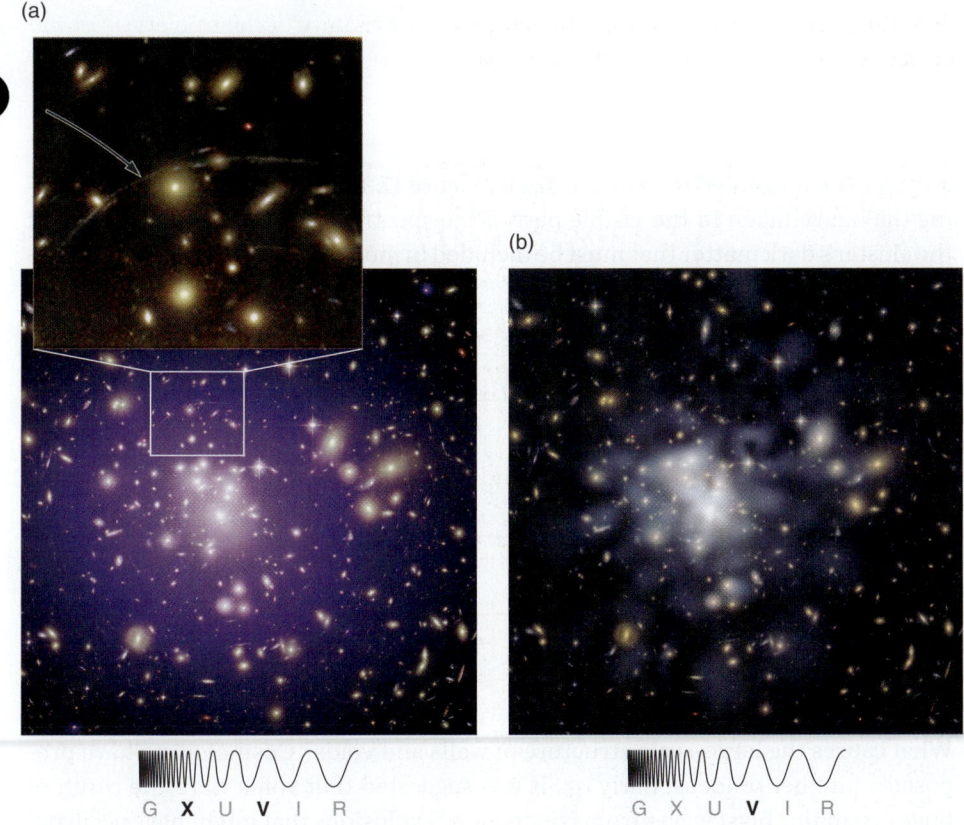

(a)

(b)

G **X** U **V** I R   G **X** U **V** I R

Figure 17.3 (a) Galaxy clusters like Abell 1689 are rich in hot, X-ray-emitting gas (shown here in purple overlying the image taken in the visible part of the spectrum). Gravitational lensing warps images of background galaxies into arcs. Astronomers use these arcs to trace the distribution of dark matter in the cluster. The inset zooms in on an arc, shown here in visible light only. (b) This is also Abell 1689, with the inferred dark matter distribution shown in blue-white.

working it out 17.1

Mass of a Cluster of Galaxies

Recall from Working It Out 15.1 that astronomers use the orbit of a star about the center of a galaxy to estimate the mass of the galaxy within the star's orbit. A similar estimation is made with groups or clusters of galaxies. In these clusters, the galaxies all orbit around a center of mass, as stars do in a binary system. The orbital velocities of the galaxies around the center of mass are measured from the Doppler shifts of the lines in their spectra. The distance between the center of mass and the orbiting galaxies (the radius of a circular orbit) is measured. The mass, then, is related to the radius and velocity by the relationship

$$M = \frac{rv^2}{G}$$

Consider a typical cluster. Suppose a smaller galaxy is orbiting the center of mass of the cluster at a speed of 1,000,000 meters per second (m/s) at a distance of about 3 Mpc, or about 3.0×10^6 parsecs (pc). The universal gravitational constant is $G = 6.673 \times 10^{-11}$ m³/(kg s²). We must convert the orbital radius (3 Mpc) to meters, too:

$$(3.0 \times 10^6 \text{ pc}) \times (3.09 \times 10^{16} \text{ m/pc}) = 9.3 \times 10^{22} \text{ m}$$

The mass of the cluster is given by

$$M = \frac{(9.3 \times 10^{22} \text{ m}) \times (1,000,000 \text{ m/s})^2}{[6.673 \times 10^{-11} \text{ m}^3/(\text{kg s}^2)]}$$

$$M = 1.4 \times 10^{45} \text{ kg}$$

We can divide this by the mass of the Sun, 2.0×10^{30} kg, to get a cluster mass of 7.0×10^{14} M_{Sun}. If we divide this cluster mass by the mass of the Milky Way Galaxy, 10^{12} M_{Sun}, we see that the cluster has the mass of about 700 Milky Way galaxies.

Astronomer Fritz Zwicky (1898–1974) made a calculation like this one in 1933 and measured more mass than was expected from the visible light. He concluded that there must be dark matter within clusters of galaxies. Later in the 1970s, Vera Rubin (1928–2016) and colleagues made rotation curves of individual galaxies and discovered dark matter in these galaxies, too.

arc is shown magnified in Figure 17.3a. In Figure 17.3b, the blue-white glow overlying the image taken in the visible part of the spectrum shows the distribution of the cluster's dark matter that must be included in models in order to produce structures that match the arcs.

CHECK YOUR UNDERSTANDING 17.1

What differentiates galaxy groups from galaxy clusters? (Choose all that apply.)
a. the volume they occupy
b. the number of galaxies
c. the total mass of the galaxies

Answers to Check Your Understanding questions are in the back of the book.

17.2 Gravity Drives the Formation of Large-Scale Structure

What causes the large-scale structure of walls and voids? Cosmologists have proposed a number of ideas. Early on, it was suggested that voids were the result of huge expanding blast waves from tremendous explosions that might have occurred in the early universe. The correct answer has turned out to be less fanciful but far more satisfying: To understand the formation of these larger structures, we need consider only the way that galaxy-sized clumps of matter fall together under the force of gravity.

Gravitational Instabilities

In Chapter 5, you learned about gravity and star formation. Star formation can begin in a molecular cloud with clumps inside it. Gravity causes those clumps to collapse faster than their surroundings—gravity can turn density variations of clouds into stars. The same gravitational instability can turn density variations of the universe into galaxies. Tiny variations (on the scale of subatomic particles) that imprinted structure on the early universe at the time of inflation provided the "clumps" or "seeds" from which galaxies and collections of galaxies grew. The physics that governs atomic nuclei, atoms, and molecules is responsible for seeding the very largest structures that we can see in our universe.

As discussed in Chapter 16, space missions such as COBE, WMAP, and Planck have revealed variations in the background radiation of about one part in 100,000. Theoretical models show that these variations are far too small to explain the structure we see in today's universe. These models indicate that, for ripples in the density of the universe to have formed today's galaxies, the variations in the cosmic microwave background (CMB) today should be at least 30 times larger than what is observed. At first glance this might seem to be irreconcilable with our understanding of the origin of structure in the universe, but it is instead a crucial result that leads us to reconsider the role of dark matter in the universe, as we will discuss shortly.

Building a Testable Model of Large-Scale Structure Formation

To build a model of the formation of large-scale structure, we need three pieces of information. First, we have to decide what universe we are going to model—what values of Ω_{mass} and Ω_Λ (that is, what densities of matter and energy) we are going to assume. These values determine how rapidly the universe expands. The more rapidly a universe expands, or the less mass it contains, the more difficult it will be for gravity to pull material together into galaxies and large-scale structures. As discussed in Chapter 16, these values have been constrained by observations, such that Ω_{mass} is about 0.3 and Ω_Λ is about 0.7. These observational conclusions are inputs to the theoretical models and simulations of the formation of large-scale structure.

Second, we need to know how large and how concentrated the early clumps were. This can be determined from structure in the CMB, which tracks density enhancements in the early universe. Additionally, models of inflation predict the structure that will emerge after the rapid expansion. These predictions are especially important to test because they tie together the large-scale structure of today's universe with our most basic ideas about what the universe was like in the briefest instant after the Big Bang. Currently, we know enough to say that the early universe was "clumpier" on galaxy-sized scales than it was on the scales of the clusters, superclusters, filaments, and voids. Therefore, smaller structures formed first and larger structures formed later—an idea that is referred to as **hierarchical clustering**.

Third, we need a complete list of the types of matter and energy that existed in the early universe. We need to know the balance between radiation, normal matter, and dark matter. We also need to make some choices about the nature of the dark matter that we use in our model. To understand dark matter at this level, we need some more information about matter in the universe.

▶❚❚ **AstroTour:** Big Bang Nucleosynthesis

what if . . .

What if the temperature of the CMB were 10 times higher: How would this change our conclusions about the temperature of the universe at the moment it became transparent? How would this change our conclusions about the amount of elements synthesized in the Big Bang?

Figure 17.4 The darker curving lines show calculated abundances of ^{2}H (blue) and ^{4}He (purple) for various possible densities of the universe (along the x-axis). The observed abundances in the actual universe are shown by the shaded horizontal bars (blue and purple). The yellow region, where the darker curves and the shaded horizontal bars cross, shows how the calculations and the observations constrain possible densities. If our *actual* density falls in this region, then the calculations agree with observation. The black vertical line shows this actual density. This line lies in the yellow region, showing that Big Bang nucleosynthesis correctly predicts the amounts of these isotopes found in the universe today.

Big Bang Nucleosynthesis

When the universe was only a few minutes old, its temperature and density were in the range for nuclear reactions to take place (see Figure 16.14). Collisions between protons in the early universe built up low-mass nuclei, including deuterium (heavy hydrogen) and isotopes of helium, lithium, beryllium, and boron in a process called **Big Bang nucleosynthesis**. These nuclear reactions determined the final chemical composition of the matter that emerged from the hot phase of the Big Bang. Even in this short time after the Big Bang, the density of the universe had fallen too low for reactions such as the triple-alpha process—which forms carbon in the interiors of stars—to occur (see Section 12.3 and Figure 12.6). Therefore, all elements more massive than boron, including most of the atoms that make up Earth and us, formed in subsequent generations of stars.

Figure 17.4 shows the observed and calculated predictions of deuterium (^{2}H) and the very stable helium isotope (^{4}He), plotted as a function of the possible present-day density of normal matter in the universe. Observations of current abundances are shown as shaded horizontal bands. Theoretical predictions, which depend on the density of normal matter in the universe, are shown as darker, thick lines. Big Bang nucleosynthesis predicts that about 24 percent of the mass of the normal matter formed in the early universe should have ended up in the form of ^{4}He. Indeed, when we look about us in the universe today, we find that about 24 percent of the mass of normal matter in the universe is in the form of ^{4}He, in complete agreement with the prediction of Big Bang nucleosynthesis. In fact, this agreement between theoretical predictions and observation provides powerful evidence that the universe began in a Big Bang.

The abundance of helium (dark purple curve in Figure 17.4) does not depend strongly on the density of normal matter, with the mass fraction varying only from 0.22 to 0.25. Unlike helium, most isotope abundances depend strongly on the density of normal matter in the universe, so comparing current abundances with models of isotope formation in the Big Bang pins down the density of the early universe. From these comparisons, astronomers can find the density of normal matter in the universe today. The location on the graph where the observations and the theoretical predictions intersect gives the average density of normal matter in the universe today: about 3.9×10^{-28} kilograms per cubic meter (kg/m^3). This value lies well within the range predicted by the observations for many different light isotopes.

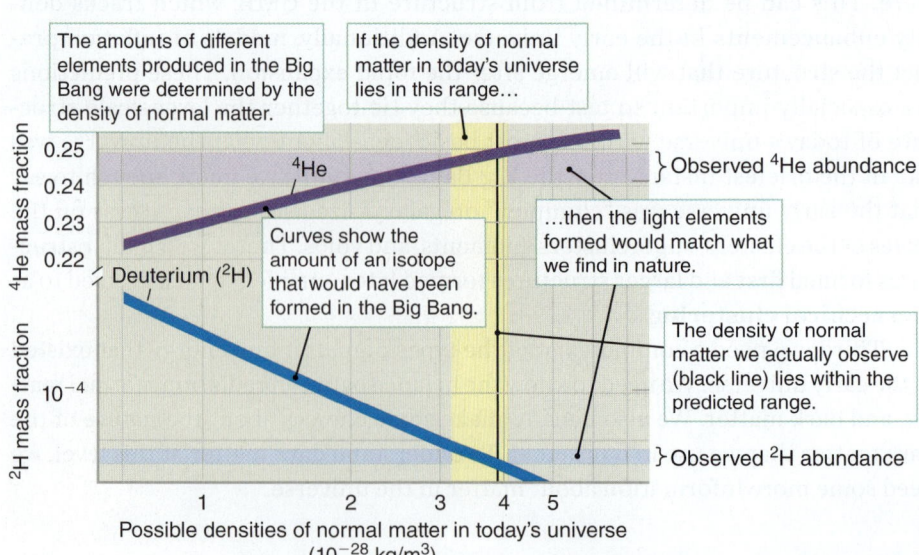

The Nature of Dark Matter

The density of normal matter inferred from nucleosynthesis is about the same as the density of normal matter in today's universe, which in turn is much less than the density of dark matter. This agreement provides a powerful constraint on the nature of dark matter. Dark matter cannot be made of neutrons and protons. If it were, the density of neutrons and protons in the early universe would have been much higher, and the resulting abundances of light elements in the universe would have been very different from what we actually observe.

Dark matter must instead be something that has no electric charge and therefore does not interact with electromagnetic radiation. Because clumps of such dark matter in the early universe did not interact with radiation or normal matter, they were not smoothed out by the pressure due to radiation. **Figure 17.5** shows how the distribution of normal matter and the distribution of dark matter in the universe differ, due to how they interact with radiation. Unlike normal matter, which smooths out over time, dark matter remains clumpy. These clumps of dark matter become the seeds of galaxy formation. Dark matter solves the problems of the formation of galaxies and clusters of galaxies by providing a source of gravitational attraction that isn't smoothed out by radiation in the early universe.

Hot and Cold Dark Matter

Dark matter in the early universe was much more densely clumped than normal matter. Within a few million years after recombination, these dark matter clumps pulled in the surrounding normal matter. Later, gravitational instabilities caused these clumps to collapse. The normal matter in the clumps went on to form visible galaxies. This story seems plausible enough, but the details of how this happened depend greatly on the properties of dark matter itself. Recall that, in Chapter 15,

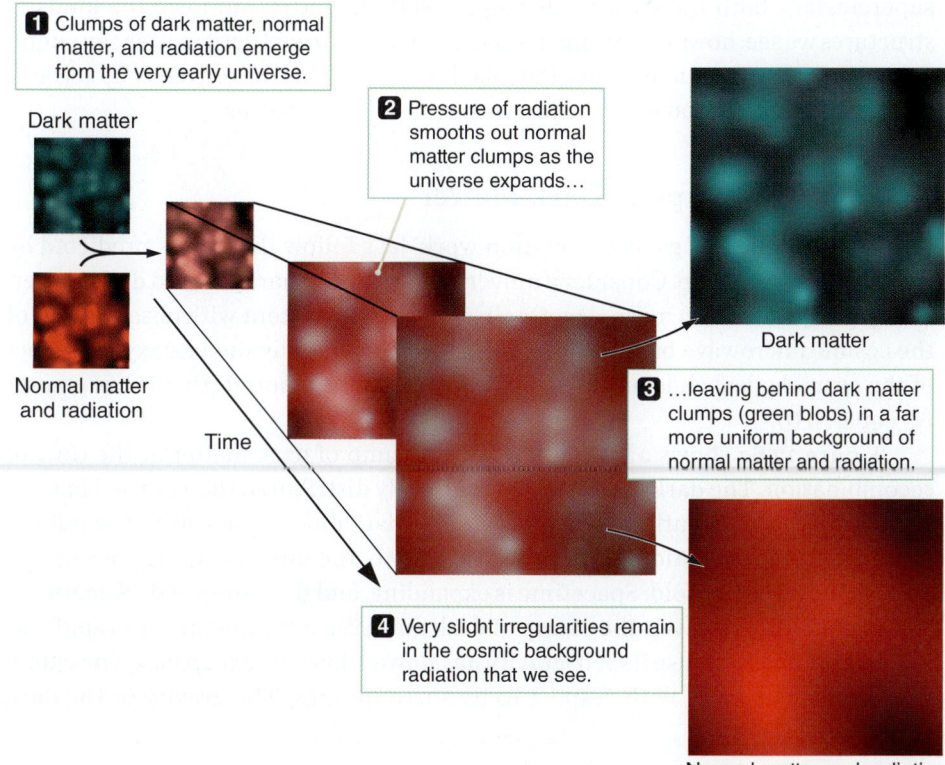

1 Clumps of dark matter, normal matter, and radiation emerge from the very early universe.

Dark matter

Normal matter and radiation

Time

2 Pressure of radiation smooths out normal matter clumps as the universe expands…

Dark matter

3 …leaving behind dark matter clumps (green blobs) in a far more uniform background of normal matter and radiation.

4 Very slight irregularities remain in the cosmic background radiation that we see.

Normal matter and radiation

Figure 17.5 The pressure of radiation in the early universe smoothed out variations in normal matter, but irregularities in dark matter survived to become the seeds of galaxy formation.

we discussed the search for massive compact halo objects (MACHOs) and weakly interacting massive particles (WIMPs) as possible explanations of dark matter and concluded that MACHOs could not explain the missing mass. Even though we do not yet know exactly what dark matter is made of, on the basis of how it behaves, we can categorize two broad classes: cold dark matter and hot dark matter.

Cold dark matter consists of feebly interacting particles that are moving about relatively slowly, like the atoms and molecules in a cold gas. Most likely, cold dark matter consists of an unknown elementary particle. There are several candidate particles for cold dark matter. One candidate is the **axion**, an as-yet-undetected particle first proposed to explain some observed properties of neutrons. Axions should have very low mass, and they would have been produced in great abundance in the Big Bang. Another candidate is the **photino**, an elementary particle related to the photon. Some theories of particle physics predict that the photino exists and has a mass about 10,000 times that of the proton. Our understanding about the particles that make up cold dark matter could soon change: Photinos might be detectable in current particle physics experiments, and experiments are under way to search for axions and photinos trapped in the dark matter halo of our galaxy.

Hot dark matter consists of particles moving so rapidly that gravity cannot confine them to the same region as the luminous matter in the galaxy. Neutrinos are one example of hot dark matter. We have seen that neutrinos interact with matter so feebly that they are able to flow freely outward from the center of the Sun. There is no question that the universe is filled with neutrinos, which might account for as much as 5 percent of its mass. Although this percentage is not high enough to account for all of the dark matter in the universe, it may still have had a noticeable effect on the formation of structure.

Slow-moving particles are more easily held by gravity than fast-moving particles, so particles of cold dark matter clump together more easily into galaxy-sized structures than do particles of hot dark matter. On the largest scales of massive superclusters, both hot dark matter and cold dark matter can form the kinds of structures we see; however, on much smaller scales, only cold dark matter can clump enough to produce structures like the galaxies we see filling the universe. To account for the formation of today's galaxies, we need cold dark matter.

Collapsing Clumps of Dark Matter

To see how models of galaxy formation work, let's follow the events predicted by the models step by step. Consider a universe made up primarily of cold dark matter, clumped together with normal matter in a manner consistent with observations of the cosmic microwave background. On the scale of an individual galaxy, the effect of the cosmological constant is so small that it can be ignored (that is, we can set Ω_Λ equal to 0).

Figure 17.6a shows a simulation of one clump of dark matter at the time of recombination. The dark matter is less uniformly distributed than normal matter, but overall the distribution of matter is still remarkably uniform. By a few million years after recombination, shown in Figure 17.6b, the universe in the simulation has expanded severalfold. Spacetime is expanding, and the clump of dark matter is also expanding. However, the clump of dark matter is not expanding as rapidly as its surroundings because its self-gravity has slowed down its expansion. The clump now stands out more with respect to its surroundings. The gravity of the dark

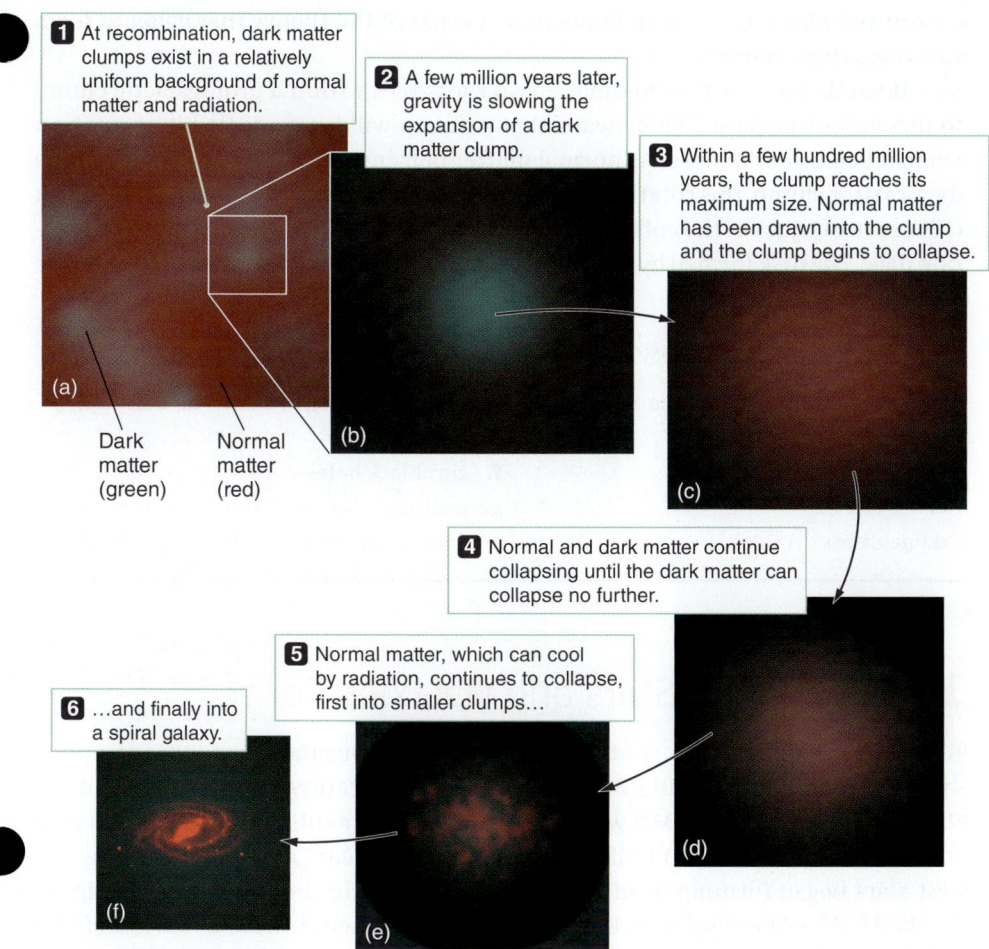

Figure 17.6 A spiral galaxy passes through roughly six stages as it forms from the collapse of a clump of cold dark matter.

1 At recombination, dark matter clumps exist in a relatively uniform background of normal matter and radiation.

2 A few million years later, gravity is slowing the expansion of a dark matter clump.

3 Within a few hundred million years, the clump reaches its maximum size. Normal matter has been drawn into the clump and the clump begins to collapse.

Dark matter (green) Normal matter (red)

4 Normal and dark matter continue collapsing until the dark matter can collapse no further.

5 Normal matter, which can cool by radiation, continues to collapse, first into smaller clumps…

6 …and finally into a spiral galaxy.

matter clump has begun to pull in normal matter as well. By the stage shown in Figure 17.6c, normal matter is clumped in much the same way as dark matter.

A ball thrown up in the air will slow, stop, and then fall back to Earth. Similarly, the clump of dark matter will stop expanding when its own self-gravity slows and then stops its initial expansion. Eventually, the clump of dark matter reaches its maximum size and begins to collapse (see Figure 17.6c). The collapse of the dark matter clump stops when the clump is about half its maximum size, however, because the particles making up the cold dark matter are moving too rapidly to be pulled in any closer (Figure 17.6d). The clump of cold dark matter is now given its shape by the orbits of its particles, in the same way that an elliptical galaxy is given its shape by the orbits of the stars it contains.

Unlike dark matter (which cannot emit radiation), the normal matter in the clump is able to radiate away energy, allowing it to cool and collapse. Small concentrations of normal matter within the dark matter collapse under their own gravity to form clumps of normal matter that range from the size of globular clusters to the size of dwarf galaxies. These clumps of normal matter then fall inward toward the center of the dark matter clump, as shown in Figure 17.6e. According to models, gas in our universe can cool quickly enough to fall in toward the center of the dark matter clump only if the clump has a mass of 10^8 to 10^{12} solar masses (M_{Sun}). This is just the range of masses of observed galaxies. This agreement between theory and observation is an important success of the theory that galaxies form from cold dark matter.

It would be nearly impossible for all the forces within a protogalactic clump to precisely cancel out. This means that a clump will have a little bit of rotation when it begins its collapse. As normal matter falls inward toward the center of the dark matter clump, this rotation forces much of the gas to settle into a rotating disk (Figure 17.6f), just as the collapsing cloud around a protostar settles into an accretion disk. The disk formed by the collapse of each protogalactic clump becomes the disk of a spiral galaxy.

CHECK YOUR UNDERSTANDING **17.2**

Cold dark matter might be composed of which of the following? (Choose all that apply.)

a. neutrinos   **e.** photinos
b. electrons   **f.** tiny black holes
c. axions   **g.** protons
d. neutrons

17.3 The First Stars and Galaxies Form

Recall from Chapter 16 that the universe first became transparent about 400,000 years after the Big Bang, when the temperature dropped low enough for atoms to form. The CMB was emitted at this moment, and we can observe it today. Between about 100 million and 200 million years after the Big Bang, the first stars began forming from the elements created in the Big Bang (see Figure 16.14). As these stars formed, they heated up until they emitted ultraviolet (UV) photons with enough energy to reionize neutral hydrogen in interstellar space. During this **reionization** stage, when the electrons were stripped from the hydrogen atoms, the hydrogen began to glow at visible wavelengths. The first stars, many of which were supermassive, died, perhaps seeding the supermassive black holes that we see today at the centers of galaxies. The reionization that began with radiation from the first stars continued with star formation in the first low-luminosity galaxies and with radiation from the first supermassive black holes. Reionization was complete about 900 million years after the Big Bang.

It is only within the past few years that astronomers have begun detecting objects from the first billion years of the universe. Many astronomers were surprised to find galaxies, quasars, and gamma-ray bursts at these early times because it was thought that they did not form until later. For example, gamma-ray bursts result from the explosive deaths of massive stars, so there must have been massive stars that had already died by 650 million years after the Big Bang. Similarly, since quasars were observed, supermassive black holes must have formed in less than 750 million years after the Big Bang. The study of these very early objects is one of the most dynamic topics in astronomy today. New telescopes are regularly detecting objects from earlier and earlier times.

The First Stars

The first stars must have formed from the elements created in the Big Bang; namely, hydrogen, helium, and tiny amounts of lithium, boron, and beryllium. In 2019, astronomers reported that they had discovered a star with this limited set of elements. Other old stars found in the halo of the Milky Way contain very low abundances of heavy elements, but not quite as low as zero. Astronomers use computer simulations that combine data from these old stars with the conditions in the early universe to figure out what happened before those old stars formed.

During the formation of the first stars, the lack of heavy elements affected star formation in multiple ways. Because there were no heavy elements, there was no dust, and also no molecular clouds full of cold, dense gas. Instead, the first stars formed inside dark matter mini-halos that were about 1 million M_{Sun} and 100 pc across (**Figure 17.7**). These mini-halos formed a few hundred million years after the Big Bang. Neutral hydrogen trapped in these mini-halos combined to form molecular hydrogen, cooling the gas. As the gas cooled, it collapsed to the center of the mini-halo, forming a tiny protostar that accreted more gas to become a star. Theoretical models predict that these first stars were likely to be hot (so their luminosity was high and peaked in the UV region) and massive (10–100 M_{Sun} for single stars and 10–40 M_{Sun} for double stars). These stars were likely singles, doubles, or small multiples, since large clusters of stars did not form.

The first stars were much more massive than today's average star. Therefore, their main-sequence lifetimes were very short—10 million years or less. These stars ended their short lives as supernovae. If the core of such a star was rotating rapidly at the time of the explosion, it might have emitted a gamma-ray burst.

Some of these stars were massive enough to become black holes. Black holes with companions can become energetic X-ray binary systems, as mass falls onto the accretion disk of the black hole as the companion evolves. Because these stars were all so massive, it is likely that both stars in a binary system (or all of them in a multiple system) would become black holes, and these black holes could then merge. In 2016, the LIGO Scientific Collaboration announced that they had observed gravitational waves emitted during a black hole merger. Some think that these merged black holes might have become the seeds for the supermassive black holes found in galaxies, but other models suggest it would take too long for these stellar black holes to build up to a mass of 1 million to 1 billion M_{Sun}.

The explosions of these massive first stars scattered heavy elements, such as carbon, iron, and oxygen, into nearby gas clouds. Some of these elements condensed into dust grains, which further cooled the clouds, so that the next generation of stars formed in a manner similar to the way stars form today. These "second-generation stars" had very low amounts of heavy elements but measurably more than the first stars. Because they formed in a cooler environment than the first stars, lower-mass stars could form. All stars less massive than about 0.8 M_{Sun} have such long lifetimes that they are still burning hydrogen on the main sequence today. These stars are not very luminous, but a few have been found in the halo of the Milky Way. Their spectra show small but measurable amounts of many of the heavier elements in the periodic table. Astronomers are very interested in studying these second-generation stars because they offer clues about the nature of the first stars and the conditions of the young Milky Way.

The First Galaxies

One model of galaxy formation proposes that the first galaxies were made up of the first systems of stars to be gravitationally bound in a dark matter halo. These stars were likely first- or second-generation stars. The properties of the first galaxies were shaped by these first star systems: their radiation, their production of heavier elements, and the black holes they spawned. The masses of these galaxies are

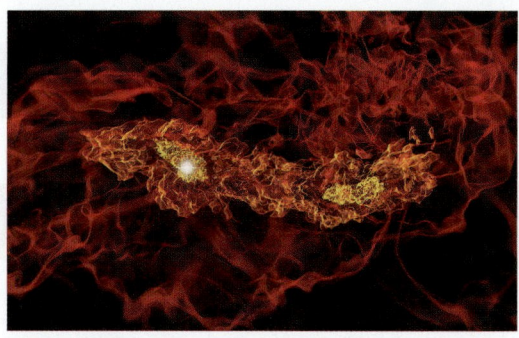

Figure 17.7 In this image from a supercomputer simulation of early star formation, two massive stars form a few hundred astronomical units apart. Each of these massive stars might have as much as 40 times the mass of the Sun. The brighter yellow regions have higher density than the purple regions.

what if . . .

What if you had time on a large telescope to search for first-generation stars: Where and how would you look for them?

(a)

(b)

Figure 17.8 (a) A standard infrared image from the Spitzer Space Telescope shows stars and some galaxies in this strip of sky. (b) In this image, the nearby stars and galaxies have been subtracted out (gray) and the remaining glow has been enhanced, showing some structure from the time when the earliest stars and galaxies were forming.

thought to have been about a hundred million M_{Sun}, and they were built up hierarchically from the merging of mini-halos.

One possible piece of evidence for this model comes from infrared observations, shown in **Figure 17.8**. Figure 17.8a includes the usual nearby stars and galaxies, but when these are all subtracted, a glow remains, as seen in Figure 17.8b. This remaining glow is not smooth, and the bright spots likely come from the first stars and galaxies, roughly 500 million years after the Big Bang.

Another piece of evidence comes from our discovery of the most distant galaxies and quasars; that is, the ones we see when they were very young. These observations of the very distant (and therefore very faint) galaxies in the early universe are difficult, requiring very large telescopes or telescopes in space. The peak of the spectrum of these galaxies has been cosmologically redshifted into the infrared (**Figure 17.9**). The images of these distant objects look like small, faint dots, with none of the detail seen in closer galaxies. Astronomers are excited by such images because just the detection of these objects contributes to an understanding of when and how the first galaxies formed. The age of these first galaxies gives us an outer limit on when galaxies formed after the Big Bang.

The first galaxies are thought to have formed by about 400 million years after the Big Bang. The heavier elements created from the first stars cooled the gas in larger dark matter halos, which then collapsed, probably to a disk, and stars formed. These youngest galaxies appear to be small, 20 times smaller than the Milky Way, which adds support for the hierarchical model of galaxy formation.

CHECK YOUR UNDERSTANDING **17.3**

Galaxy fragments in the early universe formed
a. before the first stars.
b. in dark matter mini-halos.
c. due to gravity from large dark matter halos.
d. in places where the universe was slightly less dense than average.

Figure 17.9 ★ WHAT AN ASTRONOMER SEES
An astronomer will be fascinated by a Hubble Space Telescope image like this one, which was taken over a very long time. This patch of sky was chosen because it appeared to be empty of stars in all previous images. Knowing that nearly every blob in this image is a galaxy, composed of billions or trillions of stars, will give her pause. Perhaps she will take a moment to appreciate that the size of this image on the sky is about the size of a basketball, 6 kilometers (about 3.78 miles) away. She may take a moment to think about the fact that every tiny patch of sky in every direction is as dense with galaxies as this image. Naturally, she will immediately look for the few specks that are *not* galaxies but instead are nearby stars. She will identify these from the diffraction spikes, which make an "X" across the star's image. She will focus in on the colors of the galaxies in the field and search for at least one galaxy of every Hubble type (spiral, barred spiral, elliptical, and irregular). The earliest, highest-redshifted galaxies are observable in infrared light. Contained within this image is a compact faint galaxy as it existed about 480 million years after the Big Bang. At least 100 of these small galaxies would be needed to build up a galaxy like the Milky Way.

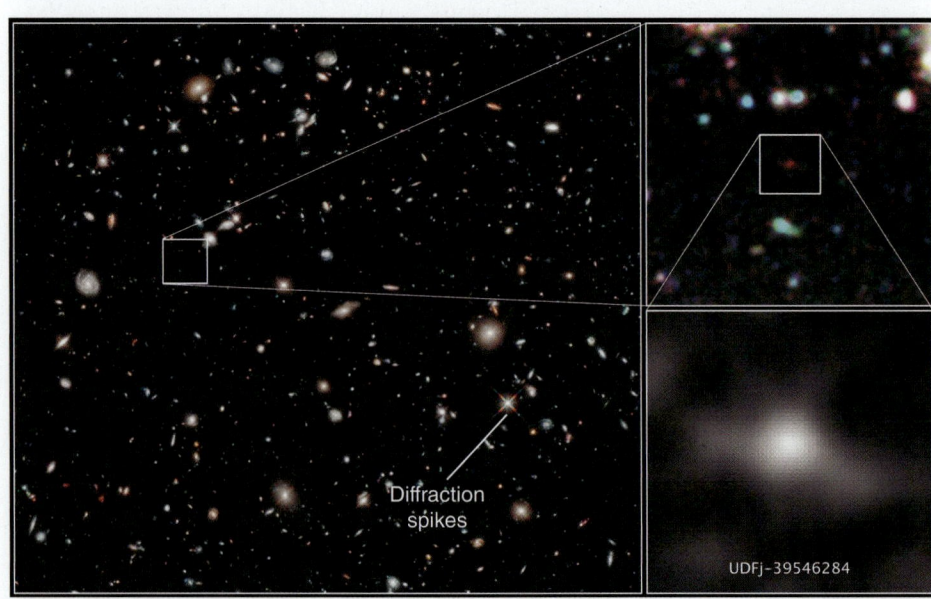

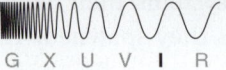

17.4 Galaxies Evolve

Galaxies continued to evolve hierarchically, with smaller fragments merging to form larger objects. The early universe was denser, so early fragments were closer together and mergers were more common.

Recall that the Local Group has small, faint dwarf galaxies orbiting the Milky Way. Streams of material from the dwarf galaxies are falling onto the Milky Way, indicating that the Milky Way is still gaining mass. Recently, the known number of these galaxies has doubled. About a dozen of these small galaxies are called **ultrafaint dwarf galaxies** because they are less than 100,000 times more luminous than the Sun. These galaxies contain mostly old, faint stars with low abundances of heavy elements, hinting that they may have contributed to building the Milky Way's halo. The ultrafaint dwarf galaxies may not have had any further star formation after the first stars, and they may be the oldest galaxies around. They may even be the fossil remains of the first galaxies or of the first mini-halos. Because they are more massive than their luminosity suggests, we have determined that even these very old galaxies are dominated by dark matter.

In a large spiral galaxy like the Milky Way, faint dwarf spheroidal galaxies (with dark matter) and the oldest globular clusters (without much dark matter) may be leftover protogalactic fragments. The gas collapsed to form a rotating disk as it cooled. A recent simulation that included dark matter, gravity, star formation, and supernova explosions was able to reproduce a Milky Way–like galaxy with a large disk and a small bulge (**Figure 17.10**). Astronomers have sorted photos of 400 spiral galaxies at various stages of evolution to understand how galaxies like the Milky Way form. Combining this information with simulations and observations of the oldest parts of the Milky Way, they have found that the oldest globular clusters are about 13.5 billion years old, but the halo itself may be only about 11.5 billion years old. Most of the stars in the Milky Way formed between 11 billion and 7 billion years ago. The disk and bulge formed at about that same time.

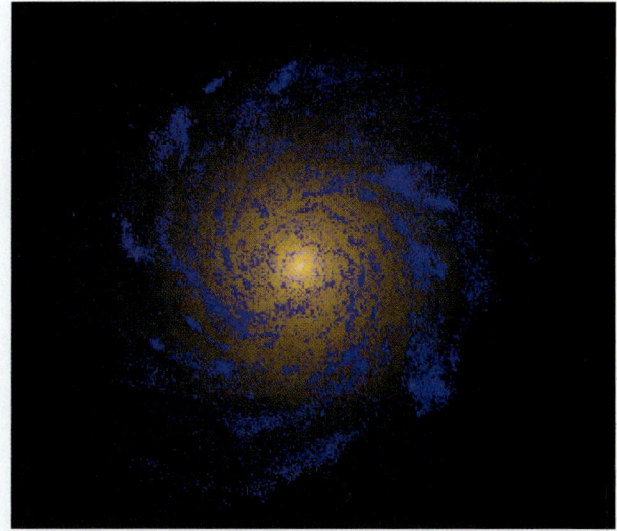
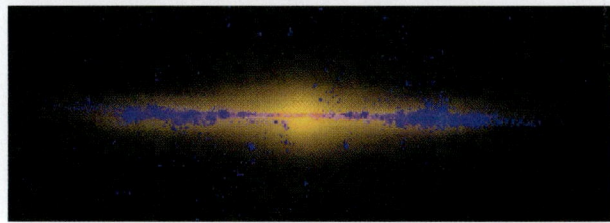

Figure 17.10 This simulation of the formation of a Milky Way–like galaxy has been able to reproduce the small bulge and big disk. Blue colors indicate more recent star formation, while older stars are redder.

The Most Distant Galaxies

Figure 17.11 shows images of galaxies throughout the history of the universe. The galaxies observed in the very early universe, before about 11 billion years ago, are so faint that no structure can be seen (for example, see Figure 17.9). Galaxies from about 11 billion years ago show visible structure, but it is much less regular than that of galaxies today. Even at 4 billion years ago, galaxies are much more irregular. Early irregular galaxies are merging galaxies, which fits with our ideas that in the early universe, when galaxies were closer together, there were more mergers. Most of today's galaxies conform to the Hubble classification, and only about 10 percent are irregular. However, 4 billion years ago, more than half of the galaxies were irregular, the number of ellipticals and S0 galaxies was about the same, and there were many fewer spirals. This difference in galaxy types at different times suggests that it took time for spirals to form. It is likely that these mergers produced spiral galaxies over time.

Hierarchical merging also may have triggered the formation of supermassive black holes at the centers of galaxies. The first supermassive black holes, which power distant quasars, could have grown from the merging of mini-halos containing stellar black holes left after the first stars. Alternatively, they could have formed through the accretion of gas from the material between the galaxies during mergers of the first galaxies or through rapid collapse from hot, dense gas at the center of the first galaxies. In nearby galaxies, the mass of the supermassive black hole and

▶❚❚ **AstroTour:** Galaxy Interactions and Mergers

Figure 17.11 A comparison of the Hubble classification of galaxies today with galaxies throughout the history of the universe. There were more peculiar galaxies in the past, indicating that it took some time for spirals to form.

G **X** U **V** I R

Figure 17.12 Chandra X-ray observations (red, orange, and yellow), combined with Hubble Space Telescope observations (blue and white) of NGC 6240 show that it has two black holes about 900 pc apart. Each white feature has a black hole at the center. The black holes will likely merge in about 100 million years.

the bulge properties are related, suggesting that the growth of the black hole and the bulge might have been linked when they were younger. Supermassive black holes could also have grown even more massive from the mergers of large galaxies. **Figure 17.12** shows a nearby galaxy with two supermassive black holes about 900 pc (roughly 3,000 light-years) apart, which are in the process of merging.

The hierarchical merging and growth of galaxies also affects star formation in the evolving galaxies. The tidal interactions between galaxies and the collisions between gas clouds trigger many regions of star formation throughout the combined system. In the early universe, star formation increased over time, peaking about 3 billion years after the Big Bang.

Merging galaxies at many different distances have been observed by the Hubble Space Telescope (**Figure 17.13a**), showing how mergers differ at various times in the history of the universe. Elliptical galaxies are now thought to result from the merger of two or more spiral galaxies. The dark matter halos of the galaxies merge, and the stars eventually take on the blob-like shape of an elliptical galaxy. Elliptical galaxies are more common in dense clusters, where mergers are more frequent. Compare the young mergers in Figure 17.13a with those of closer, older galaxies in Figure 17.13b.

Just as galaxies merge, clusters of galaxies also merge. **Figure 17.14** shows the high-speed collision and merging of two galaxy clusters in the Bullet Cluster. Optical images show the individual galaxies; hot gas is seen in X-rays (shown in red), and the distribution of the total mass can be found by observing the effects of gravitational lenses. The ordinary matter slowed down in the collision, but the dark matter (shown in blue) did not, providing evidence for dark matter in clusters. Clusters of galaxies also evolve hierarchically, growing from smaller structures to larger ones over time. As with galaxies themselves, the younger, distant clusters are

(a)

G X U **V I** R

(b)

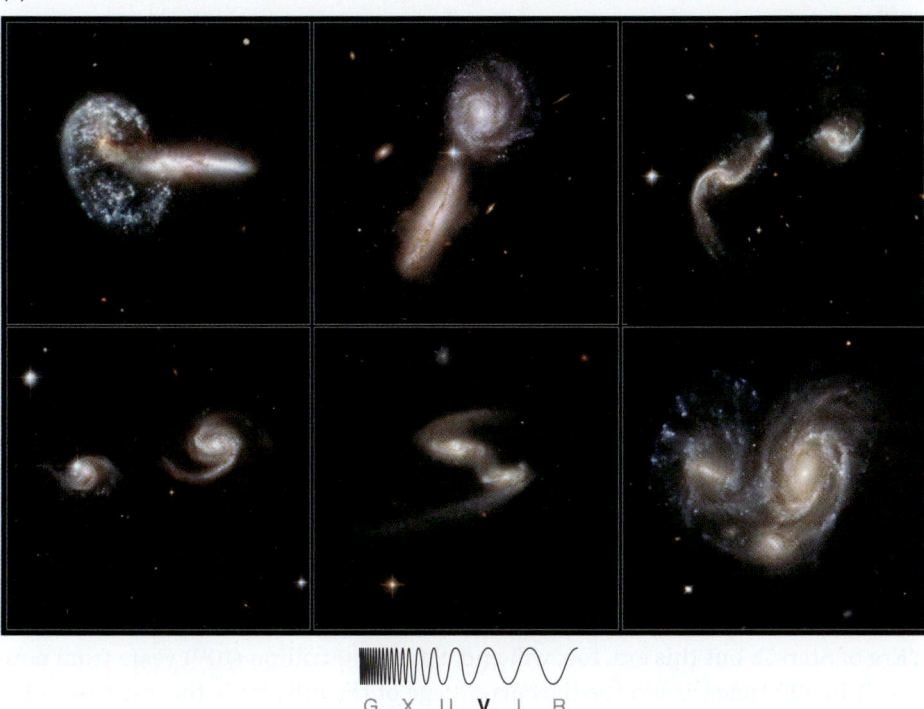

G X U **V I** R

Figure 17.13 The merging galaxies in (a) are between 730 and 2,000 Mpc from Earth, and observations show them at a younger age than the merging galaxies shown in (b), which are located closer to Earth.

G **X** U V I R

Figure 17.14 The Bullet Cluster of galaxies is an example of two giant galaxy clusters merging. In this later stage of merging, the smaller cluster of galaxies (top) appears to have passed through the larger cluster (bottom) like a bullet. The normal matter (red) slowed significantly in the interaction, while the dark matter (blue) from each cluster passed right through the dark matter from the other cluster and has traveled farther at the faster velocity.

messier than the older, nearby ones. This is additional evidence that the formation of structure in the universe was hierarchical.

Simulating Structure

Supercomputer simulations of the universe start with billions of particles of dark matter and incorporate the most recent observations of the CMB. The simulations model the formation and evolution of dark matter clumps and halos, filaments and voids, small and large galaxies, and galaxy groups and clusters. They also simulate the flow of gas made of normal matter within these structures as stars form and are used to create images of what the universe could look like at different times, if their simulation inputs are correct. These images are then compared to images of the actual universe. This comparison sets limits on the parameters of the universe: the amount of mass, for example, or the type of dark matter. If the inputs to the simulation are correct, the two sets of images should look very similar. If not, the two sets of images look very different.

In 2009, a simulation run on NASA supercomputers showed that slight variations in density after inflation led to higher-density regions that became the

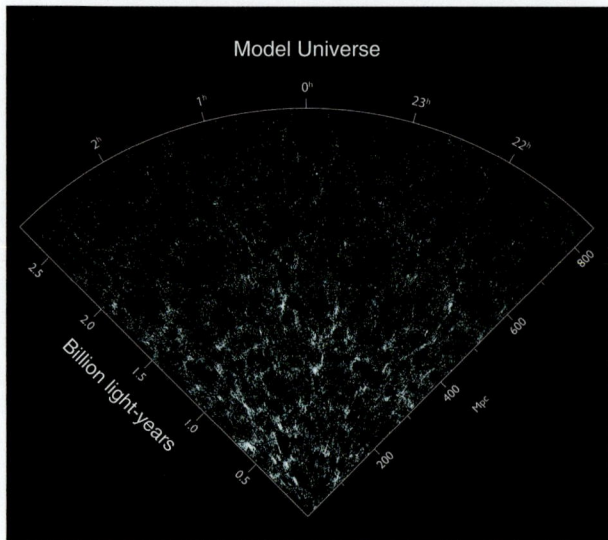

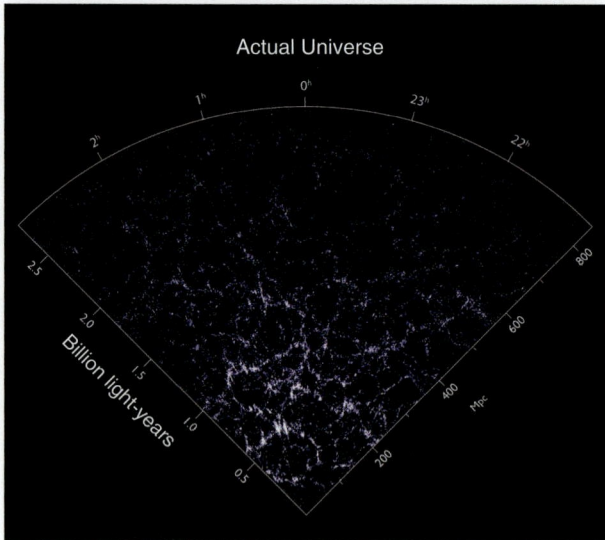

Figure 17.15 The large-scale structure of dark matter halos produced by this simulation (top) is remarkably similar to the distribution of galaxies observed in the Sloan Digital Sky Survey (bottom).

what if . . .

What if you run a computer simulation of the universe, including galaxy formation and the stars within galaxies: How would you need to change this simulation if you compared it to an infrared survey of galaxies rather than an optical survey of galaxies?

seeds for the growth of structure. These density variations are observed in the CMB (see Figure 16.9). During the first few billion years after the Big Bang, dark matter fell together into structures comparable in size to today's clusters of galaxies. The spongelike filaments, walls, and voids became well defined later. The similarities between images produced by the simulation and images of the observed universe are quite remarkable. **Figure 17.15** compares the simulated view with the observed slice of the universe from the Sloan Digital Sky Survey. These simulations constrain the combination of mass, CMB variations, types of dark matter, dark matter halos, and values for the cosmological constant in the universe. This is a very important result. This model contains assumptions consistent with observational and theoretical knowledge of the early universe and predicts the formation of large-scale structure similar to what is actually observed in today's universe.

CHECK YOUR UNDERSTANDING 17.4

Hierarchical models of structure formation are those in which
a. the smallest structures form first.
b. the largest structures form first.
c. formation of structure goes from smallest to largest to smallest.
d. formation of structure goes from largest to smallest to largest.

17.5 Astronomers Think about the Far Future

How will our universe evolve into the far future? Using well-established physics, we can calculate how existing structures in the universe will evolve over a very long time. The result of one such calculation is illustrated in **Figure 17.16**. During the first era of the universe, the Primordial Era—the first several hundred thousand years after the Big Bang and before recombination—the universe was a swarm of radiation and elementary particles. Today, we live during the second era, the Stelliferous Era ("Era of Stars"), but this era, too, will end. Some 100 trillion (10^{14}) years from now (about 10,000 times as long as the current age of the universe), the last molecular cloud will collapse to form stars, and a mere 10 trillion years later, the least massive of these stars will evolve to form white dwarfs.

After the Stelliferous Era, most of the normal matter in the universe will be locked up in brown dwarfs and degenerate stellar objects: white dwarfs, neutron stars, and black holes. During this Degenerate Era, the occasional star will still flare up as ancient substellar brown dwarfs collide, merging to form low-mass stars that burn out in a trillion years or so. However, the main source of energy during this era will come from the decay of particles like protons and neutrons and from the annihilation of particles of dark matter. Even these processes will eventually run out of fuel. In 10^{39} years, white dwarfs will have been destroyed by proton decay, and neutron stars will have been destroyed by the beta decay of neutrons.

As the Degenerate Era comes to an end, the only significant concentrations of mass left will be black holes. These will range from small ones with the masses of single stars to greedy monsters that grew during the Degenerate Era to have masses as large as those of galaxy clusters. During the period that follows—the Black Hole Era—these black holes will slowly evaporate into elementary particles through the emission of Hawking radiation (see Section 13.6). A black hole with a mass of a few solar masses will evaporate into elementary particles in 10^{65} years, and a black hole with the mass of a galaxy will evaporate in about 10^{98} years. By the time the universe

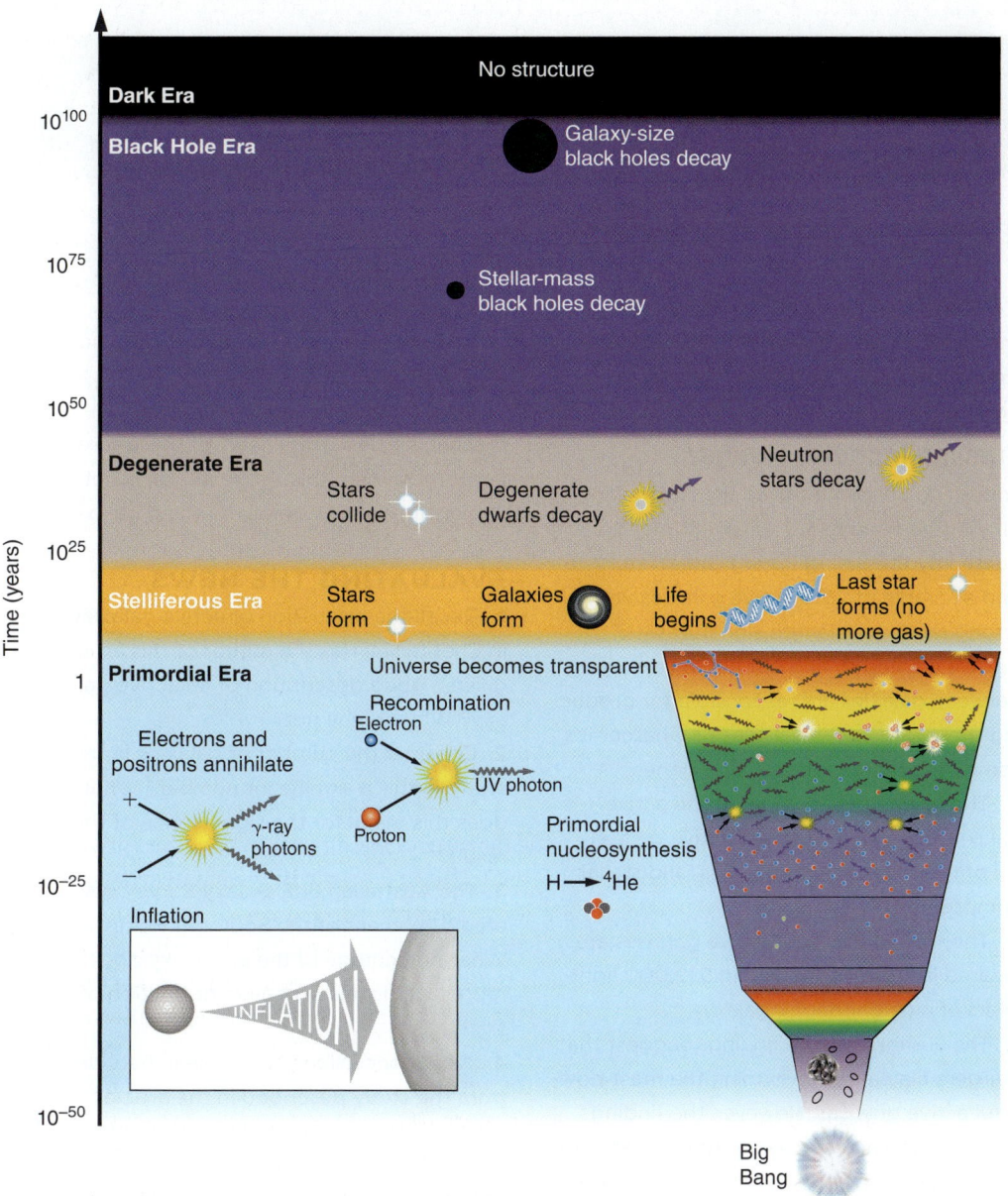

Figure 17.16 Structure in the universe will continue to evolve as it passes from the Primordial Era to the present and then to the future Dark Era. The graphic in the lower-right part of this figure is a smaller version of that shown in Figure 16.14.

reaches an age of 10^{100} years, even the largest of the black holes will be gone. A universe vastly larger than ours will contain little but photons with colossal wavelengths, neutrinos, electrons, positrons, and other waste products of black hole evaporation. The Dark Era will have arrived as the universe continues to expand forever—into the long, cold, dark night of eternity.

CHECK YOUR UNDERSTANDING 17.5

Place the following in the order in which they occur.

a. Degenerate Era

b. Primordial Era

c. Black Hole Era

d. Stelliferous Era

e. the Big Bang

reading Astronomy News

Massive black holes sidle up to other galaxies

by Stuart Gary, ABC

Star formation, supermassive black holes, and the mergers of galaxies are all related. A complete understanding of these processes requires the study of thousands of galaxies across a wide range of luminosities to find the connections.

Galaxies containing supermassive black holes tend to cluster closer to other galaxies, according to a new study.

The discovery, reported in The *Astrophysical Journal*, also reveals how gravitational changes caused by these nearby galaxies could affect star formation.

Supermassive giant black holes are millions to billions of times the mass of the Sun and lurk deep in the hearts of active galaxies.

Scientists believe galaxies interacting with, or colliding into, active galaxies could increase the rate at which gas and other matter falls into the supermassive black hole, effectively shutting down the formation of new stars.

"Astronomers have been scratching their heads to think of what mechanism can shut down star formation in some galaxies," says Dr. Heath Jones from Melbourne's Monash University, who is one of the study's authors.

"In recent years there's been ideas around whether or not these active galactic nuclei,

the supermassive black holes, actually control star formation in a galaxy . . . like a switch."

Fuelling Monster Hunger

Jones and colleagues used the Australian Six-degree-Field [6dF] Galaxy Survey conducted by the Siding Spring Observatory to find a sample of 2,178 active galaxies with strong radio wavelength signatures.

"We wanted to find out if mergers between galaxies are responsible for producing these rare active galaxies or whether it's something more random," says Jones.

The researchers compared the active galaxies to a sample of less active galaxies of the same size, brightness, and stellar composition.

They found the most active galaxies are located in clusters and within 522,000 light-years of a neighbouring galaxy.

The authors say the findings suggest that mergers play a role in forming the most powerful active galaxies. However, the findings only apply to active galaxies that are at least 200 times more powerful than their quieter counterparts.

"With the very strongest ones, there is a difference, but over the full range of radio power the differences we measure get smaller and smaller the less powerful you go," says Jones.

"What that tells us is that this is by no means the end of the story."

"Our understanding of exactly what mechanism lies at the heart of these galaxies, to give them this active nuclei, is not as straightforward as simply merging galaxies."

EVALUATING THE NEWS

1. Recall the discussion of active galaxies in Chapter 14. Has the author of this article given a good description of an active galaxy? Why or why not?

2. Consider the number of galaxies in this study. Is this a significant number? What does this mean for the significance of the result?

3. The Australian 6dF Galaxy Survey covered 41 percent of the Southern Hemisphere. What percentage of the entire sky does this represent? Is this a significant fraction of the sky?

4. What conclusions have researchers drawn from the study described in the article?

5. How will this study help astronomers understand mergers and their effects on active galaxies?

SUMMARY

Galaxies are not distributed uniformly. Instead, they are clumped into groups, clusters, superclusters, filaments, and walls. Galaxies develop due to gravitational instabilities in the presence of cold dark matter. The first stars formed in mini-halos of dark matter, whereas the first galaxies formed later, in larger dark matter halos. Over time, smaller galaxy fragments merged to form larger galaxies. Mergers still happen today. The universe will expand forever and, in an extraordinarily distant future, will eventually become cold and dark.

① Galaxies are hierarchically gathered into groups, clusters, superclusters, filaments, and walls. The walls surround voids in which very few galaxies are present.

② Cold dark matter halos provided the gravitational assist necessary to form galaxy fragments in the early universe.

③ The formation of the first stars occurred in dark matter mini-halos, rather than in clouds of dust and gas. Because of the absence of heavy elements in these dark matter halos, star formation and evolution were different than in the nearby universe.

④ Large-scale structure began with the formation of dark matter halos in the early universe. These halos caused the collapse of normal matter into stars and clusters of stars. Gravitationally bound groupings of stars formed, which were the earliest galaxies. These earliest galaxies merged to become larger galaxies, which in turn accumulated into clusters. The clusters grew hierarchically, through mergers, to become larger clusters, superclusters, and walls.

QUESTIONS AND PROBLEMS

TESTING YOUR UNDERSTANDING

1. **T/F:** Most galaxies are isolated and do not interact gravitationally with other galaxies.

2. **T/F:** Gravity caused variations in the CMB in the early universe.

3. **T/F:** The first stars were much more massive than stars forming today.

4. **T/F:** Collections of galaxies form "bottom-up"—with groups merging to form larger structures.

5. **T/F:** The universe became transparent when it cooled below the surface temperature of the Sun.

6. Place the following types of galaxy collections in order of increasing size.
   a. wall
   b. cluster
   c. group
   d. supercluster

7. It is believed that voids do not contain significant amounts of dark matter because
   a. the light from the far side of the void would not reach us.
   b. galaxies would be drawn toward the center of the voids.
   c. all the dark matter is in the walls.
   d. the walls do not rotate fast enough to require a lot of dark matter.

8. According to the definitions in Chapter 1, the concept of dark matter is classified as
   a. an idea.
   b. a law.
   c. a theory.
   d. a principle.

9. The dominant force in the formation of galaxies is
   a. gravity.
   b. angular momentum.
   c. the strong nuclear force.
   d. the electromagnetic force.

10. Our universe will
   a. expand forever.
   b. expand for a long time and then collapse.
   c. expand but gradually slow down.
   d. neither expand nor contract.

11. Following are the three major observations related to the theory of the Big Bang that have been discussed in the past few chapters. Which one led to the development of the theory (as opposed to being predicted by the theory and subsequently observed)?
   a. Hubble's law
   b. cosmic microwave background radiation
   c. the abundance of helium

12. On the largest scales, galaxies in the universe are distributed
   a. uniformly.
   b. along filaments and walls.
   c. in disconnected clumps.
   d. along lines extending radially outward from the center.

13. Galaxies in the young universe were _____ galaxies in the universe today.
   a. just like
   b. smaller and more irregular than
   c. far more numerous than
   d. larger and more prototypical than

14. The helium abundance in the current universe
   a. is due partly to the Big Bang and partly to stellar evolution.
   b. is due only to the Big Bang.
   c. is due only to stellar evolution.
   d. is unknown at large scales.

15. Galaxy formation is similar to star formation because both
   a. are the result of gravitational instabilities.
   b. are dominated by the influence of dark matter.
   c. end with the release of energy through fusion.
   d. result in the formation of a disk.

16. Dark matter cannot consist of protons, neutrons, and electrons, because if it did
   a. the abundances of isotopes would not be the same as those observed.
   b. it would have interacted with light in the early universe.
   c. stars and galaxies would be much more massive.
   d. both a and b are correct

17. Neutrinos are an example of
   a. hot dark matter.
   b. cold dark matter.
   c. charged particles.
   d. both a and c are correct

18. Dark matter clumps stop collapsing because
    a. angular momentum must be conserved.
    b. they are not affected by normal gravity.
    c. fusion begins, and radiation pressure stops the collapse.
    d. the clumps cannot radiate away any energy.

19. All stars that have been observed have elements heavier than boron. What does this imply about the first stars?
    a. They must have died before galaxies were fully formed.
    b. The first stars did not form until after galaxies formed.
    c. The first stars must have had very low masses.
    d. The first stars must have been enriched in heavy elements.

20. Giant elliptical galaxies come from
    a. the gravitational collapse of clouds of normal and dark matter.
    b. the collision of smaller elliptical galaxies.
    c. the fragmentation of large clouds of normal and dark matter.
    d. the merging of two or more spiral galaxies.

THINKING ABOUT THE CONCEPTS

21. What is the difference between a galaxy cluster and a supercluster? Is our galaxy part of either? How do we know this?

22. Figure 17.2 shows three slices of the universe at three different scales. The scale represented in panel (a) has been marked in panel (b). Resketch the pie shapes from panel (b), and add a line to represent the volume sampled in panel (c). How many billion light-years is it to the outer edge of the circle in panel (c)? ⭐

23. Suppose you could view the early universe when galaxies were first forming. How would it be different from the universe we see today?

24. Imagine that there are galaxies in the universe composed mostly of dark matter with relatively few stars or other luminous normal matter. If this were true, how might we learn of the existence of such galaxies?

25. How are the processes of star formation and galaxy formation similar? How do they differ?

26. ⭐ **WHAT AN ASTRONOMER SEES** Count the number of galaxies that you can see in the inset image in Figure 17.9. From this number, estimate the total number of galaxies visible in Figure 17.9, to the nearest power of 10. (Does the image contain 10 galaxies? 100? 1000?) This image spans only about an arcminute (1/30 of the diameter of the full Moon). Take a moment to consider that every piece of the sky looks like this, and write a sentence or two reacting to your conclusion about the number of galaxies in the observable universe. 👁

27. Why is dark matter believed to be so essential in the formation of galaxies?

28. Why does the current model of large-scale structure require that we include the effects of dark matter?

29. Why do we think that some hot dark matter exists?

30. Figure 17.15 shows simulated data in the top panel and real data in the bottom panel. These two panels are not in exact agreement. Do these differences indicate a significant problem in the simulation's ability to represent reality? Why or why not? 👁

31. How do we know that the dark matter in the universe must be composed mostly of cold—rather than hot—dark matter?

32. Describe the process of structure formation in the universe, starting at recombination (half a billion years after the Big Bang) and ending today.

33. How can we be certain that gravity, and not the other fundamental forces, is responsible for large-scale structure?

34. Previous chapters painted a fairly comprehensive picture of how and why stars form. Why, then, is it difficult to model the star formation history of a young galaxy? Is this difficulty a failure of our theories?

35. What important characteristics of the early universe are revealed by today's observed abundances of various isotopes, such as ^{2}H and ^{3}He?

APPLYING THE CONCEPTS

36. Let's get a feel for the size of a cluster of galaxies. Suppose a small galaxy is orbiting at the outer edge of a cluster of galaxies at a distance of 5 Mpc. What is this distance in meters?
    a. Make a prediction: Compare this distance to the 3 Mpc distance given in Working It Out 17.1. Do you expect your answer to be larger or smaller than the 9.3×10^{22} m found for the galaxy in the Working It Out box? About how much larger or smaller?
    b. Calculate the orbital distance in meters. Compare this orbital distance to the orbital distance of Earth from the Sun (1.5×10^{11} m). How many times larger is this cluster of galaxies than Earth's orbit?
    c. Check your work by comparing your answer to your prediction.

37. Suppose a small galaxy is orbiting a cluster of galaxies at a distance of 3 Mpc, with a speed of 1,500,000 m/s. What is the mass of the cluster?
    a. Make a prediction: This galaxy orbits at the same radius as the galaxy in the Working It Out box, but it moves faster. Do you expect to find that the cluster orbited by this galaxy is more or less massive than the cluster in Working It Out 17.1?
    b. Follow Working It Out 17.1 to calculate the mass of the cluster.
    c. Check your work by comparing your answer to your prediction.

38. Suppose a small galaxy is orbiting a cluster of galaxies at a distance of 4 Mpc, with a speed of 1,000,000 meters per second. What is the mass of the cluster?
    a. Make a prediction: This galaxy orbits at a larger radius than the galaxy in the Working It Out box, but it moves at the same speed. Do you expect to find that the cluster orbited by this galaxy is more or less massive than the cluster in Working It Out 17.1?
    b. Follow Working It Out 17.1 to calculate the mass of the cluster.
    c. Check your work by comparing your answer to your prediction.

39. Suppose a small galaxy is orbiting a cluster of galaxies at a distance of 1.5 Mpc, with a speed of 500,000 m/s. What is the mass of the cluster?
    a. Make a prediction: This galaxy orbits at a smaller radius than the galaxy in the Working It Out box, but it also moves at a slower speed. Do you expect to find that the cluster orbited by this galaxy is more or less massive than the cluster in Working It Out 17.1?
    b. Follow Working It Out 17.1 to calculate the mass of the cluster.
    c. Check your work by comparing your answer to your prediction.

40. Suppose a small galaxy is orbiting a cluster of galaxies at a distance of 1.5 Mpc, with a speed of 1,500,000 m/s. What is the mass of the cluster?
    a. Make a prediction: This galaxy orbits at a smaller radius than the galaxy in the Working It Out box, but it also moves at a greater speed. Do you expect to find that the cluster orbited by this galaxy is more or less massive than the cluster in Working It Out 17.1?
    b. Follow Working It Out 17.1 to calculate the mass of the cluster.
    c. Check your work by comparing your answer to your prediction.

exploration The Story of a Proton

Now that you have surveyed our current astronomical understanding of the universe, you are prepared to put the pieces together to make a story of how you came to be sitting in your chair, holding this book, and reading these words. It is valuable to take a moment to work your way from the Big Bang through all the intervening steps that had to occur to reach the beginning of the book, which started with looking at the sky.

1. In the Big Bang, how is a proton formed?

_____

_____

_____

2. How might that proton become part of one of the first stars?

_____

_____

_____

3. Suppose that proton later becomes part of a carbon atom in a 4-M_{Sun} star. Through what type of nebula does it pass before returning to the interstellar medium?

_____

_____

_____

4. Suppose that carbon atom then becomes part of the molecular-cloud core forming the Sun and the Solar System. What two physical processes dominate the core's collapse as the Solar System forms and that carbon atom becomes part of a planet?

_____

_____

_____

_____

5. Beginning with the Big Bang, create a timeline that traces the full history of a proton that becomes a part of the nucleus of a carbon atom on Earth.

_____

_____

_____

_____

This is in essence the astronomical story of how Earth formed. We will address what comes next—how you have come to be on Earth at this time, studying the sky—in Chapter 18.

Life in the Universe

When astronomers look for worlds that might harbor life, they look for water, as you will learn here in Chapter 18. Life as we know it depends on water because water is a terrific solvent; it can carry chemicals, minerals, and nutrients around in an organism. It has this property because it is a polar molecule—the hydrogen end of the molecule is positively charged, while the oxygen end is negatively charged. Water, therefore, is attracted to lots of different molecules. It can be attracted, in fact, to anything that carries a charge. Rub a plastic object, like a pen, a ruler, or an inflated balloon, all over your dry hair, so that the plastic object becomes electrically charged. Turn on the water from a faucet so that a thin stream falls from the tap. Take a moment to predict what will happen when you bring the charged plastic object close to the water. Then slowly bring the plastic object close to the stream of water. Sketch the result.

EXPERIMENT SETUP

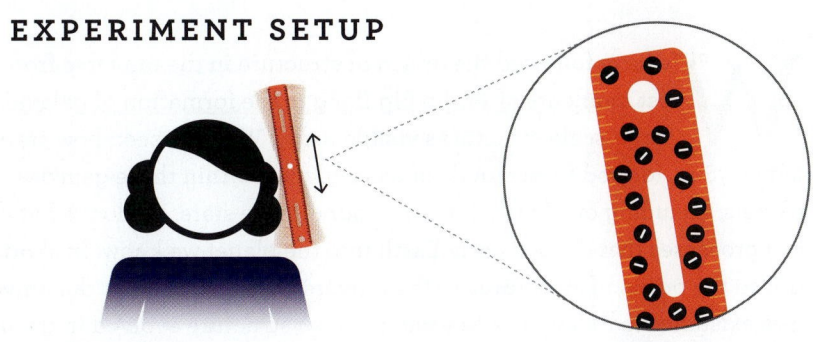

Rub a plastic object, like a pen, a ruler, or an inflated balloon, all over your hair, separating charges, so that the plastic object becomes electrically charged.

Turn on the water from a faucet so that a thin stream falls from the tap.

Take a moment to make a prediction about what will happen when you bring the charged plastic object close to the water. Then slowly bring the plastic object close to the stream of water. Sketch the result.

← SLOWLY

PREDICTION

I predict that the stream of water will
- ☐ bend away from the object.
- ☐ bend toward the object.
- ☐ fall straight down.
- ☐ yank the object out of my hand.

SKETCH OF RESULTS (in progress)

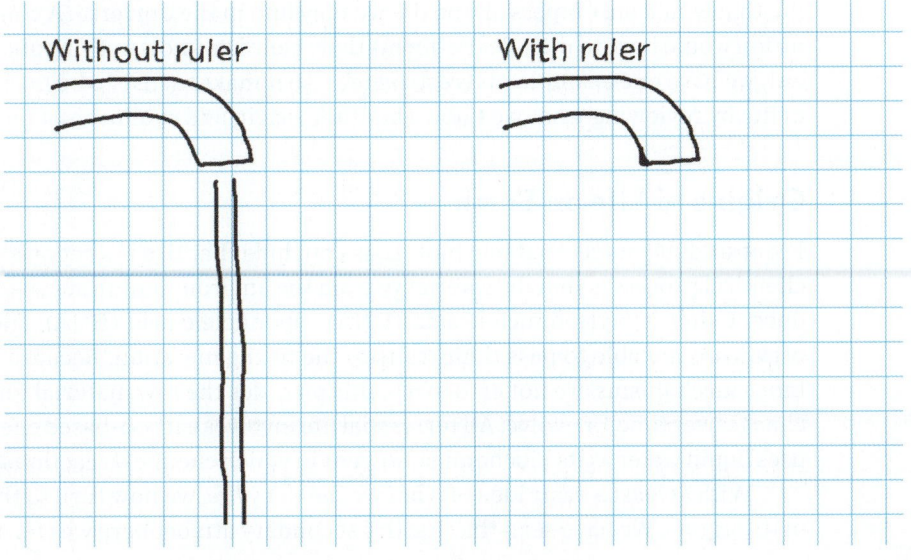

Without ruler   With ruler

18

We have followed the origin of structure in the universe from the earliest moments after the Big Bang to the formation of galaxies and other large-scale structures visible today. We have seen how stars, including our Sun, formed from clouds of gas and dust within these galaxies and how planets, including our Earth, formed around those stars. We looked at the geological processes that shaped early Earth into the planet we know. In short, we have traced the origin of structures in the universe from the instant the universe came into existence to today. No discussion of how structure evolved in the universe would be complete, however, without some consideration of the origin and evolution of *life*. Studying life on Earth gives context to our search for life elsewhere in the universe. Here in Chapter 18, we will investigate the scientific understanding of the origin and evolution of life on Earth and explore its possible existence elsewhere.

LEARNING GOALS

1. Explain our current understanding of how life began on Earth and how life evolved to reach today's complexity.

2. Describe the limits that the chemistry of known life places on possible habitats in our Solar System.

3. Define a habitable zone.

4. Explain the significance of discovering exoplanets to searches for extraterrestrial life.

5. Explain why all life on Earth will eventually end.

18.1 Life on Earth Began Early and Evolved over Time

How do we define *life*? Many scientists suggest that there is no single definition of life that would encompass all the life we may find in the universe. A complete definition would have to include life-forms that scientists know nothing about. At present, we have one example—terrestrial life—so it makes sense to begin a discussion of life by reviewing what we know about the origin and evolution of life on Earth.

Origins of Life on Earth

From studying terrestrial life, biologists conclude that **life** is a set of complex biochemical processes that draws energy from the environment to survive and reproduce. Using deoxyribonucleic acid (DNA), ribonucleic acid (RNA), and proteins, organisms are able to pass their traits to the next generation. Because the inheritance mechanisms are not error-free, this provides the raw material for species to change over time, or evolve. All terrestrial life involves carbon-based chemistry and uses liquid water as its biochemical solvent in which chemical reactions take place.

With at least a basic idea of what we mean by *life*, we now turn to the question of its origins. We have seen that Earth's secondary atmosphere was formed in part by the carbon dioxide and water vapor emitted by volcanoes; a heavy bombardment of comets and asteroids probably added large quantities of water, methane, and ammonia to the mix, too. Although these are all simple molecules, early Earth had abundant sources of energy (such as lightning and ultraviolet solar radiation) that could tear these molecules apart into fragments that could reassemble into molecules of greater mass and complexity. As rain carried the heavier organic molecules out of the atmosphere, they ended up in Earth's oceans, forming a "primordial soup."

In 1952, American chemists Harold Urey (1893–1981) and Stanley Miller (1930–2007) attempted to create conditions similar to those that they thought existed on early Earth. They placed water in a laboratory apparatus to represent the ocean and then added methane, ammonia, and hydrogen as a primitive atmosphere; electric sparks simulated lightning as a source of energy (**Figure 18.1**). Within a week, the Urey-Miller experiment yielded 11 of the 20 basic amino acids that link together to form proteins, the structural molecules of life. Other organic molecules that are components of nucleic acids, the precursors of RNA and DNA, also appeared in the mix. More recent experiments, which included carbon dioxide, nitrogen, and hydrogen sulfide as components of the atmosphere, have produced results similar to those of Urey and Miller. From laboratory experiments such as these, scientists have developed models of how life might have begun. Most of these models assume that life began in the oceans that were rich in the organic molecules necessary for life. However, the details of precisely how these simple molecules evolved into the more complex molecules of life are not yet fully understood.

Some biologists think life began in the ocean depths, where volcanic vents provided the hydrothermal energy needed to create the highly organized molecules responsible for biochemistry (**Figure 18.2**). Similar environments exist elsewhere in the Solar System. Others think that life originated in tide pools, where lightning and ultraviolet radiation supplied the energy. In both cases, short strands of molecules that could replicate (copy) themselves may have formed first, later evolving into RNA and finally into DNA, the huge molecule that serves as the biological "blueprint" for reproducing organisms.

A few scientists think that perhaps life on Earth was "seeded" from space in the form of microbes brought by meteoroids or comets. Although this hypothesis might tell us how life came to Earth, it does not explain its origin elsewhere in the Solar System or beyond. There is no scientific evidence at this time to support the seeding hypothesis.

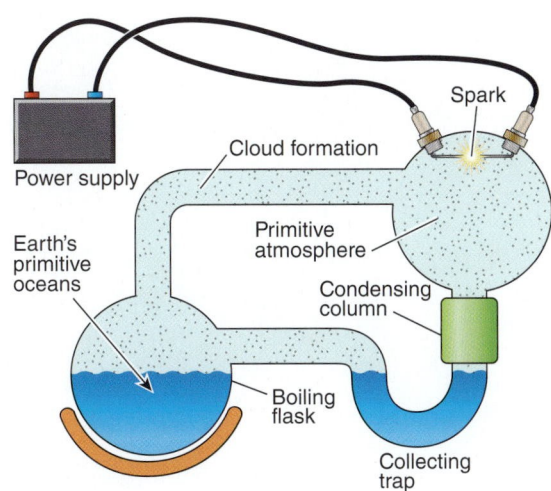

Figure 18.1 The Urey-Miller experiment was designed to simulate conditions in an early-Earth atmosphere.

Figure 18.2 (a) Life on Earth may have arisen near oceanic geothermal vents. (b) Living organisms around such vents, such as the giant tube worms shown here, rely on geothermal rather than solar energy for their survival.

(a)

(b)

The First Life

If life did indeed get its start in Earth's oceans, when did it happen? In Chapter 5, we noted that Earth formed roughly 4.6 billion years ago. We also described how young Earth was bombarded by Solar System debris for several hundred million years after its formation. These conditions might have been too harsh for life to form and evolve on Earth. Once the bombardment abated and Earth's oceans appeared, however, there were opportunities for life to appear. Carbonized material found in Greenland, in rocks dating back to 3.85 billion years ago, might be *indirect* evidence of life. Stronger and more direct evidence for early life appears in the form of stromatolites, which are masses of simple microbes such as cyanobacteria (single-celled organisms commonly known as blue-green algae). Fossilized stromatolites that date back to about 3.7 billion years ago have been found in Greenland, and some from about 3.5 billion years ago have been found in western Australia and southern Africa. Today's stromatolites (**Figure 18.3a**) share the layered structure of fossilized stromatolites (Figure 18.3b). We may never know the precise date when life first appeared on Earth, but current evidence suggests that the earliest life-forms appeared within a billion years after the formation of the Solar System and within 500 million years after the end of young Earth's catastrophic bombardment by leftover planetesimals.

The earliest organisms were extremophiles, life-forms that not only survive but thrive under extreme environmental conditions. **Extremophiles** include organisms adapted to live in subfreezing environments or in water temperatures as high as 120°C, which occur in the vicinity of deep-ocean hydrothermal vents. Other extremophiles are found under conditions of extraordinary salinity, pressure, dryness, acidity, or alkalinity. Among these early life-forms was an ancestral form of cyanobacteria. Cyanobacteria photosynthesize carbon dioxide and release oxygen as a waste product. Oxygen, however, is a highly reactive gas, so at the time, the newly released oxygen was quickly removed from Earth's atmosphere by the oxidation of surface minerals. Once the exposed minerals could no longer absorb more oxygen, atmospheric levels of oxygen began to rise. Oxygenation of Earth's atmosphere and oceans began about 2 billion years ago, and current levels were reached only about 250 million years ago (see Chapter 7). Without cyanobacteria and other photosynthesizing organisms, Earth's atmosphere would be as oxygen-free as the atmospheres of Venus and Mars.

When 20th-century biologists began peering through microscopes, it soon became clear that there were two major categories of organisms. On the basis of

Figure 18.3 (a) Stromatolites are mats of microbes that can be found in shallow tidal areas around the world. (b) Fossilized examples, like this one from Wyoming, show the layered mat structure that is common to all stromatolites, even those dating back 3.7 billion years.

(a)

(b)

cell size and internal organization, biologists identified these organisms as prokaryotes and eukaryotes. The most striking difference between the two types of cells is the lack of a nucleus in prokaryotes. In addition, prokaryotes lack the numerous internal membranes found in eukaryotes. As shown in **Figure 18.4**, a prokaryotic cell is enclosed in a membrane, and inside are found the DNA, RNA, and proteins necessary for the cell to grow and divide. A eukaryotic cell is also enclosed by a membrane. But in contrast to a prokaryote, a eukaryotic cell is typically larger, its DNA is enclosed in a membrane-bound nucleus, and many of its biochemical components are also inside membrane-bound structures (Figure 18.4).

Life Becomes More Complex

All life on Earth, whether prokaryotic or eukaryotic, shares a similar genetic code that originated from a common ancestor. DNA sequencing enables biologists to trace backward to the time when different types of life first appeared on Earth and to identify the species from which these life-forms evolved. Scientists have used DNA sequencing to establish what is known as the "phylogenetic tree of life," shown in **Figure 18.5**. This complex tree describes the evolutionary interconnectivity of all species. The tree of life is composed of two domains of prokaryotes, Bacteria and Archaea, and a third domain containing all eukaryotes, Eukarya. Comparing the DNA sequences among thousands of organisms has revealed some interesting relationships. For example, although under a microscope Archaea and Bacteria are often indistinguishable, DNA sequence comparisons show that they diverged billions of years ago. Archaea have genes and metabolic pathways that are more similar to those of Eukarya than to those of Bacteria. The first fossils of eukaryotes date from about 2 billion years ago, coincident with the rise of free oxygen in the oceans and atmosphere, although the first *multicellular* eukaryotes did not appear until a billion years after that. The phylogenetic tree places animals closest to fungi, which formed a branch of the evolutionary tree after slime molds and plants. Along the animals branch, the earliest primates branched off from other mammals about 70 million years ago, and the great apes (gorillas, chimpanzees, bonobos, orangutans, and humans) split off from the lesser apes about 20 million years ago. DNA

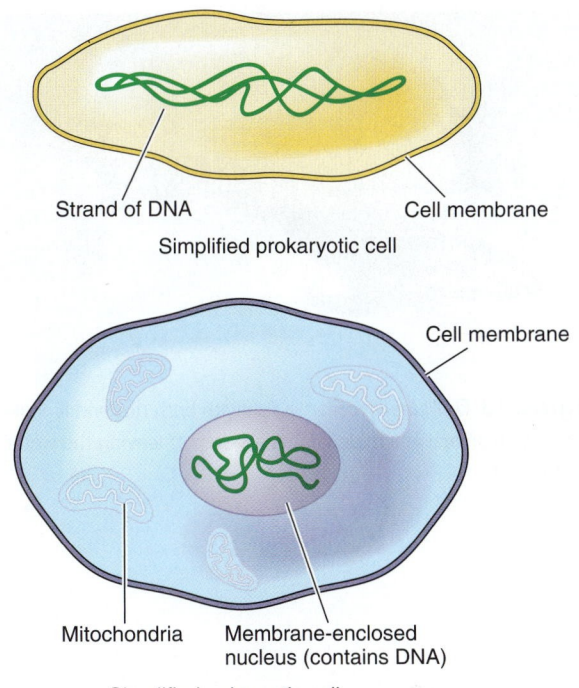

Figure 18.4 In a prokaryotic cell, its genetic material (DNA) is in the same compartment as the rest of the cell's molecules. A eukaryotic cell contains several membrane-enclosed structures, including a nucleus, which houses the cell's DNA, and mitochondria, the cell's powerhouses.

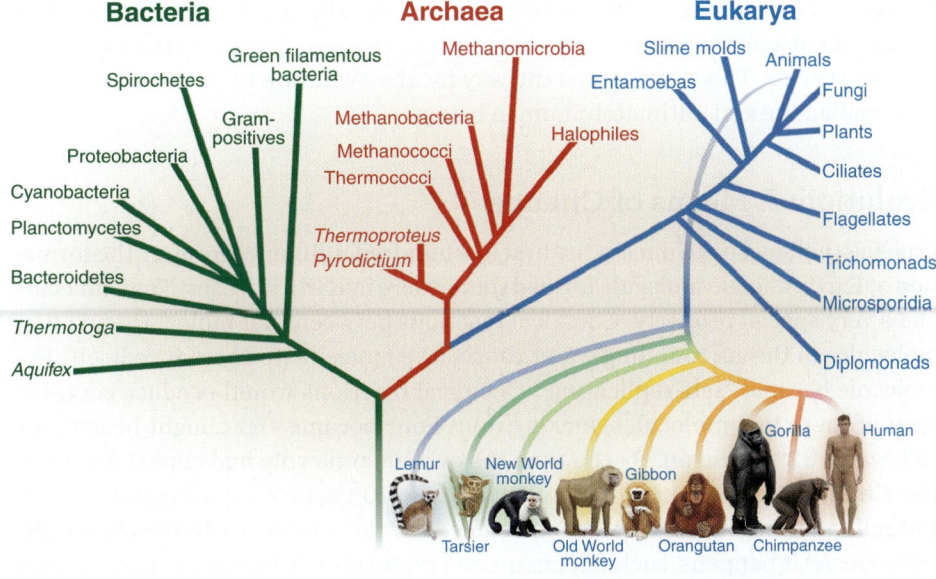

Figure 18.5 This simplified version of the phylogenetic tree has been constructed from analysis of the DNA strands of different life forms. Humans are included in the animals branch of the Eukarya. The primate branch of animals is shown as an inset at the bottom.

Figure 18.6 *Tiktaalik roseae*, a fish with both ribs and limblike fins, was an animal in a midevolutionary step of leaving the water for dry land.

what if . . .

What if life formed in *both* the ocean depths and in tide pools: Would you expect all life on Earth to be as interconnected as is shown in Figure 18.5?

what if . . .

What if we discovered self-replicating life on another planet, where the life is based on a molecule other than DNA: Would we expect life to be evolving on such a planet?

tests show that humans and chimpanzees share about 98 percent of their DNA; the two groups are believed to have evolved from a common ancestor about 6 million years ago. For comparison, all humans share 99.9 percent of their DNA.

Living creatures in Earth's oceans remained much the same—a mixture of single-celled and relatively primitive multicellular organisms—for more than 3 billion years after the first appearance of terrestrial life. Then, between 540 million and 500 million years ago, the number and diversity of biological species increased spectacularly. Biologists call this event the Cambrian explosion. The trigger of this surge in biodiversity remains unknown, but possibilities include rising oxygen levels, an increase in genetic complexity, major climate change, or a combination of all of these. The "Snowball Earth" hypothesis suggests that, before the Cambrian explosion, Earth was in a period of extreme cold between about 750 million and 550 million years ago and was covered almost entirely by ice. During this period, many organisms may have died out, thus making it easier for new species to adapt and thrive. Another possibility is that a marked increase in atmospheric oxygen (see Figure 7.5) would have been accompanied by a corresponding increase in stratospheric ozone, which, as you learned in Chapter 7, shields us from deadly solar ultraviolet radiation. With a protective ozone layer in place, life was free to leave the oceans and move to land (**Figure 18.6**).

The first plants appeared on land about 475 million years ago. Large forests and insects go back 360 million years. The age of dinosaurs began 230 million years ago and ended abruptly 65 million years ago, most likely caused by a small asteroid or comet colliding with Earth. The collision threw so much dust into the atmosphere that the sunlight was dimmed for months, causing the extinction of more than 70 percent of all existing plant and animal species. Mammals were the big winners in the aftermath. Our earliest human ancestors appeared a few million years ago, and the first civilizations occurred a mere 10,000 years ago. Our industrial society is barely more than two centuries old.

Humans are here today because of a series of events that have occurred throughout the history of the universe. Some of these events are common in the universe, such as the creation of heavy elements by earlier generations of stars and the formation of planets, including Earth. Other events in Earth's history may have been less likely to happen elsewhere, such as the formation of a planet with life-supporting conditions or the development of self-replicating molecules enclosed inside membranes that led to Earth's earliest life. A few events stand out as random, such as the impact of a piece of space debris 65 million years ago that led to the extinction of most species. This event eased the way for the evolution of advanced forms of mammalian life and, ultimately, human beings.

Evolution: A Means of Change

Imagine that, just once during the first few hundred million years after the formation of Earth, a single molecule formed somewhere in Earth's oceans. That molecule had a very special property: Chemical reactions between that molecule and other molecules in the surrounding water caused the molecule to reproduce itself. The molecule became "self-replicating." Chemical reactions would produce copies of each of these two molecules, making four. Four became eight, eight became 16, 16 became 32, and so on. By the time the original molecule had copied itself just 100 times, more than a *million trillion trillion* (10^{30}) copies would exist. That is 100 million times more of these molecules than there are stars in the observable universe. As it happens, such unconstrained replication is highly unlikely because of the limited availability of raw materials needed for reproduction and the competition from similar molecules. However, there were ample resources present to allow replication on a smaller scale—enough to create life as we know it.

The molecules of DNA (**Figure 18.7**) that make up the chromosomes in the nuclei of the cells throughout your body are direct descendants of those early self-replicating molecules that flourished in the oceans of a young Earth. Although the DNA in your body is far more elaborate than those early molecules, the fundamental rules of biochemistry remain the same.

Chemical reactions are not always error-free. Sometimes, when a molecule replicates, the new molecule is not quite a duplicate of the old one. The likelihood that a copying error will occur while a molecule is replicating increases significantly with the number of copies being made. For DNA, which contains the genetic code for entire organisms, the change in the code is called a **mutation**. Sometimes such an error leads to an organism that is no longer suited to its environment. Organisms with these changes will reproduce poorly and will not flourish. Other times a mutation is beneficial, leading to a molecule or an organism that is *better* suited to the environment than the original. Organisms with these mutations are more likely to reproduce successfully. Even if changes in the copying process crop up only once every 100,000 times that an organism reproduces its DNA, and even if only one out of 100,000 of these errors turns out to be beneficial, after only 100 generations there will still be a hundred million trillion (10^{20}) mutations that, by chance, might improve on the original's chances of survival. Copies of each "improved" molecule will inherit the change. Mutations affect the genetic **heredity** of the organism—the genetic code passed on to future generations. To summarize, *mutation* and *heredity* lead to *natural selection* and to organisms that *evolve* over time.

Cells have a membrane that encloses all of the molecular machinery and protects the structures inside from the environment. This membrane may have formed from fatty molecules that enclosed bubbles of water, along with self-replicating molecules, to make the first organic cells. These early cells were likely to be fragile, dissolving and re-forming in response to environmental factors. Over time, cells with sturdier membranes reproduced more successfully than those with weaker membranes. Biologists today are learning how these membranes contribute to the survival of cells in varied environments. For example, **thermophiles (Figure 18.8a)** are cells that thrive at high temperatures, such as those found in hot springs in Yellowstone National Park (Figure 18.8b). These cells have a larger fraction of saturated fatty acids in their membranes, which makes the membrane stable at higher temperatures. Two familiar fats, butter and olive oil, are different for the same

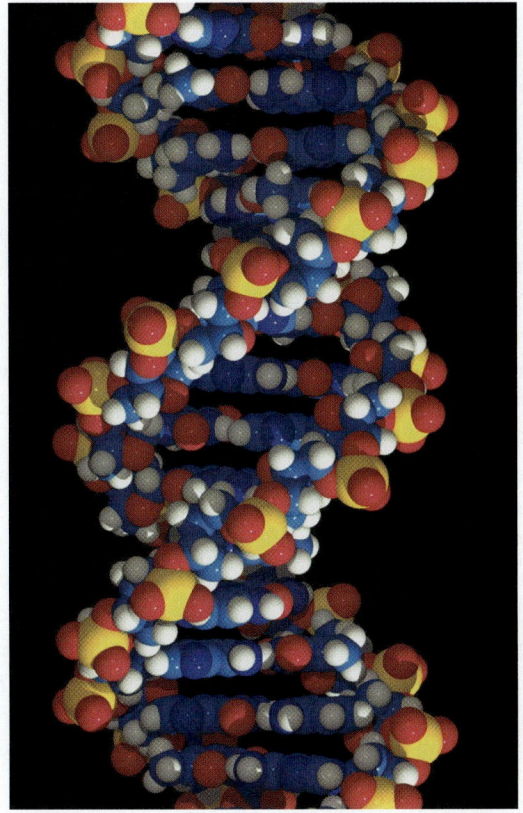

Figure 18.7 DNA is the molecular basis for heredity. This model of a DNA molecule shows the two strands that are wound together into a shape known as a double helix.

Figure 18.8 (a) Thermophiles are heat-loving microscopic organisms. (b) These organisms thrive in high-temperature environments, such as those found in hot springs and pools in Yellowstone National Park.

(a)

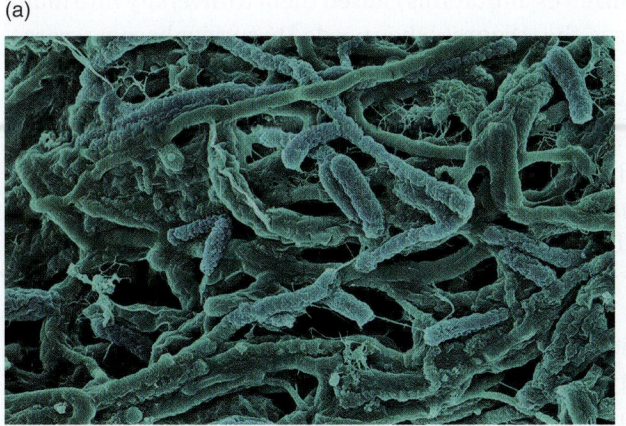

(b)

working it out 18.1

Exponential Growth

Exponential growth means that a population grows by a percentage (say, 7 percent) each time period, rather than by a fixed amount (say, increasing by 7) each time period. The difference is important because, in exponential growth, the percentage represents a larger number of individuals in each time step: The population adds more individuals with time. Biological systems, because they are self-replicating, grow by exponential growth until a resource limit is reached.

Assume a single hypothetical self-replicating molecule that makes one copy of itself each minute. The doubling time, then, is 1 minute. How many molecules will exist after an hour? We will use P_0 to represent the population at time $t = 0$. In this case, $P_0 = 1$ (a single molecule). P_F is the final population after an hour has elapsed. Population increases by 100 percent—a factor of 2—each minute for 60 minutes. So, to find the final population, we must double the initial population (multiply it by 2) 60 times. That would be tedious, but there is a shortcut. Doubling the population 60 times is the same as multiplying by 2 to the 60th power:

$$P_F = 2^{60} \times P_0$$

$$P_F = 2^{60} \times 1$$

$$P_F = 2^{60}$$

Plugging 2^{60} into the calculator gives 1.2×10^{18}. Therefore, there are a billion billion of these molecules after 1 hour.

In general, we use n to represent the number of doubling times, so the more general expression is

$$P_F = 2^n \times P_0$$

Now, suppose that a change in the copying process occurs only once every 50,000 times that a molecule reproduces itself, and one out of 200,000 of these changes is beneficial. After 100 generations, how many new molecules with these beneficial changes will form? First, we must find the final number of molecules. Again, $P_0 = 1$, but now n is 100:

$$P_F = 2^{100} \times 1$$

$$P_F = 1.3 \times 10^{30}$$

To find the number of these molecules that have changes, P_C, we must multiply by the fraction of times a change occurs. Because a change occurs one time in 50,000, we must multiply by 1/50,000:

$$P_C = 1.3 \times 10^{30} \times \frac{1}{50,000}$$

$$P_C = 2.6 \times 10^{25}$$

To find the number of molecules with beneficial changes, P_B, we must account for the fact that only 1/200,000 of these changes is beneficial. So, we multiply by 1/200,000:

$$P_B = 2.6 \times 10^{25} \times \frac{1}{200,000}$$

$$P_B = 1.3 \times 10^{20}$$

There will be roughly 10^{20} (100 million trillion) molecules in this generation that have a beneficial mutation. Because this number does not count earlier beneficial changes that themselves replicated, the total number of molecules with beneficial changes will be much larger!

reason. Butter is high in saturated fat and is a solid at room temperature, whereas olive oil is low in saturated fat and is a liquid at room temperature.

As the organisms of the early Earth continued to interact with their surroundings and make copies of themselves, mutations caused them to diversify into many different species. In some cases, the resources they needed to survive became scarce. Competition for resources, predation of one variety on another, and cooperation between organisms became important to the survival of many organisms. Some organisms were more successful under these circumstances and reproduced to become more numerous, while other organisms that were less well-suited for the conditions became less common. This process, in which better-adapted organisms reproduce and thrive while less-well-adapted ones become extinct, is called **natural selection**.

Four billion years is a long time—enough time for the combined effects of heredity and natural selection to shape the descendants of that early self-copying molecule into a huge variety of complex, competitive, successful organisms (see **Working It Out 18.1**). Among these descendants are organisms capable of thinking about their own existence and unraveling the mysteries of the stars. This evolutionary process has involved considerable randomness throughout the entire history of life on Earth. Geological processes on Earth have preserved a fossil record of the history of the organisms that existed throughout Earth's history (**Figure 18.9**). Another "Earth" would probably not produce the same organisms that we find on Earth, either today or in the fossil record.

CHECK YOUR UNDERSTANDING 18.1

Scientists think that terrestrial life probably originated in Earth's oceans because (select all that apply)

a. all the chemical pieces were in the ocean.

b. energy was available in the ocean.

c. the earliest evidence for life on Earth is from fossils of ocean-dwelling organisms.

d. the deepest parts of the ocean have hydrothermal vents.

Answers to Check Your Understanding questions are in the back of the book.

18.2 Life beyond Earth Is Possible

The story of the formation and evolution of life cannot be separated from the story of astronomy. We know that we live in a universe full of stars and that systems of planets orbit many—probably most—of those stars. The evolution of life on Earth is but one of many examples that we have encountered of the emergence of structure in an evolving universe. This point leads naturally to one of the more profound questions that we can ask about the universe: Has life arisen elsewhere? To explore this question, we need to take a closer look at the chemistry of life on Earth.

The Chemistry of Life

Recall that the early universe was composed basically of hydrogen and helium and very little else. After 9 billion years of stellar nucleosynthesis, all the heavier chemical elements essential to life were present and available in the molecular cloud that gave birth to our Solar System. As you saw in Chapters 12 and 13, those heavy elements—up to and including iron—were created during earlier generations of low-mass and high-mass stars and were then dispersed into space. At times, this dispersal was passive. For example, low-mass stars, such as dying red giants, lose their gravitational grip on their overly extended atmospheres. Along with hydrogen and helium, the newly created heavy elements are blown off into space, eventually finding their way into molecular clouds. Other dispersals were more violent. Most of the trace elements essential to our biology are more massive than iron, so they are not produced in the interiors of main-sequence stars. They are instead created within a matter of minutes during the violent supernova explosions that mark the death of high-mass stars, or in the collision of neutron stars. These elements, too, are part of the chemical mix found in molecular clouds.

All known organisms are composed of a more or less common suite of complex chemicals—very complex ones at that. Approximately two-thirds of the atoms in our bodies are hydrogen (H), about one-fourth are oxygen (O), one-tenth are carbon (C), and a few hundredths are nitrogen (N). The remaining elements, and there are several dozen of them, make up only 0.2 percent of the total inventory of the atoms in our bodies. All known living creatures are an assemblage of molecules composed almost entirely of these four elements, sometimes called CHON, along with small amounts of phosphorus and sulfur. Some of these molecules are large.

Figure 18.9 Fossils, such as this Parasaurolophus ("near crested lizard"), record the history of the evolution of life on Earth. This plant-eating dinosaur lived in North America about 75 million years ago.

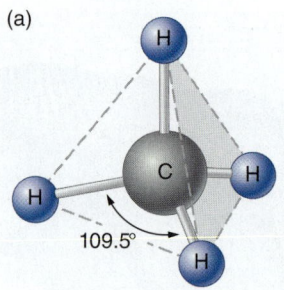

(a)

109.5°

Methane tetrahedron

(b)

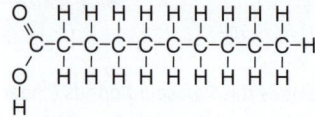

Saturated

Unsaturated

(c)

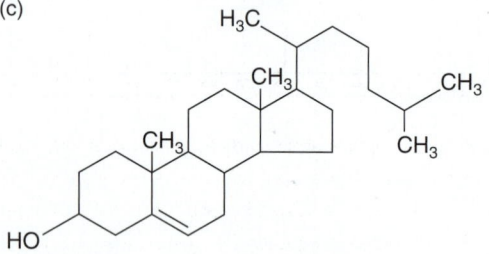

Figure 18.10 (a) Carbon is tetravalent, meaning it can bond to as many as four other atoms, as in methane. This property allows carbon to form an enormous variety of (b) long-chain molecules, like saturated and unsaturated fatty acids, as well as (c) ringed structures like cholesterol.

DNA, which is responsible for our genetic code, is made entirely from only *five* elements: CHON and phosphorus. But the DNA in each cell of our bodies is composed of combinations of *tens of billions* of atoms of these same five elements. Then there are proteins, the huge molecules responsible for the structure and function of living organisms. Proteins are long chains of smaller molecules called amino acids. Terrestrial life uses 20 specific amino acids, which also contain no more than five elements—in this case CHON plus sulfur instead of phosphorus.

The chemistry of life requires more than a mere half-dozen elements. Many others are present in smaller amounts but are essential to the complicated chemical processes of life. These include sodium, chlorine, potassium, calcium, magnesium, iron, manganese, and iodine. Trace elements, such as copper, zinc, selenium, and cobalt, also play a crucial role in biochemistry but are needed in only tiny amounts.

Terrestrial life is based on carbon. Carbon is the lightest among the tetravalent atoms, which can bond with as many as four other atoms or molecules (**Figure 18.10**). (*Tetra* means "four," and *valence* refers to an atom's ability to attach to other atoms or molecules.) If the attached molecules also contain carbon, the result can be an enormous variety of complex molecules including long chains, rings, and combinations of chains and rings. This great versatility enables carbon to form the huge array of molecules that provide the basis for the chemistry of terrestrial life.

There could be carbon-based forms of extraterrestrial life that have chemistries quite different from our own. For example, there are countless varieties of amino acids beyond the 20 used in terrestrial life. Furthermore, molecules other than RNA and DNA may be capable of self-replication.

Science-fiction writers often speculate about silicon-based life-forms because silicon, like carbon, is tetravalent—each silicon atom can bond with as many as four other atoms, so that a large number of combinations are possible. As a potential life-enabling atom, silicon has both advantages and disadvantages compared to carbon. An important advantage is that silicon-based molecules remain stable at much higher temperatures than carbon-based molecules, perhaps enabling possible silicon-based life to thrive in high-temperature environments, such as on planets that orbit close to their parent star. But silicon has a serious disadvantage; it is a larger and more massive atom than carbon, and it cannot form long chains of atoms as complex as those based on carbon. Any silicon-based life likely would be simpler than life-forms here on Earth, but it might exist in high-temperature niches somewhere in the universe.

Although carbon's unique properties make it readily adaptable to the chemistry of life on Earth, we don't know what other chemistries life might adopt. Life on Earth is highly adaptable and tenacious. Everywhere that scientists have looked on Earth, they have found life (for example, see Figures 18.2 and 18.8). When it comes to the forms that extraterrestrial life might take, nothing should surprise us.

Life within Our Solar System

The logical place to start looking for extraterrestrial life is right here in our own Solar System. Early conjectures about life in our Solar System seem naïve, considering what we now know. Two centuries ago, the eminent astronomer Sir William Herschel, discoverer of Uranus, proclaimed, "We need not hesitate to admit that the Sun is richly stored with inhabitants." In 1877, Italian astronomer Giovanni Schiaparelli (1835–1910) observed what appeared to be linear features on Mars and dubbed them *canali*, meaning "channels" in Italian (**Figure 18.11**). The famous

American observer of Mars, Percival Lowell (1855–1916), misinterpreted Schiaparelli's *canali* as "canals," suggesting that they were constructed by intelligent beings. This turned out to be incorrect, but the idea inspired a number of famous science fiction stories, including "War of the Worlds," by H. G. Wells.

During the mid-20th century, ground-based telescopes discovered that Mars possesses an atmosphere and water, both considered essential for any terrestrial-type life to begin and evolve. During the 1960s, the United States and the Soviet Union sent reconnaissance spacecraft to the Moon, Venus, and Mars, but the instrumentation carried aboard these spacecraft was more suited to learning about the physical and geological properties of these bodies than to searching for life. Serious efforts to look for signs of life—past or present—require more advanced spacecraft with specialized bio-instrumentation.

In the meantime, astronomers and biologists together were discussing how to look for life as part of a new science called **astrobiology**: the study of the origin, evolution, distribution, and future of life in the universe. Because Mercury and the Moon lacked atmospheres, astrobiologists determined they were not conducive to life. The giant planets and their moons were thought to be too remote and too cold to sustain life. Venus was far too hot, but Mars seemed just right. In the mid-1970s, two American *Viking* spacecraft were sent to Mars with detachable landers containing a suite of instruments designed to find evidence of a terrestrial type of life. When the *Viking* landers failed to find evidence of life on Mars, hopes faded for finding life on any other body orbiting our Sun.

Since that time, however, there is renewed optimism. A better understanding of the history of Mars indicates that at one time the planet was wetter and warmer, leading many scientists to believe that fossilized life or even living microbes might yet be buried under the planet's surface. The first decade of the 21st century saw a return to Mars and a continuation in the search for evidence of current or preexisting life. In 2008, NASA's *Phoenix* spacecraft landed at a far northern latitude, inside the planet's arctic circle, where specialized instruments dug into and analyzed the martian water-ice permafrost. *Phoenix* found that the martian arctic soil has a chemistry similar to that of the Antarctic dry valleys on Earth, where life exists deep below the surface at the ice-soil boundary. Minerals that form in water, such as calcium carbonate, reveal that ancient oceans were once present on the planet. However, *Phoenix* did not find direct evidence of life.

The most recent Mars lander, the Mars Science Laboratory rover *Curiosity* (**Figure 18.12**), landed in Gale Crater on Mars in 2012. This large rover—about the size of a small car—studies the rocks and soil of Mars to provide data for a better understanding of the history of the planet's climate and geology. Shortly after landing, *Curiosity* found evidence that a stream of liquid water had once flowed in the crater. The rover observed rounded, gravelly pebbles stuck together, which have been interpreted as coming from a stream roughly 30 centimeters (cm) deep and flowing at 1 meter per second (m/s). The presence of liquid, flowing water on the surface of Mars could be seen as a positive indication that life might have arisen there. Conversely, *Curiosity* has failed to find evidence of methane on Mars. Previous data had indicated that there might be methane, which is another possible indicator of life, although it is also made by geological processes. This result implies that methane-producing life is not present on Mars currently.

The *Mars Atmosphere and Volatile Evolution Mission* (*MAVEN*) arrived at Mars in September 2014. This mission's purpose is to study the upper atmosphere to learn more about the escape of carbon dioxide, hydrogen, and nitrogen from the planet's atmosphere. Data from the first martian year of the mission indicate that

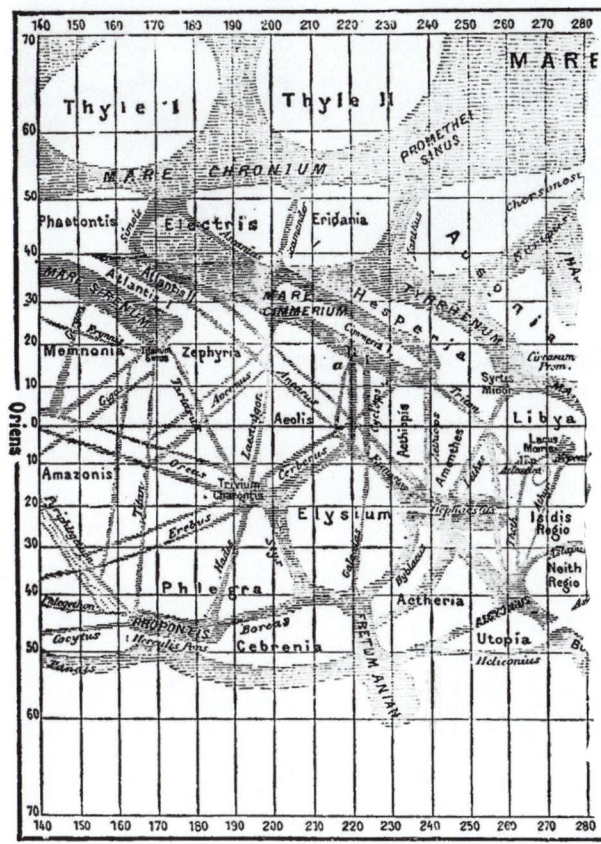

Figure 18.11 Schiaparelli made many hand-drawn maps of the surface of Mars; the streaks that he labeled *canali* ("channels") were later thought to be "canals" made by intelligent life to move water from one place to another. The idea of massive building projects by intelligent life has since been discredited, but the presence of water on Mars has been verified.

Figure 18.12 The *Curiosity* rover stitched together several images to form a "selfie," the first ever taken on Mars.

Figure 18.13 This *MRO* image shows narrow, dark, downhill streaks, about 100 meters long, which are thought to indicate liquid water flowing on Mars today. Hydrated salts and minerals, including pyroxene (the blue color), were detected by spectroscopy.

the rate of loss of these gases varies with solar activity and the solar wind. This atmospheric loss has affected the surface pressure and the existence of liquid water on Mars and has led to Mars becoming colder and drier over time.

In September 2015, NASA announced that spectroscopic observations from the orbiting *Mars Reconnaissance Orbiter* (*MRO*) indicate that there is liquid water on Mars today. Darkish streaks on Mars that change seasonally contain hydrated salts and minerals, indicating that liquid water is important to their formation (**Figure 18.13**). This water is briny (salty), so it has a much lower freezing point than non-briny water and thus could exist in a liquid state during the martian summer. Later observations have revealed subsurface ice in numerous locations, notably along the edges of eroded scarps, where the edge has been worn away to reveal the layers beneath. Future experiments will dig below the martian surface for liquid water—and fossil or living microorganisms.

NASA's instrumented robots reached the outer Solar System in the 1980s, and what they found surprised many astrobiologists. Although the outer planets themselves did not appear to be habitats for life, some of their moons became objects of special interest, because they contain significant amounts of water (**Figure 18.14**). Jupiter's moon Europa is covered with a layer of water ice that appears to overlie a great ocean of liquid water. Impacts by comet nuclei may have added a mix of organic material, another essential ingredient for life. Once thought to be a frozen and inhospitable world, Europa is now a candidate for biological exploration. Scientists using the Hubble Space Telescope observed water geysers taller than Mount Everest erupting from the icy surface. Ejected material from these geysers may make it possible to search for life on Europa without drilling down through the ice.

Saturn's moon Titan appears to be rich in organic chemicals, many of which are thought to be precursor molecules of a type that existed on Earth before life appeared here. The *Cassini* mission found additional evidence for a variety of molecules in Titan's atmosphere that might be necessary for life, as well as a liquid lake of methane on the surface and probably a liquid-water ocean under the surface. In addition, it identified another potential site for life: Saturn's tiny moon Enceladus. The spacecraft detected water-ice crystals spouting from ice volcanoes near the south pole of Enceladus. Liquid water must lie beneath its icy surface, and Enceladus therefore joins Europa as a possible habitat of extremophile life, perhaps life similar to that found near geothermal vents deep within Earth's oceans.

Efforts to find evidence of life on other bodies within our own Solar System have so far been unsuccessful, but the quest continues. The discovery of life on even one Solar System body beyond Earth would be exciting: If life arose independently *twice* in the same planetary system, then life in the universe might not be rare.

Habitable Zones

Searching for life within our own Milky Way Galaxy is a daunting task. Nevertheless, astronomers are narrowing the possibilities by searching for planets with environments conducive to the formation and evolution of life as we understand it, while eliminating planets that are clearly unsuitable.

One criterion that astrobiologists look for in planetary systems is stability. Planets in stable systems remain in nearly circular orbits that preserve relatively uniform climatological and oceanic environments. Planets in highly elliptical orbits can experience wild temperature swings that could be detrimental to the survival of life. A stable temperature that maintains the existence of water in a liquid state

Figure 18.14 The total amount of liquid water (represented by the blue spheres) on selected objects in the Solar System.

might be important. We know that liquid water was essential for the formation and evolution of life on Earth. We do not know if liquid water is an absolute requirement for life elsewhere, but it's a good starting point.

The region around a star that provides a range of temperatures in which liquid water can exist is called the star's **habitable zone**. On planets that are too close to their parent star, water would exist only as a vapor—if at all. On planets that are too far from their star, water would be permanently frozen as ice. Even if a planet is located in the habitable zone, we cannot yet determine if it is actually inhabited. We only know that liquid water could exist on the surface. Another consideration when looking for habitable planets is their size. Large planets such as Jupiter retain most of their light gases—hydrogen and helium—during formation and so become gas giants without a surface. Planets that are very small may have insufficient surface gravity to retain their atmospheric gases and so end up like our Moon. Calculating whether any particular planet is habitable in this sense is quite complicated. Many astrobiologists are working on models of the climates of known exoplanets to figure out whether they are in their star's habitable zone.

In our own Solar System, Venus, which orbits at 0.7 times Earth's distance from the Sun, has become an inferno because of its runaway greenhouse effect (see Chapter 7). Any liquid water that might once have existed on Venus has long since evaporated and been lost to space. Mars orbits about 1.5 times farther from the Sun than Earth, and nearly all of the water that we see on Mars today is frozen. The orbit of Mars is more elliptical and variable than Earth's, however, giving the planet a greater variety of climate, including long-term cycles that might occasionally permit liquid water to exist. Most astrobiologists put the habitable zone of our Solar System at about 0.9–1.4 astronomical units (AU), which includes Earth but just misses Venus and Mars. Yet this range may be too narrow because of the liquid water under ice that is seen in the moons of the outer Solar System. For example, extremophiles could be thriving in liquid water beneath the surfaces of some icy moons of Jupiter and Saturn.

Astronomers must also think about the type of star they are observing in their search for life-supporting planets. **Figure 18.15** shows that stars that are less massive than the Sun and thus cooler will have narrower habitable zones,

what if . . .

What if we find an exoplanet with a very small magnetic field and large amounts of greenhouse gases in its atmosphere: How would each of those factors increase or decrease the habitable zone distance for that exoplanet?

▶▶ **Interactive Simulation:** Habitable Zone

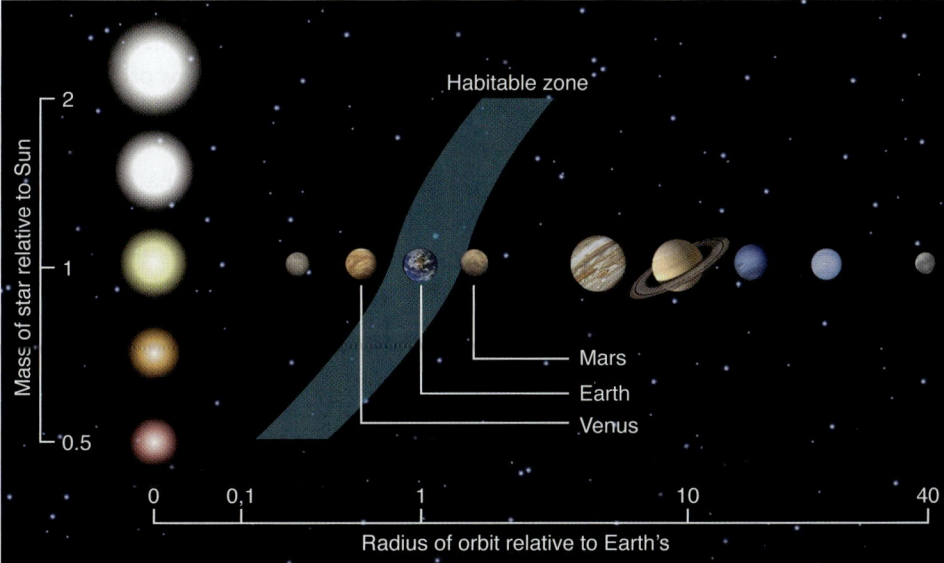

Figure 18.15 The habitable zone changes with the mass and temperature of a star. Habitable zones around hot, high-mass stars are more distant and wider than habitable zones around cooler, lower-mass stars. The Sun and Solar System are shown for comparison.

lessening the chance that life will form on an orbiting planet. Stars that are more massive than the Sun are hotter and will have a larger habitable zone. However, recall that a star's main-sequence lifetime depends on its mass. For example, a star of $2\,M_{Sun}$ would enjoy relative stability on the main sequence for only about a billion years before the helium flash incinerated everything around it. On Earth, a billion years was long enough for bacterial life to form and cover the planet but insufficient for anything more advanced to evolve. It is possible that evolution could happen at a different pace elsewhere—we have only our one terrestrial case as an example. Still, main-sequence lifetime is a sufficiently strong consideration that most astronomers prefer to focus their efforts on stars with longer lifetimes—specifically, spectral types F, G, K, and M (recall Figure 10.19 and Table 12.1).

Finally, some astronomers think about a "galactic habitable zone," referring to a star's location within the Milky Way Galaxy. Stars that are situated too far from the galactic center may not have had enough heavy elements—such as oxygen, silicon (silicates), iron, and nickel—in their protoplanetary disks to form rocky planets like Earth. The systems that form in the outer regions of the Galaxy lack heavy elements from prior episodes of star formation and death; they could form Jupiters but not Earths. Conversely, regions too close to the galactic center experience less star formation and therefore fewer opportunities to gather heavy elements into planetary environments. Perhaps more serious is the high-energy radiation environment near the galactic center (X-ray and gamma ray), which is damaging to RNA and DNA. Even so, for many stars the galactic habitable zone may not remain as a permanent home if they tend to have an orbit within the galaxy that changes their distance from the galactic center over time. In short, astronomers try to narrow down the vast numbers of stars as they conduct their search, but they acknowledge that these types of arguments—based on what worked well for planet Earth—might not be applicable when we are looking at other planetary systems.

Searching for Earth-like Planets

The search for Earth-like planets is already well under way. Currently, there are more than 100 projects from the ground and from space focused on searching for exoplanets using the methods discussed in Chapter 5. Each week, it seems, a new system makes the news, either because it is more like Earth than any other yet discovered or because it is unusual in some way.

In 2006, the European Space Agency (ESA) launched COROT, the first space telescope dedicated to exoplanet detection. COROT monitored nearby stars for transiting exoplanets. The ESA spacecraft made several discoveries, including one planet that is only 1.7 times the size of Earth. In 2013, COROT was decommissioned due to an onboard computer failure.

In 2009, NASA launched the Kepler space telescope into a solar orbit that trailed Earth in its own orbit around the Sun. This location enabled uninterrupted monitoring of Kepler's target stars, because Earth never got in the way of Kepler's cameras. Over the next several years, Kepler's photometer simultaneously and continually monitored the brightness of more than 100,000 stars, looking for transiting exoplanets. Kepler could detect planets as small as 0.8 times the size of Earth. Kepler has discovered thousands of suspected exoplanetary systems, roughly 2,500 of which have subsequently been confirmed. Many more candidates will likely be found upon closer examination of the data, and many more will likely be confirmed as follow-up observations by other ground-based or orbiting instruments proceed.

Many of these systems include multiple planets. The star Gliese 667C, for example, has seven planets in orbit around it, three of which are super-Earths in the star's habitable zone. In early 2017, data from a small ground-based telescope were followed up with 21 days of data from the Spitzer Space Telescope. Astronomers found that the red dwarf TRAPPIST-1 has seven Earth-sized planets, three of which are in the habitable zone and all of which are tidally locked to the star. This means each of the seven could potentially have liquid water somewhere on the planet. This system is particularly exciting because it is relatively nearby—40 light-years from Earth—so follow-up observations will be more detailed than those for other systems. The system is very compact, comparable in size to Jupiter and the Galilean moons (**Figure 18.16**). Interplanetary travel among these seven planets in orbit around TRAPPIST-1 would be measured in days, rather than months or years.

Determining how many of these planets are in their star's habitable zone is difficult and is a current focus for astrobiologists, who model the climate of each planet as information about its orbit becomes known. From observations so far, astronomers estimate that planets in habitable zones may be common in the Milky Way. Astronomers estimate that 10–20 percent of stars like the Sun have an Earth-sized planet in the habitable zone. There are about 40 billion Sun-like stars in the Milky Way, so this estimate implies that there could be as many as 4 billion to 8 billion Earth-sized planets in the habitable zones of their stars. Upcoming missions, such as the Transiting Exoplanet Surveying Satellite (TESS) and the James Webb Space Telescope, will seek to identify new planets and obtain information about their atmospheres.

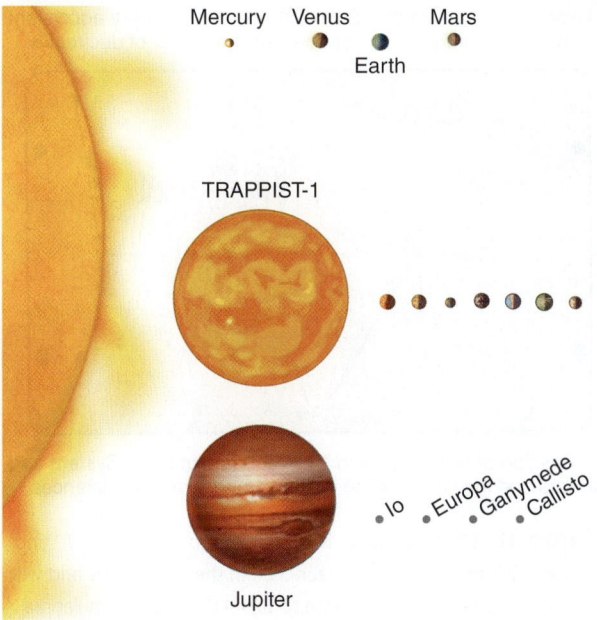

Figure 18.16 The Sun is much larger than the planets in the Solar System, compared with TRAPPIST-1 and its planets. The TRAPPIST-1 system contains seven Earth-sized planets that are all closer to the star than Mercury is to the Sun. In both size and distance, the TRAPPIST-1 system is more similar to Jupiter and its moons than it is to the Solar System.

CHECK YOUR UNDERSTANDING **18.2**

A habitable zone is the place around a star where
a. life is found on the surface of a planet.
b. heavy elements are found in the protoplanetary disk.
c. water can exist as a liquid on the surface of a planet.
d. rocky planets are found to be orbiting the star.

18.3 Scientists Search for Signs of Intelligent Life

Scientists would find any evidence of extraterrestrial life exciting, even if it were the fossil remnants of blue-green algae. But many people are interested in extraterrestrial life only if it is sufficiently advanced that we could communicate with it. Searching for signs of intelligent life in the universe requires exploring a large number of questions that are not astronomical in nature: Given the right conditions, does life always arise? Among life-forms, how common is intelligence? Or language? Or complex technology such as radio transmitters or lasers that can send messages to space? Framing the discussion of the search for intelligent life requires addressing all of these questions and more. But while scientists explore these questions, they have already sent out messages, just in case someone (or something), somewhere, is listening.

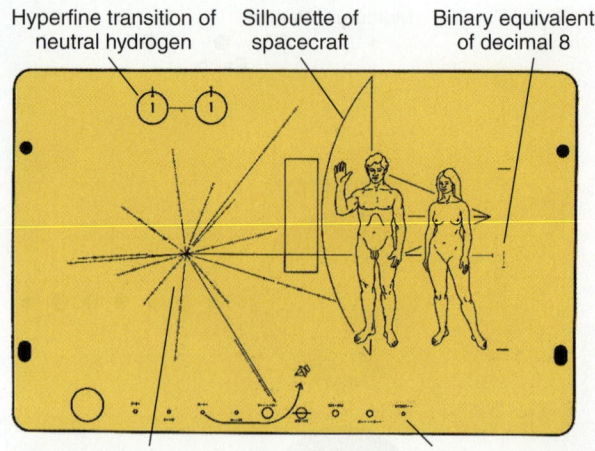

Hyperfine transition of neutral hydrogen

Silhouette of spacecraft

Binary equivalent of decimal 8

Position of Sun relative to 14 pulsars and the center of the galaxy

Planets of Solar System and binary relative distances

Figure 18.17 This plaque was attached to the *Pioneer 10* and *Pioneer 11* probes, which were launched in the early 1970s and will eventually leave the Solar System to travel through the millennia in interstellar space.

Figure 18.18 The binary-encoded message we beamed toward the globular cluster M13 in 1974 contains the binary symbols for the numbers 1–10, diagrams of hydrogen and carbon atoms, some interesting molecules, DNA, a human with description, basics of our Solar System, and basics of the Arecibo telescope.

Saying "Hello"

During the 1970s, humans made preliminary efforts to send a message to alien life. The *Pioneer 10* and *Pioneer 11* spacecraft, which will probably spend eternity drifting through interstellar space, each carry the plaque shown in **Figure 18.17**. It pictures humans and the location of Earth to any future traveler who might happen to find it. The two *Voyager* spacecraft each carry a different message prepared on an identical type of phonograph record: The message contains greetings from planet Earth in 60 languages, samples of music, animal sounds, and a message from then-President Jimmy Carter. Some politicians were concerned that scientists were dangerously advertising our location in the galaxy, even though radio signals had been broadcast into space for nearly 80 years at the time. The messages were also criticized by some philosophers, who claimed that we were making ridiculous anthropomorphic assumptions about aliens being sufficiently like us to decode these messages.

Sending messages on spacecraft is not the most efficient way to contact aliens. A somewhat more practical effort was made in 1974, when astronomers used the 300-meter-wide dish of the Arecibo radio telescope to beam a message (**Figure 18.18**) toward the globular cluster M13. It will take 25,000 years for that message to travel the distance to M13. By then, M13 will have moved, and the message will miss it entirely. In 2008, a radio telescope in Ukraine sent a message to the exoplanet Gliese 581c. The message was composed of 501 digitized images and text messages selected by users on a social networking site. It will arrive at Gliese 581c in 2029.

The Drake Equation

It took less than a billion years for life to form in Earth's oceans. It took another 3.5 billion years for modern humans to develop. Our star is now halfway through its lifetime. As noted in Section 18.2, many stars do not live long enough for life to evolve. Therefore, astronomers looking for life start by looking at planets that surround stars of about 1.2 solar masses or smaller—stars of type F, G, K, or M.

The first serious effort to search for intelligent extraterrestrial life was made by astronomer Frank Drake in 1960. Drake used what was then astronomy's most powerful radio telescope to listen for signals from two nearby stars. Although his search revealed nothing unusual, it prompted him to develop the equation that bears his name. This equation is different from other equations in this book because the values for each variable are quite uncertain. Still, it is a very useful way to organize our thoughts about whether intelligent life might exist elsewhere, because it includes the things we need to know to make an informed estimate about how many intelligent civilizations there might be. The **Drake equation** estimates the number (*N*) of intelligent, communicating civilizations that may currently exist within the Milky Way Galaxy:

$$N = R^* \times f_p \times n_e \times f_l \times f_i \times f_c \times L$$

The seven factors on the right side of the equation are the conditions that Drake thought must be met for a civilization to exist:

1. R^* is the number of stars that form in our galaxy each year that are suitable for the development of intelligent life. Astronomers consider these to be F, G, K, or M spectral-type stars because their lifetimes are sufficiently long. This is roughly seven per year and is the best understood of all these terms.

2. f_p is the fraction of those stars that form planetary systems. The abundance of exoplanets indicates that planets form as a natural by-product of star

formation and that many—perhaps most—stars have planets. We will assume that f_p is between 0.5 and 1.

3. n_e is the number of planets and moons, per stellar system, with an environment suitable for life. If we look at the Solar System, we might decide this number is about 2. Including Europa and Enceladus increases the number to 4. Earth definitely did develop life, and Mars might have. The Solar System is just one example, however, and we do not really know what this term should be, so a number between 1 and 10 seems reasonable.

4. f_l is the fraction of suitable planets and moons on which life actually arises. Remember that just a single self-replicating molecule may be enough to get the ball rolling. Some biochemists now believe that if the right chemical and environmental conditions are present, then life *will* develop, but others disagree. Estimates range from 100 percent (life always develops) to 1 percent (life is rare). Astronomers use a range of 0.01 to 1.

5. f_i is the fraction of those planets harboring life that eventually develop intelligent life. Intelligence is the kind of survival trait that might often be strongly favored by natural selection. On Earth, however, it took about 4 billion years—roughly half the expected lifetime of our star—to evolve tool-building intelligence. The correct value for f_i might be close to 0.01 or it might be closer to 1. The truth is, we just don't know.

6. f_c is the fraction of intelligent life-forms that develop technologically advanced civilizations; that is, civilizations that send communications into space. With only one example of a technological civilization to work with, f_c is hard to estimate. We will consider estimates for f_c between 0.1 and 1.

7. L is the number of years that such civilizations exist. This factor is the most difficult of all to estimate because it depends on the long-term stability of advanced civilizations. We have had a communicating, technologically advanced civilization on Earth for less than 100 years, and during that time we have developed, deployed, and used weapons with the potential to eradicate our civilization and render Earth hostile to life for many years to come. We have also so degraded our planet's ecosystems that many respectable biologists and climatologists wonder whether Earth is in danger of reducing or losing its habitability. Do all technological civilizations destroy themselves within a thousand years? Conversely, if most technological civilizations learn to use their technology for survival rather than self-destruction, might they instead survive for a million years? For L in our calculation, we will use a range between 1,000 years and 1 million years.

As illustrated in **Figure 18.19**, the conclusions we draw using the Drake equation depend a great deal on the assumptions we make. For the most pessimistic of our estimates, the Drake equation sets the number of technological civilizations currently in our galaxy at about 1. If this is correct, then we are the *only* technological civilization in the Milky Way. (Because we are here, we know that N must be at least 1.) Alternatively, we could be the only technological civilization in the Milky Way *at this time*, meaning other civilizations could have lived and died out millions or billions of years ago. When we consider that there are a trillion galaxies in the observable universe, even these pessimistic assumptions would mean that a trillion technological civilizations existed at one time or another out there somewhere. Yet, we would have to go a *very* long way (10 million parsecs or so) to find our nearest neighbors that exist now.

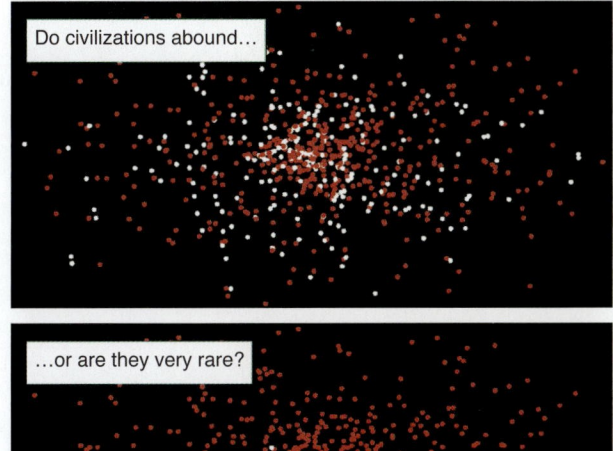

Figure 18.19 Optimistic and pessimistic estimates of the existence of intelligent civilizations in our galaxy based on the Drake equation are very different because uncertainty about the seven factors (see the text) affects the estimated prevalence of intelligent extraterrestrial life. White dots indicate stars with possible civilizations.

At the other extreme, suppose we take the most optimistic view, assuming that intelligent life arises and survives everywhere it gets the chance. The Drake equation then says that there should be roughly 15 million technological civilizations in our galaxy alone. In this case, our nearest neighbors may be "only" 50 light-years away. If scientists in that civilization are listening to the universe with their own radio telescopes, hoping to answer the question of other life in the universe for themselves, then as you read this page they may be puzzling over an episode of the original *Star Trek* television series that began in 1966.

If we did run across another technologically advanced civilization, what would it be like? Looking back at the Drake equation, we can see that unless civilizations typically live for many thousands or even millions of years, it is highly unlikely that we have neighbors nearby. In this case, any civilization that we encountered would almost certainly have been around for much longer than we have.

Technologically Advanced Civilizations

During a lunch with colleagues, the physicist Enrico Fermi (1901–1954) is reported to have asked, "If the universe is teeming with aliens . . . where is everybody?" Fermi's question—first posed in 1950 and sometimes called the Fermi paradox—remains unanswered. Consider also the following closely related question: If intelligent life-forms are common, but interstellar travel is difficult or impossible, why don't we detect their signals? We have failed so far to detect any—but it is not for lack of trying.

Drake's original project of listening for radio signals from intelligent life around two nearby stars has grown over the years into a much more elaborate program that is referred to as the Search for Extraterrestrial Intelligence, or **SETI**. Scientists from around the world have thought carefully about what strategies might be useful for finding life in the universe. Most of these have focused on the idea of using radio telescopes to listen for signals from space that bear an unambiguous signature of an intelligent source. Some have listened intently at certain key frequencies, such as the frequency of the interstellar 21-cm line from hydrogen gas. The assumption behind this approach is that if a civilization wanted to be heard, its denizens would tune their broadcasts to a channel that astronomers throughout the galaxy should be listening to. More recent searches have made use of advances in technology to record as broad a wavelength range of radio signals from space as possible. Analysts then use computers to search these databases for patterns that might suggest the signals are intelligent in origin.

Unlike much astronomical research, SETI receives its funding from private rather than government sources, and SETI researchers have found ingenious ways to continue the search for extraterrestrial civilizations with limited resources. One project, known as SETI@home, involves the use of hundreds of thousands of personal computers around the world to analyze the institute's data. SETI programs installed on personal computers worldwide download radio observations from the SETI Institute over the Internet, analyze these data while computer owners are not using their equipment, and then report the results of their searches back to the institute. Perhaps the first sign of intelligent life in the universe might be found by a computer sitting on your desk.

The SETI Institute's Allen Telescope Array (ATA) is named for Microsoft cofounder Paul Allen, who provided much of the initial financing for the project. A joint venture between the SETI Institute and the University of California, the ATA consists of a "farm" of 42 small, inexpensive radio dishes like those used to capture signals from orbiting communication satellites (**Figure 18.20**). One of the key projects of the ATA is to observe planets discovered by the Kepler mission in search of signs of intelligent life. Each dish has a diameter of 6.1 meters. All the telescopes working together have a total signal-receiving area greater than that of a 20-meter radio telescope. Just as your brain can sort out sounds coming from different directions, this array of radio telescopes is able to determine the direction a signal is coming from, allowing it to listen to many stars at the same time. Over several years, astronomers using the ATA are expected to survey as many as a million stars, hoping to find a civilization that has sent a signal in our direction.

Finding even one other nearby civilization in our galaxy would make scientists more optimistic that the universe holds a great deal more intelligent life. The likelihood of SETI's success is difficult to predict, but its potential payoff is enormous. Few discoveries would have a more profound impact than the certain knowledge that we on Earth are not alone.

Figure 18.20 The Allen Telescope Array listens for evidence of intelligent life from as many as a million stellar systems.

Science fiction is filled with tales of humans who leave Earth to "seek out new life and new civilizations." Unfortunately, these scenarios are not scientifically realistic. The distances to the stars and their planets are enormous; to explore a significant sample of stars would require extending the physical search over tens or hundreds of light-years. As we discussed in Chapter 13, special relativity limits how fast we can travel. The speed of light is the limit, and even at that rate it would take more than 4 years to get to the *nearest* star. Time dilation would favor the astronauts themselves, and they would return to Earth younger than if they had stayed at home. But suppose they visited a star 15 light-years distant. Even if they traveled at speeds close to that of light, by the time they returned to Earth, 30 years would have passed. Some science-fiction enthusiasts get around this problem by invoking "warp speed" or "hyperdrive" to travel faster than the speed of light or by using "wormholes" as shortcuts across the galaxy—but there is absolutely no evidence that any one of these shortcuts is possible. The converse, of course, is also true. It is as difficult for alien life forms to get to us as it is for us to get to them. All claims of alien life forms or "unidentified flying objects" (UFOs) visiting Earth have so far failed to stand up to scrutiny.

what if . . .

What if SETI succeeds in finding a signal from a distant civilization—one that clearly must consist of technologically advanced beings: What (if anything) should we say to them, and who should have the authority to say it?

Life as We Do Not Know It

So far, we have confined our discussion to the search for life as we know it, as it may exist now. We do this for sound scientific reasons, primarily having to do with falsifiability. Recall Chapter 1, where we defined scientific ideas as being first and foremost falsifiable. There are no tests we can apply to prove that very different forms of life did not (or will not) exist; therefore, the idea behind them is not a scientific hypothesis. The ideas of very different forms of life, or life in the distant past or future, are purely speculative and nonscientific. This does not mean these ideas are uninteresting. They are just not falsifiable, and therefore not science.

Now consider more common examples. Some people claim that aliens have already visited us: Tabloid newspapers, books, and websites are filled with tales of UFO sightings, government conspiracies and cover-ups of alien crash-landings, alleged alien abductions, and UFO religious cults. However, none of these reports meets the basic standards of science. They are not falsifiable—they lack verifiable evidence and repeatability—and we must conclude that there is no scientific evidence for any alien visitations.

When you encounter discussions of past, future, or very different alien life-forms, it is useful to ask whether the ideas are falsifiable and whether the evidence is verifiable and repeatable. If not, the discussions are speculation, not science.

CHECK YOUR UNDERSTANDING 18.3

The fact that we have not detected alien civilizations yet tells us

a. they are not there.

b. they are rare.

c. we are in a "blackout," and they are not talking to us.

d. we don't know enough yet to draw any conclusions.

18.4 The Fate of Life on Earth

In this book, we have used our understanding of physics and cosmology to look back through time and watch as structure formed throughout the universe. We have peered into the future and seen the ultimate fate of the universe, from its enormous clusters of galaxies down to its tiniest components. Now we examine our own destiny and contemplate the fate that awaits Earth, humanity, and the star on which all terrestrial life depends.

Eventually, the Sun Must Die

About 5 billion years from now, the Sun will end its long period of relative stability. Shedding its identity as the passive star that has nurtured life on Earth, the Sun will expand to become a red giant and later an asymptotic giant branch (AGB) star, swelling to hundreds of times its current size. The giant planets, orbiting outside the extended red giant atmosphere, may survive in some form. Even so, they will suffer the blistering radiation from a Sun grown thousands of times more luminous than it is today.

The terrestrial planets will not fare as well. Some—perhaps all—of the worlds of the inner Solar System will be engulfed by the expanding Sun (**Figure 18.21**). Just as an artificial satellite is slowed by drag in Earth's tenuous outer atmosphere and eventually falls to the ground in a dazzling streak of white-hot light, so, too, will a terrestrial planet caught in the Sun's atmosphere be consumed by the burgeoning star. If this is Earth's fate, our home world will leave no trace other than a slight increase in the amount of massive elements in the Sun's atmosphere. As the Sun loses more and more of its atmosphere in an AGB wind, our atoms may be expelled back into the reaches of interstellar space from which they came, perhaps to become incorporated into new generations of stars, planets, and even life itself.

Another planetary fate is possible, however. In this scenario, as the red giant Sun loses mass in a powerful wind, its gravitational grasp on the planets will weaken, and the orbits of both the inner and outer planets will spiral outward. If Earth moves out far enough, it may survive as a seared cinder, orbiting the white dwarf that the Sun will become. Barely larger than Earth and with its nuclear fuel exhausted, the

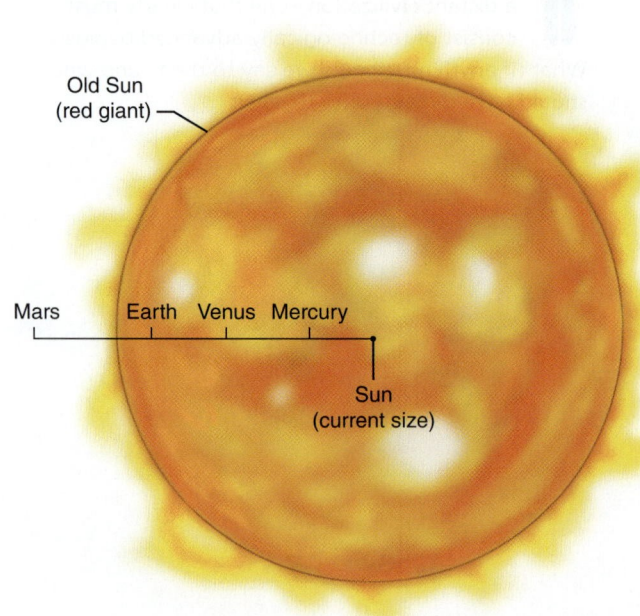

Figure 18.21 As the Sun leaves the main sequence, it will swell up to encompass some, and perhaps all, of the orbits of the terrestrial planets.

white dwarf Sun will slowly cool, eventually becoming a cold, inert sphere of degenerate carbon, orbited by what remains of its collection of planets. Thus, the ultimate outcome for Earth—consumed in the heart of the Sun or left behind as a frigid, burned rock orbiting a long-dead white dwarf—is not yet known.

The Future of Life on Earth

Life on Earth has even less of a future, because it will not survive long enough to witness the Sun's departure from the main sequence. Well before that cataclysmic event takes place, the Sun's luminosity will begin to rise. As solar luminosity increases, so will temperatures on all the planets, including Earth. The inner edge of the Sun's habitable zone will move out past Earth's orbit. Eventually, temperatures on Earth will climb so high that all animal and plant life will perish. Even the extremophiles that inhabit the oceanic depths will die as the oceans boil away. Models of the Sun's evolution are still too imprecise to predict with certainty when that fatal event will occur, but the end of all terrestrial life may be only 1 billion or 2 billion years away. That is a comfortably distant time from now, but it is well short of the Sun's departure from the main sequence.

It is far from certain, however, that the descendants of today's humanity will even be around a billion years from now. Some of the threats that await us come from beyond Earth. For the remainder of the Sun's life, the terrestrial planets, including Earth, will continue to be bombarded by asteroids and comets. Perhaps a hundred or more of these impacts will involve kilometer-sized objects, capable of causing the kind of devastation that eradicated the dinosaurs and most other species 65 million years ago. Although these events may create new surface scars, they will have little effect on the integrity of Earth itself. Earth's geological record is filled with such events, and each time they happen, life manages to recover and reorganize.

It seems likely, then, that some form of life will survive to see the Sun begin its march toward instability. However, individual species do not necessarily fare so well when faced with cosmic cataclysm. We are rapidly developing technology that could enable us to detect most threatening asteroids and modify their orbits well before they could strike Earth. However, comets are more difficult to guard against because long-period comets appear from the outer Solar System with little warning. To offer protection, various means of defending ourselves would have to be in place, ready to be used on very short notice. Although impacts from kilometer-sized objects are infrequent, smaller objects only a few dozen meters in size, carrying the punch of a several-megaton bomb, strike Earth more than once every 100 years. There have been three such events in the past 105 years: Tunguska, Sikhote-Alin, and Chelyabinsk.

We might protect ourselves from the fate of the dinosaurs, but in the long run the descendants of humanity will either leave this world or die out. Planetary systems surround other stars, and all that we know tells us that many other Earth-like planets may well exist throughout our galaxy. Colonizing other planets is currently the stuff of science fiction, but if our descendants are ultimately to survive the death of our home planet, off-Earth colonization must become science fact at some point in the future.

Although humankind may soon be capable of protecting Earth from life-threatening comet and asteroid impacts, in other ways we are our own worst enemy. Human activities are dramatically affecting the balances of atmospheric gases. As discussed in Chapter 7, the climate and ecosystem of Earth constitute a finely balanced, complex system. The record of past climatic shifts shows that the system is capable of large changes in response to even small disturbances. Drastic changes in climate and resource availability would certainly have consequences for our own survival. In addition, for the first time in human history, we possess the means to unleash nuclear, chemical, or biological disasters that could threaten the very survival of our species. In the end, the fate of humanity may depend more than anything on how well we accept responsibility for our actions and the effect we have on each other and our planet.

Figure 18.22 shows an image of Earth as taken by the *Voyager 1* spacecraft at a distance of more than 40 AU (beyond Neptune's orbit). The beams in the picture are sunlight scattered from the spacecraft. The arrow points to a tiny dot, which is Earth, the only place in the entire universe where we can confirm that life exists. Compare the size of that dot to the size of the universe. Compare the history of life on Earth to the history of the universe. Compare Earth's future with the fate of the universe. Astronomy is humbling. We occupy a tiny part of space and time. And yet we are unique, as far as we know. Think for a moment about what that means to you. This perspective may be the most important lesson the universe has to teach you.

Figure 18.22 ★ WHAT AN ASTRONOMER SEES This image from the *Voyager 1* spacecraft shows Earth from a distance of 3.8 billion miles, well past the orbit of Neptune. The streaks in this image are scattered sunlight. The "pale blue dot" in the rightmost streak is Earth. For this last "What an Astronomer Sees" image, we give you the words of Carl Sagan, describing what he saw in this image:

"We succeeded in taking that picture, and, if you look at it, you see a dot. That's here. That's home. That's us. On it, everyone you ever heard of, every human being who ever lived, lived out their lives.

The aggregate of all our joys and sufferings, thousands of confident religions, ideologies and economic doctrines, every hunter and forager, every hero and coward, every creator and destroyer of civilizations, every king and peasant, every young couple in love, every hopeful child, every mother and father, every inventor and explorer, every teacher of morals, every corrupt politician, every superstar, every supreme leader, every saint and sinner in the history of our species, lived there—on a mote of dust, suspended in a sunbeam.

The Earth is a very small stage in a vast cosmic arena. Think of the rivers of blood spilled by all those generals and emperors so that in glory and in triumph they could become the momentary masters of a fraction of a dot. Think of the endless cruelties visited by the inhabitants of one corner of the dot on scarcely distinguishable inhabitants of some other corner of the dot. How frequent their misunderstandings, how eager they are to kill one another, how fervent their hatreds. Our posturings, our imagined self-importance, the delusion that we have some privileged position in the universe, are challenged by this point of pale light ...

To my mind, there is perhaps no better demonstration of the folly of human conceits than this distant image of our tiny world. To me, it underscores our responsibility to deal more kindly and compassionately with one another and to preserve and cherish that pale blue dot, the only home we've ever known."

— *Carl Sagan, speech at Cornell University, October 13, 1994*

CHECK YOUR UNDERSTANDING 18.4

All life on Earth will eventually come to an end because
a. humans will make the planet uninhabitable.
b. asteroids will make the planet uninhabitable.
c. new species will arise that we won't recognize as life.
d. the Sun will make the planet uninhabitable.

reading Astronomy News

Tardigrades are now on the moon thanks to a crashed Israeli spacecraft

by Bonnie Burton, CNET

Searching for life—even in the Solar System—is a time-consuming process, in which following up on tantalizing hints can take decades.

Tardigrades, affectionately known as water bears, are microscopic animals that can survive in almost any environment. We already know it's possible for scientists to bring tardigrades back to life after a 30-year deep freeze. As 'extremophiles,' tardigrades can shut down their metabolism and survive in hostile conditions for long periods. But can they survive in space, more specifically on the moon?

Back in April, an Israeli spacecraft called Beresheet, which carried thousands of dehydrated tardigrades (among other cargo), crashed on the moon. Some people wondered if the water bears could survive.

One of those people was Arch Mission Foundation founder Nova Spivack. Arch Mission Foundation is a nonprofit whose main goal is to create a "backup of planet Earth."

The Israeli spacecraft was transporting Arch Mission's first lunar library, a digital archive holding the equivalent of 30 million pages of information. It also carried human DNA samples and thousands of dehydrated tardigrades. It's unknown how much of the cargo actually ended up on the moon's surface following the crash.

Based on Arch Mission's analysis of the spacecraft's path as well as the makeup of the lunar library itself, Spivack told Wired on Monday that he's confident the library, a "DVD-sized object made of thin sheets of nickel," survived the crash mostly intact.

That doesn't mean the DNA or water bears are in good shape.

"We sent enough DNA to regenerate life on Earth, if necessary," Spivack tweeted Tuesday. "Although it would require more advanced biotech than we have to do that. At least our DNA is offsite now. But note that cells and DNA cannot survive or reproduce on the moon. Yet if retrieved they could be useful."

Spivack noted that the tardigrades cannot reproduce on the moon, regardless.

"About the tardigrades in the Lunar Library: Some are sealed in epoxy with 100 million human, plant and microorganism cells," Spivack tweeted Tuesday. "Some are encapsulated onto the sticky side of a 1cm square piece of Kapton tape that is sealed inside the disc stack. They cannot reproduce on the moon."

Even though the dehydrated tardigrades can't spring to life on the moon, they could theoretically be gathered, revived and studied to teach us about their time there.

"It is not likely that cells can survive on the moon without a lot more protection from radiation," Spivack added. "However the human cells, plant cells and micro organisms we sent could be recovered, studied and their DNA extracted — perhaps to be cloned and regenerated, far in the future."

Arch Mission Foundation, Nova Spivack and SpaceIL didn't immediately respond to a request for comment.

EVALUATING THE NEWS

1. The Beresheet spacecraft was the first privately funded moon mission. What was this mission trying to accomplish?

2. Why did the Arch Mission Foundation choose tardigrades for this mission?

3. In the 20th Century, NASA tried very hard to avoid bringing Earth life (other than astronauts) to the Moon. Why would they have done that?

4. Suppose now that astrobiologists find life on the moon. What complication has been introduced by this mission?

SUMMARY

Life on Earth uses a complex carbon-based chemistry that is perpetuated by self-replicating molecules enclosed in fatty membranes. Life likely formed in Earth's oceans and then evolved chemically from simple molecules into self-replicating organisms. Life-forms that are very different from ours, including those based on silicon chemistry, cannot be ruled out. Space-based instruments have begun the search for Earth-like planets, and astronomers focus their search for extraterrestrial life on Earth-like planets in habitable zones. Although our galaxy may be teeming with intelligent life, none has yet been detected. The Drake equation organizes our thoughts about the likelihood of intelligent life in the Milky Way Galaxy. Long before the Sun ends its period of stability on the main sequence, all terrestrial life will have perished after an increasingly luminous Sun makes Earth uninhabitable.

1 Life on Earth likely began in the oceans because they were rich in the building blocks of RNA, DNA, proteins, and membranes. Life then evolved through a combination of mutation and heredity to form the complex ecosystems that exist today.

2 The chemistry of life as we know it requires liquid water. Europa and Enceladus have liquid water now, and Mars had liquid water in the past.

3 A stellar habitable zone is defined as the region around a star in which a planet might have liquid water on the surface: neither too hot nor too cold. The galactic habitable zone is not too close to the high-energy radiation that dominates the center of the Milky Way, yet it is still close enough that there is a sufficient amount of the heavy elements that make up life.

4 Scientists have found that exoplanets are common. This indicates that the term in the Drake equation representing the fraction of stars with planets is high. Extraterrestrial life has not yet been discovered, but this null result is not definitive, given the limitations of the search so far.

5 Eventually, the Sun will move off the main sequence on the Hertzsprung-Russell diagram. Life on Earth will not even last until that time, however, because as the Sun becomes more luminous, the inner edge of the Sun's habitable zone will move out past Earth's orbit.

QUESTIONS AND PROBLEMS

TESTING YOUR UNDERSTANDING

1. **T/F:** All that is required for life to begin is a solitary, self-replicating molecule.

2. **T/F:** All life on Earth evolved from a common ancestor.

3. **T/F:** Astronomers do not consider the possibility of non-carbon-based life.

4. **T/F:** There are several places other than Earth where life might exist in our Solar System.

5. **T/F:** It is likely that we will discover many civilizations that are not advanced.

6. The Urey-Miller experiment produced _____ in a laboratory jar.
   a. life
   b. RNA and DNA
   c. amino acids
   d. proteins

7. A star in the halo of the Milky Way is likely outside the galactic habitable zone because
   a. there is too much dark matter.
   b. there is too much high-energy radiation.
   c. there are not enough supernovae nearby to start star formation.
   d. there are not enough heavy elements.

8. *Natural selection* means that
   a. poorly adapted forms of life die, while well-adapted forms reproduce.
   b. well-adapted forms of life are chosen and encouraged.
   c. forms of life are designed for their environment.
   d. environments are designed for the life-forms that inhabit them.

9. Any system with heredity, mutation, and natural selection will (choose all that apply)
   a. change over time.
   b. become larger over time.
   c. become more complex over time.
   d. develop intelligence.

10. The study of life and the study of astronomy are connected because (choose all that apply)
    a. life may be quite commonplace in the universe.
    b. studying other planets may help explain why there is life on Earth.
    c. explorations of extreme environments on Earth suggest where to look for life elsewhere.
    d. life is a structure that evolved through physical processes, and life on Earth might not be unique.
    e. life elsewhere is most likely to be found by astronomers.

11. The Cambrian explosion
    a. killed the dinosaurs.
    b. produced the carbon that is now here on Earth.
    c. was a sudden increase in biodiversity.
    d. released a lot of carbon dioxide into the atmosphere.

12. The difference between prokaryotes and eukaryotes is that
   a. prokaryotes have no DNA.
   b. prokaryotes have no cell wall.
   c. prokaryotes have no nucleus.
   d. prokaryotes do not exist today.

13. In the phrase "theory of evolution," the word *theory* means that evolution
   a. is an idea that can't be tested scientifically.
   b. is an educated guess to explain natural phenomena.
   c. probably doesn't happen anymore.
   d. is a well-tested, well-corroborated scientific explanation of natural phenomena.

14. Mutations are
   a. changes to an organism's DNA.
   b. changes in the appearance of the organism.
   c. deadly to the organism's children.
   d. changes in the whole species.

15. The habitable zone is the place around a star where
   a. life has been found.
   b. atmospheres can contain oxygen.
   c. liquid water exists.
   d. liquid water can exist on the surface of a planet.

16. The length of time an intelligent, communicating civilization lasts affects _____ in the Drake equation.
   a. the value of R^*
   b. the value of f_i
   c. the value of f_c
   d. the value of L

17. Carbon is a favorable base for life because
   a. it can bond to many other atoms in long chains.
   b. it is nonreactive.
   c. it forms weak bonds that can be readily reorganized as needed.
   d. it is organic.

18. A thermophile is an organism that lives in extremely _____ conditions.
   a. salty
   b. hot
   c. cold
   d. dry

19. Life first appeared on Earth _____ of years ago.
   a. thousands
   b. hundreds of thousands
   c. millions
   d. billions

20. Astronomers think that intelligent life is more likely to be found around stars of types F, G, K, and M because
   a. those stars are hot enough to have planets and moons with liquid water.
   b. those stars are cool enough to have planets and moons with liquid water.
   c. those stars live long enough for life to begin and evolve.
   d. those stars produce no UV radiation or X-rays.

THINKING ABOUT THE CONCEPTS

21. Why do we generally talk about molecules such as DNA or RNA when discussing life on Earth? Why do we suspect that the building blocks of DNA first formed on Earth?

22. Why is silicon-based life unlikely?

23. Today, most known life enjoys moderate climates and temperatures. Compare this environment to some of the conditions in which early life developed.

24. What are the similarities and differences between prokaryotes and eukaryotes?

25. Tracing back on the evolutionary tree, what kinds of life do we find that animals are most similar to?

26. What were the general conditions needed on our planet for life to arise?

27. Figure 18.13 is a "false color" image. Why do astronomers sometimes use false color in images like these? ⊛

28. Study Figure 18.18. The four rows of white blocks at the top represent the numbers 1–10: The bottom row is a placeholder, and the top three rows are the actual counting. Explain the "rule" for the kind of counting shown here. (For example, how do three white blocks represent the number 7?) ⊛

29. The *Viking* spacecraft did not find evidence of life on Mars when it visited that planet in the late 1970s, nor did the *Phoenix* lander when it examined the martian soil in 2009. Does this imply that life never existed on Mars? Why or why not?

30. Figure 18.14 shows the relative amounts of water on Earth, Europa, and Titan. Do Europa and Titan contain more or less water than Earth? Why is that important to the search for life in the Solar System? ⊛

31. What is a habitable zone? What defines its boundaries?

32. Figure 18.15 shows the location of the habitable zone for stars of various masses. About how much more massive would the Sun have to have been for Mars to be within its habitable zone? ⊛

33. In searching for intelligent life elsewhere, why is listening with radio telescopes currently our favored method?

34. Why is it likely that life on Earth as we know it will end long before the Sun runs out of nuclear fuel?

35. ★ A WHAT AN ASTRONOMER SEES The "Pale Blue Dot" photograph in Figure 18.22 is famous. Take a moment to think back to the beginning of this course, and reflect on what you have learned. How has your concept of your place in the universe changed since then? How has your understanding of this image (and others like it) changed as you've learned about the universe beyond Earth? ⊛

APPLYING THE CONCEPTS

36. If a self-replicating molecule has begun replicating, and seven doubling times have passed, how many molecules are there?

   a. Make a prediction: Compare the number of doubling times in this problem to 60, the number of doubling times in the first part of Working It Out 18.1. Do you expect the final population to be larger or smaller than the final population in the Working It Out box? Do you expect the difference to be large or small?

   b. Calculate the final population of this molecule after seven doubling times.

   c. Check your work by comparing your answer to your prediction.

37. Suppose that an organism has a doubling time of 1 second. What will the final population be after 20 seconds?

   a. Make a prediction: Compare the number of doubling times in this problem to 60, the number of doubling times in the first part of Working It Out 18.1. Do you expect the final population to be larger or smaller than the final population in the Working It Out box? Do you expect the difference to be large or small?

   b. Calculate the final population of this organism after 20 seconds.

   c. Check your work by comparing your answer to your prediction.

38. Suppose that an organism has a doubling time of 3 seconds. After how many seconds will one organism increase to become 1,024?

   a. Make a prediction: Compare 1,024 to the final population in Working It Out 18.1. Do you expect the number of doubling times to be larger or smaller than 60?

   b. Calculate the number of doubling times, by trial and error; that is, put numbers into your calculator until you find the number for which $2^{number} = 1,024$. How many seconds will it take for one organism to become 1,024?

   c. Check your work by comparing your answer to your prediction.

39. The doubling time for a sample of the bacterium *Escherichia coli* is 20 minutes. You become infected when just 10 bacteria enter your system. How many *E. coli* bacteria are in your body 12 hours after infection begins?

   a. Make a prediction: About how many doubling times will pass in 12 hours? Consider Working It Out 18.1. Do you expect the number of *E. coli* bacteria in your body after 12 hours to be much larger, much smaller, or comparable to the number of molecules after 60 doubling times?

   b. Calculate the number of *E. coli* bacteria in your body after 12 hours. Don't forget to take into account that the initial population is 10 bacteria!

   c. Check your work by comparing your answer to your prediction.

40. As a rule of thumb, you can find the doubling time for exponential growth by dividing 70 by the rate of increase. So, if the population increases by 7 percent per year, the doubling time is 10 years ($70/7 = 10$). Earth's human population grows by about 1 percent annually. What is the doubling time for Earth's human population? How much time will pass before there are 4 times as many humans on Earth?

   a. Make a prediction: Do you expect the doubling time for Earth's population to be longer or shorter than that of a population that grows at 7 percent per year? Do you expect the time to double twice (to have 4 times as many humans) to be large or small compared to a decade?

   b. Estimate the doubling time for Earth's population, then calculate the time from now until there are 4 times as many humans on Earth as currently.

   c. Check your work by comparing your answer to your prediction.

exploration Fermi Problems and the Drake Equation

The Drake equation is a way of organizing our thoughts about whether there might be other intelligent, communicating civilizations in the galaxy. This type of thinking is very useful for getting estimates of a value, particularly when analyzing systems for which counting is not actually possible. The types of problems that can be solved in this way are often called Fermi problems after Enrico Fermi, a physicist who was mentioned in this chapter. For example, we might ask, "What is the circumference of Earth?"

You could Google this, or you could already "know" it, or you might look it up in this textbook. Alternatively, you could very carefully measure the shadow of a stick in two locations at the same time on the same day. Or you could drive around the planet. Or you could reason this way:

How many time zones are between New York and Los Angeles?
   3 time zones. You know this from traveling or from television.
How many miles is it from New York to Los Angeles?
   3,000 miles. You know this from traveling or from living in the world.
So, how many miles per time zone?
   3,000/3 = 1,000
How many time zones in the world?
   24, because there are 24 hours in a day, and each time zone marks an hour.
So, what is the circumference of Earth?
   24,000 miles, because there are 24 time zones, each 1,000 miles wide.
The measured circumference is 24,900 miles, which agrees to within 4 percent.

Following are several Fermi problems. Time yourself for an hour, and work on as many of them as possible. (You don't have to do them in order!)

1. How much has the mass of the human population on Earth increased in the past year?

2. How much energy does a horse consume in its lifetime?

3. How many pounds of potatoes are consumed in the United States annually?

4. How many cells are there in the human body?

5. If your life earnings were given to you by the hour, how much is your time worth per hour?

6. What is the weight of solid garbage thrown out by American families each year?

7. How fast does human hair grow (in feet per hour)?

8. If all the people on Earth were crowded together, how much area would we cover?

9. How many people could fit on Earth if every person occupied 1 square meter of land?

10. How much carbon dioxide (CO_2) does an automobile emit each year?

11. What is the mass of Earth?

12. What is the average annual cost of an automobile including overhead (maintenance, looking for parking, cleaning, and so forth)?

13. How much ink was used printing all the newspapers in the United States today?

PERIODIC TABLE OF THE ELEMENTS

Legend:
- 1 — Atomic number
- H — Symbol
- Hydrogen — Name
- 1.00794 — Average atomic mass

- Metals
- Metalloids
- Nonmetals

1 1A																	18 8A
1 **H** Hydrogen 1.00794	2 2A											13 3A	14 4A	15 5A	16 6A	17 7A	2 **He** Helium 4.002602
3 **Li** Lithium 6.941	4 **Be** Beryllium 9.012182											5 **B** Boron 10.811	6 **C** Carbon 12.0107	7 **N** Nitrogen 14.0067	8 **O** Oxygen 15.9994	9 **F** Fluorine 18.9984032	10 **Ne** Neon 20.1797
11 **Na** Sodium 22.98976928	12 **Mg** Magnesium 24.3050	3 3B	4 4B	5 5B	6 6B	7 7B	8	9 8B	10	11 1B	12 2B	13 **Al** Aluminum 26.9815386	14 **Si** Silicon 28.0855	15 **P** Phosphorus 30.973762	16 **S** Sulfur 32.065	17 **Cl** Chlorine 35.453	18 **Ar** Argon 39.948
19 **K** Potassium 39.0983	20 **Ca** Calcium 40.078	21 **Sc** Scandium 44.955912	22 **Ti** Titanium 47.867	23 **V** Vanadium 50.9415	24 **Cr** Chromium 51.9961	25 **Mn** Manganese 54.938045	26 **Fe** Iron 55.845	27 **Co** Cobalt 58.933195	28 **Ni** Nickel 58.6934	29 **Cu** Copper 63.546	30 **Zn** Zinc 65.38	31 **Ga** Gallium 69.723	32 **Ge** Germanium 72.64	33 **As** Arsenic 74.92160	34 **Se** Selenium 78.96	35 **Br** Bromine 79.904	36 **Kr** Krypton 83.798
37 **Rb** Rubidium 85.4678	38 **Sr** Strontium 87.62	39 **Y** Yttrium 88.90585	40 **Zr** Zirconium 91.224	41 **Nb** Niobium 92.90638	42 **Mo** Molybdenum 95.96	43 **Tc** Technetium [98]	44 **Ru** Ruthenium 101.07	45 **Rh** Rhodium 102.90550	46 **Pd** Palladium 106.42	47 **Ag** Silver 107.8682	48 **Cd** Cadmium 112.411	49 **In** Indium 114.818	50 **Sn** Tin 118.710	51 **Sb** Antimony 121.760	52 **Te** Tellurium 127.60	53 **I** Iodine 126.90447	54 **Xe** Xenon 131.293
55 **Cs** Cesium 132.9054519	56 **Ba** Barium 137.327	57 **La** Lanthanum 138.90547	72 **Hf** Hafnium 178.49	73 **Ta** Tantalum 180.94788	74 **W** Tungsten 183.84	75 **Re** Rhenium 186.207	76 **Os** Osmium 190.23	77 **Ir** Iridium 192.217	78 **Pt** Platinum 195.084	79 **Au** Gold 196.966569	80 **Hg** Mercury 200.59	81 **Tl** Thallium 204.3833	82 **Pb** Lead 207.2	83 **Bi** Bismuth 208.98040	84 **Po** Polonium [209]	85 **At** Astatine [210]	86 **Rn** Radon [222]
87 **Fr** Francium [223]	88 **Ra** Radium [226]	89 **Ac** Actinium [227]	104 **Rf** Rutherfordium [261]	105 **Db** Dubnium [262]	106 **Sg** Seaborgium [266]	107 **Bh** Bohrium [264]	108 **Hs** Hassium [277]	109 **Mt** Meitnerium [268]	110 **Ds** Darmstadtium [271]	111 **Rg** Roentgenium [272]	112 **Cn** Copernicium [285]	113 **Uut** Ununtrium [284]	114 **Fl** Flerovium [289]	115 **Uup** Ununpentium [288]	116 **Lv** Livermorium [292]	117 **Uus** Ununseptium [294]	118 **Uuo** Ununoctium [294]

6 Lanthanides

58 **Ce** Cerium 140.116	59 **Pr** Praseodymium 140.90765	60 **Nd** Neodymium 144.242	61 **Pm** Promethium [145]	62 **Sm** Samarium 150.36	63 **Eu** Europium 151.964	64 **Gd** Gadolinium 157.25	65 **Tb** Terbium 158.92535	66 **Dy** Dysprosium 162.500	67 **Ho** Holmium 164.93032	68 **Er** Erbium 167.259	69 **Tm** Thulium 168.93421	70 **Yb** Ytterbium 173.05	71 **Lu** Lutetium 174.967

7 Actinides

90 **Th** Thorium 232.03806	91 **Pa** Protactinium 231.03588	92 **U** Uranium 238.02891	93 **Np** Neptunium [237]	94 **Pu** Plutonium [244]	95 **Am** Americium [243]	96 **Cm** Curium [247]	97 **Bk** Berkelium [247]	98 **Cf** Californium [251]	99 **Es** Einsteinium [252]	100 **Fm** Fermium [257]	101 **Md** Mendelevium [258]	102 **No** Nobelium [259]	103 **Lr** Lawrencium [262]

We have used the U.S. system as well as the system recommended by the International Union of Pure and Applied Chemistry (IUPAC) to label the groups in this periodic table. The system used in the United States includes a letter and a number (1A, 2A, 3B, 4B, etc.), which is close to the system developed by Mendeleev. The IUPAC system uses numbers 1–18 and has been recommended by the American Chemical Society (ACS). Elements with atomic numbers higher than 112 have been reported but not yet fully authenticated.

PROPERTIES OF PLANETS, DWARF PLANETS, AND MOONS

Physical Data for Planets and Dwarf Planets

Planet	Equatorial Radius (km)	Equatorial Radius (R/R_{Earth})	Mass (kg)	Mass (M/M_{Earth})	Average Density*	Rotation Period (days)	Tilt of Rotation Axis (degrees, relative to orbit)	Equatorial Surface Gravity[†]	Escape Velocity (km/s)	Average Surface Temperature[‡] (K)
Mercury	2,440	0.383	3.30×10^{23}	0.055	5.427	58.65	0.01	0.378	4.3	340 (100, 725)
Venus	6,052	0.949	4.87×10^{24}	0.815	5.243	243.02[§]	177.36	0.907	10.36	737
Earth	6,378	1.000	5.97×10^{24}	1.000	5.513	1.000	23.44	1.000	11.19	288 (183, 331)
Mars	3,396	0.533	6.42×10^{23}	0.107	3.934	1.0260	25.19	0.377	5.03	210 (133, 293)
Ceres	487.3	0.075	9.47×10^{20}	0.0002	2.100	0.378	3.0	0.029	0.51	200
Jupiter	71,492	11.209	1.90×10^{27}	317.83	1.326	0.4135	3.13	2.528	59.5	165
Saturn	60,268	9.449	5.68×10^{26}	95.16	0.687	0.4440	26.73	1.065	35.5	134
Uranus	25,559	4.007	8.68×10^{25}	14.536	1.270	0.7183[§]	97.77	0.889	21.3	76
Neptune	24,764	3.883	1.02×10^{26}	17.148	1.638	0.6713	28.32	1.14	23.5	58
Pluto	1,184	0.182	1.30×10^{22}	0.0021	2.030	6.387[§]	122.53	0.083	1.23	40
Haumea	~650	0.11	4.0×10^{21}	0.0007	~3	0.163	?	0.045	0.84	<50
Makemake	750	0.12	4.18×10^{21}	0.0007	~2	0.32	?	0.048	0.8	~30
Eris	1,200	0.188	1.5×10^{22} (est.)	0.0025 (est.)	~2	>0.3?	?	0.082	~1.3	30

*Relative to the density of water, which is 1,000 kg/m^3.
[†]Relative to the surface gravity of Earth, which is 9.81 m/s^2.
[‡]Where given, values in parentheses give extremes of recorded temperatures.
[§]Venus, Uranus, and Pluto rotate opposite to the directions of their orbits. Their north poles are south of their orbital planes.

Orbital Data for Planets and Dwarf Planets

Planet	Mean Distance from Sun (A*)		Orbital Period (P, sidereal years)	Eccentricity	Inclination (degrees, relative to ecliptic)	Average Speed (km/s)
	(10^6 km)	(AU)				
Mercury	57.9	0.387	0.2408	0.2056	7.005	47.36
Venus	108.2	0.723	0.6152	0.0068	3.395	35.02
Earth	149.6	1.000	1.0000	0.0167	0.000	29.78
Mars	227.9	1.524	1.8808	0.0934	1.850	24.08
Ceres	413.7	2.765	4.6027	0.079	10.587	17.88
Jupiter	778.3	5.203	11.8626	0.0484	1.304	13.06
Saturn	1,426.7	9.537	29.4475	0.0539	2.485	9.64
Uranus	2,870.7	19.189	84.0168	0.0473	0.772	6.80
Neptune	4,495.1	30.070	164.7913	0.011	1.769	5.43
Pluto	5,906.4	39.48	247.9207	0.2488	17.14	4.72
Haumea	6,428.1	43.0	281.9	0.198	28.22	4.50
Makemake	6,789.7	45.3	305.3	0.164	29.00	4.39
Eris	10,183	68.05	561.6	0.4339	43.82	3.43

*A is the semimajor axis of the planet's elliptical orbit.

Properties of Selected Moons*

Planet	Moon	Orbital Properties		Physical Properties		
		P (days)	A (10³ km)	R (km)	M (10²⁰ kg)	Relative Density† (g/cm³)
Earth (1 moon)	Moon	27.32	384.4	1,737.5	735	3.34
Mars (2 moons)	Phobos	0.32	9.38	13.4 × 11.2 × 9.2	0.0001	1.9
	Deimos	1.26	23.46	7.5 × 6.1 × 5.2	0.00002	1.5
Jupiter (more than 78 known moons)	Metis	0.30	127.97	21.5	0.00012	3
	Amalthea	0.50	181.40	131 × 73 × 67	0.0207	0.8
	Io	1.77	421.80	1,822	893	3.53
	Europa	3.55	671.10	1,561	480	3.01
	Ganymede	7.15	1,070	2,631	1,482	1.94
	Callisto	16.69	1,883	2,410	1,080	1.83
	Himalia	250.56	11,461	85	0.067	2.6
	Pasiphae	744‡	23,624	30	0.0030	2.6
	Callirrhoe	759‡	24,102	4.3	0.00001	2.6
Saturn (more than 81 known moons)	Pan	0.58	133.58	14	0.00005	0.42
	Prometheus	0.61	139.38	74.0 × 50.0 × 34.0	0.0016	0.5
	Pandora	0.63	141.70	55 × 44 × 31	0.0014	0.5
	Mimas	0.94	185.54	198	0.38	1.15
	Enceladus	1.37	238.04	252	1.08	1.6
	Tethys	1.89	294.67	533	6.17	0.97
	Dione	2.74	377.42	562	11.0	1.48
	Rhea	4.52	527.07	764	23.1	1.23
	Titan	15.95	1,222	2,575	1,346	1.88
	Hyperion	21.28	1,501	205 × 130 × 110	0.0559	0.54
	Iapetus	79.33	3,561	736	18.1	1.08
	Phoebe	550.3‡	12,948	107	0.08	1.6
	Paaliaq	687.5	15,024	11	0.0001	2.3
Uranus (27 known moons)	Cordelia	0.34	49.80	20	0.0004	1.3
	Miranda	1.41	129.90	236	0.66	1.21
	Ariel	2.52	190.90	579	12.9	1.59
	Umbriel	4.14	264.96	585	12.2	1.46
	Titania	8.71	436.30	789	34.2	1.66
	Oberon	13.46	583.50	761	28.8	1.56
	Setebos	2,225‡	17,420	24	0.0009	1.5
Neptune (14 known moons)	Naiad	0.29	48.0	48 × 30 × 26	0.002	1.3
	Larissa	0.55	73.5	108 × 102 × 84	0.05	1.3
	Proteus	1.12	117.6	218 × 208 × 201	0.5	1.3
	Triton	5.88‡	354.8	1,353	214	2.06
	Nereid	360.13	5,513.82	170	0.3	1.5
Pluto (5 known moons)	Charon	6.39	17.54	604	15.2	1.65
Haumea (2 moons)	Namaka	18	25.66	85	0.018	~1
	Hi'iaka	49	49.88	170	0.179	~1
Eris	Dysnomia	15.8	37.4	50–125?	?	?

*Innermost, outermost, largest, and/or a few other moons are listed for each planet.
†Relative to the density of water, which is 1,000 kg/m³ or 1.00 g/cm³.
‡Irregular moon (has retrograde orbit).

NEAREST AND BRIGHTEST STARS

Stars within 12 Light-Years of Earth

Name*	Distance (ly)	Spectral Type[†]	Relative Visual Luminosity[‡]	Apparent Magnitude	Absolute Magnitude
Sun	1.58×10^{-5}	G2V	1.000	−26.74	4.83
Alpha Centauri C (Proxima Centauri)	4.243	M5.5V	0.000052	11.05	15.48
Alpha Centauri A	4.367	G2V	1.5	0.01	4.38
Alpha Centauri B	4.367	K0V	0.44	1.34	5.71
Barnard's star	5.978	M4Ve	0.00044	9.57	13.25
Luhman 16	6.516	L8	?	10.7[§]	?
CN Leonis	7.86	M5.5	0.000020	13.53	16.64
BD +36-2147	8.31	M2.0V	0.0056	7.47	10.44
Sirius A	8.60	A1V	22	−1.43	1.47
Sirius B	8.60	DA2	0.0025	8.44	11.34
BL Ceti	8.73	M5.5V	0.000059	12.61	15.40
UV Ceti	8.73	M6.0	0.000039	12.99	15.85
V1216 Sagittarii	9.60	M3.5V	0.00050	10.43	13.07
HH Andromedae	10.30	M5.5V	0.00010	12.29	14.79
Epsilon Eridani	10.475	K2V	0.28	3.73	6.19
Lacaille 9352	10.68	M1.5V	0.011	7.34	9.75
FI Virginis	11.03	M4.0V	0.00033	11.13	13.51
Wise 1506+7027	11.1	T6	?	14.3[§]	?
EZ Aquarii A	11.26	M5.0V	0.000063	13.03	15.33
EZ Aquarii B	11.26	M5e	0.000050	13.27	15.58
EZ Aquarii C	11.26		0.00001	15.07	17.37
Procyon A	11.46	F5IV-V	7.38	0.38	2.66
Procyon B	11.46	DA	0.00054	10.70	12.98
61 Cygni A	11.40	K5.0V	0.086	5.21	7.49
61 Cygni B	11.40	K7.0V	0.040	6.03	8.31
Gliese 725 A	11.50	M3.0V	0.0029	8.90	11.16
Gliese 725 B	11.50	M3.5V	0.0014	9.69	11.95

*Stars may carry many names, including common names (such as Sirius), names based on their prominence within a constellation (such as Alpha Canis Majoris, another name for Sirius), or names based on their inclusion in a catalog (such as BD +36-2147). Addition of letters (A, B, and so on) or superscripts indicates membership in a multiple-star system.

[†]Spectral types such as M3 are discussed in Chapter 10. Other letters or numbers provide additional information. For example, V after the spectral type indicates a main-sequence star, and III indicates a giant star. Stars of spectral type T are brown dwarfs.

[‡]Luminosity in this table refers only to radiation in "visual" light and is given relative to that of the Sun = 1.000.

[§]Brown dwarf, detected in IR J-band only.

(continued)

Stars within 12 Light-Years of Earth (continued)

Name*	Distance (ly)	Spectral Type[†]	Relative Visual Luminosity[‡]	Apparent Magnitude	Absolute Magnitude
Groombridge 34 A	11.62	M1.5V	0.0063	8.08	10.32
Groombridge 34 B	11.62	M3.5V	0.00041	11.06	13.30
Epsilon Indi A	11.81	K5Ve	0.15	4.69	6.89
Epsilon Indi B (brown dwarf)	11.81	T1.0			
Epsilon Indi C (brown dwarf)	11.81	T6.0			
DX Cancri	11.8	M6.0V	0.000012	14.9	17.10
Tau Ceti	11.905	G8.5V	0.46	3.49	5.68
Gliese 1061	12.04	M5.5V	0.000067	13.09	15.26

The 25 Brightest Stars in the Sky

Name	Common Name	Distance (ly)	Spectral Type	Relative Visual Luminosity*	Apparent Visual Magnitude	Absolute Visual Magnitude
Sun	Sun	1.58×10^{-5}	G2V	1.000	−26.8	4.82
Alpha Canis Majoris	Sirius	8.60	A1V	22.9	−1.47	1.42
Alpha Carinae	Canopus	313	F0II	13,800	−0.72	−5.53
Alpha Bootis	Arcturus	36.7	K1.5IIIFe-0.5	111	−0.04	−0.29
Alpha1 Centauri	Rigel Kentaurus	4.367	G2V	1.50	−0.01	4.38
Alpha Lyrae	Vega	25.04	A0Va	49.7	0.03	0.58
Alpha Aurigae	Capella	42.9	G5IIIe+G0III	132	0.08	−0.48
Beta Orionis	Rigel	860	B8Ia	40,200	0.12	−7.84
Alpha Canis Minoris	Procyon	11.4	F5IV-V	7.38	0.34	2.65
Alpha Eridani	Achernar	139	B3Vpe	1,090	0.50	−1.46
Alpha Orionis	Betelgeuse	643	M1-2Ia-Iab	9,600	0.58	−5.85
Beta Centauri	Hadar	390	B1III	8,700	0.60	−4.53
Alpha Aquilae	Altair	16.8	A7V	11.1	0.77	2.21
Alpha Tauri	Aldebaran	65.1	K5+III	151	0.85	−0.63
Alpha Virginis	Spica	250	B1III-IV+B2V	2,230	1.04	−3.55
Alpha Scorpii	Antares	604	M1.5Iab-Ib+B4Ve	11,000	1.09	−5.28
Beta Geminorum	Pollux	33.7	K0IIIb	31.0	1.15	1.09
Alpha Piscis	Fomalhaut	25.1	A3V	17.2	1.16	1.73
Alpha Cygni	Deneb	3,200	A2Ia	260,000	1.25	−8.73
Beta Crucis	Becrux	353	B0.5III	3,130	1.30	−3.92
Alpha2 Centauri	Alpha Centauri B	4.367	K1V	0.44	1.33	5.71
Alpha Leonis	Regulus	77.5	B7V	137	1.35	−0.52
Alpha Crucis	Acrux	321	B0.5IV	4,000	1.40	−4.19
Epsilon Canis Majoris	Adara	431	B2II	3,700	1.51	−4.11
Gamma Crucis	Gacrux	88.0	M3.5III	142	1.59	−0.56

Sources: Data from *The Hipparcos and Tycho Catalogues,* 1997, European Space Agency SP-1200; SIMBAD Astronomical Database (http://simbad.u-strasbg.fr/simbad); and Research Consortium on Nearby Stars (http://www.recons.org/TOP100.posted.htm).
Luminosity in this table refers only to radiation in "visual" light and is given relative to that of the Sun = 1.000.

STAR MAPS

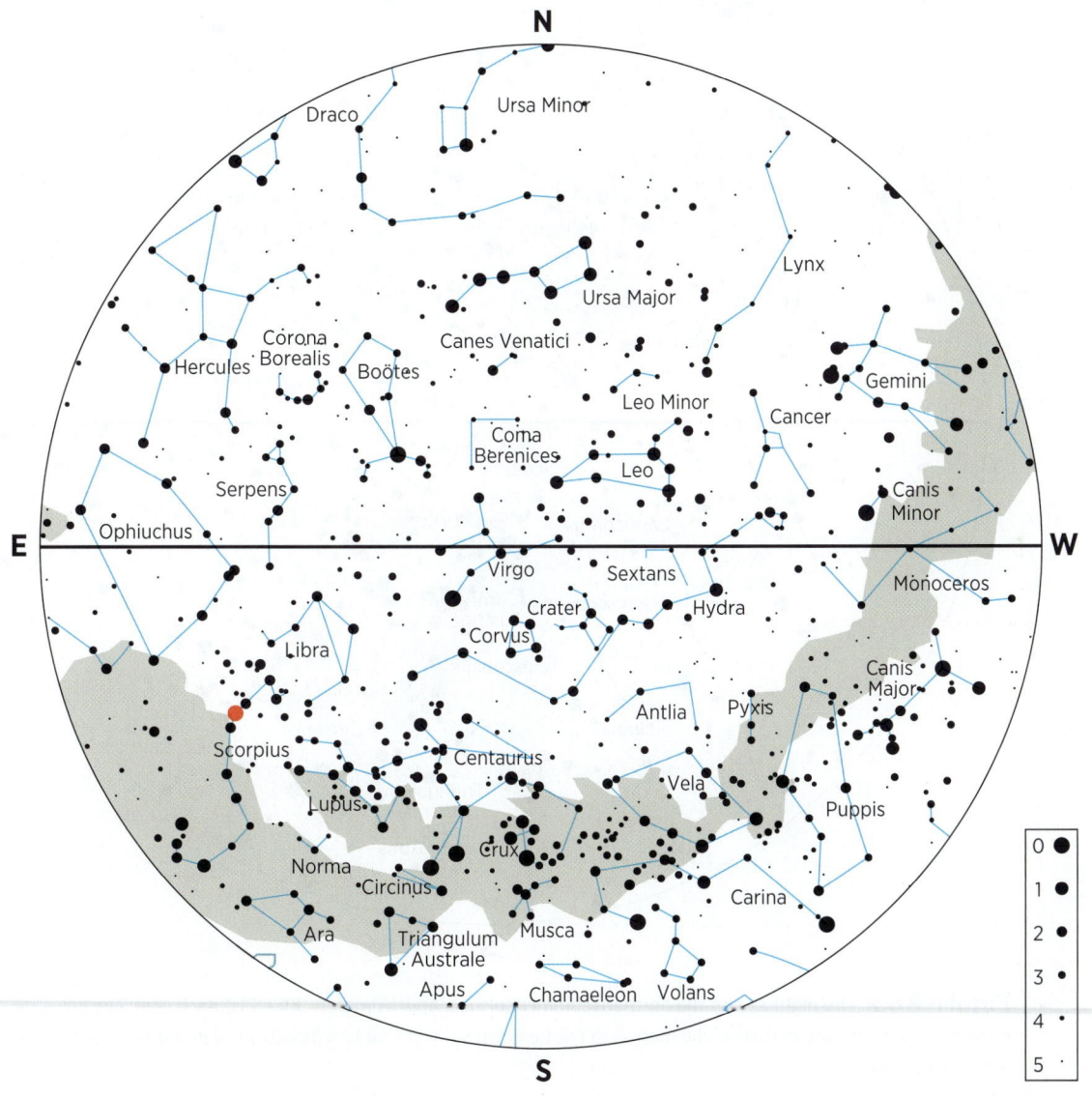

Figure A4.1 The night sky on the northern vernal equinox, approximately March 21st each year. The celestial equator runs across the middle of the map. A red dot indicates a star that is noticeably red in the sky. The shaded area is the Milky Way.

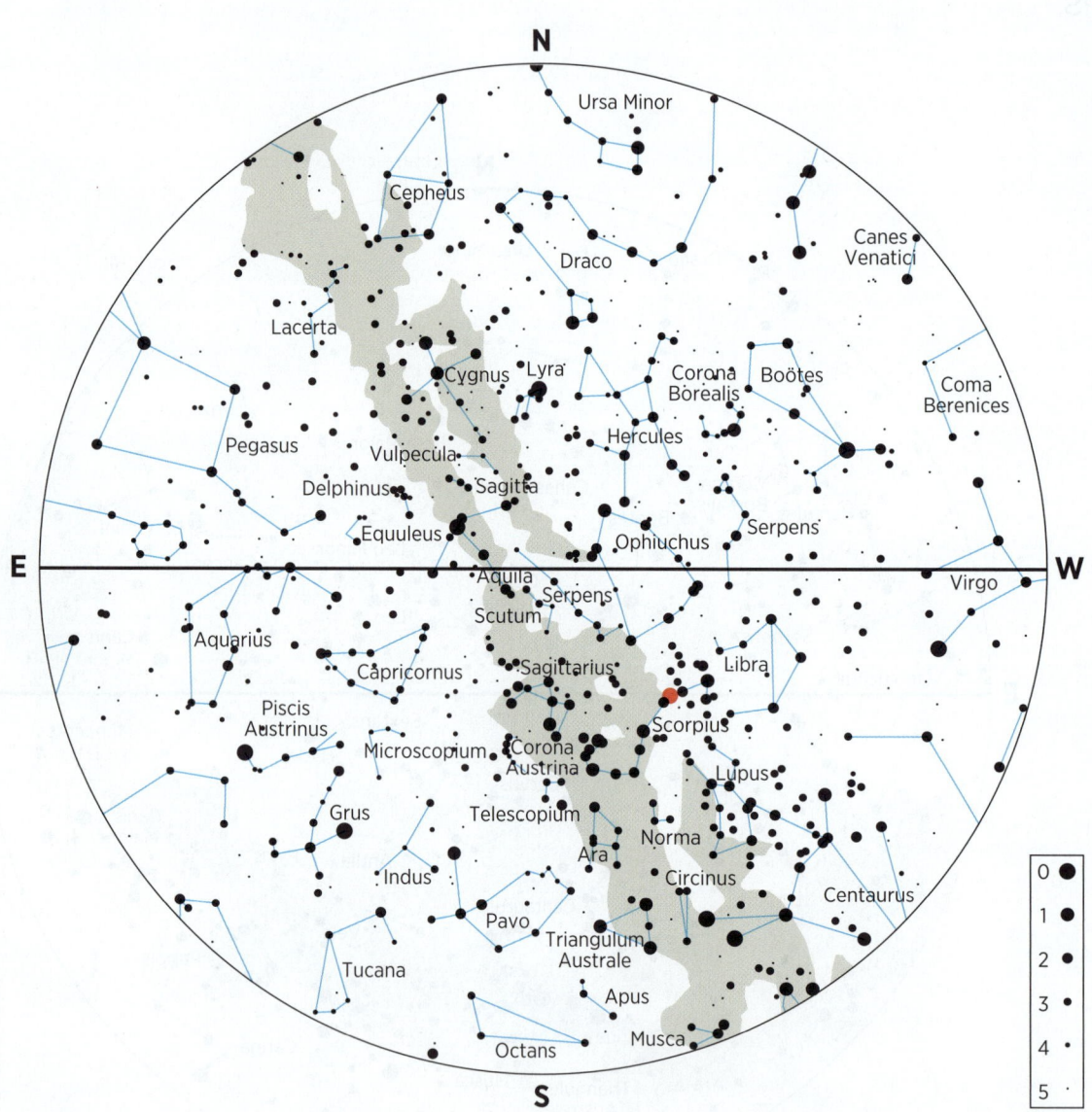

Figure A4.2 The night sky on the northern summer solstice, approximately June 21st each year. The celestial equator runs across the middle of the map. A red dot indicates a star that is noticeably red in the sky. The shaded area is the Milky Way.

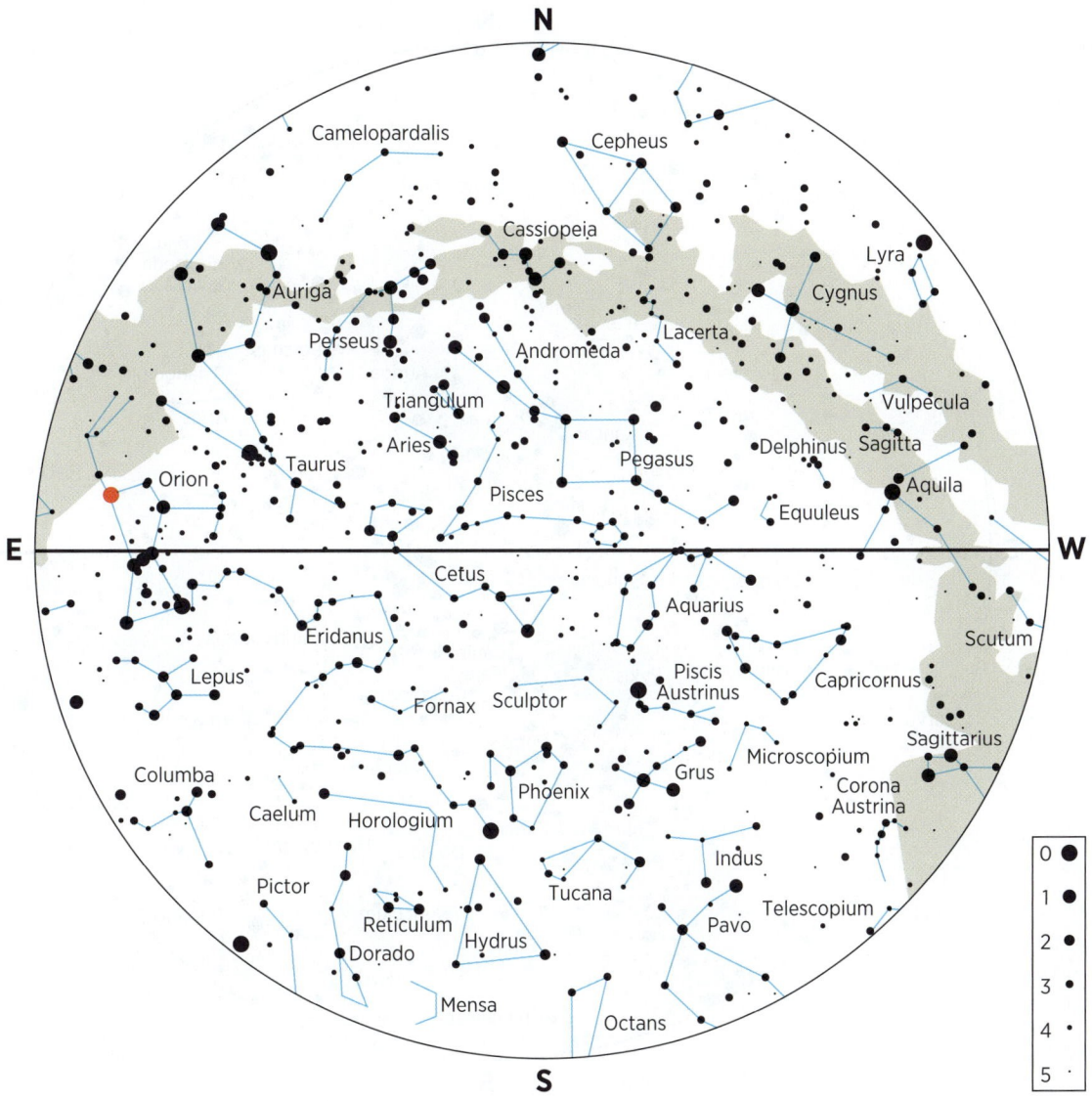

Figure A4.3 The night sky on the northern autumnal equinox, approximately September 21st each year. The celestial equator runs across the middle of the map. A red dot indicates a star that is noticeably red in the sky. The shaded area is the Milky Way.

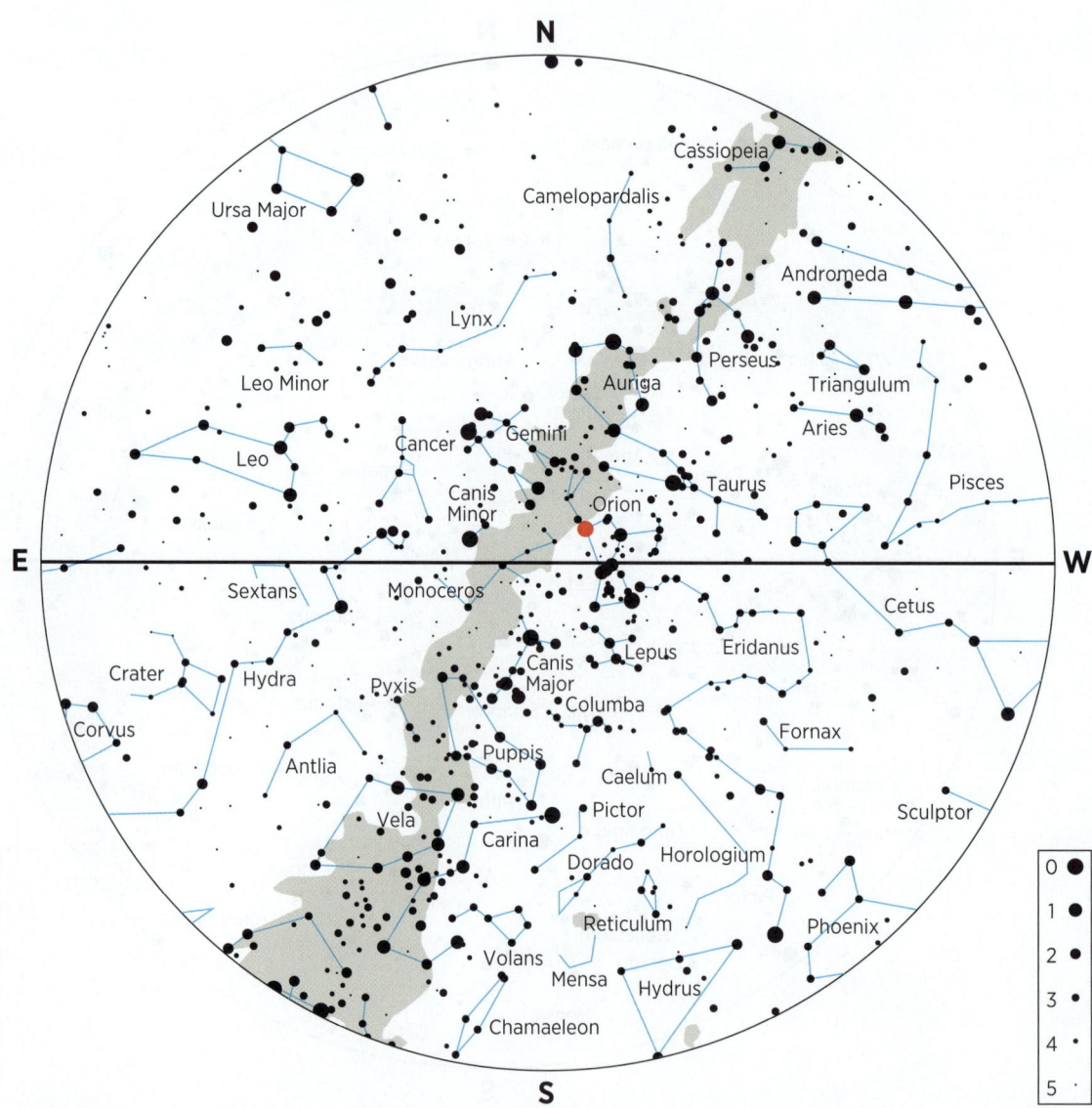

Figure A4.4 The night sky on the northern winter solstice, approximately December 21st each year. The celestial equator runs across the middle of the map. A red dot indicates a star that is noticeably red in the sky. The shaded area is the Milky Way.

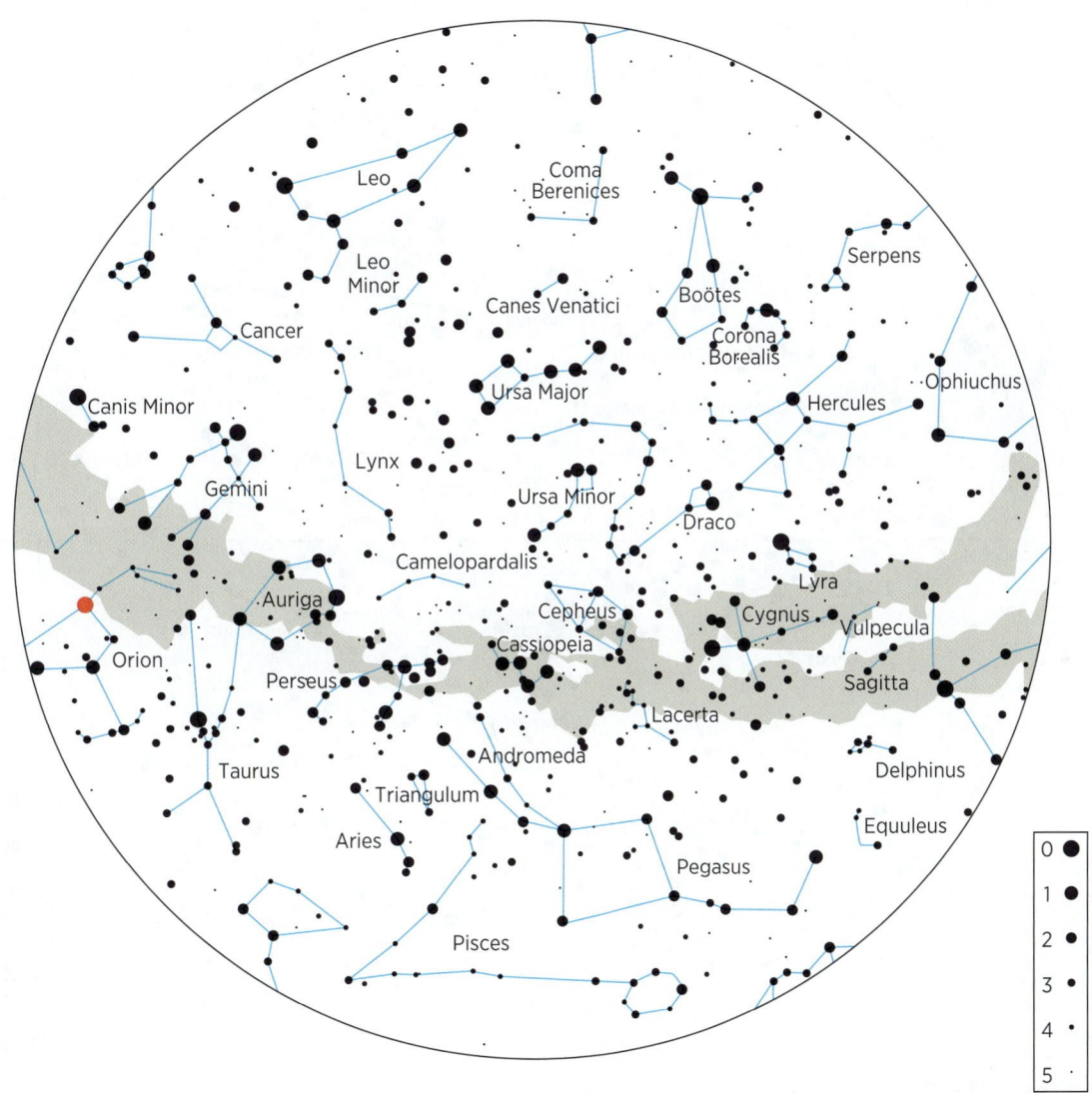

Figure A4.5 The night sky from the North Pole of Earth.

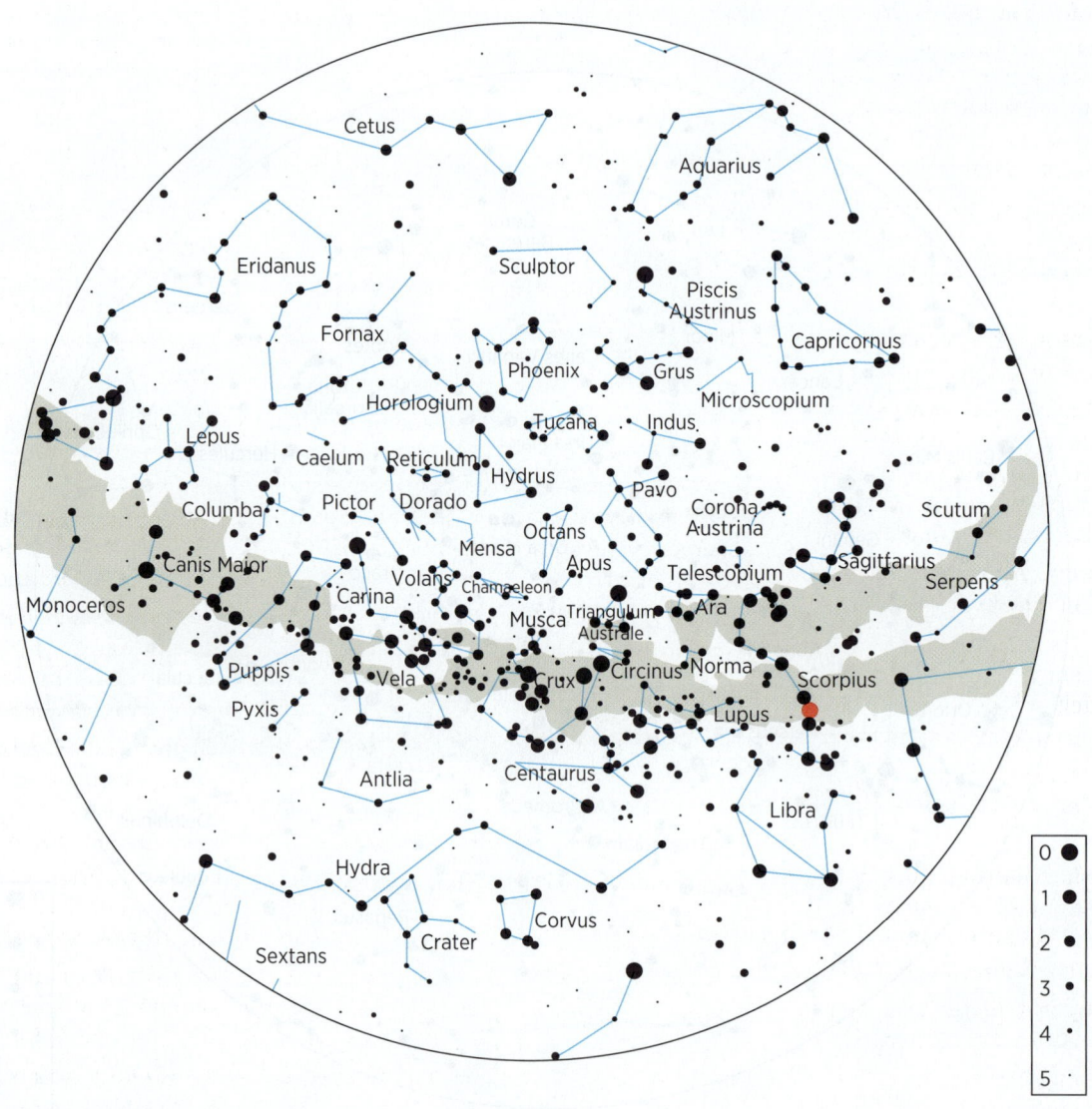

Figure A4.6 The night sky from the South Pole of Earth.

A

aberration of starlight The apparent displacement in the position of a star that is due to the finite speed of light and Earth's orbital motion around the Sun.

absolute magnitude A measure of the intrinsic brightness of a celestial object, generally a star. Specifically, the apparent brightness of an object, such as a star, if it were located at a standard distance of 10 parsecs (pc). Compare *apparent magnitude*.

absolute zero The temperature at which thermal motions cease; the lowest possible temperature, zero on the Kelvin temperature scale.

absorption The capture of electromagnetic radiation by matter. Compare *emission*.

absorption line An intensity minimum in a spectrum that is due to absorption of electromagnetic radiation at a specific wavelength, determined by the energy levels of an atom or molecule. Compare *emission line*.

absorption spectrum A spectrum showing absorption lines. Compare *emission spectrum*.

abundance A measure (by mass, number, or volume) of how much of an element is present, typically measured relative to hydrogen.

acceleration The rate at which the speed and/or direction of an object's motion is changing.

accretion disk A flat, rotating disk of gas and dust surrounding an object, such as a young stellar object, a forming planet, a collapsed star in a binary system, or a black hole.

achondrite A stony meteorite that does not contain chondrules. Compare *chondrite*.

active comet A comet nucleus that approaches close enough to the Sun to show signs of activity, such as the production of a coma and tail.

active galactic nucleus (AGN) A highly luminous, compact galactic nucleus whose luminosity may exceed that of the rest of the galaxy.

active region An area of the Sun's chromosphere that anchors bursts of intense magnetic activity.

adaptive optics Electro-optical systems that largely compensate for image distortion caused by Earth's atmosphere.

AGB See *asymptotic giant branch*.

age 1. The length of time that an object has existed. 2. The length of time since a planetary surface (for example) was last extensively modified.

AGN See *active galactic nucleus*.

albedo The fraction of electromagnetic radiation incident on a surface that is reflected by the surface.

alpha particle A ⁴He nucleus, consisting of two protons and two neutrons. Alpha particles get their name from the fact that they are given off in the type of radioactive decay referred to as "alpha decay."

altitude The location of an object above the horizon, measured by the angle formed between an imaginary line from an observer to the object and a second line from the observer to the point on the horizon directly below the object.

Amors A group of asteroids whose orbits cross the orbit of Mars but not the orbit of Earth. Compare *Apollos* and *Atens*.

amplitude In a wave, the maximum deviation from its undisturbed or relaxed position. For example, in a water wave, the amplitude is the vertical distance from the crest to the undisturbed water level.

angular momentum A conserved property of a rotating or revolving system; its value depends on the velocity and distribution of the system's mass.

angular resolution The minimum angular distance between distinguishable objects in the focal plane of an imaging device, such as a telescope.

annular solar eclipse An event that occurs when the apparent diameter of the Moon is less than that of the Sun, leaving a visible ring of light ("annulus") surrounding the dark disk of the Moon. Compare *partial solar eclipse* and *total solar eclipse*.

Antarctic Circle The circle on Earth with latitude 66.5° south, marking the northern limit where at least 1 day per year is in 24-hour daylight. Compare *Arctic Circle*.

anthropogenic climate change The modification of Earth's climate system due to human activities such as the release of greenhouse gases from the burning of fossil fuels.

anticyclonic motion The rotation of a weather system resulting from the Coriolis effect as air moves outward from a region of high atmospheric pressure. Compare *cyclonic motion*.

antimatter Matter made from antiparticles.

antiparticle An elementary particle of antimatter, identical in mass but opposite in charge and all other properties to its corresponding ordinary matter particle.

aperture The clear diameter of a telescope's objective lens or primary mirror.

aphelion (pl. aphelia) The point in a solar orbit that is farthest from the Sun. Compare *perihelion*.

Apollos A group of asteroids whose orbits cross the orbits of both Earth and Mars. Compare *Amors* and *Atens*.

apparent magnitude A measure of the brightness of a celestial object, generally a star, as it appears to us on our sky. Compare *absolute magnitude*.

apparent retrograde motion Movement of the planets with respect to the "fixed stars," in which the planets appear to move westward for a period of time before resuming their normal eastward motion. The heliocentric model explains this effect much more simply than the geocentric model.

arcminute (arcmin) A minute of arc ('), a unit used for measuring angles. An arcminute is 1/60 of a degree of arc.

arcsecond (arcsec) A second of arc ("), a unit used for measuring very small angles. An arcsecond is 1/60 of an arcminute or 1/3,600 of a degree of arc.

Arctic Circle The circle on Earth with latitude 66.5° north, marking the southern limit where at least 1 day per year is in 24-hour daylight. Compare *Antarctic Circle*.

ash The products of fusion that collect in the core of a star.

asteroid Also called *minor planet*. A primitive rocky or metallic body (planetesimal) that has survived planetary accretion. Asteroids are the parent bodies of meteoroids.

asteroid belt Also called *main asteroid belt*. The region between the orbits of Mars and Jupiter that contains most of the asteroids in our Solar System.

astrobiology An interdisciplinary science combining astronomy, biology, chemistry, geology, and physics to study life in the cosmos.

astrology The belief that the positions and aspects of stars and planets influence human affairs and characteristics, as well as terrestrial events.

astrometry Precision measurement of the position and motion of astronomical objects.

astronomical seeing A measurement of the degree to which Earth's atmosphere degrades the resolution of a telescope's view of astronomical objects.

astronomical unit (AU) The average distance from the Sun to Earth: approximately 150 million kilometers (km).

astronomy The scientific study of planets, stars, galaxies, and the universe as a whole.

astrophysics The application of physical laws to the understanding of planets, stars, galaxies, and the universe as a whole.

asymptotic giant branch (AGB) The path on the H-R diagram that goes from the horizontal branch toward higher luminosities and lower temperatures, asymptotically approaching and then rising above the red giant branch.

Atens A group of asteroids whose orbits cross the orbit of Earth but not the orbit of Mars. Compare *Amors* and *Apollos*.

atmosphere The gravitationally bound, outer gaseous envelope surrounding a planet, moon, or star.

atmospheric greenhouse effect A warming of planetary surfaces produced by atmospheric gases that transmit optical solar radiation but partially trap infrared radiation.

atmospheric probe An instrumented package designed to provide in situ measurements of the chemical and/or physical properties of a planetary atmosphere.

atmospheric window A region of the electromagnetic spectrum in which radiation is able to penetrate a planet's atmosphere.

atom The smallest piece of a chemical element that retains the properties of that element. Each atom is composed of a nucleus (neutrons and protons) surrounded by a cloud of electrons.

AU See *astronomical unit*.

aurora Emission in the upper atmosphere of a planet from atoms that have been excited by collisions with energetic particles from the planet's magnetosphere.

autumnal equinox 1. One of two points where the Sun crosses the celestial equator. 2. The day on which the Sun appears at this location, marking the first day of autumn (about September 22 in the Northern Hemisphere and about March 20 in the Southern Hemisphere). Compare *vernal equinox*.

axion A hypothetical elementary particle, first proposed to explain certain properties of the neutron and now considered a candidate for cold dark matter.

B

backlighting Illumination from behind a subject as seen by an observer. Fine material, such as human hair and dust in planetary rings, stands out best when viewed under backlighting conditions.

bar A unit of pressure. One bar is equivalent to 10^5 newtons per square meter (N/m^2)—approximately equal to Earth's atmospheric pressure at sea level.

barred spiral galaxy A spiral galaxy with a central bulge having an elongated, barlike shape.

basalt Gray to black volcanic rock, rich in iron and magnesium.

beta decay 1. The conversion of a neutron into a proton by emission of an electron (beta particle) and an antineutrino. 2. The conversion of a proton into a neutron by emission of a positron and a neutrino.

Big Bang The event that occurred 13.8 billion years ago that marks the beginning of time and the universe.

Big Bang nucleosynthesis The formation of low-mass nuclei (H, He, Li, Be, and B) during the first few minutes after the Big Bang.

Big Rip A hypothetical cosmic event in which all matter in the universe, from stars to subatomic particles, is progressively torn apart by expansion of the universe.

binary star A system in which two stars are in gravitationally bound orbits around their common center of mass.

binding energy The minimum energy required to separate an atomic nucleus into its component protons and neutrons.

biosphere The global sum of all living organisms on Earth (or any planet or moon). Compare *hydrosphere* and *lithosphere*.

bipolar outflow Material streaming away in opposite directions from either side of the accretion disk of a young star.

black hole An object so dense that its escape velocity exceeds the speed of light; a singularity in spacetime.

blackbody An object that absorbs and can reemit all electromagnetic energy it receives.

blackbody spectrum Also called *Planck spectrum*. The spectrum of electromagnetic energy emitted by a blackbody per unit area per second, which is determined only by the temperature of the object.

blueshift The Doppler shift toward shorter wavelengths of light from an approaching object. Compare *redshift*.

Bohr model A model of the atom, proposed by Niels Bohr in 1913, in which a small positively charged nucleus is surrounded by orbiting electrons, similar to a miniature solar system.

bolide A very bright, exploding meteor.

bound orbit A closed orbit in which the velocity is less than the escape velocity. Compare *unbound orbit*.

bow shock 1. The boundary at which the speed of the solar wind abruptly drops from supersonic to subsonic in its approach to a planet's magnetosphere; the boundary between the region dominated by the solar wind and the region dominated by a planet's magnetosphere. 2. The interface between strong collimated gas and dust outflow from a star and the interstellar medium.

brightness The apparent intensity of light from a luminous object, which depends on both the *luminosity* of a source and its distance. Units at the detector: watts per square meter (W/m^2).

brown dwarf A "failed" star that is not massive enough to cause hydrogen fusion in its core; an object whose mass is intermediate between the least massive stars and supermassive planets.

bulge The central region of a spiral galaxy that is similar in appearance to a small elliptical galaxy.

burning In stellar nuclear physics, the fusion of light elements into heavier ones. For example, the fusion of hydrogen into helium is often called *hydrogen burning*.

C

C See *Celsius*.

C-type asteroid An asteroid made of material that has largely been unmodified since the formation of the Solar System; the most primitive type of asteroid. Compare *M-type asteroid* and *S-type asteroid*.

caldera The summit crater of a volcano.

carbon-nitrogen-oxygen (CNO) cycle One of the ways in which hydrogen is converted to helium (hydrogen burning) in the interiors of main-sequence stars. See also *proton-proton chain* and *triple-alpha process*.

carbon star A cool red giant or asymptotic giant branch star that has an excess of carbon in its atmosphere.

carbonaceous chondrite A primitive stony meteorite that contains chondrules and is rich in carbon and volatile materials.

Cassini Division The largest gap in Saturn's rings, discovered by Jean-Dominique Cassini in 1675.

catalyst A substance that permits or encourages chemical and nuclear reactions but does not change its own chemical or nuclear properties.

CCD See *charge-coupled device*.

celestial equator The imaginary great circle that is the projection of Earth's equator onto the celestial sphere.

celestial sphere An imaginary sphere with celestial objects on its inner surface and Earth at its center. The celestial sphere has no physical existence but is a convenient tool for picturing the directions in which celestial objects are seen from the surface of Earth.

Celsius (C) Also called *centigrade scale*. The arbitrary temperature scale—defined by Anders Celsius (1701–1744)—that defines 0°C as the freezing point of water and 100°C as the boiling point of water at sea level. Unit: °C. Compare *Fahrenheit* and *Kelvin temperature scale*.

center of mass 1. The location associated with an object system at which we may regard the entire mass of the system as being concentrated. 2. The point in any isolated system that moves according to Newton's first law of motion.

centigrade scale See *Celsius*.

centripetal force A force directed toward the center of curvature of an object's curved path.

Cepheid variable An evolved high-mass star with an atmosphere that is pulsating, leading to variability in the star's luminosity and color.

Chandrasekhar limit The upper limit on the mass of an object supported by electron degeneracy pressure; approximately 1.4 solar masses (M_{Sun}).

chaos Behavior in complex, interrelated systems in which tiny differences in the initial configuration of a system result in dramatic differences in the system's later evolution.

charge-coupled device (CCD) A common type of solid-state detector of electromagnetic radiation that transforms the intensity of light directly into electric signals.

chemistry The study of the composition, structure, and properties of substances.

chondrite A stony meteorite containing chondrules. Compare *achondrite*.

chondrule A small, crystallized, spherical inclusion of rapidly cooled molten droplets found inside some meteorites.

chromatic aberration A detrimental property of a lens in which rays of different wavelengths are brought to different focal distances from the lens.

chromosphere The region in the Sun's atmosphere located between the photosphere and the corona.

circular velocity (v_{circ}) The orbital velocity needed to keep an object moving in a circular orbit.

circumpolar Referring to the part of the sky, near either celestial pole, that can always be seen above the horizon from a specific location on Earth.

circumstellar disk See *protoplanetary disk*.

classical mechanics The science of applying Newton's laws to the motion of objects.

classical planets The eight major planets of the Solar System: Mercury, Venus, Earth, Mars, Jupiter, Saturn, Uranus, and Neptune.

climate The state of an atmosphere averaged over an extended time. Compare *weather*.

closed universe A finite universe with a curved spatial structure such that the sum of the angles of a triangle always exceeds 180°. Compare *flat universe* and *open universe*.

cloud 1. In planetary astronomy, a concentration of condensed volatiles in a planetary atmosphere. 2. In the interstellar medium, a volume of dust and gas that is denser than the surroundings.

CMB See *cosmic microwave background radiation*.

CNO cycle See *carbon-nitrogen-oxygen cycle*.

cold dark matter Particles of dark matter that move slowly enough to be gravitationally bound even in the smallest galaxies. Compare *hot dark matter*.

color index The color of a celestial object, generally a star, based on the ratio of its brightness in blue light (b_B) to its brightness in "visual" (or yellow-green) light (b_V); the difference between an object's blue (B) magnitude and visual (V) magnitude, $B - V$.

coma (pl. comae) The nearly spherical cloud of gas and dust surrounding the nucleus of an active comet.

comet A complex object consisting of a small, solid, icy nucleus; an atmospheric halo; and a tail of gas and dust.

comet nucleus See *nucleus [of a comet]*.

comparative planetology The study of planets through comparison of their chemical and physical properties.

complex system An interrelated system capable of exhibiting chaotic behavior. See also *chaos*.

composite volcano A large, cone-shaped volcano formed by viscous, pasty lava flows alternating with pyroclastic (explosively generated) rock deposits. Compare *shield volcano*.

compound lens A lens made up of two or more elements of differing refractive index, the purpose of which is to minimize chromatic aberration.

confusion A condition where data points are so numerous or densely packed that they are difficult to distinguish.

conservation law A physical law stating that the amount of a particular physical quantity (such as energy or angular momentum) of an isolated system does not change over time.

conservation of angular momentum The physical law stating that the amount of angular momentum of an isolated system does not change over time.

conservation of energy The physical law stating that the amount of energy of an isolated, closed system does not change over time.

constant of proportionality The multiplicative factor by which one quantity is related to another.

constellation An imaginary image formed by patterns of stars; any of 88 defined areas on the celestial sphere used by astronomers to locate celestial objects.

constructive interference A state in which the amplitudes of two intersecting waves reinforce one another. Compare *destructive interference*.

continental drift The slow motion (centimeters per year) of Earth's continents relative to each other and to Earth's mantle. See also *plate tectonics*.

continuous radiation Electromagnetic radiation with intensity that varies smoothly over a wide range of wavelengths.

convection The transport of thermal energy from the lower (hotter) to the higher (cooler) layers of a fluid by motions within the fluid, driven by variations in buoyancy.

convective zone A region within a star through which energy is transported outward by convection. Compare *radiative zone*.

conventional greenhouse effect The solar heating of air in an enclosed space, such as a closed building or car, resulting primarily from the inability of the hot air to escape.

core 1. The innermost region of a planetary interior. 2. The innermost part of a star.

core accretion A process for forming giant planets, whereby large quantities of surrounding hydrogen and helium are gravitationally captured onto a massive rocky core.

Coriolis effect The apparent displacement of objects in a direction perpendicular to their true motion as viewed from a rotating frame of reference. On a rotating planet, different latitudes rotating at different speeds cause this effect.

corona The hot, outermost part of the Sun's atmosphere.

coronal hole A low-density region in the solar corona containing "open" magnetic field lines along which coronal material is free to stream into interplanetary space.

coronal loop An arc of plasma in the Sun's lower corona resulting from loops in the magnetic field.

coronal mass ejection An eruption on the Sun that ejects hot gas and energetic particles at much higher speeds than are typical in the solar wind.

cosmic microwave background (CMB) radiation Isotropic microwave radiation, from every direction in the sky, that has a 2.73-kelvin (K) blackbody spectrum; residual radiation from the Big Bang.

cosmic ray A very fast-moving particle (usually an atomic nucleus); these particles fill the disk of our galaxy.

cosmological constant A constant, introduced into general relativity by Einstein, that characterizes an extra, repulsive force in the universe due to the vacuum of space itself.

cosmological principle The (testable) assumption that the same physical laws that apply here and now also apply everywhere and at all times, and that there are no special locations or directions in the universe.

cosmological redshift (z) The redshift that results from the expansion of the universe rather than from the motions of galaxies or gravity (see *gravitational redshift*).

cosmology The study of the large-scale structure and evolution of the universe as a whole.

Crab Nebula The remnant of the Type II supernova explosion witnessed by Chinese astronomers in A.D. 1054.

crescent Any phase of the Moon, Mercury, or Venus in which the object appears less than half illuminated by the Sun. Compare *gibbous*.

Cretaceous-Tertiary (K-T) boundary The boundary between the Cretaceous and Tertiary periods in Earth's history; which corresponds to the time of the impact of an asteroid or comet and the extinction of the dinosaurs.

critical density The value of the mass density of the universe that, when any cosmological constant is ignored, is just barely capable of halting expansion of the universe.

crust The relatively thin, outermost, hard layer of a planet, which is chemically distinct from the interior.

cryovolcanism Low-temperature volcanism in which the magmas are composed of molten ices rather than rocky material.

cyclone See *hurricane*.

cyclonic motion The rotation of a weather system resulting from the Coriolis effect as air moves toward a region of low atmospheric pressure. Compare *anticyclonic motion*.

Cygnus X-1 A binary X-ray source and probable black hole.

D

dark energy A form of energy that permeates all of space (including the vacuum), producing a repulsive force that increases the expansion rate of the universe.

dark matter Matter in galaxies that does not emit or absorb electromagnetic radiation; it is thought to constitute most of the mass in the universe. Compare *luminous matter*.

dark matter halo The centrally condensed, greatly extended dark matter component of a galaxy that contains up to 95 percent of the galaxy's mass.

daughter product An element resulting from radioactive decay of a more massive *parent element*.

day The time for Earth to rotate around its axis. A *sidereal day* is the time to rotate relative to the fixed stars (23 hours 56 minutes). A *solar day* is the time to rotate relative to the Sun (24 hours).

decay 1. The process of a radioactive nucleus changing into its daughter product. 2. The process of an atom or molecule dropping from a higher-energy state to a lower-energy state. 3. The process of a satellite's orbit losing energy.

deep In astronomy, far from Earth or distant.

density The measure of an object's mass per unit of volume. Units: kilograms per cubic meter (kg/m^3).

destructive interference A state in which the amplitudes of two intersecting waves cancel one another. Compare *constructive interference*.

differential rotation Rotation of different parts of a system at different rates.

differentiation The process by which materials of higher density sink toward the center of a molten or fluid planetary interior.

diffraction The spreading of a wave after it passes through an opening or past the edge of an object.

diffraction limit The limit of a telescope's angular resolution caused by the bending of light waves as they pass the sharp edge of the telescope's aperture.

diffuse ring A sparsely populated planetary ring spread out both horizontally and vertically.

direct imaging A technique for detecting exoplanets by observing them directly with telescopes.

dispersion The separation of rays of light into their component wavelengths.

distance ladder A sequence of techniques for measuring cosmic distances, where each method is calibrated by using the results from other methods that have been applied to closer objects.

Doppler effect The change in wavelength of sound or light that is due to the relative motion of the source toward or away from the observer.

Doppler redshift See *redshift*.

Doppler shift The amount by which a wavelength is shifted by the Doppler effect.

Drake equation A prescription for estimating the number of intelligent civilizations that may exist elsewhere.

dust devil A small tornado-like column of air containing dust or sand.

dust tail One type of comet tail, consisting of dust particles that are pushed away from the comet's head by radiation pressure from the Sun. Compare *ion tail*.

dwarf galaxy A small galaxy with a luminosity ranging from 1 million to 1 billion solar luminosities (L_{Sun}). Compare *giant galaxy*.

dwarf planet A body with characteristics similar to those of a classical planet except that it has not cleared smaller bodies from the neighboring regions around its orbit. Compare *planet* (definition 2).

dynamic equilibrium A state in which a system is constantly changing but its configuration remains the same because one source of change is exactly balanced by another source of change. Compare *static equilibrium*.

dynamo A device that converts mechanical energy into electric energy in the form of electric currents and magnetic fields. The "dynamo effect" is thought to create magnetic fields in planets and stars by electrically charged currents of material flowing within their cores.

E

eccentricity (*e*) A measure of the departure of an ellipse from circularity; the ratio of the distance between the two foci of an ellipse to its major axis.

eclipse 1. The total or partial obscuration of one celestial body by another. 2. The total or partial obscuration of light from one celestial body as it passes through the shadow of another celestial body.

eclipse season Any time during the year when the Moon's line of nodes is sufficiently close to the Sun for eclipses to occur.

eclipsing binary A pair of stars in which the orbital plane is oriented such that each star appears to pass in front of the other as seen from Earth. Compare *spectroscopic binary and visual binary*.

ecliptic 1. The apparent annual path of the Sun against the background of stars. 2. The projection of Earth's orbital plane onto the celestial sphere.

ecliptic plane The plane of Earth's orbit around the Sun; the projection of this plane onto the celestial sphere is called the ecliptic.

effective temperature The temperature at which a black body, such as a star, appears to radiate.

ejecta 1. Material thrown outward by the impact of an asteroid or comet on a planetary surface, leaving a crater behind. 2. Material thrown outward by a stellar explosion.

electric field A field that is able to exert a force on a charged object, whether at rest or moving. Compare *magnetic field*.

electric force The force exerted on a charged particle by an electric field. Compare *magnetic force*.

electromagnetic force The force, including both electric and magnetic forces, that acts on electrically charged particles; one of four fundamental forces of nature; the force mediated by photons.

electromagnetic radiation A traveling disturbance in the electric and magnetic fields caused by accelerating electric charges; in quantum mechanics, a stream of photons; light.

electromagnetic spectrum The spectrum made up of all possible frequencies or wavelengths of electromagnetic radiation, ranging from gamma rays through radio waves and including the portion our eyes can use.

electromagnetic wave A wave consisting of oscillations in electric-field strength and magnetic-field strength.

electron (*e*$^-$) A subatomic particle that has a negative electric charge of 1.6×10^{-19} coulomb (C), a rest mass of 9.1×10^{-31} kilogram (kg), and a rest energy of 8×10^{-14} joule (J). An electron is the antiparticle of a positron. Compare *proton and neutron*.

electron-degenerate Describing the state of material compressed to the point at which electron density reaches the limit imposed by the rules of quantum mechanics.

electroweak theory The quantum theory that combines descriptions of the electromagnetic force and the weak nuclear force.

element One of 92 naturally occurring substances (such as hydrogen, oxygen, and uranium) and more than 20 human-made ones (such as plutonium). Each element is chemically defined by the specific number of protons in the nuclei of its atoms.

elementary particle One of the basic building blocks of nature that is not known to have substructure, such as the *electron* and the *quark*.

ellipse A conic section produced by the intersection of a plane with a cone when the plane is passed through the cone at an angle to the axis other than 0° or 90°. This forms an oval around two points known as *foci*. For any point on the ellipse, the sum of the distances to the foci is constant. The long axis of the ellipse is called the major axis; half of this is the *semimajor axis*. The semimajor axis of an elliptical orbit is equal to the average distance of the orbiting body from the object at the focus; for example, the average distance of a planet from the Sun. The *eccentricity* of the ellipse describes how much the ellipse deviates from a circle and is found by dividing the distance between the foci by the length of the major axis.

elliptical galaxy A galaxy of Hubble type E, with a circular to elliptical outline on the sky, containing almost no disk and a population of old stars. Compare *irregular galaxy, S0 galaxy*, and *spiral galaxy*.

emission The release of electromagnetic energy when an atom, molecule, or particle drops from a higher-energy state to a lower-energy state. Compare *absorption*.

emission line An intensity peak in a spectrum that is due to sharply defined emission of electromagnetic radiation in a narrow range of wavelengths. Compare *absorption line*.

emission spectrum The spectrum of the light emitted from an object; it may contain emission lines. Compare *absorption spectrum*.

empirical Derived directly from observations or evidence, rather than from logic or theoretical inference.

empirical science Scientific investigation that is based primarily on observations and experimental data; it is descriptive rather than based on theoretical inference.

energy The conserved quantity that gives objects and systems the ability to do work. Units: joules (J).

energy transport The transfer of energy from one location to another. In stars, energy transport is carried out mainly by radiation or convection.

entropy A measure of the disorder of a system, related to the number of ways a system can be rearranged without its appearance being affected.

equator The imaginary great circle on the surface of a body midway between its poles that divides the body into northern and southern hemispheres. The equatorial plane passes through the center of the body and is perpendicular to its rotation axis. Compare *meridian*.

equilibrium The state of an object in which physical processes balance each other so that its properties or conditions remain constant.

equinox Literally, "equal night." 1. One of two positions on the ecliptic where it intersects the celestial equator. 2. Either of the two times of year (the *autumnal equinox* and *vernal equinox*) when the Sun is at one of these two positions. At this time, night and day are of the same length everywhere on Earth. Compare *solstice*.

equivalence principle The principle stating that there is no difference between a frame of reference that is freely floating through space and one that is freely falling within a gravitational field.

erosion The degradation of a planet's surface topography by the mechanical action of wind and/or water.

escape velocity (v_{esc}) The minimum velocity needed for an object to achieve a parabolic trajectory and thus permanently leave the gravitational grasp of another mass.

eternal inflation The idea that a universe might inflate forever. In such a universe, quantum effects could randomly cause regions to slow their expansion, eventually stop inflating, and experience an explosion resembling our Big Bang.

event Something that happens at a particular place in space at a particular time.

event horizon The effective "surface" of a black hole. Nothing inside this surface—not even light—can escape from the black hole.

evolutionary track The path that a star follows across the H-R diagram as it evolves through its lifetime.

excited state An energy level of a particular atom, molecule, or particle that is higher than its ground state. Compare *ground state*.

exoplanet Also called *extrasolar planet*. A planet orbiting a star other than the Sun.

extrasolar planet See *exoplanet*.

extremophiles Microbes that are adapted to tolerate extreme conditions, such as high or low temperatures.

F

F See *Fahrenheit*.

fact A piece of information; an observation or measurement.

Fahrenheit (F) The arbitrary temperature scale—defined by Daniel Gabriel Fahrenheit (1686–1736)—that defines 32°F as the melting point of water and 212°F as the boiling point of water at sea level. Unit: °F. Compare *Celsius* and *Kelvin temperature scale*.

falsified Proven wrong.

fault A fracture in the crust of a planet or moon along which blocks of material can slide.

filter An instrument component that transmits a limited wavelength range of electromagnetic radiation. For the optical range, such components are typically made of different kinds of glass and take on the hue of the light they transmit.

first quarter Moon The phase of the Moon in which only its western half, as viewed from Earth, is illuminated by the Sun. This phase occurs about 1 week after a new Moon. Compare *third quarter Moon*.

fissure A fracture in the planetary lithosphere from which magma emerges.

flat rotation curve A rotation curve of a spiral galaxy in which rotation rates do not decline in the outer part of the galaxy but remain relatively constant to the outermost points.

flat universe An infinite universe whose spatial structure obeys Euclidean geometry, such that the sum of the angles of a triangle always equals 180°. Compare *closed universe* and *open universe*.

flatness problem The surprising result that the sum of $\Omega_{mass} + \Omega_\Lambda$ is extremely close to unity in the present-day universe; equivalent to saying that it is surprising the universe is so close to being exactly flat.

fluid A liquid or a gas; a substance that flows to take on the shape of its container.

flux The total amount of energy passing through each square meter of a surface each second. Units: watts per square meter (W/m²).

flux tube A strong magnetic field contained within a tubelike structure. Flux tubes are found in the solar atmosphere and connecting Jupiter to its moon Io.

flyby A spacecraft that first approaches and then continues flying past a planet or moon. Flybys can visit multiple objects, but they remain in the vicinity of their targets only briefly. Compare *orbiter*.

focal length The optical distance between a telescope's objective lens or primary mirror and the plane (called the *focal plane*) on which the light from a distant object is focused.

focal plane The plane, perpendicular to the optical axis of a lens or mirror, in which an image is formed.

focus (pl. foci) 1. One of two points that define an ellipse. 2. A point in the focal plane of a telescope.

force A push or a pull on an object.

frame of reference A coordinate system within which an observer measures positions and motions.

free fall The motion of an object when the only force acting on it is gravity.

frequency The number of times per second that a periodic process occurs. Unit: hertz (Hz), 1/s.

full Moon The phase of the Moon in which its near side, as viewed from Earth, is fully illuminated by the Sun. This phase occurs about 2 weeks after a *new Moon*.

G

galaxy A gravitationally bound system that consists of stars and star clusters, gas, dust, and dark matter; typically greater than 1,000 light-years across and recognizable as a discrete, single object.

galaxy cluster A large, gravitationally bound collection of galaxies containing hundreds to thousands of members; typically about 10 million to 15 million light-years (3–5 Mpc) across. Compare *galaxy group* and *supercluster*.

galaxy group A small, gravitationally bound collection of galaxies containing several to a hundred members; typically 4 million to 6 million light-years across. Compare *galaxy cluster* and *supercluster*.

gamma rays Electromagnetic radiation with higher frequencies and photon energies and shorter wavelengths than all other types.

gamma-ray burst (GRB) An intense burst of gamma rays from a particular region of the sky. The systems that produce these bursts have not yet been confirmed, but it is thought that the short-period bursts come from a merger of two dense objects, whereas the long-period gamma-ray bursts are created in the collapse of a rapidly spinning object.

gas giant A giant planet formed mostly of hydrogen and helium. In our Solar System, Jupiter and Saturn are the gas giants. Compare *ice giant*.

gauss A basic unit of magnetic flux density.

general relativistic time dilation The verified prediction that time passes more slowly in a gravitational field than in the absence of a gravitational field. Compare *time dilation*.

general relativity See *general theory of relativity*.

general theory of relativity Sometimes referred to as simply *general relativity*. Einstein's theory explaining gravity as the distortion of spacetime by massive objects, such that particles travel on the shortest path between two events in spacetime. This theory deals with all types of motion. Compare *special theory of relativity*.

geocentric A coordinate system having the center of Earth as its origin. Compare *heliocentric*.

geodesic The path an object will follow through spacetime in the absence of external forces.

geometry A branch of mathematics that deals with points, lines, angles, and shapes.

giant galaxy A galaxy with luminosity greater than about 1 billion solar luminosities (L_{Sun}). Compare *dwarf galaxy*.

giant molecular cloud An interstellar cloud composed primarily of molecular gas and dust, having hundreds of thousands of solar masses.

giant planet One of the largest planets in the Solar System (Saturn, Jupiter, Uranus, or Neptune), typically 10 times the size and many times the mass of any *terrestrial planet* and lacking a solid surface.

gibbous Any phase of the Moon, Mercury, or Venus in which the object appears more than half illuminated by the Sun. Compare *crescent*.

global circulation The overall, planet-wide circulation pattern of a planet's atmosphere.

globular cluster A spherically symmetric, highly condensed group of stars that contains tens of thousands to a million members. Compare *open cluster*.

gluon The particle that carries (or, equivalently, mediates) interactions due to the strong nuclear force.

gossamer ring An extremely tenuous planetary ring found beyond Jupiter's main ring.

grand unified theory (GUT) A unified quantum theory that combines the strong nuclear, weak nuclear, and electromagnetic forces but does not include gravity.

granite Rock that is cooled from magma and is relatively rich in silicon and oxygen.

grating An optical surface containing many narrow, closely and equally spaced parallel grooves or slits that spectrally disperse reflected or transmitted light.

gravitational force Force due to the gravitational interaction between two or more objects.

gravitational lens A massive object that gravitationally focuses the light of a more distant object to produce multiple brighter, magnified, possibly distorted images.

gravitational lensing The bending of light by gravity; can be used to detect exoplanets.

gravitational potential energy The stored energy in an object that is due solely to its position within a gravitational field.

gravitational redshift The shifting to longer wavelengths of radiation from an object deep within a gravitational well.

gravitational wave A wave in the fabric of spacetime emitted by accelerating masses.

gravity 1. The mutually attractive force between massive objects. 2. An effect arising from the bending of spacetime by massive objects. 3. One of four fundamental forces of nature.

great circle Any circle on a sphere that has as its center the center of the sphere. The celestial equator, the meridian, and the ecliptic are all great circles on the sphere of the sky, as is any circle drawn through the zenith.

Great Red Spot The giant, oval, brick-red anticyclone seen in Jupiter's southern hemisphere.

greenhouse effect See *atmospheric greenhouse effect* and *conventional greenhouse effect*.

greenhouse gas One of a group of atmospheric gases, such as carbon dioxide, that are transparent to visible radiation but absorb infrared radiation.

Gregorian calendar The modern calendar: a modification of the Julian calendar decreed by Pope Gregory XIII in 1582. By this time, the less accurate Julian calendar had developed an error of 10 days over the 13 centuries since its inception.

ground state The lowest possible energy state for a system or part of a system, such as an atom, molecule, or particle. Compare *excited state*.

GUT See *grand unified theory*.

H

H II region A region of interstellar gas that has been ionized by ultraviolet radiation from nearby hot massive stars.

H-R diagram The Hertzsprung-Russell diagram, which is a plot of the luminosities versus the surface temperatures of stars. The evolving properties of stars are plotted as tracks across the H-R diagram.

habitable zone The distance from its star at which a planet must be located in order to have a temperature suitable for life; often assumed to be temperatures at which water exists in a liquid state.

Hadley circulation A simplified, and therefore uncommon, atmospheric global circulation that carries thermal energy directly from the equator to the polar regions of a planet.

half-life The time it takes for half a sample of a particular radioactive parent element to decay to a daughter product.

halo The spherically symmetric, low-density distribution of stars and dark matter that defines the outermost regions of a galaxy.

harmonic law See *Kepler's third law*.

Hawking radiation Radiation from a black hole.

Hayashi track The path that a protostar follows on the H-R diagram as it contracts toward the main sequence.

head The part of a comet that includes both the nucleus and the inner part of the coma.

heat death The possible eventual fate of an open universe, in which entropy has triumphed and all energy- and structure-producing processes have come to an end.

heavy element Also called *massive element*. 1. In astronomy, any element more massive than helium. 2. In other sciences (and sometimes also in astronomy), any of the most massive elements in the periodic table, such as uranium and plutonium.

Heisenberg uncertainty principle The physical limitation that the product of uncertainties of the position and the momentum of a particle cannot be smaller than a well-defined value, Planck's constant (h).

heliocentric A coordinate system having the center of the Sun as its origin. Compare *geocentric*.

helioseismology The use of solar oscillations to study the interior of the Sun.

helium flash The runaway explosive burning of helium in the degenerate helium core of a red giant star.

Herbig-Haro (HH) object A glowing, rapidly moving knot of gas and dust that is excited by bipolar outflows in very young stars.

heredity The process by which the characteristics of one generation of organisms are passed on to future generations.

hertz (Hz) A unit of frequency equivalent to cycles per second.

Hertzsprung-Russell diagram See *H-R diagram*.

HH object See *Herbig-Haro object*.

hierarchical clustering The "bottom-up" process of forming large-scale structure. Small-scale structure first produces groups of galaxies, which in turn form clusters, which then form superclusters.

high-mass star A star with a main-sequence mass greater than about 8 solar masses (M_{Sun}). Compare *low-mass star*.

high-velocity star A star belonging to the halo found near the Sun, distinguished from disk stars by moving far faster and often in the direction opposite to the rotation of the disk and its stars.

homogeneous In cosmology, describing a universe in which observers at any location would observe the same properties.

horizon The boundary that separates the sky from the ground.

horizon problem The puzzling observation that the cosmic background radiation is so uniform in all directions, despite the fact that widely separated regions should have been "over the horizon" from each other in the early universe.

horizontal branch A region on the H-R diagram defined by stars burning helium to carbon in a stable core.

hot dark matter Particles of dark matter that move so fast that gravity cannot confine them to the volume occupied by a galaxy's normal luminous matter. Compare *cold dark matter*.

hot Jupiter A large, Jupiter-like extrasolar planet located very close to its parent star.

hot spot A place where hot plumes of mantle material rise near the surface of a planet.

Hubble constant (H_0) The constant of proportionality relating the recession velocities of galaxies to their distances. See also *Hubble time*.

Hubble time An estimate of the age of the universe from the inverse of the Hubble constant, $1/H_0$.

Hubble's law The law stating that the speed at which a galaxy is moving away from us is proportional to the distance of that galaxy.

hurricane Also called *cyclone* or *typhoon*. A large tropical cyclonic system circulating counterclockwise in the Northern Hemisphere and clockwise in the Southern Hemisphere. Hurricanes can extend outward from their center to more than 600 kilometers (km) and generate winds in excess of 300 kilometers per hour (km/h).

hydrogen burning The release of energy from the nuclear fusion of four hydrogen atoms into a single helium atom.

hydrogen fusion The nuclear process of turning hydrogen into helium.

hydrogen shell burning The fusion of hydrogen in a shell surrounding a stellar core that may be either degenerate or fusing more massive elements.

hydrosphere The portion of Earth that is largely liquid water. Compare *biosphere* and *lithosphere*.

hydrostatic equilibrium The condition in which the weight bearing down at a particular point within an object is balanced by the pressure within the object.

hypothesis A well-thought-out idea, based on scientific principles and knowledge, that leads to testable predictions. Compare *theory*.

Hz See *hertz*.

I

ice The solid form of a volatile material; sometimes the *volatile material* itself, regardless of its physical form.

ice giant A giant planet formed mostly of the condensed form of volatile substances (ices). In our Solar System, Uranus and Neptune are the ice giants. Compare *gas giant*.

ideal gas law The relationship of pressure (P) to number density of particles (n) and temperature (T) expressed as $P = nkT$, where k is Boltzmann's constant.

igneous activity The formation and action of molten rock (magma).

impact A collision between two bodies.

impact crater The scar of the impact left on a solid planetary or moon surface by collision with another object. Compare *secondary crater*.

impact cratering A process involving collisions between solid planetary objects.

index of refraction (n) The ratio of the speed of light in a vacuum (c) to the speed of light in an optical medium (v): $n = c/v$.

inert Nonreactive.

inert gas A gaseous element that combines with other elements only under conditions of extreme temperature and pressure. Examples include helium, neon, and argon.

inertia The tendency for objects to retain their state of motion.

inertial frame of reference 1. A frame of reference that is not accelerating. 2. In general relativity, a frame of reference that is falling freely in a gravitational field.

infinite Limitless; extending without end.

inflation An extremely brief phase of ultra-rapid expansion of the very early universe. After inflation, the standard Big Bang models of expansion apply.

infrared (IR) radiation Electromagnetic radiation with frequencies, photon energies, and wavelengths between those of visible light and microwaves. Compare *ultraviolet radiation*.

instability strip A region of the H-R diagram containing stars that pulsate with a periodic variation in luminosity.

integration time The time interval during which photons are collected in a detecting device.

intensity (I) (of light) The amount of radiant energy emitted per second per unit area. Units: watts per square meter (W/m²).

intercloud gas A low-density region of the interstellar medium that fills the space between interstellar clouds.

interference The interaction of two sets of waves producing high and low intensity, depending on whether their amplitudes reinforce (*constructive interference*) or cancel (*destructive interference*).

interferometer Also called *interferometric array*. A group or array of separate but linked optical or radio telescopes whose overall separation determines the angular resolution of the system.

interferometric array See *interferometer*.

interstellar cloud A discrete, high-density region of the interstellar medium made up mostly of atomic or molecular hydrogen and dust.

interstellar dust Small particles or grains of matter (0.01–10 micrometers [μm]), primarily carbon and silicates, distributed throughout interstellar space.

interstellar extinction The dimming of visible and ultraviolet light by interstellar dust.

interstellar gas The tenuous gas, far less dense than air, comprising 99 percent of the matter in the interstellar medium.

interstellar medium The gas and dust that fill the space between the stars within a galaxy.

inverse square law The rule that a quantity or effect diminishes with the square of the distance from the source.

ion An atom or molecule that has lost or gained one or more electrons.

ion tail One type of comet tail, consisting of ionized gas; particles in the ion tail are pushed directly away from the comet's head in the antisunward direction at high speeds by the solar wind. Compare *dust tail*.

ionization The process by which electrons are stripped free from an atom or molecule, resulting in free electrons and a positively charged atom or molecule.

ionosphere A layer high in Earth's atmosphere in which most of the atoms are ionized by solar radiation.

IR See *infrared radiation*.

iron meteorite A meteorite composed primarily of iron-nickel alloys. Compare *stony-iron meteorite* and *stony meteorite*.

irregular galaxy A galaxy without regular or symmetric appearance. Compare *elliptical galaxy*, *S0 galaxy*, and *spiral galaxy*.

irregular moon A moon that has been captured by a planet. Some irregular moons revolve in the opposite direction from the rotation of the planet, and many are in distant, unstable orbits. Compare *regular moon*.

isotopes Forms of the same element with differing numbers of neutrons.

isotropic In cosmology, describing a universe where all observers measure the same properties, regardless of the direction in which they are looking.

J

J See *joule*.

jansky (Jy) The basic unit of flux density. Units: watts per square meter per hertz or (W/m²)/Hz.

jet 1. A stream of gas and dust ejected from a comet nucleus by solar heating. 2. A collimated linear feature of bright emission extending from a protostar or active galactic nucleus.

joule (J) A unit of energy or work. 1 J = 1 kg m²/s².

Jy See *jansky*.

K

K See *kelvin*.

K-T boundary See *Cretaceous-Tertiary boundary*.

KBO See *Kuiper Belt object*.

kelvin (K) The basic unit of the Kelvin scale of temperature.

Kelvin temperature scale The temperature scale—defined by William Thomson, better known as Lord Kelvin (1824–1907)—that uses Celsius-sized degrees but defines its zero point (that is, 0 K) as absolute zero instead of the melting point of water. Compare *Celsius* and *Fahrenheit*.

Kepler's first law A rule of planetary motion, inferred by Johannes Kepler, stating that planets move in orbits of elliptical shapes with the Sun at one focus.

Kepler's laws The three rules of planetary motion inferred by Johannes Kepler from the data acquired by Tycho Brahe.

Kepler's second law Also called *law of equal areas*. A rule of planetary motion, inferred by Johannes Kepler, stating that a line drawn from the Sun to a planet sweeps out equal areas in equal times as the planet orbits the Sun.

Kepler's third law Also called *harmonic law*. A rule of planetary motion inferred by Johannes Kepler that describes the relationship between the period of a planet's orbit and its distance from the Sun. The law states that the square of the period of a planet's orbit, measured in years, is equal to the cube of the semimajor axis of the planet's orbit, measured in astronomical units: $(P_{years})^2 = (A_{AU})^3$.

kinetic energy (E_K) The energy of an object that is due to its motions, expressed as $E_K = \frac{1}{2} mv^2$. Units: joules (J).

Kirkwood gap A gap in the main asteroid belt related to orbital resonances with Jupiter.

Kuiper Belt A disk-shaped population of comet nuclei extending from Neptune's orbit to perhaps several thousand astronomical units (AU) from the Sun. The highly populated innermost part of the Kuiper Belt has an outer edge approximately 50 AU from the Sun.

Kuiper Belt object (KBO) Also called *trans-Neptunian object*. An icy planetesimal (comet nucleus) that orbits within the Kuiper Belt beyond the orbit of Neptune.

L

Lagrangian equilibrium point One of five points of equilibrium in a system consisting of two massive objects in nearly circular orbit around a common center of mass. Only two Lagrangian points (L_4 and L_5) represent stable equilibrium. A third smaller body located at one of the five points will move in lockstep with the center of mass of the larger bodies.

lander An instrumented spacecraft designed to land on a planet or moon. Compare *rover*.

large-scale structure Observable aggregates on the largest scales in the universe, including galaxy groups, clusters, and superclusters.

latitude The angular distance north (+) or south (−) from the equatorial plane of a nearly spherical body.

law of equal areas See *Kepler's second law*.

law of gravitation See *universal law of gravitation*.

leap year A year that contains 366 days. Leap years occur every fourth year when the year is divisible by 4, correcting for the accumulated excess time in a normal year, which is approximately 365¼ days long.

length contraction The relativistic compression of moving objects in the direction of their motion.

Leonids A November meteor shower associated with the dust debris left by comet Tempel-Tuttle. Compare *Perseids*.

libration The apparent wobble of an orbiting body that is tidally locked to its companion (such as Earth's Moon) resulting from the fact that its orbit is elliptical rather than circular.

life A biochemical process in which living organisms can reproduce, evolve, and sustain themselves by drawing energy from their environment. All terrestrial life involves carbon-based chemistry, assisted by the self-replicating molecules ribonucleic acid (RNA) and deoxyribonucleic acid (DNA).

light All electromagnetic radiation, which comprises the entire electromagnetic spectrum.

light travel time How long it would take for light to travel a given distance.

light-day The distance that light travels in 1 day—about 26 billion kilometers (km).

light-year (ly) The distance that light travels in 1 year—about 9 trillion kilometers (km).

limb The outer edge of the visible disk of a planet, moon, or the Sun.

limb darkening The darker appearance caused by increased atmospheric absorption near the limb of a planet or star.

line of nodes 1. A line defined by the intersection of two orbital planes. 2. The line defined by the intersection of Earth's equatorial plane and the plane of the ecliptic.

lithosphere The solid, brittle part of Earth (or any planet or moon), including the crust and the upper part of the mantle. Compare *biosphere* and *hydrosphere*.

lithospheric plate A separate piece of Earth's lithosphere capable of moving independently. See also *continental drift* and *plate tectonics*.

Local Group The small group of galaxies of which the Milky Way and the Andromeda galaxies are members.

long-period comet A comet with an orbital period of greater than 200 years. Compare *short-period comet*.

longitudinal wave A wave that oscillates parallel to the direction of the wave's propagation. Compare *transverse wave*.

look-back time The time that it has taken for the light from an astronomical object to reach Earth.

low-mass star A star with a main-sequence mass of less than about 8 solar masses (M_{Sun}). Compare *high-mass star*.

luminosity The total flux emitted by an object. Unit: watts (W). See also *brightness*.

luminosity class A spectral classification based on stellar size, from the largest supergiants to the smallest white dwarfs.

luminosity-temperature-radius relationship A relationship among these three properties of stars, indicating that if any two are known, the third can be calculated.

luminous Shining.

luminous matter Also called *normal matter*. Matter in galaxies—including stars, gas, and dust—that emits electromagnetic radiation. Compare *dark matter*.

lunar eclipse An event that occurs when the Moon is partially or entirely in Earth's shadow. Compare *solar eclipse*.

lunar tide A tide on Earth that is due to the differential gravitational pull of the Moon. Compare *solar tide*. See also *tide* (definition 2).

ly See *light-year*.

M

M-type asteroid An asteroid that was once part of the metallic core of a larger, differentiated body that has since been broken into pieces; made mostly of iron and nickel. Compare *C-type asteroid* and *S-type asteroid*.

MACHO Literally, "massive compact halo object." These objects include the dark matter candidates small main-sequence stars, planets, brown dwarfs, white dwarfs, neutron stars, and black holes. Compare *WIMP*.

magma Molten rock, often containing dissolved gases and solid minerals.

magnetic field A field that is able to exert a force on a moving electric charge. Compare *electric field*.

magnetic force A force associated with, or caused by, the relative motion of charges. Compare *electric force*. See also *electromagnetic force*.

magnetosphere The region surrounding a planet that is filled with relatively intense magnetic fields and plasmas.

magnitude A system used by astronomers to describe the brightness or luminosity of stars. The brighter the star, the smaller its magnitude.

main asteroid belt See *asteroid belt*.

main sequence The strip on the H-R diagram where most stars are found. Main-sequence stars are fusing hydrogen to helium in their cores.

main-sequence lifetime The amount of time a star spends on the main sequence, fusing hydrogen into helium in its core.

main-sequence turnoff The location on the H-R diagram of a single-aged stellar population (such as a star cluster) where stars have just evolved off the main sequence. The position of the main-sequence turnoff is determined by the age of the stellar population.

mantle The solid portion of a rocky planet that lies between the crust and the core.

mare (pl. maria) A dark region on the Moon composed of basaltic lava flows.

mass 1. Inertial mass: the property of matter that resists changes in motion. 2. Gravitational mass: the property of matter defined by its attractive force on other objects. According to general relativity, the two are equivalent.

mass-luminosity relationship An empirical relationship between the luminosity (L) and mass (M) of main-sequence stars expressed as a power law: for example, $L \propto M^{3.5}$.

mass transfer The loss of mass from one member of a binary star system to its companion. Mass transfer occurs when one of the stars evolves to the point that it overfills its Roche lobe, so that its outer layers are pulled toward its binary companion.

massive Containing mass.

massive element See *heavy element*.

matter 1. Objects made of particles that have mass, such as protons, neutrons, and electrons. 2. Anything that occupies space and has mass.

Maunder minimum The time period from 1645 to 1715, when very few sunspots were observed.

medium The material through which a wave travels.

megabar A unit of pressure equal to 1 million bars.

meridian The imaginary arc in the sky that runs from the horizon at due north through the zenith to the horizon at due south. The meridian divides the observer's sky into eastern and western halves. Compare *equator*.

mesosphere The layer of Earth's atmosphere immediately above the *stratosphere*, extending from an altitude of 50 kilometers (km) to about 90 km.

meteor The incandescent trail produced by a small piece of interplanetary debris as it travels through the atmosphere at very high speeds. Compare *meteorite* and *meteoroid*.

meteor shower A larger-than-normal display of meteors that occurs when Earth passes through the orbit of a disintegrating comet and sweeps up some of its debris.

meteorite A meteoroid that survives to reach a planet's surface. Compare *meteor* and *meteoroid*.

meteoroid A small cometary or asteroid fragment ranging in size from 100 micrometers (μm) to 100 meters. When it enters a planetary atmosphere, the meteoroid creates a meteor, which is an atmospheric phenomenon. Compare *meteor* and *meteorite*. See also *planetesimal* and *zodiacal dust*.

microlensing Gravitational lensing by relatively small objects like planets (rather than galaxies).

micrometer (μm) Also called *micron*. 10^{-6} meter; a unit of length used for the wavelength of electromagnetic radiation. Compare *nanometer*.

micron See *micrometer*.

microwave radiation Electromagnetic radiation with frequencies, photon energies, and wavelengths between those of infrared radiation and radio waves.

Milky Way Galaxy The galaxy in which our Sun and Solar System reside.

minor planet See *asteroid*.

minute of arc See *arcminute*.

μm See *micrometer*.

model 1. A representation (often mathematical) of objects and the interaction between them. 2. In computing, a simulation to reproduce the behavior of a system in one, two, or three dimensions.

modern physics Usually, the physical principles, including relativity and quantum mechanics, that have been developed since James Maxwell's equations were published.

molecular cloud An interstellar cloud composed primarily of molecular hydrogen.

molecular-cloud core A dense clump within a molecular cloud that forms as the cloud collapses and fragments. Protostars form from molecular-cloud cores.

molecule Generally, the smallest particle of a substance that retains its chemical properties and is composed of two or more atoms. A very few types of molecules, such as helium, are composed of single atoms.

momentum The product of the mass and velocity of a particle. Units: kilograms times meters per second (kg m/s).

moon A less massive satellite orbiting a more massive object. Moons are found around planets, dwarf planets, asteroids, and Kuiper Belt objects.

multiverse A collection of parallel universes that together comprise all that exists.

mutation In biology, an imperfect reproduction of self-replicating material.

N

N See *newton*.

nadir The point on the celestial sphere located directly below an observer, opposite the *zenith*.

nanometer (nm) One billionth (10^{-9}) of a meter; a unit of length used for the wavelength of light. Compare *micrometer*.

natural selection The process by which forms of structure—ranging from molecules to whole organisms—that are best adapted to their environments become more common than less-well-adapted forms.

NCP See *north celestial pole*.

neap tide An especially weak tide that occurs near the time of a first or third quarter Moon, when lunar tides and solar tides are at right angles and partially cancel each other. Compare *spring tide*. See also *tide* (definition 2).

near-Earth asteroid An asteroid whose orbit brings it close to the orbit of Earth. See also *near-Earth object*.

near-Earth object (NEO) An asteroid, comet, or large meteoroid whose orbit intersects Earth's orbit.

nebula (pl. nebulae) A cloud of interstellar gas and dust, either illuminated by stars (a bright nebula) or seen in silhouette against a brighter background (a dark nebula).

nebular hypothesis The first plausible theory of the formation of the Solar System, proposed by Immanuel Kant in 1734. Kant hypothesized that the Solar System formed from the collapse of an interstellar cloud of rotating gas.

NEO See *near-Earth object*.

neutrino A very low mass, electrically neutral particle emitted during beta decay. Neutrinos interact with matter only very feebly and so can penetrate through great quantities of matter.

neutrino cooling The process in which thermal energy is carried out of the center of a star by neutrinos rather than by electromagnetic radiation or convection.

neutron A subatomic particle that has no net electric charge and a rest mass and rest energy nearly equal to those of a proton. Compare *electron* and *proton*.

neutron star The neutron-degenerate remnant left behind by a Type II supernova.

new Moon The phase of the Moon in which it is between Earth and the Sun, and from Earth we see only the side of the Moon not being illuminated by the Sun. Compare *full Moon*.

newton (N) The force required to accelerate a 1-kilogram (kg) mass at a rate of 1 meter per second per second (m/s^2). Units: kilograms times meters per second squared (kg m/s^2).

Newton's first law of motion The law, formulated by Isaac Newton, stating that an object will remain at rest or will continue moving along a straight line at a constant speed until an unbalanced force acts on it.

Newton's laws See *Newton's first law of motion, Newton's second law of motion*, and *Newton's third law of motion*.

Newton's second law of motion The law, formulated by Isaac Newton, stating that if an unbalanced force acts on a body, the body will have an acceleration proportional to the unbalanced force and inversely proportional to the object's mass: $a = F/m$. The acceleration will be in the direction of the unbalanced force.

Newton's third law of motion The law, formulated by Isaac Newton, stating that for every force there is an equal and opposite force.

nm See *nanometer*.

normal matter See *luminous matter*.

north celestial pole (NCP) The northward projection of Earth's rotation axis onto the celestial sphere. Compare *south celestial pole*.

North Pole The location in the Northern Hemisphere where Earth's rotation axis intersects the surface of Earth. Compare *South Pole*.

nova (pl. novae) A stellar explosion that results from runaway nuclear fusion in a layer of material on the surface of a white dwarf in a binary system.

nuclear burning Release of energy by fusion of low-mass elements.

nuclear fusion The combination of two less massive atomic nuclei into a single more massive atomic nucleus.

nucleosynthesis The formation of more massive atomic nuclei from less massive nuclei, either in the Big Bang (Big Bang nucleosynthesis) or in the interiors of stars (stellar nucleosynthesis).

nucleus (pl. nuclei) 1. The dense, central part of an atom. 2. The central core of a galaxy, comet, or other diffuse object.

nucleus [of a comet] A primitive planetesimal composed of ices and refractory materials that has survived planetary accretion; the "heart" of a comet, containing nearly the entire mass of the comet; a "dirty snowball."

O

objective lens The primary optical element in a telescope or camera that produces an image of an object.

oblate Flattened, as of an otherwise spherical planet or star caused by its rapid rotation. The radius is larger at the equator than at the poles.

obliquity The inclination of a celestial body's equator to its orbital plane.

observable universe The part of the universe that can be observed, because light has had time to travel to Earth.

observational uncertainty The fact that real measurements are never perfect; all observations are uncertain by some amount.

Occam's razor The principle that the simplest hypothesis is the most likely; named after William of Occam (ca. 1285–1349), the medieval English cleric to whom the idea is attributed.

ocean A vast expanse of liquid, not necessarily water.

Oort Cloud A spherical distribution of comet nuclei stretching from beyond the Kuiper Belt to more than 50,000 astronomical units (AU) from the Sun.

opacity A measure of how effectively a material blocks the electromagnetic radiation going through it.

open cluster A loosely bound group of a few dozen to a few thousand stars that formed together in the disk of a spiral galaxy. Compare *globular cluster*.

open universe An infinite universe with a negatively curved spatial structure (much like the surface of a saddle) such that the sum of the angles of a triangle is always less than 180°. Compare *closed universe* and *flat universe*.

orbit The path taken by one object moving around another object under the influence of their mutual gravitational or electrical attraction.

orbital resonance A situation in which the orbital periods of two objects are related by a ratio of small integers.

orbiter A spacecraft that is placed in orbit around a planet or moon. Compare *flyby*.

organic Describing a substance, not necessarily of biological origin, that contains the element carbon.

P

P wave See *primary wave*.

pair production The creation of a particle-antiparticle pair from a source of electromagnetic energy.

paleomagnetism The record of Earth's magnetic field as preserved in rocks.

parallax 1. The apparent shift in the position of one object relative to another object, caused by the changing perspective of the observer. 2. In astronomy, the displacement in the apparent position of a nearby star caused by the changing location of Earth in its orbit.

parent element A radioactive element that decays to form more stable *daughter products*.

parsec (pc) The distance to a star with a parallax of 1 arcsecond, using a base of 1 astronomical unit (AU). One parsec is approximately 3.26 light-years.

partial lunar eclipse An event that occurs when the Moon passes through the penumbra of Earth's shadow. Compare *total lunar eclipse*.

partial solar eclipse An event that occurs when Earth passes through the penumbra of the Moon's shadow, so that the Moon blocks only a portion of the Sun's disk. Compare *annular solar eclipse* and *total solar eclipse*.

pc See *parsec*.

peculiar velocity The motion of a galaxy relative to the overall expansion of the universe.

penumbra (pl. penumbrae) 1. The outer part of a shadow, where the source of light is only partially blocked. 2. The region surrounding the *umbra* of a sunspot. The penumbra is cooler and darker than the surrounding surface of the Sun but not as cool or dark as the umbra of the sunspot. Compare *umbra*.

perihelion (pl. perihelia) The point in a solar orbit that is closest to the Sun. Compare *aphelion*.

period The time it takes for a regularly repetitive process to complete one cycle.

period-luminosity relationship The relationship between the period of variability of a pulsating variable star, such as a Cepheid or RR Lyrae variable, and the luminosity of the star. Longer-period Cepheid or RR Lyrae variables are more luminous than their shorter-period cousins.

Perseids A prominent August meteor shower associated with the dust debris left by comet Swift-Tuttle. Compare *Leonids*.

perturb Move an astronomical object from its undisturbed path or location.

phase One of the various appearances of the sunlit surface of the Moon or a planet caused by the change in viewing location of Earth relative to both the Sun and the object. Examples include *crescent* and *gibbous* phases.

photino An elementary particle related to the photon. One of the leading candidates for cold dark matter.

photochemical reaction A chemical reaction driven by the absorption of electromagnetic radiation.

photodissociation Breaking apart of molecules into smaller fragments or individual atoms by the action of photons.

photoelectric effect An effect whereby electrons are emitted from a substance illuminated by photons above a certain critical frequency.

photometry The process of measuring the brightness of a source of light, generally over a specific range of wavelength.

photon Also called *quantum of light*. A discrete unit or particle of electromagnetic radiation. The energy of a photon is equal to Planck's constant (h) multiplied by the frequency (f) of its electromagnetic radiation: $E_{photon} = h \times f$. The photon is the particle that mediates the electromagnetic force.

photosphere The apparent surface of the Sun as seen in visible light.

physical law A broad statement that predicts a particular aspect of how the physical universe behaves and that is supported by many empirical tests. See also *theory*.

pixel The smallest picture element in a digital image array.

Planck era The early time, just after the Big Bang, when the universe as a whole must be described with quantum mechanics.

Planck spectrum See *blackbody spectrum*.

Planck's constant (h) The constant of proportionality between the energy of a photon and the frequency of the photon. This constant determines how much energy a single photon of a given frequency or wavelength has. Value: $h = 6.63 \times 10^{-34}$ joule-second.

planet 1. A large body that orbits the Sun or another star and shines only by light reflected from the Sun or star. 2. In the Solar System, a body that orbits the Sun, has sufficient mass for self-gravity to overcome rigid-body forces so that it assumes a spherical shape, and has cleared smaller bodies from the neighborhood around its orbit. Compare *dwarf planet*.

planet migration The theory that a planet can move from its formation distance around its parent star to a different distance though gravitational interactions with other bodies or loss of orbital energy from interaction with gas in the protoplanetary disk.

planetary nebula The expanding shell of material ejected by a dying asymptotic giant branch star. A planetary nebula glows from fluorescence caused by intense ultraviolet light coming from the hot, stellar remnant at its center.

planetary system A system of planets and other smaller objects in orbit around a star.

planetesimal A primitive body of rock and ice, 100 meters or more in diameter, that combines with others to form a planet. Compare *meteoroid* and *zodiacal dust*.

plasma A gas that is composed largely of charged particles but also may include some neutral atoms.

plate tectonics The geological theory concerning the motions of lithospheric plates, which in turn provides the theoretical basis for continental drift.

positron A positively charged subatomic particle; the antiparticle of the *electron*.

potential energy Energy that is stored in a system.

power The rate at which work is done or at which energy is delivered. Unit: watts (W) or joules per second (J/s).

precession of the equinoxes The slow change in orientation between the ecliptic plane and the celestial equator caused by the wobbling of Earth's axis.

pressure Force per unit area. Units: newtons per square meter (N/m^2) or bars.

primary atmosphere An atmosphere, composed mostly of hydrogen and helium, that forms at the same time as its host planet. Compare *secondary atmosphere*.

primary mirror The principal optical mirror in a reflecting telescope. The primary mirror determines the telescope's light-gathering power and resolution. Compare *secondary mirror*.

primary wave Also called *P wave*. A longitudinal seismic wave, in which the oscillations involve compression and decompression parallel to the direction of travel (that is, a pressure wave). Compare *secondary wave*.

principle A general idea or sense about how the universe is that guides us in constructing new scientific theories. Principles can be testable theories.

prograde motion 1. Rotational or orbital motion of a moon that is in the same sense as the planet it orbits. 2. The counterclockwise orbital motion of Solar System objects as seen from above Earth's orbital plane. Compare *retrograde motion*.

prominence An archlike projection above the solar photosphere often associated with a sunspot.

proportional Describing two things whose ratio is a constant.

proton (p or p^+) A subatomic particle that has a positive electric charge of 1.6×10^{-19} coulomb (C), a rest mass of 1.67×10^{-27} kilogram (kg), and a rest energy of 1.5×10^{-10} joule (J). Compare *electron* and *neutron*.

proton-proton chain One of the ways in which hydrogen burning can take place; the most important path for hydrogen burning in low-mass stars such as the Sun. See also *carbon-nitrogen-oxygen cycle* and *triple-alpha process*.

protoplanetary disk Also called *circumstellar disk*. The remains of the accretion disk around a young star, from which a planetary system may form.

protostar A young stellar object that derives its luminosity from the conversion of gravitational energy to thermal energy rather than from nuclear reactions in its core.

pulsar A rapidly rotating neutron star that beams radiation into space in two searchlight-like beams. To a distant observer, the star appears to flash on and off or "pulse," earning its name.

pulsating variable star A variable star that undergoes periodic radial pulsations.

Q

QCD See *quantum chromodynamics*.

QED See *quantum electrodynamics*.

quantized Describing a quantity that exists as discrete, irreducible units.

quantum chromodynamics (QCD) The quantum mechanical theory describing the strong nuclear force and its mediation by gluons. Compare *quantum electrodynamics*.

quantum efficiency The fraction of photons falling on a detector that actually produces a response in the detector.

quantum electrodynamics (QED) The quantum theory describing the electromagnetic force and its mediation by photons. Compare *quantum chromodynamics*.

quantum mechanics The branch of physics that deals with the quantized and probabilistic behavior of atoms and subatomic particles.

quantum of light See *photon*.

quark The building block of protons and neutrons.

quasar Short for *"quasi-stellar radio source."* The most luminous of the active galactic nuclei, seen only at great distances from our galaxy.

R

radial velocity The component of velocity that is directed toward or away from the observer.

radian The angle at the center of a circle subtended by an arc equal to the length of the circle's radius. Therefore, 2π radians equals $360°$, and 1 radian equals approximately $57.3°$.

radiant The direction in the sky from which the meteors in a meteor shower seem to come.

radiation 1. Energy that has been emitted as particles or waves. 2. Light of all wavelengths.

radiation belt A toroidal ring of high-energy particles surrounding a planet.

radiative transfer The transport of energy from one location to another by electromagnetic radiation.

radiative zone A region within a star through which energy is transported outward by radiation. Compare *convective zone*.

radio galaxy A type of elliptical galaxy with an active galactic nucleus at its center that has very strong emission (10^{35} to 10^{38} watts [W]) in the radio portion of the electromagnetic spectrum. Compare *Seyfert galaxy*.

radio telescope An instrument for detecting and measuring radio frequency emissions from celestial sources.

radio waves Electromagnetic radiation with lower frequencies and photon energies and longer wavelengths than all other types.

radioisotope A radioactive element.

radiometric dating Use of radioactive decay to measure the ages of materials such as minerals.

ratio The relationship in quantity or size between two or more things.

ray 1. A beam of electromagnetic radiation. 2. A bright streak emanating from a young impact crater.

recession Moving away, receding.

recombination 1. The combining of positive ions and electrons to form neutral atoms. 2. An event early in the evolution of the universe in which hydrogen and helium nuclei combined with electrons to form neutral atoms. The removal of electrons caused the universe to become transparent to electromagnetic radiation.

red giant A low-mass star that has evolved beyond the main sequence and is now fusing hydrogen in a shell surrounding a degenerate helium core.

red giant branch A region on the H-R diagram defined by low-mass stars evolving from the main sequence toward the horizontal branch.

reddening The effect by which stars and other objects, when viewed through interstellar dust, appear redder than they actually are, caused by the fact that blue light is more strongly absorbed and scattered than red light.

redshift Also called *Doppler redshift*. The shift toward longer wavelengths of light by any of several effects, including Doppler shift, gravitational redshift, or cosmological redshift. Compare *blueshift*.

reflecting telescope A type of telescope that uses mirrors to collect and focus incoming electromagnetic radiation and form an image in its focal plane. The size of a reflecting telescope is defined by the diameter of the primary mirror. Compare *refracting telescope*.

reflection The redirection of a beam of light that is incident on, but does not cross, the surface between two media having different refractive indices. If the surface is flat and smooth, the angle of incidence equals the angle of reflection. Compare *refraction*.

refracting telescope A type of telescope that uses objective lenses to collect and focus light. Compare *reflecting telescope*.

refraction The redirection or bending of a beam of light when it crosses the boundary between two media having different refractive indices. Compare *reflection*.

refractory material Material that remains solid at high temperatures. Compare *volatile material*.

regular moon A moon that formed together with the planet it orbits. Compare *irregular moon*.

reionization The process of ultraviolet photons stripping electrons from neutral hydrogen in the early universe after the Big Bang.

relative humidity The amount of water vapor held by a volume of air at a given temperature (stated as a percentage) compared to the total amount of water that could be held by the same volume of air at the same temperature.

relative motion The difference in motion between two individual frames of reference.

relativistic Describing physical processes that take place in systems traveling at nearly the speed of light or located in the vicinity of very strong gravitational fields.

relativistic beaming The effect created when material moving at nearly the speed of light beams the radiation it emits in the direction of its motion.

remote sensing The use of images, spectra, radar, or other techniques to measure the properties of an object from a distance.

resolution The ability of a telescope to separate two point sources of light. Resolution is determined by the telescope's aperture and the wavelength of light it receives.

rest wavelength The wavelength of light we see coming from an object at rest with respect to the observer.

retrograde motion 1. Rotation or orbital motion of a moon that is in the opposite sense to the rotation of the planet it orbits. 2. The clockwise orbital motion of Solar System objects as seen from above Earth's orbital plane. Compare *prograde motion*.

revolve In astronomy, this indicates that one object orbits another.

ring An aggregation of small particles orbiting a planet or star. The rings of the four giant planets of the Solar System are composed variously of silicates, organic materials, and ices.

ring arc A discontinuous, higher-density region within an otherwise continuous, narrow ring.

ringlet A narrowly confined concentration of ring particles.

Roche limit The distance at which a planet's tidal forces exceed the self-gravity of a smaller object, such as a moon, asteroid, or comet, causing the object to break apart.

Roche lobe The hourglass-shaped or figure eight–shaped volume of space surrounding two stars, which constrains material that is gravitationally bound by one or the other.

rotate In astronomy, this indicates that an object spins about an axis.

rotation curve A plot showing how the orbital velocity of stars and gas in a galaxy changes with radial distance from the galaxy's center.

rover A remotely controlled instrumented vehicle designed to traverse and explore the surface of a terrestrial planet or moon. Compare *lander*.

RR Lyrae variable A variable giant star whose regularly timed pulsations are good predictors of its luminosity. RR Lyrae stars are used for distance measurements to globular clusters.

S

S-type asteroid An asteroid made of material that has been modified from its original state, likely as the outer part of a larger, differentiated body that has since broken into pieces. Compare *C-type asteroid* and *M-type asteroid*.

S wave See *secondary wave*.

S0 galaxy A galaxy with a bulge and a disk-like spiral but a smooth elliptical appearance. Compare *elliptical galaxy*, *irregular galaxy*, and *spiral galaxy*.

satellite 1. An object in orbit around a more massive body. 2. A moon.

scale factor (R_U) A dimensionless number proportional to the distance between two points in space. The scale factor increases as the universe expands.

scattering The random change in the direction of travel of photons, caused by their interactions with molecules or dust particles.

Schwarzschild radius The distance from the center of a nonrotating, spherical black hole at which the escape velocity equals the speed of light.

scientific method The formal procedure—including hypothesis, prediction, and experiment or observation—used to test (attempt to falsify) the validity of scientific hypotheses and theories.

scientific notation The standard expression of numbers with one digit (which can be zero) to the left of the decimal point and multiplied by 10 to the exponent required to give the number its correct value. Example: $2.99 \times 10^8 = 299{,}000{,}000$.

SCP See *south celestial pole*.

second law of thermodynamics The law stating that the entropy or disorder of an isolated system always increases as the system evolves.

second of arc See *arcsecond*.

secondary atmosphere An atmosphere that formed—as a result of volcanism, comet impacts, or another process—sometime after its host planet formed. Compare *primary atmosphere*.

secondary crater A crater formed from ejected material thrown from an *impact crater*.

secondary mirror A small mirror placed on the optical axis of a reflecting telescope that returns the beam back through a small hole in the *primary mirror*, thereby shortening the mechanical length of the telescope.

secondary wave Also called *S wave*. A transverse seismic wave, which involves the sideways motion of material. Compare *primary wave*.

seeing The reduction in quality of observed images resulting from atmospheric distortions.

seismic wave A vibration due to an earthquake, a large explosion, or an impact on the surface that travels through a planet's interior.

seismometer An instrument that measures the amplitude and frequency of seismic waves.

self-gravity The gravitational attraction among all the parts of the same object.

semimajor axis Half of the longer axis of an ellipse.

SETI The Search for Extraterrestrial Intelligence project, which uses advanced technology combined with radio telescopes to search for evidence of intelligent life elsewhere in the universe.

Seyfert galaxy A type of spiral galaxy with an active galactic nucleus at its center; first discovered in 1943 by Carl Seyfert. Compare *radio galaxy*.

shepherd moon A moon that orbits a planet close to rings and gravitationally confines the orbits of the ring particles.

shield volcano A volcano formed by highly fluid lava flowing from a single source and spreading out from that source. Compare *composite volcano*.

short-period comet A comet with an orbital period of less than 200 years. Compare *long-period comet*.

sidereal day The length of time Earth takes to rotate to the same position relative to the stars: 23 hours 56 minutes. Compare *solar day*.

sidereal period An object's orbital or rotational period measured with respect to the stars. Compare *synodic period*.

silicate One of the family of minerals composed of silicon and oxygen in combination with other elements.

singularity The point where a mathematical expression or equation becomes meaningless, such as the denominator of a fraction approaching zero. See also *black hole*.

solar abundance The relative amount of an element detected in the atmosphere of the Sun, expressed as the ratio of the number of atoms of that element to the number of hydrogen atoms.

solar day The 24-hour period of Earth's axial rotation that brings the Sun back to the same local meridian where the rotation started. Compare *sidereal day*.

solar eclipse An event that occurs when the Sun is partially or entirely blocked by the Moon. Compare *lunar eclipse*.

solar flare An explosion on the Sun's surface associated with complex sunspot groups and strong magnetic fields.

solar maximum (pl. maxima) The time, occurring about every 11 years, when the Sun is at its peak activity, meaning that sunspot activity and related phenomena (such as prominences, flares, and coronal mass ejections) are at their peak.

solar neutrino problem The historical observation that only about a third as many neutrinos as predicted by theory seemed to be coming from the Sun.

Solar System The gravitationally bound system made up of the Sun, planets, dwarf planets, moons, asteroids, comets, and Kuiper Belt objects, along with their associated gas and dust.

solar tide A tide on Earth that is due to the differential gravitational pull of the Sun. Compare *lunar tide*. See also *tide* (definition 2).

solar wind The stream of charged particles emitted by the Sun that flows at high speeds through interplanetary space.

solstice Literally, "Sun standing still." 1. Either of the two most northerly and southerly points on the ecliptic. 2. Either of the two times of year (the *summer solstice* and *winter solstice*) when the Sun is at one of these two positions. Compare *equinox*.

south celestial pole (SCP) The southward projection of Earth's rotation axis onto the celestial sphere. Compare *north celestial pole*.

South Pole The location in the Southern Hemisphere where Earth's rotation axis intersects the surface of Earth. Compare *North Pole*.

spacetime The four-dimensional continuum in which we live, that we experience as three spatial dimensions plus time.

special relativity See *special theory of relativity*.

special theory of relativity Sometimes referred to as simply *special relativity*. Einstein's theory explaining how the fact that the speed of light is a constant affects nonaccelerating frames of reference. Compare *general theory of relativity*.

spectral type A classification system for stars that is based on the presence and relative strength of absorption lines in their spectra. Spectral type is related to the surface temperature of a star.

spectrograph A device that spreads out the light from an object into its component wavelengths. See also *spectrometer*.

spectrometer A device that spreads out the light from an object into its component wavelengths, in which the spectrum is generally recorded digitally by electronic means.

spectroscopic binary A pair of stars whose existence and properties are revealed only by the Doppler shifts of their spectral lines. Most spectroscopic binaries are close pairs. Compare *eclipsing binary* and *visual binary*.

spectroscopic parallax Use of the spectroscopically determined luminosity and the observed brightness of a star to determine the star's distance.

spectroscopic radial velocity method A technique for detecting exoplanets by examining Doppler shifts in the light from stars.

spectroscopy The study of electromagnetic radiation from an object in terms of its component wavelengths.

spectrum (pl. spectra) 1. The intensity of electromagnetic radiation as a function of wavelength. 2. Waves sorted by wavelength.

speed The rate of change of an object's position with time, without regard to the direction of movement. Units: meters per second (m/s) or kilometers per hour (km/h). Compare *velocity*.

spherically symmetric Describing an object whose properties depend only on distance from the object's center, so that the object has the same form viewed from any direction.

spin-orbit resonance A relationship between the orbital and rotation periods of an object such that the ratio of their periods can be expressed by simple integers.

spiral density wave A stable, spiral-shaped change in the local gravity of a galactic disk that can be produced by periodic gravitational kicks from neighboring galaxies or from nonspherical bulges and bars in spiral galaxies.

spiral galaxy A galaxy of Hubble type S, with a discernible disk in which large spiral patterns exist. Compare *elliptical galaxy*, *irregular galaxy*, and *S0 galaxy*.

spoke One of several narrow radial features seen occasionally in Saturn's B Ring. Spokes appear dark in backscattered light and bright in forward-scattering light, indicating that they are composed of tiny particles. Their origin is not well understood.

sporadic meteor A meteor that is not associated with a specific meteor shower.

spreading center A zone from which two tectonic plates diverge.

spring tide An especially strong tide that occurs near the time of a new or full Moon, when lunar tides and solar tides reinforce each other. Compare *neap tide*. See also *tide* (definition 2).

stable equilibrium A state in which the system returns to its former condition after a small disturbance. Compare *unstable equilibrium*.

standard candle An object whose luminosity either is known or can be predicted in a distance-independent way, so its brightness can be used to determine its distance via the inverse square law of radiation.

standard model The theory of particle physics that combines electroweak theory with quantum mechanical theory of the strong nuclear force to describe the structure of known forms of matter.

star A luminous ball of gas that is held together by gravity and normally powered by nuclear reactions in its interior.

star cluster A group of stars that all formed at the same time and in the same general location.

static equilibrium A state in which the forces within a system are all in balance so that the system does not change. Compare *dynamic equilibrium*.

Stefan-Boltzmann constant (σ) The proportionality constant that relates the flux emitted by an object to the fourth power of its absolute temperature. Value: 5.67×10^{-8} W/(m^2 K^4) (W = watts, m = meters, K = kelvins).

Stefan-Boltzmann law The law stating that the amount of electromagnetic energy emitted from the surface of a body, summed over the energies of all photons of all wavelengths emitted, is proportional to the fourth power of the temperature of the body.

stellar mass loss The loss of mass from the outermost parts of a star's atmosphere during the course of its evolution.

stellar occultation An event in which a planet or other Solar System body moves between the observer and a star, eclipsing the light emitted by that star.

stellar population A group of stars with similar ages, chemical compositions, and dynamic properties.

stereoscopic vision The way the brains of humans and some animals combine the different information from two eyes to perceive the distances to surrounding objects.

stony-iron meteorite A meteorite composed of a mixture of silicate minerals and iron-nickel alloys. Compare *iron meteorite* and *stony meteorite*.

stony meteorite A meteorite composed primarily of silicate minerals, similar to those found on Earth. Compare *iron meteorite* and *stony-iron meteorite*.

stratosphere The atmospheric layer immediately above the *troposphere*. On Earth it extends upward to an altitude of 50 kilometers (km).

strength 1. Magnitude of a force. 2. In spectroscopy, brightness of an emission line or depth of an absorption line.

string theory The theory that conceives of particles as strings in 10 dimensions of space and time; the current contender for a theory of everything.

strong nuclear force The attractive short-range force between protons and neutrons that holds atomic nuclei together. One of the four fundamental forces of nature. Compare *weak nuclear force*.

subduction zone A region where two tectonic plates converge, with one plate sliding under the other and being drawn downward into the interior.

subgiant A giant star smaller and lower in luminosity than normal giant stars of the same spectral type. Subgiants evolve to become giants.

subgiant branch A region of the H-R diagram defined by stars that have left the main sequence but have not yet reached the red giant branch.

sublimation The process in which a solid becomes a gas without first becoming a liquid.

subsonic Moving within a medium at a speed slower than the speed of sound in that medium. Compare *supersonic*.

summer solstice 1. One of two points where the Sun is at its greatest distance from the celestial equator. 2. The day on which the Sun appears at this location, marking the first day of summer (about June 21 in the Northern Hemisphere and about December 21 in the Southern Hemisphere). Compare *winter solstice*.

sungrazer A comet whose perihelion is within a few solar diameters of the surface of the Sun.

sunspot A cooler, transitory region on the solar surface, produced when loops of magnetic flux break through the surface of the Sun.

sunspot cycle The approximate 11-year cycle during which sunspot activity increases and then decreases. This is one-half of a full 22-year cycle, in which the magnetic polarity of the Sun first reverses and then returns to its original configuration.

supercluster A very large, gravitationally bound collection of galaxy clusters and galaxy groups containing tens of thousands to hundreds of thousands of galaxies; typically more than 100 million light-years (more than 30 Mpc) across. Compare *galaxy cluster* and *galaxy group*.

superluminal motion The appearance (though not the reality) that a jet is moving faster than the speed of light.

supermassive black hole A black hole of 1,000 solar masses (M_{Sun}) or more that resides in the center of a galaxy and whose gravity powers active galactic nuclei.

supernova (pl. supernovae) A stellar explosion resulting in the release of tremendous amounts of energy, including the high-speed ejection of matter into the interstellar medium. See also *Type Ia supernova* and *Type II supernova*.

supersonic Moving within a medium at a speed faster than the speed of sound in that medium. Compare *subsonic*.

superstring theory See *string theory*.

surface The outermost layer of something. In astronomy, this is often defined to be the visible surface.

surface brightness The amount of electromagnetic radiation emitted or reflected per unit area.

surface wave A seismic wave that travels on the surface of a planet or moon.

symmetry 1. The property that an object has if the object is unchanged by rotation or reflection about a particular point, line, or plane. 2. In theoretical physics, the correspondence of different aspects of physical laws or systems, such as the symmetry between matter and antimatter.

synchronous rotation The case in which the period of rotation of a body on its axis equals the period of revolution in its orbit around another body; a special type of spin-orbit resonance.

synchrotron radiation Radiation from electrons moving at close to the speed of light as they spiral in a strong magnetic field; named because this kind of radiation was first identified on Earth in particle accelerators called synchrotrons.

synodic period An object's orbital or rotational period measured with respect to the Sun. Compare *sidereal period*.

T

T Tauri star A young stellar object that has dispersed enough of the material surrounding it to be seen in visible light.

tail A stream of gas and dust swept away from the coma of a comet by the solar wind and by radiation pressure from the Sun.

tectonism Deformation of the lithosphere of a planet.

telescope The basic tool of astronomers. Working over the entire range from gamma rays to radio, astronomical telescopes collect and concentrate electromagnetic radiation from celestial objects.

temperature A measure of the average kinetic energy of the atoms or molecules in a gas, solid, or liquid.

terrestrial planet An Earth-like planet, made of rock and metal and having a solid surface. In our Solar System, the terrestrial planets are Mercury, Venus, Earth, and Mars. Compare *giant planet*.

test A prediction of a hypothesis that can be measured or observed.

theoretical model A detailed description of the properties of a particular object or system in terms of known physical laws or theories. Often, a computer calculation of predicted properties based on such a description.

theory A well-developed idea or group of ideas that are tied solidly to known physical laws and make testable predictions about the world. A very well-tested theory may be called a physical law or simply a fact. Compare *hypothesis*.

theory of everything (TOE) A theory that unifies all four fundamental forces of nature: strong nuclear, weak nuclear, electromagnetic, and gravitational forces.

thermal conduction The transfer of energy in which the thermal energy of particles is transferred to adjacent particles by collisions or other interactions. Conduction is the most important way that thermal energy is transported in solid matter.

thermal energy The energy that resides in the random motion of atoms, molecules, and particles, by which we measure their temperature.

thermal equilibrium The state in which the rates of thermal-energy emission and absorption by an object are equal.

thermal motion The random motion of atoms, molecules, and particles that gives rise to thermal radiation.

thermal radiation Electromagnetic radiation resulting from the random motion of the charged particles in every substance.

thermophile A type of extremophile that tolerates high temperatures (between 45°C and 122°C).

thermosphere The layer of Earth's atmosphere at altitudes greater than 90 kilometers (km), above the *mesosphere*. Near its top, at an altitude of 600 km, the temperature can reach 1000 K.

third quarter Moon The phase of the Moon in which only its eastern half, as viewed from Earth, is illuminated by the Sun. This phase occurs about 1 week after a full Moon. Compare *first quarter Moon*.

tidal bulge A distortion of a body resulting from tidal stresses.

tidal heating The process of heating a body by repeated gravitational stretching and contraction.

tidal locking Synchronous rotation of an object caused by internal friction as the object rotates through its tidal bulge.

tidal stress Stress due to differences in the gravitational force of one mass on different parts of another mass.

tide 1. The deformation of a mass due to differential gravitational effects of one mass on another because of the extended size of the masses. 2. On Earth, the rise and fall of the oceans as Earth rotates through a tidal bulge caused by the Moon and Sun.

time dilation The relativistic "stretching" of time. Compare *general relativistic time dilation*.

TNO Trans-Neptunian object; see *Kuiper Belt object*.

TOE See *theory of everything*.

topographic relief The differences in elevation from point to point on a planetary surface.

tornado A violent rotating column of air, typically 75 meters across with 200-kilometer-per-hour (km/h) winds. Some tornadoes can be more than 3 km across, and winds up to 500 km/h have been observed.

torus A three-dimensional, doughnut-shaped ring.

total lunar eclipse An event that occurs when the Moon passes through the umbra of Earth's shadow. Compare *partial lunar eclipse*.

total solar eclipse An event that occurs when Earth passes through the umbra of the Moon's shadow, so that the Moon completely blocks the disk of the Sun. Compare *annular solar eclipse* and *partial solar eclipse*.

transform fault The actively slipping segment of a fracture zone between lithospheric plates.

transit method A technique for detecting exoplanets by observing a star's decrease in brightness when a planet passes in front of it.

trans-Neptunian object (TNO) See *Kuiper Belt object*.

transverse wave A wave that oscillates perpendicular to the direction of the wave's propagation. Compare *longitudinal wave*.

triple-alpha process The nuclear fusion reaction that combines three helium nuclei (alpha particles) together into a single nucleus of carbon. See also *carbon-nitrogen-oxygen cycle* and *proton-proton chain*.

Trojan asteroid One of a group of asteroids orbiting in the L_4 and L_5 Lagrangian points of Jupiter's orbit.

tropical year The time between one crossing of the vernal equinox and the next. Because of the precession of the equinoxes, a tropical year is slightly shorter than the time it takes for Earth to orbit once about the Sun.

Tropics The region on Earth between latitudes 23.5° south and 23.5° north, in which the Sun appears directly overhead twice during the year.

tropopause The top of a planet's *troposphere*.

troposphere The convection-dominated layer of a planet's atmosphere. On Earth, the atmospheric region closest to the ground within which most weather phenomena take place.

tuning fork diagram The two-pronged diagram showing Hubble's classification of galaxies into elliptical, S0, spiral, and barred spiral types.

turbulence The random motion of blobs of gas within a larger cloud of gas.

Type Ia supernova A supernova explosion in which no trace of hydrogen is seen in the ejected material. Most Type Ia supernovae are thought to be the result of runaway carbon burning in a white dwarf star onto which material is being deposited by a binary companion.

Type II supernova A supernova explosion in which the degenerate core of an evolved massive star suddenly collapses and rebounds.

typhoon See *hurricane*.

U

ultrafaint dwarf galaxy A dwarf galaxy of very low luminosity.

ultraviolet (UV) radiation Electromagnetic radiation with frequencies, photon energies, and wavelengths between those of visible light and X-rays. Compare *infrared radiation*.

umbra (pl. umbrae) 1. The darkest part of a shadow, where the source of light is completely blocked. 2. The darkest, innermost part of a sunspot. Compare *penumbra*.

unbalanced force The nonzero net force acting on a body.

unbound orbit An orbit in which the velocity is greater than the escape velocity. Compare *bound orbit*.

uncertainty A description of the accuracy of a measurement, sometimes expressed as a percentage, more often as an interval. The uncertainty gives the range over which one

might expect to obtain measurements if the experiment were repeated multiple times.

uncertainty principle See *Heisenberg uncertainty principle*.

unified model of active galactic nuclei A model in which many different types of activity in the nuclei of galaxies are all explained by accretion of matter around a supermassive black hole.

uniform circular motion Motion in a circular path at a constant speed.

unit A fundamental quantity of measurement; for example, metric units or English units.

universal gravitational constant (G) The constant of proportionality in the universal law of gravity. Value: $G = 6.673 \times 10^{-11}$ newtons times meters squared per kilogram squared (N m²/kg²), equivalently 6.673×10^{-11} meters cubed per kilogram second squared [m³/(kg s²)].

universal law of gravitation The law stating that the gravitational force between any two objects is proportional to the product of their masses and inversely proportional to the square of the distance between them: $F \propto (m_1 m_2 / r^2)$.

universe All of space and everything contained therein.

unresolved Images overlapping to the extent that they cannot be distinguished from one another.

unstable equilibrium A state in which a small disturbance will cause a system to move away from equilibrium. Compare *stable equilibrium*.

UV See *ultraviolet radiation*.

V

vacuum A region of space that contains very little matter. In quantum mechanics and general relativity, however, even a perfect vacuum has physical properties.

vaporize Turn to gas.

variable star A star with varying luminosity. Many periodic variables are found within the instability strip on the H-R diagram.

velocity The rate and direction of change of an object's position with time. Units: meters per second (m/s) or kilometers per hour (km/h). Compare *speed*.

vernal equinox 1. One of two points where the Sun crosses the celestial equator. 2. The day on which the Sun appears at this location, marking the first day of spring (about March 20 in the Northern Hemisphere and about September 22 in the Southern Hemisphere). Compare *autumnal equinox*.

virtual particle A particle that, according to quantum mechanics, comes into existence only momentarily. According to theory, fundamental forces are mediated by the exchange of virtual particles.

visible spectrum The small part of the electromagnetic spectrum that the human eye can perceive, with frequencies, photon energies, and wavelengths between those of ultraviolet and infrared radiation.

visual binary A pair of stars in which both stars can be seen individually from Earth. Compare *eclipsing binary* and *spectroscopic binary*.

void A region in space containing little or no matter. Examples include regions in cosmological space that are largely empty of galaxies.

volatile material Sometimes called *ice*. Material that remains gaseous at moderate temperature. Compare *refractory material*

volcanism The occurrence of volcanic activity on a planet or moon.

vortex (pl. vortices) Any circulating fluid system. Specifically, 1. an atmospheric anticyclone or cyclone or 2. a whirlpool or eddy.

W

W See *watt*.

waning The changing phases of the Moon as it becomes less fully illuminated between full Moon and new Moon, as seen from Earth. Compare *waxing*.

watt (W) A measure of *power*. Units: joules per second (J/s).

wave A disturbance moving along a surface or passing through a space or a medium.

wavefront The imaginary surface of an electromagnetic wave, either plane or spherical, oriented perpendicular to the direction of travel.

wavelength The distance on a wave between two adjacent points having identical characteristics; the distance a wave travels in one period. Unit: meter.

waxing The changing phases of the Moon as it becomes more fully illuminated between new Moon and full Moon, as seen from Earth. Compare *waning*.

weak nuclear force The force underlying some forms of radioactivity and certain interactions between subatomic particles, responsible for radioactive beta decay and for the initial proton-proton interactions that lead to nuclear fusion in the Sun and other stars. One of the four fundamental forces of nature. Compare *strong nuclear force*.

weather The state of an atmosphere at any given time and place. Compare *climate*.

weathering The first step in the process of erosion, where rocks begin to break down by various physical or chemical processes.

weight 1. The force equal to the mass of an object multiplied by the local acceleration due to gravity. 2. In general relativity, the force equal to the mass of an object multiplied by the acceleration of the frame of reference in which the object is observed.

white dwarf The stellar remnant left at the end of the evolution of a low-mass star. A typical white dwarf has a mass of 0.6 solar mass (M_{Sun}) and a size about equal to that of Earth; it is made of nonburning, electron-degenerate carbon.

Wien's law A relationship describing how the peak wavelength, and therefore the color, of electromagnetic radiation from a glowing blackbody changes with temperature.

WIMP Literally, "weakly interacting massive particle." A hypothetical massive particle that interacts through gravity but not with electromagnetic radiation and is a candidate for dark matter. Compare *MACHO*.

wind The natural movement of air, both locally and on a global scale, in response to variations in temperature from place to place.

winter solstice 1. One of two points where the Sun is at its greatest distance from the celestial equator. 2. The day on which the Sun appears at this location, marking the first day of winter (about December 21 in the Northern Hemisphere and about June 21 in the Southern Hemisphere). Compare *summer solstice*.

X

X-ray binary A system in which mass from an evolving star spills over onto a collapsed companion, such as a neutron star or black hole. The material falling in is heated to such high temperatures that it glows brightly in the X-ray region.

X-ray burster A binary star system that releases a burst of X-ray emission when material falls from the main-sequence star onto the evolved star. The system's X-ray luminosity increases by a factor of about 100 during a burst.

X-rays Electromagnetic radiation with frequencies, photon energies, and wavelengths between those of ultraviolet light and gamma rays.

Y

year The time it takes Earth to make one revolution around the Sun. A solar year is measured from equinox to equinox. A sidereal year, Earth's true orbital period, is measured relative to the stars.

Z

Zeeman effect An interaction between atomic electrons and an external magnetic field that causes an atomic emission line to split into three lines: one at the original wavelength, one slightly redshifted, and one slightly blueshifted. The amount by which the lines are shifted can be used to find the strength of the magnetic field.

zenith The point on the celestial sphere located directly overhead from an observer. Compare *nadir*.

zero-age main sequence The strip on the H-R diagram plotting where stars of all masses in a cluster begin their lives.

zodiac The constellations that lie along the plane of the ecliptic.

zodiacal dust Particles of cometary and asteroidal debris less than 100 micrometers (μm) in size that orbit the inner Solar System close to the plane of the ecliptic. Compare *meteoroid* and *planetesimal*.

zodiacal light A band of light in the night sky caused by sunlight reflected by zodiacal dust.

zonal wind The planet-wide circulation of air that moves in directions parallel to the planet's equator.

Chapter 1

CHECK YOUR UNDERSTANDING

1.1. Smallest to largest: Earth, Sun, Solar System, Milky Way Galaxy, Local Group, Laniakea Supercluster, Universe

1.2. c

1.3. a

WHAT IF

p. 5. Earth, Solar System, Andromeda Galaxy, Local Group, Virgo Supercluster, Laniakea Supercluster

p. 6. We'd expect a much lower abundance of elements heavier than boron, like carbon, oxygen, etc. There was less time for earlier stars to fuse heavy elements and recycle them to the ISM.

p. 9. We would not know what the physical laws are elsewhere in the universe. It would then be difficult to interpret observations. We would then likely try to find a metaphysical law that explains what physics applies where.

p. 10. Of course it is convenient if a theory is falsifiable immediately. But nature is not always governed by our convenience. What is crucial is that a theory is falsifiable in principle.

TESTING YOUR UNDERSTANDING

6. b, d, a, c, e

8. d

11. d

14. a

17. b

THINKING ABOUT THE CONCEPTS

25. The distance between the Milky Way Galaxy and the Andromeda Galaxy would be placed after the figure showing the diameter of the Milky Way Galaxy.

29. A *hypothesis* is an idea that might explain some physical occurrence. A *theory* is a hypothesis that has been rigorously tested.

34. You would observe the same large-scale structure in the Universe as we see from Earth's viewpoint. You would observe stars grouped into galaxies, galaxies grouped into galaxy clusters, and galaxy clusters grouped into superclusters.

APPLYING THE CONCEPTS

36. (a) You should get a smaller number of light-hours than of light-minutes.

(b) 8.3 light-minutes $\times \dfrac{1\,\text{light-hour}}{60\,\text{light-minutes}} = 0.14$ light-hour.

(c) 0.14 is less than 8.3; the answer matches the prediction.

38. (a) $86{,}400 = 8.64 \times 10^4$ and $0.0123 = 1.23 \times 10^{-2}$ in scientific notation. (b) 86,400 is greater than 1, and its exponent in scientific notation is positive. 0.0123 is less than 1, and its exponent in scientific notation is negative.

Chapter 2

CHECK YOUR UNDERSTANDING

2.1. a

2.2. b

2.3. a

2.4. b, c

WHAT IF

p. 25. This would mean that a "day," from sunrise to sunrise, would last 1 full year. Any given location on Earth would always see the same half of the sky during its 6 months of night. It would have been much harder to figure out that Earth orbits the Sun.

p. 32. Mars' rotation axis is more tilted relative to the ecliptic than Earth's. Therefore, we expect greater seasonal variations than we see on Earth.

p. 36. If one face always points toward Earth, you would see a square at full moon, with the square becoming a rectangle of decreasing width, disappearing when closer to the Sun than Earth.

p. 38. You'd expect to see more eclipses, one from each Moon in each eclipse season, separated by about ⅓ of a month (10 days). In some eclipse seasons, the Sun would move away from the node before the second moon could pass in front of it; this means there will be fewer than twice as many eclipses.

TESTING YOUR UNDERSTANDING

6. a

13. b

16. d

19. b

20. d

THINKING ABOUT THE CONCEPTS

22. It is the June solstice. The Northern Hemisphere of Earth is at its greatest tilt toward the Sun, so the Sun spends the longest time above the horizon on this day if you are in the Northern Hemisphere. The Southern Hemisphere is at its greatest tilt away from the Sun, so the Sun spends the shortest time above the horizon on this day if you are in the Southern Hemisphere.

25. On the vernal equinox, the Sun rises directly in the east, travels along the celestial equator, and sets directly in the west.

28. In order for the Moon to appear full, it has to be on the opposite side of the sky from the Sun. Consequently, a full Moon would not be directly overhead while the Sun is still in the sky.

APPLYING THE CONCEPTS

36. (a) $s = \dfrac{d}{t}$, so $t = \dfrac{d}{s}$.

(b) Inserting units: $t = \dfrac{\text{kilometers}}{\text{kilometers/hour}} = \text{hours}$

38. (a) 27 days. (b) The Moon's circumference is smaller than Earth's, but its rotation period is much longer, so you might predict that a point on the Moon's equator would move more slowly than a point on the Earth's equator.

(c) $27\ \text{days} \times \dfrac{24\ \text{hours}}{1\ \text{day}} = 648\ \text{hours}$.

(d) Speed $= \dfrac{11{,}000\ \text{km}}{648\ \text{h}} = 17\ \text{km/h}$.

(e) This is much slower than Earth's rotation speed of 1,670 km/h.

Chapter 3

CHECK YOUR UNDERSTANDING

3.1. (a) Closest, farthest. (b) Sun, nothing. (c) Larger.

3.2. All three apples would have the same acceleration.

3.3. $b = e < c < d < a$

3.4. a

3.5. $c > b > d > a$

WHAT IF

p. 48. This means that the Southern Hemisphere gets a double whammy. In southern summer, it is both tilted toward the Sun AND closer. Just the opposite occurs in winter. The Northern Hemisphere experiences offsetting effects (that cancel out). Seasonal weather variations will be more extreme in the Southern Hemisphere than in the Northern Hemisphere.

p. 49. The period squared is $4 \times 4 = 16$. But the semimajor axis cubed is only $2 \times 2 \times 2 = 8$. So these supposed data do not satisfy the relationship that $P^2 = A^3$.

p. 57. We would observe many differences! This would mean that objects with different masses would fall at different rates, because the mass would not cancel out when the gravitational force equals mass times acceleration. Your hair, for example, would not fall at the same rate as the rest of you. In spacecraft, all the objects would not freely fall together, so items like wrenches would fly out of an astronaut's hands. Orbits would depend on the planet's mass, and Kepler's third law would not apply, as written.

p. 61. Ignore air resistance. In that case, you would feel weightless while falling, just like an astronaut in orbit!

TESTING YOUR UNDERSTANDING

7. c

11. d

16. c

17. d

20. a

THINKING ABOUT THE CONCEPTS

21. Mars appears larger in the retrograde part of the loop because Earth and Mars are closer together at that time. Venus must also appear larger when it is closer to Earth.

24. No, they would not be in the same reference frame, because they are in motion with respect to each other.

30. Your weight on a planet (or moon) is the gravitational force of that body acting on you. That gravitational force is directly proportional to the mass of the body. So, on the less massive Moon, that gravitational force acting on you would be less than it is on Earth.

APPLYING THE CONCEPTS

37. (a) You might predict that the acceleration would be small. 5 km/h is not very fast, and 1 hour seems like a long time for something to reach this speed. (b) Kilometers per hour per hour (km/h²). (c) 5 km/h². (d) This acceleration is much less than the acceleration due to gravity.

39. (a) $F = ma$, so $a = \dfrac{F}{m}$

    (b) $a = \dfrac{(100\ \text{N})}{(50\ \text{kg})} = 2\ \dfrac{\text{N}}{\text{kg}} = 2\ \dfrac{\text{kg m/s}^2}{\text{kg}} = 2\ \text{m/s}^2$

Chapter 4

CHECK YOUR UNDERSTANDING

4.1. $e > c > b > d > a$

4.2. a, b, c

4.3.

a. aperture	3. diameter
b. resolution	4. ability to distinguish objects that appear close together in the sky
c. focal length	2. distance from lens to focal plane
d. chromatic aberration	6. rainbow-making effect
e. diffraction	7. smearing effect due to a sharp edge
f. interferometer	1. several telescopes connected to act as one
g. adaptive optics	5. computer-controlled active focusing

WHAT IF

p. 75. The speed of light from an object at any given time would depend on the direction Earth is traveling. It would be smallest in the direction of Earth's motion. This smallest direction would rotate by 360° over the course of a year.

p. 78. Sunlight is basically optical, so you would still see the objects you currently see. We rarely encounter X-rays in our daily life, but your eyes would see a flash when your doctor takes an X-ray. Infrared radiation is all over the place. The ground is an intense source of infrared radiation. People and trees emit infrared radiation. You could see at night, but you would be "blinded" by the radiation from the ground.

p. 83. You would have 4 times the light-gathering power! You'd see better in the dark, and you'd have better angular resolution.

p. 89. Ultraviolet and infrared astronomy would have developed much earlier. Gamma ray, X-ray, and microwave astronomy would not need to be done from space telescopes.

TESTING YOUR UNDERSTANDING

8. a

12. d

14. a

15. b

19. b

THINKING ABOUT THE CONCEPTS

22. A light-year is a measure of distance.

25. Large refractors are impractical to build because the large lenses have to be perfectly shaped on both sides, and it is difficult to adequately support the weight of a large heavy lens without blocking any light that is sent through it.

31. The Earth's atmosphere scatters and absorbs some wavelengths of light, preventing them from passing through to the ground. Turbulence in the atmosphere also causes blurring of incoming light.

APPLYING THE CONCEPTS

36. (a) Radio waves appear to span wavelengths from about 2 cm to beyond 1 km.

    (b) $\lambda = \dfrac{c}{f} = \dfrac{3 \times 10^8\ \text{m/s}}{7.90 \times 10^5\ \text{Hz}} = 380\ \text{m}.$

    (c) 380 m falls within the radio range.

38. (a) For a given diameter, a longer wavelength should yield a poorer resolution (a greater number of arcseconds of separation).

    (b) $\theta_{\text{NIR}} = (2.06 \times 10^5) \left[\dfrac{(1000 \times 10^{-9}\ \text{m})}{1\ \text{m}} \right]$ arc-seconds = 0.21 arcsecond for the near-IR resolution;

    $\theta_{\text{V}} = (2.06 \times 10^5) \left[\dfrac{(400 \times 10^{-9}\ \text{m})}{1\ \text{m}} \right]$ arcseconds = 0.08 arcsecond for the violet resolution.

    (c) The longer-wavelength near-IR light resulted in a larger separation in arcseconds, which is a poorer resolution.

Chapter 5

CHECK YOUR UNDERSTANDING

5.1. b

5.2. b

5.3. c

5.4. a

5.5. b

5.6. a

WHAT IF

p. 107. A disk surrounding the binary should be in the same plane. The disk may be sparser, since most of the angular momentum can be stored in the orbits of the binary system. Each star could have its own disk as well, with the planes of the disks aligned with the orbital plane of the stars.

p. 110. The planet is still too hot to retain a primary atmosphere. Little secondary atmosphere implies both little volcanism and few comets.

p. 118 (top). It would tell us that star formation is boring; that is it is highly deterministic with very few variables. It also would imply that planets do not move very far from their place of birth.

p. 118 (bottom). One would conclude that solar systems are quite rare; fewer than 10 million of them in our galaxy. More stars may have planets, but the planets would be of very low mass and/or very far from their central stars.

TESTING YOUR UNDERSTANDING

7. d

9. c

15. a

17. d

19. d

THINKING ABOUT THE CONCEPTS

30. The inner planets were close to the Sun. The atmospheres were heated enough that the speeds of the atmospheric gas particles exceeded the escape velocities of the planets, and the atmospheres escaped.

31. The leftover debris still remains in the solar system, as asteroids and other small bodies.

35. Many of these stars are probably orbited by planets. The process of star formation naturally leads to planet formation, and astronomers are finding enough exoplanets to support this hypothesis.

APPLYING THE CONCEPTS

36. (a) Since the planet has a shorter orbital period than Earth, you might expect it to be closer to its star than Earth is to the Sun. (b) $(P_{\text{years}})^2 = (A_{\text{AU}})^3$, so $(A_{\text{AU}}) = (P_{\text{years}})^{2/3} = \left[(200\ \text{days}) \left(\dfrac{1\ \text{year}}{365\ \text{days}} \right) \right]^{2/3} = 0.7\ \text{AU}.$ (c) This is smaller than the orbit of Earth, smaller than the orbit of Mars, and larger than the orbit of Mercury. (d) This planet is probably receiving

more heat from its star than Earth receives from the Sun. It is probably not in its star's habitable zone.

38. (a) This planet has a higher surface temperature than Earth's 288 K. You might expect it to have a shorter peak wavelength and to emit more light.

(b) $\lambda_{\text{peak}} = \dfrac{2.9 \times 10^6\ \text{nm K}}{400\ \text{K}} = 7{,}250$ nm. This is in the infrared region of the electromagnetic spectrum.

(c) $I = \sigma T^4 = \left[5.67 \times 10^{-8}\dfrac{\text{W}}{\text{m}^2\,\text{K}^4}\right](400\ \text{K})^4 = 1450$ W/m^2.

(d) $L = I(4\pi r^2) = \left(1450\ \dfrac{\text{W}}{\text{m}^2}\right)(4\pi)(12.8 \times 10^6\ \text{m})^2 = 3.0 \times 10^{18}$ W.

(e) As predicted, this planet has a shorter peak wavelength and a greater luminosity than Earth.

Chapter 6

CHECK YOUR UNDERSTANDING

6.1. b
6.2. c, d, e
6.3. c
6.4. c
6.5. c

WHAT IF

p. 129. No. Plate tectonics, volcanism, and long-term erosion would erase such features. Only recent craters would remain.

p. 138. Earth's magnetic field protects the surface from both cosmic rays and the solar wind. Without it, humans could not survive. So yes, we would care.

p. 145. Hawaii would have volcanoes more like those on Mars, with a fixed hot spot producing a huge shield volcano. There would be fewer (perhaps only one), larger, higher islands.

p. 147. It would have smoothed the surface and decreased the prominence of these features.

TESTING YOUR UNDERSTANDING

7. d
11. a
14. c
15. d
19. c

THINKING ABOUT THE CONCEPTS

22. This surface appears to have many craters, so it has probably not been altered much since the early period of heavy bombardment. It is likely an older surface.

27. The Moon's core is cooler than Earth's mainly because the Moon is a smaller body. The volume of a body helps it retain its heat, while it radiates heat (cools) through its surface area. If a body's surface area to volume ratio is relatively high, like the Moon's, it will cool relatively quickly.

33. In tectonism, it is the shifting of these plates with respect to each other that causes these features and phenomena.

APPLYING THE CONCEPTS

36. (a) Since ¼ of the original radium remains, not many half-lives have occurred. With a half-life for radium of 1,620 years, you might easily predict that the sample should be less than 5,000 years old.

(b) $\dfrac{P_F}{P_O} = \left(\dfrac{1}{2}\right)^N = \left(\dfrac{1}{4}\right)$, so $N = 2$.

(c) $N \times$ half-life $= 2 \times 1{,}620$ years $= 3{,}240$ years.

(d) This answer is in thousands of years, as predicted.

38. (a) You might predict that a medium-dense object, such as a rock, made this crater. So you might predict an answer on the order of thousands of kilograms per cubic meter.

(b) $D = \dfrac{m}{V} = \dfrac{300 \times 10^6\ \text{kg}}{(4/3)\pi(25\ \text{m})^3} = 4{,}600$ kg/m^3, on the order of the density of a rock. (c) A rocklike composition seems reasonable for a meteorite.

Chapter 7

CHECK YOUR UNDERSTANDING

7.1. a
7.2. c
7.3. c
7.4. c
7.5. d

WHAT IF

p. 164. Oxygen, because it is so reactive that it would oxidize the surface material and disappear from the atmosphere without a source such as photosynthesis.

p. 168. Both cosmic rays and charged particles in the solar wind constitute a major radiation hazard. It is particularly dangerous during solar flares and other solar activity.

p. 172. Volcanism and cometary (and asteroid) impacts would restore some atmosphere over time.

p. 173. Since the rotation and inclination of Mars are similar to those of Earth, we expect similar zonal winds to Earth, with one major difference: The larger differences in surface temperatures on Mars would create higher wind speeds.

p. 177. Earth's surface will be much hotter! If a hotter ocean releases stored CO$_2$, then there is a positive feedback loop that can raise temperatures catastrophically. Rising oceans will wipe out coastal areas where more than half the world's population resides.

TESTING YOUR UNDERSTANDING

7. d
9. a
14. c
17. a, b, c
19. c

THINKING ABOUT THE CONCEPTS

22. The secondary atmospheres were created through volcanism, impacts, and the contributions of life.

29. The atmosphere of Venus is so thick that it blocks most wavelengths of light, and the atmospheric pressure on Venus is so great that it quickly destroys any probes that we send to the surface of Venus.

33. The overall trend is a gradual rise in temperature over time, maybe around 0.2C° over 20 years on average. The rise after the year 1980 is steeper than the rise before the year 1951.

APPLYING THE CONCEPTS

36. (a) An albedo greater than 0 means that the planet is reflecting some of the radiation that it receives from the Sun, instead of absorbing all of it. It should, therefore, be cooler than the black-body temperature. A higher albedo should lead to a greater amount of reflected radiation and a cooler planet.

(b) $T = 279\ \text{K} \times \left[\dfrac{1 - a}{d_{\text{AU}}{}^2}\right]^{1/4} = 279\ \text{K} \times \left[\dfrac{1 - 0.1}{1^2}\right]^{1/4} = 272$ K.

(c) $T = 279\ \text{K} \times \left[\dfrac{1 - a}{d_{\text{AU}}{}^2}\right]^{1/4} = 279\ \text{K} \times \left[\dfrac{1 - 0.9}{1^2}\right]^{1/4} = 157$ K.

(d) Both of these temperatures are less than 279 K, with the greater albedo leading to a lower temperature.

40. (a) Since we are talking about moving the Earth farther from the Sun to achieve a lower temperature, you would predict that we should calculate a distance greater than 1 AU.

(b) $d = \dfrac{\sqrt{1 - a}}{(T/279)^2} = \dfrac{\sqrt{1 - 0.306}}{(251/279)^2} = 1.03$ AU.

(c) This would put the Earth seemingly slightly farther from the Sun than it is now. The difference is

$(0.03\ \text{AU}) \times \dfrac{1.496 \times 10^8\ \text{km}}{\text{AU}} = 4.488 \times 10^6$ km.

This is $\dfrac{4.488 \times 10^6\ \text{km}}{6{,}371\ \text{km}} = 704$ times as long as

Earth's radius and $\dfrac{4.488 \times 10^6\ \text{km}}{385{,}000\ \text{km}} = 11.7$ times

as long as the Moon's orbital radius around the Earth. This does not seem like a feasible solution.

Chapter 8

CHECK YOUR UNDERSTANDING

8.1. b
8.2. d
8.3. b
8.4. a
8.5. a

WHAT IF

p. 191. Its rotation would be tidally locked, and the side facing the Sun would be much hotter.

p. 196. Since more mass means more particles, each particle will have to "fall" less to achieve the same energy release. The contraction would be slower.

p. 200. Radio emission from the planet could be dwarfed by radio emission from the star itself. This could work only if the radio resolution allows the planet to be resolved from the star.

p. 203. The ring would be within the plane of the Moon's orbit. The ring would be short-lived because there are no shepherding moons. As it dissipates, the ring material would, for a time, be near Earth, and it would scatter radiation from the Sun, leading to a brighter night sky.

TESTING YOUR UNDERSTANDING

6. b
9. c
12. d
18. c
20. d

THINKING ABOUT THE CONCEPTS

22. Hydrogen and helium are by far the most abundant elements in the Sun and in the Universe as a whole.
24. Jupiter's Great Red Spot is considered an anticyclone because it is rotating counterclockwise and it is a high-pressure region.
34. Saturn's moons move through its rings. A single moon will exert gravitational forces on ring particles just inside and just outside its orbit, clearing a gap in the rings.

APPLYING THE CONCEPTS

36. (a) Saturn should be less dense than water, so we should find a density less than 1,000 kg/m³.

    (b) $M_{Saturn} = 95\, M_{Earth} \times \dfrac{5.972 \times 10^{24}\ \text{kg}}{1\, M_{Earth}} = 5.673 \times 10^{26}$ kg.

    (c) $V_{Saturn} = \frac{4}{3}\pi R_{Saturn}^3 = \frac{4}{3}\pi(60{,}270 \times 10^3\ \text{m})^3 = 9.170 \times 10^{23}\ \text{m}^3$; $D_{Saturn} = \dfrac{M_{Saturn}}{V_{Saturn}} = \dfrac{5.673 \times 10^{26}\ \text{kg}}{9.170 \times 10^{23}\ \text{m}^3} = 619$ kg/m³.

    (d) This is significantly lower than the density of water.
38. (a) Since Neptune is much farther from the Sun than Uranus, you might predict that it would occult the star for longer since it is moving more slowly around the Sun. (b) $S = \dfrac{d}{t}$, so $t = \dfrac{d}{s}$.

    (c) $\dfrac{48{,}600\ \text{km}}{18\ \text{km/s}}\ (2700\ \text{s}) \times \dfrac{\text{min}}{60\ \text{s}} = 45$ min.

    (d) This is longer than the 37 minutes that Saturn took to occult the star, so it matches your prediction.

Chapter 9

CHECK YOUR UNDERSTANDING

9.1. b, e
9.2. b, d
9.3. b
9.4. c
9.5. b
9.6. b

WHAT IF

p. 214. All of the dwarf planets would qualify, as would many asteroids, comet nuclei, and any spherical rock orbiting the Sun.

p. 221. Answers will vary, but an active or possibly active moon would be far more interesting than an inactive one.

p. 227 (top). The tails will separate; two tails will become four. Absent a collision in the Kuiper Belt, we'd expect to see both again at the usual time.

p. 227 (bottom). (1) Comet tails point away from the Sun, not toward it. (2) Just after sunset, a full Moon cannot be high in the sky.

p. 230. Early detection is important. Despite popular movies, exploding the asteroid is impractical. Deflection is the right course of action. The most promising approach is an explosion on one side of the asteroid. Hitting the asteroid with high-momentum projectiles (i.e., shooting at it) is also possible, but more difficult.

TESTING YOUR UNDERSTANDING

7. a
10. a
13. a
16. d
18. a, b, c

THINKING ABOUT THE CONCEPTS

21. During the heavy bombardment period, all Solar System bodies would have been heavily impacted and would have been covered in craters. A smooth surface implies that some process has erased craters in that region in more recent times.
25. The gravitational force of Jupiter tidally heats Io's interior as Io rotates and orbits Jupiter. This keeps the interior molten and drives volcanic activity.
30. Comets that originate in the Kuiper Belt are short-period comets and will typically orbit the Sun in a prograde direction, the same as the planets. Comets that originate in the Oort Cloud are long-period comets and can enter the inner Solar System from any direction, including orbiting the Sun in a retrograde direction.

APPLYING THE CONCEPTS

39. (a) The semimajor axis of Ceres is larger than the semimajor axis of Earth, so you might predict that the orbital period of Ceres will be longer than that of Earth. (b) $P[year]^2 = A[AU]^3$, so $P[year] = A[AU]^{3/2} = (2.768)^{3/2} = 4.6$ years. (c) This is longer than Earth's orbital period, as predicted.
40. (a) Since Makemake is so far from the Sun, it is probably a mix of ice and rock. So you might predict that its density would fall between the density of ice (917 kg/m³) and the average density of rock (3,000 kg/m³).

    (b) $V = \left(\dfrac{4}{3}\right)\pi R^3 = \left(\dfrac{4}{3}\right)\pi(750 \times 10^3\ \text{m})^3 = 1.77 \times 10^{18}\ \text{m}^3$; $D = \dfrac{M}{V} = \dfrac{4.18 \times 10^{21}\ \text{kg}}{1.77 \times 10^{18}\ \text{m}^3} = 2{,}367$ kg/m³.

(c) This density is between those of ice and rock, as predicted.

Chapter 10

CHECK YOUR UNDERSTANDING

10.1. a
10.2. a
10.3. c
10.4. b

WHAT IF

p. 246. We expect the distances to be varied. Sagittarius is a pattern on the sky, but not a physical cluster of stars. If the distances were similar, the stars in Sagittarius would be a physical cluster of stars, all close to each other in space.

p. 256. (1) The elemental abundances are different OR (2) something (a magnetic field maybe) affects the strength of the spectral lines.

p. 260. As long as the periods of the three stars are the same, the measured velocities will determine their orbital radii. But Kepler's third law is no longer valid, although it may be a fair approximation.

p. 262. It would look about the same, extending from top left to bottom right, as long as the temperature axis still ran from hottest at left to coolest at right.

TESTING YOUR UNDERSTANDING

10. b
11. a
13. c
15. c
20. d

THINKING ABOUT THE CONCEPTS

23. If we were on Mars, Jupiter, or Saturn, we could still measure stellar parallax. It would just take longer to measure the parallax of a given star, because it would take longer for our planet to reach the other side of the Sun.
28. The presence of most of a continuous spectrum tells us that there is a white light source. The cool gas between us and the white light source is inferred by the absorption features in the spectrum. If the cool gas cloud were behind the white light source, it would not absorb any of the white source's light on its way to us, so the resulting spectrum would be a continuous one.
32. In order to directly measure the mass of a body, we need to be able to measure the gravitational effect that it has on another body. A companion star gives us this opportunity.

APPLYING THE CONCEPTS

36. (a) Sirius has a smaller parallax than Proxima Centauri, so you would expect it to be farther away than Proxima Centauri. From this parallax, you might predict a few parsecs for the distance to Sirius. Since a light-year is smaller than a parsec, Sirius should be more light-years away than parsecs.

(b) $d[pc] = \dfrac{1}{p[arcsec]} = \dfrac{1}{0.379 \text{ arcsec}} = 2.639 \text{ pc}$

$\times \left(\dfrac{3.26 \text{ ly}}{1 \text{ pc}}\right) = 8.6 \text{ ly}.$

(c) Sirius is farther than Proxima Centauri, as predicted, and is on the order of a few parsecs away.

38. (a) Barnard's star is farther away than Proxima Centauri, so you would expect its parallax to be smaller than that of Proxima Centauri.

(b) $5.978 \text{ ly} \times \dfrac{1 \text{ pc}}{3.26 \text{ ly}} = 1.83 \text{ pc}$;

$p[arcsec] = \dfrac{1}{d[pc]} = \left(\dfrac{1}{1.83}\right) = 0.546 \text{ arcsec}.$

(c) This is a smaller parallax than that of Proxima Centauri, as predicted.

Chapter 11

CHECK YOUR UNDERSTANDING

11.1. b

11.2. c

11.3. c

11.4. a

WHAT IF

p. 276. It wouldn't vary at all, since Earth is transparent to neutrinos.

p. 278. Variations in the core are more likely to be seen at the surface, and lower opacity means that the Sun's surface will be much hotter.

p. 284. Instead of emitting UV and soft X-rays, it would emit higher-energy X-rays.

p. 291. There would be no sunspots and no solar prominences.

TESTING YOUR UNDERSTANDING

8. c

13. b

15. c

17. d

19. c

THINKING ABOUT THE CONCEPTS

23. (a) In part (a) the protons are moving more slowly than in part (b), so the electric repulsion between the protons does not allow them to get as close to each other. (b) The blue arrows are larger in part (b) because they represent faster proton speeds.

26. Hydrogen fusion requires very high temperature and very high pressure.

33. The number of neutrinos predicted by core hydrogen fusion in the Sun was not being observed. After the discovery was made that neutrinos oscillate among three different types, the predicted number was found.

APPLYING THE CONCEPTS

36. (a) If the Sun used its fuel up faster, you might predict a shorter lifetime for the Sun.

(b) $M_{\text{fusion}} = 2 \times 10^{29} \text{ kg}$. $M_{\text{year}} = (1,200 \times 10^9 \text{ kg/s})$ $(3.16 \times 10^7 \text{ s/year}) = 4 \times 10^{19} \text{ kg/year}$. Lifetime =

$\dfrac{M_{\text{fusion}}}{M_{\text{year}}} = \dfrac{(2 \times 10^{29} \text{ kg})}{(4 \times 10^{19} \text{ kg/year})} = 5 \times 10^9 \text{ years} =$

5 billion years. (c) This is half the currently predicted lifetime of the Sun; as predicted, it is less.

38. (a) The mass of this star is 60% of the Sun's mass, but its luminosity is 10%, so you might predict a significantly longer lifetime for this star.

(b) $M_{\text{fusion}} = (0.1)(0.6\,M_{\text{Sun}}) = (0.1)(0.6)(2 \times 10^{30} \text{ kg})$ $= 1.2 \times 10^{29} \text{ kg}$. $M_{\text{year}} = (0.1)(600 \times 10^9 \text{ kg/s})$ $(3.16 \times 10^7 \text{ s/year}) = 2 \times 10^{18} \text{ kg/year}$. Lifetime =

$\dfrac{M_{\text{fusion}}}{M_{\text{year}}} = \dfrac{(1.2 \times 10^{29} \text{ kg})}{(2 \times 10^{18} \text{ kg/year})} = 60 \times 10^9 \text{ years} =$

60 billion years. (c) This is significantly longer than the Sun's lifetime of 10 billion years.

Chapter 12

CHECK YOUR UNDERSTANDING

12.1. a

12.2. a

12.3. d

12.4. c, f, h, d, i, a, e, g, b

12.5. c, a, f, d, h, b, e, g

12.6. b

WHAT IF

p. 300. Jupiter would glow in the infrared and deep red, fueled by its own gravitational collapse. But the deep red light would be faint, compared to the reflected light from the Sun. So it would look about the same.

p. 302. This would have no effect on the orbit of the Moon. Because the gravitational acceleration depends on $1/r^2$, the gravitational acceleration on Earth's surface would increase by a factor of 100^2, or $10,000$.

p. 304. The inner planets would be absorbed into the Sun. The outer planets would absorb and emit much more energy because of the Sun's increasing luminosity.

p. 314. The two stars would evolve independently. When mass is lost from the more massive star, the binary may become more loosely bound.

p. 316. No. The lifetime of a $0.8\text{-}M_{\text{sun}}$ star is nearly 20 billion years. Even if it formed when the universe did, the system won't form a supernova until after the Sun has become a red giant.

TESTING YOUR UNDERSTANDING

6. c

9. c

13. a

15. d

19. a

THINKING ABOUT THE CONCEPTS

30. The degenerate core is not able to expand to cool itself.

33. In this nebula, the "fur hood" was probably released first, because this layer has expanded farther from the central star.

35. Novae were probably named "new stars" because some of them would have been invisible to the naked eye previous to their flare-ups, and then they became visible.

APPLYING THE CONCEPTS

36. (a) An F5 main sequence star is more massive than our Sun, so you would expect it to have a shorter lifetime than that our Sun. (b) Lifetime =

$(1.0 \times 10^{10}) \times \left(\dfrac{M_{\text{MS}}}{M_{\text{Sun}}}\right)^{-2.5} \text{ years} = (1.0 \times 10^{10}) \times$

$(1.4)^{-2.5} \text{ years} = 4.3 \times 10^9 \text{ years}$. (c) This is shorter than the lifetime of our Sun, as predicted.

38. (a) A main-sequence star that is twice as massive as the Sun would be larger and hotter, so you would predict that it would be more luminous than our Sun.

(b) $\dfrac{L_{\text{MS}}}{L_{\text{Sun}}} = \left(\dfrac{M_{\text{MS}}}{M_{\text{Sun}}}\right)^{3.5} = (2.0)^{3.5} = 11$ times the luminosity of the Sun. (c) This is more luminous than our Sun, as predicted.

Chapter 13

CHECK YOUR UNDERSTANDING

13.1. c

13.2. b

13.3. d

13.4. a

13.5. a, b, d, e, f

13.6. d

WHAT IF

p. 329. Possibly the white dwarf was originally a star with mass between 4 and 8 solar masses, which evolved faster than its companion and transferred mass to it.

p. 332. You would expect to see neutrinos first, since they escape the star more easily.

p. 338. The nebula will have expanded and grown more diffuse and less luminous. The pulsar's rotation will have slowed considerably.

p. 342 (top). The mass would decrease slightly, due to $E = mc^2$, where E is the energy emitted. For chemical reactions, this change in measured mass is very, very small.

p. 342 (bottom). The astronaut will have aged differently because of time dilation. Since the astronaut accelerates to a higher speed when leaving Earth, her clock will run more slowly than an Earth-bound clock.

p. 344. Once you throw the ball, you might expect it not to accelerate any more from your perspective. But once you've thrown the ball, it is on a different orbit than yours and you will see it accelerate. If you were on a circular orbit, the ball would now be on a slightly elliptical orbit, so it would accelerate relative to you. From the perspective of Earth, both you and the ball are accelerating.

p. 348. If the stars are massive and close together, they will produce a larger signal.

p. 351. Hawking radiation would be very weak. The black hole might accrete some material from the interstellar medium and emit radiation, but that will also be very weak. The best possibility for detection is by gravitational lensing of background stars.

TESTING YOUR UNDERSTANDING

7. b
10. b
13. c
17. d
20. c

THINKING ABOUT THE CONCEPTS

22. The core of a high-mass star doesn't shrink enough to become degenerate before the pressure and temperature become high enough for helium fusion to begin.

24. Cepheid variables are higher up on the H-R diagram because they have higher luminosity. Because they have higher luminosity, they are visible for longer distances.

26. The material from a supernova will mix with the interstellar medium to become part of a future generation of stars. Additionally, shock waves from supernovae can disrupt molecular clouds, causing clumps to collapse to begin star formation.

APPLYING THE CONCEPTS

37. Figure 13.11 shows a log plot. Oxygen lies on the y-axis at 10^{-3}, which means that there are 10^{-3} oxygen atoms for every hydrogen atom in the Solar System. The number of oxygen atoms is 1/1000 the number of hydrogen atoms.

39. (a) We know that the more massive a black hole is, the larger its Schwarzschild radius, so you would predict that a $6\text{-}M_{\text{Sun}}$ black hole would have a larger Schwarzschild radius than a $3\text{-}M_{\text{Sun}}$ black hole. Since it has twice the mass, you might even predict twice the radius.

(b) $R_s = (3 \text{ km}) \left(\dfrac{M_{\text{BH}}}{M_{\text{Sun}}} \right) = (3 \text{ km})(6) = 18 \text{ km}$.

(c) This is twice the Schwarzschild radius of a $3\text{-}M_{\text{Sun}}$ black hole, as predicted.

Chapter 14

CHECK YOUR UNDERSTANDING

14.1. a
14.2. a
14.3. c
14.4. a, b, c

WHAT IF

p. 360. Not necessarily. For example, imagine you live on the two-dimensional surface of an infinitely long cylinder. That surface is homogeneous (the same everywhere), but not isotropic (the same in every direction). No location is special.

p. 367. The distances to galaxies would also be 10% larger.

p. 370. In that case, Earth WOULD be in a special place in the universe. The universe would not be homogeneous.

p. 374. This system will be emitting gravitational waves; it's similar to other binary black hole systems that have been detected (although much more massive). But the system will probably be too far away to be detectable with current technology.

TESTING YOUR UNDERSTANDING

8. b
11. d
12. a
16. c
20. c

THINKING ABOUT THE CONCEPTS

22. A *standard candle* is an object with a known luminosity. If we can measure the brightness of a standard candle, we can calculate its distance. Globular clusters, which trace the halo of our galaxy, contain standard candles called RR Lyrae stars. Finding the distances to these stars allows us to estimate the size of the galaxy's halo.

30. At the center of a galaxy with an AGN, evidence shows that there must be a supermassive black hole with an accretion disk and a dusty torus.

35. Nine of the interactions appear to be between two spiral galaxies, one appears to be between two elliptical galaxies, and two appear to be between a spiral galaxy and an elliptical galaxy.

APPLYING THE CONCEPTS

36. (a) Comparing 230 million to 4.6 billion, you might predict that the Sun has made tens of orbits.

(b) Number of orbits $= \dfrac{\text{age of Sun}}{\text{time for Sun to orbit}} = \dfrac{4.6 \times 10^9 \text{ years}}{230 \times 10^6 \text{ years}} = 20$ orbits.

(c) 20 orbits is in the range of our prediction.

38. (a) Given that this galaxy has a lower recessional velocity than the galaxy in Working It Out 14.1, you might predict that this galaxy is closer.

(b) $d_G = \dfrac{v_r}{H_0} = \dfrac{2{,}000 \text{ km/s}}{70 \text{ (km/s)/Mpc}} = 29$ Mpc.

(c) As predicted, this is a smaller distance than that found in Working It Out 14.1.

Chapter 15

CHECK YOUR UNDERSTANDING

15.1. a
15.2. b
15.3. d
15.4. c

WHAT IF

p. 390. If the distribution of WIMPs is uniform, we should see more WIMPs coming from the direction of the Sun's motion in the galaxy, as the Sun moves through them. But we should also see a small variation annually as Earth orbits the Sun.

p. 391. There are no short-lived stars, such as O or B types, which implies that there has been no recent star formation. This, in turn, indicates that we do not expect to see much dust and gas in the arms.

p. 396. The lack of heavy elements indicates that this is an ancient star, nearly as old as the universe. Its main-sequence lifetime, therefore, must be longer than the current age of the universe, so it must have a mass less than about 0.7 M_{sun}. Ancient stars like these are found in the halo, so it will be moving fast relative to the Sun as it passes through the disk of the Milky Way.

TESTING YOUR UNDERSTANDING

6. c
11. b
14. d
16. c
20. a

THINKING ABOUT THE CONCEPTS

21. Regular matter emits, reflects, and absorbs electromagnetic radiation. Dark matter does not appear to interact with electromagnetic radiation at all.

24. A face-on galaxy would not be a good candidate to study the rotation curve of galaxies. This is because we need to be able to measure the Doppler shifts of light being emitted by stars that are moving toward or away from us. Observing a face-on galaxy, we mostly see motions that are perpendicular to our line of sight.

29. Radio observations at 21 cm detect neutral hydrogen gas. The spiral arms of a galaxy are mostly made up of neutral hydrogen gas. Detecting the Doppler shift of this emission then allows us to map the motions of the spiral arms. The observations show that the outer regions of spiral galaxies are rotating faster than would be expected by taking only luminous matter into account. This suggests the presence of dark matter.

APPLYING THE CONCEPTS

36. (a) A parsec is larger than a light-year, so you would expect the number of parsecs to be fewer than the number of light-years. An astronomical unit is smaller than a light-year, so you would expect the number of astronomical units to be more than the number of light-years.

(b) $160{,}000 \text{ ly} \times \dfrac{1 \text{ pc}}{3.26 \text{ ly}} = 49{,}000 \text{ pc}$.

(c) $160{,}000 \text{ ly} \times \dfrac{6.324 \times 10^4 \text{ AU}}{1 \text{ ly}} = 10{,}000{,}000{,}000 \text{ AU}$.

(d) We calculated fewer parsecs than light-years and more astronomical units than light-years, as predicted.

37. (a) The circumference of the orbit should be larger than 33,000 light-years by a factor of 2π, or about 6. (b) $C = 2\pi R = (2\pi)(33{,}000 \text{ ly}) = 200{,}000 \text{ ly}$.

(c) This orbit is about 6 times greater than the distance from the center of the galaxy, as predicted.

Chapter 16

CHECK YOUR UNDERSTANDING

16.1. a
16.2. b
16.3. c
16.4. b, c
16.5. a, b
16.6. c

WHAT IF

p. 412. They would be blue-shifted and related by a Hubble law with a negative slope.

p. 416. The density would have to be so high that it blocks microwave photons from passing through it. It must also be so large that its gas has not had time to come to equilibrium with the CMB.

p. 421. The universe would have been in rapid expansion for a long time. Very early, highly redshifted objects might have formed, but galaxy formation would have ceased billions of years ago.

p. 424. Yes, it is consistent under certain circumstances. Today, antiparticles would not be apparent because they interact so weakly. Possibly in the early universe, annihilation could have occurred just as for electron-positron pairs.

p. 428 (top). Without inflation, the universe (1) would be chaotic with large variations in temperature, (2) would have curvature that would grow more rapidly as the universe expands, and (3) would not develop fluctuations in density.

p. 428 (bottom). This is too close to be a parallel universe. We can see the galaxy. Something weird is happening.

TESTING YOUR UNDERSTANDING

6. b
9. a
11. a, c, e
15. a
19. b

THINKING ABOUT THE CONCEPTS

24. *Critical density* for the Universe is how dense the Universe would need to be for the expansion to slowly approach zero over a very long time.
29. The four fundamental forces are gravity, the electromagnetic force, the strong nuclear force, and the weak nuclear force. The electromagnetic force depends on electric charge.
33. *Hubble time* is the amount of time that has passed since the Universe began expanding.

APPLYING THE CONCEPTS

36. (a) A galaxy that is 60 Mpc away should be moving away from us about twice as fast as a galaxy that is 30 Mpc away. (b) $v_r = H_0 d_G = 70$ (km/s)/Mpc $\times$ 60 Mpc = 4,200 km/s. (c) This is twice as fast as the galaxy that was 30 Mpc away, as predicted.

38. (a) You would expect to calculate about ⅔ the number of kilometers (9.3×10^{20}), since 20 Mpc is ⅔ of 30 Mpc.

(b) $20 \text{ Mpc} \times \dfrac{3.086 \times 10^{19} \text{ km}}{1 \text{ Mpc}} = 6.172 \times 10^{20}$ km.

(c) This is ⅔ the number of kilometers to the 30 Mpc galaxy, as predicted.

Chapter 17

CHECK YOUR UNDERSTANDING

17.1. b, c
17.2. c, e
17.3. c
17.4. a
17.5. e, b, d, a, c

WHAT IF

p. 440. It proves dark matter exists in voids; it shows galaxy-like structures can form in voids; it shows that something in voids prevented galaxies from lighting up.

p. 444. If the temperature of the CMB were 10 times higher today, it would NOT change the temperature at recombination (3000 K), but recombination would be more recent than in the real universe. If, in addition, the density of today's universe is the same, then at the temperature of primordial nucleosynthesis, which is more recent, the density in the universe would be lower. This means collisions would be less frequent, and much less helium (and other elements) would be synthesized.

p. 449. Perhaps look in concentrations of protogalaxies for bright objects or supernovae. If possible, correlate an optical/IR search with X-ray observations.

p. 454. You'd have to correct for dust formation and its effect on galaxy observability; you need to model star-forming regions; and you must use good IR emission models for the stars.

TESTING YOUR UNDERSTANDING

7. b
11. a
15. a
18. d
20. d

THINKING ABOUT THE CONCEPTS

29. We have observed neutrinos, which are a form of hot dark matter and interact very weakly with other matter.
30. No, the slight differences exist because the simulation shows one possible outcome. Very small changes in different factors will make the outcome look different each time. The simulation is very close to observation and shows the same general structure.

31. Cold dark matter moves more slowly than hot dark matter and would be more easily clumped together by gravity to form galaxies.

APPLYING THE CONCEPTS

36. (a) You would expect to calculate a greater number of meters for the galaxy orbiting at 5 Mpc than for the galaxy orbiting at 3 Mpc, by almost a factor of 2.

(b) $5 \text{ Mpc} \times \dfrac{3.086 \times 10^{22} \text{ m}}{1 \text{ Mpc}} = 1.5 \times 10^{23}$ m. This

is $\dfrac{1.5 \times 10^{23} \text{ m}}{1.5 \times 10^{11} \text{ m}} = 10^{12}$ times larger than Earth's

orbit around the Sun.

(c) This is almost twice as far as the galaxy orbiting at 3 Mpc, as predicted.

38. (a) If this galaxy orbits at the same radius but at a greater speed than the previous galaxy, you would predict that the galaxy cluster orbited by this galaxy is more massive than

the previous galaxy cluster. (b) $M = \dfrac{rv^2}{G} =$

$\dfrac{(4.0 \times 10^6 \text{ pc})(3.09 \times 10^{16} \text{ m/pc})(1 \times 10^6 \text{ m/s})^2}{6.673 \times 10^{-11} \text{ m}^3/(\text{kg s}^2)} =$

1.9×10^{45} kg

(c) This is a greater mass than the first galaxy cluster by at least 2 times, as predicted.

Chapter 18

CHECK YOUR UNDERSTANDING

18.1. a, b, c, d
18.2. c
18.3. d
18.4. d

WHAT IF

p. 466 (top). No. One would imagine that they would be based upon different self-replicating chemicals, different amino acids, etc.

p. 466 (bottom). Yes. It is hard to imagine life forming with "just the right" molecular structure for survival. In addition, as conditions change, life has to change to optimally thrive.

p. 473. A very small magnetic field means no shielding from the particles in the stellar wind. This could mean that the inner part of the habitable zone is less habitable. Greenhouse gases in the atmosphere would raise the planet's temperature, so that the entire habitable zone would move farther away from the star.

p. 479. This is a completely open-ended discussion question, based purely on opinion.

TESTING YOUR UNDERSTANDING

7. d
11. c
13. d
15. d
20. c

THINKING ABOUT THE CONCEPTS

22. Silicon is unable to form long complex chains of atoms like carbon can, so any silicon-based life would have to be very simple.

27. Astronomers use false-color images to easily view emissions in certain wavelengths that are not visible to the human eye. These emissions in other wavelengths can give us information about energies, temperatures, and compositions of objects, among other characteristics.

35. Answers will vary. You might feel like a smaller part of the Universe than before. You might feel more "at home," having a better knowledge of the Universe. You might feel the importance of preserving our unique lives and planet.

APPLYING THE CONCEPTS

36. (a) Seven doublings is significantly fewer than 60 doublings, so you would expect a much smaller population than 1 billion molecules this time.

(b) $P_F = 2^7 = 128$ molecules. (c) This is a much smaller population than 1 billion molecules, as predicted.

38. (a) 1,024 is much less than 1.2 billion, so you would expect many fewer doubling times for this population. (b) $2^{10} = 1,024$; and (1,024 doublings) $\times$ (3 s/doubling) = 3,072 seconds. (c) 10 doublings is many fewer than 60, as predicted.

Photos

FRONT MATTER

Pages ix–xxviii (in order of appearance): W. W. Norton & Company, Inc.; (trees): Andrew Bret Wallis/Getty Images; W. W. Norton & Company, Inc.; (moon): Deanna Truesdale/EyeEm/Getty Images; W. W. Norton & Company, Inc.; (trees): Anthony Qualkinbush/EyeEm/Getty Images; W. W. Norton & Company, Inc.; Richard Sharrocks/Getty Images; W. W. Norton & Company, Inc.; NASA/GSFC/Arizona State University; W. W. Norton & Company, Inc.; Jordana Meilleur/Alamy Stock Photo; Daniel Pludowski/EyeEm/Getty Images; Orchidpoet/Getty Images; W. W. Norton & Company, Inc.; Knickohr/Getty Images; W. W. Norton & Company, Inc.; Hadafee/Shutterstock; W. W. Norton & Company, Inc.; Stewart Bremner/Getty Images; Teemu Tretjako/Alamy Stock Photo; W. W. Norton & Company, Inc.; W. W. Norton & Company, Inc.; Yuji Kotani/Getty Images; W. W. Norton & Company, Inc.; NASA; Stacy Palen; George Blumenthal.

CHAPTER 1

Page 2: W.W. Norton & Company, Inc.; **Page 3**: Andrew Bret Wallis/Getty Images; **Page 6**: Private Collection/© Look and Learn/Bridgeman Images; **Page 7 (top)**: Courtesy Douglas Finkbeiner http://halpha.skymaps.info and Swinburne Astronomy Productions; **(center)**: Michael J. Tuttle (NASM/NASA); **(bottom)**: NASA/JPL-Caltech/MSSS; **Page 8**: Caltech/MIT/LIGO Lab; **Page 10**: GL Archive/Alamy Stock Photo; **Page 12**: Matheisl/Getty Images

CHAPTER 2

Page 20: Deanna Truesdale/EyeEm/Getty Images; **Page 21 (left)**: W.W. Norton & Company, Inc.; **(right)**: Anthony Qualkinbush/EyeEm/Getty Images; **Page 28 (top)**: Pekka Parviainen/Science Source; **(bottom)**: David Nunuk/Science Source; **Page 35**: Castleski/Shutterstock; **Page 37**: Martin Ratcliffe; **Page 38 (a)**: B.O'Kane/Alamy Stock Photo; **(b)**: Anthony Ayiomamitis (TWAN)

CHAPTER 3

Page 44: W.W. Norton & Company, Inc.; **Page 47**: Tunc Tezel; **Page 51**: Jaime Cooper/SSPL/Getty Images; **Page 65**: ESA/Hubble

CHAPTER 4

Page 70: Richard Sharrocks/Getty Images; **Page 76**: Stockbyte/Getty Images; **Page 80 (a)**: Science Source; **(b)**: © Hulton-Deutsch Collection/Corbis via Getty Images; **Page 81**: NASA and Ball Aerospace; **Page 82**: ASU Physics Instructional Resource Team. © 2009 Arizona Board of Regents. Used with permission.

CHAPTER 5

Page 96: W.W. Norton & Company, Inc.; **Page 97**: NASA, ESA, and M. Livio and the Hubble 20th Anniversary Team (STScI); **Page 101**: Jeff Hester and Paul Scowen (Arizona State University), Bradford Smith (University of Hawaii), Roger Thompson (University of Arizona), and NASA; **Page 104 (a, left)**: STScI, photo: Karl Stapelfeldt (JPL); **(a, right)**: D. Padgett (IPAC/Caltech), W. Brandner (IPAC), K. Stapelfeldt (JPL) and NASA; **(b)**: Science Source; **(bottom)**: R. Koenig/picture alliance/blickwinkel/R/Newscom; **Page 105**: John Schults/Reuters/Newscom; **Page 107 (a)**: Karl Stapelfeldt/JPL/NASA; **(b)**: NASA/JPL-Caltech/UIU; **Page 111**: NASA/Johns Hopkins University Applied Physics Laboratory/Carnegie Institution of Washington; **Page 114**: Sally Anderson Wildlife/Alamy Stock Photo; **Page 116**: NRC Canada, C. Marois and Keck Observatory; **Page 117**: NASA, ESA and P. Kalas (University of California, Berkeley and Seti Institute)

CHAPTER 6

Page 124: NASA/GSFC/Arizona State University; **Page 126 (Mercury inset)**: NASA/Johns Hopkins University Applied Physics Laboratory/Carnegie Institution of Washington; **(Venus inset)**: NASA-JPL, **(Earth inset)**: PlanetObserver/Science Source, **(Moon inset)**: Science Source; **Page 127**: NASA; **Page 128 (Mars inset)**: NASA/USG; **(b)**: NASA/JSC; **(bottom)**: Photograph by D. J. Roddy and K. A. Zeller, USGS, Flagstaff, AZ; **Page 129**: NASA/JPL; **Page 130 (all)**: NASA/JPL/Caltech; **Page 137**: Marmaduke St. John/Alamy Stock Photo; **Page 138**: Duckeesue/Shutterstock; **Page 142 (a)**: NASA; **(b)**: NSSDC/NASA; **(bottom)**: NASA/Magellan Image/JPL; **Page 143**: NASA/JPL; **Page 144 (inset a)**: Ana Phelps/Shutterstock; **(inset b)**: Jukurae/Shutterstock; **(bottom)**: NASA/JSC/Arizona State University; **Page 145 (top)**: NASA/JSC; **(bottom)**: NASA/MOLA/Science Team; **Page 146**: NASA/JPL/ASU; **Page 147**: NASA/JPL/University of Arizona; **Page 148 (a–b)**: NASA/JPL-Caltech/MSSS and PSI; **(bottom)**: NASA/JPL-Caltech/University of Arizona/Texas A&M University

CHAPTER 7

Page 154: W.W. Norton & Company, Inc.; **Page 159 (all)**: NASA/NSSDC/GSFC; **Page 164 (Cyanobcteria inset)**: Aurélien Celette, Mona Lisa Production/Science Source; **(Stegosaurus inset)**: Sabena Jane Blackbird/Alamy Stock Photo; **(Multi-cellular inset)**: P&R Fotos/agefotostock/Alamy Stock Photo; **(Arthropod inset)**: John Cancalosi/Nature Picture Library/Alamy Stock Photo; **(Eusthenopteron inset)**: Universal Images Group North America LLC/DeAgostini/Alamy Stock Photo; **(Fossilised flower inset)**: The Natural History Museum/Alamy Stock Photo; **(Eukaryotes inset)**: Biophoto Associates/Science Source; **Page 165**: Paranyu Pithayarungsarit/Getty Images; **Page 168 (a)**: NASA; **(b)**: LOOK Die Bildagentur der Fotografen GmbH/Alamy Stock Photo; **Page 171**: NASA/NSSD; **Page 172**: NASA/JPL; **Page 173 (left)**: NASA; **(right)**: NASA/STSc

CHAPTER 8

Page 182: Jordana Meilleur/Alamy Stock Photo; **Page 185 (all)**: NASA/JPL; **Page 188**: NASA, ESA, and the Hubble Heritage Team (STScI/AURA); **Page 189 (top)**: NASA/JPL/University of Arizona; **(bottom)**: NASA/JPL-Caltech/SwRI/MSSS/Jason Major; **Page 190 (top)**: NASA Pictures/Alamy Stock Photo; **(bottom)**: NASA/JPL/Space Science Institute; **Page 191 (a)**: NASA/JPL/Space Science Institute; **(inset a)**: NASA/JPL/Space Science Institute; **(b)**: Courtesy of Cassini Imaging Central Laboratory for Operation; **(bottom a–b)**: NASA, L. Sromovsky, and P. Fry (University of Wisconsin-Madison); **Page 194 (a)**: NASA/JPL; **(bottom)**: NASA, ESA, L. Sromovsky and P. Fry (University of Wisconsin), H. Hammel (Space Science Institute), and K. Rages (SETI Institute); **Page 199 (top)**: SNC Art and More/Shutterstock; **(bottom)**: CSIRo (G. A. Dulk, Y. Leblanc, R. Sault, and R. W. Hunstead); **Page 200 (top)**: Nick Schneider, University of Colorado, Boulder; **(a)**: J. Clarke (University of Michigan), NASA; **(a inset top)**: NASA/JPL-Caltech/SwRI/ASI/INAF/JIRA; **(a inset bottom)**: J. Clarke (University of Michigan), NASA; **(b)**: Courtesy of John Clark, Boston University and NASA/STSc; **Page 203 (a)**: NASA/JPL; **Page 204 (top)**: NASA/JPL/Space Science Institute; **(bottom)**: NASA and The Hubble Heritage Team (STScI/AURA); **Page 205**: NASA/JPL/Caltech; **Page 209**: NASA/JPL

CHAPTER 9

Page 210: Daniel Pludowski/EyeEm/Getty Images; **Page 212 (from top to bottom)**: NASA/JPL/USGS; NASA/Johns Hopkins University Applied Physics Laboratory/Southwest Research Institute; NASA Visualization Technology Applications and Development (VTAD); NASA; Hubble Space Telescope, NASA; NASA/JPL-Caltech/UCLA/MPS/DLR/IDA; **Page 213 (a–b)**: NASA/JPL-Caltech/UCLA/MPS/DLR/IDA/PSI; **Page 214**: NASA/JHUAPL/SwR; **Page 215 (all but noted)**: NASA/JPL/

Caltech; **(Pluto and Charon)**: NASA-JHUAPL-SWRI; **Page 216 (all)**: NASA/JPL/Caltech; **Page 217 (top a–b)**: NASA/JPL/Space Science Institute; **(bottom a–b)**: NASA/JPL/Caltech; **Page 218 (a)**: NASA/JPL; **(b)**: NASA/JPL-Caltech/SETI Institute; **Page 219 (top a–d)**: NASA/JPL/Space Science Institute; **(bottom a)**: NASA/JPL; **(bottom b)**: NASA/JPL/ESA/University of Arizona; **Page 220 (top)**: ESA/NASA/JPL/University of Arizona; **(bottom)**: NASA; **Page 222 (top)**: NASA/JPL/Caltech; **(top inset)**: NASA/JPL/Caltech; **(center)**: NEAR Project, NLR, JHUAPL, Goddard SVS, NASA; **(bottom a–b)**: NASA/JPL/University of Arizona; **(bottom b)**: ESA/DLR/FU Berlin (G. Neukum)/Science Source; **Page 223**: NASA/Johns Hopkins Applied Physics Laboratory/Southwest Research Institute, National Optical Astronomy Observatory; **Page 225**: Courtesy of Terry Acomb; **Page 226 (a)**: S. Deiries/ESO; **(b)** ©Don Goldman; **Page 227**: NASA/NSSDC/GSF; **Page 228 (top a–b)**: NASA/JPL-Caltech/UMD; **(bottom)**: NASA/JPL-Caltech/UMD; **Page 229 (top)**: ESA/Rosetta/MPS for OSIRIS Team MPS /UPD /LAM /IAA/SSO/INTA/UPM/DASP/IDA; **(bottom)**: Science Source; **Page 231 (a)**: AP Photo/AP Video; **(b)**: Xinhua/RIA Xinhua News Agency/Newscom; **Page 232 (a)**: RGB Ventures/SuperStock/Alamy Stock Photo; **(b)**: Noel Power/Shutterstock; **p. 233 (a)**: John Cancalosi/Alamy Stock Photo; **(b)**: Collection and photo courtesy Thomas Witzke; **(c)**: Aerolite Meteorites; **(d)**: Dr. Svend Buhl; **Page 234**: NASA/JPL/Cornell

CHAPTER 10

Page 240: Orchidpoet/Getty Images; **Page 254**: Chris Stoughton; **Page 260**: Image taken by the D. McKinnon, Charles Sturt University Remote Telescope (12" F/6.7, Starlight Xpress HX9 CCD, exposure 0.01s); **Page 263**: Gaia Collaboration, A&A, 616, A10, 2018, reproduced with permission © ESO

CHAPTER 11

Page 270: W.W. Norton & Company, Inc.; Page 280: Hinode JAXA/NASA/PPARC; **Page 282**: NASA/Bill Ingalls; **Page 283 (top)**: Nigel Sharp, NOAO/NSO/Kitt Peak FTS/AURA/NSF; **(a)**: © Stefan Seip, www.astromeeting.de; **(b–c)**: © Laura Kay; **Page 284**: Science Source; **Page 285**: SOHO/ESA/NASA; **Page 286 (a–b)**: NASA/SDO/Solar Dynamics Observatory; **Page 288 (top, all)**: SOHO-EIT Consortium, ESA, NASA; **(a–b)**: SOHO/ESA/NASA; **Page 289 (a–c)**: NASA SDO; **(bottom)**: SOHO/ESA and NASA

CHAPTER 12

Page 296: Knickohr/Getty Images; **Page 310 (top)**: NASA and The Hubble Heritage Team (STScI/AURA); **(a)**: Dr. Raghvendra Sahai (JPL) and Dr. Arsen R. Hajian (USNO), NASA and The Hubble Heritage Team (STScI/AURA); **(b)**: J. P. Harrington and K. J. Borkowski (University of Maryland), HST, NASA; **(c)**: Bruce Balick (University of Washington), Vincent Icke (Leiden University, The Netherlands), Garrelt Mellema (Stockholm University), and NASA; **(d)**: NASA, A. Fruchter and the ERO team

CHAPTER 13

Page 324: W. W. Norton & Company, Inc.; **Page 330 (top)**: N. Smith, J. A. Morse (U. Colorado) et al., NASA; **(a–b)**: Anglo-Australian Observatory, photographs by David Malin; **Page 331**: NASA, ESA, and R. Kirshner (Harvard-Smithsonian Center for Astrophysics and Gordon and Betty Moore Foundation), and P. Challis (Harvard-Smithsonian Center for Astrophysics); **Page 334 (a)**: N. Levenson (Johns Hopkins), S. Snowden (USRA/GSFC); **(b)**: NASA/Hubble Heritage; **(c)**: Jeff Hester, Arizona State University, and NASA; **Page 337 (a)**: ESO; **(b)**: NASA and ESA, Acknowledgment: J. Hester (ASU) and M. Weisskopf (NASA/MSFC); **(inset)**: Hubble Space Telescope; **(c)**: NASA/CXC/SAO; **Page 338**: Event Horizon Telescope Collaboration; **Page 348**: Caltech/MIT/LIGO Lab; **Page 351**: ESA/Hubble

CHAPTER 14

Page 358: Hadafee/Shutterstock; **Page 361 (top, all but b-2)**: Australian Astronomical Observatory, photograph by David Malin; **(b-2)**: CFHT/Coelum - J.-C. Cuillandre/G. Anselmi; **(bottom all)**: NASA/JPL-Caltech; **Page 362 (top all)**: NASA/JPL-Caltech; **(bottom a)**: Australian Astronomical Observatory, photograph by David Malin; **(bottom b)**: NASA and The Hubble Heritage Team (AURA/STScI); **Page 363 (a)**: NOAO/AURA/NSF/Science Source; **(b)**: Anglo-Australian Telescope photograph by David Malin; **Page 364 (top)**: NOAO/AURA/NSF; **(bottom)**: NASA; **Page 365**: NASA/Swift/Stefan Immler (GSFC) and Erin Grand (UMCP); **Page 367 (a–b)**: WIYN/NOAO/NSF, WIYN Consortium, Inc.; **Page 368 (a–b)**: Courtesy of Jeff Hester; **Page 372**: Courtesy of Dr. Michael Brotherton; Weblink: http://www.mikebrotherton.com/wp-content/uploads/2008/05/figure1.jpg; **Page 374 (all)**: F. Owen, NRAO, with J. Biretta, STScI, and J. Eilek, NMIMT; **Page 375**: NASA, ESA, the Hubble Heritage (AURA/STScI)-ESA/Hubble Collaboration, and A. Evans (University of Virginia, Charlottesville/NRAO/Stony Brook University); **Page 376 (top all)**: NASA/CXC/MIT/F.K. Baganoff et al./E. Slawik; **(bottom)**: References and Credit: UCLA Galactic Center Group - W.M. Keck Observatory Laser Team. These images/animations were created by Prof. Andrea Ghez and her research team at UCLA and are from data sets obtained with the W. M. Keck Telescopes; **Page 377**: David A. Aguilar; **Page 381**: NASA, ESA, and The Hubble Heritage Team (STScI/AURA)

CHAPTER 15

Page 382 (a): Stewart Bremner/Getty Images; **(b)**: Teemu Tretjako/Alamy Stock Photo; **Page 387**: NASA, ESA and the Hubble Heritage (STScI/AURA)-ESA/Hubble Collaboration Acknowledgment: M. West (ESO, Chile); **Page 391 (top a)**: NASA/Swift/Stefan Immler (GSFC) and Erin Grand (UMCP); **(top b)**: Bill Schoening, Vanessa Harvey/REU program/NOAO/AURA/NSF; **(bottom a)**: Todd Boroson/NOAO/AURA/NSF; **(bottom b)**: Richard Rand, University of New Mexico; **Page 393 (a)**: Jasonfang/Getty Images; **(b)**: C. Howk, University of Notre Dame, B. Savage, University of Wisconsin, N.S. Sharp (NOAO)/WIYN/NOAO/NSF; **Page 394 (top)**: Lambda/GSFC/NASA;

(bottom left): Lynette Cook, all rights reserved; **(bottom right)**: NOAO/AURA/NSF; **Page 395 (a)**: Courtesy of Jeff Hester; **(b)**: The Electronic Universe Project; **Page 397**: ESA; **Page 400 (a)**: ASPERA; **(b)**: Philippe Widling/Design Pics/AgeFotoStock; **Page 402 (a–b)**: Anglo-Australian Observatory, photographs by David Malin

CHAPTER 16

Page 408: W. W. Norton & Company, Inc.; **Page 416**: GRANGER — All rights reserved; **Page 417 (a)**: Greg Balkin/Shutterstock; **(b)**: Roger Kisby/Cavan Images/Shutterstock; **p. 418 (a–c)**: NASA COBE Science Team; **(d)**: ESA and the Planck Collaboration; **Page 434 (a–b)**: Courtesy of the author

CHAPTER 17

Page 436: W. W. Norton & Company, Inc.; **Page 439**: Andrew Z. Colvin; **Page 440 (a–b)**: Max Tegmark and the Sloan Digital Sky Survey; **(c)**: M. Blanton and the Sloan Digital Sky Survey; Page 441 (a): X-ray: NASA/CXC/MIT/E.-H Peng. Optical: NASA/STScI; **(a inset)**: NASA, N. Benitez (JHU), T. Broadhurst (The Hebrew University), H. Ford (JHU), M. Clampin (STScI), G. Hartig (STScI), G. Illingworth (UCO/Lick Observatory), the ACS Science Team and ESA; **(b)**: NASA, ESA, D. Coe (NASA JPL/California Institute of Technology, and Space Telescope Science Institute), N. Benitez (Institute of Astrophysics of Andalusia, Spain), T. Broadhurst (University of the Basque Country, Spain), and H. Ford; **Page 449**: Ralf Kaehler, Matthew Turk, and Tom Abel; **Page 450 (top a–b)**: NASA/JPL-Caltech/GSFC; **(bottom)**: NASA, ESA, G. Illingworth (University of California, Santa Cruz), R. Bouwens (University of California, Santa Cruz, and Leiden University), and the HUDF09 Team; **Page 451**: Guedes, Javiera; Callegari, Simone; Madau, Piero; Mayer, Lucio The Astrophysical Journal, Volume 742, Issue 2, article id. 76 (2011); **Page 452 (top)**: Illustration Credit: NASA, ESA, and M. Kornmesser (ESO); **(bottom)**: X-ray (NASA/CXC/SAO/P. Green et al.), Optical (Carnegie Obs./Magellan/W.Baade Telescope/J.S. Mulchaey et al.); **Page 453 (a)**: NASA, ESA, J. Lotz (STScI), M. Davis (University of California, Berkeley), and A. Koekemoer (STScI); **(b)**: NASA, ESA, the Hubble Heritage (STScI/AURA)-ESA/Hubble Collab-

oration, and A. Evans (University of Virginia, Charlottesville/NRAO/Stony Brook University); **(right)**: NASA/CXC/CfA/M. Markevitch et al., and NASA/STScI, Magellan/U. Arizona/D.Clowe et al.; Lensing Map: NASA/STScI, ESO WFI, Magellan/U.Arizona/D.Clowe et al.; **Page 454 (top)**: Sloan Digital Sky Survey; **(bottom)**: Courtesy Joel Primack and George Blumenthal

CHAPTER 18

Page 460: Yuji Kotani/Getty Images; **Page 463 (a)**: NOAA PMEL EOI Program; **(b)**: Woods Hole Oceanographic Institution; **Page 464 (a)**: Associate Professor Warwick Hillier (dec); **(b)**: James St. John; **Page 466**: Science History Images/Alamy Stock Photo; **Page 467 (top)**: Kenneth E. Ward/ Biografx/ Science Source; **(a)**: Ted Kinsman/Science Source; **(b)**: National Park Service; **Page 469**: ImageBROKER/Alamy Stock Photo; **Page 471 (top)**: Popular Science Monthly, Volume 35, 1889/ University of Toronto via archive.org; **(bottom)**: NASA/JPL-Caltech/MSSS; **Page 472**: NASA/JPL-Caltech/Univ. of Arizona; **Page 476 (both)**: NASA; **Page 479**: Seti/NASA; **Page 482**: Seti/NASA

Figures

Figure 1.10: Figure 1 from B. P. Abott et al., "Observation of Gravitational Waves from a Binary Black Hole Merger," *Physical Review Letters*, Vol. 166, No. 6, 2016. Used with permission of the authors.
Figure 7.19: Figure: Global temperature and CO_2 variations over the last 800,000 years of Earth's history. From "Past atmospheric composition and greenhouse gases," iceandclimate.nbiku.dk. Reprinted with permission of the Niels Bohr Institute, University of Copenhagen, Denmark, http://www.iceandclimate.nbi.ku.dk
Figure 7.20: Figure: Global Average Temperature 1850–2018 from "Global Temperature Report for 2018," www.berkeleyearth.org
Figure 13.28: Masses in the Stellar Graveyard. Adapted from LIGO/Frank Elavsky/Aaron Geller/Northwestern University. Used with permission.
Figure 18.21: Figure: Size comparison: Sun now and when it will be a Red Giant by Ramses Ramirez. Courtesy of Ramses Ramirez, Cornell.

Page numbers followed by *f* and *t* indicate figures and tables, respectively.

Units and Values

Quantity	Fundamental Unit	Values
Length	meters (m)	radius of Sun (R_{Sun}) = 6.96265×10^8 m astronomical unit (AU) = 1.49598×10^{11} m 1 AU = 149,598,000 km light-year (ly) = 9.4605×10^{15} m 1 ly = 6.324×10^4 AU 1 parsec (pc) = 3.262 ly 1 pc = 3.0857×10^{16} m 1 m = 3.281 feet
Volume	cubic meters (m^3)	1 m^3 = 1,000 liters = 264.2 gallons
Mass	kilograms (kg)	1 kg = 1,000 grams mass of Earth (M_{Earth}) = 5.9736×10^{24} kg mass of Sun (M_{Sun}) = 1.9891×10^{30} kg
Time	seconds (s)	1 hour (h) = 60 minutes (min) = 3,600 s solar day (noon to noon) = 86,400 s sidereal day (Earth rotation period) = 86,164.1 s tropical year (equinox to equinox) = 365.24219 days = 3.15569×10^7 s sidereal year (Earth orbital period) = 365.25636 days = 3.15581×10^7 s
Speed	meters per second (m/s)	1 m/s = 2.237 miles/h 1 km/s = 1,000 m/s = 3,600 km/h c = 3.00×10^8 m/s = 300,000 km/s
Acceleration	meters per second squared (m/s^2)	gravitational acceleration on Earth (g) = 9.81 m/s^2
Energy	joules (J)	1 J = 1 kg m^2/s^2
Power	watts (W)	1 W = 1 J/s solar luminosity (L_{Sun}) = 3.827×10^{26} W
Force	newtons (N)	1 N = 1 kg m/s^2 1 pound (lb) = 4.448 N 1 N = 0.22481 lb
Pressure	newtons per square meter (N/m^2)	atmospheric pressure at sea level = 1.013×10^5 N/m^2 = 1.013 bar
Temperature	kelvins (K)	absolute zero = 0 K = −273.15°C = −459.67°F

Sources: Data from the National Space Science Data Center (2002); *Observer's Handbook 2002* (Royal Astronomical Society of Canada, 2001); and National Institute of Standards and Technology (2002).

Norton Digital Resources
digital.wwnorton.com/universe4

The registration code, available at no extra cost in every new copy of this book, provides access to the following online resources:

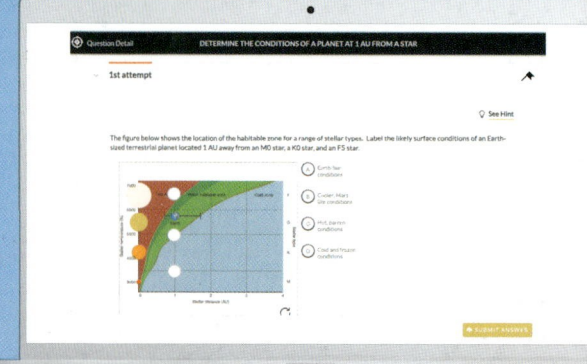

 **AT PLAY IN THE COSMOS** THE VIDEOGAME

At Play in the Cosmos is a videogame that spans the introductory astronomy course and invites you to experience the excitement of scientific discovery. You will apply concepts from the course in order to fly and repair your ship, make decisions based on scientific data, and complete missions.

smartwork

Smartwork is Norton's online tutorial and homework system. Questions in Smartwork allow you to test your understanding of course concepts. Questions provide hints, answer-specific feedback, and ebook links when you need help.

Ebook

The ebook gives you the ability to highlight, bookmark, and take notes while viewing on any computer or mobile device. The ebook is linked to AstroTour animations, Interactive Simulations, and Astronomy in Action videos, which provide demonstrations of key concepts in an easy-to-understand and memorable way.

To register or purchase a registration code, go to
digital.wwnorton.com/universe4

Cover design: DeMarinis Design LLC

Cover image: IC 410, The Tadpole Nebula in Auriga. Rolf Geissinger/Stocktrek Images/Alamy Images

ISBN 978-0-393-53383-5

9 780393 533835

90000

 **NORTON**
WWNORTON.COM/EDUCATOR